COLLEGE ALGEBRA

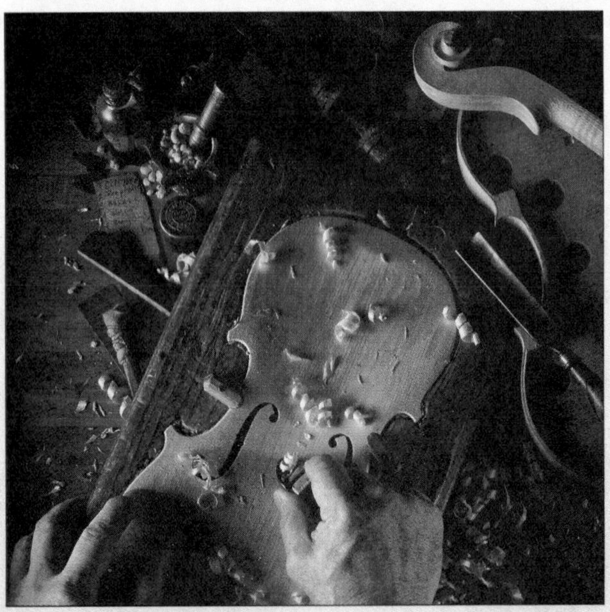

ABOUT THE COVER

The College Algebra course is an opportunity to learn about the beauty and practical power of mathematics. The course may also be a stepping stone to the study of calculus.

The violin, with its sound hole in the shape of an integral sign, has become a symbol of lead author James Stewart's calculus textbook series, which includes *Calculus, Fourth Edition, Calculus: Early Transcendentals, Fourth Edition,* and *Calculus: Concepts and Contexts.*

The cover of this edition, which shows violin making as a "work in progress," reflects both the Stewart authorship and the foundational importance of the College Algebra course.

ABOUT THE AUTHORS

James Stewart was educated at the University of Toronto and Stanford University, did research at the University of London, and now teaches at McMaster University. His research field is harmonic analysis.

He is the author of a best-selling calculus textbook series published by Brooks/Cole, including *Calculus, 4th Ed., Calculus: Early Transcendentals, 4th Ed.,* and *Calculus: Concepts and Contexts*, as well as a series of high-school mathematics textbooks.

A talented violinist, Stewart was concertmaster of the McMaster Symphony Orchestra for eight years and played professionally in the Hamilton Philharmonic Orchestra. One of his greatest pleasures is playing string quartets.

Lothar Redlin grew up on Vancouver Island, received a Bachelor of Science degree from the University of Victoria, and a Ph.D. from McMaster University in 1978. After completing his education, he did research and taught at the University of Washington, the University of Waterloo, and California State University, Long Beach.

He is currently Professor of Mathematics at The Pennsylvania State University, Abington College. His research field is topology.

Saleem Watson received his Bachelor of Science degree from Andrews University in Michigan. He did his graduate studies at Dalhousie University and McMaster University, where he received his Ph.D. in 1978. After completing his education, he did research at the Mathematics Institute of the University of Warsaw in Poland. He subsequently taught and did research at McMaster University and The Pennsylvania State University.

He is currently Professor of Mathematics at California State University, Long Beach. His research field is functional analysis.

The authors have also published *Precalculus, Third Edition* (Brooks/Cole, 1998).

COLLEGE ALGEBRA

Third Edition

James Stewart
McMaster University

Lothar Redlin
The Pennsylvania State University

Saleem Watson
California State University, Long Beach

Australia • Canada • Mexico • Singapore • Spain • United Kingdom • United States

Publisher: *Gary W. Ostedt*
Marketing Communications: *Samantha Cabaluna*
Assistant Editor: *Carol Ann Benedict*
Production Editor: *Tom Novack*
Production Service: *Luana Richards*
Manuscript Editor: *Luana Richards*
Permissions Editor: *Mary Kay Hancharick*
Interior Design: *John Edeen*
Cover Design: *Roy Neuhaus*
Cover Photo: *Mark Tomalty/Masterfile*
Interior Illustration: *Brian Betsill*
Page Layout: *Stephanie Kuhns*
Photo Researcher: *Terry Powell*
Print Buyer: *Vena Dyer*
Typesetting: *Better Graphics, Inc.*
Cover Printing: *Phoenix Color Corporation*
Printing and Binding: *World Color Corporation, Taunton*

For more information, contact:
BROOKS/COLE
511 Forest Lodge Road
Pacific Grove, CA 93950 USA
www.brookscole.com

Printed in United States of America

10 9 8 7 6 5 4 3 2

Library of Congress Cataloging-in-Publication Data

Stewart, James
College algebra / James Stewart, Lothar Redlin, Saleem Watson.—3rd ed.
p. cm.
Includes index.
ISBN 0-534-37352-6 (alk. paper)
1. Algebra. I. Redlin, L. II. Watson, Saleem. III. Title
QA152.2.S74 2000
512.9—dc21 99-052141

To Timothy Kottsieper

PREFACE

The art of teaching is the art of assisting discovery.
MARK VAN DOREN

For many students a College Algebra course represents the first opportunity to discover the beauty and practical power of mathematics. Thus, College Algebra instructors are faced with the challenge of teaching the technical skills of algebra while at the same time imparting an understanding of the concepts that underlie the subject. In this edition, as in the first two editions, our aim is to provide instructors and students with the tools needed to meet this challenge.

Good teaching comes in different forms, and each instructor brings unique strengths to the classroom. But certainly all instructors are committed to the goal of encouraging *conceptual understanding.* To implement this goal, instructors use different approaches. Some use *technology* to help students become active learners; others use the *rule of four,* "topics should be presented geometrically, numerically, algebraically, and verbally," as well as an expanded emphasis on applications to promote conceptual reasoning; still others use *group learning,* extended *projects,* or *writing exercises* as a way of encouraging students to explore their own understanding of a given concept; and all present College Algebra as a *problem-solving* endeavor. In this book we have used all these methods of presenting College Algebra, as enhancements to a central core of fundamental skills. These methods are tools to be utilized by instructors and their students to navigate their own course of action toward the goal of conceptual understanding.

In writing this third edition, our premise has been that conceptual understanding and technical skills go hand in hand, each reinforcing the other. Both are necessary for success, not only in subsequent mathematics and science courses, but also in today's increasingly technology-centered career world.

Many of the changes in the structure of this book are a result of our own experiences in teaching College Algebra. We have benefited from the insightful comments of our colleagues who have used the text and critiqued it section by section. We have also benefited from the sharp insight of our reviewers.

SPECIAL FEATURES

Exercise Sets The most important way to foster conceptual understanding is through the problems that the instructor assigns. To that end we have provided a wide selection of exercises, many of them new to this edition. Each exercise set is carefully graded, progressing from basic conceptual exercises and skill-development problems to more challenging problems requiring synthesis of previously learned material with new concepts.

Focus on Modeling

In addition to the applied problems that occur throughout the text, we have concluded several of the chapters with sections entitled *Focus on Modeling.* The first such section, after Chapter 2, introduces the basic idea of modeling a real-life situation using linear equations. Other sections present ways in which polynomial, exponential, and logarithmic functions, systems of inequalities, and probability can all be used to model familiar phenomena from the sciences and from everyday life.

Focus on Problem Solving

We are committed to helping students develop mathematical thinking rather than "memorizing all the rules," an endless and pointless task that many students are not taught to avoid. Thus, an emphasis on problem solving is integrated throughout the text. In addition, we have concluded several chapters with sections entitled *Focus on Problem Solving,* each of which highlights a particular problem-solving technique. The first such section, following Chapter 1, gives a general introduction to the principles of problem solving. Each *Focus* section includes problems of a varied nature that encourage students to apply their problem-solving skills. In selecting these problems, we have kept in mind the following advice from David Hilbert: "A mathematical problem should be difficult in order to entice us, yet not inaccessible lest it mock our efforts."

Projects

One way of engaging students and making them active learners is to have them work (perhaps in groups) on extended projects that give a feeling of substantial accomplishment when completed. We have included three types of projects: *Discovery Projects* encourage discovery through algebraic and geometric pattern recognition; the one on page 43 involves connecting factoring formulas to areas and volumes. *Laboratory Projects* involve technology; the one on page 486 uses repeated matrix multiplication to predict population distribution within a species. The *Writing Project* on page 150 compares modern algebraic methods to ancient ones.

Graphing Calculators and Computers

Calculator and computer technology provides a powerful way of visualizing mathematics and affects not only how a topic is taught but also what is emphasized. We have integrated the use of graphing calculators in numerous subsections, examples, and exercises (see, for instance, pages 263–266, 314, and 396). One example of how technology enhances understanding is that we can use it to draw accurate graphs of families of functions, so that students can see how varying a parameter affects the graph of a function (see pages 323 and 229, Exercises 53–58). These graphing calculator subsections, examples, and exercises, all marked with the special symbol, are optional and may be omitted without loss of continuity.

The availability of graphing calculators makes it not less important but far more important to clearly understand the concepts that underlie the image the calculator produces on its screen. Accordingly, all our calculator-oriented subsections are preceded by sections in which students must sketch graphs by hand and analyze them, so that they can understand precisely what the graphing device is doing when they later use it to simplify the routine, mechanical part of their work. Thus, we treat the calculator as an aid to understanding, an extension of pencil and paper, not as the central feature of the course.

Mathematical Vignettes

Throughout this book we make use of the margins to provide short biographies of interesting mathematicians as well as applications of mathematics to the "real world." The biographies often include a key insight that the mathematician discovered and which is relevant to College Algebra. (See, for instance, the vignettes on Viète, page 154; Salt Lake City, page 79; and radiocarbon dating, page 416.) They serve to enliven the material and show that mathematics is an important, vital activity, and that even at this elementary level, it is fundamental to everyday life. A new series of vignettes, entitled *Mathematics in the Modern World,* emphasizes the central role of mathematics in current advances in technology and the sciences (see pages 134, 174, and 318, for example).

Real-World Applications

We have included substantial applied problems that we believe will capture the attention of students. These are integrated throughout the text in both examples and exercises. Applications from engineering, physics, chemistry, business, biology, environmental studies, and other fields show how mathematics is used to model real-life situations. We have carefully chosen applications that show the relevance of algebra to our daily lives and its remarkable power as a problem-solving tool. (See, for instance, the examples and exercises on energy expended in bird flight, pages 176–177; the food price index, page 266; terminal velocity of a skydiver, page 395; and establishing time of death, page 432, Exercise 30.) We have used real-world data where appropriate.

Discovery, Writing, and Group Learning

Each exercise set ends with a block of exercises labeled *Discovery·Discussion.* These exercises are designed to encourage the students to experiment, preferably in groups, with the concepts developed in the section, and then to write out what they have learned, rather than simply look for "the answer."

Check Your Answer

The *Check Your Answer* feature is used wherever possible to emphasize the importance of looking back to check whether an answer is reasonable. (See, for instance, page 56.)

Review Sections and Chapter Tests

Each chapter ends with an extensive review section, including a *Chapter Test* designed to help the students gauge their progress. Brief answers to odd-numbered exercises in each section (including the review exercises), and to all questions in the Chapter Tests, are given in the back of the book.

The review material in each chapter now begins with a *Concept Check,* designed to get the students to think about and explain in their own words the ideas presented in the chapter. These can be used as writing exercises, in a classroom discussion setting, or for personal study.

MAJOR CHANGES FOR THE THIRD EDITION

- Chapter 1, which is a review chapter, now includes a section on basic equations.
- Chapter 2 is a new chapter that contains the basic material on the coordinate plane and graphs of equations. These are review topics for many College Algebra students. This reorganization permits the use of graphical methods when solving equations, and also makes early introduction of the graphing calculator possible.
- Chapter 3 on equations and inequalities has been rewritten to include graphical solution methods to parallel the algebraic methods.
- Chapter 4 on functions has been extensively rewritten with an eye to streamlining the sections on applied functions and extreme values. A new section dealing with the average rate of change of a function has been added. The new *Focus on Modeling* that follows the chapter contains a thorough introduction to setting up and using applied functions in optimization problems.
- Chapter 5 on polynomial and rational functions has been reorganized into five sections. The graphing calculator material has been integrated as subsections of the appropriate main sections. The section on rational functions has been clarified, and now makes more use of transformations of functions.
- Chapter 6 is followed by a new *Focus on Modeling* that covers the use of exponential functions in population growth.
- Chapter 7 now begins with the substitution and elimination methods for both linear and nonlinear systems of equations. It also includes the concept of reduced echelon form to clarify the Gauss-Jordan elimination process. Many of the Gaussian elimination exercises have been simplified from a computational point of view.
- Chapter 9 has been reorganized so that the material on arithmetic sequences and series appears together in one section, as does the material on geometric sequences and series.
- Chapter 10 on counting and probability has been streamlined as requested by users of this chapter.
- *Focus on Modeling* sections have been introduced after Chapters 2, 4, 6, 7, and 10.
- *Focus on Problem Solving* sections now appear after Chapters 1, 3, 5, 8, and 9.
- Several new vignettes have been added and existing ones updated. The new vignettes include a series entitled *Mathematics in the Modern World,* designed to showcase some of the latest developments in the applications of mathematics.
- The Projects, one per chapter, and the *Discovery·Discussion* exercises at the end of each section, are new.

MATHEMATICS IN THE MODERN WORLD

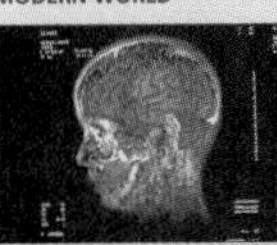

LOOKING INSIDE YOUR HEAD

How would you like to take a look inside your head? The idea may not be especially appealing to you, but doctors often need to do just that. If they can look without invasive surgery, all the better. An X-ray doesn't really give a look inside, it simply gives a "graph" of the density of tissue the X-rays must pass through. So an X-ray is a "flattened" view in one direction. Suppose you get an X-ray view from many different directions—can these "graphs" be used to reconstruct the three-dimensional inside view? This is a purely mathematical problem and was solved by mathematicians a long time ago. However, reconstructing the inside view requires thousands of tedious computations. Today, mathematics and high-speed computers make it possible to "look inside" by a process called Computer Aided Tomography (or CAT scan). Mathematicians continue to search for better ways of using mathematics to reconstruct images. One of the latest techniques, called magnetic resonance imaging (MRI), combines molecular biology and mathematics for a clear "look inside."

SHIFTING GRAPHS OF EQUATIONS

If h and k are positive real numbers, then replacing x by $x - h$ or by $x + h$ and replacing y by $y - k$ or by $y + k$ has the following effect(s) on the graph of any equation in x and y.

Replacement	How the graph is shifted
1. x replaced by $x - h$	Right h units
2. x replaced by $x + h$	Left h units
3. y replaced by $y - k$	Upward k units
4. y replaced by $y + k$	Downward k units

For example, consider the ellipse with equation

$$\frac{x^2}{a^2} + \frac{y^2}{b^2} = 1$$

which is shown in Figure 1. If we shift it so that its center is at the point (h, k) instead of at the origin, then its equation becomes

$$\frac{(x-h)^2}{a^2} + \frac{(y-k)^2}{b^2} = 1$$

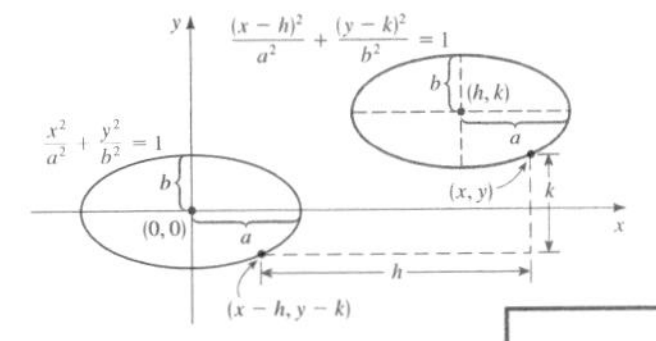

EXAMPLE 1 ■ Sketching the Graph of a Shifted Ellipse

Sketch the graph of the ellipse

$$\frac{(x+1)^2}{4} + \frac{(y-2)^2}{9} = 1$$

and determine the coordinates of the foci.

SOLUTION

The ellipse

$$\frac{(x+1)^2}{4} + \frac{(y-2)^2}{9} = 1 \quad \text{Shifted ell}$$

◀ Summary boxes organize and clarify topics and highlight key ideas.

◀ Mathematical vignettes provide applications of algebra to the real world or give short biographies of mathematicians, contemporary as well as historical. A series of vignettes entitled *Mathematics in the Modern World* (new to this edition) emphasize the central role of mathematics in current advances in technology and the sciences.

CHECK YOUR ANSWER

$x = 4$:

$$\text{LHS} = \frac{1}{4+1} + \frac{1}{4-2} = \frac{1}{5} + \frac{1}{2} = \frac{7}{10}$$

$$\text{RHS} = \frac{4+3}{4^2-4-2} = \frac{7}{10}$$

LHS = RHS ✓

equation by the LCD to get

$$(x+1)(x-2)\left(\frac{1}{x+1} + \frac{1}{x-2}\right) = (x+1)(x-2)\left(\frac{x+3}{x^2-x-2}\right) \quad \text{Multiply by LCD}$$

$$(x-2) + (x+1) = x + 3 \quad \text{Expand}$$

$$2x - 1 = x + 3 \quad \text{Subtract } x \text{ and add 1}$$

$$x = 4 \quad \text{Solve}$$

The solution is $x = 4$. ■

It is always important to check your answer, even if you never make a mistake in your calculations. This is because you sometimes end up with **extraneous solutions**, which are potential solutions that do not satisfy the original equation. The next example shows how this can happen.

EXAMPLE 4 ■ An Equation with No Solution

Solve the equation $2 + \frac{5}{x-4} = \frac{x+1}{x-4}$.

SOLUTION

First, we multiply each side by the common denominator, which is $x - 4$.

$$(x-4)\left(2 + \frac{5}{x-4}\right) = (x-4)\left(\frac{x+1}{x-4}\right) \quad \text{Multiply by } x - 4$$

$$2(x-4) + 5 = x + 1 \quad \text{Expand}$$

$$2x - 8 + 5 = x + 1 \quad \text{Distributive Property}$$

$$2x - 3 = x + 1 \quad \text{Simplify}$$

$$2x = x + 4 \quad \text{Add 3}$$

$$x = 4 \quad \text{Subtract } x$$

CHECK YOUR ANSWER

$x = 4$:

$$\text{LHS} = 2 + \frac{5}{4-4} = 2 + \frac{5}{0}$$

$$\text{RHS} = \frac{4+1}{4-4} = \frac{5}{0}$$

Impossible—can't divide by 0. LHS and RHS are undefined, so $x = 4$ is not a solution. ✕

But now if we try to substitute $x = 4$ back into the original equation, we would be dividing by 0, which is impossible. So, this equation has *no solution*. ■

The first step in the preceding solution, multiplying by $x - 4$, had the effect of multiplying by 0. (Do you see why?) Multiplying each side of an equation by an expression that contains the variable may introduce extraneous solutions. That is why it is important to check every answer.

Example titles clarify the purpose of examples. ▶

Author notes provide step-by-step comments on solutions. ▶

Check Your Answer emphasizes the importance of looking back at your work to check whether an answer is reasonable. ▶

This *Warning* symbol points out situations where many students often make the same mistake. ▶

Graphing calculator examples, exercises, and subsections are identified with the graphing calculator icon. ▶

Families of curves can be studied using the graphing calculator, so that students can see how varying a parameter in an equation affects the graph. ▶

With a graphing calculator we can quickly draw the graphs of many functions at once, on the same viewing screen. This allows us to see how changing a value in the definition of the functions affects the shape of its graph. In the next example we apply this principle to a family of third-degree polynomials.

EXAMPLE 8 ■ A Family of Polynomials

Sketch the family of polynomials $P(x) = x^3 - cx^2$ for $c = 0, 1, 2$, and 3. How does changing the value of c affect the graph?

SOLUTION

The polynomials

$$P_0(x) = x^3 \qquad P_1(x) = x^3 - x^2$$
$$P_2(x) = x^3 - 2x^2 \qquad P_3(x) = x^3 - 3x^2$$

are graphed in Figure 13. We see that increasing the value of c causes the graph to develop an increasingly deep "valley" to the right of the y-axis, creating a local maximum at the origin and a local minimum at a point in quadrant IV. This local minimum moves lower and farther to the right as c increases. To see why this happens, factor $P(x) = x^2(x - c)$. The polynomial P has zeros at 0 and c, and the larger c gets, the farther to the right the minimum between 0 and c will be. ■

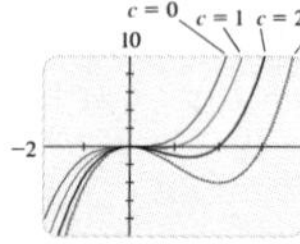

FIGURE 13
A family of polynomials $P(x) = x^3 - cx^2$

5.1 EXERCISES

1–8 ■ Sketch the graph of the function by transforming the graph of an appropriate function of the form $y = x^n$. Indicate all x- and y-intercepts on each graph.

1. $P(x) = x^3 - 8$
2. $P(x) = (x - 2)^3$
4. $P(x) = -2(x - 1)^3$
6. $P(x) = 3x^5 - 9$
8. $P(x) = 3x^4 - 27$

...nction. Make sure your graph ...e proper end behavior.

15. $P(x) = (x - 1)^2(x - 3)$
16. $P(x) = \frac{1}{4}(x + 1)^3(x - 3)$
17. $P(x) = \frac{1}{12}(x + 2)^2(x - 3)^2$
18. $P(x) = (x - 1)^2(x + 2)^3$
19. $P(x) = x^3(x + 2)(x - 3)^2$ **20.** $P(x) = (x^2 - 2x - 3)^2$
21. $P(x) = x^3 - x^2 - 6x$ **22.** $P(x) = x^3 + 2x^2 - 8x$
23. $P(x) = -x^3 + x^2 + 12x$ **24.** $P(x) = -2x^3 - x^2 + x$
25. $P(x) = x^4 - 3x^3 + 2x^2$ **26.** $P(x) = x^5 - 9x^3$
27. $P(x) = x^3 + x^2 - x - 1$
28. $P(x) = x^3 + 3x^2 - 4x - 12$
29. $P(x) = 2x^3 - x^2 - 18x + 9$
30. $P(x) = \frac{1}{8}(2x^4 + 3x^3 - 16x - 24)^2$
31. $P(x) = x^4 - 2x^3 - 8x + 16$
32. $P(x) = x^4 - 2x^3 + 8x - 16$
33. $P(x) = x^4 - 3x^2 - 4$ **34.** $P(x) = x^6 - 2x^3 + 1$

boardwalk before veering off onto the sand if he wishes to reach his umbrella in exactly 4 min 45 s?

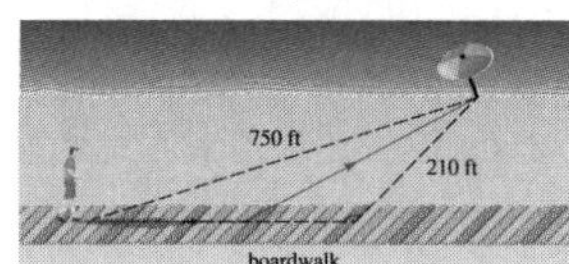

65. Grain is falling from a chute onto the ground, forming a conical pile whose diameter is always three times its height. How high is the pile (to the nearest hundredth of a foot) when it contains 1000 ft³ of grain?

66. A spherical tank has a capacity of 750 gallons. Using the fact that one gallon is about 0.1337 ft³, find the radius of the tank (to the nearest hundredth of a foot).

67. A city lot has the shape of a right triangle whose hypotenuse is 7 ft longer than one of the other sides. The perimeter of the lot is 392 ft. How long is each side of the lot?

68. Two television monitors sitting beside each other on a shelf in an appliance store have the same screen height. One has a conventional screen, which is 5 in. wider than it is high. The other has a wider, high-definition screen, which is 2.4 times as wide as it is high. The diagonal measure of the wider screen is 14 in. more than the diagonal measure of the smaller. What are the heights of the screens? (Assume that the height is greater than 10 in.)

69. One method for determining the depth of a well is to drop a stone into it and then measure the time it takes until the splash is heard. If d is the depth of the well (in feet) and t_1 the time (in seconds) it takes for the stone to fall, then $d = 16t_1^2$, so $t_1 = \sqrt{d}/4$. Now if t_2 is the time it takes for the sound to travel back up, then $d = 1090t_2$ because the speed of sound is 1090 ft/s. So $t_2 = d/1090$. Thus, the total time elapsed between dropping the stone and hearing the splash is $t_1 + t_2 = \sqrt{d}/4 + d/1090$. How deep is the well if this total time is 3 s?

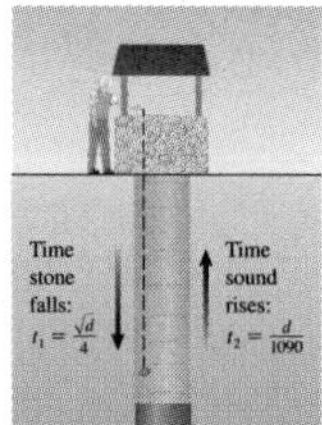

70–73 ■ Solve the equation for the variable x. The constants a and b represent positive real numbers.

70. $x^4 + 5ax^2 + 4a^2 = 0$ **71.** $a^3x^3 + b^3 = 0$

72. $\sqrt{x + a} + \sqrt{x - a} = \sqrt{2}\sqrt{x + 6}$

73. $\sqrt{x} + a\sqrt[3]{x} + b\sqrt[6]{x} + ab = 0$

DISCOVERY · DISCUSSION

74. Solving an Equation in Different Ways We have learned several different ways to solve an equation in this section. Some equations can be tackled by more than one method. For example, the equation $x - \sqrt{x} - 2 = 0$ is of quadratic type: We can solve it by letting $\sqrt{x} = u$ and $x = u^2$, and factoring. Or we could solve for $\sqrt{x}$, square, and then solve the resulting quadratic equation. Solve the following equations using both methods indicated, and show that you get the same final answers.

(a) $x - \sqrt{x} - 2 = 0$ quadratic type; solve for the radical and square

(b) $\dfrac{12}{(x - 3)^2} + \dfrac{10}{x - 3} + 1 = 0$ quadratic type; multiply by LCD

◀ Real-world applications show the relevance of algebra to everyday life and indicate its remarkable problem-solving power.

◀ *Discovery · Discussion* exercises encourage students to experiment with, discuss, and write about the concepts they have learned.

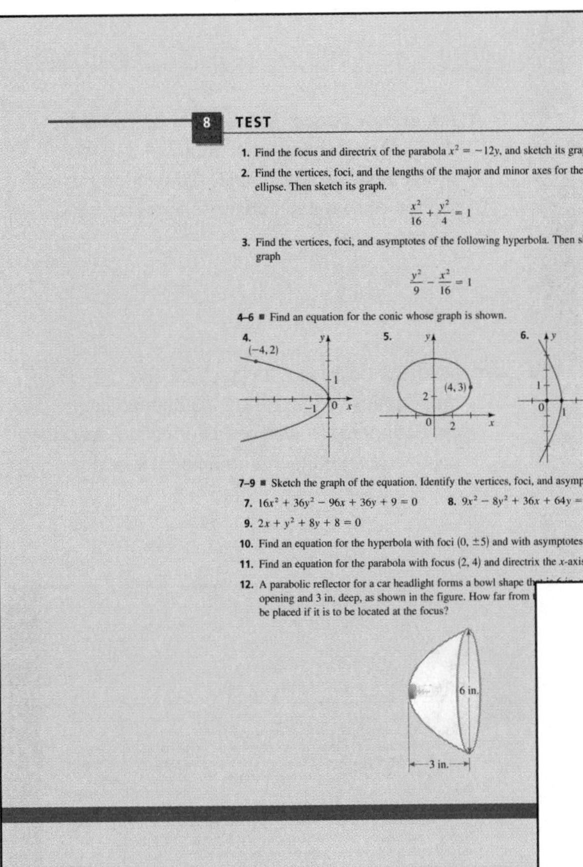

8 TEST

1. Find the focus and directrix of the parabola $x^2 = -12y$, and sketch its graph.
2. Find the vertices, foci, and the lengths of the major and minor axes for the following ellipse. Then sketch its graph.
$$\frac{x^2}{16} + \frac{y^2}{4} = 1$$
3. Find the vertices, foci, and asymptotes of the following hyperbola. Then sketch its graph
$$\frac{y^2}{9} - \frac{x^2}{16} = 1$$

4–6 ■ Find an equation for the conic whose graph is shown.

7–9 ■ Sketch the graph of the equation. Identify the vertices, foci, and asymptotes, if any.

7. $16x^2 + 36y^2 - 96x + 36y + 9 = 0$
8. $9x^2 - 8y^2 + 36x + 64y = 92$
9. $2x + y^2 + 8y + 8 = 0$
10. Find an equation for the hyperbola with foci $(0, \pm 5)$ and with asymptotes $y = \pm\frac{3}{4}x$.
11. Find an equation for the parabola with focus (2, 4) and directrix the x-axis.
12. A parabolic reflector for a car headlight forms a bowl shape th... opening and 3 in. deep, as shown in the figure. How far from t... be placed if it is to be located at the focus?

◀ Each chapter ends with a review section containing a *Concept Check,* extensive review exercises, and a *Chapter Test.*

278 CHAPTER 4 FUNCTIONS

Laboratory Project

ITERATION AND CHAOS

The **iterates** of a function f at a point x_0 are $f(x_0)$, $f(f(x_0))$, $f(f(f(x_0)))$, and so on. We write

$x_1 = f(x_0)$ The first iterate

$x_2 = f(f(x_0))$ The second iterate

$x_3 = f(f(f(x_0)))$ The third iterate

For example, if $f(x) = x^2$, then the iterates of f at 2 are $x_1 = 4$, $x_2 = 16$, $x_3 = 256$, and so on. (Check this.) Iterates can be described graphically as in Figure 1. Start with x_0 on the x-axis, move vertically to the graph of f, then horizontally to the line $y = x$, then vertically to the graph of f, and so on. The x-coordinates of the points on the graph of f are the iterates of f at x_0.

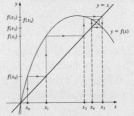

FIGURE 1

n	x_n
0	0.1
1	0.234
2	0.46603
3	0.64700
4	0.59382
5	0.62712
6	0.60799
7	0.61968
8	0.61276
9	0.61694
10	0.61444
11	0.61595
12	0.61505

Iterates are important in studying the **logistic function**
$$f(x) = kx(1 - x)$$
which models the population of a species with limited potential for growth (such as rabbits on an island or fish in a pond). In this model the maximum population that the environment can support is 1 (that is, 100%); if we start with a fraction of that population, say 0.1 (10%), then the iterates of f at 0.1 give the population after each time interval (days, months, or years, depending on the species). The constant k depends on the rate of growth of the species being modeled; it is called the **growth constant**. For example, for $k = 2.6$ and $x_0 = 0.1$ the iterates shown in the table give the population of the species for the first 12 time intervals. The population seems to be stabilizing around 0.615 (that is, 61.5% of maximum).

In the three graphs in Figure 2 we plot the iterates of f at 0.1 for different values of the growth constant k. For $k = 2.6$ the population appears to stabilize at a

Discovery, Laboratory, and *Writing Projects* ▶ are designed to help students become active learners by providing them with an extended problem, often involving experimentation and a written report, that requires them to use their algebra skills in a substantial fashion.

PRINCIPLES OF PROBLEM SOLVING

There are no hard and fast rules that will ensure success in solving problems. However, it is possible to outline some general steps in the problem-solving process and to give principles that are useful in solving certain problems. These steps and principles are just common sense made explicit. They have been adapted from George Polya's insightful book *How To Solve It.*

George Polya (1887–1985) is famous among mathematicians for his ideas on problem solving. His lectures on problem solving at Stanford University attracted overflow crowds whom he held on the edges of their seats, leading them to discover solutions for themselves. He was able to do this because of his deep insight into the psychology of problem solving. His well-known book *How To Solve It* has been translated into 15 languages. He said that Euler (see page 165) was unique among great mathematicians because he explained *how* he found his results. Polya often said to his students and colleagues, "Yes, I see that your proof is correct, but how did you discover it?" In the preface to *How To Solve It*, Polya writes, "A great discovery solves a great problem but there is a grain of discovery in the solution of any problem. Your problem may be modest; but if it challenges your curiosity and brings into play your inventive faculties, and if you solve it by your own means, you may experience the tension and enjoy the triumph of discovery."

1 UNDERSTAND THE PROBLEM

The first step is to read the problem and make sure that you understand it. Ask yourself the following questions:

What is the unknown?
What are the given quantities?
What are the given conditions?

For many problems it is useful to

draw a diagram

and identify the given and required quantities on the diagram.

Usually it is necessary to

introduce suitable notation

In choosing symbols for the unknown quantities, we often use letters such as a, b, c, m, n, x, and y, but in some cases it helps to use initials as suggestive symbols, for instance, V for volume or t for time.

2 THINK OF A PLAN

Find a connection between the given information and the u
to calculate the unknown. It often helps to ask yourself exp
the given to the unknown?" If you don't see a connection
ing ideas may be helpful in devising a plan.

■ **Try to recognize something familiar**

Relate the given situation to previous knowledge. Look at
recall a more familiar problem that has a similar unknown.

■ **Try to recognize patterns**

Certain problems are solved by recognizing that some kin
The pattern could be geometric, or numerical, or algebraic.

◀ *Focus on Problem Solving* sections include many new problems that are related to the problem-solving principle discussed in the section or to the material of the preceding chapters.

◀ Mathematical vignettes provide short biographies of interesting mathematicians, contemporary as well as historical, or describe applications of algebra to the real world.

Focus on Modeling sections show how algebra is used to model important phenomena from the sciences and from everyday life. ▶

PRINCIPLES OF MODELING

A model is a representation of an object or process. For example, a toy Ferrari is a *model* of the actual car; a road map is a model of the streets and highways in a city. A model usually represents just one aspect of the original thing. The toy Ferrari is not an actual car, but it does represent what a real Ferrari looks like; a road map does not contain the actual streets in a city, but it does represent the relationship of the streets to each other.

A **mathematical model** is a mathematical representation of an object or process. Often a mathematical model is an equation that describes a certain phenomenon. In Example 11 of Section 2.4 we found that the equation $T = -10h + 20$ models the atmospheric temperature T at elevation h. We then used this equation to predict the temperature at a certain height. Figure 1 illustrates the process of mathematical modeling.

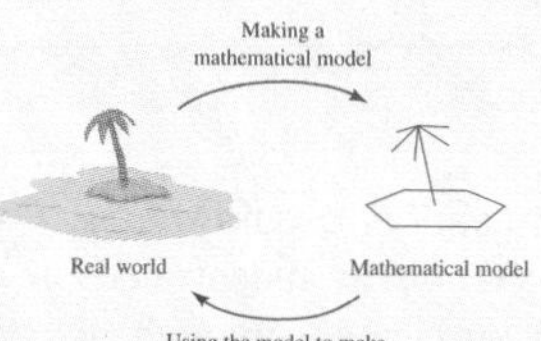

FIGURE 1

Mathematical models are useful because they enable us to isolate critical aspects of the thing we are studying and then to predict how it will behave. Models are used extensively in engineering, industry, and manufacturing. For example, engineers use computer models of skyscrapers to predict their strength and how they would behave in an earthquake. Aircraft manufacturers use elaborate mathematical models to predict the aerodynamic properties of a new design before the aircraft is actually built.

How are mathematical models developed? How are they used to predict the behavior of a process? In the next few pages and in subsequent *Focus on Modeling* sections, we explain how mathematical models can be constructed from real-world data, and we describe some of their applications.

LINEAR EQUATIONS AS MODELS

Consider the table on page 120, which gives the winning Olympic pole vaults in this century.

119

ACKNOWLEDGMENTS

We thank the following people for their thoughtful and constructive comments.

Reviewers of the First Edition

Barry W. Brunson, *Western Kentucky University;* Gay Ellis, *Southwest Missouri State University;* Martha Ann Larkin, *Southern Utah University;* Franklin A. Michello, *Middle Tennessee State University;* Kathryn Wetzel, *Amarillo College.*

Reviewers of the Second Edition

David Watson, *Rutgers University;* Floyd Downs, *Arizona State University at Tempe;* Muserref Wiggins, *University of Akron;* Marjorie Kreienbrink, *University of Wisconsin;* Richard Dodge, *Jackson Community College;* Christine Panoff, *University of Michigan at Flint;* Arnold Volbach, *University of Houston, University Park;* Keith Oberlander, *Pasadena City College;* Tom Walsh, *City College of San Francisco;* and George Wang, *Whittier College.*

Reviewers of the Third Edition

Christine Oxley, *Indiana University;* Linda B. Hutchison, *Belmont University;* David Rollins, *University of Central Florida;* and Max Warshauer, *Southwest Texas State University.*

We have benefited greatly from the suggestions and comments of our colleagues who have used our books in previous editions. We extend special thanks in this regard to Linda Byun, Bruce Chaderjian, David Gau, YongHee Kim, Daniel Martinez, David McKay, Robert Mena, Kent Merryfield, Viet Ngo, Marilyn Oba, Robert Valentini, and Derming Wang, of California State University, Long Beach; to Karen Gold, Betsy Huttenlock, Cecilia McVoy, Mike McVoy, Samir Ouzomgi, and Ralph Rush, of The Pennsylvania State University, Abington College; and to Gloria Dion, of Educational Testing Service, Princeton, New Jersey.

We thank Luana Richards, manager of our production service, for her excellent work and tireless attention to detail. Without her energy this edition could not have been completed on time. At TECH·arts we thank Brian Betsill for his elegant graphics, Stephanie Kuhns for her careful page layout, and Kathi Townes for her invaluable advice and assistance. We also thank Phyllis Panman for checking the accuracy of examples, exercises, and answers, and for helping with the proofreading process. At Brooks/Cole, our thanks go to Production Editor Tom Novack, Assistant Editor Carol Ann Benedict, Director of Marketing Edward N. Kelly, Marketing Manager Caroline Croley, Marketing Communications Project Manager Samantha Cabaluna, National Sales Manager Lucas R. Tomasso, and Art Director Vernon Boes. They have all done an outstanding job.

We are especially grateful to mathematics publisher Gary W. Ostedt. His extensive experience, keen editorial insight, and skillful management were invaluable resources in guiding this third edition to its completion.

ANCILLARIES FOR COLLEGE ALGEBRA, THIRD EDITION

For the Instructor

PRINTED

Instructor's Solutions Manual
by John Banks, San Jose City College

- Solutions to all even-numbered text exercises

Printed Test Items
by Steve Harmon, El Paso Community College

- Text specific
- Over 3000 multiple-choice and short-answer test items

SOFTWARE

Thomson Learning Testing Tools CD

- Available in Windows and Mac platforms
- Over 3000 multiple-choice and short-answer test items
- Algorithmic test generation
- Print version also available

Technical Support

- Toll-free technical support: (800) 423-0563 or E-mail: *support@kdc.com*

TUTORIAL SOFTWARE

Interactive Algebra Plus, Instructor's Version
This text-specific tutorial software provides explanations of concepts, along with carefully graded, step-by-step examples. Exercises are algorithmically generated and correspond to every chapter and section of the text, so that students have unlimited practice with concepts and skills. *Interactive Algebra Plus* also contains course management features that enable instructors to make assignments, give tests, and track student progress.

VIDEO

Text-Specific Video Tutorial Series
by Barbara Brown, Anoka-Ramsey Community College

- Provides additional examples and explanations for each chapter of the text

For the Student

PRINTED

Student Solutions Manual
by John Banks, San Jose City College
0-534-37361-5

- Solutions to all odd-numbered text exercises

Study Guide
by John Banks, San Jose City College
0-534-37360-7

- Detailed explanations
- Worked-out practice problems

TUTORIAL SOFTWARE

Interactive Algebra Plus, Student Version

- Provides unlimited practice with the concepts and skills of *College Algebra*
- Offers additional exercises, quizzes, and tests
- Hints are provided when you answer incorrectly, and the on-screen calculator is always available

RELATED PRODUCTS

Mastering Mathematics: How to Be a Great Math Student, 3/e © 1998
by Richard Manning Smith, Bryant College
0-534-34947-1

This practical guide will help you

- Avoid mental blocks during math exams
- Identify and improve areas of weakness
- Get the most out of class time
- Study more effectively
- Overcome a perceived low math ability
- Be successful on math tests
- Get back on track when feeling lost

Explorations in College Algebra Using the TI-82/83 with Integrated Appendix Notes for the TI-85 © 1997
by Deborah J. Cochener and Bonnie M. Hodge, Austin Peay State University
0-534-34228-0

- This unique and user-friendly workbook improves both understanding and retention of college algebra concepts using the graphing calculator
- By integrating technology into mathematics, the authors help readers develop problem-solving and critical-thinking skills

TO THE STUDENT

This textbook was written for you to use as a guide to mastering algebra. Here are some suggestions to help you get the most out of your course.

First of all, you should read the appropriate section of text *before* you attempt your homework problems. Reading a mathematics text is quite different from reading a novel, a newspaper, or even another textbook. You may find that you have to reread a passage several times before you understand it. Pay special attention to the examples, and work them out yourself with pencil and paper as you read. With this kind of preparation you will be able to do your homework much more quickly and with more understanding.

Don't make the mistake of trying to memorize every single rule or fact you may come across. Mathematics does not consist simply of memorization. Mathematics is a *problem-solving art*, not just a collection of facts. To master the subject you must solve problems—lots of problems. Do as many of the exercises as you can. Be sure to write your solutions in a logical, step-by-step fashion. Don't give up on a problem if you can't solve it right away. Try to understand the problem more clearly—reread it thoughtfully and relate it to what you have learned from your teacher and from the examples in the text. Struggle with it until you solve it. Once you have done this a few times you will begin to understand what mathematics is really all about.

Answers to the odd-numbered exercises, as well as all the answers to each chapter test, appear at the back of the book. If your answer differs from the one given, don't immediately assume that you are wrong. There may be a calculation that connects the two answers and makes both correct. For example, if you get $1/(\sqrt{2} - 1)$ but the answer given is $1 + \sqrt{2}$, your answer *is* correct, because you can multiply both numerator and denominator of your answer by $\sqrt{2} + 1$ to change it to the given answer.

The symbol ⊘ is used to warn against committing an error. We have placed this symbol in the margin to point out situations where we have found that many of our students make the same mistake.

CALCULATORS AND CALCULATIONS

Calculators are essential in most mathematics and science subjects. They free us from performing routine tasks, so we can focus more clearly on the concepts we are studying. Calculators are powerful tools but their results need to be interpreted with care. In what follows, we describe the features a calculator suitable for a college algebra course should have, and we give guidelines for interpreting the results of its calculations.

SCIENTIFIC AND GRAPHING CALCULATORS

For this course you will need a *scientific* calculator—one that has, as a minimum, the usual arithmetic operations ($+$, $-$, $\times$, $\div$) and exponential and logarithmic function (e^x, 10^x, $\ln x$, log). In addition, a memory and at least some degree of programmability will be useful. Many scientific calculators can perform operations on matrices and determinants, which are studied in Chapter 7.

Your instructor may recommend or require that you purchase a *graphing* calculator. This book has optional sections and exercises that require the use of a graphing calculator or a computer with graphing software. These special sections and exercises are indicated by the symbol .

It is important to realize that, because of limited resolution, a graphing calculator gives only an *approximation* to the graph of a function. It can plot only a finite number of points and then connect them to form a *representation* of the graph. In many of our examples we point out that we must be careful when interpreting graphs produced by calculators.

CALCULATIONS AND SIGNIFICANT FIGURES

Most of the applied examples and exercises in this book involve approximate values. For example, one exercise states that the moon has a radius of 1074 miles. This does not mean that the moon's radius is exactly 1074 miles but simply that this is the radius rounded to the nearest mile.

One simple method for specifying the accuracy of a number is to state how many **significant digits** it has. The significant digits in a number are the ones from the first nonzero digit to the last nonzero digit (reading from left to right). Thus, 1074 has four significant digits, 1070 has three, 1100 has two, and 1000 has one significant digit. This rule may sometimes lead to ambiguities. For example, if a distance is 200 km to the nearest kilometer, then the number 200 really has three significant digits, not just one. This ambiguity is avoided if we use scientific notation—that is, if we express the number as a multiple of a power of 10:

$$2.00 \times 10^2$$

When working with approximate values, students often make the mistake of giving a final answer with *more* significant digits than the original data. This is incorrect because you cannot "create" precision by using a calculator. The final result can be no more accurate than the measurements given in the problem. For example, suppose we are told that the two shorter sides of a right triangle are measured to be 1.25 and 2.33 inches long. By the Pythagorean Theorem, we find, using a calculator, that the hypotenuse has length

$$\sqrt{1.25^2 + 2.33^2} \approx 2.644125564 \text{ in.}$$

But since the given lengths were expressed to three significant digits, the answer cannot be any more accurate. We can therefore say only that the hypotenuse is 2.64 in. long, rounding to the nearest hundredth.

In general, the final answer should be expressed with the same accuracy as the *least*-accurate measurement given in the statement of the problem. The following rules make this principle more precise.

RULES FOR WORKING WITH APPROXIMATE DATA

1. When multiplying or dividing, round off the final result so that it has as many *significant digits* as the given value with the fewest number of significant digits.
2. When adding or subtracting, round off the final result so that it has its last significant digit in the *decimal place* in which the least-accurate given value has its last significant digit.
3. When taking powers or roots, round off the final result so that it has the same number of *significant digits* as the given value.

As an example, suppose that a rectangular table top is measured to be 122.64 in. by 37.3 in. We express its area and perimeter as follows:

$$\text{Area} = \text{length} \times \text{width} = 122.64 \times 37.3 \approx 4570 \text{ in}^2$$ Three significant digits

$$\text{Perimeter} = 2(\text{length} + \text{width}) = 2(122.64 + 37.3) \approx 319.9 \text{ in.}$$ Tenths digit

Note that in the formula for the perimeter, the value 2 is an exact value, not an approximate measurement. It therefore does not affect the accuracy of the final result. In general, if a problem involves only exact values, we may express the final answer with as many significant digits as we wish.

Note also that to make the final result as accurate as possible, *you should wait until the last step to round off your answer*. If necessary, use the memory feature of your calculator to retain the results of intermediate calculations.

ABBREVIATIONS

cm	centimeter
dB	decibel
F	farad
ft	foot
g	gram
gal	gallon
h	hour
H	henry
Hz	Hertz
in.	inch
J	Joule
kcal	kilocalorie
kg	kilogram
km	kilometer
kPa	kilopascal
L	liter
lb	pound
M	mole of solute per liter of solution
m	meter
MHz	megahertz
mi	mile
min	minute
mL	milliliter
mm	millimeter
N	Newton
qt	quart
oz	ounce
s	second
Ω	ohm
V	volt
W	watt
yd	yard
yr	year
°C	degree Celsius
°F	degree Fahrenheit
K	Kelvin
$\Rightarrow$	implies
$\Leftrightarrow$	is equivalent to

CONTENTS

4 FUNCTIONS 208

POLYNOMIAL AND RATIONAL FUNCTIONS 310

6 EXPONENTIAL AND LOGARITHMIC FUNCTIONS 378

7 SYSTEMS OF EQUATIONS AND INEQUALITIES 446

8 CONIC SECTIONS 532

9 SEQUENCES AND SERIES 580

10 COUNTING AND PROBABILITY 640

COLLEGE ALGEBRA

1 BASIC ALGEBRA

Algebra has been an essential element of scientific progress in every age, from the construction of the pyramids to the development of the space shuttle.

One learns by doing the thing; for though you think you know it, you have no certainty until you try.

SOPHOCLES

In this first chapter we review the basic ideas from algebra that we will need throughout this book.

1.1 WHAT IS ALGEBRA?

The word *algebra* comes from the ninth century Arabic book *Hisâb al-Jabr w'al-Muqabala,* written by al-Khowarizmi. The title refers to transposing and combining terms, two processes used in solving equations. In Latin translations the title was shortened to *Aljabr*, from which we get the word *algebra*. The author's name itself made its way into the English language in the form of our word *algorithm.*

The power of algebra stems from its ability to say many things at once.

Let's begin with a simple example. Consider the following list of facts about addition:

$$3 + 4 = 4 + 3$$

$$5 + 7 = 7 + 5$$

$$11 + 2 = 2 + 11$$

$$51032 + 873 = 873 + 51032$$

The order in which we add two numbers appears to make no difference—we get the same answer in either case. A little thought will convince you that this is *always* true. One way to express this fact is to complete the list to include all possible pairs of numbers; however, this would take forever because there are infinitely many pairs of numbers. A better way to express this fact is to say that when adding two numbers we have

$$(\text{first number}) + (\text{second number}) = (\text{second number}) + (\text{first number})$$

This sentence covers all possible cases at once. If we use letters such as $a, b, c, \ldots$ to stand for numbers, we can write this even more briefly as

$$a + b = b + a$$

This is the shortest way of saying what we mean.

This example illustrates the basic idea in all of algebra. Instead of working with specific numbers, as we do in arithmetic, we work with letters that can stand for any number. The crucial advantage in writing things this way is that we can see patterns more easily. This, in turn, helps us write formulas and solve problems. We illustrate these key ideas with some examples.

WRITING FORMULAS

We begin with examples that illustrate how to write formulas in algebra.

EXAMPLE 1 ■ Finding a Formula for Average

Find a formula for the average of three numbers.

SOLUTION

To find the average of three numbers, we add them and divide by 3. For example, if your scores on three tests in a college algebra course are 78 and 81 and 93, then your average test score is

Intuitively, the average is where the numbers "balance."

$$\frac{78 + 81 + 93}{3} = 84$$

We see that the formula for the average A of any three numbers a, b, and c is

$$A = \frac{a + b + c}{3}$$ ■

EXAMPLE 2 ■ Finding a Formula for Weekly Pay

Find a formula for a worker's weekly pay in terms of the hourly wage and the number of hours worked.

SOLUTION

If you are paid \$6 an hour and you work 20 hours in one week, then your pay (which we will call P) for that week is

$$P = 6 \times 20$$

If you work only 18 hours the following week, then your pay for that week is $P = 6 \times 18$. The general rule is that your pay is \$6 times the number of hours worked. Letting H stand for the number of hours worked, we have

$$P = 6H$$

The formula would be different for a worker who makes \$10 or \$14 an hour. But if we let W stand for the hourly wage, we can write

$$P = WH$$ ■

The simple formula in Example 2 gives the pattern for calculating the weekly pay of any worker, with any hourly wage, for any number of hours worked. For example, the weekly pay of a worker who is paid \$8 per hour and has worked 32 hours is

$$P = WH = 8 \times 32 = \$256$$

The ability to discover and use formulas as we have done here is fundamental to understanding what algebra is all about.

SOLVING PROBLEMS

The methods of algebra enable us to solve many types of problems. The next examples use the formulas we have just discovered.

MATHEMATICS IN THE MODERN WORLD

Although we are often unaware of its presence, mathematics permeates nearly every aspect of life in the modern world. With the advent of modern technology, mathematics has steadily been playing a greater role in our everyday life. In fact, you were probably awakened this morning by a digital alarm clock, made a phone call that used digital transmission, sent an e-mail message over the Internet, drove a car with digitally controlled fuel injection, listened to music on a CD player, then fell asleep in a room whose temperature is controlled by a digital thermostat. In each of these activities mathematics is crucially involved. In general, a property such as the intensity or frequency of sound, oxygen level in the exhaust emission from a car, the colors in an image, or the temperature in your bedroom is transformed into sequences of numbers by sophisticated mathematical algorithms. These numerical data, which usually consist of many millions of bits (the digits 0 and 1), are then transmitted and reinterpreted. Dealing with such huge amounts of data was not feasible until the invention of computers, machines whose logical processes were invented by mathematicians.

(continued next page)

EXAMPLE 3 ■ Solving an Algebra Problem Involving Averages

If your scores on the first two college algebra tests are 76 and 87, what score must you earn on the third test so that your average for the three tests is 85?

SOLUTION

The formula we discovered in Example 1 tells us that the average of three numbers is

$$A = \frac{a+b+c}{3}$$

Since we want $A = 85$ and we know that $a = 76$ and $b = 87$, we substitute all these values in our formula. Thus

$$85 = \frac{76+87+c}{3}$$

$$= \frac{163+c}{3}$$

To find the necessary test score on the third test, we solve for c:

$$85 \times 3 = \frac{163+c}{3} \times 3 \qquad \text{Multiply each side by 3}$$

$$255 = 163 + c$$

$$92 = c \qquad \text{Subtract 163 from each side}$$

You must earn a test score of 92 on the third test to attain an average test score of 85. ■

EXAMPLE 4 ■ Solving an Algebra Problem Involving Pay

James makes \$5.50 an hour working at a car wash. He wants to earn enough money to buy a portable CD player that costs \$77 (including tax). How many hours does he need to work to earn this amount?

(continued from page 5)

The contributions of mathematics in the modern world are not limited to technological advances. The logical processes of mathematics are now used to analyze complex problems in the social, political, and life sciences in new and surprising ways. Advances in mathematics continue to be made, some of the most exciting of these just within the past decade.

In other *Mathematics in the Modern World*, we will describe in more detail how mathematics affects us in our everyday activities.

SOLUTION

First let's write down what we know. James's hourly wage is $W = \$5.50$ and the CD player costs \$77. So, the amount of pay he needs is $P = \$77$. From the formula in Example 2 we know that

$$P = WH$$

We now substitute the values we know for P and W and solve for H:

$$77 = 5.50H$$

$$H = \frac{77}{5.50} \qquad \text{Divide each side by 5.50}$$

$$H = 14$$

James needs to work 14 hours in order to earn \$77. ■

You are probably familiar with the formulas for the area and volume of geometric shapes; these are listed on the inside back cover of this book. For example, the formula for the area A of a rectangle is

$$A = lw$$

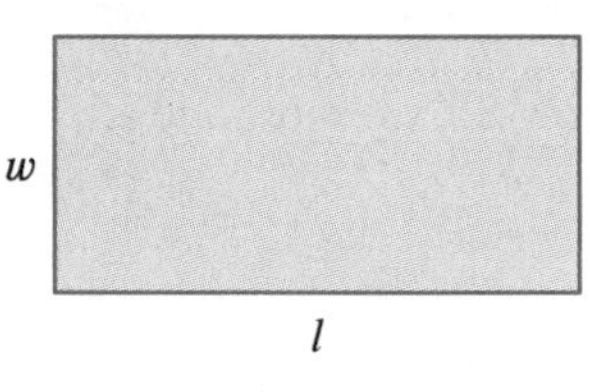

Area = lw

where l is the length and w is the width. In the next example we use this formula to derive another formula.

EXAMPLE 5 ■ An Algebra Problem Involving Area

(a) Find a formula for the area of a rectangle whose length is twice its width.

(b) A manufacturing company requires a rectangular sheet of metal whose length is twice its width and whose total area is 50 ft^2. What are the dimensions of this sheet?

SOLUTION

FIGURE 1

(a) If x is the width of the rectangle, then its length is $2x$ (see Figure 1). So, from the formula for the area of a rectangle we get the formula

$$A = lw = (2x)x = 2x^2$$

(b) We know that $A = 50$, and we want to find x. Using the formula in part (a), we get

$$50 = 2x^2$$

$$25 = x^2 \qquad \text{Divide each side by 2}$$

$$5 = x \qquad \text{Take the square root of each side}$$

Thus, the width of the sheet is 5 ft and its length is 10 ft. ■

The problems we have solved in the preceding examples are simple, but the methods are far-reaching. In fact, algebra applies to a wide variety of problems and is an indispensable tool in the study of any exact science. In algebra we use letters to stand for numbers so that we can talk concisely about numbers, see patterns (or formulas), and manipulate these patterns to solve problems.

1.1 EXERCISES

1–8 ■ The list of facts illustrates a general principle about numbers. State the principle in algebraic notation.

1. $3 \cdot 7 = 7 \cdot 3, \quad 8 \cdot 4 = 4 \cdot 8, \quad 5 \cdot 9 = 9 \cdot 5$

2. $(2 + 3) + 4 = 2 + (3 + 4)$
$(4 + 6) + 2 = 4 + (6 + 2)$

3. $(2 \cdot 3) \cdot 4 = 2 \cdot (3 \cdot 4)$
$(4 \cdot 6) \cdot 2 = 4 \cdot (6 \cdot 2)$

4. $6(2 + 3) = 6 \cdot 2 + 6 \cdot 3$
$2(3 + 5) = 2 \cdot 3 + 2 \cdot 5$

5. $(5 - 2)(5 + 2) = 5^2 - 2^2$
$(6 - 4)(6 + 4) = 6^2 - 4^2$

6. $(3 + 5)^2 = 3^2 + 2(3 \cdot 5) + 5^2$
$(4 + 7)^2 = 4^2 + 2(4 \cdot 7) + 7^2$

7. $(2 \cdot 3)^2 = 2^2 \cdot 3^2,$
$(5 \cdot 7)^2 = 5^2 \cdot 7^2$

8. $(5 - 3) = -(3 - 5),$
$(9 - 4) = -(4 - 9)$

9–20 ■ Write an algebraic formula for the given quantity.

9. The average of two numbers

10. The average of four numbers

11. The sum of a number and twice its square

12. The area of a rectangle whose length is 4 ft more than its width

13. The number of days in w weeks

14. The number of cents in n quarters

15. The product of two consecutive integers

16. The sum of three consecutive integers

17. The sum of the squares of two numbers

18. The distance in miles that a car travels in t hours at a speed of r miles per hour

19. The time it takes an airplane to travel d miles if its speed is r miles per hour

20. The speed of a boat that travels d miles in t hours

21–30 ■ Write an algebraic formula for the given quantity. You may need to consult the formulas for area and volume listed on the inside back cover of this book.

21. The area of a square of side x

22. The volume of a cube of side x

23. The volume of a box with a square base of side x and height $2x$

24. The surface area of a cube of side x

25. The surface area of a box of dimensions l, w, and h

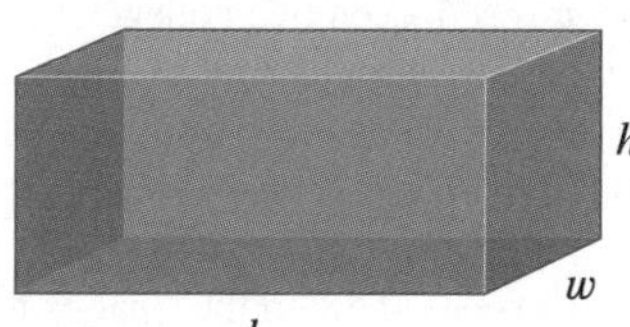

26. The area of what remains when a circle is cut out of a square, as shown in the figure

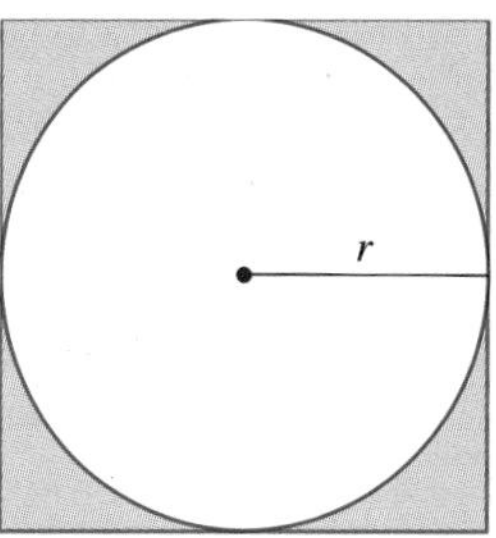

27. The length of the race track shown in the figure

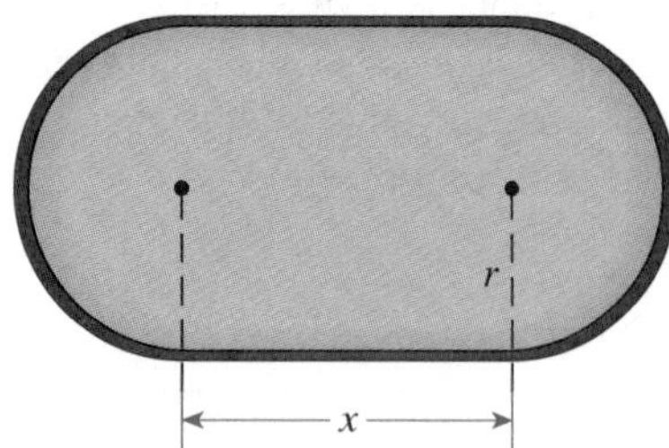

28. The area enclosed by the race track shown in the preceding figure

29. The area of a triangle whose base is twice its height

30. The volume of what remains of a solid ball of radius R after a smaller solid ball of radius r is removed from inside the larger ball, as shown in the figure

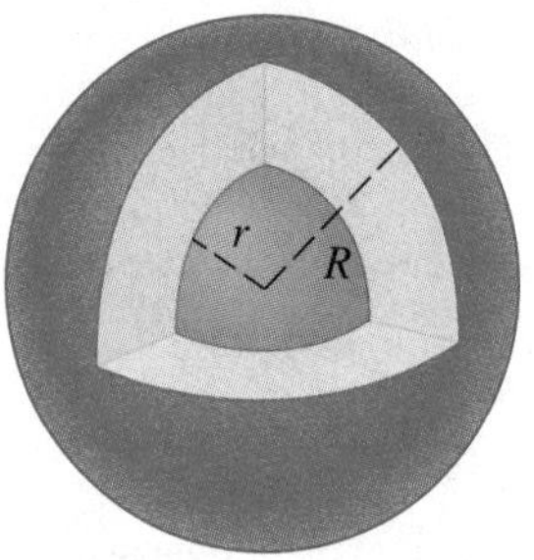

31. Suppose that your scores on two college algebra tests are 79 and 83, and the only test left to take is the final exam. This exam, however, counts double (this means that your test score on the final exam counts twice in the formula for average).

(a) If you earn a score of 88 on the final exam, what is your average test score for the course?

(b) Write a formula for a student's average test score in terms of the two test scores a and b and the test score f on the final exam.

(c) What test score do you need on the final exam for your average test score for the course to be 85?

32. Suppose that you walk at a constant rate of 4 feet per second.

(a) If you walk for 20 seconds, how far will you have walked?

(b) Find a formula for the distance d you walk in t seconds.

(c) How long do you have to walk to cover a distance of one mile?

33. Maria can mow a lawn at a rate of 150 square feet per minute.

(a) What area can she mow in half an hour?

(b) Find a formula for the area she can mow in T minutes.

(c) How long would it take Maria to mow a lawn that is 80 ft wide and 120 ft long?

(d) If she wants to mow the lawn in part (c) in one hour, what would her rate of mowing have to be?

34. A fruit crate has square ends and is twice as long as it is wide (see the figure).

(a) Find the volume of the box if its width is 20 inches.

(b) Find a formula for the volume V of the box in terms of its width x.

(c) If the box is to have volume 6750 in^3, what dimensions must the box have?

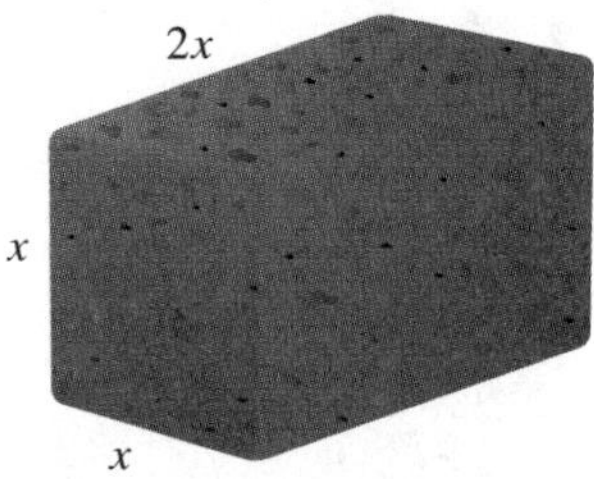

35. In many universities a student is given grade points for each credit unit according to the following scale:

A	4 points
B	3 points
C	2 points
D	1 point
F	0 point

For example, a grade of A in a 3-unit course earns $4 \times 3 = 12$ grade points and a grade of B in a 5-unit course earns $3 \times 5 = 15$ grade points. A student's grade point average (GPA) for these two courses is the total number of grade points earned divided by the number of units; in this case the GPA is $(12 + 15)/8 = 3.375$.

(a) Find a formula for the GPA of a student who earns a grade of A in a units of course work, B in b units, C in c units, D in d units, and F in f units.

(b) Find the GPA of a student who has earned a grade of A in two 3-unit courses, B in one 4-unit course, and C in three 3-unit courses.

1.2 REAL NUMBERS

The different types of real numbers were invented to meet specific needs. For example, natural numbers are needed for counting, negative numbers for describing debt or below-zero temperatures, rational numbers for concepts like "half a gallon of milk," and irrational numbers for measuring certain distances, like the diagonal of a square.

Let's review the types of numbers that make up the real number system. We start with the **natural numbers**:

$$1, 2, 3, 4, \ldots$$

The **integers** consist of the natural numbers together with their negatives and 0:

$$\ldots, -3, -2, -1, 0, 1, 2, 3, 4, \ldots$$

We construct the **rational numbers** by taking ratios of integers. Thus, any rational number r can be expressed as

$$r = \frac{m}{n} \qquad \text{where } m \text{ and } n \text{ are integers and } n \neq 0$$

Examples are

$$\tfrac{1}{2} \qquad -\tfrac{3}{7} \qquad 46 = \tfrac{46}{1} \qquad 0.17 = \tfrac{17}{100}$$

(Recall that division by 0 is always ruled out, so expressions like $\frac{3}{0}$ and $\frac{0}{0}$ are undefined.) There are also real numbers, such as $\sqrt{2}$, that cannot be expressed as a ratio of integers and are therefore called **irrational numbers**. It can be shown, with varying degrees of difficulty, that these numbers are also irrational:

$$\sqrt{3} \qquad \sqrt{5} \qquad \sqrt[3]{2} \qquad \pi \qquad \frac{3}{\pi^2}$$

The set of all real numbers is usually denoted by the symbol $\mathbb{R}$. When we use the word *number* without qualification, we will mean "real number." Figure 1 is a diagram of the types of real numbers that we work with in this book.

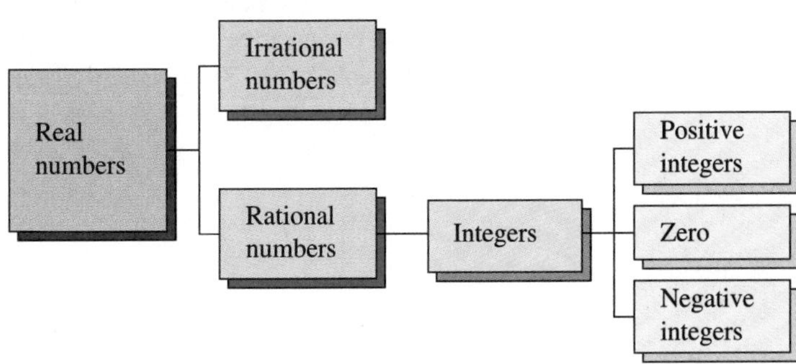

FIGURE 1
The real number system

Every real number has a decimal representation. If the number is rational, then its corresponding decimal is repeating. For example,

$$\tfrac{1}{2} = 0.5000\ldots = 0.5\overline{0} \qquad \tfrac{2}{3} = 0.66666\ldots = 0.\overline{6}$$

$$\tfrac{157}{495} = 0.3171717\ldots = 0.3\overline{17} \qquad \tfrac{9}{7} = 1.285714285714\ldots = 1.\overline{285714}$$

(The bar indicates that the sequence of digits repeats forever.) If the number is irrational, the decimal representation is nonrepeating:

$$\sqrt{2} = 1.414213562373095\ldots \qquad \pi = 3.141592653589793\ldots$$

If we stop the decimal expansion of any number at a certain place, we get an approximation to the number. For instance, we can write

$$\pi \approx 3.14159265$$

where the symbol $\approx$ is read "is approximately equal to." The more decimal places we retain, the better the approximation we get.

A repeating decimal such as

$$x = 3.5474747\ldots$$

is a rational number. To convert it to a ratio of two integers, we write

$$\begin{aligned} 1000x &= 3547.47474747\ldots \\ 10x &= 35.47474747\ldots \\ \hline 990x &= 3512.0 \end{aligned}$$

Thus, $x = \frac{3512}{990}$. (The idea is to multiply x by appropriate powers of 10, and then subtract to eliminate the repeating part.)

The real numbers can be represented by points on a line, as shown in Figure 2. The positive direction (toward the right) is indicated by an arrow. We choose an arbitrary reference point O, called the **origin**, which corresponds to the real number 0. Given any convenient unit of measurement, each positive number x is represented by the point on the line a distance of x units to the right of the origin, and each negative number $-x$ is represented by the point x units to the left of the origin. Thus, every real number is represented by a point on the line, and every point P on the line corresponds to exactly one real number. The number associated with the point P is called the coordinate of P, and the line is then called a **coordinate line**, or a **real number line**, or simply a **real line**. Often we identify the point with its coordinate and think of a number as being a point on the real line.

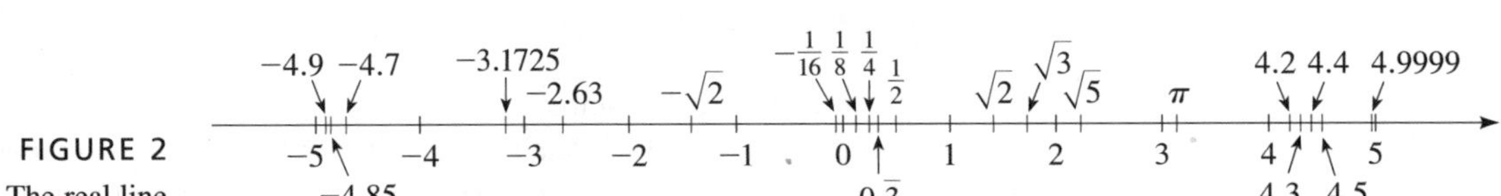

FIGURE 2
The real line

The real numbers are *ordered*. We say that ***a* is less than *b*** and write $a < b$ if $b - a$ is a positive number. Geometrically, this means that a lies to the left of b on the number line. (Equivalently, we can say that ***b* is greater than *a*** and write $b > a$.) The symbol $a \leqslant b$ (or $b \geqslant a$) means that either $a < b$ or $a = b$ and is read "a is less than or equal to b." For instance, the following are true inequalities (see Figure 3):

$$7 < 7.4 < 7.5 \qquad -\pi < -3 \qquad \sqrt{2} < 2 \qquad 2 \leqslant 2$$

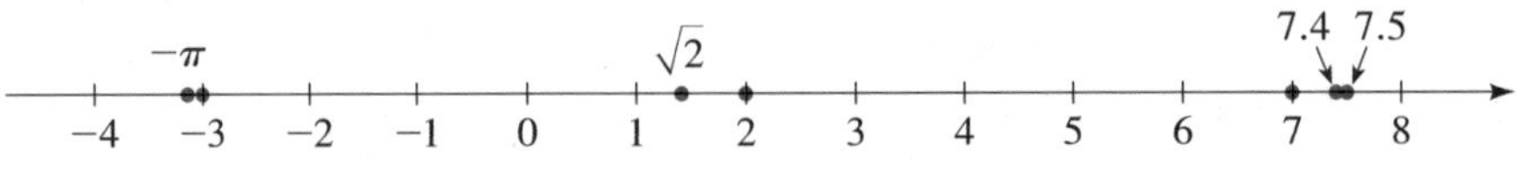

FIGURE 3

PROPERTIES OF REAL NUMBERS

In combining real numbers using the familiar operations of addition and multiplication, we use the following properties of real numbers.

PROPERTIES OF REAL NUMBERS

Property	Example	Description
Commutative Properties		
$a + b = b + a$	$7 + 3 = 3 + 7$	When we add two numbers, order doesn't matter.
$ab = ba$	$3 \cdot 5 = 5 \cdot 3$	When we multiply two numbers, order doesn't matter.
Associative Properties		
$(a + b) + c = a + (b + c)$	$(2 + 4) + 7 = 2 + (4 + 7)$	When we add three numbers, it doesn't matter which two we add first.
$(ab)c = a(bc)$	$(3 \cdot 7) \cdot 5 = 3 \cdot (7 \cdot 5)$	When we multiply three numbers, it doesn't matter which two we multiply first.
Distributive Property		
$a(b + c) = ab + ac$ $(b + c)a = ab + ac$	$2 \cdot (3 + 5) = 2 \cdot 3 + 2 \cdot 5$ $(3 + 5) \cdot 2 = 2 \cdot 3 + 2 \cdot 5$	When we multiply a number by a sum of two numbers, we get the same result as multiplying the number by each of the terms and then adding the results.

From our experience with numbers we know intuitively that these properties are valid. To reacquaint yourself with these properties, calculate the expressions on both sides of the equal sign in each example in the table. The operations in the parentheses are to be performed first. Thus,

$$(2 + 4) + 7 = 6 + 7 = 13 \qquad \text{and} \qquad 2 + (4 + 7) = 2 + 11 = 13$$

The Distributive Property is crucial because it describes the way addition and multiplication interact with each other.

The Distributive Property applies whenever we multiply a number by a sum. Figure 4 explains why this property works for the case in which all the numbers are positive integers, but the property is true for any real numbers a, b, and c.

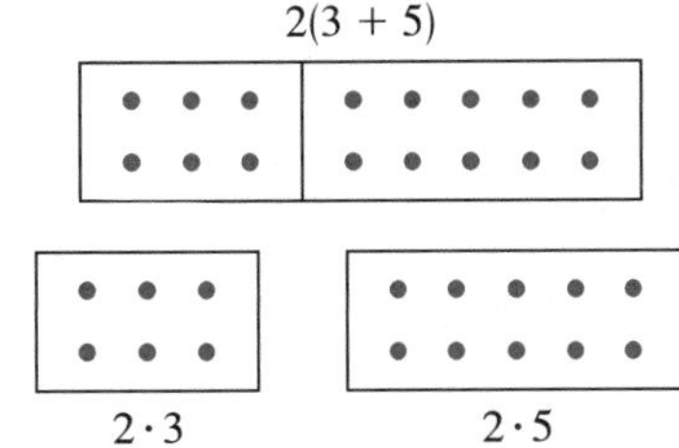

FIGURE 4
The Distributive Property

EXAMPLE 1 ■ Using the Properties of the Real Numbers

Let x, y, z, and w be real numbers.

(a) $(x + y)(2zw) = (2zw)(x + y)$ — Commutative Property for multiplication

(b) $(x + y)(z + w) = (x + y)z + (x + y)w$ — Distributive Property (with $a = x + y$)

$= (zx + zy) + (wx + wy)$ — Distributive Property

$= zx + zy + wx + wy$ — Associative Property of addition

In the last step we removed the parentheses because, according to the Associative Property, the order of addition doesn't matter. ■

⊘ Don't make the mistake of assuming that $-a$ is a negative number. Whether $-a$ is negative or positive depends on the value of a. For example, if $a = 5$, then $-a = -5$, a negative number, but if $a = -5$, then $-a = -(-5) = 5$ (Property 2), a positive number.

The number 0 is special for addition; it is called the **additive identity** because $a + 0 = a$ for any real number a. Every real number a has a **negative**, $-a$, that satisfies $a + (-a) = 0$. **Subtraction** is the operation that undoes addition; to subtract a number from another, we simply add the negative of that number. By definition

$$a - b = a + (-b)$$

To combine real numbers involving negatives, we use the following properties.

PROPERTIES OF NEGATIVES

Property	Example
1. $(-1)a = -a$	$(-1)5 = -5$
2. $-(-a) = a$	$-(-5) = 5$
3. $(-a)b = a(-b) = -(ab)$	$(-5)7 = 5(-7) = -(5 \cdot 7)$
4. $(-a)(-b) = ab$	$(-4)(-3) = 4 \cdot 3$
5. $-(a + b) = -a - b$	$-(3 + 5) = -3 - 5$
6. $-(a - b) = b - a$	$-(5 - 8) = 8 - 5$

Property 6 states the intuitive fact that $a - b$ and $b - a$ are negatives of each other. Property 5 is often used with more than two terms:

$$-(a + b + c) = -a - b - c$$

EXAMPLE 2 ■ Using Properties of Negatives

Let x, y, and z be real numbers.

(a) $-(x + 2) = -x - 2$ — Property 5

(b) $-(x + y - z) = -x - y - (-z)$ — Property 5

$= -x - y + z$ — Property 2 ■

The number 1 is special for multiplication; it is called the **multiplicative identity** because $a \cdot 1 = a$ for any real number a. Every nonzero real number a has an **inverse**, $1/a$, that satisfies $a \cdot (1/a) = 1$. **Division** is the operation that undoes multiplication; to divide by a number, we multiply by the inverse of that number. If $b \neq 0$, then, by definition.

$$a \div b = a \cdot \frac{1}{b}$$

We write $a \cdot (1/b)$ as simply a/b. We refer to a/b as the **quotient** of a and b or as the **fraction** a over b; a is the **numerator** and b is the **denominator** (or **divisor**). To combine real numbers using the operation of division, we use the following properties.

PROPERTIES OF FRACTIONS

Property	Example	Description
1. $\frac{a}{b} \cdot \frac{c}{d} = \frac{ac}{bd}$	$\frac{2}{3} \cdot \frac{5}{7} = \frac{2 \cdot 5}{3 \cdot 7} = \frac{10}{21}$	To multiply fractions, multiply numerators and denominators.
2. $\frac{a}{b} \div \frac{c}{d} = \frac{a}{b} \cdot \frac{d}{c}$	$\frac{2}{3} \div \frac{5}{7} = \frac{2}{3} \cdot \frac{7}{5} = \frac{14}{15}$	To divide fractions, invert the divisor and multiply.
3. $\frac{a}{c} + \frac{b}{c} = \frac{a+b}{c}$	$\frac{2}{5} + \frac{7}{5} = \frac{2+7}{5} = \frac{9}{5}$	To add fractions with the same denominator, add the numerators.
4. $\frac{a}{b} + \frac{c}{d} = \frac{ad+bc}{bd}$	$\frac{2}{5} + \frac{3}{7} = \frac{2 \cdot 7 + 3 \cdot 5}{35} = \frac{29}{35}$	To add fractions with different denominators, find a common denominator. Then add the numerators.
5. $\frac{ac}{bc} = \frac{a}{b}$	$\frac{2 \cdot 5}{3 \cdot 5} = \frac{2}{3}$	Cancel numbers that are common factors in numerator and denominator.
6. If $\frac{a}{b} = \frac{c}{d}$, then $ad = bc$.	$\frac{2}{3} = \frac{6}{9}$, so $2 \cdot 9 = 3 \cdot 6$	Cross multiply.

When adding fractions with different denominators, we don't usually use Property 4. Instead we rewrite the fractions so that they have the smallest possible common denominator (often smaller than the product of the denominators), and then we use Property 3. This denominator is the **L**east **C**ommon **D**enominator (LCD) described in the next example.

EXAMPLE 3 ■ Using the LCD to Add Fractions

Evaluate: $\dfrac{5}{36} + \dfrac{7}{120}$

SOLUTION

Factoring each denominator into prime factors gives

$$36 = 2^2 \cdot 3^2 \qquad \text{and} \qquad 120 = 2^3 \cdot 3 \cdot 5$$

We find the least common denominator (LCD) by forming the product of all the factors that occur in these factorizations, using the highest power of each factor. Thus, the LCD is $2^3 \cdot 3^2 \cdot 5 = 360$. So

$$\begin{aligned} \frac{5}{36} + \frac{7}{120} &= \frac{5 \cdot 10}{36 \cdot 10} + \frac{7 \cdot 3}{120 \cdot 3} && \text{Property 5} \\ &= \frac{50}{360} + \frac{21}{360} \\ &= \frac{71}{360} && \text{Property 3} \end{aligned}$$

SETS AND INTERVALS

In the discussion that follows, we need to use set notation. A **set** is a collection of objects, and these objects are called the **elements** of the set. If S is a set, the notation $a \in S$ means that a is an element of S, and $b \notin S$ means that b is not an element of S. For example, if Z represents the set of integers, then $-3 \in Z$ but $\pi \notin Z$.

Some sets can be described by listing their elements within braces. For instance, the set A that consists of all positive integers less than 7 can be written as

$$A = \{1, 2, 3, 4, 5, 6\}$$

We could also write A in **set-builder notation** as

$$A = \{x \mid x \text{ is an integer and } 0 < x < 7\}$$

which is read "A is the set of all x such that x is an integer and $0 < x < 7$."

If S and T are sets, then their **union** $S \cup T$ is the set that consists of all elements that are in S *or* T (or in both). The **intersection** of S and T is the set $S \cap T$ consisting of all elements that are in both S *and* T. In other words, $S \cap T$ is the common part of S and T. The **empty set**, denoted by $\varnothing$, is the set that contains no element.

EXAMPLE 4 ■ Union and Intersection of Sets

If $S = \{1, 2, 3, 4, 5\}$, $T = \{4, 5, 6, 7\}$, and $V = \{6, 7, 8\}$, find the sets $S \cup T$, $S \cap T$, and $S \cap V$.

SOLUTION

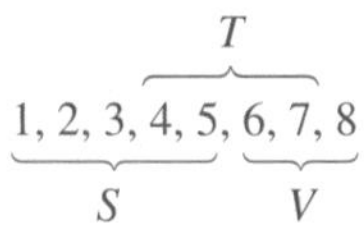

$$\begin{aligned} S \cup T &= \{1, 2, 3, 4, 5, 6, 7\} && \text{All elements in } S \text{ or } T \\ S \cap T &= \{4, 5\} && \text{Elements common to both } S \text{ and } T \\ S \cap V &= \varnothing && S \text{ and } V \text{ have no element in common} \end{aligned}$$

Certain sets of real numbers, called **intervals**, occur frequently in calculus and correspond geometrically to line segments. For example, if $a < b$, then the **open interval** from a to b consists of all numbers between a and b and is denoted by the symbol (a, b). Using set-builder notation, we can write

$$(a, b) = \{x \mid a < x < b\}$$

FIGURE 5
The open interval (a, b)

Note that the endpoints, a and b, are excluded from this interval. This fact is indicated by the parentheses ( ) in the interval notation and the open circles on the graph of the interval in Figure 5.

The **closed interval** from a to b is the set

$$[a, b] = \{x \mid a \leqslant x \leqslant b\}$$

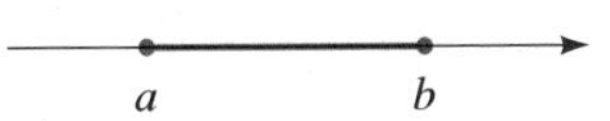

FIGURE 6
The closed interval $[a, b]$

Here the endpoints of the interval are included. This is indicated by the square brackets [ ] in the interval notation and the solid circles on the graph of the interval in Figure 6. It is also possible to include only one endpoint in an interval, as shown in the table of intervals below.

We also need to consider infinite intervals, such as

$$(a, \infty) = \{x \mid x > a\}$$

This does not mean that ∞ ("infinity") is a number. The notation (a, ∞) stands for the set of all numbers that are greater than a, so the symbol ∞ simply indicates that the interval extends indefinitely far in the positive direction.

The following table lists the nine possible types of intervals. When these intervals are discussed, we will always assume that $a < b$.

Notation	Set description	Graph
(a, b)	$\{x \mid a < x < b\}$	a b
$[a, b]$	$\{x \mid a \leqslant x \leqslant b\}$	a b
$[a, b)$	$\{x \mid a \leqslant x < b\}$	a b
$(a, b]$	$\{x \mid a < x \leqslant b\}$	a b
(a, ∞)	$\{x \mid a < x\}$	a
$[a, \infty)$	$\{x \mid a \leqslant x\}$	a
$(-\infty, b)$	$\{x \mid x < b\}$	b
$(-\infty, b]$	$\{x \mid x \leqslant b\}$	b
$(-\infty, \infty)$	$\mathbb{R}$ (set of all real numbers)	

NO SMALLEST OR LARGEST NUMBER IN AN OPEN INTERVAL

Any interval contains infinitely many numbers—every point on the graph of an interval corresponds to a real number. In the closed interval $[0, 1]$, the smallest number is 0 and the largest is 1, but the open interval $(0, 1)$ contains no smallest or largest number. To see this, note that 0.01 is close to zero, but 0.001 is closer, 0.0001 closer yet, and so on. So we can always find a number in the interval $(0, 1)$ closer to zero than any given number. Since 0 itself is not in the interval, the interval contains no smallest number. Similarly, 0.99 is close to 1, but 0.999 is closer, 0.9999 closer yet, and so on. Since 1 itself is not in the interval, the interval has no largest number.

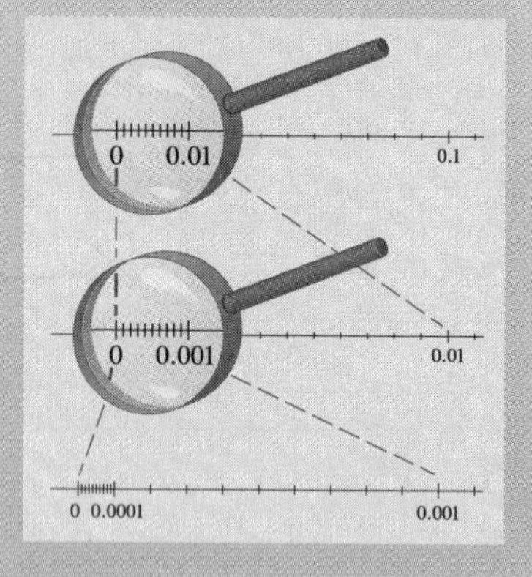

EXAMPLE 5 ■ Graphing Intervals

Express each interval in terms of inequalities, and then graph the interval.

(a) $[-1, 2) = \{x \mid -1 \leq x < 2\}$

(b) $[1.5, 4] = \{x \mid 1.5 \leq x \leq 4\}$

(c) $(-3, \infty) = \{x \mid -3 < x\}$

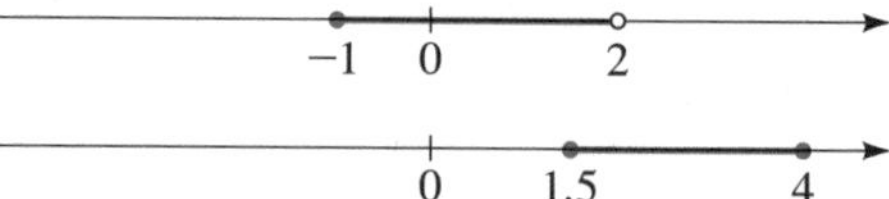

■

EXAMPLE 6 ■ Finding Unions and Intersections of Intervals

Graph each set.

(a) $(1, 3) \cap [2, 7]$ (b) $(1, 3) \cup [2, 7]$

SOLUTION

(a) The intersection of two intervals consists of the numbers that are in both intervals. Therefore

$$\begin{aligned}(1, 3) \cap [2, 7] &= \{x \mid 1 < x < 3 \text{ and } 2 \leq x \leq 7\}\\ &= \{x \mid 2 \leq x < 3\}\\ &= [2, 3)\end{aligned}$$

This set is illustrated in Figure 7.

(b) The union of two intervals consists of the numbers that are in either one interval or the other (or both). Therefore

$$\begin{aligned}(1, 3) \cup [2, 7] &= \{x \mid 1 < x < 3 \text{ or } 2 \leq x \leq 7\}\\ &= \{x \mid 1 < x \leq 7\}\\ &= (1, 7]\end{aligned}$$

This set is illustrated in Figure 8.

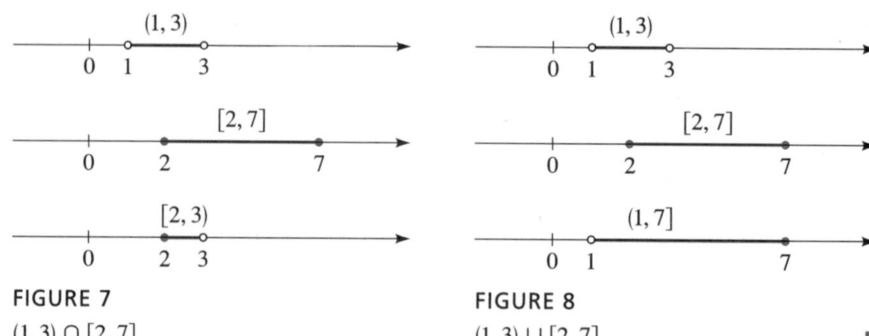

FIGURE 7
$(1, 3) \cap [2, 7]$

FIGURE 8
$(1, 3) \cup [2, 7]$

■

ABSOLUTE VALUE AND DISTANCE

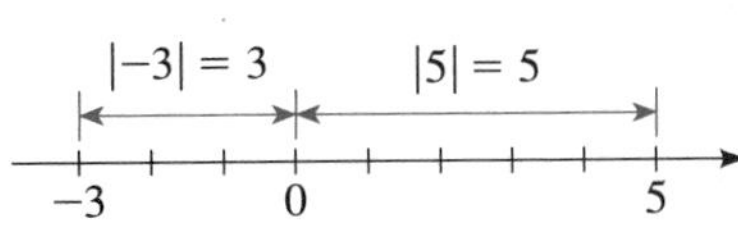

FIGURE 9

The **absolute value** of a number a, denoted by $|a|$, is the distance from a to 0 on the real number line (see Figure 9). Distance is always positive or zero, so we have $|a| \geqslant 0$ for every number a. Remembering that $-a$ is positive when a is negative, we have the following definition.

DEFINITION OF ABSOLUTE VALUE

If a is a real number, then the **absolute value** of a is

$$|a| = \begin{cases} a & \text{if } a \geqslant 0 \\ -a & \text{if } a < 0 \end{cases}$$

EXAMPLE 7 ■ Evaluating Absolute Values of Numbers

(a) $|3| = 3$

(b) $|-3| = -(-3) = 3$

(c) $|0| = 0$

(d) $|\sqrt{2} - 1| = \sqrt{2} - 1 \qquad (\text{since } \sqrt{2} > 1 \Rightarrow \sqrt{2} - 1 > 0)$

(e) $|3 - \pi| = -(3 - \pi) = \pi - 3 \qquad (\text{since } \pi > 3 \Rightarrow 3 - \pi < 0)$ ■

When working with absolute values, we use the following properties.

PROPERTIES OF ABSOLUTE VALUE

Property	Example	Description
1. $\lvert a\rvert \geqslant 0$	$\lvert -3\rvert = 3 \geqslant 0$	The absolute value of a number is always positive or zero.
2. $\lvert a\rvert = \lvert -a\rvert$	$\lvert 5\rvert = \lvert -5\rvert$	A number and its negative have the same absolute value.
3. $\lvert ab\rvert = \lvert a\rvert\lvert b\rvert$	$\lvert -2 \cdot 5\rvert = \lvert -2\rvert\lvert 5\rvert$	The absolute value of a product is the product of the absolute values.
4. $\left\lvert \dfrac{a}{b}\right\rvert = \dfrac{\lvert a\rvert}{\lvert b\rvert}$	$\left\lvert \dfrac{12}{-3}\right\rvert = \dfrac{\lvert 12\rvert}{\lvert -3\rvert}$	The absolute value of a quotient is the quotient of the absolute values.

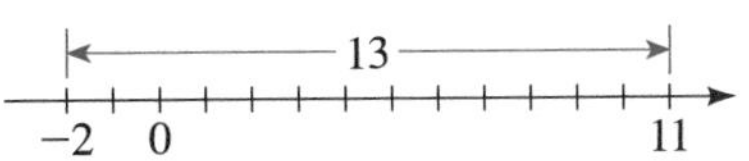

FIGURE 10

What is the distance on the real line between the numbers -2 and 11? From Figure 10 we see that the distance is 13. We arrive at this by finding either $|11 - (-2)| = 13$ or $|(-2) - 11| = 13$. From this observation we make the following definition (see Figure 11).

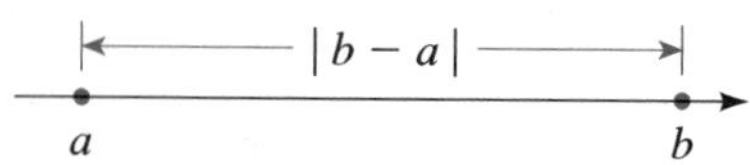

FIGURE 11

Length of a line segment = $|b - a|$

DISTANCE BETWEEN POINTS ON THE REAL LINE

If a and b are real numbers, then the **distance** between the points a and b on the real line is

$$d(a, b) = |b - a|$$

From Property 6 of negatives it follows that $|b - a| = |a - b|$. This confirms that, as we would expect, the distance from a to b is the same as the distance from b to a.

EXAMPLE 8 ■ Distance between Points on the Real Line

The distance between the numbers -8 and 2 is

$$d(a, b) = |-8 - 2| = |-10| = 10$$

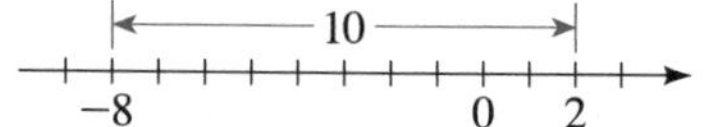

FIGURE 12

We can check this calculation geometrically, as shown in Figure 12. ■

1.2 EXERCISES

1–6 ■ State the property of real numbers being used.

1. $3x + 4y = 4y + 3x$

2. $c(a + b) = (a + b)c$

3. $(x + 2y) + 3z = x + (2y + 3z)$

4. $2(A + B) = 2A + 2B$

5. $(5x + 1)3 = 15x + 3$

6. $(x + a)(x + b) = (x + a)x + (x + a)b$

7–12 ■ Use properties of real numbers to write the expression without parentheses.

7. $3(x + y)$ **8.** $(a - b)8$

9. $4(2m)$ **10.** $\frac{4}{3}(-6y)$

11. $-\frac{5}{2}(2x - 4y)$ **12.** $(3a)(b + c - 2d)$

13–16 ■ Perform the indicated operations.

13. (a) $\dfrac{4}{13} + \dfrac{3}{13}$ (b) $\dfrac{3}{10} + \dfrac{7}{15}$

14. (a) $\dfrac{7}{45} + \dfrac{2}{25}$ (b) $\dfrac{5}{14} - \dfrac{1}{21} + 1$

15. (a) $\dfrac{2}{5} \div \dfrac{9}{10}$ (b) $\left(4 \div \dfrac{1}{2}\right) - \dfrac{1}{2}$

16. (a) $\left(\dfrac{1}{8} - \dfrac{1}{9}\right) \div \dfrac{1}{72}$ (b) $\left(2 \div \dfrac{2}{3}\right) - \left(\dfrac{2}{3} \div 2\right)$

17–20 ■ State whether each inequality is true or false.

17. (a) $-6 < -10$ (b) $\sqrt{2} > 1.41$

18. (a) $\dfrac{10}{11} < \dfrac{12}{13}$ (b) $-\dfrac{1}{2} < -1$

19. (a) $-\pi > -3$ (b) $8 \leq 9$

20. (a) $1.1 > 1.\overline{1}$ (b) $8 \leq 8$

21–22 ■ Write each statement in terms of inequalities.

21. (a) x is positive
(b) t is less than 4

(c) a is greater than or equal to π
(d) x is less than $\frac{1}{3}$ and is greater than -5
(e) The distance from p to 3 is at most 5

22. (a) y is negative
(b) z is greater than 1
(c) b is at most 8
(d) w is positive and is less than or equal to 17
(e) y is at least 2 units from π

23–26 ■ Find the indicated set if

$$A = \{1, 2, 3, 4, 5, 6, 7\} \qquad B = \{2, 4, 6, 8\}$$
$$C = \{7, 8, 9, 10\}$$

23. (a) $A \cup B$ (b) $A \cap B$
24. (a) $B \cup C$ (b) $B \cap C$
25. (a) $A \cup C$ (b) $A \cap C$
26. (a) $A \cup B \cup C$ (b) $A \cap B \cap C$

27–28 ■ Find the indicated set if

$$A = \{x \mid x \geqslant -2\} \qquad B = \{x \mid x < 4\}$$
$$C = \{x \mid -1 < x \leqslant 5\}$$

27. (a) $B \cup C$ (b) $B \cap C$
28. (a) $A \cap C$ (b) $A \cap B$

29–34 ■ Express the interval in terms of inequalities, and then graph the interval.

29. $(-3, 0)$
30. $(2, 8]$
31. $[2, 8)$
32. $[-6, -\frac{1}{2}]$
33. $[2, \infty)$
34. $(-\infty, 1)$

35–40 ■ Express the inequality in interval notation, and then graph the corresponding interval.

35. $x \leqslant 1$
36. $1 \leqslant x \leqslant 2$
37. $-2 < x \leqslant 1$
38. $x \geqslant -5$
39. $x > -1$
40. $-5 < x < 2$

41–46 ■ Graph the set.

41. $(-2, 0) \cup (-1, 1)$
42. $(-2, 0) \cap (-1, 1)$
43. $[-4, 6] \cap [0, 8)$
44. $[-4, 6) \cup [0, 8)$
45. $(-\infty, -4) \cup (4, \infty)$
46. $(-\infty, 6] \cap (2, 10)$

47–52 ■ Evaluate each expression.

47. (a) $|100|$ (b) $|-73|$
48. (a) $|\sqrt{5} - 5|$ (b) $|10 - \pi|$
49. (a) $\big||-6| - |-4|\big|$ (b) $\dfrac{-1}{|-1|}$
50. (a) $\big|2 - |-12|\big|$ (b) $-1 - \big|1 - |-1|\big|$
51. (a) $|(-2) \cdot 6|$ (b) $|(-\frac{1}{3})(-15)|$
52. (a) $\left|\dfrac{-6}{24}\right|$ (b) $\left|\dfrac{7 - 12}{12 - 7}\right|$

53–54 ■ Find the distance between the given numbers.

53. (a) 2 and 17
(b) -3 and 21
(c) $\frac{11}{8}$ and $-\frac{3}{10}$

54. (a) $\frac{7}{15}$ and $-\frac{1}{21}$
(b) -38 and -57
(c) -2.6 and -1.8

55–56 ■ Express each repeating decimal as a fraction. (See the margin note on page 10.)

55. (a) $0.\overline{7}$ (b) $0.\overline{28}$ (c) $0.\overline{57}$
56. (a) $5.\overline{23}$ (b) $1.3\overline{7}$ (c) $2.1\overline{35}$

57. Suppose an automobile's fuel consumption is 28 mi/gal in city driving and 34 mi/gal in highway driving. If x denotes the number of city miles and y the number of highway miles, then the total miles this car can travel on a 15-gallon tank of fuel must satisfy the inequality

$$\tfrac{1}{28}x + \tfrac{1}{34}y \leqslant 15$$

Use this inequality to answer the following questions. (Assume the car has a full tank of fuel.)
(a) Can the car travel 165 city miles and 230 highway miles before running out of gas?
(b) If the car has been driven 280 miles in the city, how many highway miles can it be driven before running out of fuel?

DISCOVERY • DISCUSSION

58. **Sums and Products of Rational and Irrational Numbers** Explain why the sum, the difference, and the product of two rational numbers are rational numbers. Is the product of two irrational numbers necessarily irrational? What about the sum?

59. Limiting Behavior of Reciprocals Complete the tables. What happens to the size of the fraction $1/x$ as x gets large? As x gets small?

x	$1/x$
1	
2	
10	
100	
1000	

x	$1/x$
1.0	
0.5	
0.1	
0.01	
0.001	

60. Irrational Numbers and Geometry Using the following figure, explain how to locate the point $\sqrt{2}$ on a number line. Can you locate $\sqrt{5}$ by a similar method? What about $\sqrt{6}$? List some other irrational numbers that can be located this way.

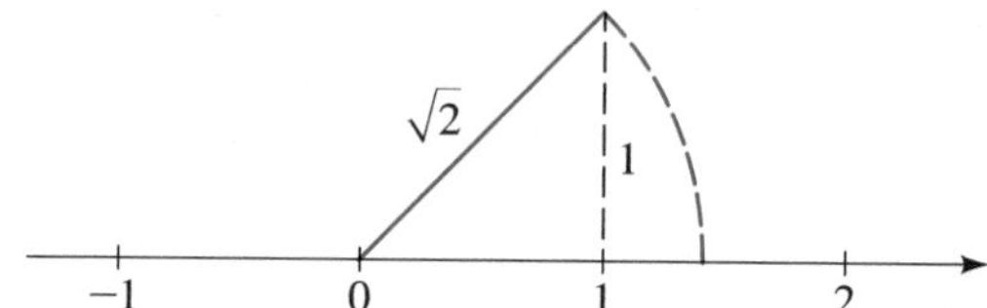

1.3 EXPONENTS AND RADICALS

In this section we give meaning to expressions such as $a^{m/n}$ in which the exponent m/n is a rational number. To do this, we need to recall some facts about integer exponents, radicals, and nth roots.

INTEGER EXPONENTS

A product of identical numbers is usually written in exponential notation. For example, $5 \cdot 5 \cdot 5$ is written as 5^3. In general, we have the following definition.

EXPONENTIAL NOTATION

If a is any real number and n is a positive integer, then the ***n*th power** of a is

$$a^n = \underbrace{a \cdot a \cdot \cdots \cdot a}_{n \text{ factors}}$$

The number a is called the **base** and n is called the **exponent**.

EXAMPLE 1 ■ Exponential Notation

(a) $\left(\frac{1}{2}\right)^5 = \left(\frac{1}{2}\right)\left(\frac{1}{2}\right)\left(\frac{1}{2}\right)\left(\frac{1}{2}\right)\left(\frac{1}{2}\right) = \frac{1}{32}$

(b) $(-3)^4 = (-3) \cdot (-3) \cdot (-3) \cdot (-3) = 81$

(c) $-3^4 = -(3 \cdot 3 \cdot 3 \cdot 3) = -81$ ■

Note the distinction between $(-3)^4$ and -3^4. In $(-3)^4$ the exponent applies to -3, but in -3^4 the exponent applies only to 3.

We can state several useful rules for working with exponential notation. To discover the rule for multiplication, we multiply 5^4 by 5^2:

$$5^4 \cdot 5^2 = \underbrace{(5 \cdot 5 \cdot 5 \cdot 5)}_{4 \text{ factors}}\underbrace{(5 \cdot 5)}_{2 \text{ factors}} = \underbrace{5 \cdot 5 \cdot 5 \cdot 5 \cdot 5 \cdot 5}_{6 \text{ factors}} = 5^6 = 5^{4+2}$$

It appears that *to multiply two powers of the same base, we add their exponents.* In general, we can confirm this by considering any real number a and any positive integers m and n:

$$a^m a^n = \underbrace{(a \cdot a \cdot \cdots \cdot a)}_{m \text{ factors}}\underbrace{(a \cdot a \cdot \cdots \cdot a)}_{n \text{ factors}} = \underbrace{a \cdot a \cdot a \cdot \cdots \cdot a}_{m + n \text{ factors}} = a^{m+n}$$

Thus, we have shown that

$$a^m a^n = a^{m+n}$$

whenever m and n are positive integers.

We would like this rule to be true even when m and n are 0 or negative integers. Thus, for instance, we must have

$$2^0 \cdot 2^3 = 2^{0+3} = 2^3$$

But this can happen only if $2^0 = 1$. Likewise, we want to have

$$5^4 \cdot 5^{-4} = 5^{4+(-4)} = 5^{4-4} = 5^0 = 1$$

and this will be true if $5^{-4} = 1/5^4$. These observations lead to the following definition.

ZERO AND NEGATIVE EXPONENTS

If $a \neq 0$ is any real number and n is a positive integer, then

$$a^0 = 1 \qquad \text{and} \qquad a^{-n} = \frac{1}{a^n}$$

EXAMPLE 2 ■ Zero and Negative Exponents

(a) $\left(\frac{4}{7}\right)^0 = 1$

(b) $x^{-1} = \dfrac{1}{x^1} = \dfrac{1}{x}$

(c) $(-2)^{-3} = \dfrac{1}{(-2)^3} = \dfrac{1}{-8} = -\dfrac{1}{8}$ ■

Familiarity with the following rules is essential for our work with exponents and bases. In the table the bases a and b are real numbers, and the exponents m and n are integers.

LAWS OF EXPONENTS

Law	Description
1. $a^m a^n = a^{m+n}$	To multiply two powers of the same number, add the exponents.
2. $\dfrac{a^m}{a^n} = a^{m-n}$	To divide two powers of the same number, subtract the exponents.
3. $(a^m)^n = a^{mn}$	To raise a power to a new power, multiply the exponents.
4. $(ab)^n = a^n b^n$	To raise a product to a power, raise each factor to the power.
5. $\left(\dfrac{a}{b}\right)^n = \dfrac{a^n}{b^n}$	To raise a quotient to a power, raise both numerator and denominator to the power.

We have seen how to prove Law 1 whenever m and n are positive integers, but it can also be proved for any integer exponents by using the definition of negative exponents. We now give the proofs of Laws 3 and 4. (You are asked to prove Laws 2 and 5 in Exercise 76.)

■ **Proof of Law 3** If m and n are positive integers, we have

$$(a^m)^n = (\underbrace{a \cdot a \cdot \cdots \cdot a}_{m \text{ factors}})^n$$

$$= \underbrace{(\underbrace{a \cdot a \cdot \cdots \cdot a}_{m \text{ factors}})(\underbrace{a \cdot a \cdot \cdots \cdot a}_{m \text{ factors}}) \cdots (\underbrace{a \cdot a \cdot \cdots \cdot a}_{m \text{ factors}})}_{n \text{ groups of factors}}$$

$$= \underbrace{a \cdot a \cdot \cdots \cdot a}_{mn \text{ factors}} = a^{mn}$$

The cases for which $m \leqslant 0$ or $n \leqslant 0$ can be proved using the definition of negative exponents. □

■ **Proof of Law 4** If m and n are positive integers, we have

$$(ab)^n = \underbrace{(ab)(ab) \cdots (ab)}_{n \text{ factors}} = \underbrace{(a \cdot a \cdot \cdots \cdot a)}_{n \text{ factors}} \cdot \underbrace{(b \cdot b \cdot \cdots \cdot b)}_{n \text{ factors}} = a^n b^n$$

Here we have used the Commutative and Associative Properties repeatedly. If

$m \leqslant 0$ or $n \leqslant 0$, Law 4 can be proved using the definition of negative exponents. □

EXAMPLE 3 ■ Using Laws of Exponents

(a) $x^4x^7 = x^{4+7} = x^{11}$ Law 1

(b) $y^4y^{-7} = y^{4-7} = y^{-3} = \dfrac{1}{y^3}$ Law 1

(c) $\dfrac{c^9}{c^5} = c^{9-5} = c^4$ Law 2

(d) $(b^4)^5 = b^{4\cdot 5} = b^{20}$ Law 3

(e) $(3x)^3 = 3^3x^3 = 27x^3$ Law 4

(f) $\left(\dfrac{x}{2}\right)^5 = \dfrac{x^5}{2^5} = \dfrac{x^5}{32}$ Law 5 ■

EXAMPLE 4 ■ Simplifying Expressions with Exponents

Simplify: (a) $(2a^3b^2)(3ab^4)^3$ (b) $\left(\dfrac{x}{y}\right)^3\left(\dfrac{y^2x}{z}\right)^4$

SOLUTION

(a)
$$\begin{aligned}(2a^3b^2)(3ab^4)^3 &= (2a^3b^2)[3^3a^3(b^4)^3] && \text{Law 4}\\ &= (2a^3b^2)(27a^3b^{12}) && \text{Law 3}\\ &= (2)(27)a^3a^3b^2b^{12} && \text{Group factors with the same base}\\ &= 54a^6b^{14} && \text{Law 1}\end{aligned}$$

(b)
$$\begin{aligned}\left(\frac{x}{y}\right)^3\left(\frac{y^2x}{z}\right)^4 &= \frac{x^3}{y^3}\,\frac{(y^2)^4x^4}{z^4} && \text{Laws 5 and 4}\\ &= \frac{x^3}{y^3}\,\frac{y^8x^4}{z^4} && \text{Law 3}\\ &= (x^3x^4)\left(\frac{y^8}{y^3}\right)\frac{1}{z^4} && \text{Group factors with the same base}\\ &= \frac{x^7y^5}{z^4} && \text{Laws 1 and 2}\end{aligned}$$ ■

When simplifying an expression, you will find that many different methods will lead to the same result; you should feel free to use any of the rules of exponents to arrive at your own method. We now give two additional laws that are useful in simplifying expressions with negative exponents.

LAWS OF EXPONENTS

6. $\left(\frac{a}{b}\right)^{-n} = \left(\frac{b}{a}\right)^{n}$	To raise a fraction to a negative power, invert the fraction and change the sign of the exponent.
7. $\frac{a^{-n}}{b^{-m}} = \frac{b^{m}}{a^{n}}$	To move a number raised to a power from numerator to denominator or from denominator to numerator, change the sign of the exponent.

■ **Proof of Law 7** Using the definition of negative exponents and then Property 2 of fractions, we have

$$\frac{a^{-n}}{b^{-m}} = \frac{1/a^{n}}{1/b^{m}} = \frac{1}{a^{n}} \cdot \frac{b^{m}}{1} = \frac{b^{m}}{a^{n}} \qquad \square$$

(You are asked to prove Law 6 in Exercise 76.)

EXAMPLE 5 ■ Simplifying Expressions with Negative Exponents

Eliminate negative exponents and simplify each expression.

(a) $\dfrac{6st^{-4}}{2s^{-2}t^{2}}$ (b) $\left(\dfrac{y}{3z^{2}}\right)^{-2}$

SOLUTION

(a) We use Law 7, which allows us to move a number raised to a power from the numerator to the denominator (or vice versa) by changing the sign of the exponent.

$$\frac{6st^{-4}}{2s^{-2}t^{2}} = \frac{6ss^{2}}{2t^{4}t^{2}} \qquad \text{Law 7}$$

$$= \frac{3s^{3}}{t^{6}} \qquad \text{Law 1}$$

(b) We use Law 6, which allows us to change the sign of the exponent of a fraction by inverting the fraction.

$$\left(\frac{y}{3z^{2}}\right)^{-2} = \left(\frac{3z^{2}}{y}\right)^{2} \qquad \text{Law 6}$$

$$= \frac{9z^{4}}{y^{2}} \qquad \text{Laws 5 and 4}$$

■

SCIENTIFIC NOTATION

Exponential notation is used by scientists as a compact way of writing very large numbers and very small numbers. For example, the nearest star beyond the sun, Proxima Centauri, is approximately 40,000,000,000,000 km away. The mass of a hydrogen atom is about 0.00000000000000000000000166 g. Scientists usually write such numbers in a more convenient way, called *scientific notation*. A positive number x is said to be written in **scientific notation** if it is expressed as follows:

$$x = a \times 10^n \qquad \text{where } 1 \leqslant a < 10 \text{ and } n \text{ is an integer}$$

For instance, when we state that the distance to the star Proxima Centauri is 4×10^{13} km, the positive exponent 13 indicates that the decimal point should be moved 13 places to the *right*:

$$4 \times 10^{13} = \underbrace{40,000,000,000,000}_{13 \text{ places}}$$

When we state that the mass of a hydrogen atom is 1.66×10^{-24} g, the exponent -24 indicates that the decimal point should be moved 24 places to the *left*:

$$1.66 \times 10^{-24} = \underbrace{0.00000000000000000000000166}_{24 \text{ places}}$$

EXAMPLE 6 ■ Writing Numbers in Scientific Notation

(a) $56920 = 5.692 \times 10^4$ (b) $0.000093 = 9.3 \times 10^{-5}$ ■

To use scientific notation on a calculator, use a key labeled EE or EXP or EEX to enter the exponent. For example, to enter the number 3.629×10^{15} on a TI-83 calculator, we enter

3.629 2nd EE 15

and the display reads

3.629E15

Scientific notation is often used on a calculator to display a very large or very small number. For instance, if we use a calculator to square the number 1,111,111, the display panel may show (depending on the calculator model) the approximation

1.234568 12 or 1.234568 E12

Here the final digits indicate the power of 10, and we interpret the result as

$$1.234568 \times 10^{12}$$

EXAMPLE 7 ■ Using Scientific Notation

If $a \approx 0.00046$, $b \approx 1.697 \times 10^{22}$, and $c \approx 2.91 \times 10^{-18}$, use a calculator to approximate the quotient ab/c.

SOLUTION

We could enter the data using scientific notation, or we could use laws of exponents as follows:

$$\frac{ab}{c} \approx \frac{(4.6 \times 10^{-4})(1.697 \times 10^{22})}{2.91 \times 10^{-18}}$$

$$= \frac{(4.6)(1.697)}{2.91} \times 10^{-4+22+18}$$

$$\approx 2.7 \times 10^{36}$$

We state the answer correct to two significant figures because the least accurate of the given numbers is stated to two significant figures. ■

RADICALS

We know what 2^n means whenever n is an integer. To give meaning to a power, such as $2^{4/5}$, whose exponent is a rational number, we need to discuss radicals.

The symbol $\sqrt{}$ means "the positive square root of." Thus

$$\sqrt{a} = b \qquad \text{means} \qquad b^2 = a \quad \text{and} \quad b \geqslant 0$$

It is true that the number 9 has two square roots, 3 and -3, but the notation $\sqrt{9}$ is reserved for the *positive* square root of 9 (sometimes called the *principal square root* of 9). If we want the negative root, we *must* write $-\sqrt{9}$, which is -3.

Since $a = b^2 \geqslant 0$, the symbol $\sqrt{a}$ makes sense only when $a \geqslant 0$. For instance,

$$\sqrt{9} = 3 \qquad \text{because} \qquad 3^2 = 9 \quad \text{and} \quad 3 \geqslant 0$$

Square roots are special cases of nth roots. The nth root of x is the number that, when raised to the nth power, gives x.

DEFINITION OF nth ROOT

If n is any positive integer, then the **principal nth root** of a is defined as follows:

$$\sqrt[n]{a} = b \qquad \text{means} \qquad b^n = a$$

If n is even, we must have $a \geqslant 0$ and $b \geqslant 0$.

Thus

$$\sqrt[4]{81} = 3 \qquad \text{because} \qquad 3^4 = 81 \quad \text{and} \quad 3 \geqslant 0$$

$$\sqrt[3]{-8} = -2 \qquad \text{because} \qquad (-2)^3 = -8$$

But $\sqrt{-8}$, $\sqrt[4]{-8}$, and $\sqrt[6]{-8}$ are not defined. (For instance, $\sqrt{-8}$ is not defined because the square of every real number is nonnegative.)

Notice that

$$\sqrt{4^2} = \sqrt{16} = 4 \qquad \text{but} \qquad \sqrt{(-4)^2} = \sqrt{16} = 4 = |-4|$$

Thus, the equation $\sqrt{a^2} = a$ is not always true; it is true only when $a \geqslant 0$. However, we can always write $\sqrt{a^2} = |a|$. This last equation is true not only for square roots, but for any even root. This and other rules used in working with nth roots are listed in the following box. In each property we assume that all the given roots exist.

PROPERTIES OF nth ROOTS

Property	Example
1. $\sqrt[n]{ab} = \sqrt[n]{a}\,\sqrt[n]{b}$	$\sqrt[3]{-8 \cdot 27} = \sqrt[3]{-8}\,\sqrt[3]{27} = (-2)(3) = -6$
2. $\sqrt[n]{\dfrac{a}{b}} = \dfrac{\sqrt[n]{a}}{\sqrt[n]{b}}$	$\sqrt[4]{\dfrac{16}{81}} = \dfrac{\sqrt[4]{16}}{\sqrt[4]{81}} = \dfrac{2}{3}$
3. $\sqrt[m]{\sqrt[n]{a}} = \sqrt[mn]{a}$	$\sqrt{\sqrt[3]{729}} = \sqrt[6]{729} = 3$
4. $\sqrt[n]{a^n} = a$ if n is odd	$\sqrt[3]{(-5)^3} = -5,\ \sqrt[5]{2^5} = 2$
5. $\sqrt[n]{a^n} = \lvert a \rvert$ if n is even	$\sqrt[4]{(-3)^4} = \lvert -3 \rvert = 3$

EXAMPLE 8 ■ Simplifying Expressions Involving nth Roots

(a) $\sqrt[3]{x^4} = \sqrt[3]{x^3 x}$ Factor out the largest cube

$= \sqrt[3]{x^3}\,\sqrt[3]{x}$ Property 1

$= x\,\sqrt[3]{x}$ Property 4

(b) $\sqrt[4]{81x^8y^4} = \sqrt[4]{81}\,\sqrt[4]{x^8}\,\sqrt[4]{y^4}$ Property 1

$= 3\,\sqrt[4]{(x^2)^4}\,|y|$ Property 5

$= 3x^2\,|y|$ Property 5 ■

It is frequently useful to combine like radicals in an expression such as $2\sqrt{3} + 5\sqrt{3}$. This can be done by using the Distributive Property. Thus

$$2\sqrt{3} + 5\sqrt{3} = (2 + 5)\sqrt{3} = 7\sqrt{3}$$

The next example further illustrates this process.

EXAMPLE 9 ■ Combining Radicals

(a) $10\sqrt[3]{4} + 7\sqrt[3]{4} - 2\sqrt[3]{4} = (10 + 7 - 2)\sqrt[3]{4}$ Distributive Property

$= 15\sqrt[3]{4}$

Avoid making the following error:

$$\sqrt{a+b} \neq \sqrt{a} + \sqrt{b}$$

For instance, if we let $a = 9$ and $b = 16$, then we see the error:

$$\sqrt{9+16} \stackrel{?}{=} \sqrt{9} + \sqrt{16}$$
$$\sqrt{25} \stackrel{?}{=} 3 + 4$$
$$5 \stackrel{?}{=} 7 \quad \text{Wrong!}$$

(b) $\sqrt{32} + \sqrt{200} = \sqrt{16 \cdot 2} + \sqrt{100 \cdot 2}$ Factor out the largest squares

$= \sqrt{16}\sqrt{2} + \sqrt{100}\sqrt{2}$ Property 1

$= 4\sqrt{2} + 10\sqrt{2} = 14\sqrt{2}$ Distributive Property

(c) If $b > 0$, then

$b\sqrt{25b} - \sqrt{b^3} = b\sqrt{25}\sqrt{b} - \sqrt{b^2}\sqrt{b}$ Property 1

$= 5b\sqrt{b} - b\sqrt{b}$ Property 5

$= (5b - b)\sqrt{b}$ Distributive Property

$= 4b\sqrt{b}$ ■

RATIONAL EXPONENTS

To define what is meant by a *rational exponent* or, equivalently, a *fractional exponent* such as $a^{1/3}$, we need to use radicals. In order to give meaning to the symbol $a^{1/n}$ in a way that is consistent with the Laws of Exponents, we would have to have

$$(a^{1/n})^n = a^{(1/n)n} = a^1 = a$$

So, by the definition of nth root,

$$a^{1/n} = \sqrt[n]{a}$$

In general, we define rational exponents as follows.

DEFINITION OF RATIONAL EXPONENTS

For any rational exponent m/n in lowest terms, where m and n are integers and $n > 0$, we define

$$a^{m/n} = (\sqrt[n]{a})^m$$

or equivalently

$$a^{m/n} = \sqrt[n]{a^m}$$

If n is even, then we require that $a \geqslant 0$.

With this definition it can be proved that *the Laws of Exponents also hold for rational exponents.*

EXAMPLE 10 ■ Using the Definition of Rational Exponents

(a) $4^{1/2} = \sqrt{4} = 2$

(b) $64^{1/3} = \sqrt[3]{64} = 4$

(c) $4^{3/2} = (\sqrt{4})^3 = 2^3 = 8$ Alternative solution: $4^{3/2} = \sqrt{4^3} = \sqrt{64} = 8$

(d) $(125)^{-1/3} = \dfrac{1}{125^{1/3}} = \dfrac{1}{\sqrt[3]{125}} = \dfrac{1}{5}$

(e) $\dfrac{1}{\sqrt[3]{x^4}} = \dfrac{1}{x^{4/3}} = x^{-4/3}$ ■

EXAMPLE 11 ■ Using the Laws of Exponents with Rational Exponents

(a) $a^{1/3}a^{7/3} = a^{8/3}$ Law 1

(b) $\dfrac{a^{2/5}a^{7/5}}{a^{3/5}} = a^{2/5+7/5-3/5} = a^{6/5}$ Laws 1 and 2

(c) $(2a^3b^4)^{3/2} = 2^{3/2}(a^3)^{3/2}(b^4)^{3/2}$ Law 4

$= (\sqrt{2})^3 a^{3(3/2)} b^{4(3/2)}$ Law 3

$= 2\sqrt{2}\, a^{9/2} b^6$

(d) $\left(\dfrac{2x^{3/4}}{y^{1/3}}\right)^3 \left(\dfrac{y^4}{x^{-1/2}}\right) = \dfrac{2^3(x^{3/4})^3}{(y^{1/3})^3} \cdot (y^4 x^{1/2})$ Laws 5, 4, and 7

$= \dfrac{8x^{9/4}}{y} \cdot y^4 x^{1/2}$ Law 3

$= 8x^{11/4}y^3$ Laws 1 and 2 ■

EXAMPLE 12 ■ Simplifying by Writing Radicals as Rational Exponents

(a) $(2\sqrt{x})(3\sqrt[3]{x}) = (2x^{1/2})(3x^{1/3})$ Definition of rational exponents

$= 6x^{1/2+1/3} = 6x^{5/6}$ Law 1

(b) $\sqrt{x\sqrt{x}} = (xx^{1/2})^{1/2}$ Definition of rational exponents

$= (x^{3/2})^{1/2}$ Law 1

$= x^{3/4}$ Law 3 ■

It is often useful to eliminate the radical in a denominator by multiplying both numerator and denominator by an appropriate expression. This procedure is called **rationalizing the denominator**. If the denominator is of the form $\sqrt{a}$, we multiply numerator and denominator by $\sqrt{a}$. In doing so we multiply the given quantity by 1, so we do not change its value. For instance,

$$\frac{1}{\sqrt{a}} = \frac{1}{\sqrt{a}} \cdot 1 = \frac{1}{\sqrt{a}} \cdot \frac{\sqrt{a}}{\sqrt{a}} = \frac{\sqrt{a}}{a}$$

Note that the denominator in the last fraction contains no radical. In general, if the denominator is of the form $\sqrt[n]{a^m}$ with $m < n$, then multiplying numerator and denominator by $\sqrt[n]{a^{n-m}}$ will rationalize the denominator, because (for $a > 0$)

$$\sqrt[n]{a^m}\,\sqrt[n]{a^{n-m}} = \sqrt[n]{a^{m+n-m}} = \sqrt[n]{a^n} = a$$

EXAMPLE 13 ■ Rationalizing Denominators

(a) $\dfrac{2}{\sqrt{3}} = \dfrac{2}{\sqrt{3}} \cdot \dfrac{\sqrt{3}}{\sqrt{3}} = \dfrac{2\sqrt{3}}{3}$

(b) $\dfrac{1}{\sqrt[3]{x^2}} = \dfrac{1}{\sqrt[3]{x^2}}\,\dfrac{\sqrt[3]{x}}{\sqrt[3]{x}} = \dfrac{\sqrt[3]{x}}{\sqrt[3]{x^3}} = \dfrac{\sqrt[3]{x}}{x}$

(c) $\sqrt[7]{\dfrac{1}{a^2}} = \dfrac{1}{\sqrt[7]{a^2}} = \dfrac{1}{\sqrt[7]{a^2}}\,\dfrac{\sqrt[7]{a^5}}{\sqrt[7]{a^5}} = \dfrac{\sqrt[7]{a^5}}{\sqrt[7]{a^7}} = \dfrac{\sqrt[7]{a^5}}{a}$ ■

1.3 EXERCISES

1–10 ■ Evaluate each number.

1. (a) $(-2)^4$ (b) -2^4 (c) $(-2)^0$

2. (a) $\left(\frac{1}{2}\right)^4 4^{-2}$ (b) $\left(\frac{1}{4}\right)^{-2}$ (c) $\left(\frac{1}{4}\right)^0 2^{-1}$

3. (a) $2^4 5^{-2}$ (b) $\dfrac{10^7}{10^4}$ (c) $(2^3 \cdot 2^2)^2$

4. (a) $\sqrt{64}$ (b) $\sqrt[3]{-64}$ (c) $\sqrt[5]{-32}$

5. (a) $\sqrt{\frac{4}{9}}$ (b) $\sqrt[4]{256}$ (c) $\sqrt[6]{\frac{1}{64}}$

6. (a) $\sqrt{7}\,\sqrt{28}$ (b) $\sqrt[3]{3}\,\sqrt[3]{9}$ (c) $\sqrt[4]{24}\,\sqrt[4]{54}$

7. (a) $\dfrac{\sqrt{72}}{\sqrt{2}}$ (b) $\dfrac{\sqrt{48}}{\sqrt{3}}$ (c) $\sqrt{\frac{9}{25}}$

8. (a) $9^{7/2}$ (b) $(-32)^{2/5}$ (c) $(-125)^{-1/3}$

9. (a) $\left(\frac{4}{9}\right)^{-1/2}$ (b) $\left(-\frac{27}{8}\right)^{2/3}$ (c) $\left(\frac{25}{64}\right)^{3/2}$

10. (a) $1024^{-0.1}$ (b) $3^{2/7}3^{5/7}$ (c) $3^{1/2}9^{1/4}$

11–14 ■ Simplify the expression.

11. $\sqrt[3]{108} - \sqrt[3]{32}$

12. $\sqrt{8} + \sqrt{50}$

13. $\sqrt{245} - \sqrt{125}$

14. $\sqrt[3]{54} - \sqrt[3]{16}$

15–32 ■ Simplify the expression and eliminate any negative exponent(s).

15. a^9a^{-5}

16. $(3y^2)(4y^5)$

17. $(12x^2y^4)\left(\frac{1}{2}x^5y\right)$

18. $(6y)^3$

19. $\dfrac{x^9(2x)^4}{x^3}$

20. $\dfrac{a^{-3}b^4}{a^{-5}b^5}$

21. $b^4\left(\frac{1}{3}b^2\right)(12b^{-8})$

22. $(2s^3t^{-1})\left(\frac{1}{4}s^6\right)(16t^4)$

23. $(rs)^3(2s)^{-2}(4r)^4$

24. $(2u^2v^3)^3(3u^3v)^{-2}$

25. $\dfrac{(6y^3)^4}{2y^5}$

26. $\dfrac{(2x^3)^2(3x^4)}{(x^3)^4}$

27. $\dfrac{(x^2y^3)^4(xy^4)^{-3}}{x^2y}$

28. $\left(\dfrac{c^4d^3}{cd^2}\right)\left(\dfrac{d^2}{c^3}\right)^3$

29. $\dfrac{(xy^2z^3)^4}{(x^3y^2z)^3}$

30. $\left(\dfrac{xy^{-2}z^{-3}}{x^2y^3z^{-4}}\right)^{-3}$

31. $\left(\dfrac{q^{-1}rs^{-2}}{r^{-5}sq^{-8}}\right)^{-1}$

32. $(3ab^2c)\left(\dfrac{2a^2b}{c^3}\right)^{-2}$

33–48 ■ Simplify the expression and eliminate any negative exponent(s). Assume that all letters denote positive numbers.

33. $x^{2/3}x^{1/5}$

34. $(-2a^{3/4})(5a^{3/2})$

35. $(4b)^{1/2}(8b^{2/5})$

36. $(8x^6)^{-2/3}$

37. $(c^2d^3)^{-1/3}$

38. $(4x^6y^8)^{3/2}$

39. $(y^{3/4})^{2/3}$

40. $(a^{2/5})^{-3/4}$

41. $(2x^4y^{-4/5})^3(8y^2)^{2/3}$

42. $(x^{-5}y^3z^{10})^{-3/5}$

43. $\left(\dfrac{x^6y}{y^4}\right)^{5/2}$

44. $\left(\dfrac{-2x^{1/3}}{y^{1/2}z^{1/6}}\right)^4$

45. $\left(\dfrac{3a^{-2}}{4b^{-1/3}}\right)^{-1}$

46. $\dfrac{(y^{10}z^{-5})^{1/5}}{(y^{-2}z^{3})^{1/3}}$

47. $\dfrac{(9st)^{3/2}}{(27s^3t^{-4})^{2/3}}$

48. $\left(\dfrac{a^2b^{-3}}{x^{-1}y^2}\right)^3\left(\dfrac{x^{-2}b^{-1}}{a^{3/2}y^{1/3}}\right)$

49–56 ■ Simplify the expression. Assume the letters denote any real numbers.

49. $\sqrt[4]{x^4}$

50. $\sqrt[3]{x^3y^6}$

51. $\sqrt[3]{x^3y}$

52. $\sqrt{x^4y^4}$

53. $\sqrt[5]{a^6b^7}$

54. $\sqrt[3]{a^2b}\ \sqrt[3]{a^4b}$

55. $\sqrt[3]{\sqrt{64x^6}}$

56. $\sqrt[4]{x^4y^2z^2}$

57–60 ■ Rationalize the denominator.

57. (a) $\dfrac{1}{\sqrt{6}}$ (b) $\sqrt{\dfrac{x}{3y}}$ (c) $\sqrt{\dfrac{3}{20}}$

58. (a) $\sqrt{\dfrac{x^5}{2}}$ (b) $\sqrt{\dfrac{2}{3}}$ (c) $\sqrt{\dfrac{1}{2x^3y^5}}$

59. (a) $\dfrac{1}{\sqrt[3]{x}}$ (b) $\dfrac{1}{\sqrt[5]{x^2}}$ (c) $\dfrac{1}{\sqrt[7]{x^3}}$

60. (a) $\dfrac{1}{\sqrt[3]{x^2}}$ (b) $\dfrac{1}{\sqrt[4]{x^3}}$ (c) $\dfrac{1}{\sqrt[3]{x^4}}$

61–62 ■ Write each number in scientific notation.

61. (a) 69,300,000
(b) 0.000028536
(c) 129,540,000

62. (a) 7,259,000,000
(b) 0.0000000014
(c) 0.0007029

63–64 ■ Write each number in ordinary decimal notation.

63. (a) 3.19×10^5
(b) 2.670×10^{-8}
(c) 7.1×10^{14}

64. (a) 8.55×10^{-3}
(b) 6×10^{12}
(c) 6.257×10^{-10}

65–66 ■ Write the number indicated in each statement in scientific notation.

65. (a) A light-year, the distance that light travels in one year, is about 5,900,000,000,000 mi.
(b) The diameter of an electron is about 0.0000000000004 cm.
(c) A drop of water contains more than 33 billion billion molecules.

66. (a) The distance from the earth to the sun is about 93 million miles.
(b) The mass of an oxygen molecule is about 0.000000000000000000000053 g.
(c) The mass of the earth is about 5,970,000,000,000,000,000,000,000 kg.

67–72 ■ Use scientific notation, the Laws of Exponents, and a calculator to perform the indicated operations. State your answer correct to the number of significant digits indicated by the given data.

67. $(7.2 \times 10^{-9})(1.806 \times 10^{-12})$

68. $(1.062 \times 10^{24})(8.61 \times 10^{19})$

69. $\dfrac{1.295643 \times 10^9}{(3.610 \times 10^{-17})(2.511 \times 10^6)}$

70. $\dfrac{(73.1)(1.6341 \times 10^{28})}{0.0000000019}$

71. $\dfrac{(0.0000162)(0.01582)}{(594621000)(0.0058)}$

72. $\dfrac{(3.542 \times 10^{-6})^9}{(5.05 \times 10^4)^{12}}$

73. The speed of light is about 186,000 mi/s. Use the information in Exercise 66(a) to find how long it takes for a light ray from the sun to reach the earth.

74. Police use the formula $s = \sqrt{30fd}$ to estimate the speed s (in mi/h) at which a car is traveling if it skids d feet after the brakes are applied suddenly. The number f is the coefficient of friction of the road, which is a measure of the "slipperiness" of the road. The table gives some typical estimates for f.

	Tar	Concrete	Gravel
Dry	1.0	0.8	0.2
Wet	0.5	0.4	0.1

(a) If a car skids 65 ft on wet concrete, how fast was it moving when the brakes were applied?
(b) If a car is traveling at 50 mi/h, how far will it skid on wet tar?

75. Due to the curvature of the earth, the maximum distance D that you can see from the top of a tall building of height h is estimated by the formula

$$D = \sqrt{2rh + h^2}$$

where $r = 3960$ mi is the radius of the earth and D and h are also measured in miles. How far can you see from the observation deck of the Toronto CN Tower, 1135 ft above the ground?

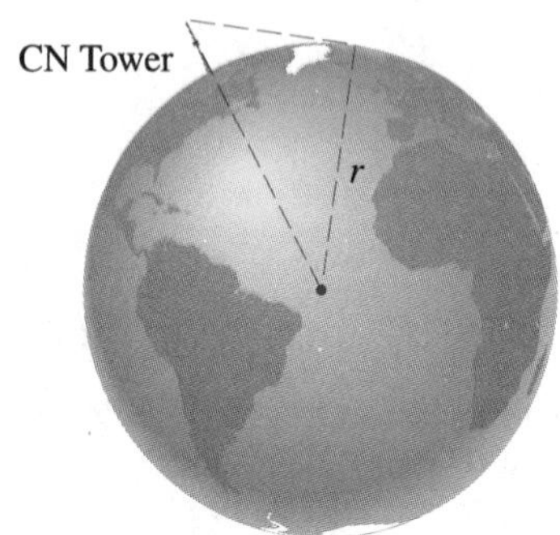

76. Prove the given Laws of Exponents for the case in which m and n are positive integers and $m > n$.

(a) Law 2 (b) Law 5 (c) Law 6

DISCOVERY • DISCUSSION

77. Limiting Behavior of Powers Complete the following tables. What happens to the nth root of 2 as n gets large? What about the nth root of $\frac{1}{2}$?

n	$2^{1/n}$
1	
2	
5	
10	
100	

n	$\left(\frac{1}{2}\right)^{1/n}$
1	
2	
5	
10	
100	

Construct a similar table for $n^{1/n}$. What happens to the nth root of n as n gets large?

78. Comparing Roots Without using a calculator, determine which number is larger in each pair.
(a) $2^{1/2}$ or $2^{1/3}$
(b) $\left(\frac{1}{2}\right)^{1/2}$ or $\left(\frac{1}{2}\right)^{1/3}$
(c) $7^{1/4}$ or $4^{1/3}$
(d) $\sqrt[3]{5}$ or $\sqrt{3}$

79. Easy Powers That Look Hard Calculate these expressions in your head. Use the Laws of Exponents to help you.
(a) $\dfrac{18^5}{9^5}$ (b) $20^6 \cdot (0.5)^6$

80. Distances between Powers Which pair of numbers is closer together?

$$10^{10} \text{ and } 10^{50} \qquad \text{or} \qquad 10^{100} \text{ and } 10^{101}$$

81. Relativity The theory of relativity states that as an object travels with velocity v, its rest mass m_0 changes to a mass m given by the formula

$$m = \frac{m_0}{\sqrt{1 - \dfrac{v^2}{c^2}}}$$

where $c \approx 3.0 \times 10^8$ m/s is the speed of light. By what factor is the rest mass of a spaceship multiplied if the ship travels at one-tenth the speed of light? At one-half the speed of light? At 90% of the speed of light? How does the mass of the spaceship change as it travels at a speed very close to the speed of light? Do we need to know the actual value of the speed of light to answer these questions?

1.4 ALGEBRAIC EXPRESSIONS

Algebraic expressions such as

$$2x^2 - 3x + 4 \qquad ax + b$$

$$\frac{y-1}{y^2+2} \qquad \frac{cx^2y + dy^2z}{\sqrt{x^2+y^2+z^2}}$$

are obtained by starting with variables such as x, y, and z and constants such as 2, -3, a, b, c, and d, and combining them using addition, subtraction, multiplication, division, and roots. A **variable** is a letter that can represent any number in a given set of numbers, whereas a **constant** represents a fixed (or specific) number. The **domain** of a variable is the set of numbers that the variable is permitted to have. For instance, in the expression $\sqrt{x}$ the domain of x is $\{x \mid x \geqslant 0\}$, whereas in the expression $2/(x - 3)$ the domain of x is $\{x \mid x \neq 3\}$.

The simplest types of algebraic expressions use only addition, subtraction, and multiplication. Such expressions are called **polynomials**. The general form of a polynomial of degree n (where n is a nonnegative integer) in the variable x is

$$a_n x^n + a_{n-1}x^{n-1} + \cdots + a_1 x + a_0$$

where $a_0, a_1 \ldots, a_n$ are constants and $a_n \neq 0$. The **degree** of a polynomial is the highest power of the variable. Any polynomial is a sum of **terms** of the form ax^k, called **monomials**, where a is a constant and k is a nonnegative integer. A **binomial** is a sum of two monomials, a **trinomial** is the sum of three monomials, and so on. Thus, $2x^2 - 3x + 4$, $ax + b$, and $x^4 + 2x^3$ are polynomials of degree 2, 1, and 4, respectively; the first is a trinomial, the other two are binomials.

We **add** and **subtract** polynomials using the properties of real numbers that were discussed in Section 1.2. The idea is to combine **like terms** (that is, terms with the same variables raised to the same powers) using the Distributive Property. For instance,

$ac + bc = (a + b)c$

$$5x^7 + 3x^7 = (5 + 3)x^7 = 8x^7$$

In subtracting polynomials we have to remember that if a minus sign precedes an expression in parentheses, then the sign of every term within the parentheses is changed when we remove the parentheses:

$$-(b + c) = -b - c$$

[This is simply a case of the Distributive Property, $a(b + c) = ab + ac$, with $a = -1$.]

EXAMPLE 1 ■ Adding and Subtracting Polynomials

(a) Find the sum $(x^3 - 6x^2 + 2x + 4) + (x^3 + 5x^2 - 7x)$.

(b) Find the difference $(x^3 - 6x^2 + 2x + 4) - (x^3 + 5x^2 - 7x)$.

SOLUTION

(a) $(x^3 - 6x^2 + 2x + 4) + (x^3 + 5x^2 - 7x)$

$= (x^3 + x^3) + (-6x^2 + 5x^2) + (2x - 7x) + 4$ Group like terms

$= 2x^3 - x^2 - 5x + 4$ Combine like terms

(b) $(x^3 - 6x^2 + 2x + 4) - (x^3 + 5x^2 - 7x)$

$= x^3 - 6x^2 + 2x + 4 - x^3 - 5x^2 + 7x$ Distributive Property

$= (x^3 - x^3) + (-6x^2 - 5x^2) + (2x + 7x) + 4$ Group like terms

$= -11x^2 + 9x + 4$ Combine like terms ■

To find the **product** of polynomials or other algebraic expressions, we need to use the Distributive Property repeatedly. In particular, using it three times on the product of two binomials, we get

$$(a + b)(c + d) = a(c + d) + b(c + d) = ac + ad + bc + bd$$

This says that we multiply the two factors by multiplying each term in one factor by each term in the other factor and adding these products. Schematically we have

The acronym **FOIL** helps us remember that the product of two binomials is the sum of the products of the **F**irst terms, the **O**uter terms, the **I**nner terms, and the **L**ast terms.

$$(a + b)(c + d) = \underset{\text{F}}{ac} + \underset{\text{O}}{ad} + \underset{\text{I}}{bc} + \underset{\text{L}}{bd}$$

In general, we can multiply any two polynomials by using the Distributive Property and the Laws of Exponents.

EXAMPLE 2 ■ Multiplying Polynomials

(a) $(2x + 1)(3x - 5) = 6x^2 - 10x + 3x - 5$ Distributive Property

$= 6x^2 - 7x - 5$ Combine like terms

(b) $(x^2 - 3)(x^3 + 2x + 1) = x^2(x^3 + 2x + 1) - 3(x^3 + 2x + 1)$ Distributive Property

$= x^5 + 2x^3 + x^2 - 3x^3 - 6x - 3$ Distributive Property

$= x^5 - x^3 + x^2 - 6x - 3$ Combine like terms ■

The next example shows that the methods we have used for multiplying polynomials also apply to other algebraic expressions.

EXAMPLE 3 ■ Multiplying Algebraic Expressions

(a) $\sqrt{x}\,(x^2 + 2x + \sqrt{x}) = x^2\sqrt{x} + 2x\sqrt{x} + \sqrt{x}\,\sqrt{x}$ Distributive Property

$= x^{5/2} + 2x^{3/2} + x$ Laws of Exponents

(b) $(1 + \sqrt{x})(2 - 3\sqrt{x}) = 2 - 3\sqrt{x} + 2\sqrt{x} - 3(\sqrt{x})^2$ Distributive Property

$= 2 - \sqrt{x} - 3x$ Combine like terms ■

We can also consider polynomials in two or more variables. For instance,

$$4x^2y^3 - 2xy + 6$$

is a polynomial in the variables x and y. The techniques for combining such polynomials are similar to those for polynomials in one variable.

EXAMPLE 4 ■ Multiplying Polynomials with More Than One Variable

Find the product $(x^2 - xy + y^2)(x - y)$.

SOLUTION

$$\begin{aligned}(x^2 - xy + y^2)(x - y) &= (x^2 - xy + y^2)x - (x^2 - xy + y^2)y && \text{Distributive Property}\\ &= x^3 - x^2y + xy^2 - x^2y + xy^2 - y^3 && \text{Distributive Property}\\ &= x^3 - 2x^2y + 2xy^2 - y^3 && \text{Combine like terms}\end{aligned}$$

■

Certain types of products occur so frequently that you should memorize them. You can verify the following formulas by performing the multiplications.

See the Discovery Project on page 43 for a geometric interpretation of some of these formulas.

SPECIAL PRODUCT FORMULAS

1. $(A - B)(A + B) = A^2 - B^2$
2. $(A + B)^2 = A^2 + 2AB + B^2$
3. $(A - B)^2 = A^2 - 2AB + B^2$
4. $(A + B)^3 = A^3 + 3A^2B + 3AB^2 + B^3$
5. $(A - B)^3 = A^3 - 3A^2B + 3AB^2 - B^3$

The key idea in using these formulas (or any other formula in algebra) is the **Principle of Substitution**: We may substitute any algebraic expression for any letter in a formula. For example, to find $(x^2 + y^3)^2$ we use Product Formula 2:

$$(A + B)^2 = A^2 + 2AB + B^2$$

We substitute x^2 for A and y^2 for B to get

$$(x^2 + y^3)^2 = (x^2)^2 + 2(x^2)(y^3) + (y^3)^2$$

This type of substitution is valid because every algebraic expression (in this case, x^2 or y^3) represents a number.

EXAMPLE 5 ■ Using the Special Product Formulas

Use the Special Product Formulas to find each product.

(a) $(3x + 5)^2$ (b) $(2x - \sqrt{y})(2x + \sqrt{y})$ (c) $(x^2 - 2)^3$

SOLUTION

(a) Product Formula 2, with $A = 3x$ and $B = 5$, gives

$$(3x + 5)^2 = (3x)^2 + 2(3x)(5) + 5^2 = 9x^2 + 30x + 25$$

(b) Using Product Formula 1 with $A = 2x$ and $B = \sqrt{y}$, we have

$$(2x - \sqrt{y})(2x + \sqrt{y}) = (2x)^2 - (\sqrt{y})^2 = 4x^2 - y$$

(c) Substituting $A = x^2$ and $B = 2$ in Product Formula 5, we get

$$\begin{aligned}(x^2 - 2)^3 &= (x^2)^3 - 3(x^2)^2(2) + 3(x^2)(2)^2 - 2^3 \\ &= x^6 - 6x^4 + 12x^2 - 8\end{aligned}$$

■

FACTORING

We used the Distributive Property to expand algebraic expressions. We sometimes need to reverse this process (again using the Distributive Property) by **factoring** an expression as a product of simpler ones. For example, we can write

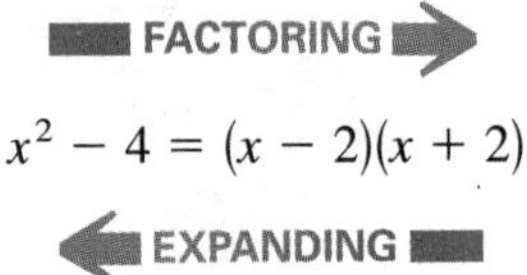

$$x^2 - 4 = (x - 2)(x + 2)$$

We say that $x - 2$ and $x + 2$ are **factors** of $x^2 - 4$. The easiest type of factoring occurs when the terms have a common factor.

EXAMPLE 6 ■ Factoring Out Common Factors

Factor each expression.

(a) $3x^2 - 6x$ (b) $8x^4y^2 + 6x^3y^3 - 2xy^4$

CHECK YOUR ANSWER

Multiplying gives

$$3x(x - 2) = 3x^2 - 6x \quad ✓$$

SOLUTION

(a) The greatest common factor of the terms $3x^2$ and $-6x$ is $3x$, so we have

$$3x^2 - 6x = 3x(x - 2)$$

(b) We note that

8, 6, and -2 have the greatest common factor 2

x^4, x^3, and x have the greatest common factor x

y^2, y^3, and y^4 have the greatest common factor y^2

CHECK YOUR ANSWER

Multiplying gives

$2xy^2(4x^3 + 3x^2y - y^2) = 8x^4y^2 + 6x^3y^3 - 2xy^4$ ✓

So the greatest common factor of the three terms in the polynomial is $2xy^2$, and we have

$$8x^4y^2 + 6x^3y^3 - 2xy^4 = (2xy^2)(4x^3) + (2xy^2)(3x^2y) + (2xy^2)(-y^2)$$
$$= 2xy^2(4x^3 + 3x^2y - y^2)$$ ■

To factor a second-degree polynomial, or **quadratic**, of the form $x^2 + bx + c$, we note that

$$(x + r)(x + s) = x^2 + (r + s)x + rs$$

so we need to choose numbers r and s so that $r + s = b$ and $rs = c$.

EXAMPLE 7 ■ Factoring $x^2 + bx + c$ by Trial and Error

Factor: $x^2 + 7x + 12$

SOLUTION

In this case, $rs = 12$, so r and s must be factors of 12 and their sum must be 7. We enumerate the trial factors of 12:

r	1	2	3
s	12	6	4
Sum	13	8	7

CHECK YOUR ANSWER

Multiplying gives

$(x + 3)(x + 4) = x^2 + 7x + 12$ ✓

Thus, $r = 3$ and $s = 4$ are the factors of 12 whose sum is 7. The factorization is

$$x^2 + 7x + 12 = (x + 3)(x + 4)$$ ■

To factor a quadratic of the form $ax^2 + bx + c$ with $a \neq 1$, we look for factors of the form $px + r$ and $qx + s$:

$$ax^2 + bx + c = (px + r)(qx + s) = pqx^2 + (ps + qr)x + rs$$

factors of a

$$ax^2 + bx + c = (\underset{\uparrow}{\overset{\downarrow}{p}}x + r)(\overset{\downarrow}{q}x + \underset{\uparrow}{s})$$

factors of c

Therefore, we try to find numbers p, q, r, and s such that

$$pq = a \qquad rs = c \qquad ps + qr = b$$

If these numbers are all integers, then we will have a limited number of possibilities to try for p, q, r, and s.

EXAMPLE 8 ■ Factoring $ax^2 + bx + c$ by Trial and Error

Factor: $6x^2 + 7x - 5$

SOLUTION

We can factor 6 as $6 \cdot 1$ or $3 \cdot 2$, and -5 as $-5 \cdot 1$ or $5 \cdot (-1)$. By trying these possibilities, we arrive at the factorization

$$6x^2 + 7x - 5 = (3x + 5)(2x - 1)$$ ■

CHECK YOUR ANSWER

Multiplying gives

$$(3x + 5)(2x - 1) = 6x^2 + 7x - 5$$ ✓

Some special algebraic expressions can be factored using the following formulas. The first three are simply the Special Product Formulas written backward.

FACTORING FORMULAS

Formula	Name
1. $A^2 - B^2 = (A - B)(A + B)$	Difference of squares
2. $A^2 + 2AB + B^2 = (A + B)^2$	Perfect square
3. $A^2 - 2AB + B^2 = (A - B)^2$	Perfect square
4. $A^3 - B^3 = (A - B)(A^2 + AB + B^2)$	Difference of cubes
5. $A^3 + B^3 = (A + B)(A^2 - AB + B^2)$	Sum of cubes

EXAMPLE 9 ■ Factoring Differences of Squares

Factor each polynomial.

(a) $4x^2 - 25$ (b) $9x^4 - y^6$ (c) $(x + y)^2 - x^2$

SOLUTION

(a) Using the formula for a difference of squares with $A = 2x$ and $B = 5$, we have

$$4x^2 - 25 = (2x)^2 - 5^2$$
$$= (2x - 5)(2x + 5)$$

(b) We recognize that $9x^4 = (3x^2)^2$ and $y^6 = (y^3)^2$ and so the expression is a difference of squares. We may use the formula for the difference of squares with $A = 3x^2$ and $B = y^3$. We have

$$9x^4 - y^6 = (3x^2)^2 - (y^3)^2$$
$$= (3x^2 - y^3)(3x^2 + y^3)$$

MATHEMATICS IN THE MODERN WORLD

UNBREAKABLE CODES

If you read spy novels, you know about secret codes, and how the hero "breaks" the code. Today secret codes have a much more common use. Most of the information stored on computers is coded to prevent unauthorized use. For example, your banking records, medical records, and school records are coded. Many cellular and cordless phones code the signal carrying your voice so no one can listen in. Fortunately, because of recent advances in mathematics, today's codes are "unbreakable."

Modern codes are based on a simple principle: Factoring is much harder than multiplying. For example, try multiplying 78 and 93; now try factoring 9991. It takes a long time to factor 9991 because it is a product of two primes 97×103, so to factor it we had to find one of these primes. Now imagine trying to factor a number N that is the

(continued)

product of two primes p and q, each about 200 digits long. Even the fastest computers would take many millions of years to factor such a number! But the same computer would take less than a second to multiply two such numbers. This fact was used by Ted Rivset, Adi Shamir, and Leonard Adelman in the 1970s to devise the RSA code. Their code uses an extremely large number to encode a message but requires us to know its factors to decode it. As you can see, such a code is practically unbreakable.

The RSA code is an example of a "public key encryption" code. In such codes, anyone can code a message using a publicly known procedure based on N, but to decode the message they must know p and q, the factors of N. When the RSA code was developed, it was thought that a carefully selected 80-digit number would provide an unbreakable code. But interestingly, recent advances in the study of factoring have made much larger numbers necessary.

(c) We use the formula for the difference of squares with $A = x + y$ and $B = x$.

$$\begin{aligned}(x + y)^2 - x^2 &= [(x + y) - x][(x + y) + x] \\ &= y(2x + y)\end{aligned}$$

■

EXAMPLE 10 ■ Factoring Perfect Squares

Factor each trinomial: (a) $x^2 + 6x + 9$ (b) $4x^2 - 4xy + y^2$

SOLUTION

(a) Using Formula 2 with $A = x$ and $B = 3$, we get

$$x^2 + 6x + 9 = x^2 + 2(3x) + 3^2 = (x + 3)^2$$

(b) Here we use $A = 2x$ and $B = y$ in Formula 3.

$$\begin{aligned}4x^2 - 4xy + y^2 &= (2x)^2 - 2(2x)y + y^2 \\ &= (2x - y)^2\end{aligned}$$

■

EXAMPLE 11 ■ Factoring Differences and Sums of Cubes

Factor each polynomial: (a) $27x^3 - 1$ (b) $x^6 + 8$

SOLUTION

(a) Using the formula for a difference of cubes with $A = 3x$ and $B = 1$, we get

$$\begin{aligned}27x^3 - 1 &= (3x)^3 - 1^3 = (3x - 1)[(3x)^2 + (3x)(1) + 1^2] \\ &= (3x - 1)(9x^2 + 3x + 1)\end{aligned}$$

(b) Using the formula for a sum of cubes with $A = x^2$ and $B = 2$, we have

$$x^6 + 8 = (x^2)^3 + 2^3 = (x^2 + 2)(x^4 - 2x^2 + 4)$$

■

When we factor an expression, the result can sometimes be factored further. In general, we first factor out common factors, then inspect the result to see if it can be factored by any of the other methods of this section. We repeat this process until we have factored the expression completely.

EXAMPLE 12 ■ Factoring an Expression Completely

Factor each expression completely.

(a) $2x^4 - 8x^2$ (b) $x^5y^2 - xy^6$

SOLUTION

(a) We first factor out the power of x with the smallest exponent.

$$2x^4 - 8x^2 = 2x^2(x^2 - 4) \qquad \text{Common factor is } 2x^2$$

$$= 2x^2(x - 2)(x + 2) \qquad \text{Factor } x^2 - 4 \text{ as a difference of squares}$$

(b) We first factor out the powers of x and y with the smallest exponents.

$$x^5y^2 - xy^6 = xy^2(x^4 - y^4) \qquad \text{Common factor is } xy^2$$

$$= xy^2(x^2 + y^2)(x^2 - y^2) \qquad \text{Factor } x^4 - y^4 \text{ as a difference of squares}$$

$$= xy^2(x^2 + y^2)(x + y)(x - y) \qquad \text{Factor } x^2 - y^2 \text{ as a difference of squares}$$

■

In the next example we factor out variables with fractional exponents. This type of factoring occurs in calculus.

EXAMPLE 13 ■ Factoring Expressions with Fractional Exponents

Factor each expression.

(a) $3x^{3/2} - 9x^{1/2} + 6x^{-1/2}$ (b) $(1 + x)^{-2/3}x + (1 + x)^{1/3}$

SOLUTION

(a) Factor out the power of x with the *smallest exponent*, that is, $x^{-1/2}$.

$$3x^{3/2} - 9x^{1/2} + 6x^{-1/2} = 3x^{-1/2}(x^2 - 3x + 2) \qquad \text{Factor out } 3x^{-1/2}$$

$$= 3x^{-1/2}(x - 1)(x - 2) \qquad \text{Factor the quadratic } x^2 - 3x + 2$$

(b) Factor out the power of $1 + x$ with the *smallest exponent*, that is, $(1 + x)^{-2/3}$.

$$(1 + x)^{-2/3}x + (1 + x)^{1/3} = (1 + x)^{-2/3}[x + (1 + x)] \qquad \text{Factor out } (1 + x)^{-2/3}$$

$$= (1 + x)^{-2/3}(1 + 2x)$$

■

CHECK YOUR ANSWER

To see that you have factored correctly, multiply using the Laws of Exponents.

(a) $3x^{-1/2}(x^2 - 3x + 2)$
$= 3x^{3/2} - 9x^{1/2} + 6x^{-1/2}$ ✓

(b) $(1 + x)^{-2/3}[x + (1 + x)]$
$= (1 + x)^{-2/3}x + (1 + x)^{1/3}$ ✓

Polynomials with at least four terms can sometimes be factored by grouping terms. The following example illustrates the idea.

EXAMPLE 14 ■ Factoring by Grouping

Factor each polynomial.

(a) $x^3 + x^2 + 4x + 4$ (b) $x^3 - 2x^2 - 3x + 6$

SOLUTION

(a) $x^3 + x^2 + 4x + 4 = (x^3 + x^2) + (4x + 4)$ Group terms

$= x^2(x + 1) + 4(x + 1)$ Factor out common factors

$= (x^2 + 4)(x + 1)$ Factor out $x + 1$ from each term

(b) $x^3 - 2x^2 - 3x + 6 = (x^3 - 2x^2) - (3x - 6)$ Group terms

$= x^2(x - 2) - 3(x - 2)$ Factor out common factors

$= (x^2 - 3)(x - 2)$ Factor out $x - 2$ from each term ■

1.4 EXERCISES

1–38 ■ Perform the indicated operations and simplify.

1. $(3x^2 + x + 1) + (2x^2 - 3x - 5)$

2. $(3x^2 + x + 1) - (2x^2 - 3x - 5)$

3. $(x^3 + 6x^2 - 4x + 7) - (3x^2 + 2x - 4)$

4. $3(x - 1) + 4(x + 2)$

5. $8(2x + 5) - 7(x - 9)$

6. $4(x^2 - 3x + 5) - 3(x^2 - 2x + 1)$

7. $2(2 - 5t) + t^2(t - 1) - (t^4 - 1)$

8. $5(3t - 4) - (t^2 + 2) - 2t(t - 3)$

9. $\sqrt{x}\,(x - \sqrt{x})$

10. $x^{3/2}(\sqrt{x} - 1/\sqrt{x})$

11. $\sqrt[3]{y}\,(y^2 - 1)$

12. $(4x - 1)(3x + 7)$

13. $(3t - 2)(7t - 5)$

14. $(t + 6)(t + 5) - 3(t + 4)$

15. $(x + 2y)(3x - y)$

16. $(4x - 3y)(2x + 5y)$

17. $(1 - 2y)^2$

18. $(3x + 4)^2$

19. $(2x - 5)(x^2 - x + 1)$

20. $(x^2 + 3)(5x - 6)$

21. $x(x - 1)(x + 2)$

22. $(1 + 2x)(x^2 - 3x + 1)$

23. $(2x^2 + 3y^2)^2$

24. $(x^{1/2} + y^{1/2})(x^{1/2} - y^{1/2})$

25. $(x^2 - a^2)(x^2 + a^2)$

26. $\left(\sqrt{h^2 + 1} + 1\right)\left(\sqrt{h^2 + 1} - 1\right)$

27. $(1 + a^3)^3$

28. $(x - 1)(x^2 + x + 1)$

29. $\left(\sqrt{a} - \dfrac{1}{b}\right)\left(\sqrt{a} + \dfrac{1}{b}\right)$

30. $\left(c + \dfrac{1}{c}\right)^2$

31. $(x^2 + x - 2)(x^3 - x + 1)$

32. $(1 + x + x^2)(1 - x + x^2)$

33. $(1 + x^{4/3})(1 - x^{2/3})$

34. $(x^{3/2} - x + 1)(x^2 + x^{1/2} - 2)$

35. $(1 - b)^2(1 + b)^2$

36. $(1 + x - x^2)^2$

37. $(3x^2y + 7xy^2)(x^2y^3 - 2y^2)$

38. $(x^4y - y^5)(x^2 + xy + y^2)$

39–88 ■ Factor the expression completely.

39. $2x + 12x^3$

40. $8x^5 + 4x^3$

41. $6y^4 - 15y^3$

42. $5ab - 8abc$

43. $x^2 + 7x + 6$

44. $x^2 - x - 6$

45. $x^2 - 2x - 8$

46. $x^2 - 14x + 48$

47. $y^2 - 8y + 15$

48. $z^2 + 6z - 16$

49. $2x^2 + 5x + 3$

50. $2x^2 + 7x - 4$

51. $9x^2 - 36$

52. $8x^2 + 10x + 3$

53. $6x^2 - 5x - 6$

54. $6 + 5t - 6t^2$

55. $(x - 1)(x + 2)^2 - (x - 1)^2(x + 2)$

56. $(x + 1)^3x - 2(x + 1)^2x^2 + x^3(x + 1)$

57. $y^4(y + 2)^3 + y^5(y + 2)^4$

58. $n(x - y) + (n - 1)(y - x)$

59. $(a^2 - 1)b^2 - 4(a^2 - 1)$

60. $(a + b)^2 - (a - b)^2$

61. $t^3 + 1$

62. $4t^2 - 9s^2$

63. $4t^2 - 12t + 9$

64. $x^3 - 27$

65. $x^3 + 2x^2 + x$

66. $3x^3 - 27x$

67. $4x^2 + 4xy + y^2$

68. $4r^2 - 12rs + 9s^2$

69. $x^4 + 2x^3 - 3x^2$

70. $x^6 + 64$

71. $8x^3 - 125$

72. $x^4 + 2x^2 + 1$

73. $x^4 + x^2 - 2$

74. $x^3 + 3x^2 - x - 3$

75. $y^3 - 3y^2 - 4y + 12$

76. $y^3 - y^2 + y - 1$

77. $2x^3 + 4x^2 + x + 2$

78. $3x^3 + 5x^2 - 6x - 10$

79. $(3 + x^2)^2 - (1 + x^2)^2$

80. $\left(1 + \frac{1}{x}\right)^2 - \left(1 - \frac{1}{x}\right)^2$

81. $x^{5/2} - x^{1/2}$

82. $3x^{-1/2} + 4x^{1/2} + x^{3/2}$

83. $x^{-3/2} + 2x^{-1/2} + x^{1/2}$

84. $(x - 1)^{7/2} - (x - 1)^{3/2}$

85. $(x^2 + 1)^{1/2} + 2(x^2 + 1)^{-1/2}$

86. $x^{-1/2}(x + 1)^{1/2} + x^{1/2}(x + 1)^{-1/2}$

87. $(a^2 + 1)^2 - 7(a^2 + 1) + 10$

88. $(a^2 + 2a)^2 - 2(a^2 + 2a) - 3$

89–92 ■ Factor the expression completely. (This type of expression arises in calculus when using the "product rule.")

89. $3x^2(4x - 12)^2 + x^3(2)(4x - 12)(4)$

90. $5(x^2 + 4)^4(2x)(x - 2)^4 + (x^2 + 4)^5(4)(x - 2)^3$

91. $3(2x - 1)^2(2)(x + 3)^{1/2} + (2x - 1)^3(\frac{1}{2})(x + 3)^{-1/2}$

92. $\frac{1}{3}(x + 6)^{-2/3}(2x - 3)^2 + (x + 6)^{1/3}(2)(2x - 3)(2)$

93. (a) Show that $ab = \frac{1}{2}[(a + b)^2 - (a^2 + b^2)]$.

(b) Show that $(a^2 + b^2)^2 - (a^2 - b^2)^2 = 4a^2b^2$.

(c) Show that

$$(a^2 + b^2)(c^2 + d^2) = (ac + bd)^2 + (ad - bc)^2$$

(d) Factor completely: $4a^2c^2 - (a^2 - b^2 + c^2)^2$.

(e) Factor $x^4 + 3x^2 + 4$. [*Hint*: Write the expression as $(x^4 + 4x^2 + 4) - x^2$ and use the Difference of Squares Formula.]

94. Verify each formula algebraically.

(a) Special Product Formulas 1 and 2

(b) Special Product Formulas 3 and 4

(c) Factoring Formula 4

(d) Factoring Formula 5

DISCOVERY • DISCUSSION

95. The Power of Algebraic Formulas Use the Difference of Squares Formula to factor $17^2 - 16^2$. Notice that it is easy to calculate the factored form in your head, but not so easy to calculate the original form in this way. Evaluate each expression in your head:

(a) $528^2 - 527^2$ (b) $122^2 - 120^2$

(c) $1020^2 - 1010^2$

Now use Special Product Formula 1 to evaluate these products in your head:

(d) $49 \cdot 51$ (e) $998 \cdot 1002$

96. Volume of a Cylindrical Shell Using the formula for the volume of a cylinder given on the inside back cover of this book, explain why the volume of the cylindrical shell shown in the figure is

$$V = \pi R^2 h - \pi r^2 h$$

Factor this to show that

$$V = 2\pi \cdot \text{average radius} \cdot \text{height} \cdot \text{thickness}$$

Use the "unrolled" diagram to explain why this makes sense geometrically.

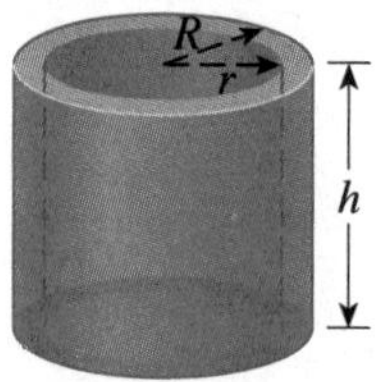

97. Differences of Even Powers

(a) Factor the expressions completely: $A^4 - B^4$ and $A^6 - B^6$.

(b) Verify that $18{,}335 = 12^4 - 7^4$ and that $2{,}868{,}335 = 12^6 - 7^6$.

(c) Use the results of parts (a) and (b) to factor the integers 18,335 and 2,868,335. Show that in both of these factorizations, all the factors are prime numbers.

98. Factoring $A^n - 1$ Verify the factoring formulas in the list by expanding and simplifying the right-hand side in each case.

$$A^2 - 1 = (A - 1)(A + 1)$$
$$A^3 - 1 = (A - 1)(A^2 + A + 1)$$
$$A^4 - 1 = (A - 1)(A^3 + A^2 + A + 1)$$

Based on the pattern displayed in this list, how do you think $A^5 - 1$ would factor? Verify your conjecture. Now generalize the pattern you have observed to obtain a factorization formula for $A^n - 1$, where n is a positive integer.

VISUALIZING A FORMULA

Many of the Special Product Formulas that we learned in this section can be "seen" as geometrical facts about length, area, and volume. For example, the figure shows how the formula for the square of a binomial can be interpreted as a fact about areas of squares and rectangles.

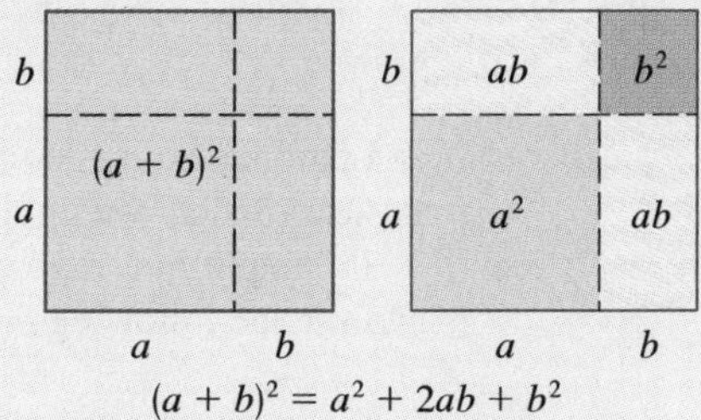

$$(a + b)^2 = a^2 + 2ab + b^2$$

In the figure, a and b represent lengths, a^2, b^2, ab, and $(a + b)^2$ represent areas. The ancient Greeks always interpreted algebraic formulas in terms of geometric figures as we have done here.

1. Explain how the figure verifies the formula $a^2 - b^2 = (a + b)(a - b)$.

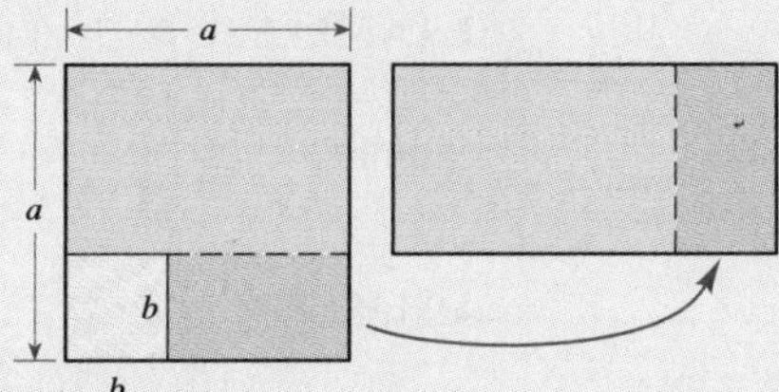

2. Find a figure that verifies the formula $(a - b)^2 = a^2 - 2ab + b^2$.
3. Explain how the figure verifies the formula
$(a + b)^3 = a^3 + 3a^2b + 3ab^2 + b^3$.

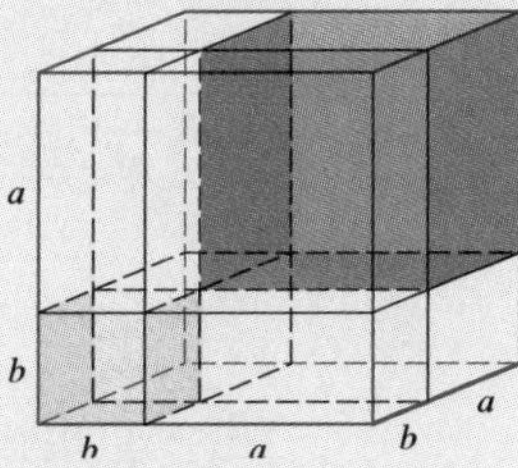

4. Is it possible to draw a geometric figure that verifies the formula for $(a + b)^4$? Explain.
5. (a) Expand $(a + b + c)^2$.
(b) Make a geometric figure that verifies the formula you found in part (a).

1.5 FRACTIONAL EXPRESSIONS

A quotient of two algebraic expressions is called a **fractional expression**. We assume that all fractions are defined; that is, *we deal only with values of the variables such that the denominators are not zero.*

A common type of fractional expression occurs when both the numerator and denominator are polynomials. This is called a **rational expression**. For instance,

$$\frac{4x^3 + 2x + 5}{x + 3}$$

is a rational expression whose denominator is 0 when $x = -3$. So, in dealing with this expression, we implicitly assume that $x \neq -3$.

In simplifying rational expressions we factor both numerator and denominator and use the following property of fractions:

$$\frac{AC}{BC} = \frac{A}{B}$$

This says that we can **cancel** common factors from numerator and denominator.

EXAMPLE 1 ■ Simplifying Fractional Expressions by Cancellation

Simplify: (a) $\dfrac{x^2 - 1}{x^2 + x - 2}$ (b) $\dfrac{2x^3 + 5x^2 - 3x}{6 - x - x^2}$

SOLUTION

We can't cancel the x^2's in $\dfrac{x^2 - 1}{x^2 + x - 2}$ because x^2 is not a factor.

(a) $$\frac{x^2 - 1}{x^2 + x - 2} = \frac{(x - 1)(x + 1)}{(x - 1)(x + 2)} \quad \text{Factor}$$

$$= \frac{x + 1}{x + 2} \quad \text{Cancel common factors}$$

(b) $$\frac{2x^3 + 5x^2 - 3x}{6 - x - x^2} = \frac{x(2x^2 + 5x - 3)}{-(x^2 + x - 6)} = \frac{x(2x - 1)(x + 3)}{-(x - 2)(x + 3)} \quad \text{Factor}$$

$$= -\frac{x(2x - 1)}{x - 2} \quad \text{Cancel common factors}$$ ■

When multiplying fractional expressions, we use the following property of fractions:

$$\frac{A}{B} \cdot \frac{C}{D} = \frac{AC}{BD}$$

Diophantus lived in Alexandria about 250 A.D. His book *Arithmetica* is considered the first book on algebra. In it he gives methods for finding integer solutions of algebraic equations. *Arithmetica* was read and studied for more than a thousand years. Fermat (see page 578) made some of his most important discoveries while studying this book. Diophantus' major contribution is the use of symbols to stand for the unknowns in a problem. Although his symbolism is not as simple as what we use today, it was a major advance over writing everything in words. In Diophantus' notation the equation

$$x^5 - 7x^2 + 8x - 5 = 24$$

is written

$$\Delta K^{\gamma}\alpha\zeta\theta \pitchfork \Delta^{\gamma}\eta\dot{M}\varepsilon\iota\kappa\delta$$

Our modern algebraic notation did not come into common use until the 17th century.

This says that to multiply two fractions we multiply their numerators and multiply their denominators.

EXAMPLE 2 ■ Multiplying Fractional Expressions

Perform the indicated multiplication and simplify: $\dfrac{x^2 + 2x - 3}{x^2 + 8x + 16} \cdot \dfrac{3x + 12}{x - 1}$

SOLUTION

We first factor.

$$\frac{x^2 + 2x - 3}{x^2 + 8x + 16} \cdot \frac{3x + 12}{x - 1} = \frac{(x - 1)(x + 3)}{(x + 4)^2} \cdot \frac{3(x + 4)}{x - 1} \qquad \text{Factor}$$

$$= \frac{3(x - 1)(x + 3)(x + 4)}{(x - 1)(x + 4)^2} \qquad \text{Property of fractions}$$

$$= \frac{3(x + 3)}{x + 4} \qquad \text{Cancel common factors}$$

■

When dividing fractional expressions, we use the following property of fractions:

$$\frac{A}{B} \div \frac{C}{D} = \frac{A}{B} \cdot \frac{D}{C}$$

This says that to divide a fraction by another fraction we invert the divisor and multiply.

EXAMPLE 3 ■ Dividing Fractional Expressions

Perform the indicated division and simplify: $\dfrac{x - 4}{x^2 - 4} \div \dfrac{x^2 - 3x - 4}{x^2 + 5x + 6}$

SOLUTION

$$\frac{x - 4}{x^2 - 4} \div \frac{x^2 - 3x - 4}{x^2 + 5x + 6} = \frac{x - 4}{x^2 - 4} \cdot \frac{x^2 + 5x + 6}{x^2 - 3x - 4} \qquad \text{Invert and multiply}$$

$$= \frac{(x - 4)(x + 2)(x + 3)}{(x - 2)(x + 2)(x - 4)(x + 1)} \qquad \text{Factor}$$

$$= \frac{x + 3}{(x - 2)(x + 1)} \qquad \text{Cancel common factors}$$

■

 Avoid making the following error:

$$\frac{A}{B+C} \neq \frac{A}{B} + \frac{A}{C}$$

For instance, if we let $A = 2$, $B = 1$, and $C = 1$, then we see the error:

$$\frac{2}{1+1} \stackrel{?}{=} \frac{2}{1} + \frac{2}{1}$$

$$\frac{2}{2} \stackrel{?}{=} 2 + 2$$

$$1 \stackrel{?}{=} 4 \quad \text{Wrong!}$$

In adding and subtracting rational expressions, we first find a common denominator and then use the following property of fractions:

$$\frac{A}{C} + \frac{B}{C} = \frac{A+B}{C}$$

Although any common denominator will work, it is best to use the **least common denominator** (LCD) as explained in Section 1.2. The LCD is found by factoring each denominator and taking the product of the distinct factors, using the highest power that appears in any of the factors.

EXAMPLE 4 ■ Adding and Subtracting Fractional Expressions

Perform the indicated operations and simplify.

(a) $\dfrac{3}{x-1} + \dfrac{x}{x+2}$ (b) $\dfrac{1}{x^2-1} - \dfrac{2}{(x+1)^2}$

SOLUTION

(a) Here the LCD is simply the product $(x-1)(x+2)$, so we have

$$\frac{3}{x-1} + \frac{x}{x+2} = \frac{3(x+2)}{(x-1)(x+2)} + \frac{x(x-1)}{(x-1)(x+2} \quad \text{Write fractions using LCD}$$

$$= \frac{3x + 6 + x^2 - x}{(x-1)(x+2)} \quad \text{Add fractions}$$

$$= \frac{x^2 + 2x + 6}{(x-1)(x+2)} \quad \text{Combine terms in numerator}$$

(b) The LCD of $x^2 - 1 = (x-1)(x+1)$ and $(x+1)^2$ is $(x-1)(x+1)^2$, so we have

$$\frac{1}{x^2-1} - \frac{2}{(x+1)^2} = \frac{1}{(x-1)(x+1)} - \frac{2}{(x+1)^2} \quad \text{Factor}$$

$$= \frac{(x+1) - 2(x-1)}{(x-1)(x+1)^2} \quad \text{Combine fractions using LCD}$$

$$= \frac{x + 1 - 2x + 2}{(x-1)(x+1)^2} \quad \text{Distributive Property}$$

$$= \frac{3 - x}{(x-1)(x+1)^2} \quad \text{Combine terms in numerator}$$

■

In the next example we simplify a **compound fraction**, which contains a fraction in both the numerator and the denominator.

EXAMPLE 5 ■ Simplifying a Compound Fraction

Simplify: $\dfrac{\dfrac{x}{y} + 1}{1 - \dfrac{y}{x}}$

SOLUTION 1

We combine the terms in the numerator into a single fraction. We do the same in the denominator. Then we invert and multiply.

$$\frac{\dfrac{x}{y} + 1}{1 - \dfrac{y}{x}} = \frac{\dfrac{x + y}{y}}{\dfrac{x - y}{x}} = \frac{x + y}{y} \cdot \frac{x}{x - y}$$

$$= \frac{x(x + y)}{y(x - y)}$$

SOLUTION 2

We find the LCD of all the fractions in the expression, then multiply numerator and denominator by it. In this example the LCD of all the fractions is xy. Thus

$$\frac{\dfrac{x}{y} + 1}{1 - \dfrac{y}{x}} = \frac{\dfrac{x}{y} + 1}{1 - \dfrac{y}{x}} \cdot \frac{xy}{xy}$$ Multiply numerator and denominator by xy

$$= \frac{x^2 + xy}{xy - y^2}$$ Simplify

$$= \frac{x(x + y)}{y(x - y)}$$ Factor

■

The remaining examples show situations in calculus that require the ability to work with fractional expressions.

EXAMPLE 6 ■ Simplifying a Compound Fraction

Simplify: $\dfrac{\dfrac{1}{(a + h)^2} - \dfrac{1}{a^2}}{h}$

SOLUTION

As in the first solution to Example 5, we begin by combining the fractions in the numerator using a common denominator.

$$\frac{\dfrac{1}{(a+h)^2}-\dfrac{1}{a^2}}{h}=\frac{\dfrac{a^2-(a+h^2)}{(a+h)a^2}}{h}$$ Combine fractions in the numerator

$$=\frac{a^2-(a+h)^2}{(a+h)^2a^2}\cdot\frac{1}{h}$$ Property 2 of fractions (invert divisor and multiply)

$$=\frac{a^2-(a^2+2ah+h^2)}{(a+h)^2a^2}\cdot\frac{1}{h}$$ Product Formula 2

$$=\frac{-2ah-h^2}{h(a+h)^2a^2}$$ Subtract

$$=\frac{h(-2a-h)}{h(a+h)^2a^2}$$ Factor out h

$$=-\frac{2a+h}{(a+h)^2a^2}$$ Property 5 of fractions (cancel common factors) ■

EXAMPLE 7 ■ Simplifying a Compound Fraction

Simplify: $$\frac{(1+x^2)^{1/2}-x^2(1+x^2)^{-1/2}}{1+x^2}$$

SOLUTION 1

Factor $(1+x^2)^{-1/2}$ from the numerator.

$$\frac{(1+x^2)^{1/2}-x^2(1+x^2)^{-1/2}}{1+x^2}=\frac{(1+x^2)^{-1/2}[(1+x^2)-x^2]}{1+x^2}$$

$$=\frac{(1+x^2)^{-1/2}}{1+x^2}=\frac{1}{(1+x^2)^{3/2}}$$

SOLUTION 2

Since $(1+x^2)^{-1/2}=1/(1+x^2)^{1/2}$ is a fraction, we may clear all fractions by multiplying numerator and denominator by $(1+x^2)^{1/2}$.

$$\frac{(1+x^2)^{1/2}-x^2(1+x^2)^{-1/2}}{1+x^2}=\frac{(1+x^2)^{1/2}-x^2(1+x^2)^{-1/2}}{1+x^2}\cdot\frac{(1+x^2)^{1/2}}{(1+x^2)^{1/2}}$$

$$=\frac{(1+x^2)-x^2}{(1+x^2)^{3/2}}=\frac{1}{(1+x^2)^{3/2}}$$ ■

If a fraction has a denominator of the form $A + B\sqrt{C}$, we may rationalize the denominator by multiplying numerator and denominator by the **conjugate radical** $A - B\sqrt{C}$. This is effective because, by Product Formula 1 in Section 1.4, the product of the denominator and its conjugate radical does not contain a radical:

$$\left(A + B\sqrt{C}\right)\left(A - B\sqrt{C}\right) = A^2 - B^2C$$

EXAMPLE 8 ■ Rationalizing the Denominator

Rationalize the denominator: $\dfrac{1}{1 + \sqrt{2}}$

SOLUTION

We multiply both the numerator and the denominator by the conjugate radical of $1 + \sqrt{2}$, which is $1 - \sqrt{2}$.

Product Formula 1
$(a + b)(a - b) = a^2 - b^2$

$$\begin{aligned}
\frac{1}{1 + \sqrt{2}} &= \frac{1}{1 + \sqrt{2}} \cdot \frac{1 + \sqrt{2}}{1 + \sqrt{2}} && \text{Multiply numerator and denominator by the conjugate radical} \\
&= \frac{1 - \sqrt{2}}{1^2 - \left(\sqrt{2}\right)^2} && \text{Product Formula 1} \\
&= \frac{1 - \sqrt{2}}{1 - 2} = \frac{1 - \sqrt{2}}{-1} = \sqrt{2} - 1
\end{aligned}$$

■

EXAMPLE 9 ■ Rationalizing a Numerator

Rationalize the numerator: $\dfrac{\sqrt{4 + h} - 2}{n}$

SOLUTION

We multiply numerator and denominator by the conjugate radical $\sqrt{4 + h} + 2$.

$$\begin{aligned}
\frac{\sqrt{4 + h} - 2}{h} &= \frac{\sqrt{4 + h} - 2}{h} \cdot \frac{\sqrt{4 + h} + 2}{\sqrt{4 + h} + 2} && \text{Multiply numerator and denominator by the conjugate radical} \\
&= \frac{\left(\sqrt{4 + h}\,\right)^2 - 2^2}{h\left(\sqrt{4 + h} + 2\right)} && \text{Product Formula 1} \\
&= \frac{4 + h - 4}{h\left(\sqrt{4 + h} + 2\right)} \\
&= \frac{h}{h\left(\sqrt{4 + h} + 2\right)} = \frac{1}{\sqrt{4 + h} + 2} && \text{Property 5 of fractions (cancel common factors)}
\end{aligned}$$

■

Don't make the mistake of applying properties of multiplication to the operation of addition. Many of the common errors in algebra involve doing just that. The following table states several properties of multiplication and illustrates the error in applying them to addition.

Correct multiplication property	**Common error with addition**
$(a \cdot b)^2 = a^2 \cdot b^2$	$(a + b)^2 \neq a^2 + b^2$
$\sqrt{a \cdot b} = \sqrt{a}\,\sqrt{b} \quad (a, b \geq 0)$	$\sqrt{a + b} \neq \sqrt{a} + \sqrt{b}$
$\sqrt{a^2 \cdot b^2} = a \cdot b \quad (a, b \geq 0)$	$\sqrt{a^2 + b^2} \neq a + b$
$\dfrac{1}{a} \cdot \dfrac{1}{b} = \dfrac{1}{a \cdot b}$	$\dfrac{1}{a} + \dfrac{1}{b} \neq \dfrac{1}{a + b}$
$\dfrac{ab}{a} = b$	$\dfrac{a + b}{a} \neq b$
$a^{-1} \cdot b^{-1} = (a \cdot b)^{-1}$	$a^{-1} + b^{-1} \neq (a + b)^{-1}$

To verify that the formulas in the right-hand column are wrong, simply substitute numbers for a and b and calculate each side. For example, if we take $a = 2$ and $b = 2$ in the fourth property, we find that the left-hand side is

$$\frac{1}{a} + \frac{1}{b} = \frac{1}{2} + \frac{1}{2} = 1$$

whereas the right-hand side is

$$\frac{1}{a + b} = \frac{1}{2 + 2} = \frac{1}{4}$$

Since $1 \neq \frac{1}{4}$, the stated formula is wrong. You should similarly convince yourself of the error in each of the other formulas. (See Exercise 82.)

1.5 EXERCISES

1–52 ■ Simplify the expression.

1. $\dfrac{x - 2}{x^2 - 4}$

2. $\dfrac{x^2 - x - 2}{x^2 - 1}$

3. $\dfrac{x^2 + 6x + 8}{x^2 + 5x + 4}$

4. $\dfrac{x^2 - x - 12}{x^2 + 5x + 6}$

5. $\dfrac{y^2 + y}{y^2 - 1}$

6. $\dfrac{y^2 - 3y - 18}{2y^2 + 5y + 3}$

7. $\dfrac{2x^3 - x^2 - 6x}{2x^2 - 7x + 6}$

8. $\dfrac{1 - x^2}{x^3 - 1}$

9. $\dfrac{t - 3}{t^2 + 9} \cdot \dfrac{t + 3}{t^2 - 9}$

10. $\dfrac{x^2 - x - 6}{x^2 + 2x} \cdot \dfrac{x^3 + x^2}{x^2 - 2x - 3}$

11. $\dfrac{x^2 + 7x + 12}{x^2 + 3x + 2} \cdot \dfrac{x^2 + 5x + 6}{x^2 + 6x + 9}$

12. $\dfrac{x^2 + 2xy + y^2}{x^2 - y^2} \cdot \dfrac{2x^2 - xy - y^2}{x^2 - xy - 2y^2}$

13. $\dfrac{2x^2+3x+1}{x^2+2x-15} \div \dfrac{x^2+6x+5}{2x^2-7x+3}$

14. $\dfrac{4y^2-9}{2y^2+9y-18} \div \dfrac{2y^2+y-3}{y^2+5y-6}$

15. $\dfrac{\dfrac{x^3}{x+1}}{\dfrac{x}{x^2+2x+1}}$

16. $\dfrac{\dfrac{2x^2-3x-2}{x^2-1}}{\dfrac{2x^2+5x+2}{x^2+x-2}}$

17. $\dfrac{x/y}{z}$

18. $\dfrac{x}{y/z}$

19. $\dfrac{1}{x+5}+\dfrac{2}{x-3}$

20. $\dfrac{1}{x+1}+\dfrac{1}{x-1}$

21. $\dfrac{1}{x+1}-\dfrac{1}{x+2}$

22. $\dfrac{x}{x-4}-\dfrac{3}{x+6}$

23. $\dfrac{x}{(x+1)^2}+\dfrac{2}{x+1}$

24. $\dfrac{5}{2x-3}-\dfrac{3}{(2x-3)^2}$

25. $u+1+\dfrac{u}{u+1}$

26. $\dfrac{2}{a^2}-\dfrac{3}{ab}+\dfrac{4}{b^2}$

27. $\dfrac{1}{x^2}+\dfrac{1}{x^2+x}$

28. $\dfrac{1}{x}+\dfrac{1}{x^2}+\dfrac{1}{x^3}$

29. $\dfrac{2}{x+3}-\dfrac{1}{x^2+7x+12}$

30. $\dfrac{x}{x^2-4}+\dfrac{1}{x-2}$

31. $\dfrac{1}{x+3}+\dfrac{1}{x^2-9}$

32. $\dfrac{x}{x^2+x-2}-\dfrac{2}{x^2-5x+4}$

33. $\dfrac{2}{x}+\dfrac{3}{x-1}-\dfrac{4}{x^2-x}$

34. $\dfrac{x}{x^2-x-6}-\dfrac{1}{x+2}-\dfrac{2}{x-3}$

35. $\dfrac{1}{x^2+3x+2}-\dfrac{1}{x^2-2x-3}$

36. $\dfrac{1}{x+1}-\dfrac{2}{(x+1)^2}+\dfrac{3}{x^2-1}$

37. $\dfrac{\dfrac{x}{y}-\dfrac{y}{x}}{\dfrac{1}{x^2}-\dfrac{1}{y^2}}$

38. $x-\dfrac{y}{\dfrac{x}{y}+\dfrac{y}{x}}$

39. $\dfrac{1+\dfrac{1}{c-1}}{1-\dfrac{1}{c-1}}$

40. $1+\dfrac{1}{1+\dfrac{1}{1+x}}$

41. $\dfrac{\dfrac{5}{x-1}-\dfrac{2}{x+1}}{\dfrac{x}{x-1}+\dfrac{1}{x+1}}$

42. $\dfrac{\dfrac{a-b}{a}-\dfrac{a+b}{b}}{\dfrac{a-b}{b}+\dfrac{a+b}{a}}$

43. $\dfrac{x^{-2}-y^{-2}}{x^{-1}+y^{-1}}$

44. $\dfrac{x^{-1}+y^{-1}}{(x+y)^{-1}}$

45. $\dfrac{1}{1+a^n}+\dfrac{1}{1+a^{-n}}$

46. $\dfrac{\left(a+\dfrac{1}{b}\right)^m\left(a-\dfrac{1}{b}\right)^n}{\left(b+\dfrac{1}{a}\right)^m\left(b-\dfrac{1}{a}\right)^n}$

47. $\dfrac{\dfrac{1}{a+h}-\dfrac{1}{a}}{h}$

48. $\dfrac{(x+h)^{-3}-x^{-3}}{h}$

49. $\dfrac{\dfrac{1-(x+h)}{2-(x+h)}-\dfrac{1-x}{2+x}}{h}$

50. $\dfrac{(x+h)^3-7(x+h)-(x^3-7x)}{h}$

51. $\sqrt{1+\left(\dfrac{x}{\sqrt{1-x^2}}\right)^2}$

52. $\sqrt{1+\left(x^3-\dfrac{1}{4x^3}\right)^2}$

53–58 ■ Simplify the expression. (This type of expression arises in calculus when using the "quotient rule.")

53. $\dfrac{3(x+2)^2(x-3)^2-(x+2)^3(2)(x-3)}{(x-3)^4}$

54. $\dfrac{2x(x+6)^4-x^2(4)(x+6)^3}{(x+6)^8}$

55. $\dfrac{2(1+x)^{1/2}-x(1+x)^{-1/2}}{x+1}$

56. $\dfrac{(1-x^2)^{1/2}+x^2(1-x^2)^{-1/2}}{1-x^2}$

57. $\dfrac{3(1+x)^{1/3}-x(1+x)^{-2/3}}{(1+x)^{2/3}}$

58. $\dfrac{(7-3x)^{1/2}+\frac{3}{2}x(7-3x)^{-1/2}}{7-3x}$

59–62 ■ Rationalize the denominator.

59. $\dfrac{2}{3+\sqrt{5}}$

60. $\dfrac{1}{\sqrt{x}+1}$

61. $\dfrac{2}{\sqrt{2}+\sqrt{7}}$

62. $\dfrac{y}{\sqrt{3}+\sqrt{y}}$

63–68 ■ Rationalize the numerator.

63. $\dfrac{1-\sqrt{5}}{3}$

64. $\dfrac{\sqrt{3}+\sqrt{5}}{2}$

65. $\dfrac{\sqrt{r}+\sqrt{2}}{5}$

66. $\dfrac{\sqrt{x}-\sqrt{x+h}}{h\sqrt{x}\sqrt{x+h}}$

67. $\sqrt{x^2+1}-x$

68. $\sqrt{x+1}-\sqrt{x}$

69–78 ■ State whether the given equation is true for all values of the variables. (Disregard any value that makes a denominator zero.)

69. $\dfrac{16+a}{16}=1+\dfrac{a}{16}$

70. $\dfrac{b}{b-c}=1-\dfrac{b}{c}$

71. $\dfrac{2}{4+x}=\dfrac{1}{2}+\dfrac{2}{x}$

72. $\dfrac{x+1}{y+1}=\dfrac{x}{y}$

73. $\dfrac{x}{x+y}=\dfrac{1}{1+y}$

74. $2\left(\dfrac{a}{b}\right)=\dfrac{2a}{2b}$

75. $\dfrac{-a}{b}=-\dfrac{a}{b}$

76. $\dfrac{1+x+x^2}{x}=\dfrac{1}{x}+1+x$

77. $\dfrac{x^2+1}{x^2+x-1}=\dfrac{1}{x-1}$

78. $\dfrac{x^2-1}{x-1}=x+1$

79. If two electrical resistors with resistances R_1 and R_2 are connected in parallel (see the figure), then the total resistance R is given by

$$R=\frac{1}{\dfrac{1}{R_1}+\dfrac{1}{R_2}}$$

(a) Simplify the expression for R.

(b) If $R_1 = 10$ ohms and $R_2 = 20$ ohms, what is the total resistance R?

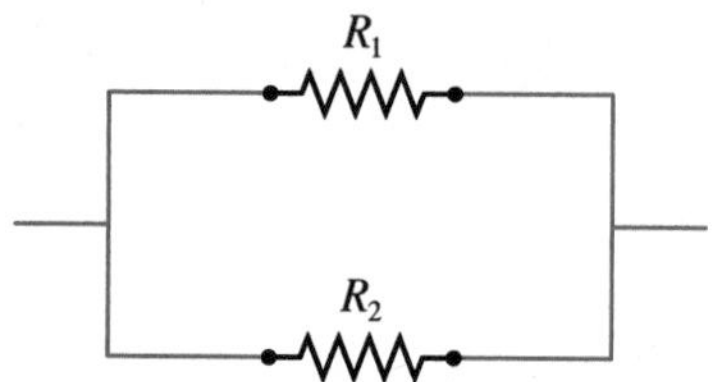

DISCOVERY • DISCUSSION

80. Limiting Behavior of a Rational Expression The rational expression

$$\frac{x^2-9}{x-3}$$

is not defined for $x = 3$. Complete the tables and determine what value the expression approaches as x gets closer and closer to 3. Why is this reasonable? Factor the numerator of the expression and simplify to see why.

x	$\dfrac{x^2-9}{x-3}$
2.80	
2.90	
2.95	
2.99	
2.999	

x	$\dfrac{x^2-9}{x-3}$
3.20	
3.10	
3.05	
3.01	
3.001	

81. Is This Rationalization? In the expression

$$\frac{2}{\sqrt{x}}$$

we would eliminate the radical if we were to square both numerator and denominator. Is this the same thing as rationalizing the denominator?

82. Algebraic Errors The left-hand column in the table lists some common algebraic errors. In each case, give an example using numbers that shows that the formula is not valid. An example of this type, which shows that a statement is false, is called a *counterexample*.

Algebraic error	Counterexample
$\dfrac{1}{a}+\dfrac{1}{b}\;\cancel{=}\;\dfrac{1}{a+b}$	$\dfrac{1}{2}+\dfrac{1}{2}\neq\dfrac{1}{2+2}$
$(a+b)^2\;\cancel{=}\;a^2+b^2$	
$\sqrt{a^2+b^2}\;\cancel{=}\;a+b$	
$\dfrac{a+b}{a}\;\cancel{=}\;b$	
$(a^3+b^3)^{1/3}\;\cancel{=}\;a+b$	
$a^m/a^n\;\cancel{=}\;a^{m/n}$	
$a^{-1/n}\;\cancel{=}\;\dfrac{1}{a^n}$	

1.6 BASIC EQUATIONS

An equation is a statement that two mathematical expressions are equal. For example,

$$3 + 5 = 8$$

is an equation. Most equations that we study in algebra contain variables, which are symbols (usually letters) that stand for numbers. In the equation

$$4x + 7 = 19$$

the letter x is the variable. We think of x as the "unknown" in the equation, and our goal is to find the value of x that makes the equation true. The values of the unknown that make the equation true are called the **solutions** or **roots** of the equation, and the process of finding the solutions is called **solving the equation**.

Two equations with exactly the same solutions are called **equivalent equations**. To solve an equation, we try to find a simpler, equivalent equation in which the variable stands alone on one side of the "equal" sign. Here are the properties that we use to solve an equation. (In these properties, A, B, and C stand for any algebraic expressions, and the symbol $\Leftrightarrow$ means "is equivalent to.")

PROPERTIES OF EQUALITY

Property	Description
1. $A = B \Leftrightarrow A + C = B + C$	Adding the same quantity to both sides of an equation gives an equivalent equation.
2. $A = B \Leftrightarrow CA = CB \quad (C \neq 0)$	Multiplying both sides of an equation by the same nonzero quantity gives an equivalent equation

These properties require that you *perform the same operation on both sides of an equation* when solving it. Thus, if we say "*add* -7" when solving an equation, that is just a short way of saying "*add* -7 to each side of the equation."

This is how we use the properties of equality to solve the equation $4x + 7 = 19$:

$$4x + 7 + (-7) = 19 + (-7) \qquad \text{Add } -7$$

$$4x = 12 \qquad \text{Simplify}$$

$$\tfrac{1}{4} \cdot 4x = \tfrac{1}{4} \cdot 12 \qquad \text{Multiply by } \tfrac{1}{4}$$

$$x = 3 \qquad \text{Simplify}$$

So the solution of this equation is $x = 3$. To verify this, we check our answer by subsituting $x = 3$ to make sure that this value of x does indeed make the equation true:

$$\boxed{x = 3}$$
$$\downarrow$$
$$4(3) + 7 \stackrel{?}{=} 19$$
$$19 = 19 \qquad \text{Correct!}$$

Because it is so important to CHECK YOUR ANSWER, we do this in many of the examples throughout the book. In these checks, LHS stands for "left-hand side" of the original equation, and RHS stands for "right-hand side."

LINEAR EQUATIONS

The simplest type of equation is a **linear equation**, or first-degree equation, which is an equation in which each term is either a constant or a nonzero multiple of the variable. This means that it is equivalent to an equation of the form $ax + b = 0$. Here a and b represent real numbers with $a \neq 0$, and x is the unknown variable that we are solving for. The equation in the following example is linear.

Linear Equations

$4x - 5 = 3$

$2x = \frac{1}{2}x - 5$

Nonlinear Equations

$x^2 + 2x = 8$

$\sqrt{x} - \dfrac{3}{x} = 6x - 1$

EXAMPLE 1 ■ Solving a Linear Equation

Solve the equation $7x - 4 = 3x + 8$.

SOLUTION

We solve this by changing it to an equivalent equation with all terms that have the variable x on one side and all constant terms on the other.

$$\begin{aligned} 7x - 4 &= 3x + 8 && \\ (7x - 4) + 4 &= (3x + 8) + 4 && \text{Add 4} \\ 7x &= 3x + 12 && \text{Simplify} \\ 7x - 3x &= (3x + 12) - 3x && \text{Subtract } 3x \\ 4x &= 12 && \text{Simplify} \\ \tfrac{1}{4} \cdot 4x &= \tfrac{1}{4} \cdot 12 && \text{Multiply by } \tfrac{1}{4} \\ x &= 3 && \text{Simplify} \end{aligned}$$

CHECK YOUR ANSWER

$x = 3$:

$$\begin{aligned} \text{LHS} &= 7(3) - 4 \\ &= 17 \end{aligned} \qquad \begin{aligned} \text{RHS} &= 3(3) + 8 \\ &= 17 \end{aligned}$$

LHS = RHS ✓

■

Euclid (circa 300 B.C.) taught in Alexandria. His *Elements* is the most widely influential scientific book in history. For 2000 years it was the standard introduction to geometry in the schools, and for many generations it was considered the best way to develop logical reasoning. Abraham Lincoln, for instance, studied the *Elements* as a way to sharpen his mind. The story is told that King Ptolemy once asked Euclid if there was a faster way to learn geometry than through the *Elements*. Euclid replied that there is "no royal road to geometry"—meaning by this that mathematics does not respect wealth or social status. Euclid was revered in his own time and was referred to by the title "The Geometer" or "The Writer of the *Elements*." The greatness of the *Elements* stems from its precise, logical, and systematic treatment of geometry. For dealing with equality, Euclid lists the following rules, which he calls "common notions."

1. Things that are equal to the same thing are equal to each other.
2. If equals are added to equals, the sums are equal.
3. If equals are subtracted from equals, the remainders are equal.
4. Things that coincide with one another are equal.
5. The whole is greater than the part.

When a linear equation involves fractions, solving the equation is usually easier if we first multiply each side by the lowest common denominator (LCD) of the fractions, as we see in the following examples.

EXAMPLE 2 ■ Solving an Equation That Involves Fractions

Solve the equation $\dfrac{x}{6} + \dfrac{2}{3} = \dfrac{3}{4}x$.

SOLUTION

The LCD of the denominators 6, 3, and 4 is 12, so we first multiply each side of the equation by 12 to clear the denominators.

$$12 \cdot \left(\frac{x}{6} + \frac{2}{3}\right) = 12 \cdot \frac{3}{4}x \qquad \text{Multiply by LCD}$$

$$2x + 8 = 9x \qquad \text{Distributive Property}$$

$$8 = 7x \qquad \text{Subtract } 2x$$

$$\frac{8}{7} = x \qquad \text{Divide by 7}$$

The solution is $x = \frac{8}{7}$.

CHECK YOUR ANSWER

$x = \frac{8}{7}$:

$$\text{LHS} = \frac{\frac{8}{7}}{6} + \frac{2}{3} = \frac{4}{21} + \frac{2}{3} = \frac{4 + 14}{21} = \frac{18}{21} = \frac{6}{7}$$

$$\text{RHS} = \frac{3}{4}\left(\frac{8}{7}\right) = \frac{24}{28} = \frac{6}{7}$$

LHS = RHS ✓ ■

In the next example, we solve an equation that doesn't look like a linear equation, but it does simplify to an equivalent linear equation when we multiply by the LCD.

EXAMPLE 3 ■ An Equation Involving Fractional Expressions

Solve the equation $\dfrac{1}{x+1} + \dfrac{1}{x-2} = \dfrac{x+3}{x^2 - x - 2}$.

SOLUTION

The LCD of the fractional expressions is $(x + 1)(x - 2) = x^2 - x - 2$. So as long as $x \neq -1$ and $x \neq 2$, we can multiply both sides of the

CHECK YOUR ANSWER

$x = 4$:

$$\text{LHS} = \frac{1}{4+1} + \frac{1}{4-2} = \frac{1}{5} + \frac{1}{2} = \frac{7}{10}$$

$$\text{RHS} = \frac{4+3}{4^2-4-2} = \frac{7}{10}$$

LHS = RHS ✓

equation by the LCD to get

$$(x+1)(x-2)\left(\frac{1}{x+1} + \frac{1}{x-2}\right) = (x+1)(x-2)\left(\frac{x+3}{x^2-x-2}\right) \qquad \text{Multiply by LCD}$$

$$(x-2) + (x+1) = x+3 \qquad \text{Expand}$$

$$2x - 1 = x + 3 \qquad \text{Subtract } x \text{ and add 1}$$

$$x = 4 \qquad \text{Solve}$$

The solution is $x = 4$. ■

It is always important to check your answer, even if you never make a mistake in your calculations. This is because you sometimes end up with **extraneous solutions**, which are potential solutions that do not satisfy the original equation. The next example shows how this can happen.

EXAMPLE 4 ■ An Equation with No Solution

Solve the equation $2 + \dfrac{5}{x-4} = \dfrac{x+1}{x-4}$.

SOLUTION

First, we multiply each side by the common denominator, which is $x - 4$.

$$(x-4)\left(2 + \frac{5}{x-4}\right) = (x-4)\left(\frac{x+1}{x-4}\right) \qquad \text{Multiply by } x-4$$

$$2(x-4) + 5 = x + 1 \qquad \text{Expand}$$

$$2x - 8 + 5 = x + 1 \qquad \text{Distributive Property}$$

$$2x - 3 = x + 1 \qquad \text{Simplify}$$

$$2x = x + 4 \qquad \text{Add 3}$$

$$x = 4 \qquad \text{Subtract } x$$

CHECK YOUR ANSWER

$x = 4$:

$$\text{LHS} = 2 + \frac{5}{4-4} = 2 + \frac{5}{0}$$

$$\text{RHS} = \frac{4+1}{4-4} = \frac{5}{0}$$

Impossible—can't divide by 0. LHS and RHS are undefined, so $x = 4$ is not a solution. ✕

But now if we try to substitute $x = 4$ back into the original equation, we would be dividing by 0, which is impossible. So, this equation has *no solution*. ■

The first step in the preceding solution, multiplying by $x - 4$, had the effect of multiplying by 0. (Do you see why?) Multiplying each side of an equation by an expression that contains the variable may introduce extraneous solutions. That is why it is important to check every answer.

SOLVING EQUATIONS USING RADICALS

Linear equations only have variables to the first power. Now let's consider some equations that involve squares, cubes, and other powers of the variable. Such equations will be studied more extensively in Chapter 3. Here we just consider basic equations that can be simplified into the form $X^n = a$. These can be solved by taking radicals of both sides of the equation.

In Section 1.3 we saw that if n is an even number and a is positive, then a has two nth roots, namely $\sqrt[n]{a}$ and $-\sqrt[n]{a}$, but if n is even and a is negative, then the nth root of a does not exist. If n is odd, then the nth root of a exists and is unique, whether a is positive or negative. For example, this means that

The equation $x^5 = 32$ has only one real solution: $x = \sqrt[5]{32} = 2$.

The equation $x^4 = 16$ has two real solutions: $x = \pm\sqrt[4]{16} = \pm 2$.

The equation $x^5 = -32$ has only one real solution: $x = \sqrt[5]{-32} = -2$.

The equation $x^4 = -16$ has no real solutions because $\sqrt[4]{-16}$ does not exist.

Let's use these facts and the other properties of nth roots to solve simple equations involving powers of x.

EXAMPLE 5 ■ Solving Simple Quadratic Equations

Solve each equation: (a) $x^2 - 5 = 0$ (b) $(x - 4)^2 = 5$

SOLUTION

(a)

$$\begin{aligned} x^2 - 5 &= 0 \\ x^2 &= 5 && \text{Add 5} \\ x &= \pm\sqrt{5} && \text{Take the square root} \end{aligned}$$

The solutions are $x = \sqrt{5}$ and $x = -\sqrt{5}$.

(b) We can take the square root of each side of this equation as well.

$$\begin{aligned} (x - 4)^2 &= 5 \\ x - 4 &= \pm\sqrt{5} && \text{Take the square root} \\ x &= 4 \pm \sqrt{5} && \text{Add 4} \end{aligned}$$

The solutions are $x = 4 + \sqrt{5}$ and $x = 4 - \sqrt{5}$.

You should check that each answer satisifies the original equation. ■

EXAMPLE 6 ■ Solving Equations Using *n*th Roots

Find all real solutions for each equation.

(a) $x^3 = -8$ (b) $16x^4 = 81$

SOLUTION

(a) Since every real number has exactly one real cube root, we can solve this equation by taking the cube root of each side.

$$(x^3)^{1/3} = (-8)^{1/3}$$
$$x = -2$$

If n is even, the equation $x^n = c\ (c > 0)$ has two solutions, $x = c^{1/n}$ and $x = -c^{1/n}$.

(b) Here we must remember that if n is even, then every positive real number has *two* real nth roots, a positive one and a negative one.

$$x^4 = \tfrac{81}{16} \qquad \text{Divide by 16}$$
$$(x^4)^{1/4} = \pm\left(\tfrac{81}{16}\right)^{1/4} \qquad \text{Take the fourth root}$$
$$x = \pm\tfrac{3}{2}$$

■

The next example shows how to solve an equation that involves a fractional power of the variable.

EXAMPLE 7 ■ Solving an Equation with a Fractional Exponent

Solve the equation $5x^{2/3} - 2 = 43$.

SOLUTION

The idea is to first isolate the term with the fractional exponent, then raise both sides of the equation to the *reciprocal* of that exponent.

If n is even, the equation $x^{n/m} = c$ has two solutions, $x = c^{m/n}$ and $x = -c^{m/n}$.

$$5x^{2/3} - 2 = 43$$
$$5x^{2/3} = 45 \qquad \text{Add 2}$$
$$x^{2/3} = 9 \qquad \text{Divide by 5}$$
$$x = \pm 9^{3/2} \qquad \text{Raise both sides to } \tfrac{3}{2} \text{ power}$$
$$x = \pm 27 \qquad \text{Simplify}$$

The solutions are $x = 27$ and $x = -27$.

CHECK YOUR ANSWER:

$x = 27$:

$$\begin{aligned} \text{LHS} &= 5(27)^{2/3} - 2 \\ &= 5(9) - 2 \\ &= 43 \\ \text{RHS} &= 43 \\ \text{LHS} &= \text{RHS} \quad \checkmark \end{aligned}$$

$x = -27$:

$$\begin{aligned} \text{LHS} &= 5(-27)^{2/3} - 2 \\ &= 5(9) - 2 \\ &= 43 \\ \text{RHS} &= 43 \\ \text{LHS} &= \text{RHS} \quad \checkmark \end{aligned}$$

■

SOLVING FOR ONE VARIABLE IN TERMS OF OTHERS

Many formulas in the sciences involve several variables, and it is often necessary to express one of the variables in terms of the others. In the next example we solve for a variable in Newton's Law of Gravity.

EXAMPLE 8 ■ Solving for One Variable in Terms of Others

This is Newton's Law of Gravity. It gives the gravitational force F between two masses m and M that are a distance r apart. The constant G is the universal gravitational constant.

Solve for the variable M in the equation

$$F = G\frac{mM}{r^2}$$

SOLUTION

Although this equation involves more than one variable, we solve it as usual by isolating M on one side and treating the other variables as we would numbers.

$$F = \left(\frac{Gm}{r^2}\right)M \qquad \text{Factor } M \text{ from RHS}$$

$$\left(\frac{r^2}{Gm}\right)F = \left(\frac{r^2}{Gm}\right)\left(\frac{Gm}{r^2}\right)M \qquad \text{Multiply by reciprocal of } \frac{Gm}{r^2}$$

$$\frac{r^2F}{Gm} = M \qquad \text{Simplify}$$

The solution is $M = \dfrac{r^2F}{Gm}$. ■

EXAMPLE 9 ■ Solving for One Variable in Terms of Others

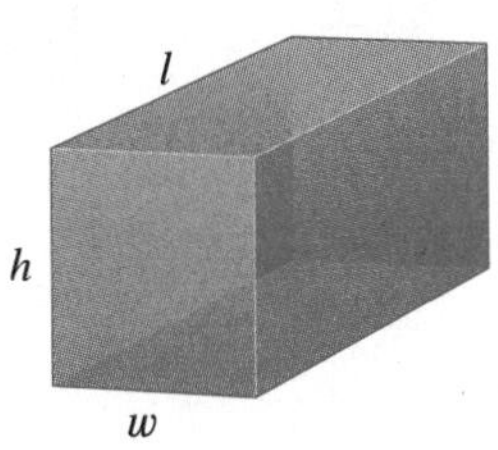

FIGURE 1
A closed rectangular box

The surface area A of the closed rectangular box shown in Figure 1 can be calculated from the length l, the width w, and the height h according to the formula

$$A = 2lw + 2wh + 2lh$$

Solve for w in terms of the other variables in this equation.

SOLUTION

Although this equation involves more than one variable, we solve it as usual by isolating w on one side, treating the other variables as we would numbers.

$$A = (2lw + 2wh) + 2lh \quad \text{Collect terms involving } w$$

$$A - 2lh = 2lw + 2wh \quad \text{Subtract } 2lh$$

$$A - 2lh = (2l + 2h)w \quad \text{Factor } w \text{ from RHS}$$

$$\frac{A - 2lh}{2l + 2h} = w \quad \text{Divide by } 2l + 2h$$

The solution is $w = \dfrac{A - 2lh}{2l + 2h}$. ■

1.6 EXERCISES

1–6 ■ Determine whether the given value is a solution of the equation.

1. $2x - 3 = x + 1$
(a) $x = 4$ (b) $x = \frac{3}{2}$

2. $4(x - 1) - (2 - x) = 5(x - 2) + 4$
(a) $x = -3$ (b) $x = 0$

3. $\dfrac{1}{x} - \dfrac{1}{x + 3} = \dfrac{1}{6}$
(a) $x = -3$ (b) $x = 3$

4. $1 - [2 - (3 - x)] = 4x - (6 + x)$
(a) $x = 2$ (b) $x = 22$

5. $ax - 2b = 0 \quad (a \neq 0, b \neq 0)$
(a) $x = 0$ (b) $x = \dfrac{2b}{a}$

6. $\dfrac{ax - b}{bx - a} = \dfrac{a}{b} \quad (a \neq b, a > 0, b > 0)$
(a) $x = 0$ (b) $x = 1$

7–42 ■ Solve the equation.

7. $3x - 5 = 7$

8. $4x + 12 = 28$

9. $x - 3 = 2x + 6$

10. $4x + 7 = 9x - 13$

11. $-7w = 15 - 2w$

12. $5t - 13 = 12 - 5t$

13. $\frac{1}{2}y - 2 = \frac{1}{3}y$

14. $\dfrac{z}{5} = \dfrac{3}{10}z + 7$

15. $2(1 - x) = 3(1 + 2x) + 5$

16. $5(x + 3) + 9 = -2(x - 2) - 1$

17. $4\left(y - \frac{1}{2}\right) - y = 6(5 - y)$

18. $\dfrac{2}{3}y + \dfrac{1}{2}(y - 3) = \dfrac{y + 1}{4}$

19. $\dfrac{1}{x} = \dfrac{4}{3x} + 1$

20. $\dfrac{2x - 1}{x + 2} = \dfrac{4}{5}$

21. $\dfrac{2}{t + 6} = \dfrac{3}{t - 1}$

22. $\dfrac{1}{t - 1} + \dfrac{t}{3t - 2} = \dfrac{1}{3}$

23. $r - 2[1 - 3(2r + 4)] = 61$

24. $(t - 4)^2 = (t + 4)^2 + 32$

25. $\sqrt{3}\,x + \sqrt{12} = \dfrac{x + 5}{\sqrt{3}}$

26. $\frac{2}{3}x - \frac{1}{4} = \frac{1}{6}x - \frac{1}{9}$

27. $\frac{2}{x} - 5 = \frac{6}{x} + 4$

28. $\frac{6}{x-3} = \frac{5}{x+4}$

29. $\frac{3}{x+1} - \frac{1}{2} = \frac{1}{3x+3}$

30. $\frac{4}{x-1} + \frac{2}{x+1} = \frac{35}{x^2-1}$

31. $\frac{2x-7}{2x+4} = \frac{2}{3}$

32. $\frac{12x-5}{6x+3} = 2 - \frac{5}{x}$

33. $x - \frac{1}{3}x - \frac{1}{2}x - 5 = 0$

34. $2x - \frac{x}{2} + \frac{x+1}{4} = 6x$

35. $\frac{1}{z} - \frac{1}{2z} - \frac{1}{5z} = \frac{10}{z+1}$

36. $\dfrac{1}{1 - \dfrac{3}{2+w}} = 60$

37. $\dfrac{u}{u - \dfrac{u+1}{2}} = 4$

38. $\frac{1}{3-t} + \frac{4}{3+t} + \frac{16}{9-t^2} = 0$

39. $\frac{x}{2x-4} - 2 = \frac{1}{x-2}$

40. $\frac{1}{x+3} + \frac{5}{x^2-9} = \frac{2}{x-3}$

41. $\frac{3}{x+4} = \frac{1}{x} + \frac{6x+12}{x^2+4x}$

42. $\frac{1}{x} - \frac{2}{2x+1} = \frac{1}{2x^2+x}$

43–66 ■ Find all real solutions of the equation.

43. $x^2 = 49$

44. $x^2 = 18$

45. $x^2 - 24 = 0$

46. $x^2 - 7 = 0$

47. $8x^2 - 64 = 0$

48. $5x^2 - 125 = 0$

49. $x^2 + 16 = 0$

50. $6x^2 + 100 = 0$

51. $(x+2)^2 = 4$

52. $3(x-5)^2 = 15$

53. $x^3 = 27$

54. $x^5 + 32 = 0$

55. $x^4 - 16 = 0$

56. $64x^6 = 27$

57. $x^4 + 64 = 0$

58. $(x-1)^3 + 8 = 0$

59. $(x+2)^4 - 81 = 0$

60. $(x+1)^4 + 16 = 0$

61. $3(x-3)^3 = 375$

62. $4(x+2)^5 = 1$

63. $\sqrt[3]{x} = 5$

64. $x^{4/3} - 16 = 0$

65. $2x^{5/3} + 64 = 0$

66. $6x^{2/3} - 216 = 0$

67–70 ■ Find the solution of the equation correct to two decimals.

67. $2.15x - 4.63 = x + 1.19$

68. $3.95 - x = 2.32x + 2.00$

69. $3.16(x + 4.63) = 4.19(x - 7.24)$

70. $\frac{0.26x - 1.94}{3.03 - 2.44x} = 1.76$

71–84 ■ Solve the equation for the indicated variable.

71. $PV = nRT$; for R

72. $F = G\frac{mM}{r^2}$; for m

73. $\frac{1}{R} = \frac{1}{R_1} + \frac{1}{R_2}$; for R_1

74. $P = 2l + 2w$; for w

75. $\frac{ax+b}{cx+d} = 2$; for x

76. $a - 2[b - 3(c - x)] = 6$; for x

77. $a^2x + (a-1) = (a+1)x$; for x

78. $\frac{a+1}{b} = \frac{a-1}{b} + \frac{b+1}{a}$; for a

79. $V = \frac{1}{3}\pi r^2 h$; for r

80. $F = G\frac{mM}{r^2}$; for r

81. $a^2 + b^2 = c^2$; for b

82. $A = P\left(1 + \frac{i}{100}\right)^2$; for i

83. $V = \frac{4}{3}\pi r^3$; for r

84. $x^4 + y^4 + z^4 = 100$; for x

85. The speed that a sailboat is capable of sailing is determined by three factors: its total length L, the surface area A of its sails, and its displacement V (the volume of water it displaces), as shown in the sketch.

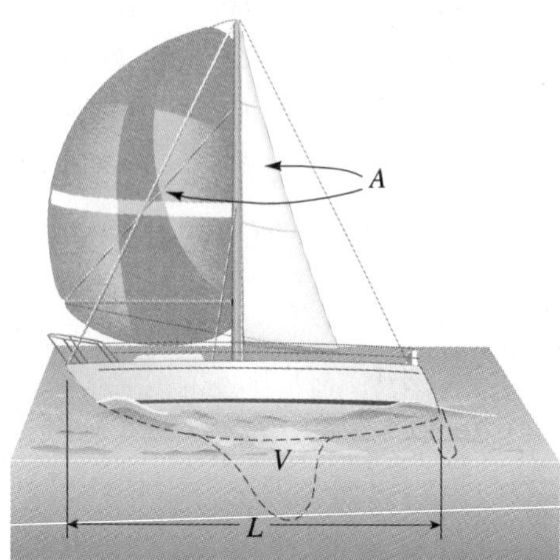

In general, a sailboat is capable of greater speed if it is longer, has a larger sail area, or displaces less water. To make sailing races fair, only boats in the same "class" can qualify to race together. For a certain race a boat is considered to qualify if

$$0.30L + 0.38A^{1/2} - 3V^{1/3} \leq 16$$

where L is measured in feet, A in square feet, and V in cubic feet. Use this inequality to answer the following questions.

(a) A sailboat has length 60 ft, sail area 3400 ft^2, and displacement 650 ft^3. Does this boat qualify for the race?

(b) Suppose a sailboat has length 65 ft and displaces 600 ft^3. What is the largest possible sail area that could be used and still allow the boat to qualify for this race?

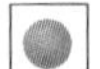

DISCOVERY • DISCUSSION

86. A Family of Equations The equation

$$3x + k - 5 = kx - k + 1$$

is really a **family of equations**, because for each value of k, we get a different equation with the unknown x. The letter k is called a **parameter** for this family. What value should we pick for k to make the given value of x a solution of the resulting equation?

(a) $x = 0$ **(b)** $x = 1$ **(c)** $x = 2$

87. Proof That $0 = 1$? The following steps appear to give equivalent equations, which seem to prove that $1 = 0$. Find the error.

$x = 1$	Given
$x^2 = x$	Multiply by x
$x^2 - x = 0$	Subtract x
$x(x - 1) = 0$	Factor
$\dfrac{x(x-1)}{x-1} = \dfrac{0}{x-1}$	Divide by $x - 1$
$x = 0$	Simplify
$1 = 0$	Given $x = 1$

88. Volumes of Solids The sphere, cylinder, and cone shown here all have the same radius r and the same volume V. Express the heights of the cylinder and cone in terms of r only, not V. (Use the volume formulas given on the inside back cover of this book.)

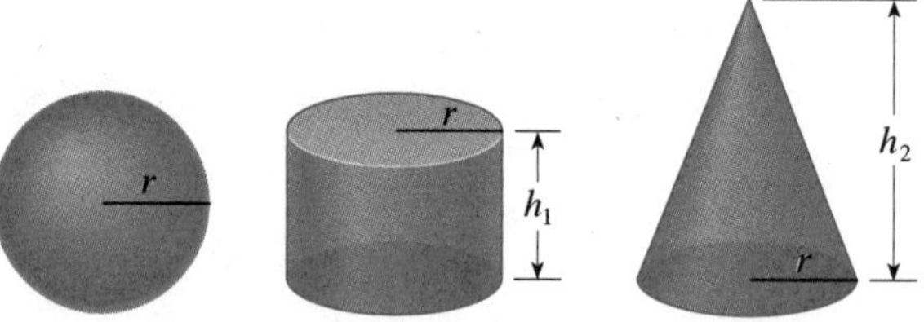

1 REVIEW

CONCEPT CHECK

1. Define each term in your own words. (Check by referring to the definition in the text.)
   (a) An integer
   (b) A rational number
   (c) An irrational number
   (d) A real number

2. State each of these properties of real numbers.
   (a) Commutative Property
   (b) Associative Property
   (c) Distributive Property

3. Explain what is meant by the union and intersection of sets. What notation is used for these ideas?

4. What is an open interval? What is a closed interval? What notation is used for these intervals?

5. What is the absolute value of a number?

6. (a) In the expression a^x, which is the base and which is the exponent?
   (b) What does a^x mean if $x = n$, a positive integer?
   (c) What if $x = 0$?
   (d) What if x is a negative integer: $x = -n$, where n is a positive integer?
   (e) What if $x = m/n$, a rational number?
   (f) State the Laws of Exponents.

7. (a) What does $\sqrt[n]{a} = b$ mean?
   (b) Why is $\sqrt{a^2} = |a|$?
   (c) How many real nth roots does a positive real number have if n is odd? If n is even?

8. Explain how the procedure of rationalizing the denominator works.

9. What is the difference between a variable and a constant in an algebraic expression?

10. State the Special Product Formulas for $(a + b)^2$, $(a - b)^2$, $(a + b)^3$, and $(a - b)^3$.

11. State each factoring formula.
    (a) Difference of squares
    (b) Difference of cubes
    (c) Sum of cubes

12. What is a solution of an equation?

13. Describe how you would solve the general linear equation $ax + b = 0$ for the unknown x.

14. Describe how you would solve the equation $x^n = a$, where $a > 0$.

EXERCISES

1–4 ■ State the property of real numbers being used.

1. $x + 5 = 5 + x$
2. $(a + b)(a - b) = (a - b)(a + b)$
3. $A(x + y) = Ax + Ay$
4. $(A + 1)(x + y) = (A + 1)x + (A + 1)y$

5–6 ■ Express the interval in terms of inequalities, and then graph the interval.

5. $(-1, 3]$
6. $(-\infty, 4]$

7–8 ■ Express the inequality in interval notation, and then graph the corresponding interval.

7. $x > 2$
8. $1 \leq x \leq 6$

9–18 ■ Evaluate the expression.

9. $\big|3 - |-9|\big|$
10. $1 - \big|1 - |-1|\big|$
11. $2^{-3} - 3^{-2}$
12. $\sqrt[3]{-125}$
13. $216^{-1/3}$
14. $64^{2/3}$
15. $\dfrac{\sqrt{242}}{\sqrt{2}}$
16. $\sqrt[4]{4}\,\sqrt[4]{324}$
17. $2^{1/2}8^{1/2}$
18. $\sqrt{2}\,\sqrt{50}$

19–26 ■ Write the expression as a power of x.

19. $\dfrac{1}{x^2}$
20. $x\sqrt{x}$
21. $x^2x^m(x^3)^m$
22. $((x^m)^2)^n$

23. $x^a x^b x^c$

24. $((x^a)^b)^c$

25. $x^{c+1}(x^{2c-1})^2$

26. $\dfrac{(x^2)^n x^5}{x^n}$

27–36 ■ Simplify the expression.

27. $(2x^3y)^2(3x^{-1}y^2)$

28. $(a^2)^{-3}(a^3b)^2(b^3)^4$

29. $\dfrac{x^4(3x)^2}{x^3}$

30. $\left(\dfrac{r^2s^{4/3}}{r^{1/3}s}\right)^6$

31. $\sqrt[3]{(x^3y)^2y^4}$

32. $\sqrt{x^2y^4}$

33. $\dfrac{x}{2+\sqrt{x}}$

34. $\dfrac{\sqrt{x}+1}{\sqrt{x}-1}$

35. $\dfrac{8r^{1/2}s^{-3}}{2r^{-2}s^4}$

36. $\left(\dfrac{ab^2c^{-3}}{2a^3b^{-4}}\right)^{-2}$

37. Write the number 78,250,000,000 in scientific notation.

38. Write the number 2.08×10^{-8} in ordinary decimal notation.

39. If $a \approx 0.00000293$, $b \approx 1.582 \times 10^{-14}$, and $c \approx 2.8064 \times 10^{12}$, use a calculator to approximate the number ab/c.

40. If your heart beats 80 times per minute and you live to be 90 years old, estimate the number of times your heart beats during your lifetime. State your answer in scientific notation.

41–60 ■ Factor the expression.

41. $12x^2y^4 - 3xy^5 + 9x^3y^2$

42. $x^2 - 9x + 18$

43. $x^2 + 3x - 10$

44. $6x^2 + x - 12$

45. $4t^2 - 13t - 12$

46. $x^4 - 2x^2 + 1$

47. $25 - 16t^2$

48. $2y^6 - 32y^2$

49. $x^6 - 1$

50. $y^3 - 2y^2 - y + 2$

51. $x^{-1/2} - 2x^{1/2} + x^{3/2}$

52. $a^4b^2 + ab^5$

53. $4x^3 - 8x^2 + 3x - 6$

54. $8x^3 + y^6$

55. $(x^2+2)^{5/2} + 2x(x^2+2)^{3/2} + x^2\sqrt{x^2+2}$

56. $3x^3 - 2x^2 + 18x - 12$

57. $a^2y - b^2y$

58. $ax^2 + bx^2 - a - b$

59. $(x+1)^2 - 2(x+1) + 1$

60. $(a+b)^2 + 2(a+b) - 15$

61–84 ■ Perform the indicated operations.

61. $(2x+1)(3x-2) - 5(4x-1)$

62. $(2y-7)(2y+7)$

63. $(2a^2-b)^2$

64. $(1+x)(2-x) - (3-x)(3+x)$

65. $(x-1)(x-2)(x-3)$

66. $(2x+1)^3$

67. $\sqrt{x}\left(\sqrt{x}+1\right)\left(2\sqrt{x}-1\right)$

68. $x^3(x-6)^2 + x^4(x-6)$

69. $x^2(x-2) + x(x-2)^2$

70. $\dfrac{x^3+2x^2+3x}{x}$

71. $\dfrac{x^2-2x-3}{2x^2+5x+3}$

72. $\dfrac{t^3-1}{t^2-1}$

73. $\dfrac{x^2+2x-3}{x^2+8x+16} \cdot \dfrac{3x+12}{x-1}$

74. $\dfrac{x^3/(x-1)}{x^2/(x^3-1)}$

75. $\dfrac{x^2-2x-15}{x^2-6x+5} \div \dfrac{x^2-x-12}{x^2-1}$

76. $x - \dfrac{1}{x+1}$

77. $\dfrac{1}{x-1} - \dfrac{x}{x^2+1}$

78. $\dfrac{2}{x} + \dfrac{1}{x-2} + \dfrac{3}{(x-2)^2}$

79. $\dfrac{1}{x-1} - \dfrac{2}{x^2-1}$

80. $\dfrac{1}{x+2} + \dfrac{1}{x^2-4} - \dfrac{2}{x^2-x-2}$

81. $\dfrac{\dfrac{1}{x} - \dfrac{1}{2}}{x-2}$

82. $\dfrac{\dfrac{1}{x} - \dfrac{1}{x+1}}{\dfrac{1}{x} + \dfrac{1}{x+1}}$

83. $\dfrac{3(x+h)^2 - 5(x+h) - (3x^2-5x)}{h}$

84. $\dfrac{\sqrt{x+h}-\sqrt{x}}{h}$ (rationalize the numerator)

85–104 ■ Find all real solutions of the equation.

85. $3x + 12 = 24$

86. $5x - 7 = 42$

87. $7x - 6 = 4x + 9$

88. $8 - 2x = 14 + x$

89. $\frac{1}{3}x - \frac{1}{2} = 2$

90. $\frac{2}{3}x + \frac{3}{5} = \frac{1}{5} - 2x$

91. $2(x + 3) - 4(x - 5) = 8 - 5x$

92. $\dfrac{x - 5}{2} - \dfrac{2x + 5}{3} = \dfrac{5}{6}$

93. $\dfrac{x + 1}{x - 1} = \dfrac{2x - 1}{2x + 1}$

94. $\dfrac{x}{x + 2} - 3 = \dfrac{1}{x + 2}$

95. $x^2 = 144$

96. $x^3 - 27 = 0$

97. $5x^3 - 15 = 0$

98. $6x^4 + 15 = 0$

99. $(x + 1)^3 = -64$

100. $(x + 2)^2 - 2 = 0$

101. $\sqrt[3]{x} = -3$

102. $x^{2/3} - 4 = 0$

103. $4x^{3/4} - 500 = 0$

104. $(x - 2)^{1/5} = 2$

105–111 ■ State whether the given equation is true for all values of the variables. (Disregard any value that makes a denominator 0.)

105. $(x + y)^3 = x^3 + y^3$

106. $\dfrac{1 + \sqrt{a}}{1 - a} = \dfrac{1}{1 - \sqrt{a}}$

107. $\dfrac{12 + y}{y} = \dfrac{12}{y} + 1$

108. $\sqrt[3]{a + b} = \sqrt[3]{a} + \sqrt[3]{b}$

109. $\sqrt{a^2} = a$

110. $\dfrac{1}{x + 4} = \dfrac{1}{x} + \dfrac{1}{4}$

111. $x^3 + y^3 = (x + y)(x^2 + xy + y^2)$

112. If $m > n > 0$ and $a = 2mn$, $b = m^2 - n^2$, $c = m^2 + n^2$, show that $a^2 + b^2 = c^2$.

113. If $t = \dfrac{1}{2}\left(x^3 - \dfrac{1}{x^3}\right)$ and $x > 0$, show that

$$\sqrt{1 + t^2} = \frac{1}{2}\left(x^3 + \frac{1}{x^3}\right)$$

114. Assume $a = b$. What is wrong with the following argument?

$$a = b$$
$$a^2 = ab$$
$$a^2 - b^2 = ab - b^2$$
$$(a + b)(a - b) = b(a - b)$$
$$a + b = b$$

Now set $a = b = 1$: $\quad 2 = 1$

1 TEST

1. Write an algebraic expression for the indicated quantity.
   (a) The number of hours H in n days
   (b) The product P of three consecutive integers, where the middle one is n.

2. (a) Graph the intervals $[-3, 2]$ and $(4, \infty)$ on a real number line.
   (b) Express the inequalities $x < 5$ and $-2 \leqslant x \leqslant 1$ in interval notation.
   (c) Find the distance between -22 and 31 on the number line.

3. Evaluate each expression.
   (a) $(-3)^4$ (b) 2^{-4} (c) $\dfrac{5^{18}}{5^{12}}$

4. Evaluate each expression.
   (a) $\left(\dfrac{2}{3}\right)^{-1}$ (b) $\dfrac{\sqrt{32}}{\sqrt{8}}$ (c) $16^{-3/4}$

5. Express $\dfrac{(x^2)^a(\sqrt{x})^b}{x^{a+b}x^{a-b}}$ as a power of x.

6. Simplify each expression.
   (a) $\sqrt{200} - \sqrt{8}$ (b) $(2a^3b^2)(3ab^4)^3$
   (c) $\left(\dfrac{x^2y^{-3}}{y^5}\right)^{-4}$ (d) $\left(\dfrac{2x^{1/4}}{y^{1/3}x^{1/6}}\right)^3$

7. Simplify each expression.
   (a) $\dfrac{x^2 + 3x + 2}{x^2 - x - 2}$ (b) $\dfrac{x^2}{x^2 - 4} - \dfrac{x + 1}{x + 2}$ (c) $\dfrac{\dfrac{y}{x} - \dfrac{x}{y}}{\dfrac{1}{y} - \dfrac{1}{x}}$

8. Write each number in scientific notation.
   (a) 325,000,000,000 (b) 0.000008931

9. Perform the indicated operations and simplify.
   (a) $4(3 - x) - 3(x + 5)$ (b) $(x - 5)(2x + 3)$ (c) $(\sqrt{x} + \sqrt{y})(\sqrt{x} - \sqrt{y})$
   (d) $(3t + 4)^2$ (e) $(2 - x^2)^3$

10. Factor each expression completely.
    (a) $9x^2 - 25$ (b) $6x^2 + 7x - 5$ (c) $x^3 - 4x^2 - 3x + 12$
    (d) $x^4 + 27x$ (e) $3x^{3/2} - 9x^{1/2} + 6x^{-1/2}$ (f) $x^3y - 4xy$

11. Rationalize the denominator and simplify: $\dfrac{\sqrt{10}}{\sqrt{5} - 2}$

12. Solve for x.
    (a) $2x + 7 = 12 + \frac{5}{2}x$ (b) $\dfrac{2x}{x + 1} = \dfrac{2x - 1}{x}$
    (c) $3(x - 2)^2 - 16 = 0$ (d) $x^5 = -243$ (e) $3x^{2/3} - 12 = 0$

13. Solve for w in the equation $V = 4wh^2 + 2wh$.

PRINCIPLES OF PROBLEM SOLVING

There are no hard and fast rules that will ensure success in solving problems. However, it is possible to outline some general steps in the problem-solving process and to give principles that are useful in solving certain problems. These steps and principles are just common sense made explicit. They have been adapted from George Polya's insightful book *How To Solve It.*

1 UNDERSTAND THE PROBLEM

The first step is to read the problem and make sure that you understand it. Ask yourself the following questions:

What is the unknown?
What are the given quantities?
What are the given conditions?

For many problems it is useful to

draw a diagram

and identify the given and required quantities on the diagram.

Usually it is necessary to

introduce suitable notation

In choosing symbols for the unknown quantities, we often use letters such as a, b, c, m, n, x, and y, but in some cases it helps to use initials as suggestive symbols, for instance, V for volume or t for time.

2 THINK OF A PLAN

Find a connection between the given information and the unknown that enables you to calculate the unknown. It often helps to ask yourself explicitly: "How can I relate the given to the unknown?" If you don't see a connection immediately, the following ideas may be helpful in devising a plan.

■ Try to recognize something familiar

Relate the given situation to previous knowledge. Look at the unknown and try to recall a more familiar problem that has a similar unknown.

■ Try to recognize patterns

Certain problems are solved by recognizing that some kind of pattern is occurring. The pattern could be geometric, or numerical, or algebraic. If you can see regularity or repetition in a problem, then you might be able to guess what the pattern is and then prove it.

■ **Use analogy**

Try to think of an analogous problem, that is, a similar or related problem, but one that is easier than the original. If you can solve the similar, simpler problem, then it might give you the clues you need to solve the original, more difficult one. For instance, if a problem involves very large numbers, you could first try a similar problem with smaller numbers. Or if the problem is in three-dimensional geometry, you could look for something similar in two-dimensional geometry. Or if the problem you start with is a general one, you could first try a special case.

■ **Introduce something extra**

You may sometimes need to introduce something new—an auxiliary aid—to make the connection between the given and the unknown. For instance, in a problem for which a diagram is useful, the auxiliary aid could be a new line drawn in the diagram. In a more algebraic problem the aid could be a new unknown that relates to the original unknown.

■ **Take cases**

You may sometimes have to split a problem into several cases and give a different argument for each case. For instance, we often have to use this strategy in dealing with absolute value.

■ **Work backward**

Sometimes it is useful to imagine that your problem is solved and work backward, step by step, until you arrive at the given data. Then you may be able to reverse your steps and thereby construct a solution to the original problem. This procedure is commonly used in solving equations. For instance, in solving the equation $3x - 5 = 7$, we suppose that x is a number that satisfies $3x - 5 = 7$ and work backward. We add 5 to each side of the equation and then divide each side by 3 to get $x = 4$. Since each of these steps can be reversed, we have solved the problem.

■ **Establish subgoals**

In a complex problem it is often useful to set subgoals (in which the desired situation is only partially fulfilled). If you can attain or accomplish these subgoals, then you may be able to build on them to reach your final goal.

■ **Indirect reasoning**

Sometimes it is appropriate to attack a problem indirectly. In using **proof by contradiction** to prove that P implies Q, we assume that P is true and Q is false and try

George Polya (1887–1985) is famous among mathematicians for his ideas on problem solving. His lectures on problem solving at Stanford University attracted overflow crowds whom he held on the edges of their seats, leading them to discover solutions for themselves. He was able to do this because of his deep insight into the psychology of problem solving. His well-known book *How To Solve It* has been translated into 15 languages. He said that Euler (see page 165) was unique among great mathematicians because he explained *how* he found his results. Polya often said to his students and colleagues, "Yes, I see that your proof is correct, but how did you discover it?" In the preface to *How To Solve It*, Polya writes, "A great discovery solves a great problem but there is a grain of discovery in the solution of any problem. Your problem may be modest; but if it challenges your curiosity and brings into play your inventive faculties, and if you solve it by your own means, you may experience the tension and enjoy the triumph of discovery."

to see why this cannot happen. Somehow we have to use this information and arrive at a contradiction to what we absolutely know is true.

■ **Mathematical induction**

In proving statements that involve a positive integer n, it is frequently helpful to use the Principle of Mathematical Induction, which is discussed in Section 10.5.

3 CARRY OUT THE PLAN

In Step 2, a plan was devised. In carrying out that plan, you must check each stage of the plan and write the details that prove each stage is correct.

4 LOOK BACK

Having completed your solution, it is wise to look back over it, partly to see if any errors have been made and partly to see if you can discover an easier way to solve the problem. Looking back also familiarizes you with the method of solution, and this may be useful for solving a future problem. Descartes said, "Every problem that I solved became a rule which served afterwards to solve other problems."

We illustrate some of these principles of problem solving with an example. Further illustrations of these principles will be presented at the end of selected chapters.

PROBLEM ■ Average Speed

A driver sets out on a journey. For the first half of the distance she drives at the leisurely pace of 30 mi/h; during the second half she drives 60 mi/h. What is her average speed on this trip?

PRELIMINARY THOUGHTS It is tempting to take the average of the speeds and say that the average speed for the entire trip is

$$\frac{30 + 60}{2} = 45 \text{ mi/h}$$

But is this simple-minded approach really correct?

Try a special case

Let's look at an easily calculated special case. Suppose that the total distance traveled is 120 mi. Since the first 60 mi is traveled at 30 mi/h, it takes 2 h. The second 60 mi is traveled at 60 mi/h, so it takes one hour. Thus, the total time is $2 + 1 = 3$ hours and the average speed is

$$\frac{120}{3} = 40 \text{ mi/h}$$

So our guess of 45 mi/h was wrong.

SOLUTION

Understand the problem

We need to look more carefully at the meaning of average speed. It is defined as

$$\text{average speed} = \frac{\text{distance traveled}}{\text{time elapsed}}$$

Introduce notation

State what is given

Let d be the distance traveled on each half of the trip. Let t_1 and t_2 be the times taken for the first and second halves of the trip. Now we can write down the information we have been given. For the first half of the trip, we have

(1) $$30 = \frac{d}{t_1}$$

and for the second half, we have

(2) $$60 = \frac{d}{t_2}$$

Identify the unknown

Now we identify the quantity we are asked to find:

$$\text{average speed for entire trip} = \frac{\text{total distance}}{\text{total time}} = \frac{2d}{t_1 + t_2}$$

Connect the given with the unknown

To calculate this quantity, we need to know t_1 and t_2, so we solve Equations 1 and 2 for these times:

$$t_1 = \frac{d}{30} \qquad t_2 = \frac{d}{60}$$

Now we have the ingredients needed to calculate the desired quantity:

$$\text{average speed} = \frac{2d}{t_1 + t_1} = \frac{2d}{\dfrac{d}{30} + \dfrac{d}{60}}$$

$$= \frac{60(2d)}{60\left(\dfrac{d}{30} + \dfrac{d}{60}\right)} \qquad \text{Multiply numerator and denominator by 60}$$

$$= \frac{120d}{2d + d} = \frac{120d}{3d} = 40$$

So, the average speed for the entire trip is 40 mi/h. ■

PROBLEMS

1. A man drives from home to work at a speed of 50 mi/h. The return trip from work to home is traveled at the more leisurely pace of 30 mi/h. What is the man's average speed for the round-trip?

2. An old car has to travel a 2-mile route, uphill and down. Because it is so old, the car can climb the first mile—the ascent—no faster than an average speed of 15 mi/h. How fast does the car have to travel the second mile—on the descent it can go faster, of course—in order to achieve an average speed of 30 mi/h for the trip?

3. A car and a van are parked 120 mi apart on a straight road. The drivers start driving toward each other at noon, each at a speed of 40 mi/h. A fly starts from the front bumper of the van at noon and flies to the bumper of the car, then immediately back to the bumper of the van, back to the car, and so on, until the car and the van meet. If the fly flies at a speed of 100 mi/h, what is the total distance it travels?

4. Which price is better for the buyer, a 40% discount or two successive discounts of 20%?

5. Use a calculator to find the value of the expression

$$\sqrt{3 + 2\sqrt{2}} - \sqrt{3 - 2\sqrt{2}}$$

The number looks very simple. Show that the calculated value is correct.

6. Use a calculator to evaluate

$$\frac{\sqrt{2} + \sqrt{6}}{\sqrt{2 + \sqrt{3}}}$$

Show that the calculated value is correct.

7. An amoeba propagates by simple division; each split takes 3 minutes to complete. When such an amoeba is put into a glass container with a nutrient fluid, the container is full of amoebas in one hour. How long would it take for the container to be filled if we start with not one amoeba, but two?

8. Two runners start running laps at the same time, from the same starting position. George runs a lap in 50 s; Sue runs a lap in 30 s. When will the runners next be side by side?

9. Player A has a higher batting average than player B for the first half of the baseball season. Player A also has a higher batting average than player B for the second half of the season. Is it necessarily true that player A has a higher batting average than player B for the entire season?

10. A woman starts at a point P on the earth's surface and walks 1 mi south, then 1 mi east, then 1 mi north, and finds herself back at P, the starting point. Describe all points P for which this is possible (there are infinitely many).

Don't feel bad if you don't solve these problems right away. Problems 2 and 7 were sent to Albert Einstein by his friend Wertheimer. Einstein (and his friend Bucky) enjoyed the problems and wrote back to Wertheimer. Here is part of his reply:

> Your letter gave us a lot of amusement. The first intelligence test fooled both of us (Bucky and me). Only on working it out did I notice that no time is available for the downhill run! Mr. Bucky was also taken in by the second example, but I was not. Such drolleries show us how stupid we are!

(See *Mathematical Intelligencer*, Spring 1990, page 41)

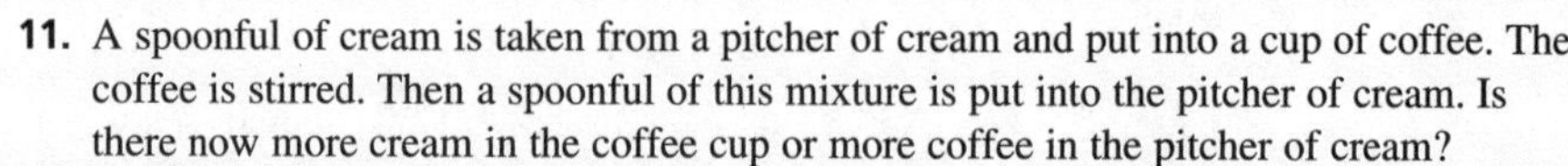

11. A spoonful of cream is taken from a pitcher of cream and put into a cup of coffee. The coffee is stirred. Then a spoonful of this mixture is put into the pitcher of cream. Is there now more cream in the coffee cup or more coffee in the pitcher of cream?

12. An ice cube is floating in a cup of water, full to the brim, as shown in the sketch. As the ice melts, what happens? Does the cup overflow, or does the water level drop, or does it remain the same? (You need to know Archimedes' principle: A floating object displaces a volume of water whose weight equals the weight of the object.)

13. An extended family consists of the members

{Baby, Son, Daughter, Mother, Father, Uncle, Aunt, Grandma, Grandpa}

listed in increasing order of age. If x and y are members of this family, define $x + y$ to be the older of x and y, and define $x \cdot y$ to be the younger of x and y.

(a) Find (Baby + Uncle) + Mother.

(b) Find Father $\cdot$ (Grandpa + Aunt).

(c) Show that the operations + and $\cdot$ that we have defined satisfy the same Commutative, Associative, and Distributive Properties as do the real numbers with regular addition and multiplication.

(d) Show that the operations also satisfy the property

$$x + (y \cdot z) = (x + y) \cdot (x + z)$$

Do the operations of regular addition and multiplication with real numbers also satisfy this property?

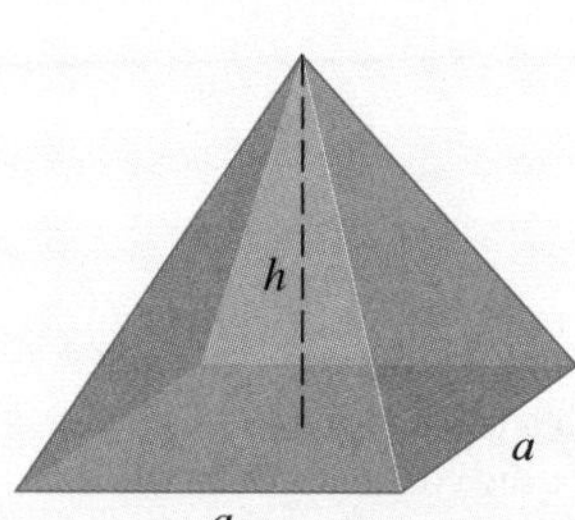

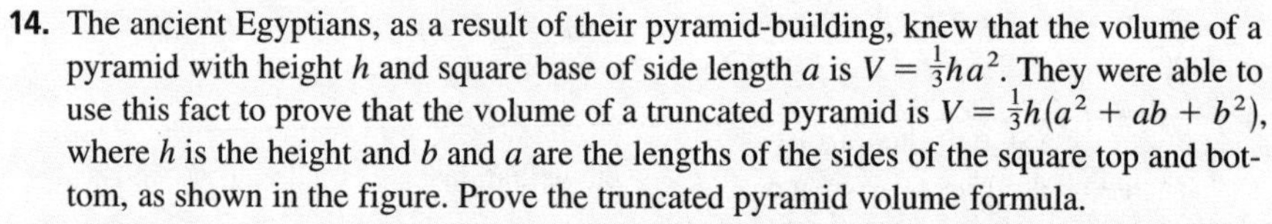

14. The ancient Egyptians, as a result of their pyramid-building, knew that the volume of a pyramid with height h and square base of side length a is $V = \frac{1}{3}ha^2$. They were able to use this fact to prove that the volume of a truncated pyramid is $V = \frac{1}{3}h(a^2 + ab + b^2)$, where h is the height and b and a are the lengths of the sides of the square top and bottom, as shown in the figure. Prove the truncated pyramid volume formula.

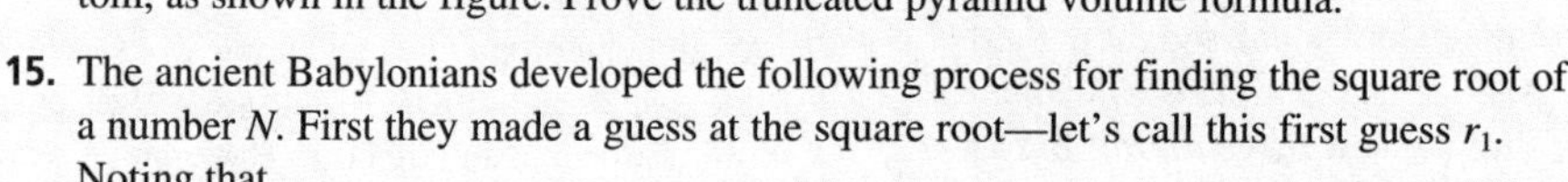

15. The ancient Babylonians developed the following process for finding the square root of a number N. First they made a guess at the square root—let's call this first guess r_1. Noting that

$$r_1 \cdot \left(\frac{N}{r_1}\right) = N$$

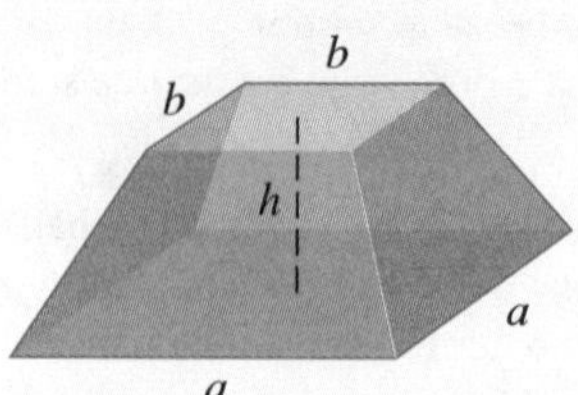

they concluded that the actual square root must be somewhere between r_1 and N/r_1, so their next guess for the square root, r_2, was the average of these two numbers:

$$r_2 = \frac{1}{2}\left(r_1 + \frac{N}{r_1}\right)$$

Continuing in this way, their next approximation was given by

$$r_3 = \frac{1}{2}\left(r_2 + \frac{N}{r_2}\right)$$

and so on. In general, once we have the nth approximation to the square root of N, we find the $(n + 1)$st using

$$r_{n+1} = \frac{1}{2}\left(r_n + \frac{N}{r_n}\right)$$

Use this procedure to find $\sqrt{72}$, correct to two decimal places.

16. This problem is adapted from a tenth century Arab manuscript. In his will, a nobleman leaves half of his horses to his oldest son, a third to his second son, and a ninth to his youngest. At his death he has 17 horses, so none of these bequests results in a whole number of horses for any of the sons. The executor of the will solves this dilemma by adding one of his own horses to the estate, making a total of 18. Now, according to the provisions of the will, he gives 9 to the oldest son, 6 to the second, and 2 to the third. Thus, each son inherits a little more than he was originally entitled to, and one horse is left over, which the executor takes back again. Everybody is happy, but something seems wrong. What, in fact, *is* wrong?

17. A piece of wire is bent as shown in the figure. You can see that one cut through the wire produces four pieces and two parallel cuts produce seven pieces. How many pieces will be produced by 142 parallel cuts? Write a formula for the number of pieces produced by n parallel cuts.

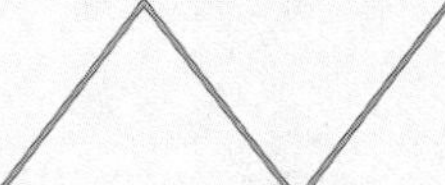
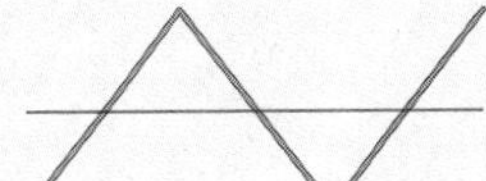
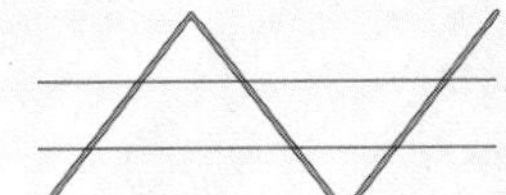

18. Find the last digit in the number 3^{459}. [*Hint:* Calculate the first few powers of 3, and look for a pattern.]

19. How many digits does the number $8^{15} \cdot 5^{37}$ have? [*Hint:* Express the number as a multiple of a power of 10.]

20. Use the techniques of solving a simpler problem and looking for a pattern to evaluate the number

$$3999999999999^2$$

21. Suppose $x + y = 1$ and $x^2 + y^2 = 4$. Find $x^3 + y^3$.

22. A red ribbon is tied tightly around the earth at the equator. How much more ribbon would you need if you raised the ribbon 1 ft above the equator everywhere? (You don't need to know the radius of the earth to solve this problem.)

2

COORDINATES AND GRAPHS

Coordinates are used to describe the location of a sailboat in the ocean or a satellite in space.

The function of mathematics in providing answers is often less important than its function in providing understanding.

BEN NOBLE

In this chapter we study the coordinate plane. In the coordinate plane, we can draw a graph of an equation, thus allowing us to "see" the relationship of the variables in the equation. We can also reverse this process and describe a geometrical figure (such as a circle or a line) by an algebraic equation. Thus, the coordinate plane is the link between algebra and geometry that enables us to use the techniques of algebra to solve geometric problems and the techniques of geometry to solve algebraic problems, thereby giving us a deeper understanding of both fields.

2.1 THE COORDINATE PLANE

The Cartesian plane is named in honor of the French mathematician René Descartes (1596–1650), although another Frenchman, Pierre Fermat (1601–1665), also invented the principles of coordinate geometry at the same time. (See their biographies on pages 76 and 578.)

Just as points on a line can be identified with real numbers to form the coordinate line, points in a plane can be identified with ordered pairs of numbers to form the **coordinate plane** or **Cartesian plane**. To do this, we draw two perpendicular real lines that intersect at 0 on each line. Usually one line is horizontal with positive direction to the right and is called the ***x*-axis**; the other line is vertical with positive direction upward and is called the ***y*-axis**. The point of intersection of the x-axis and the y-axis is the **origin *O***, and the two axes divide the plane into four **quadrants**, labeled I, II, III, and IV in Figure 1. (The points *on* the coordinate axes are not assigned to any quadrant.)

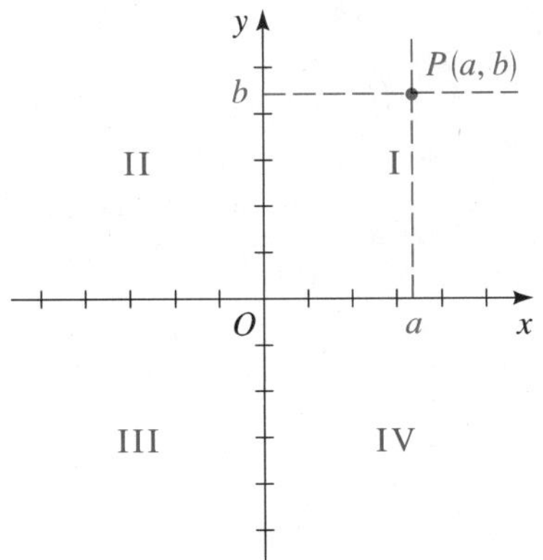

FIGURE 1

Although the notation for a point (a, b) is the same as the notation for an open interval (a, b), the context should make clear which meaning is intended.

Any point P in the coordinate plane can be located by a unique **ordered pair** of numbers (a, b), as shown in Figure 1. The first number a is called the ***x*-coordinate** of P; the second number b is called the ***y*-coordinate** of P. We can think of the coordinates of P as its "address," because they specify its location in the plane.

René Descartes (1596–1650) was born in the town of La Haye in southern France. From an early age Descartes liked mathematics because of "the certainty of its results and the clarity of its reasoning." He believed that in order to arrive at truth, one must begin by doubting everything, including one's own existence; this led him to formulate perhaps the most well-known sentence in all of philosophy: "I think, therefore I am." In his book *Discourse on Method* he described what is now called the Cartesian plane. This idea of combining algebra and geometry enabled mathematicians for the first time to "see" the equations they were studying. The philosopher John Stuart Mill called this invention "the greatest single step ever made in the progress of the exact sciences." Descartes liked to get up late and spend the morning in bed thinking and writing. He invented the coordinate plane while lying in bed watching a fly crawl on the ceiling, reasoning that he could describe the exact location of the fly by knowing its distance from two perpendicular walls. In 1649 Descartes became the tutor of Queen Christina of Sweden. She liked her lessons at 5 o'clock in the morning when, she said, her mind

(continued)

Several points are labeled with their coordinates in Figure 2.

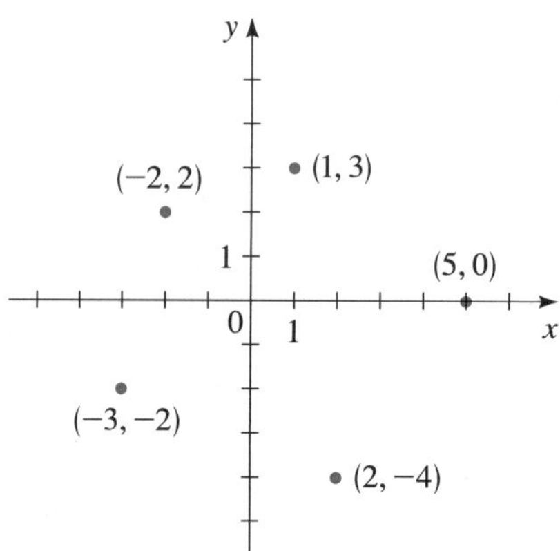

FIGURE 2

EXAMPLE 1 ■ Graphing Regions in the Coordinate Plane

Describe and sketch the regions given by each set.

(a) $\{(x, y) \mid x \geqslant 0\}$ (b) $\{(x, y) \mid y = 1\}$ (c) $\{(x, y) \mid -1 < y < 1\}$

SOLUTION

(a) The points whose x-coordinates are 0 or positive lie on the y-axis or to the right of it, as shown in Figure 3(a).

(b) The set of all points with y-coordinate 1 is a horizontal line one unit above the x-axis, as in Figure 3(b).

(c) The given region consists of those points in the plane whose y-coordinates lie between -1 and 1. Thus, the region consists of all points that lie between (but not on) the horizontal lines $y = 1$ and $y = -1$. These lines are shown as broken lines in Figure 3(c) to indicate that the points on these lines do not lie in the set.

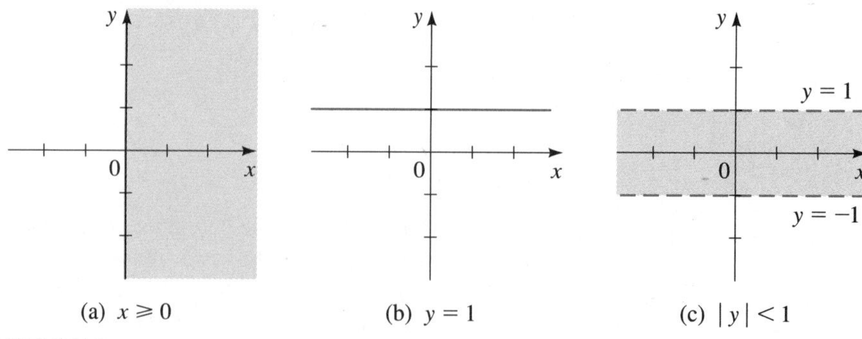

(a) $x \geqslant 0$ (b) $y = 1$ (c) $|y| < 1$

FIGURE 3

■

was sharpest. However, the change from his usual habits and the ice-cold library where they studied proved too much for him. In February 1650, after just two months of this, he caught pneumonia and died.

We now find a formula for the distance $d(A, B)$ between two points $A(x_1, y_1)$ and $B(x_2, y_2)$ in the plane. Recall from Section 1.2 that the distance between points a and b on a number line is $d(a, b) = |b - a|$. So, from Figure 4 we see that the distance between the points $A(x_1, y_1)$ and $C(x_2, y_1)$ on a horizontal line must be $|x_2 - x_1|$, and the distance between $B(x_2, y_2)$ and $C(x_2, y_1)$ on a vertical line must be $|y_2 - y_1|$. Since triangle ABC is a right triangle, the Pythagorean Theorem gives

$$\begin{aligned} d(A, B) &= \sqrt{|x_2 - x_1|^2 + |y_2 - y_1|^2} \\ &= \sqrt{(x_2 - x_1)^2 + (y_2 - y_1)^2} \end{aligned}$$

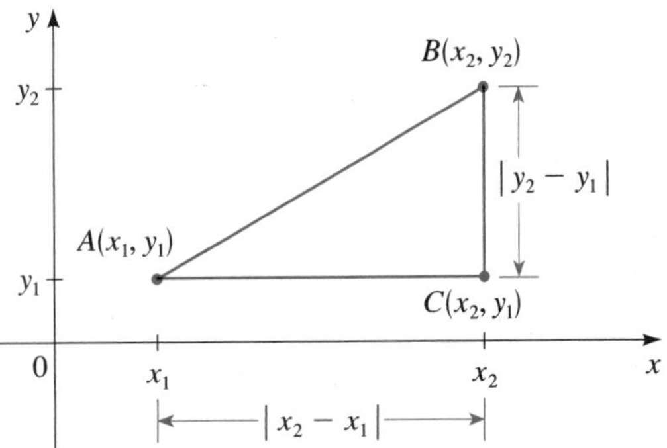

FIGURE 4

DISTANCE FORMULA

The distance between the points $A(x_1, y_1)$ and $B(x_2, y_2)$ in the plane is

$$d(A, B) = \sqrt{(x_2 - x_1)^2 + (y_2 - y_1)^2}$$

EXAMPLE 2 ■ Finding the Distance between Two Points

Find the distance between the points $A(2, 5)$ and $B(4, -1)$.

SOLUTION

Using the distance formula, we have

$$\begin{aligned} d(A, B) &= \sqrt{(4 - 2)^2 + (-1 - 5)^2} \\ &= \sqrt{2^2 + (-6)^2} \\ &= \sqrt{4 + 36} = \sqrt{40} \approx 6.32 \end{aligned}$$

See Figure 5. ■

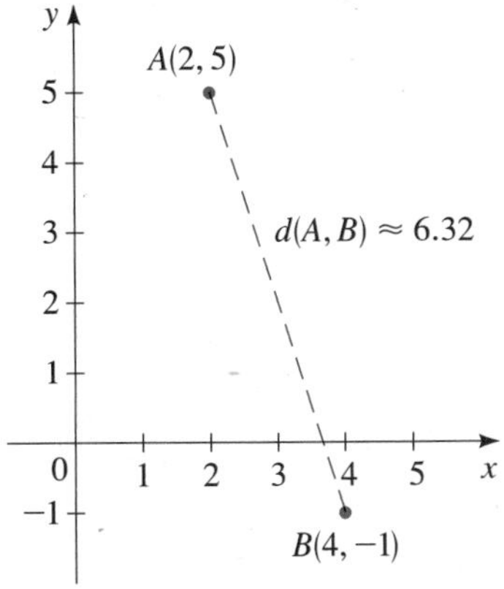

FIGURE 5

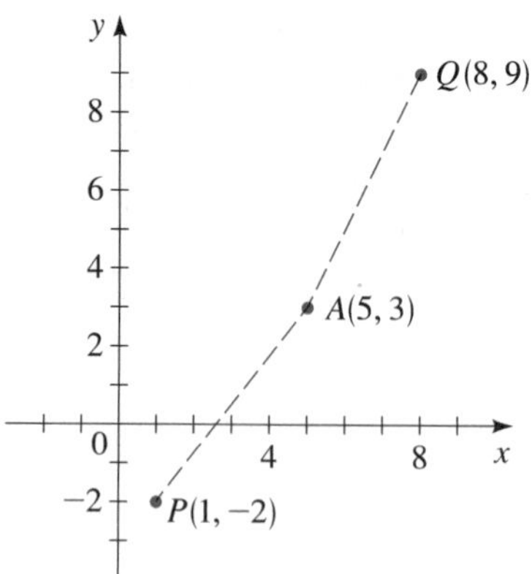

FIGURE 6

EXAMPLE 3 ■ Applying the Distance Formula

Which of the points $P(1, -2)$ or $Q(8, 9)$ is closer to the point $A(5, 3)$?

SOLUTION

By the distance formula, we have

$$d(P, A) = \sqrt{(5-1)^2 + [3-(-2)]^2} = \sqrt{4^2 + 5^2} = \sqrt{41}$$

$$d(Q, A) = \sqrt{(5-8)^2 + (3-9)^2} = \sqrt{(-3)^2 + (-6)^2} = \sqrt{45}$$

This shows that $d(P, A) < d(Q, A)$, so P is closer to A (see Figure 6). ■

Now let's find the coordinates (x, y) of the midpoint M of the line segment that joins the point $A(x_1, y_1)$ to the point $B(x_2, y_2)$. In Figure 7 notice that triangles APM and MQB are congruent because $d(A, M) = d(M, B)$ and the corresponding angles are equal.

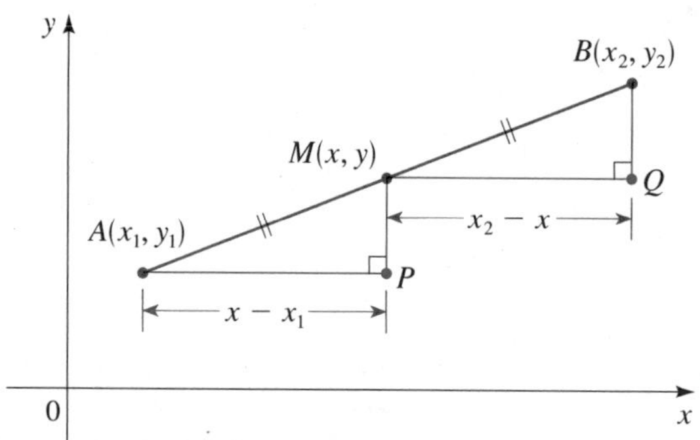

FIGURE 7

It follows that $d(A, P) = d(M, Q)$ and so

$$x - x_1 = x_2 - x$$

Solving this equation for x, we get

$$2x = x_1 + x_2$$

$$x = \frac{x_1 + x_2}{2}$$

Similarly, $$y = \frac{y_1 + y_2}{2}$$

MIDPOINT FORMULA

The midpoint of the line segment from $A(x_1, y_1)$ to $B(x_2, y_2)$ is

$$\left(\frac{x_1 + x_2}{2}, \frac{y_1 + y_2}{2}\right)$$

The coordinates of a point in the xy-plane uniquely determine its location. We can think of the coordinates as the "address" of the point. In Salt Lake City, Utah, the addresses of most buildings are in fact expressed as coordinates. The city is divided into quadrants with Main Street as the vertical (North-South) axis and S. Temple Street as the horizontal (East-West) axis. An address such as

1760 W 2100 S

indicates a location 17.6 blocks west of Main Street and 21 blocks south of S. Temple Street. (This is the address of the main post office in Salt Lake City.) With this logical system it is possible for someone unfamiliar with the city to locate any address immediately, as easily as one locates a point in the coordinate plane.

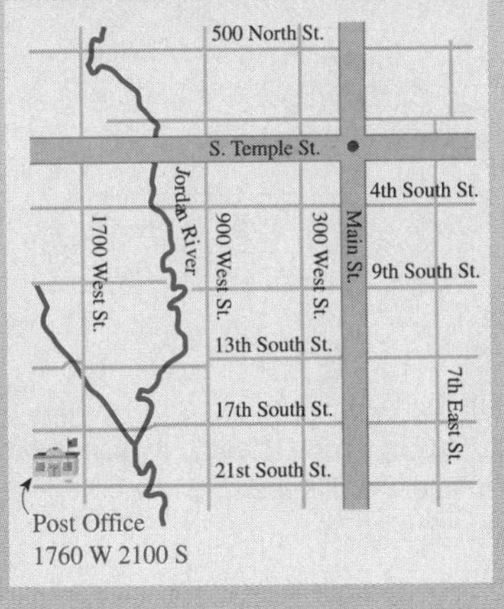

EXAMPLE 4 ■ Finding the Midpoint

The midpoint of the line segment that joins the points $(-2, 5)$ and $(4, 9)$ is

$$\left(\frac{-2+4}{2}, \frac{5+9}{2}\right) = (1, 7)$$

See Figure 8.

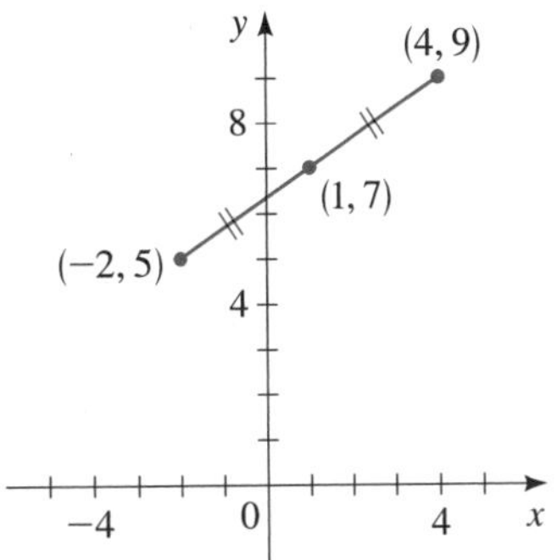

FIGURE 8

■

EXAMPLE 5 ■ Applying the Midpoint Formula

Show that the quadrilateral with vertices $P(1, 2)$, $Q(4, 4)$, $R(5, 9)$, and $S(2, 7)$ is a parallelogram by proving that its two diagonals bisect each other.

SOLUTION

If the two diagonals have the same midpoint, then they must bisect each other. The midpoint of the diagonal PR is

$$\left(\frac{1+5}{2}, \frac{2+9}{2}\right) = \left(3, \frac{11}{2}\right)$$

and the midpoint of the diagonal QS is

$$\left(\frac{4+2}{2}, \frac{4+7}{2}\right) = \left(3, \frac{11}{2}\right)$$

so each diagonal bisects the other, as shown in Figure 9. (A theorem from elementary geometry states that the quadrilateral is therefore a parallelogram.)

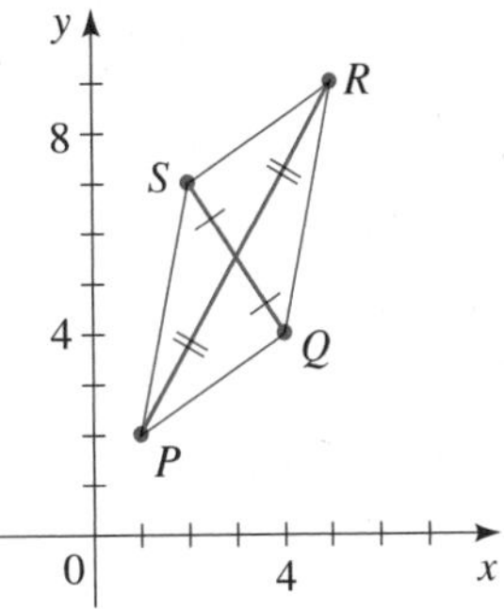

FIGURE 9

■

2.1 EXERCISES

1. Plot the given points in a coordinate plane:

   $(2, 3), (-2, 3), (4, 5), (4, -5), (-4, 5), (-4, -5)$

2. Find the coordinates of the points shown in the figure.

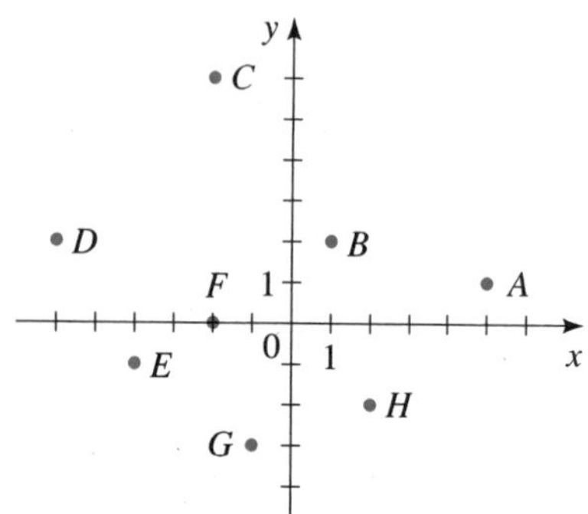

3–8 ■ A pair of points is given.

(a) Plot the points in a coordinate plane.
(b) Find the distance between them.
(c) Find the midpoint of the segment that joins them.

3. $(2, 3), \quad (5, 2)$
4. $(2, -1), \quad (4, 3)$
5. $(6, -2), \quad (-1, 3)$
6. $(1, -6), \quad (-1, -3)$
7. $(3, 4), \quad (-3, -4)$
8. $(5, 0), \quad (0, 6)$
9. Draw the rectangle with vertices $A(1, 3)$, $B(5, 3)$, $C(1, -3)$, and $D(5, -3)$ on a coordinate plane. Find the area of the rectangle.
10. Draw the parallelogram with vertices $A(1, 2)$, $B(5, 2)$, $C(3, 6)$, and $D(7, 6)$ on a coordinate plane. Find the area of the parallelogram.
11. Plot the points $A(1, 0)$, $B(5, 0)$, $C(4, 3)$, and $D(2, 3)$ on a coordinate plane. Draw the segments AB, BC, CD, and DA. What kind of quadrilateral is $ABCD$, and what is its area?
12. Plot the points $P(5, 1)$, $Q(0, 6)$, and $R(-5, 1)$ on a coordinate plane. Where must the point S be located so that the quadrilateral $PQRS$ is a square? Find the area of this square.

13–24 ■ Sketch the region given by the set.

13. $\{(x, y) \mid x \leqslant 0\}$
14. $\{(x, y) \mid y \geqslant 0\}$
15. $\{(x, y) \mid x = 3\}$
16. $\{(x, y) \mid y = -2\}$
17. $\{(x, y) \mid 1 < x < 2\}$
18. $\{(x, y) \mid 0 \leqslant y \leqslant 4\}$
19. $\{(x, y) \mid xy < 0\}$
20. $\{(x, y) \mid xy > 0\}$
21. $\{(x, y) \mid x \geqslant 1 \text{ and } y < 3\}$
22. $\{(x, y) \mid y > 1 \text{ or } y < -1\}$
23. $\{(x, y) \mid -1 < x \leqslant 2\}$
24. $\{(x, y) \mid -2 < x < 2 \text{ and } y \geqslant 3\}$
25. Which of the points $A(6, 7)$ or $B(-5, 8)$ is closer to the origin?
26. Which of the points $C(-6, 3)$ or $D(3, 0)$ is closer to the point $E(-2, 1)$?
27. Which of the points $P(3, 1)$ or $Q(-1, 3)$ is closer to the point $R(-1, -1)$?
28. (a) Show that the points $(7, 3)$ and $(3, 7)$ are the same distance from the origin.
    (b) Show that the points (a, b) and (b, a) are the same distance from the origin.
29. Show that the triangle with vertices $A(0, 2)$, $B(-3, -1)$, and $C(-4, 3)$ is isosceles.
30. Find the area of the triangle shown in the figure.

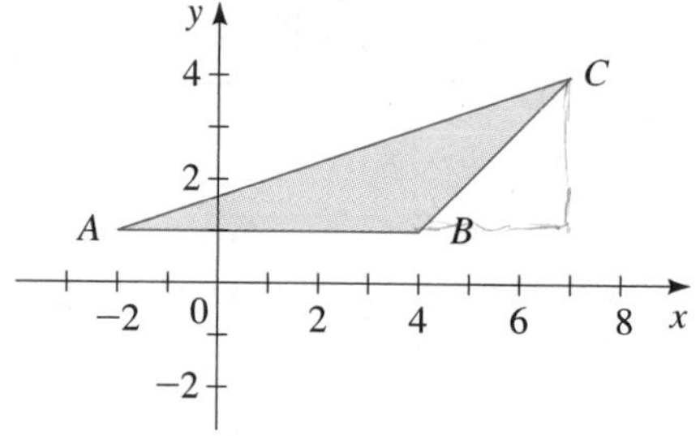

31. Refer to triangle ABC in the figure.
    (a) Show that triangle ABC is a right triangle by using the converse of the Pythagorean Theorem.
    (b) Find the area of triangle ABC.

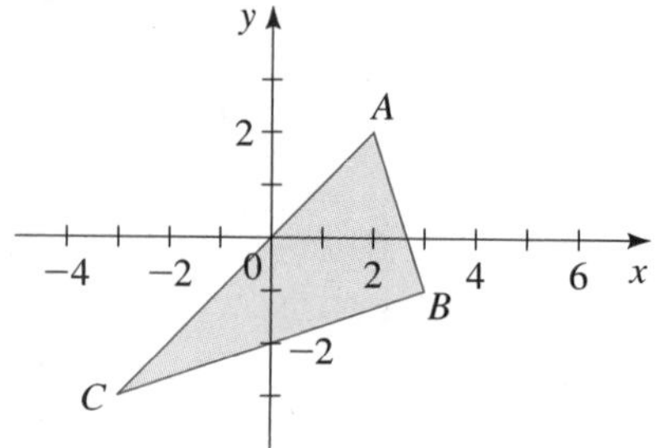

32. Show that the triangle with vertices $A(6, -7)$, $B(11, -3)$, and $C(2, -2)$ is a right triangle by using the converse of the Pythagorean Theorem. Find the area of the triangle.

33. Show that the points $A(-2, 9)$, $B(4, 6)$, $C(1, 0)$, and $D(-5, 3)$ are the vertices of a square.

34. Show that the points $A(-1, 3)$, $B(3, 11)$, and $C(5, 15)$ are collinear by showing that $d(A, B) + d(B, C) = d(A, C)$.

35. Find a point on the y-axis that is equidistant from the points $(5, -5)$ and $(1, 1)$.

36. Find the lengths of the medians of the triangle with vertices $A(1, 0)$, $B(3, 6)$, and $C(8, 2)$. (A *median* is a line segment from a vertex to the midpoint of the opposite side.)

37. Find the point that is one-fourth of the distance from the point $P(-1, 3)$ to the point $Q(7, 5)$ along the segment PQ.

38. Plot the points $P(-2, 1)$ and $Q(12, -1)$. Which (if either) of the points $A(5, -7)$ and $B(6, 7)$ lies on the perpendicular bisector of the segment PQ?

39. Plot the points $P(-1, -4)$, $Q(1, 1)$, and $R(4, 2)$ on a coordinate plane. Where should the point S be located so that the figure $PQRS$ is a parallelogram?

40. If $M(6, 8)$ is the midpoint of the line segment AB, and if A has coordinates $(2, 3)$, find the coordinates of B.

41. (a) Sketch the parallelogram with vertices $A(-2, -1)$, $B(4, 2)$ $C(7, 7)$, and $D(1, 4)$.

(b) Find the midpoints of the diagonals of this parallelogram.

(c) From part (b) show that the diagonals bisect each other.

42. The point M in the figure is the midpoint of the line segment AB. Show that M is equidistant from the vertices of triangle ABC.

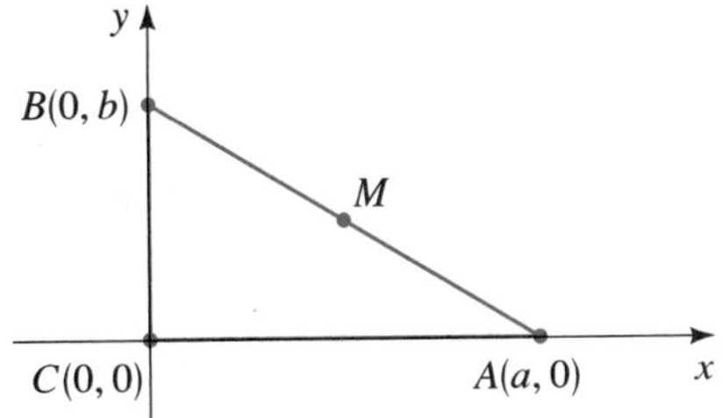

DISCOVERY • DISCUSSION

43. Shifting the Coordinate Plane Suppose that each point in the coordinate plane is shifted 3 units to the right and 2 units upward.

(a) The point $(5, 3)$ is shifted to what new point?

(b) The point (a, b) is shifted to what new point?

(c) What point is shifted to $(3, 4)$?

(d) Triangle ABC in the figure has been shifted to triangle $A'B'C'$. Find the coordinates of the points A', B', and C'.

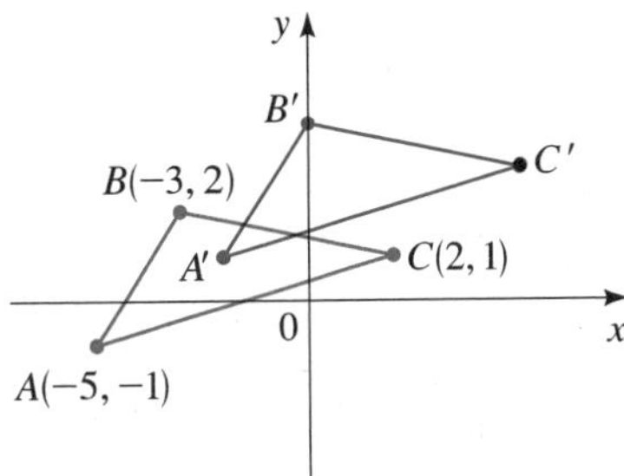

44. Reflecting in the Coordinate Plane Suppose that the y-axis acts as a mirror that reflects each point to the right of it into a point to the left of it.

(a) The point $(3, 7)$ is reflected to what point?

(b) The point (a, b) is reflected to what point?

(c) What point is reflected to $(-4, -1)$?

(d) Triangle ABC in the figure is reflected to triangle $A'B'C'$. Find the coordinates of the points A', B', and C'.

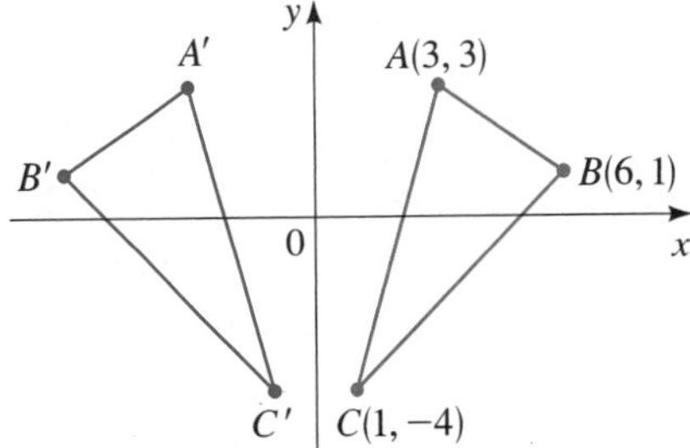

45. Completing a Line Segment Plot the points $M(6, 8)$ and $A(2, 3)$ on a coordinate plane. If M is the midpoint of the line segment AB, find the coordinates of B. Write a brief description of the steps you took in finding B, and your reasons for taking them.

46. Completing a Parallelogram Plot the points $P(-1, -4)$, $Q(1, 1)$, and $R(4, 2)$ on a coordinate plane. Where should the point S be located so that the figure $PQRS$ is a parallelogram? Write a brief description of the steps you took and your reasons for taking them.

VISUALIZING DATA

When scientists analyze data, they look for a trend or pattern from which they can draw a conclusion about the process they are studying. It's often hard to look at lists of numbers and see any kind of pattern. One of the best ways to reveal a hidden pattern in data is to draw a graph. For instance, a biologist measures the levels of three different enzymes (call them A, B, and C) in 20 blood samples taken from expectant mothers, yielding the data shown in the table (enzyme levels in mg/dL).

Sample	A	B	C	Sample	A	B	C
1	1.3	1.7	49	11	2.2	0.6	25
2	2.6	6.8	22	12	1.5	4.8	32
3	0.9	0.6	53	13	3.1	1.9	20
4	3.5	2.4	15	14	4.1	3.1	10
5	2.4	3.8	25	15	1.8	7.5	31
6	1.7	3.3	30	16	2.9	5.8	18
7	4.0	6.7	12	17	2.1	5.1	30
8	3.2	4.3	17	18	2.7	2.5	20
9	1.3	8.4	45	19	1.4	2.0	39
10	1.4	5.8	47	20	0.8	2.3	56

The biologist wishes to determine whether there is a relationship between the serum levels of these enzymes, so she decides to make some **scatter plots** of the data. The scatter plot in Figure 1 shows the levels of enzymes A and B. Each point represents the results for one sample—for instance, sample 1 had 1.3 mg/dL of enzyme A and 1.7 mg/dL of enzyme B, so we plot the point $(1.3, 1.7)$ to represent this pair of data. Similarly, Figure 2 shows the levels of enzymes A and C. From Figure 1 the biologist sees that there is no obvious relationship between enzymes A and B, but from Figure 2 it appears that when the level of enzyme A goes up, the level of enzyme C goes down.

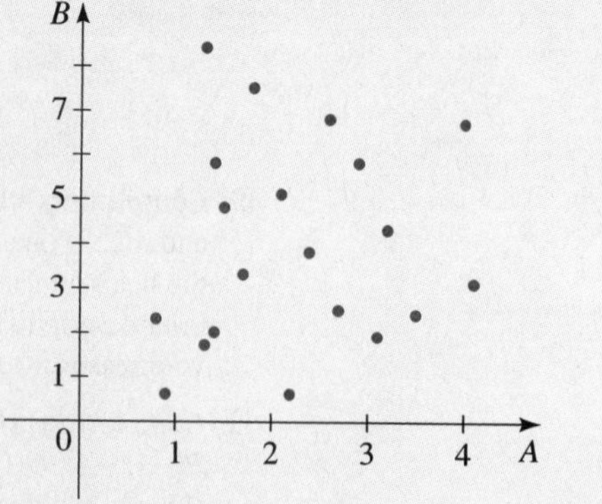

FIGURE 1

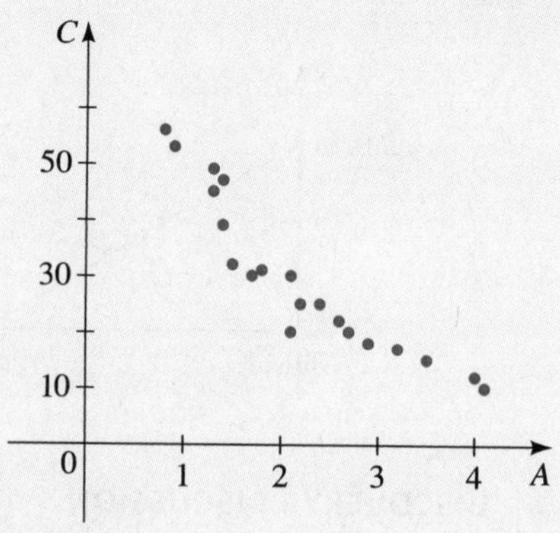

FIGURE 2

1. Make a scatter plot of enzyme B and C levels in the blood samples. Do you detect any relationship from your graph?

2. For each scatter plot, determine whether there is a relationship between the two variables in the graphs. If there is, describe the relationship; that is, explain what happens to y as x increases.

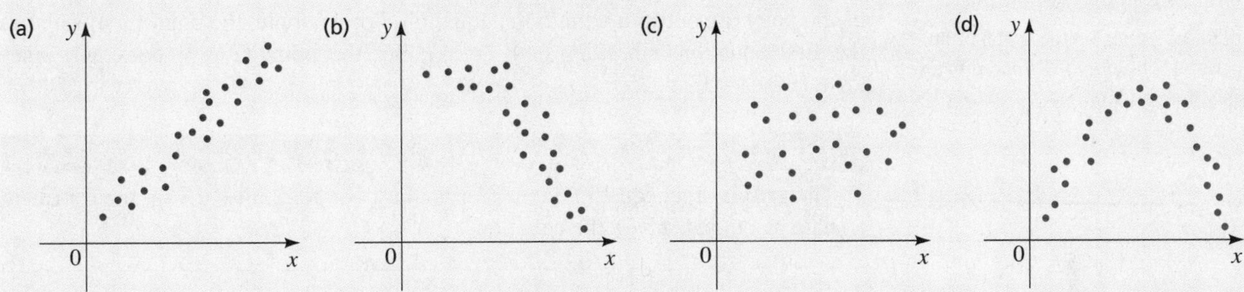

3. In each scatter plot below, the value of y increases as x increases. Explain how the relationship between x and y differs in the two cases.

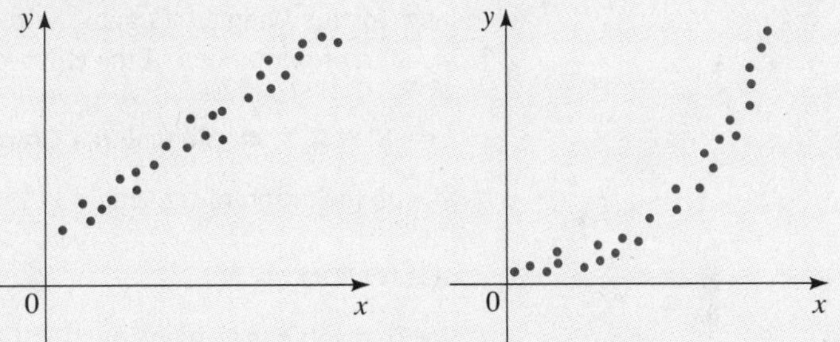

4. For the data given in the following table, make three scatter plots; one for A and B, one for B and C, and one for A and C. Determine whether any of these pairs of variables are related. If so, describe the relationship.

Sample	A	B	C
1	58	4.1	51.7
2	39	5.2	15.4
3	15	7.6	2.0
4	30	6.0	7.3
5	46	4.3	34.2
6	59	3.9	72.4
7	22	6.3	4.1
8	7	8.1	0.5
9	41	4.7	22.6
10	62	3.7	96.3
11	10	7.9	1.3
12	6	8.3	0.2

2.2 GRAPHS OF EQUATIONS

Suppose we have an equation involving the variables x and y, such as

$$x^2 + y^2 = 25 \qquad \text{or} \qquad x = y^2 \qquad \text{or} \qquad y = \frac{2}{x}$$

Fundamental Principle of Analytic Geometry

A point (x, y) lies on the graph of an equation if and only if its coordinates satisfy the equation.

A point (x, y) **satisfies** the equation if the equation is true when the coordinates of the point are substituted into the equation. For example, the point $(3, 4)$ satisfies the first equation, since $3^2 + 4^2 = 25$, but the point $(2, -3)$ does not, since $2^2 + (-3)^2 = 13 \neq 25$.

THE GRAPH OF AN EQUATION

The **graph** of an equation in x and y is the set of all points (x, y) in the coordinate plane that satisfy the equation.

The graphs of most of the equations that we will encounter are curves. For example, we will see later in this section that the first equation, $x^2 + y^2 = 25$, represents a circle; we will determine the shapes represented by the other equations later in this chapter. Graphs help us understand equations because they give us visual representations of the equations.

EXAMPLE 1 ■ Sketching a Graph by Plotting Points

Sketch the graph of the equation $2x - y = 3$.

SOLUTION

We first solve the given equation for y to get

$$y = 2x - 3$$

This helps us calculate the y-coordinates in the following table.

x	$y = 2x - 3$	(x, y)
-1	-5	$(-1, -5)$
0	-3	$(0, -3)$
1	-1	$(1, -1)$
2	1	$(2, 1)$
3	3	$(3, 3)$
4	5	$(4, 5)$

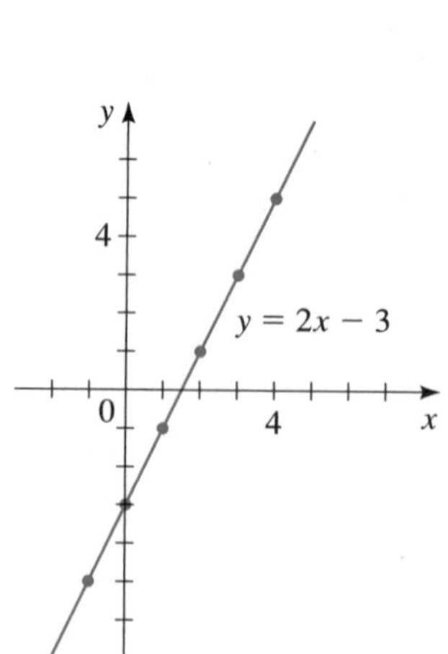

FIGURE 1

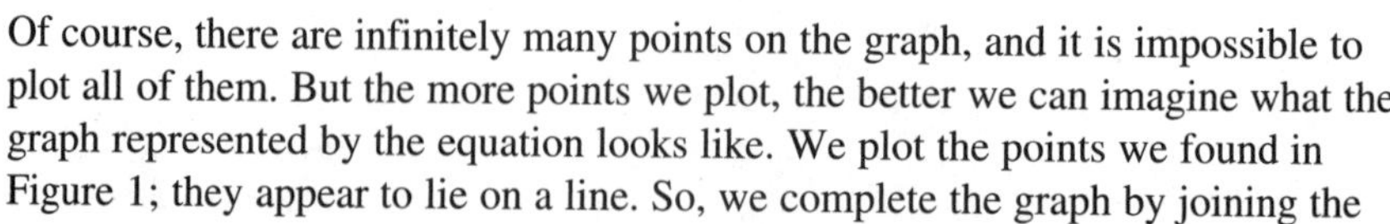

Of course, there are infinitely many points on the graph, and it is impossible to plot all of them. But the more points we plot, the better we can imagine what the graph represented by the equation looks like. We plot the points we found in Figure 1; they appear to lie on a line. So, we complete the graph by joining the points by a line. (In Section 2.4 we verify that the graph of this equation is indeed a line.) ■

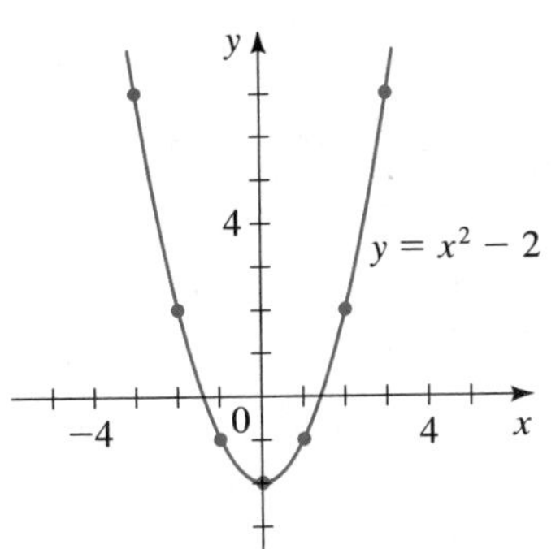

FIGURE 2

A detailed discussion of parabolas and their geometric properties is presented in Chapter 8.

EXAMPLE 2 ■ Sketching a Graph by Plotting Points

Sketch the graph of the equation $y = x^2 - 2$.

SOLUTION

We find some of the points that satisfy the equation in the following table. In Figure 2 we plot these points and then connect them by a smooth curve. A curve with this shape is called a *parabola*.

x	$y = x^2 - 2$	(x, y)
−3	7	(−3, 7)
−2	2	(−2, 2)
−1	−1	(−1, −1)
0	−2	(0, −2)
1	−1	(1, −1)
2	2	(2, 2)
3	7	(3, 7)

■

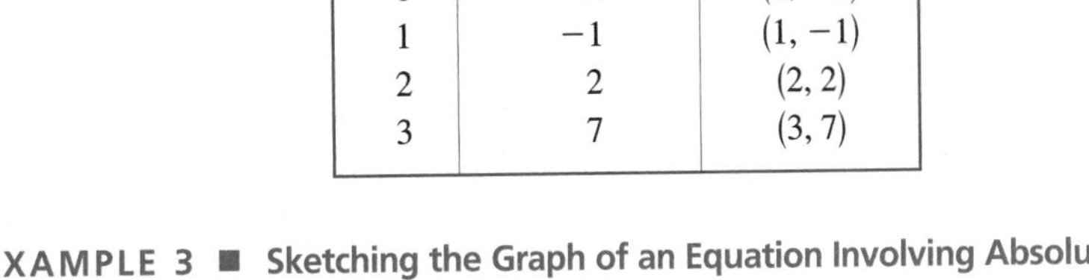

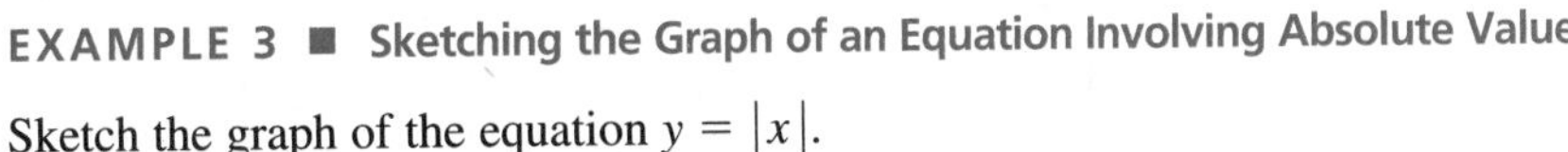

EXAMPLE 3 ■ Sketching the Graph of an Equation Involving Absolute Value

Sketch the graph of the equation $y = |x|$.

SOLUTION

We make a table of values:

x	$y = \|x\|$	(x, y)
−3	3	(−3, 3)
−2	2	(−2, 2)
−1	1	(−1, 1)
0	0	(0, 0)
1	1	(1, 1)
2	2	(2, 2)
3	3	(3, 3)

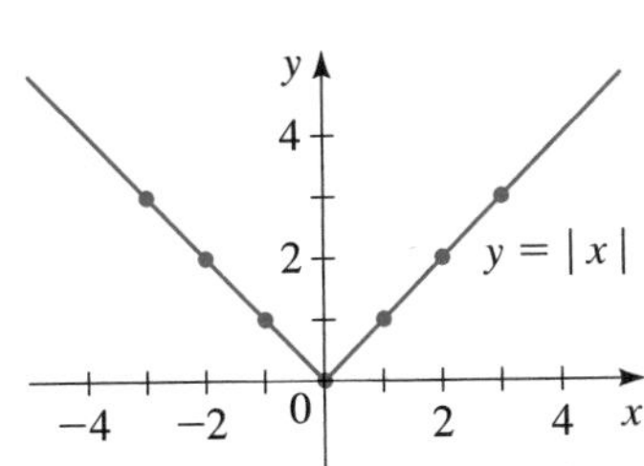

FIGURE 3

In Figure 3 we plot these points and use them to sketch the graph of the equation. ■

The x-coordinates of the points where a graph intersects the x-axis are called the **x-intercepts** of the graph and are obtained by setting $y = 0$ in the equation of the graph. The y-coordinates of the points where a graph intersects the y-axis are called the **y-intercepts** of the graph and are obtained by setting $x = 0$ in the equation of the graph.

DEFINITION OF INTERCEPTS

Intercepts	How to find them	Where they are on the graph
x-intercepts: The x-coordinates of points where the graph of an equation intersects the x-axis	Set $y = 0$ and solve for x	y, 0, x
y-intercepts: The y-coordinates of points where the graph of an equation intersects the y-axis	Set $x = 0$ and solve for y	y, 0, x

EXAMPLE 4 ■ Finding Intercepts

Find the x- and y-intercepts of the graph of the equation $y = x^2 - 2$.

SOLUTION

To find the x-intercepts, we set $y = 0$ and solve for x. Thus

$$0 = x^2 - 2 \qquad \text{Set } y = 0$$

$$x^2 = 2 \qquad \text{Add 2 to each side}$$

$$x = \pm\sqrt{2} \qquad \text{Take the square root}$$

The x-intercepts are $\sqrt{2}$ and $-\sqrt{2}$.

To find the y-intercepts, we set $x = 0$ and solve for y. Thus

$$y = 0^2 - 2 \qquad \text{Set } x = 0$$

$$y = -2$$

The y-intercept is -2.

The graph of this equation was sketched in Example 2. It is repeated in Figure 4 with the x- and y-intercepts labeled. ■

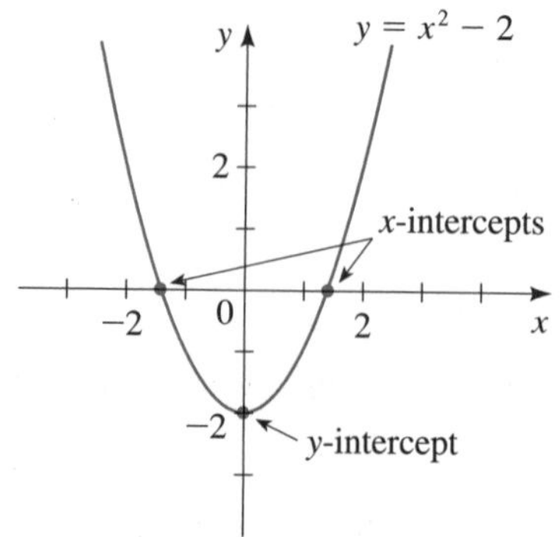

FIGURE 4

CIRCLES

So far we have discussed how to find the graph of an equation in x and y. The converse problem is to find an equation of a graph, that is, an equation that represents a given curve in the xy-plane. Such an equation is satisfied by the coordinates of the points on the curve and by no other point. This is the other half of the fundamental principle of analytic geometry as formulated by Descartes and Fermat. The idea is that if a geometric curve can be represented by an algebraic equation, then the rules of algebra can be used to analyze the curve.

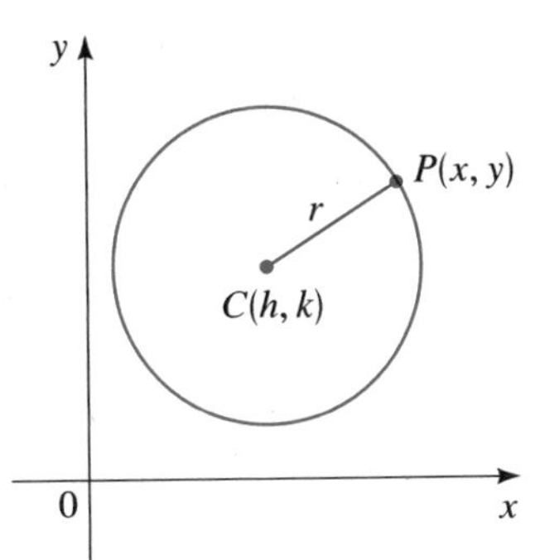

FIGURE 5

As an example of this type of problem, let's find the equation of a circle with radius r and center (h, k). By definition, the circle is the set of all points $P(x, y)$ whose distance from the center $C(h, k)$ is r (see Figure 5). Thus, P is on the circle if and only if $d(P, C) = r$. From the distance formula we have

$$\sqrt{(x-h)^2 + (y-k)^2} = r$$

$$(x-h)^2 + (y-k)^2 = r^2 \qquad \text{Square each side}$$

This is the desired equation.

EQUATION OF A CIRCLE

An equation of the circle with center (h, k) and radius r is

$$(x-h)^2 + (y-k)^2 = r^2$$

This is called the **standard form** for the equation of the circle. If the center of the circle is the origin $(0, 0)$, then the equation is

$$x^2 + y^2 = r^2$$

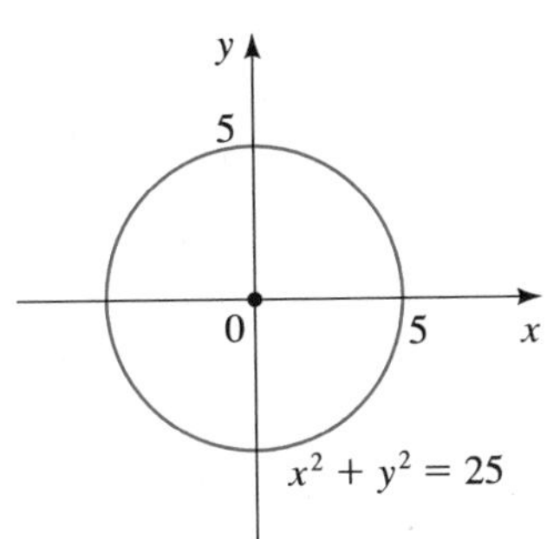

FIGURE 6

EXAMPLE 5 ■ Graphing a Circle

Graph each equation.

(a) $x^2 + y^2 = 25$

(b) $(x-2)^2 + (y+1)^2 = 25$

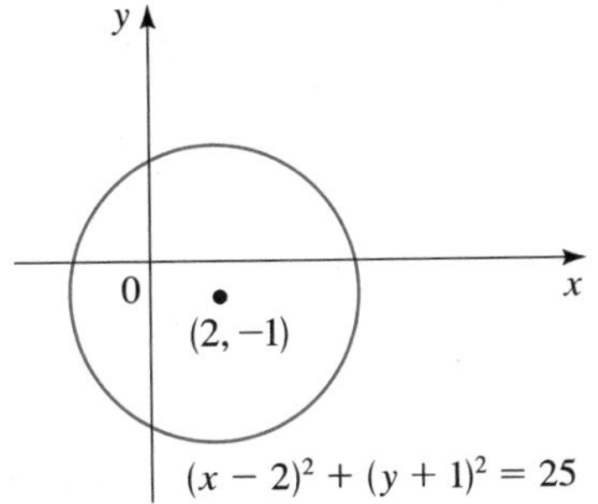

FIGURE 7

SOLUTION

(a) Rewriting the equation as $x^2 + y^2 = 5^2$, we see that this is an equation of the circle of radius 5 centered at the origin. Its graph is shown in Figure 6.

(b) Rewriting the equation as $(x-2)^2 + (y+1)^2 = 5^2$, we see that this is an equation of the circle of radius 5 centered at $(2, -1)$. Its graph is shown in Figure 7. ■

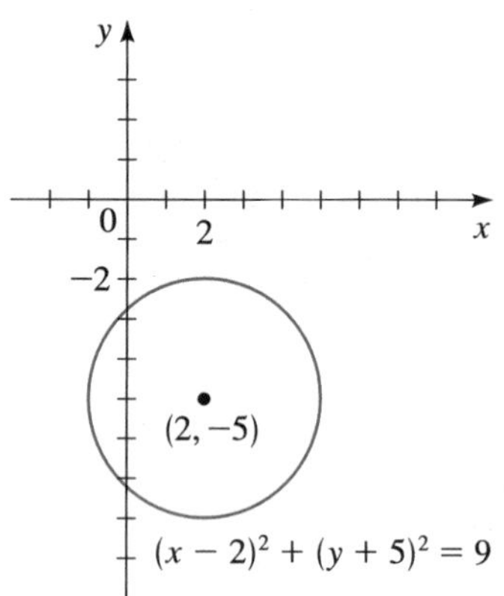

FIGURE 8

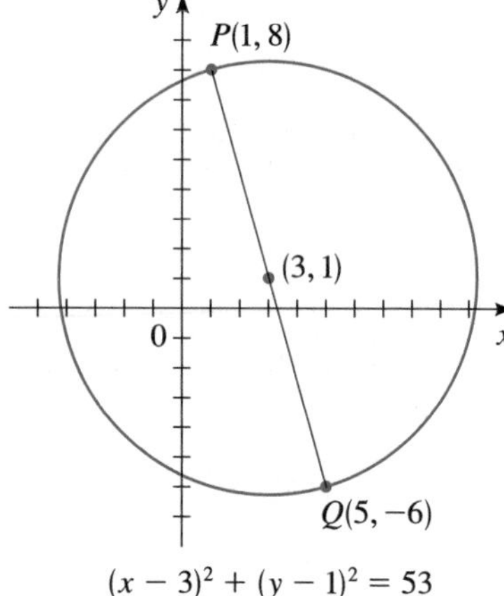

FIGURE 9

EXAMPLE 6 ■ Finding an Equation of a Circle

(a) Find an equation of the circle with radius 3 and center $(2, -5)$.

(b) Find an equation of the circle that has the points $P(1, 8)$ and $Q(5, -6)$ as the endpoints of a diameter.

SOLUTION

(a) Using the equation of a circle with $r = 3$, $h = 2$, and $k = -5$, we obtain

$$(x - 2)^2 + (y + 5)^2 = 9$$

Its graph is shown in Figure 8.

(b) We first observe that the center is the midpoint of the diameter PQ, so by the Midpoint Formula the center is

$$\left(\frac{1 + 5}{2}, \frac{8 - 6}{2}\right) = (3, 1)$$

The radius r is the distance from P to the center, so by the Distance Formula

$$r^2 = (3 - 1)^2 + (1 - 8)^2 = 2^2 + (-7)^2 = 53$$

Therefore, the equation of the circle is

$$(x - 3)^2 + (y - 1)^2 = 53$$

Its graph is shown in Figure 9. ■

Let's expand the equation of the circle in the preceding example.

$$(x - 3)^2 + (y - 1)^2 = 53 \qquad \textbf{Standard form}$$

$$x^2 - 6x + 9 + y^2 - 2y + 1 = 53 \qquad \text{Expand the squares}$$

$$x^2 - 6x + y^2 - 2y = 43 \qquad \text{Subtract 10 to get } \textbf{expanded form}$$

Completing the square is used in many contexts in algebra. In Section 3.3, we use completing the square to solve quadratic equations.

Suppose we are given the equation of a circle in expanded form. Then to find its center and radius we must put the equation back in standard form. That means we must reverse the steps in the preceding calculation, and to do that we need to know what to add to an expression like $x^2 - 6x$ to make it a perfect square—that is, we need to "complete the square." It turns out that to **complete the square**, we must *add the square of half the coefficient of the variable*. For example, to complete the square for $x^2 - 6x$, we must add the square of half of -6:

$$x^2 - 6x + \left(\frac{1}{2} \cdot (-6)\right)^2 = x^2 - 6x + 9 = (x - 3)^2$$

In general, to make $X^2 + aX$ a perfect square, add $(a/2)^2$.

EXAMPLE 7 ■ Identifying an Equation of a Circle

Sketch the graph of the equation $x^2 + y^2 + 2x - 6y + 7 = 0$ by first showing that it represents a circle and then finding its center and radius.

SOLUTION

We first group the x-terms and y-terms. Then we complete the square within each grouping. That is, we complete the square for $x^2 + 2x$ by adding $\left(\frac{1}{2} \cdot 2\right)^2 = 1$, and we complete the square for $y^2 - 6y$ by adding $\left[\frac{1}{2} \cdot (-6)\right]^2 = 9$.

$$(x^2 + 2x \qquad) + (y^2 - 6y \qquad) = -7 \qquad \text{Group terms}$$

$$(x^2 + 2x + 1) + (y^2 - 6y + 9) = -7 + 1 + 9 \qquad \text{Complete the square by adding 1 and 9 to each side}$$

$$(x + 1)^2 + (y - 3)^2 = 3$$

⊘ We must add the same numbers to *each side* to maintain equality.

Comparing this equation with the standard equation of a circle, we see that $h = -1$, $k = 3$, and $r = \sqrt{3}$, so the given equation represents a circle with center $(-1, 3)$ and radius $\sqrt{3}$. The circle is sketched in Figure 10.

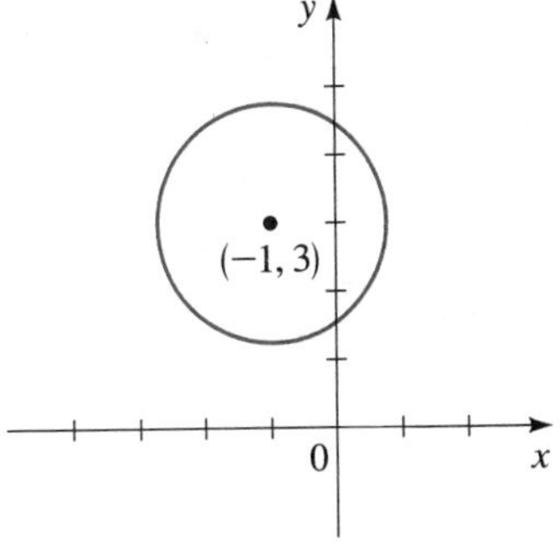

FIGURE 10
$x^2 + y^2 + 2x - 6y + 7 = 0$

■

SYMMETRY

Figure 11 shows the graph of $y = x^2$. Notice that the part of the graph to the left of the y-axis is the mirror image of the part to the right of the y-axis. The reason is that if the point (x, y) is on the graph, then so is $(-x, y)$, and these points are reflections of each other about the y-axis. In this situation we say the graph is **symmetric with**

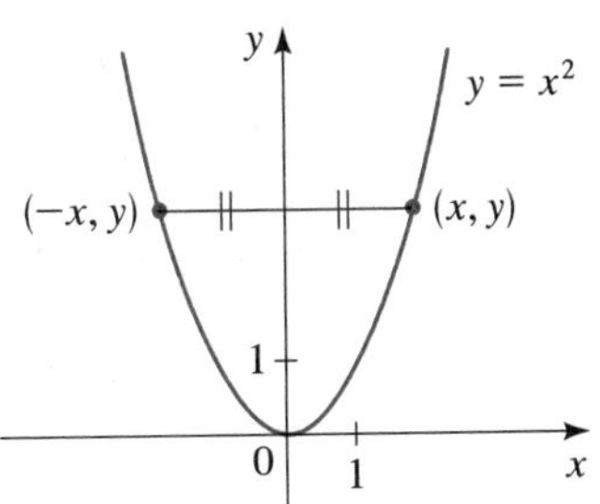

FIGURE 11

respect to the y-axis. Similarly, we say a graph is **symmetric with respect to the x-axis** if whenever the point (x, y) is on the graph, then so is $(x, -y)$. A graph is **symmetric with respect to the origin** if whenever (x, y) is on the graph, so is $(-x, -y)$.

DEFINITION OF SYMMETRY

Type of symmetry	How to test for symmetry	What the graph looks like (figures in this section)	Geometric meaning
Symmetry with respect to the x-axis	The equation is unchanged when y is replaced by $-y$	y; (x, y); 0; x; $(x, -y)$ (Figures 6, 12, 14)	Graph is unchanged when reflected in the x-axis
Symmetry with respect to the y-axis	The equation is unchanged when x is replaced by $-x$	y; $(-x, y)$; (x, y); 0; x (Figures 2, 3, 4, 6, 11, 14)	Graph is unchanged when reflected in the y-axis
Symmetry with respect to the origin	The equation is unchanged when x is replaced by $-x$ and y by $-y$	y; (x, y); 0; x; $(-x, -y)$ (Figures 6, 13, 14)	Graph is unchanged when rotated 180° about the origin

The remaining examples in this section show how symmetry helps us sketch the graphs of equations.

EXAMPLE 8 ■ Using Symmetry to Sketch a Graph

Test the equation $x = y^2$ for symmetry and sketch the graph.

SOLUTION

If y is replaced by $-y$ in the equation $x = y^2$, we get

$$x = (-y)^2 = y^2$$

and so the equation is unchanged. Therefore, the graph is symmetric about the x-axis. But changing x to $-x$ gives the equation $-x = y^2$, which is not the same as the original equation, so the graph is not symmetric about the y-axis.

We use the symmetry about the x-axis to sketch the graph by first plotting points just for $y > 0$ and then reflecting the graph in the x-axis, as shown in Figure 12.

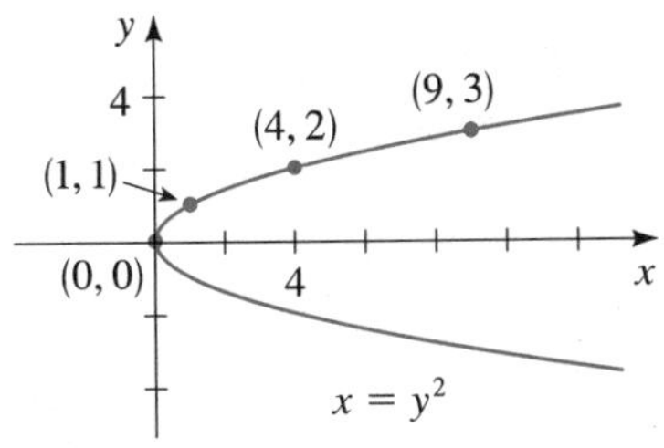

FIGURE 12

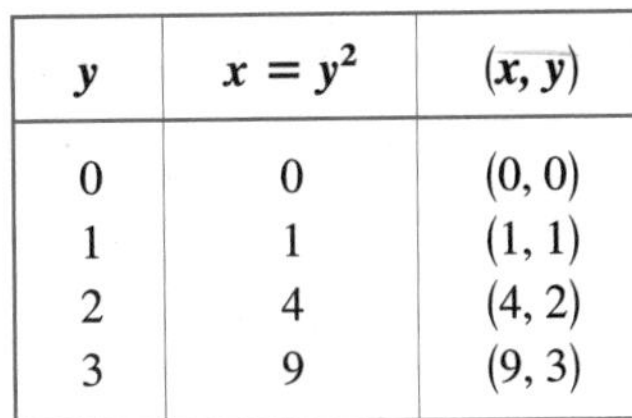

y	$x = y^2$	(x, y)
0	0	(0, 0)
1	1	(1, 1)
2	4	(4, 2)
3	9	(9, 3)

■

EXAMPLE 9 ■ Using Symmetry to Sketch a Graph

Test the equation $y = x^3 - 9x$ for symmetry and sketch its graph.

SOLUTION

If we replace x by $-x$ and y by $-y$ in the equation, we get

$$-y = (-x)^3 - 9(-x)$$

$$-y = -x^3 + 9x$$

$$y = x^3 - 9x$$

and so the equation is unchanged. This means that the graph is symmetric with respect to the origin. We sketch it by first plotting points for $x > 0$ and then using symmetry about the origin (see Figure 13).

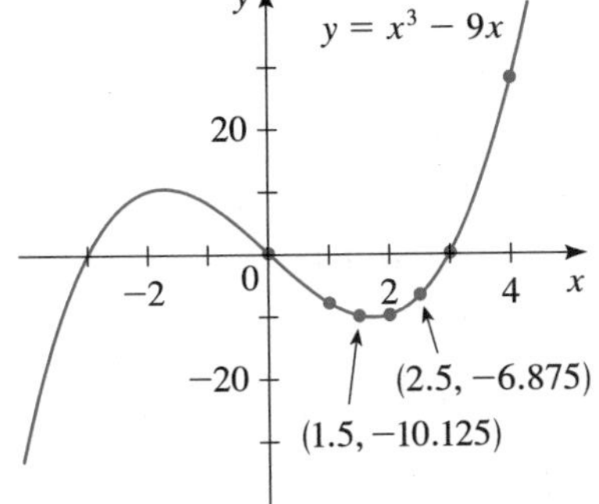

FIGURE 13

x	$y = x^3 - 9x$	(x, y)
0	0	(0, 0)
1	−8	(1, −8)
1.5	−10.125	(1.5, −10.125)
2	−10	(2, −10)
2.5	−6.875	(2.5 −6.875)
3	0	(3, 0)
4	28	(4, 28)

■

EXAMPLE 10 ■ A Circle That Has All Three Types of Symmetry

Test the equation of the circle $x^2 + y^2 = 4$ for symmetry.

SOLUTION

The equation $x^2 + y^2 = 4$ remains unchanged when x is replaced by $-x$ and y is replaced by $-y$, since $(-x)^2 = x^2$ and $(-y)^2 = y^2$, so the circle exhibits all three types of symmetry. It is symmetric with respect to the x-axis, the y-axis, and the origin, as shown in Figure 14.

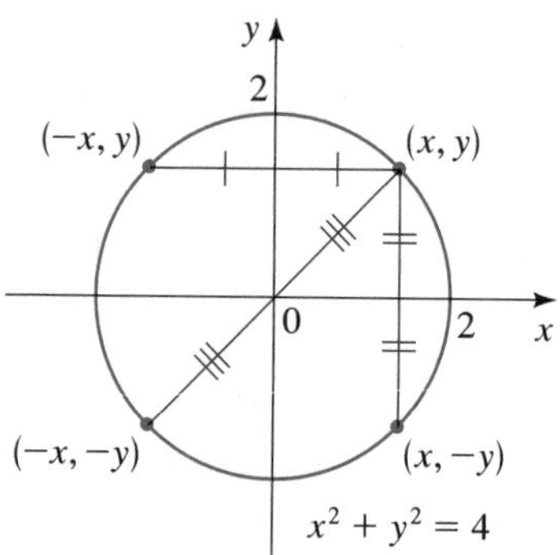

FIGURE 14

■

2.2 EXERCISES

1–6 ■ Determine whether the given points are on the graph of the equation.

1. $y = 2x + 3$; $(0, 0), (\frac{1}{2}, 4), (1, 4)$

2. $y = \sqrt{x + 1}$; $(1, 0), (0, 1), (3, 2)$

3. $2y - x + 1 = 0$; $(0, 0), (1, 0), (-1, -1)$

4. $y(x^2 + 1) = 1$; $(1, 1), \left(1, \frac{1}{2}\right), \left(-1, \frac{1}{2}\right)$

5. $x^2 + xy + y^2 = 4$; $(0, -2), (1, -2), (2, -2)$

6. $x^2 + y^2 - 1 = 0$; $(0, 1), \left(\frac{1}{\sqrt{2}}, \frac{1}{\sqrt{2}}\right), \left(\frac{\sqrt{3}}{2}, \frac{1}{2}\right)$

7–14 ■ Find the x- and y-intercepts of the graph of the equation.

7. $y = x - 3$

8. $y = x^2 - 5x + 6$

9. $y = x^2 - 9$

10. $y - 2xy + 2x = 1$

11. $x^2 + y^2 = 4$

12. $y = \sqrt{x + 1}$

13. $xy = 5$

14. $x^2 - xy + y = 1$

15–40 ■ Make a table of values and sketch the graph of the equation. Find x- and y-intercepts and test for symmetry.

15. $y = x$

16. $y = -x$

17. $y = x - 1$

18. $y = 2x + 5$

19. $3x - y = 5$

20. $x + y = 3$

21. $y = 1 - x^2$

22. $y = x^2 + 2$

23. $4y = x^2$

24. $8y = x^3$

25. $y = x^2 - 9$

26. $y = 9 - x^2$

27. $xy = 2$

28. $x + y^2 = 4$

29. $y = \sqrt{x}$

30. $x^2 + y^2 = 9$

31. $y = \sqrt{4 - x^2}$

32. $y = -\sqrt{4 - x^2}$

33. $y = |x|$

34. $x = |y|$

35. $y = 4 - |x|$

36. $y = |4 - x|$

37. $x = y^3$

38. $y = x^3 - 1$

39. $y = x^4$

40. $y = 16 - x^4$

41–46 ■ Test the equation for symmetry.

41. $y = x^4 + x^2$

42. $x = y^4 - y^2$

43. $x^2y^2 + xy = 1$

44. $x^4y^4 + x^2y^2 = 1$

45. $y = x^3 + 10x$

46. $y = x^2 + |x|$

47–50 ■ Complete the graph using the given symmetry property.

47. Symmetric with respect to the y-axis

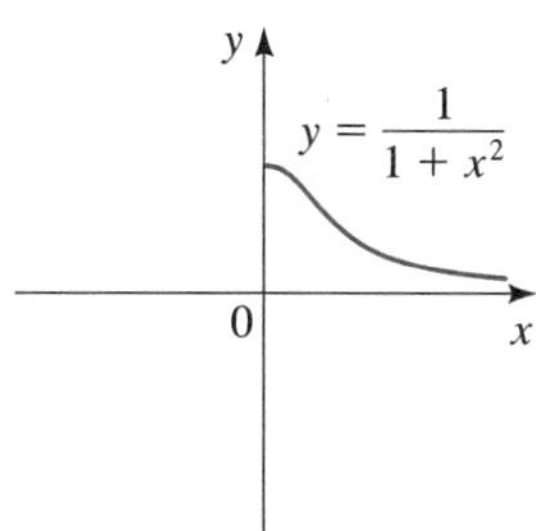

48. Symmetric with respect to the x-axis

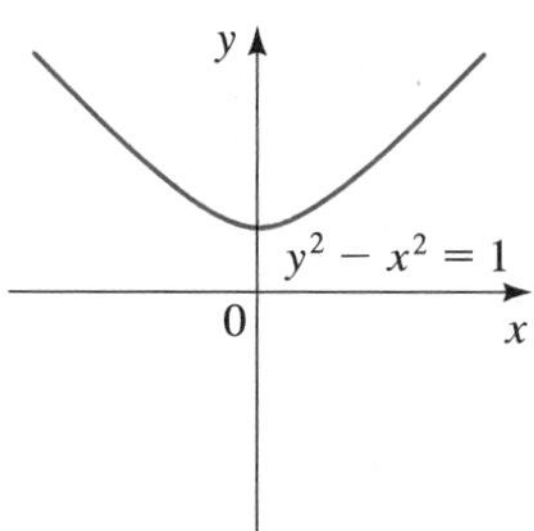

49. Symmetric with respect to the origin

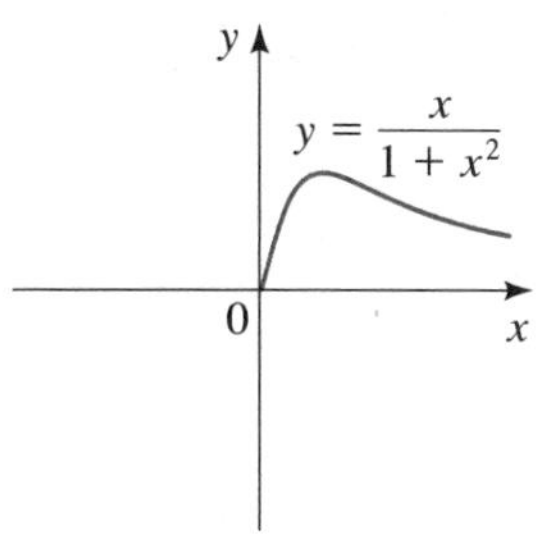

50. Symmetric with respect to the origin

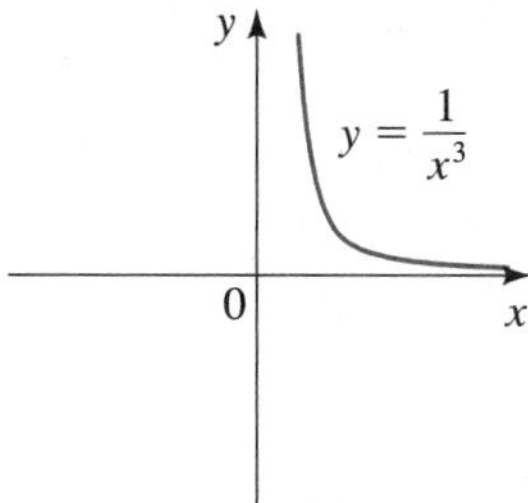

51–58 ■ Find an equation of the circle that satisfies the given conditions.

51. Center $(2, -1)$; radius 3

52. Center $(-1, -4)$; radius 8

53. Center at the origin; passes through $(4, 7)$

54. Center $(-1, 5)$; passes through $(-4, -6)$

55. Endpoints of a diameter are $P(-1, 1)$ and $Q(5, 5)$

56. Endpoints of a diameter are $P(-1, 3)$ and $Q(7, -5)$

57. Center $(7, -3)$; tangent to the x-axis

58. Circle lies in the first quadrant, tangent to both x- and y-axes; radius 5

59–60 ■ Find the equation of the circle shown in the figure.

59.

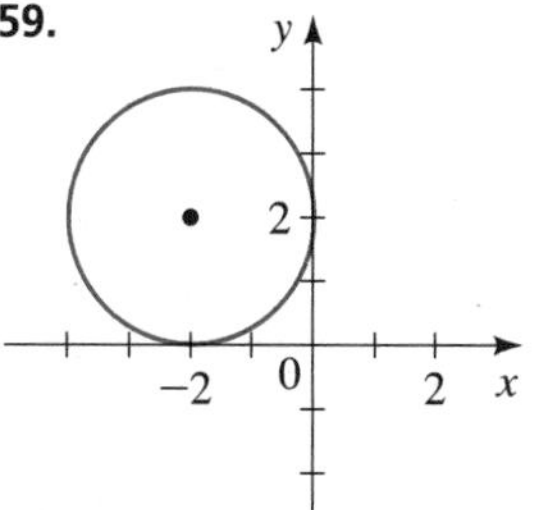

60.

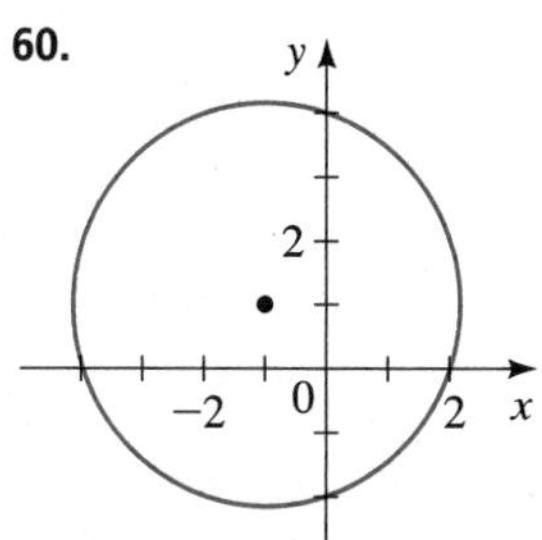

61–68 ■ Show that the equation represents a circle, and find the center and radius of the circle.

61. $x^2 + y^2 - 2x + 4y + 1 = 0$

62. $x^2 + y^2 - 2x - 2y = 2$

63. $x^2 + y^2 - 4x + 10y + 13 = 0$

64. $x^2 + y^2 + 6y + 2 = 0$

65. $x^2 + y^2 + x = 0$

66. $x^2 + y^2 + 2x + y + 1 = 0$

67. $x^2 + y^2 - \frac{1}{2}x + \frac{1}{2}y = \frac{1}{8}$

68. $x^2 + y^2 + \frac{1}{2}x + 2y + \frac{1}{16} = 0$

69–72 ■ Sketch the graph of the equation.

69. $x^2 + y^2 + 4x - 10y = 21$

70. $4x^2 + 4y^2 + 2x = 0$

71. $x^2 + y^2 + 6x - 12y + 45 = 0$

72. $x^2 + y^2 - 16x + 12y + 200 = 0$

73–76 ■ Sketch the region given by the set.

73. $\{(x, y) \mid x^2 + y^2 \leqslant 1\}$

74. $\{(x, y) \mid x^2 + y^2 > 4\}$

75. $\{(x, y) \mid 1 \leqslant x^2 + y^2 < 9\}$

76. $\{(x, y) \mid 2x < x^2 + y^2 \leqslant 4\}$

77. Find the area of the region that lies outside the circle $x^2 + y^2 = 4$ but inside the circle

$$x^2 + y^2 - 4y - 12 = 0$$

78. Sketch the region in the coordinate plane that satisfies both the inequalities $x^2 + y^2 \leqslant 9$ and $y \geqslant |x|$. What is the area of this region?

DISCOVERY • DISCUSSION

79. Circle, Point, or Empty Set? Complete the squares in the general equation $x^2 + ax + y^2 + by + c = 0$ and simplify the result as much as possible. Under what conditions on the coefficients a, b, and c does this equation represent a circle? a single point? the empty set? In the case that the equation does represent a circle, find its center and radius.

80. Do the Circles Intersect?

(a) Find the radius of each circle in the pair, and the distance between their centers; then use this information to determine whether the circles intersect.

(i) $(x-2)^2 + (y-1)^2 = 9$; $(x-6)^2 + (y-4)^2 = 16$

(ii) $x^2 + (y-2)^2 = 4$; $(x-5)^2 + (y-14)^2 = 9$

(iii) $(x-3)^2 + (y+1)^2 = 1$; $(x-2)^2 + (y-2)^2 = 25$

(b) How can you tell, just by knowing the radii of two circles and the distance between their centers, whether the circles intersect? Write a short paragraph describing how you would decide this and draw graphs to illustrate your answer.

81. Making a Graph Symmetric The graph shown in the figure is not symmetric about the x-axis, the y-axis, or the origin. Add more line segments to the graph so that it exhibits the indicated symmetry. In each case, add as little as possible.

(a) Symmetry about the x-axis.

(b) Symmetry about the y-axis.

(c) Symmetry about the origin.

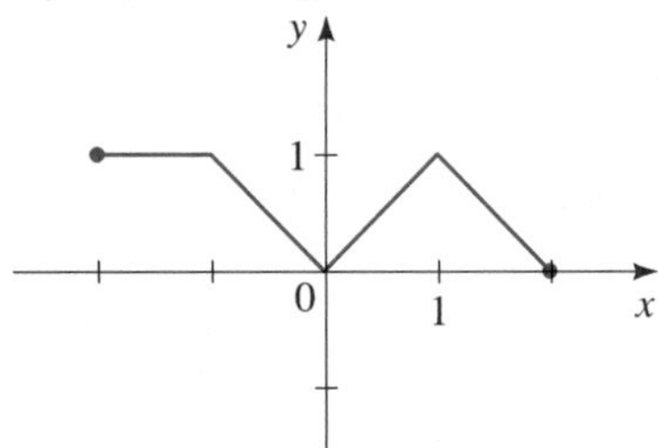

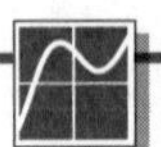

2.3 GRAPHING CALCULATORS AND COMPUTERS

In this section we make use of graphing calculators or computers to graph more complicated equations and to solve more complex problems. We begin by describing how these devices work and how to recognize and avoid some common pitfalls of using such devices.

HOW TO USE GRAPHING DEVICES

A graphing calculator or computer displays a rectangular portion of the graph of an equation in a **display window** or **viewing screen**, which we call a **viewing rectangle**. The default screen often gives an incomplete or misleading picture, so it is important to choose the viewing rectangle with care. If we choose the x-values to range from a minimum value of Xmin $= a$ to a maximum of Xmax $= b$ and the y-values to range from a minimum of Ymin $= c$ to a maximum of Ymax $= d$, then the portion of the graph displayed lies in the rectangle

$$[a, b] \times [c, d] = \{(x, y) \mid a \leq x \leq b, c \leq y \leq d\}$$

shown in Figure 1. We refer to this as the $[a, b]$ *by* $[c, d]$ *viewing rectangle.*

(a, d) $y = d$ (b, d)
$x = a$ $x = b$
(a, c) $y = c$ (b, c)

FIGURE 1
The viewing rectangle $[a, b]$ by $[c, d]$

The graphing device draws the graph of an equation much as you would. It plots points of the form (x, y) for a certain number of values of x, equally spaced between a and b. If the equation is not defined for an x-value, or if the corresponding y-value lies outside the viewing rectangle, the device ignores this value and moves on to the next x-value. The machine connects each point to the preceding plotted point to form a representation of the graph of the equation.

Alan Turing (1912–1954) was at the center of two pivotal events of the 20th century—World War II and the invention of computers. At the age of 23 Turing made his mark on mathematics by solving an important problem in the foundations of mathematics that was posed by David Hilbert at the 1928 International Congress of Mathematicians (see page 498). In this research he invented a theoretical machine, now called a Turing machine, which was the inspiration for modern digital computers. During World War II Turing was in charge of the British effort to decipher secret German codes. His complete success in this endeavor played a decisive role in the Allies' victory. To carry out the numerous logical steps required to break a coded message. Turing developed decision procedures similar to modern computer programs. After the war he helped develop the first electronic computers in Britain. He also did pioneering work on artificial intelligence and computer models of biological processes. At the age of 42 Turing died of cyanide poisoning under mysterious circumstances.

EXAMPLE 1 ■ Drawing Graphs in Different Viewing Rectangles

Draw the graph of the equation $y = x^2 + 3$ in each viewing rectangle.

(a) $[-2, 2]$ by $[-2, 2]$ (b) $[-4, 4]$ by $[-4, 4]$

(c) $[-10, 10]$ by $[-5, 30]$ (d) $[-50, 50]$ by $[-100, 1000]$

SOLUTION

For part (a) select the range by setting Xmin $= -2$, Xmax $= 2$, Ymin $= -2$, and Ymax $= 2$. The resulting graph is shown in Figure 2(a). The display window is blank! A moment's thought provides the explanation. Notice that $x^2 \geqslant 0$ for all x, so $y = x^2 + 3 \geqslant 3$ for all x. This means that the graph of the equation lies totally outside the viewing rectangle $[-2, 2]$ by $[-2, 2]$.

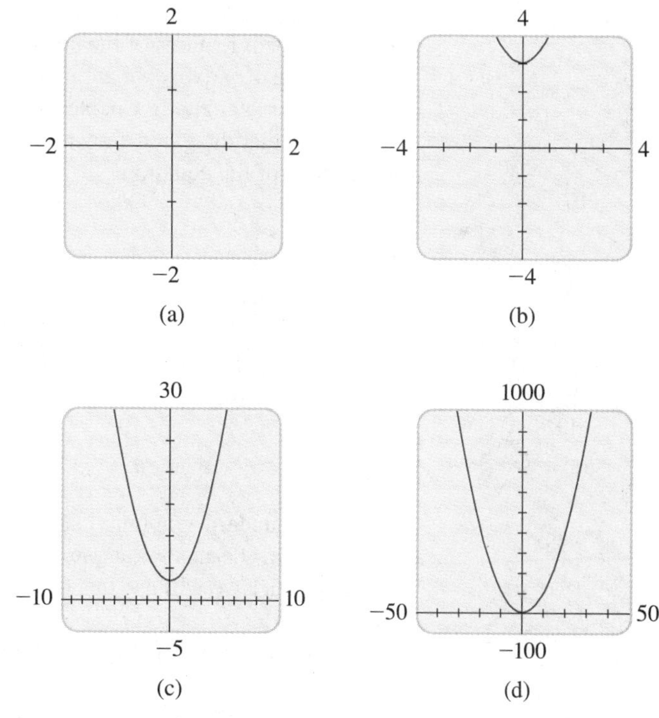

FIGURE 2 Graphs of $y = x^2 + 3$

The graphs for the viewing rectangles in parts (b), (c), and (d) are shown in Figure 2, parts (b), (c), and (d). Observe that we get more complete pictures in parts (c) and (d), but the graph in part (d) doesn't show clearly that the y-intercept is 3. ■

EXAMPLE 2 ■ Graphing a Cubic Equation

Graph the equation $y = x^3 - 49x$.

SOLUTION

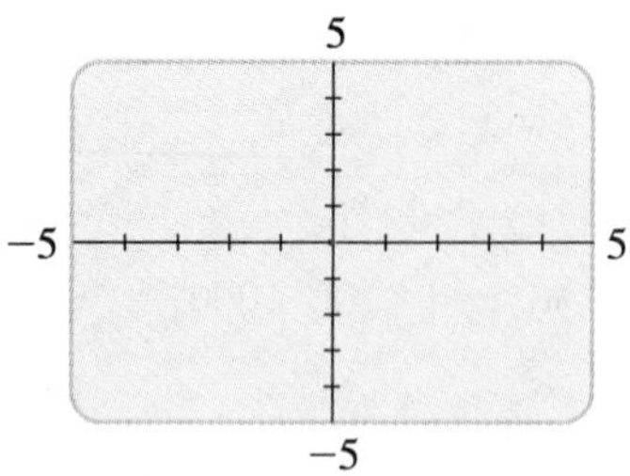

FIGURE 3

Let's experiment with different viewing rectangles. If we start with the viewing rectangle $[-5, 5]$ by $[-5, 5]$ we get the graph in Figure 3. On most calculators the screen appears to be blank, but it is not quite blank because the point $(0, 0)$ has been plotted. It turns out that for all the other x-values that the calculator chooses between -5 and 5, the value of y is greater than 5 or less than -5, so the corresponding point on the graph lies outside the viewing rectangle.

If we use the zoom-out feature to change the viewing rectangle to $[-10, 10]$ by $[-10, 10]$, then we get the picture shown in Figure 4(a). The graph appears to consist of vertical lines, but we know that it can't be correct. If we look carefully while the graph is being drawn, we see that the graph leaves the screen and reappears during the graphing process. This indicates that we need to see more of the graph in the vertical direction, so we change the viewing rectangle to $[-10, 10]$ by $[-100, 100]$. The resulting graph is shown in Figure 4(b). It still doesn't reveal all the main features of the equation, so we try $[-10, 10]$ by $[-200, 200]$ in Figure 4(c). Now we are more confident that we have arrived at an appropriate viewing rectangle. In Chapter 5, when we discuss third-degree polynomials, we will see that the graph shown in Figure 4(c) does indeed reveal all the main features of the equation.

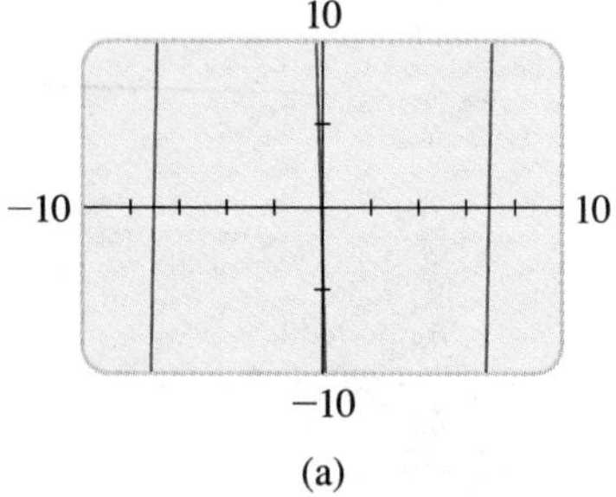

(a)

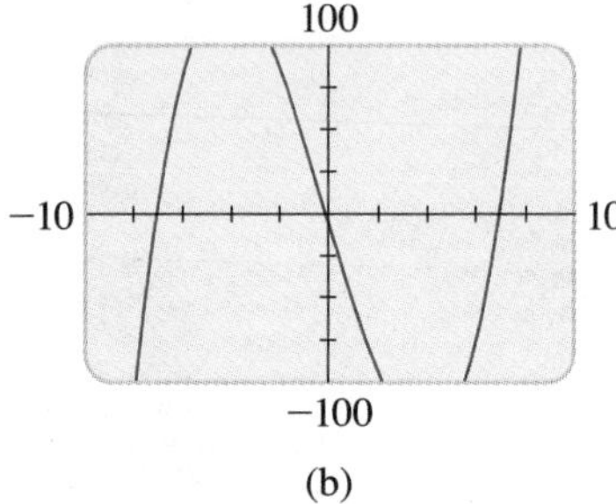

(b)

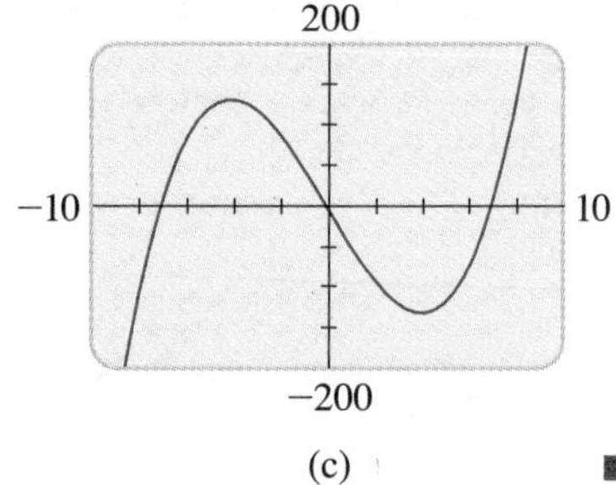

(c) ■

FIGURE 4
Graphs of $y = x^3 - 49x$

All standard graphing calculators can show more than one graph on the same screen. In the next example we use this feature to see if two graphs intersect. In Section 3.1 we will use this technique to solve equations.

EXAMPLE 3 ■ Two Graphs on the Same Screen

Graph the equations $y = 3x^2 - 6x + 1$ and $y = 0.23x - 2.25$ together in the viewing rectangle $[-1, 3]$ by $[-2.5, 1.5]$. Do the graphs intersect in this viewing rectangle?

SOLUTION

Figure 5(a) shows the essential features of both graphs. One is a parabola and the other is a line. It looks as if the graphs intersect near the point $(1, -2)$. However, if we zoom in on the area around this point as shown in Figure 5(b), we see that although the graphs almost touch, they don't actually intersect.

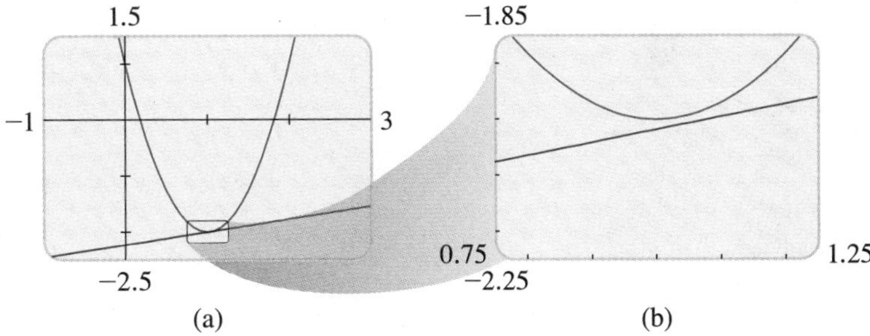

FIGURE 5
Graphs of $y = 3x^2 - 6x + 1$ and $y = 0.23x - 2.25$

■

You can see from Examples 1, 2, and 3 that the choice of a viewing rectangle makes a big difference in the appearance of a graph. If you want an overview of the essential features of a graph, you must choose a relatively large viewing rectangle to obtain a global view of the graph. If you want to investigate the details of a graph, you must zoom in to a small viewing rectangle that shows just the feature of interest.

Graphing calculators easily graph equations in which y can be isolated on one side of the equal sign. The next example shows how to graph equations that don't have this property.

EXAMPLE 4 ■ Graphing a Circle

Graph the circle $x^2 + y^2 = 1$.

SOLUTION

We know from Section 2.2 how to graph this equation by hand; it is a circle with center the origin and radius 1. But graphing a circle with a graphing calculator is not a straightforward task, because we cannot isolate y on one side of the equal sign. If we try to solve the equation of the circle for y, we have

$$y^2 = 1 - x^2$$

$$y = \pm\sqrt{1 - x^2}$$

Therefore, we regard the circle as being described by the graphs of *two* equations:

$$y = \sqrt{1 - x^2} \qquad \text{and} \qquad y = -\sqrt{1 - x^2}$$

The first equation represents the top half of the circle (because $y \geq 0$), and the second represents the bottom half of the circle (because $y \leq 0$). If we graph the first equation in the viewing rectangle $[-2, 2]$ by $[-2, 2]$, we get the semicircle shown in Figure 6(a). The graph of the second equation is the semicircle in Figure 6(b). Graphing these semicircles together on the same viewing screen, we get the full circle in Figure 6(c).

The graph in Figure 6(c) looks somewhat flattened. Most graphing calculators allow you to set the scales on the axes so that circles really look like circles. On the TI-82 and TI-83, from the Zoom menu, choose "ZSquare" to set the scales appropriately. (On the TI-85 the command is "Zsq.")

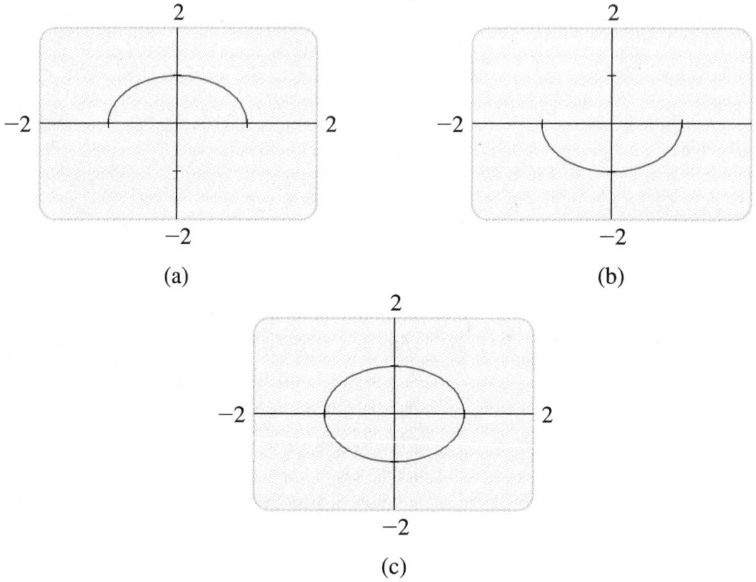

FIGURE 6
Graphing the equation $x^2 + y^2 = 1$

■

We have already discovered one of the pitfalls of using graphing calculators and computers: Examples 1 and 2 showed that the use of an inappropriate viewing rectangle can give a misleading representation of the graph of an equation. We also saw how to remedy the situation—we included the crucial parts of the graph by changing to a larger viewing rectangle. Another pitfall is illustrated in the next example.

EXAMPLE 5 ■ Extraneous Vertical Lines in Graphs

Draw the graph of the equation $y = \dfrac{1}{1 - x}$.

SOLUTION

Figure 7(a) shows the graph produced with viewing rectangle $[-9, 9]$ by $[-9, 9]$. In connecting successive points on the graph, the calculator produced a steep line segment from the top to the bottom of the screen. That line segment should not be part of the graph. Notice that the right side of the equation $y = 1/(1 - x)$ is not defined for $x = 1$. Sometimes you can get rid of the extraneous vertical line by experimenting with changes in scale. Here, for example, when you change to the smaller viewing rectangle $[-5, 5]$ by $[-5, 5]$, you obtain the much better graph in Figure 7(b).

Another way to avoid the extraneous line is to change the graphing mode on the calculator so that the dots are not connected. Alternatively, we could zoom in using the Zoom Decimal mode.

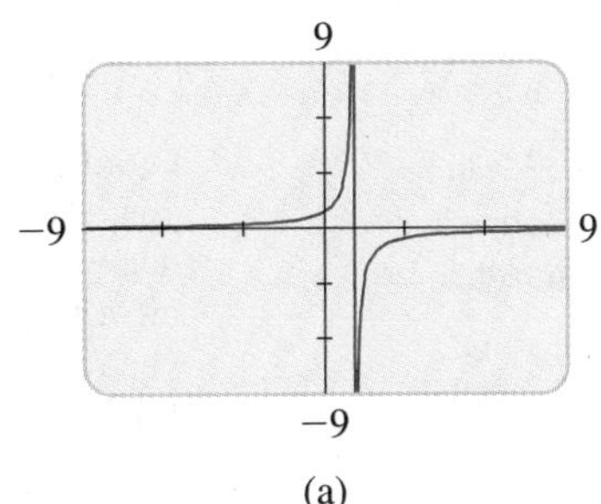

(a)

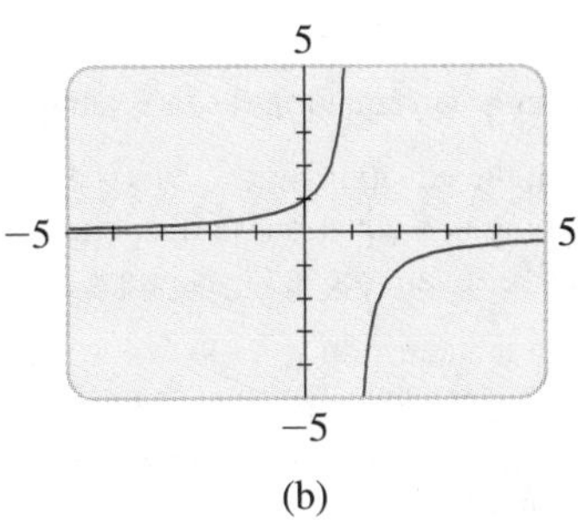

(b)

FIGURE 7

Graphs of $y = \frac{1}{1 - x}$

2.3 EXERCISES

1–8 ■ Use a graphing calculator or computer to decide which viewing rectangle (a)–(d) produces the most appropriate graph of the equation.

1. $y = x^2 - 16$
(a) $[-10, 10]$ by $[-10, 10]$
(b) $[-5, 5]$ by $[-20, 10]$
(c) $[-2, 2]$ by $[-2, 2]$
(d) $[0, 4]$ by $[-16, 0]$

2. $y = \sqrt{8 - x}$
(a) $[-2, 2]$ by $[-2, 2]$
(b) $[-4, 4]$ by $[-1, 4]$
(c) $[-10, 10]$ by $[-10, 10]$
(d) $[-20, 20]$ by $[-20, 20]$

3. $y = x^4 + 2$
(a) $[-2, 2]$ by $[-2, 2]$
(b) $[0, 4]$ by $[0, 4]$
(c) $[-8, 8]$ by $[-4, 40]$
(d) $[-40, 40]$ by $[-80, 800]$

4. $y = x^2 + 7x + 6$
(a) $[-5, 5]$ by $[-5, 5]$
(b) $[0, 10]$ by $[-20, 100]$
(c) $[-15, 8]$ by $[-20, 100]$
(d) $[-10, 3]$ by $[-100, 20]$

5. $y = 100 - x^2$
(a) $[-4, 4]$ by $[-4, 4]$
(b) $[-10, 10]$ by $[-10, 10]$
(c) $[-15, 15]$ by $[-30, 110]$
(d) $[-4, 4]$ by $[-30, 110]$

6. $y = 2x^2 - 1000$
(a) $[-10, 10]$ by $[-10, 10]$
(b) $[-10, 10]$ by $[-100, 100]$
(c) $[-10, 10]$ by $[-1000, 1000]$
(d) $[-25, 25]$ by $[-1200, 200]$

7. $y = 10 + 25x - x^3$
(a) $[-4, 4]$ by $[-4, 4]$
(b) $[-10, 10]$ by $[-10, 10]$
(c) $[-20, 20]$ by $[-100, 100]$
(d) $[-100, 100]$ by $[-200, 200]$

8. $y = \sqrt{8x - x^2}$
(a) $[-4, 4]$ by $[-4, 4]$
(b) $[-5, 5]$ by $[0, 100]$
(c) $[-10, 10]$ by $[-10, 40]$
(d) $[-2, 10]$ by $[-2, 6]$

9–24 ■ Determine an appropriate viewing rectangle for the equation and use it to draw the graph.

9. $y = 100x^2$

10. $y = -100x^2$

11. $y = 4 + 6x - x^2$

12. $y = 0.3x^2 + 1.7x - 3$

13. $y = \sqrt[4]{256 - x^2}$

14. $y = \sqrt{12x - 17}$

15. $y = 0.01x^3 - x^2 + 5$

16. $y = x(x + 6)(x - 9)$

17. $y = \dfrac{1}{x^2 + 25}$

18. $y = \dfrac{x}{x^2 + 25}$

19. $y = x^4 - 4x^3$

20. $y = x^3 + \dfrac{1}{x}$

21. $y = 1 + |x - 1|$

22. $y = 2x - |x^2 - 5|$

23. $y = \dfrac{|x + 1|}{x + 1}$

24. $y = x^{1/3} - 8x^{-1/3}$

25. Graph the circle $x^2 + y^2 = 9$ by solving for y and graphing two equations as in Example 4.

26. Graph the circle $(y-1)^2 + x^2 = 1$ by solving for y and graphing two equations as in Example 4.

27. Graph the equation $4x^2 + 2y^2 = 1$ by solving for y and graphing two equations corresponding to the negative and positive square roots. (This graph is called an *ellipse*.)

28. Graph the equation $y^2 - 9x^2 = 1$ by solving for y and graphing the two equations corresponding to the positive and negative square roots. (This graph is called a *hyperbola*.)

29–32 ■ Do the graphs intersect in the given viewing rectangle? If they do, how many points of intersection are there?

29. $y = -3x^2 + 6x - \frac{1}{2}$, $y = \sqrt{7 - \frac{7}{12}x^2}$; $[-4, 4]$ by $[-1, 3]$

30. $y = \sqrt{49 - x^2}$, $y = \frac{1}{5}(41 - 3x)$; $[-8, 8]$ by $[-1, 8]$

31. $y = 6 - 4x - x^2$, $y = 3x + 18$; $[-6, 2]$ by $[-5, 20]$

32. $y = x^3 - 4x$, $y = x + 5$; $[-4, 4]$ by $[-15, 15]$

DISCOVERY • DISCUSSION

33. Equation Notation on Graphing Calculators When you enter the following equations into your calculator, how does what you see on the screen differ from the usual way of writing the equations? (Check your user's manual if you're not sure.)

(a) $y = |x|$ (b) $y = \sqrt[5]{x}$

(c) $y = \dfrac{x}{x-1}$ (d) $y = x^3 + \sqrt[3]{x+2}$

34. Enter Equations Carefully A student wishes to graph the equations

$$y = x^{1/3} \qquad \text{and} \qquad y = \frac{x}{x+4}$$

on the same screen, so he enters the following information into his calculator:

$$Y_1 = X\text{^}1/3 \qquad Y_2 = X/X + 4$$

The calculator graphs two lines instead of the equations he wanted. What went wrong? Why are the graphs lines?

2.4 LINES

In this section we find equations for straight lines lying in a coordinate plane. We first need a way to measure the "steepness" of a line, or how quickly it rises (or falls) as we move from left to right. We define *run* to be the distance we move to the right and *rise* to be the corresponding distance that the line rises (or falls). The *slope* of a line is the ratio of rise to run:

$$\text{slope} = \frac{\text{rise}}{\text{run}}$$

Figure 1 shows situations where slope is important. Carpenters use the term *pitch* for the slope of a roof or a staircase; the term *grade* is used for the slope of a road.

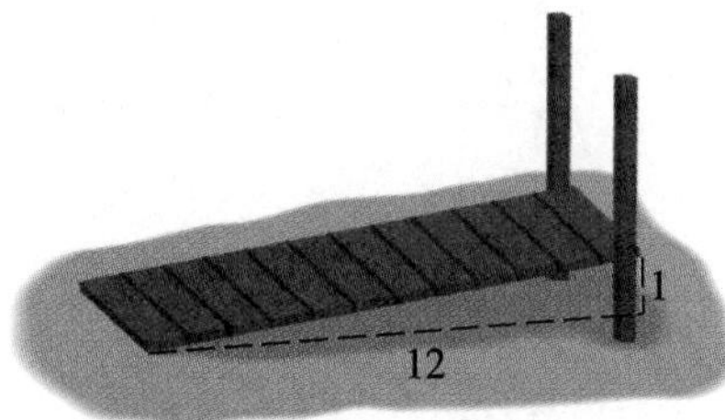

Slope of a ramp

Slope $= \frac{1}{12}$

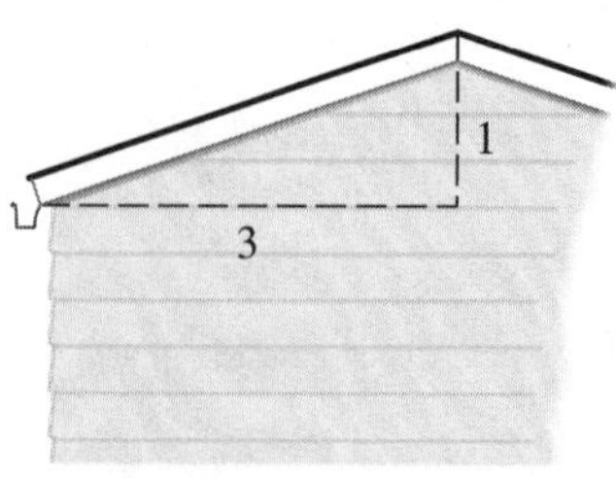

Pitch of a roof

Slope $= \frac{1}{3}$

Grade of a road

Slope $= \frac{8}{100}$

FIGURE 1

If a line lies in a coordinate plane, then the **run** is the change in the x-coordinate and the **rise** is the corresponding change in the y-coordinate between any two points on the line (see Figure 2). This gives us the following definition of slope.

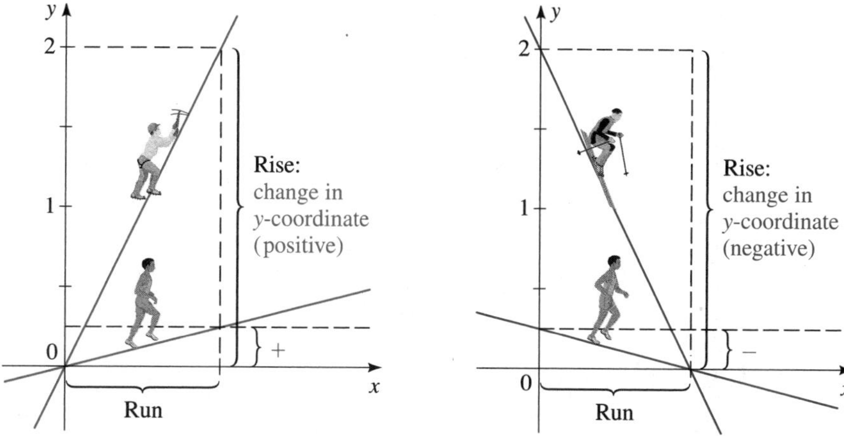

FIGURE 2

SLOPE OF A LINE

The **slope** m of a nonvertical line that passes through the points $A(x_1, y_1)$ and $B(x_2, y_2)$ is

$$m = \frac{\text{rise}}{\text{run}} = \frac{y_2 - y_1}{x_2 - x_1}$$

The slope of a vertical line is not defined.

The slope is independent of which two points are chosen on the line. We can see that this is true from the similar triangles in Figure 3:

$$\frac{y_2 - y_1}{x_2 - x_1} = \frac{y_2' - y_1'}{x_2' - x_1'}$$

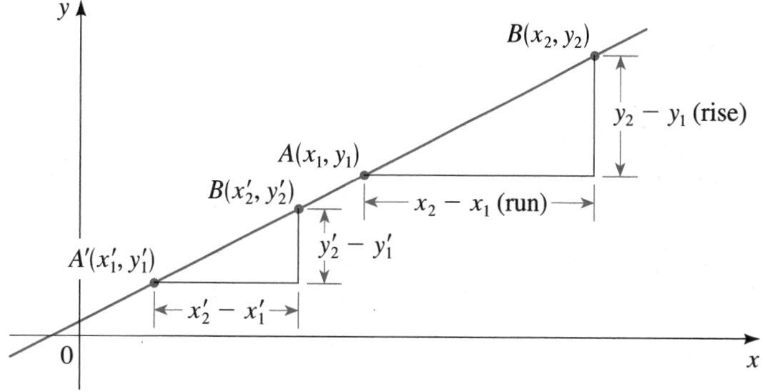

FIGURE 3

Figure 4 shows several lines labeled with their slopes. Notice that lines with positive slope slant upward to the right, whereas lines with negative slope slant downward to the right. The steepest lines are those for which the absolute value of the slope is the largest; a horizontal line has slope zero.

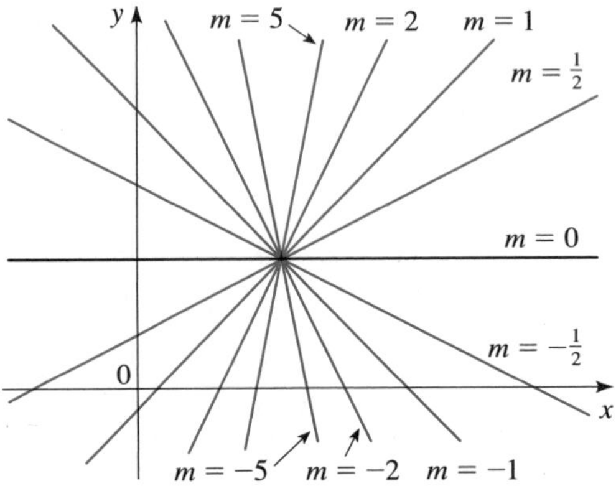

FIGURE 4
Lines with various slopes

EXAMPLE 1 ■ Finding the Slope of a Line through Two Points

Find the slope of the line that passes through the points $P(2, 1)$ and $Q(8, 5)$.

SOLUTION

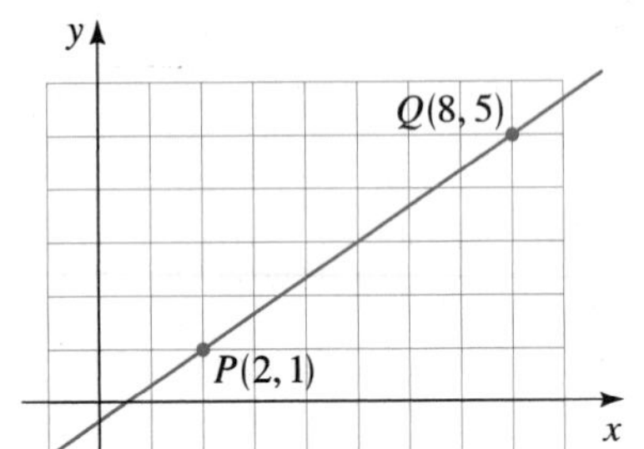

FIGURE 5

Since any two different points determine a line, only one line passes through these two points. From the definition, the slope is

$$m = \frac{y_2 - y_1}{x_2 - x_1} = \frac{5 - 1}{8 - 2} = \frac{4}{6} = \frac{2}{3}$$

This says that for every 3 units we move to the right, the line rises 2 units. The line is drawn in Figure 5. ■

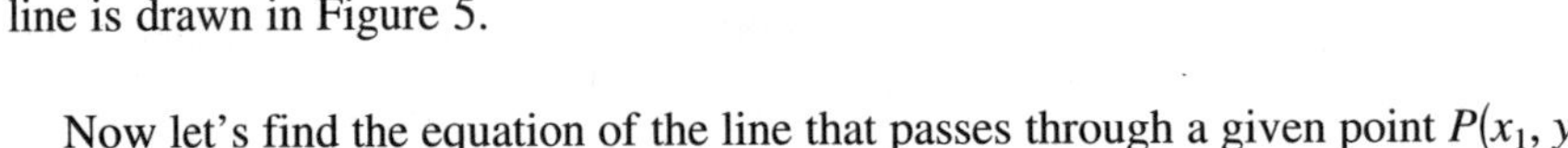

Now let's find the equation of the line that passes through a given point $P(x_1, y_1)$ and has slope m. A point $P(x, y)$ with $x \neq x_1$ lies on this line if and only if the slope of the line through P_1 and P is equal to m (see Figure 6), that is,

$$\frac{y - y_1}{x - x_1} = m$$

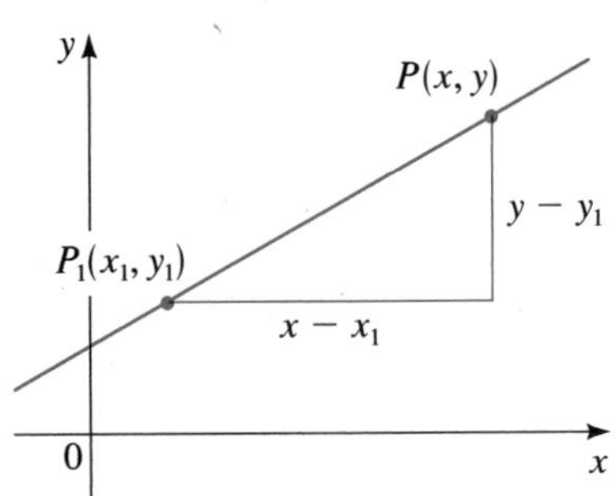

FIGURE 6

This equation can be rewritten in the form $y - y_1 = m(x - x_1)$; note that the equation is also satisfied when $x = x_1$ and $y = y_1$. Therefore, it is an equation of the given line.

POINT-SLOPE FORM OF THE EQUATION OF A LINE

An equation of the line that passes through the point (x_1, y_1) and has slope m is

$$y - y_1 = m(x - x_1)$$

EXAMPLE 2 ■ Finding the Equation of a Line with Given Point and Slope

(a) Find an equation of the line through $(1, -3)$ with slope $-\frac{1}{2}$.
(b) Sketch the line.

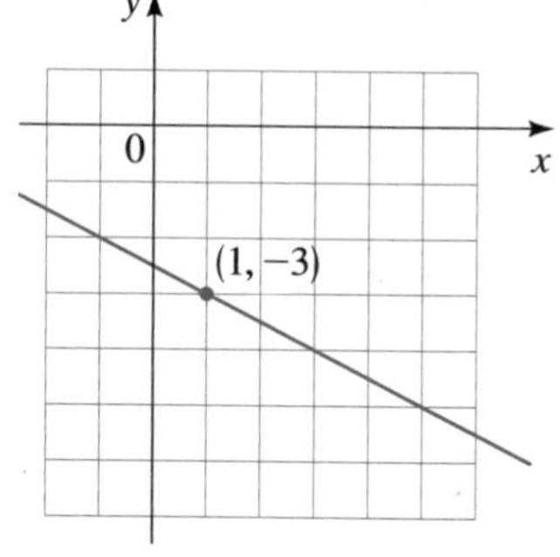

FIGURE 7

SOLUTION

(a) Using the point-slope form with $m = -\frac{1}{2}$, $x_1 = 1$, and $y_1 = -3$, we obtain an equation of the line as

$$y + 3 = -\tfrac{1}{2}(x - 1) \qquad \text{From point-slope equation}$$
$$2y + 6 = -x + 1 \qquad \text{Multiply by 2}$$
$$x + 2y + 5 = 0 \qquad \text{Rearrange}$$

(b) The fact that the slope is $-\frac{1}{2}$ tells us that when we move to the right 2 units, the line drops 1 unit. This enables us to sketch the line in Figure 7. ■

EXAMPLE 3 ■ Finding the Equation of a Line through Two Given Points

Find an equation of the line through the points $(-1, 2)$ and $(3, -4)$.

SOLUTION

The slope of the line is

$$m = \frac{-4 - 2}{3 - (-1)} = -\frac{6}{4} = -\frac{3}{2}$$

Using the point-slope form with $x_1 = -1$ and $y_1 = 2$, we obtain

$$y - 2 = -\tfrac{3}{2}(x + 1) \qquad \text{From point-slope equation}$$
$$2y - 4 = -3x - 3 \qquad \text{Multiply by 2}$$
$$3x + 2y - 1 = 0 \qquad \text{Rearrange}$$
■

Suppose a nonvertical line has slope m and y-intercept b (see Figure 8). This means the line intersects the y-axis at the point $(0, b)$, so the point-slope form of the equation of the line, with $x = 0$ and $y = b$, becomes

$$y - b = m(x - 0)$$

This simplifies to $y = mx + b$, which is called the **slope-intercept form** of the equation of a line.

FIGURE 8

SLOPE-INTERCEPT FORM OF THE EQUATION OF A LINE

An equation of the line that has slope m and y-intercept b is

$$y = mx + b$$

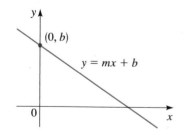

EXAMPLE 4 ■ Lines in Slope-Intercept Form

(a) Find the equation of the line with slope 3 and y-intercept -2.
(b) Find the slope and y-intercept of the line $3y - 2x = 1$.

SOLUTION

(a) Since $m = 3$ and $b = 2$, from the slope-intercept form of the equation of a line we get

$$y = 3x - 2$$

(b) We first write the equation in the form $y = mx + b$:

$$\begin{aligned} 3y - 2x &= 1 \\ 3y &= 2x + 1 && \text{Add } 2x \\ y &= \tfrac{2}{3}x + \tfrac{1}{3} && \text{Divide by 3} \end{aligned}$$

From the slope-intercept form of the equation of a line, we see that the slope is $m = \frac{2}{3}$ and the y-intercept is $b = \frac{1}{3}$. ■

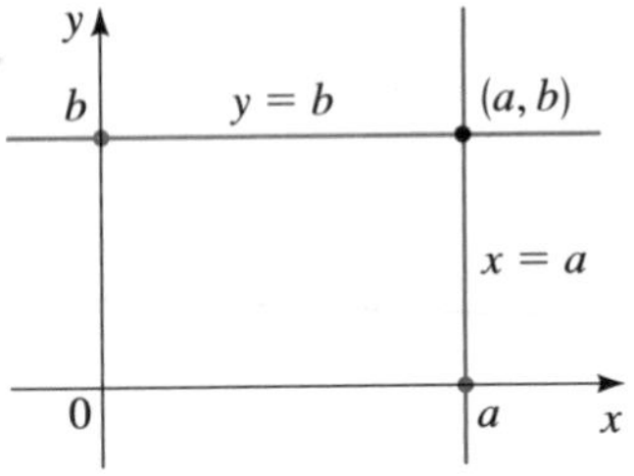

If a line is horizontal, its slope is $m = 0$, so its equation is $y = b$, where b is the y-intercept (see Figure 9). A vertical line does not have a slope, but we can write its equation as $x = a$, where a is the x-intercept, because the x-coordinate of every point on the line is a.

FIGURE 9

VERTICAL AND HORIZONTAL LINES

An equation of the vertical line through (a, b) is $x = a$.

An equation of the horizontal line through (a, b) is $y = b$.

EXAMPLE 5 ■ Vertical and Horizontal Lines

(a) The graph of the equation $x = 3$ is a vertical line with x-intercept 3.

(b) The graph of the equation $y = -2$ is a horizontal line with y-intercept -2.

The lines are graphed in Figure 10. ■

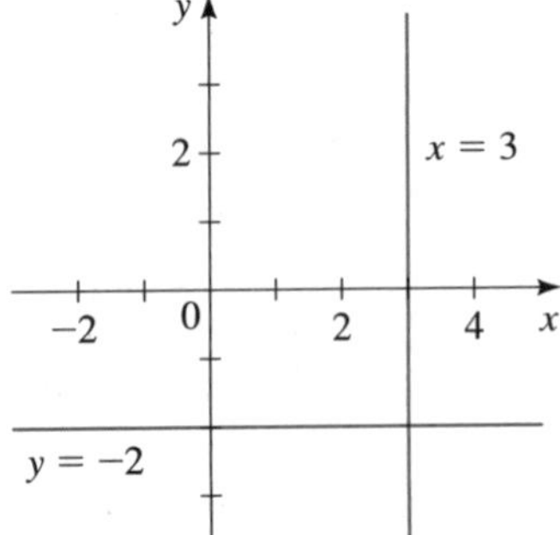

FIGURE 10

A **linear equation** is an equation of the form

$$Ax + By + C = 0$$

where A, B, and C are constants and A and B are not both 0. The equation of a line is a linear equation:

- A nonvertical line has the equation $y = mx + b$ or $-mx + y - b = 0$, which is a linear equation with $A = -m$, $B = 1$, and $C = -b$.
- A vertical line has the equation $x = a$ or $x - a = 0$, which is a linear equation with $A = 1$, $B = 0$, and $C = -a$.

Conversely, the graph of a linear equation is a line:

- If $B \neq 0$, the equation becomes

$$y = -\frac{A}{B}x - \frac{C}{B}$$

  and this is the slope-intercept form of the equation of a line (with $m = -A/B$ and $b = -C/B$).

- If $B = 0$, the equation becomes $Ax + C = 0$, or $x = -C/A$, which represents a vertical line.

We have proved the following.

GENERAL EQUATION OF A LINE

The graph of every **linear equation**

$$Ax + By + C = 0 \qquad (A, B \text{ not both zero})$$

is a line. Conversely, every line is the graph of a linear equation.

EXAMPLE 6 ■ Graphing a Linear Equation

Sketch the graph of the equation $2x - 3y - 12 = 0$.

SOLUTION 1

Since the equation is linear, its graph is a line. To draw the graph, it is enough to find any two points on the line. The intercepts are the easiest points to find.

x-intercept: Substitute $y = 0$, to get $2x - 12 = 0$, so $x = 6$

y-intercept: Substitute $x = 0$, to get $-3y - 12 = 0$, so $y = -4$

With these points we can sketch the graph in Figure 11.

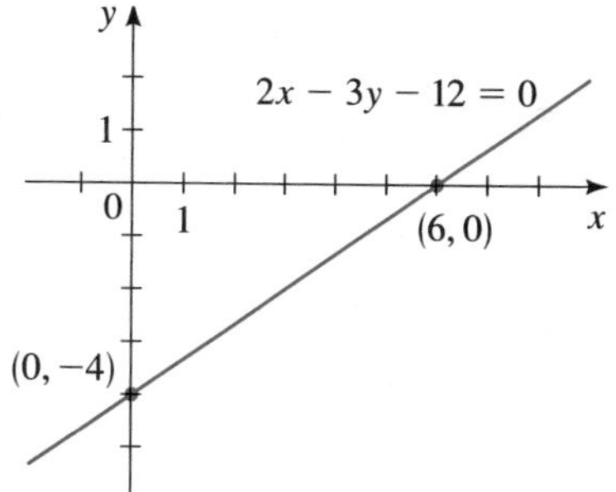

FIGURE 11

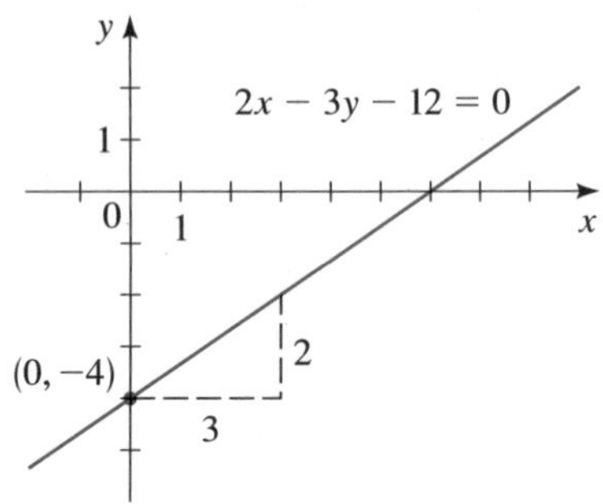

FIGURE 12

SOLUTION 2

We write the equation in slope-intercept form:

$$\begin{aligned} 2x - 3y - 12 &= 0 \\ 2x - 3y &= 12 && \text{Add 12} \\ -3y &= -2x + 12 && \text{Subtract } 2x \\ y &= \tfrac{2}{3}x - 4 && \text{Divide by } -3 \end{aligned}$$

This equation is in the form $y = mx + b$, so the slope is $m = \frac{2}{3}$ and the y-intercept is $b = -4$. To sketch the graph, we plot the y-intercept, and then move 3 units to the right and 2 units up as shown in Figure 12. ■

PARALLEL AND PERPENDICULAR LINES

Since slope measures the steepness of a line, it seems reasonable that parallel lines should have the same slope. In fact, we can prove this.

PARALLEL LINES

Two nonvertical lines are parallel if and only if they have the same slope.

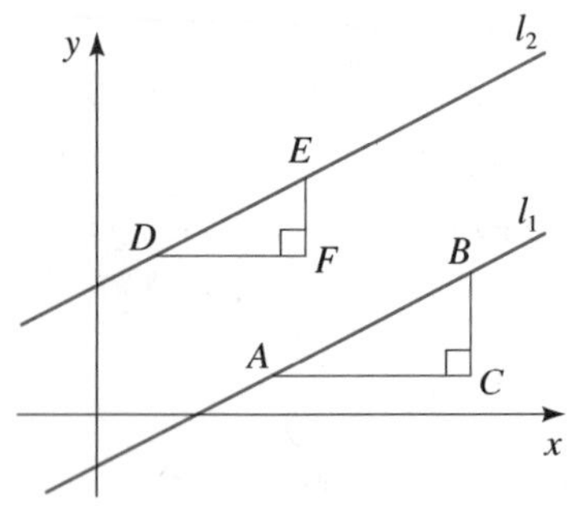

FIGURE 13

■ **Proof** Let the lines l_1 and l_2 in Figure 13 have slopes m_1 and m_2. If the lines are parallel, then the right triangles ABC and DEF are similar, so

$$m_1 = \frac{d(B, C)}{d(A, C)} = \frac{d(E, F)}{d(D, F)} = m_2$$

Conversely, if the slopes are equal, then the triangles will be similar, so $\angle BAC = \angle EDF$ and the lines are parallel. □

EXAMPLE 7 ■ Finding the Equation of a Line Parallel to a Given Line

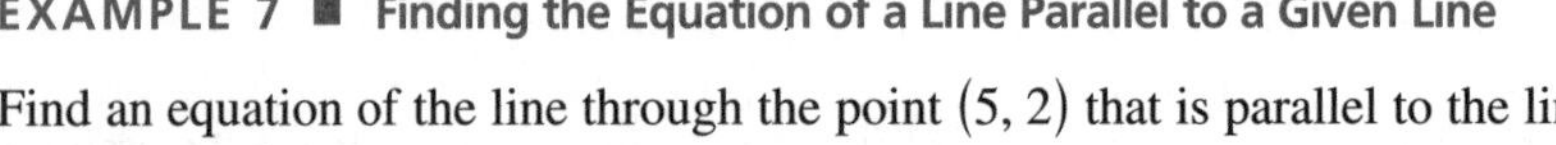

Find an equation of the line through the point $(5, 2)$ that is parallel to the line $4x + 6y + 5 = 0$.

SOLUTION

First we write the equation of the given line in slope-intercept form.

$$\begin{aligned} 4x + 6y + 5 &= 0 \\ 6y &= -4x - 5 && \text{Subtract } 4x + 5 \\ y &= -\tfrac{2}{3}x - \tfrac{5}{6} && \text{Divide by 6} \end{aligned}$$

So the line has slope $m = -\frac{2}{3}$. Since the required line is parallel to the given line, it also has slope $m = -\frac{2}{3}$. From the point-slope form of the equation of a line, we get

$$y - 2 = -\tfrac{2}{3}(x - 5) \qquad \text{Slope } m = -\tfrac{2}{3}, \text{ point } (5, 2)$$

$$3y - 6 = -2x + 10 \qquad \text{Multiply by 3}$$

$$2x + 3y - 16 = 0 \qquad \text{Rearrange}$$

Thus, the equation of the required line is $2x + 3y - 16 = 0$. ■

The condition for perpendicular lines is not as obvious as that for parallel lines.

PERPENDICULAR LINES

Two lines with slopes m_1 and m_2 are perpendicular if and only if $m_1m_2 = -1$, that is, their slopes are negative reciprocals:

$$m_2 = -\frac{1}{m_1}$$

Also, a horizontal line (slope 0) is perpendicular to a vertical line (no slope).

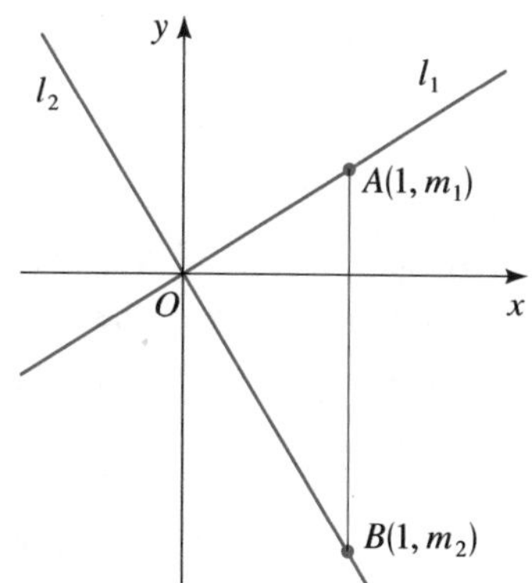

FIGURE 14

■ **Proof** In Figure 14 we show two lines intersecting at the origin. (If the lines intersect at some other point, we consider lines parallel to these that intersect at the origin. These lines have the same slopes as the original lines.)

If the lines l_1 and l_2 have slopes m_1 and m_2, then their equations are $y = m_1x$ and $y = m_2x$. Notice that $A(1, m_1)$ lies on l_1 and $B(1, m_2)$ lies on l_2. By the Pythagorean Theorem and its converse, $OA \perp OB$ if and only if

$$[d(O, A)]^2 + [d(O, B)]^2 = [d(A, B)]^2$$

By the Distance Formula, this becomes

$$(1^2 + m_1^2) + (1^2 + m_2^2) = (1 - 1)^2 + (m_2 - m_1)^2$$

$$2 + m_1^2 + m_2^2 = m_2^2 - 2m_1m_2 + m_1^2$$

$$2 = -2m_1m_2$$

$$m_1m_2 = -1$$

□

MATHEMATICS IN THE MODERN WORLD

CHANGING WORDS, SOUND, AND PICTURES INTO NUMBERS

Pictures, sound, and text are routinely transmitted from one place to another via the Internet, fax machines, or modems. How can such things be transmitted through telephone wires? The key to doing this is to change them into numbers or bits (the digits 0 or 1). It's easy to see how to change text to numbers. For example, we could use the correspondence A = 00000001, B = 00000010, C = 00000011, D = 00000100, E = 00000101, and so on. The word "BED" then becomes 000000100000010100000100. By reading the digits in groups of eight, it is possible to translate this number back to the word "BED."

Changing sound to bits is more complicated. A sound wave can be graphed on an oscilloscope or a computer. The graph is then broken down mathematically into simpler components corresponding to the different frequencies of the original sound. (A branch of mathematics called Fourier analysis is used here). The intensity of each component is a number, and the original sound can be reconstructed from these numbers. For example, music is stored on a CD as a sequence of bits; it may look like 101010001010 010100101010100000101111101010 00101011. . . . (One second of music requires 1.5 million bits!) The CD player reconstructs the music from the numbers on the CD.

Changing pictures into numbers involves expressing the color and brightness of each dot (or pixel)

(continued)

EXAMPLE 8 ■ Perpendicular Lines

Show that the points $P(3, 3)$, $Q(8, 17)$, and $R(11, 5)$ are the vertices of a right triangle.

SOLUTION

The slopes of the lines containing PR and QR are, respectively,

$$m_1 = \frac{5-3}{11-3} = \frac{1}{4} \qquad \text{and} \qquad m_2 = \frac{5-17}{11-8} = -4$$

Since $m_1 m_2 = -1$, these lines are perpendicular and so PQR is a right triangle. It is sketched in Figure 15. ■

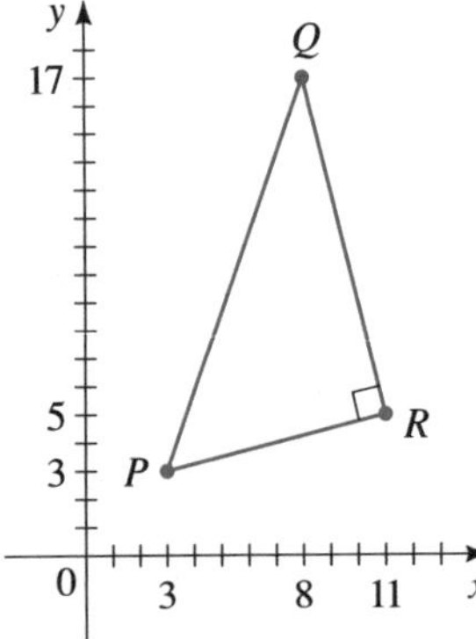

FIGURE 15

EXAMPLE 9 ■ Finding an Equation of a Line Perpendicular to a Given Line

Find an equation of the line that is perpendicular to the line $4x + 6y + 5 = 0$ and passes through the origin.

SOLUTION

In Example 7 we found that the slope of the line $4x + 6y + 5 = 0$ is $-\frac{2}{3}$. Thus, the slope of a perpendicular line is the negative reciprocal, that is, $\frac{3}{2}$. Since the required line passes through $(0, 0)$, the point-slope form gives

$$y - 0 = \tfrac{3}{2}(x - 0)$$

$$y = \tfrac{3}{2}x$$

■

APPLICATIONS OF LINEAR EQUATIONS; RATES OF CHANGE

Linear equations and their graphs are often used in applications to model the relationship between two quantities. We have seen that the slope measures the steepness of a line. In the next two examples we see that the slope in a linear model can be interpreted as a **rate of change**.

into a number. This is done very efficiently using a branch of mathematics called wavelet theory. The FBI now uses wavelets as a compact way of storing the millions of fingerprints they need on file.

EXAMPLE 10 ■ Slope as Rate of Change

A dam is built on a river to create a reservoir. The water level w in the reservoir is given by the equation

$$w = 4.5t + 28$$

where t is the number of years since the dam was constructed, and w is measured in feet.

(a) Sketch a graph of this equation.

(b) What do the slope and w-intercept of this graph represent?

SOLUTION

(a) This equation is linear, so its graph is a line. Since two points determine a line, we plot two points that lie on the graph and draw a line through them.

When $t = 0$, then $w = 4.5(0) + 28 = 28$, so $(0, 28)$ is on the line.

When $t = 2$, then $w = 4.5(2) + 28 = 37$, so $(2, 37)$ is on the line.

The line determined by these points is shown in Figure 16.

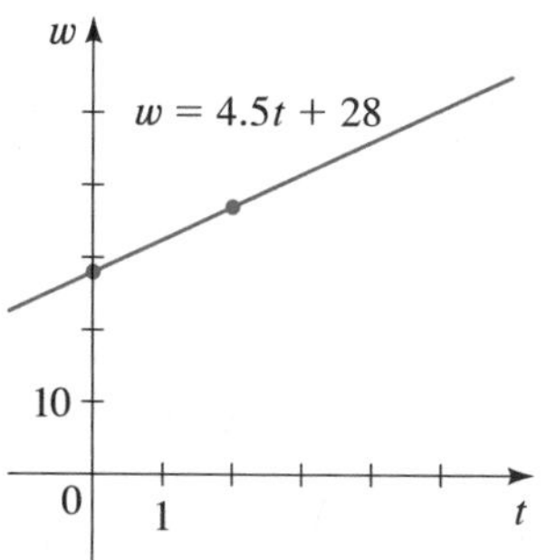

FIGURE 16

(b) The slope is $m = 4.5$; it represents the rate of change of water level with respect to time. This means that the water level *increases* 4.5 ft per year. The w-intercept is 28, and occurs when $t = 0$, so it represents the water level when the dam was constructed. ■

EXAMPLE 11 ■ Linear Relationship between Temperature and Elevation

(a) As dry air moves upward, it expands and cools. If the ground temperature is 20°C and the temperature at a height of 1 km is 10°C, express the temperature T (in °C) in terms of the height h (in kilometers). (Assume that the relationship between T and h is linear.)

(b) Draw the graph of the linear equation. What does its slope represent?

(c) What is the temperature at a height of 2.5 km?

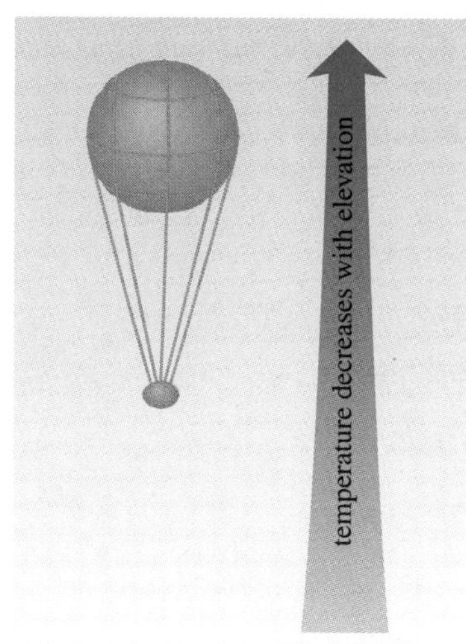

Weather balloons are used to measure temperature at high elevations.

SOLUTION

(a) Because we are assuming a linear relationship between T and h, the equation must be of the form

$$T = mh + b$$

where m and b are constants. When $h = 0$, we are given that $T = 20$, so

$$20 = m(0) + b$$

$$b = 20$$

Thus, we have

$$T = mh + 20$$

When $h = 1$, we have $T = 10$ and so

$$10 = m(1) + 20$$

$$m = 10 - 20 = -10$$

The required expression is

$$T = -10h + 20$$

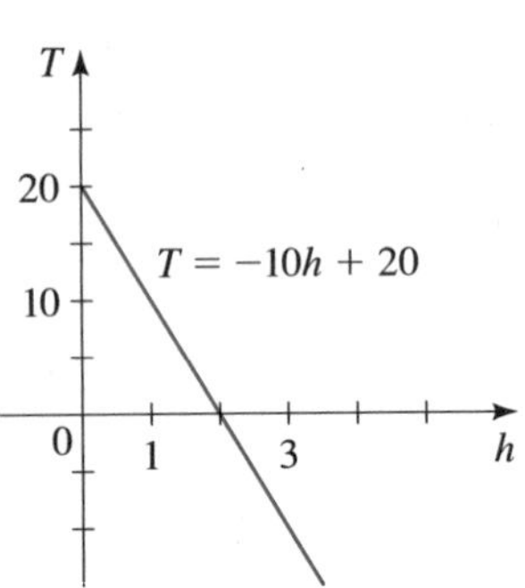

FIGURE 17

(b) The graph is sketched in Figure 17. The slope is $m = -10°\text{C/km}$, and this represents the rate of change of temperature with respect to distance above the ground.

(c) At a height of $h = 2.5$ km, the temperature is

$$T = -10(2.5) + 20 = -25 + 20 = 5°\text{C}$$ ■

In the supply and demand equations, most economists express p, the price, in terms of y, the amount produced or sold. For clarity, we have chosen to express y in terms of p.

Economists model supply and demand for a commodity by linear equations. For example, for a certain commodity we might have

Supply equation: $y = 8p - 10$

Demand equation: $y = -3p + 15$

where p is the price of the commodity. In the supply equation, y (the *amount produced*) increases as the price increases because if the price is high, more people will produce the commodity. The demand equation indicates that y (the *amount sold*) decreases as the price increases. The **equilibrium point** is the point of intersection of the graphs of the supply and demand equations; at that point the amount produced equals the amount sold.

EXAMPLE 12 ■ Supply and Demand for Wheat

An economist models the market for wheat by the following equations.

Supply: $y = 8.33p - 14.58$

Demand: $y = -1.39p + 23.35$

Here p is the price per bushel (in dollars), and y is the number of bushels produced and sold (in millions).

(a) At what point is the price so low that no wheat is produced?

(b) At what point is the price so high that no wheat is sold?

(c) Draw the graph of the supply and demand lines in the same viewing rectangle and find the equilibrium point. Estimate the equilibrium price and the quantities produced and sold at equilibrium.

SOLUTION

(a) If no wheat is produced, then $y = 0$ in the supply equation.

$$0 = 8.33p - 14.58 \qquad \text{Set } y = 0 \text{ in the supply equation}$$

$$p = 1.75 \qquad \text{Solve for } p$$

So, at the low price of \$1.75 per bushel, the production of wheat halts completely.

(b) If no wheat is sold, then $y = 0$ in the demand equation.

$$0 = -1.39p + 23.35 \qquad \text{Set } y = 0 \text{ in the demand equation}$$

$$p = 16.80 \qquad \text{Solve for } p$$

So, at the high price of \$16.80 per bushel, no wheat is sold.

(c) Figure 18 shows the graphs of the supply and demand equations in the same viewing rectangle. By moving the cursor to their point of intersection, we find that the lines intersect (approximately) at the point (3.9, 17.9). Thus, the equilibrium point occurs when the price is \$3.90 per bushel and 17.9 million bushels are produced and sold. ■

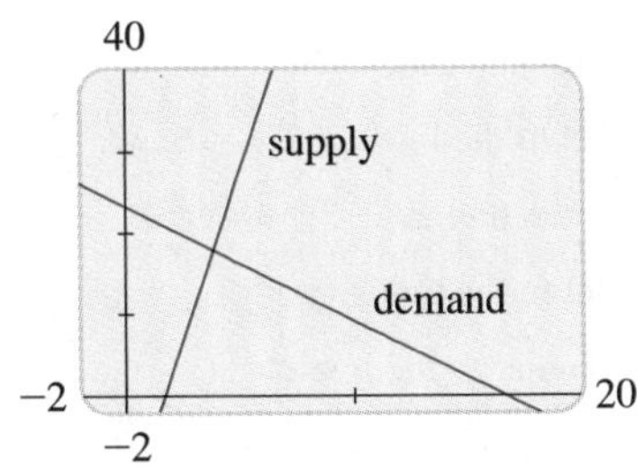

FIGURE 18

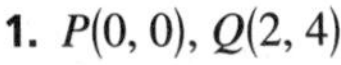

2.4 EXERCISES

1–8 ■ Find the slope of the line through P and Q.

1. $P(0, 0)$, $Q(2, 4)$
2. $P(0, 0)$, $Q(2, -6)$
3. $P(2, 2)$, $Q(10, 0)$
4. $P(1, 2)$, $Q(3, 3)$
5. $P(2, 4)$, $Q(4, 12)$
6. $P(2, -5)$, $Q(-4, 3)$
7. $P(1, -3)$, $Q(-1, 6)$
8. $P(-1, -4)$, $Q(6, 0)$
9. Find the slopes of the lines l_1, l_2, l_3, and l_4 in the figure at the right.

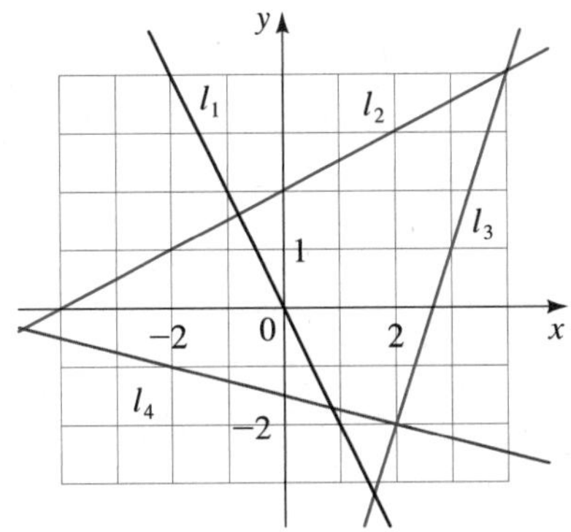

10. (a) Sketch lines through (0, 0) with slopes 1, 0, $\frac{1}{2}$, 2, and -1.

(b) Sketch lines through (0, 0) with slopes $\frac{1}{3}$, $\frac{1}{2}$, $-\frac{1}{3}$, and 3.

11–14 ■ Find an equation for the line whose graph is sketched.

11.

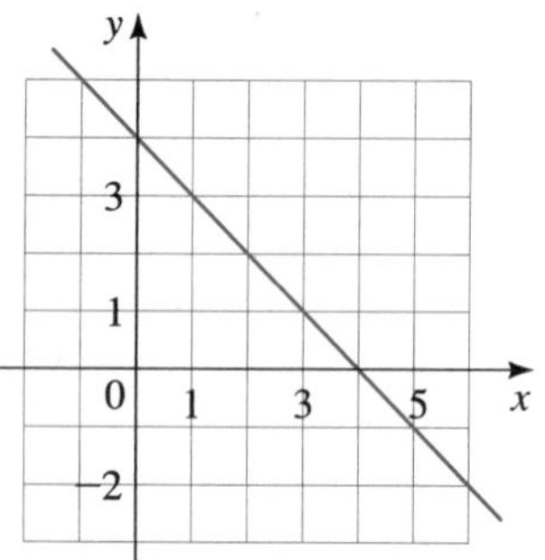

12.

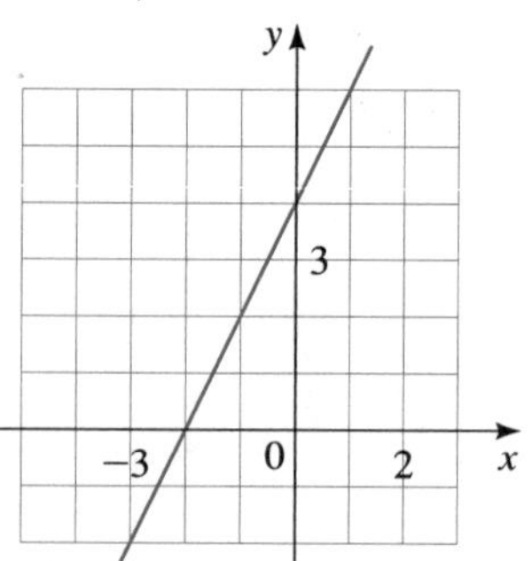

13.

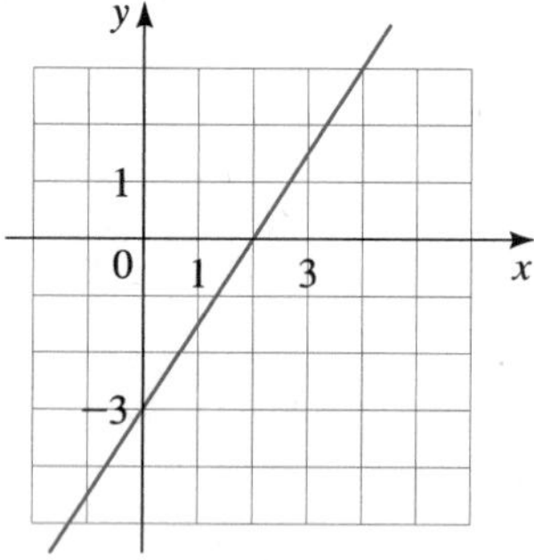

14.

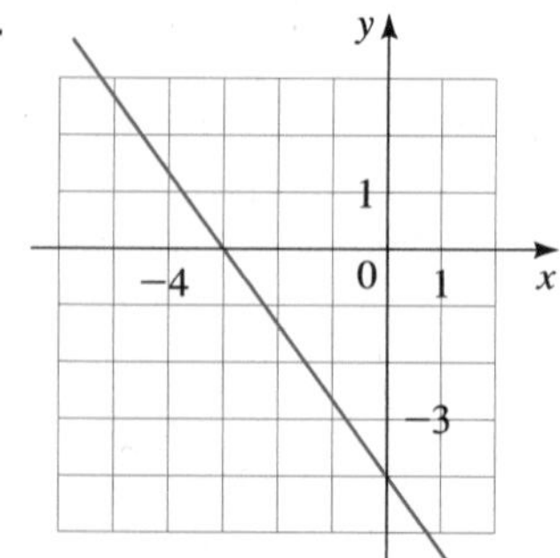

15–34 ■ Find an equation of the line that satisfies the given conditions.

15. Through (2, 3); slope 1

16. Through $(-2, 4)$; slope -1

17. Through (1, 7); slope $\frac{2}{3}$

18. Through $(-3, -5)$; slope $-\frac{7}{2}$

19. Through (2, 1) and (1, 6)

20. Through $(-1, -2)$ and (4, 3)

21. Slope 3; y-intercept -2

22. Slope $\frac{2}{5}$; y-intercept 4

23. x-intercept 1; y-intercept -3

24. x-intercept -8; y-intercept 6

25. Through (4, 5); parallel to the x-axis

26. Through (4, 5); parallel to the y-axis

27. Through $(1, -6)$; parallel to the line $x + 2y = 6$

28. y-intercept 6; parallel to the line $2x + 3y + 4 = 0$

29. Through $(-1, 2)$; parallel to the line $x = 5$

30. Through (2, 6); perpendicular to the line $y = 1$

31. Through $(-1, -2)$; perpendicular to the line $2x + 5y + 8 = 0$

32. Through $\left(\frac{1}{2}, -\frac{2}{3}\right)$; perpendicular to the line $4x - 8y = 1$

33. Through (1, 7); parallel to the line passing through (2, 5) and $(-2, 1)$

34. Through $(-2, -11)$; perpendicular to the line passing through (1, 1) and $(5, -1)$

35. (a) Sketch the line with slope $\frac{3}{2}$ that passes through the point $(-2, 1)$.
(b) Find an equation for this line.

36. (a) Sketch the line with slope -2 that passes through the point $(4, -1)$.
(b) Find an equation for this line.

37–38 ■ Use a graphing device to graph the given family of lines in the same viewing rectangle. What do the lines have in common?

37. $y = 2 + m(x - 3)$ for $m = 0, \pm 0.5, \pm 1, \pm 2, \pm 10$

38. $y = 1.3x + b$ for $b = 0, \pm 1, \pm 2.8$

39–50 ■ Find the slope and y-intercept of the line and draw its graph.

39. $x + y = 3$

40. $3x - 2y = 12$

41. $x + 3y = 0$

42. $2x - 5y = 0$

43. $\frac{1}{2}x - \frac{1}{3}y + 1 = 0$

44. $-3x - 5y + 30 = 0$

45. $y = 4$

46. $4y + 8 = 0$

47. $3x - 4y = 12$

48. $x = -5$

49. $3x + 4y - 1 = 0$

50. $4x + 5y = 10$

51. Show that $A(1, 1)$, $B(7, 4)$, $C(5, 10)$, and $D(-1, 7)$ are vertices of a parallelogram.

52. Show that $A(-3, -1)$, $B(3, 3)$, and $C(-9, 8)$ are vertices of a right triangle.

53. Show that $A(1, 1)$, $B(11, 3)$, $C(10, 8)$, and $D(0, 6)$ are vertices of a rectangle.

54. Use slopes to determine whether the given points are collinear (lie on a line).

(a) $(1, 1)$, $(3, 9)$, $(6, 21)$

(b) $(-1, 3)$, $(1, 7)$, $(4, 15)$

55. Find an equation of the perpendicular bisector of the line segment joining the points $A(1, 4)$ and $B(7, -2)$.

56. Find the area of the triangle formed by the coordinate axes and the line $2y + 3x - 6 = 0$.

57. (a) Show that if the x- and y-intercepts of a line are nonzero numbers a and b, then the equation of the line can be written in the form

$$\frac{x}{a} + \frac{y}{b} = 1$$

This is called the **two-intercept form** of the equation of a line.

(b) Use part (a) to find an equation of the line whose x-intercept is 6 and whose y-intercept is -8.

58. (a) Find an equation for the line tangent to the circle $x^2 + y^2 = 25$ at the point $(3, -4)$. (See the figure.)

(b) At what other point on the circle will a tangent line be parallel to the tangent line in part (a)?

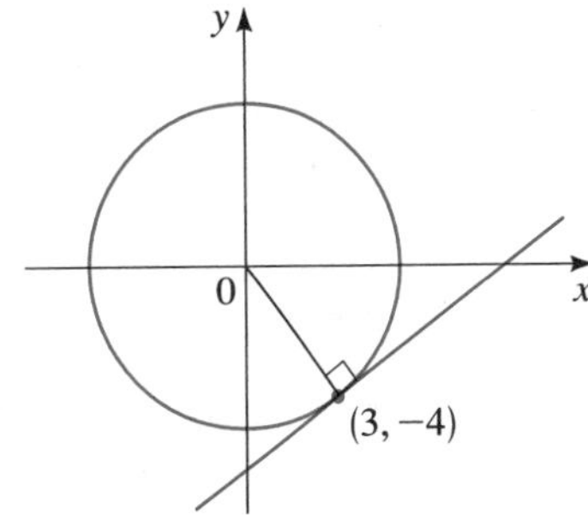

59. West of Albuquerque, New Mexico, Route 40 eastbound is straight and makes a steep descent towards the city. The highway has a 6% grade, which means that its slope is $-\frac{6}{100}$. Driving on this road you notice from elevation signs that you have descended a distance of 1000 ft. What is the change in your horizontal distance?

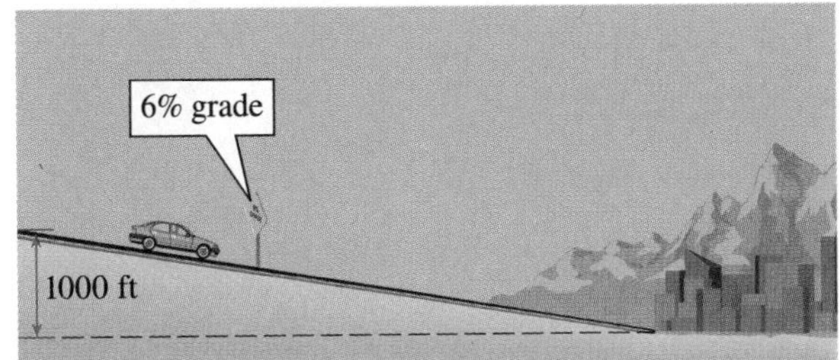

60. Some scientists believe that the average surface temperature of the world has been rising steadily. The average surface temperature is given by

$$T = 0.02t + 8.50$$

where T is temperature in °C and t is years since 1900.

(a) What do the slope and T-intercept represent?

(b) Use the equation to predict the average global surface temperature in 2100.

61. If the recommended adult dosage for a drug is D (in mg), then to determine the appropriate dosage c for a child of age a, pharmacists use the equation

$$c = 0.0417D(a + 1)$$

Suppose the dosage for an adult is 200 mg.

(a) Find the slope. What does it represent?

(b) What is the dosage for a newborn?

62. The manager of a weekend flea market knows from past experience that if she charges x dollars for a rental space at the flea market, then the number y of spaces she can rent is given by the equation $y = 200 - 4x$.

(a) Sketch a graph of this linear equation. (Remember that the rental charge per space and the number of spaces rented must both be nonnegative quantities.)

(b) What do the slope, the y-intercept, and the x-intercept of the graph represent?

63. A small-appliance manufacturer finds that if he produces x toaster ovens in a month his production cost is given by the equation $y = 6x + 3000$ (where y is measured in dollars).

(a) Sketch a graph of this linear equation.

(b) What do the slope and y-intercept of the graph represent?

64. The relationship between the Fahrenheit (F) and Celsius (C) temperature scales is given by the equation $F = \frac{9}{5}C + 32$.

(a) Complete the table to compare the two scales at the given values.

C	F
$-30°$	
$-20°$	
$-10°$	
$0°$	
	$50°$
	$68°$
	$86°$

(b) Find the temperature at which the scales agree. [*Hint:* Suppose that a is the temperature at which the scales agree. Set $F = a$ and $C = a$. Then solve for a.]

65. Biologists have observed that the chirping rate of crickets of a certain species is related to temperature, and the relationship appears to be very nearly linear. A cricket produces 120 chirps per minute at 70°F and 168 chirps per minute at 80°F.

(a) Find the linear equation that relates the temperature t and the number of chirps per minute n.

(b) If the crickets are chirping at 150 chirps per minute, estimate the temperature.

66. A small business buys a computer for \$4000. After 4 years the value of the computer is expected to be \$200. For accounting purposes, the business uses *linear depreciation* to assess the value of the computer at a given time. This means that if V is the value of the computer at time t, then a linear equation is used to relate V and t.

(a) Find a linear equation that relates V and t.

(b) Find the depreciated value of the computer 3 years from the date of purchase.

67. At the surface of the ocean, the water pressure is the same as the air pressure above the water, 15 lb/in^2. Below the surface, the water pressure increases by 4.34 lb/in^2 for every 10 ft of descent.

(a) Find an equation for the relationship between pressure and depth below the ocean surface.

(b) At what depth is the pressure 100 lb/in^2?

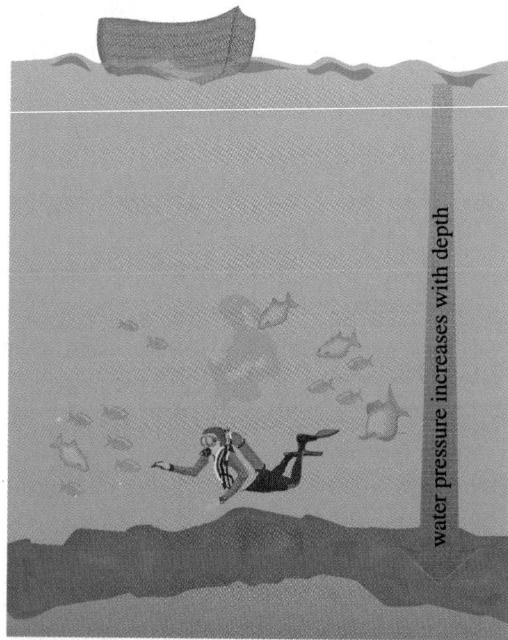

68. Jason and Debbie leave Detroit at 2:00 P.M. and drive at a constant speed, traveling west on I-90. They pass Ann Arbor, 40 mi from Detroit, at 2:50 P.M.

(a) Express the distance traveled in terms of the time elapsed.

(b) Draw the graph of the equation in part (a).

(c) What is the slope of this line? What does it represent?

69. The monthly cost of driving a car depends on the number of miles driven. Lynn found that in May her driving cost was \$380 for 480 mi and in June her cost was \$460 for 800 mi.

(a) Express the monthly cost C in terms of the distance driven d, assuming that a linear relationship gives a suitable model.

(b) Use part (a) to predict the cost of driving 1500 mi per month.

(c) Draw the graph of the linear equation. What does the slope of the line represent?

(d) What does the y-intercept of the graph represent?

(e) Why is a linear relationship a suitable model for this situation?

70. The manager of a furniture factory finds that it costs \$2200 to manufacture 100 chairs in one day and \$4800 to produce 300 chairs in one day.

(a) Assuming that the relationship between cost and the number of chairs produced is linear, find an equation that expresses this relationship. Then graph the equation.

(b) What is the slope of the line in part (a), and what does it represent?

(c) What is the y-intercept of this line, and what does it represent?

71–72 ■ Equations for supply and demand are given.

(a) Draw the graphs of the two equations in an appropriate viewing rectangle.

(b) Estimate the equilibrium point from the graph.

(c) Estimate the price and the amount of the commodity produced and sold at equilibrium.

71. Supply: $y = 0.45p + 4$
Demand: $y = -0.65p + 28$

72. Supply: $y = 8.5p + 45$
Demand: $y = -0.6p + 300$

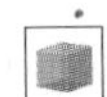

DISCOVERY • DISCUSSION

73. What Does the Slope Mean? Suppose that the graph of the outdoor temperature over a certain period of time is a line. How is the weather changing if the slope of the line is positive? If it's negative? If it's zero?

74. Collinear Points Suppose you are given the coordinates of three points in the plane, and you want to see whether they lie on the same line. How can you do this using slopes? Using the Distance Formula? Can you think of another method?

2 REVIEW

CONCEPT CHECK

1. (a) Describe the coordinate plane.
(b) How do you locate points in the coordinate plane?

2. State each formula.
(a) The Distance Formula
(b) The Midpoint Formula

3. Given an equation, what is its graph?

4. How do you find the x-intercepts and y-intercepts of a graph?

5. Write an equation of the circle with center (h, k) and radius r.

6. Explain the meaning of each type of symmetry. How do you test for it?
(a) Symmetry with respect to the x-axis
(b) Symmetry with respect to the y-axis
(c) Symmetry with respect to the origin

7. Define the slope of a line.

8. Write each form of the equation of a line.
(a) The point-slope form
(b) The slope-intercept form

9. (a) What is the equation of a vertical line?
(b) What is the equation of a horizontal line?

10. What is the general equation of a line?

11. Given lines with slopes m_1 and m_2, explain how you can tell if the lines are
(a) parallel
(b) perpendicular

EXERCISES

1–2 ■ Two points P and Q are given.

(a) Plot P and Q on a coordinate plane.
(b) Find the distance from P to Q.
(c) Find the midpoint of the segment PQ.
(d) Sketch the line determined by P and Q, and find its equation in slope-intercept form.
(e) Sketch the circle that passes through Q and has center P, and find the equation of this circle.

1. $P(2, 0)$, $Q(-5, 12)$ **2.** $P(7, -1)$, $Q(2, -11)$

3–4 ■ Sketch the region given by the set.

3. $\{(x, y) \mid -4 < x < 4 \text{ and } -2 < y < 2\}$

4. $\{(x, y) \mid x \geq 4 \text{ or } y \geq 2\}$

5. Which of the points $A(4, 4)$ or $B(5, 3)$ is closer to the point $C(-1, -3)$?

6. Find an equation of the circle that has center $(2, -5)$ and radius $\sqrt{2}$.

7. Find an equation of the circle that has center $(-5, -1)$ and passes through the origin.

8. Find an equation of the circle that contains the points $P(2, 3)$ and $Q(-1, 8)$ and has the midpoint of the segment PQ as its center.

9–12 ■ Determine whether the equation represents a circle, a point, or has no graph. If the equation is that of a circle, find its center and radius.

9. $x^2 + y^2 + 2x - 6y + 9 = 0$

10. $2x^2 + 2y^2 - 2x + 8y = \frac{1}{2}$

11. $x^2 + y^2 + 72 = 12x$

12. $x^2 + y^2 - 6x - 10y + 34 = 0$

13–22 ■ Test the equation for symmetry and sketch its graph.

13. $y = 2 - 3x$ **14.** $2x - y + 1 = 0$

15. $x + 3y = 21$ **16.** $x = 2y + 12$

17. $\dfrac{x}{2} - \dfrac{y}{7} = 1$ **18.** $\dfrac{x}{4} + \dfrac{y}{5} = 0$

19. $y = 16 - x^2$ **20.** $8x + y^2 = 0$

21. $x = \sqrt{y}$ **22.** $y = -\sqrt{1 - x^2}$

23–26 ■ Use a graphing device to graph the equation in an appropriate viewing rectangle.

23. $y = x^2 - 6x$

24. $y = \sqrt{5 - x}$

25. $y = x^3 - 4x^2 - 5x$

26. $\dfrac{x^2}{4} + y^2 = 1$

27. Find an equation for the line that passes through the points $(-1, -6)$ and $(2, -4)$.

28. Find an equation for the line that passes through the point $(6, -3)$ and has slope $-\frac{1}{2}$.

29. Find an equation for the line that has x-intercept 4 and y-intercept 12.

30. Find an equation for the line that passes through the point $(1, 7)$ and is perpendicular to the line $x - 3y + 16 = 0$.

31. Find an equation for the line that passes through the origin and is parallel to the line $3x + 15y = 22$.

32. Find an equation for the line that passes through the point $(5, 2)$ and is parallel to the line passing through $(-1, -3)$ and $(3, 2)$.

33. Hooke's Law states that if a weight w is attached to a hanging spring, then the stretched length s of the spring is linearly related to w. For a particular spring we have

$$s = 0.3w + 2.5$$

where s is measured in inches and w in pounds.

(a) What do the slope and s-intercept in this equation represent?
(b) How long is the spring when a 5-lb weight is attached?

34. Margarita is hired by an accounting firm at a salary of \$60,000 per year. Three years later her annual salary has increased to \$70,500. Assume her salary increases linearly.

(a) Find an equation that relates her annual salary S and the number of years t that she has worked for the firm.
(b) What do the slope and S-intercept of her salary equation represent?
(c) What will her salary be after 12 years with the firm?

35–36 ■ Find equations for the circle and the line in the figure.

35.

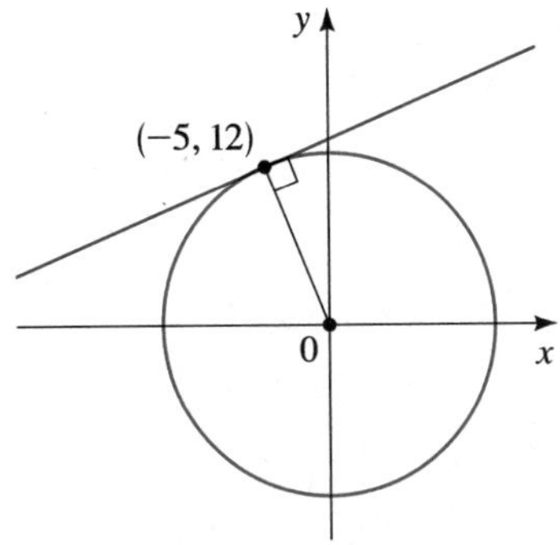

36.

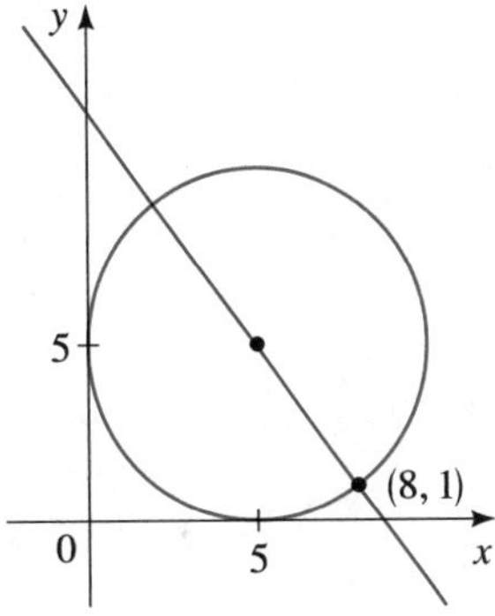

2 TEST

1. (a) Plot the points $P(0, 3)$, $Q(3, 0)$, and $R(6, 3)$ in the coordinate plane. Where must the point S be located so that $PQRS$ is a square?
   (b) Find the area of $PQRS$.

2. (a) Sketch the graph of $y = x^2 - 4$.
   (b) Find the x- and y-intercepts of the graph.
   (c) Is the graph symmetric about the x-axis, the y-axis, or the origin?

3. Let $A(-7, 4)$ and $B(5, -12)$ be points in the plane.
   (a) Find the length of the segment AB.
   (b) Find the midpoint of the segment AB.
   (c) Find an equation of the line that passes through the points A and B.
   (d) Find an equation of the perpendicular bisector of AB.
   (e) Find an equation of the circle for which AB is a diameter.

4. Find the center and radius of each circle and sketch its graph.
   (a) $x^2 + (y - 4)^2 = 9$
   (b) $x^2 + y^2 - 6x + 10y + 9 = 0$

5. Write the linear equation

$$3x - 4y + 24 = 0$$

   in slope-intercept form. What are the slope and y-intercept? Sketch the graph.

6. Find an equation for the line with the given property.
   (a) Passing through $(-2, 3)$ and parallel to the line $2x + 6y = 17$
   (b) Having x-intercept -3 and y-intercept 12

7. A manufacturer finds that the cost C of producing x blenders is

$$C = 7x + 17{,}500$$

   (a) How much does it cost to produce 2500 blenders?
   (b) What do the slope and C-intercept of the cost equation represent?

PRINCIPLES OF MODELING

A model is a representation of an object or process. For example, a toy Ferrari is a *model* of the actual car; a road map is a model of the streets and highways in a city. A model usually represents just one aspect of the original thing. The toy Ferrari is not an actual car, but it does represent what a real Ferrari looks like; a road map does not contain the actual streets in a city, but it does represent the relationship of the streets to each other.

A **mathematical model** is a mathematical representation of an object or process. Often a mathematical model is an equation that describes a certain phenomenon. In Example 11 of Section 2.4 we found that the equation $T = -10h + 20$ models the atmospheric temperature T at elevation h. We then used this equation to predict the temperature at a certain height. Figure 1 illustrates the process of mathematical modeling.

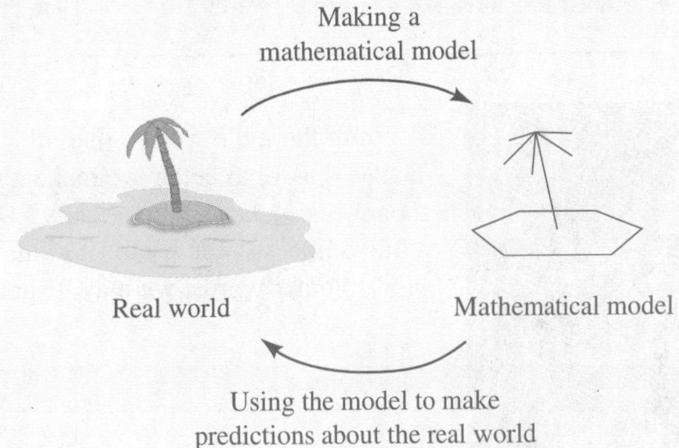

FIGURE 1

Mathematical models are useful because they enable us to isolate critical aspects of the thing we are studying and then to predict how it will behave. Models are used extensively in engineering, industry, and manufacturing. For example, engineers use computer models of skyscrapers to predict their strength and how they would behave in an earthquake. Aircraft manufacturers use elaborate mathematical models to predict the aerodynamic properties of a new design before the aircraft is actually built.

How are mathematical models developed? How are they used to predict the behavior of a process? In the next few pages and in subsequent *Focus on Modeling* sections, we explain how mathematical models can be constructed from real-world data, and we describe some of their applications.

LINEAR EQUATIONS AS MODELS

Consider the table on page 120, which gives the winning Olympic pole vaults in this century.

Year	Gold medalist(s)	Height (ft)	Year	Gold medalist	Height (ft)
1900	Irving Baxter, USA	10.83	1956	Robert Richards, USA	14.96
1904	Charles Dvorak, USA	11.48	1960	Don Bragg, USA	15.42
1908	A. Gilbert, E. Cook, USA	12.17	1964	Fred Hansen, USA	16.73
1912	Harry Babcock, USA	12.96	1968	Bob Seagren, USA	17.71
1920	Frank Foss, USA	13.42	1972	W. Nordwig, E. Germany	18.04
1924	Lee Barnes, USA	12.96	1976	Tadeusz Slusarski, Poland	18.04
1928	Sabin Carr, USA	13.77	1980	W. Kozakiewicz, Poland	18.96
1932	William Miller, USA	14.15	1984	Pierre Quinon, France	18.85
1936	Earle Meadows, USA	14.27	1988	Sergei Bubka, USSR	19.77
1948	Guinn Smith, USA	14.10	1992	M. Tarassob, Unified Team	19.02
1952	Robert Richards, USA	14.92	1996	Jean Jalfione, France	19.42

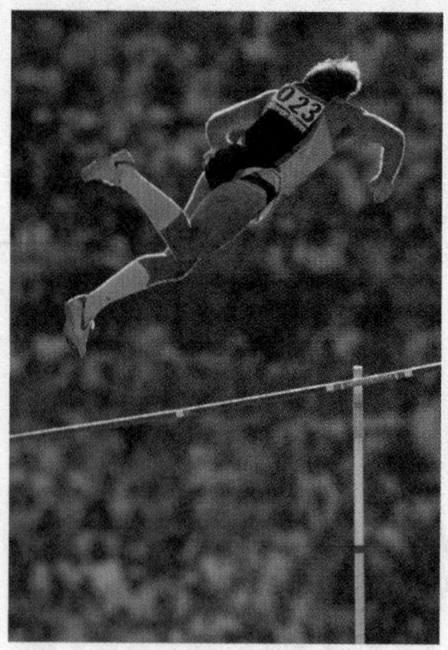

From the table we see that athletes generally pole-vault higher as time goes on. To see this trend better we make a **scatter plot**, as shown in Figure 2. How can we model this data mathematically? Since it appears that the points lie more or less along a line, we can try to fit a line visually to approximate the points in the scatter plot. In other words, we may be able to use a linear equation as our model.

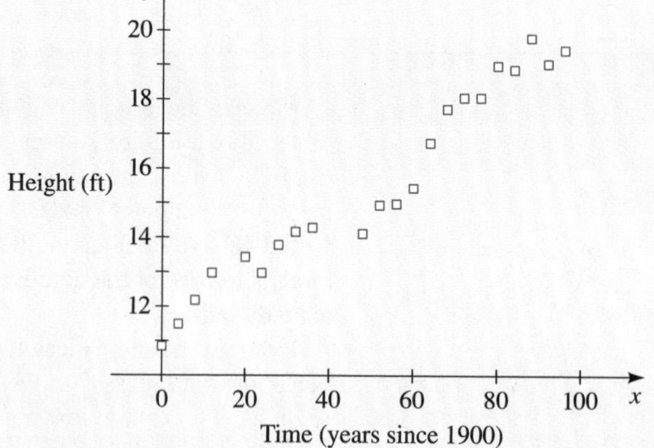

FIGURE 2
Scatter plot of pole-vault data

WHAT IS THE LINE OF BEST FIT?

How do we find a line that best fits the data? It seems reasonable to choose the line that is as close as possible to all the data points in the scatter plot. So, we want the

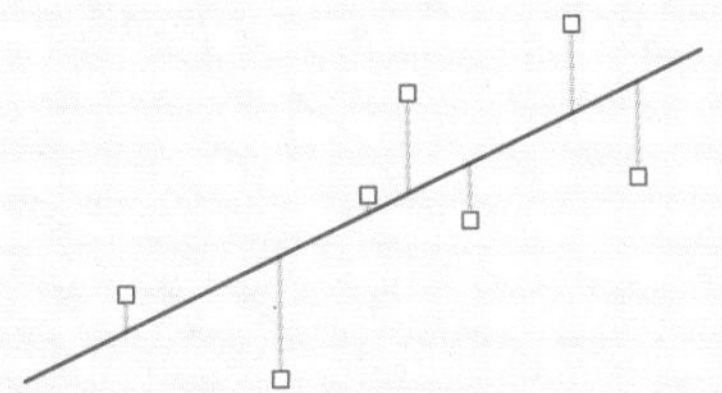

FIGURE 3
Distances from the points to the line

line for which the sum of the distances from the data points to the line is smallest (see Figure 3).

For technical reasons it is better to find the line where the sum of the squares of these distances is smallest. The resulting line is called the **regression line** or the **least squares line**. The formula for the regression line is found using calculus.* Fortunately, this formula is programmed into most graphing calculators (such as the TI-83, TI-85, or TI-86). Using a calculator, we find that the regression line for the pole-vault data is

$$y = 0.0891x + 11.09 \qquad \text{Model}$$

In Figure 4 the regression line is graphed, together with the scatter plot.

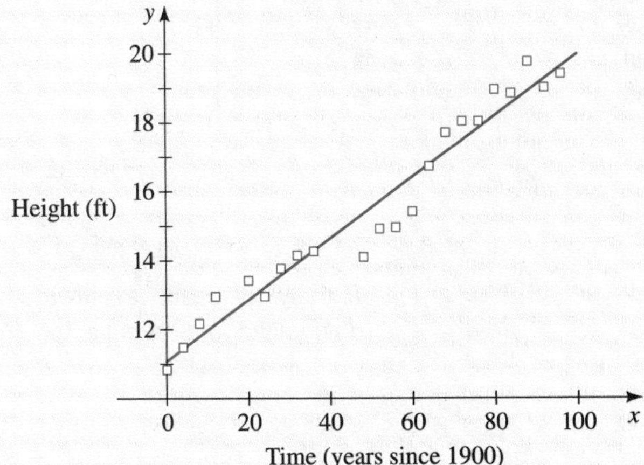

FIGURE 4
Scatter plot and regression line

What does the model predict for the Olympic pole-vault record in the year 2000? In the year 2000, $x = 100$ so the regression line gives

$$y = 0.0891(100) + 11.09 \approx 19.99 \text{ ft}$$

(Look up the actual record for 2000, and compare with this prediction.) Such predictions are reasonable for points close to our measured data, but we can't predict too far away from the measured data. Is it reasonable to use this model to predict the record 100 years from now?

ANOTHER EXAMPLE

When laboratory rats are exposed to asbestos fibers, some of them develop lung tumors. The following table lists the results of several such experiments by different scientists. Using a graphing calculator we get the regression line

$$y = 0.0177x + 0.54047 \qquad \text{Model}$$

*This formula is discussed in Exercise 53 on page 463.

The scatter plot and the regression line are graphed in Figure 5. Is it reasonable to use a linear model for this data?

Asbestos exposure (fibers/mL)	Percent that develop lung tumors
50	2
400	6
500	5
900	10
1100	26
1600	42
1800	37
2000	28
3000	50

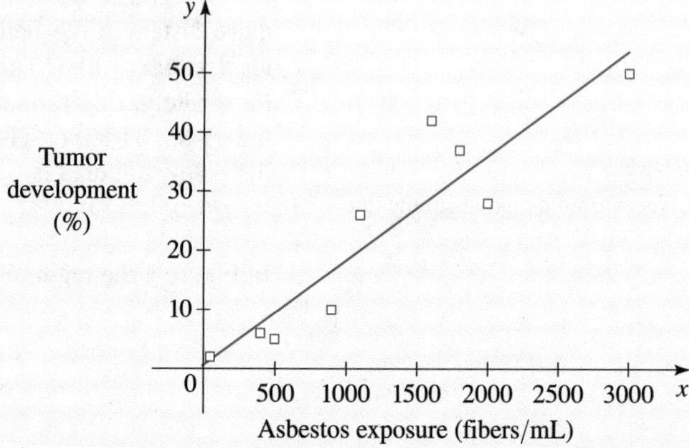

FIGURE 5 Scatter plot and regression line for asbestos-tumor data

HOW GOOD IS THE FIT?

For any given data it is always possible to find the regression line, even if the data do not tend to lie along a line. Consider the three scatter plots in Figure 6.

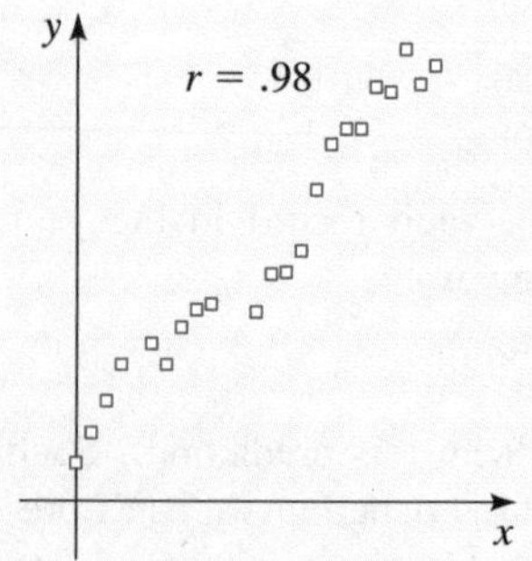

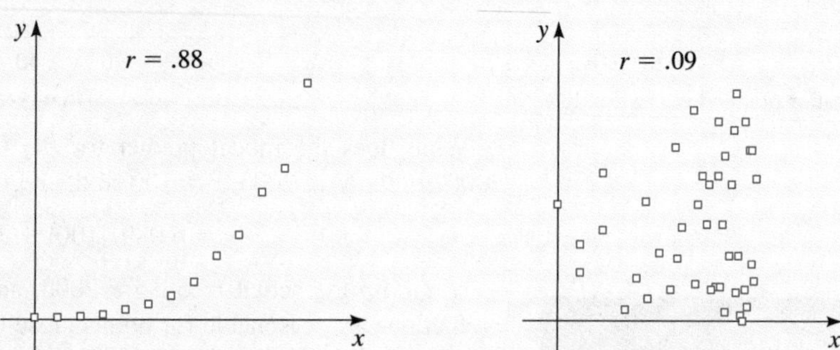

FIGURE 6

The data in the first scatter plot appear to lie along a line, in the second they appear to lie along a parabola, and in the third there appears to be no discernible trend. We can easily find the regression lines for each scatter plot using a graphing calculator. But how well do these lines represent the data? The calculator gives a **correlation coefficient** r, which is a statistical measure of how well the data lie along the regression line, or how well the two variables are **correlated**. The correlation coefficient is a number between -1 and 1. A correlation coefficient r close to 1 or -1 indicates strong correlation and a coefficient close to 0 indicates very little correlation; the slope of the line determines whether the correlation coefficient is positive or negative. Also, the more data points we have, the more meaningful the correlation coefficient will be. Using a calculator we find that the correlation coefficient between asbestos fibers and lung tumors in rats is $r = .92$. We can reasonably conclude that the presence of asbestos and the risk of lung tumors in rats are related. Can we conclude that asbestos *causes* lung tumors in rats?

If two variables are correlated, it does not necessarily mean that a change in one variable *causes* a change in the other. For example, the mathematician John Allen Paulos points out that shoe size is strongly correlated to mathematics scores among school children. Does this mean that big feet cause high math scores? Certainly not—both shoe size and math skills increase independently as children get older. So it is important not to jump to conclusions: Correlation and causation are not the same thing. Correlation is a useful tool in bringing important cause and effect relationships to light, but to prove causation, we must explain the mechanism by which one variable affects the other. For example, the link between smoking and lung cancer was observed as a correlation long before science found the mechanism by which smoking causes lung cancer.

PROBLEMS

Year	CO_2 level (ppm)
1980	338.5
1982	341.0
1984	344.3
1986	347.0
1988	351.3
1990	354.0
1992	356.3
1994	358.9
1996	362.7

1. The table lists average carbon dioxide (CO_2) levels in the atmosphere, measured in parts per million (ppm) at Mauna Loa Observatory from 1980 to 1996.

(a) Make a scatter plot of the data.

(b) Find and graph the regression line.

(c) Use the linear model in part (b) to estimate the CO_2 level in the atmosphere in 1998. Compare your answer with the actual CO_2 level of 366.7 measured in 1998.

2. Biologists have observed that the chirping rate of crickets of a certain species appears to be related to temperature. The table shows the chirping rates for various temperatures.

(a) Make a scatter plot of the data.

(b) Find and graph the regression line.

(c) Use the linear model in part (b) to estimate the chirping rate at 100°F.

Temperature (°F)	Chirping rate (chirps/min)
50	20
55	46
60	79
65	91
70	113
75	140
80	173
85	198
90	211

Income	Ulcer rate
$4,000	14.1
$6,000	13.0
$8,000	13.4
$12,000	12.4
$16,000	12.0
$20,000	12.5
$30,000	10.5
$45,000	9.4
$60,000	8.2

3. The table shows (lifetime) peptic ulcer rates (per 100 population) for various family incomes as reported by the 1989 National Health Interview Survey.

(a) Make a scatter plot of the data.

(b) Find and graph the regression line.

(c) Estimate the peptic ulcer rate for an income level of $25,000 according to the linear model in part (b).

(d) Estimate the peptic ulcer rate for an income level of $80,000 according to the linear model in part (b).

4. The table lists the relative abundance of mosquitoes (as measured by the mosquito positive rate) versus the flow rate (measured as a percentage of maximum flow) of canal networks in Saga City, Japan.

(a) Make a scatter plot of the data.

(b) Find and graph the regression line.

(c) Use the linear model in part (b) to estimate the mosquito positive rate if the canal flow is 70% of maximum.

Flow rate (%)	Mosquito positive rate (%)
0	22
10	16
40	12
60	11
90	6
100	2

5. The tables show the results of experiments measuring the average number of mosquito bites per 15 minutes versus temperature and wind speed.

(a) Make a scatter plot for each set of data.

Temperature (°C)	Bites/15 min
10	0.03
15	7.2
20	46.2
25	65.5
30	37.8

Wind speed (km/h)	Bites/15 min
0.5	59.3
2	35.7
4	24.8
6	21.9
8	12.0
14	4.8

(b) Find and graph the regression line for the temperature-bites data and for the wind-bites data.

(c) Find the correlation coefficient for each set of data. Are linear models appropriate for these data?

6. The table shows (lifetime) peptic ulcer rates (per 100 population) for various education levels as reported by the 1989 National Health Interview Survey.

(a) Make a scatter plot of the data.

(b) Find and graph the regression line.

(c) Use the linear model in part (b) to estimate the ulcer rate for a person with 20 years of education.

Education level (years)	Ulcer rate
8	12
11	12.5
12	11
15	10
16	8
18	9

7. Audiologists study the intelligibility of spoken sentences under different noise levels. Intelligibility, the MRT score, is measured as the percent of a spoken sentence that the listener can decipher at a certain noise level in decibels (dB). The table shows the results of one such test.

(a) Make a scatter plot of the data.

(b) Find and graph the regression line.

(c) Find the correlation coefficient. Is a linear model appropriate?

(d) Use the linear model in part (b) to estimate the intelligibility of a sentence at a 94-dB noise level.

Noise level (dB)	MRT score (%)
80	99
84	91
88	84
92	70
96	47
100	23
104	11

Year	IQ index
1940	85
1950	89
1960	90
1970	95
1980	97
1990	100

8. Average IQ scores in the United States have risen dramatically since the tests were first introduced, as shown in the table at the left.

(a) Make a scatter plot of the data.

(b) Find and graph the regression line.

(c) Use the linear model you found in part (b) to predict the IQ index in the year 2000.

9. The average life expectancy in the United States has been rising steadily over the past few decades, as shown in the table.

(a) Make a scatter plot of the data.

(b) Find and graph the regression line.

(c) Use the linear model you found in part (b) to predict the life expectancy in the year 2000.

(d) Search the Internet or your campus library to find the actual 2000 average life expectancy. Compare to your answer in part (c).

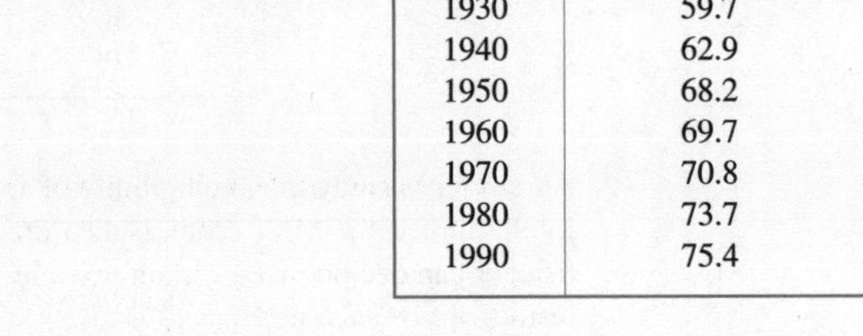

Year	Life expectancy
1920	54.1
1930	59.7
1940	62.9
1950	68.2
1960	69.7
1970	70.8
1980	73.7
1990	75.4

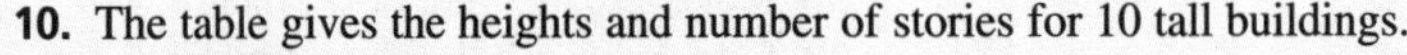

10. The table gives the heights and number of stories for 10 tall buildings.

(a) Make a scatter plot of the data.

(b) Find and graph the regression line.

(c) What is the slope of your regression line? What does its value indicate?

Building	Height (ft)	Stories
Empire State Building, New York	1250	102
One Liberty Place, Philadelphia	945	61
Canada Trust Tower, Toronto	863	51
One World Trade Center, New York	1368	110
Sears Tower, Chicago	1450	110
Petronas Tower I, Malaysia	1483	88
Commerzbank Tower, Germany	850	60
Palace of Culture and Science, Poland	758	42
Republic Plaza, Singapore	919	66
Transamerica Pyramid, San Francisco	853	48

11. The table gives U.S. teachers' average pay and the high school graduation rate by state for the 1995–1996 academic year.

(a) Make a scatter plot of the data.

(b) Find and graph the regression line.

(c) Calculate the correlation coefficient r. With 50 data points, a correlation coefficient of .3 or greater indicates strong correlation. Do you think that teachers' salaries affect the graduation rate?

State	Teachers' average pay ($)	Graduation rate (%)	State	Teachers' average pay ($)	Graduation rate (%)
Alabama	32,818	57.8	Montana	30,617	82.8
Alaska	51,738	64.7	Nebraska	32,688	82.9
Arizona	33,850	58.4	Nevada	37,093	65.4
Arkansas	30,576	74.9	New Hampshire	36,640	74.9
California	43,725	65.3	New Jersey	50,442	82.9
Colorado	37,052	71.9	New Mexico	30,152	63.4
Connecticut	50,730	73.5	New York	49,034	62.0
Delaware	42,439	65.8	North Carolina	33,315	62.4
Florida	34,475	57.8	North Dakota	28,230	89.0
Georgia	37,378	55.0	Ohio	38,977	70.6
Hawaii	38,377	74.8	Oklahoma	30,606	73.0
Idaho	32,775	79.6	Oregon	42,150	66.6
Illinois	43,873	80.0	Pennsylvania	47,650	76.3
Indiana	39,682	70.1	Rhode Island	44,300	71.4
Iowa	34,040	85.3	South Carolina	33,608	54.4
Kansas	36,811	75.8	South Dakota	27,341	86.6
Kentucky	34,525	68.1	Tennessee	35,340	63.4
Louisiana	29,650	57.9	Texas	33,648	58.4
Maine	34,349	72.4	Utah	32,950	78.4
Maryland	41,739	73.9	Vermont	36,299	89.9
Massachusetts	43,930	75.8	Virginia	36,654	75.5
Michigan	49,277	69.6	Washington	38,788	72.2
Minnesota	39,106	85.3	West Virginia	33,398	76.1
Mississippi	29,547	56.8	Wisconsin	39,899	80.4
Missouri	33,975	71.2	Wyoming	32,022	77.8

12. The table gives the gold medal times in the men's and women's 100-m freestyle Olympic swimming event. The data are graphed in the scatter plot, with the women's times in red and the men's in blue.

(a) Find the regression lines for the men's data and the women's data.

(b) Sketch both regression lines on the same graph. When do these lines predict that the women will overtake the men in this event? Does this conclusion seem reasonable?

MEN

Year	Gold medalist	Time (s)
1908	C. Daniels, USA	65.6
1912	D. Kahanamoku, USA	63.4
1920	D. Kahanamoku, USA	61.4
1924	J. Weissmuller, USA	59.0
1928	J. Weissmuller, USA	58.6
1932	Y. Miyazaki, Japan	58.2
1936	F. Csik, Hungary	57.6
1948	W. Ris, USA	57.3
1952	C. Scholes, USA	57.4
1956	J. Henricks, Australia	55.4
1960	J. Devitt, Australia	55.2
1964	D. Schollander, USA	53.4
1968	M. Wenden, Australia	52.2
1972	M. Spitz, USA	51.22
1976	J. Montgomery, USA	49.99
1980	J. Woithe, E. Germany	50.40
1984	R. Gaines, USA	49.80
1988	M. Biondi, USA	48.63
1992	A. Popov, Russia	49.02
1996	A. Popov, Russia	48.74

WOMEN

Year	Gold medalist(s)	Time (s)
1912	F. Durack, Australia	82.2
1920	E. Bleibtrey, USA	73.6
1924	E. Lackie, USA	72.4
1928	A. Osipowich, USA	71.0
1932	H. Madison, USA	66.8
1936	H. Mastenbroek, Holland	65.9
1948	G. Andersen, Denmark	66.3
1952	K. Szoke, Hungary	66.8
1956	D. Fraser, Australia	62.0
1960	D. Fraser, Australia	61.2
1964	D. Fraser, Australia	59.5
1968	J. Henne, USA	60.0
1972	S. Nielson, USA	58.59
1976	K. Ender, E. Germany	55.65
1980	B. Krause, E. Germany	54.79
1984	(Tie) C. Steinseifer, USA	55.92
	N. Hogshead, USA	55.92
1988	K. Otto, E. Germany	54.93
1992	Z. Yong, China	54.64
1996	L. Jingyi, China	54.50

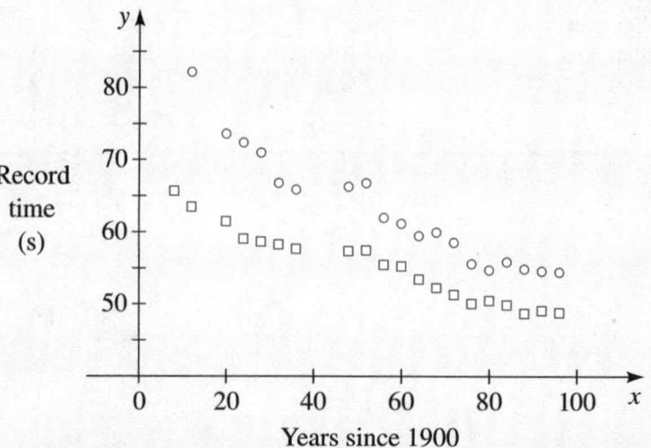

13. Do you think that shoe size and height are correlated? Find out by surveying the shoe sizes and heights of people in your class. (Of course, the data for men and women should be separate.) Find the correlation coefficient.

14. In this problem you will determine a linear demand equation that describes the demand for candy bars in your class. (See page 110 for a description of supply and demand equations.) Survey your classmates to determine what price they would be willing to pay for a candy bar. Your survey form might look like the sample below.

(a) Make a table of the number of respondents who answered "yes" at each price level.
(b) Make a scatter plot of your data.
(c) Find and graph the regression line $y = mp + b$, which gives the number of respondents y who would buy a candy bar if the price were p cents. This is the demand equation. Why is the slope m negative?
(d) What is the p-intercept of the demand equation? What does this intercept tell you about pricing candy bars?

Would you buy a candy bar from the vending machine in the hallway if the price is as indicated?

Price	Yes or No
30¢	
40¢	
50¢	
60¢	
70¢	
80¢	
90¢	
$1.00	
$1.10	
$1.20	

3

EQUATIONS AND INEQUALITIES

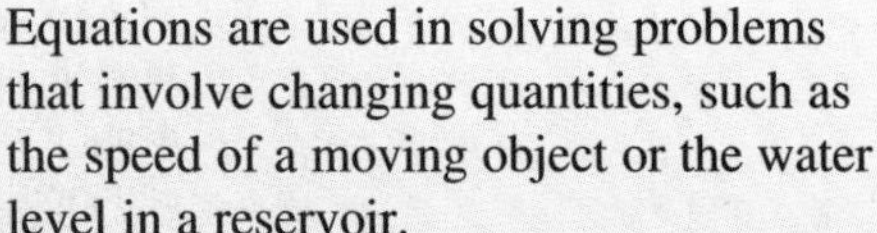

Equations are used in solving problems that involve changing quantities, such as the speed of a moving object or the water level in a reservoir.

I keep the subject constantly before me, and wait till the first dawnings open little by little into the full light.

ISAAC NEWTON

In this chapter we study equations and inequalities. We introduce algebraic techniques for solving more complicated equations than the simple ones we reviewed in Section 1.6. We also introduce the graphical method for solving equations and inequalities. Solving equations that arise in real life was the original motivation for inventing algebra. In this chapter we consider a variety of applications of equations and inequalities.

3.1 ALGEBRAIC AND GRAPHICAL SOLUTIONS OF EQUATIONS

"Algebra is a merry science," Uncle Jakob would say. "We go hunting for a little animal whose name we don't know, so we call it x. When we bag our game we pounce on it and give it its right name."

ALBERT EINSTEIN

In Section 1.6 we learned how to solve basic equations. To solve an equation like

$$3x - 5 = 0$$

we used the **algebraic method**. This means we used the rules of algebra to isolate x on one side of the equation. We view x as an *unknown* and we use the rules of algebra to hunt it down. Here are the steps in the solution:

$$\begin{aligned} 3x - 5 &= 0 \\ 3x &= 5 && \text{Add 5} \\ x &= \tfrac{5}{3} && \text{Divide by 3} \end{aligned}$$

So the solution is $x = \frac{5}{3}$.

We can also solve this equation by the **graphical method**. In this method we view x as a *variable* and sketch the graph of the equation

$$y = 3x - 5$$

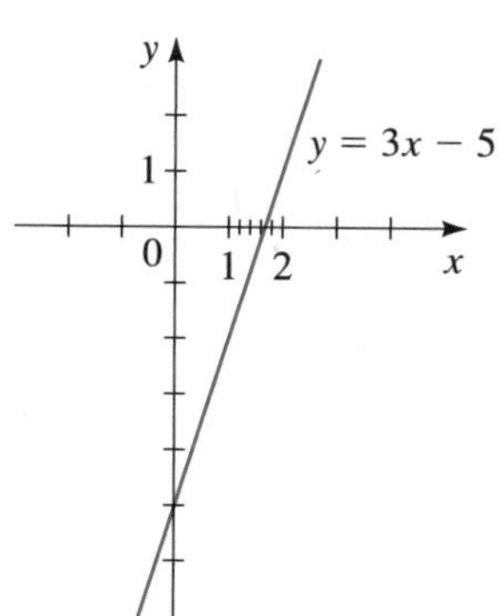

FIGURE 1

Different values for x give different values for y. Our goal is to find the value of x for which $y = 0$. From the graph in Figure 1 we see that $y = 0$ when $x \approx 1.7$. Thus, the solution is $x \approx 1.7$. Note that from the graph we obtain an approximate solution.

We summarize these methods in the following box.

SOLVING AN EQUATION

Algebraic Method	Graphical Method
Use the rules of algebra to isolate the unknown x on one side of the equation.	Move all terms to one side and set equal to y. Sketch the graph to find the value of x where $y = 0$.
Example: $2x = 6 - x$ $3x = 6$ Add x $x = 2$ Divide by 3	**Example:** $2x = 6 - x$ $0 = 6 - 3x$. Set $y = 6 - 3x$ and graph.
The solution is $x = 2$.	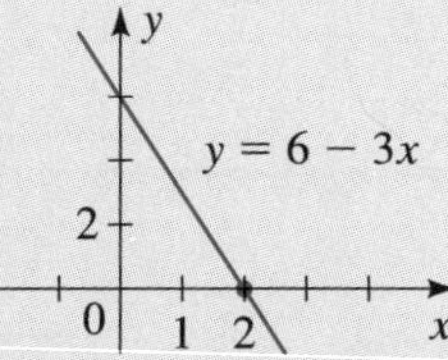
	From the graph the solution is $x \approx 2$.

The advantage of the algebraic method is that it gives exact answers. However, for many equations it is difficult or impossible to algebraically isolate x. In that case the graphical method is preferable. Graphing calculators and computer graphing programs make the graphical method practical and easily accessible.

EXAMPLE 1 ■ Solving an Equation Algebraically and Graphically

Solve the equation algebraically and graphically: $x^2 - 7 = 0$

SOLUTION 1: ALGEBRAIC

We isolate x^2 on one side of the equal sign and take square roots.

$$x^2 - 7 = 0$$

$$x^2 = 7 \qquad \text{Add 7}$$

$$x = \pm\sqrt{7} \qquad \text{Take square roots}$$

The solutions are $x = \sqrt{7}$ and $x = -\sqrt{7}$.

SOLUTION 2: GRAPHICAL

We graph the equation $y = x^2 - 7$ and determine the x-intercepts from the graph. From Figure 2 we see that the graph crosses the x-axis at $x \approx 2.6$ and $x \approx -2.6$.

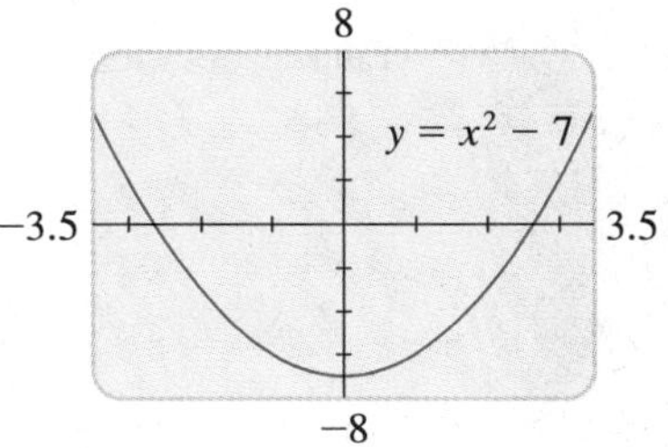

FIGURE 2

■

EXAMPLE 2 ■ Another Graphical Method

Solve the equation algebraically and graphically: $5 - 3x = 8x - 20$

SOLUTION 1: ALGEBRAIC

$$5 - 3x = 8x - 20$$

$$-3x = 8x - 25 \qquad \text{Subtract 5}$$

$$-11x = -25 \qquad \text{Subtract } 8x$$

$$x = \frac{-25}{-11} = 2\tfrac{3}{11} \qquad \text{Divide by } -11 \text{ and simplify}$$

SOLUTION 2: GRAPHICAL

We could move all terms to one side of the equal sign, set the result equal to y, and graph the resulting equation. But to avoid all this algebra, we graph two equations instead:

$$y_1 = 5 - 3x \qquad \text{and} \qquad y_2 = 8x - 20$$

The solution of the original equation will be the value of x that makes y_1 equal to y_2; that is, the solution is the x-coordinate of the intersection point of the two graphs. Using the trace feature on a graphing calculator, we see from Figure 3 that the solution is $x \approx 2.27$.

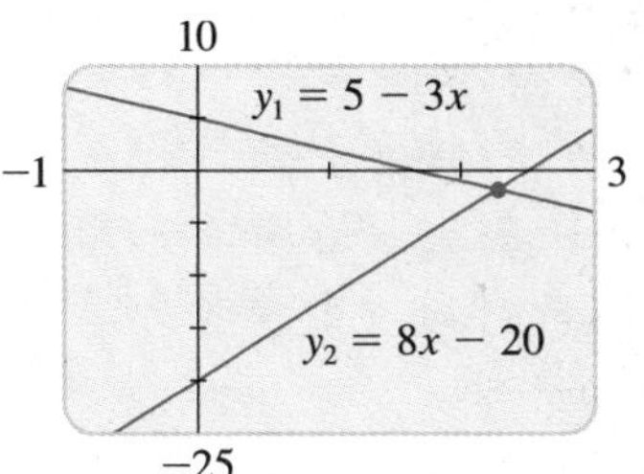

FIGURE 3

■

In the next example we use the graphical method to solve an equation that is extremely difficult to solve algebraically.

EXAMPLE 3 ■ Solving an Equation Graphically

Find all solutions, correct to two decimals: $x^5 - 2x^4 + 5x - 18 = 0$.

SOLUTION

After some experimentation we find that an appropriate viewing rectangle is $[-2, 3]$ by $[-90, 50]$. The graph of

$$y = x^5 - 2x^4 + 5x - 18$$

is shown in Figure 4. There appears to be just one x-intercept. Zooming in to this point, and using the trace feature on the calculator, we see in Figure 5 that the only solution is $x \approx 2.26$.

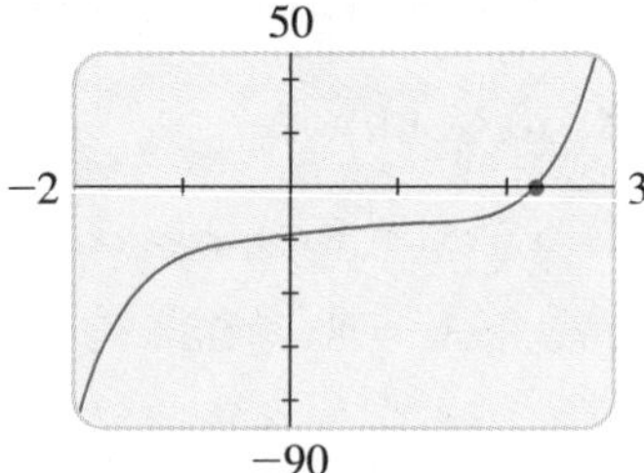

FIGURE 4
$y = x^5 - 2x^4 + 5x - 18$

FIGURE 5 ■

How can we be sure that we have found all the solutions in Example 3? If we look at the graph in a very wide viewing rectangle, we will still see just one solution, so it seems likely that no other solutions exist. In Chapter 5 we will learn some algebraic facts about polynomial graphs that will enable us to prove that this equation has just one solution.

EXAMPLE 4 ■ Solving an Equation in an Interval

Solve the equation

$$x^3 - 6x^2 + 9x = \sqrt{x}$$

in the interval $[1, 6]$.

SOLUTION

We are asked to find all solutions x that satisfy $1 \leq x \leq 6$, so we will graph the equation in a viewing rectangle for which the x-values are restricted to this interval.

$$x^3 - 6x^2 + 9x = \sqrt{x}$$

$$x^3 - 6x^2 + 9x - \sqrt{x} = 0 \qquad \text{Subtract } \sqrt{x}$$

MATHEMATICS IN THE MODERN WORLD

WEATHER PREDICTION

Modern meteorologists do much more than predict tomorrow's weather. They research long-term weather patterns, depletion of the ozone layer, global warming, and other effects of human activity on the weather. But daily weather prediction is still a major part of meteorology; its value is measured by the innumerable human lives saved each year through accurate prediction of hurricanes, blizzards, and other catastrophic weather phenomena. At the turn of the century mathematicians proposed to model weather with equations that used the current values of hundreds of atmospheric variables. Although this model worked in principle, it was impossible to predict future weather patterns with it because of the difficulty of measuring all the variables accurately and solving all the equations. Today, new mathematical models combined with high-speed computer simulations have vastly improved weather prediction. As a result, many human as well as economic disasters have been averted. Mathematicians at the National Oceanographic and Atmospheric Administration (NOAA) are continually researching better methods of weather prediction.

Figure 6 shows the graph of the equation $y = x^3 - 6x^2 + 9x - \sqrt{x}$ in the viewing rectangle $[1, 6]$ by $[-2, 10]$. There are two x-intercepts in this viewing rectangle; zooming in we see that the solutions are $x \approx 2.18$ and $x \approx 3.7$.

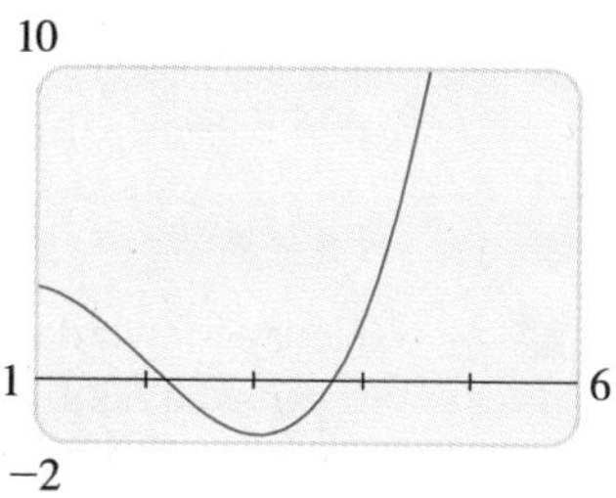

FIGURE 6

■

The equation in Example 4 actually has four solutions. You are asked to find the other two in Exercise 27.

3.1 EXERCISES

1–14 ■ Solve the equation both algebraically and graphically.

1. $2x - 9 = 5$

2. $3x + 7 = 31$

3. $x - 4 = 5x + 12$

4. $6x + 15 = -3x$

5. $\frac{1}{2}x - 3 = 6 + 2x$

6. $\frac{3}{4}x + 2(x + 1) = -\frac{5}{2}x + 14$

7. $\dfrac{2}{x} + \dfrac{1}{2x} = 7$

8. $\dfrac{4}{x + 2} - \dfrac{6}{2x} = \dfrac{5}{2x + 4}$

9. $x^2 - 32 = 0$

10. $x^3 + 16 = 0$

11. $16x^4 = 625$

12. $2x^5 - 243 = 0$

13. $(x - 5)^4 - 80 = 0$

14. $6(x + 2)^5 = 64$

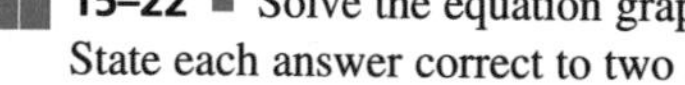

15–22 ■ Solve the equation graphically in the given interval. State each answer correct to two decimals.

15. $x^2 - 7x + 12 = 0$; $[0, 6]$

16. $x^2 - 0.75x + 0.125 = 0$; $[-2, 2]$

17. $x^3 - 6x^2 + 11x - 6 = 0$; $[-1, 4]$

18. $16x^3 + 16x^2 = x + 1$; $[-2, 2]$

19. $x - \sqrt{x + 1} = 0$; $[-1, 5]$

20. $1 + \sqrt{x} = \sqrt{1 + x^2}$; $[-1, 5]$

21. $x^{1/3} - x = 0$; $[-3, 3]$

22. $x^{1/2} + x^{1/3} - x = 0$; $[-1, 5]$

23–26 ■ Find all real solutions of the equation, correct to two decimals.

23. $x^3 - 2x^2 - x - 1 = 0$

24. $x^4 - 8x^2 + 2 = 0$

25. $x(x - 1)(x + 2) = \frac{1}{6}x$

26. $x^4 = 16 - x^3$

27. In Example 4 we found two solutions of the equation $x^3 - 6x^2 + 9x = \sqrt{x}$, the solutions that lie between 1 and 6. Find two more solutions, correct to two decimals.

DISCOVERY • DISCUSSION

28. Algebraic and Graphical Solution Methods Write a short essay comparing the algebraic and graphical methods for solving equations. Make up your own examples to illustrate the advantages and disadvantages of each method.

29. How Many Solutions? This exercise deals with the family of equations $x^3 - 3x = k$.

(a) Draw the graphs of $y_1 = x^3 - 3x$ and $y_2 = k$ in the same viewing rectangle, in the cases $k = -4, -2, 0, 2$, and 4. How many solutions of the equation $x^3 - 3x = k$ are there in each case? Find the solutions correct to two decimals.

(b) For what ranges of values of k does the equation have one solution? two solutions? three solutions?

3.2 MODELING WITH EQUATIONS

Many problems in the sciences, economics, finance, medicine, and numerous other fields can be translated into algebra problems; this is one reason that algebra is so useful. In this section we use equations as mathematical models to solve real-life problems.

EXAMPLE 1 ■ Annual Salary

Henry is awarded a 5% annual salary increase at the beginning of 1999. He also received a \$2500 bonus for completing a special project in August 1999. If his total pay in 1999 was \$40,300, what was his salary in 1998?

SOLUTION

We are asked to find Henry's salary in 1998. So we let

Identify the variable

$$x = \text{Henry's salary in 1998}$$

Then we translate all the information given in the problem into the language of algebra.

Express all unknown quantities in terms of the variable

In Words	In Algebra
Henry's salary in 1998	x
Salary increase of 5%	$0.05x$
Henry's salary in 1999	$x + 0.05x = 1.05x$

Now we set up the model.

Set up the model

$$\text{Henry's 1999 salary} + \text{amount of the bonus} = \text{Henry's total 1999 pay}$$

$$1.05x + 2500 = 40{,}300$$

Solve

$$1.05x = 37{,}800 \qquad \text{Subtract 2500}$$

$$x = \frac{37{,}800}{1.05} \qquad \text{Divide by 1.05}$$

$$= 36{,}000 \qquad \text{Simplify}$$

Henry's salary in 1998 was \$36,000.

CHECK YOUR ANSWER

$$\begin{aligned} \text{salary in 1999} &= \text{salary in 1998} + 5\% \text{ raise} + \text{bonus} \\ &= \$36{,}000 + 0.05 \times \$36{,}000 + \$2500 \\ &= \$40{,}300 \quad \checkmark \end{aligned}$$

■

We now adapt the problem-solving principles given on pages 67–70 to the process of modeling real-life problems with equations. The following guidelines should help you set up the equation that expresses the English statement of a problem in the language of algebra. These guidelines are referred to in the margin notes for the examples in this section.

GUIDELINES FOR MODELING WITH EQUATIONS

1. IDENTIFY THE VARIABLE. Identify the quantity that the problem asks you to find. This quantity can usually be determined by a careful reading of the question posed at the end of the problem. **Introduce notation** for the variable (call it x or some other letter).

2. EXPRESS ALL UNKNOWN QUANTITIES IN TERMS OF THE VARIABLE. Read each sentence in the problem again, and express all the quantities mentioned in the problem in terms of the variable you defined in Step 1. To organize this information, it is sometimes helpful to **draw a diagram** or **make a table** (see Examples 4, 6, 7, and 9 in this section).

3. SET UP THE MODEL. Find the crucial fact in the problem that gives a relationship between the expressions you listed in Step 2. **Set up an equation** (or **model**) that expresses this relationship.

4. SOLVE THE EQUATION AND CHECK YOUR ANSWER. Solve the equation, check your answer, and express it as a sentence that answers the question posed in the problem.

EXAMPLE 2 ■ An Unknown Sum of Money

Harriet gives her two grandsons gifts of cash for Christmas. John, the younger one, asks his brother Michael how much he received. Michael is unwilling to state the amount, but he says, "If I had gotten \$10 more and then doubled that amount, then I'd have \$50 more than what I actually got." How much did Michael get from his grandmother?

SOLUTION

We are asked to find the amount of Michael's gift. So let

Identify the variable

$$x = \text{the amount of Michael's gift}$$

Now we translate all the information in the problem into the language of algebra.

Express all unknown quantities in terms of the variable

In Words	In Algebra
Michael's gift	x
\$10 more than the gift	$x + 10$
Double that amount	$2(x + 10)$
\$50 more than the gift	$x + 50$

We now set up the model.

Set up the model

double · \$10 more than the gift = \$50 more than the gift

Solve

$$2(x + 10) = x + 50$$

$$2x + 20 = x + 50 \quad \text{Expand}$$

$$2x = x + 30 \quad \text{Subtract 20}$$

$$x = 30 \quad \text{Subtract } x$$

Michael received \$30 from his grandmother. (You should check that this answer satisfies the riddle.) ■

EXAMPLE 3 ■ Interest on an Investment

Mary inherits \$100,000 and invests it in two certificates of deposit. One certificate pays 6% and the other pays $4\frac{1}{2}\%$ simple interest annually. If Mary's total interest is \$5025 per year, how much money is invested at each rate?

SOLUTION

The problem asks for the amount she has invested at each rate. So let

Identify the variable

$$x = \text{the amount invested at 6\%}$$

Since Mary's total inheritance is \$100,000, it follows that she invested $100{,}000 - x$ at $4\frac{1}{2}\%$. We translate all the information given into the language of algebra.

Express all unknown quantities in terms of the variable

In Words	In Algebra
Amount invested at 6%	x
Amount invested at $4\frac{1}{2}\%$	$100{,}000 - x$
Interest earned at 6%	$0.06x$
Interest earned at $4\frac{1}{2}\%$	$0.045(100{,}000 - x)$

Now we use the fact that Mary's total interest is \$5025 to set up the model.

Set up the model

$$\text{interest at 6\%} + \text{interest at } 4\tfrac{1}{2}\% = \text{total interest}$$

$$0.06x + 0.045(100{,}000 - x) = 5025$$

Solve

$$0.06x + 4500 - 0.045x = 5025 \qquad \text{Multiply}$$

$$0.015x + 4500 = 5025 \qquad \text{Combine the } x \text{ terms}$$

$$0.015x = 525 \qquad \text{Subtract 4500}$$

$$x = \frac{525}{0.015} = 35{,}000 \qquad \text{Divide by 0.015}$$

So Mary has invested \$35,000 at 6% and the remaining \$65,000 at $4\frac{1}{2}$%.

CHECK YOUR ANSWER

total interest = 6% of \$35,000 + $4\frac{1}{2}$% of \$65,000

= \$2100 + \$2925 = \$5025 ✓

■

EXAMPLE 4 ■ Dimensions of a Poster

A poster has a rectangular printed area 100 cm by 140 cm, and a blank strip of uniform width around the four edges. The perimeter of the poster is $1\frac{1}{2}$ times the perimeter of the printed area. What is the width of the blank strip, and what are the dimensions of the poster?

In a problem such as this, which involves geometry, it is essential to draw a diagram like the one shown in Figure 1.

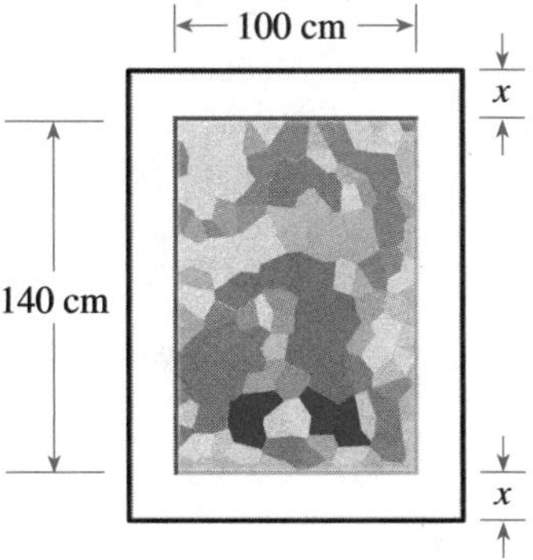

FIGURE 1

SOLUTION

We are asked to find the width of the blank strip. So let

Identify the variable

$$x = \text{the width of the blank strip}$$

Then we translate the information in Figure 1 into the language of algebra:

Express all unknown quantities in terms of the variable

In Words	In Algebra
Width of blank strip	x
Perimeter of printed area	$2(100) + 2(140) = 480$
Width of poster	$100 + 2x$
Length of poster	$140 + 2x$
Perimeter of poster	$2(100 + 2x) + 2(140 + 2x)$

Now we use the fact that the perimeter of the poster is $1\frac{1}{2}$ times the perimeter of the printed area to set up the model.

Set up the model

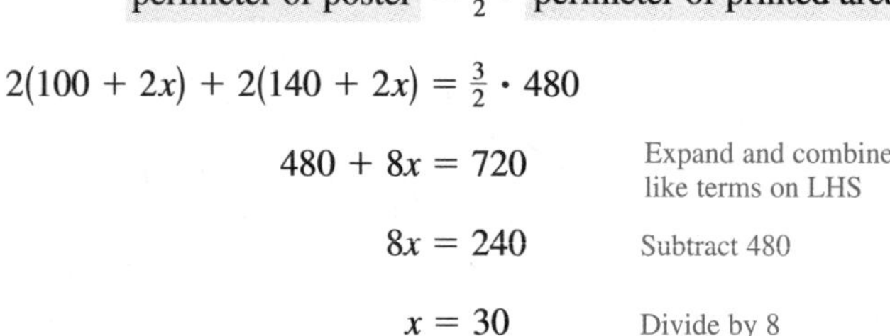

$$\text{perimeter of poster} = \tfrac{3}{2} \cdot \text{perimeter of printed area}$$

$$2(100 + 2x) + 2(140 + 2x) = \tfrac{3}{2} \cdot 480$$

Solve

$$480 + 8x = 720 \qquad \text{Expand and combine like terms on LHS}$$

$$8x = 240 \qquad \text{Subtract 480}$$

$$x = 30 \qquad \text{Divide by 8}$$

The blank strip is 30 cm wide, so the dimensions of the poster are

$$100 + 30 + 30 = 160 \text{ cm wide}$$

by

$$140 + 30 + 30 = 200 \text{ cm wide}$$

■

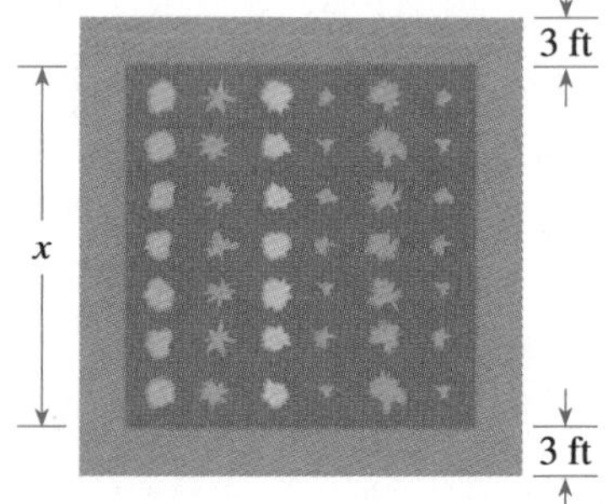

FIGURE 2

EXAMPLE 5 ■ Dimensions of a Garden

A square garden has a walkway 3 ft wide around its outer edge, as shown in Figure 2. If the area of the entire garden, including the walkway, is 18,000 ft^2, what are the dimensions of the planted area?

SOLUTION

We are asked to find the length and width of the planted area. So let

Identify the variable

$$x = \text{the length of the planted area}$$

Next we translate the information from Figure 2 into the language of algebra.

Express all unknown quantities in terms of the variable

In Words	In Algebra
Length of planted area	x
Length of entire garden	$x + 6$
Area of entire garden	$(x + 6)^2$

We now set up the model.

Set up the model

$$\text{area of entire garden} = 18{,}000 \text{ ft}^2$$

$$(x+6)^2 = 18{,}000$$

Solve

$$x + 6 = \sqrt{18{,}000} \qquad \text{Take square roots}$$

$$x = \sqrt{18{,}000} - 6 \qquad \text{Subtract 6}$$

$$x \approx 128$$

The planted area of the garden is about 128 ft by 128 ft. ■

EXAMPLE 6 ■ Determining the Height of a Building Using Similar Triangles

A man 6 ft tall wishes to find the height of a certain four-story building. He measures its shadow and finds it to be 28 ft long, while his own shadow is $3\frac{1}{2}$ ft long. How tall is the building?

SOLUTION

The problem asks for the height of the building. So let

Identify the variable

$$x = \text{the height of the building}$$

We use the fact that the triangles in Figure 3 are similar. Recall that for any pair of similar triangles the ratios of corresponding sides are equal. Now we translate these observations into the language of algebra.

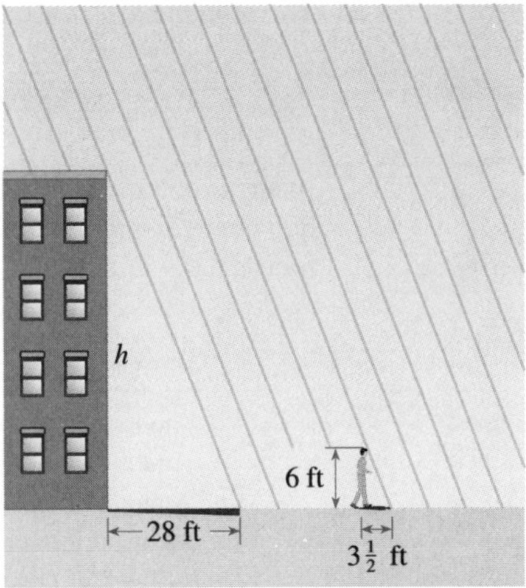

FIGURE 3

Express all unknown quantities in terms of the variable

In Words	In Algebra
Height of building	h
Ratio of height to base in large triangle	$\frac{h}{28}$
Ratio of height to base in small triangle	$\frac{6}{3.5}$

Since the large and small triangles are similar, we get the equation

Set up the model

$$\text{ratio of height to base in large triangle} = \text{ratio of height to base in small triangle}$$

$$\frac{h}{28} = \frac{6}{3.5}$$

Solve

$$h = \frac{6 \cdot 28}{3.5} = 48$$

The building is 48 ft tall. ■

EXAMPLE 7 ■ Mixtures and Concentration

A manufacturer of soft drinks advertises their orange soda as "naturally flavored," although it contains only 5% orange juice. A new federal regulation stipulates that to be called "natural" a drink must contain at least 10% fruit juice. How much pure orange juice must this manufacturer add to 900 gal of orange soda to conform to the new regulation?

SOLUTION

The problem asks for the amount of pure orange juice to be added. So let

Identify the variable

x = the amount (in gallons) of pure orange juice to be added

In any problem of this type—in which two different substances are to be mixed—drawing a diagram helps us organize the given information (see Figure 4). We then translate the information in the figure into the language of algebra.

Express all quantities in terms of the variable

In Words	In Algebra
Amount of orange juice to be added	x
Amount of the mixture	$900 + x$
Amount of orange juice in the first vat	$0.05(900) = 45$
Amount of orange juice in the second vat	$1 \cdot x = x$
Amount of orange juice in the mixture	$0.10(900 + x)$

Volume	900 gallons	x gallons	$900 + x$ gallons
Amount of orange juice	5% of 900 gallons $= 45$ gallons	100% of x gallons $= x$ gallons	10% of $900 + x$ gallons $= 0.1(900 + x)$ gallons

FIGURE 4

To set up the model, we use the fact that the total amount of orange juice in the mixture is equal to the orange juice in the first two vats.

Set up the model

amount of orange juice in the first vat	+	amount of orange juice in the second vat	=	amount of orange juice in the mixture

Solve

$$45 + x = 0.1(900 + x) \quad \text{From Figure 4}$$

$$45 + x = 90 + 0.1x \quad \text{Multiply}$$

$$0.9x = 45 \quad \text{Subtract } 0.1x \text{ and } 45$$

$$x = \frac{45}{0.9} = 50 \quad \text{Divide by } 0.9$$

The manufacturer should add 50 gal of pure orange juice to the soda.

CHECK YOUR ANSWER

amount of juice before mixing = 5% of 900 gal + 50 gal pure juice
= 45 gal + 50 gal = 95 gal

amount of juice after mixing = 10% of 950 gal = 95 gal

Amounts are equal. ✓

■

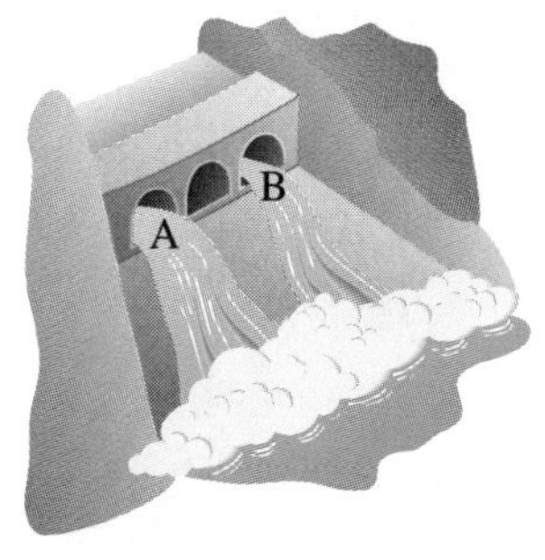

EXAMPLE 8 ■ Time Needed to Do a Job

Because of an anticipated heavy rainstorm, the water level in a reservoir must be lowered by 1 ft. Opening spillway A lowers the level by this amount in 4 hours, whereas opening the smaller spillway B does the job in 6 hours. How long will it take to lower the water level by 1 ft if both spillways are opened?

SOLUTION

We are asked to find the time needed to lower the level by 1 ft if both spillways are open. So let

Identify the variable

x = the time (in hours) it takes to lower the water level by 1 ft if both spillways are open

Finding an equation relating x to the other quantities in this problem is not easy. Certainly x is not simply $4 + 6$, because that would mean that together the two spillways require longer to lower the water level than either spillway alone. Instead, *we look at the fraction of the job that can be done in one hour by each spillway.*

Express all unknown quantities in terms of the variable

In Words	In Algebra
Time it takes to lower level 1 ft with A and B together	x h
Distance A lowers level in 1 h	$\frac{1}{4}$ ft
Distance B lowers level in 1 h	$\frac{1}{6}$ ft
Distance A and B together lower levels in 1 h	$\frac{1}{x}$ ft

Now we set up the model.

Set up the model

fraction done by A + fraction done by B = fraction done by both

$$\frac{1}{4} + \frac{1}{6} = \frac{1}{x}$$

Solve

$$3x + 2x = 12 \qquad \text{Multiply by the LCD, } 12x$$

$$5x = 12 \qquad \text{Add}$$

$$x = \frac{12}{5} \qquad \text{Divide by 5}$$

It will take $2\frac{2}{5}$ hours, or 2 h 24 min to lower the water level by 1 ft if both spillways are open. ■

The next example deals with distance, rate (speed), and time. The formula to keep in mind here is

$$\text{distance} = \text{rate} \times \text{time}$$

where the rate is either the constant speed or average speed of a moving object. For example, driving at 60 mi/h for 4 hours takes you a distance of $60 \cdot 4 = 240$ mi.

EXAMPLE 9 ■ Distance, Rate, and Time

Bill left his house at 2:00 P.M. and rode his bicycle down Main Street at a speed of 12 mi/h. When his friend Mary arrived at his house at 2:10 P.M., Bill's mother told her the direction in which Bill went, and Mary cycled after him at a speed of 16 mi/h. At what time did Mary catch up with Bill?

SOLUTION

We are asked to find the time that it took Mary to catch up with Bill. Let

Identify the variable

t = the time (in hours) it took Mary to catch up with Bill

Because Bill had a 10-minute, or $\frac{1}{6}$-hour head start, he cycled for $t + \frac{1}{6}$ hours.

In problems involving motion, it is often helpful to organize the information in a table, using the formula distance = rate × time. First we fill in the "Rate" column in the table, since we are told the speeds at which Mary and Bill cycled. Then we fill in the "Time" column, since we know what their travel times were in terms of t. Finally, we multiply these columns to calculate the entries in the "Distance" column.

Express all unknown quantities in terms of the variable

	Distance (mi)	Rate (mi/h)	Time (h)
Mary	$16t$	16	t
Bill	$12\left(t + \frac{1}{6}\right)$	12	$t + \frac{1}{6}$

At the instant Mary caught up with Bill, they had both cycled the same distance. We use this fact to set up the model for this problem.

Set up the model

distance traveled by Mary = distance traveled by Bill

Solve

$$16t = 12\left(t + \tfrac{1}{6}\right) \quad \text{From table}$$

$$16t = 12t + 2 \quad \text{Distributive Property}$$

$$4t = 2 \quad \text{Subtract } 12t$$

$$t = \tfrac{1}{2} \quad \text{Divide by 4}$$

Mary caught up with Bill after cycling for half an hour, that is, at 2:40 P.M.

CHECK YOUR ANSWER ■ Bill traveled for $\frac{1}{2} + \frac{1}{6} = \frac{2}{3}$ h, so

$$\text{distance Bill traveled} = 12 \text{ mi/h} \times \tfrac{2}{3} \text{ h} = 8 \text{ mi}$$

$$\text{distance Mary traveled} = 16 \text{ mi/h} \times \tfrac{1}{2} \text{ h} = 8 \text{ mi}$$

Distances are equal. ✓

■

3.2 EXERCISES

1–12 ■ Express the given quantity in terms of the indicated variable.

1. The sum of three consecutive integers; n = the first integer of the three

2. The sum of three consecutive integers; n = the middle integer of the three

3. The sum of the squares of two consecutive integers; n = the first integer of the two

4. The interest obtained after 2 years on an investment at 7% simple interest per year; x = the number of dollars invested.

5. The average of three test scores if the first two scores are 78 and 82; s = the third test score

6. The perimeter (in cm) of a rectangle that is 5 cm longer than it is wide; w = width of the rectangle (in cm)

7. The area (in in^2) of a rectangle that is 50 in. long; w = width of the rectangle (in inches)

8. The time (in hours) it takes to travel a given distance at 55 mi/h; d = the given distance (in mi)

9. The distance (in miles) traveled when driving at a certain speed for 2 h, then driving 15 mi/h faster for another hour; s = initial speed (in mi/h)

10. The concentration (in oz/gal) of salt in a mixture of 3 gal of brine containing 25 oz of salt, to which has been added some pure water; x = volume of pure water added (in gal)

11. The average age of three sisters if the second was born 3 years after the first and the third was born 2 years after the second; a = age of firstborn (in years)

12. The value (in cents) of the change in a purse that contains twice as many nickels as pennies, four more dimes than nickels, and as many quarters as dimes and nickels combined; p = number of pennies

13–67 ■ Use the principles described in this section to answer the question posed.

13. An executive in an engineering firm earns a monthly salary plus a Christmas bonus of \$8500. If she earns a total of \$97,300 per year, what is her monthly salary?

14. A woman earns 15% more than her husband. Together they make \$69,875 per year. What is the husband's annual salary?

15. Craig is saving to buy a vacation home. He inherits some money from a wealthy uncle, then combines this with the \$22,000 he has already saved and doubles the total in a lucky investment. He ends up with \$134,000, just enough to buy a cabin on the lake. How much did he inherit?

16. A father is four times as old as his daughter. In 6 years, he will be three times as old as she is. How old is the daughter now?

17. The oldest child in a family of four children is twice as old as the youngest. The two middle children are 10 and 11 years old. If the average age of the children is $10\frac{1}{2}$ years, how old is the youngest child?

18. A movie star, unwilling to give his age, posed the following riddle to a gossip columnist. "Seven years ago, I was eleven times as old as my daughter. Now I am four times as old as she is." How old is the star?

19. Find three consecutive integers whose sum is 336.

20. Find four consecutive odd integers whose sum is 272.

21. A room is $1\frac{1}{2}$ times as long as it is wide. Its perimeter is 80 feet. How wide is the room?

22. Helen earns \$7.50 per hour at her job, but if she works more than 35 hours in a week she is paid $1\frac{1}{2}$ times her regular salary for the overtime hours worked. One week her gross pay was \$352.50. How many overtime hours did she work that week?

23. A rectangular garden is 25 ft wide. If its area is 1125 ft^2, what is the length of the garden?

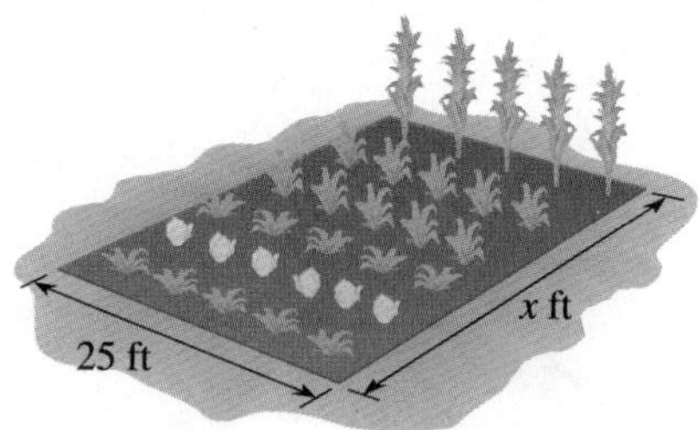

24. During his major league career, Hank Aaron hit 31 more home runs than Babe Ruth hit during his career. Together they hit 1459 home runs. How many home runs did Babe Ruth hit?

25. Phyllis invested \$12,000, a portion earning a simple interest rate of $4\frac{1}{2}$% per year and the rest earning a rate of 4% per year. After one year the total interest earned on these investments was \$525. How much money did she invest at each rate?

26. If Ben invests \$4000 at 4% interest per year, how much additional money must he invest at $5\frac{1}{2}\%$ annual interest to ensure that the interest he receives each year is $4\frac{1}{2}\%$ of the total amount invested?

27. A plumber and his assistant work together to replace the pipes in an old house. The plumber charges \$45 per hour for his own labor and \$25 per hour for his assistant's labor. The plumber works twice as long as his assistant on this job, and the labor charge on the final bill is \$4025. How long did the plumber and his assistant work on this job?

28. A change purse contains an equal number of pennies, nickels, and dimes. The total value of the coins is \$1.44. How many coins of each type does the purse contain?

29. Mary has \$3.00 in nickels, dimes, and quarters. If she has twice as many dimes as quarters and five more nickels than dimes, how many coins of each type does she have?

30. A merchant blends tea that sells for \$3.00 a pound with tea that sells for \$2.75 a pound to produce 80 lb of a mixture that sells for \$2.90 a pound. How many pounds of each type of tea does the merchant use in the blend?

31. Al paints with watercolors on a sheet of paper 20 in. wide by 15 in. high. He then places this sheet on a mat so that a uniformly wide strip of the mat shows all around the picture. The perimeter of the mat is 102 in. How wide is the strip of the mat showing around the picture?

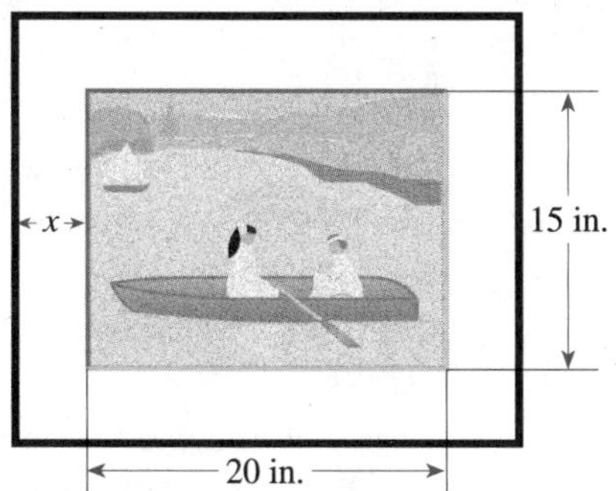

32. Dr. Plath has been raising fruit flies for use in a biology experiment. She has noticed that the number of new baby flies in her colony each day is always half the previous day's total population, but that 200 adult flies die each day. If after three days she has 3100 fruit flies, how many did she start with?

33. A restaurant is open five days a week, from Tuesday through Saturday. During a certain week, gross receipts averaged \$1200 per day. If the restaurant sold \$650 in meals on Tuesday, \$550 on Wednesday, and \$300 on Thursday, and if Saturday's gross receipts were twice those of Friday, how much gross income did the restaurant earn on Saturday?

34. Rochelle's psychology professor gives a grade of A for an average of 90 or better and a grade of B for an average of 80 or better in her course. Rochelle has obtained scores of 82, 75, and 71 on her midterm examinations. If the final exam counts twice as much as a midterm, what score must she make on her final exam to earn a grade of B? to earn an A? (Assume that the maximum possible score on each test is 100.)

35. Light travels at about 3.0×10^8 m/s. The distance from the sun to the earth is about 1.5×10^{11} m. How long does it take for the sun's light to reach the earth?

36. The fuel consumption for William's car is 30 mi/gal on the highway and 25 mi/gal in the city. On a vacation trip of 400 mi he used 14 gal of gasoline. How many highway miles did he drive on this trip?

37. The figure shows a lever system, similar to a seesaw that you might find in a children's playground. For the system to balance the product of the weight and its distance from the fulcrum must be the same on each side; that is

$$w_1 x_1 = w_2 x_2$$

This equation is called the **law of the lever**, and was first discovered by Archimedes (see page 538).

A woman and her son are playing on a seesaw. The boy is at one end, 8 ft from the fulcrum. If the son weighs 100 lb and the mother weighs 125 lb, where should the woman sit so that the seesaw is balanced?

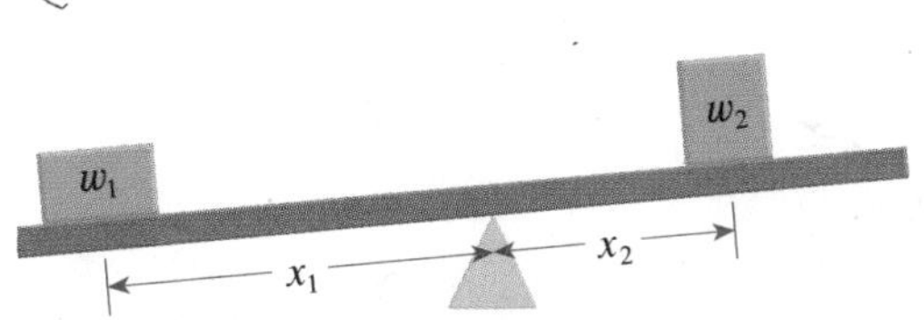

38. A plank 30 ft long rests on top of a flat-roofed building, with 5 ft of the plank projecting over the edge, as shown in the figure. A worker weighing 240 lb sits on one end of the plank. What is the largest weight that can be hung on the projecting end of the plank if it is to remain in balance? (Use the law of the lever stated in Exercise 37.)

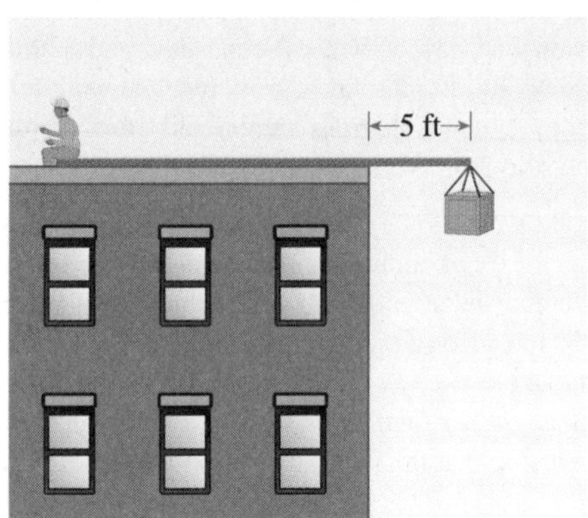

39. A pasture is twice as long as it is wide. Its area is 115,200 ft^2. How wide is the pasture?

40. A square plot of land has a building 60 ft long and 40 ft wide at one corner. The rest of the land outside the building forms a parking lot. If the parking lot has area 12,000 ft^2, what are the dimensions of the entire plot of land?

41. Find the length x in the figure, if the shaded area is 144 cm^2.

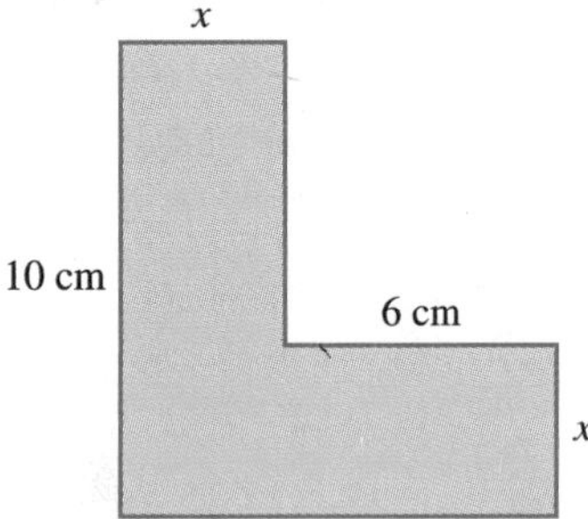

42. Find the length y in the figure, if the shaded area is 120 in^2.

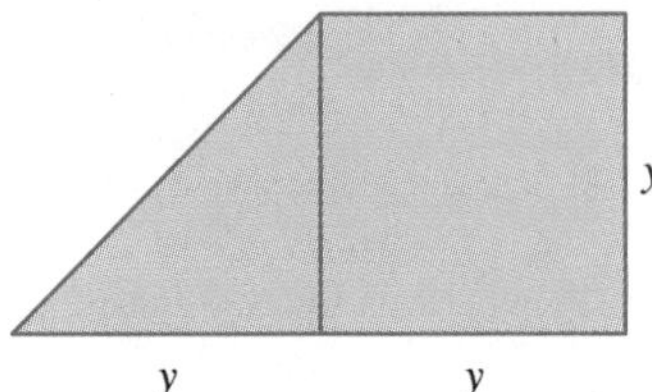

43. A man is walking away from a lamppost with a light source 6 m above the ground. The man is 2 m tall. How far from the lamppost is the man when his shadow is 5 m long? [*Hint:* Use similar triangles.]

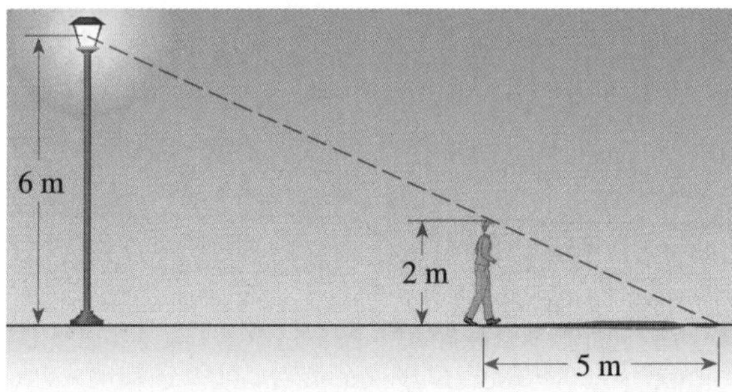

44. A woodcutter determines the height of a tall tree by first measuring a smaller one 125 ft away, then moving so that his eyes are in the line of sight along the tops of the trees, and measuring how far he is standing from the small tree (see the figure). Suppose the small tree is 20 ft tall, the man is 25 ft from the small tree, and his eye level is 5 ft above the ground. How tall is the taller tree?

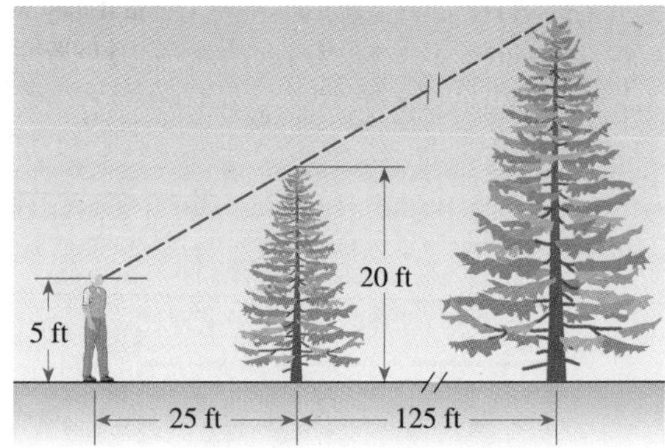

45. What quantity of a 60% acid solution must be mixed with a 30% solution to produce 300 mL of a 50% solution?

46. A pot contains 6 L of brine at a concentration of 120 g/L. How much of the water should be boiled off to increase the concentration to 200 g/L?

47. A jeweler has five rings, each weighing 18 g, made of an alloy of 10% silver and 90% gold. He decides to melt down the rings and add enough silver to reduce the gold content to 75%. How much silver should he add?

48. A health clinic uses a solution of bleach to sterilize petri dishes in which cultures are grown. The sterilization tank contains 100 gal of a solution of 2% ordinary household bleach mixed with pure distilled water. New research indicates that the concentration of bleach should be 5% for complete sterilization. How much of the solution should be drained and replaced with bleach to increase the bleach content to the recommended level?

49. The radiator in a car is filled with a solution of 60% antifreeze and 40% water. The manufacturer of the antifreeze suggests that, for summer driving, optimal cooling of the engine is obtained with only 50% antifreeze. If the capacity of the radiator is 3.6 L, how much coolant should be drained and replaced with water to reduce the antifreeze concentration to the recommended level?

50. A bottle contains 750 mL of fruit punch with a concentration of 50% pure fruit juice. Jill drinks 100 mL of the punch and then refills the bottle with an equal amount of a cheaper brand of punch. If the concentration of juice in the bottle is now reduced to 48%, what was the concentration in the punch that Jill added?

51. Candy and Tim share a paper route. It takes Candy 70 min to deliver all the papers, whereas Tim takes 80 min. How long does it take them when they work together?

52. Stan and Hilda can mow the lawn in 40 min if they work together. If Hilda works twice as fast as Stan, how long would it take Stan to mow the lawn alone?

53. Betty and Karen have been hired to paint the houses in a new development. Working together the women can paint a house in two-thirds the time that it takes Karen working alone. Betty takes 6 h to paint a house alone. How long does it take Karen to paint a house working alone?

54. Next-door neighbors Bob and Jim use hoses from both houses to fill Bob's swimming pool. They know it takes 18 h using both hoses. They also know that Bob's hose, used alone, takes 20% less time than Jim's hose alone. How much time is required to fill the pool by each hose alone?

55. A commercial jet took off from Kansas City for San Francisco, a distance of 2550 km, at a speed of 800 km/h. At the same time a private jet, traveling at 900 km/h, left San Francisco for Kansas City. How long after takeoff will the jets pass each other?

56. After robbing a bank in Dodge City, the robber gallops off at 14 mi/h. Ten minutes later the marshal leaves in hot pursuit at 16 mi/h. How long does it take the marshal to catch up with the bank robber?

57. Wilma drove at an average speed of 50 mi/h from her home in Boston to visit her sister in Buffalo. She stayed in Buffalo 10 h, and on the trip back averaged 45 mi/h. She returned home 29 h after leaving. How many miles is Buffalo from Boston?

58. A coast guard boat that patrols a river separating two countries cruises at 20 knots (nautical miles per hour) in still water. The river flows at 4 knots. For each patrol the captain of the boat is ordered to travel upstream and then return. His trip takes exactly 6 h. How far upstream does the boat travel?

59. Wendy took a trip from Davenport to Omaha, a distance of 300 mi. She traveled part of the way by bus, which arrived at the train station just in time for Wendy to complete her journey by train. The bus averaged 40 mi/h and the train 60 mi/h. The entire trip took $5\frac{1}{2}$ h. How long did Wendy spend on the train?

60. Two cyclists, 90 mi apart, start riding toward each other at the same time. One cycles twice as fast as the other. If they meet 2 h later, at what average speed is each cyclist traveling?

61. A pilot flew a jet from Montreal to Los Angeles, a distance of 2500 mi. On the return trip the average speed was 20% faster than the outbound speed. The round-trip took 9 h 10 min. What was the speed from Montreal to Los Angeles?

62. A woman driving a car 14 ft long is passing a truck 30 ft long. The truck is traveling at 50 mi/h. How fast must the woman drive her car so that she can pass the truck completely in 6 s, from the position shown in figure (a) to the position shown in figure (b)? [*Hint:* Use feet and seconds instead of miles and hours.]

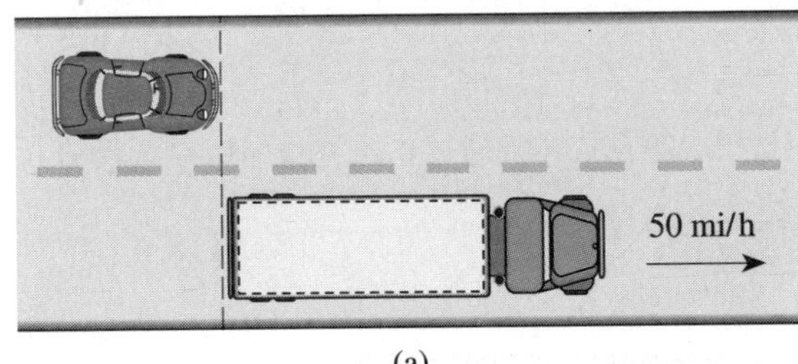

(a)

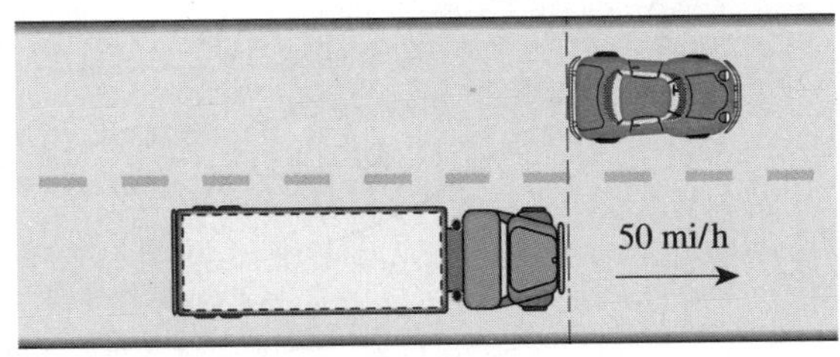

(b)

63. A rectangular parcel of land is 50 ft wide. The length of a diagonal between opposite corners is 10 ft more than the length of the parcel. What is the length of the parcel?

64. The hypotenuse of a right triangle is 3 cm longer than one of the legs, and the other leg is 10 cm long. How long is each side of the triangle? (Express your answer correct to two decimals.)

65. A running track has the shape shown in the figure, with straight sides and semicircular ends. If the length of the track is 440 yd and the two straight parts are each 110 yd long, what is the radius of the semicircular parts (to the nearest yard)?

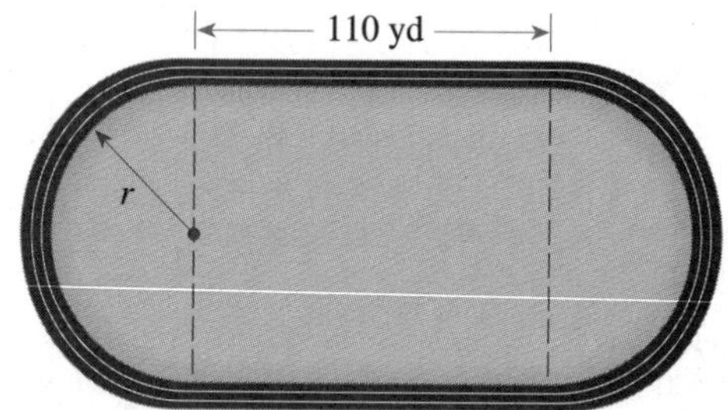

66. A storage bin for corn consists of a cylindrical section made of wire mesh, surmounted by a conical tin roof, as shown in the figure. The height of the roof is one-third the entire height of the structure. If the total volume of the structure is 1400π ft^3 and its radius is 10 ft, what is its height? [*Hint:* Use the volume formulas listed on the inside back cover of this book.]

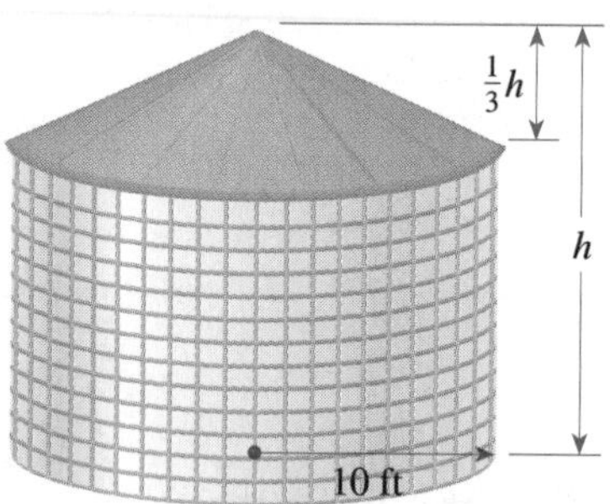

67. This problem is taken from a Chinese mathematics textbook called *Chui-chang suan-shu*, or *Nine Chapters on the Mathematical Art*, which was written about 250 B.C.

> A 10-ft-long stem of bamboo is broken in such a way that its tip touches the ground 3 ft from the base of the stem, as shown in the figure. What is the height of the break?

[*Hint:* Use the Pythagorean Theorem.]

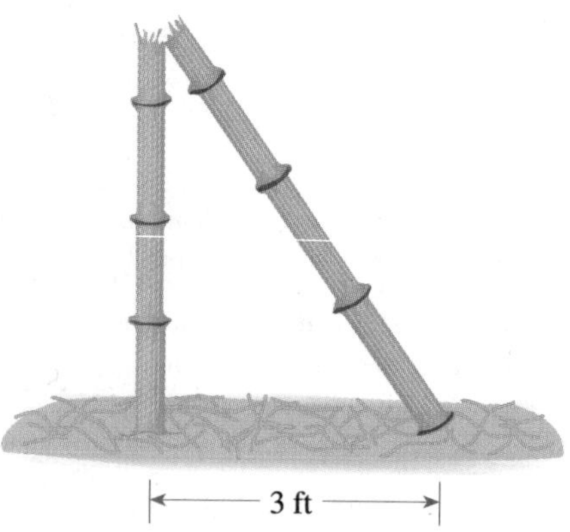

DISCOVERY • DISCUSSION

68. Historical Research Read the biographical notes on Pythagoras (page 306), Euclid (page 55), and Thales (page 304). Choose one of these mathematicians and find out more about him from the library. Write a short essay on your findings. Include both biographical information and a description of the mathematics for which he is famous.

Writing Project

EQUATIONS THROUGH THE AGES

Equations have been used to solve problems throughout recorded history, in every civilization. Here is a problem from ancient Babylon (ca. 2000 B.C.).

> I found a stone but did not weigh it. After I added a seventh, and then added an eleventh of the result, I weighed it and found it weighed 1 mina. What was the original weight of the stone?

The answer given on the cuneiform tablet was $\frac{2}{3}$ mina, 8 sheqel, and $22\frac{1}{2}$ se, where 1 mina = 60 sheqel, and 1 sheqel = 180 se.

In ancient Egypt, knowing how to solve word problems was a highly prized secret. The Rhind Papyrus (ca. 1850 B.C.) contains many such problems (see page 513). Problem 32 in the Papyrus states

> A quantity, its third, its quarter, added together become 2. What is the quantity?

The answer in Egyptian notation is $1 + \overline{4} + \overline{76}$, where the bar indicates "reciprocal," much like our notation 4^{-1}.

The Greek mathematician Diophantus (ca. 250 A.D., see page 45) wrote the book *Arithmetica*, which contains many word problems and equations. The Indian mathematician Bhaskara (12th century A.D.) and the Chinese mathematician Chang Ch'iu-Chien (6th century A.D.) also studied and wrote about equations. Of course, equations continue to be important today.

1. Solve the Babylonian problem and show that their answer was correct.
2. Solve the Egyptian problem and show that their answer was correct.
3. The ancient Egyptians and Babylonians used equations to solve practical problems. From the examples given here, do you think that they may have enjoyed posing and solving word problems just for fun?
4. Consider the following problem from 6th century China.

   > If a rooster is worth 5 coins, a hen 3 coins, and three chicks together one coin, how many roosters, hens, and chicks, totalling 100, can be bought for 100 coins?

   This problem has several answers. Use trial and error to find at least one answer. Is this a practical problem or more of a riddle? Write a short essay to support your opinion.
5. Write a short essay explaining how equations affect your own life in today's world.

3.3 QUADRATIC EQUATIONS

In Section 1.6 we learned how to solve linear equations, which are first-degree equations, like $2x + 1 = 5$ or $4 - 3x = 2$. In this section we learn how to algebraically solve quadratic equations, which are second-degree equations, like $x^2 + 2x - 3 = 0$ or $2x^2 + 3 = 5x$. We will see in the next section that applied problems often lead to quadratic equations.

QUADRATIC EQUATIONS

A **quadratic equation** is an equation of the form

$$ax^2 + bx + c = 0$$

where a, b, and c are real numbers with $a \neq 0$.

SOLVING QUADRATIC EQUATIONS BY FACTORING

Some quadratic equations can be solved by factoring and using the following basic property of real numbers.

ZERO-PRODUCT PROPERTY

$$AB = 0 \quad \text{if and only if} \quad A = 0 \quad \text{or} \quad B = 0$$

This means that if we can factor the left-hand side of a quadratic (or other) equation, then we can solve it by setting each factor equal to 0 in turn. This method works only when the right-hand side of the equation is 0.

EXAMPLE 1 ■ Solving a Quadratic Equation by Factoring

Solve the equation $x^2 + 5x = 24$.

SOLUTION

We must first rewrite the equation so that the right-hand side is 0.

$$x^2 + 5x = 24$$

$$x^2 + 5x - 24 = 0 \qquad \text{Subtract 24}$$

$$(x - 3)(x + 8) = 0 \qquad \text{Factor}$$

$$x - 3 = 0 \quad \text{or} \quad x + 8 = 0 \qquad \text{Zero-Product Property}$$

$$x = 3 \qquad\qquad x = -8 \qquad \text{Solve}$$

The solutions are $x = 3$ and $x = -8$. ■

CHECK YOUR ANSWERS

$x = 3$:

$(3)^2 + 5(3) = 9 + 15 = 24$ ✓

$x = -8$:

$(-8)^2 + 5(-8) = 64 - 40 = 24$ ✓

Do you see why one side of the equation must be 0 in Example 1? Factoring the equation as $x(x + 5) = 24$ does not help us find the solutions, since 24 can be factored in infinitely many ways, such as $6 \cdot 4$, $\frac{1}{2} \cdot 48$, $\left(-\frac{2}{3}\right) \cdot (-60)$, and so on.

SOLVING QUADRATIC EQUATIONS BY COMPLETING THE SQUARE

As we saw in Section 1.6, Example 5(b), if a quadratic equation is of the form $(x \pm a)^2 = c$, then we can solve it by taking the square root of each side. In an equation of this form the left-hand side is a *perfect square*: the square of a linear expression in x. So, if a quadratic equation does not factor readily, then we can solve it using the technique of **completing the square**. This means that we add a constant to an expression to make it a perfect square. For example, to make $x^2 - 6x$ a perfect square we must add 9, since $x^2 - 6x + 9 = (x - 3)^2$.

Completing the Square

Area of blue region is

$$x^2 + 2\left(\frac{b}{2}\right)x = x^2 + bx$$

Add a small square of area $(b/2)^2$ to "complete" the square.

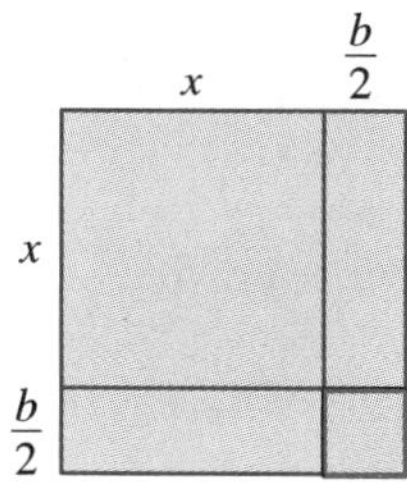

COMPLETING THE SQUARE

To make $x^2 + bx$ a perfect square, add $\left(\frac{b}{2}\right)^2$, the square of half the coefficient of x. This gives the perfect square

$$x^2 + bx + \left(\frac{b}{2}\right)^2 = \left(x + \frac{b}{2}\right)^2$$

EXAMPLE 2 ■ Solving Quadratic Equations by Completing the Square

Solve each equation.

(a) $x^2 - 8x + 13 = 0$ (b) $3x^2 - 12x + 6 = 0$

SOLUTION

(a) $x^2 - 8x + 13 = 0$

$x^2 - 8x = -13$	Subtract 13
$x^2 - 8x + 16 = -13 + 16$	Complete the square: add $\left(\frac{-8}{2}\right)^2 = 16$
$(x - 4)^2 = 3$	Perfect square
$x - 4 = \pm\sqrt{3}$	Take square root
$x = 4 \pm \sqrt{3}$	Add 4

(b) After subtracting 6 from each side of the equation, we must factor the coefficient of x^2 (the 3) from the left side to put the equation in the correct form for completing the square.

$3x^2 - 12x + 6 = 0$

$3x^2 - 12x = -6$	Subtract 6
$3(x^2 - 4x \quad) = -6$	Factor 3 from LHS

When completing the square, make sure the coefficient of x^2 is 1. If it isn't, you must factor this coefficient from both terms that contain x:

$$ax^2 + bx = a\left(x^2 + \frac{b}{a}x \quad\right)$$

Then complete the square inside the parentheses. Remember that the term added inside the parentheses is multiplied by a.

François Viète (1540–1603) had a successful political career before taking up mathematics late in life. He became one of the most famous French mathematicians of the 16th century. Viète introduced a new level of abstraction in algebra by using letters to stand for *known* quantities in an equation. Before Viète's time, each equation had to be solved on its own. For instance, the quadratic equations

$$3x^2 + 2x + 8 = 0$$

and $\quad 5x^2 - 6x + 4 = 0$

had to be solved separately for the unknown x. Viète's idea was to consider all quadratic equations at once by writing

$$ax^2 + bx + c = 0$$

where a, b, and c are known quantities. Thus, he made it possible to write a *formula* (in this case, the quadratic formula) involving a, b, and c that can be used to solve all such equations in one fell swoop. Viète's mathematical genius proved valuable during a war between France and Spain. To communicate with their troops, the Spaniards used a complicated code which Viète managed to decipher. Unaware of Viète's accomplishment, King Philip II of Spain protested to the Pope, claiming that the French were using witchcraft to read his messages.

Now we complete the square by adding $(-2)^2 = 4$ inside the parentheses. Since everything inside the parentheses is multiplied by 3, this means that we are actually adding $3 \cdot 4 = 12$ to the left side of the equation. Thus, we must add 12 to the right side as well.

$3(x^2 - 4x + 4) = -6 + 3 \cdot 4$	Complete the square: add 4
$3(x - 2)^2 = 6$	Perfect square
$(x - 2)^2 = 2$	Divide by 3
$x - 2 = \pm\sqrt{2}$	Take square root
$x = 2 \pm \sqrt{2}$	Add 2 ■

THE QUADRATIC FORMULA

We can use the technique of completing the square to derive a formula for the roots of the general quadratic equation $ax^2 + bx + c = 0$. First, we divide each side of the equation by a and move the constant to the right side, giving

$$x^2 + \frac{b}{a}x = -\frac{c}{a}$$

We now complete the square by adding $(b/2a)^2$ to each side of the equation:

$x^2 + \frac{b}{a}x + \left(\frac{b}{2a}\right)^2 = -\frac{c}{a} + \left(\frac{b}{2a}\right)^2$	Complete the square: Add $\left(\frac{b}{2a}\right)^2$
$\left(x + \frac{b}{2a}\right)^2 = \frac{-4ac + b^2}{4a^2}$	Perfect square
$x + \frac{b}{2a} = \pm\sqrt{\frac{-4ac + b^2}{4a^2}} = \pm\frac{\sqrt{b^2 - 4ac}}{2a}$	Take square root
$x = \frac{-b^2 \pm \sqrt{b^2 - 4ac}}{2a}$	Subtract $\frac{b}{2a}$

This is the quadratic formula.

THE QUADRATIC FORMULA

The roots of the quadratic equation $ax^2 + bx + c = 0$, where $a \neq 0$, are

$$x = \frac{-b \pm \sqrt{b^2 - 4ac}}{2a}$$

The quadratic formula could be used to solve the equations in Examples 1 and 2. You should carry out the details of these calculations.

EXAMPLE 3 ■ Using the Quadratic Formula

Find all solutions of each equation.

(a) $3x^2 - 5x - 1 = 0$ (b) $4x^2 + 12x + 9 = 0$ (c) $x^2 + 2x + 2 = 0$

SOLUTION

(a) Using the quadratic formula with $a = 3$, $b = -5$, and $c = -1$, we get

$$x = \frac{-(-5) \pm \sqrt{(-5)^2 - 4(3)(-1)}}{2(3)} = \frac{5 \pm \sqrt{37}}{6}$$

If approximations are desired, we use a calculator and obtain

$$x = \frac{5 + \sqrt{37}}{6} \approx 1.8471 \quad \text{and} \quad x = \frac{5 - \sqrt{37}}{6} \approx -0.1805$$

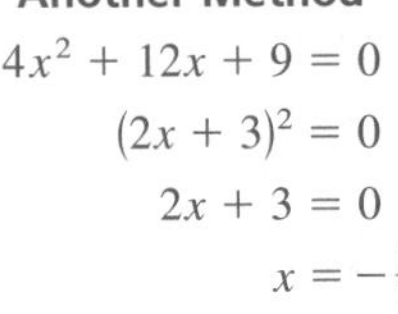
Another Method
$4x^2 + 12x + 9 = 0$
$(2x + 3)^2 = 0$
$2x + 3 = 0$
$x = -\frac{3}{2}$

(b) Using the quadratic formula with $a = 4$, $b = 12$, and $c = 9$ gives

$$x = \frac{-12 \pm \sqrt{(12)^2 - 4 \cdot 4 \cdot 9}}{2 \cdot 4} = \frac{-12 \pm 0}{8} = -\frac{3}{2}$$

This equation has only one solution, $x = -\frac{3}{2}$.

(c) Using the quadratic formula with $a = 1$, $b = 2$, and $c = 2$ gives

$$x = \frac{-2 \pm \sqrt{2^2 - 4 \cdot 2}}{2} = \frac{-2 \pm \sqrt{-4}}{2} = \frac{-2 \pm 2\sqrt{-1}}{2} = -1 \pm \sqrt{-1}$$

Since the square of any real number is nonnegative, $\sqrt{-1}$ is undefined in the real number system. The equation has no real solution. ■

In the next section we study the complex number system, in which the square roots of negative numbers do exist. The equation in Example 3(c) does have solutions in the complex number system.

EXAMPLE 4 ■ Solving a Quadratic Equation Graphically

Solve the equation $2x^2 + 3x - 7 = 0$.

SOLUTION

We graph the equation

$$y = 2x^2 + 3x - 7$$

In Example 4 we can also use the quadratic formula to get

$$x = \frac{-3 \pm \sqrt{3^2 - 4 \cdot 2 \cdot (-7)}}{2 \cdot 2} = \frac{-3 \pm \sqrt{65}}{4}$$

You can check that, correct to two decimals, these are the solutions we obtained using the graphing calculator.

in the viewing rectangle $[-5, 5]$ by $[-15, 15]$ (see Figure 1). The solutions occur at the x-intercepts. Moving the cursor to the x-intercepts, we find that $x \approx -2.8$ and $x \approx 1.3$. Using the zoom-in feature to get a closer look at the x-intercepts, we read the more accurate approximations.

$$x \approx -2.77 \qquad \text{and} \qquad x \approx 1.27$$

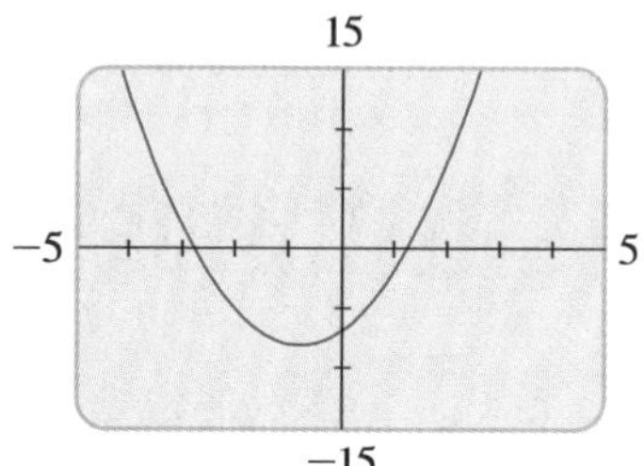

FIGURE 1
$y = 2x^2 + 3x - 7$

■

THE DISCRIMINANT

The quantity $b^2 - 4ac$ that appears under the square root sign in the quadratic formula is called the *discriminant* of the equation $ax^2 + bx + c = 0$ and is given the symbol D. If $D < 0$, then $\sqrt{b^2 - 4ac}$ is undefined, and the quadratic equation has no real solution, as in Example 3(c). If $D = 0$, then the equation has only one real solution, as in Example 3(b). Finally, if $D > 0$, then the equation has two distinct real solutions, as in Example 3(a). The following box summarizes these observations.

THE DISCRIMINANT

The **discriminant** of the general quadratic $ax^2 + bx + c = 0$ $(a \neq 0)$ is $D = b^2 - 4ac$.

1. If $D > 0$, then the equation has two distinct real solutions.
2. If $D = 0$, then the equation has exactly one real solution.
3. If $D < 0$, then the equation has no real solution.

EXAMPLE 5 ■ Using the Discriminant

Use the discriminant to determine how many real solutions each equation has.

(a) $x^2 + 4x - 1 = 0$ (b) $4x^2 - 12x + 9 = 0$ (c) $\frac{1}{3}x^2 - 2x + 4 = 0$

SOLUTION

(a) The discriminant is $D = 4^2 - 4(1)(-1) = 20 > 0$, so the equation has two distinct real solutions.

(b) The discriminant is $D = (-12)^2 - 4 \cdot 4 \cdot 9 = 0$, so the equation has exactly one real solution.

(c) The discriminant is $D = (-2)^2 - 4\left(\frac{1}{3}\right)4 = -\frac{4}{3} < 0$, so the equation has no real solution. ■

The graphs in Figure 2 show why the equations in Example 5 have two, one, and no real solutions, respectively.

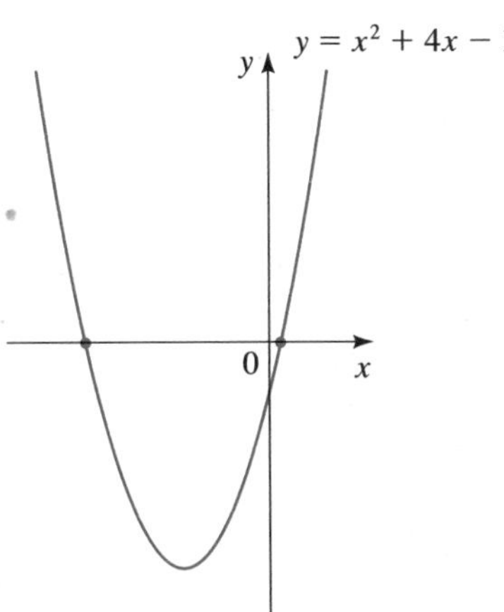

(a) Two real solutions

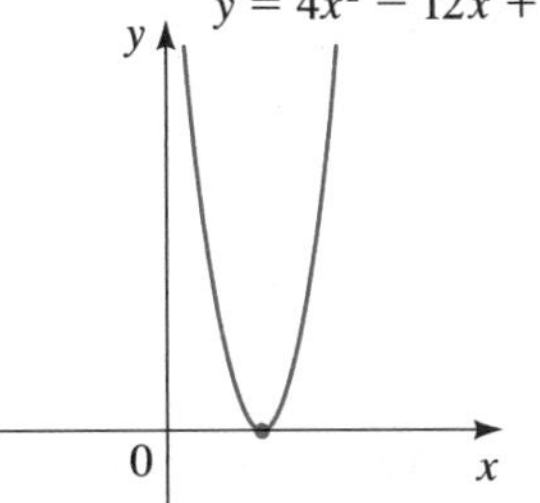

(b) One real solution

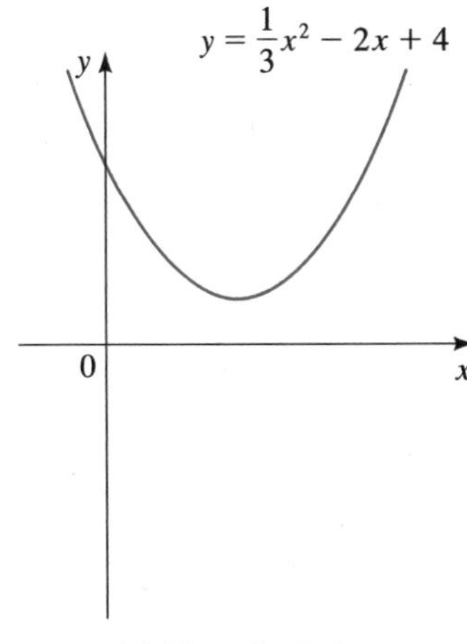

(c) No real solution

FIGURE 2 The two solutions are where the graph intersects the x-axis.

MODELING WITH QUADRATIC EQUATIONS

Let's look at some real-life problems that can be modeled by quadratic equations. The principles discussed in Section 3.2 for setting up equations as models are useful here as well.

EXAMPLE 6 ■ Dimensions of a Building Lot

A rectangular building lot is 8 ft longer than it is wide and has an area of 2900 ft^2. Find the dimensions of the lot.

SOLUTION

We are asked to find the width and length of the lot. So let

Identify the variable

$$w = \text{width of lot}$$

Then we translate the information given in the problem into the language of algebra (see Figure 3).

Express all unknown quantities in terms of the variable

In Words	In Algebra
Width of lot	w
Length of lot	$w + 8$

Now we set up the model.

Set up the model

$$\text{width of lot} \cdot \text{length of lot} = \text{area of lot}$$

$$w(w + 8) = 2900$$

Solve

$$w^2 + 8w = 2900 \qquad \text{Expand}$$

$$w^2 + 8w - 2900 = 0 \qquad \text{Subtract 2900}$$

$$(w - 50)(w + 58) = 0 \qquad \text{Factor}$$

$$w = 50 \quad \text{or} \quad w = -58 \qquad \text{Zero-Product Property}$$

Since the width of the lot must be a positive number, we conclude that $w = 50$ ft. The length of the lot is $w + 8 = 50 + 8 = 58$ ft.

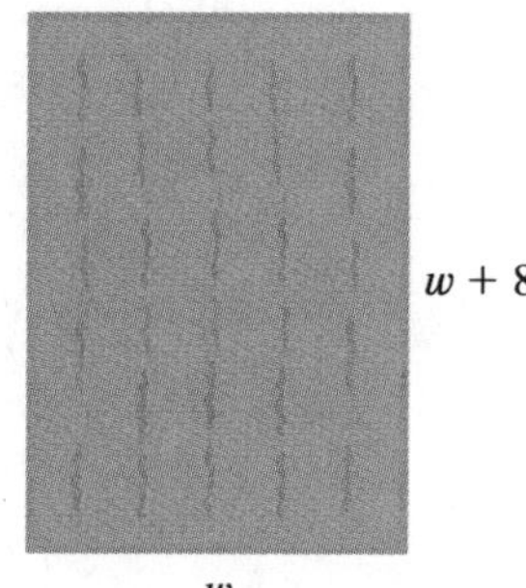

FIGURE 3

■

EXAMPLE 7 ■ A Distance-Rate-Time Problem

A jet flew from New York to Los Angeles, a distance of 4200 km. The speed for the return trip was 100 km/h faster than the outbound speed. If the total trip took 13 hours, what was the speed from New York to Los Angeles?

SOLUTION

We are asked for the speed of the jet from New York to Los Angeles. So let

Identify the variable

$$s = \text{speed from New York to Los Angeles}$$

Then $\qquad s + 100 =$ speed from Los Angeles to New York

Now we organize the information in a table. We fill in the "Distance" column first, since we know that the cities are 4200 km apart. Then we fill in the "Rate" column, since we have expressed both speeds (rates) in terms of the variable s. Finally, we calculate the entries for the "Time" column, using

$$\text{time} = \frac{\text{distance}}{\text{rate}}$$

Express all unknown quantities in terms of the variable

	Distance (km)	**Rate (km/h)**	**Time (h)**
N.Y. to L.A.	4200	s	$\frac{4200}{s}$
L.A. to N.Y.	4200	$s + 100$	$\frac{4200}{s + 100}$

The total trip took 13 hours, so we have the model

Set up the model

$$\text{time from N.Y. to L.A.} + \text{time from L.A. to N.Y.} = \text{total time}$$

$$\frac{4200}{s} + \frac{4200}{s + 100} = 13$$

Multiplying by the common denominator, $s(s + 100)$, we get

$$4200(s + 100) + 4200s = 13s(s + 100)$$

$$8400s + 420{,}000 = 13s^2 + 1300s$$

$$0 = 13s^2 - 7100s - 420{,}000$$

Although this equation does factor, with numbers this large it is probably quickest to use the quadratic formula and a calculator.

Solve

$$s = \frac{7100 \pm \sqrt{(7100)^2 - 4(13)(-420{,}000)}}{2(13)}$$

$$= \frac{7100 \pm 8500}{26}$$

$$s = 600 \qquad \text{or} \qquad s = \frac{-700}{13} \approx -53.8$$

Since s represents speed, we reject the negative answer and conclude that the jet's speed from New York to Los Angeles was 600 km/h. ■

EXAMPLE 8 ■ The Path of a Projectile

This formula depends on the fact that acceleration due to gravity is constant near the earth's surface. Here we neglect the effect of air resistance.

An object thrown or fired straight upward at an initial speed of v_0 ft/s will reach a height of h feet after t seconds, where h and t are related by the formula

$$h = -16t^2 + v_0 t$$

Suppose that a bullet is shot straight upward with an initial speed of 800 ft/s. Its path is shown in Figure 4.

(a) When does the bullet fall back to ground level?

(b) When does it reach a height of 6400 ft?

(c) When does it reach a height of 2 mi?

(d) How high is the highest point the bullet reaches?

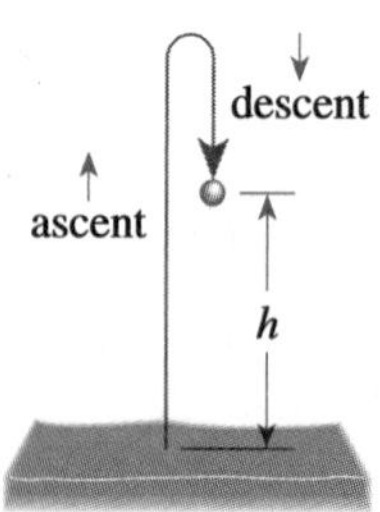

FIGURE 4

SOLUTION

Since the initial speed in this case is $v_0 = 800$ ft/s, the formula is

$$h = -16t^2 + 800t$$

(a) Ground level corresponds to $h = 0$, so we must solve the equation

$$0 = -16t^2 + 800t$$
$$= -16t(t - 50)$$

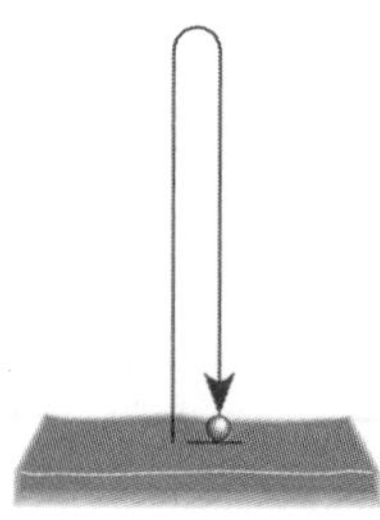

Thus, $t = 0$ or $t = 50$. This means the bullet starts $(t = 0)$ at ground level and returns to ground level after 50 s.

(b) Setting $h = 6400$ gives the equation

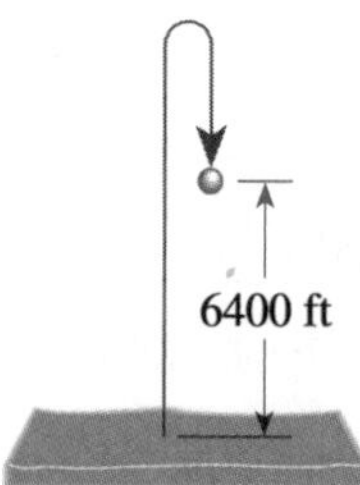

$$6400 = -16t^2 + 800t$$
$$16t^2 - 800t + 6400 = 0$$
$$t^2 - 50t + 400 = 0 \qquad \text{Divide by 16}$$
$$(t - 10)(t - 40) = 0 \qquad \text{Factor}$$
$$t = 10 \quad \text{or} \quad t = 40$$

The bullet reaches 6400 ft after 10 s (on its ascent) and again after 40 s (on its descent to earth).

(c) Two miles is $2 \times 5280 = 10{,}560$ ft.

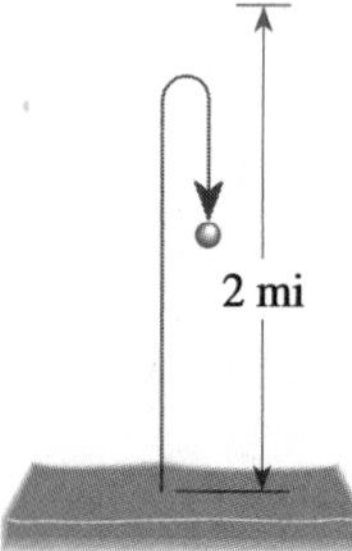

$$10{,}560 = -16t^2 + 800t$$
$$16t^2 - 800t + 10{,}560 = 0$$
$$t^2 - 50t + 660 = 0 \qquad \text{Divide by 16}$$

The discriminant of this equation is $D = (-50)^2 - 4(660) = -140$, which is negative. Thus, the equation has no real solution. The bullet never reaches a height of 2 mi.

(d) Each height the bullet reaches is attained twice, once on its ascent and once on its descent. The only exception is the highest point of its path, which is reached only once. This means that for the highest value of h, the equation

$$h = -16t^2 + 800t$$

or

$$16t^2 - 800t + h = 0$$

has only one solution for t. This in turn means that the discriminant of the equation is 0, and so

$$D = (-800)^2 - 4(16)h = 0$$

$$640{,}000 - 64h = 0$$

$$h = 10{,}000$$

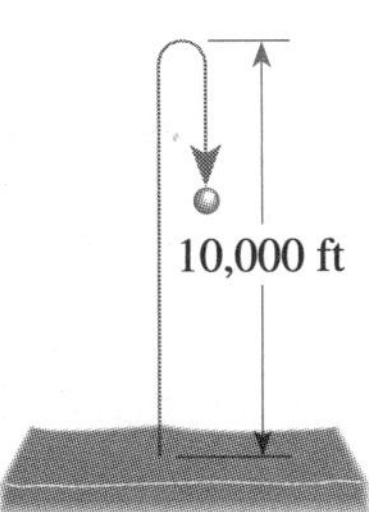

The maximum height reached is 10,000 ft. ■

3.3 EXERCISES

1–10 ■ Solve the equation by factoring.

1. $x^2 - x - 6 = 0$
2. $x^2 + 2x - 8 = 0$
3. $x^2 - 4x + 4 = 0$
4. $x^2 + 6x + 8 = 0$
5. $2y^2 + 7y + 3 = 0$
6. $4w^2 = 4w + 3$
7. $6x^2 + 5x = 4$
8. $3x^2 + 1 = 4x$
9. $x^2 = 5(x + 100)$
10. $a^2x^2 + 2ax + 1 = 0 \quad (a > 0)$

11–20 ■ Solve the equation by completing the square.

11. $x^2 + 2x - 2 = 0$
12. $x^2 - 4x + 2 = 0$
13. $x^2 - 6x - 9 = 0$
14. $x^2 + x - \frac{3}{4} = 0$
15. $x^2 + 22x + 21 = 0$
16. $x^2 - 18x = 19$
17. $2x^2 + 8x + 1 = 0$
18. $3x^2 - 6x - 1 = 0$
19. $4x^2 - x = 0$
20. $x^2 = \frac{3}{4}x - \frac{1}{8}$

21–40 ■ Find all real solutions of the equation.

21. $x^2 - 2x - 8 = 0$
22. $2x^2 + x - 3 = 0$
23. $x^2 + 12x - 27 = 0$
24. $8x^2 - 6x - 9 = 0$
25. $3x^2 + 6x - 5 = 0$
26. $x^2 - 6x + 1 = 0$
27. $2y^2 - y - \frac{1}{2} = 0$
28. $\theta^2 - \frac{3}{2}\theta + \frac{9}{16} = 0$
29. $4x^2 + 16x - 9 = 0$
30. $0 = x^2 - 4x + 1$
31. $3 + 5z + z^2 = 0$
32. $w^2 = 3(w - 1)$
33. $x^2 - \sqrt{5}\,x + 1 = 0$
34. $\sqrt{6}\,x^2 + 2x - \sqrt{\frac{3}{2}} = 0$
35. $10y^2 - 16y + 5 = 0$
36. $25x^2 + 70x + 49 = 0$
37. $3x^2 + 2x + 2 = 0$
38. $5x^2 - 7x + 5 = 0$
39. $ax^2 - (2a + 1)x + (a + 1) = 0 \quad (a \neq 0)$
40. $bx^2 + 2x + \dfrac{1}{b} = 0 \quad (b \neq 0)$

41–44 ■ Use the quadratic formula and a calculator to find all real solutions, correct to three decimals.

41. $x^2 - 0.011x - 0.064 = 0$
42. $2.232x^2 - 4.112x = 6.219$
43. $x^2 - 2.450x + 1.500 = 0$

44. $x^2 - 2.450x + 1.501 = 0$

45–48 ■ Solve the equation for the indicated variable.

45. $h = \frac{1}{2}gt^2 + v_0t$; for t **46.** $S = \dfrac{n(n+1)}{2}$; for n

47. $A = 2\pi r^2 + 2\pi rh$; for r

48. $\dfrac{1}{s+a} + \dfrac{1}{s+b} = \dfrac{1}{c}$; for s

49–52 ■ Solve the equation graphically, correct to two decimals.

49. $2x^2 + 19x - 11 = 0$ **50.** $0.24x^2 - 4.66x - 10.51 = 0$

51. $1.33x^2 + 0.26x - 0.65 = 0$

52. $8.3x^2 + 0.4x + 1.5 = 0$

53–58 ■ Use the discriminant to determine the number of real solutions of the equation. Do not solve the equation.

53. $x^2 - 6x + 1 = 0$ **54.** $x^2 = 6x - 9$

55. $x^2 + 2.20x + 1.21 = 0$ **56.** $x^2 + 2.21x + 1.21 = 0$

57. $x^2 + rx - s = 0 \quad (s > 0)$

58. $x^2 - rx + s = 0 \quad (s > 0, r > 2\sqrt{s})$

59–60 ■ Find all values of k that ensure that the given equation has exactly one solution.

59. $4x^2 + kx + 25 = 0$ **60.** $kx^2 + 36x + k = 0$

61. Find two numbers whose sum is 55 and whose product is 684.

62. The sum of the squares of two consecutive even integers is 1252. Find the integers.

63. A rectangular garden is 10 ft longer than it is wide. Its area is 875 ft^2. What are its dimensions?

64. A rectangular bedroom is 7 ft longer than it is wide. Its area is 228 ft^2. What is the width of the room?

65. A farmer has a rectangular garden plot surrounded by 200 ft of fence. Find the length and width of the garden if its area is 2400 ft^2.

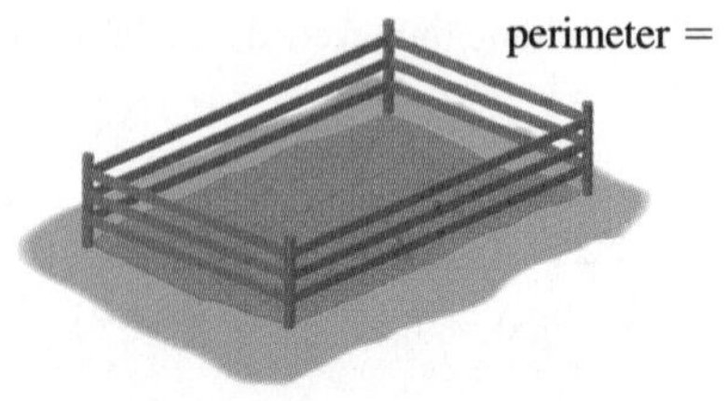

66. Find the length x if the shaded area is 160 in^2.

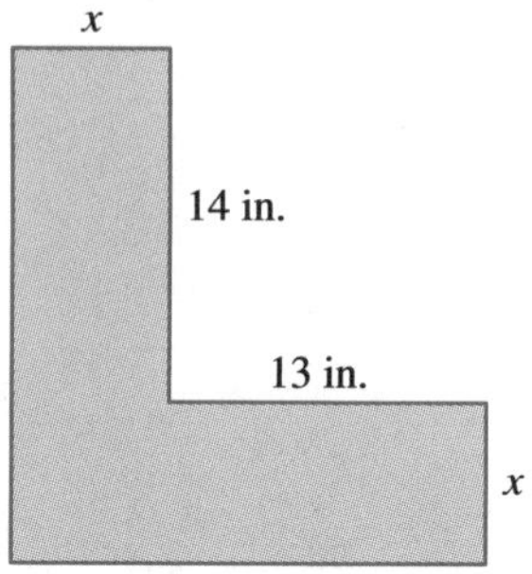

67. Find the length x if the shaded area is 1200 cm^2.

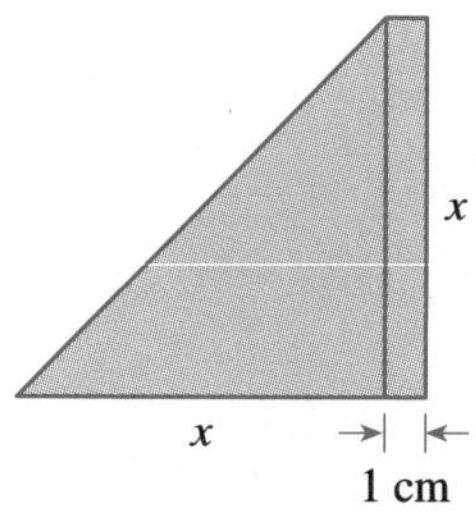

68. A small-appliance manufacturer finds that the profit P (in dollars) generated by producing x microwave ovens per week is given by the formula $P = \frac{1}{10}x(300 - x)$ provided that $0 \leqslant x \leqslant 200$. How many ovens must be manufactured in a given week to generate a profit of \$1250?

69. A box with a square base and no top is to be made from a square piece of carboard by cutting 4-in. squares from each corner and folding up the sides, as shown in the figure. The box is to hold 100 in^3. How big a piece of cardboard is needed?

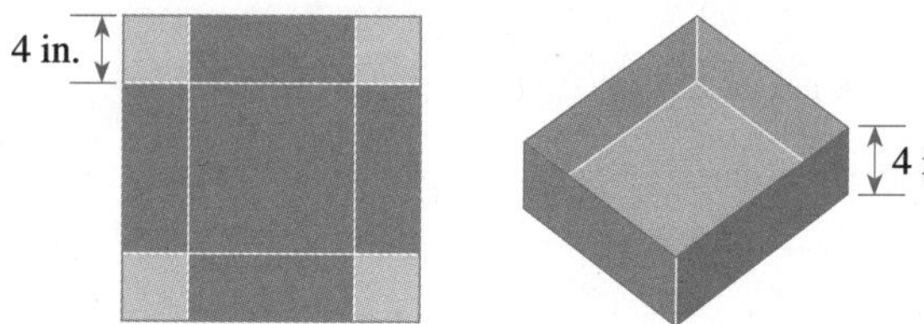

70. A cylindrical can has a volume of 40π cm^3 and is 10 cm tall. What is its diameter? [*Hint:* Use the volume formula listed on the inside back cover of this book.]

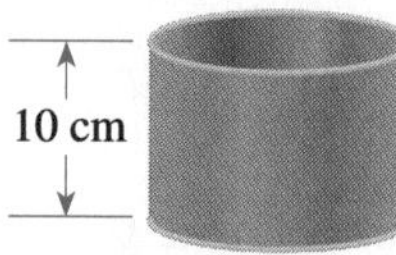

71. A parcel of land is 6 ft longer than it is wide. Each diagonal from one corner to the opposite corner is 174 ft long. What are the dimensions of the parcel?

72. A flagpole is secured on opposite sides by two guy wires, each of which is 5 ft longer than the pole. The distance between the points where the wires are fixed to the ground is equal to the length of one guy wire. How tall is the flagpole (to the nearest inch)?

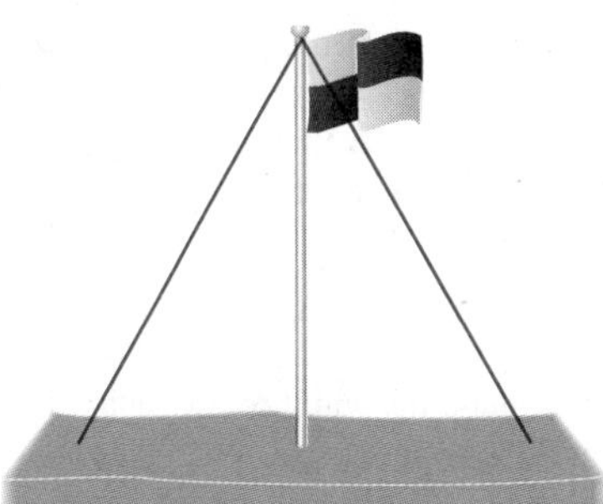

73–74 ■ Suppose an object is dropped from a height h_0 above the ground. Then its height after t seconds is given by $h = -16t^2 + h_0$, where h is measured in feet. Use this information to solve the problem.

73. If a ball is dropped from 288 ft above the ground, how long does it take to reach ground level?

74. A ball is dropped from the top of a building 96 ft tall.

(a) How long will it take to fall half the distance to ground level?

(b) How long will it take to fall to ground level?

75–76 ■ Use the formula $h = -16t^2 + v_0 t$ discussed in Example 8.

75. A ball is thrown straight upward at an initial speed of 40 ft/s.

(a) When does the ball reach a height of 24 ft?

(b) When does it reach a height of 48 ft?

(c) What is the greatest height reached by the ball?

(d) When does the ball reach the highest point of its path?

(e) When does the ball hit the ground?

76. How fast would a ball have to be thrown upward to reach a maximum height of 100 ft? [*Hint:* Use the discriminant of the equation $16t^2 - v_0 t + h = 0$.]

77. The fish population in a certain lake rises and falls according to the formula

$$F = 1000(30 + 17t - t^2)$$

Here F is the number of fish at time t, where t is measured in years since January 1, 1992, when the fish population was first estimated.

(a) On what date will the fish population again be the same as on January 1, 1992?

(b) By what date will all the fish in the lake have died?

78. A wire 360 in. long is cut into two pieces. One piece is formed into a square and the other into a circle. If the two figures have the same area, what are the lengths of the two pieces of wire (to the nearest tenth of an inch)?

79. A salesman drives from Ajax to Barrington, a distance of 120 mi, at a steady speed. He then increases his speed by 10 mi/h to drive the 150 mi from Barrington to Collins. If the second leg of his trip took 6 min more time than the first leg, how fast was he driving between Ajax and Barrington?

80. Kiran drove from Tortula to Cactus, a distance of 250 mi. She increased her speed by 10 mi/h for the 360-mi trip from Cactus to Dry Junction. If the total trip took 11 h, what was her speed from Tortula to Cactus?

81. It took a crew 2 h 40 min to row 6 km upstream and back again. If the rate of flow of the stream was 3 km/h, what was the rowing rate of the crew in still water?

82. A factory is to be built on a lot measuring 180 ft by 240 ft. A local building code specifies that a lawn of uniform width and equal in area to the factory must surround the factory. What must the width of this lawn be, and what are the dimensions of the factory?

83. Henry and Irene working together can wash all the windows of their house in 1 h 48 min. Working alone, it takes Henry $1\frac{1}{2}$ h more than Irene to do the job. How long does it take each person working alone to wash all the windows?

84. Jack, Kay, and Lynn deliver advertising flyers in a small town. If each person works alone, it takes Jack 4 h to deliver all the flyers, and it takes Lynn 1 h longer than it takes Kay. Working together, they can deliver all the flyers in 40% of the time it takes Kay working alone. How long does it take Kay to deliver all the flyers alone?

85. If an imaginary line segment is drawn between the centers of the earth and the moon, then the net gravitational force F acting on an object situated on this line segment is

$$F = \frac{-K}{x^2} + \frac{0.012K}{(239 - x)^2}$$

where $K > 0$ is a constant and x is the distance of the object from the center of the earth, measured in thousands of miles. How far from the center of the earth is the "dead spot" where no net gravitational force acts upon the object? (Express your answer to the nearest thousand miles.)

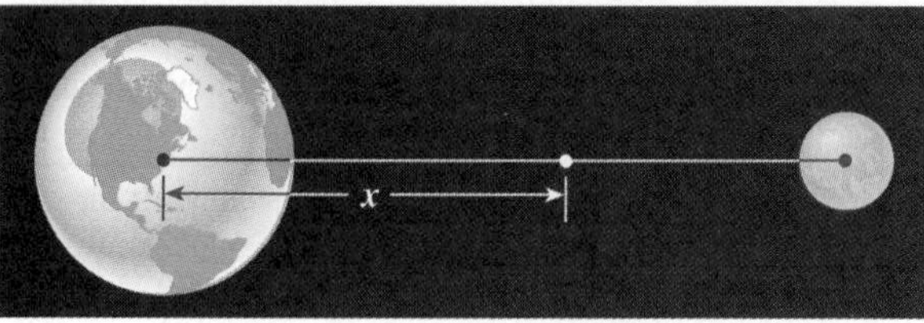

DISCOVERY • DISCUSSION

86. **Relationship between Roots and Coefficients** The quadratic formula gives us the roots of a quadratic equation from its coefficients. We can also obtain the coefficients from the roots. For example, find the roots of the equation $x^2 - 9x + 20 = 0$ and show that the product of the roots is the constant term 20 and the sum of the roots is 9, the negative of the coefficient of x. Show that the same relationship between roots and coefficients holds for the following equations:

$$x^2 - 2x - 8 = 0$$

$$x^2 + 4x + 2 = 0$$

Use the quadratic formula to prove that in general, if the equation $x^2 + bx + c = 0$ has roots r_1 and r_2, then $c = r_1 r_2$ and $b = -(r_1 + r_2)$.

87. **A Babylonian Quadratic Equation** The ancient Babylonians knew how to solve quadratic equations. Here is a problem from a cuneiform tablet found in a Babylonian school dating back to about 2000 B.C.

> I have a reed, I know not its length. I broke from it one cubit, and it fit 60 times along the length of my field. I restored to the reed what I had broken off, and it fit 30 times along the width of my field. The area of my field is 375 square nindas. What was the original length of the reed?

Solve this problem, using the fact that 1 ninda = 12 cubits.

3.4 COMPLEX NUMBERS

In Section 3.3 we saw that if the discriminant of a quadratic equation is negative, the equation has no real solution. For example, the equation

$$x^2 + 4 = 0$$

has no real solution. If we try to solve this equation, we get $x^2 = -4$, so

$$x = \pm\sqrt{-4}$$

But this is impossible, since the square of any real number is positive. [For example, $(-2)^2 = 4$, a positive number.] Thus, negative numbers don't have real square roots.

See the note on Cardano, page 351, for an example of how complex numbers are used to find real solutions of polynomial equations.

To make it possible to solve *all* quadratic equations, mathematicians invented an expanded number system, called the *complex number system*. First they defined the new number

$$i = \sqrt{-1}$$

This means $i^2 = -1$. A complex number is then a number of the form $a + bi$, where a and b are real numbers.

Leonhard Euler (1707–1783) was born in Basel, Switzerland, the son of a pastor. At age 13 his father sent him to the University at Basel to study theology, but Euler soon decided to devote himself to the sciences. Besides theology he studied mathematics, medicine, astronomy, physics, and oriental languages. It is said that Euler could calculate as effortlessly as "men breathe or as eagles fly." One hundred years before Euler, Fermat (see page 578) had conjectured that $2^{2^n} + 1$ is a prime number for all n. The first five of these numbers are 5, 17, 257, 65537, and 4,294,967,297. It's easy to show that the first four are prime. The fifth was also thought to be prime until Euler, with his phenomenal calculating ability, showed that it is the product $641 \times 6{,}700{,}417$ and so is not prime. Euler published more than any other mathematician in history. His collected works comprise 75 large volumes. Although he was blind for the last 17 years of his life, he continued to work and publish. In his writings he popularized the use of the symbols π, e, and i, which you will find in this textbook. One of Euler's most lasting contributions is the development of complex numbers.

DEFINITION OF COMPLEX NUMBERS

A **complex number** is an expression of the form

$$a + bi$$

where a and b are real numbers and $i^2 = -1$. The **real part** of this complex number is a and the **imaginary part** is b. Two complex numbers are **equal** if and only if their real parts are equal and their imaginary parts are equal.

Note that both the real and imaginary parts of a complex number are real numbers.

EXAMPLE 1 ■ Complex Numbers

The following are examples of complex numbers.

$3 + 4i$	Real part 3, imaginary part 4
$\frac{1}{2} - \frac{2}{3}i$	Real part $\frac{1}{2}$, imaginary part $-\frac{2}{3}$
$6i$	Real part 0, imaginary part 6
-7	Real part -7, imaginary part 0

■

A number such as $6i$, which has real part 0, is called a **pure imaginary number**. A real number like -7 can be thought of as a complex number with imaginary part 0.

In the complex number system every quadratic equation has solutions. The numbers $2i$ and $-2i$ are solutions of $x^2 = -4$ because

$$(2i)^2 = 2^2 i^2 = 4(-1) = -4 \qquad \text{and} \qquad (-2i)^2 = (-2)^2 i^2 = 4(-1) = -4$$

Although we use the term *imaginary* in this context, imaginary numbers should not be thought of as any less "real" (in the ordinary rather than the mathematical sense of that word) than negative numbers or irrational numbers. All numbers (except possibly the positive integers) are creations of the human mind—the numbers -1 and $\sqrt{2}$ as well as the number i. We study complex numbers because they complete, in a useful and elegant fashion, the study of the solutions of polynomial equations. In fact, imaginary numbers are useful not only in algebra and mathematics, but in the other sciences as well. To give just one example, in electrical theory the *reactance* of a circuit is a quantity whose measure is an imaginary number.

ARITHMETIC OPERATIONS ON COMPLEX NUMBERS

Complex numbers are added, subtracted, multiplied, and divided just as we would any number of the form $a + b\sqrt{c}$. The only difference we must keep in mind is that $i^2 = -1$. Thus, the following calculations are valid.

$$\begin{aligned}(a + bi)(c + di) &= ac + (ad + bc)i + bdi^2 && \text{Multiply and collect like terms}\\ &= ac + (ad + bc)i + bd(-1) && i^2 = -1\\ &= (ac - bd) + (ad + bc)i && \text{Combine real and imaginary parts}\end{aligned}$$

We therefore define the sum, difference, and product of complex numbers as follows.

ADDING, SUBTRACTING, AND MULTIPLYING COMPLEX NUMBERS

Definition	Description
Addition	
$(a + bi) + (c + di) = (a + c) + (b + d)i$	To add complex numbers, add the real parts and the imaginary parts.
Subtraction	
$(a + bi) - (c + di) = (a - c) + (b - d)i$	To subtract complex numbers, subtract the real parts and the imaginary parts.
Multiplication	
$(a + bi) \cdot (c + di) = (ac - bd) + (ad + bc)i$	Multiply complex numbers like binomials, using $i^2 = -1$.

EXAMPLE 2 ■ Adding, Subtracting, and Multiplying Complex Numbers

Express the following in the form $a + bi$.

(a) $(3 + 5i) + (4 - 2i)$

(b) $(3 + 5i) - (4 - 2i)$

(c) $(3 + 5i)(4 - 2i)$

(d) i^{23}

SOLUTION

(a) According to the definition, we add the real parts and we add the imaginary parts.

$$(3 + 5i) + (4 - 2i) = (3 + 4) + (5 - 2)i = 7 + 3i$$

(b) $(3 + 5i) - (4 - 2i) = (3 - 4) + [5 - (-2)]i = -1 + 7i$

(c) $(3 + 5i)(4 - 2i) = [3 \cdot 4 - 5(-2)] + [3(-2) + 5 \cdot 4]i = 22 + 14i$

(d) $i^{23} = i^{20+3} = (i^2)^{10}i^3 = (-1)^{10}i^2 i = (1)(-1)i = -i$ ■

Complex Conjugates

Number	Conjugate
$3 + 2i$	$3 - 2i$
$1 - i$	$1 + i$
$4i$	$-4i$
5	5

Division of complex numbers is much like rationalizing the denominator of a radical expression, which we considered in Section 1.3. For the complex number $z = a + bi$ we define its **complex conjugate** to be $\overline{z} = a - bi$. Note that

$$z \cdot \overline{z} = (a + bi)(a - bi) = a^2 + b^2$$

So the product of a complex number and its conjugate is always a nonnegative real number. We use this property to divide complex numbers.

DIVIDING COMPLEX NUMBERS

To simplify the quotient $\dfrac{a + bi}{c + di}$, multiply the numerator and the denominator by the complex conjugate of the denominator:

$$\frac{a + bi}{c + di} = \left(\frac{a + bi}{c + di}\right)\left(\frac{c - di}{c - di}\right) = \frac{(ac + bd) + (bc - ad)i}{c^2 + d^2}$$

Rather than memorize this entire formula, it's best just to remember the first step and then mutiply out the numerator and the denominator as usual.

EXAMPLE 3 ■ Dividing Complex Numbers

Express the following in the form $a + bi$.

(a) $\dfrac{3 + 5i}{1 - 2i}$ (b) $\dfrac{7 + 3i}{4i}$

SOLUTION

We multiply both numerator and denominator by the complex conjugate of the denominator to make the new denominator a real number.

(a) The complex conjugate of $1 - 2i$ is $\overline{1 - 2i} = 1 + 2i$.

$$\frac{3 + 5i}{1 - 2i} = \left(\frac{3 + 5i}{1 - 2i}\right)\left(\frac{1 + 2i}{1 + 2i}\right) = \frac{-7 + 11i}{5} = -\frac{7}{5} + \frac{11}{5}i$$

(b) The complex conjugate of $4i$ is $-4i$. Therefore

$$\frac{7 + 3i}{4i} = \left(\frac{7 + 3i}{4i}\right)\left(\frac{-4i}{-4i}\right) = \frac{12 - 28i}{16} = \frac{3}{4} - \frac{7}{4}i$$

■

SQUARE ROOTS OF NEGATIVE NUMBERS

Just as every positive real number r has two square roots ($\sqrt{r}$ and $-\sqrt{r}$), every negative number has two square roots as well. If $-r$ is a negative number, then its square roots are $\pm i\sqrt{r}$, because $(i\sqrt{r})^2 = i^2 r = -r$ and $(-i\sqrt{r})^2 = i^2 r = -r$.

SQUARE ROOTS OF NEGATIVE NUMBERS

If $-r$ is negative, then the **principal square root** of $-r$ is

$$\sqrt{-r} = i\sqrt{r}$$

The two square roots of $-r$ are $i\sqrt{r}$ and $-i\sqrt{r}$.

We usually write $i\sqrt{b}$ instead of $\sqrt{b}\,i$ to avoid confusion with $\sqrt{bi}$.

EXAMPLE 4 ■ Square Roots of Negative Numbers

(a) $\sqrt{-1} = i\sqrt{1} = i$ (b) $\sqrt{-16} = i\sqrt{16} = 4i$ (c) $\sqrt{-3} = i\sqrt{3}$ ■

Special care must be taken when performing calculations involving square roots of negative numbers. Although $\sqrt{a} \cdot \sqrt{b} = \sqrt{ab}$ when a and b are positive, this is *not* true when both are negative. For example,

$$\sqrt{-2} \cdot \sqrt{-3} = i\sqrt{2} \cdot i\sqrt{3} = i^2\sqrt{6} = -\sqrt{6}$$

but $$\sqrt{(-2)(-3)} = \sqrt{6}$$

so $$\sqrt{-2} \cdot \sqrt{-3} \neq \sqrt{(-2)(-3)}$$

When multiplying radicals of negative numbers, express them first in the form $i\sqrt{r}$ (where $r > 0$) to avoid possible errors of this type.

EXAMPLE 5 ■ Using Square Roots of Negative Numbers

Evaluate $(\sqrt{12} - \sqrt{-3})(3 + \sqrt{-4})$ and express in the form $a + bi$.

SOLUTION

$$\begin{aligned}(\sqrt{12} - \sqrt{-3})(3 + \sqrt{-4}) &= (\sqrt{12} - i\sqrt{3})(3 + i\sqrt{4}) \\ &= (2\sqrt{3} - i\sqrt{3})(3 + 2i) \\ &= (6\sqrt{3} + 2\sqrt{3}) + i(2 \cdot 2\sqrt{3} - 3\sqrt{3}) \\ &= 8\sqrt{3} + i\sqrt{3}\end{aligned}$$ ■

IMAGINARY ROOTS OF QUADRATIC EQUATIONS

We have already seen that, if $a \neq 0$, the solutions of the quadratic equation $ax^2 + bx + c = 0$ are

$$x = \frac{-b \pm \sqrt{b^2 - 4ac}}{2a}$$

If $b^2 - 4ac < 0$, then the equation has no real solution. But in the complex number system, this equation will always have solutions, because negative numbers have square roots in this expanded setting.

EXAMPLE 6 ■ Quadratic Equations with Imaginary Solutions

Solve each equation: (a) $x^2 + 9 = 0$ (b) $x^2 + 4x + 5 = 0$

SOLUTION

(a) The equation $x^2 + 9 = 0$ means $x^2 = -9$, so

$$x = \pm\sqrt{-9} = \pm i\sqrt{9} = \pm 3i$$

The solutions are therefore $3i$ and $-3i$.

(b) By the quadratic formula we have

$$x = \frac{-4 \pm \sqrt{4^2 - 4 \cdot 5}}{2}$$

$$= \frac{-4 \pm \sqrt{-4}}{2}$$

$$= \frac{-4 \pm 2i}{2} = \frac{2(-2 \pm i)}{2} = -2 \pm i$$

The two solutions of *any* quadratic equation that has real coefficients are complex conjugates of each other. To understand why this is true, think about the $\pm$ sign in the quadratic formula.

So, the solutions are $-2 + i$ and $-2 - i$. ■

EXAMPLE 7 ■ Complex Conjugates as Solutions of a Quadratic

Show that the solutions of the equation

$$4x^2 - 24x + 37 = 0$$

are complex conjugates of each other.

SOLUTION

We use the quadratic formula to get

$$x = \frac{24 \pm \sqrt{(24)^2 - 4(4)(37)}}{2(4)}$$

$$= \frac{24 \pm \sqrt{-16}}{8} = \frac{24 \pm 4i}{8} = 3 \pm \frac{1}{2}i$$

So, the solutions are $3 + \frac{1}{2}i$ and $3 - \frac{1}{2}i$, and these are complex conjugates. ■

3.4 EXERCISES

1–6 ■ Find the real and imaginary part of the complex number.

1. $3 - 5i$

2. $\dfrac{2 + 4i}{3}$

3. $6i$

4. $\frac{1}{2}$

5. $\sqrt{2} + \sqrt{-3}$

6. $\dfrac{2 + 4i}{\sqrt{-16}}$

7–46 ■ Evaluate the expression and write the result in the form $a + bi$.

7. $(4 + 3i) + (5 - 2i)$

8. $(7 - 6i) + (-3 + 7i)$

9. $\left(7 - \frac{1}{2}i\right) + \left(5 + \frac{3}{2}i\right)$

10. $(-4 + i) - (2 - 5i)$

11. $(-12 + 8i) - (7 + 4i)$

12. $6i - (4 - i)$

13. $4(-1 + 2i)$

14. $2i\left(\frac{1}{2} - i\right)$

15. $(7 - i)(4 + 2i)$

16. $(5 - 3i)(1 + i)$

17. $(3 - 4i)(5 - 12i)$

18. $\left(\frac{2}{3} + 12i\right)\left(\frac{1}{6} + 24i\right)$

19. $(6 + 5i)(2 - 3i)$

20. $(-2 + i)(3 - 7i)$

21. $\dfrac{1}{i}$

22. $\dfrac{1}{1 + i}$

23. $\dfrac{2 - 3i}{1 - 2i}$

24. $\dfrac{5 - i}{3 + 4i}$

25. $\dfrac{26 + 39i}{2 - 3i}$

26. $\dfrac{25}{4 - 3i}$

27. $\dfrac{10i}{1 - 2i}$

28. $(2 - 3i)^{-1}$

29. $\dfrac{4 + 6i}{3i}$

30. $\dfrac{-3 + 5i}{15i}$

31. $\dfrac{1}{1 + i} - \dfrac{1}{1 - i}$

32. $\dfrac{(1 + 2i)(3 - i)}{2 + i}$

33. i^3

34. $(2i)^4$

35. i^{100}

36. i^{1002}

37. $\sqrt{-25}$

38. $\sqrt{\dfrac{-9}{4}}$

39. $\sqrt{-3}\sqrt{-12}$

40. $\sqrt{\frac{1}{3}}\sqrt{-27}$

41. $(3 - \sqrt{-5})(1 + \sqrt{-1})$

42. $\dfrac{1 - \sqrt{-1}}{1 + \sqrt{-1}}$

43. $\dfrac{2 + \sqrt{-8}}{1 + \sqrt{-2}}$

44. $(\sqrt{3} - \sqrt{-4})(\sqrt{6} - \sqrt{-8})$

45. $\dfrac{\sqrt{-36}}{\sqrt{-2}\sqrt{-9}}$

46. $\dfrac{\sqrt{-7}\sqrt{-49}}{\sqrt{28}}$

47–60 ■ Find all solutions of the equation and express them in the form $a + bi$.

47. $x^2 + 9 = 0$

48. $9x^2 + 4 = 0$

49. $x^2 - 4x + 5 = 0$

50. $x^2 + 2x + 2 = 0$

51. $x^2 + x + 1 = 0$

52. $x^2 - 3x + 3 = 0$

53. $2x^2 - 2x + 1 = 0$

54. $2x^2 + 3 = 2x$

55. $t + 3 + \dfrac{3}{t} = 0$

56. $z + 4 + \dfrac{12}{z} = 0$

57. $6x^2 + 12x + 7 = 0$

58. $4x^2 - 16x + 19 = 0$

59. $\frac{1}{2}x^2 - x + 5 = 0$

60. $x^2 + \frac{1}{2}x + 1 = 0$

61–68 ■ Recall that the symbol $\overline{z}$ represents the complex conjugate of z. If $z = a + bi$ and $w = c + di$, prove each statement.

61. $\overline{z} + \overline{w} = \overline{z + w}$

62. $\overline{zw} = \overline{z} \cdot \overline{w}$

63. $(\overline{z})^2 = \overline{z^2}$

64. $\overline{\overline{z}} = z$

65. $z + \overline{z}$ is a real number

66. $z - \overline{z}$ is a pure imaginary number

67. $z \cdot \overline{z}$ is a real number

68. $z = \overline{z}$ if and only if z is real

DISCOVERY • DISCUSSION

69. Complex Conjugate Roots Suppose that the equation $ax^2 + bx + c = 0$ has real coefficients and complex roots. Why must the roots be complex conjugates of each other? (Think about how you would find the roots using the quadratic formula.)

70. Powers of i Calculate the first 12 powers of i, that is, $i, i^2, i^3, \ldots, i^{12}$. Do you notice a pattern? Explain how you would calculate any whole number power of i, using the pattern you have discovered. Use this procedure to calculate i^{4446}.

3.5 OTHER EQUATIONS

So far we have learned how to solve linear and quadratic equations. In this section we study other types of equations, including those that involve higher powers, fractional expressions, and radicals.

POLYNOMIAL EQUATIONS

Some equations can be solved by factoring and using the Zero-Product Property, which says that if a product equals 0, then at least one of the factors must equal 0.

EXAMPLE 1 ■ Solving an Equation by Factoring

Solve the equation $x^5 = 9x^3$.

SOLUTION

We bring all terms to one side and then factor.

$$x^5 - 9x^3 = 0 \qquad \text{Subtract } 9x^3$$

$$x^3(x^2 - 9) = 0 \qquad \text{Factor } x^3$$

$$x^3(x - 3)(x + 3) = 0 \qquad \text{Difference of squares}$$

$$x^3 = 0 \quad \text{or} \quad x - 3 = 0 \quad \text{or} \quad x + 3 = 0 \qquad \text{Zero-Product Property}$$

$$x = 0 \qquad x = 3 \qquad x = -3 \qquad \text{Solve}$$

The solutions are $x = 0$, $x = 3$, and $x = -3$. You should check that each of these satisfies the original equation. ■

To divide each side of the equation in Example 1 by the common factor x^3 would be wrong, because in doing so we would lose the solution $x = 0$. Never divide both sides of an equation by an expression containing the variable, unless you know that the expression cannot equal 0.

EXAMPLE 2 ■ Factoring by Grouping

Solve the equation $x^3 + 3x^2 - 4x - 12 = 0$.

SOLUTION

The left-hand side of the equation can be factored by grouping the terms in pairs.

$$(x^3 + 3x^2) - (4x + 12) = 0 \qquad \text{Group terms}$$

$$x^2(x + 3) - 4(x + 3) = 0 \qquad \text{Factor } x^2 \text{ and } 4$$

$$(x^2 - 4)(x + 3) = 0 \qquad \text{Factor } x + 3$$

$$(x - 2)(x + 2)(x + 3) = 0 \qquad \text{Difference of squares}$$

Solving $(x-2)(x+2)(x+3)=0$, we obtain

$x-2=0$	or	$x+2=0$	or	$x+3=0$	Zero-Product Property
$x=2$		$x=-2$		$x=-3$	Solve

The solutions are $x=2$, -2, and -3. ■

EXAMPLE 3 ■ An Equation Involving Fractional Expressions

Solve the equation $\dfrac{3}{x}+\dfrac{5}{x+2}=2$.

SOLUTION

To simplify the equation, we multiply each side by the common denominator.

$$\left(\frac{3}{x}+\frac{5}{x+2}\right)x(x+2)=2x(x+2) \qquad \text{Multiply by LCD } x(x+2)$$

$$3(x+2)+5x=2x^2+4x \qquad \text{Expand}$$

$$8x+6=2x^2+4x \qquad \text{Expand LHS}$$

$$0=2x^2-4x-6 \qquad \text{Subtract } 8x+6$$

$$0=x^2-2x-3 \qquad \text{Divide both sides by 2}$$

$$0=(x-3)(x+1) \qquad \text{Factor}$$

$$x-3=0 \quad \text{or} \quad x+1=0 \qquad \text{Zero-Product Property}$$

$$x=3 \qquad\qquad x=-1 \qquad \text{Solve}$$

CHECK YOUR ANSWERS

$x=3$:

$\text{LHS}=\dfrac{3}{3}+\dfrac{5}{3+2}$

$=1+1=2$

$\text{RHS}=2$

$\text{LHS}=\text{RHS}$ ✓

$x=-1$:

$\text{LHS}=\dfrac{3}{-1}+\dfrac{5}{-1+2}$

$=-3+5=2$

$\text{RHS}=2$

$\text{LHS}=\text{RHS}$ ✓

We must check our answers because multiplying by an expression that contains the variable can introduce extraneous solutions (see the *Warning* on page 56). From *Check Your Answers* we see that the solutions are $x=3$ and -1. ■

EQUATIONS INVOLVING RADICALS

When you solve an equation that involves radicals, you must be especially careful to check your final answers. The next example demonstrates why.

EXAMPLE 4 ■ An Equation Involving a Radical

Solve the equation $2x=1-\sqrt{2-x}$.

SOLUTION

To eliminate the square root, we first isolate it on one side of the equal sign, then square.

$$\begin{aligned} 2x - 1 &= -\sqrt{2 - x} && \text{Subtract 1} \\ (2x - 1)^2 &= 2 - x && \text{Square each side} \\ 4x^2 - 4x + 1 &= 2 - x && \text{Expand LHS} \\ 4x^2 - 3x - 1 &= 0 && \text{Add } -2 + x \\ (4x + 1)(x - 1) &= 0 && \text{Factor} \\ 4x + 1 = 0 \quad &\text{or} \quad x - 1 = 0 && \text{Zero-Product Property} \\ x = -\tfrac{1}{4} \quad & \quad\quad x = 1 && \text{Solve} \end{aligned}$$

CHECK YOUR ANSWERS

$x = -\frac{1}{4}$:

$\text{LHS} = 2\left(-\frac{1}{4}\right) = -\frac{1}{2}$

$\text{RHS} = 1 - \sqrt{2 - \left(-\frac{1}{4}\right)}$

$= 1 - \sqrt{\frac{9}{4}}$

$= 1 - \frac{3}{2} = -\frac{1}{2}$

$\text{LHS} = \text{RHS}$ ✓

$x = 1$:

$\text{LHS} = 2(1) = 2$

$\text{RHS} = 1 - \sqrt{2 - 1}$

$= 1 - 1 = 0$

$\text{LHS} \neq \text{RHS}$ ✗

The values $x = -\frac{1}{4}$ and $x = 1$ are only potential solutions. We must check them to see if they satisfy the original equation. From *Check Your Answers* we see that $x = -\frac{1}{4}$ is a solution but $x = 1$ is not. The only solution is

$$x = -\tfrac{1}{4}$$ ■

When we solve an equation, we may end up with one or more **extraneous solutions**, that is, potential solutions that do not satisfy the original equation. In Example 4, the value $x = 1$ is an extraneous solution. Extraneous solutions may be introduced when we square each side of an equation because the operation of squaring can turn a false equation into a true one. For example, $-1 \neq 1$, but $(-1)^2 = 1^2$. Thus, the squared equation may be true for more values of the variable than the original equation.  That is why you must always check your answers to make sure that each satisfies the original equation.

EQUATIONS OF QUADRATIC TYPE

An equation of the form $aw^2 + bw + c = 0$, where w is an algebraic expression, is an equation of **quadratic type**. We solve equations of quadratic type by substituting for the algebraic expression, as we see in the next three examples.

EXAMPLE 5 ■ An Equation of Quadratic Type

Solve the equation $\left(1 + \dfrac{1}{x}\right)^2 - 6\left(1 + \dfrac{1}{x}\right) + 8 = 0$.

SOLUTION

We could solve this equation by multiplying it out first. But it's easier to think of the expression $1 + \frac{1}{x}$ as the unknown in this equation, and give it a new name w.

MATHEMATICS IN THE MODERN WORLD

ERROR-CORRECTING CODES

The pictures sent back by the *Pathfinder* spacecraft from the surface of Mars on July 4, 1997, were astoundingly clear. But few watching these pictures were aware of the complex mathematics used to accomplish that feat. The distance to Mars is enormous, and the background noise (or static) is many times stronger than the original signal emitted by the spacecraft. So, when scientists receive the signal, it is full of errors. To get a clear picture, the errors must be found and corrected. This same problem of errors is routinely encountered in transmitting bank records when you use an ATM machine, or voice when you are talking on the telephone.

To understand how errors are found and corrected, we must first understand that to transmit pictures, sound, or text we transform them into bits (the digits 0 or 1; see page 108). To help the receiver recognize errors, the message is "coded" by inserting additional bits. For example, suppose you want to transmit the message "10100." A very simple-minded code is as follows: Send each digit a million times. The person receiv-

(continued)

This turns the equation into a quadratic equation in the new variable w.

$$\left(1 + \frac{1}{x}\right)^2 - 6\left(1 + \frac{1}{x}\right) + 8 = 0$$

$$w^2 - 6w + 8 = 0 \qquad \text{Let } w = 1 + \frac{1}{x}$$

$$(w - 4)(w - 2) = 0 \qquad \text{Factor}$$

$$w - 4 = 0 \quad \text{or} \quad w - 2 = 0 \qquad \text{Zero-Product Property}$$

$$w = 4 \qquad\qquad w = 2 \qquad \text{Solve}$$

Now we change these values of w back into the corresponding values of x.

$$1 + \frac{1}{x} = 4 \qquad\qquad 1 + \frac{1}{x} = 2 \qquad w = 1 + \frac{1}{x}$$

$$\frac{1}{x} = 3 \qquad\qquad \frac{1}{x} = 1 \qquad \text{Subtract 1}$$

$$x = \frac{1}{3} \qquad\qquad x = 1 \qquad \text{Take reciprocals}$$

The solutions are $x = \frac{1}{3}$ and $x = 1$. ■

EXAMPLE 6 ■ A Fourth-Degree Equation of Quadratic Type

Find all solutions of the equation $x^4 - 8x^2 + 8 = 0$.

SOLUTION

If we set $w = x^2$, then we get a quadratic equation in the new variable w:

$$(x^2)^2 - 8x^2 + 8 = 0 \qquad \text{Write } x^4 \text{ as } (x^2)^2$$

$$w^2 - 8w + 8 = 0 \qquad \text{Let } w = x^2$$

$$w = \frac{-(-8) \pm \sqrt{(-8)^2 - 4 \cdot 8}}{2} = 4 \pm 2\sqrt{2} \qquad \text{Quadratic formula}$$

$$x^2 = 4 \pm 2\sqrt{2} \qquad w = x^2$$

$$x = \pm\sqrt{4 \pm 2\sqrt{2}} \qquad \text{Take square roots}$$

So, there are four solutions:

$$\sqrt{4 + 2\sqrt{2}}, \quad \sqrt{4 - 2\sqrt{2}}, \quad -\sqrt{4 + 2\sqrt{2}}, \quad -\sqrt{4 - 2\sqrt{2}}$$

Using a calculator, we obtain the approximations $x \approx 2.61, 1.08, -2.61, -1.08$. ■

ing the message reads it in blocks of a million digits. If the first block is mostly 1's, he concludes that you are probably trying to transmit a 1, and so on. To say that this code is not efficient is a bit of an understatement; it requires sending a million times more data than the original message. Another method inserts "check digits." For example, for each block of eight digits insert a ninth digit; the inserted digit is 0 if there is an even number of 1's in the block and 1 if there is an odd number. So, if a single digit is wrong (a 0 changed to a 1, or vice versa), the check digits allow us to recognize that an error has occurred. This method does not tell us where the error is, so we can't correct it. Modern error-correcting codes use interesting mathematical algorithms that require inserting relatively few digits but which allow the receiver to not only recognize, but also correct, errors. The first error-correcting code was developed in the 1940s by Richard Hamming at MIT. It is interesting to note that the English language has a built-in error-correcting mechanism; to test it, try reading this error-laden sentence: Gve mo libty ox giv ne deth.

EXAMPLE 7 ■ An Equation Involving Fractional Powers

Find all solutions of the equation $x^{1/3} + x^{1/6} - 2 = 0$.

SOLUTION

This equation is of quadratic type because if we let $w = x^{1/6}$, then $w^2 = x^{1/3}$.

$$x^{1/3} + x^{1/6} - 2 = 0$$

$$w^2 + w - 2 = 0 \qquad \text{Let } w = x^{1/6}$$

$$(w - 1)(w + 2) = 0 \qquad \text{Factor}$$

$$w - 1 = 0 \quad \text{or} \quad w + 2 = 0 \qquad \text{Zero-Product Property}$$

$$w = 1 \qquad w = -2 \qquad \text{Solve}$$

$$x^{1/6} = 1 \qquad x^{1/6} = -2 \qquad w = x^{1/6}$$

$$x = 1^6 = 1 \qquad x = (-2)^6 = 64 \qquad \text{Take the 6th power}$$

From *Check Your Answers* we see that $x = 1$ is a solution but $x = 64$ is not. The only solution is $x = 1$.

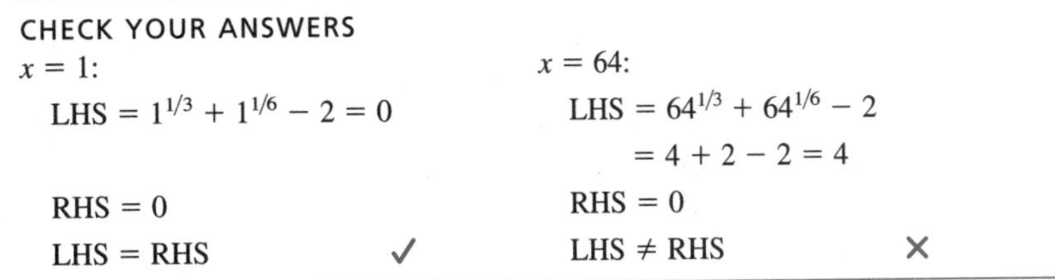

CHECK YOUR ANSWERS

$x = 1$:

LHS $= 1^{1/3} + 1^{1/6} - 2 = 0$

RHS $= 0$

LHS = RHS ✓

$x = 64$:

LHS $= 64^{1/3} + 64^{1/6} - 2$
$= 4 + 2 - 2 = 4$

RHS $= 0$

LHS ≠ RHS ✗

■

APPLICATIONS

Many real-life problems can be modeled with the types of equations we have studied in this section.

EXAMPLE 8 ■ Dividing a Lottery Jackpot

A group of people come forward to claim a $1,000,000 lottery jackpot, which the winners are to share equally. Before the jackpot is divided, three more winning ticket holders show up. As a result, each person's share is reduced by $75,000. How many winners were in the original group?

SOLUTION

We are asked for the number of people in the original group. So let

Identify the variable

$$x = \text{number of winners in the original group}$$

We translate the information in the problem as follows:

Express all unknown quantities in terms of the variable

In Words	In Algebra
Number of winners in original group	x
Number of winners in final group	$x + 3$
Winnings per person, originally	$\dfrac{1{,}000{,}000}{x}$
Winnings per person, finally	$\dfrac{1{,}000{,}000}{x+3}$

Now we set up the model.

Set up the model

winnings per person, originally	−	\$75,000	=	winnings per person, finally

$$\frac{1{,}000{,}000}{x} - 75{,}000 = \frac{1{,}000{,}000}{x+3}$$

Solve

$$1{,}000{,}000(x+3) - 75{,}000x(x+3) = 1{,}000{,}000x \qquad \text{Multiply by LCD } x(x+3)$$

$$40(x+3) - 3x(x+3) = 40x \qquad \text{Divide by 25,000}$$

$$x^2 + 3x - 40 = 0 \qquad \text{Expand, simplify, and divide by 3}$$

$$(x+8)(x-5) = 0 \qquad \text{Factor}$$

$$x + 8 = 0 \quad \text{or} \quad x - 5 = 0 \qquad \text{Zero-Product Property}$$

$$x = -8 \qquad x = 5 \qquad \text{Solve}$$

Since we can't have a negative number of people, we conclude that there were five winners in the original group.

CHECK YOUR ANSWER

$$\text{winnings per person, originally} = \frac{\$1{,}000{,}000}{5} = \$200{,}000$$

$$\text{winnings per person, finally} = \frac{\$1{,}000{,}000}{8} = \$125{,}000$$

$$\$200{,}000 - \$75{,}000 = \$125{,}000 \quad ✓$$

■

EXAMPLE 9 ■ Energy Expended in Bird Flight

Ornithologists have determined that some species of birds tend to avoid flights over large bodies of water during daylight hours, because air generally rises over land and falls over water in the daytime, so flying over water requires more energy. A bird is released from point A on an island, 5 mi from B, the nearest point

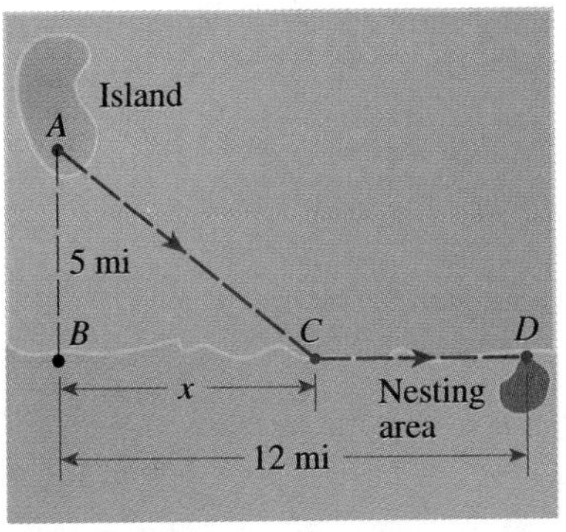

FIGURE 1

on a straight shoreline. The bird flies to a point C on the shoreline and then flies along the shoreline to its nesting area D, as shown in Figure 1. Suppose the bird has 170 kcal of energy reserves. It uses 10 kcal/mi flying over land and 14 kcal/mi flying over water.

(a) Where should the point C be located so that the bird uses exactly 170 kcal of energy during its flight?

(b) Does the bird have enough energy reserves to fly directly from A to D?

SOLUTION

Identify the variable

(a) We are asked to find the location of C. So let

$$x = \text{the distance from } B \text{ to } C$$

From the figure, and from the fact that

$$\text{energy used} = \text{energy per mile} \times \text{miles flown}$$

we determine the following:

Express all unknown quantities in terms of the variable

In Words	In Algebra	
Distance from B to C	x	
Distance flown over water (from A to C)	$\sqrt{x^2 + 25}$	Pythagorean Theorem
Distance flown over land (from B to C)	$12 - x$	
Energy used over water	$14\sqrt{x^2 + 25}$	
Energy used over land	$10(12 - x)$	

Now we set up the model.

Set up the model

$$\text{total energy used} = \text{energy used over water} + \text{energy used over land}$$

$$170 = 14\sqrt{x^2 + 25} + 10(12 - x)$$

To solve this equation, we eliminate the square root by first bringing all other terms to the left of the equal sign and then squaring each side.

Solve

$$170 - 10(12 - x) = 14\sqrt{x^2 + 25} \quad \text{Isolate the square-root term on RHS}$$

$$50 + 10x = 14\sqrt{x^2 + 25} \quad \text{Simplify LHS}$$

$$(50 + 10x)^2 = (14)^2(x^2 + 25) \quad \text{Square each side}$$

$$2500 + 1000x + 100x^2 = 196x^2 + 4900 \quad \text{Expand}$$

$$0 = 96x^2 - 1000x + 2400 \quad \text{Move all terms to RHS}$$

This equation could be factored, but because the numbers are so large it is easier to use the quadratic formula:

$$x = \frac{1000 \pm \sqrt{(-1000)^2 - 4(96)(2400)}}{2(96)}$$

$$= \frac{1000 \pm 280}{192} = 6\tfrac{2}{3} \quad \text{or} \quad 3\tfrac{3}{4}$$

Point C should be either $6\frac{2}{3}$ mi or $3\frac{3}{4}$ mi from B so that the bird uses exactly 170 kcal of energy during its flight.

(b) By the Pythagorean Theorem, the length of the route directly from A to D is $\sqrt{5^2 + 12^2} = 13$ mi, so the energy the bird requires for that route is $14 \times 13 = 182$ kcal. This is more energy than the bird has available, so it can't use this route. ■

3.5 EXERCISES

1–46 ■ Find all real solutions of the equation.

1. $x^4 = 64x^2$

2. $x^5 = 64x^2$

3. $x^3 - 6x = 0$

4. $x^6 - 16x^2 = 0$

5. $x^3 - 3x^2 + 2x = 0$

6. $x^4 - x^3 - 2x^2 = 0$

7. $x^4 + 4x^3 + 2x^2 = 0$

8. $(x - 2)^5 - 9(x - 2)^3 = 0$

9. $x^3 - 5x^2 - 2x + 10 = 0$

10. $2x^3 + x^2 - 18x - 9 = 0$

11. $x^3 - x^2 + x - 1 = x^2 + 1$

12. $7x^3 - x + 1 = x^3 + 3x^2 + x$

13. $\dfrac{1}{x-1} + \dfrac{1}{x+2} = \dfrac{5}{4}$

14. $\dfrac{10}{x} - \dfrac{12}{x-3} + 4 = 0$

15. $\dfrac{x^2}{x+100} = 50$

16. $1 + \dfrac{2x}{(x+3)(x+4)} = \dfrac{2}{x+3} + \dfrac{4}{x+4}$

17. $\dfrac{x+5}{x-2} = \dfrac{5}{x+2} + \dfrac{28}{x^2-4}$

18. $\dfrac{x}{2x+7} - \dfrac{x+1}{x+3} = 1$

19. $\dfrac{1}{x-1} - \dfrac{2}{x^2} = 0$

20. $\dfrac{x + \frac{2}{x}}{3 + \frac{4}{x}} = 5x$

21. $(x + 5)^2 - 3(x + 5) - 10 = 0$

22. $\left(\dfrac{x+1}{x}\right)^2 + 4\left(\dfrac{x+1}{x}\right) + 3 = 0$

23. $\left(\dfrac{1}{x+1}\right)^2 - 2\left(\dfrac{1}{x+1}\right) - 8 = 0$

24. $\left(\dfrac{x}{x+2}\right)^2 = \dfrac{4x}{x+2} - 4$

25. $x^4 - 13x^2 + 40 = 0$

26. $x^4 - 5x^2 + 4 = 0$

27. $2x^4 + 4x^2 + 1 = 0$

28. $x^6 - 2x^3 - 3 = 0$

29. $x^{4/3} - 5x^{2/3} + 6 = 0$

30. $\sqrt{x} - 3\sqrt[4]{x} - 4 = 0$

31. $4(x + 1)^{1/2} - 5(x + 1)^{3/2} + (x + 1)^{5/2} = 0$

32. $x^{1/2} + 3x^{-1/2} = 10x^{-3/2}$

33. $x^{1/2} - 3x^{1/3} = 3x^{1/6} - 9$

34. $x - 5\sqrt{x} + 6 = 0$

35. $\dfrac{1}{x^3} + \dfrac{4}{x^2} + \dfrac{4}{x} = 0$

36. $4x^{-4} - 16x^{-2} + 4 = 0$

37. $\sqrt{2x+1} + 1 = x$

38. $x - \sqrt{9 - 3x} = 0$

39. $\sqrt{5 - x} + 1 = x - 2$

40. $2x + \sqrt{x + 1} = 8$

41. $\sqrt{\sqrt{x - 5} + x} = 5$

42. $\sqrt[3]{4x^2 - 4x} = x$

43. $x^2\sqrt{x + 3} = (x + 3)^{3/2}$

44. $\sqrt{11 - x^2} - \dfrac{2}{\sqrt{11 - x^2}} = 1$

45. $\sqrt{x + \sqrt{x + 2}} = 2$

46. $\sqrt{1 + \sqrt{x + \sqrt{2x + 1}}} = \sqrt{5 + \sqrt{x}}$

47–56 ■ Find all solutions, real and complex, of the equation.

47. $x^3 = 1$

48. $x^4 - 16 = 0$

49. $x^3 + x^2 + x = 0$

50. $x^4 + x^3 + x^2 + x = 0$

51. $x^4 - 6x^2 + 8 = 0$

52. $x^3 + 3x^2 + 9x + 27 = 0$

53. $x^6 - 9x^3 + 8 = 0$

54. $x^6 + 9x^4 - 4x^2 - 36 = 0$

55. $\sqrt{x^2 + 1} + \dfrac{8}{\sqrt{x^2 + 1}} = \sqrt{x^2 + 9}$

56. $1 - \sqrt{x^2 + 7} = 6 - x^2$

57. A social club charters a bus at a cost of \$900 to take a group of members on an excursion to Atlantic City. At the last minute, five people in the group decide not to go. This raises the transportation cost per person by \$2. How many people originally intended to take the trip?

58. A group of friends decides to buy a vacation home for \$120,000, sharing the cost equally. If they could find one more person to join them, each person's contribution would drop by \$6000. How many people are in the group?

59. A large pond is stocked with fish. The fish population P is modeled by the formula $P = 3t + 10\sqrt{t} + 140$, where t is the number of days since the fish were first introduced into the pond. How many days will it take for the fish population to reach 500?

60. If F is the focal length of a convex lens and an object is placed at a distance x from the lens, then its image will be at a distance y from the lens, where F, x, and y are related by the *lens equation*

$$\frac{1}{F} = \frac{1}{x} + \frac{1}{y}$$

Suppose that a lens has a focal length of 4.8 cm, and that the image of an object is 4 cm closer to the lens than the object itself. How far from the lens is the object?

61. A large plywood box has a volume of 180 ft^3. Its length is 9 ft greater than its height, and its width is 4 ft less than its height. What are the dimensions of the box?

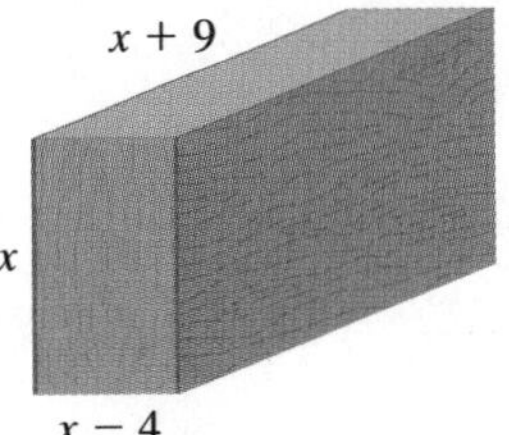

62. A jeweler has three small solid spheres made of gold, of radius 2 mm, 3 mm, and 4 mm. He decides to melt these down and make just one sphere out of them. What will the radius of this larger sphere be?

63. The town of Foxton lies 10 mi north of an abandoned east-west road that runs through Grimley, as shown in the figure. The point on the abandoned road closest to Foxton is 40 mi from Grimley. County officials are about to build a new road connecting the two towns. They have determined that restoring the old road would cost \$100,000 per mile, while building a new road would cost \$200,000 per mile. How much of the abandoned road should be used (as indicated in the figure) if the officials intend to spend exactly \$6.8 million? Would it cost less than this amount to build a new road connecting the towns directly?

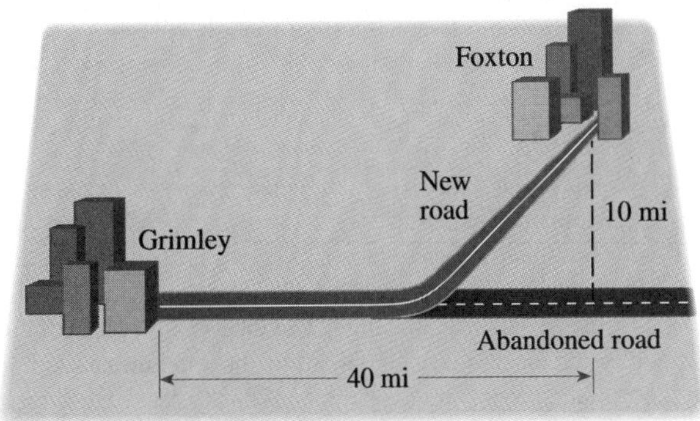

64. A boardwalk is parallel to and 210 ft inland from a straight shoreline. A sandy beach lies between the boardwalk and the shoreline. A man is standing on the boardwalk, exactly 750 ft across the sand from his beach umbrella, which is right at the shoreline. The man walks 4 ft/s on the boardwalk and 2 ft/s on the sand. How far should he walk on the boardwalk before veering off onto the sand if he wishes to reach his umbrella in exactly 4 min 45 s?

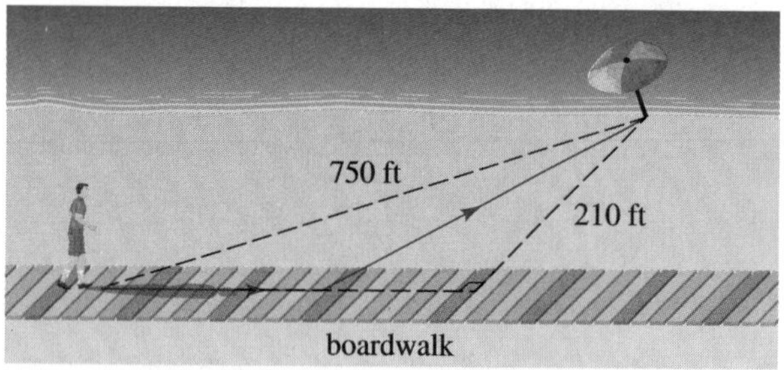

65. Grain is falling from a chute onto the ground, forming a conical pile whose diameter is always three times its height. How high is the pile (to the nearest hundredth of a foot) when it contains 1000 ft^3 of grain?

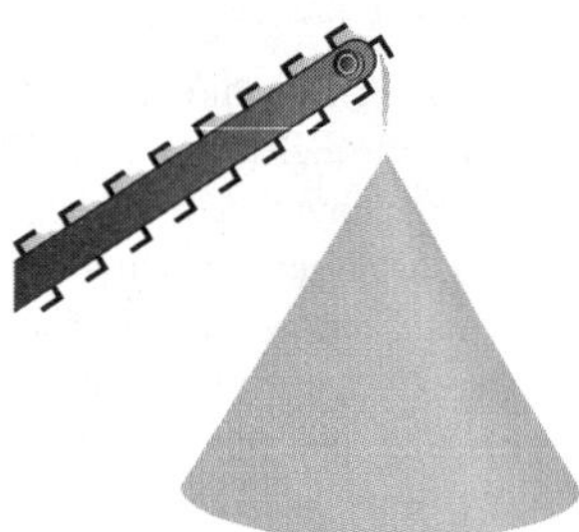

66. A spherical tank has a capacity of 750 gallons. Using the fact that one gallon is about 0.1337 ft^3, find the radius of the tank (to the nearest hundredth of a foot).

67. A city lot has the shape of a right triangle whose hypotenuse is 7 ft longer than one of the other sides. The perimeter of the lot is 392 ft. How long is each side of the lot?

68. Two television monitors sitting beside each other on a shelf in an appliance store have the same screen height. One has a conventional screen, which is 5 in. wider than it is high. The other has a wider, high-definition screen, which is 2.4 times as wide as it is high. The diagonal measure of the wider screen is 14 in. more than the diagonal measure of the smaller. What are the heights of the screens? (Assume that the height is greater than 10 in.)

69. One method for determining the depth of a well is to drop a stone into it and then measure the time it takes until the splash is heard. If d is the depth of the well (in feet) and t_1 the time (in seconds) it takes for the stone to fall, then $d = 16t_1^2$, so $t_1 = \sqrt{d}/4$. Now if t_2 is the time it takes for the sound to travel back up, then $d = 1090t_2$ because the speed of sound is 1090 ft/s. So $t_2 = d/1090$. Thus, the total time elapsed between dropping the stone and hearing the splash is $t_1 + t_2 = \sqrt{d}/4 + d/1090$. How deep is the well if this total time is 3 s?

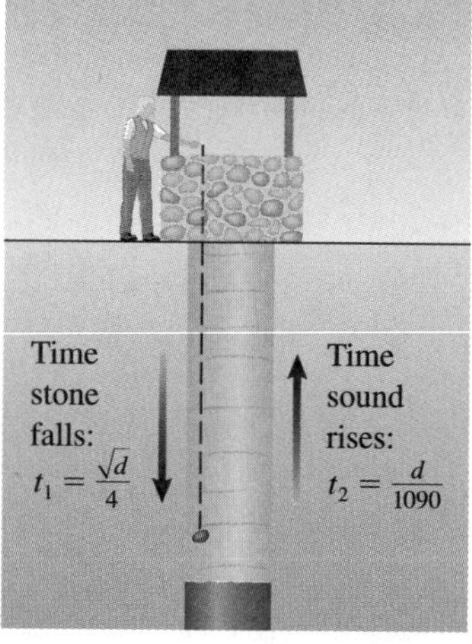

70–73 ■ Solve the equation for the variable x. The constants a and b represent positive real numbers.

70. $x^4 + 5ax^2 + 4a^2 = 0$

71. $a^3x^3 + b^3 = 0$

72. $\sqrt{x + a} + \sqrt{x - a} = \sqrt{2}\sqrt{x + 6}$

73. $\sqrt{x} + a\sqrt[3]{x} + b\sqrt[6]{x} + ab = 0$

DISCOVERY • DISCUSSION

74. **Solving an Equation in Different Ways** We have learned several different ways to solve an equation in this section. Some equations can be tackled by more than one method. For example, the equation $x - \sqrt{x} - 2 = 0$ is of quadratic type: We can solve it by letting $\sqrt{x} = u$ and $x = u^2$, and factoring. Or we could solve for $\sqrt{x}$, square, and then solve the resulting quadratic equation. Solve the following equations using both methods indicated, and show that you get the same final answers.

(a) $x - \sqrt{x} - 2 = 0$ quadratic type; solve for the radical and square

(b) $\dfrac{12}{(x-3)^2} + \dfrac{10}{x-3} + 1 = 0$ quadratic type; multiply by LCD

3.6 LINEAR INEQUALITIES

Some problems in algebra lead to **inequalities** instead of equations. An inequality looks just like an equation, except that in the place of the equal sign is one of the symbols, $<$, $>$, $\leqslant$, or $\geqslant$. An inequality is **linear** if each term is a constant or a multiple of the variable. Here is an example of a linear inequality:

$$4x + 7 \leqslant 19$$

x	$4x + 7 \leqslant 19$
1	$11 \leqslant 19$ ✓
2	$15 \leqslant 19$ ✓
3	$19 \leqslant 19$ ✓
4	$23 \leqslant 19$ ✗
5	$27 \leqslant 19$ ✗

The table in the margin shows that some numbers satisfy the inequality and some numbers don't.

To **solve** an inequality that contains a variable means to find all values of the variable that make the inequality true. Unlike an equation, an inequality generally has infinitely many solutions, which form an interval or a union of intervals on the real line. The following illustration shows how an inequality differs from its corresponding equation:

		Solution	Graph
Equation:	$4x + 7 = 19$	$x = 3$	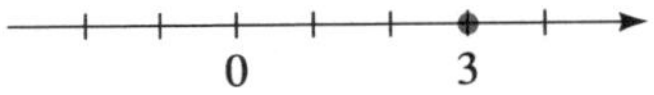
Inequality:	$4x + 7 \leqslant 19$	$x \leqslant 3$	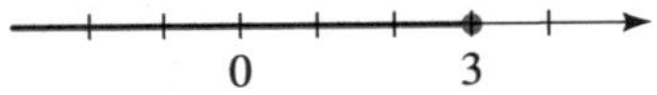

To solve inequalities, we use the following rules to isolate the variable on one side of the inequality sign. These rules tell us when two inequalities are *equivalent* (the symbol $\Leftrightarrow$ means "is equivalent to"). In these rules the symbols A, B, and C stand for real numbers or algebraic expressions. Here we state the rules for inequalities involving the symbol $\leqslant$, but they apply to all four inequality symbols.

RULES FOR INEQUALITIES

Rule	Description
1. $A \leqslant B \quad \Leftrightarrow \quad A + C \leqslant B + C$	**Adding** the same quantity to each side of an inequality gives an equivalent inequality.
2. $A \leqslant B \quad \Leftrightarrow \quad A - C \leqslant B - C$	**Subtracting** the same quantity from each side of an inequality gives an equivalent inequality.
3. If $C > 0$, then $A \leqslant B \quad \Leftrightarrow \quad CA \leqslant CB$	**Multiplying** each side of an inequality by the same *positive* quantity gives an equivalent inequality.
4. If $C < 0$, then $A \leqslant B \quad \Leftrightarrow \quad CA \geqslant CB$	**Multiplying** each side of an inequality by the same *negative* quantity *reverses the direction* of the inequality.
5. If $A > 0$ and $B > 0$, then $A \leqslant B \quad \Leftrightarrow \quad \dfrac{1}{A} \geqslant \dfrac{1}{B}$	**Taking reciprocals** of each side of an inequality involving *positive* quantities *reverses the direction* of the inequality.
6. If $A \leqslant B$ and $C \leqslant D$, then $A + C \leqslant B + D$	Inequalities can be added.

Pay special attention to Rules 3 and 4. Rule 3 says that we can multiply (or divide) each side of an inequality by a *positive* number, but Rule 4 says that if we multiply each side of an inequality by a *negative* number, then we reverse the direction of the inequality. For example, if we start with the inequality

$$3 < 5$$

and multiply by 2, we get

$$6 < 10$$

but if we multiply by -2, we get

$$-6 > -10$$

EXAMPLE 1 ■ Solving a Linear Inequality

Solve the inequality $3x - 1 \leq 5$.

SOLUTION

As when solving an equation, we need to simplify this inequality to an equivalent one in which the variable x is isolated on one side of the $\leq$ sign.

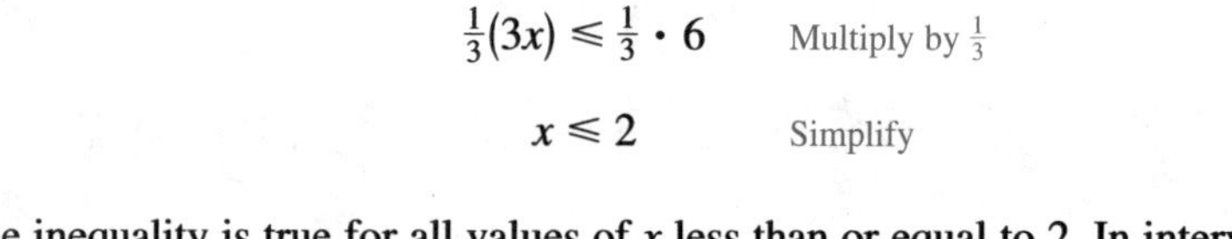

$$3x - 1 + 1 \leq 5 + 1 \qquad \text{Add 1 to each side}$$

$$3x \leq 6 \qquad \text{Simplify}$$

$$\tfrac{1}{3}(3x) \leq \tfrac{1}{3} \cdot 6 \qquad \text{Multiply by } \tfrac{1}{3}$$

$$x \leq 2 \qquad \text{Simplify}$$

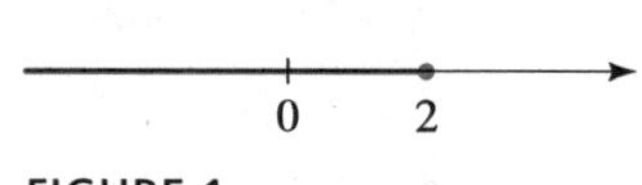

FIGURE 1

The inequality is true for all values of x less than or equal to 2. In interval notation this is the set $(-\infty, 2]$. The solution set is sketched in Figure 1. ■

EXAMPLE 2 ■ Solving a Linear Inequality

Solve the inequality $3x < 9x + 4$ and sketch the solution set.

SOLUTION

$$3x < 9x + 4$$

$$3x - 9x < 9x + 4 - 9x \qquad \text{Subtract } 9x$$

$$-6x < 4 \qquad \text{Simplify}$$

$$\left(-\tfrac{1}{6}\right)(-6x) > -\tfrac{1}{6}(4) \qquad \text{Multiply by } -\tfrac{1}{6} \text{ (or divide by } -6\text{)}$$

$$x > -\tfrac{2}{3} \qquad \text{Simplify}$$

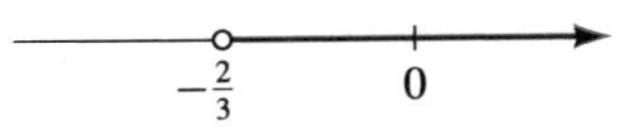

FIGURE 2

The solution set consists of all numbers greater than $-\frac{2}{3}$. In other words the solution of the inequality is the interval $\left(-\frac{2}{3}, \infty\right)$. It is graphed in Figure 2. ■

EXAMPLE 3 ■ Solving a Pair of Simultaneous Inequalities

Solve the inequalities $4 \leq 3x - 2 < 13$.

SOLUTION

The solution set consists of all values of x that satisfy both inequalities. Using Rules 1 and 3, we see that the following inequalities are equivalent:

$$4 \leq 3x - 2 < 13$$

$$6 \leq 3x < 15 \qquad \text{Add 2}$$

$$2 \leq x < 5 \qquad \text{Divide by 3}$$

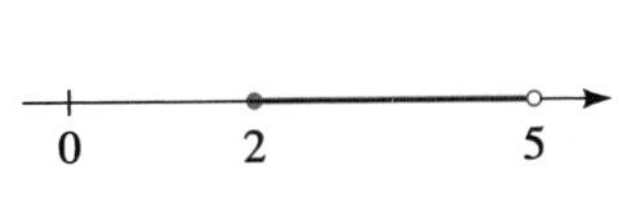

FIGURE 3

Therefore, the solution set is $[2, 5)$, as shown in Figure 3. ■

SOLVING INEQUALITIES GRAPHICALLY

Just as we solve equations graphically, we can also solve inequalities graphically.

EXAMPLE 4 ■ Solving an Inequality Graphically

Solve the inequality $3x - 2 \geq \frac{1}{2}x + 3$.

SOLUTION

We begin by graphing the equations $y_1 = 3x - 2$ and $y_2 = \frac{1}{2}x + 3$ in the same viewing rectangle, as shown in Figure 4. We are looking for those values of x for which y_1 is greater than or equal to y_2; that is, we must determine where the graph of y_1 lies above the graph of y_2. From the graph we see that this occurs for $x \geq 2$, so the solution is the interval $[2, \infty)$. ■

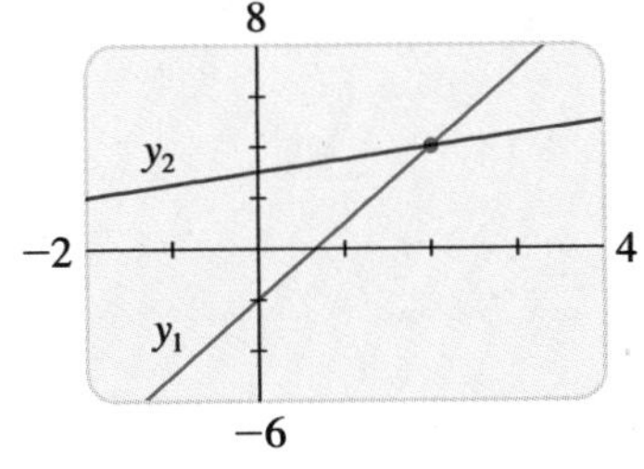

FIGURE 4
$y_1 = 3x - 2$
$y_2 = \frac{1}{2}x + 3$

Another way to solve the inequality in Example 4 is to move all terms to one side of the inequality symbol to get $\frac{5}{2}x - 5 \geq 0$, then graph $y = \frac{5}{2}x - 5$ and determine where the graph lies above the x-axis. Try this yourself to see that you get the same answer.

MODELING WITH INEQUALITIES

Modeling real-life problems frequently leads to inequalities because we are often interested in determining when one quantity is more (or less) than another.

EXAMPLE 5 ■ Carnival Tickets

A carnival has two plans for tickets.

Plan A: \$5 entrance fee and 25¢ each ride

Plan B: \$2 entrance fee and 50¢ each ride

How many rides would you have to take for plan A to be less expensive than plan B?

SOLUTION

We are asked for the number of rides for which plan A is less expensive than plan B. So let

Identify the variable

$$x = \text{number of rides}$$

The information in the problem may be organized as follows

Express all unknown quantities in terms of the variable

In Words	In Algebra
Number of rides	x
Cost with Plan A	$5 + 0.25x$
Cost with Plan B	$2 + 0.50x$

Now we set up the model.

Set up the model

$$\text{cost with plan A} < \text{cost with plan B}$$

$$5 + 0.25x < 2 + 0.50x$$

Solve

$$3 + 0.25x < 0.50x \qquad \text{Subtract 2}$$

$$3 < 0.25x \qquad \text{Subtract } 0.25x$$

$$12 < x \qquad \text{Divide by } 0.25$$

So if you plan to take *more than* 12 rides, plan A is less expensive. ■

EXAMPLE 6 ■ Relationship between Fahrenheit and Celsius Scales

The instructions on a box of film indicate that the box should be stored at a temperature between 5°C and 30°C. What range of temperatures does this correspond to on the Fahrenheit scale?

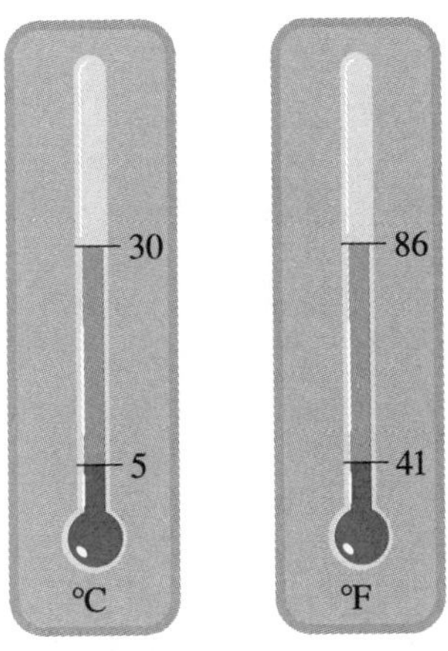

SOLUTION

The relationship between degrees Celsius (C) and degrees Fahrenheit (F) is given by the equation $C = \frac{5}{9}(F - 32)$. Expressing the statement on the box in terms of inequalities, we have

$$5 < C < 30$$

so the corresponding Fahrenheit temperatures satisfy the inequalities

$$5 < \tfrac{5}{9}(F - 32) < 30$$

$$\tfrac{9}{5} \cdot 5 < F - 32 < \tfrac{9}{5} \cdot 30 \qquad \text{Multiply by } \tfrac{9}{5}$$

$$9 < F - 32 < 54 \qquad \text{Simplify}$$

$$9 + 32 < F < 54 + 32 \qquad \text{Add 32}$$

$$41 < F < 86 \qquad \text{Simplify}$$

The film should be stored at a temperature between 41°F and 86°F. ■

3.6 EXERCISES

1–6 ■ Let $S = \{-1, 0, \frac{1}{2}, \sqrt{2}, 2\}$. Use substitution to determine which elements of S satisfy the inequality.

1. $x + 1 \geq 0$

2. $x - 2 < 0$

3. $2x + 10 > 8$

4. $4x + 1 \leq 2x$

5. $\dfrac{1}{x} \leq \dfrac{1}{2}$

6. $x^2 + 2 < 4$

7–36 ■ Solve the inequality. Express the solution using interval notation and graph the solution set.

7. $3x \leq 12$

8. $-4x > 16$

9. $20 < -4x$

10. $-1 \leq 7x$

11. $2x - 5 > 3$

12. $3x + 11 < 5$

13. $7 - x \geq 5$

14. $5 - 3x \leq -16$

15. $2x + 1 < 0$

16. $0 < 5 - 2x$

17. $3x + 11 \leq 6x + 8$

18. $6 - x \geq 2x + 9$

19. $1 - x \leq 2$

20. $4 - 3x \geq 6$

21. $\frac{1}{2}x - \frac{2}{3} > 2$

22. $\frac{2}{5}x + 1 < \frac{1}{5} - 2x$

23. $\frac{3}{2} - \frac{1}{2}x \leq 1 + \frac{1}{4}x$

24. $3\left(1 - \frac{3}{4}x\right) > 5 - \frac{1}{4}x$

25. $4 - 3x \leq -(1 + 8x)$

26. $2(7x - 3) \leq 12x + 16$

27. $2 \leq x + 5 < 4$

28. $5 \leq 3x - 4 \leq 14$

29. $-1 < 2x - 5 < 7$

30. $1 < 3x + 4 \leq 16$

31. $0 \leq 1 - x < 1$

32. $-5 \leq 3 - 2x \leq 9$

33. $-2 < 8 - 2x \leq -1$

34. $-3 \leq 3x + 7 \leq \frac{1}{2}$

35. $\dfrac{2}{3} \geq \dfrac{2x - 3}{12} > \dfrac{1}{6}$

36. $-\dfrac{1}{2} < \dfrac{4 - 3x}{5} \leq \dfrac{1}{4}$

37. A car rental company offers two plans for renting a car.
Plan A: \$30 per day and 10¢ per mile
Plan B: \$50 per day with free unlimited mileage
For what range of miles will plan B save you money?

38. A telephone company offers two long-distance plans.
Plan A: \$25 per month and 5¢ per minute
Plan B: \$5 per month and 12¢ per minute
For how many minutes of long-distance calls would plan B be financially advantageous?

39. Use the relationship between C and F given in Example 6 to find the interval on the Fahrenheit scale corresponding to the temperature range $20 \leq C \leq 30$.

40. What interval on the Celsius scale corresponds to the temperature range $50 \leq F \leq 95$?

41. A charter airline finds that on its Saturday flights from Philadelphia to London, all 120 seats will be sold if the ticket price is \$200. However, for each \$3 increase in ticket price, the number of seats sold decreases by one.

(a) Find a formula for the number of seats sold if the ticket price is P dollars.

(b) Over a certain period, the number of seats sold for this flight ranged between 90 and 115. What was the corresponding range of ticket prices?

42. As dry air moves upward, it expands and in so doing cools at a rate of about 1°C for each 100 m rise, up to about 12 km.

(a) If the ground temperature is 20°C, write a formula for the temperature at height h.

(b) What range of temperatures can be expected if a plane takes off and reaches a maximum height of 5 km?

43. It is estimated that the annual cost of driving a certain new car is given by the formula

$$C = 0.35m + 2200$$

where m represents the number of miles driven per year and C is the cost in dollars. Jane has purchased such a car, and decides to budget between \$6400 and \$7100 for next year's driving costs. What is the corresponding range of miles that she can drive her new car?

44. A coffee merchant sells a customer 3 lb of Hawaiian Kona at \$6.50 per pound. His scale is accurate to within ±0.03 lb. By how much could the customer have been overcharged or undercharged because of possible inaccuracy in the scale?

45–46 ■ Solve the inequality for x, assuming that a, b, and c are positive constants.

45. $a(bx - c) \geqslant bc$

46. $a \leqslant bx + c < 2a$

47–48 ■ Solve the inequality for x, assuming that a, b, and c are negative constants.

47. $ax + b < c$

48. $\dfrac{ax + b}{c} \leqslant b$

49. Show that if $a < b$, then $a < \dfrac{a + b}{2} < b$.

50. Show that if $0 < a < b$, then $a^2 < b^2$.

51. Suppose that a, b, c, and d are positive numbers such that

$$\frac{a}{b} < \frac{c}{d}$$

Show that

$$\frac{a}{b} < \frac{a + c}{b + d} < \frac{c}{d}$$

DISCOVERY • DISCUSSION

52. Ordering the Points in the Cartesian Plane Given any two different real numbers x and y, we know that either $x < y$ or $y < x$. We can think of the real numbers as being *totally ordered* by the relation $<$, in the sense that we can always tell which number in any pair is smaller—that is, which comes earlier on the real line. The words in a dictionary are also totally ordered. Given two words, the one whose first letter comes earlier in the alphabet will be earlier in the dictionary; if the first letter is the same, we apply the alphabetic rule to the second letter, and so on. This is called the *lexicographic* order. Thus, we can say that cat < dog, man < mountain, or pantry < pastry. A similar **lexicographic order** can be defined for the points in the coordinate plane. We'll say that $(a, b) < (c, d)$ if $a < c$, or if $a = c$ and $b < d$. Thus, $(1, 4) < (3, 2)$ and $(5, 0) < (5, 1)$.

(a) List the following points in lexicographic order, from smallest to largest.

$(2, 5)$ $(4, -2)$ $(2, 0)$ $(0, 2)$ $(2, -3)$ $(-1, 4)$ $(-2, 5)$

(b) Graph the following sets of points in the coordinate plane.

$$S = \{(x, y) \mid (0, 3) < (x, y)\}$$
$$T = \{(x, y) \mid (-2, 2) < (x, y) \leqslant (4, 0)\}$$

3.7 NONLINEAR INEQUALITIES

To solve inequalities involving squares and other powers of the variable, we use factoring, together with the following principle.

THE SIGN OF A PRODUCT OR QUOTIENT

If a product or a quotient has an *even* number of *negative* factors, then its value is *positive*.

If a product or a quotient has an *odd* number of *negative* factors, then its value is *negative*.

EXAMPLE 1 ■ A Quadratic Inequality

Solve the inequality $x^2 - 5x + 6 \leq 0$.

SOLUTION

First we factor the left side.

$$(x - 2)(x - 3) \leq 0$$

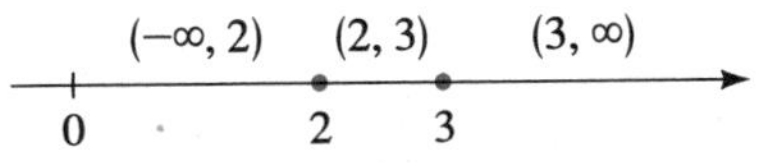

$(-\infty, 2)$ $(2, 3)$ $(3, \infty)$

0 2 3

FIGURE 1

We know that the corresponding equation $(x - 2)(x - 3) = 0$ has the solutions 2 and 3. As shown in Figure 1, the numbers 2 and 3 divide the real line into three intervals: $(-\infty, 2)$, $(2, 3)$, and $(3, \infty)$. On each of these intervals we determine the signs of the factors using **test values**. We choose a number inside each interval and check the sign of the factors $x - 2$ and $x - 3$ at the value selected. For instance, if we use the test value $x = 1$ for the interval $(-\infty, 2)$ shown in Figure 2, then substitution in the factors $x - 2$ and $x - 3$ gives

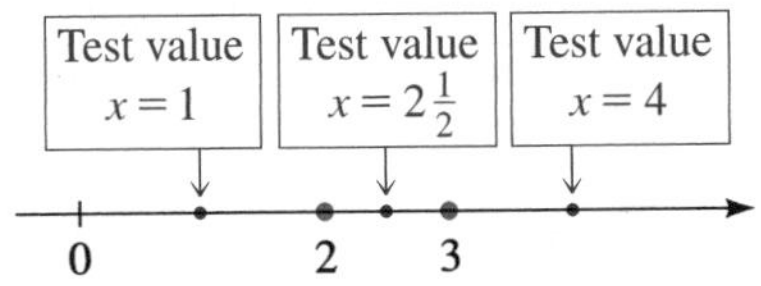

Test value $x = 1$ | Test value $x = 2\frac{1}{2}$ | Test value $x = 4$

0 2 3

FIGURE 2

$$x - 2 = 1 - 2 = -1 < 0$$

and

$$x - 3 = 1 - 3 = -2 < 0$$

So both factors are negative on this interval. (The factors $x - 2$ and $x - 3$ change sign only at 2 and 3, respectively, so they maintain their signs over the length of each interval. That is why using a single test value on each interval is sufficient.)

Using the test values $x = 2\frac{1}{2}$ and $x = 4$ for the intervals $(2, 3)$ and $(3, \infty)$ (see Figure 2), respectively, we construct the sign table on page 188. The final row of the table is obtained from the fact that the expression in the last row is the product of the two factors.

Interval	$(-\infty, 2)$	$(2, 3)$	$(3, \infty)$
Sign of $x - 2$ Sign of $x - 3$	$-$ $-$	$+$ $-$	$+$ $+$
Sign of $(x - 2)(x - 3)$	$+$	$-$	$+$

If you prefer, you can represent this information on a real number line, as in the following sign diagram. The vertical lines indicate the points at which the real line is divided into intervals:

		2		3	
Sign of $x - 2$	$-$		$+$		$+$
Sign of $x - 3$	$-$		$-$		$+$
Sign of $(x - 2)(x - 3)$	$+$		$-$		$+$

We read from the table or the diagram that $(x - 2)(x - 3)$ is negative on the interval $(2, 3)$. Thus, the solution of the inequality $(x - 2)(x - 3) \leqslant 0$ is

$$\{x \mid 2 \leqslant x \leqslant 3\} = [2, 3]$$

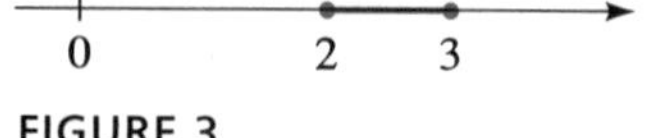

FIGURE 3

We have included the endpoints 2 and 3 because we seek values of x such that the product is either less than *or equal to* zero. The solution is illustrated in Figure 3. ■

We use the following guidelines to solve an inequality that can be factored.

GUIDELINES FOR SOLVING NONLINEAR INEQUALITIES

1. MOVE ALL TERMS TO ONE SIDE. If necessary, rewrite the inequality so that all nonzero terms appear on one side of the inequality sign. If the nonzero side of the inequality involves quotients, bring them to a common denominator.

2. FACTOR. Factor the nonzero side of the inequality.

3. FIND THE INTERVALS. Use the factorization to find all solutions of the equation corresponding to the given inequality. These numbers will divide the real line into intervals. List the intervals determined by these numbers.

4. MAKE A TABLE OR DIAGRAM. Use test values to make a table or diagram of the signs of each factor on each interval. In the last row of the table determine the sign of the product (or quotient) of these factors.

5. SOLVE. Determine the solution of the inequality from the last row of the sign table. Be sure to check whether the inequality is satisfied by some or all of the endpoints of the intervals (this may happen if the inequality involves $\leqslant$ or $\geqslant$).

The factoring technique described in these guidelines works only if all nonzero terms appear on one side of the inequality symbol. If the inequality is not written in this form, first rewrite it, as indicated in Step 1. This technique is illustrated in the examples that follow.

EXAMPLE 2 ■ Solving an Inequality by Factoring

Solve $x^2 + 3x > 4$.

SOLUTION

First we move all nonzero terms to one side of the inequality sign and factor the resulting expression.

$$x^2 + 3x - 4 > 0 \qquad \text{Subtract 4}$$

$$(x - 1)(x + 4) > 0 \qquad \text{Factor}$$

The corresponding equation $(x - 1)(x + 4) = 0$ has solutions -4 and 1, so we obtain the three intervals $(-\infty, -4)$, $(-4, 1)$, and $(1, \infty)$. On each interval the product keeps a constant sign, as shown in the following diagram. The signs in the diagram could be obtained by using, for example, the test values -5, 0, and 2, respectively, for the three intervals.

	$(-\infty, -4)$	$(-4, 1)$	$(1, \infty)$
Sign of $x - 1$	−	−	+
Sign of $x + 4$	−	+	+
Sign of $(x - 1)(x + 4)$	+	−	+

We read from the diagram that the solution set is

$$\{x \mid x < -4 \text{ or } x > 1\} = (-\infty, -4) \cup (1, \infty)$$

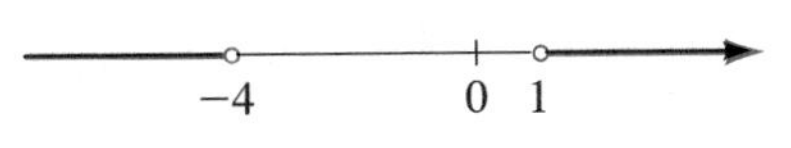

FIGURE 4

The solution is illustrated in Figure 4. ■

In the next example we solve an inequality involving a quotient instead of a product.

EXAMPLE 3 ■ An Inequality Involving a Quotient

Solve $\dfrac{1 + x}{1 - x} \geq 1$.

SOLUTION

First we move all nonzero terms to the left side, and then we simplify using a common denominator.

$$\frac{1+x}{1-x} \geqslant 1$$

$$\frac{1+x}{1-x} - 1 \geqslant 0 \qquad \text{Subtract 1}$$

$$\frac{1+x}{1-x} - \frac{1-x}{1-x} \geqslant 0 \qquad \text{Common denominator } 1-x$$

$$\frac{1+x-1+x}{1-x} \geqslant 0 \qquad \text{Combine the fractions}$$

$$\frac{2x}{1-x} \geqslant 0 \qquad \text{Simplify}$$

The numerator is zero when $x = 0$ and the denominator is zero when $x = 1$, so we construct the following sign diagram using these values to define intervals on the real line.

	$(-\infty, 0)$	0 to 1	$(1, \infty)$
Sign of $2x$	$-$	$+$	$+$
Sign of $1-x$	$+$	$+$	$-$
Sign of $\frac{2x}{1-x}$	$-$	$+$	$-$

From the diagram we see that the solution set is $\{x \mid 0 \leqslant x < 1\} = [0, 1)$. We include the endpoint 0 because the original inequality requires the quotient to be greater than *or equal to* 1. However, we do not include the other endpoint 1, since the quotient in the inequality is not defined at 1. Always check the endpoints of solution intervals to determine whether they satisfy the original inequality.

The solution set $[0, 1)$ is illustrated in Figure 5. ■

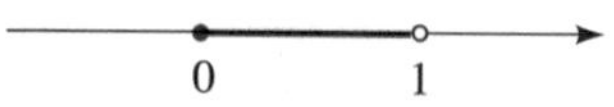

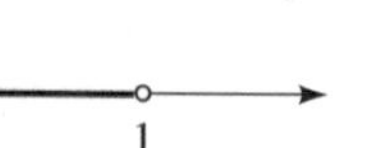

FIGURE 5

EXAMPLE 4 ■ Solving an Inequality with Three Factors

Solve the inequality $x < \dfrac{2}{x-1}$.

SOLUTION

After moving all nonzero terms to one side of the inequality, we use a common denominator to combine the terms.

$$x - \frac{2}{x-1} < 0 \qquad \text{Subtract } \frac{2}{x-1}$$

$$\frac{x(x-1)}{x-1} - \frac{2}{x-1} < 0 \qquad \text{Common denominator } x-1$$

$$\frac{x^2 - x - 2}{x-1} < 0 \qquad \text{Combine the fractions}$$

$$\frac{(x+1)(x-2)}{x-1} < 0 \qquad \text{Factor the numerator}$$

The factors in this quotient change sign at -1, 1, and 2, so we must examine the intervals $(-\infty, -1)$, $(-1, 1)$, $(1, 2)$, and $(2, \infty)$. Using test values, we get the following sign diagram.

	$(-\infty, -1)$	$(-1, 1)$	$(1, 2)$	$(2, \infty)$
Sign of $x + 1$	$-$	$+$	$+$	$+$
Sign of $x - 2$	$-$	$-$	$-$	$+$
Sign of $x - 1$	$-$	$-$	$+$	$+$
Sign of $\frac{(x+1)(x-2)}{x-1}$	$-$	$+$	$-$	$+$

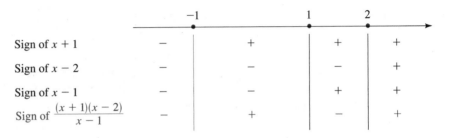

Since the quotient must be negative, the solution is

$$(-\infty, -1) \cup (1, 2)$$

as illustrated in Figure 6. ■

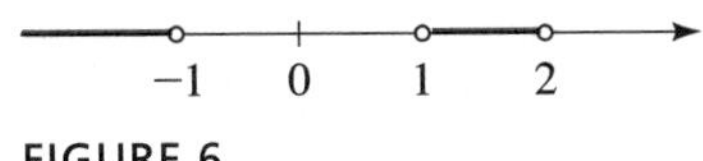

FIGURE 6

SOLVING INEQUALITIES GRAPHICALLY

Graphs give us another way to understand the solution of an inequality.

EXAMPLE 5 ■ Solving an Inequality Graphically

Solve the inequality $3.7x^2 + 1.3x - 1.9 \leq 2.0 - 1.4x$.

SOLUTION

We graph the equations

$$y_1 = 3.7x^2 + 1.3x - 1.9$$

$$y_2 = 2.0 - 1.4x$$

in the same viewing rectangle in Figure 7. We are interested in those values of x

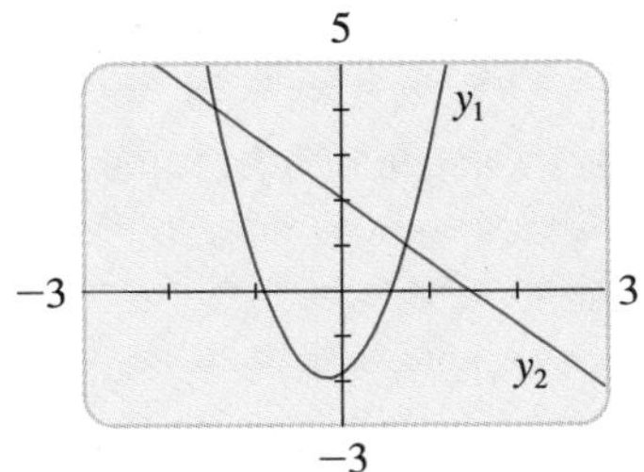

FIGURE 7
$y_1 = 3.7x^2 + 1.3x - 1.9$
$y_2 = 2.0 - 1.4x$

for which $y_1 \leqslant y_2$; these are points for which the graph of y_2 lies on or above the graph of y_1. To determine the appropriate interval, we look for the x-coordinates of points where the graphs intersect. We conclude that the solution is (approximately) the interval $[-1.45, 0.72]$. ■

EXAMPLE 6 ■ Solving an Inequality Graphically

Solve the inequality $x^3 - 5x^2 \geqslant -8$.

SOLUTION

We write the inequality as

$$x^3 - 5x^2 + 8 \geqslant 0$$

and then graph the equation

$$y = x^3 - 5x^2 + 8$$

in the viewing rectangle $[-6, 6]$ by $[-15, 15]$, as shown in Figure 8. The solution of the inequality consists of those intervals on which the graph lies on or above the x-axis. By moving the cursor to the x-intercepts we find that, correct to one decimal place, the solution is $[-1.1, 1.5] \cup [4.6, \infty)$.

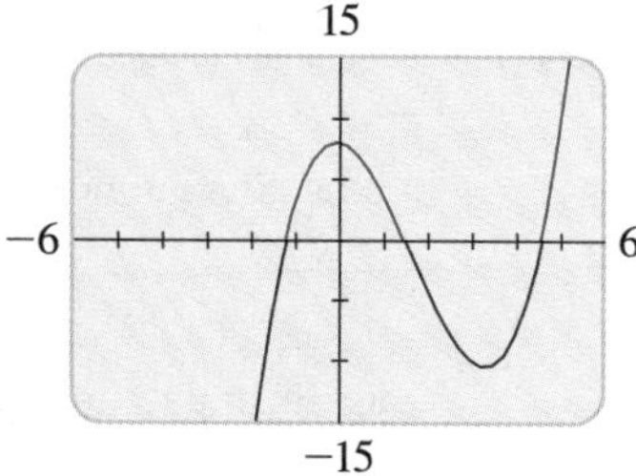

FIGURE 8
$y = x^3 - 5x^2 + 8$

■

MODELING WITH NONLINEAR INEQUALITIES

Now let's consider applied problems that lead to nonlinear inequalities.

EXAMPLE 7 ■ The Path of a Projectile

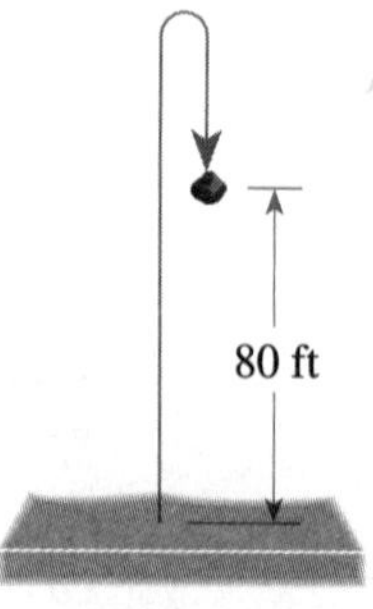

A stone thrown straight up into the air at 96 ft/s reaches a height of h feet after t seconds, where h and t are related by the formula

$$h = 96t - 16t^2$$

During what time interval will the stone be at least 80 ft above the ground?

SOLUTION

We are asked to determine the times t for which $h \geqslant 80$. Thus, we must solve the inequality:

$$96t - 16t^2 \geqslant 80$$

$$6t - t^2 \geqslant 5 \qquad \text{Divide by 6}$$

$$0 \geqslant t^2 - 6t + 5 \qquad \text{Subtract } 6t - t^2$$

$$0 \geqslant (t - 1)(t - 5) \qquad \text{Factor}$$

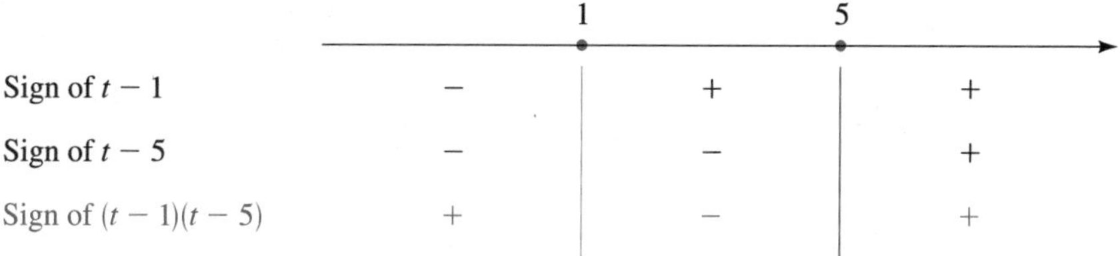

1
5
Sign of $t - 1$ − + +
Sign of $t - 5$ − − +
Sign of $(t - 1)(t - 5)$ + − +

From the sign diagram the solution is the interval $[1, 5]$. Thus, the stone is at least 80 ft high between 1 and 5 seconds after it is thrown. ■

EXAMPLE 8 ■ Concert Tickets

A group of students decide to attend a concert. The cost of chartering a bus to take them to the concert is \$450, which is to be shared equally among the students. The concert promoters offer discounts to groups arriving by bus. Tickets normally cost \$50 each but are reduced by 10¢ per ticket for each person in the group (up to the maximum capacity of the bus). How many students must be in the group for the total cost per student to be less than \$54?

SOLUTION

We are asked for the number of students in the group. So let

Identify the variable

$$x = \text{number of students in the group}$$

The information in the problem may be organized as follows.

Express all unknown quantities in terms of the variable

In Words	In Algebra
Number of students in group	x
Bus cost per student	$\frac{450}{x}$
Ticket cost per student	$50 - 0.10x$

Now we set up the model.

Set up the model

$$\boxed{\text{bus cost per student}} + \boxed{\text{ticket cost per student}} < 54$$

$$\frac{450}{x} + (50 - 0.10x) < 54$$

Solve

$$\frac{450}{x} - 4 - 0.10x < 0 \qquad \text{Subtract 54}$$

$$\frac{450 - 4x - 0.10x^2}{x} < 0 \qquad \text{Common denominator}$$

$$\frac{4500 - 40x - x^2}{x} < 0 \qquad \text{Multiply by 10}$$

$$\frac{(90 + x)(50 - x)}{x} < 0 \qquad \text{Factor numerator}$$

	$(-\infty, -90)$	$(-90, 0)$	$(0, 50)$	$(50, \infty)$
Sign of $90 + x$	−	+	+	+
Sign of $50 - x$	+	+	+	−
Sign of x	−	−	+	+
Sign of $\frac{(90 + x)(50 - x)}{x}$	+	−	+	−

The sign diagram shows that the solution of the inequality is the set $(-90, 0) \cup (50, \infty)$. Because we cannot have a negative number of students, it follows that the group must have more than 50 students for the total cost per person to be less than \$54. ■

3.7 EXERCISES

1–40 ■ Solve the inequality. Express the solution in interval form and graph the solution set on the real number line.

1. $(x - 2)(x - 5) > 0$

2. $(3x + 1)(x - 1) \geq 0$

3. $x^2 - 3x - 18 \leq 0$

4. $x^2 + 5x + 6 > 0$

5. $2x^2 + x \geq 1$

6. $x^2 < x + 2$

7. $3x^2 - 3x < 2x^2 + 4$

8. $5x^2 + 3x \geq 3x^2 + 2$

9. $x^2 > 3(x + 6)$

10. $x^2 + 2x > 3$

11. $x^2 < 4$

12. $x^2 \geq 9$

13. $-2x^2 \leq 4$

14. $(x + 2)(x - 1)(x - 3) \leq 0$

15. $x(x^2 - 4) \geq 0$

16. $x^3 > x$

17. $\dfrac{x - 3}{x + 1} \geq 0$

18. $\dfrac{2x + 6}{x - 2} < 0$

19. $\dfrac{4x}{2x + 3} > 2$

20. $-2 < \dfrac{x + 1}{x - 3}$

21. $\dfrac{2x+1}{x-5} \leqslant 3$

22. $\dfrac{3+x}{3-x} \geqslant 1$

23. $\dfrac{4}{x} < x$

24. $\dfrac{x}{x+1} > 3x$

25. $\dfrac{x^2-4}{x^2+4} \geqslant 0$

26. $\dfrac{x}{x+2} \leqslant \dfrac{1}{x}$

27. $1 + \dfrac{2}{x+1} \leqslant \dfrac{2}{x}$

28. $\dfrac{3}{x-1} - \dfrac{4}{x} \geqslant 1$

29. $\dfrac{1}{1-x} \leqslant \dfrac{3}{x}$

30. $\dfrac{(x-1)^2}{(x+1)(x+2)} > 0$

31. $\dfrac{x^2+2x-3}{x^2-7x+6} > 0$

32. $\dfrac{x^2-16}{x^4-16} < 0$

33. $\dfrac{x-3}{2x+5} \geqslant 1$

34. $\dfrac{1}{x} + \dfrac{1}{x+1} < \dfrac{2}{x+2}$

35. $\dfrac{6}{x-1} - \dfrac{6}{x} \geqslant 1$

36. $\dfrac{x}{2} \geqslant \dfrac{5}{x+1} + 4$

37. $\dfrac{x+2}{x+3} < \dfrac{x-1}{x-2}$

38. $\dfrac{1}{x+1} + \dfrac{1}{x+2} \leqslant 0$

39. $x^4 > x^2$

40. $x^5 > x^2$

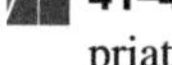 **41–48** ■ Find the solutions of the inequality by drawing appropriate graphs. State each answer correct to two decimals.

41. $x^2 - 3x - 10 \leqslant 0$

42. $0.5x^2 + 0.875x \leqslant 0.25$

43. $x^3 + 11x \leqslant 6x^2 + 6$

44. $16x^3 + 24x^2 > -9x - 1$

45. $x^{1/3} < x$

46. $\sqrt{0.5x^2+1} \leqslant 2|x|$

47. $(x+1)^2 < (x-1)^2$

48. $(x+1)^2 \leqslant x^3$

49. Using calculus it can be shown that if a ball is thrown upward with an initial velocity of 16 ft/s from the top of a building 128 ft high, then its height h above the ground t seconds later will be

$$h = 128 + 16t - 16t^2$$

During what time interval will the ball be at least 32 ft above the ground?

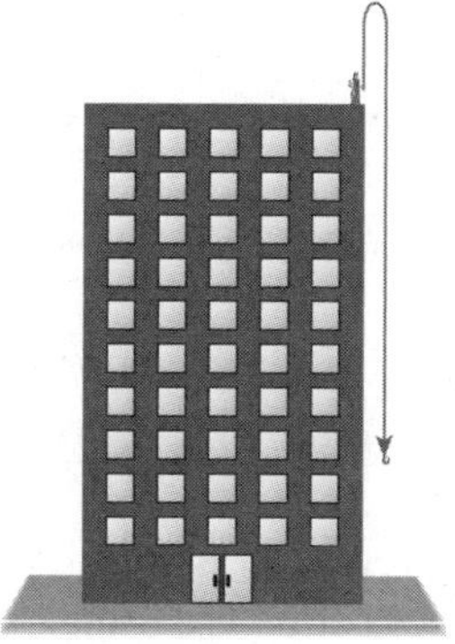

50. The gravitational force F exerted by the earth on an object having a mass of 100 kg is given by the equation

$$F = \frac{4{,}000{,}000}{d^2}$$

where d is the distance (in km) of the object from the center of the earth, and the force F is measured in newtons (N). For what distances will the gravitational force exerted by the earth on this object be between 0.0004 N and 0.01 N?

51. In the vicinity of a bonfire, the temperature T in °C at a distance of x meters from the center of the fire was given by

$$T = \frac{600{,}000}{x^2 + 300}$$

At what range of distances from the fire center was the temperature less than 500°C?

52. The gas mileage g (measured in mi/gal) for a particular vehicle, driven at v mi/h, is given by the formula $g = 10 + 0.9v - 0.01v^2$, as long as v is between 10 mi/h and 75 mi/h. For what range of speeds is the vehicle's mileage 30 mi/gal or better?

53. For a certain model of car the distance d required to stop the vehicle if it is traveling at v mi/h is given by the following formula:

$$d = v + \frac{v^2}{20}$$

where d is measured in feet. Kerry wants her stopping distance not to exceed 240 ft. At what range of speeds can she travel?

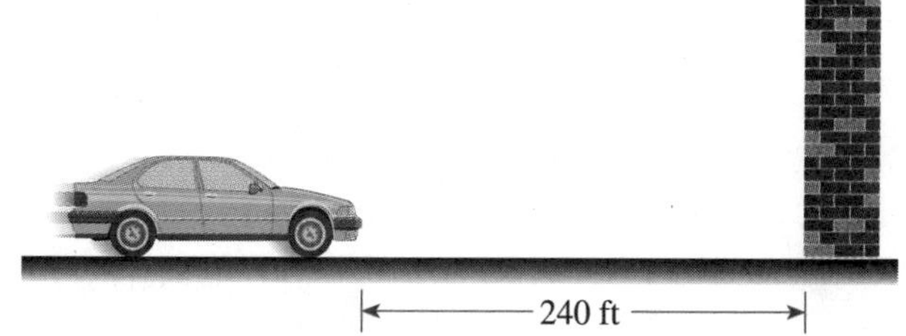

54. If a manufacturer sells x units of a certain product, his revenue R and cost C (in dollars) are given by:

$$R = 20x$$

$$C = 2000 + 8x + 0.0025x^2$$

Use the fact that

$$\text{profit} = \text{revenue} - \text{cost}$$

to determine how many units he should sell to enjoy a profit of at least \$2400.

55. A gardener has 120 ft of deer-resistant fence. She wants to enclose a rectangular vegetable garden in her backyard, and she wants the area enclosed to be at least 800 ft^2. What range of values is possible for the length of her garden?

56. A riverboat theater offers bus tours to groups on the following basis. Hiring the bus costs the group \$360, to be shared equally by the group members. Theater tickets, normally \$30 each, are discounted by 25¢ times the number of people in the group. How many members must be in the group so that the cost of the theater tour (bus fare plus theater ticket) is less than \$39 per person?

57–60 ■ Determine the values of the variable for which the expression is defined as a real number.

57. $\sqrt{16 - 9x^2}$

58. $\sqrt{3x^2 - 5x + 2}$

59. $\left(\dfrac{1}{x^2 - 5x - 14}\right)^{1/2}$

60. $\sqrt[4]{\dfrac{1 - x}{2 + x}}$

61. Solve $\dfrac{x^2 + (a - b)x - ab}{x + c} \leq 0$, where $0 < a < b < c$.

DISCOVERY • DISCUSSION

62. Do Powers Preserve Order? If $a < b$, is $a^2 < b^2$? (Check both positive and negative values for a and b.) If $a < b$, is $a^3 < b^3$? Based on your observations, state a general rule about the relationship between a^n and b^n when $a < b$ and n is a positive integer.

63. What's Wrong Here? It is tempting to try to solve an inequality like an equation. For instance, we might try to solve $1 < 3/x$ by multiplying both sides by x, to get $x < 3$, so the solution would be $(-\infty, 3)$. But that's wrong; for example, $x = -1$ lies in this interval but does not satisfy the original inequality. Explain why this method doesn't work (think about the *sign* of x). Then solve the inequality correctly.

3.8 ABSOLUTE VALUE

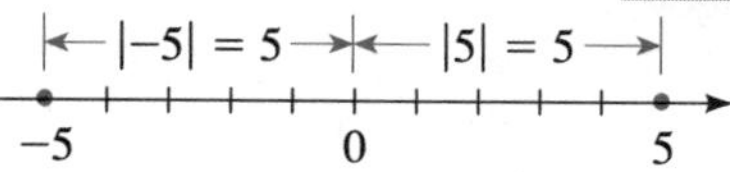

FIGURE 1

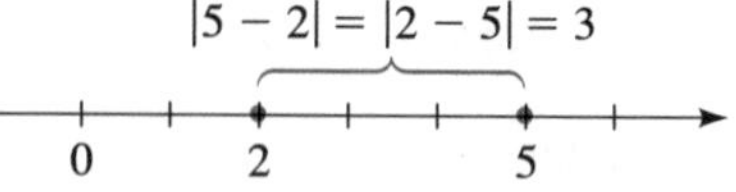

FIGURE 2

Recall from Section 1.2 that the absolute value of a number a is given by

$$|a| = \begin{cases} a & \text{if } a \geq 0 \\ -a & \text{if } a < 0 \end{cases}$$

and that it represents the distance from a to the origin on the real number line (see Figure 1). More generally, $|x - a|$ is the distance between x and a on the real number line (see Figure 2).

ABSOLUTE VALUE EQUATIONS

We use the following property to solve equations that involve absolute value.

$$|x| = C \quad \text{is equivalent to} \quad x = \pm C$$

This property says that to solve an absolute value equation, we must solve *two* separate equations. For example, the equation $|x| = 5$ is equivalent to the two equations $x = 5$ and $x = -5$.

EXAMPLE 1 ■ Solving an Absolute Value Equation

Solve the equation $|2x - 5| = 3$.

SOLUTION

The equation $|2x - 5| = 3$ is equivalent to two equations:

$$2x - 5 = 3 \quad \text{or} \quad 2x - 5 = -3$$

$$2x = 8 \qquad 2x = 2 \qquad \text{Add 5}$$

$$x = 4 \qquad x = 1 \qquad \text{Divide by 2}$$

The solutions are 1 and 4. ■

CHECK YOUR ANSWERS

$x = 1$:

$\text{LHS} = |2 \cdot 1 - 5|$

$= |-3| = 3 = \text{RHS}$ ✓

$x = 4$:

$\text{LHS} = |2 \cdot 4 - 5|$

$= |3| = 3 = \text{RHS}$ ✓

EXAMPLE 2 ■ Solving an Absolute Value Equation

Solve the equation $3|x - 7| + 5 = 14$.

SOLUTION

We first isolate the absolute value on one side of the equal sign.

$$3|x - 7| + 5 = 14$$

$$3|x - 7| = 9 \qquad \text{Subtract 5}$$

$$|x - 7| = 3 \qquad \text{Divide by 3}$$

$$x - 7 = 3 \quad \text{or} \quad x - 7 = -3 \qquad \text{Take cases}$$

$$x = 10 \qquad x = 4 \qquad \text{Add 7}$$

The solutions are 4 and 10. ■

ABSOLUTE VALUE INEQUALITIES

We use the following properties to solve inequalities that involve absolute value.

PROPERTIES OF ABSOLUTE VALUE INEQUALITIES

Inequality	Equivalent form	Graph
1. $\lvert x\rvert < c$	$-c < x < c$	$-c$ 0 c
2. $\lvert x\rvert \le c$	$-c \le x \le c$	$-c$ 0 c
3. $\lvert x\rvert > c$	$x < -c$ or $c < x$	$-c$ 0 c
4. $\lvert x\rvert \ge c$	$x \le -c$ or $c \le x$	$-c$ 0 c

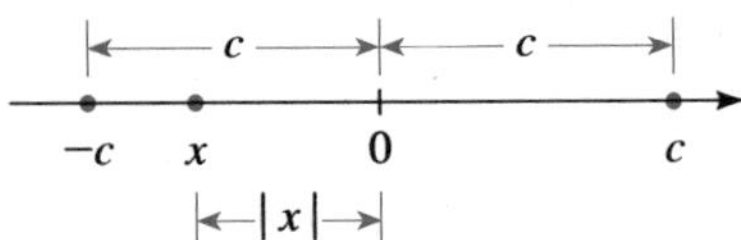

FIGURE 3

These properties can be proved using the definition of absolute value. To prove Property 1, for example, we see that the inequality $|x| < c$ says that the distance from x to 0 is less than c, and from Figure 3 you can see that this is true if and only if x is between c and $-c$.

EXAMPLE 3 ■ Solving an Absolute Value Inequality

Solve the inequality $|x - 5| < 2$.

SOLUTION 1

The inequality $|x - 5| < 2$ is equivalent to

$$-2 < x - 5 < 2 \qquad \text{Property 1}$$

$$3 < x < 7 \qquad \text{Add 5}$$

The solution set is the open interval $(3, 7)$.

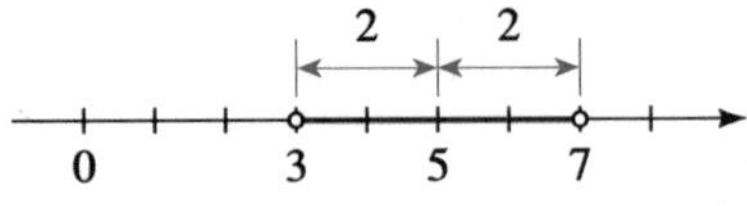

FIGURE 4

SOLUTION 2

Geometrically, the solution set consists of all numbers x whose distance from 5 is less than 2. From Figure 4 we see that this is the interval $(3, 7)$. ■

EXAMPLE 4 ■ Solving an Absolute Value Inequality

Solve the inequality $|3x + 2| \geq 4$.

SOLUTION

By Property 4 the inequality $|3x + 2| \geq 4$ is equivalent to

$$3x + 2 \geq 4 \quad \text{or} \quad 3x + 2 \leq -4$$

$$3x \geq 2 \qquad\qquad 3x \leq -6 \qquad \text{Subtract 2}$$

$$x \geq \tfrac{2}{3} \qquad\qquad x \leq -2 \qquad \text{Divide by 3}$$

So the solution set is

$$\{x \mid x \leq -2 \quad \text{or} \quad x \geq \tfrac{2}{3}\} = (-\infty, -2] \cup \left[\tfrac{2}{3}, \infty\right)$$

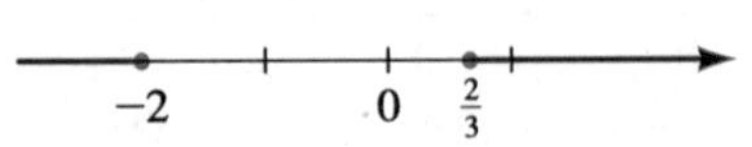

FIGURE 5

The set is graphed in Figure 5. ■

3.8 EXERCISES

1–16 ■ Solve the equation.

1. $|3x| = 15$

2. $4|x| = 24$

3. $|x - 3| = 2$

4. $|2x - 3| = 7$

5. $|x + 4| = 0.5$

6. $|x - 4| = -3$

7. $|4x + 7| = 9$

8. $|\frac{1}{2}x - 2| = 1$

9. $4 - |3x + 6| = 1$

10. $|5 - 2x| + 6 = 14$

11. $3|x + 5| + 6 = 15$

12. $20 + |2x - 4| = 15$

13. $8 + 5|\frac{1}{3}x - \frac{5}{6}| = 33$

14. $|\frac{3}{5}x + 2| - \frac{1}{2} = 4$

15. $|x - 1| = |3x + 2|$

16. $|x + 3| = |2x + 1|$

17–40 ■ Solve the inequality. Express the answer using interval notation.

17. $|x| < 2$

18. $|x| \geqslant 4$

19. $|x - 5| \leqslant 3$

20. $|x - 9| > 9$

21. $|x + 1| \geqslant 1$

22. $|x + 4| \leqslant 0$

23. $|x + 5| \geqslant 2$

24. $|x + 1| \geqslant 3$

25. $|2x - 3| \leqslant 0.4$

26. $|5x - 2| < 6$

27. $\left|\frac{x - 2}{3}\right| < 2$

28. $\left|\frac{x + 1}{2}\right| \geqslant 4$

29. $|x + 6| < 0.001$

30. $|x - a| < d$

31. $4|x + 2| - 3 < 13$

32. $3 - |2x + 4| \leqslant 1$

33. $8 - |2x - 1| \geqslant 6$

34. $7|x + 2| + 5 > 4$

35. $\frac{1}{2}|4x + \frac{1}{3}| > \frac{5}{6}$

36. $2|\frac{1}{2}x + 3| + 3 \leqslant 51$

37. $1 \leqslant |x| \leqslant 4$

38. $0 < |x - 5| \leqslant \frac{1}{2}$

39. $\frac{1}{|x + 7|} > 2$

40. $\frac{1}{|2x - 3|} \leqslant 5$

41–44 ■ Solve the equation graphically. Express the answers correct to two decimals.

41. $|3x| + x = 16$

42. $|x^2 - 16| = 9$

43. $|2x - 1| = \frac{1}{2}x^2$

44. $||x - 2| - 10| = 5$

45–48 ■ Solve the inequality graphically. Express the answers using interval notation, with the interval endpoints accurate to one decimal.

45. $|2x - 4| \leqslant |3 - x|$

46. $|x^2 - 25| \geqslant |x^2 - 4|$

47. $x^2 - 2x + \frac{5}{4} > |x - 1|$

48. $||x - 3| - 12| < 6$

DISCOVERY • DISCUSSION

49. Using Distances to Solve Absolute Value Inequalities Recall that $|a - b|$ is the distance between a and b on the number line. For any number x, what do $|x - 1|$ and $|x - 3|$ represent? Use this interpretation to solve the inequality $|x - 1| < |x - 3|$ geometrically. In general, if $a < b$, what is the solution of the inequality $|x - a| < |x - b|$?

3 REVIEW

CONCEPT CHECK

1. What is a solution or root of an equation?

2. How do you solve an equation
(a) algebraically?
(b) graphically?

3. Write the general form of each type of equation.
(a) A linear equation
(b) A quadratic equation

4. What are the three ways to solve a quadratic equation?

5. State the Zero-Product Property.

6. Describe the process of completing the square.

7. State the quadratic formula.

8. What is the discriminant of a quadratic equation?

9. For the complex number $a + bi$, what is the real part and what is the imaginary part?

10. Describe how to perform the following arithmetic operations with complex numbers.
(a) Addition (b) Subtraction
(c) Multiplication (d) Division

11. Define the complex conjugate of the complex number $a + bi$ and state one of its properties.

12. How do you solve an equation involving radicals? Why is it important to check your answers when solving equations of this type?

13. What is an equation of quadratic type? Give an example, and use it to show how you solve such equations.

14. State the rules for working with inequalities.

15. How do you solve
(a) a linear inequality?
(b) a nonlinear inequality?

16. (a) How do you solve an equation involving an absolute value?
(b) How do you solve an inequality involving an absolute value?

EXERCISES

1–4 ■ Solve the equation algebraically and graphically.

1. $5x + 11 = 36$
2. $3 - x = 5 + 3x$
3. $4x^2 = 49$
4. $5x^4 - 16 = 0$

5–24 ■ Find all real solutions of the equation.

5. $\dfrac{x+1}{x-1} = \dfrac{3x}{3x-6}$
6. $(x + 2)^2 = (x - 4)^2$
7. $x^2 - 9x + 14 = 0$
8. $x^2 + 24x + 144 = 0$
9. $2x^2 + x = 1$
10. $3x^2 + 5x - 2 = 0$
11. $4x^3 - 25x = 0$
12. $x^3 - 2x^2 - 5x + 10 = 0$
13. $3x^2 + 4x - 1 = 0$
14. $x^2 - 3x + 9 = 0$
15. $\dfrac{1}{x} + \dfrac{2}{x-1} = 3$
16. $\dfrac{x}{x-2} + \dfrac{1}{x+2} = \dfrac{8}{x^2-4}$
17. $x^4 - 8x^2 - 9 = 0$
18. $x - 4\sqrt{x} = 32$
19. $x^{-1/2} - 2x^{1/2} + x^{3/2} = 0$
20. $\left(1 + \sqrt{x}\right)^2 - 2\left(1 + \sqrt{x}\right) - 15 = 0$
21. $|x - 7| = 4$
22. $|3x| = 18$
23. $|2x - 5| = 9$
24. $4|3 - x| + 3 = 15$

25–28 ■ Solve the equation graphically.

25. $x^2 - 4x = 2x + 7$
26. $\sqrt{x+4} = x^2 - 5$
27. $x^4 - 9x^2 = x - 9$
28. $\big||x + 3| - 5\big| = 2$

29. The owner of a store sells raisins for \$3.20 per pound and nuts for \$2.40 per pound. He decides to mix the raisins and nuts and sell 50 lb of the mixture for \$2.72 per pound. What quantities of raisins and nuts should he use?

30. Anthony leaves Kingstown at 2:00 P.M. and drives to Queensville, 160 mi distant, at 45 mi/h. At 2:15 P.M. Helen leaves Queensville and drives to Kingstown at 40 mi/h. At what time do they pass each other on the road?

31. A woman cycles 8 mi/h faster than she runs. Every morning she cycles 4 mi and runs $2\frac{1}{2}$ mi, for a total of one hour of exercise. How fast does she run?

32. The approximate distance d (in feet) that drivers travel after noticing that they must come to a sudden stop is given by the following formula, where x is the speed of the car (in mi/h):

$$d = x + \frac{x^2}{20}$$

If a car travels 75 ft before stopping, what must its speed have been before the brakes were applied?

33. The hypotenuse of a right triangle has length 20 cm. The sum of the lengths of the other two sides is 28 cm. Find the lengths of the other two sides of the triangle.

34. Abbie paints twice as fast as Beth and three times as fast as Cathie. If it takes them 60 min to paint a living room with all three working together, how long would it take Abbie if she works alone?

35. A rectangular swimming pool is 8 ft deep everywhere and twice as long as it is wide. If the pool holds 8464 ft³ of water, what are its dimensions?

36. A homeowner wishes to fence in three adjoining garden plots, one for each of her children, as shown in the figure. If each plot is to be 80 ft^2 in area, and she has 88 ft of fencing material at hand, what dimensions should each plot have?

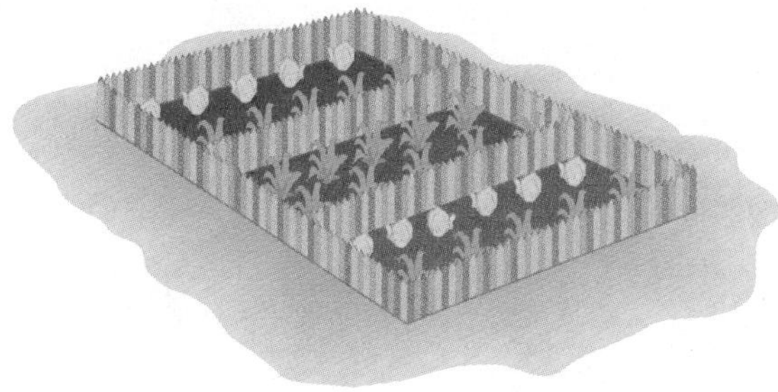

37–46 ■ Evaluate the expression and write the result in the form $a + bi$.

37. $(3 - 5i) - (6 + 4i)$

38. $(-2 + 3i) + \left(\frac{1}{2} - i\right)$

39. $(2 + 7i)(6 - i)$

40. $3(5 - 2i)\dfrac{i}{5}$

41. $\dfrac{2 - 3i}{2 + 3i}$

42. $\dfrac{2 + i}{4 - 3i}$

43. i^{45}

44. $(3 - i)^3$

45. $\left(1 - \sqrt{-3}\right)\left(2 + \sqrt{-4}\right)$

46. $\sqrt{-5} \cdot \sqrt{-20}$

47–54 ■ Find all real and imaginary solutions of the equation.

47. $x^2 + 16 = 0$

48. $x^2 = -12$

49. $x^2 + 6x + 10 = 0$

50. $2x^2 - 3x + 2 = 0$

51. $x^4 - 256 = 0$

52. $x^3 - 2x^2 + 4x - 8 = 0$

53. $x^2 + 4x = (2x + 1)^2$

54. $x^3 = 125$

55–68 ■ Solve the inequality. Express the solution using interval notation and graph the solution set on the real number line.

55. $3x - 2 > -11$

56. $12 - x \geqslant 7x$

57. $-1 < 2x + 5 \leqslant 3$

58. $3 - x \leqslant 2x - 7$

59. $x^2 + 4x - 12 > 0$

60. $x^2 \leqslant 1$

61. $\dfrac{2x + 5}{x + 1} \leqslant 1$

62. $2x^2 \geqslant x + 3$

63. $\dfrac{x - 4}{x^2 - 4} \leqslant 0$

64. $\dfrac{5}{x^3 - x^2 - 4x + 4} < 0$

65. $|x - 5| \leqslant 3$

66. $|x - 4| < 0.02$

67. $|2x + 1| \geqslant 1$

68. $|x - 1| < |x - 3|$
[*Hint:* Interpret the quantities as distances.]

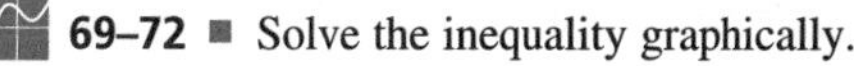

69–72 ■ Solve the inequality graphically.

69. $4x - 3 \geqslant x^2$

70. $x^3 - 4x^2 - 5x > 2$

71. $x^4 - 4x^2 < \frac{1}{2}x - 1$

72. $|x^2 - 16| - 10 \geqslant 0$

73. For what values of x is the algebraic expression defined as a real number?

(a) $\sqrt{24 - x - 3x^2}$

(b) $\dfrac{1}{\sqrt[4]{x - x^4}}$

74. The volume of a sphere is given by $V = \frac{4}{3}\pi r^3$, where r is the radius. Find the interval of values of the radius so that the volume is between 8 ft^3 and 12 ft^3, inclusive.

3 TEST

1. Solve the equation $3x - 7 = \frac{1}{2}x$ by the indicated method. Give reasons for each step in your solution.
   (a) Solve the equation algebraically.
   (b) Solve the equation graphically.

2. Bill drove from Ajax to Bixby at an average speed of 50 mi/h. On the way back, he drove at 60 mi/h. The total trip took $4\frac{2}{5}$ h of driving time. Find the distance between these two cities.

3. Calculate and write the result in the form $a + bi$:
   (a) $(6 - 2i) - (7 - \frac{1}{2}i)$
   (b) $(1 + i)(3 - 2i)$
   (c) $\dfrac{5 + 10i}{3 - 4i}$
   (d) i^{50}
   (e) $\left(2 - \sqrt{-2}\right)\left(\sqrt{8} + \sqrt{-4}\right)$

4. Find all solutions, real and imaginary, of each equation.
   (a) $x^2 - x - 12 = 0$
   (b) $2x^2 + 4x + 3 = 0$
   (c) $\sqrt{3 - \sqrt{x + 5}} = 2$
   (d) $x^{1/2} - 3x^{1/4} + 2 = 0$
   (e) $x^4 - 16x^2 = 0$
   (f) $3|x - 4| - 10 = 0$

5. Solve the equation graphically: $x^3 - 9x - 1 = 0$

6. A rectangular parcel of land is 70 ft longer than it is wide. Each diagonal between opposite corners is 130 ft. What are the dimensions of the parcel?

7. Solve each inequality. Sketch the solution on a real number line, and write the answer using interval notation.
   (a) $-1 \leq 5 - 2x < 10$
   (b) $x(x - 1)(x - 2) > 0$
   (c) $|x - 3| < 2$
   (d) $\dfrac{2x + 5}{x + 1} \leq 1$

8. Solve the inequality graphically: $x^2 - 2 \leq |x + 1|$

9. A bottle of medicine is to be stored at a temperature between 5°C and 10°C. What range does this correspond to on the Fahrenheit scale? [*Note:* The Fahrenheit (F) and Celsius (C) scales satisfy the relation $C = \frac{5}{9}(F - 32)$.]

10. For what values of x is the expression $\sqrt{4x - x^2}$ defined as a real number?

FOCUS ON PROBLEM SOLVING

One way to show that a statement is true is to show that its opposite leads to a contradiction. This type of **indirect reasoning** is an important problem-solving tool. The two examples we give here are problems that were solved more than 2000 years ago but continue to have a profound influence in mathematics.

IRRATIONAL NUMBERS EXIST

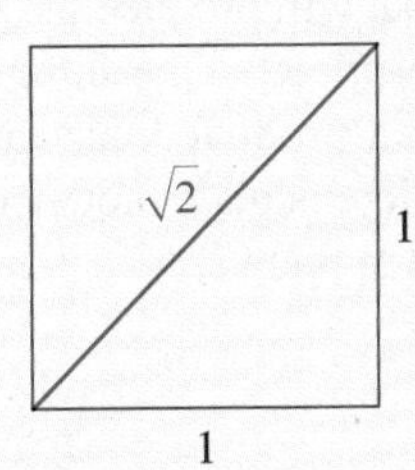

The Pythagoreans were fascinated by the beauty of the natural numbers: 1, 2, 3, ... They hoped that the length of every interval constructed in geometry would be a ratio of two natural numbers so that the natural numbers would describe all of geometry.

But what is the length of the diagonal of a square of side 1? The Pythagoreans knew (by the Pythagorean Theorem) that this length is $\sqrt{2}$. The question was, Is $\sqrt{2}$ the ratio of two natural numbers? One Pythagorean, Hippasus, is reputed to have proved, during a sea voyage, that $\sqrt{2}$ is in fact *not* rational. His colleagues were so upset by this discovery that they hurled him overboard! His proof, nevertheless, is a classic use of indirect reasoning. Here is the proof.

Suppose $\sqrt{2}$ is rational, so that

$$\sqrt{2} = \frac{a}{b}$$

where a and b are natural numbers with no factor in common. Then

$$a = \sqrt{2}\,b \qquad \text{Multiply by } b$$

$$a^2 = 2b^2 \qquad \text{Square each side}$$

This means that a^2 is an even number and therefore a is also an even number, say $a = 2m$. From the preceding equation we get

$$(2m)^2 = 2b^2 \qquad \text{Substitute } a = 2m$$

$$4m^2 = 2b^2$$

$$2m^2 = b^2 \qquad \text{Divide by 2}$$

Thus, b is also even. So a and b have 2 as a common factor, and this contradicts our assumption that a and b have no factor in common. The assumption that $\sqrt{2}$ is rational therefore leads to a contradiction; therefore, $\sqrt{2}$ must be irrational.

THERE ARE INFINITELY MANY PRIMES

A prime number has no factor other than 1 and itself. The first primes are

$$2, 3, 5, 7, 11, 13, 17, 19, 23, 29, 31, \ldots$$

How do we find the next prime? No formula is known that will produce the primes, and no one has found a pattern for the location of the primes. One thing we do know is that infinitely many primes exist. Euclid gave a proof for this fact over 2000 years ago by a brilliant use of indirect reasoning. Here is his famous proof.

Suppose there is only a finite number of primes. We list them as

$$p_1, p_2, p_3, \ldots, p_n$$

Then the number

$$N = p_1 \cdot p_2 \cdot p_3 \cdot \cdots \cdot p_n + 1$$

is not divisible by any prime (Why?) and so is itself prime. But N is not on our list of primes (Why?). This contradicts our assumption that the list contains all the primes. Thus, there are infinitely many primes.

Eratosthenes (circa 276–195 B.C.) was a renowned Greek geographer, mathematician, and astronomer. He accurately calculated the circumference of the earth by an ingenious method. He is most famous, however, for his method for finding primes, now called the *sieve of Eratosthenes*. The method consists of listing the integers, beginning with 2 (the first prime), and then crossing out all the multiples of 2, which are not prime. The next number remaining on the list is 3 (the second prime), so we again cross out all multiples of it. The next remaining number is 5 (the third prime number), and we cross out all multiples of it, and so on. In this way all numbers that are not prime are crossed out, and the remaining numbers are the primes.

	2	3	4	5	6	7	8	9	10
11	12	13	14	15	16	17	18	19	20
21	22	23	24	25	26	27	28	29	30
31	32	33	34	35	36	37	38	39	40
41	42	43	44	45	46	47	48	49	50
51	52	53	54	55	56	57	58	59	60
61	62	63	64	65	66	67	68	69	70
71	72	73	74	75	76	77	78	79	80
81	82	83	84	85	86	87	88	89	90
91	92	93	94	95	96	97	98	99	100

PROBLEMS

1. (a) Prove that $\sqrt{6}$ is an irrational number.

(b) Prove that $\sqrt{2} + \sqrt{3}$ is an irrational number.

2. Prove that it's possible to raise an irrational number to an irrational power and get a rational result. [*Hint:* The number $a = \sqrt{2}^{\sqrt{2}}$ is either rational or irrational. If a is rational, we are done. If a is irrational, consider $a^{\sqrt{2}}$.]

3. The ancient Egyptians considered rectangles whose perimeters and areas are numerically the same to be special. Find all rectangles whose perimeter (in feet) and whose area (in ft^2) are the same integer.

4. Each letter in the following multiplication represents a different digit. Find the value of each letter.

$$\begin{array}{r} ABCDE \\ 4 \\ \hline EDCBA \end{array}$$

LARGE PRIME NUMBERS

The search for large primes fascinates many people. As of this writing, the largest known prime number is

$$2^{3,021,377} - 1$$

It was discovered in 1998 by Roland Clarkson, a 19-year-old student at California State University, Dominguez Hills, using a 200-MHz Pentium computer. In decimal notation this number contains 909,526 digits. If it were written in full, it would occupy about 150 pages of this book. Clarkson was working with a large Internet group known as GIMPS (the Great Internet Mersenne Prime Search). Numbers of the form $2^p - 1$, where p is prime, are called Mersenne numbers and are more easily checked for primality than others. That is why the largest known primes are of this form.

5. Prove that at any party there are two people who know the same number of people. (Assume that if person A knows person B, then B knows A. Assume also that everyone knows himself or herself.)

6. Of nine eggs, eight have exactly the same weight. The ninth egg weighs less than the other eight. How can you determine which is the lighter egg with exactly two weighings on a balance?

7. Of 12 similar coins, one is a counterfeit. It is not known whether the counterfeit coin is lighter or heavier than a genuine coin. Using a balance three times, how can the counterfeit be identified and in the process be determined to be lighter or heavier than a genuine coin?

8. Augustus DeMorgan, the famous 19th century logician, once stated that he was x years old in the year x^2. He died at age 65. In what year did he die?

9. Suppose that each point in the coordinate plane is colored either red or blue. Show that there must always be two points of the same color that are exactly one unit apart.

10. Suppose that each point (x, y) in the plane, both of whose coordinates are rational numbers, represents a tree. If you are standing at the point $(0, 0)$, how far could you see in this forest?

11. A thousand points are graphed in the coordinate plane. Explain why it is possible to draw a straight line in the plane so that half of the points are on one side of the line and half are on the other. [*Hint:* Consider the slopes of the lines determined by each *pair* of points.]

12. Sketch the region in the plane consisting of all points (x, y) such that

$$|x| + |y| \leq 1$$

Bhaskara (born 1114) was an Indian mathematician, astronomer, and astrologer. Among his many accomplishments was an ingenious proof of the Pythagorean Theorem (see Problem 16). His important mathematical book *Lilavati* [*The Beautiful*] consists of algebra problems posed in the form of stories to his daughter Lilavati. Many of the problems begin "Oh beautiful maiden, suppose . . ." The story is told that using astrology, Bhaskara had determined that great misfortune would befall his daughter if she married at any time other than at a certain hour of a certain day. On her wedding day, as she was anxiously watching the water clock, a pearl fell unnoticed from her headdress. It stopped the flow of water in the clock, causing her to miss the opportune moment for marriage. Bhaskara's *Lilavati* was written to console her.

13. Graph the equation

$$x^2y - y^3 - 5x^2 + 5y^2 = 0$$

[*Hint:* Factor.]

14. Find the area of the region between the two concentric circles shown in the figure.

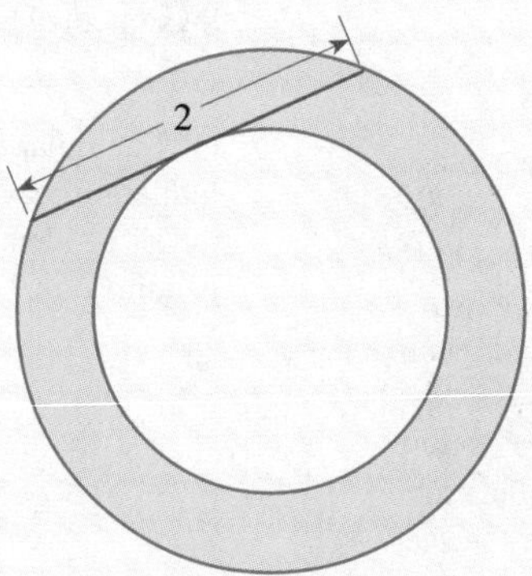

15. In a singles "knock-out" tennis tournament, players are eliminated as soon as they lose a match. A prearranged schedule determines who plays whom initially, and the winner of each match advances to the next round, as shown in the following sample.

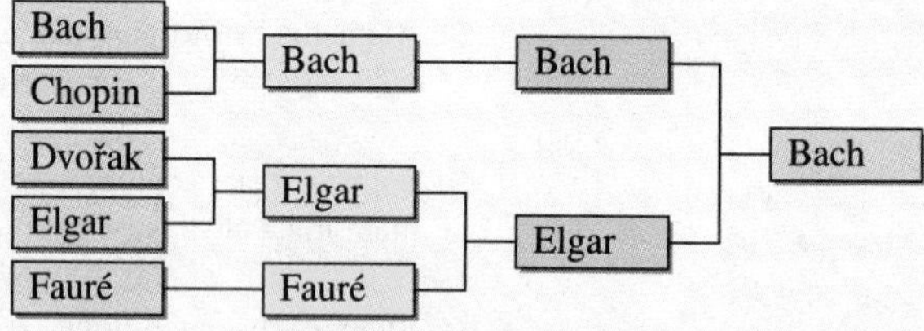

If an odd number of players participates in any round, then the round will have at least one *bye*, that is, a player who doesn't play but advances automatically to the next round. Suppose n players participate in the tournament.

(a) For what values of n is it possible to avoid a bye?

(b) Show that exactly $n - 1$ matches must be played, no matter how the schedule is arranged. [*Hint:* There's an easy way to see this. Consider the losers of matches.]

16. The Indian mathematician Bhaskara sketched the two figures shown here and wrote below them, "Behold!" Explain how his sketches prove the Pythagorean Theorem.

 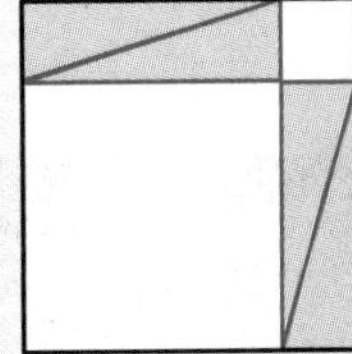

17. If the lengths of the sides of a right triangle, in increasing order, are a, b, and c, show that $a^3 + b^3 < c^3$.

4

FUNCTIONS

A function is a rule that describes how one quantity depends on another. For example, the price of a commodity is a function of demand; the biomass of the rain forest is a function of time.

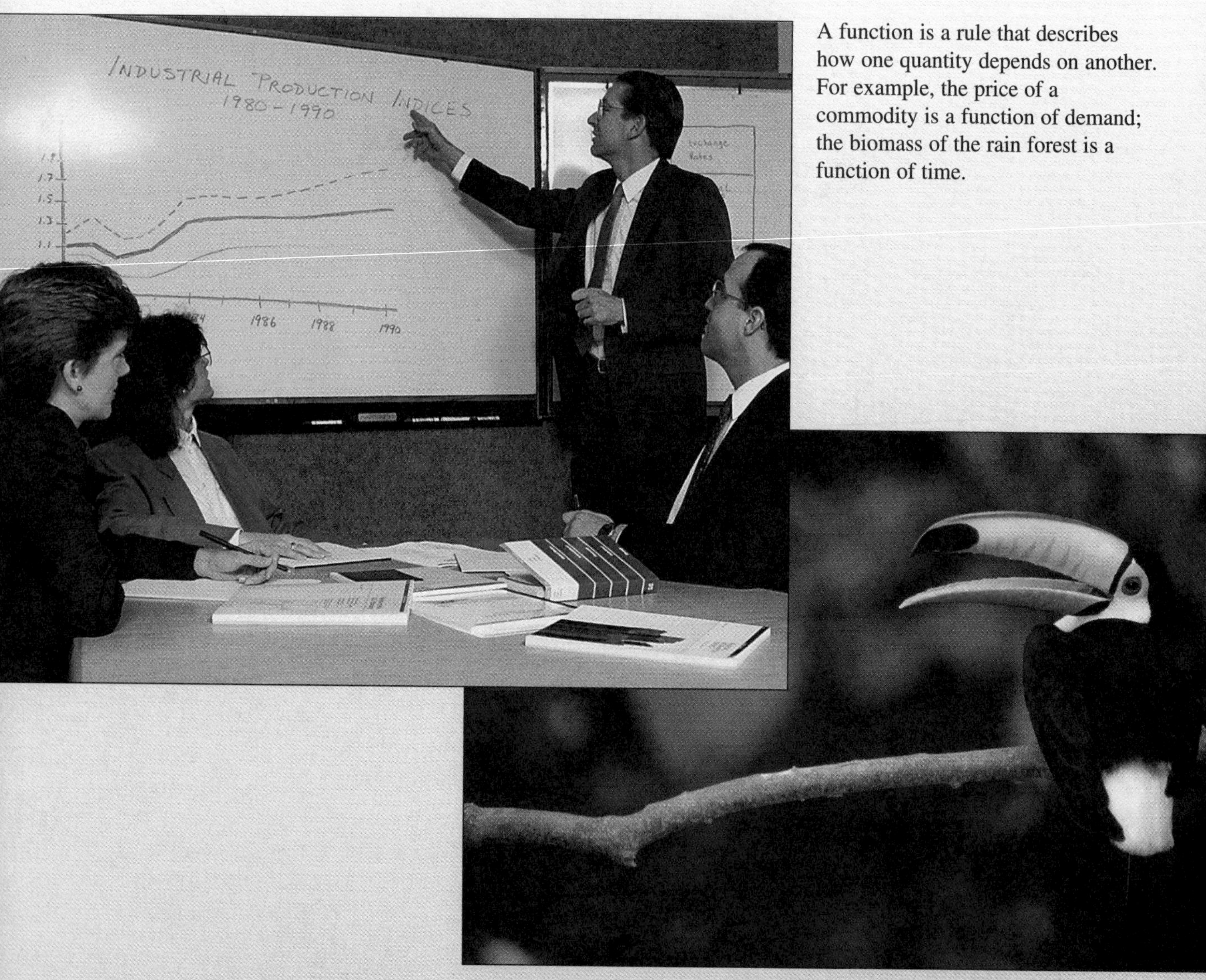

That flower of modern mathematical thought—the notion of a function.

THOMAS J. MCCORMACK

One of the most basic and important ideas in all of mathematics is the concept of a *function*. In this chapter we study functions, their graphs, and some of their applications.

4.1 WHAT IS A FUNCTION?

In this section we explore the idea of a function and then give the mathematical definition of function.

FUNCTIONS ALL AROUND US

In nearly every physical phenomenon, we observe that one quantity depends on another. For example, your height depends on your age, the temperature depends on the date, the cost of mailing a package depends on its weight (see Figure 1). We use the term *function* to describe this dependence of one quantity on another. That is, we say the following:

- Height is a function of age
- Temperature is a function of date
- Cost of mailing a package is a function of weight

The U.S. Post Office uses a simple rule to determine the cost of mailing a package based on its weight. But it's not so easy to describe the rule that relates height to age or temperature to date.

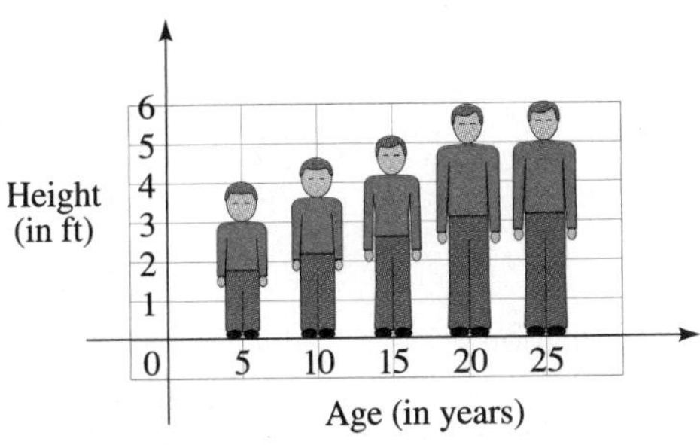

Height is a function of age.

°F
80
60
40
20
0
5
10
15
20
25
30
Date
Daily high temperature
Columbia, MO, April 1995

Temperature is a function of date.

w (ounces)	Postage (dollars)
$0 < w \leq 1$	0.33
$1 < w \leq 2$	0.55
$2 < w \leq 3$	0.77
$3 < w \leq 4$	0.99
$4 < w \leq 5$	1.21
$5 < w \leq 6$	1.43

Postage is a function of weight.

FIGURE 1

Can you think of other functions? Here are some more examples:

- The area of a circle is a function of its radius
- The number of bacteria in a culture is a function of time
- The weight of an astronaut is a function of her elevation
- The price of a commodity is a function of the demand for that commodity

The area A of a circle depends on its radius r. The rule that describes this dependence is given by the formula $A = \pi r^2$. The number N of bacteria in a culture depends on the time t (see Table 1). The rule that connects N and t in this case is given by the formula $N = 50 \cdot 2^t$. The weight w of an astronaut depends on her elevation h. Physicists use the rule $w = w_0 R^2/(R + h)^2$, where R is the radius of the earth. The price p of a commodity (wheat, for instance) depends on the demand y for that commodity (see Example 12, Section 2.4).

t (min)	*N* (bacteria)
0	50
1	100
2	200
3	400
4	800
5	1600
6	3200

TABLE 1

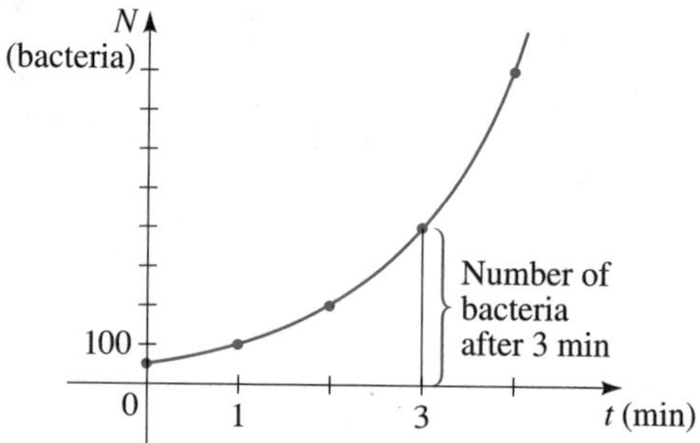

FIGURE 2

We can visualize a function by sketching its *graph*. For instance, to see how the number N of bacteria in a culture depends on the time t, we graph the data given in Table 1. The height of the graph above the horizontal axis represents the number of bacteria at a given time t—the higher the graph, the more bacteria. The shape of the graph in Figure 2 indicates how quickly the bacteria population grows.

Even when a precise rule or formula describing a function is not available, we can still describe the function by a graph. For example, when you turn on a hot water faucet, the temperature of the water depends on how long the water has been running. So we can say

- Temperature of water from the faucet is a function of time

Figure 3 shows a rough graph of the temperature T of the water as a function of the time t that has elapsed since the faucet was turned on. The graph shows that the initial temperature of the water is close to room temperature. When the water from the hot water tank reaches the faucet, the water's temperature T increases quickly. In

the next phase, T is constant at the temperature of the water in the tank. When the tank is drained, T decreases to the temperature of the cold water supply.

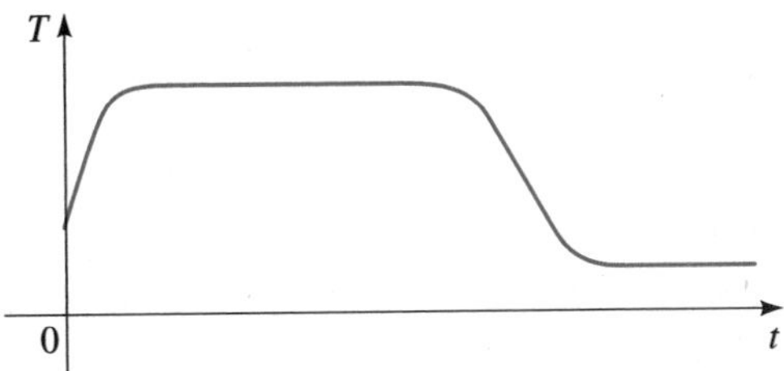

FIGURE 3
Graph of water temperature T as a function of time t

You can see why scientists routinely use functions in their work. For instance, a biologist observes that the number of bacteria in a culture increases with time. To predict the size of the bacteria culture in the future, the biologist tries to find the rule or function that relates the number of bacteria to the time. A physicist observes that the weight of an astronaut depends on her elevation. The physicist then tries to discover the rule or function that relates these quantities. An economist observes that the price of a certain commodity depends on the demand. The economist then tries to find the rule or function that relates these quantities. The process of finding a function to describe a real-world phenomenon is called *modeling*.

In the *Focus on Modeling*, pages 296–309, we learn how to find functions that model real-life data.

DEFINITION OF FUNCTION

Before we give the formal definition of a function, let's examine one more example. In the following list the numbers on the right are related to those on the left.

$$\begin{aligned} 1 &\to 1 \\ 2 &\to 4 \\ 3 &\to 9 \\ 4 &\to 16 \end{aligned}$$

Can you discover the rule that relates these numbers? The rule is "square the number." Thus, if x is any number, we have

$$x \to x^2$$

In order to talk about this rule, we need to give it a name. Let's call it f. So, in this example, f is the rule "square the number." When we apply the rule f to the number x, we get x^2. We express this mathematically as

$$f(x) = x^2$$

When we write $f(2)$, we mean "apply the rule f to the number 2." Applying the rule gives $f(2) = 2^2 = 4$. Similarly, $f(3) = 3^2 = 9$ and $f(4) = 4^2 = 16$.

The following definition concisely captures these ideas.

DEFINITION OF FUNCTION

A **function** f is a rule that assigns to each element x in a set A exactly one element, called $f(x)$, in a set B.

We have previously used letters to stand for numbers. Here we do something quite different. We use letters to represent *rules*. If the letter f represents a function, then the notation $f(x)$ means "apply the rule f to the number x."

We usually consider functions for which the sets A and B are sets of real numbers. The symbol $f(x)$ is read "f of x" or "f at x" and is called the **value of f at x**, or the **image of x under f**. The set A is called the **domain** of the function. The **range** of f is the set of all possible values of $f(x)$ as x varies throughout the domain, that is,

$$\{f(x) \mid x \in A\}$$

The symbol that represents an arbitrary number in the domain of a function f is called an **independent variable**. The symbol that represents a number in the range of f is called a **dependent variable**. For instance, in the bacteria example, t is the independent variable and N is the dependent variable.

The $\boxed{\sqrt{\ }}$ key on your calculator is a good example of a function as a machine. First you input x into the display. Then you press the key labeled $\boxed{\sqrt{\ }}$. If $x < 0$, then x is not in the domain of this function; that is, x is not an acceptable input and the calculator will indicate an error. If $x \geqslant 0$, then an approximation to $\sqrt{x}$ appears in the display, correct to a certain number of decimal places. [Thus, the $\boxed{\sqrt{\ }}$ key on your calculator is not quite the same as the exact mathematical function f defined by $f(x) = \sqrt{x}$.]

It's helpful to think of a function as a **machine** (see Figure 4). If x is in the domain of the function f, then when x enters the machine, it is accepted as an input and the machine produces an output $f(x)$ according to the rule of the function. Thus, we can think of the domain as the set of all possible inputs and the range as the set of all possible outputs.

FIGURE 4 Machine diagram of f

Another way to picture a function is by an **arrow diagram** as in Figure 5. Each arrow connects an element of A to an element of B. The arrow indicates that $f(x)$ is associated with x, $f(a)$ is associated with a, and so on.

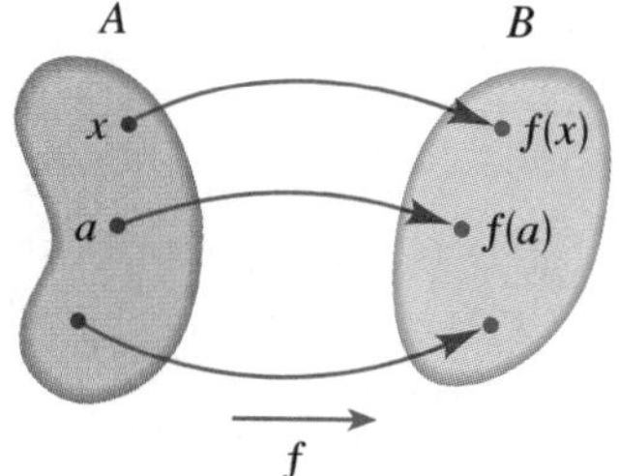

FIGURE 5 Arrow diagram of f

EXAMPLE 1 ■ The Squaring Function

The squaring function assigns to each real number x its square x^2. It is defined by

$$f(x) = x^2$$

(a) Evaluate $f(3)$, $f(-2)$, and $f(\sqrt{5})$.
(b) Find the domain and range of f.
(c) Draw a machine diagram for f.

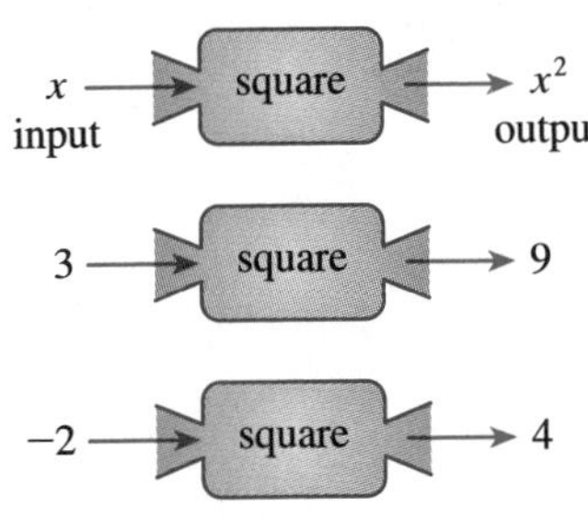

FIGURE 6
Machine diagram

SOLUTION

(a) The values of f are found by substituting for x in $f(x) = x^2$.

$$f(3) = 3^2 = 9 \qquad f(-2) = (-2)^2 = 4 \qquad f(\sqrt{5}) = (\sqrt{5})^2 = 5$$

(b) The domain of f is the set $\mathbb{R}$ of all real numbers. The range of f consists of all values of $f(x)$, that is, all numbers of the form x^2. Since $x^2 \geqslant 0$ for all real numbers x, we can see that the range of f is $\{y \mid y \geqslant 0\} = [0, \infty)$.

(c) A machine diagram for this function is shown in Figure 6. ■

EVALUATING A FUNCTION

In the definition of a function the independent variable x plays the role of a "placeholder." For example, the function $f(x) = 3x^2 + x - 5$ can be thought of as

$$f(\square) = 3 \cdot \square^2 + \square - 5$$

To evaluate f at a number, we substitute the number for the placeholder.

EXAMPLE 2 ■ Evaluating a Function

Let $f(x) = 3x^2 + x - 5$. Evaluate each function value.

(a) $f(-2)$ (b) $f(0)$ (c) $f(4)$ (d) $f\left(\frac{1}{2}\right)$

SOLUTION

To evaluate f at a number, we substitute the number for x in the definition of f.

(a) $f(-2) = 3 \cdot (-2)^2 + (-2) - 5 = 5$

(b) $f(0) = 3 \cdot 0^2 + 0 - 5 = -5$

(c) $f(4) = 3 \cdot 4^2 + 4 - 5 = 47$

(d) $f\left(\frac{1}{2}\right) = 3 \cdot \left(\frac{1}{2}\right)^2 + \frac{1}{2} - 5 = -\frac{15}{4}$ ■

EXAMPLE 3 ■ A Piecewise Defined Function

Evaluate the following function at $x = -2, 1, 2,$ and 3.

$$f(x) = \begin{cases} 1 - x & \text{if } x \leqslant 1 \\ x^2 & \text{if } x > 1 \end{cases}$$

SOLUTION

Remember that a function is a rule. Here is how we apply the rule for this particular function: First look at the value of the input x. If it happens that $x \leqslant 1$, then the value of $f(x)$ is $1 - x$. On the other hand, if $x > 1$, then the value of $f(x)$ is x^2.

Since $-2 \leqslant 1$, we have $f(-2) = 1 - (-2) = 3$.

Since $1 \leqslant 1$, we have $f(1) = 1 - 1 = 0$.

Since $2 > 1$, we have $f(2) = 2^2 = 4$.

Since $3 > 1$, we have $f(3) = 3^2 = 9$. ■

EXAMPLE 4 ■ Evaluating a Function

Expressions like the one in part (d) of Example 4 occur frequently in calculus; they are called *difference quotients*, and they represent the average change in the value of f between $x = a$ and $x = a + h$.

If $f(x) = 2x^2 + 3x - 1$, evaluate the following.

(a) $f(a)$ (b) $f(-a)$

(c) $f(a + h)$ (d) $\dfrac{f(a+h) - f(a)}{h}$, $h \neq 0$

SOLUTION

(a) $f(a) = 2a^2 + 3a - 1$

(b) $f(-a) = 2(-a)^2 + 3(-a) - 1 = 2a^2 - 3a - 1$

(c)
$$\begin{aligned} f(a+h) &= 2(a+h)^2 + 3(a+h) - 1 \\ &= 2(a^2 + 2ah + h^2) + 3(a+h) - 1 \\ &= 2a^2 + 4ah + 2h^2 + 3a + 3h - 1 \end{aligned}$$

(d) Using the results from parts (c) and (a), we have

$$\begin{aligned} \frac{f(a+h) - f(a)}{h} &= \frac{(2a^2 + 4ah + 2h^2 + 3a + 3h - 1) - (2a^2 + 3a - 1)}{h} \\ &= \frac{4ah + 2h^2 + 3h}{h} = 4a + 2h + 3 \end{aligned}$$

■

THE DOMAIN OF A FUNCTION

The domain of a function may be stated explicitly. For example, if we write

$$f(x) = x^2 \qquad 0 \leqslant x \leqslant 5$$

then the domain is the set of all real numbers x for which $0 \leqslant x \leqslant 5$. If the function is given by an algebraic expression and the domain is not stated explicitly, then by convention *the domain is the set of all real numbers for which the expression is defined as a real number.* For example, the function

$$f(x) = \frac{1}{x - 4}$$

is not defined at $x = 4$, so its domain is $\{x \mid x \neq 4\}$. The function

$$f(x) = \sqrt{x}$$

is not defined for negative x, so its domain is $\{x \mid x \geqslant 0\}$.

EXAMPLE 5 ■ Finding Domains of Functions

Find the domain of each function.

(a) $f(x) = \dfrac{1}{x^2 - x}$ (b) $g(x) = \sqrt{9 - x^2}$ (c) $h(t) = \dfrac{t}{\sqrt{t + 1}}$

SOLUTION

(a) The function is not defined when the denominator is 0. Since

$$f(x) = \frac{1}{x^2 - x} = \frac{1}{x(x - 1)}$$

we see that $f(x)$ is not defined when $x = 0$ or $x = 1$. Thus, the domain of f is

$$\{x \mid x \neq 0, x \neq 1\}$$

The domain may also be written in interval notation as

$$(\infty, 0) \cup (0, 1) \cup (1, \infty)$$

(b) We can't take the square root of a negative number, so we must have $9 - x^2 \geqslant 0$. Using the methods of Section 3.7, we can solve this inequality to find that $-3 \leqslant x \leqslant 3$. Thus, the domain of g is

$$\{x \mid -3 \leqslant x \leqslant 3\} = [-3, 3]$$

(c) We can't take the square root of a negative number, and we can't divide by 0, so we must have $t + 1 > 0$, that is, $t > -1$. So the domain of h is

$$\{t \mid t > -1\} = (-1, \infty)$$

■

FOUR WAYS TO REPRESENT A FUNCTION

To help us understand what a function is, we have used machine and arrow diagrams. To describe a specific function, we can use the following four ways:

- verbally (by a description in words)
- algebraically (by an explicit formula)
- visually (by a graph)
- numerically (by a table of values)

A single function may be represented in all four ways, and it is often useful to go from one representation to another to gain insight into the function. However, certain functions are described more naturally by one method than by the other. An example of a verbal description is

$$P(t) \quad \text{is} \quad \text{“the population of the world at time } t\text{”}$$

The function P can also be described numerically by giving a table of values (see Table 1 on page 438). A useful representation of the area of a circle as a function of its radius is the algebraic formula

$$A(r) = \pi r^2$$

The graph produced by a seismograph (see the box) is a visual representation of the vertical acceleration function $a(t)$ of the ground during an earthquake. As a final example, consider the function $C(w)$, which is described verbally as "the cost of mailing a first-class letter with weight w." The most convenient way of describing this function is numerically—that is, using a table of values.

We will be using all four representations of functions throughout this book. We summarize them in the following box.

FOUR WAYS TO REPRESENT A FUNCTION

Verbal

Using words:

$P(t)$ is "the population of the world at time t"

Relation of population P and time t

Algebraic

Using a formula:

$$A(r) = \pi r^2$$

Area of a circle

Visual

Using a graph:

Source: Calif. Dept. of Mines and Geology

Vertical acceleration during an earthquake

Numerical

Using a table of values:

w (ounces)	$C(w)$ (dollars)
$0 < w \leq 1$	0.33
$1 < w \leq 2$	0.55
$2 < w \leq 3$	0.77
$3 < w \leq 4$	0.99
$4 < w \leq 5$	1.21
⋮	⋮

Cost of mailing a first-class letter

4.1 EXERCISES

1–4 ■ Express the rule in function notation. [For example, the rule "square, then subtract 5" is expressed as the function $f(x) = x^2 - 5$.]

1. Multiply by 3, then add 1

2. Subtract 5, then divide by 7

3. Add 2, then square

4. Square, add 1, then take the square root

5–8 ■ Express the function (or rule) in words.

5. $f(x) = \dfrac{x}{3} - 5$

6. $g(x) = \dfrac{x - 5}{3}$

7. $h(x) = 2x^2 - 3$

8. $k(x) = \sqrt{2x + 1}$

9–10 ■ Draw a machine diagram for the function.

9. $f(x) = \sqrt{x}$

10. $f(x) = \dfrac{2}{x}$

11–12 ■ Complete the table.

11. $f(x) = 2x^2 + 1$

x	$f(x)$
-1	
0	
1	
2	
3	

12. $g(x) = |2x - 3|$

x	$g(x)$
-2	
0	
1	
3	
5	

13–20 ■ Evaluate the function at the indicated values.

13. $f(x) = 2x + 1$;
$f(1), f(-2), f(\frac{1}{2}), f(a), f(-a), f(a + b)$

14. $f(x) = x^2 + 2x$;
$f(0), f(3), f(-3), f(a), f(-x), f\left(\dfrac{1}{a}\right)$

15. $g(x) = \dfrac{1 - x}{1 + x}$;
$g(2), g(-2), g(\frac{1}{2}), g(a), g(a - 1), g(-1)$

16. $h(t) = t + \dfrac{1}{t}$;
$h(1), h(-1), h(2), h(\frac{1}{2}), h(x), h\left(\dfrac{1}{x}\right)$

17. $f(x) = 2x^2 + 3x - 4$;
$f(0), f(2), f(-2), f(\sqrt{2}), f(x + 1), f(-x)$

18. $f(x) = x^3 - 4x^2$;
$f(0), f(1), f(-1), f(\frac{3}{2}), f\left(\dfrac{x}{2}\right), f(x^2)$

19. $f(x) = 2|x - 1|$;
$f(-2), f(0), f(\frac{1}{2}). f(2), f(x + 1), f(x^2 + 2)$

20. $f(x) = \dfrac{|x|}{x}$;
$f(-2), f(-1), f(0), f(5), f(x^2), f\left(\dfrac{1}{x}\right)$

21–24 ■ Evaluate the piecewise defined function at the indicated values.

21. $f(x) = \begin{cases} x^2 & \text{if } x < 0 \\ x + 1 & \text{if } x \geqslant 0 \end{cases}$
$f(-2), f(-1), f(0), f(1), f(2)$

22. $f(x) = \begin{cases} 5 & \text{if } x \leqslant 2 \\ 2x - 3 & \text{if } x > 2 \end{cases}$
$f(-3), f(0), f(2), f(3), f(5)$

23. $f(x) = \begin{cases} x^2 + 2x & \text{if } x \leqslant -1 \\ x & \text{if } x > -1 \end{cases}$
$f(-4), f(-\frac{3}{2}), f(-1), f(0), f(1)$

24. $f(x) = \begin{cases} 3x & \text{if } x < 0 \\ x + 1 & \text{if } 0 \leqslant x \leqslant 2 \\ (x - 2)^2 & \text{if } x > 2 \end{cases}$
$f(-5), f(0), f(1), f(2), f(5)$

25–28 ■ Use the function to evaluate the indicated expressions and simplify.

25. $f(x) = x^2 + 1$; $f(x + 2), f(x) + f(2)$

26. $f(x) = 3x - 1$; $f(2x), 2f(x)$

27. $f(x) = x + 4$; $f(x^2), (f(x))^2$

28. $f(x) = 6x - 18$; $f\left(\dfrac{x}{3}\right), \dfrac{f(x)}{3}$

29–34 ■ Find $f(a)$, $f(a + h)$, and $\dfrac{f(a + h) - f(a)}{h}$, where $h \neq 0$.

29. $f(x) = 3x + 2$

30. $f(x) = x^2 + 1$

31. $f(x) = 5$

32. $f(x) = \dfrac{1}{x + 1}$

33. $f(x) = 3 - 5x + 4x^2$

34. $f(x) = x^3$

35–52 ■ Find the domain of the function.

35. $f(x) = 2x$

36. $f(x) = x^2 + 1$

37. $f(x) = 2x, \quad -1 \leq x \leq 5$

38. $f(x) = x^2 + 1, \quad 0 \leq x \leq 5$

39. $f(x) = \dfrac{1}{x - 3}$

40. $f(x) = \dfrac{1}{3x - 6}$

41. $f(x) = \dfrac{x + 2}{x^2 - 1}$

42. $f(x) = \dfrac{x^4}{x^2 + x - 6}$

43. $f(x) = \sqrt{x - 5}$

44. $f(x) = \sqrt[4]{x + 9}$

45. $f(t) = \sqrt[3]{t - 1}$

46. $g(x) = \sqrt{7 - 3x}$

47. $h(x) = \sqrt{2x - 5}$

48. $G(x) = \sqrt{x^2 - 9}$

49. $g(x) = \dfrac{\sqrt{2 + x}}{3 - x}$

50. $g(x) = \dfrac{\sqrt{x}}{2x^2 + x - 1}$

51. $g(x) = \sqrt[4]{x^2 - 6x}$

52. $g(x) = \sqrt{x^2 - 2x - 8}$

53. The graph gives the weight of a certain person as a function of age. Describe in words how this person's weight has varied over time. What do you think happened when this person was 30 years old?

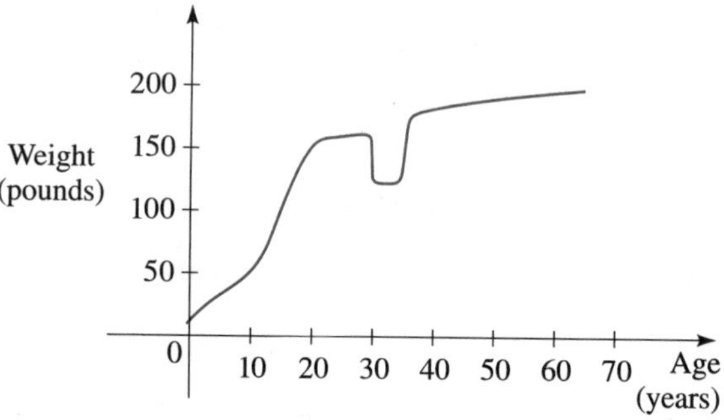

54. The graph gives a salesman's distance from his home as a function of time on a certain day. Describe in words what the graph indicates about his travels on this day.

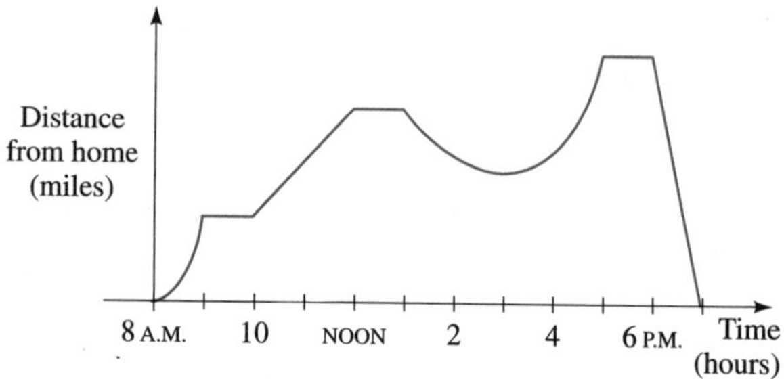

55. You put some ice cubes in a glass, fill the glass with cold water, and then let the glass sit on a table. Sketch a rough graph of the temperature of the water as a function of the elapsed time.

56. A home owner mows the lawn every Wednesday afternoon. Sketch a rough graph of the height of the grass as a function of time over the course of a four-week period beginning on a Sunday.

57. When a football is kicked from a tee, its height depends on the time elapsed since the kickoff. Sketch a rough graph of the height of the football as a function of time.

58. Sketch a rough graph of the number of hours of daylight as a function of the time of year in the Northern Hemisphere.

59. The number of Christmas cards sold by a greeting-card store depends on the time of year. Sketch a rough graph of

the number of Christmas cards sold as a function of the time of year.

60. You place a frozen pie in an oven and bake it for an hour. Then you take it out and let it cool before eating it. Sketch a rough graph of the temperature of the pie as a function of time.

61. The figure shows the power consumption in San Francisco for September 19, 1996 (P is measured in megawatts; t is measured in hours starting at midnight).

(a) What was the power consumption at 6 A.M.? At 6 P.M.?

(b) When was the power consumption the lowest?

(c) When was the power consumption the highest?

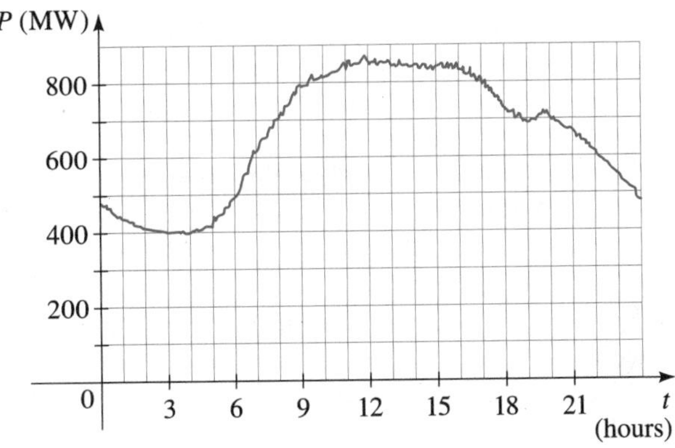

Source: Pacific Gas & Electric

62. The graph shows the vertical acceleration of the ground from the 1994 Northridge earthquake in Los Angeles, as measured by a seismograph. (Here t represents the time in seconds.)

(a) At what time t did the earthquake first make noticeable movements of the earth?

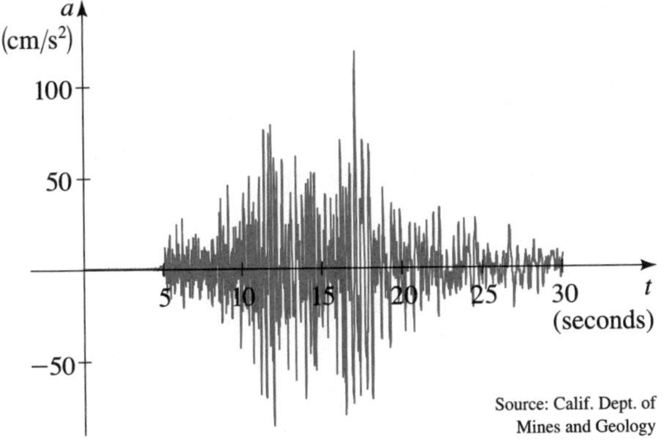

Source: Calif. Dept. of Mines and Geology

(b) At what time t did the earthquake seem to end?

(c) At what time t was the maximum intensity of the earthquake reached?

63. Temperature readings T (in °F) were recorded every 2 hours from midnight to noon in Atlanta, Georgia, on March 18, 1996. The time t was measured in hours from midnight. Sketch a rough graph of T as a function of t.

t	T
0	58
2	57
4	53
6	50
8	51
10	57
12	61

64. The population P (in thousands) of San Jose, California, from 1984 to 1994 is shown in the table. (Midyear estimates are given.) Draw a rough graph of P as a function of time t.

t	P
1984	695
1986	716
1988	733
1990	782
1992	800
1994	817

DISCOVERY • DISCUSSION

65. Examples of Functions At the beginning of this section we discussed three examples of everyday, ordinary functions: Height is a function of age, temperature is a function of date, and postage cost is a function of weight. Give three other examples of functions from everyday life.

66. Four Ways to Represent a Function In the table on page 216 we represented four different functions verbally, algebraically, visually, and numerically. Think of a function that can be represented all four ways, and write the four representations.

4.2 GRAPHS OF FUNCTIONS

The most important way to visualize a function is through its graph. In this section we investigate in more detail the concept of graphing functions.

GRAPHING FUNCTIONS

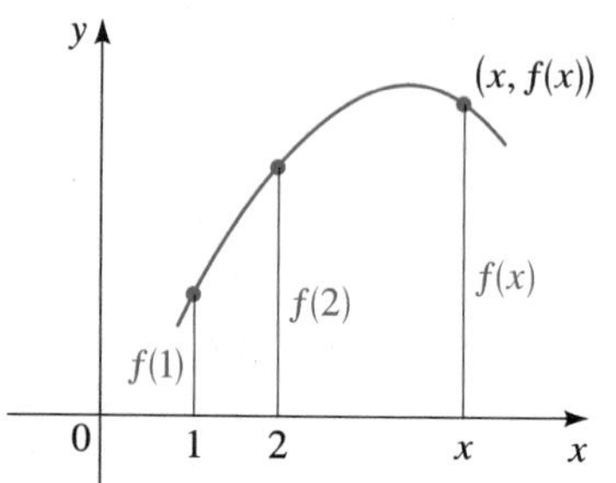

FIGURE 1
The graph of f

The height of the graph above the point x is the value of $f(x)$.

THE GRAPH OF A FUNCTION

If f is a function with domain A, then the **graph** of f is the set of ordered pairs

$$\{(x, f(x)) \mid x \in A\}$$

In other words, the graph of f is the set of all points (x, y) such that $y = f(x)$; that is, the graph of f is the graph of the equation $y = f(x)$.

The graph of a function f gives us a useful picture of the behavior or "life history" of a function. Since the y-coordinate of any point (x, y) on the graph is $y = f(x)$, we can read the value of $f(x)$ from the graph as being the height of the graph above the point x (see Figure 1).

A function f of the form $f(x) = mx + b$ is called a **linear function** because its graph is the graph of the equation $y = mx + b$, which represents a line with slope m and y-intercept b. A special case of a linear function occurs when the slope is $m = 0$. The function $f(x) = b$, where b is a given number, is called a **constant function** because all its values are the same number, namely, b. Its graph is the horizontal line $y = b$. Figure 2 shows the graphs of the constant function $f(x) = 3$ and the linear function $f(x) = 2x + 1$.

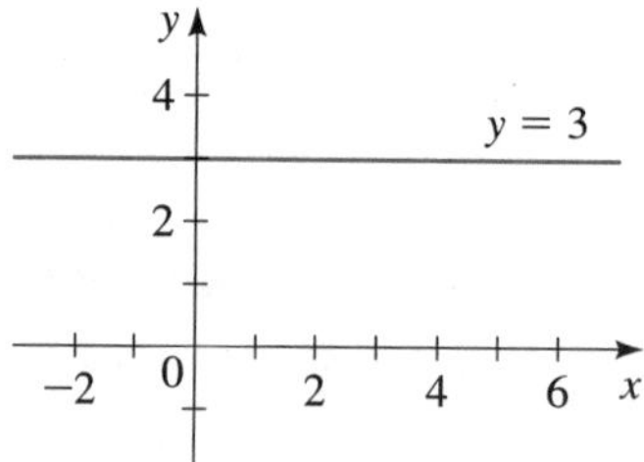

The constant function $f(x) = 3$

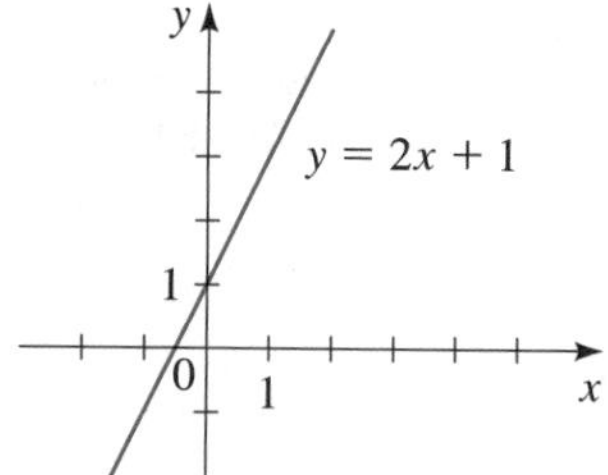

The linear function $f(x) = 2x + 1$

FIGURE 2

EXAMPLE 1 ■ Graphing Functions

Sketch the graphs of the following functions.

(a) $f(x) = x^2$ (b) $g(x) = x^3$ (c) $h(x) = \sqrt{x}$

SOLUTION

We first make a table of values. Then we plot the points given by the table and join them by a smooth curve to obtain the graph. The graphs are sketched in Figure 3.

x	$f(x) = x^2$
0	0
$\pm\frac{1}{2}$	$\frac{1}{4}$
± 1	1
± 2	4
± 3	9

x	$f(x) = x^3$
0	0
$\frac{1}{2}$	$\frac{1}{8}$
1	1
2	8
$-\frac{1}{2}$	$-\frac{1}{8}$
-1	-1
-2	-8

x	$f(x) = \sqrt{x}$
0	0
1	1
2	$\sqrt{2}$
3	$\sqrt{3}$
4	2
5	$\sqrt{5}$

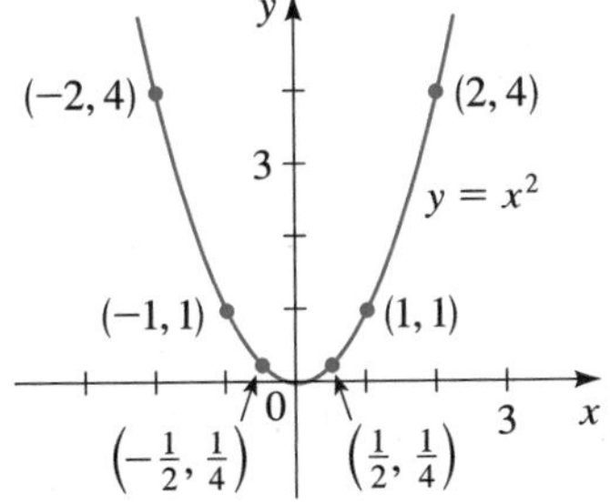

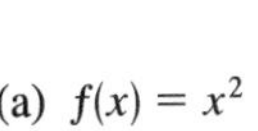
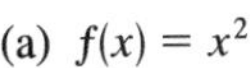

(a) $f(x) = x^2$

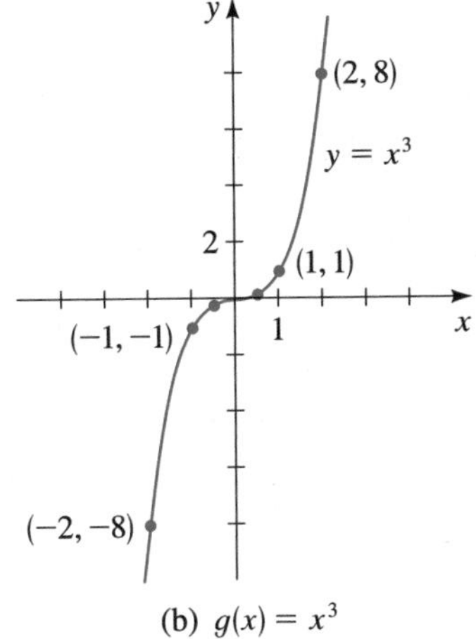

(b) $g(x) = x^3$

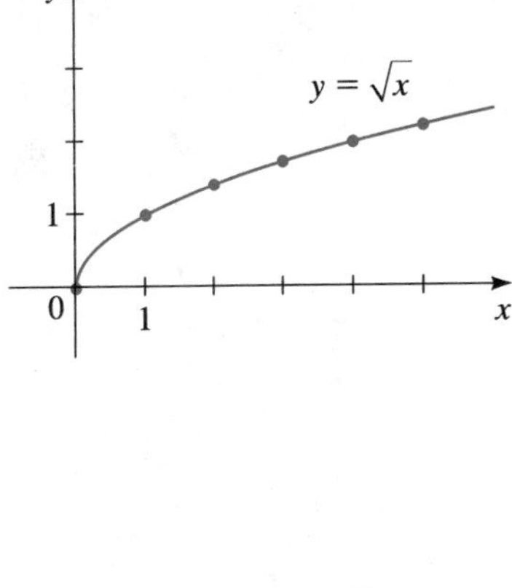

(c) $h(x) = \sqrt{x}$

FIGURE 3

■

EXAMPLE 2 ■ Graph of a Piecewise Defined Function

Sketch the graph of the function

$$f(x) = \begin{cases} x^2 & \text{if } x \leq 1 \\ 2x + 1 & \text{if } x > 1 \end{cases}$$

SOLUTION

If $x \leq 1$, then $f(x) = x^2$, so the part of the graph to the left of $x = 1$ coincides with the graph of $y = x^2$, which we sketched in Figure 3. If $x > 1$, then $f(x) = 2x + 1$,

On many graphing calculators the graph in Figure 4 can be produced by using the logical functions in the calculator. For example, on the TI-83 the following equation gives the required graph:

$$y = (x \leqslant 1)x\text{^}2 + (x > 1)(2x + 1)$$

so the part of the graph to the right of $x = 1$ coincides with the line $y = 2x + 1$, which we graphed in Figure 2. This enables us to sketch the graph in Figure 4.

The solid dot at $(1, 1)$ indicates that this point is included in the graph; the open dot at $(1, 3)$ indicates that this point is excluded from the graph.

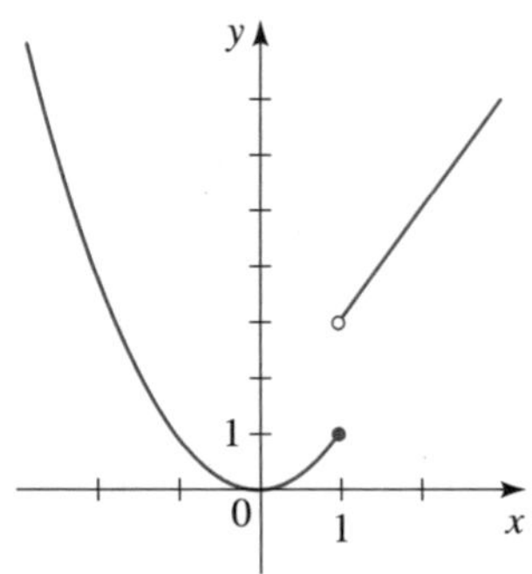

FIGURE 4

$$f(x) = \begin{cases} x^2 & \text{if } x \leqslant 1 \\ 2x + 1 & \text{if } x > 1 \end{cases}$$

■

EXAMPLE 3 ■ Graph of the Absolute Value Function

Sketch the graph of the absolute value function $f(x) = |x|$.

SOLUTION

Recall that

$$|x| = \begin{cases} x & \text{if } x \geqslant 0 \\ -x & \text{if } x < 0 \end{cases}$$

Using the same method as in Example 2, we note that the graph of f coincides with the line $y = x$ to the right of the y-axis and coincides with the line $y = -x$ to the left of the y-axis (see Figure 5).

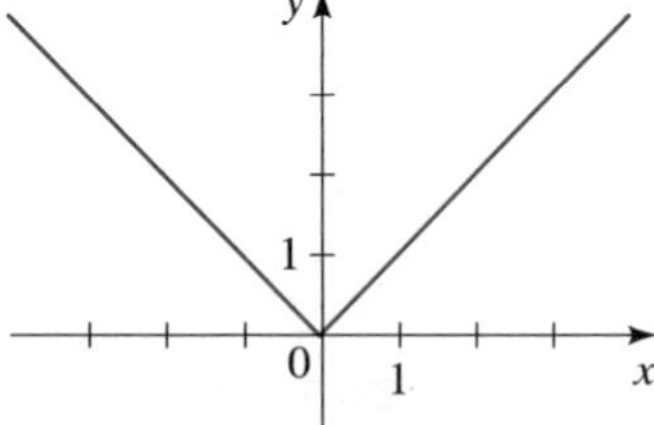

FIGURE 5
Graph of $f(x) = |x|$

■

The **greatest integer function** is defined by

$$[\![x]\!] = \text{greatest integer less than or equal to } x$$

For example, $[\![2]\!] = 2$, $[\![2.3]\!] = 2$, $[\![1.999]\!] = 1$, $[\![0.002]\!] = 0$, $[\![-3.5]\!] = -4$, $[\![-0.5]\!] = -1$.

EXAMPLE 4 ■ Graph of the Greatest Integer Function

Sketch the graph of $f(x) = [\![x]\!]$.

SOLUTION

The table shows the values of f for some values of x. Note that $f(x)$ is constant between consecutive integers so the graph between integers is a horizontal line segment as shown in Figure 6.

x	$[\![x]\!]$
$\vdots$	$\vdots$
$-2 \leq x < -1$	-2
$-1 \leq x < 0$	-1
$0 \leq x < 1$	0
$1 \leq x < 2$	1
$2 \leq x < 3$	2
$\vdots$	$\vdots$

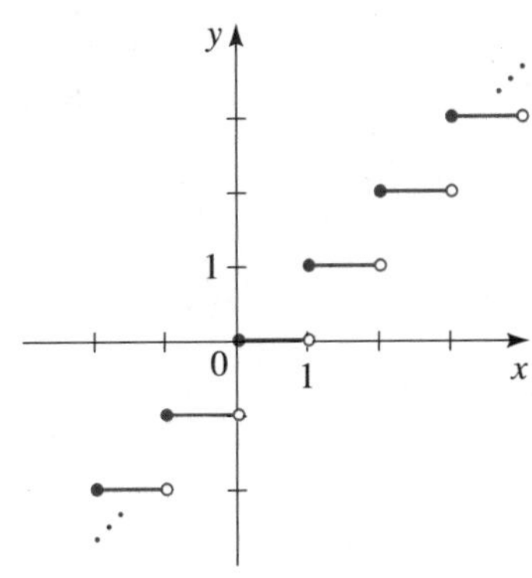

FIGURE 6
The greatest integer function, $y = [\![x]\!]$ ■

The greatest integer function is an example of a **step function**. The next example gives a real-world example of a step function.

EXAMPLE 5 ■ The Cost Function for Long-Distance Phone Calls

The cost of a long-distance daytime phone call from Toronto to New York City is 69 cents for the first minute and 58 cents for each additional minute (or part of a minute). Draw the graph of the cost C (in dollars) of the phone call as a function of time t (in minutes).

SOLUTION

Let $C(t)$ be the cost for t minutes. Since $t > 0$, the domain of the function is $(0, \infty)$. From the given information, we have

$$C(t) = 0.69 \qquad \text{if } 0 < t \leq 1$$

$$C(t) = 0.69 + 0.58 = 1.27 \qquad \text{if } 1 < t \leq 2$$

$$C(t) = 0.69 + 2(0.58) = 1.85 \qquad \text{if } 2 < t \leq 3$$

$$C(t) = 0.69 + 3(0.58) = 2.43 \qquad \text{if } 3 < t \leq 4$$

and so on. The graph is shown in Figure 7. ■

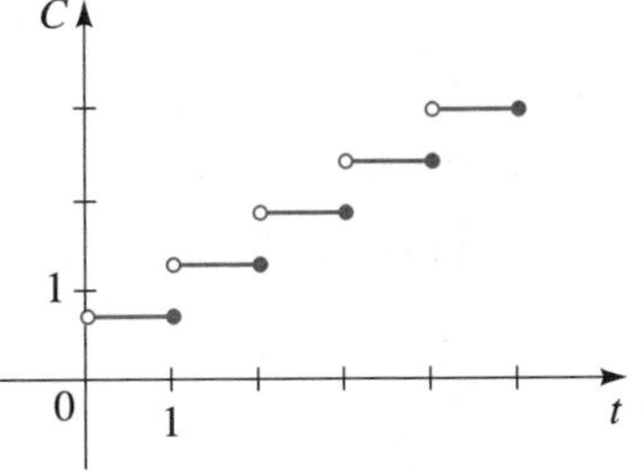

FIGURE 7
Cost of a long-distance call

To understand how the equation of a function relates to its graph, it's helpful to graph a **family of functions**, that is, a collection of functions whose equations are related. Usually, the equations differ only in a single constant called a *parameter*.

In the next example, we graph a family of **power functions** $f(x) = x^n$, when n is a positive integer. In this case, n is the parameter.

EXAMPLE 6 ■ A Family of Power Functions

(a) Graph the functions $f(x) = x^n$ for $n = 2$, 4, and 6 in the viewing rectangle $[-2, 2]$ by $[-1, 3]$.
(b) Graph the functions $f(x) = x^n$ for $n = 1$, 3, and 5 in the viewing rectangle $[-2, 2]$ by $[-2, 2]$.
(c) What conclusions can you draw from these graphs?

SOLUTION

The graphs for parts (a) and (b) are shown in Figure 8.

(c) We see that the general shape of the graph of $f(x) = x^n$ depends on whether n is even or odd.

If n is even, the graph of $f(x) = x^n$ is similar to the parabola $y = x^2$.
If n is odd, the graph of $f(x) = x^n$ is similar to that of $y = x^3$.

Notice from part (c) that as n increases the graph of $y = x^n$ becomes flatter near 0 and steeper when $x > 1$. When $0 < x < 1$, the lower powers of x are the "bigger" functions. But when $x > 1$, the higher powers of x are the dominant functions.

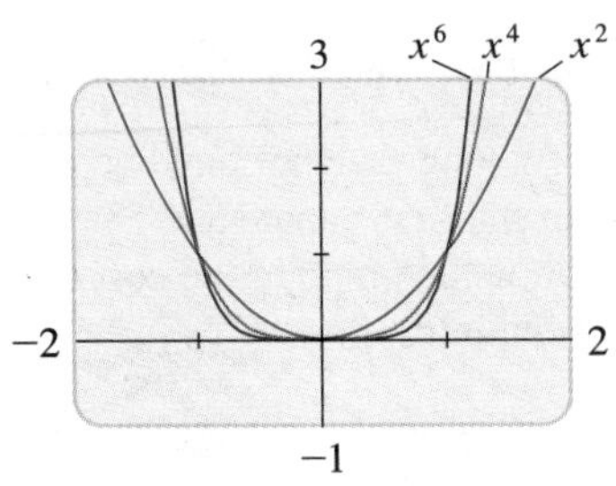

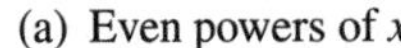
(a) Even powers of x

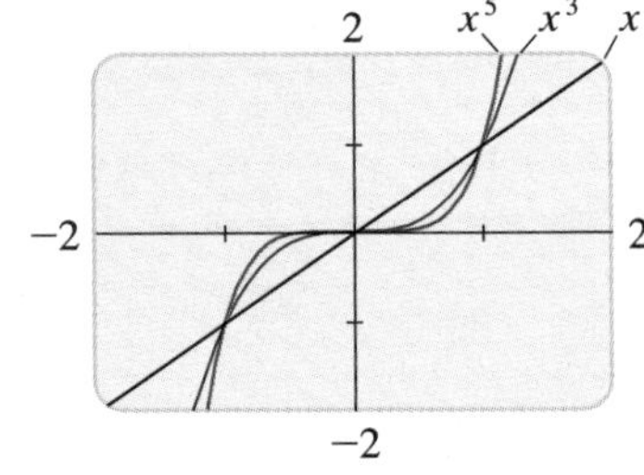

(b) Odd powers of x ■

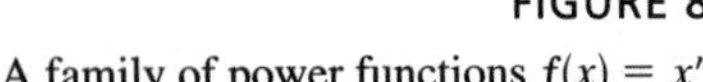
FIGURE 8
A family of power functions $f(x) = x^n$

FINDING DOMAIN AND RANGE FROM THE GRAPH

The graph of a function helps us picture the domain and range of the function on the x-axis and y-axis as shown in Figure 9.

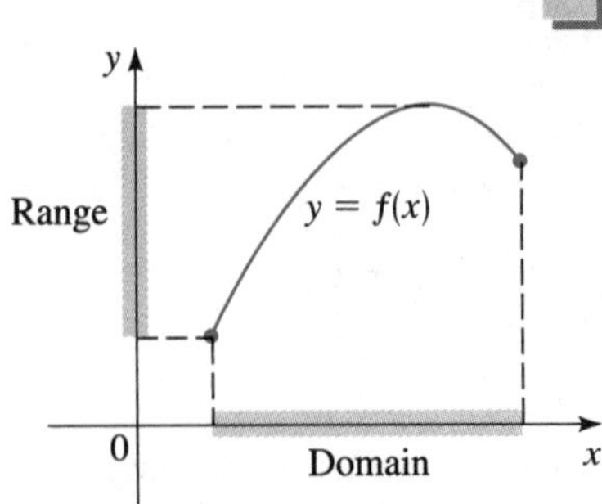

FIGURE 9
Domain and range of f

EXAMPLE 7 ■ Finding the Domain and Range from a Graph

(a) Sketch the graph of $f(x) = \sqrt{4 - x^2}$.
(b) Find the domain and range of f.

SOLUTION

(a) We must graph the equation $y = \sqrt{4 - x^2}$. Because we are taking the positive square root, we know that $y \geq 0$. Squaring each side of the equation,

Donald Knuth was born in Milwaukee in 1938 and is now Professor of Computer Science at Stanford University. While still a graduate student at Caltech, he started writing a monumental series of books entitled *The Art of Computer Programming*. President Carter awarded him the National Medal of Science in 1979. When Knuth was a high school student, he became fascinated with graphs of functions and laboriously drew many hundreds of them because he wanted to see the behavior of a great variety of functions. (Now, with progress in computers and programming, it is far easier to use computers and graphing calculators to do this.) Knuth is also famous for his invention of $\TeX$, a system of computer-assisted typesetting. This system was used in the preparation of the manuscript for this textbook. He has also written a novel entitled *Surreal Numbers: How Two Ex-Students Turned On to Pure Mathematics and Found Total Happiness.*

we get

$$y^2 = 4 - x^2 \qquad \text{Square both sides}$$

$$x^2 + y^2 = 4 \qquad \text{Add } x^2$$

which we recognize as the equation of a circle with center the origin and radius 2. But, since $y \geq 0$, the graph of f consists of just the upper half of this circle. See Figure 10.

(b) From the graph in Figure 10 we see that the domain is the closed interval $[-2, 2]$ and the range is $[0, 2]$.

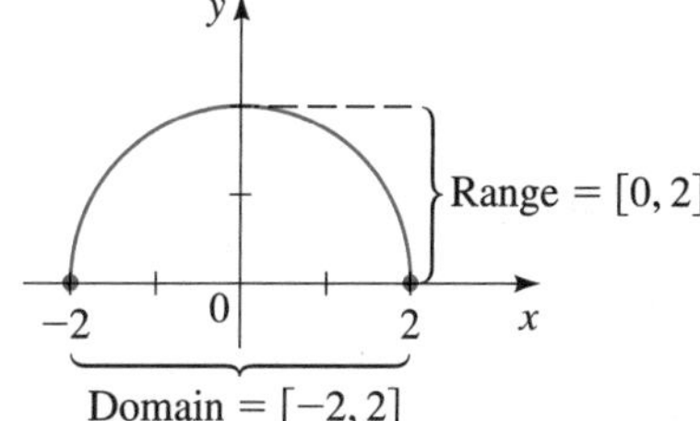

FIGURE 10
Graph of $f(x) = \sqrt{4 - x^2}$

■

THE VERTICAL LINE TEST

The graph of a function is a curve in the *xy*-plane. But the question arises: Which curves in the *xy*-plane are graphs of functions? This is answered by the following test.

THE VERTICAL LINE TEST

A curve in the coordinate plane is the graph of a function if and only if no vertical line intersects the curve more than once.

We can see from Figure 11 why the Vertical Line Test is true. If each vertical line $x = a$ intersects a curve only once at (a, b), then exactly one functional value is defined by $f(a) = b$. But if a line $x = a$ intersects the curve twice, at (a, b) and at

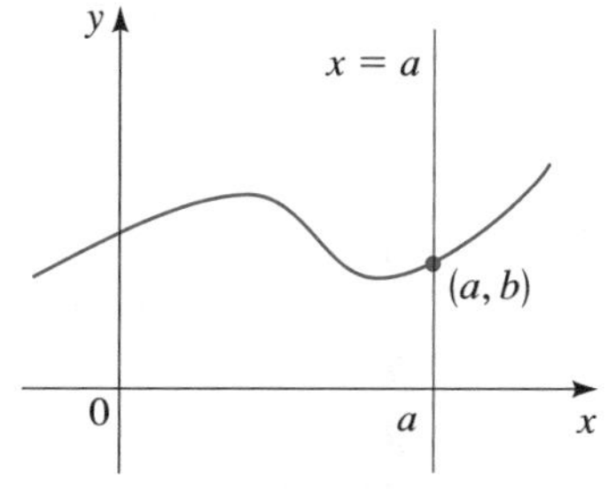

Graph of a function

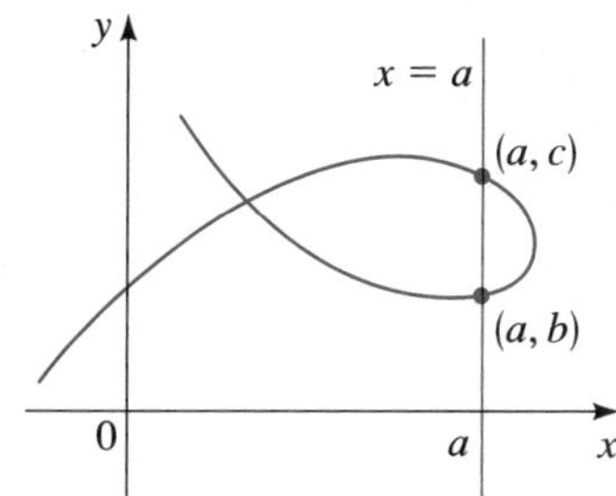

Not a graph of a function

FIGURE 11
Vertical Line Test

(a, c), then the curve can't represent a function because a function cannot assign two different values to a.

EXAMPLE 8 ■ Using the Vertical Line Test

Using the Vertical Line Test, we see that the curves in parts (b) and (c) of Figure 12 represent functions, whereas those in parts (a) and (d) do not.

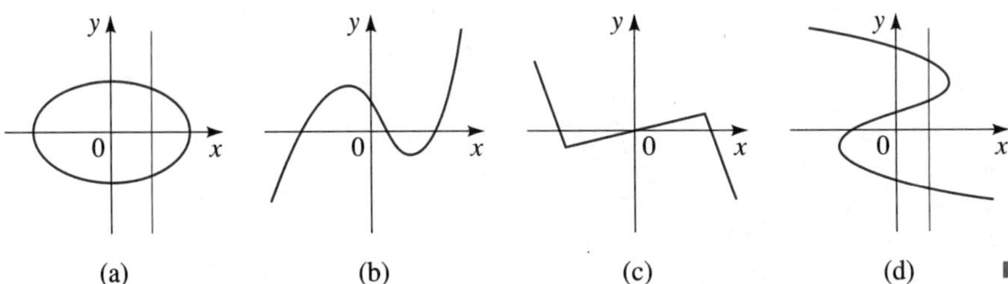

FIGURE 12 ■

EXAMPLE 9 ■ Equations That Define Functions

Does the equation define y as a function of x?

(a) $y - x^2 = 2$ (b) $x^2 + y^2 = 4$

SOLUTION

(a) Solving for y in terms of x gives

$$y - x^2 = 2$$

$$y = x^2 + 2 \qquad \text{Add } x^2$$

The last equation is a rule that gives one value of y for each value of x, so it defines y as a function of x. We can write the function as $f(x) = x^2 + 2$.

(b) We try to solve for y in terms of x:

$$x^2 + y^2 = 4$$

$$y^2 = 4 - x^2 \qquad \text{Subtract } x^2$$

$$y = \pm\sqrt{4 - x^2} \qquad \text{Take square roots}$$

The last equation gives two values of y for a given value of x. Thus, the equation does not define y as a function of x. ■

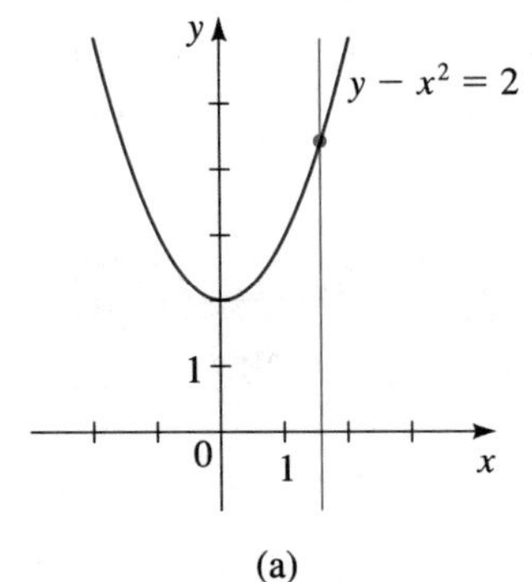

(a)

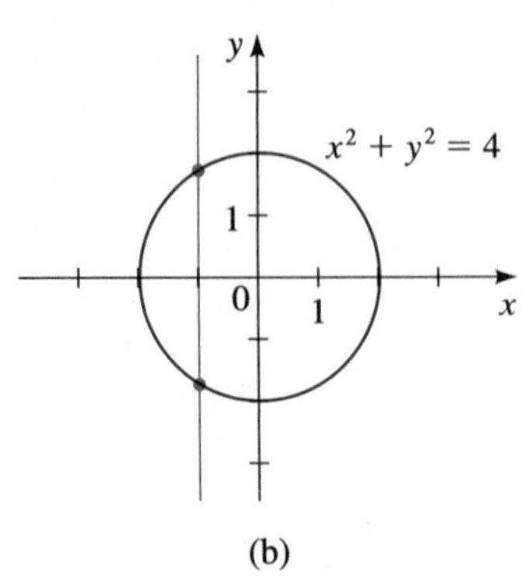

(b)

FIGURE 13

The graphs of the equations in Example 9 are shown in Figure 13. The Vertical Line Test shows graphically that the equation in Example 9(a) defines a function but the equation in Example 9(b) does not.

The following table shows the graphs of some functions that you will see frequently in this book.

SOME FUNCTIONS AND THEIR GRAPHS

Linear functions
$f(x) = mx + b$

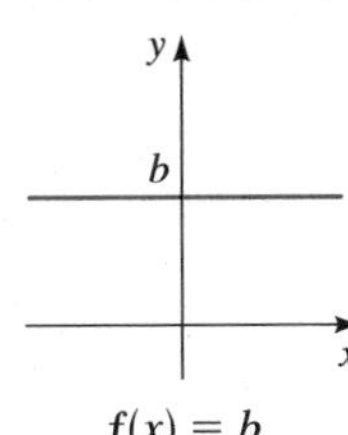

$f(x) = b$

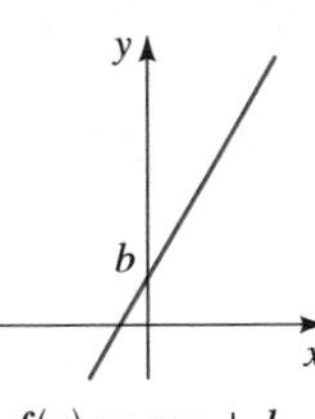

$f(x) = mx + b$

Power functions
$f(x) = x^n$

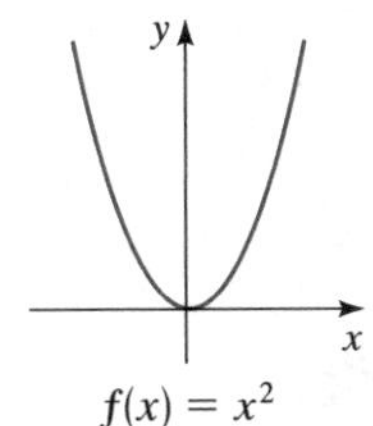

$f(x) = x^2$

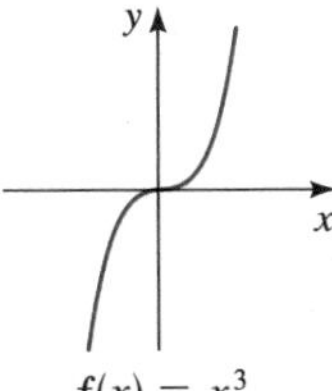

$f(x) = x^3$

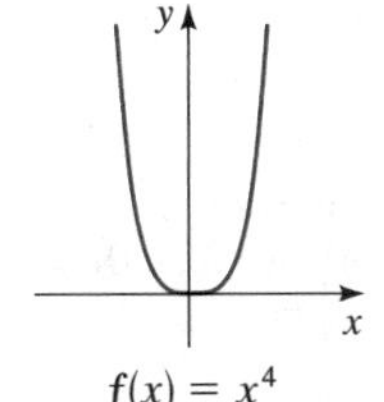

$f(x) = x^4$

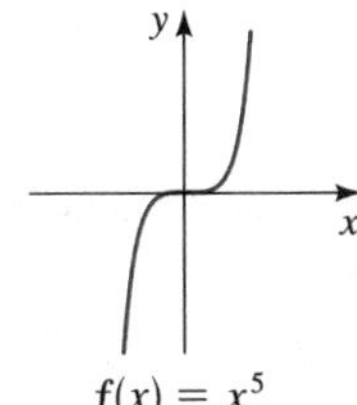

$f(x) = x^5$

Root functions
$f(x) = \sqrt[n]{x}$

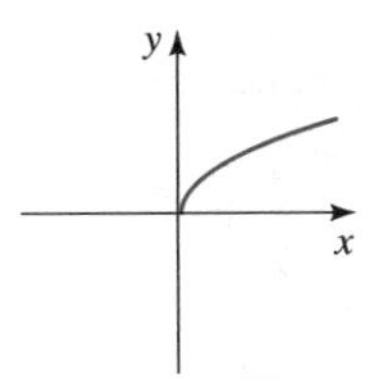

$f(x) = \sqrt{x}$

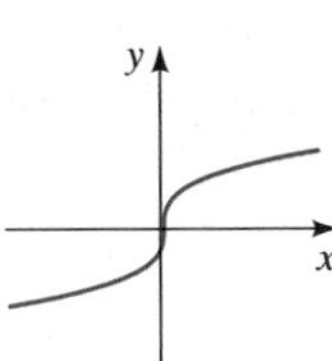

$f(x) = \sqrt[3]{x}$

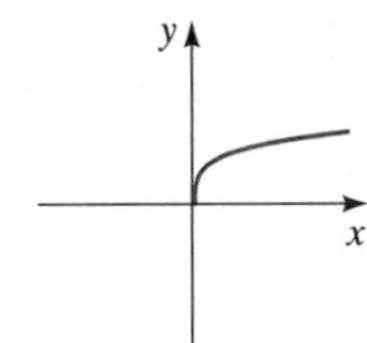

$f(x) = \sqrt[4]{x}$

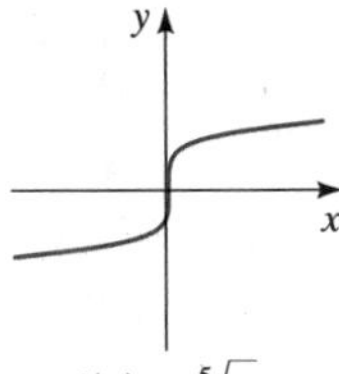

$f(x) = \sqrt[5]{x}$

Reciprocal functions
$f(x) = 1/x^n$

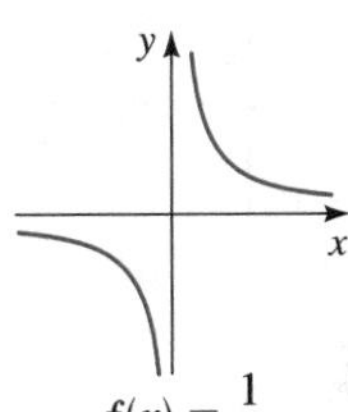

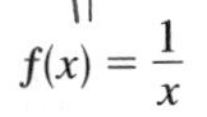

$f(x) = \dfrac{1}{x}$

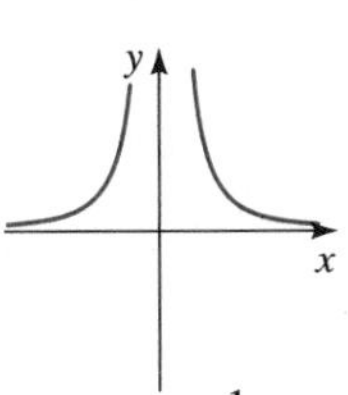

$f(x) = \dfrac{1}{x^2}$

Absolute value function
$f(x) = |x|$

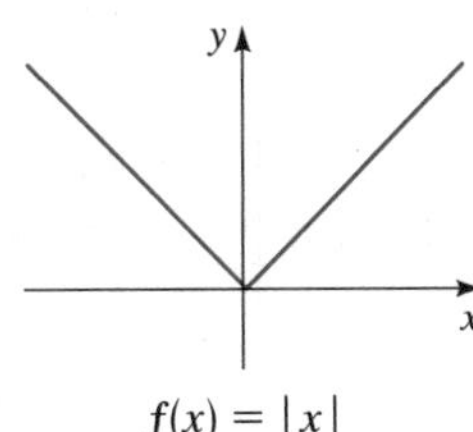

$f(x) = |x|$

Greatest integer function
$f(x) = [\![x]\!]$

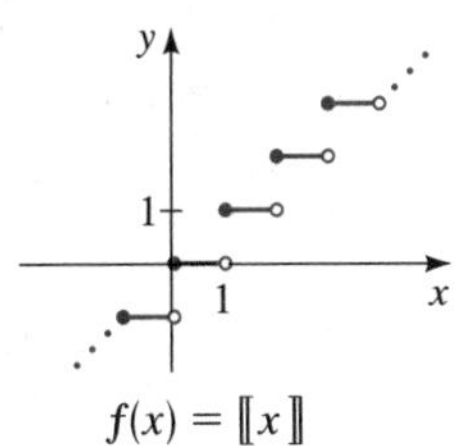

$f(x) = [\![x]\!]$

4.2 EXERCISES

1. The graph of a function h is given.

(a) State the values of $h(-2)$, $h(0)$, $h(2)$, and $h(3)$.

(b) State the domain and range of h.

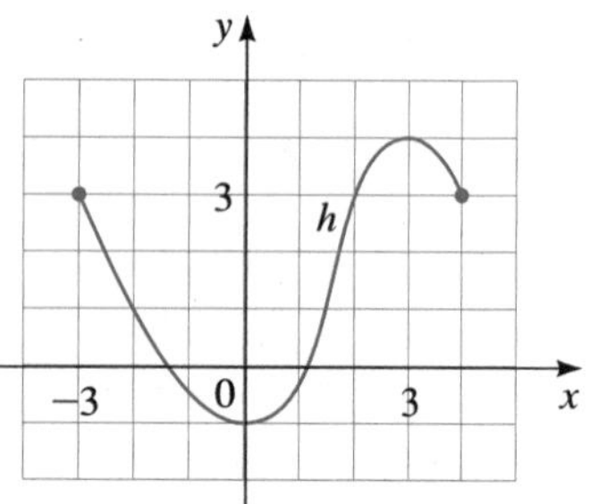

2. The graph of a function g is given.

(a) State the values of $g(-4)$, $g(-2)$, $g(0)$, $g(2)$, and $g(4)$.

(b) State the domain and range of g.

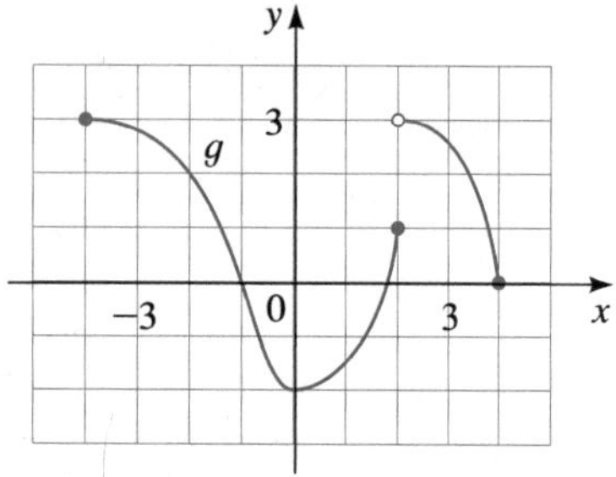

3. Graphs of the functions f and g are given.

(a) Which is larger, $f(0)$ or $g(0)$?

(b) Which is larger, $f(-3)$ or $g(-3)$?

(c) For which values of x is $f(x) = g(x)$?

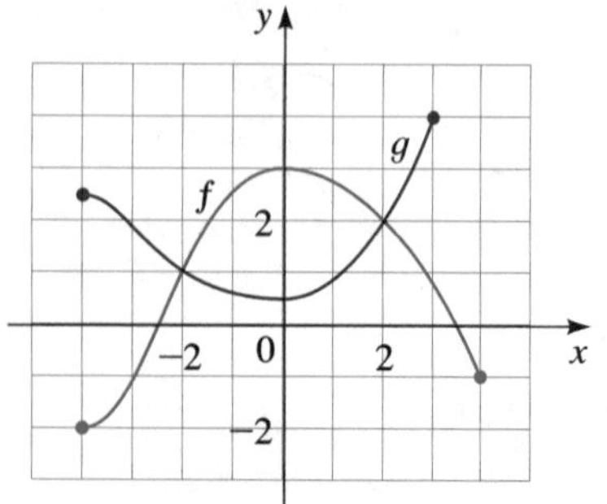

4. The graph of a function f is given.

(a) Estimate $f(0.5)$ to the nearest tenth.

(b) Estimate $f(3)$ to the nearest tenth.

(c) Find all the numbers x in the domain of f so that $f(x) = 1$.

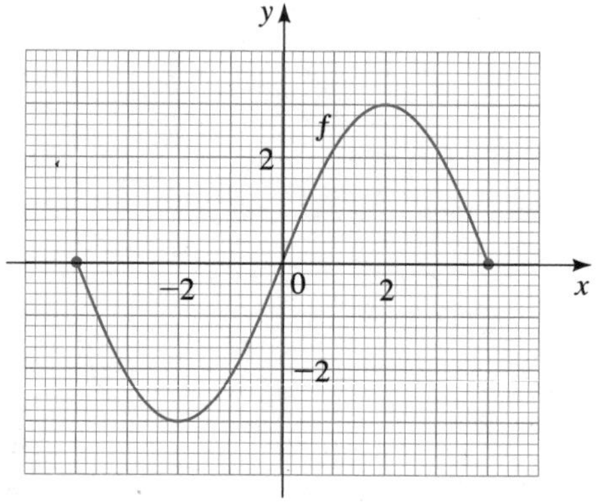

5–6 ■ Determine whether the curve is the graph of a function of x.

5. (a)

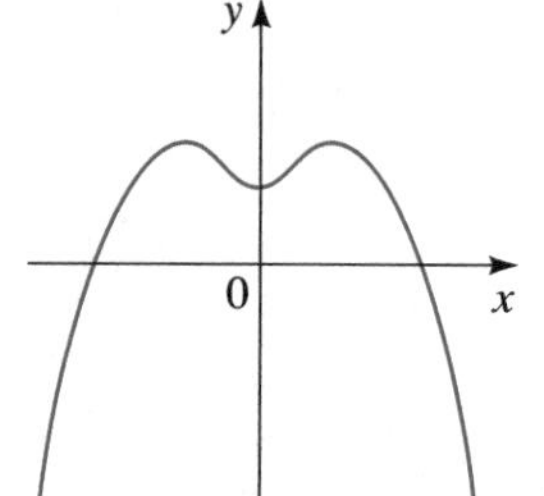

(b)

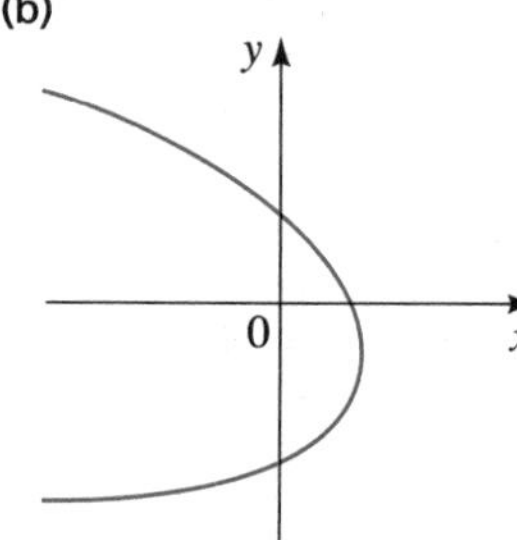

(c)

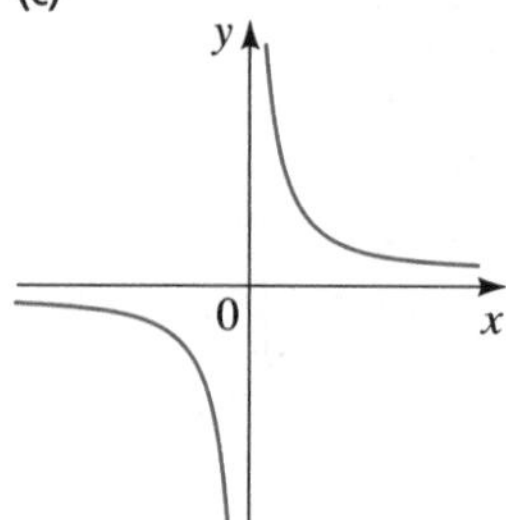

(d)

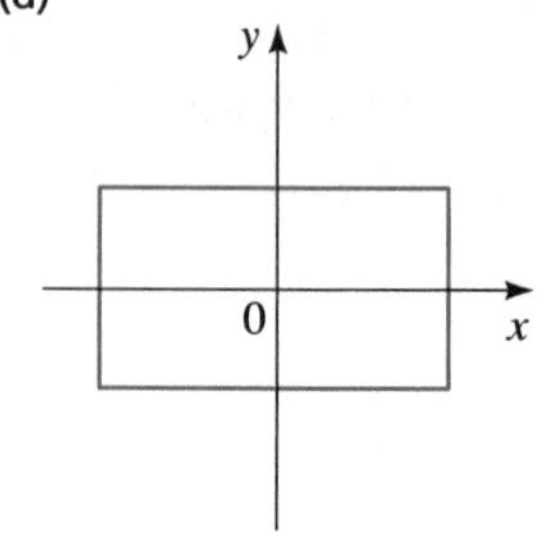

6. (a)

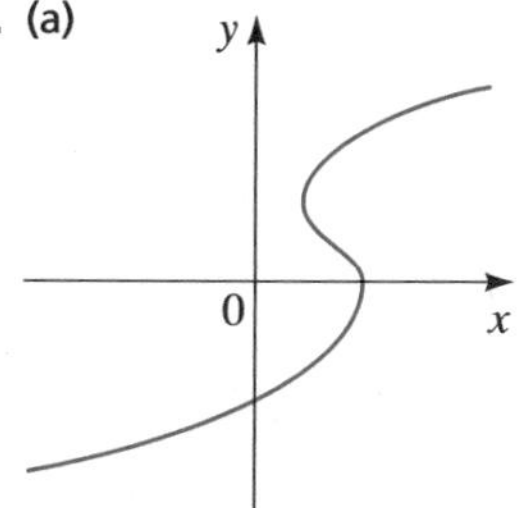

(b)

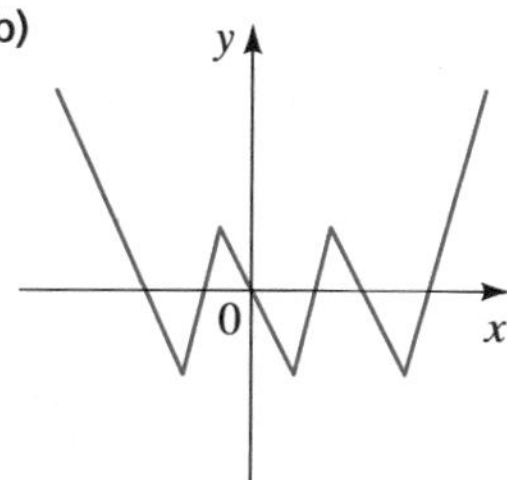

(c)

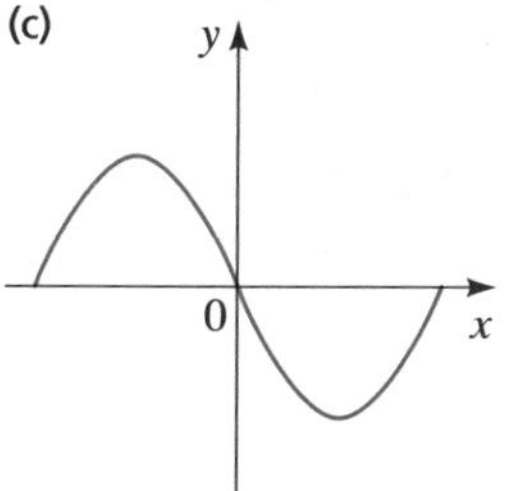

(d)

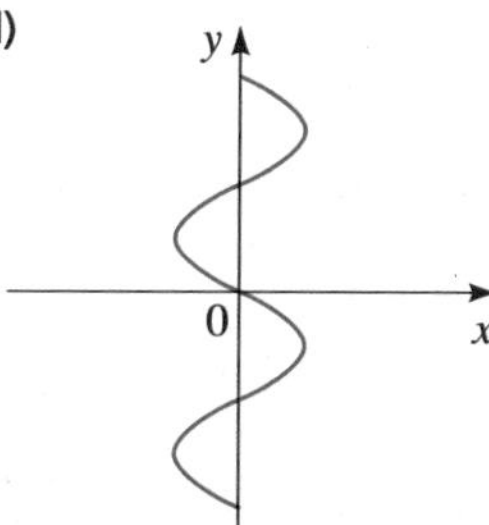

7–10 ■ Determine whether the curve is the graph of a function x. If it is, state the domain and range of the function.

7.

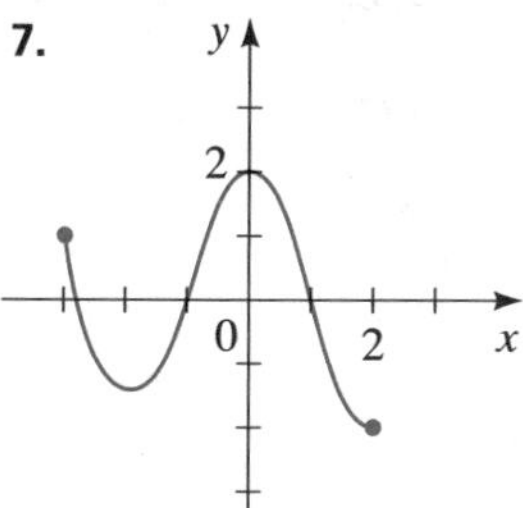

8.

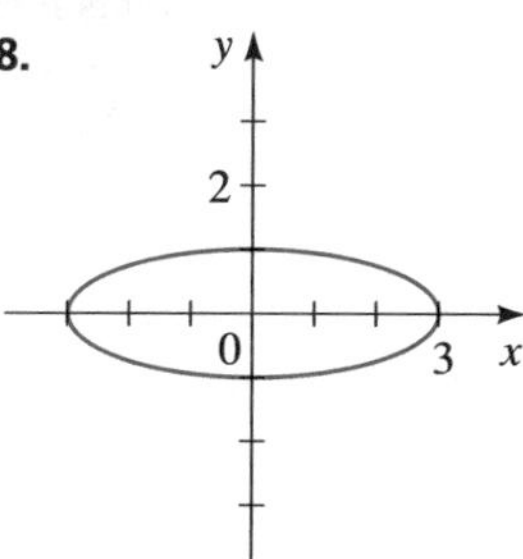

9.

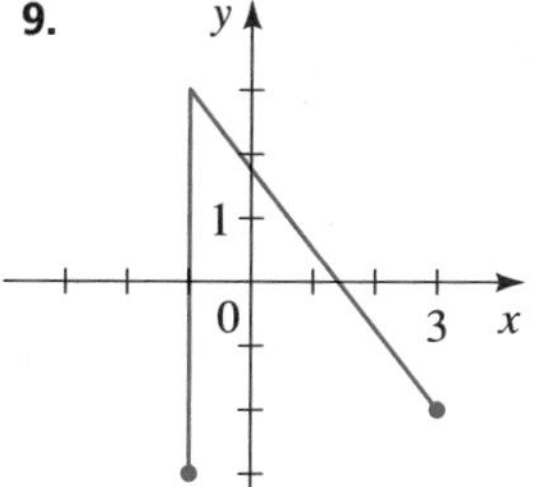

10.

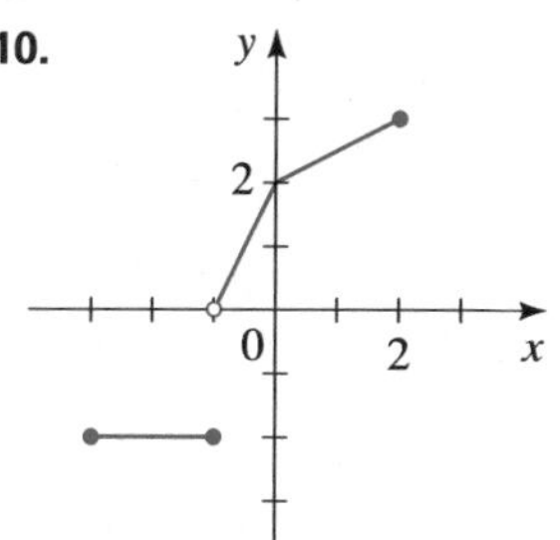

11–18 ■ A function f is given.
(a) Sketch the graph of f.
(b) Find the domain of f.

11. $f(x) = 1 - x$

12. $f(x) = \frac{1}{2}(x + 1)$

13. $f(x) = x^2 - 4x$

14. $f(x) = x^2 - 4x + 4$

15. $f(x) = \sqrt{9 - x}$

16. $f(x) = \sqrt{2x + 6}$

17. $f(x) = \sqrt{16 - x^2}$

18. $f(x) = -\sqrt{25 - x^2}$

19–40 ■ Sketch the graph of the function.

19. $f(x) = 3$

20. $f(x) = -5$

21. $f(x) = 2x + 3$

22. $f(x) = 6 - 3x$

23. $f(x) = -x + 4, \quad -1 \leq x \leq 4$

24. $f(x) = \dfrac{x + 3}{2}, \quad -2 \leq x \leq 2$

25. $f(x) = -x^2$

26. $f(x) = x^2 - 4$

27. $g(x) = x^3 - 8$

28. $g(x) = 4x^2 - x^4$

29. $g(x) = \sqrt{-x}$

30. $g(x) = \sqrt{6 - 2x}$

31. $F(x) = \dfrac{1}{x}$

32. $F(x) = \dfrac{2}{x + 4}$

33. $H(x) = |2x|$

34. $H(x) = |x + 1|$

35. $G(x) = |x| + x$

36. $G(x) = |x| - x$

37. $f(x) = |2x - 2|$

38. $f(x) = \dfrac{x}{|x|}$

39. $g(x) = \dfrac{2}{x^2}$

40. $g(x) = \dfrac{|x|}{x^2}$

41–52 ■ Determine whether the equation defines y as a function of x. (See Example 9.)

41. $x^2 + 2y = 4$

42. $3x + 7y = 21$

43. $x = y^2$

44. $x^2 + (y - 1)^2 = 4$

45. $x + y^2 = 9$

46. $x^2 + y = 9$

47. $x^2y + y = 1$

48. $\sqrt{x} + y = 12$

49. $2|x| + y = 0$

50. $2x + |y| = 0$

51. $x = y^3$

52. $x = y^4$

53–58 ■ A family of functions is given. In parts (a) and (b) graph all the given members of the family in the viewing rectangle indicated. In part (c) state the conclusions you can make from your graphs.

53. $f(x) = x^2 + c$
(a) $c = 0, 2, 4, 6;\quad [-5, 5]$ by $[-10, 10]$
(b) $c = 0, -2, -4, -6;\quad [-5, 5]$ by $[-10, 10]$
(c) How does the value of c affect the graph?

54. $f(x) = (x - c)^2$
(a) $c = 0, 1, 2, 3$; $[-5, 5]$ by $[-10, 10]$
(b) $c = 0, -1, -2, -3$; $[-5, 5]$ by $[-10, 10]$
(c) How does the value of c affect the graph?

55. $f(x) = (x - c)^3$
(a) $c = 0, 2, 4, 6$; $[-10, 10]$ by $[-10, 10]$
(b) $c = 0, -2, -4, -6$; $[-10, 10]$ by $[-10, 10]$
(c) How does the value of c affect the graph?

56. $f(x) = cx^2$
(a) $c = 1, \frac{1}{2}, 2, 4$; $[-5, 5]$ by $[-10, 10]$
(b) $c = 1, -1, -\frac{1}{2}, -2$; $[-5, 5]$ by $[-10, 10]$
(c) How does the value of c affect the graph?

57. $f(x) = x^c$
(a) $c = \frac{1}{2}, \frac{1}{4}, \frac{1}{6}$; $[-1, 4]$ by $[-1, 3]$
(b) $c = 1, \frac{1}{3}, \frac{1}{5}$; $[-3, 3]$ by $[-2, 2]$
(c) How does the value of c affect the graph?

58. $f(x) = 1/x^n$
(a) $n = 1, 3$; $[-3, 3]$ by $[-3, 3]$
(b) $n = 2, 4$; $[-3, 3]$ by $[-3, 3]$
(c) How does the value of n affect the graph?

59–72 ■ Sketch the graph of the piecewise defined function.

59. $f(x) = \begin{cases} 0 & \text{if } x < 2 \\ 1 & \text{if } x \geqslant 2 \end{cases}$

60. $f(x) = \begin{cases} 1 & \text{if } x \leqslant 1 \\ x + 1 & \text{if } x > 1 \end{cases}$

61. $f(x) = \begin{cases} 3 & \text{if } x < 2 \\ x - 1 & \text{if } x \geqslant 2 \end{cases}$

62. $f(x) = \begin{cases} 1 - x & \text{if } x < -2 \\ 5 & \text{if } x \geqslant -2 \end{cases}$

63. $f(x) = \begin{cases} x & \text{if } x \leqslant 0 \\ x + 1 & \text{if } x > 0 \end{cases}$

64. $f(x) = \begin{cases} 2x + 3 & \text{if } x < -1 \\ 3 - x & \text{if } x \geqslant -1 \end{cases}$

65. $f(x) = \begin{cases} -1 & \text{if } x < -1 \\ 1 & \text{if } -1 \leqslant x \leqslant 1 \\ -1 & \text{if } x > 1 \end{cases}$

66. $f(x) = \begin{cases} -1 & \text{if } x < -1 \\ x & \text{if } -1 \leqslant x \leqslant 1 \\ 1 & \text{if } x > 1 \end{cases}$

67. $f(x) = \begin{cases} 2 & \text{if } x \leqslant -1 \\ x^2 & \text{if } x > -1 \end{cases}$

68. $f(x) = \begin{cases} 1 - x^2 & \text{if } x \leqslant 2 \\ x & \text{if } x > 2 \end{cases}$

69. $f(x) = \begin{cases} 0 & \text{if } |x| \leqslant 2 \\ 3 & \text{if } |x| > 2 \end{cases}$

70. $f(x) = \begin{cases} x^2 & \text{if } |x| \leqslant 1 \\ 1 & \text{if } |x| > 1 \end{cases}$

71. $f(x) = \begin{cases} 4 & \text{if } x < -2 \\ x^2 & \text{if } -2 \leqslant x \leqslant 2 \\ -x + 6 & \text{if } x > 2 \end{cases}$

72. $f(x) = \begin{cases} -x & \text{if } x \leqslant 0 \\ 9 - x^2 & \text{if } 0 < x \leqslant 3 \\ x - 3 & \text{if } x > 3 \end{cases}$

73–74 ■ Use a graphing device to draw the graph of the piecewise defined function. (See the margin note on page 222.)

73. $f(x) = \begin{cases} x + 2 & \text{if } x \leqslant -1 \\ x^2 & \text{if } x > -1 \end{cases}$

74. $f(x) = \begin{cases} 2x - x^2 & \text{if } x > 2 \\ (x - 1)^3 & \text{if } x \leqslant 2 \end{cases}$

75. A taxi company charges $2.00 for the first mile (or part of a mile) and 20 cents for each succeeding tenth of a mile (or part). Express the cost C (in dollars) of a ride as a function of the distance x traveled (in miles) for $0 < x < 2$, and sketch the graph of this function.

76. The domestic postage rate for first-class letters weighing 12 oz or less is 33 cents for a letter weighing 1 oz or less and 22 cents for each additional ounce (or part of an ounce). Express the postage P as a function of the weight x of a letter, with $0 < x \leqslant 12$.

77–80 ■ Find a function whose graph is the given curve.

77. The line segment joining the points $(-2, 1)$ and $(4, -6)$

78. The line segment joining the points $(-3, -2)$ and $(6, 3)$

79. The bottom half of the parabola $x + (y - 1)^2 = 0$

80. The top half of the circle $(x - 1)^2 + y^2 = 1$

DISCOVERY • DISCUSSION

81. When Does a Graph Represent a Function? For every integer n, the graph of the equation $y = x^n$ is the graph of a function, namely $f(x) = x^n$. Explain why the graph of $x = y^2$ is *not* the graph of a function of x. Is the graph of $x = y^3$ the graph of a function of x? If so, of what function of x is it the graph? Determine for what integers n the graph of $x = y^n$ is the graph of a function of x.

82. Step Functions In Example 5 and Exercises 75 and 76 we are given functions whose graphs consist of horizontal line segments. Such functions are often called *step functions*, because their graphs look like stairs. Give some other examples of step functions that arise in everyday life.

83. Stretched Step Functions Sketch graphs of the functions $f(x) = [\![x]\!]$, $g(x) = [\![2x]\!]$, and $h(x) = [\![3x]\!]$ on separate graphs. How are the graphs related? If n is a positive integer, what does the graph of $k(x) = [\![nx]\!]$ look like?

84. Graph of the Absolute Value of a Function

(a) Draw the graphs of the functions $f(x) = x^2 + x - 6$ and $g(x) = |x^2 + x - 6|$. How are the graphs of f and g related?

(b) Draw the graphs of the functions $f(x) = x^4 - 6x^2$ and $g(x) = |x^4 - 6x^2|$. How are the graphs of f and g related?

(c) In general, if $g(x) = |f(x)|$, how are the graphs of f and g related? Draw graphs to illustrate your answer.

4.3 APPLIED FUNCTIONS: VARIATION

Mathematical models are discussed in more detail in *Focus on Modeling*, which begins on page 296.

When scientists or applied mathematicians talk about a *mathematical model* for a real-world phenomenon, they often mean a function that describes, at least approximately, the dependence of one physical quantity on another. For instance, the model may describe the population of an animal species as a function of time, or the pressure of a gas as a function of its volume. In this section we study a kind of modeling that occurs frequently in the sciences, called *variation*.

DIRECT VARIATION

Two types of mathematical models occur so often in various sciences that they are given special names. The first is called *direct variation* and occurs when one quantity is a constant multiple of another, so we use a function of the form $f(x) = kx$ to model this dependence.

DIRECT VARIATION

If the quantities x and y are related by an equation

$$y = kx$$

for some constant $k \neq 0$, we say that y **varies directly as** x, or y **is directly proportional to** x, or simply y **is proportional to** x. The constant k is called the **constant of proportionality**.

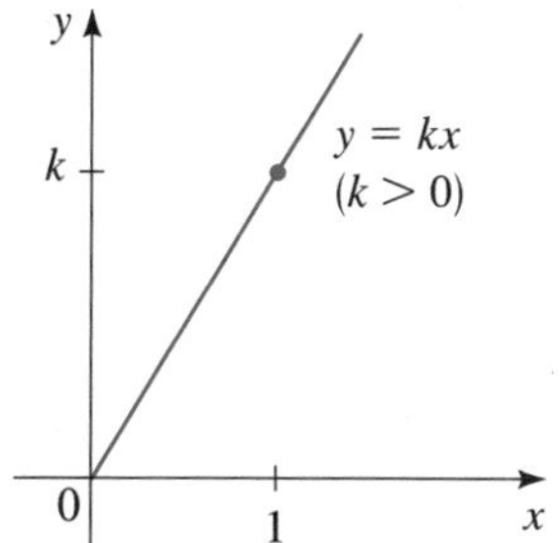

FIGURE 1
Direct variation

Recall that a linear function is a function of the form $f(x) = mx + b$ and its graph is a line with slope m and y-intercept b. So if y varies directly as x, then the equation $y = kx$ or, equivalently, $f(x) = kx$, shows that y is a linear function of x. The graph of this function is a line with slope k, the constant of proportionality (see Figure 1). The y-intercept is $b = 0$, so this line passes through the origin.

EXAMPLE 1 ■ Direct Variation

During a thunderstorm you see the lightning before you hear the thunder because light travels much faster than sound. The distance between you and the storm varies directly as the time interval between the lightning and the thunder.

(a) Suppose that the thunder from a storm 5400 ft away takes 5 s to reach you. Determine the constant of proportionality and write the equation for the variation.

(b) Sketch the graph of this equation. What does the constant of proportionality represent?

(c) If the time interval between the lightning and thunder is now 8 s, how far away is the storm?

SOLUTION

(a) Let d be the distance from you to the storm and let t be the length of the time interval. We are given that d varies directly as t, so

$$d = kt$$

where k is a constant. To find k, we use the fact that $t = 5$ when $d = 5400$. Substituting these values in the equation, we get

$$5400 = k(5) \qquad \text{Substitute}$$

$$k = \frac{5400}{5} = 1080 \qquad \text{Solve for } k$$

Substituting this value of k in the equation for d, we obtain

$$d = 1080t$$

as the equation for d as a function of t.

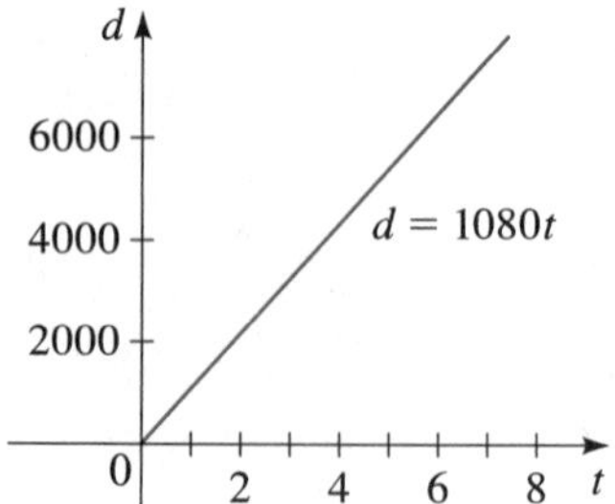

FIGURE 2

(b) The graph of the equation $d = 1080t$ is a line through the origin with slope 1080 and is shown in Figure 2. The constant $k = 1080$ is the approximate speed of sound (in ft/s).

(c) When $t = 8$, we have

$$d = 1080 \cdot 8 = 8640$$

So, the storm is 8640 ft $\approx$ 1.6 mi away. ■

INVERSE VARIATION

Another function that is frequently used in mathematical modeling is the function $f(x) = k/x$, where k is a constant.

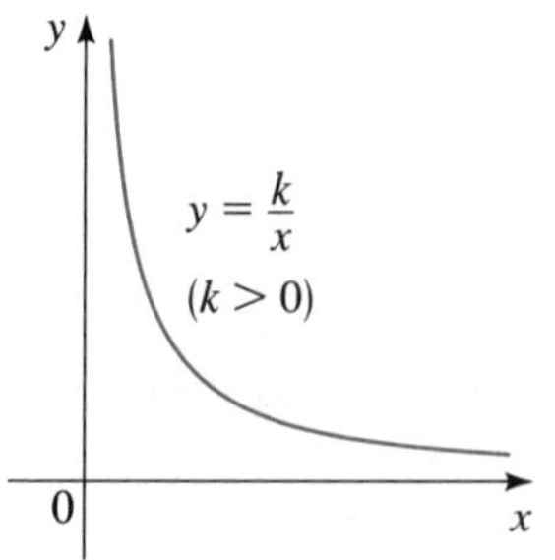

FIGURE 3
Inverse variation

INVERSE VARIATION

If the quantities x and y are related by the equation

$$y = \frac{k}{x}$$

for some constant $k \neq 0$, we say that y **is inversely proportional to** x, or y **varies inversely as** x.

The graph of the function $f(x) = k/x$ for $x > 0$ is shown in Figure 3 for the case $k > 0$. It gives a picture of what happens when y is inversely proportional to x.

EXAMPLE 2 ■ Inverse Variation

Boyle's law states that when a sample of gas is compressed at a constant temperature, the pressure of the gas is inversely proportional to the volume of the gas.

(a) Suppose the pressure of a sample of air that occupies 0.106 m^3 at 25°C is 50 kPa. Find the constant of proportionality, and write the equation that expresses the inverse proportionality.
(b) If the sample expands to a volume of 0.3 m^3, find the new pressure.

SOLUTION

(a) Let P be the pressure of the sample of gas and let V be its volume. Then, by the definition of inverse proportionality, we have

$$P = \frac{k}{V}$$

where k is a constant. To find k we use the fact that $P = 50$ when $V = 0.106$. Substituting these values in the equation, we get

$$50 = \frac{k}{0.106} \qquad \text{Substitute}$$

$$k = (50)(0.106) = 5.3 \qquad \text{Solve for } k$$

Putting this value of k in the equation for P, we have

$$P = \frac{5.3}{V}$$

(b) When $V = 0.3$, we have

$$P = \frac{5.3}{0.3} \approx 17.7$$

So, the new pressure is about 17.7 kPa. ■

JOINT VARIATION

A physical quantity often depends on more than one other quantity. If one quantity is proportional to two or more other quantities, we call this relationship *joint variation.*

> **JOINT VARIATION**
>
> If the quantities x, y, and z are related by the equation
>
> $$z = kxy$$
>
> where k is a nonzero constant, we say that z **varies jointly as** x and y, or z **is jointly proportional to** x and y.

In the sciences, functions of three or more variables are common, and any combination of the different types of proportionality that we have discussed is possible. For example, if

$$z = k\frac{x}{y}$$

we say that z **is proportional to** x and **inversely proportional to** y.

EXAMPLE 3 ■ Newton's Law of Gravitation

Newton's Law of Gravitation says that two objects with masses m_1 and m_2 attract each other with a force F that is jointly proportional to their masses and inversely proportional to the square of the distance r between the objects. Express Newton's Law of Gravitation as an equation.

SOLUTION

Using the definitions of joint and inverse proportionality, and the traditional notation G for the constant of proportionality, we have

$$F = G\frac{m_1 m_2}{r^2}$$

■

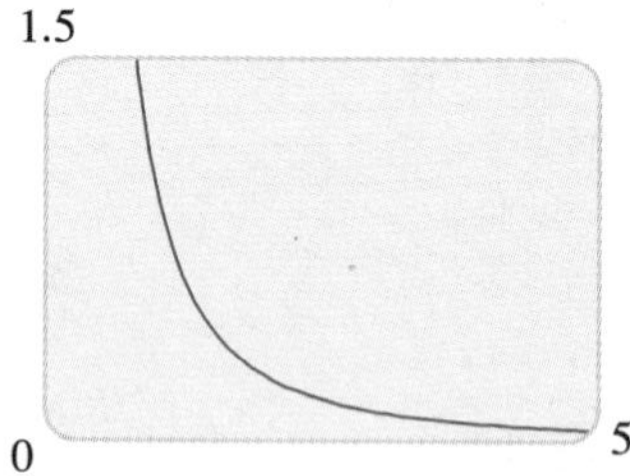

FIGURE 4

Graph of $F = \frac{1}{r^2}$

If m_1 and m_2 are fixed masses, then the gravitational force between them is $F = C/r^2$ (where $C = Gm_1m_2$ is a constant). Figure 4 shows the graph of this function for $r > 0$ with $C = 1$. Observe how the gravitational attraction decreases with increasing distance.

4.3 EXERCISES

1–12 ■ Write an equation that expresses the statement.

1. R varies directly as t.
2. P is directly proportional to u.
3. v is inversely proportional to z.
4. w is jointly proportional to m and n.
5. y is proportional to s and inversely proportional to t.
6. P varies inversely as T.
7. z is proportional to the square root of y.
8. A is proportional to the square of t and inversely proportional to the cube of x.
9. V is jointly proportional to l, w, and h.
10. S is jointly proportional to the squares of r and θ.
11. R is proportional to i and inversely proportional to P and t.
12. A is jointly proportional to the square roots of x and y.

13–22 ■ Express the statement as a formula. Use the given information to find the constant of proportionality.

13. y is directly proportional to x. If $x = 4$, then $y = 72$.
14. z varies inversely as t. If $t = 3$, then $z = 5$.
15. M varies directly as x and inversely as y. If $x = 2$ and $y = 6$, then $M = 5$.
16. S varies jointly as p and q. If $p = 4$ and $q = 5$, then $S = 180$.
17. W is inversely proportional to the square of r. If $r = 6$, then $W = 10$.
18. t is jointly proportional to x and y and inversely proportional to r. If $x = 2$, $y = 3$, and $r = 12$, then $t = 25$.
19. C is jointly proportional to l, w, and h. If $l = w = h = 2$, then $C = 128$.
20. H is jointly proportional to the squares of l and w. If $l = 2$ and $w = \frac{1}{3}$, then $H = 36$.
21. s is inversely proportional to the square root of t. If $s = 100$, then $t = 25$.
22. M is jointly proportional to a, b, and c, and inversely proportional to d. If a and d have the same value, and if b and c are both 2, then $M = 128$.

23. Hooke's law states that the force needed to keep a spring stretched x units beyond its natural length is directly proportional to x. Here the constant of proportionality is called the **spring constant**.
    (a) Write Hooke's law as an equation.
    (b) If a spring has a natural length of 10 cm and a force of 40 N is required to maintain the spring stretched to a length of 15 cm, find the spring constant.
    (c) What force is needed to keep the spring stretched to a length of 14 cm?

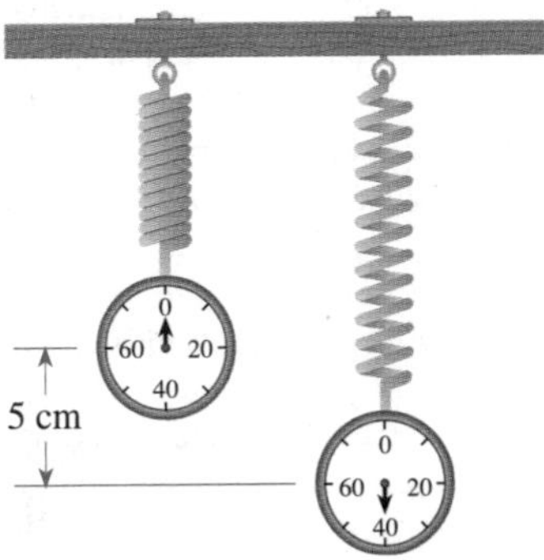

24. The period of a pendulum (the time elapsed during one complete swing of the pendulum) varies directly with the square root of the length of the pendulum.
    (a) Express this relationship by writing an equation.
    (b) In order to double the period, how would we have to change the length l?

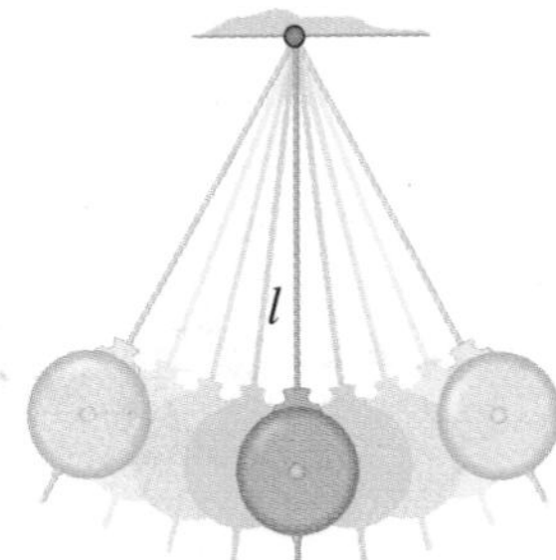

25. The cost C of printing a magazine is jointly proportional to the number of pages p in the magazine and the number of magazines printed m.
    (a) Write an equation that expresses this joint variation.
    (b) Find the constant of proportionality if the printing cost is $60,000 for 4000 copies of a 120-page magazine.

(c) How much would the printing cost be for 5000 copies of a 92-page magazine?

26. The pressure P of a sample of gas is directly proportional to the temperature T and inversely proportional to the volume V.
(a) Write an equation that expresses this variation.
(b) Find the constant of proportionality if 100 L of gas exerts a pressure of 33.2 kPa at a temperature of 400 K (absolute temperature measured on the Kelvin scale).
(c) If the temperature is increased to 500 K and the volume is decreased to 80 L, what is the pressure of the gas?

27. The resistance R of a wire varies directly as its length L and inversely as the square of its diameter d.
(a) Write an equation that expresses this joint variation.
(b) Find the constant of proportionality if a wire 1.2 m long and 0.005 m in diameter has a resistance of 140 ohms.
(c) Find the resistance of a wire made of the same material that is 3 m long and has a diameter of 0.008 m.

28. Kepler's Third Law of planetary motion states that the square of the period T of a planet (the time it takes for the planet to make a complete revolution about the sun) is directly proportional to the cube of its average distance d from the sun.
(a) Express Kepler's Third Law as an equation.
(b) Find the constant of proportionality by using the fact that for our planet the period is about 365 days and the average distance is about 93 million miles.
(c) The planet Neptune is about 2.79×10^9 mi from the sun. Find the period of Neptune.

29. The cost of a sheet of gold foil is proportional to its area. If a rectangular sheet measuring 15 cm by 20 cm costs \$75, how much would a 3 cm by 5 cm sheet cost?

30. In the short growing season of the Canadian arctic territory of Nunavut, some gardeners find it possible to grow gigantic cabbages in the midnight sun. Assume that the final size of a cabbage is proportional to the amount of nutrients it receives, and inversely proportional to the number of other cabbages surrounding it. A cabbage that received 20 oz of nutrients and had 12 other cabbages around it grew to 30 lb. What size would it grow to if it received 10 oz of nutrients and had only 5 cabbage "neighbors"?

31. The value of a building lot on Galiano Island is jointly proportional to its area and the quantity of water produced by a well on the property. A 200 ft by 300 ft lot has a well producing 10 gallons of water per minute, and is valued at \$48,000. What is the value of a 400 ft by 400 ft lot if the well on the lot produces 4 gallons of water per minute?

32. The heat experienced by a hiker at a campfire is proportional to the amount of wood on the fire, and inversely proportional to the cube of his distance from the fire. If he is 20 ft from the fire, and someone doubles the amount of wood burning, how far from the fire would he have to be so that he feels the same heat as before?

DISCOVERY • DISCUSSION

33. Is Proportionality Everything? A great many laws of physics and chemistry are expressible as proportionalities. Give at least one example of a function that occurs in the sciences that is *not* a proportionality.

4.4 AVERAGE RATE OF CHANGE: INCREASING AND DECREASING FUNCTIONS

Functions are often used to model changing quantities. In this section we learn how to determine the rate at which the values of a function change as the variable changes.

AVERAGE RATE OF CHANGE

We are all familiar with the concept of speed: If you drive a distance of 120 miles in 2 hours, then your average speed, or rate of travel, is $\frac{120 \text{ mi}}{2 \text{ h}} = 60$ mi/h.

Now suppose you take a car trip and record the distance that you travel every few minutes. The distance s you have traveled is a function of the time t:

$$s(t) = \text{total distance traveled at time } t$$

We graph the function s as shown in Figure 1. The graph shows that you have traveled a total of 50 miles after 1 hour, 75 miles after 2 hours, 140 miles after 3 hours, and so on. To find your *average* speed between any two points on the trip, we divide the distance traveled by the time elapsed. Let's calculate your average speed between 1:00 P.M. and 4:00 P.M. The time elapsed is $4 - 1 = 3$ hours. To find the distance you traveled, we subtract the distance at 1:00 P.M. from the distance at 4:00 P.M., that is, $200 - 50 = 150$ mi. Thus, your average speed is

$$\text{average speed} = \frac{\text{distance traveled}}{\text{time elapsed}} = \frac{150 \text{ mi}}{3 \text{ h}} = 50 \text{ mi/h}$$

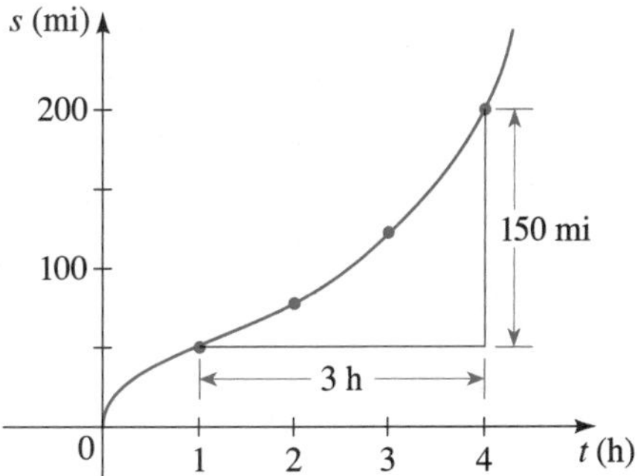

FIGURE 1
Average speed

The average speed we have just calculated can be expressed using function notation:

$$\text{average speed} = \frac{s(4) - s(1)}{4 - 1} = \frac{200 - 50}{3} = 50 \text{ mi/h}$$

Note that the average speed is different over different time intervals. For example, between 2:00 P.M. and 3:00 P.M. we find that

$$\text{average speed} = \frac{s(3) - s(2)}{3 - 2} = \frac{140 - 75}{1} = 65 \text{ mi/h}$$

Finding average rates of change is important in many contexts. For instance, we may be interested in knowing how quickly the air temperature is dropping as a storm approaches, or how fast revenues are increasing from the sale of a new product. So we need to know how to determine the average rate of change of the functions that model these quantities. In fact, the concept of average rate of change can be defined for any function.

AVERAGE RATE OF CHANGE

The **average rate of change** of the function $y = f(x)$ between $x = a$ and $x = b$ is

$$\text{average rate of change} = \frac{\text{change in } y}{\text{change in } x} = \frac{f(b) - f(a)}{b - a}$$

The average rate of change is the slope of the **secant line** between $x = a$ and $x = b$ on the graph of f, that is, the line that passes through $(a, f(a))$ and $(b, f(b))$.

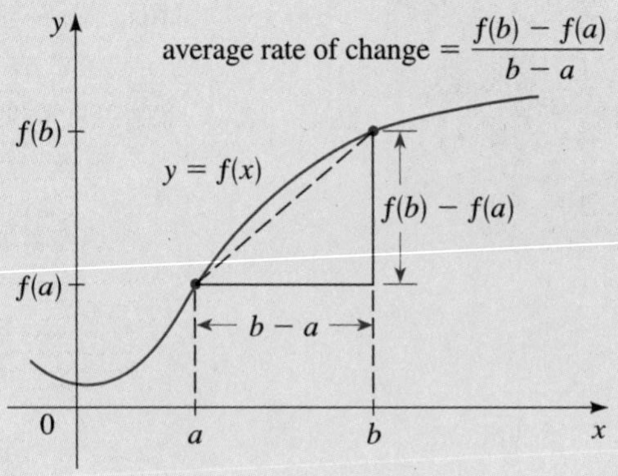

EXAMPLE 1 ■ Calculating the Average Rate of Change

For the function $f(x) = x^2 + 4$, find the average rate of change of the function between the following points:

(a) $x = 2$ and $x = 6$
(b) $x = 5$ and $x = 10$
(c) $x = a$ and $x = a + h$ $\quad (h \neq 0)$

SOLUTION

(a) Average rate of change $= \dfrac{f(6) - f(2)}{6 - 2}$ — Definition

$= \dfrac{(6^2 + 4) - (2^2 + 4)}{6 - 2}$ — Use $f(x) = x^2 + 4$

$= \dfrac{40 - 8}{4} = \dfrac{32}{4}$ — Simplify

$= 8$

(b) Average rate of change $= \dfrac{f(10) - f(5)}{10 - 5}$ Definition

$= \dfrac{(10^2 + 4) - (5^2 + 4)}{10 - 5}$ Use $f(x) = x^2 + 4$

$= \dfrac{104 - 29}{5} = \dfrac{75}{5}$ Simplify

$= 15$

(c) Average rate of change $= \dfrac{f(a + h) - f(a)}{(a + h) - a}$

$= \dfrac{[(a + h)^2 + 4] - (a^2 + 4)}{h}$ Use $f(x) = x^2 + 4$

$= \dfrac{[a^2 + 2ah + h^2 + 4] - (a^2 + 4)}{h}$ Expand

$= \dfrac{2ah + h^2}{h}$ Simplify numerator

$= \dfrac{(2a + h)h}{h}$ Factor

$= 2a + h$ Cancel h ■

The average rate of change calculated in Example 1(c) is known as a *difference quotient*. In calculus we use difference quotients to calculate instantaneous rates of change.

EXAMPLE 2 ■ Average Speed of a Falling Object

If an object is dropped from a height of 3000 ft, its distance above the ground (in feet) after t seconds is given by $h(t) = 3000 - 16t^2$. Find the object's average speed for the following times.

(a) Between 1 s and 2 s

(b) Between 4 s and 5 s

SOLUTION

(a) Average rate of change $= \dfrac{h(2) - h(1)}{2 - 1}$ Definition of average rate

$= \dfrac{(3000 - 16(2)^2) - (3000 - 16(1)^2)}{2 - 1}$ Use $h(t) = 3000 - 16t^2$

$= \dfrac{-48}{1}$ Simplify

$= -48$ ft/s

The negative sign indicates that height is decreasing; this makes sense because the object is falling.

(b) $$\text{Average rate of change} = \frac{h(5) - h(4)}{2 - 1} \quad \text{Definition of average rate}$$

$$= \frac{(3000 - 16(5)^2) - (3000 - 16(4)^2)}{5 - 4} \quad \text{Use } h(t) = 3000 - 16t^2$$

$$= \frac{-144}{1} \quad \text{Simplify}$$

$$= -144 \text{ ft/s}$$

The object is falling much faster during the 4- to 5-s period than the 1- to 2-s period. ■

EXAMPLE 3 ■ Average Rate of Temperature Change

The table gives the outdoor temperatures observed by a science student on a spring day. Draw a graph of the data, and find the average rate of change of temperature between the following times:

(a) 8:00 A.M. and 9:00 A.M.
(b) 1:00 P.M. and 3:00 P.M.
(c) 4:00 P.M. and 7:00 P.M.

Time	Temperature (°F)
8:00 A.M.	38
9:00 A.M.	40
10:00 A.M.	44
11:00 A.M.	50
12:00 NOON	56
1:00 P.M.	62
2:00 P.M.	66
3:00 P.M.	67
4:00 P.M.	64
5:00 P.M.	58
6:00 P.M.	55
7:00 P.M.	51

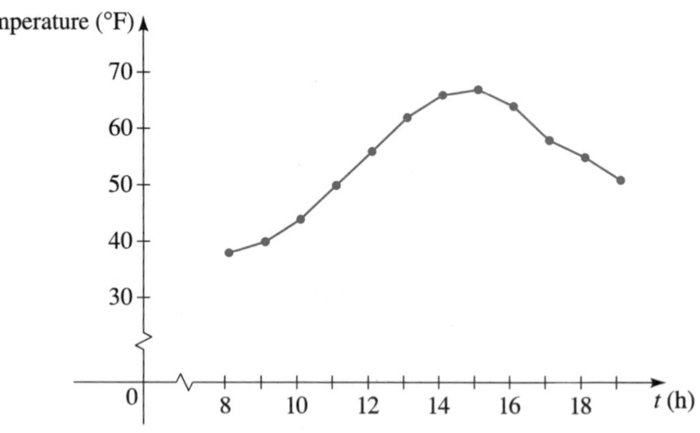

FIGURE 2

SOLUTION

A graph of the temperature data is shown in Figure 2. Let t represent time, measured in hours since midnight (so that 2:00 P.M., for example, corresponds to $t = 14$). Define the function F by

$$F(t) = \text{temperature at time } t$$

MATHEMATICS IN THE MODERN WORLD

COMPUTERS

For centuries machines have been designed to perform specific tasks. For example, a washing machine washes clothes, a weaving machine weaves cloth, an adding machine adds numbers, and so on. The computer has changed all that. The computer is a machine that does nothing—until it is given instructions on what to do. So your computer can play games, draw pictures, or calculate π to a million decimal places; it all depends on what program (or instructions) you give the computer. The computer can do all this because it is able to accept instructions and logically change those instructions based on incoming data. This versatility makes computers useful in nearly every aspect of human endeavor. The idea of a computer was described theoretically in the 1940s by the mathematician Allan Turing (see page 95) in what he called a *universal machine*. In 1945 the mathematician John Von Neumann, extending Turing's ideas, built one of the first electronic computers. Mathematicians continue to develop new theoretical bases for the design of computers. The heart of the computer is the "chip," which is capable of processing logical instructions. To get an idea of the chip's complexity, consider that the Pentium chip has over 3.5 million logic circuits!

(a) $$\text{Average rate of change} = \frac{\text{temperature at 9 A.M.} - \text{temperature at 8 A.M.}}{9-8} = \frac{F(9)-F(8)}{9-8} = \frac{40-38}{9-8} = 2$$

The average rate of change was 2°F per hour.

(b) $$\text{Average rate of change} = \frac{\text{temperature at 3 P.M.} - \text{temperature at 1 P.M.}}{15-13} = \frac{F(15)-F(13)}{15-13} = \frac{67-62}{2} = 2.5$$

The average rate of change was 2.5°F per hour.

(c) $$\text{Average rate of change} = \frac{\text{temperature at 7 P.M.} - \text{temperature at 4 P.M.}}{19-16} = \frac{F(19)-F(16)}{19-16} = \frac{51-64}{3} \approx -4.3$$

The average rate of change was about −4.3°F per hour during this time period. The negative sign indicates that the temperature was dropping. ■

EXAMPLE 4 ■ A Function with Constant Rate of Change

Let $f(x) = 3x - 5$. Find the average rate of change of f between the following points.

(a) $x = 0$ and $x = 1$ (b) $x = 3$ and $x = 7$ (c) $x = a$ and $x = a + h$

What conclusion can you draw from your answers?

SOLUTION

(a) $$\text{Average rate of change} = \frac{f(1)-f(0)}{1-0} = \frac{(3\cdot 1-5)-(3\cdot 0-5)}{1} = \frac{(-2)-(-5)}{1} = 3$$

$$\text{(b) Average rate of change} = \frac{f(7) - f(3)}{7 - 3}$$

$$= \frac{(3 \cdot 7 - 5) - (3 \cdot 3 - 5)}{4}$$

$$= \frac{16 - 4}{4} = 3$$

$$\text{(c) Average rate of change} = \frac{f(a + h) - f(a)}{(a + h) - a}$$

$$= \frac{[3(a + h) - 5] - [3a - 5]}{h}$$

$$= \frac{3a + 3h - 5 - 3a + 5}{h}$$

$$= \frac{3h}{h} = 3$$

It appears that the average rate of change is always 3 for this function. In fact, part (c) proves that the rate of change between any two arbitrary points $x = a$ and $x = a + h$ is 3. ■

As Example 4 indicates, for a linear function $f(x) = mx + b$, the average rate of change between any two points is the slope m of the line. Why is this true? (Think about the definitions of the slope and the average rate of change.)

INCREASING AND DECREASING FUNCTIONS

It is very useful to know where the graph of a function rises and where it falls. The graph shown in Figure 3 rises, falls, then rises again as we move from left to right: It rises from A to B, falls from B to C, and rises again from C to D. The function f is said to be *increasing* when its graph rises and *decreasing* when its graph falls. We have the following definition.

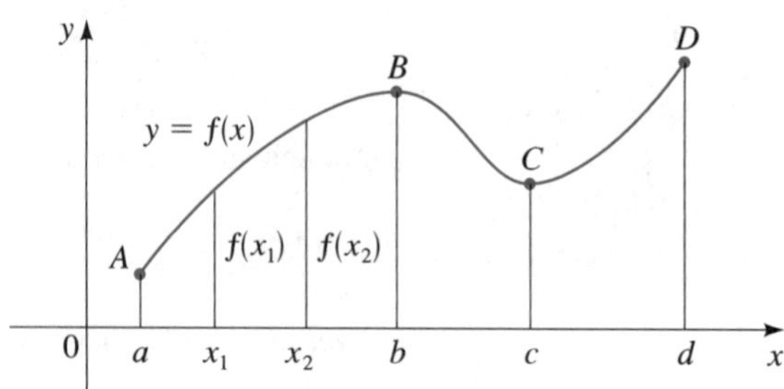

FIGURE 3

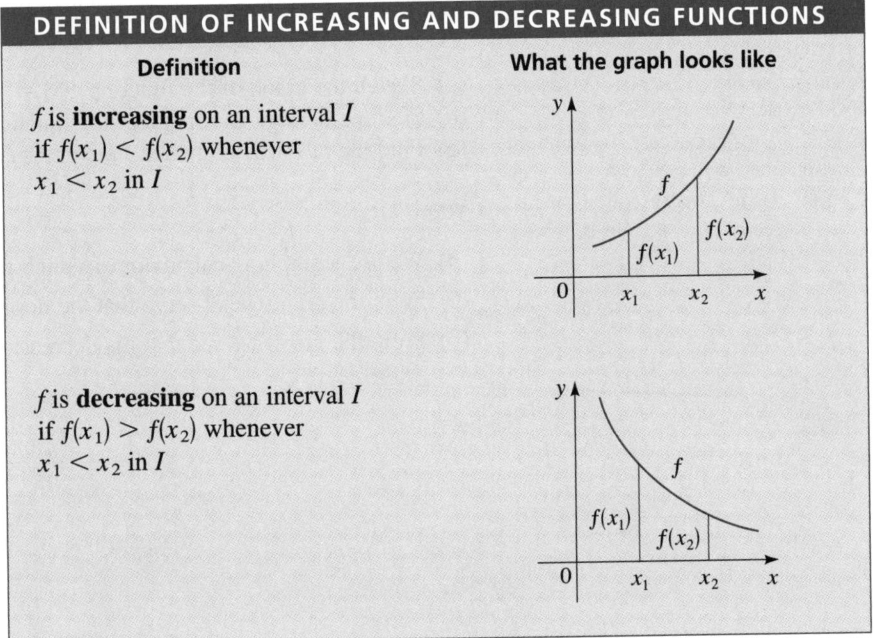

DEFINITION OF INCREASING AND DECREASING FUNCTIONS

Definition	What the graph looks like
f is **increasing** on an interval I if $f(x_1) < f(x_2)$ whenever $x_1 < x_2$ in I	
f is **decreasing** on an interval I if $f(x_1) > f(x_2)$ whenever $x_1 < x_2$ in I	

EXAMPLE 5 ■ Intervals on which a Function Increases or Decreases

State the intervals on which the function whose graph is shown in Figure 4 is increasing or decreasing.

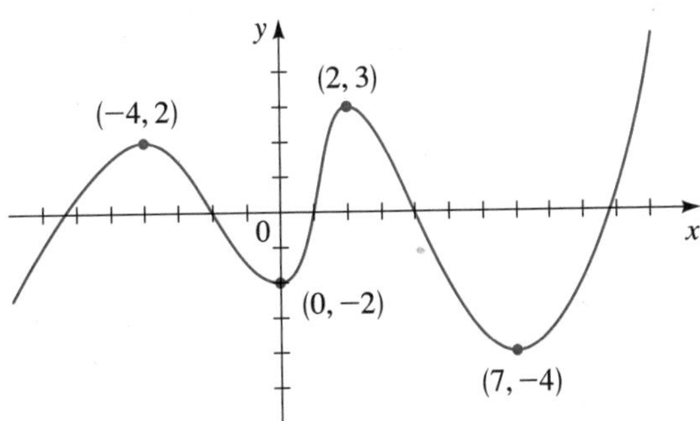

FIGURE 4

SOLUTION

The function is increasing on $(-\infty, -4]$, $[0, 2]$, and $[7, \infty)$. It is decreasing on $[-4, 0]$ and $[2, 7]$. ■

EXAMPLE 6 ■ Using a Graph to Find Intervals where a Function Increases and Decreases

Some graphing calculators, such as the TI-82 do not evaluate $x^{2/3}$ [entered as x^(2/3)] for negative x. To graph a function like $f(x) = x^{2/3}$, we enter it as y_1 = (x^(1/3))^2 because these calculators correctly evaluate powers of the form x^(1/n).

(a) Sketch the graph of the function $f(x) = x^{2/3}$.
(b) Find the domain and range of the function.
(c) Find the intervals on which f increases and decreases.

SOLUTION

(a) We use a graphing calculator to sketch the graph in Figure 5.
(b) From the graph we observe that the domain of f is $\mathbb{R}$ and the range is $[0, \infty)$.
(c) From the graph we see that f is decreasing on $(-\infty, 0]$ and increasing on $[0, \infty)$.

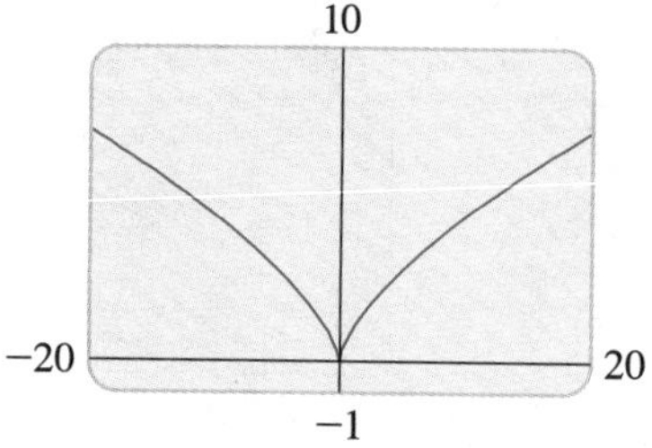

FIGURE 5
Graph of $f(x) = x^{2/3}$

■

The graphs in Figure 6 show that if a function is increasing on an interval, then the average rate of change between any two points is positive, whereas if a function is decreasing on an interval, then the average rate of change between any two points is negative.

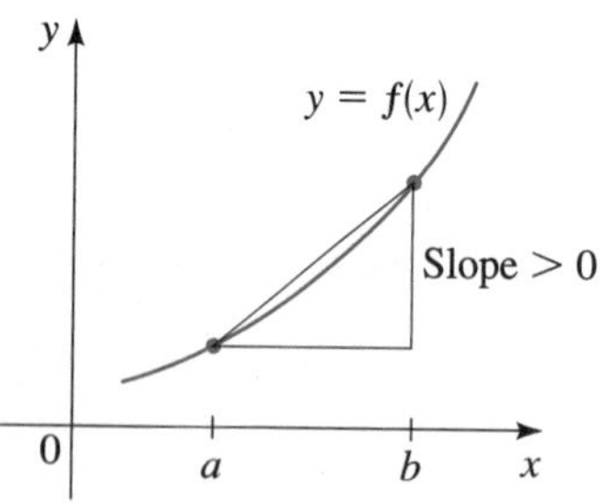

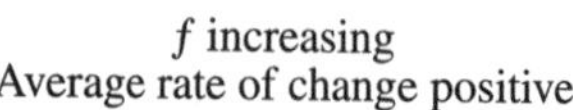

f increasing
Average rate of change positive

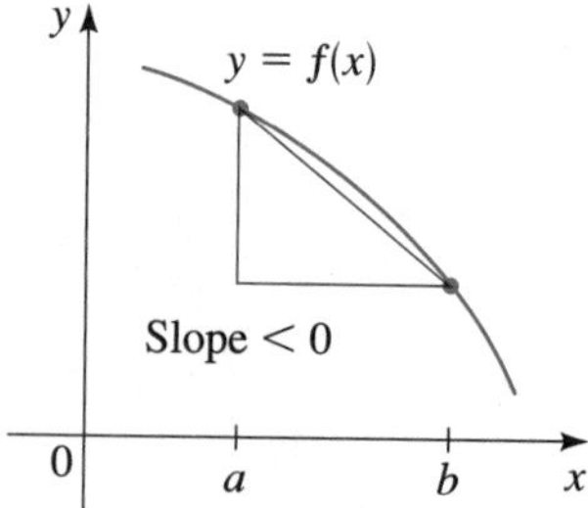

f decreasing
Average rate of change negative

FIGURE 6

4.4 EXERCISES

1–12 ■ A function is given. Determine the average rate of change of the function between the given values of the variable.

1. $f(x) = 3x - 2;\quad x = 2, x = 3$

2. $g(x) = 5 + \frac{1}{2}x;\quad x = 1, x = 5$

3. $h(t) = t^2 + 2t;\quad t = -1, t = 4$

4. $f(z) = 1 - 3z^2;\quad z = -2, z = 0$

5. $f(x) = x^3 - 4x^2;\quad x = 0, x = 10$

6. $f(x) = x + x^4;\quad x = -1, x = 3$

7. $f(x) = 3x^2;\quad x = 2, x = 2 + h$

8. $f(x) = 4 - x^2;\quad x = 1, x = 1 + h$

9. $g(x) = \dfrac{1}{x};\quad x = 1, x = a$

10. $g(x) = \dfrac{2}{x + 1};\quad x = 0, x = h$

11. $f(t) = \dfrac{2}{t};\quad t = a, t = a + h$

12. $f(t) = \sqrt{t};\quad t = a, t = a + h$

13–14 ■ A linear function is given. (a) Find the average rate of change of the function between $x = a$ and $x = a + h$. (b) Show that the average rate of change is the same as the slope of the line.

13. $f(x) = \frac{1}{2}x + 3$ **14.** $g(x) = -4x + 2$

15–18 ■ The graph of a function is given. Determine the average rate of change of the function between the indicated values of the variable.

15.

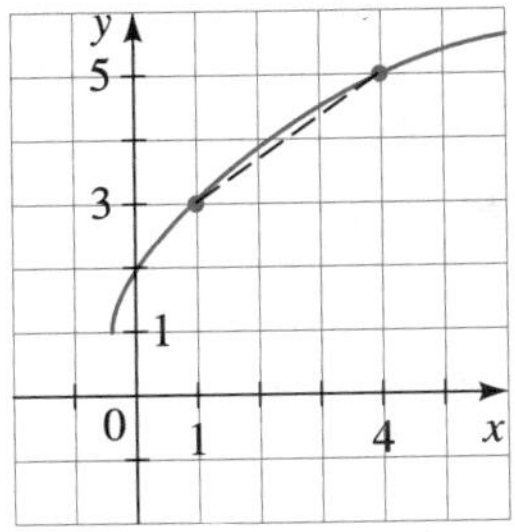

16.

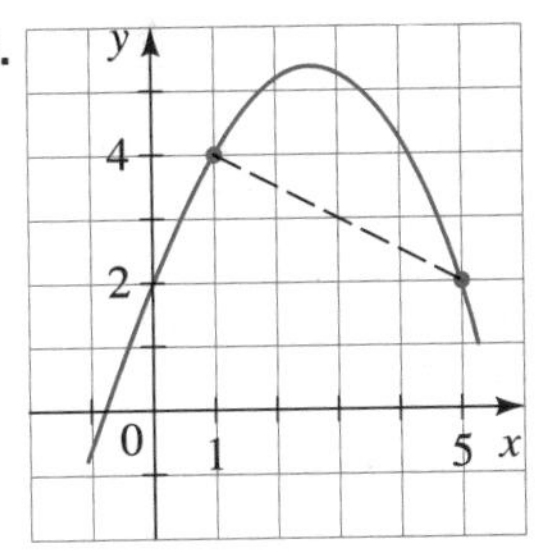

17.

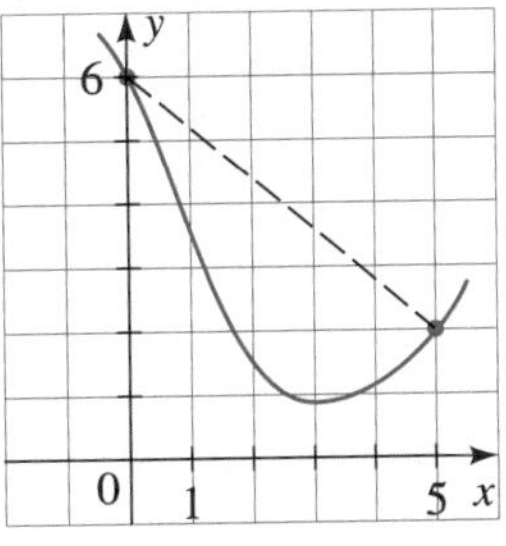

18.

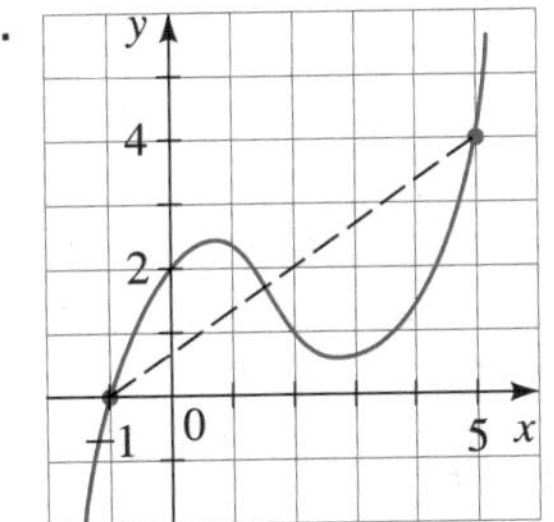

19. The table gives the population in a small coastal community for the period 1990–1999. Figures shown are for January 1 in each year.

(a) What was the average rate of change of population between 1991 and 1994?

(b) What was the average rate of change of population between 1995 and 1997?

(c) For what period of time was the population increasing?

(d) For what period of time was the population decreasing?

Year	Population
1990	624
1991	856
1992	1,336
1993	1,578
1994	1,591
1995	1,483
1996	994
1997	826
1998	801
1999	745

20. A man is running around a circular track 200 m in circumference. An observer uses a stopwatch to time each lap, obtaining the data in the following table.

(a) What was the man's average speed (rate) between 68 s and 152 s?

(b) What was the man's average speed between 263 s and 412 s?

(c) Calculate the man's speed for each lap. Is he slowing down, speeding up, or neither?

Time (s)	Distance (m)
32	200
68	400
108	600
152	800
203	1000
263	1200
335	1400
412	1600

21. The table shows the number of CD players sold in a small electronics store in the years 1989–1999.

(a) What was the average rate of change of sales between 1989 and 1999?

(b) What was the average rate of change of sales between 1989 and 1990?

(c) What was the average rate of change of sales between 1990 and 1992?

(d) Between which two successive years did CD player sales *increase* most quickly? *decrease* most quickly?

Year	CD players sold
1989	512
1990	520
1991	413
1992	410
1993	468
1994	510
1995	590
1996	607
1997	732
1998	612
1999	584

22. Between 1980 and 2000, a rare book collector purchased books for his collection at the rate of 40 books per year. Use this information to complete the following table. (Note that not every year is given in the table.)

Year	Number of books
1980	420
1981	460
1982	
1985	
1990	
1992	
1995	
1997	
1998	
1999	
2000	1220

23–26 ■ The graph of a function is given. Determine the intervals on which the function is increasing and on which it is decreasing.

23.

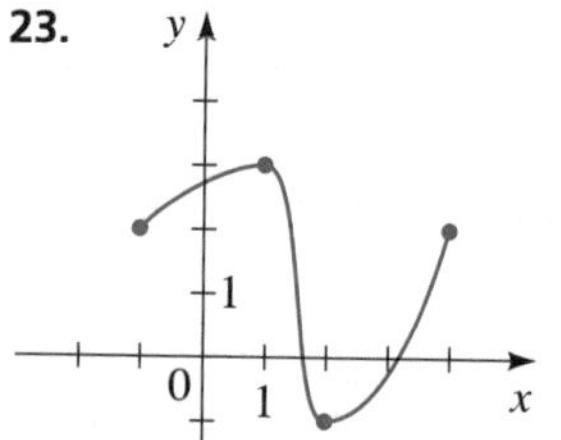

24.

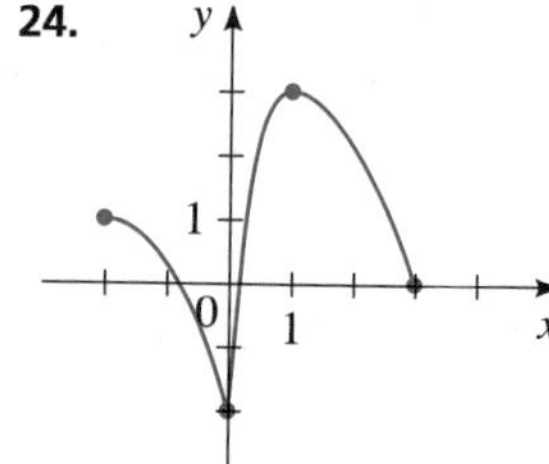

25.

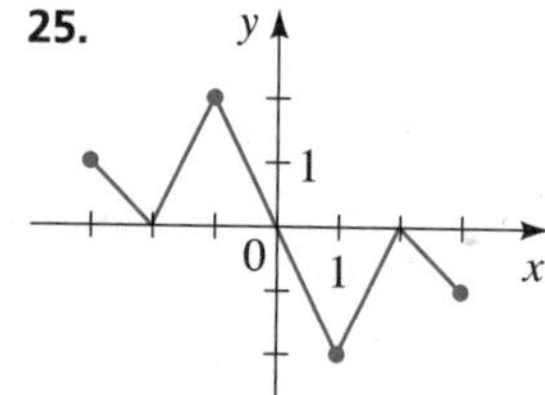

26.

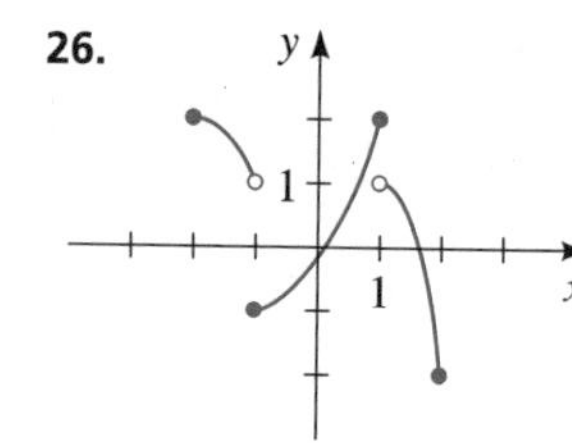

27–34 ■ A function f is given.

(a) Use a graphing device to draw the graph of f.

(b) State approximately the intervals on which f is increasing and on which f is decreasing.

27. $f(x) = x^{2/5}$

28. $f(x) = 4 - x^{2/3}$

29. $f(x) = x^2 - 5x$

30. $f(x) = x^3 - 4x$

31. $f(x) = 2x^3 - 3x^2 - 12x$

32. $f(x) = x^4 - 16x^2$

33. $f(x) = x^3 + 2x^2 - x - 2$

34. $f(x) = x^4 - 4x^3 + 2x^2 + 4x - 3$

DISCOVERY • DISCUSSION

35. **Changing Rates of Change: Concavity** The two tables and graphs give the distances traveled by a racing car during two different 10-s portions of a race. In each case, calculate the average speed at which the car is traveling between the observed data points. Is the speed increasing or decreasing? In other words, is the car *accelerating* or *decelerating* on each of these intervals? How does the shape of the graph tell you whether the car is accelerating or decelerating? (The first graph is said to be *concave up* and the second graph *concave down*.)

(a)

Time (s)	Distance (ft)
0	0
2	34
4	70
6	196
8	490
10	964

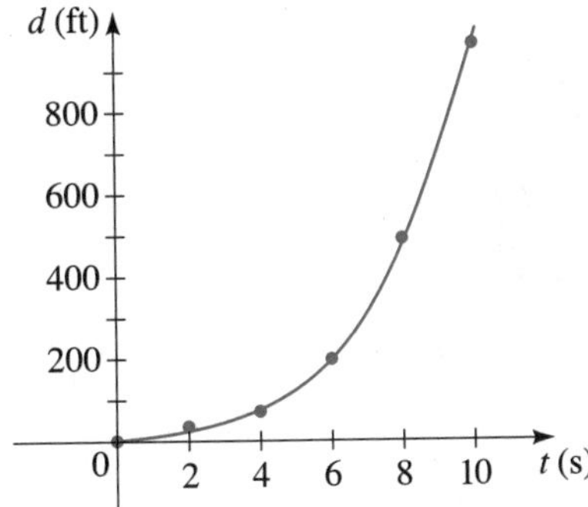

(b)

Time (s)	Distance (ft)
30	5208
32	5734
34	6022
36	6204
38	6352
40	6448

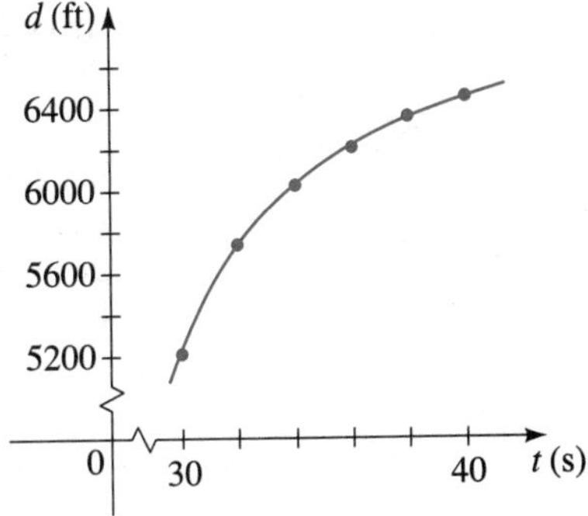

36. **Functions That Are Always Increasing or Decreasing** Sketch rough graphs of functions that are defined for all real numbers and that exhibit the indicated behavior (or explain why it is impossible).
(a) f is always increasing, and $f(x) > 0$ for all x
(b) f is always decreasing, and $f(x) > 0$ for all x
(c) f is always increasing, and $f(x) < 0$ for all x
(d) f is always decreasing, and $f(x) < 0$ for all x

4.5 TRANSFORMATIONS OF FUNCTIONS

In this section we study how certain transformations of a function affect its graph. This will give us a better understanding of how to graph functions. The transformations we study are shifting, reflecting, and stretching.

VERTICAL SHIFTING

Adding a constant to a function shifts its graph vertically: upward if the constant is positive and downward if it is negative.

EXAMPLE 1 ■ Vertical Shifts of Graphs

Sketch the graph of each function.

(a) $f(x) = x^2 + 3$ (b) $g(x) = x^2 - 2$

SOLUTION

Recall that the graph of the function f is the same as the graph of the equation $y = f(x)$. It is often convenient to express a function in equation form, particularly when discussing its graph. For instance, we may refer to the function $f(x) = x^2$ by the equation $y = x^2$, as we do in Example 1.

(a) We start with the graph of the function

$$y = x^2$$

(from Example 1(a) in Section 4.2). The equation

$$y = x^2 + 3$$

indicates that the y-coordinate of a point on the graph of f is 3 more than the y-coordinate of the corresponding point on the graph of $y = x^2$. This means that we obtain the graph of $f(x) = x^2 + 3$ simply by shifting the graph of $y = x^2$ upward 3 units, as shown in Figure 1.

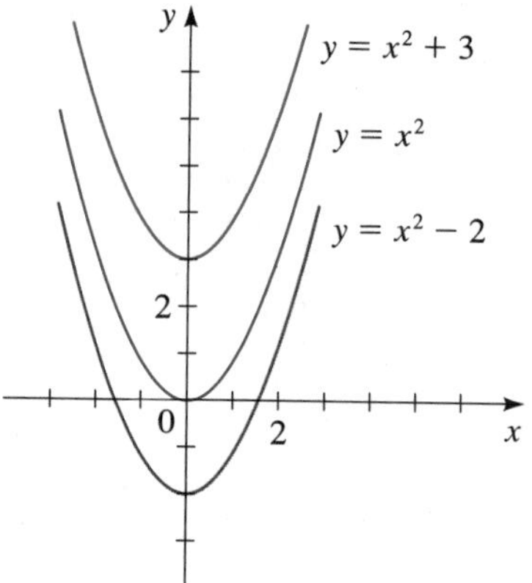

FIGURE 1

(b) Similarly, we get the graph of $g(x) = x^2 - 2$ by shifting the parabola $y = x^2$ downward 2 units (see Figure 1). ■

In general, suppose we know the graph of $y = f(x)$. How do we obtain from it the graphs of

$$y = f(x) + c \quad \text{and} \quad y = f(x) - c \quad (c > 0)$$

The equation $y = f(x) + c$ tells us that the y-coordinate of each point on its graph is c units above the y-coordinate of the corresponding point on the graph of $y = f(x)$. So, we obtain the graph of $y = f(x) + c$ simply by shifting the graph of $y = f(x)$ upward c units. Similarly, we obtain the graph of $y = f(x) - c$ by shifting the graph of $y = f(x)$ downward c units.

We summarize these observations in the following box.

VERTICAL SHIFTS OF GRAPHS

Equation	How to obtain the graph	What the graph looks like
$y = f(x) + c$ $(c > 0)$	Shift graph of $y = f(x)$ upward c units	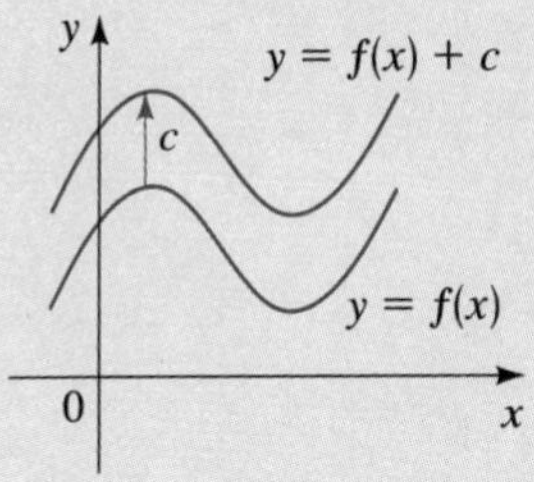
$y = f(x) - c$ $(c > 0)$	Shift graph of $y = f(x)$ downward c units	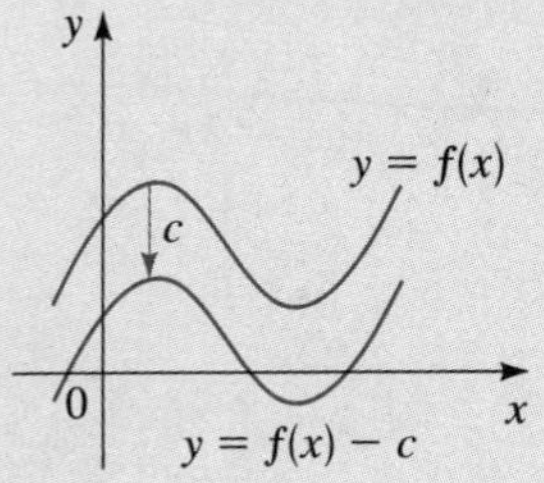

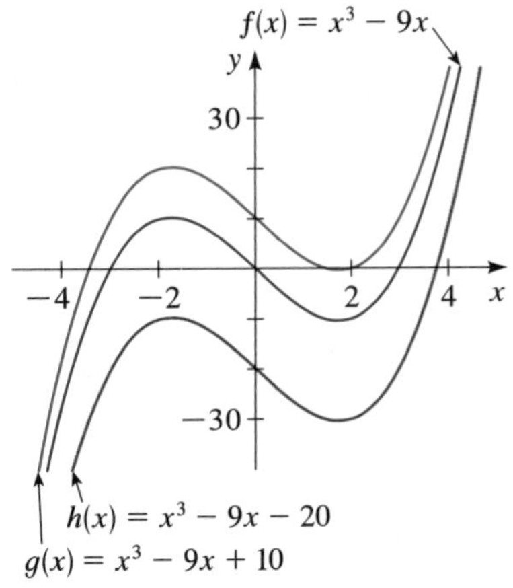

FIGURE 2

EXAMPLE 2 ■ Vertical Shifts of Graphs

Sketch the graph of each function.

(a) $f(x) = x^3 - 9x$ (b) $g(x) = x^3 - 9x + 10$

(c) $h(x) = x^3 - 9x - 20$

SOLUTION

The graphs of f, g, and h are sketched in Figure 2.

(a) The graph of f was sketched in Example 9 in Section 2.2. It is sketched again in Figure 2.

(b) To obtain the graph of g, we shift the graph of f upward 10 units.

(c) To obtain the graph of h, we shift the graph of f downward 20 units. ■

HORIZONTAL SHIFTING

Now we consider transformations of functions that shift the graph horizontally.

Suppose we know the graph of $y = f(x)$. How do we use it to obtain the graphs of

$$y = f(x + c) \qquad \text{and} \qquad y = f(x - c) \qquad (c > 0)$$

The value of $f(x - c)$ at x is the same as the value of $f(x)$ at $x - c$. Since $x - c$ is c units to the left of x, it follows that the graph of $y = f(x - c)$ is just the graph of

$y = f(x)$ shifted to the right c units. Similar reasoning shows that the graph of $y = f(x + c)$ is the graph of $y = f(x)$ shifted to the left c units. The following box summarizes these facts.

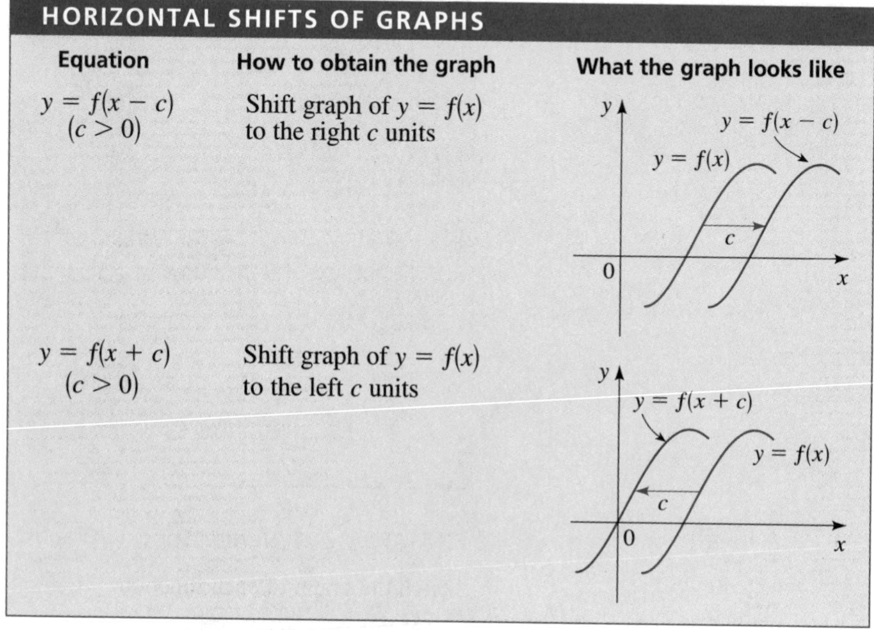

HORIZONTAL SHIFTS OF GRAPHS

Equation	How to obtain the graph	What the graph looks like
$y = f(x - c)$ $(c > 0)$	Shift graph of $y = f(x)$ to the right c units	
$y = f(x + c)$ $(c > 0)$	Shift graph of $y = f(x)$ to the left c units	

EXAMPLE 3 ■ Horizontal Shifts of Graphs

Sketch the graph of each function.

(a) $f(x) = (x + 4)^2$ (b) $g(x) = (x - 2)^2$

SOLUTION

We start with the graph of $y = x^2$, then move it to the left 4 units to get the graph of f and to the right 2 units to obtain the graph of g. See Figure 3.

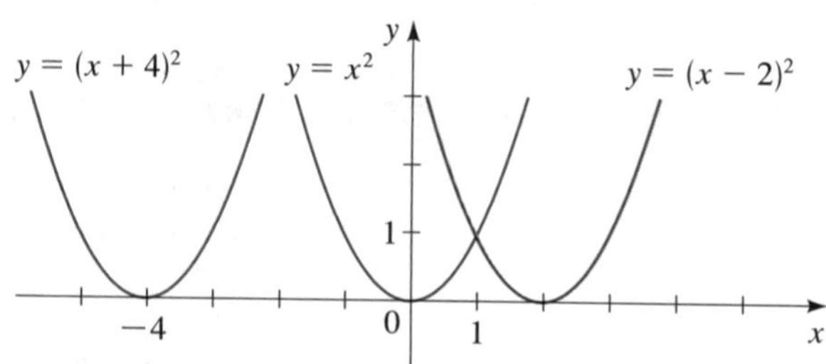

FIGURE 3

■

EXAMPLE 4 ■ Combining Horizontal and Vertical Shifts

Sketch the graph of the function $f(x) = \sqrt{x - 3} + 4$.

SOLUTION

We start with the graph of the square root function $y = \sqrt{x}$ (Example 1(c) in Section 4.2). We move it to the right 3 units to get the graph of $y = \sqrt{x - 3}$. Then we move the resulting graph upward 4 units to obtain the graph of $f(x) = \sqrt{x - 3} + 4$ shown in Figure 4.

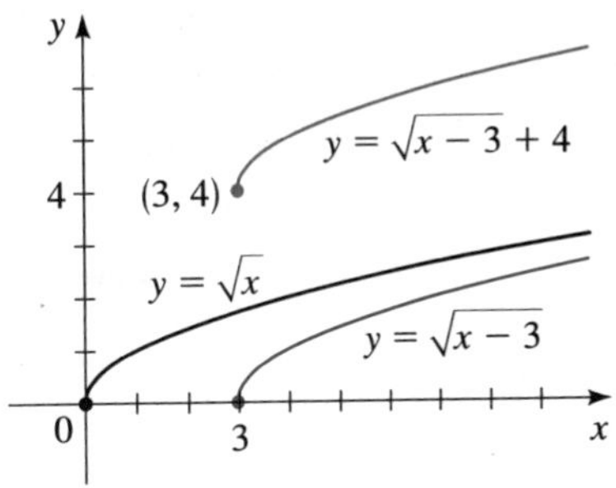

FIGURE 4

■

REFLECTING GRAPHS

Suppose we know the graph of $y = f(x)$. How do we use it to obtain the graphs of $y = -f(x)$ and $y = f(-x)$? The y-coordinate of each point on the graph of $y = -f(x)$ is simply the negative of the y-coordinate of the corresponding point on the graph of $y = f(x)$. So the desired graph is the reflection of the graph of $y = f(x)$ in the x-axis. On the other hand, the value of $y = f(-x)$ at x is the same as the value of $y = f(x)$ at $-x$ and so the desired graph here is the reflection of the graph of $y = f(x)$ in the y-axis. The following box summarizes these observations.

REFLECTING GRAPHS

Equation	How to obtain the graph	What the graph looks like
$y = -f(x)$	Reflect the graph of $y = f(x)$ in the x-axis	y; $y = f(x)$; 0; x; $y = -f(x)$
$y = f(-x)$	Reflect the graph of $y = f(x)$ in the y-axis	y; $y = f(x)$; 0; x; $y = f(-x)$

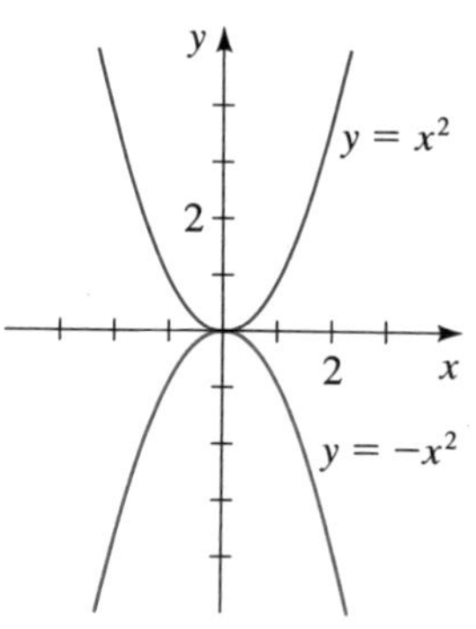

FIGURE 5

EXAMPLE 5 ■ Reflecting Graphs

Sketch the graph of each function.

(a) $f(x) = -x^2$ (b) $g(x) = \sqrt{-x}$

SOLUTION

(a) We start with the graph of $y = x^2$. The graph of $f(x) = -x^2$ is the graph of $y = x^2$ reflected in the x-axis (see Figure 5).

(b) We start with the graph of $y = \sqrt{x}$ (Example 1(c) in Section 4.2). The graph of $g(x) = \sqrt{-x}$ is the graph of $y = \sqrt{x}$ reflected in the y-axis (see Figure 6). Note that the domain of the function $g(x) = \sqrt{-x}$ is $\{x \mid x \leq 0\}$.

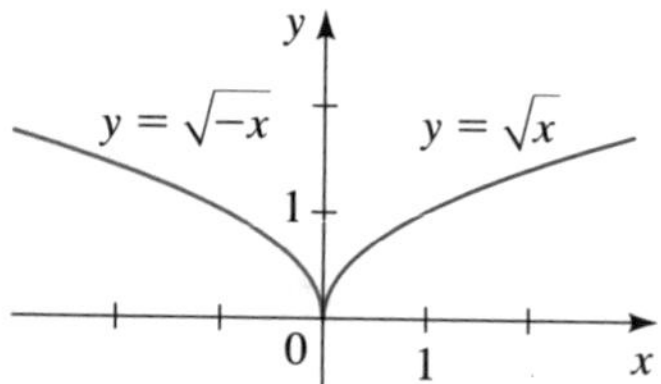

FIGURE 6

■

VERTICAL STRETCHING AND SHRINKING

Suppose we know the graph of $y = f(x)$. How do we use it to obtain the graph of $y = af(x)$? The y-coordinate of $y = af(x)$ at x is the same as the corresponding y-coordinate of $y = f(x)$ multiplied by a. Multiplying the y-coordinates by a has the effect of vertically stretching or shrinking the graph by a factor of a.

VERTICAL STRETCHING AND SHRINKING OF GRAPHS

Equation	How to obtain the graph	What the graph looks like
$y = af(x)$ $(a > 1)$	Stretch the graph of $y = f(x)$ vertically by a factor of a	
$y = af(x)$ $(0 < a < 1)$	Shrink the graph of $y = f(x)$ vertically by a factor of a	

EXAMPLE 6 ■ Vertical Stretching and Shrinking of Graphs

Sketch the graph of each function.

(a) $f(x) = 3x^2$ (b) $g(x) = \frac{1}{3}x^2$

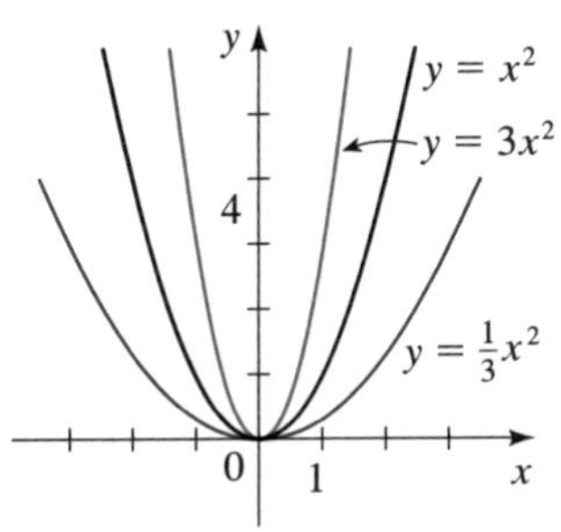

FIGURE 7

SOLUTION

(a) We start with the graph of $y = x^2$. The graph of $f(x) = 3x^2$ is the graph of $y = x^2$ stretched vertically by a factor of 3. The result is the narrower parabola in Figure 7. The graph is obtained by multiplying the y-coordinate of each point on the graph of $y = x^2$ by 3.

(b) The graph of $g(x) = \frac{1}{3}x^2$ is the graph of $y = x^2$ shrunk vertically by a factor of $\frac{1}{3}$. The result is the wider parabola in Figure 7. The graph is obtained by multiplying the y-coordinate of each point on the graph of $y = x^2$ by $\frac{1}{3}$. ■

We illustrate the effect of combining shifts, reflections, and stretching in the following example.

EXAMPLE 7 ■ Combining Shifting, Stretching, and Reflecting

Sketch the graph of the function $f(x) = 1 - 2(x - 3)^2$.

SOLUTION

Starting with the graph of $y = x^2$, we first shift to the right 3 units to get the graph of $y = (x - 3)^2$. Then we reflect in the x-axis and stretch by a factor of 2 to get the graph of $y = -2(x - 3)^2$. Finally, we shift upward 1 unit to get the graph of $f(x) = 1 - 2(x - 3)^2$ shown in Figure 8.

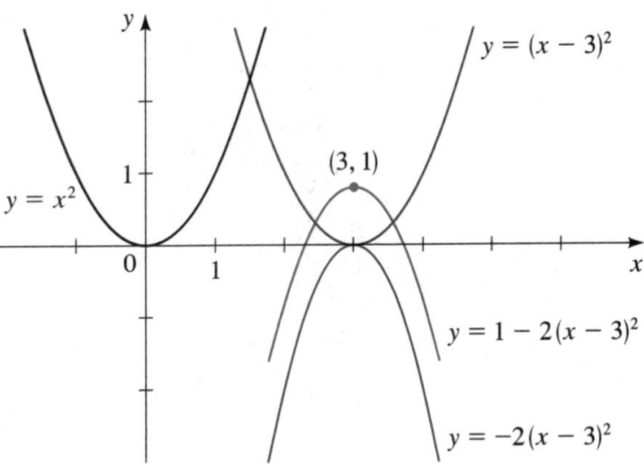

FIGURE 8

■

EVEN AND ODD FUNCTIONS

If a function f satisfies $f(-x) = f(x)$ for every number x in its domain, then f is called an **even function**. For instance, the function $f(x) = x^2$ is even because

$$f(-x) = (-x)^2 = (-1)^2x^2 = x^2 = f(x)$$

The geometric significance of an even function is that its graph is symmetric with respect to the y-axis (see Figure 9). This means that if we have plotted the graph of f for $x \geqslant 0$, then we can obtain the entire graph simply by reflecting this portion in the y-axis.

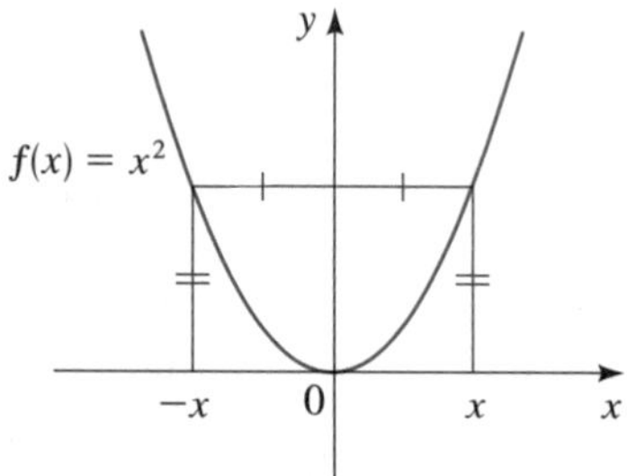

FIGURE 9
$f(x) = x^2$ is an even function.

If f satisfies $f(-x) = -f(x)$ for every number x in its domain, then f is called an **odd function**. For example, the function $f(x) = x^3$ is odd because

$$f(-x) = (-x)^3 = (-1)^3x^3 = -x^3 = -f(x)$$

The graph of an odd function is symmetric about the origin (see Figure 10). If we have plotted the graph of f for $x \geqslant 0$, then we can obtain the entire graph by rotating this portion through 180° about the origin. (This is equivalent to reflecting first in the x-axis and then in the y-axis.)

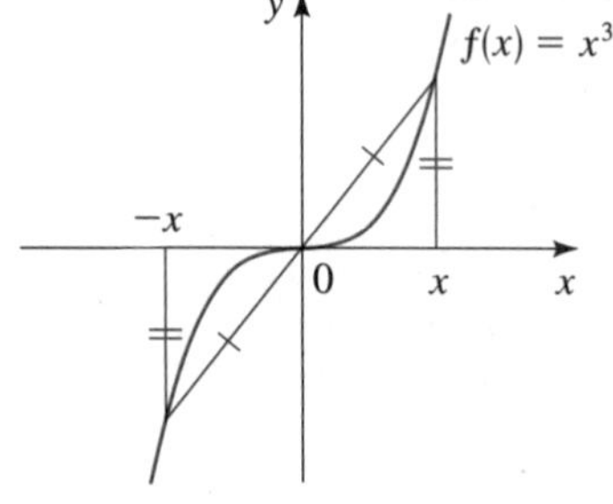

FIGURE 10
$f(x) = x^3$ is an odd function.

EVEN AND ODD FUNCTIONS

Definition	Symmetry of graph of f	What the graph looks like
f is **even** if $f(-x) = f(x)$ for all x in the domain of f	Graph of f is symmetric with respect to the y-axis	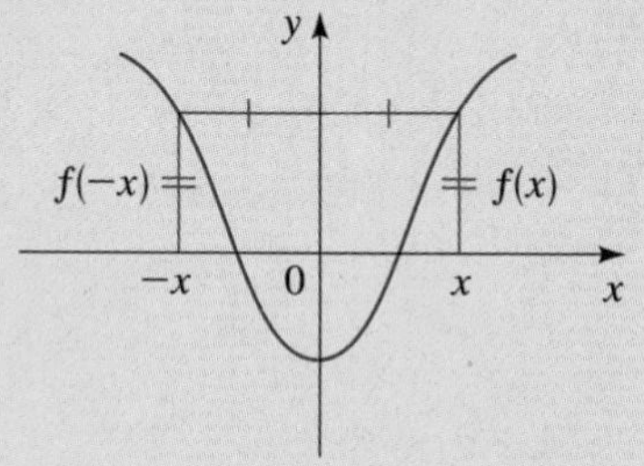
f is **odd** if $f(-x) = -f(x)$ for all x in the domain of f	Graph of f is symmetric with respect to the origin	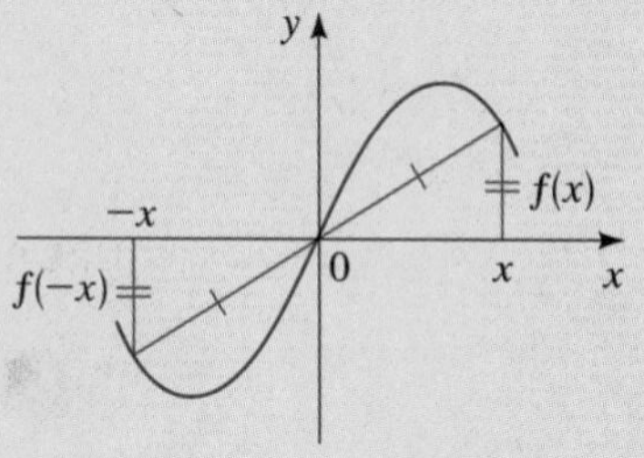

Sonya Kovalevsky (1850–1891) is considered the most important woman mathematician of the 19th century. She was born in Moscow to an aristocratic family. While a child, she was exposed to the principles of calculus in a very unusual fashion—her bedroom was temporarily wallpapered with the pages of a calculus book. She later wrote that she "spent many hours in front of that wall, trying to understand it." Since Russian law forbade women from studying in universities, she entered a marriage of convenience, which allowed her to travel to Germany and obtain a doctorate in mathematics from the University of Göttingen. She eventually was awarded a full professorship at the University of Stockholm, where she taught for eight years before dying in an influenza epidemic at the age of 41. Her research was instrumental in helping put the ideas and applications of functions and calculus on a sound and logical foundation. She received many accolades and prizes for her research work.

EXAMPLE 8 ■ Even and Odd Functions

Determine whether the functions are even, odd, or neither even nor odd.

(a) $f(x) = x^5 + x$

(b) $g(x) = 1 - x^4$

(c) $h(x) = 2x - x^2$

SOLUTION

(a)
$$\begin{aligned} f(-x) &= (-x)^5 + (-x) \\ &= -x^5 - x = -(x^5 + x) \\ &= -f(x) \end{aligned}$$

Therefore, f is an odd function.

(b) $g(-x) = 1 - (-x)^4 = 1 - x^4 = g(x)$

So g is even.

(c) $h(-x) = 2(-x) - (-x)^2 = -2x - x^2$

Since $h(-x) \neq h(x)$ and $h(-x) \neq -h(x)$, we conclude that h is neither even nor odd. ■

The graphs of the functions in Example 8 are shown in Figure 11. The graph of f was drawn by plotting points for $x \geqslant 0$ and rotating 180° about the origin. The graph of g was drawn by plotting points for $x \geqslant 0$ and reflecting in the y-axis. Notice that the graph of h is not symmetric either about the y-axis or the origin.

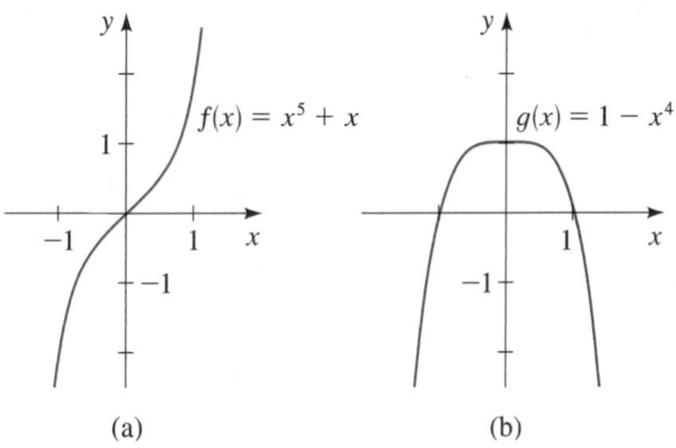

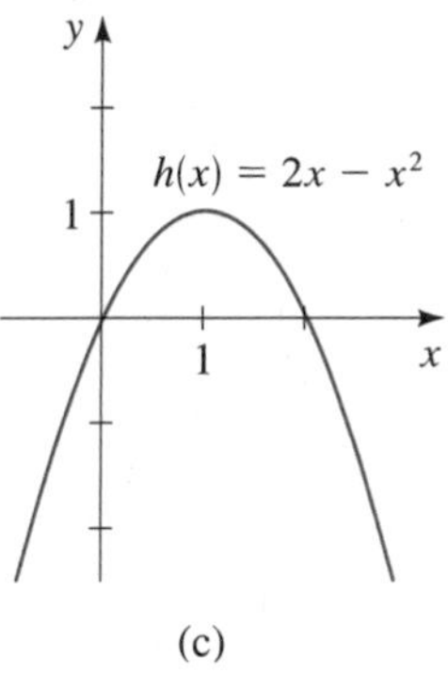

FIGURE 11 (a) (b) (c)

4.5 EXERCISES

1–8 ■ Suppose the graph of f is given. Describe how the graph of each function can be obtained from the graph of f.

1. (a) $y = f(x) - 4$ (b) $y = f(x - 4)$

2. (a) $y = f(x + 5)$ (b) $y = f(x) + 5$

3. (a) $y = 3f(x)$ (b) $y = \frac{1}{3}f(x)$

4. (a) $y = -f(x)$ (b) $y = f(-x)$

5. (a) $y = -f(x) + 5$ (b) $y = f(-x) + 5$

6. (a) $y = -4f(x)$ (b) $y = -\frac{1}{4}f(x)$

7. (a) $y = f(x - 2) - 3$ (b) $y = 2f(x - 3)$

8. (a) $y = \frac{1}{2}f(x) + 10$ (b) $y = \frac{1}{2}f(x + 10)$

9. The graph of f is given. Sketch the graphs of the following functions.
(a) $y = f(x - 2)$ (b) $y = f(x) - 2$
(c) $y = 2f(x)$ (d) $y = -f(x) + 3$
(e) $y = f(-x)$ (f) $y = \frac{1}{2}f(x - 1)$

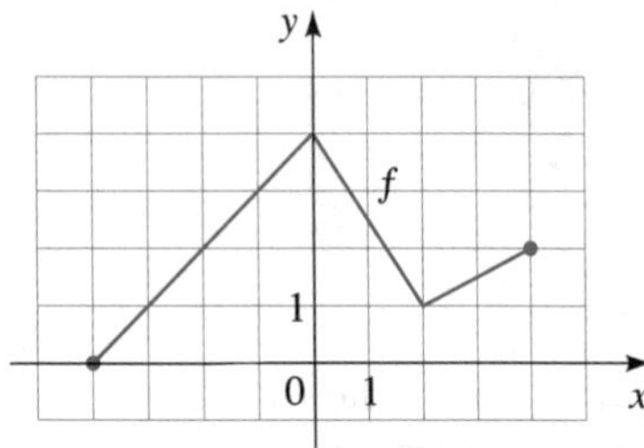

10. The graph of $y = f(x)$ is given. Match each equation with its graph.
(a) $y = f(x - 4)$ (b) $y = f(x) + 3$
(c) $y = \frac{1}{3}f(x)$ (d) $y = -f(x + 4)$
(e) $y = 2f(x + 6)$

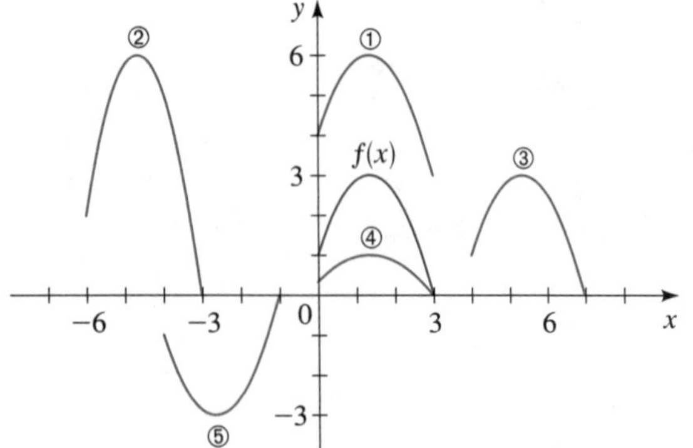

11. (a) Sketch the graph of $f(x) = \dfrac{1}{x}$ by plotting points.
(b) Use the graph of f to sketch the graphs of the following functions.
(i) $y = -\dfrac{1}{x}$ (ii) $y = \dfrac{1}{x - 1}$
(iii) $y = \dfrac{2}{x + 2}$ (iv) $y = 1 + \dfrac{1}{x - 3}$

12. (a) Sketch the graph of $g(x) = \sqrt[3]{x}$ by plotting points.
(b) Use the graph of g to sketch the graphs of the following functions.
(i) $y = \sqrt[3]{x - 2}$ (ii) $y = \sqrt[3]{x + 2} + 2$
(iii) $y = 1 - \sqrt[3]{x}$

13–16 ■ Explain how the graph of g is obtained from the graph of f.

13. (a) $f(x) = x^2, \quad g(x) = (x + 2)^2$
(b) $f(x) = x^2, \quad g(x) = x^2 + 2$

14. (a) $f(x) = x^3, \quad g(x) = (x - 4)^3$
(b) $f(x) = x^3, \quad g(x) = x^3 - 4$

15. (a) $f(x) = \sqrt{x}, \quad g(x) = 2\sqrt{x}$
(b) $f(x) = \sqrt{x}, \quad g(x) = \frac{1}{2}\sqrt{x - 2}$

16. (a) $f(x) = |x|, \quad g(x) = 3|x| + 1$
(b) $f(x) = |x|, \quad g(x) = -|x + 1|$

17–32 ■ Sketch the graph of the function, not by plotting points, but by starting with the graph of a standard function and applying transformations.

17. $f(x) = (x - 2)^2$ **18.** $f(x) = (x + 7)^2$

19. $f(x) = -(x + 1)^2$ **20.** $f(x) = 1 - x^2$

21. $f(x) = x^3 + 2$ **22.** $f(x) = -x^3$

23. $y = 1 + \sqrt{x}$ **24.** $y = 2 - \sqrt{x + 1}$

25. $y = \frac{1}{2}\sqrt{x + 4} - 3$ **26.** $y = 3 - 2(x - 1)^2$

27. $y = 5 + (x + 3)^2$ **28.** $y = \frac{1}{3}x^3 - 1$

29. $y = |x| - 1$

30. $y = |x - 1|$

31. $y = |x + 2| + 2$

32. $y = 2 - |x|$

33–36 ■ Graph the functions on the same screen using the given viewing rectangle. How is each graph related to the graph in part (a)?

33. Viewing rectangle $[-8, 8]$ by $[-2, 8]$
(a) $y = \sqrt[4]{x}$ (b) $y = \sqrt[4]{x + 5}$
(c) $y = 2\sqrt[4]{x + 5}$ (d) $y = 4 + 2\sqrt[4]{x + 5}$

34. Viewing rectangle $[-8, 8]$ by $[-6, 6]$
(a) $y = |x|$ (b) $y = -|x|$
(c) $y = -3|x|$ (d) $y = -3|x - 5|$

35. Viewing rectangle $[-4, 6]$ by $[-4, 4]$
(a) $y = x^6$ (b) $y = \frac{1}{3}x^6$
(c) $y = -\frac{1}{3}x^6$ (d) $y = -\frac{1}{3}(x - 4)^6$

36. Viewing rectangle $[-6, 6]$ by $[-4, 4]$
(a) $y = \dfrac{1}{\sqrt{x}}$ (b) $y = \dfrac{1}{\sqrt{x + 3}}$
(c) $y = \dfrac{1}{2\sqrt{x + 3}}$ (d) $y = \dfrac{1}{2\sqrt{x + 3}} - 3$

37–44 ■ Determine whether the function f is even, odd, or neither. If f is even or odd, use symmetry to sketch its graph.

37. $f(x) = x^{-2}$ **38.** $f(x) = x^{-3}$

39. $f(x) = x^2 + x$ **40.** $f(x) = x^4 - 4x^2$

41. $f(x) = x^3 - x$ **42.** $f(x) = 3x^3 + 2x^2 + 1$

43. $f(x) = 1 - \sqrt[3]{x}$ **44.** $f(x) = x + \dfrac{1}{x}$

45. The graphs of $f(x) = x^2 - 4$ and $g(x) = |x^2 - 4|$ are shown. Explain how the graph of g is obtained from the graph of f.

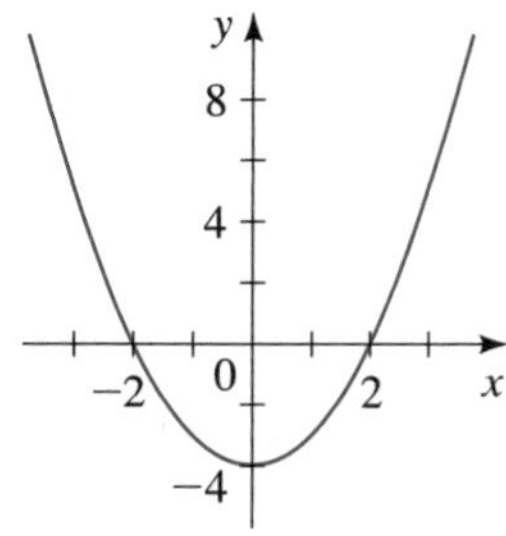

$f(x) = x^2 - 4$

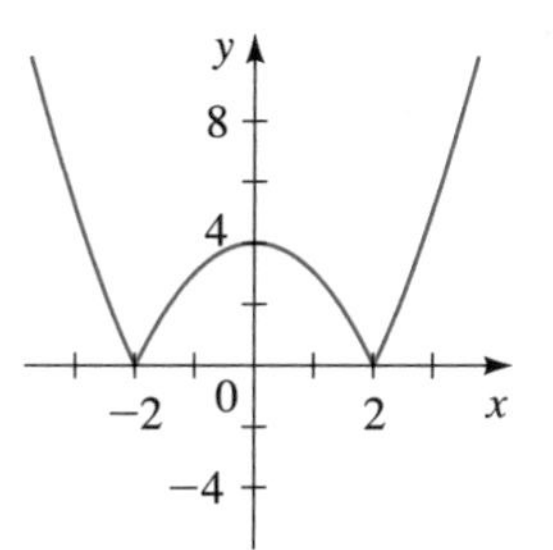

$g(x) = |x^2 - 4|$

46. The graph of $f(x) = x^4 - 4x^2$ is shown. Use this graph to sketch the graph of $g(x) = |x^4 - 4x^2|$.

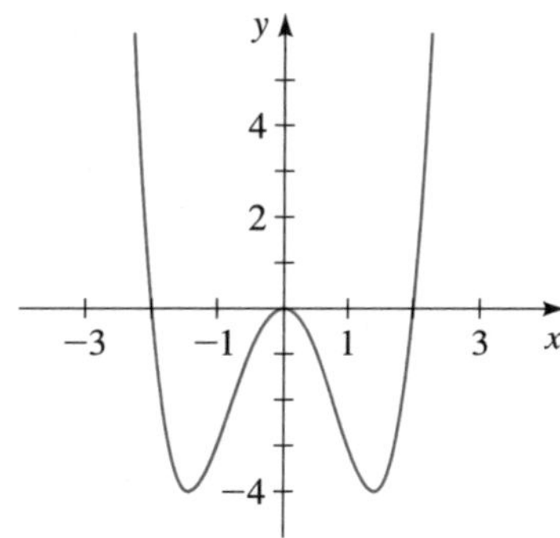

47–48 ■ Sketch the graph of each function.

47. (a) $f(x) = 4x - x^2$ (b) $g(x) = |4x - x^2|$

48. (a) $f(x) = x^3$ (b) $g(x) = |x^3|$

DISCOVERY • DISCUSSION

49. Sums of Even and Odd Functions If f and g are both even functions, is $f + g$ necessarily even? If both are odd, is their sum necessarily odd? What can you say about the sum if one is odd and one is even? In each case, prove your answer.

50. Products of Even and Odd Functions Answer the same questions as in Exercise 49, except this time consider the *product* of f and g instead of the sum.

51. Even and Odd Power Functions What must be true about the integer n if the function $f(x) = x^n$ is an even function? If it is an odd function? Why do you think the names "even" and "odd" were chosen for these function properties?

52. Horizontal Shrinking and Stretching of Graphs In this section we learned that to graph $y = af(x)$, we must stretch or shrink the graph of $y = f(x)$ vertically by the factor a. In this exercise we learn how the factor a affects the graph of $y = f(ax)$. In parts (a) and (b), let $f(x) = \sqrt{2x - x^2}$.

(a) Graph $y = f(x)$, $y = f(2x)$, and $y = f\left(\frac{1}{2}x\right)$ in the viewing rectangle $[-5, 5]$ by $[-4, 4]$. How are the graphs related to the graph of f?

(b) Based on your observations in part (a), describe how the graph of $y = f(ax)$ is related to the graph of

$y = f(x)$. Your description should consider the cases $a > 1$ and $0 < a < 1$.

(c) The graph of a function is given. Use the principles that you discovered in part (b) to graph $y = g(2x)$ and $y = g\left(\frac{1}{2}x\right)$.

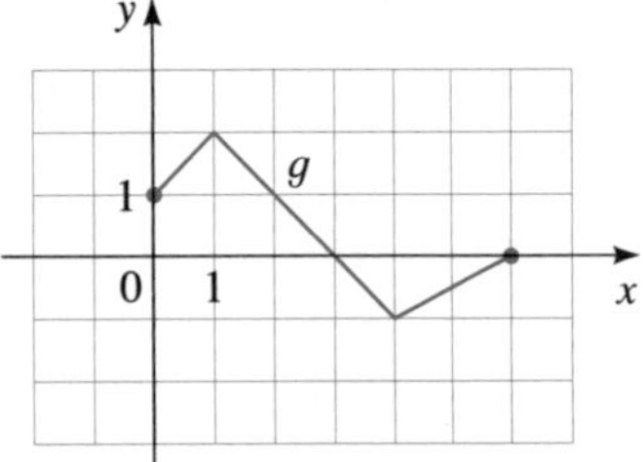

(d) The graph of a function is given. Use the principles that you discovered in part (b) to graph $y = h(3x)$ and $y = h\left(\frac{1}{3}x\right)$.

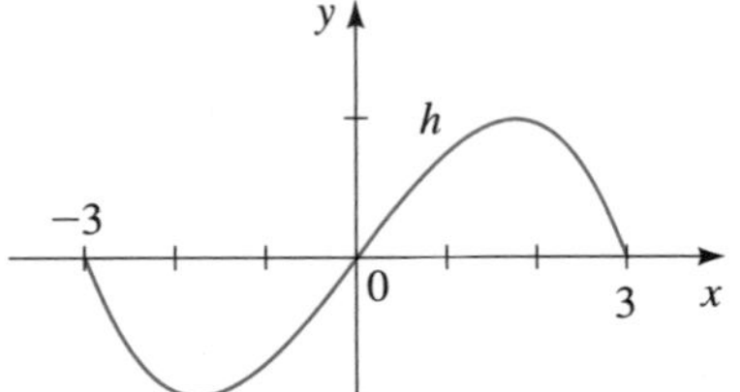

(e) Use the graph of $f(x) = [\![x]\!]$ described in Section 4.2 to graph the functions $y = [\![2x]\!]$ and $y = [\![\frac{1}{4}x]\!]$.

4.6 EXTREME VALUES OF FUNCTIONS

An extreme value of a function is the largest or smallest value of the function on some interval. Finding points where functions achieve extreme values is important in many applications. For example, if the function represents the profit in a business, we would be interested in the maximum values; if a function represents the amount of material to be used in a manufacturing process, we would be interested in the minimum values. We begin by giving an algebraic method for finding extreme values of quadratic functions and then use graphing devices to find extreme values of other functions.

EXTREME VALUES OF QUADRATIC FUNCTIONS

A **quadratic function** is a function f of the form

$$f(x) = ax^2 + bx + c$$

where a, b, and c are real numbers and $a \neq 0$.

In particular, if we take $a = 1$ and $b = c = 0$, we get the simple quadratic function $f(x) = x^2$ whose graph is the parabola that we drew in Example 1 of Section 4.2. In fact, the graph of any quadratic function is a **parabola**; it can be obtained from the graph of $y = x^2$ by the transformations given in Section 4.5.

Completing the square is discussed in Section 3.3.

To graph a quadratic function, we first complete the square to put the function in a more convenient form for graphing.

EXAMPLE 1 ■ Graphing a Quadratic Function by Completing the Square

Sketch the graph of the quadratic function $f(x) = 2x^2 - 12x + 23$.

SOLUTION

Since the coefficient of x^2 is not 1, we must factor this coefficient from the terms involving x before we complete the square.

$$\begin{aligned} f(x) &= 2x^2 - 12x + 23 \\ &= 2(x^2 - 6x) + 23 && \text{Factor 2 from the } x\text{-terms} \\ &= 2(x^2 - 6x + 9) + 23 - 2 \cdot 9 && \text{Complete the square: Add 9 inside parentheses, subtract } 2 \cdot 9 \text{ outside} \\ &= 2(x - 3)^2 + 5 \end{aligned}$$

This form of the function tells us that we get the graph of f by taking the parabola $y = x^2$, shifting it to the right 3 units, stretching it by a factor of 2, and moving it upward 5 units. Notice that the lowest point of the parabola is at $(3, 5)$ and the parabola opens upward. We sketch the graph in Figure 1 after noting that the y-intercept is $f(0) = 23$.

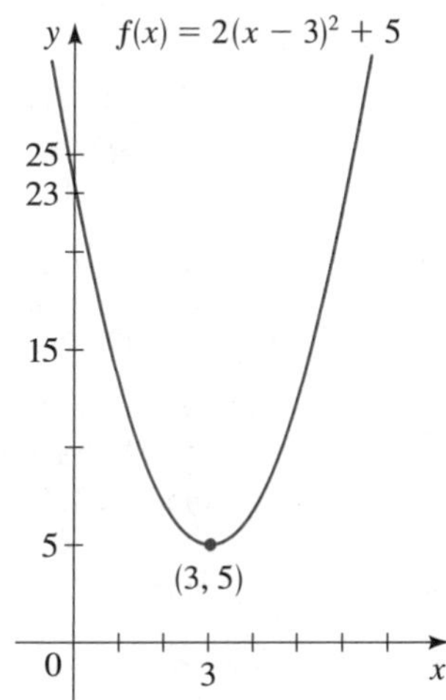

FIGURE 1

■

If we start with a general quadratic function $f(x) = ax^2 + bx + c$ and complete the square as in Example 1, we arrive at an expression in the standard form $f(x) = a(x - h)^2 + k$. We know from Section 4.5 that the graph of this function is a parabola obtained from the graph of $y = x^2$ by a horizontal shift, a vertical stretch, and then a vertical shift. The *vertex*, that is, the highest or lowest point of the parabola, is the point (h, k). If $a > 0$, the parabola opens upward. In this case, the vertex (h, k) is the lowest point on the parabola, so the *minimum* value of the function is $f(h) = k$. If $a < 0$, the parabola opens downward. In this case, the vertex (h, k) is the highest point on the parabola, so the *maximum* value of the function is $f(h) = k$.

STANDARD FORM OF A QUADRATIC FUNCTION

A quadratic function $f(x) = ax^2 + bx + c$ can be expressed in the **standard form**

$$f(x) = a(x - h)^2 + k$$

by completing the square. The graph of f is a parabola with **vertex** (h, k); the parabola opens upward if $a > 0$ or downward if $a < 0$.

If $a > 0$, then the **minimum value** of f occurs at $x = h$ and this value is $f(h) = k$

If $a < 0$, then the **maximum value** of f occurs at $x = h$ and this value is $f(h) = k$

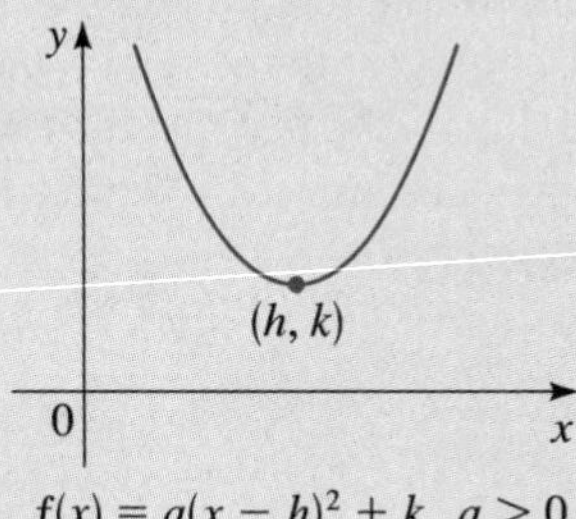

$f(x) = a(x - h)^2 + k,\ a > 0$ $\quad$ $f(x) = a(x - h)^2 + k,\ a < 0$

EXAMPLE 2 ■ Expressing a Quadratic Function in Standard Form

Consider the quadratic function $f(x) = 5x^2 - 30x + 49$.

(a) Express f in standard form.
(b) Sketch the graph of f.
(c) Find the minimum value of f.

SOLUTION

(a) To express this quadratic function in standard form, we complete the square.

$$\begin{aligned} f(x) &= 5x^2 - 30x + 49 \\ &= 5(x^2 - 6x) + 49 && \text{Factor 5 from the } x\text{-terms} \\ &= 5(x^2 - 6x + 9) + 49 - 5 \cdot 9 && \text{Complete the square: Add 9 inside parentheses, subtract } 5 \cdot 9 \text{ outside} \\ &= 5(x - 3)^2 + 4 \end{aligned}$$

(b) The graph is a parabola that has its vertex at $(3, 4)$ and opens upward, as sketched in Figure 2.

(c) Since the coefficient of x^2 is positive, f has a minimum value. The minimum value is $f(3) = 4$. ■

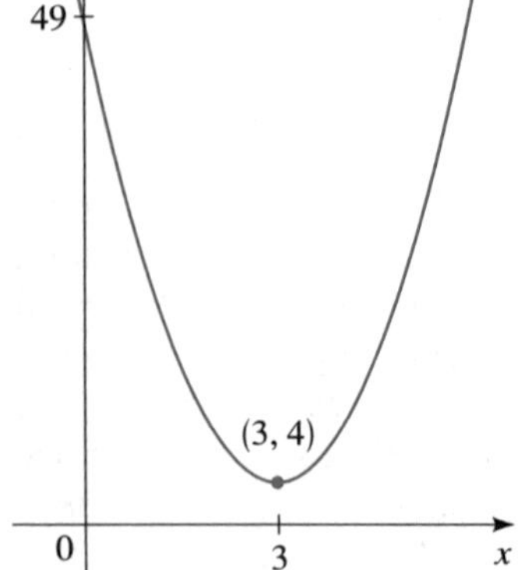

FIGURE 2

EXAMPLE 3 ■ Expressing a Quadratic Function in Standard Form

Consider the quadratic function $f(x) = -x^2 + x + 2$.

(a) Express f in standard form.
(b) Sketch the graph of f.
(c) Find the maximum value of f.

SOLUTION

(a) To express this quadratic function in standard form, we complete the square.

$$\begin{aligned} y &= -x^2 + x + 2 \\ &= -(x^2 - x) + 2 && \text{Factor } -1 \text{ from the } x\text{-terms} \\ &= -\left(x^2 - x + \tfrac{1}{4}\right) + 2 - (-1)\tfrac{1}{4} && \text{Complete the square: Add } \tfrac{1}{4} \text{ inside parentheses, subtract } (-1)\tfrac{1}{4} \text{ outside} \\ &= -\left(x - \tfrac{1}{2}\right)^2 + \tfrac{9}{4} \end{aligned}$$

(b) From the standard form we see that the graph is a parabola that opens downward and has vertex $\left(\frac{1}{2}, \frac{9}{4}\right)$. As an aid to sketching the graph, we find the intercepts. The y-intercept is $f(0) = 2$. To find the x-intercepts, we set $f(x) = 0$ and factor the resulting equation.

$$\begin{aligned} -x^2 + x + 2 &= 0 \\ -(x^2 - x - 2) &= 0 \\ -(x - 2)(x + 1) &= 0 \end{aligned}$$

Thus, the x-intercepts are $x = 2$ and $x = -1$. The graph of f is sketched in Figure 3.

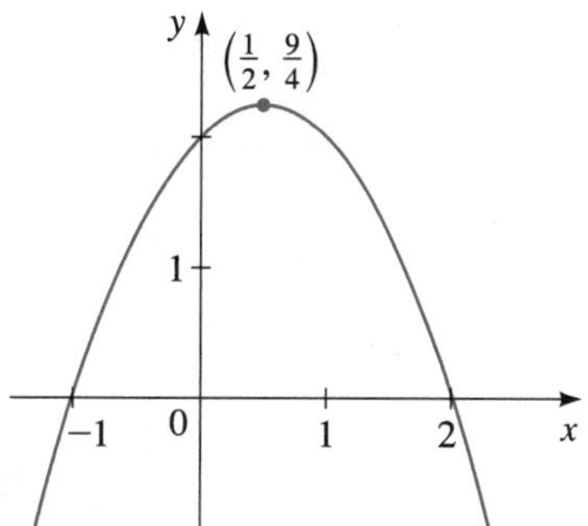

FIGURE 3
Graph of $f(x) = -x^2 + x + 2$

(c) Since the coefficient of x^2 is negative, f has a maximum value, which is $f\left(\frac{1}{2}\right) = \frac{9}{4}$. ■

Expressing a quadratic function in standard form helps us sketch its graph as well as find its maximum or minimum value. If we are interested only in finding the maximum or minimum value, then a formula is available for doing so. This formula is obtained by completing the square for the general quadratic function as follows:

$$f(x) = ax^2 + bx + c$$

$$= a\left(x^2 + \frac{b}{a}x\right) + c \qquad \text{Factor } a \text{ from the } x\text{-terms}$$

$$= a\left(x^2 + \frac{b}{a}x + \frac{b^2}{4a^2}\right) + c - \frac{b^2}{4a} \qquad \text{Complete the square: Add } \frac{b^2}{4a^2} \text{ inside parentheses, subtract } a\left(\frac{b^2}{4a^2}\right) \text{ outside}$$

$$= a\left(x + \frac{b}{2a}\right)^2 + c - \frac{b^2}{4a}$$

This equation is in standard form with $h = -b/(2a)$ and $k = c - b^2/(4a)$. Since the maximum or minimum value occurs at $x = h$, we have the following result.

MAXIMUM OR MINIMUM VALUE OF A QUADRATIC FUNCTION

The maximum or minimum value of a quadratic function $f(x) = ax^2 + bx + c$ occurs at

$$x = -\frac{b}{2a}$$

If $a > 0$, then the **minimum value** is $f\left(-\frac{b}{2a}\right)$.

If $a < 0$, then the **maximum value** is $f\left(-\frac{b}{2a}\right)$.

EXAMPLE 4 ■ Finding Maximum and Minimum Values of Quadratic Functions

Find the maximum or minimum value of each quadratic function.

(a) $f(x) = x^2 + 4x$ (b) $g(x) = -2x^2 + 4x - 5$

SOLUTION

(a) This is a quadratic function with $a = 1$ and $b = 4$. Thus, the maximum or minimum value occurs at

$$x = -\frac{b}{2a} = -\frac{4}{2 \cdot 1} = -2$$

Since $a > 0$, the function has the minimum value

$$f(-2) = (-2)^2 + 4(-2) = -4$$

(b) This is a quadratic function with $a = -2$ and $b = 4$. Thus, the maximum or minimum value occurs at

$$x = -\frac{b}{2a} = -\frac{4}{2 \cdot (-2)} = 1$$

Since $a < 0$, the function has the maximum value

$$f(1) = -2(1)^2 + 4(1) - 5 = -3$$

■

Many real-world problems involve finding a maximum or minimum value for a function that models a given situation. In the next example we find the maximum value of a quadratic function that models the gas mileage for a car.

EXAMPLE 5 ■ Maximum Gas Mileage for a Car

Most cars get their best gas mileage when traveling at a relatively modest speed. The gas mileage M for a certain new car is modeled by the function

$$M(s) = -\frac{1}{28}s^2 + 3s - 31, \qquad 15 \leqslant s \leqslant 70$$

where s is the speed in mi/h and M is measured in mi/gal. What is the car's best gas mileage, and at what speed is it attained?

SOLUTION

The function M is a quadratic function with $a = -\frac{1}{28}$ and $b = 3$. Thus its maximum value occurs when

$$s = -\frac{b}{2a} = -\frac{3}{-2\left(\frac{1}{28}\right)} = 42$$

The maximum is $M(42) = -\frac{1}{28}(42)^2 + 3(42) - 31 = 32$. Thus, the car's best gas mileage is 32 mi/gal, when it is traveling at 42 mi/h. ■

 USING GRAPHING DEVICES TO FIND EXTREME VALUES

The methods we have discussed apply to finding extreme values of quadratic functions only. We now show how to locate extreme values of any function that can be graphed with a calculator or computer.

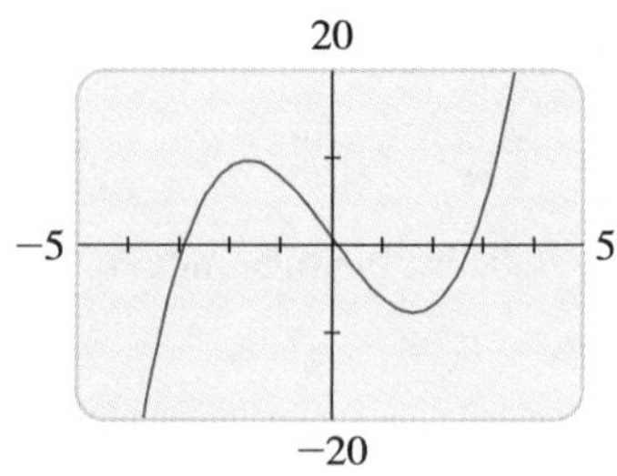

FIGURE 4

Graph of $f(x) = x^3 - 8x + 1$

EXAMPLE 6 ■ Peaks and Valleys of a Graph

Draw the graph of the function $f(x) = x^3 - 8x + 1$ in the viewing rectangle $[-5, 5]$ by $[-20, 20]$. Comment on the peaks and valleys of the graph.

SOLUTION

The graph is shown in Figure 4. A high point of the graph occurs for a value of x between -3 and 0. In fact, if we restrict our attention to the viewing rectangle $[-3, 0]$ by $[0, 15]$, as in Figure 5(a), we see that there is a *highest* point on the restricted graph. Similarly, when we confine our attention to the viewing rectangle $[0, 3]$ by $[-15, 0]$ in Figure 5(b), we see that there is also a *lowest* point on the restricted graph.

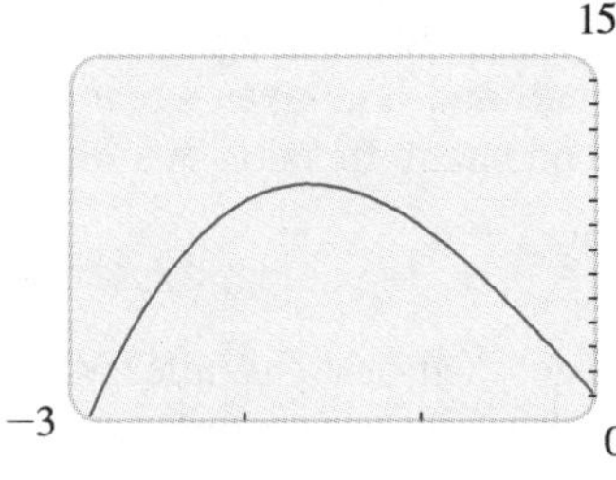

(a) f restricted to $[-3, 0]$ by $[0, 15]$

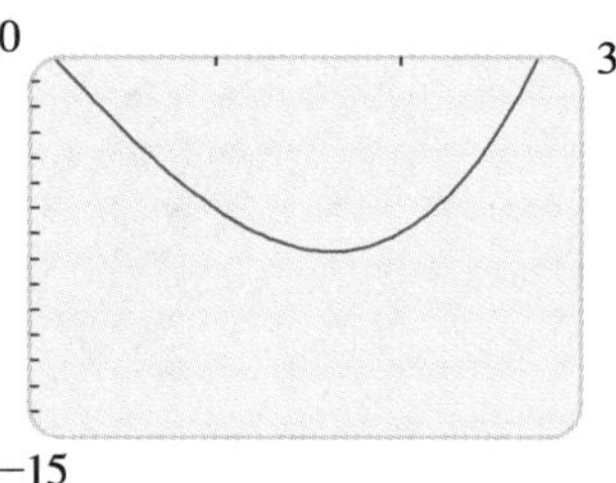

(b) f restricted to $[0, 3]$ by $[-15, 0]$ ■

FIGURE 5

In general, if there is a viewing rectangle such that the point $(a, f(a))$ is the highest point on the graph of f *within* the viewing rectangle (not on the edge), then the number $f(a)$ is called a **local maximum value** of f (see Figure 6). Notice that $f(a) \geqslant f(x)$ for all numbers x that are close to a.

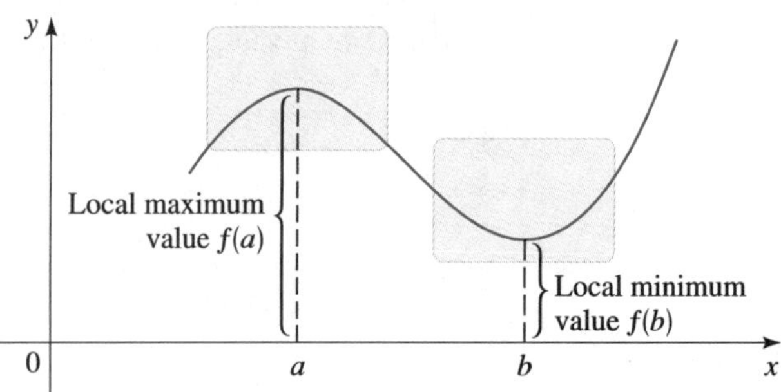

FIGURE 6

Similarly, if there is a viewing rectangle such that the point $(b, f(b))$ is the lowest point on the graph of f within the viewing rectangle, then the number $f(b)$ is called a **local minimum value** of f. In this case, $f(b) \leqslant f(x)$ for all numbers x that are close to b.

EXAMPLE 7 ■ Finding Local Maxima and Minima from a Graph

Find the approximate local maximum and minimum values of the function $f(x) = x^3 - 8x + 1$ of Example 6.

SOLUTION

Let's first find the local maximum value. In Figure 5(a) the maximum point lies in the rectangle $[-2, -1]$ by $[9, 10]$, so we draw the graph restricted to this viewing rectangle in Figure 7(a). By moving the cursor close to the maximum point, we see that the y-coordinates don't change very much in the vicinity of the maximum. The maximum value is about 9.7, and since the distance between the scale marks is 0.1, this estimate for the maximum value is accurate to within 0.1.

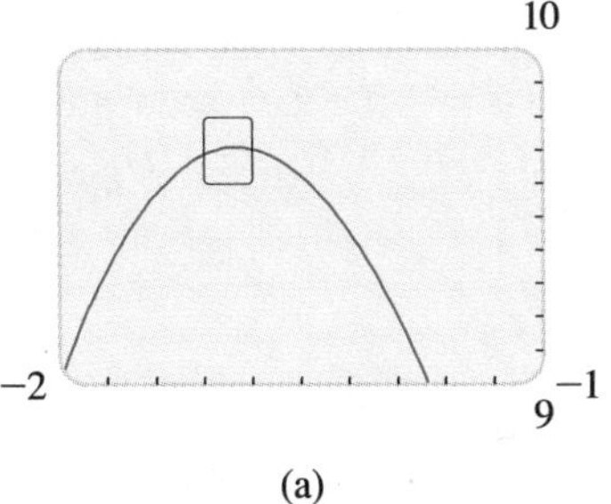

(a)

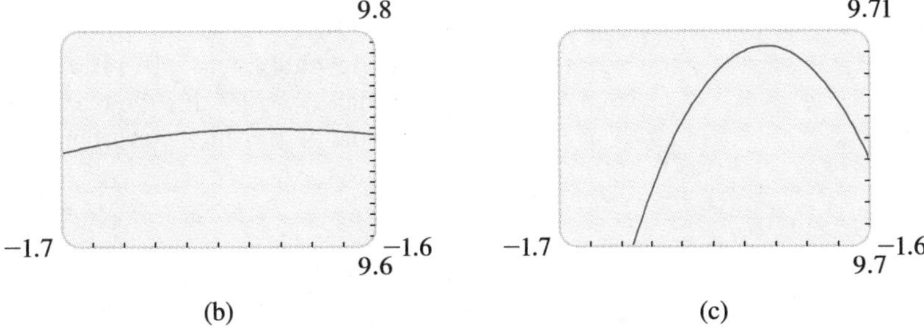

(b) (c)

FIGURE 7
Steps in finding the local maximum of $f(x) = x^3 - 8x + 1$

If we want an even more accurate estimate, we zoom in to a smaller rectangle containing the maximum point. Figure 7(b) shows the graph restricted to the viewing rectangle $[-1.7, -1.6]$ by $[9.6, 9.8]$. The curve in Figure 7(b) looks quite flat. It is easier to pinpoint the maximum value if we use a viewing rectangle that is much wider than tall. So we adjust the viewing rectangle to $[-1.7, -1.6]$ by $[9.7, 9.71]$ in Figure 7(c). By moving the cursor along the curve and observing how the y-coordinates change, we can estimate the maximum value quite accurately. It appears that the local maximum value is about 9.709, and this value occurs when x is about -1.633.

We locate the local minimum in a similar fashion. By zooming in to the rectangles shown in Figure 8, we find that the local minimum value is about -7.709, and this value occurs when $x \approx 1.633$.

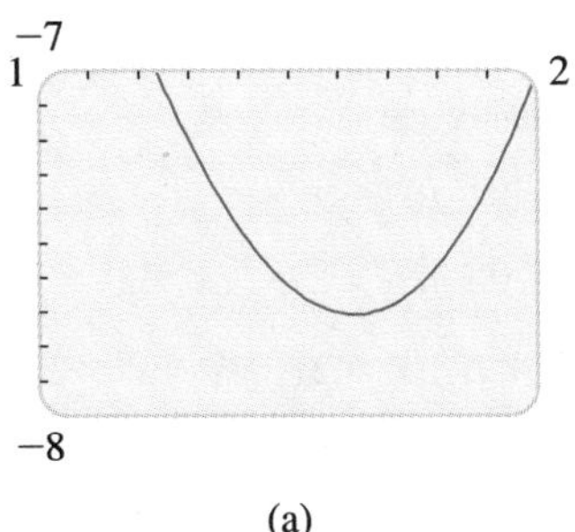

(a)

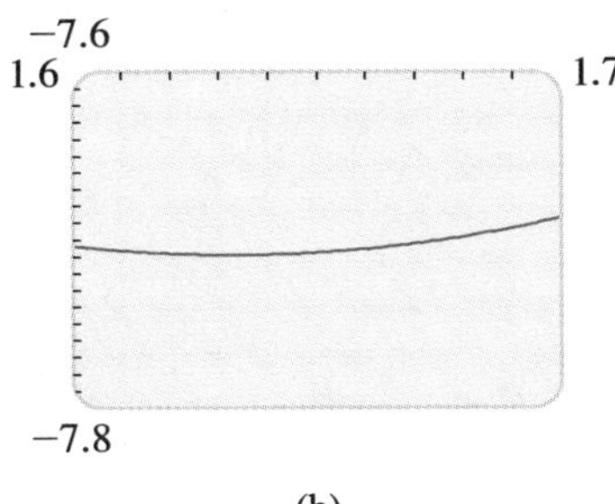

(b)

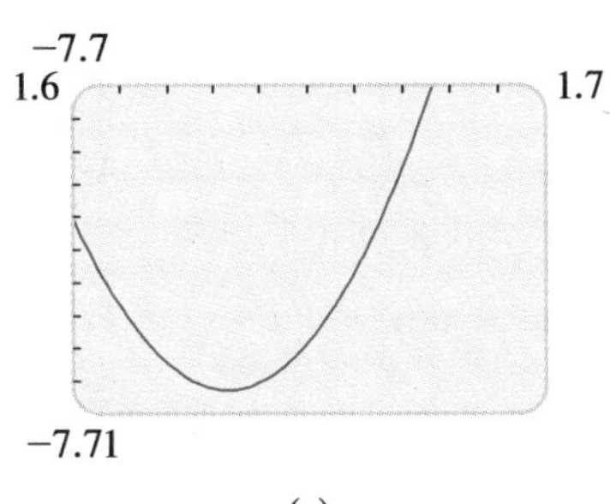

(c)

FIGURE 8 Steps in finding the local minimum of $f(x) = x^3 - 8x + 1$ ■

EXAMPLE 8 ■ A Model for the Food Price Index

A model for the food price index (the price of a representative "basket" of foods) between 1984 and 1994 is given by the function

$$I(t) = -0.0113t^3 + 0.0681t^2 + 0.198t + 99.1$$

where t is measured in years since midyear 1984, so $0 \leq t \leq 10$, and $I(t)$ is scaled so that $I(3) = 100$. Estimate the time when food was most expensive during the period 1984–1994.

SOLUTION

The graph of I as a function of t is shown in Figure 9(a). We see that the maximum occurs near the point (5, 100.5), so we zoom in on the viewing rectangle [4.5, 5.5] by [100, 101], as shown in part (b). But that graph is very flat, so we change the view to [4.5, 5.5] by [100.35, 100.4] in part (c). Using the trace feature on the calculator, we see that the maximum value of I is about 100.4, and it occurs when t is about 5.15, which corresponds to August 1989.

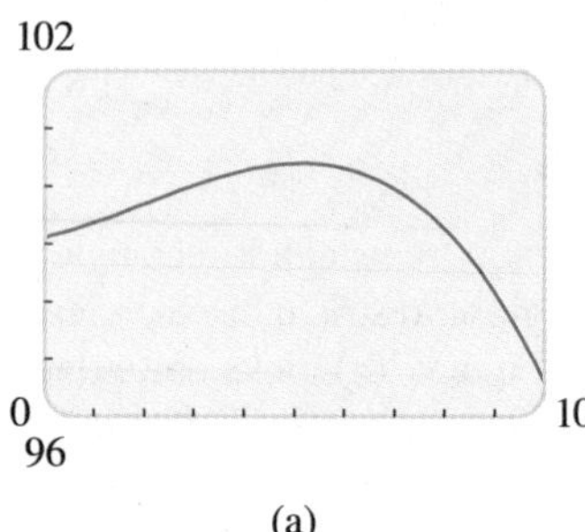

(a)

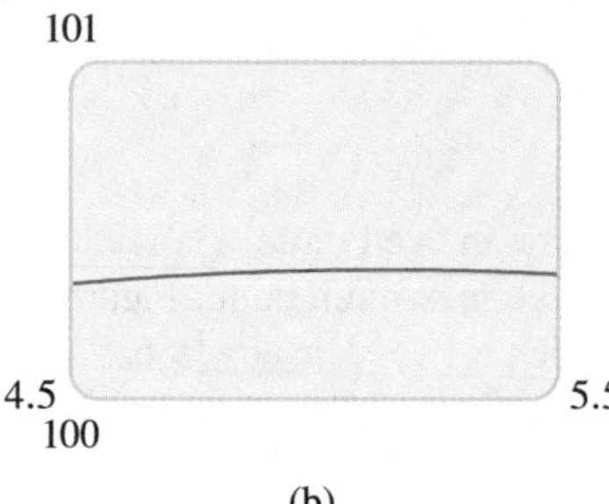

(b)

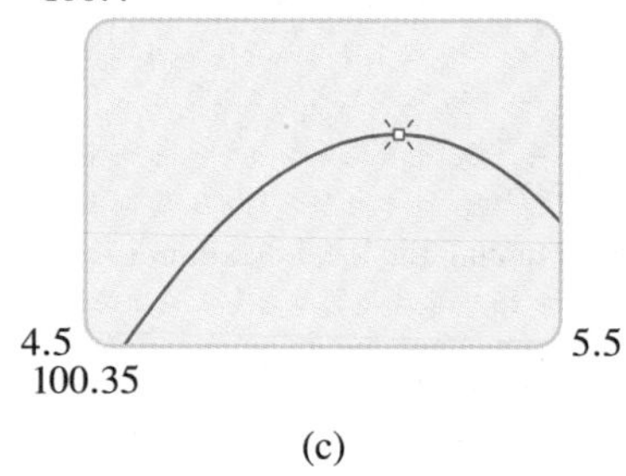

(c)

FIGURE 9

■

4.6 EXERCISES

1–14 ■ Sketch the graph of the given parabola and state the coordinates of its vertex and its intercepts.

1. $y = x^2 - 8x$ **2.** $y = x^2 + 6x$

3. $y = 2x^2 - 6x$ **4.** $y = -x^2 + 10x$

5. $y = x^2 + 4x + 1$ **6.** $y = x^2 - 2x + 2$

7. $y = x^2 + 6x + 8$ **8.** $y = -x^2 - 4x + 4$

9. $y = 2x^2 + 4x + 3$ **10.** $y = -3x^2 + 6x - 2$

11. $y = 2x^2 - 20x + 57$ **12.** $y = 2x^2 + x - 6$

13. $y = -4x^2 - 16x + 3$ **14.** $y = 6x^2 + 12x - 5$

15–24 ■ A quadratic function is given.
(a) Express the quadratic function in standard form.
(b) Sketch its graph.
(c) Find its maximum or minimum value.

15. $f(x) = 2x - x^2$ **16.** $f(x) = x + x^2$

17. $f(x) = x^2 + 2x - 1$ **18.** $f(x) = x^2 - 8x + 8$

19. $f(x) = -x^2 - 3x + 3$ **20.** $f(x) = 1 - 6x - x^2$

21. $g(x) = 3x^2 - 12x + 13$ **22.** $g(x) = 2x^2 + 8x + 11$

23. $h(x) = 1 - x - x^2$ **24.** $h(x) = 3 - 4x - 4x^2$

25–34 ■ Find the maximum or minimum value of the function.

25. $f(x) = x^2 + x + 1$

26. $f(x) = 1 + 3x - x^2$

27. $f(t) = 100 - 49t - 7t^2$

28. $f(t) = 10t^2 + 40t + 113$

29. $f(s) = s^2 - 1.2s + 16$

30. $g(x) = 100x^2 - 1500x$

31. $h(x) = \frac{1}{2}x^2 + 2x - 6$

32. $f(x) = -\dfrac{x^2}{3} + 2x + 7$

33. $f(x) = 3 - x - \frac{1}{2}x^2$

34. $g(x) = 2x(x - 4) + 7$

35. Find a function whose graph is a parabola with vertex $(1, -2)$ and that passes through the point $(4, 16)$.

36. Find a function whose graph is a parabola with vertex $(3, 4)$ and that passes through the point $(1, -8)$.

37–38 ■ Find the domain and range of the function.

37. $f(x) = -x^2 + 4x - 3$

38. $f(x) = x^2 - 2x - 3$

39. If a ball is thrown directly upward with a velocity of 40 ft/s, its height (in feet) after t seconds is given by $y = 40t - 16t^2$. What is the maximum height attained by the ball?

40. A ball is thrown across a playing field. Its path is given by the equation $y = -0.005x^2 + x + 5$, where x is the distance the ball has traveled horizontally, and y is its height above ground level, both measured in feet.

(a) What is the maximum height attained by the ball?

(b) How far has it traveled horizontally when it hits the ground?

41. A manufacturer finds that the revenue generated by selling x units of a certain commodity is given by the function $R(x) = 80x - 0.4x^2$, where the revenue $R(x)$ is measured in dollars. What is the maximum revenue, and how many units should be manufactured to obtain this maximum?

42. A soft-drink vendor at a popular beach analyzes his sales records, and finds that if he sells x cans of soda pop in one day, his profit (in dollars) is given by

$$P(x) = -0.001x^2 + 3x - 1800$$

What is his maximum profit per day, and how many cans must he sell for maximum profit?

43. The effectiveness of a television commercial depends on how many times a viewer watches it. After some experiments an advertising agency found that if the effectiveness E is measured on a scale of 0 to 10, then

$$E(n) = \tfrac{2}{3}n - \tfrac{1}{90}n^2$$

where n is the number of times a viewer watches a given commercial. For a commercial to have maximum effectiveness, how many times should a viewer watch it?

44. When a certain drug is taken orally, the concentration of the drug in the patient's bloodstream after t minutes is given by $C(t) = 0.06t - 0.0002t^2$, where $0 \leq t \leq 240$ and the concentration is measured in mg/L. When is the maximum serum concentration reached, and what is that maximum concentration?

 **45–46** ■ A quadratic function is given.

(a) Use a graphing device to find the maximum or minimum value of the quadratic function f, correct to two decimal places.

(b) Find the exact maximum or minimum value of f, and compare with your answer to part (a).

45. $f(x) = x^2 + 1.79x - 3.21$

46. $f(x) = 1 + x - \sqrt{2}\,x^2$

 **47–54** ■ Find the local maximum and minimum values of the function and the value of x at which each occurs. State each answer correct to two decimal places.

47. $f(x) = x^3 - x$

48. $f(x) = 3 + x + x^2 - x^3$

49. $g(x) = x^4 - 2x^3 - 11x^2$

50. $g(x) = x^5 - 8x^3 + 20x$

51. $U(x) = x\sqrt{6 - x}$

52. $U(x) = x\sqrt{x - x^2}$

53. $V(x) = \dfrac{1 - x^2}{x^3}$

54. $V(x) = \dfrac{1}{x^2 + x + 1}$

 **55.** A fish swims at a speed v relative to the water, against a current of 5 mi/h. Using a mathematical model of energy expenditure, it can be shown that the total energy E required to swim a distance of 10 mi is given by

$$E(v) = 2.73v^3 \frac{10}{v - 5}$$

Biologists believe that migrating fish try to minimize the total energy required to swim a fixed distance. Find the value of v that minimizes energy required.

NOTE: This result has been verified; migrating fish swim against a current at a speed 50% greater than the speed of the current.

56. A highway engineer wants to estimate the maximum number of cars that can safely travel a particular highway at a given speed. She assumes that each car is 17 ft long, travels at a speed s, and follows the car in front of it at the "safe following distance" for that speed. She finds that the number N of cars that can pass a given point per minute is modeled by the function

$$N(s) = \frac{88s}{17 + 17\left(\frac{s}{20}\right)^2}$$

At what speed can the greatest number of cars travel the highway safely?

57. Between 0°C and 30°C, the volume V (in cubic centimeters) of 1 kg of water at a temperature T is given by the formula

$$V = 999.87 - 0.06426T + 0.0085043T^2 - 0.0000679T^3$$

Find the temperature at which the volume of 1 kg of water is a minimum.

58. When a foreign object lodged in the trachea (windpipe) forces a person to cough, the diaphragm thrusts upward causing an increase in pressure in the lungs. At the same time, the trachea contracts, causing the expelled air to move faster and increasing the pressure on the foreign object. According to a mathematical model of coughing, the velocity v of the airstream through an average-sized person's trachea is related to the radius r of the trachea (in centimeters) by the function

$$v(r) = 3.2(1 - r)r^2 \qquad \tfrac{1}{2} \leqslant r \leqslant 1$$

Determine the value of r for which v is a maximum.

59. Find the maximum value of the function $f(x) = 3 + 4x^2 - x^4$. [*Hint:* Let $t = x^2$.]

DISCOVERY • DISCUSSION

60. Maxima and Minima In Example 5 we saw a real-world situation in which the maximum value of a function is important. Name several other everyday situations in which a maximum or minimum value is important.

61. Minimizing a Distance When we seek a minimum or maximum value of a function, it is sometimes easier to work with a simpler function instead.

(a) Suppose $g(x) = \sqrt{f(x)}$, where $f(x) \geqslant 0$ for all x. Explain why the local minima and maxima of f and g occur at the same values of x.

(b) Let $g(x)$ be the distance between the point $(3, 0)$ and the point (x, x^2) on the graph of the parabola $y = x^2$. Express g as a function of x.

(c) Find the minimum value of the function g that you found in part (b). Use the principle described in part (a) to simplify your work.

4.7 COMBINING FUNCTIONS

Two functions f and g can be combined to form new functions $f + g$, $f - g$, fg, and f/g in a manner similar to the way we add, subtract, multiply, and divide real numbers. For example, we define the function $f + g$ by

$$(f + g)(x) = f(x) + g(x)$$

The sum of f and g is defined by

$$(f + g)(x) = f(x) + g(x)$$

The name of the new function is "$f + g$." So this + sign stands for the operation of addition of *functions*. The + sign on the right side, however, stands for addition of the *numbers* $f(x)$ and $g(x)$.

The new function $f + g$ is called the **sum** of the functions f and g; its value at x is $f(x) + g(x)$. Of course, the sum on the right-hand side makes sense only if both $f(x)$ and $g(x)$ are defined, that is, if x belongs to the domain of f and also to the domain of g. So, if the domain of f is A and the domain of g is B, then the domain of $f + g$ is the intersection of these domains, that is, $A \cap B$. Similarly, we can define the **difference** $f - g$, the **product** fg, and the **quotient** f/g of the functions f and g. Their domains are $A \cap B$, but in the case of the quotient we must remember not to divide by 0.

ALGEBRA OF FUNCTIONS

Let f and g be functions with domains A and B. Then the functions $f + g$, $f - g$, fg, and f/g are defined as follows.

$$(f + g)(x) = f(x) + g(x) \qquad \text{Domain } A \cap B$$

$$(f - g)(x) = f(x) - g(x) \qquad \text{Domain } A \cap B$$

$$(fg)(x) = f(x)g(x) \qquad \text{Domain } A \cap B$$

$$\left(\frac{f}{g}\right)(x) = \frac{f(x)}{g(x)} \qquad \text{Domain } \{x \in A \cap B \mid g(x) \neq 0\}$$

EXAMPLE 1 ■ Combining Functions

Let $f(x) = x^3$ and $g(x) = \sqrt{x}$.

(a) Find the functions $f + g$, $f - g$, fg, and f/g and their domains.
(b) Find $(f + g)(4)$, $(f - g)(4)$, $(fg)(3)$, and $(f/g)(1)$.

SOLUTION

(a) The domain of f is $\mathbb{R}$ and the domain of g is $\{x \mid x \geq 0\}$. We have

$$(f + g)(x) = f(x) + g(x) = x^3 + \sqrt{x} \qquad \text{Domain } \{x \mid x \geq 0\}$$

$$(f - g)(x) = f(x) - g(x) = x^3 - \sqrt{x} \qquad \text{Domain } \{x \mid x \geq 0\}$$

$$(fg)(x) = f(x)g(x) = x^3\sqrt{x} \qquad \text{Domain } \{x \mid x \geq 0\}$$

$$\left(\frac{f}{g}\right)(x) = \frac{f(x)}{g(x)} = \frac{x^3}{\sqrt{x}} \qquad \text{Domain } \{x \mid x > 0\}$$

Note that in the domain of f/g we exclude 0 because $g(0) = 0$.

(b) $(f + g)(4) = f(4) + g(4) = 4^3 + \sqrt{4} = 66$

$(f - g)(4) = f(4) - g(4) = 4^3 - \sqrt{4} = 62$

$(fg)(3) = f(3)g(3) = 3^3\sqrt{3} = 27\sqrt{3}$

$$\left(\frac{f}{g}\right)(1) = \frac{f(1)}{g(1)} = \frac{1^3}{\sqrt{1}} = 1$$

■

EXAMPLE 2 ■ Finding Domains of Combinations of Functions

If $f(x) = \sqrt{x}$ and $g(x) = \sqrt{4 - x^2}$, find the functions $f + g$, $f - g$, fg, and f/g and their domains.

SOLUTION

The domain of $f(x) = \sqrt{x}$ is $[0, \infty)$. The domain of $g(x) = \sqrt{4 - x^2}$ is the interval $[-2, 2]$ (from Example 7 in Section 4.2). The intersection of the domains of f and g is

$$[0, \infty) \cap [-2, 2] = [0, 2]$$

Thus, we have

$$(f + g)(x) = \sqrt{x} + \sqrt{4 - x^2} \qquad \text{Domain } \{x \mid 0 \leqslant x \leqslant 2\}$$

$$(f - g)(x) = \sqrt{x} - \sqrt{4 - x^2} \qquad \text{Domain } \{x \mid 0 \leqslant x \leqslant 2\}$$

$$(fg)(x) = \sqrt{x}\,\sqrt{4 - x^2} = \sqrt{4x - x^3} \qquad \text{Domain } \{x \mid 0 \leqslant x \leqslant 2\}$$

$$\left(\frac{f}{g}\right)(x) = \frac{\sqrt{x}}{\sqrt{4 - x^2}} = \sqrt{\frac{x}{4 - x^2}} \qquad \text{Domain } \{x \mid 0 \leqslant x < 2\}$$

Notice that the domain of f/g is the interval $[0, 2)$ because we must exclude the points where $g(x) = 0$, that is, $x = \pm 2$. ■

The graph of the function $f + g$ can be obtained from the graphs of f and g by **graphical addition**. This means that we add corresponding y-coordinates, as illustrated in the next example.

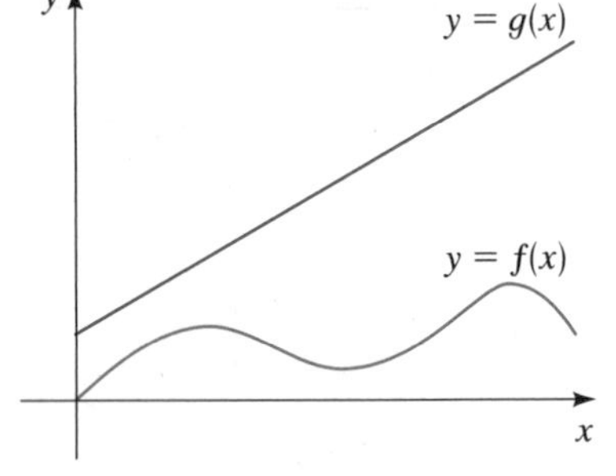

FIGURE 1

EXAMPLE 3 ■ Using Graphical Addition

The graphs of f and g are shown in Figure 1. Use graphical addition to graph the function $f + g$.

SOLUTION

In Figure 2 we obtain the graph of $f + g$ by "graphically adding" the value of $f(x)$ to $g(x)$. This is implemented by copying the line segment PQ on top of PR to obtain the point S on the graph of $f + g$.

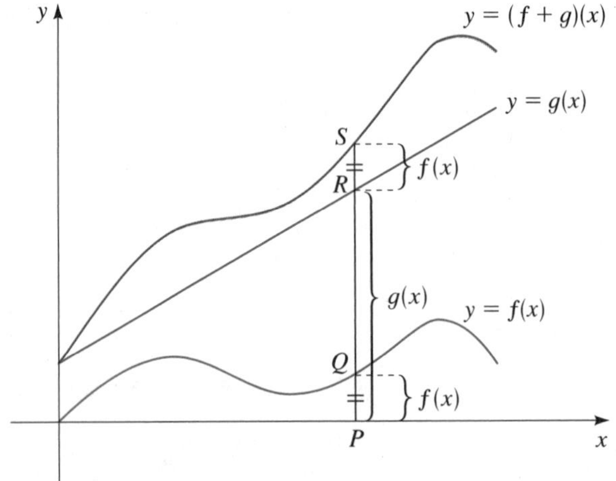

FIGURE 2
Graphical addition

■

COMPOSITION OF FUNCTIONS

Now let's consider a very important way of combining two functions to get a new function. Suppose $f(x) = \sqrt{x}$ and $g(x) = x^2 + 1$. We may define a function h as

$$h(x) = f(g(x)) = f(x^2 + 1) = \sqrt{x^2 + 1}$$

The function h is made up of the functions f and g in an interesting way: Given a number x, we first apply to it the function g, then apply f to the result. In this case, f is the rule "take the square root," g is the rule "square, then add 1," and h is the rule "square, then add 1, then take the square root." In other words, we get the rule h by applying the rule g and then the rule f.

In general, given any two functions f and g, we start with a number x in the domain of g and find its image $g(x)$. If this number $g(x)$ is in the domain of f, we can then calculate the value of $f(g(x))$. The result is a new function $h(x) = f(g(x))$ obtained by substituting g into f. It is called the *composition* (or *composite*) of f and g and is denoted by $f \circ g$ ("f composed with g").

COMPOSITION OF FUNCTIONS

Given two functions f and g, the **composite function** $f \circ g$ (also called the **composition** of f and g) is defined by

$$(f \circ g)(x) = f(g(x))$$

The domain of $f \circ g$ is the set of all x in the domain of g such that $g(x)$ is in the domain of f. In other words, $(f \circ g)(x)$ is defined whenever both $g(x)$ and $f(g(x))$ are defined. We can picture $f \circ g$ using a machine diagram (Figure 3) or an arrow diagram (Figure 4).

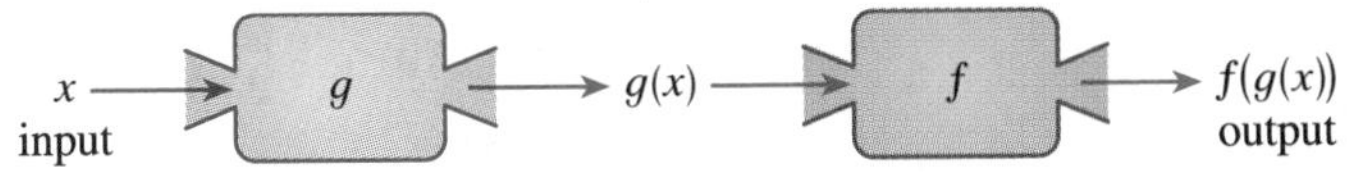

FIGURE 3
The $f \circ g$ machine is composed of the g machine (first) and then the f machine.

x input → g → $g(x)$ → f → $f(g(x))$ output

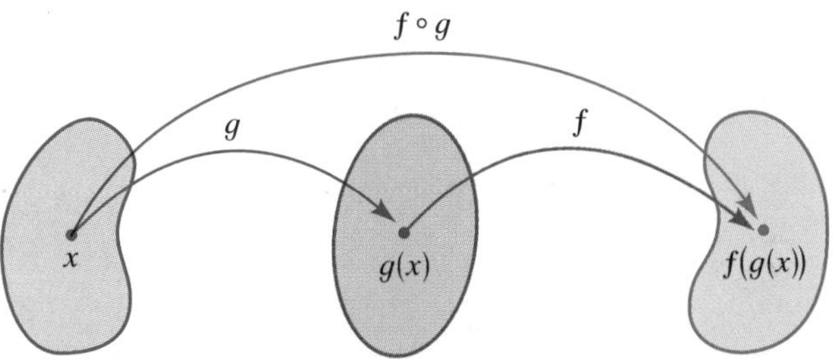

FIGURE 4
Arrow diagram for $f \circ g$

EXAMPLE 4 ■ Finding the Composition of Functions

Let $f(x) = x^2$ and $g(x) = x - 3$.

(a) Find the functions $f \circ g$ and $g \circ f$ and their domains.
(b) Find $(f \circ g)(5)$ and $(g \circ f)(7)$.

SOLUTION

In Example 4, f is the rule "square" and g is the rule "subtract 3." The function $f \circ g$ *first* subtracts 3 and *then* squares; the function $g \circ f$ *first* squares and *then* subtracts 3.

(a) We have

$$\begin{aligned}(f \circ g)(x) &= f(g(x)) && \text{Definition of } f \circ g\\ &= f(x-3) && \text{Definition of } g\\ &= (x-3)^2 && \text{Definition of } f\end{aligned}$$

and

$$\begin{aligned}(g \circ f)(x) &= g(f(x)) && \text{Definition of } g \circ f\\ &= g(x^2) && \text{Definition of } f\\ &= x^2 - 3 && \text{Definition of } g\end{aligned}$$

The domains of both $f \circ g$ and $g \circ f$ are $\mathbb{R}$.

(b) We have

$$(f \circ g)(5) = f(g(5)) = f(2) = 2^2 = 4$$
$$(g \circ f)(7) = g(f(7)) = g(49) = 49 - 3 = 46$$

■

You can see from Example 4 that, in general, $f \circ g \neq g \circ f$. Remember that the notation $f \circ g$ means that the function g is applied first and then f is applied second.

EXAMPLE 5 ■ Finding the Composition of Functions

If $f(x) = \sqrt{x}$ and $g(x) = \sqrt{2 - x}$, find the following functions and their domains.

(a) $f \circ g$ (b) $g \circ f$ (c) $f \circ f$ (d) $g \circ g$

SOLUTION

(a)

$$\begin{aligned}(f \circ g)(x) &= f(g(x)) && \text{Definition of } f \circ g\\ &= f\left(\sqrt{2-x}\right) && \text{Definition of } g\\ &= \sqrt{\sqrt{2-x}} && \text{Definition of } f\\ &= \sqrt[4]{2-x}\end{aligned}$$

The domain of $f \circ g$ is $\{x \mid 2 - x \geqslant 0\} = \{x \mid x \leqslant 2\} = (-\infty, 2]$.

The graphs of f and g of Example 5, as well as $f \circ g$, $g \circ f$, $f \circ f$, and $g \circ g$, are shown. These graphs indicate that the operation of composition can produce functions quite different from the original functions.

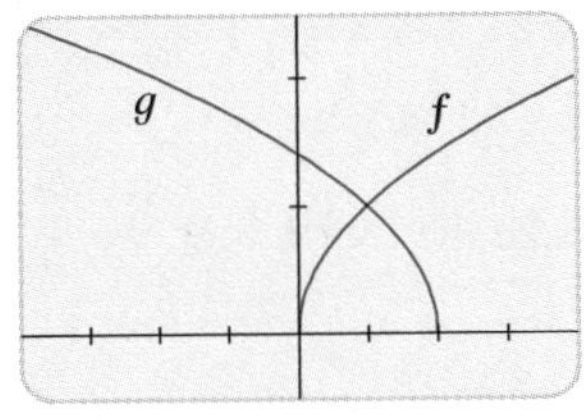

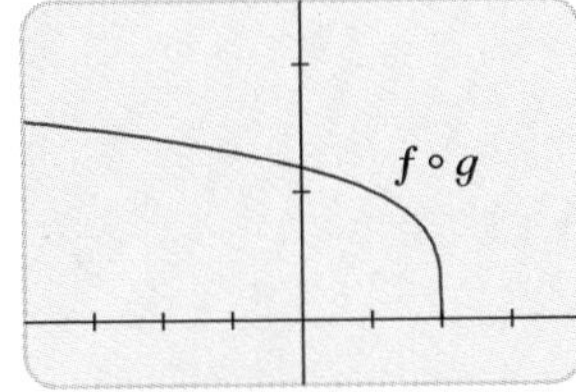

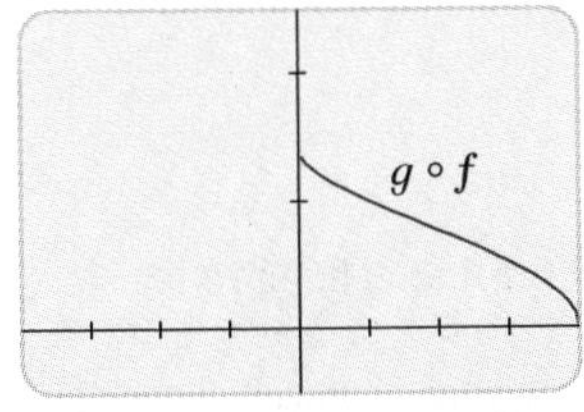

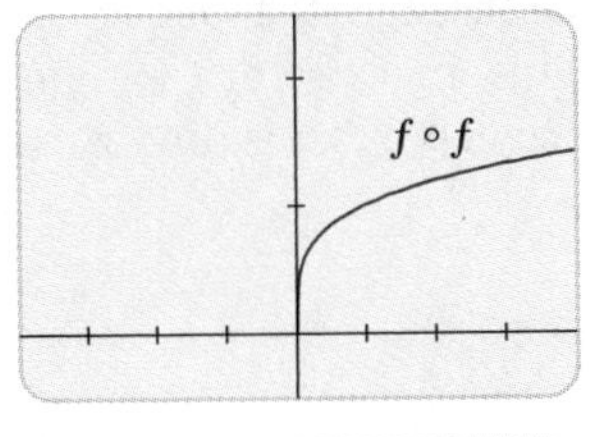

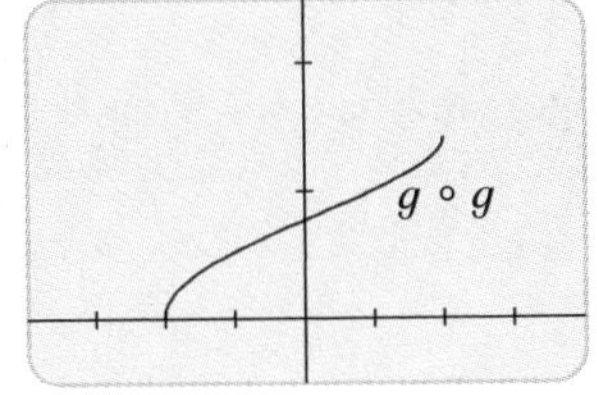

(b)

$$\begin{aligned}(g \circ f)(x) &= g(f(x)) && \text{Definition of } g \circ f\\ &= g(\sqrt{x}) && \text{Definition of } f\\ &= \sqrt{2-\sqrt{x}} && \text{Definition of } g\end{aligned}$$

For $\sqrt{x}$ to be defined, we must have $x \geqslant 0$. For $\sqrt{2-\sqrt{x}}$ to be defined, we must have $2-\sqrt{x} \geqslant 0$, that is, $\sqrt{x} \leqslant 2$, or $x \leqslant 4$. Thus, we have $0 \leqslant x \leqslant 4$, so the domain of $g \circ f$ is the closed interval $[0, 4]$.

(c)

$$\begin{aligned}(f \circ f)(x) &= f(f(x)) && \text{Definition of } f \circ f\\ &= f(\sqrt{x}) && \text{Definition of } f\\ &= \sqrt{\sqrt{x}} && \text{Definition of } f\\ &= \sqrt[4]{x}\end{aligned}$$

The domain of $f \circ f$ is $[0, \infty)$.

(d)

$$\begin{aligned}(g \circ g)(x) &= g(g(x)) && \text{Definition of } g \circ g\\ &= g(\sqrt{2-x}) && \text{Definition of } g\\ &= \sqrt{2-\sqrt{2-x}} && \text{Definition of } g\end{aligned}$$

This expression is defined when both $2 - x \geqslant 0$ and $2 - \sqrt{2-x} \geqslant 0$. The first inequality means $x \leqslant 2$, and the second is equivalent to $\sqrt{2-x} \leqslant 2$, or $2 - x \leqslant 4$, or $x \geqslant -2$. Thus, $-2 \leqslant x \leqslant 2$, so the domain of $g \circ g$ is $[-2, 2]$. ■

It is possible to take the composition of three or more functions. For instance, the composite function $f \circ g \circ h$ is found by first applying h, then g, and then f as follows:

$$(f \circ g \circ h)(x) = f(g(h(x)))$$

EXAMPLE 6 ■ A Composition of Three Functions

Find $f \circ g \circ h$ if $f(x) = x/(x + 1)$, $g(x) = x^{10}$, and $h(x) = x + 3$.

SOLUTION

$$\begin{aligned}(f \circ g \circ h)(x) &= f(g(h(x))) && \text{Definition of } f \circ g \circ h\\ &= f(g(x+3)) && \text{Definition of } h\\ &= f((x+3)^{10}) && \text{Definition of } g\\ &= \frac{(x+3)^{10}}{(x+3)^{10}+1} && \text{Definition of } f\end{aligned}$$

■

So far we have used composition to build complicated functions from simpler ones. But in calculus it is useful to be able to "decompose" a complicated function into simpler ones, as shown in the following example.

EXAMPLE 7 ■ Recognizing a Composition of Functions

Given $F(x) = \sqrt[4]{x + 9}$, find functions f and g such that $F = f \circ g$.

SOLUTION

Since the formula for F says to first add 9 and then take the fourth root, we let

$$g(x) = x + 9 \qquad \text{and} \qquad f(x) = \sqrt[4]{x}$$

Then

$$\begin{aligned}(f \circ g)(x) &= f(g(x)) \\ &= f(x + 9) \\ &= \sqrt[4]{x + 9} \\ &= F(x)\end{aligned}$$

■

EXAMPLE 8 ■ An Application of Composition of Functions

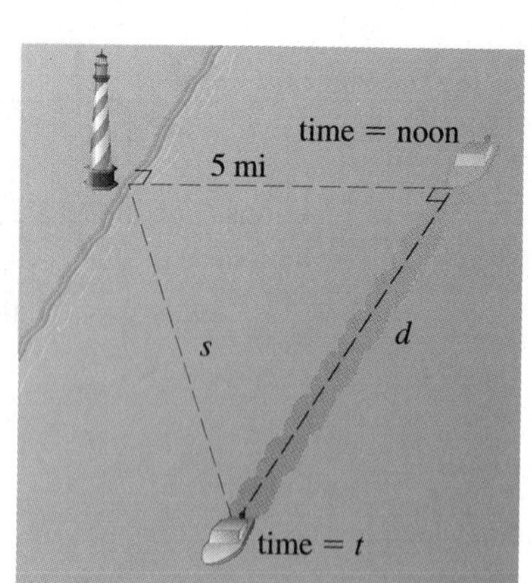

FIGURE 5

A ship is traveling at 20 mi/h parallel to a straight shoreline. The ship is 5 mi from shore. It passes a lighthouse at noon.

(a) Express the distance s between the lighthouse and the ship as a function of d, the distance the ship has traveled since noon; that is, find f so that $s = f(d)$.
(b) Express d as a function of t, the time elapsed since noon; that is, find g so that $d = g(t)$.
(c) Find $f \circ g$. What does this function represent?

SOLUTION

We first draw a diagram as in Figure 5.

(a) We can relate the distances s and d by the Pythagorean Theorem. Thus, s can be expressed as a function of d by

$$s = f(d) = \sqrt{25 + d^2}$$

(b) Since the ship is traveling at 20 mi/h, the distance d it has traveled is a function of t as follows:

distance = rate × time

$$d = g(t) = 20t$$

(c) We have

$$\begin{aligned}(f \circ g)(t) &= f(g(t)) && \text{Definition of } f \circ g \\ &= f(20t) && \text{Definition of } g \\ &= \sqrt{25 + (20t)^2} && \text{Definition of } f\end{aligned}$$

The function $f \circ g$ gives the distance of the ship from the lighthouse as a function of time. ■

4.7 EXERCISES

1–6 ■ Find $f + g$, $f - g$, fg, and f/g and their domains.

1. $f(x) = x^2 - x, \quad g(x) = x + 5$

2. $f(x) = x^3 + 2x^2, \quad g(x) = 3x^2 - 1$

3. $f(x) = \sqrt{1 + x}, \quad g(x) = \sqrt{1 - x}$

4. $f(x) = \sqrt{9 - x^2}, \quad g(x) = \sqrt{x^2 - 1}$

5. $f(x) = \dfrac{2}{x}, \quad g(x) = -\dfrac{2}{x + 4}$

6. $f(x) = \dfrac{1}{x + 1}, \quad g(x) = \dfrac{x}{x + 1}$

7–10 ■ Find the domain of the function.

7. $f(x) = \sqrt{x} + \sqrt{1 - x}$

8. $g(x) = \sqrt{x + 1} - \dfrac{1}{x}$

9. $h(x) = (x + 1)^2(2x - 8)^{1/4}$

10. $k(x) = \dfrac{\sqrt{x + 3}}{x - 1}$

11–12 ■ Use graphical addition to sketch the graph of $f + g$.

11.

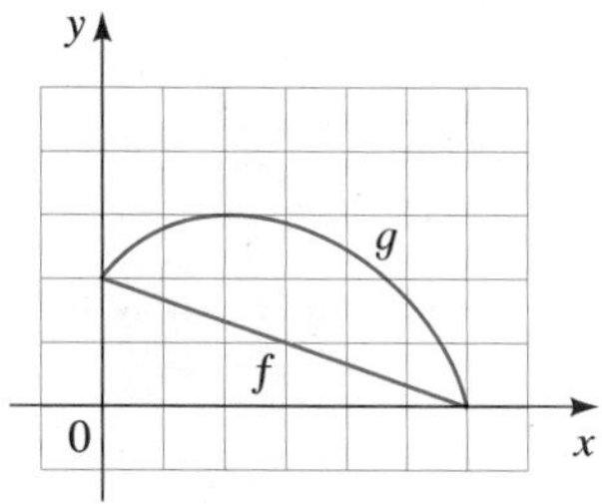

12.

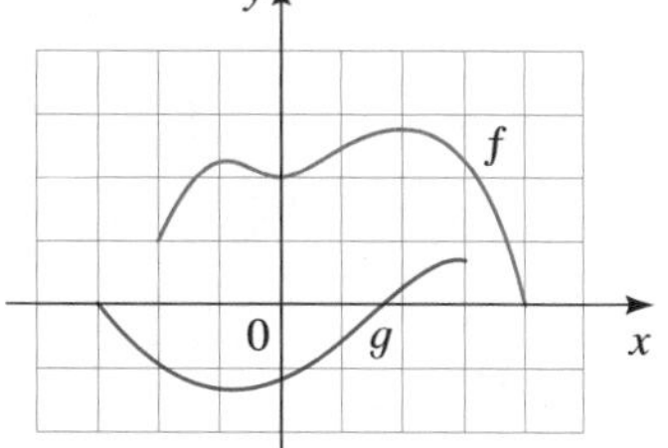

13–16 ■ Draw the graphs of f, g, and $f + g$ on a common screen to illustrate graphical addition.

13. $f(x) = \sqrt{1 + x}, \quad g(x) = \sqrt{1 - x}$

14. $f(x) = x^2, \quad g(x) = \sqrt{x}$

15. $f(x) = x^2, \quad g(x) = x^3$

16. $f(x) = \sqrt[4]{1 - x}, \quad g(x) = \sqrt{1 - \dfrac{x^2}{9}}$

17–22 ■ Use $f(x) = 3x - 5$ and $g(x) = 2 - x^2$ to evaluate the expression.

17. (a) $f(g(0))$ (b) $g(f(0))$

18. (a) $f(f(4))$ (b) $g(g(3))$

19. (a) $(f \circ g)(-2)$ (b) $(g \circ f)(-2)$

20. (a) $(f \circ f)(-1)$ (b) $(g \circ g)(2)$

21. (a) $(f \circ g)(x)$ (b) $(g \circ f)(x)$

22. (a) $(f \circ f)(x)$ (b) $(g \circ g)(x)$

23–28 ■ Use the given graphs of f and g to evaluate the expression.

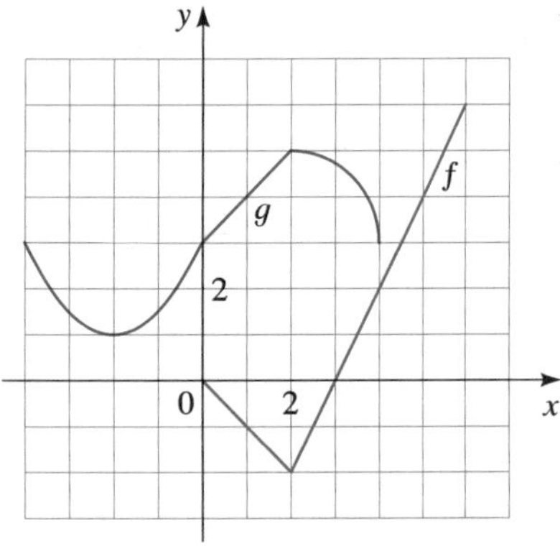

23. $f(g(2))$

24. $g(f(0))$

25. $(g \circ f)(4)$

26. $(f \circ g)(0)$

27. $(g \circ g)(-2)$

28. $(f \circ f)(4)$

29–40 ■ Find the functions $f \circ g$, $g \circ f$, $f \circ f$, and $g \circ g$ and their domains.

29. $f(x) = 2x + 3, \quad g(x) = 4x - 1$

30. $f(x) = 6x - 5, \quad g(x) = \dfrac{x}{2}$

31. $f(x) = x^2, \quad g(x) = x + 1$

32. $f(x) = x^3 + 2, \quad g(x) = \sqrt[3]{x}$

33. $f(x) = \dfrac{1}{x}, \quad g(x) = 2x + 4$

34. $f(x) = x^2, \quad g(x) = \sqrt{x - 3}$

35. $f(x) = |x|, \quad g(x) = 2x + 3$

36. $f(x) = x - 4, \quad g(x) = |x + 4|$

37. $f(x) = \dfrac{x}{x + 1}, \quad g(x) = 2x - 1$

38. $f(x) = \dfrac{1}{\sqrt{x}}, \quad g(x) = x^2 - 4x$

39. $f(x) = \sqrt[3]{x}, \quad g(x) = \sqrt[4]{x}$

40. $f(x) = \dfrac{2}{x}, \quad g(x) = \dfrac{x}{x + 2}$

41–44 ■ Find $f \circ g \circ h$.

41. $f(x) = x - 1, \quad g(x) = \sqrt{x}, \quad h(x) = x - 1$

42. $f(x) = \dfrac{1}{x}, \quad g(x) = x^3, \quad h(x) = x^2 + 2$

43. $f(x) = x^4 + 1, \quad g(x) = x - 5, \quad h(x) = \sqrt{x}$

44. $f(x) = \sqrt{x}, \quad g(x) = \dfrac{x}{x - 1}, \quad h(x) = \sqrt[3]{x}$

45–50 ■ Express the function in the form $f \circ g$.

45. $F(x) = (x - 9)^5$

46. $F(x) = \sqrt{x} + 1$

47. $G(x) = \dfrac{x^2}{x^2 + 4}$

48. $G(x) = \dfrac{1}{x + 3}$

49. $H(x) = |1 - x^3|$

50. $H(x) = \sqrt{1 + \sqrt{x}}$

51–54 ■ Express the function in the form $f \circ g \circ h$.

51. $F(x) = \dfrac{1}{x^2 + 1}$

52. $F(x) = \sqrt[3]{\sqrt{x} - 1}$

53. $G(x) = (4 + \sqrt[3]{x})^9$

54. $G(x) = \dfrac{2}{(3 + \sqrt{x})^2}$

55. A stone is dropped in a lake, creating a circular ripple that travels outward at a speed of 60 cm/s. Express the area of this circle as a function of time t (in seconds).

56. A spherical balloon is being inflated. If the radius of the balloon is increasing at a rate of 1 cm/s, express the volume of the balloon as a function of time t (in seconds).

57. An airplane is flying at a speed of 350 mi/h at an altitude of one mile. The plane passes directly above a radar station at time $t = 0$.

(a) Express the distance s between the plane and the radar station as a function of d.

(b) Express the horizontal distance d (in miles) that the plane has flown as a function of t (in hours).

(c) Use composition to express s as a function of t.

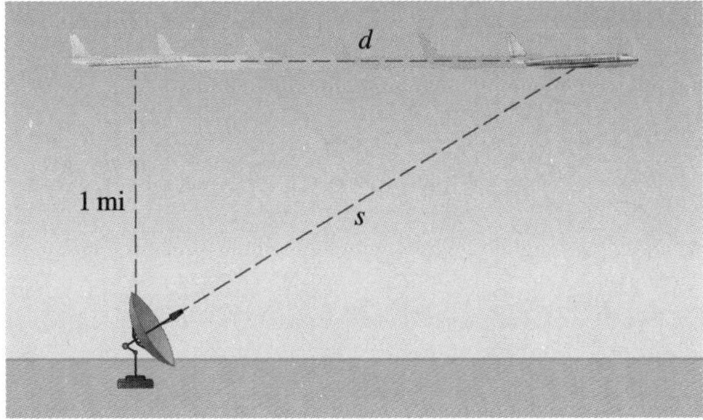

DISCOVERY • DISCUSSION

58. Compound Interest A savings account earns 5% interest compounded annually. If you invest x dollars in such an account, then the amount $A(x)$ of the investment after one year is the initial investment plus 5%; that is, $A(x) = x + 0.05x = 1.05x$. Find $A \circ A$, $A \circ A \circ A$, and $A \circ A \circ A \circ A$. What do these compositions represent? Find a formula for what you get when you compose n copies of A.

59. Composing Linear Functions The graphs of the functions $f(x) = m_1x + b_1$ and $g(x) = m_2x + b_2$ are lines with slopes m_1 and m_2, respectively. Is the graph of $f \circ g$ a line? If so, what is its slope?

60. Solving an Equation for an Unknown Function If $g(x) = 2x + 1$ and $h(x) = 4x^2 + 4x + 7$, find a function f such that $f \circ g = h$. (Think about what operations you would have to perform on the formula for g to end up with the formula for h.)

Now suppose that

$$f(x) = 3x + 5 \qquad \text{and} \qquad h(x) = 3x^2 + 3x + 2$$

Use the same sort of reasoning to find a function g such that $f \circ g = h$.

61. Compositions of Odd and Even Functions Suppose that $h = f \circ g$. If g is an even function, is h necessarily even? If g is odd, is h odd? What if g is odd and f is odd? What if g is odd and f is even?

ITERATION AND CHAOS

The **iterates** of a function f at a point x_0 are $f(x_0)$, $f(f(x_0))$, $f(f(f(x_0)))$, and so on. We write

$$x_1 = f(x_0) \qquad \text{The first iterate}$$

$$x_2 = f(f(x_0)) \qquad \text{The second iterate}$$

$$x_3 = f(f(f(x_0))) \qquad \text{The third iterate}$$

For example, if $f(x) = x^2$, then the iterates of f at 2 are $x_1 = 4$, $x_2 = 16$, $x_3 = 256$, and so on. (Check this.) Iterates can be described graphically as in Figure 1. Start with x_0 on the x-axis, move vertically to the graph of f, then horizontally to the line $y = x$, then vertically to the graph of f, and so on. The x-coordinates of the points on the graph of f are the iterates of f at x_0.

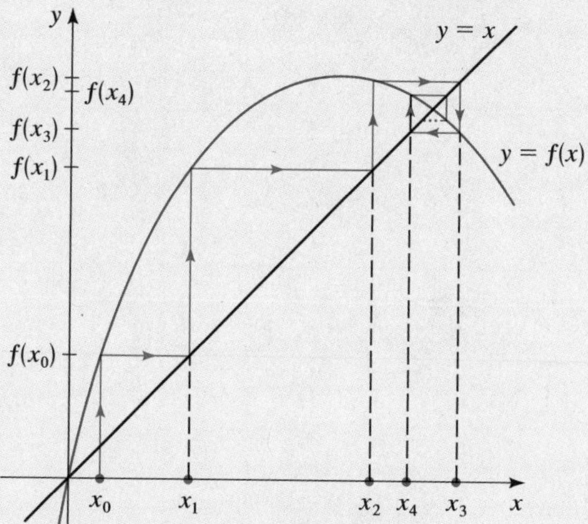

FIGURE 1

n	x_n
0	0.1
1	0.234
2	0.46603
3	0.64700
4	0.59382
5	0.62712
6	0.60799
7	0.61968
8	0.61276
9	0.61694
10	0.61444
11	0.61595
12	0.61505

Iterates are important in studying the **logistic function**

$$f(x) = kx(1 - x)$$

which models the population of a species with limited potential for growth (such as rabbits on an island or fish in a pond). In this model the maximum population that the environment can support is 1 (that is, 100%); if we start with a fraction of that population, say 0.1 (10%), then the iterates of f at 0.1 give the population after each time interval (days, months, or years, depending on the species). The constant k depends on the rate of growth of the species being modeled; it is called the **growth constant**. For example, for $k = 2.6$ and $x_0 = 0.1$ the iterates shown in the table give the population of the species for the first 12 time intervals. The population seems to be stabilizing around 0.615 (that is, 61.5% of maximum).

In the three graphs in Figure 2 we plot the iterates of f at 0.1 for different values of the growth constant k. For $k = 2.6$ the population appears to stabilize at a value 0.615 of maximum, for $k = 3.1$ the population appears to oscillate between two values, and for $k = 3.8$ no obvious pattern emerges. This latter situation is described mathematically by the word **chaos**.

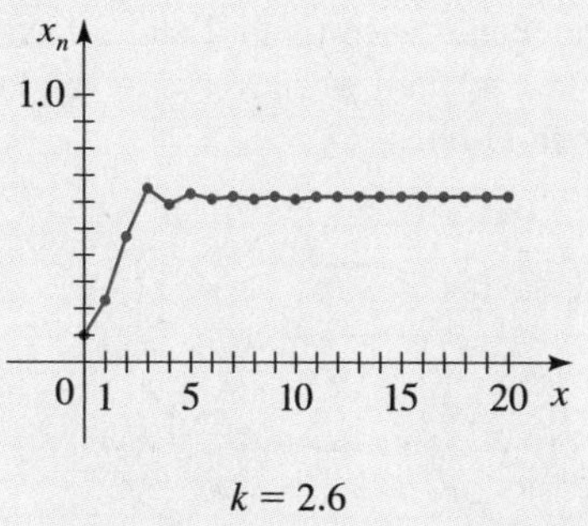

$k = 2.6$

$k = 3.1$

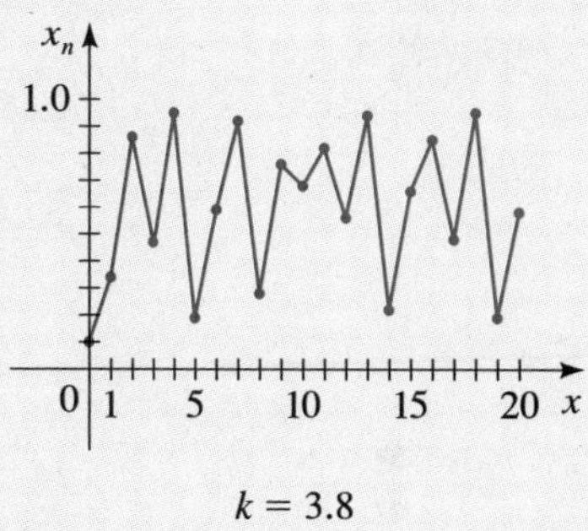

$k = 3.8$

FIGURE 2

1. Use the graphical procedure illustrated in Figure 1 to find the first five iterates of $f(x) = 2x(1 - x)$ at $x = 0.1$.

2. Find the iterates of $f(x) = x^2$ at $x = 1$.

3. Find the iterates of $f(x) = 1/x$ at $x = 2$.

4. Find the first six iterates of $f(x) = 1/(1 - x)$ at $x = 2$. What is 1000th iterate of f at 2?

5. Find the first 10 iterates of the logistic function at $x = 0.1$ for the given value of k. Does the population appear to stabilize, oscillate, or is it chaotic?
(a) $k = 2.1$ (b) $k = 3.2$ (c) $k = 3.9$

6. It is easy to find iterates using a graphing calculator. The following steps show how to find the iterates of $f(x) = kx(1 - x)$ at 0.1 for $k = 3$ on a TI-85 calculator. (The procedure can be adapted for any graphing calculator.)

y1 = K * x * (1 − x)	Enter f as Y1 on the graph list
3 → K	Store 3 in the variable K
0.1 → x	Store 0.1 in the variable x
y1 → x	Evaluate f at x and store result back in x
0.27	Hit ENTER and obtain first iterate
0.5913 0.72499293 0.59813454435	Keep hitting ENTER to re-execute the command and obtain successive iterates

Use a graphing calculator to experiment with how the value of k affects the iterates of $f(x) = kx(1 - x)$ at 0.1. Find several different values of k that make the iterates stabilize at one value, oscillate between two values, and exhibit chaos. (Use values of k between 1 and 4.) Can you find a value of k that makes the iterates oscillate between *four* values?

4.8 ONE-TO-ONE FUNCTIONS AND THEIR INVERSES

Let's compare the functions f and g whose arrow diagrams are shown in Figure 1. Note that f never takes on the same value twice (any two numbers in A have different images), whereas g does take on the same value twice (both 2 and 3 have the same image, 4). In symbols, $g(2) = g(3)$ but $f(x_1) \neq f(x_2)$ whenever $x_1 \neq x_2$. Functions that have this latter property are called *one-to-one*.

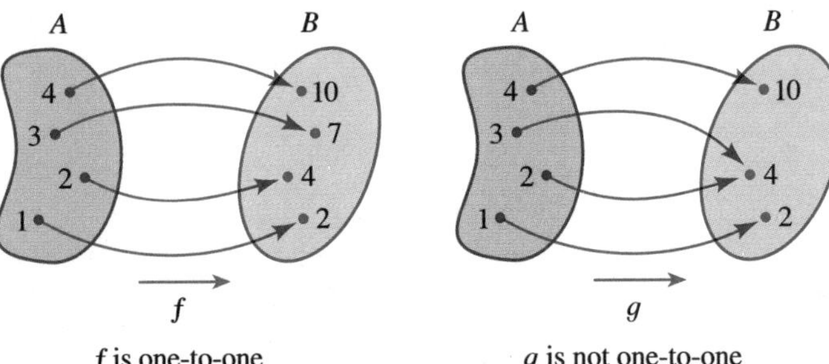

FIGURE 1 f is one-to-one g is not one-to-one

DEFINITION OF A ONE-TO-ONE FUNCTION

A function within domain A is called a **one-to-one function** if no two elements of A have the same image, that is,

$$f(x_1) \neq f(x_2) \qquad \text{whenever } x_1 \neq x_2$$

An equivalent way of writing the condition for a one-to-one function is this:

$$\text{If } f(x_1) = f(x_2), \text{ then } x_1 = x_2.$$

If a horizontal line intersects the graph of f at more than one point, then we see from Figure 2 that there are numbers $x_1 \neq x_2$ such that $f(x_1) = f(x_2)$. This means that f is not one-to-one. Therefore, we have the following geometric method for determining whether a function is one-to-one.

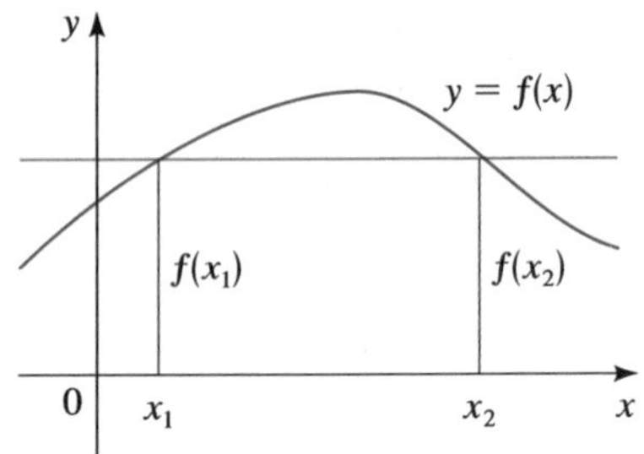

FIGURE 2
This function is not one-to-one because $f(x_1) = f(x_2)$.

HORIZONTAL LINE TEST

A function is one-to-one if and only if no horizontal line intersects its graph more than once.

EXAMPLE 1 ■ Deciding whether a Function Is One-to-One

Is the function $f(x) = x^3$ one-to-one?

SOLUTION 1

If $x_1 \neq x_2$, then $x_1^3 \neq x_2^3$ (two different numbers cannot have the same cube). Therefore, $f(x) = x^3$ is one-to-one.

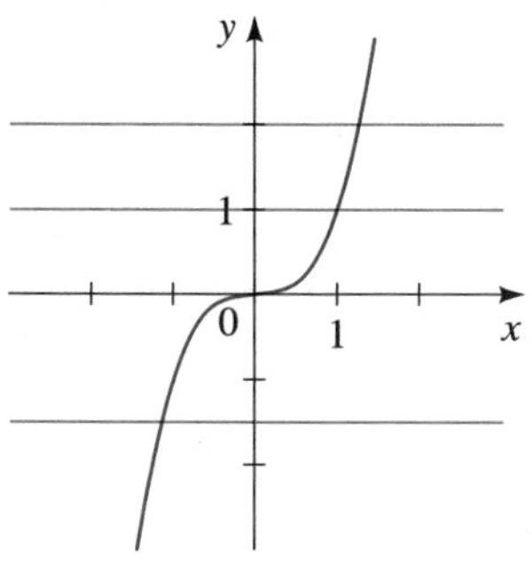

FIGURE 3
$f(x) = x^3$ is one-to-one.

SOLUTION 2

From Figure 3 we see that no horizontal line intersects the graph of $f(x) = x^3$ more than once. Therefore, by the Horizontal Line Test, f is one-to-one. ■

Notice that the function f of Example 1 is increasing and is also one-to-one. In fact, it can be proved that every increasing function and every decreasing function is one-to-one.

EXAMPLE 2 ■ **Deciding whether a Function Is One-to-One**

Is the function $g(x) = x^2$ one-to-one?

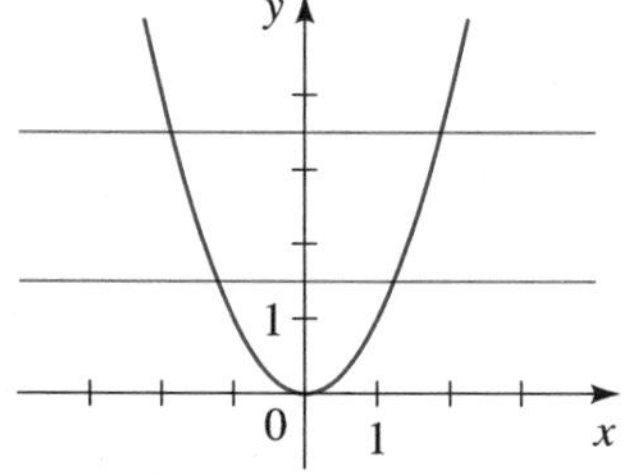

FIGURE 4
$g(x) = x^2$ is not one-to-one.

SOLUTION 1

This function is not one-to-one because, for instance,

$$g(1) = 1 = g(-1)$$

and so 1 and -1 have the same image.

SOLUTION 2

From Figure 4 we see that there are horizontal lines that intersect the graph of g more than once. Therefore, by the Horizontal Line Test, g is not one-to-one. ■

Although the function g in Example 2 is not one-to-one, it is possible to restrict its domain so that the resulting function is one-to-one. In fact, if we define

$$h(x) = x^2, \qquad x \geqslant 0$$

then h is one-to-one, as you can see from Figure 5 and the Horizontal Line Test.

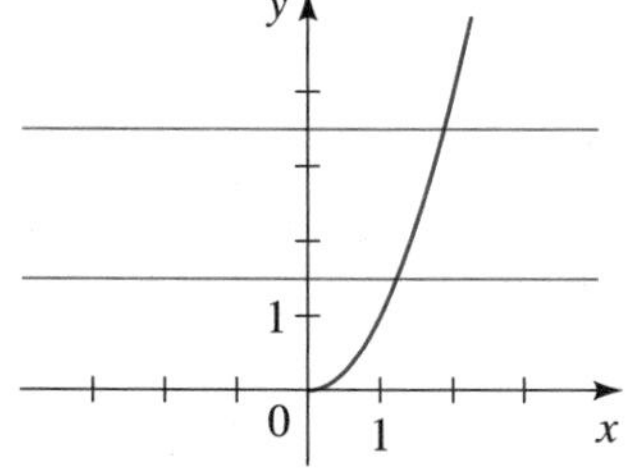

FIGURE 5
$h(x) = x^2$ $(x \geqslant 0)$ is one-to-one.

EXAMPLE 3 ■ **Showing That a Function Is One-to-One**

Show that the function $f(x) = 3x + 4$ is one-to-one.

SOLUTION

Suppose there are numbers x_1 and x_2 such that $f(x_1) = f(x_2)$. Then

$$3x_1 + 4 = 3x_2 + 4$$

$$3x_1 = 3x_2$$

$$x_1 = x_2$$

Therefore, f is one-to-one. ■

THE INVERSE OF A FUNCTION

One-to-one functions are important because they are precisely the functions that possess inverse functions according to the following definition.

Don't mistake the -1 in f^{-1} for an exponent.

$$f^{-1} \quad \textit{does not mean} \quad \frac{1}{f(x)}$$

The reciprocal $1/f(x)$ is written as $(f(x))^{-1}$.

DEFINITION OF THE INVERSE OF A FUNCTION

Let f be a one-to-one function with domain A and range B. Then its **inverse function** f^{-1} has domain B and range A and is defined by

$$f^{-1}(y) = x \quad \Leftrightarrow \quad f(x) = y$$

for any y in B.

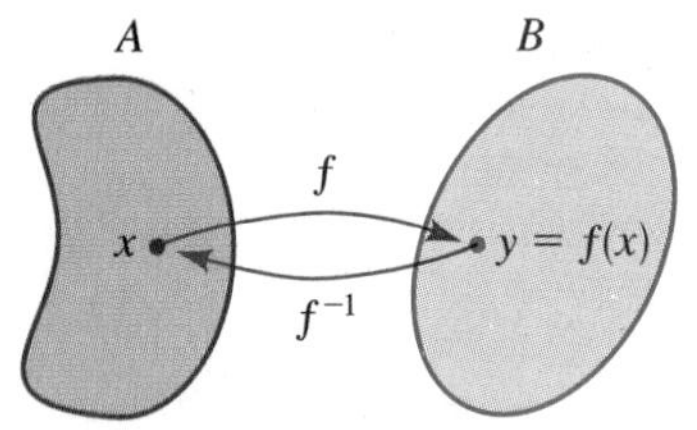

FIGURE 6

This definition says that if f takes x into y, then f^{-1} takes y back into x. (If f were not one-to-one, then f^{-1} would not be defined uniquely.) The arrow diagram in Figure 6 indicates that f^{-1} reverses the effect of f. From the definition we have

$$\text{domain of } f^{-1} = \text{range of } f$$

$$\text{range of } f^{-1} = \text{domain of } f$$

EXAMPLE 4 ■ Finding f^{-1} for Specific Values

If $f(1) = 5$, $f(3) = 7$, and $f(8) = -10$, find $f^{-1}(5)$, $f^{-1}(7)$, and $f^{-1}(-10)$.

SOLUTION

From the definition of f^{-1} we have

$$f^{-1}(5) = 1 \quad \text{because} \quad f(1) = 5$$

$$f^{-1}(7) = 3 \quad \text{because} \quad f(3) = 7$$

$$f^{-1}(-10) = 8 \quad \text{because} \quad f(8) = -10$$

Figure 7 shows how f^{-1} reverses the effect of f in this case.

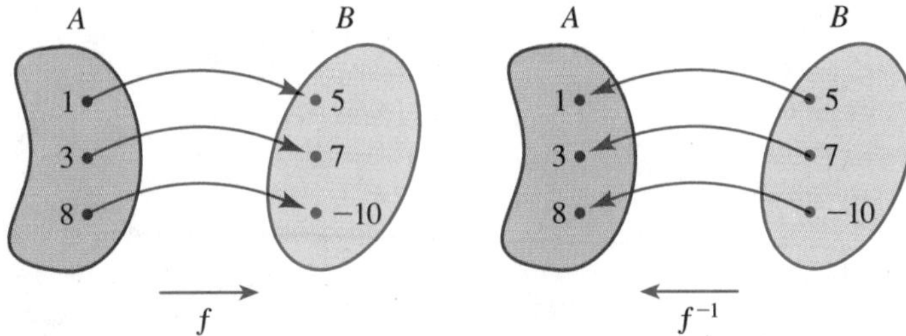

FIGURE 7

■

By definition the inverse function f^{-1} undoes what f does: If we start with x, apply f, and then apply f^{-1}, we arrive back at x, where we started. Similarly, f undoes what f^{-1} does. In general, any function that reverses the effect of f in this way must be the inverse of f. These observations are expressed precisely as follows.

PROPERTY OF INVERSE FUNCTIONS

Let f be a one-to-one function with domain A and range B. The inverse function f^{-1} satisfies the following cancellation properties.

$$f^{-1}(f(x)) = x \qquad \text{for every } x \text{ in } A$$

$$f(f^{-1}(x)) = x \qquad \text{for every } x \text{ in } B$$

Conversely, any function f^{-1} satisfying these equations is the inverse of f.

These properties indicate that f is the inverse function of f^{-1}, so we say that f and f^{-1} are *inverses of each other.*

EXAMPLE 5 ■ Verifying That Two Functions Are Inverses

Show that $f(x) = x^3$ and $g(x) = x^{1/3}$ are inverses of each other.

SOLUTION

Note that the domain and range of both f and g is $\mathbb{R}$. We have

$$g(f(x)) = g(x^3) = (x^3)^{1/3} = x$$

$$f(g(x)) = f(x^{1/3}) = (x^{1/3})^3 = x$$

So, by the Property of Inverse Functions, f and g are inverses of each other. These equations simply say that the cube function and the cube root function, when composed, cancel each other. ■

Now let's examine how we compute inverse functions. We first observe from the definition of f^{-1} that

$$y = f(x) \quad \Leftrightarrow \quad f^{-1}(y) = x$$

So, if $y = f(x)$ and if we are able to solve this equation for x in terms of y, then we must have $x = f^{-1}(y)$. If we then interchange x and y, we have $y = f^{-1}(x)$, which is the desired equation.

HOW TO FIND THE INVERSE OF A ONE-TO-ONE FUNCTION

1. Write $y = f(x)$.
2. Solve this equation for x in terms of y (if possible).
3. Interchange x and y. The resulting equation is $y = f^{-1}(x)$.

Note that Steps 2 and 3 can be reversed. In other words, we can interchange x and y first and then solve for y in terms of x.

In Example 6 note how f^{-1} reverses the effect of f. The function f is the rule "multiply by 3, then subtract 2," whereas f^{-1} is the rule "add 2, then divide by 3."

EXAMPLE 6 ■ Finding the Inverse of a Function

Find the inverse of the function $f(x) = 3x - 2$.

SOLUTION

First we write $y = f(x)$.

$$y = 3x - 2$$

Then we solve this equation for x:

$$3x = y + 2 \qquad \text{Add 2}$$

$$x = \frac{y + 2}{3} \qquad \text{Divide by 3}$$

Finally, we interchange x and y:

$$y = \frac{x + 2}{3}$$

Therefore, the inverse function is $f^{-1}(x) = \dfrac{x + 2}{3}$.

CHECK YOUR ANSWER

We use the Inverse Function Property.

$$\begin{aligned} f^{-1}(f(x)) &= f^{-1}(3x - 2) \\ &= \frac{(3x - 2) + 2}{3} \\ &= \frac{3x}{3} = x \end{aligned}$$

$$\begin{aligned} f(f^{-1}(x)) &= f\left(\frac{x + 2}{3}\right) \\ &= 3\left(\frac{x + 2}{3}\right) - 2 \\ &= x + 2 - 2 = x \quad ✓ \end{aligned}$$

■

In Example 7 note how f^{-1} reverses the effect of f. The function f is the rule "take the fifth power, subtract 3, then divide by 2," whereas f^{-1} is the rule "multiply by 2, add 3, then take the fifth root."

EXAMPLE 7 ■ Finding the Inverse of a Function

Find the inverse of the function $f(x) = \dfrac{x^5 - 3}{2}$.

SOLUTION

We first write $y = (x^5 - 3)/2$ and solve for x.

$$y = \frac{x^5 - 3}{2} \qquad \text{Equation defining function}$$

$$2y = x^5 - 3 \qquad \text{Multiply by 2}$$

$$x^5 = 2y + 3 \qquad \text{Add 3}$$

$$x = (2y + 3)^{1/5} \qquad \text{Take fifth roots}$$

Then we interchange x and y to get $y = (2x + 3)^{1/5}$. Therefore, the inverse function is $f^{-1}(x) = (2x + 3)^{1/5}$. ■

CHECK YOUR ANSWER

We use the Inverse Function Property.

$$\begin{aligned} f^{-1}(f(x)) &= f^{-1}\left(\frac{x^5 - 3}{2}\right) \\ &= \left[2\left(\frac{x^5 - 3}{2}\right) + 3\right]^{1/5} \\ &= (x^5 - 3 + 3)^{1/5} \\ &= (x^5)^{1/5} = x \end{aligned}$$

$$\begin{aligned} f(f^{-1}(x)) &= f((2x + 3)^{1/5}) \\ &= \frac{[(2x + 3)^{1/5}]^5 - 3}{2} \\ &= \frac{2x + 3 - 3}{2} \\ &= \frac{2x}{2} = x \quad ✓ \end{aligned}$$

The principle of interchanging x and y to find the inverse function also gives us a method for obtaining the graph of f^{-1} from the graph of f. If $f(a) = b$, then $f^{-1}(b) = a$. Thus, the point (a, b) is on the graph of f if and only if the point (b, a) is on the graph of f^{-1}. But we get the point (b, a) from the point (a, b) by reflecting in the line $y = x$ (see Figure 8).

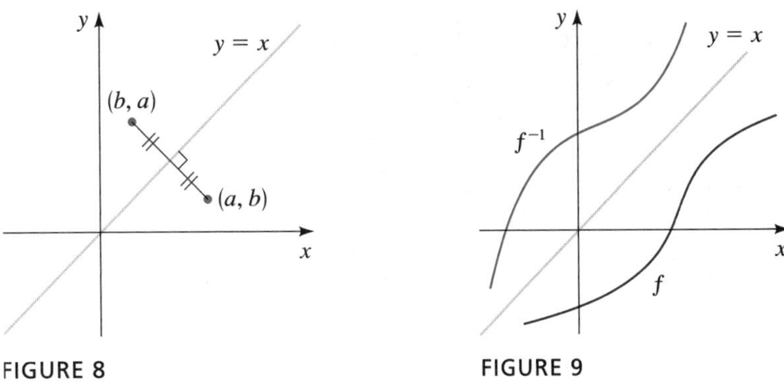

FIGURE 8

FIGURE 9

Therefore, as Figure 9 illustrates, the following is true.

> The graph of f^{-1} is obtained by reflecting the graph of f in the line $y = x$.

EXAMPLE 8 ■ Finding the Inverse of a Function

(a) Sketch the graph of $f(x) = \sqrt{x - 2}$.
(b) Use the graph of f to sketch the graph of f^{-1}.
(c) Find an equation for f^{-1}.

SOLUTION

(a) Using the transformations from Section 4.5, we sketch the graph of $y = \sqrt{x - 2}$ by plotting the graph of the function $y = \sqrt{x}$ (Example 1(c) in Section 4.2) and moving it to the right 2 units.

(b) The graph of f^{-1} is obtained from the graph of f in part (a) by reflecting it in the line $y = x$, as shown in Figure 10.

(c) Solve $y = \sqrt{x - 2}$ for x, noting that $y \geqslant 0$.

$$\sqrt{x - 2} = y$$

$$x - 2 = y^2 \qquad \text{Square each side}$$

$$x = y^2 + 2, \quad y \geqslant 0 \qquad \text{Add 2}$$

Interchange x and y:

$$y = x^2 + 2, \qquad x \geqslant 0$$

Thus

$$f^{-1}(x) = x^2 + 2, \qquad x \geqslant 0$$

This expression shows that the graph of f^{-1} is the right half of the parabola $y = x^2 + 2$ and, from the graph shown in Figure 10, this seems reasonable.

In Example 8 note how f^{-1} reverses the effect of f. The function f is the rule "subtract 2, then take the square root"; f^{-1} is the rule "square, then add 2."

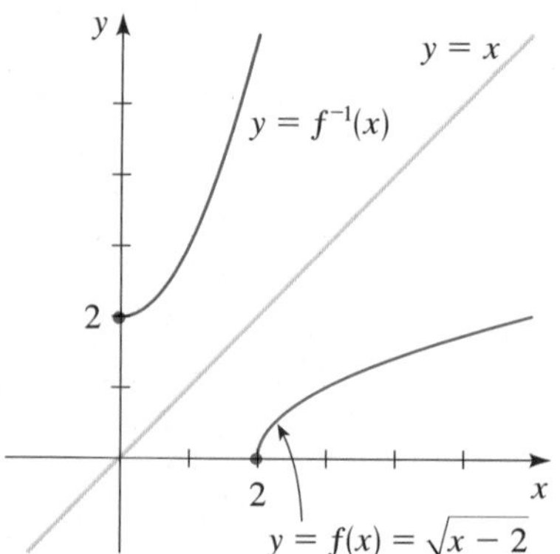

FIGURE 10

■

4.8 EXERCISES

1–6 ■ The graph of a function f is given. Determine whether f is one-to-one.

1.

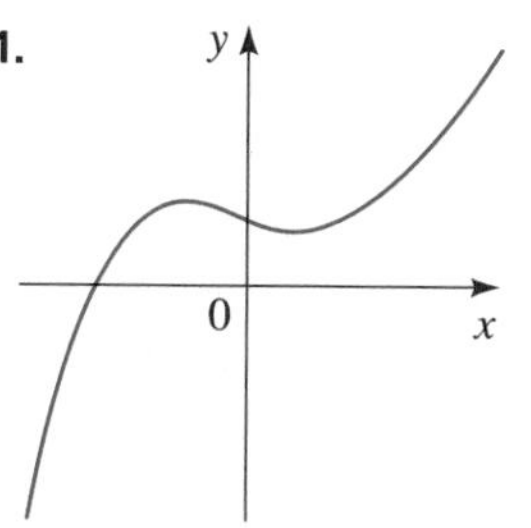

2.

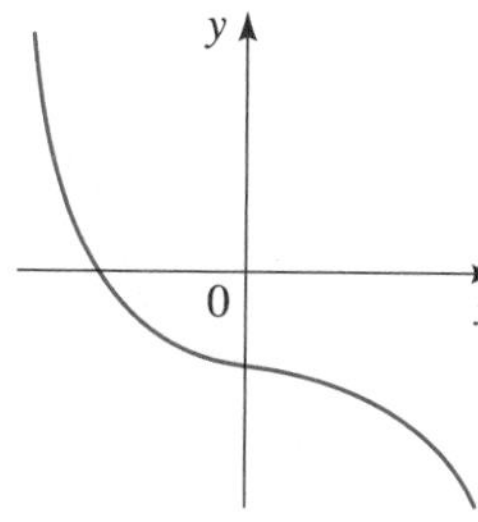

3.

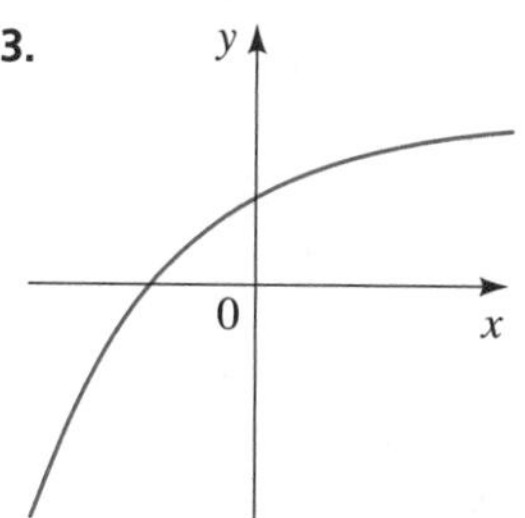

4.

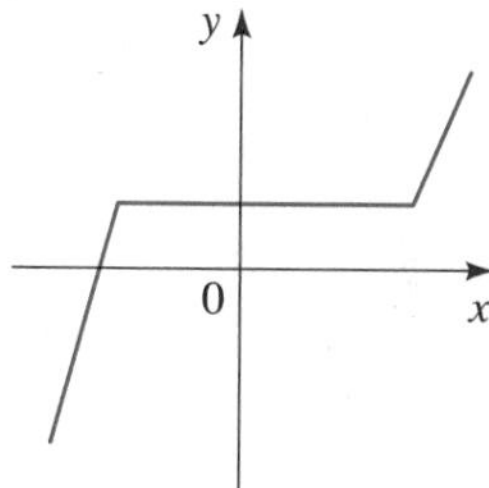

5.

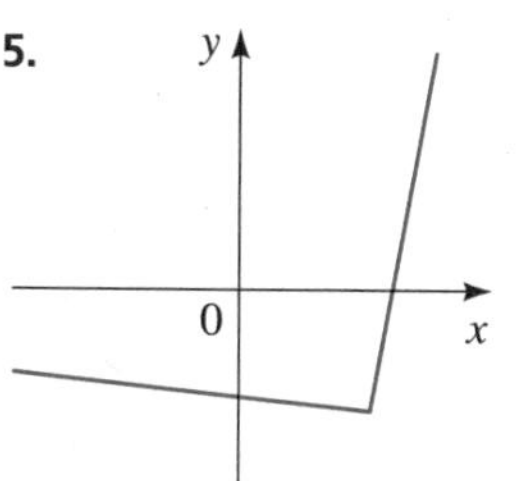

6.

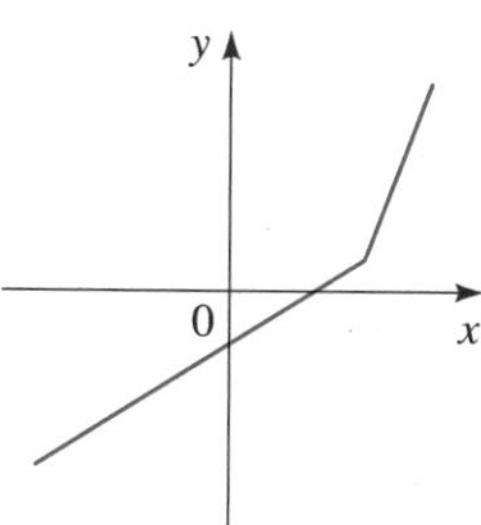

7–16 ■ Determine whether the function is one-to-one.

7. $f(x) = 7x - 3$

8. $f(x) = x^2 - 2x + 5$

9. $g(x) = \sqrt{x}$

10. $g(x) = |x|$

11. $h(x) = x^3 + 1$

12. $h(x) = \sqrt[3]{x}$

13. $f(x) = x^4 + 5$

14. $f(x) = x^4 + 5, \quad 0 \leqslant x \leqslant 2$

15. $f(x) = \dfrac{1}{x^2}$

16. $f(x) = \dfrac{1}{x}$

17–20 ■ Assume f is a one-to-one function.

17. (a) If $f(2) = 7$, find $f^{-1}(7)$.
(b) If $f^{-1}(3) = -1$, find $f(-1)$.

18. (a) If $f(5) = 18$, find $f^{-1}(18)$.
(b) If $f^{-1}(4) = 2$, find $f(2)$.

19. If $f(x) = 5 - 2x$, find $f^{-1}(3)$.

20. If $g(x) = x^2 + 4x$ with $x \geqslant -2$, find $g^{-1}(5)$.

21–30 ■ Use the Property of Inverse Functions to show that f and g are inverses of each other.

21. $f(x) = x + 3, \quad g(x) = x - 3$

22. $f(x) = 2x, \quad g(x) = \dfrac{x}{2}$

23. $f(x) = 2x - 5; \quad g(x) = \dfrac{x + 5}{2}$

24. $f(x) = \dfrac{3 - x}{4}; \quad g(x) = 3 - 4x$

25. $f(x) = \dfrac{1}{x}, \quad g(x) = \dfrac{1}{x}$

26. $f(x) = x^5, \quad g(x) = \sqrt[5]{x}$

27. $f(x) = x^2 - 4, \quad x \geqslant 0;$
$g(x) = \sqrt{x + 4}, \quad x \geqslant -4$

28. $f(x) = x^3 + 1; \quad g(x) = (x - 1)^{1/3}$

29. $f(x) = \dfrac{1}{x - 1}, \quad x \neq 1;$

$g(x) = \dfrac{1}{x} + 1, \quad x \neq 0$

30. $f(x) = \sqrt{4 - x^2}, \quad 0 \leqslant x \leqslant 2;$
$g(x) = \sqrt{4 - x^2}, \quad 0 \leqslant x \leqslant 2$

31–50 ■ Find the inverse function of f.

31. $f(x) = 2x + 1$

32. $f(x) = 6 - x$

33. $f(x) = 4x + 7$

34. $f(x) = 3 - 5x$

35. $f(x) = \dfrac{x}{2}$

36. $f(x) = \dfrac{1}{x^2} \quad (x > 0)$

37. $f(x) = \dfrac{1}{x + 2}$

38. $f(x) = \dfrac{x - 2}{x + 2}$

39. $f(x) = \dfrac{1 + 3x}{5 - 2x}$

40. $f(x) = 5 - 4x^3$

41. $f(x) = \sqrt{2 + 5x}$

42. $f(x) = x^2 + x, \quad x \geqslant -\frac{1}{2}$

43. $f(x) = 4 - x^2, \quad x \geq 0$

44. $f(x) = \sqrt{2x - 1}$

45. $f(x) = 4 + \sqrt[3]{x}$

46. $f(x) = (2 - x^3)^5$

47. $f(x) = 1 + \sqrt{1 + x}$

48. $f(x) = \sqrt{9 - x^2}, \quad 0 \leq x \leq 3$

49. $f(x) = x^4, \quad x \geq 0$

50. $f(x) = 1 - x^3$

51–54 ■ A function f is given.

(a) Sketch the graph of f

(b) Use the graph of f to sketch the graph of f^{-1}.

(c) Find f^{-1}.

51. $f(x) = 3x - 6$

52. $f(x) = 16 - x^2, \quad x \geq 0$

53. $f(x) = \sqrt{x + 1}$

54. $f(x) = x^3 - 1$

55–60 ■ Draw the graph of f and use it to determine whether the function is one-to-one.

55. $f(x) = x^3 - x$

56. $f(x) = x^3 + x$

57. $f(x) = \dfrac{x + 12}{x - 6}$

58. $f(x) = \sqrt{x^3 - 4x + 1}$

59. $f(x) = |x| - |x - 6|$

60. $f(x) = x \cdot |x|$

61–64 ■ The given function is not one-to-one. Restrict its domain so that the resulting function *is* one-to-one. Find the inverse of the function with the restricted domain. (There is more than one correct answer.)

61. $f(x) = 4 - x^2$

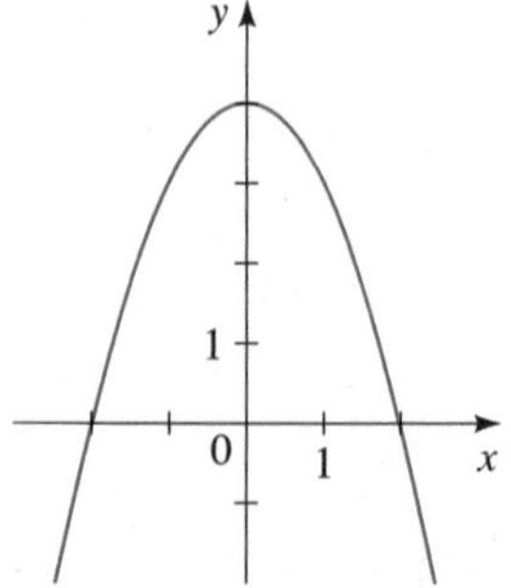

62. $g(x) = (x - 1)^2$

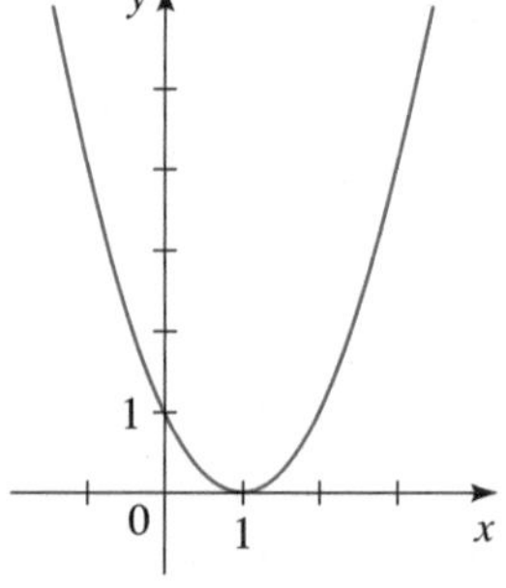

63. $h(x) = (x + 2)^2$

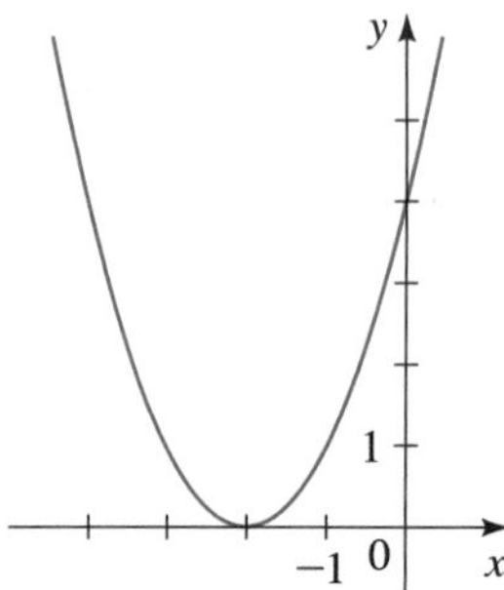

64. $k(x) = |x - 3|$

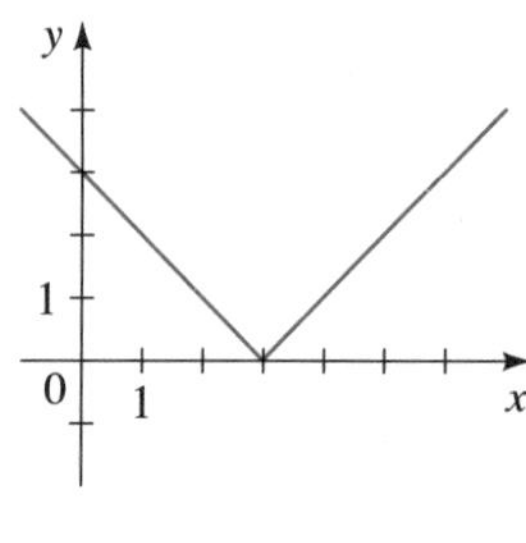

65–66 ■ Use the graph of f to sketch the graph of f^{-1}.

65.

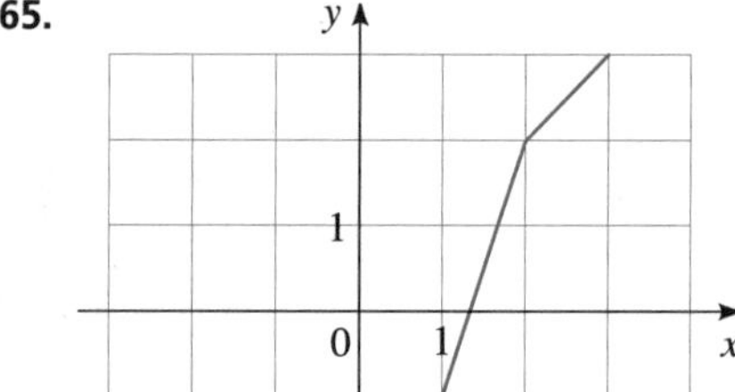

66.

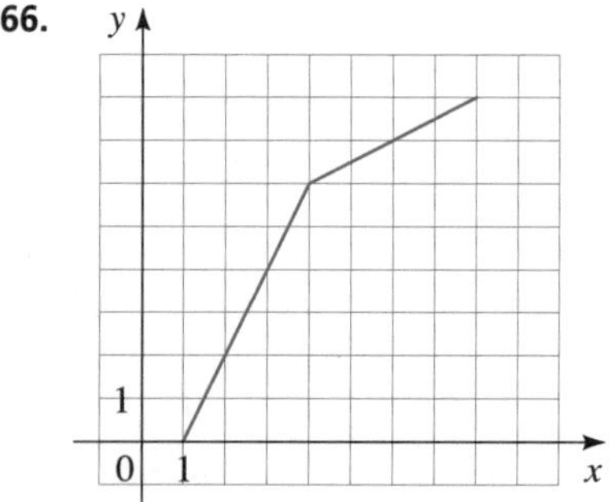

DISCOVERY • DISCUSSION

67. The Inverse of a Price Function Marcello's Pizza charges a base price of \$7 for a large pizza, plus \$2 for each topping. Thus, if you order a large pizza with x toppings, the price of your pizza is given by the function $f(x) = 7 + 2x$. Find f^{-1}. What does the function f^{-1} represent?

68. Determining when a Linear Function Has an Inverse For the linear function $f(x) = mx + b$ to be one-to-one, what must be true about its slope? If it is one-to-one, find its inverse. Is the inverse linear? If so, what is its slope?

69. Finding an Inverse "In Your Head" In the margin notes in this section we pointed out that the inverse of a function can be found by simply reversing the operations that make up the function. For instance, in Example 6 we saw that the inverse of

$$f(x) = 3x - 2 \qquad \text{is} \qquad f^{-1}(x) = \frac{x+2}{3}$$

because the "reverse" of "multiply by 3 and subtract 2" is "add 2 and divide by 3." Use the same procedure to find the inverse of the following functions.

(a) $f(x) = \dfrac{2x+1}{5}$

(b) $f(x) = 3 - \dfrac{1}{x}$

(c) $f(x) = \sqrt{x^3 + 2}$

Now consider another function:

$$f(x) = \frac{3x-2}{x+7}$$

Is it possible to use the same sort of simple reversal of operations to find the inverse of this function? If so, do it. If not, explain what is different about this function that makes this task difficult.

70. The Identity Function The function $I(x) = x$ is called the **identity function**. Show that for any function f we have $f \circ I = f$, $I \circ f = f$, and $f \circ f^{-1} = f^{-1} \circ f = I$. (This means that the identity function I behaves for functions and composition just like the number 1 behaves for real numbers and multiplication.)

71. Solving an Equation for an Unknown Function In Exercise 60 of Section 4.7 you were asked to solve equations in which the unknowns were functions. Now that we know about inverses and the identity function (see Exercise 70), we can use algebra to solve such equations. For instance, to solve $f \circ g = h$ for the unknown function f, we perform the following steps:

$f \circ g = h$	Problem: Solve for f
$f \circ g \circ g^{-1} = h \circ g^{-1}$	Compose with g^{-1} on the right
$f \circ I = h \circ g^{-1}$	$g \circ g^{-1} = I$
$f = h \circ g^{-1}$	$f \circ I = f$

So the solution is $f = h \circ g^{-1}$. Use this technique to solve the equation $f \circ g = h$ for the indicated unknown function.

(a) Solve for f, where $g(x) = 2x + 1$ and $h(x) = 4x^2 + 4x + 7$

(b) Solve for g, where $f(x) = 3x + 5$ and $h(x) = 3x^2 + 3x + 2$

4 REVIEW

CONCEPT CHECK

1. Define each concept in your own words. (Check by referring to the definition in the text.)
(a) Function
(b) Domain and range of a function
(c) Graph of a function
(d) Independent and dependent variables

2. Give an example of each type of function.
(a) Constant function
(b) Linear function
(c) Quadratic function

3. Sketch by hand, on the same axes, the graphs of the following functions.
(a) $f(x) = x$ (b) $g(x) = x^2$
(c) $h(x) = x^3$ (d) $j(x) = x^4$

4. (a) State the Vertical Line Test.
(b) State the Horizontal Line Test.

5. Write an equation that expresses each relationship.
(a) y is directly proportional to x.
(b) y is inversely proportional to x.
(c) z is jointly proportional to x and y.

6. How is the average rate of change of the function f between two points defined?

7. Define each concept in your own words.
   (a) Increasing function
   (b) Decreasing function
   (c) Constant function

8. Suppose the graph of f is given. Write an equation for each graph that is obtained from the graph of f as follows.
   (a) Shift 3 units upward.
   (b) Shift 3 units downward.
   (c) Shift 3 units to the right.
   (d) Shift 3 units to the left.
   (e) Reflect in the x-axis.
   (f) Reflect in the y-axis.
   (g) Stretch vertically by a factor of 3.
   (h) Shrink vertically by a factor of $\frac{1}{3}$.

9. (a) What is an even function? What symmetry does its graph possess? Give an example of an even function.
   (b) What is an odd function? What symmetry does its graph possess? Give an example of an odd function.

10. Write the standard form of a quadratic function.

11. What does it mean to say that $f(3)$ is a local maximum value of f?

12. Suppose that f has domain A and g has domain B.
   (a) What is the domain of $f + g$?
   (b) What is the domain of fg?
   (c) What is the domain of f/g?

13. How is the composite function $f \circ g$ defined?

14. (a) What is a one-to-one function?
   (b) How can you tell from the graph of a function whether it is one-to-one?
   (c) Suppose f is a one-to-one function with domain A and range B. How is the inverse function f^{-1} defined? What is the domain of f^{-1}? What is the range of f^{-1}?
   (d) If you are given a formula for f, how do you find a formula for f^{-1}?
   (e) If you are given the graph of f, how do you find the graph of f^{-1}?

EXERCISES

1. If $f(x) = x^2 - x + 1$, find $f(0)$, $f(2)$, $f(-2)$, $f(a)$, $f(-a)$, $f(x+1)$, $f(2x)$, and $2f(x) - 2$.

2. If $f(x) = 1 + \sqrt{x-1}$, find $f(5)$, $f(9)$, $f(a+1)$, $f(-x)$, $f(x^2)$, and $[f(x)]^2$.

3. The graph of a function is given.
   (a) Find $f(-2)$ and $f(2)$.
   (b) Find the domain of f.
   (c) Find the range of f.
   (d) On what intervals is f increasing? On what intervals is f decreasing?
   (e) Is f one-to-one?

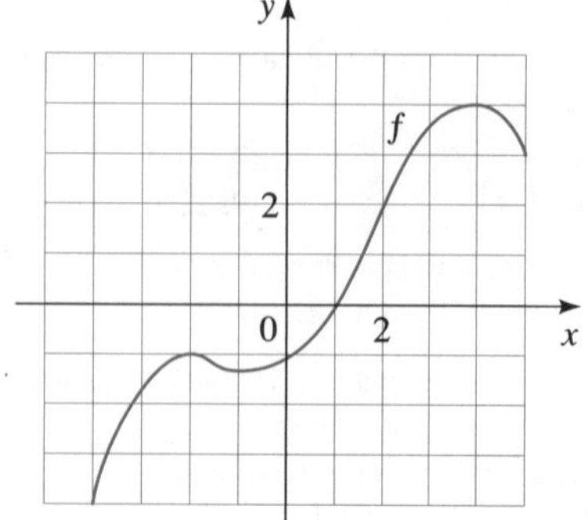

4. Which of the following figures are graphs of functions? Which of the functions are one-to-one?

(a)

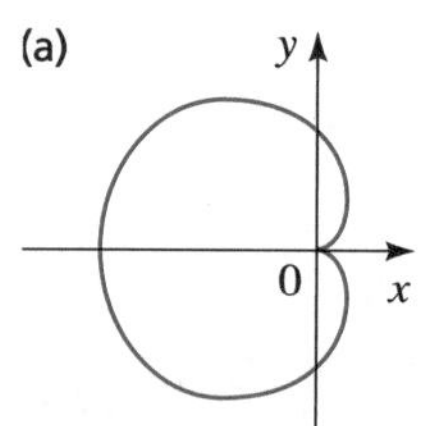

(b)

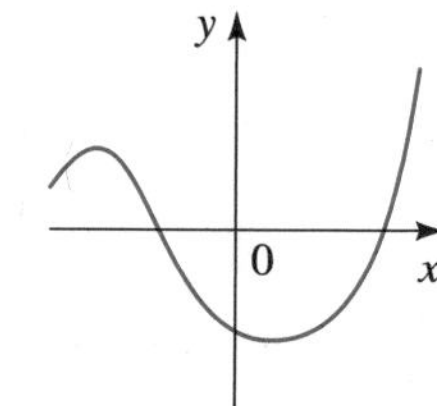

(c)

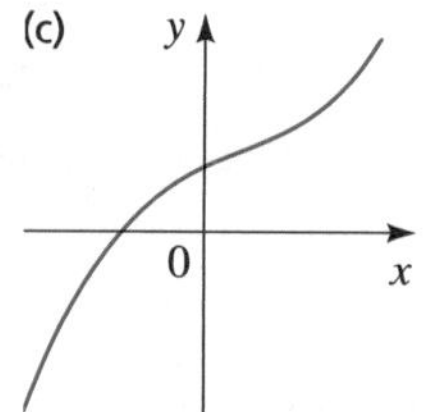

(d)

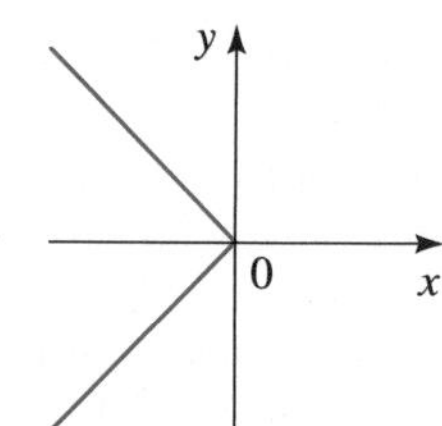

5–6 ■ Find the domain and range of the function.

5. $f(x) = \sqrt{x+3}$

6. $F(t) = t^2 + 2t + 5$

7–14 ■ Find the domain of the function.

7. $f(x) = 7x + 15$

8. $f(x) = \dfrac{2x + 1}{2x - 1}$

9. $f(x) = \sqrt{x + 4}$

10. $f(x) = 3x - \dfrac{2}{\sqrt{x + 1}}$

11. $f(x) = \dfrac{1}{x} + \dfrac{1}{x + 1} + \dfrac{1}{x + 2}$

12. $g(x) = \dfrac{2x^2 + 5x + 3}{2x^2 - 5x - 3}$

13. $h(x) = \sqrt{4 - x} + \sqrt{x^2 - 1}$

14. $f(x) = \dfrac{\sqrt[3]{2x + 1}}{\sqrt[3]{2x} + 2}$

15–32 ■ Sketch the graph of the function.

15. $f(x) = 1 - 2x$

16. $f(x) = \frac{1}{3}(x - 5), \quad 2 \leqslant x \leqslant 8$

17. $f(t) = 1 - \frac{1}{2}t^2$

18. $g(t) = t^2 - 2t$

19. $f(x) = x^2 - 6x + 6$

20. $f(x) = 3 - 8x - 2x^2$

21. $y = 1 - \sqrt{x}$

22. $y = -|x|$

23. $y = \frac{1}{2}x^3$

24. $y = \sqrt{x + 3}$

25. $h(x) = \sqrt[3]{x}$

26. $H(x) = x^3 - 3x^2$

27. $g(x) = \dfrac{1}{x^2}$

28. $G(x) = \dfrac{1}{(x - 3)^2}$

29. $f(x) = \begin{cases} 1 - x & \text{if } x < 0 \\ 1 & \text{if } x \geqslant 0 \end{cases}$

30. $f(x) = \begin{cases} 1 - 2x & \text{if } x \leqslant 0 \\ 2x - 1 & \text{if } x > 0 \end{cases}$

31. $f(x) = \begin{cases} x + 6 & \text{if } x < -2 \\ x^2 & \text{if } x \geqslant -2 \end{cases}$

32. $f(x) = \begin{cases} -x & \text{if } x < 0 \\ x^2 & \text{if } 0 \leqslant x < 2 \\ 1 & \text{if } x \geqslant 2 \end{cases}$

33. Determine which viewing rectangle produces the most appropriate graph of the function $f(x) = 6x^3 - 15x^2 + 4x - 1$.
(i) $[-2, 2]$ by $[-2, 2]$
(ii) $[-8, 8]$ by $[-8, 8]$
(iii) $[-4, 4]$ by $[-12, 12]$
(iv) $[-100, 100]$ by $[-100, 100]$

34. Determine which viewing rectangle produces the most appropriate graph of the function $f(x) = \sqrt{100 - x^3}$.
(i) $[-4, 4]$ by $[-4, 4]$
(ii) $[-10, 10]$ by $[-10, 10]$
(iii) $[-10, 10]$ by $[-10, 40]$
(iv) $[-100, 100]$ by $[-100, 100]$

35–38 ■ Draw the graph of the function in an appropriate viewing rectangle.

35. $f(x) = x^2 + 25x + 173$

36. $f(x) = 1.1x^3 - 9.6x^2 - 1.4x + 3.2$

37. $y = \dfrac{x}{\sqrt{x^2 + 16}}$

38. $y = |x(x + 2)(x + 4)|$

39. Find, approximately, the domain of the function $f(x) = \sqrt{x^3 - 4x + 1}$.

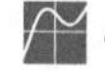

40. Find, approximately, the range of the function $f(x) = x^4 - x^3 + x^2 + 3x - 6$.

41. Suppose that M varies directly as z, and $M = 120$ when $z = 15$. Write an equation that expresses this variation.

42. Suppose that z is inversely proportional to y, and that $z = 12$ when $y = 16$. Write an equation that expresses z in terms of y.

43. The intensity of illumination I from a light varies inversely as the square of the distance d from the light.
(a) Write this statement as an equation.
(b) Determine the constant of proportionality if it is known that a lamp has an intensity of 1000 candles at a distance of 8 m.
(c) What is the intensity of this lamp at a distance of 20 m?

44. The frequency of a vibrating string under constant tension is inversely proportional to its length. If a violin string 12 inches long vibrates 440 times per second, to what length would it have to be shortened to vibrate 660 times per second?

45. The terminal velocity of a parachutist is directly proportional to the square root of his weight. A 160-lb parachutist attains a terminal velocity of 9 mi/h. What is the terminal velocity for a parachutist weighing 240 lb?

46. The maximum range of a projectile is directly proportional to the square of its velocity. A baseball pitcher can throw a ball at 60 mi/h, with a maximum range of 242 ft. What would his maximum range be if he could throw the ball at 70 mi/h?

47–50 ■ Find the average rate of change of the function between the given points.

47. $f(x) = x^2 + 3x; \quad x = 0, x = 2$

48. $f(x) = \dfrac{1}{x - 2}; \quad x = 4, x = 8$

49. $f(x) = \dfrac{1}{x}; \quad x = 3, x = 3 + h$

50. $f(x) = (x + 1)^2; \quad x = a, x = a + h$

51–52 ■ Draw a graph of the function f, and determine the intervals on which f is increasing and on which f is decreasing.

51. $f(x) = x^3 - 4x^2$

52. $f(x) = |x^4 - 16|$

53. Suppose the graph of f is given. Describe how the graphs of the following functions can be obtained from the graph of f.

(a) $y = f(x) + 8$ (b) $y = f(x + 8)$
(c) $y = 1 + 2f(x)$ (d) $y = f(x - 2) - 2$
(e) $y = f(-x)$ (f) $y = -f(-x)$
(g) $y = -f(x)$ (h) $y = f^{-1}(x)$

54. The graph of f is given. Draw the graphs of the following functions.

(a) $y = f(x - 2)$ (b) $y = -f(x)$
(c) $y = 3 - f(x)$ (d) $y = \frac{1}{2}f(x) - 1$
(e) $y = f^{-1}(x)$ (f) $y = f(-x)$

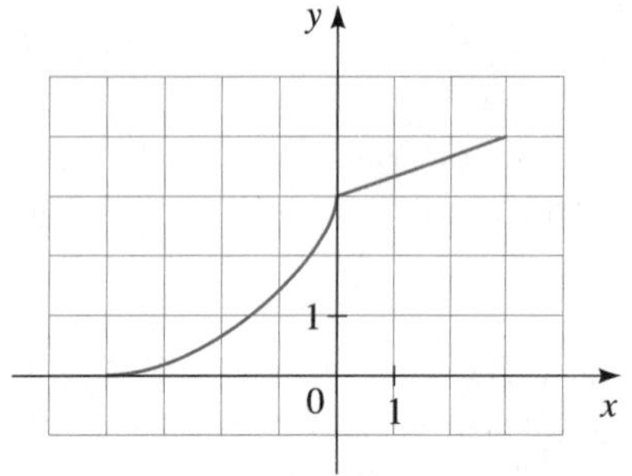

55. Determine whether f is even, odd, or neither.

(a) $f(x) = 2x^5 - 3x^2 + 2$ (b) $f(x) = x^3 - x^7$

(c) $f(x) = \dfrac{1 - x^2}{1 + x^2}$ (d) $f(x) = \dfrac{1}{x + 2}$

56. Determine whether the function in the figure is even, odd, or neither.

(a)

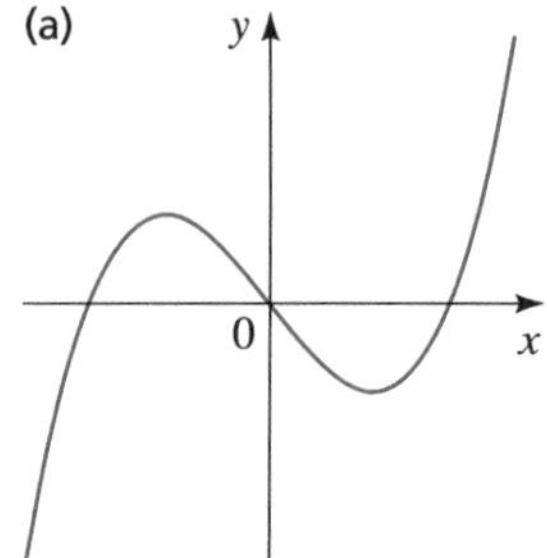

(b)

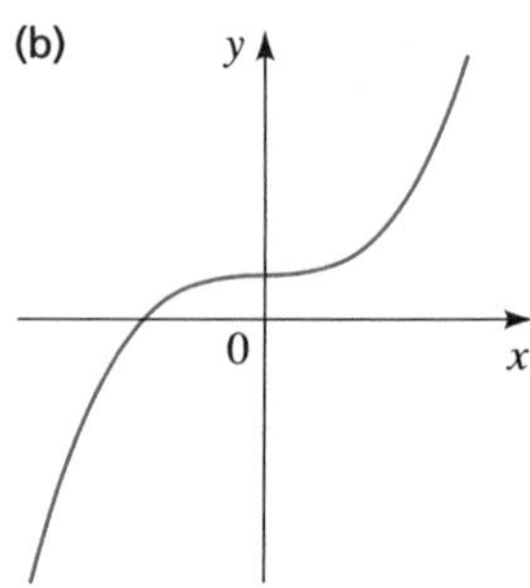

(c)

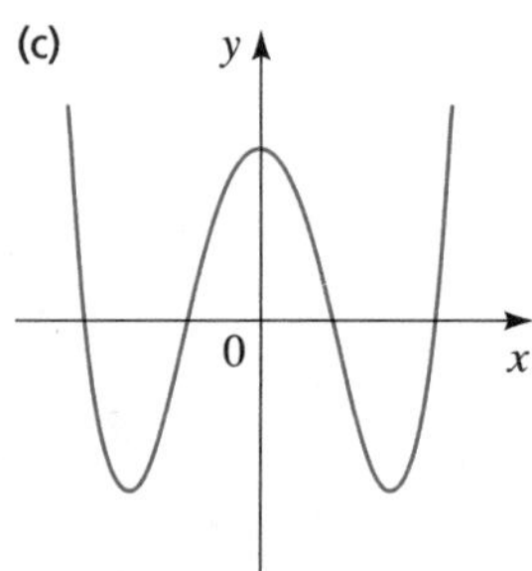

57. Express the quadratic function $f(x) = x^2 + 4x + 1$ in standard form.

58. Express the quadratic function $f(x) = -2x^2 + 12x + 12$ in standard form.

59. Find the minimum value of the function $g(x) = 2x^2 + 4x - 5$.

60. Find the maximum value of the function $f(x) = 1 - x - x^2$.

61. A stone is thrown upward from the top of a building. Its height (in feet) above the ground after t seconds is given by $h(t) = -16t^2 + 48t + 32$. What maximum height does it reach?

62. The profit P (in dollars) generated by selling x units of a certain commodity is given by

$$P(x) = -1500 + 12x - 0.0004x^2$$

What is the maximum profit, and how many units must be sold to generate it?

63–64 ■ Find the local maximum and minimum values of the function and the values of x at which they occur. State each answer correct to two decimal places.

63. $f(x) = 3.3 + 1.6x - 2.5x^3$

64. $f(x) = x^{2/3}(6 - x)^{1/3}$

65. If $f(x) = x^2 - 3x + 2$ and $g(x) = 4 - 3x$, find the following functions.
(a) $f + g$ (b) $f - g$ (c) fg
(d) f/g (e) $f \circ g$ (f) $g \circ f$

66. If $f(x) = 1 + x^2$ and $g(x) = \sqrt{x - 1}$, find the following functions.
(a) $f \circ g$ (b) $g \circ f$ (c) $(f \circ g)(2)$
(d) $(f \circ f)(2)$ (e) $f \circ g \circ f$ (f) $g \circ f \circ g$

67–68 ■ Find the functions $f \circ g$, $g \circ f$, $f \circ f$, and $g \circ g$ and their domains.

67. $f(x) = 3x - 1, \quad g(x) = 2x - x^2$

68. $f(x) = \sqrt{x}, \quad g(x) = \dfrac{2}{x - 4}$

69. Find $f \circ g \circ h$, where $f(x) = \sqrt{1 - x}$, $g(x) = 1 - x^2$, and $h(x) = 1 + \sqrt{x}$.

70. If $T(x) = \dfrac{1}{\sqrt{1 + \sqrt{x}}}$, find functions f, g, and h such that $f \circ g \circ h = T$.

71–76 ■ Determine whether the function is one-to-one.

71. $f(x) = 3 + x^3$

72. $g(x) = 2 - 2x + x^2$

73. $h(x) = \dfrac{1}{x^4}$

74. $r(x) = 2 + \sqrt{x + 3}$

75. $p(x) = 3.3 + 1.6x - 2.5x^3$

76. $q(x) = 3.3 + 1.6x + 2.5x^3$

77–80 ■ Find the inverse of the function.

77. $f(x) = 3x - 2$

78. $f(x) = \dfrac{2x + 1}{3}$

79. $f(x) = (x + 1)^3$

80. $f(x) = 1 + \sqrt[5]{x - 2}$

81. (a) Sketch the graph of the function

$$f(x) = x^2 - 4, \quad x \geqslant 0$$

(b) Use part (a) to sketch the graph of f^{-1}.
(c) Find an equation for f^{-1}.

82. (a) Show that the function $f(x) = 1 + \sqrt[4]{x}$ is one-to-one.
(b) Sketch the graph of f.
(c) Use part (b) to sketch the graph of f^{-1}.
(d) Find an equation for f^{-1}.

4 TEST

1. Which of the following are graphs of functions? If the graph is that of a function, is it one-to-one?

(a)

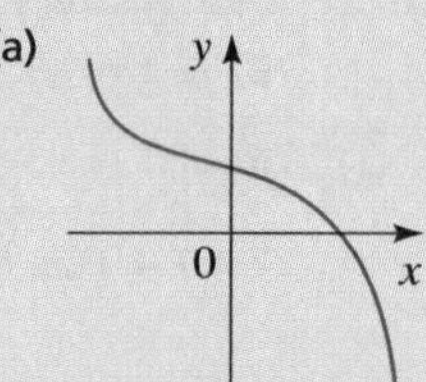

(b)

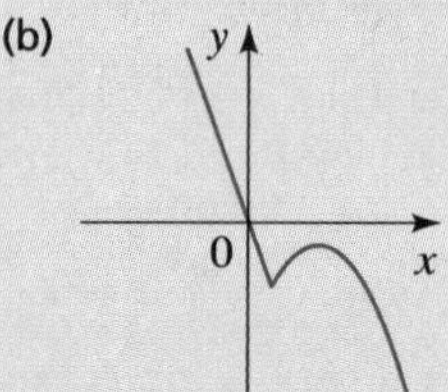

(c)

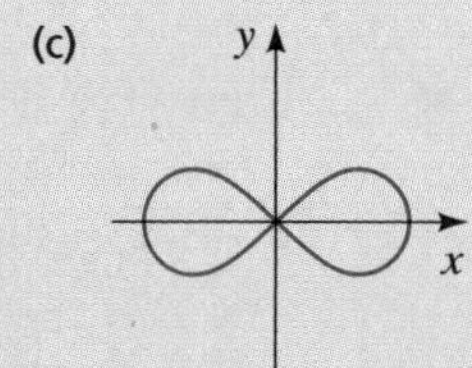

(d)

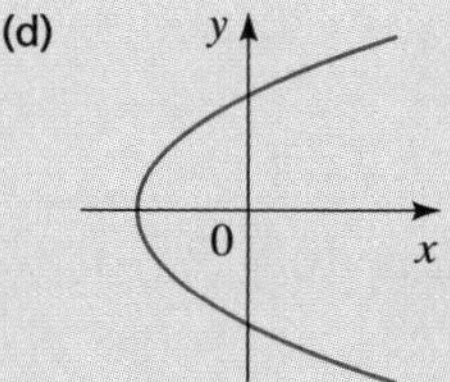

2. Let $f(x) = \dfrac{\sqrt{x}}{x - 1}$.
   (a) Evaluate $f(4)$, $f(6)$, and $f(a + 1)$.
   (b) Find the domain of f.

3. The maximum weight M that can be supported by a beam is jointly proportional to its width w and the square of its height h, and inversely proportional to its length L.
   (a) Write an equation that expresses this proportionality.
   (b) Determine the constant of proportionality if a beam 4 in. wide, 6 in. high, and 12 ft long can support a weight of 4800 lb.
   (c) If a 10-ft beam made of the same material is 3 in. wide and 10 in. high, what is the maximum weight it can support?

4. Determine the average rate of change for the function $f(t) = t^2 + 3t$ between $t = 2$ and $t = 2 + h$.

5. (a) Sketch the graph of the function $f(x) = x^3$.
   (b) Use part (a) to graph the function $g(x) = (x - 1)^3 - 2$.

6. (a) How is the graph of $y = f(x - 3) + 2$ obtained from the graph of f?
(b) How is the graph of $y = f(-x)$ obtained from the graph of f?

7. (a) Sketch the graph of the function $f(x) = 2x^2 - 8x + 13$.
(b) What is the minimum value of f?

8. Let $f(x) = \begin{cases} 1 - x^2 & \text{if } x \leq 0 \\ 2x + 1 & \text{if } x > 0 \end{cases}$
(a) Evaluate $f(-2)$ and $f(1)$.
(b) Sketch the graph of f.

9. If $f(x) = x^2 + 2x - 1$ and $g(x) = 2x - 3$, find the following functions.
(a) $f \circ g$
(b) $g \circ f$
(c) $f(g(2))$
(d) $g(f(2))$
(e) $g \circ g \circ g$

10. (a) If $f(x) = \sqrt{3 - x}$, find the inverse function f^{-1}.
(b) Sketch the graphs of f and f^{-1} on the same coordinate axes.

11. The graph of a function f is given.
(a) Find the domain and range of f.
(b) Sketch the graph of f^{-1}.
(c) Find the average rate of change of f between $x = 2$ and $x = 6$.

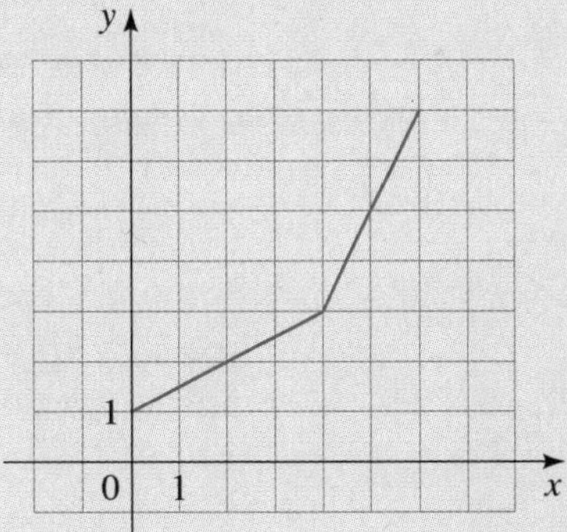

12. Let $f(x) = 3x^4 - 14x^2 + 5x - 3$.
(a) Draw the graph of f in an appropriate viewing rectangle.
(b) Is f one-to-one?
(c) Find the local maximum and minimum values of f and the values of x at which they occur. State each answer correct to two decimal places.
(d) Find the approximate range of f.
(e) Find the approximate intervals on which f is increasing and on which f is decreasing.

FOCUS ON MODELING

Recall that a **mathematical model** is a mathematical representation of an object or a process. The purpose of making a mathematical model is to be able to analyze and predict properties of the thing we are modeling. In the *Principles of Modeling* following Chapter 2 we used linear equations as mathematical models of real-life phenomena. Here we will use *functions* to construct mathematical models and then use these models to find maximum or minimum values of the quantity we are modeling.

FUNCTIONS AS MODELS

Let's see how we can use functions to model real-world situations.

EXAMPLE 1 ■ Modeling the Volume of a Box

A breakfast cereal company manufactures boxes to package their product. The prototype box has the following shape: Its width is three times its depth and its height is five times its depth. Find a function that models the volume of the box in terms of its depth.

SOLUTION

EXPRESS THE MODEL IN WORDS The model we want is a function that gives the *volume* of any such box. We know that the volume of a rectangular box is given by

$$\text{volume} = \text{width} \times \text{depth} \times \text{height}$$

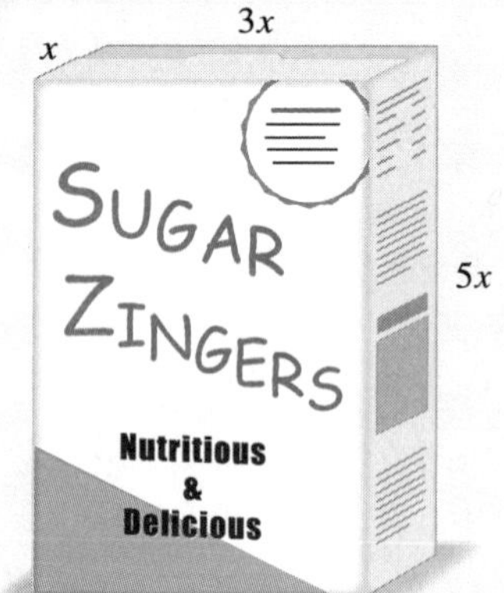

FIGURE 1

CHOOSE THE VARIABLE There are three variables—width, depth, and height. Since the function we want depends on the depth, we let

$$x = \text{depth of the box}$$

Then we express the other dimensions of the box in terms of x (see Figure 1).

In Words	In Algebra
Depth	x
Width	$3x$
Height	$5x$

SET UP THE MODEL The model is a function V that gives the volume of the box in terms of its depth x.

$$\text{volume} = \text{width} \times \text{depth} \times \text{height}$$

$$V(x) = x \cdot 3x \cdot 5x$$

$$V(x) = 15x^3 \qquad \text{Simplify}$$

The volume of the box is modeled by the function $V(x) = 15x^3$. ■

In the next example we use a function model to solve a real-life problem.

EXAMPLE 2 ■ Fencing a Garden

A gardener has 140 feet of fencing for her rectangular vegetable garden. Find the dimensions of the biggest area she can fence.

THINKING ABOUT THE PROBLEM Let's experiment with the problem. If the gardener fences a 10-ft-wide plot, then the length must be

$$\frac{140 - 2(10)}{2} = 60$$

So the area is $A = w \cdot l = 10 \cdot 60 = 600 \text{ ft}^2$. If she fences a 20-ft-wide plot, then its length must be 50 ft. In this case, the area is $A = 20 \cdot 50 = 1000 \text{ ft}^2$. The table and Figure 2 present various choices for fencing the garden.

Width	Length	Area
10	60	600
20	50	1000
30	40	1200
40	30	1200
50	20	1000
60	10	600

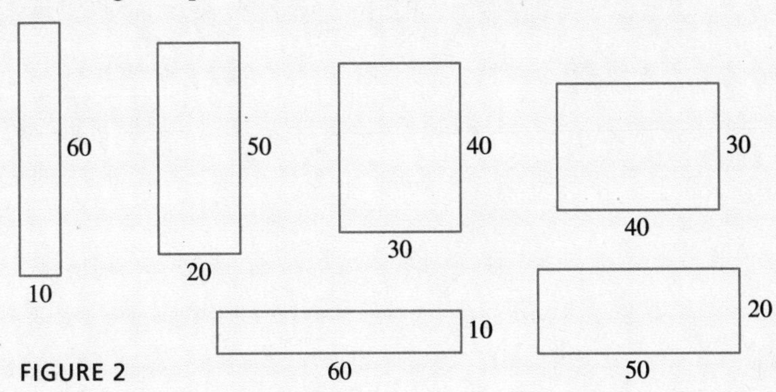

FIGURE 2

The table indicates that as the width increases, the fenced area increases, then decreases.

SOLUTION

EXPRESS THE MODEL IN WORDS The model we want is a function that gives the *area* of the garden for any rectangular shape. We know that

$$\text{area} = \text{width} \times \text{length}$$

CHOOSE THE VARIABLE There are two variables—width and length. Since the function we want can have only one variable, we let

$$x = \text{width of the garden}$$

Next we must express the length in terms of x. The perimeter is fixed at 140 ft, so the length is determined once we choose the width. If we let the length be l as shown in Figure 3, then

$$2x + 2l = 140 \qquad \text{so} \qquad l = \frac{140 - 2x}{2} = 70 - x$$

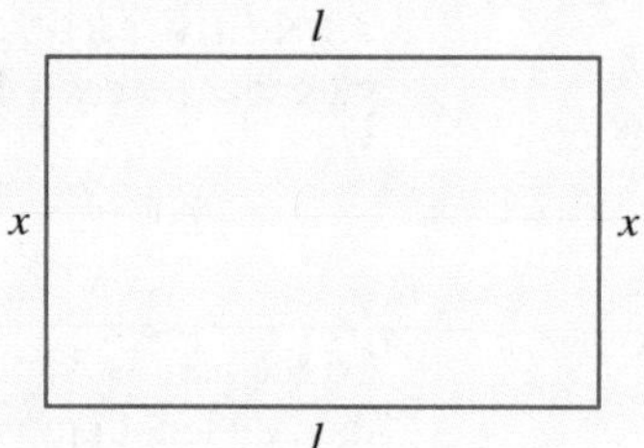

FIGURE 3

We summarize these facts as follows.

In Words	In Algebra
Width of garden	x
Length of garden	$70 - x$

SET UP THE MODEL The model is a function A that represents the area of the field for any width x.

$$\text{area} = \text{width} \times \text{length}$$

$$A(x) = x(70 - x)$$

$$A(x) = 70x - x^2 \qquad \text{Expand}$$

USE THE MODEL We use the model to find the dimensions of the garden with largest area. Note that $A(x) = 70x - x^2$ is a quadratic function with $a = -1$ and $b = 70$. From Section 4.6 we know that the maximum occurs at

$$x = -\frac{b}{2a} = -\frac{70}{2(-1)} = 35$$

STATE THE ANSWER The maximum-area garden that can be fenced with 140 ft of fencing has width 35 ft and length $70 - x = 35$ ft. ■

In Example 2 we can also find the maximum of the function A by sketching its graph. The graph in Figure 4 shows that the maximum area occurs when $x = 35$. The graph helps us *see* how the area of the garden changes as the width varies.

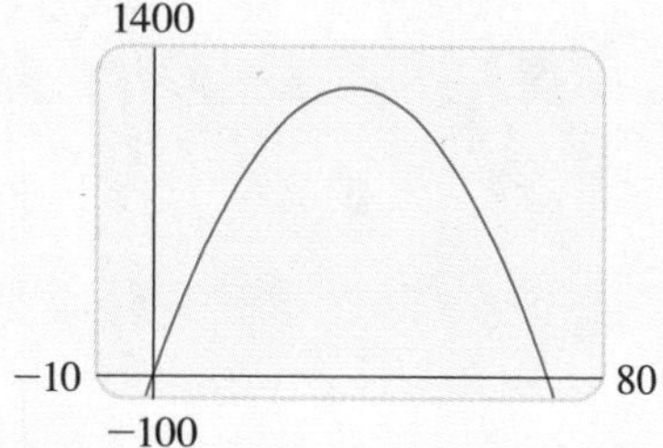

FIGURE 4
$A(x) = 70x - x^2$

The steps used in Example 2 are typical of how we model with functions. They are explained in the following box.

GUIDELINES FOR MODELING WITH FUNCTIONS

1. EXPRESS THE MODEL IN WORDS. Identify the quantity you want to model and express it as a function of the other quantities in the problem.

2. CHOOSE THE VARIABLE. Identify all the variables used to express the function in Step 1. Assign a symbol, such as x, to one variable and express the other variables in terms of this symbol.

3. SET UP THE MODEL. Express the function in the language of algebra, by writing it as a function of the single variable chosen in Step 2.

4. USE THE MODEL. Use the function to answer the question posed in the problem. To find a maximum or minimum, use the algebraic or graphical methods described in Section 4.6.

5. STATE THE ANSWER. Write your answer as a complete sentence.

EXAMPLE 3 ■ Maximizing Revenue from Ticket Sales

A hockey team plays in an arena with a seating capacity of 15,000 spectators. With the ticket price set at \$14, average attendance at recent games has been 9500. A market survey indicates that for each dollar the ticket price is lowered, the average attendance increases by 1000. What price maximizes revenue from ticket sales, and what is the maximum revenue?

THINKING ABOUT THE PROBLEM With a ticket price of \$14, the revenue is $9500 \times \$14 = \$133{,}000$. If the ticket price is lowered to \$13, attendance increases to $9500 + 1000 = 10{,}500$, so the revenue becomes $10{,}500 \times \$13 = \$136{,}500$. The table shows the revenue for several ticket prices.

Price	Attendance	Revenue
\$15	8,500	\$127,500
\$14	9,500	\$133,000
\$13	10,500	\$136,500
\$12	11,500	\$138,000
\$11	12,500	\$137,500
\$10	13,500	\$135,000
\$ 9	14,500	\$130,500

Note that as the ticket price is lowered, revenue increases, but if ticket price is lowered too much, revenue decreases. The maximum revenue appears to be generated when the ticket price is about \$12.

SOLUTION

EXPRESS THE MODEL IN WORDS The model we want is a function that gives the *revenue* for any ticket price. We know that

$$\text{revenue} = \text{ticket price} \times \text{attendance}$$

CHOOSE THE VARIABLE There are two variables in this problem—ticket price and attendance. Since the function we want can have only one variable, we let

$$x = \text{ticket price in dollars}$$

Next, we must express the attendance in terms of x. We can do this because we know that ticket price determines attendance. The following table shows how to express the quantities in the problem in terms of x.

In Words	In Algebra
Ticket price	x
Amount by which price is lowered	$14 - x$
Increase in attendance	$1000(14 - x)$
Attendance	$9500 + 1000(14 - x)$ $= 23{,}500 - 1000x$

SET UP THE MODEL Our model is a function R that represents the revenue for a given ticket price x.

$$\text{revenue} = \text{ticket price} \times \text{attendance}$$

$$R(x) = x(23{,}500 - 1000x)$$

$$R(x) = 23{,}500x - 1000x^2 \qquad \text{Expand}$$

USE THE MODEL We use the model to find which ticket price generates the greatest revenue. Since

$$R(x) = 23{,}500x - 1000x^2$$

is a quadratic function with $a = -1000$ and $b = 23{,}500$, the maximum occurs at

$$x = -\frac{b}{2a} = -\frac{23{,}500}{2(-1000)} = 11.75$$

STATE THE ANSWER A ticket price of \$11.75 yields the maximum revenue. At this price the revenue is

$$R(11.75) = 23{,}500(11.75) - 1000(11.75)^2 = \$138{,}062.50$$ ■

EXAMPLE 4 ■ Minimizing the Metal in a Can

A manufacturer makes a metal can that holds 1 L (liter) of oil. What radius minimizes the amount of metal in the can?

THINKING ABOUT THE PROBLEM To use the least amount of metal, we must minimize the surface area of the can, that is, the area of the top, the bottom, and the sides. The top and bottom of the can are circles, so together their area is $\pi r^2 + \pi r^2 = 2\pi r^2$. The area of the sides is the circumference times the height, that is, $2\pi rh$ (see Figure 5).

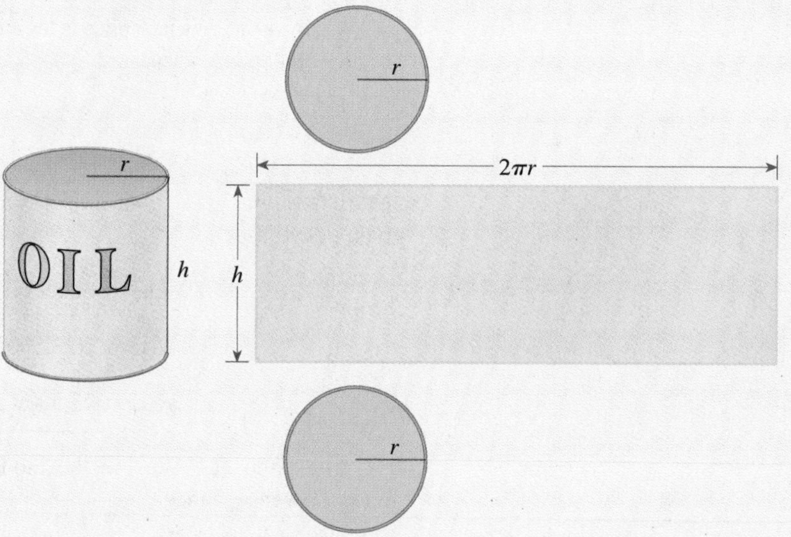

FIGURE 5

The radius and height of the can must be chosen so that the volume is exactly 1 L, or 1000 cm^3. If we want a small radius, say $r = 3$ cm, then the height must be just tall enough so that the total volume is 1000 cm^3. In other words, we must have

$$\pi(3)^2 h = 1000 \qquad \text{Volume of can is } \pi r^2 h$$

$$h = \frac{1000}{9\pi} \approx 35.4 \text{ cm} \qquad \text{Solve for } h$$

Now that we know the radius and height, we can find the surface area of the can.

$$\text{surface area} = 2\pi(3)^2 + 2\pi(3)(35.4) \approx 729.1 \text{ cm}^2$$

If we want a different radius, we can find the corresponding height and surface area in a similar fashion.

SOLUTION

EXPRESS THE MODEL IN WORDS The model we want is a function that gives the *surface area* of the can. We know that for a cylindrical can

$$\text{surface area} = \text{area of top and bottom} + \text{area of sides}$$

To find these areas, we need to know the radius and height of the can.

CHOOSE THE VARIABLE There are two variables in this problem—radius and height. Since the function we want can have only one variable, we let

$$r = \text{radius of can}$$

Next, we express the height h in terms of the radius r. Since the volume of a cylindrical can is $V = \pi r^2 h$ and the volume must be 1000 cm^3, we have

$$\pi r^2 h = 1000 \qquad \text{Volume of can is 1000 cm}^3$$

$$h = \frac{1000}{\pi r^2} \qquad \text{Solve for } h$$

We can now write expressions for the areas of the top, bottom, and sides of the can in terms of r.

In Words	In Algebra
Radius of can	r
Height of can	$\dfrac{1000}{\pi r^2}$
Area of top and bottom of can	$2\pi r^2$
Area of sides of can ($2\pi rh$)	$2\pi r\left(\dfrac{1000}{\pi r^2}\right)$

SET UP THE MODEL We set up the model, which is a function S that represents the surface area of the can as a function of r.

$$\text{surface area} = \text{area of top and bottom} + \text{area of sides}$$

$$S(r) = 2\pi r^2 + 2\pi r\left(\frac{1000}{\pi r^2}\right)$$

$$S(r) = 2\pi r^2 + \frac{2000}{r} \qquad \text{Simplify}$$

USE THE MODEL We use the model to find the minimum surface area of the can, that is, the minimum value of the function S. To do this, we graph S as shown in Figure 6 and note that the graph has a minimum. By zooming in on the minimum point and using the trace feature, we see that the minimum value of S is about 554 and this value occurs when the radius is about 5.4 cm.

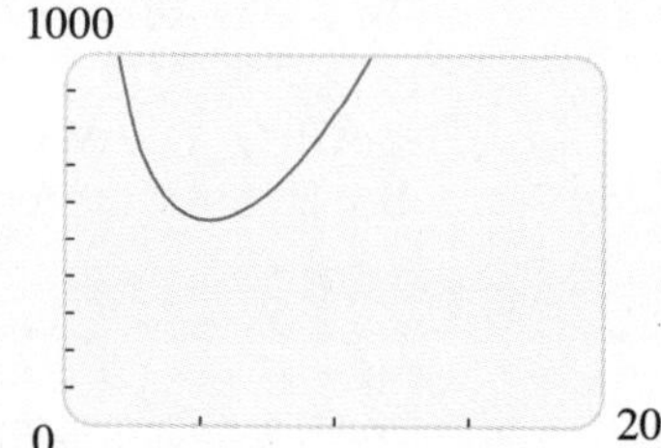

FIGURE 6
Graph of $S = 2\pi r^2 + \dfrac{2000}{r}$

STATE THE ANSWER To minimize the amount of metal in the can, the radius should be about 5.4 cm. ■

PROBLEMS

1. A rectangular building lot is three times as long as it is wide. Find a function that models its area in terms of its width.
2. A poster is 10 inches longer than it is wide. Find a function that models its area in terms of its width.
3. A rectangular box has a square base. Its height is half the width of the base. Find a function that models its volume in terms of its width.
4. The height of a cylinder is four times its radius. Find a function that models the volume of the cylinder in terms of its radius.
5. A rectangle has a perimeter of 20 ft. Find a function that models its area in terms of the length of one of its sides.
6. A rectangle has an area of 16 m^2. Find a function that models its perimeter in terms of the length of one of its sides.
7. Find a function that models the area of an equilateral triangle in terms of the length of one of its sides.
8. Find a function that models the surface area of a cube in terms of its volume.
9. Find a function that models the radius of a circle in terms of its area.
10. Find a function that models the area of a circle in terms of its circumference.
11. A rectangular box with a volume of 60 ft^3 has a square base. Find a function that models its surface area in terms of the length of one side of its base.

Thales of Miletus (circa 625–547 B.C.) is the legendary founder of Greek geometry. It is said that he calculated the height of a Greek column by comparing the length of the shadow of his staff with that of the column. Using properties of similar triangles, he argued that the ratio of the height h of the column to the height h' of his staff was equal to the ratio of the length s of the column's shadow to the length s' of the staff's shadow:

$$\frac{h}{h'} = \frac{s}{s'}$$

Since three of these quantities are known, Thales was able to calculate the height of the column.

(continued)

According to legend, Thales used a similar method to find the height of the Great Pyramid in Egypt, a feat that impressed Egypt's king. Plutarch wrote that "although he [the king of Egypt] admired you [Thales] for other things, yet he particularly liked the manner by which you measured the height of the pyramid without any trouble or instrument." The principle Thales used, the fact that ratios of corresponding sides of similar triangles are equal, is the foundation of trigonometry.

12. A woman 5 ft tall is standing near a street lamp that is 12 ft tall, as shown in the figure. Find a function that models the length L of her shadow in terms of her distance d from the base of the lamp.

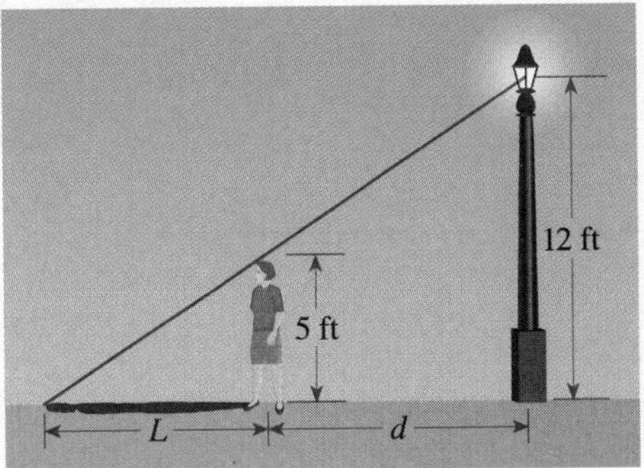

13. Two ships leave port at the same time. One sails south at 15 mi/h and the other sails east at 20 mi/h. Find a function that models the distance between the ships in terms of the time (in hours) elapsed since their departure.

14. The sum of two positive numbers is 60. Find a function that models their product in terms of one of the numbers.

15. An isosceles triangle has a perimeter of 8 cm. Find a function that models its area in terms of the length its base.

16. A right triangle has one leg twice as long as the other. Find a function that models its perimeter in terms of the length of the shorter leg.

17. A rectangle is inscribed in a semicircle of radius 10, as shown in the figure. Find a function that models the area A of the rectangle in terms of its height h.

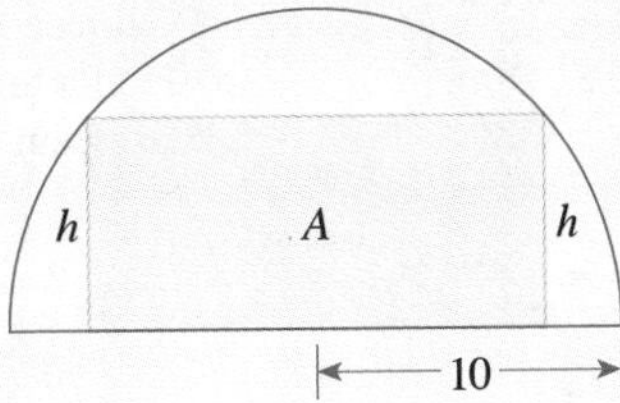

18. The volume of a cone is 100 in^3. Find a function that models the height of the cone in terms of its radius.

Pythagoras (circa 580–500 B.C.) founded a school in Croton in southern Italy, devoted to the study of arithmetic, geometry, music, and astronomy. The Pythagoreans, as they were called, were a secret society with peculiar rules and initiation rites. They wrote nothing down, and were not to reveal to anyone what they had learned from the Master. Although women were barred by law from attending public meetings, Pythagoras allowed women in his school, and his most famous student was Theano (whom he later married).

According to Aristotle, the Pythagoreans were convinced that "the principles of mathematics are the principles of all things." Their motto was "Everything is Number," by which they meant *whole* numbers. The outstanding contribution of Pythagoras is the theorem that bears his name: In a right triangle the area of the square on the hypotenuse is equal to the sum of the areas of the square on the other two sides.

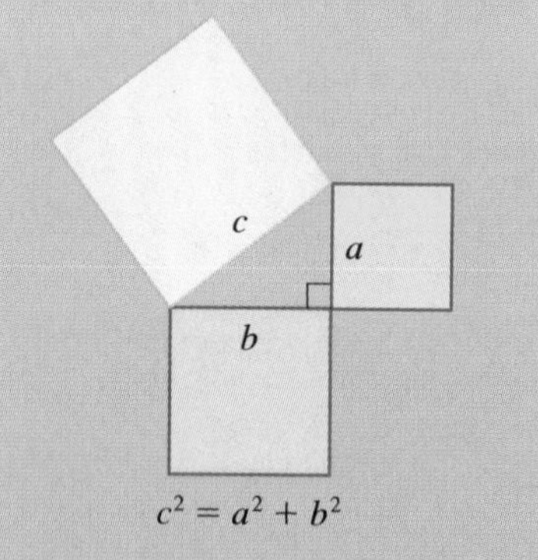

19. Consider the following problem: Find two numbers whose sum is 19 and whose product is as large as possible.

(a) Experiment with the problem by making a table like the one below, showing the product of different pairs of numbers that add up to 19. Based on the evidence in your table, estimate the answer to the problem.

First number	Second number	Product
1	18	18
2	17	34
3	16	48
⋮	⋮	⋮

(b) Find a function that models the product in terms of one of the two numbers.

(c) Use your model to solve the problem, and compare with your answer to part (a).

20. Find two positive numbers whose sum is 100 and the sum of whose squares is a minimum.

21. Find two numbers whose sum is -24 and whose product is a maximum.

22. Among all rectangles that have a perimeter of 20 ft, find the dimensions of the one with the largest area.

23. Consider the following problem: A farmer has 2400 ft of fencing and wants to fence off a rectangular field that borders a straight river. He does not need a fence along the river. (See the figure.) What are the dimensions of the field of largest area that he can fence?

(a) Experiment with the problem by drawing several diagrams illustrating the situation, as in Example 2. Calculate the area of each configuration, and use your results to estimate the dimensions of the largest possible field.

(b) Find a function that models the area of the field in terms of one of its sides.

(c) Use your model to solve the problem, and compare with your answer to part (a).

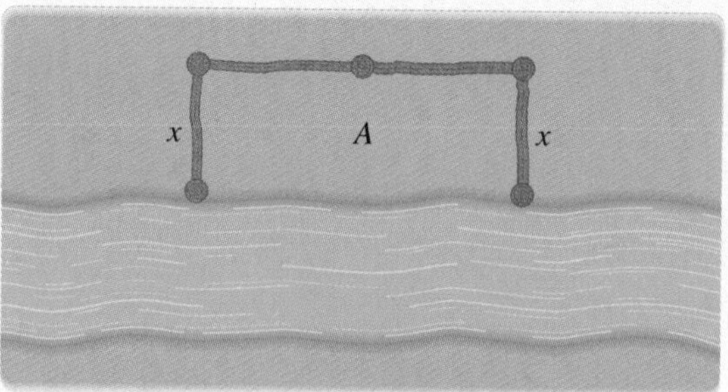

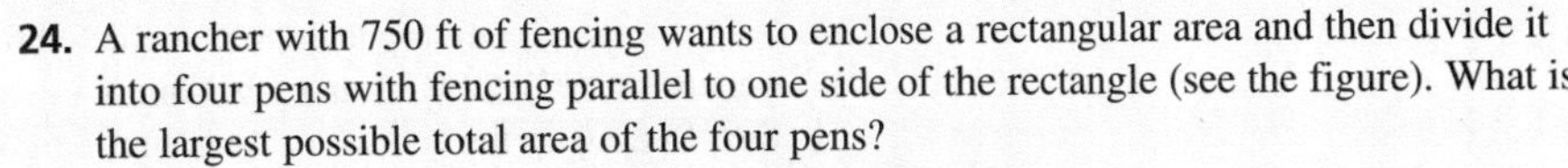
24. A rancher with 750 ft of fencing wants to enclose a rectangular area and then divide it into four pens with fencing parallel to one side of the rectangle (see the figure). What is the largest possible total area of the four pens?

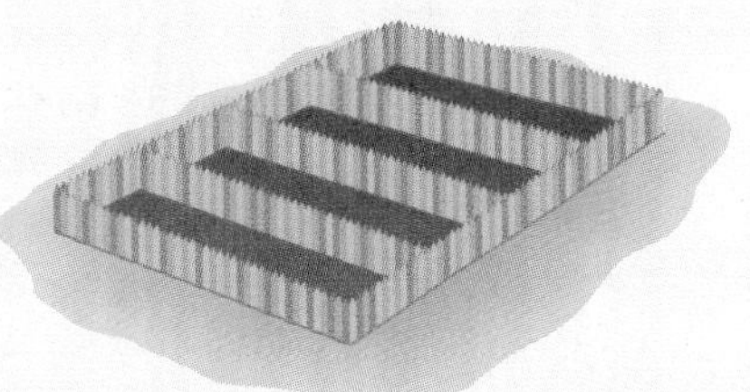

25. A property owner wants to fence a garden plot adjacent to a road, as shown in the figure. The fencing next to the road must be sturdier and costs \$5 per foot, while the other fencing costs just \$3 per foot. If the garden is to have an area of 600 ft^2, what dimensions should she choose to minimize the cost of the fence?

26. A wire 10 cm long is cut into two pieces, one of length x and the other of length $10 - x$, as shown in the figure. Each piece is bent into the shape of a square. What value of x minimizes the total area enclosed by the two squares?

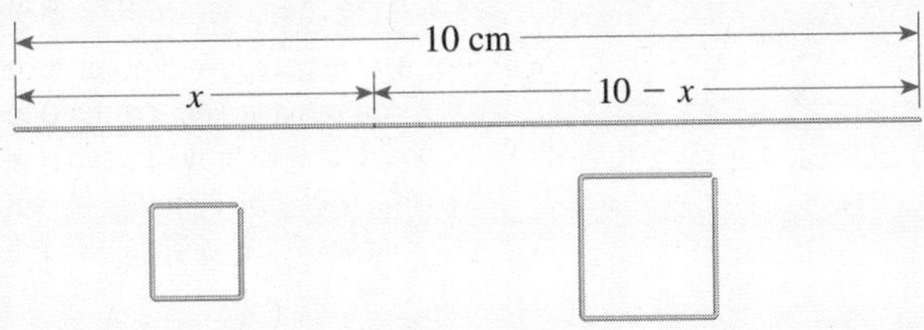

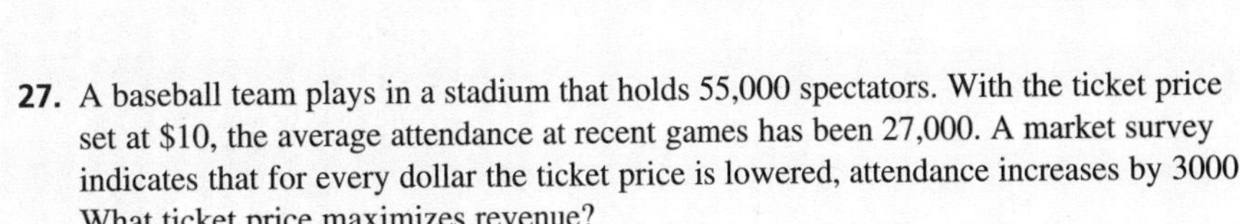
27. A baseball team plays in a stadium that holds 55,000 spectators. With the ticket price set at \$10, the average attendance at recent games has been 27,000. A market survey indicates that for every dollar the ticket price is lowered, attendance increases by 3000. What ticket price maximizes revenue?

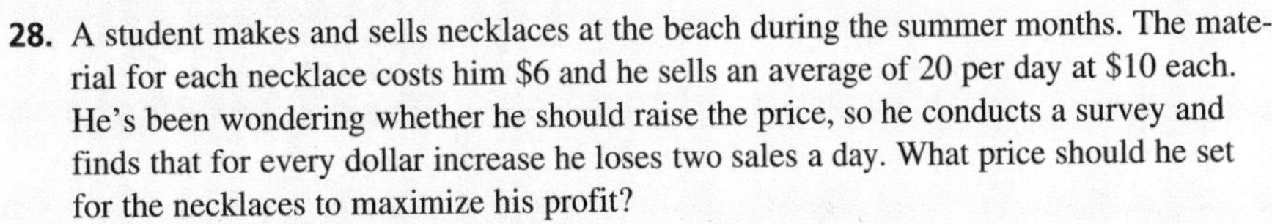
28. A student makes and sells necklaces at the beach during the summer months. The material for each necklace costs him \$6 and he sells an average of 20 per day at \$10 each. He's been wondering whether he should raise the price, so he conducts a survey and finds that for every dollar increase he loses two sales a day. What price should he set for the necklaces to maximize his profit?

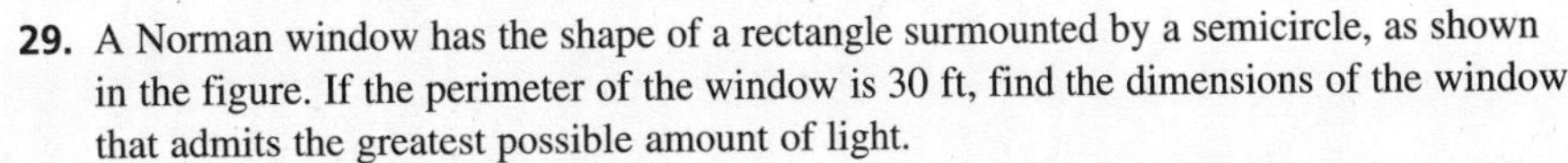
29. A Norman window has the shape of a rectangle surmounted by a semicircle, as shown in the figure. If the perimeter of the window is 30 ft, find the dimensions of the window that admits the greatest possible amount of light.

30. A box with an open top is to be constructed from a rectangular piece of cardboard with dimensions 12 in. by 20 in. by cutting out equal squares of side x at each corner and then folding up the sides, as shown in the figure. Find the largest volume that such a box can have.

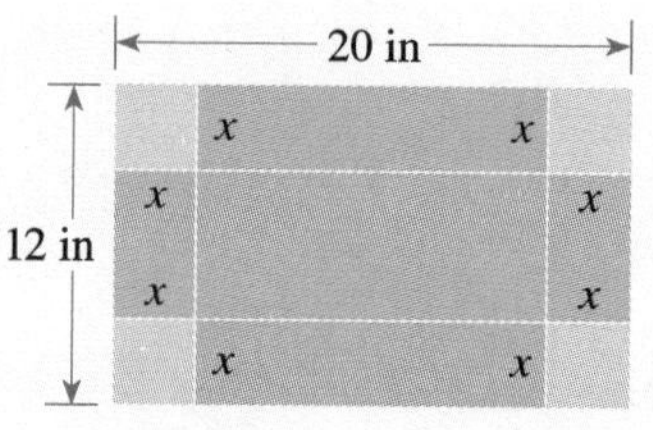

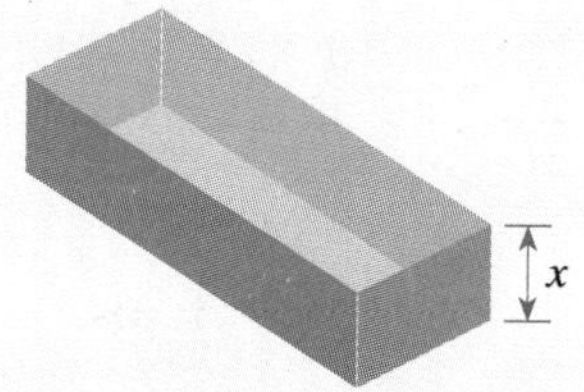

 **31.** An open box with a square base must have a volume of 12 ft^3. Find the dimensions of the box that minimize the amount of material used.

32. Find the dimensions that give the largest area for the rectangle shown in the figure. Its base is on the x-axis and its other two vertices are above the x-axis, lying on the parabola $y = 8 - x^2$.

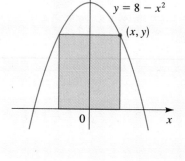

 **33.** A rancher wants to build a rectangular pen with an area of 100 m^2. Find the dimensions of the pen that require the minimum amount of fencing.

 **34.** A man stands at a point A on the bank of a straight river, 2 mi wide. To reach point B, 7 mi downstream on the opposite bank, he first rows his boat to point P on the opposite bank and then walks the remaining distance x to B, as shown in the figure. He can row at a speed of 2 mi/h and walk at a speed of 5 mi/h. Where should he land so that he reaches B as soon as possible?

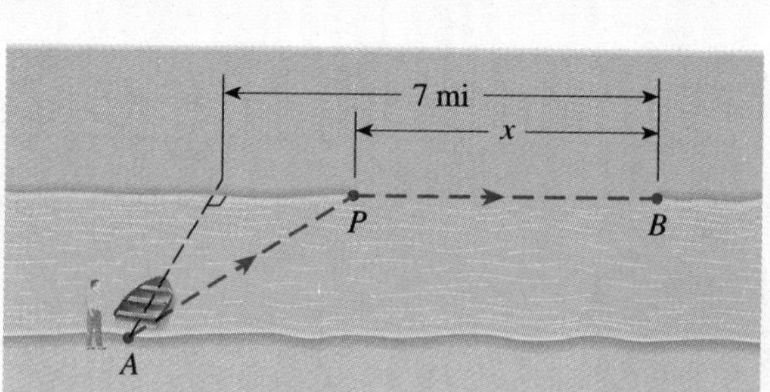

35. A bird is released from point A on an island, 5 mi from the nearest point B on a straight shoreline. The bird flies to a point C on the shoreline, and then flies along the shoreline to its nesting area D (see the figure). Suppose the bird requires 10 kcal/mi of energy to fly over land and 14 kcal/mi to fly over water (see Example 9 in Section 3.5). If the bird instinctively chooses a path that minimizes its energy expenditure, to what point does it fly?

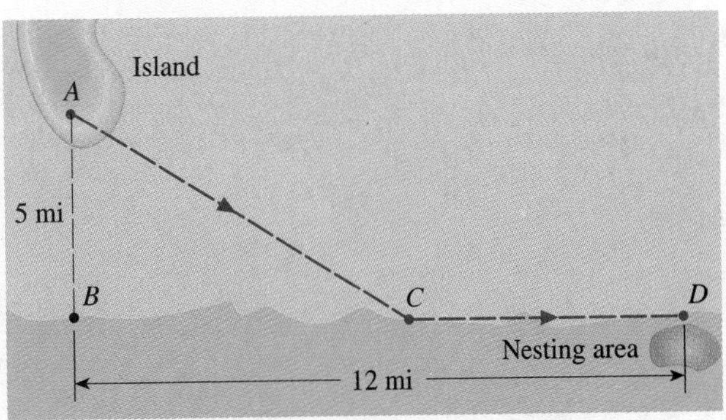

36. A kite frame is to be made from six pieces of wood. The four pieces that form its border have been cut to the lengths indicated in the figure. Let x be as shown in the figure.

(a) Show that the area of the kite is given by the function

$A(x) = x(\sqrt{25 - x^2} + \sqrt{144 - x^2})$.

(b) How long should each of the two crosspieces be to maximize the area of the kite?

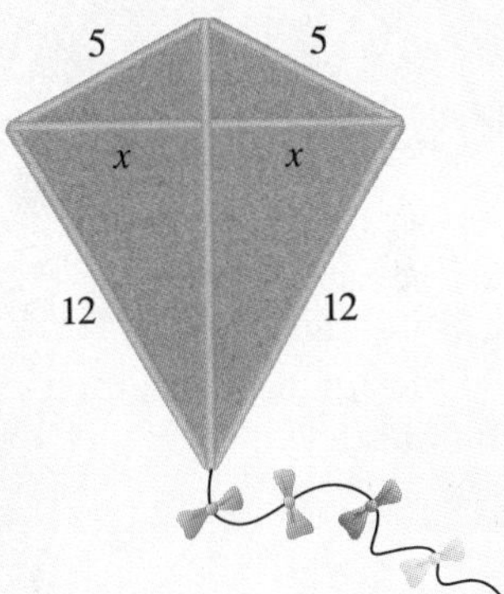

5

POLYNOMIAL AND RATIONAL FUNCTIONS

Many real-life situations can be modeled by polynomial or rational functions. The volume of a silo is a polynomial function of the radius; the observed pitch of a train whistle is a rational function of the speed of the train.

Each problem that I solved became a rule, which served afterwards to solve other problems.

RENÉ DESCARTES

In this chapter we study functions defined by polynomial expressions. Polynomials are constructed using just addition, subtraction, multiplication, and taking powers, so their values are easy to calculate. For this reason virtually all the functions used in mathematics and the sciences are evaluated numerically by using polynomial approximations. That is why polynomial functions play such an important role in mathematics. We will also study rational functions, which are quotients of polynomial functions.

5.1 POLYNOMIAL FUNCTIONS AND THEIR GRAPHS

Before we work with polynomials, we must agree on some terminology.

POLYNOMIAL FUNCTIONS

A **polynomial of degree n** is a function of the form

$$P(x) = a_n x^n + a_{n-1}x^{n-1} + \cdots + a_1 x + a_0$$

where $a_n \neq 0$. The numbers $a_0, a_1, a_2, \ldots, a_n$ are called the **coefficients** of the polynomial. The number a_0 is the **constant coefficient** or **constant term**. The number a_n, the coefficient of the highest power, is the **leading coefficient**, and the term $a_n x^n$ is the **leading term**.

For example, the following are polynomial functions.

$P(x) = 3$	degree 0
$Q(x) = 4x - 7$	degree 1
$R(x) = x^2 + x$	degree 2
$S(x) = 2x^3 - 6x^2 - 10$	degree 3

If a polynomial consists of just a single term, then it is called a **monomial**. For example, $P(x) = x^3$ and $Q(x) = -6x^5$ are monomials.

GRAPHS OF POLYNOMIALS

The graphs of polynomials of degree 0 or 1 are lines (Section 2.4), and the graphs of polynomials of degree 2 are parabolas (Section 4.6). The greater the degree of the polynomial, the more complicated its graph can be. However, the graph of a polynomial function is always a smooth curve; that is, it has no breaks or corners (see Figure 1). The proof of this fact requires calculus.

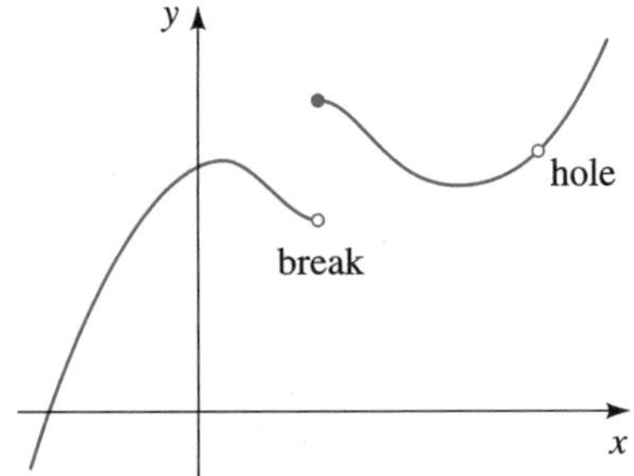

Not the graph of a polynomial function

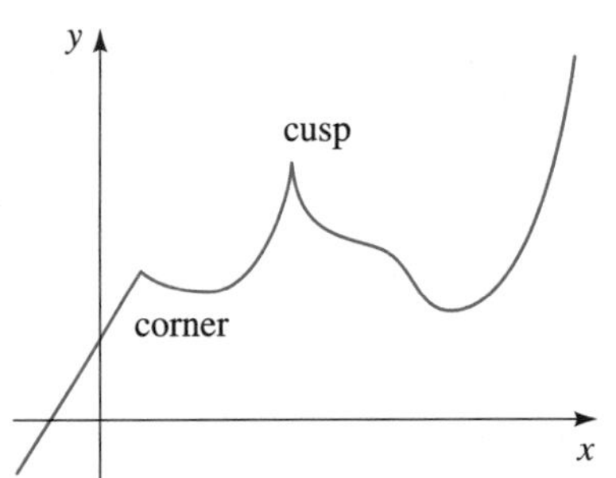

Not the graph of a polynomial function

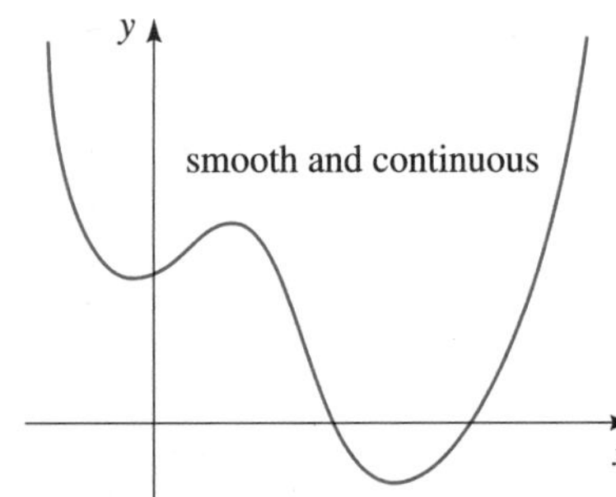

Graph of a polynomial function

FIGURE 1

The simplest polynomial functions are the monomials $P(x) = x^n$, whose graphs are shown in Figure 2. As the figure suggests, the graph of $P(x) = x^n$ has the same general shape as $y = x^2$ when n is even, and the same general shape as $y = x^3$ when n is odd. However, as the degree n becomes larger, the graphs become flatter around the origin and steeper elsewhere.

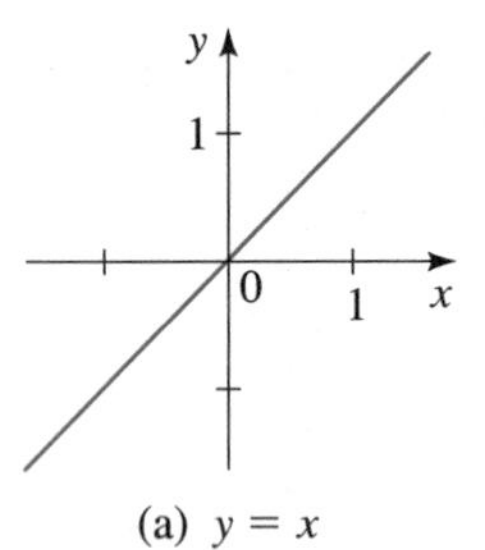

(a) $y = x$

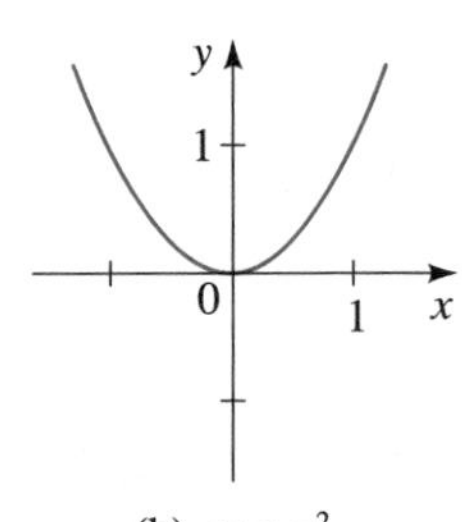

(b) $y = x^2$

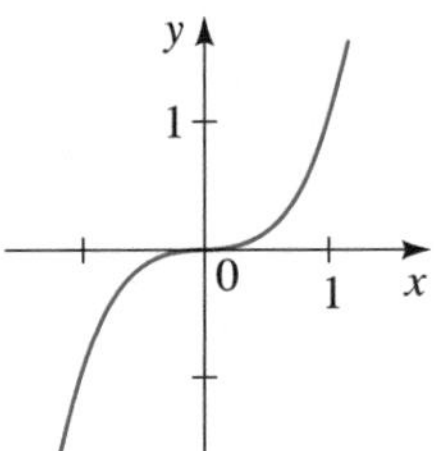

(c) $y = x^3$

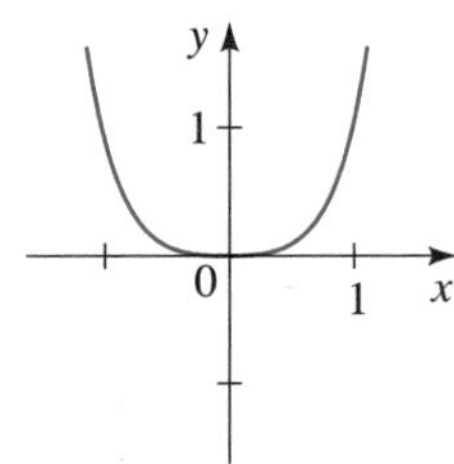

(d) $y = x^4$

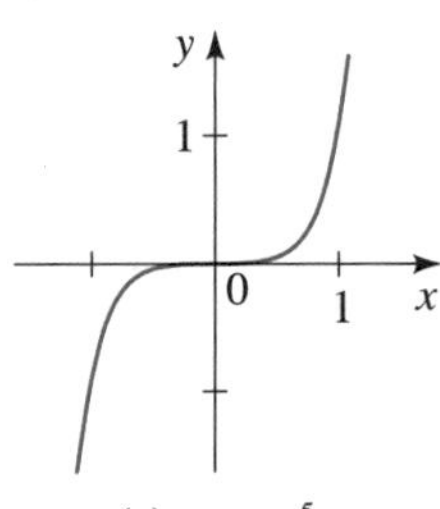

(e) $y = x^5$

FIGURE 2

EXAMPLE 1 ■ Transformations of Monomials

Sketch the graphs of the following functions.

(a) $P(x) = -x^3$ (b) $Q(x) = (x - 2)^4$ (c) $R(x) = -2x^5 + 4$

SOLUTION

We use the graphs in Figure 2 and transform them using the techniques of Section 4.5.

(a) The graph of $P(x) = -x^3$ is the reflection of the graph of $y = x^3$ in the x-axis, as shown in Figure 3(a).

(b) The graph of $Q(x) = (x - 2)^4$ is the graph of $y = x^4$ shifted to the right 2 units, as shown in Figure 3(b).

(c) We begin with the graph of $y = x^5$. The graph of $y = -2x^5$ is obtained by stretching the graph vertically and reflecting it in the x-axis (see the dashed blue graph in Figure 3(c)). Finally, the graph of $R(x) = -2x^5 + 4$ is obtained by shifting upward 4 units (see the red graph in Figure 3(c)).

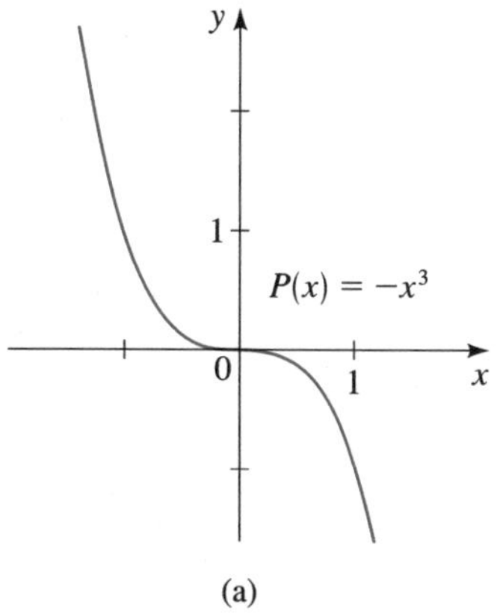

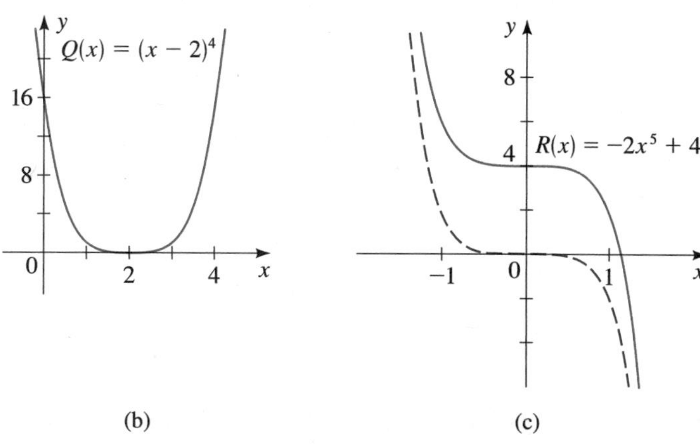

FIGURE 3

■

END BEHAVIOR OF POLYNOMIALS

The **end behavior** of a polynomial is a description of what happens as x becomes large in the positive or negative direction. To describe end behavior, we use the following notation:

$x \to \infty$	means	"x becomes large in the positive direction"
$x \to -\infty$	means	"x becomes large in the negative direction"

For example, the monomial $y = x^2$ in Figure 2(b) has the following end behavior:

$$y \to \infty \quad \text{as} \quad x \to \infty \qquad \text{and} \qquad y \to \infty \quad \text{as} \quad x \to -\infty$$

The monomial $y = x^3$ in Figure 2(c) has the end behavior

$$y \to \infty \quad \text{as} \quad x \to \infty \qquad \text{and} \qquad y \to -\infty \quad \text{as} \quad x \to -\infty$$

For any polynomial, *the end behavior is determined by the term that contains the highest power of x*, because when x is large, the other terms are relatively insignificant in size. The following box shows the four possible types of end behavior, based on the highest power and the sign of its coefficient.

END BEHAVIOR OF POLYNOMIALS

The polynomial $P(x) = a_n x^n + a_{n-1}x^{n-1} + \cdots + a_1 x + a_0$ has the same end behavior as the monomial $Q(x) = a_n x^n$, so its end behavior is determined by the degree n and the sign of the leading coefficient a_n.

$y = P(x)$ has odd degree

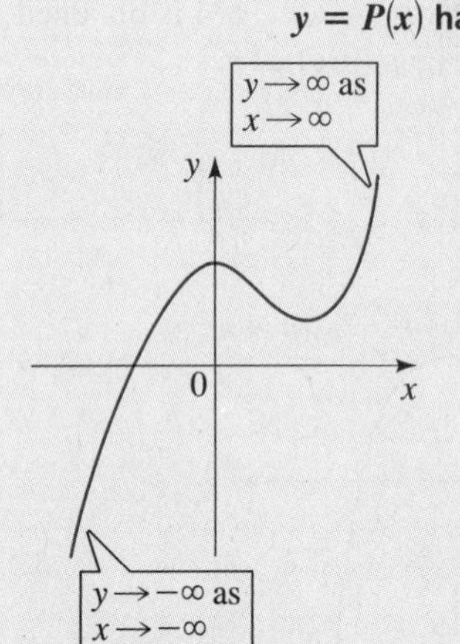

Leading coefficient positive

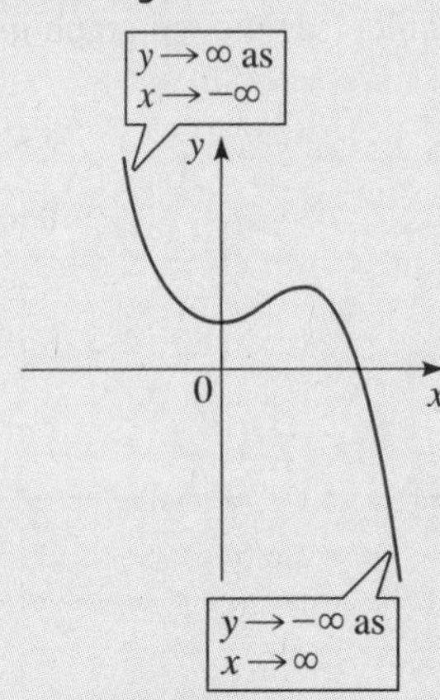

Leading coefficient negative

$y = P(x)$ has even degree

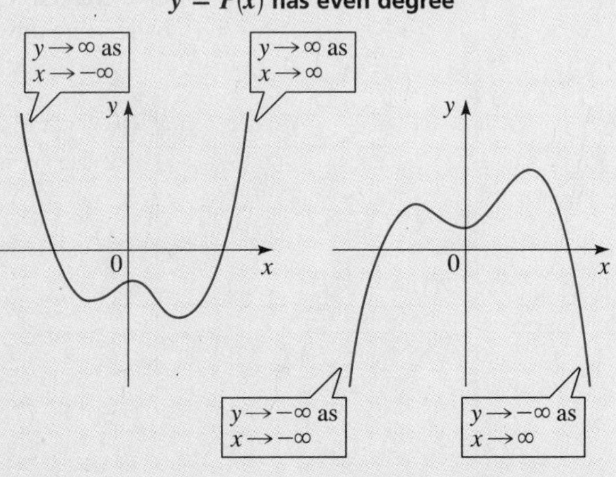

Leading coefficient positive Leading coefficient negative

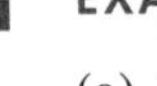

EXAMPLE 2 ■ End Behavior of a Polynomial

(a) Determine the end behavior of the polynomial $P(x) = 3x^5 - 5x^3 + 2x$.

(b) Confirm that P and its leading term $Q(x) = 3x^5$ have the same end behavior by graphing them together on a large viewing rectangle.

(c) Compare the graphs of P and Q on a small viewing rectangle.

SOLUTION

(a) Since P has odd degree and positive leading coefficient, it has the following end behavior:

$$y \to \infty \quad \text{as} \quad x \to \infty \qquad \text{and} \qquad y \to -\infty \quad \text{as} \quad x \to -\infty$$

(b) On the large viewing rectangle in Figure 4(a), the graphs of P and Q look almost the same. This confirms that they have the same end behavior.

(c) On the small viewing rectangle in Figure 4(b), the graphs of P and Q look quite different. For instance, P has five x-intercepts, while Q has only one. ■

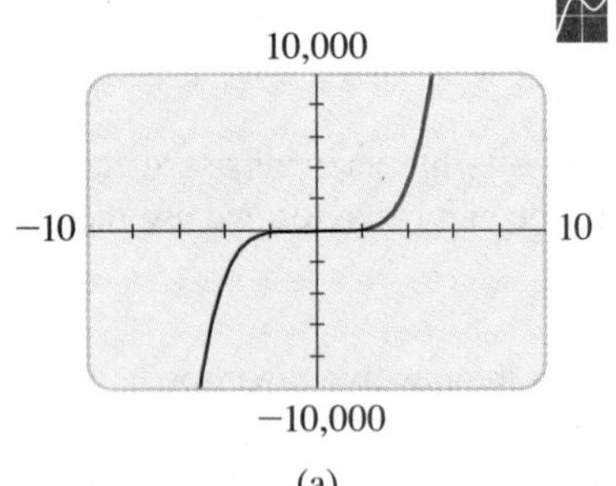

(a)

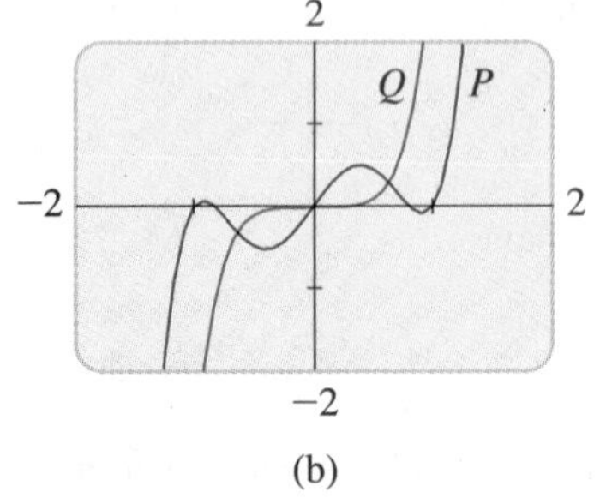

(b)

FIGURE 4

$P(x) = 3x^5 - 5x^3 + 2x$
$Q(x) = 3x^5$

To see algebraically why P and Q in Example 2 have the same end behavior, factor P as follows and compare with Q.

$$P(x) = 3x^5\left(1 - \frac{5}{3x^2} + \frac{2}{3x^4}\right)$$

$$Q(x) = 3x^5$$

When x is large, the terms $5/3x^2$ and $2/3x^4$ are close to 0 (see Exercise 59 on page 20). So for large x, we have

$$P(x) \approx 3x^5(1 - 0 - 0)$$
$$= 3x^5 = Q(x)$$

x	$P(x)$	$Q(x)$
15	2,261,280	2,278,125
30	72,765,060	72,900,000
50	936,875,100	937,500,000

So, when x is large, P and Q have approximately the same values. We can also see this numerically by making a table like the one in the margin.

By the same reasoning we can show that the end behavior of *any* polynomial is determined by its leading term.

USING ZEROS TO GRAPH POLYNOMIALS

If $y = P(x)$ is a polynomial and if c is a number such that $P(c) = 0$, then we say that c is a **zero** of P. In other words, the zeros of P are the solutions of the polynomial equation $P(x) = 0$. When we deal with polynomial equations, we often refer to solutions as **roots**. To find the roots, we factor the polynomial and then use the Zero-Product Property, as in Section 3.3. Note that if $P(c) = 0$, then the graph of $y = P(x)$ has an x-intercept at $x = c$, so the x-intercepts of the graph are the zeros of the function.

ZEROS OF POLYNOMIALS

If P is a polynomial and if c is a number such that $P(c) = 0$, then we say that c is a **zero** of P. The following are equivalent ways of saying the same thing.

1. c is a zero of P
2. $x = c$ is a root of the equation $P(x) = 0$
3. $x - c$ is a factor of $P(x)$

For example, to find the zeros of $P(x) = x^2 + x - 6$, we factor P to get

$$P(x) = (x - 2)(x + 3)$$

From this factored form we easily see that

1. 2 is a zero of P
2. $x = 2$ is a root of the equation $x^2 + x - 6 = 0$
3. $x - 2$ is a factor of $x^2 + x - 6$
4. $x = 2$ is an x-intercept of the graph of P

The same facts are true for the other zero, -3.

The following theorem has many important consequences. (See, for instance, the Discovery Project on page 345.) Here we use it to help us graph polynomial functions.

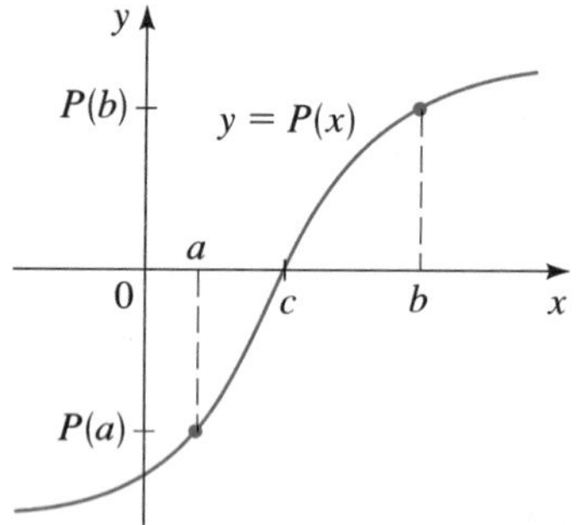

FIGURE 5

INTERMEDIATE VALUE THEOREM FOR POLYNOMIALS

If P is a polynomial function and $P(a)$ and $P(b)$ have opposite signs, then there exists at least one value c between a and b for which $P(c) = 0$.

The proof of this theorem requires calculus, but Figure 5 shows why it is intuitively plausible.

One important consequence of this theorem is that between any two successive zeros, the values of a polynomial are either all positive or all negative. That is, between two successive zeros the graph of a polynomial lies *entirely above* or *entirely below* the x-axis. To see why, suppose c_1 and c_2 are successive zeros of P. If P has both positive and negative values between c_1 and c_2, then by the Intermediate Value Theorem P must have another zero between c_1 and c_2. But that's not possible because c_1 and c_2 are successive zeros.

So to sketch the graph of P, we first find all the zeros of P. Then we choose **test points** between (and to the right and left of) successive zeros to determine whether $P(x)$ is positive or negative on each interval determined by the zeros. For example, the polynomial

$$P(x) = (x + 2)(x - 1)(x - 2)$$

has zeros at $x = -2$, 1, and 2. These determine the intervals $(-\infty, -2)$, $(-2, 1)$, $(1, 2)$, and $(2, \infty)$. Using test points in these intervals, we get the information in the following sign diagram (see Section 3.7).

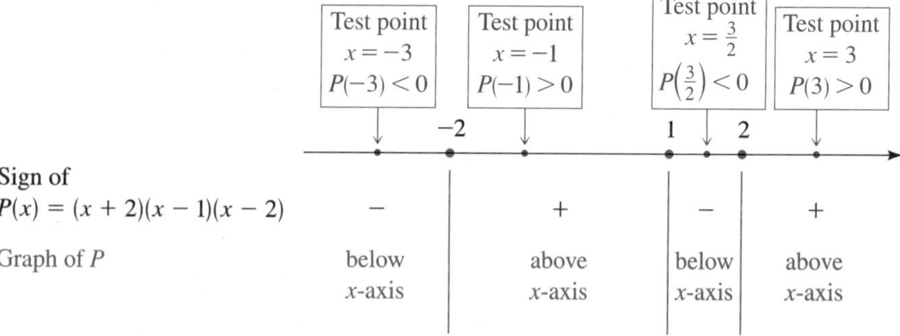

Plotting a few additional points and connecting them with a smooth curve helps us complete the graph in Figure 6.

	x	$P(x)$
Test point →	-3	-20
	-2	0
Test point →	-1	6
	0	4
	1	0
Test point →	$\frac{3}{2}$	$-\frac{7}{8}$
	2	0
Test point →	3	10

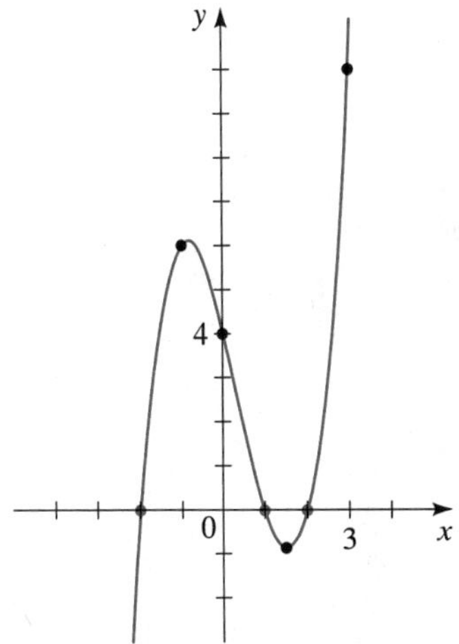

FIGURE 6
$P(x) = (x + 2)(x - 1)(x - 2)$

GUIDELINES FOR GRAPHING POLYNOMIAL FUNCTIONS

1. ZEROS. Factor the polynomial to find all its real zeros; these are the x-intercepts of the graph.

2. TEST POINTS. Make a table of values for the polynomial. Include test points to determine whether the graph of the polynomial lies above or below the x-axis on the intervals determined by the zeros. Include the y-intercept in the table.

3. END BEHAVIOR. Determine the end behavior of the polynomial.

4. GRAPH. Plot the intercepts and other points you found in the table. Sketch a smooth curve that passes through these points and exhibits the required end behavior.

EXAMPLE 3 ■ Finding Zeros and Graphing a Polynomial Function

Let $P(x) = x^3 - 5x^2 + 6x$.

(a) Find the zeros of P. (b) Sketch the graph of P.

SOLUTION

(a) To find the zeros, we factor completely.

$$\begin{aligned} P(x) &= x^3 - 2x^2 - 3x \\ &= x(x^2 - 2x - 3) && \text{Factor } x \\ &= x(x - 3)(x + 1) && \text{Factor the quadratic} \end{aligned}$$

Thus, the zeros are $x = 0$, $x = 3$, and $x = -1$.

(b) The x-intercepts are $x = 0$, $x = 3$, and $x = -1$. The y-intercept is $P(0) = 0$. We make a table of values of $P(x)$, making sure we choose test points between (and to the right and left of) successive zeros.

Since P is of odd degree and its leading coefficient is positive, it has the following end behavior:

$$y \to \infty \quad \text{as} \quad x \to \infty \qquad \text{and} \qquad y \to -\infty \quad \text{as} \quad x \to -\infty$$

We plot the points in the table and connect them by a smooth curve to complete the graph, as shown in Figure 7.

	x	$P(x)$
Test point →	-2	-10
	-1	0
Test point →	$-\frac{1}{2}$	$\frac{7}{8}$
	0	0
Test point →	1	-4
	2	-6
	3	0
Test point →	4	20

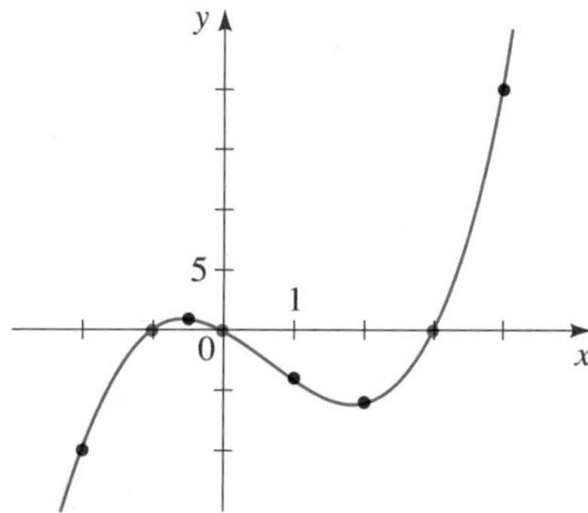

FIGURE 7
$P(x) = x^3 - 2x^2 - 3x$

■

MATHEMATICS IN THE MODERN WORLD

AUTOMOTIVE DESIGN

Computer-aided design (CAD) has completely changed the way car companies design and manufacture cars. Before the 1980s automotive engineers would build a full-scale "nuts and bolts" model of a proposed new car; this was really the only way to tell whether the design was feasible. Today automotive engineers build a mathematical model, one that exists only in the memory of a computer. The model incorporates all the main design features of the car. Certain polynomial curves, called *splines*, are used in shaping the body of the car. The resulting "mathematical car" can be tested for structural stability, handling, aerodynamics, suspension response, and more. All this testing is done before a prototype is built. As you can imagine, CAD saves car manufacturers millions of dollars each year. More importantly, CAD gives automotive engineers far more flexibility in design; desired changes can be created and tested within seconds. With the help of computer graphics designers can see how good the "mathematical car" looks before they build the real one. Moreover, the mathematical car can be viewed from any perspective; it can be

(continued)

EXAMPLE 4 ■ Finding Zeros and Graphing a Polynomial Function

Let $P(x) = -2x^4 - x^3 + 3x^2$.

(a) Find the zeros of P.
(b) Sketch the graph of P.

SOLUTION

(a) To find the zeros, we factor completely.

$$\begin{aligned} P(x) &= -2x^4 - x^3 + 3x^2 \\ &= -x^2(2x^2 + x - 3) && \text{Factor } -x^2 \\ &= -x^2(2x + 3)(x - 1) && \text{Factor the quadratic} \end{aligned}$$

Thus, the zeros are $x = 0$, $x = -\frac{3}{2}$, and $x = 1$.

(b) The x-intercepts are $x = 0$, $x = -\frac{3}{2}$, and $x = 1$. The y-intercept is $P(0) = 0$. We make a table of values of $P(x)$, making sure we choose test points between (and to the right and left of) successive zeros.

moved, rotated, or seen from the inside. These manipulations of the car on the computer monitor translate mathematically into solving large systems of linear equations.

Since P is of even degree and its leading coefficient is negative, it has the following end behavior:

$$y \to -\infty \quad \text{as} \quad x \to \infty \qquad \text{and} \qquad y \to -\infty \quad \text{as} \quad x \to -\infty$$

We plot the points from the table and connect the points by a smooth curve to complete the graph in Figure 8.

Tables of values are most easily calculated using a programmable calculator.

x	$P(x)$
−2	−12
−1.5	0
−1	2
−0.5	0.75
0	0
0.5	0.5
1	0
1.5	−6.75

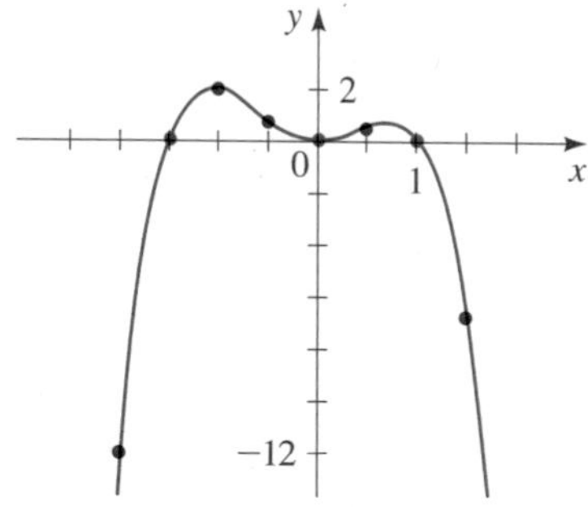

FIGURE 8
$P(x) = -2x^4 - x^3 + 3x^2$

■

EXAMPLE 5 ■ Finding Zeros and Graphing a Polynomial Function

Let $P(x) = x^3 - 2x^2 - 4x + 8$

(a) Find the zeros of P. (b) Sketch the graph of P.

SOLUTION

(a) To find the zeros, we factor completely.

$$\begin{aligned} P(x) &= x^3 - 2x^2 - 4x + 8 \\ &= x^2(x-2) - 4(x-2) && \text{Group and factor} \\ &= (x^2-4)(x-2) && \text{Factor } x-2 \\ &= (x+2)(x-2)(x-2) && \text{Difference of squares} \\ &= (x+2)(x-2)^2 && \text{Simplify} \end{aligned}$$

Thus, the zeros are $x = -2$ and $x = 2$.

(b) The x-intercepts are $x = -2$ and $x = 2$. The y-intercept is $P(0) = 8$. The table gives additional values of $P(x)$.

Since P is of odd degree and its leading coefficient is positive, it has the following end behavior:

$$y \to \infty \quad \text{as} \quad x \to \infty \qquad \text{and} \qquad y \to -\infty \quad \text{as} \quad x \to -\infty$$

We connect the points by a smooth curve to complete the graph in Figure 9.

x	$P(x)$
−3	−25
−2	0
−1	9
0	8
1	3
2	0
3	5

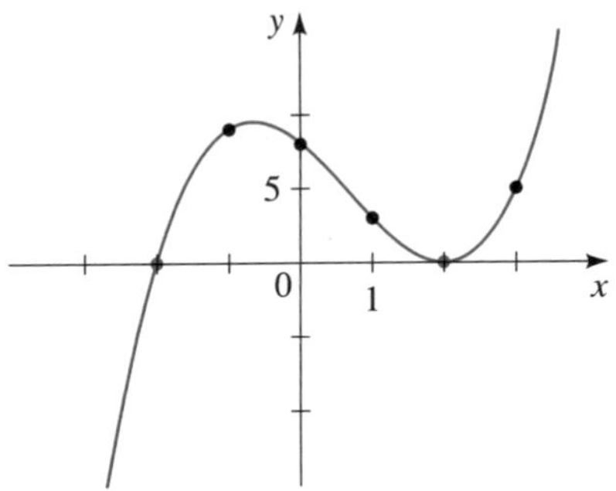

FIGURE 9
$P(x) = x^3 - 2x^2 - 4x + 8$ ■

Although $x = 2$ is a zero of the polynomial in Example 5, the graph does not cross the x-axis at the x-intercept 2. This is because the factor $(x - 2)^2$ corresponding to that zero is raised to an even power, so it doesn't change sign as we test points on either side of 2. In the same way, the graph does not cross the x-axis at $x = 0$ in Example 4.

In general, suppose c is a zero of P and the corresponding factor $(x - c)^m$ occurs *exactly m* times in the factorization of P. Then by considering test points on either side of the x-intercept c, we conclude that the graph crosses the x-axis at c if m is odd and does not cross the x-axis if m is even. Moreover, it can be shown using calculus that near $x = c$ the graph has the same general shape as $y = A(x - c)^m$.

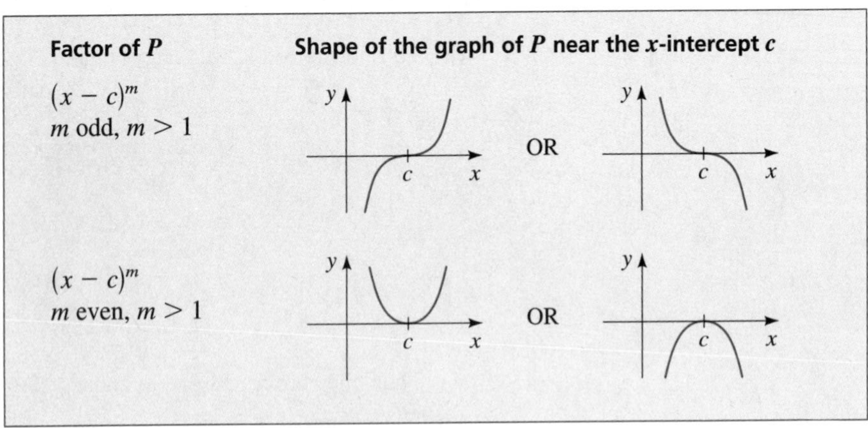

EXAMPLE 6 ■ Graphing a Polynomial Function Using Its Zeros

Graph the polynomial $P(x) = x^4(x - 2)^3(x + 1)^2$.

SOLUTION

The zeros of P are -1, 0, and 2, so the x-intercepts are at $x = -1$, 0, and 2. These zeros correspond to the factors $(x + 1)^2$, x^4, and $(x - 2)^3$. So the graph crosses the x-axis at the x-intercept 2 (the corresponding factor has odd exponent 3), but does not cross at the x-intercepts 0 and -1 (the corresponding factors have even exponents 4 and 2). Near the x-intercepts -1, 0, and 2, the graph has the same general shape as $y = A(x + 1)^2$, $y = Ax^4$, and $y = A(x - 2)^3$, respectively.

Since P is a polynomial of degree 9 and has positive leading coefficient, it has the following end behavior:

$$y \to \infty \quad \text{as} \quad x \to \infty \qquad \text{and} \qquad y \to -\infty \quad \text{as} \quad x \to -\infty$$

With this information and a table of values, we sketch the graph in Figure 10.

x	$P(x)$
-1.3	-9.2
-1	0
-0.5	-3.9
0	0
1	-4
2	0
2.3	8.2

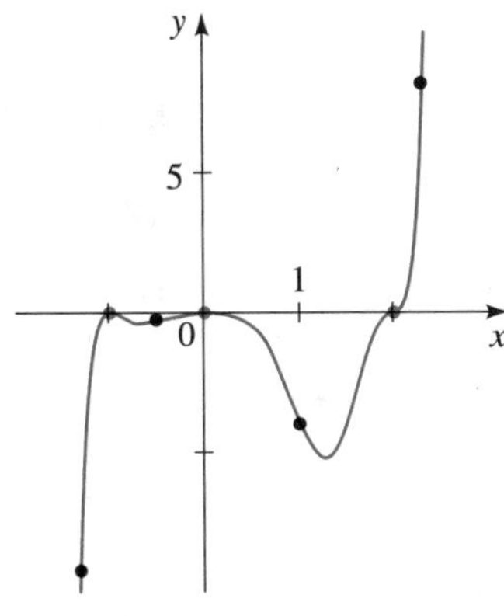

FIGURE 10
$P(x) = x^4(x - 2)^3(x + 1)^2$

■

LOCAL EXTREMA OF POLYNOMIALS

Recall from Section 4.6 that if the point $(a, f(a))$ is the highest point on the graph of f within some viewing rectangle, then $f(a)$ is a local maximum value of f, and if $(b, f(b))$ is the lowest point on the graph of f within a viewing rectangle, then $f(b)$ is a local minimum value (see Figure 11). We say that such a point $(a, f(a))$ is a **local**

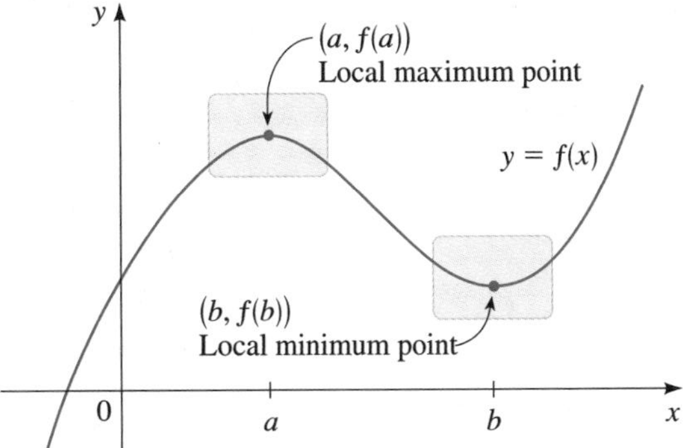

FIGURE 11

maximum point on the graph and that $(b, f(b))$ is a **local minimum point**. The set of all local maximum and minimum points on the graph of a function is called its **local extrema**.

For a polynomial function the number of local extrema must be less than the degree, as the following principle indicates. (A proof of this principle requires calculus.)

LOCAL EXTREMA OF POLYNOMIALS

If $P(x) = a_n x^n + a_{n-1}x^{n-1} + \cdots + a_1 x + a_0$ is a polynomial of degree n, then the graph of P has at most $n - 1$ local extrema.

A polynomial of degree n may in fact have less than $n - 1$ local extrema. For example, $P(x) = x^5$ (graphed in Figure 2) has *no* local extrema, even though it is of degree 5. The preceding principle tells us only that a polynomial of degree n can have no more than $n - 1$ local extrema.

EXAMPLE 7 ■ The Number of Local Extrema

Determine how many local extrema each polynomial has.

(a) $P_1(x) = x^4 + x^3 - 16x^2 - 4x + 48$

(b) $P_2(x) = x^5 + 3x^4 - 5x^3 - 15x^2 + 4x - 15$

(c) $P_3(x) = 7x^4 + 3x^2 - 10x$

SOLUTION

The graphs are shown in Figure 12.

(a) P_1 has two local minimum points and one local maximum point, for a total of three local extrema.

(b) P_2 has two local minimum points and two local maximum points, for a total of four local extrema.

(c) P_3 has just one local extremum, a local minimum.

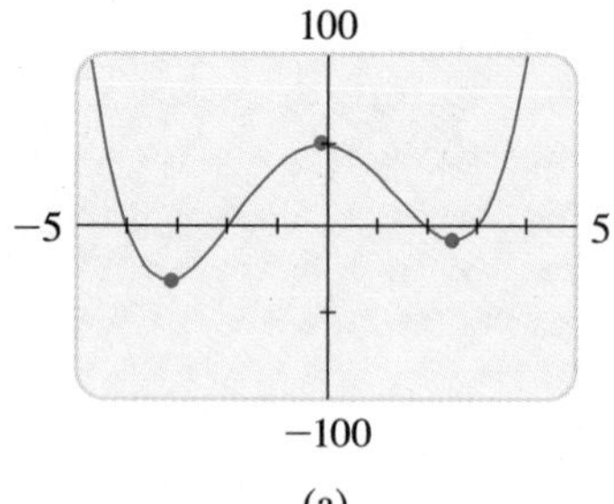

(a)

$P_1(x) = x^4 + x^3 - 16x^2 - 4x + 48$

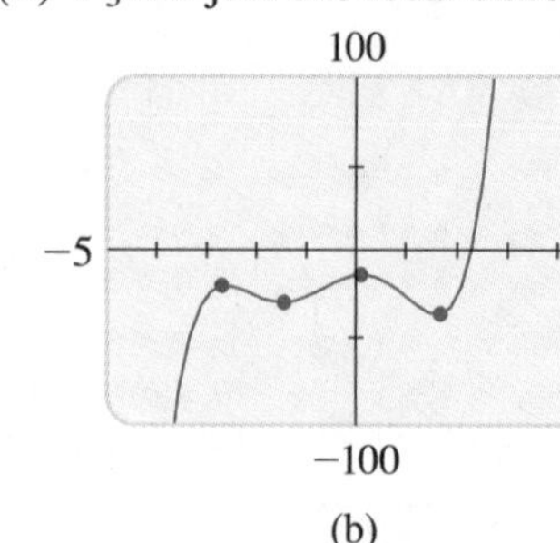

(b)

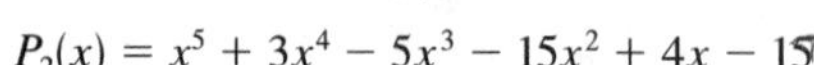

$P_2(x) = x^5 + 3x^4 - 5x^3 - 15x^2 + 4x - 15$

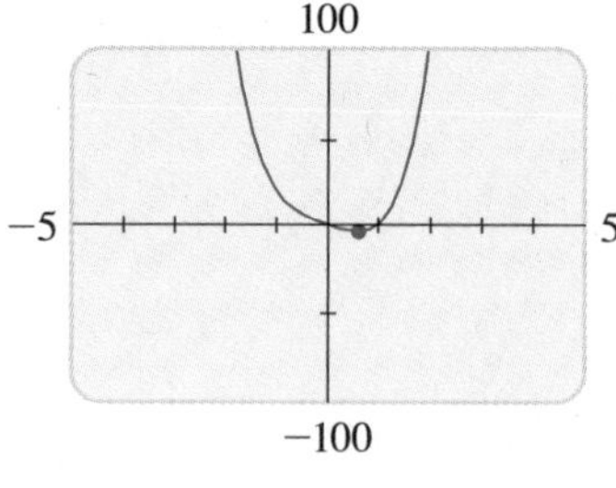

(c)

$P_3(x) = 7x^4 + 3x^2 - 10x$

FIGURE 12

■

With a graphing calculator we can quickly draw the graphs of many functions at once, on the same viewing screen. This allows us to see how changing a value in the definition of the functions affects the shape of its graph. In the next example we apply this principle to a family of third-degree polynomials.

EXAMPLE 8 ■ A Family of Polynomials

Sketch the family of polynomials $P(x) = x^3 - cx^2$ for $c = 0, 1, 2$, and 3. How does changing the value of c affect the graph?

SOLUTION

The polynomials

$$P_0(x) = x^3 \qquad P_1(x) = x^3 - x^2$$

$$P_2(x) = x^3 - 2x^2 \qquad P_3(x) = x^3 - 3x^2$$

are graphed in Figure 13. We see that increasing the value of c causes the graph to develop an increasingly deep "valley" to the right of the y-axis, creating a local maximum at the origin and a local minimum at a point in quadrant IV. This local minimum moves lower and farther to the right as c increases. To see why this happens, factor $P(x) = x^2(x - c)$. The polynomial P has zeros at 0 and c, and the larger c gets, the farther to the right the minimum between 0 and c will be. ■

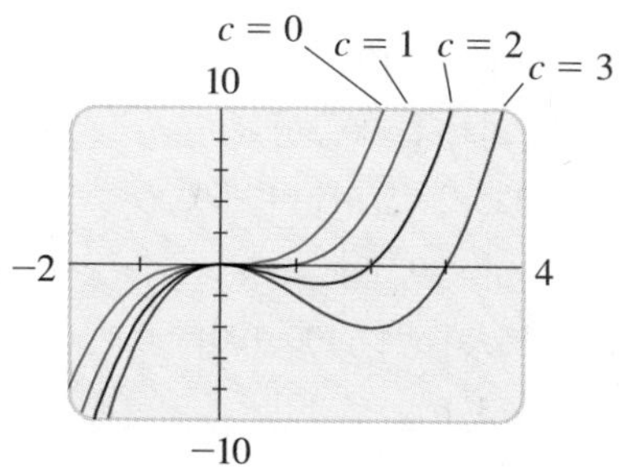

FIGURE 13
A family of polynomials
$P(x) = x^3 - cx^2$

5.1 EXERCISES

1–8 ■ Sketch the graph of the function by transforming the graph of an appropriate function of the form $y = x^n$. Indicate all x- and y-intercepts on each graph.

1. $P(x) = x^3 - 8$

2. $P(x) = (x - 2)^3$

3. $P(x) = -x^4 + 16$

4. $P(x) = -2(x - 1)^3$

5. $P(x) = -(x - 1)^4 + 1$

6. $P(x) = 3x^5 - 9$

7. $P(x) = 4(x - 2)^5 - 4$

8. $P(x) = 3x^4 - 27$

9–34 ■ Sketch the graph of the function. Make sure your graph shows all intercepts and exhibits the proper end behavior.

9. $P(x) = (x - 3)(x + 1)$

10. $P(x) = (x - 1)(x + 1)(x - 2)$

11. $P(x) = x(x - 2)(x + 1)$

12. $P(x) = (2x - 1)(x + 1)(x + 3)$

13. $P(x) = (x - 3)(x + 2)(3x - 2)$

14. $P(x) = \frac{1}{5}x(x - 5)^2$

15. $P(x) = (x - 1)^2(x - 3)$

16. $P(x) = \frac{1}{4}(x + 1)^3(x - 3)$

17. $P(x) = \frac{1}{12}(x + 2)^2(x - 3)^2$

18. $P(x) = (x - 1)^2(x + 2)^3$

19. $P(x) = x^3(x + 2)(x - 3)^2$

20. $P(x) = (x^2 - 2x - 3)^2$

21. $P(x) = x^3 - x^2 - 6x$

22. $P(x) = x^3 + 2x^2 - 8x$

23. $P(x) = -x^3 + x^2 + 12x$

24. $P(x) = -2x^3 - x^2 + x$

25. $P(x) = x^4 - 3x^3 + 2x^2$

26. $P(x) = x^5 - 9x^3$

27. $P(x) = x^3 + x^2 - x - 1$

28. $P(x) = x^3 + 3x^2 - 4x - 12$

29. $P(x) = 2x^3 - x^2 - 18x + 9$

30. $P(x) = \frac{1}{8}(2x^4 + 3x^3 - 16x - 24)^2$

31. $P(x) = x^4 - 2x^3 - 8x + 16$

32. $P(x) = x^4 - 2x^3 + 8x - 16$

33. $P(x) = x^4 - 3x^2 - 4$

34. $P(x) = x^6 - 2x^3 + 1$

35–40 ■ Match the polynomial function with one of the graphs I–VI. Give reasons for your choice.

35. $P(x) = x(x^2 - 4)$ **36.** $Q(x) = -x^2(x^2 - 4)$

37. $R(x) = -x^5 + 5x^3 - 4x$ **38.** $S(x) = \frac{1}{2}x^6 - 2x^4$

39. $T(x) = x^4 + 2x^3$ **40.** $U(x) = -x^3 + 2x^2$

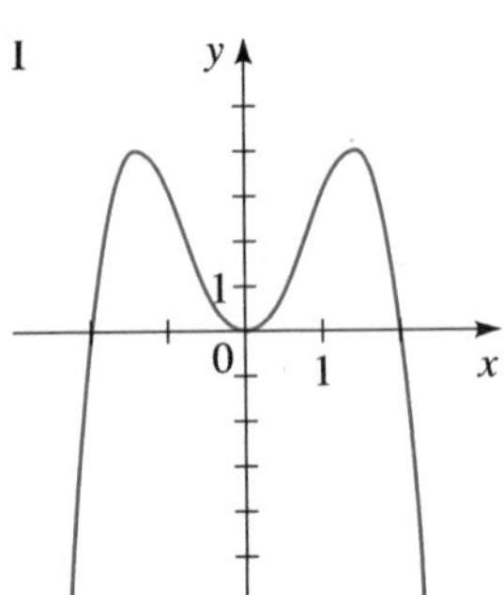

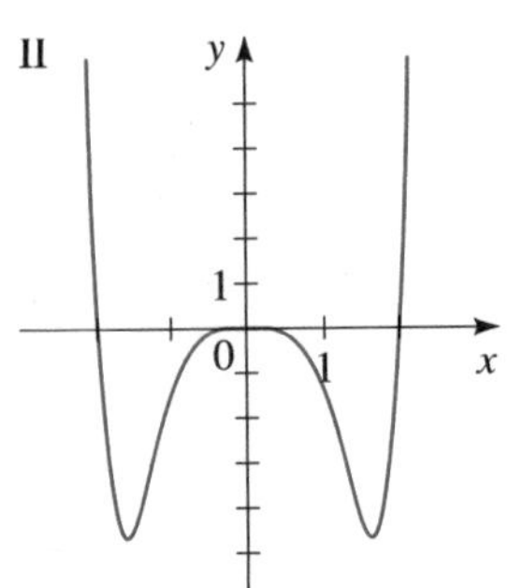

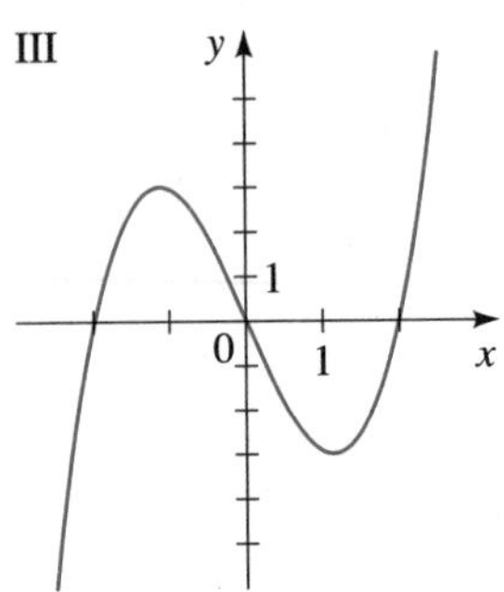

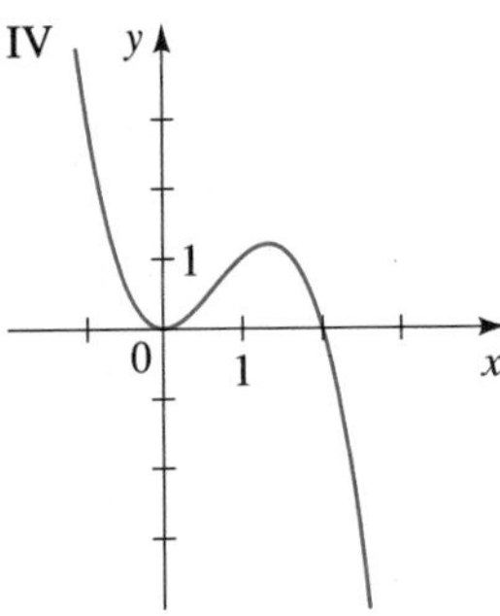

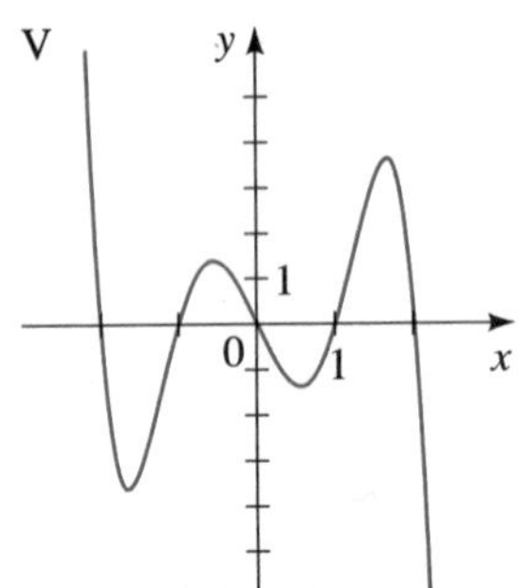

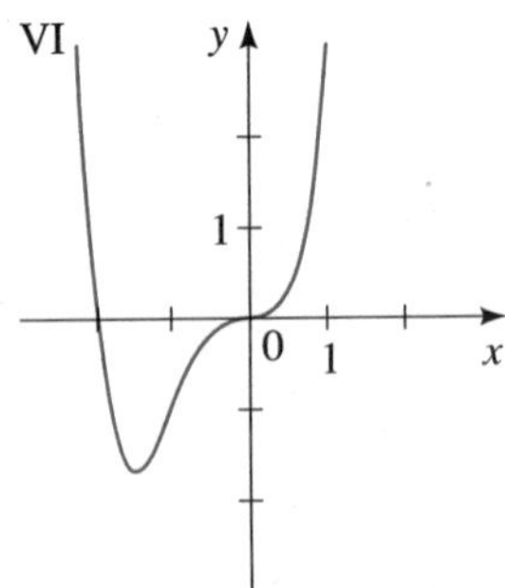

41–46 ■ Determine the end behavior of P. Compare the graphs of P and Q on a large and small viewing rectangle, as in Example 2.

41. $P(x) = 3x^3 - x^2 + 5x + 1; \quad Q(x) = 3x^3$

42. $P(x) = -\frac{1}{8}x^3 + \frac{1}{4}x^2 + 12x; \quad Q(x) = -\frac{1}{8}x^3$

43. $P(x) = x^4 - 7x^2 + 5x + 5; \quad Q(x) = x^4$

44. $P(x) = -x^5 + 2x^2 + x; \quad Q(x) = -x^5$

45. $P(x) = x^{11} - 9x^9; \quad Q(x) = x^{11}$

46. $P(x) = 2x^2 - x^{12}; \quad Q(x) = -x^{12}$

47–54 ■ Graph the polynomial in the given viewing rectangle. Find the coordinates of all local extrema. State each answer correct to two decimal places.

47. $y = -x^2 + 8x, \quad [-4, 12]$ by $[-50, 30]$

48. $y = x^3 - 3x^2, \quad [-2, 5]$ by $[-10, 10]$

49. $y = x^3 - 12x + 9, \quad [-5, 5]$ by $[-30, 30]$

50. $y = 2x^3 - 3x^2 - 12x - 32, \quad [-5, 5]$ by $[-60, 30]$

51. $y = x^4 + 4x^3, \quad [-5, 5]$ by $[-30, 30]$

52. $y = x^4 - 18x^2 + 32, \quad [-5, 5]$ by $[-100, 100]$

53. $y = 3x^5 - 5x^3 + 3, \quad [-3, 3]$ by $[-5, 10]$

54. $y = x^5 - 5x^2 + 6, \quad [-3, 3]$ by $[-5, 10]$

55–64 ■ Graph the polynomial and determine how many local maxima and minima it has.

55. $y = -2x^2 + 3x + 5$ **56.** $y = x^3 + 12x$

57. $y = x^3 - x^2 - x$ **58.** $y = 6x^3 + 3x + 1$

59. $y = x^4 - 5x^2 + 4$

60. $y = 1.2x^5 + 3.75x^4 - 7x^3 - 15x^2 + 18x$

61. $y = (x - 2)^5 + 32$

62. $y = (x^2 - 2)^3$

63. $y = x^8 - 3x^4 + x$

64. $y = \frac{1}{3}x^7 - 17x^2 + 7$

65–70 ■ Graph the family of polynomials in the same viewing rectangle, using the given values of c. Explain how changing the value of c affects the graph.

65. $P(x) = cx^3; \quad c = 1, 2, 5, \frac{1}{2}$

66. $P(x) = (x - c)^4; \quad c = -1, 0, 1, 2$

67. $P(x) = x^4 + c; \quad c = -1, 0, 1, 2$

68. $P(x) = x^3 + cx; \quad c = 2, 0, -2, -4$

69. $P(x) = x^4 - cx; \quad c = 0, 1, 8, 27$

70. $P(x) = x^c; \quad c = 1, 3, 5, 7$

71. (a) On the same coordinate axes, sketch graphs (as accurately as possible) of the functions

$$y = x^3 - 2x^2 - x + 2 \quad \text{and} \quad y = -x^2 + 5x + 2$$

(b) Based on your sketch in part (a), at how many points do the two graphs appear to intersect?

(c) Find the coordinates of all intersection points.

72. Portions of the graphs of $y = x^2$, $y = x^3$, $y = x^4$, $y = x^5$, and $y = x^6$ are plotted in the figures. Determine which function belongs to each graph.

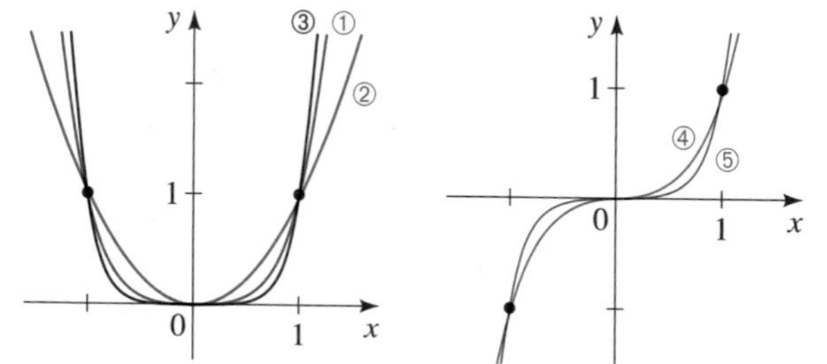

73. Recall that a function f is *odd* if $f(-x) = -f(x)$ or *even* if $f(-x) = f(x)$ for all real x.

(a) Show that if f and g are both odd, then so is the function $f + g$.

(b) Show that if f and g are both even, then so is the function $f + g$.

(c) Show that if f is odd and g is even, and neither is the constant function 0, then the function $f + g$ is neither even nor odd.

(d) Show that a polynomial $P(x)$ that contains only odd powers of x is an odd function.

(e) Show that a polynomial $P(x)$ that contains only even powers of x is an even function.

(f) Show that if a polynomial $P(x)$ contains both odd and even powers of x, then it is neither an odd nor an even function.

(g) Express the function

$$P(x) = x^5 + 6x^3 - x^2 - 2x + 5$$

as the sum of an odd function and an even function.

74. An open box is to be constructed from a piece of cardboard 20 cm by 40 cm by cutting squares of side length x from each corner and folding up the sides, as shown in the figure.

(a) Express the volume V of the box as a function of x.

(b) What is the domain of V? (Use the fact that length and volume must be positive.)

(c) Draw a graph of the function V and use it to estimate the maximum volume for such a box.

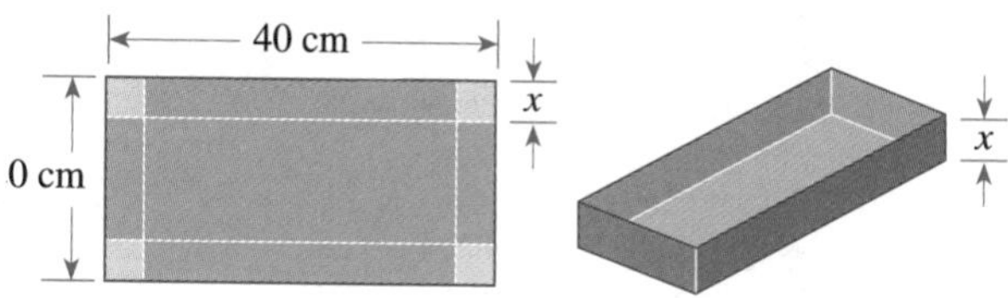

75. A cardboard box has a square base, with each edge of the base having length x inches, as shown in the figure. The total length of all 12 edges of the box is 144 in.

(a) Show that the volume of the box is given by the function $V(x) = 2x^2(18 - x)$.

(b) What is the domain of V? (Use the fact that length and volume must be positive.)

(c) Draw a graph of the function V and use it to estimate the maximum volume for such a box.

76. A market analyst working for a small-appliance manufacturer finds that if the firm produces and sells x blenders annually, the total profit (in dollars) is

$$P(x) = 8x + 0.3x^2 - 0.0013x^3 - 372$$

Graph the function P in an appropriate viewing rectangle and use the graph to answer the following questions.

(a) When just a few blenders are manufactured, the firm loses money (profit is negative). [For example, $P(10) = -263.3$, so the firm loses \$263.30 if it produces and sells only 10 blenders.] How many blenders must the firm produce to break even?

(b) Does profit increase indefinitely as more blenders are produced and sold? If not, what is the largest possible profit the firm could have?

77. The rabbit population on a small island is observed to be given by the function

$$P(t) = 120t - 0.4t^4 + 1000$$

where t is the time (in months) since observations of the island began.

(a) When is the maximum population attained, and what is that maximum population?

(b) When does the rabbit population disappear from the island?

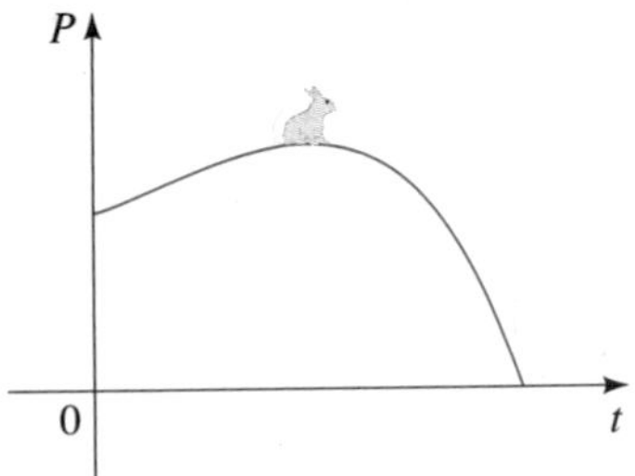

78. (a) Graph the function $P(x) = (x - 1)(x - 3)(x - 4)$ and find all local extrema, correct to the nearest tenth.

(b) Graph the function

$$Q(x) = (x - 1)(x - 3)(x - 4) + 5$$

and use your answers to part (a) to find all local extrema, correct to the nearest tenth.

(c) If $a < b < c$, explain why the function

$$P(x) = (x - a)(x - b)(x - c)$$

must have two local extrema.

(d) If $a < b < c$ and d is any real number, explain why the function $Q(x) = (x - a)(x - b)(x - c) + d$ must have two local extrema.

79. (a) How many x-intercepts and how many local extrema does the polynomial $P(x) = x^3 - 4x$ have?

(b) How many x-intercepts and how many local extrema does the polynomial $Q(x) = x^3 + 4x$ have?

(c) If $a > 0$, how many x-intercepts and how many local extrema does each of the polynomials $P(x) = x^3 - ax$ and $Q(x) = x^3 + ax$ have? Explain your answer.

DISCOVERY • DISCUSSION

80. **Graphs of Large Powers** Graph the functions $y = x^2$, $y = x^3$, $y = x^4$, and $y = x^5$, for $-1 \leqslant x \leqslant 1$, on the same coordinate axes. What do you think the graph of $y = x^{100}$ would look like on this same interval? What about $y = x^{101}$? Make a table of values to confirm your answers.

81. **Maximum Number of Local Extrema** What is the smallest possible degree that the polynomial whose graph is shown can have? Explain.

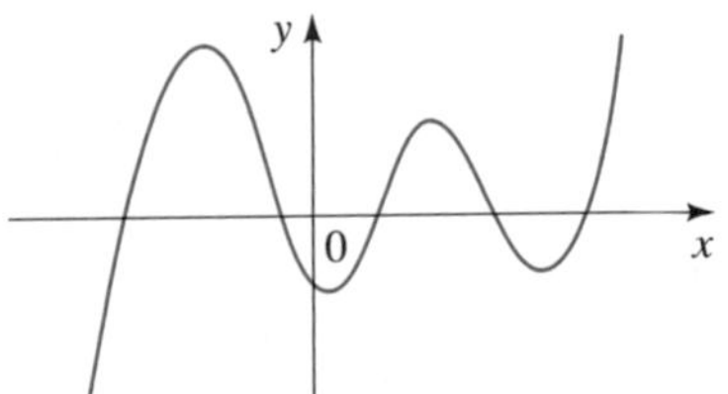

82. **Possible Number of Local Extrema** Is it possible for a third-degree polynomial to have exactly one local extremum? Can a fourth-degree polynomial have exactly two local extrema? How many local extrema can polynomials of third, fourth, fifth, and sixth degree have? (Think about the end behavior of such polynomials.) Now give an example of a polynomial that has six local extrema.

83. **Impossible Situation?** Is it possible for a polynomial to have two local maxima and no local minimum? Explain.

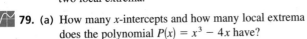

5.2 DIVIDING POLYNOMIALS

So far in this chapter we have been studying polynomial functions *graphically*. In this section we begin to study polynomials *algebraically*. Most of our work will be concerned with factoring polynomials, and to factor, we need to know how to divide polynomials.

LONG DIVISION OF POLYNOMIALS

Long division for polynomials is very much like long division for numbers. For example, to divide $6x^2 - 26x + 12$ (the **dividend**) by $x - 4$ (the **divisor**), we arrange our work as follows.

$$\begin{array}{r|l} & \text{quotient} \downarrow \\ & 6x - 2 \\ \text{divisor} \rightarrow & x - 4\overline{)6x^2 - 26x + 12} \leftarrow \text{dividend} \\ & \underline{6x^2 - 24x} \quad \text{Multiply divisor by } 6x \\ & -2x + 12 \quad \text{Subtract, then ``bring down'' 12} \\ & \underline{-2x + 8} \quad \text{Multiply divisor by } -2 \\ & 4 \leftarrow \text{remainder} \quad \text{Subtract} \end{array}$$

The division process ends when the last line is of lesser degree than the divisor. The last line then contains the **remainder**, and the top line contains the **quotient**. The result of the division can be interpreted in either of two ways.

$$\frac{6x^2 - 26x + 12}{x - 4} = 6x - 2 + \frac{4}{x - 4}$$

or

$$6x^2 - 26x + 12 = (x - 4)(6x - 2) + 4$$

We summarize what happens in any long division problem in the following theorem.

DIVISION ALGORITHM

If $P(x)$ and $D(x)$ are polynomials, with $D(x) \neq 0$, then there exist unique polynomials $Q(x)$ and $R(x)$ such that

$$P(x) = D(x) \cdot Q(x) + R(x)$$

where $R(x)$ is either 0 or of degree less than the degree of $D(x)$. The polynomials $P(x)$ and $D(x)$ are called the **dividend** and **divisor**, respectively, $Q(x)$ is the **quotient**, and $R(x)$ is the **remainder**.

EXAMPLE 1 ■ Long Division of Polynomials

Let $P(x) = 8x^4 + 6x^2 - 3x + 1$ and $D(x) = 2x^2 - x + 2$. Find polynomials $Q(x)$ and $R(x)$ such that $P(x) = D(x) \cdot Q(x) + R(x)$.

MATHEMATICS IN THE MODERN WORLD

FAIR DIVISION OF ASSETS

Dividing an asset fairly among a number of people is of great interest to mathematicians. Problems of this nature include dividing the national budget, disputed land, or assets in divorce cases. In 1994 Brams and Taylor found a mathematical way of dividing things fairly. Their solution has been applied to division problems in political science, legal proceedings, and other areas. To understand the problem, consider the following example. Suppose persons A and B want to divide a property fairly between them. To divide it *fairly* means that both A and B must be satisfied with the outcome of the division. Solution: A gets to divide the property into two pieces, then B gets to choose the piece he wants. Since both A and B had a part in the division process, each should be satisfied. The situation becomes much more complicated if three or more people are involved (and that's where mathematics comes in). Dividing things fairly involves much more than simply cutting things in half; it must take into account the *relative worth* each person attaches to the thing being divided. A story from the Bible illustrates this clearly. Two women appear before King Solomon, each claiming to be the mother of the same newborn baby. King Solomon's solution is to divide the baby in half! The real mother, who attaches far more worth to the baby than anyone,

(continued)

SOLUTION

We use long division after first inserting the term $0x^3$ into the dividend to ensure that the columns line up correctly.

$$
\begin{array}{r}
4x^2 + 2x \phantom{{}- 3x + 1} \\
2x^2 - x + 2\overline{)\,8x^4 + 0x^3 + 6x^2 - 3x + 1} \\
\underline{8x^4 - 4x^3 + 8x^2} \phantom{{}- 3x + 1} \\
4x^3 - 2x^2 - 3x \phantom{{}+ 1} \\
\underline{4x^3 - 2x^2 + 4x} \phantom{{}+ 1} \\
-7x + 1
\end{array}
$$

The process is complete at this point because $-7x + 1$ is of lesser degree than the divisor $2x^2 - x + 2$. From the long division table we see that $Q(x) = 4x^2 + 2x$ and $R(x) = -7x + 1$, so

$$8x^4 + 6x^2 - 3x + 1 = (2x^2 - x + 2)(4x^2 + 2x) + (-7x + 1)$$ ■

SYNTHETIC DIVISION

Synthetic division is a quick method of dividing polynomials; it can be used when the divisor is of the form $x - c$. In synthetic division we write only the essential part of the long division table. Compare these long division and synthetic division tables, in which we divide $2x^3 - 7x^2 + 5$ by $x - 3$:

Long Division

$$
\begin{array}{r}
2x^2 - x - 3 \phantom{{}+ 5} \\
x - 3\overline{)\,2x^3 - 7x^2 + 0x + 5} \\
\underline{2x^3 - 6x^2} \phantom{{}+ 0x + 5} \\
-x^2 + 0x \phantom{{}+ 5} \\
\underline{-x^2 + 3x} \phantom{{}+ 5} \\
-3x + 5 \\
\underline{-3x + 9} \\
-4
\end{array}
\leftarrow \text{remainder}
$$

Synthetic Division

$$
\begin{array}{r|rrrr}
3 & 2 & -7 & 0 & 5 \\
 & & 6 & -3 & -9 \\
\hline
 & 2 & -1 & -3 & -4
\end{array}
$$

quotient: $2 \quad -1 \quad -3$; remainder: -4

Note that in synthetic division we abbreviate $2x^3 - 7x^2 + 5$ by writing only the coefficients: $2 \quad -7 \quad 0 \quad 5$, and instead of $x - 3$, we simply write 3. (Writing 3 instead of -3 allows us to add instead of subtract, but this changes the sign of all the numbers that appear in the gold boxes.)

Here is how you get the synthetic division table in practice. Start by writing the divisor and dividend:

$$
\begin{array}{r|rrrr}
3 & 2 & -7 & 0 & 5 \\
\hline
\end{array}
$$

immediately gives up her claim to the baby in order to save its life.

Mathematical solutions to fair-division problems have recently been applied in an international treaty, the Convention on the Law of the Sea. If a country wants to develop a portion of the sea floor, it is required to divide the portion into two parts, one part to be used by itself, the other by a consortium that will preserve it for later use by a less developed country. The consortium gets first pick.

Bring down the 2, multiply $3 \cdot 2 = 6$, and write the result in the middle row. Then add:

$$\begin{array}{r|rrrr} 3 & 2 & -7 & 0 & 5 \\ & & 6 & & \\ \hline & \boxed{2} & -1 & & \end{array}$$

Repeat this process of multiplying and then adding until the table is complete.

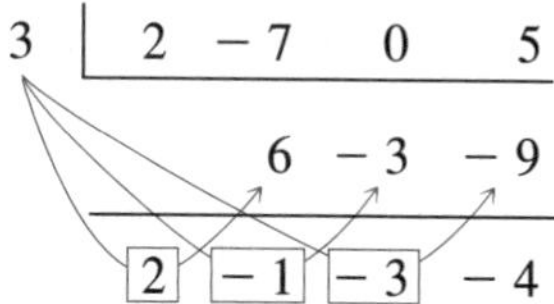

SYNTHETIC DIVISION

To divide $a_nx^n + a_{n-1}x^{n-1} + \cdots + a_1x + a_0$ by $x - c$, we proceed as follows:

$$\begin{array}{c|cccccccc} c & a_n & a_{n-1} & a_{n-2} & a_{n-3} & \cdots & a_2 & a_1 & a_0 \\ & & cb_{n-1} & cb_{n-2} & cb_{n-3} & \cdots & cb_2 & cb_1 & cb_0 \\ \hline & b_{n-1} & b_{n-2} & b_{n-3} & b_{n-4} & \cdots & b_1 & b_0 & r \end{array}$$

Here $b_{n-1} = a_n$, and each number in the bottom row is obtained by adding the numbers above it. The remainder is r and the quotient is

$$b_{n-1}x^{n-1} + b_{n-2}x^{n-2} + \cdots + b_1x + b_0$$

THE REMAINDER AND FACTOR THEOREMS

The next theorem shows how synthetic division can be used to evaluate polynomials easily.

REMAINDER THEOREM

If the polynomial $P(x)$ is divided by $x - c$, then the remainder is the value $P(c)$.

■ **Proof** If the divisor in the Division Algorithm is of the form $x - c$ for some real number c, then the remainder must be a constant (since the degree of the remainder is less than the degree of the divisor). If we call this constant r, then

$$P(x) = (x - c) \cdot Q(x) + r$$

Setting $x = c$ in this equation, we get $P(c) = (c - c) \cdot Q(x) + r = 0 + r = r$, that is, $P(c)$ is the remainder r. ☐

EXAMPLE 2 ■ Using the Remainder Theorem to Find the Value of a Polynomial

Let $P(x) = 3x^5 + 5x^4 - 4x^3 + 7x + 3$.

(a) Find the quotient and remainder when $P(x)$ is divided by $x + 2$.
(b) Use the Remainder Theorem to find $P(-2)$.

SOLUTION

(a) Since $x + 2 = x - (-2)$, the synthetic division table for this problem takes the following form.

$$\begin{array}{r|rrrrrr} -2 & 3 & 5 & -4 & 0 & 7 & 3 \\ & & -6 & 2 & 4 & -8 & 2 \\ \hline & 3 & -1 & -2 & 4 & -1 & 5 \end{array}$$

The quotient is $3x^4 - x^3 - 2x^2 + 4x - 1$ and the remainder is 5. Thus

$$3x^5 + 5x^4 - 4x^3 + 7x + 3 = (x + 2)(3x^4 - x^3 - 2x^2 + 4x - 1) + 5$$

or

$$\frac{3x^5 + 5x^4 - 4x^3 + 7x + 3}{x + 2} = 3x^4 - x^3 - 2x^2 + 4x - 1 + \frac{5}{x + 2}$$

(b) By the Remainder Theorem, $P(-2)$ is the remainder when $P(x)$ is divided by $x - (-2) = x + 2$. From part (a) the remainder is 5, so $P(-2) = 5$. ■

The next theorem says that *zeros* of polynomials correspond to *factors*; we used this fact in Section 5.1 to graph polynomials.

FACTOR THEOREM

c is a zero of P if and only if $x - c$ is a factor of $P(x)$.

■ **Proof** If $P(x)$ factors as $P(x) = (x - c) \cdot Q(x)$, then

$$P(c) = (c - c) \cdot Q(c) = 0 \cdot Q(c) = 0$$

Conversely, if $P(c) = 0$, then by the Remainder Theorem

$$P(x) = (x - c) \cdot Q(x) + 0 = (x - c) \cdot Q(x)$$

so $x - c$ is a factor of $P(x)$. ☐

EXAMPLE 3 ■ Factoring a Polynomial Using the Factor Theorem

Let $P(x) = x^3 - 7x + 6$. Show that $P(1) = 0$, and use this fact to factor $P(x)$ completely.

$$\begin{array}{r|rrrr} 1 & 1 & 0 & -7 & 6 \\ & & 1 & 1 & -6 \\ \hline & 1 & 1 & -6 & 0 \end{array}$$

SOLUTION

Substituting, we see that $P(1) = 1^3 - 7 \cdot 1 + 6 = 0$. By the Factor Theorem, this means that $x - 1$ is a factor of $P(x)$. Using synthetic or long division (shown in the margin), we see that

$$\begin{array}{r} x^2 + x - 6 \\ x - 1 \overline{) x^3 + 0x^2 - 7x + 6} \\ x^3 - x^2 \\ \hline x^2 - 7x \\ x^2 - x \\ \hline -6x + 6 \\ -6x + 6 \\ \hline 0 \end{array}$$

$$\begin{aligned} P(x) &= x^3 - 7x + 6 \\ &= (x - 1)(x^2 + x - 6) && \text{See margin} \\ &= (x - 1)(x - 2)(x + 3) && \text{Factor the quadratic } x^2 + x - 6 \end{aligned}$$

■

EXAMPLE 4 ■ Finding a Polynomial with Specified Zeros

Find a polynomial $P(x)$ of degree 4 that has zeros -3, 0, 1, and 5.

SOLUTION

By the Factor Theorem, $x - (-3)$, $x - 0$, $x - 1$, and $x - 5$ must all be factors of the desired polynomial, so let

$$P(x) = (x + 3)(x - 0)(x - 1)(x - 5) = x^4 - 3x^3 - 13x^2 + 15x$$

Since $P(x)$ is to have degree 4, any other solution of the problem must be a constant multiple of the polynomial we have chosen (because multiplication by any polynomial other than a constant will increase the degree). ■

The polynomial P of Example 4 is graphed in Figure 1. Note that the zeros of P correspond to the x-intercepts of the graph.

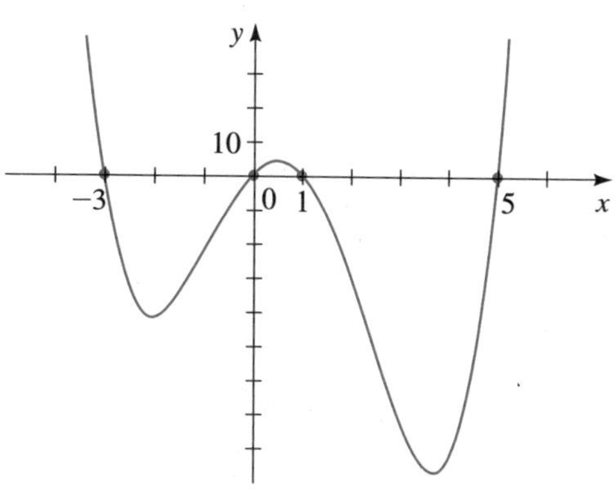

FIGURE 1

$P(x) = (x + 3)x(x - 1)(x - 5)$ has zeros -3, 0, 1, and 5

5.2 EXERCISES

1–10 ■ Find the quotient and remainder using long division.

1. $\dfrac{x^2 + 4x - 8}{x + 3}$

2. $\dfrac{x^3 - x^2 - 2x + 6}{x - 2}$

3. $\dfrac{x^3 + 6x + 5}{x - 4}$

4. $\dfrac{x^3 + 3x^2 + 4x + 3}{3x + 6}$

5. $\dfrac{x^3 + 6x + 3}{x^2 - 2x + 2}$

6. $\dfrac{3x^4 - 5x^3 - 20x - 5}{x^2 + x + 3}$

7. $\dfrac{6x^3 + 2x^2 + 22x}{2x^2 + 5}$

8. $\dfrac{9x^2 - x + 5}{3x^2 - 7x}$

9. $\dfrac{x^6 + x^4 + x^2 + 1}{x^2 + 1}$

10. $\dfrac{2x^5 - 7x^4 - 13}{4x^2 - 6x + 8}$

11–24 ■ Find the quotient and remainder using synthetic division.

11. $\dfrac{x^2 - 5x + 4}{x - 3}$

12. $\dfrac{x^2 - 5x + 4}{x - 1}$

13. $\dfrac{3x^2 + 5x}{x - 6}$

14. $\dfrac{4x^2 - 3}{x + 5}$

15. $\dfrac{x^3 + 2x^2 + 2x + 1}{x + 2}$

16. $\dfrac{3x^3 - 12x^2 - 9x + 1}{x - 5}$

17. $\dfrac{x^3 - 8x + 2}{x + 3}$

18. $\dfrac{x^4 - x^3 + x^2 - x + 2}{x - 2}$

19. $\dfrac{x^5 + 3x^3 - 6}{x - 1}$

20. $\dfrac{x^3 - 9x^2 + 27x - 27}{x - 3}$

21. $\dfrac{2x^3 + 3x^2 - 2x + 1}{x - \frac{1}{2}}$

22. $\dfrac{6x^4 + 10x^3 + 5x^2 + x + 1}{x + \frac{2}{3}}$

23. $\dfrac{x^3 - 27}{x - 3}$

24. $\dfrac{x^4 - 16}{x + 2}$

25–37 ■ Use synthetic division and the Remainder Theorem to evaluate $P(c)$.

25. $P(x) = 4x^2 + 12x + 5, \quad c = -1$

26. $P(x) = 2x^2 + 9x + 1, \quad c = \frac{1}{2}$

27. $P(x) = x^3 + 3x^2 - 7x + 6, \quad c = 2$

28. $P(x) = x^3 - x^2 + x + 5, \quad c = -1$

29. $P(x) = x^3 + 2x^2 - 7, \quad c = -2$

30. $P(x) = 2x^3 - 21x^2 + 9x - 200, \quad c = 11$

31. $P(x) = 5x^4 + 30x^3 - 40x^2 + 36x + 14, \quad c = -7$

32. $P(x) = 6x^5 + 10x^3 + x + 1, \quad c = -2$

33. $P(x) = x^7 - 3x^2 - 1, \quad c = 3$

34. $P(x) = -2x^6 + 7x^5 + 40x^4 - 7x^2 + 10x + 112,$ $c = -3$

35. $P(x) = 3x^3 + 4x^2 - 2x + 1, \quad c = \frac{2}{3}$

36. $P(x) = x^3 - x + 1, \quad c = \frac{1}{4}$

37. $P(x) = x^3 + 2x^2 - 3x - 8, \quad c = 0.1$

38. Let

$$P(x) = 6x^7 - 40x^6 + 16x^5 - 200x^4 - 60x^3 - 69x^2 + 13x - 139$$

Calculate $P(7)$ by **(a)** using synthetic division and **(b)** substituting $x = 7$ into the polynomial and evaluating directly.

39–42 ■ Use the Factor Theorem to show that $x - c$ is a factor of $P(x)$ for the given value(s) of c.

39. $P(x) = x^3 - 3x^2 + 3x - 1, \quad c = 1$

40. $P(x) = x^3 + 2x^2 - 3x - 10, \quad c = 2$

41. $P(x) = 2x^3 + 7x^2 + 6x - 5, \quad c = \frac{1}{2}$

42. $P(x) = x^4 + 3x^3 - 16x^2 - 27x + 63, \quad c = 3, -3$

43–44 ■ Show that the given value(s) of c are zeros of $P(x)$, and find all other zeros of $P(x)$.

43. $P(x) = x^3 - x^2 - 11x + 15, \quad c = 3$

44. $P(x) = 3x^4 - x^3 - 21x^2 - 11x + 6, \quad c = \frac{1}{3}, -2$

45–48 ■ Find a polynomial of the specified degree that has the given zeros.

45. Degree 3; zeros $-1, 1, 3$

46. Degree 4; zeros $-2, 0, 2, 4$

47. Degree 4; zeros $-1, 1, 3, 5$

48. Degree 5; zeros $-2, -1, 0, 1, 2$

49. Find a polynomial of degree 3 that has zeros 1, -2, and 3, and in which the coefficient of x^2 is 3.

50. Find a polynomial of degree 4 that has integer coefficients and zeros 1, -1, 2, and $\frac{1}{2}$.

DISCOVERY • DISCUSSION

51. Impossible Division? Suppose you were asked to solve the following two problems on a test:

A. Find the remainder when $6x^{1000} - 17x^{562} + 12x + 26$ is divided by $x + 1$.

B. Is $x - 1$ a factor of $x^{567} - 3x^{400} + x^9 + 2$?

Obviously, it's impossible to solve these problems by dividing, because the polynomials are of such large degree. Use one or more of the theorems in this section to solve these problems *without* actually dividing.

52. Nested Form of a Polynomial Expand Q to prove that the polynomials P and Q are the same.

$$P(x) = 3x^4 - 5x^3 + x^2 - 3x + 5$$

$$Q(x) = (((3x - 5)x + 1)x - 3)x + 5$$

Try to evaluate $P(2)$ and $Q(2)$ in your head, using the forms given. Which is easier? Now write the polynomial $R(x) = x^5 - 2x^4 + 3x^3 - 2x^2 + 3x + 4$ in "nested" form, like the polynomial Q. Use the nested form to find $R(3)$ in your head.

Do you see how calculating with the nested form follows the same arithmetic steps as calculating the value of a polynomial using synthetic division?

5.3 REAL ZEROS OF POLYNOMIALS

The Factor Theorem tells us that finding the zeros of a polynomial is really the same thing as factoring it into linear factors. In this section we study some algebraic methods that help us find the real zeros of a polynomial, and thereby factor the polynomial. We begin with the *rational* zeros of a polynomial.

RATIONAL ZEROS OF POLYNOMIALS

To help us understand the next theorem, let's consider the polynomial

$$P(x) = (x - 2)(x - 3)(x + 4) \qquad \text{Factored form}$$

$$= x^3 - x^2 - 14x + 24 \qquad \text{Expanded form}$$

From the factored form we see that the zeros of P are 2, 3, and -4. When the polynomial is expanded, the constant 24 is obtained by multiplying $(-2) \times (-3) \times 4$. This means that the zeros of the polynomial are all factors of the constant term. The following generalizes this observation.

RATIONAL ZEROS THEOREM

If the polynomial $P(x) = a_n x^n + a_{n-1}x^{n-1} + \cdots + a_1 x + a_0$ has integer coefficients, then every rational zero of P is of the form

$$\frac{p}{q}$$

where p is a factor of the constant coefficient a_0

and q is a factor of the leading coefficient a_n.

Evariste Galois (1811–1832) is one of the very few mathematicians to have an entire theory named in his honor. Not yet 21 when he died, he completely settled the central problem in the theory of equations by describing a criterion that reveals whether a polynomial equation can be solved by algebraic operations. Galois was one of the greatest mathematicians in the world at that time, although no one knew it but him. He repeatedly sent his work to the eminent mathematicians Cauchy and Poisson, who either lost his letters or did not understand his ideas. Galois wrote in a terse style and included few details, which probably played a role in his failure to pass the entrance exams at the Ecole Polytechnique in Paris. A political radical, Galois spent several months in prison for his revolutionary activities. His brief life came to a tragic end when he was killed in a duel over a love affair. The night before his duel, fearing he would die, Galois wrote down the essence of his ideas and entrusted them to his friend Auguste Chevalier. He concluded by writing ". . . there will, I hope, be people who will find it to their advantage to decipher all this mess." The mathematician Camille Jordan did just that, 14 years later.

■ **Proof** If p/q is a rational zero, in lowest terms, of the polynomial P, then we have

$$a_n\left(\frac{p}{q}\right)^n + a_{n-1}\left(\frac{p}{q}\right)^{n-1} + \cdots + a_1\left(\frac{p}{q}\right) + a_0 = 0$$

$$a_np^n + a_{n-1}p^{n-1}q + \cdots + a_1pq^{n-1} + a_0q^n = 0 \qquad \text{Multiply by } q^n$$

$$p(a_np^{n-1} + a_{n-1}p^{n-2}q + \cdots + a_1q^{n-1}) = -a_0q^n \qquad \text{Subtract } a_0q^n \text{ and factor LHS}$$

Now p is a factor of the left side, so it must be a factor of the right as well. Since p/q is in lowest terms, p and q have no factor in common, and so p must be a factor of a_0. A similar proof shows that q is a factor of a_n. □

If $a_n = 1$ in the Rational Zeros Theorem, then q must be 1 or -1, so in this case, any rational zero p/q is in fact an *integer* factor of a_0. Thus, if the highest-power term in a polynomial is just x^n, all its rational zeros are integers, not fractions.

EXAMPLE 1 ■ Using the Rational Zeros Theorem to Factor a Polynomial

Factor the polynomial $P(x) = 2x^3 + x^2 - 13x + 6$.

SOLUTION

By the Rational Zeros Theorem, if p/q is a zero of $P(x)$, then p divides 6 and q divides 2, so p/q is of the form

$$\frac{\text{factor of 6}}{\text{factor of 2}}$$

The factors of 6 are $\pm1, \pm2, \pm3, \pm6$, and the factors of 2 are $\pm1, \pm2$. Thus, the possible values of p/q are

$$\pm\frac{1}{1}, \quad \pm\frac{2}{1}, \quad \pm\frac{3}{1}, \quad \pm\frac{6}{1}, \quad \pm\frac{1}{2}, \quad \pm\frac{2}{2}, \quad \pm\frac{3}{2}, \quad \pm\frac{6}{2}$$

Simplifying the fractions and eliminating duplicates, we get the following list of possible values of p/q:

$$\pm1, \quad \pm2, \quad \pm3, \quad \pm6, \quad \pm\frac{1}{2}, \quad \pm\frac{3}{2}$$

Now we check which of these *possible* zeros actually *are* zeros by substituting them, one at a time, into the polynomial P until we find one that makes $P(x) = 0$. We have

$$P(1) = 2(1)^3 + (1)^2 - 13(1) + 6 = -4 \qquad 1 \text{ is } \textit{not} \text{ a zero of } P$$

$$P(2) = 2(2)^3 + (2)^2 - 13(2) + 6 = 0 \qquad 2 \textit{ is} \text{ a zero of } P$$

Since $x = 2$ is a zero of P, it follows that $x - 2$ is a factor of $P(x)$. Using synthetic division (shown in the margin), we obtain the following factorization:

2	2	1	−13	6
		4	10	−6
	2	5	−3	0

$$\begin{aligned} P(x) &= 2x^3 + x^2 - 13x + 6 \\ &= (x-2)(2x^2 + 5x - 3) && \text{See margin} \\ &= (x-2)(2x-1)(x+3) && \text{Factor } 2x^2 + 5x - 3 \end{aligned}$$

■

The following box explains how we use the Rational Zeros Theorem with synthetic division to factor a polynomial.

FINDING THE RATIONAL ZEROS OF A POLYNOMIAL

1. POSSIBLE ZEROS. List all possible rational zeros using the Rational Zeros Theorem.

2. DIVIDE. Use synthetic division to evaluate the polynomial at each of the candidates for rational zeros that you found in Step 1. When the remainder is 0, note the quotient you have obtained.

3. REPEAT. Repeat Steps 1 and 2 for the quotient. Stop when you reach a quotient that is quadratic or factors easily, and use the quadratic formula or factor to find the remaining zeros.

EXAMPLE 2 ■ Using the Rational Zeros Theorem and the Quadratic Formula

Let $P(x) = x^4 - 5x^3 - 5x^2 + 23x + 10$.

(a) Find the zeros of P.
(b) Sketch the graph of P.

SOLUTION

1	1	−5	−5	23	10
		1	−4	−9	14
	1	−4	−9	14	24

2	1	−5	−5	23	10
		2	−6	−22	2
	1	−3	−11	1	12

5	1	−5	−5	23	10
		5	0	−25	−10
	1	0	−5	−2	0

(a) The leading coefficient of P is 1, so all the rational zeros are integers: They are divisors of the constant term 10. Thus, the possible candidates are

$$\pm 1, \quad \pm 2, \quad \pm 5, \quad \pm 10$$

Using synthetic division (see the margin) we find that 1 and 2 are not zeros, but that 5 is a zero and that P factors as

$$x^4 - 5x^3 - 5x^2 + 23x + 10 = (x-5)(x^3 - 5x - 2)$$

We now try to factor the quotient $x^3 - 5x - 2$. Its possible zeros are the divisors of -2, namely,

$$\pm 1, \quad \pm 2$$

Since we already know that 1 and 2 are not zeros of the original polynomial P, we don't need to try them again. Checking the remaining candidates -1 and -2, we see that -2 is a zero (see the margin), and P factors as

$$\begin{array}{r|rrrr} -2 & 1 & 0 & -5 & -2 \\ & & -2 & 4 & 2 \\ \hline & 1 & -2 & -1 & 0 \end{array}$$

$$\begin{aligned} x^4 - 5x^3 - 5x^2 + 23x + 10 &= (x - 5)(x^3 - 5x - 2) \\ &= (x - 5)(x + 2)(x^2 - 2x - 1) \end{aligned}$$

Now we use the quadratic formula to obtain the two remaining zeros of P:

$$x = \frac{2 \pm \sqrt{(-2)^2 - 4(1)(-1)}}{2} = 1 \pm \sqrt{2}$$

The solutions of the equation are $5, -2, 1 + \sqrt{2}$, and $1 - \sqrt{2}$.

(b) Now that we know the zeros of P, we can use the methods of Section 5.1 to sketch the graph. If we want to use a graphing calculator instead, knowing the zeros allows us to choose an appropriate viewing rectangle—one that is wide enough to contain all the x-intercepts of P. Numerical approximations to the zeros of P are

$$5, \qquad -2, \qquad 2.4, \qquad \text{and} \qquad -0.4$$

So in this case we choose the rectangle $[-3, 6]$ by $[-50, 50]$ and draw the graph shown in Figure 1.

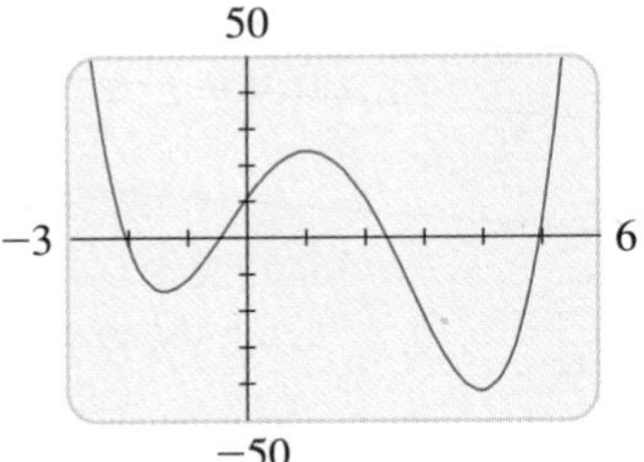

FIGURE 1
$P(x) = x^4 - 5x^3 - 5x^2 + 23x + 10$

■

DESCARTES' RULE OF SIGNS AND UPPER AND LOWER BOUNDS FOR ROOTS

In some cases, the following rule—discovered by the French philosopher and mathematician René Descartes around 1637 (see page 76)—is helpful in eliminating candidates from lengthy lists of possible rational roots. To describe this rule, we need the concept of *variation in sign*. If $P(x)$ is a polynomial with real coefficients, written with descending powers of x (and omitting powers with coefficient 0), then a **variation in sign** occurs whenever adjacent coefficients have opposite signs. For example

$$P(x) = 5x^7 - 3x^5 - x^4 + 2x^2 + x - 3$$

has three variations in sign.

Polynomial	Variations in sign
$x^2 + 4x + 1$	0
$2x^3 + x - 6$	1
$x^4 - 3x^2 - x + 4$	2

DESCARTES' RULE OF SIGNS

Let P be a polynomial with real coefficients.

1. The number of positive real zeros of $P(x)$ is either equal to the number of variations in sign in $P(x)$ or is less than that by an even whole number.
2. The number of negative real zeros of $P(x)$ is either equal to the number of variations in sign in $P(-x)$ or is less than that by an even whole number.

EXAMPLE 3 ■ Using Descartes' Rule

Use Descartes' Rule of Signs to determine the possible number of positive and negative real zeros of the polynomial

$$P(x) = 3x^6 + 4x^5 + 3x^3 - x - 3$$

SOLUTION

The polynomial has one variation in sign and so it has one positive zero. Now

$$\begin{aligned} P(-x) &= 3(-x)^6 + 4(-x)^5 + 3(-x)^3 - (-x) - 3 \\ &= 3x^6 - 4x^5 - 3x^3 + x - 3 \end{aligned}$$

So, $P(-x)$ has three variations in sign. Thus, $P(x)$ has either three or one negative zero(s), making a total of either two or four real zeros. ■

We say that a is a **lower bound** and b is an **upper bound** for the roots of a polynomial equation if every real root c of the equation satisfies $a \leq c \leq b$. The next theorem helps us find such bounds for any polynomial equation.

THE UPPER AND LOWER BOUNDS THEOREM

Let P be a polynomial with real coefficients.

1. If we divide $P(x)$ by $x - b$ (with $b > 0$) using synthetic division, and if the row that contains the quotient and remainder has no negative entry, then b is an upper bound for the real zeros of P.
2. If we divide $P(x)$ by $x - a$ (with $a < 0$) using synthetic division, and if the row that contains the quotient and remainder has entries that are alternately nonpositive and nonnegative, then a is a lower bound for the real zeros of P.

A proof of this theorem is suggested in Exercises 90 and 91. The phrase "alternately nonpositive and nonnegative" simply means that the signs of the numbers alternate, with 0 considered to be positive or negative as required.

EXAMPLE 4 ■ Upper and Lower Bounds for Zeros of a Polynomial

Show that all the real zeros of the polynomial $P(x) = x^4 - 3x^2 + 2x - 5$ lie between -3 and 2.

SOLUTION

We divide $P(x)$ by $x - 2$ and $x + 3$ using synthetic division.

$$\begin{array}{r|rrrrr} 2 & 1 & 0 & -3 & 2 & -5 \\ & & 2 & 4 & 2 & 8 \\ \hline & 1 & 2 & 1 & 4 & 3 \end{array} \quad \leftarrow \text{All entries positive}$$

$$\begin{array}{r|rrrrr} -3 & 1 & 0 & -3 & 2 & -5 \\ & & -3 & 9 & -18 & 48 \\ \hline & 1 & -3 & 6 & -16 & 43 \end{array} \quad \leftarrow \text{Entries alternate in sign}$$

By the Upper and Lower Bounds Theorem, -3 is a lower bound and 2 is an upper bound for the zeros. Since neither -3 nor 2 is a zero (the remainders are not 0 in the division table), all the real zeros lie between these numbers. ■

EXAMPLE 5 ■ Factoring a Fifth-Degree Polynomial

Factor completely the polynomial

$$P(x) = 2x^5 + 5x^4 - 8x^3 - 14x^2 + 6x + 9$$

SOLUTION

The possible rational zeros of P are $\pm\frac{1}{2}$, ± 1, $\pm\frac{3}{2}$, ± 3, $\pm\frac{9}{2}$, and ± 9. We check the positive candidates first, beginning with the smallest.

$$\begin{array}{r|rrrrrr} \frac{1}{2} & 2 & 5 & -8 & -14 & 6 & 9 \\ & & 1 & 3 & -\frac{5}{2} & -\frac{33}{4} & -\frac{9}{8} \\ \hline & 2 & 6 & -5 & -\frac{33}{2} & -\frac{9}{4} & \frac{63}{8} \end{array} \quad \leftarrow \tfrac{1}{2} \text{ is not a zero}$$

$$\begin{array}{r|rrrrrr} 1 & 2 & 5 & -8 & -14 & 6 & 9 \\ & & 2 & 7 & -1 & -15 & -9 \\ \hline & 2 & 7 & -1 & -15 & -9 & 0 \end{array} \quad \leftarrow P(1) = 0$$

So 1 is a zero, and $P(x) = (x - 1)(2x^4 + 7x^3 - x^2 - 15x - 9)$. We continue by factoring the quotient.

$$\begin{array}{r|rrrrr} 1 & 2 & 7 & -1 & -15 & -9 \\ & & 2 & 9 & 8 & -7 \\ \hline & 2 & 9 & 8 & -7 & -16 \end{array} \quad \leftarrow 1 \text{ is not a zero}$$

$$\begin{array}{r|rrrrr} \frac{3}{2} & 2 & 7 & -1 & -15 & -9 \\ & & 3 & 15 & 21 & 9 \\ \hline & 2 & 10 & 14 & 6 & 0 \end{array} \quad \leftarrow P(\tfrac{3}{2}) = 0, \text{ all entries nonnegative}$$

We see that $\frac{3}{2}$ is both a zero and an upper bound for the zeros of $P(x)$, so we don't need to check any further for positive zeros, because all the remaining candidates are greater than $\frac{3}{2}$.

$$\begin{aligned} P(x) &= (x-1)(x-\tfrac{3}{2})(2x^3+10x^2+14x+6) \\ &= (x-1)(2x-3)(x^3+5x^2+7x+3) \end{aligned}$$

Factor 2 from last factor, multiply into second factor

By Descartes' Rule of Signs, $x^3 + 5x^2 + 7x + 3$ has no positive zero, so the only possible rational zeros are -1 and -3.

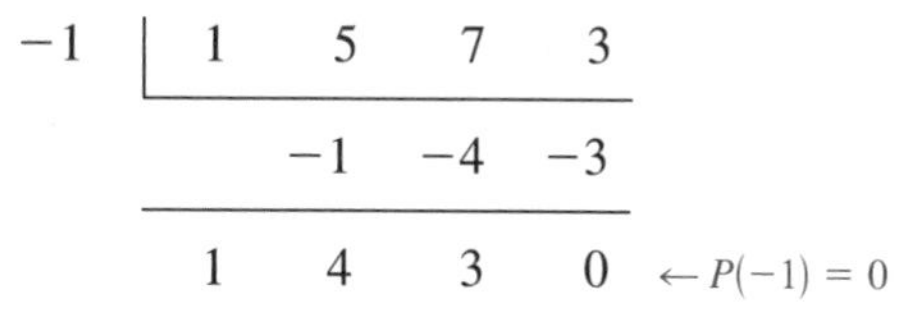

Therefore $$\begin{aligned} P(x) &= (x-1)(2x-3)(x+1)(x^2+4x+3) \\ &= (x-1)(2x-3)(x+1)^2(x+3) \end{aligned}$$

Factor quadratic

This means that the zeros of P are $1, \frac{3}{2}, -1$, and -3. ■

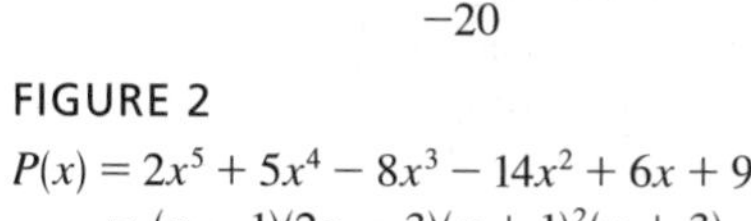

FIGURE 2
$P(x) = 2x^5 + 5x^4 - 8x^3 - 14x^2 + 6x + 9$
$= (x-1)(2x-3)(x+1)^2(x+3)$

The graph of the polynomial in Example 5 is shown in Figure 2.

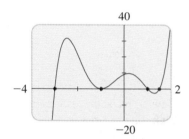

USING ALGEBRA AND GRAPHING DEVICES TO SOLVE POLYNOMIAL EQUATIONS

In Section 3.1 we used graphing devices to solve equations graphically. We can now use the algebraic techniques we've learned to select an appropriate viewing rectangle when solving a polynomial equation graphically.

EXAMPLE 6 ■ Solving a Fourth-Degree Equation Graphically

Find all real solutions of the following equation, correct to the nearest tenth.

$$3x^4 + 4x^3 - 7x^2 - 2x - 3 = 0$$

SOLUTION

To solve the equation graphically, we graph

$$P(x) = 3x^4 + 4x^3 - 7x^2 - 2x - 3$$

First we use the Upper and Lower Bounds Theorem to find two numbers between which all the solutions must lie. This allows us to choose a viewing rectangle that is certain to contain all the x-intercepts of P. We use synthetic division and proceed by trial and error.

We use the Upper and Lower Bounds Theorem to see where the roots can be found.

To find an upper bound, we try the whole numbers, 1, 2, 3, . . . as potential candidates. We see that 2 is an upper bound for the roots.

2	3	4	−7	−2	−3	
		6	20	26	48	
	3	10	13	24	45	← All positive

Now we look for a lower bound, trying the numbers −1, −2, and −3 as potential candidates. We see that −3 is a lower bound for the roots.

−3	3	4	−7	−2	−3	
		−9	15	−24	78	
	3	−5	8	−26	75	← Entries alternate in sign

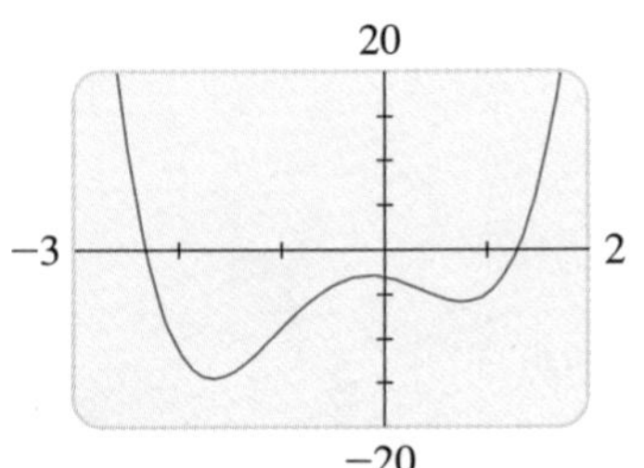

FIGURE 3

$y = 3x^4 + 4x^3 - 7x^2 - 2x - 3$

Thus, all the roots lie between −3 and 2. So the viewing rectangle $[-3, 2]$ by $[-20, 20]$ contains all the x-intercepts of P. The graph in Figure 3 has two x-intercepts, one between −3 and −2 and the other between 1 and 2. Zooming in, we find that the solutions of the equation, to the nearest tenth, are −2.3 and 1.3. ■

EXAMPLE 7 ■ Determining the Size of a Fuel Tank

A fuel tank consists of a cylindrical center section 4 ft long and two hemispherical end sections, as shown in Figure 4. If the tank has a volume of 100 ft^3, what is the radius r shown in the figure, correct to the nearest hundredth of a foot?

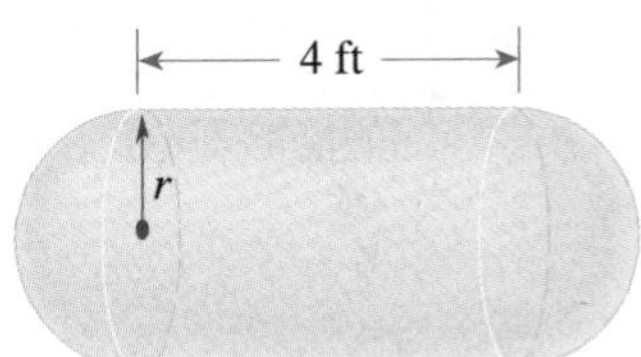

FIGURE 4

SOLUTION

Using the volume formula listed on the inside back cover of this book, we see that the volume of the cylindrical section of the tank is

Volume of a cylinder: $V = \pi r^2 h$

$$\pi \cdot r^2 \cdot 4$$

The two hemispherical parts together form a complete sphere whose volume is

Volume of a sphere: $V = \frac{4}{3}\pi r^3$

$$\tfrac{4}{3}\pi r^3$$

Because the total volume of the tank is 100 ft^3, we get the following equation:

$$\tfrac{4}{3}\pi r^3 + 4\pi r^2 = 100$$

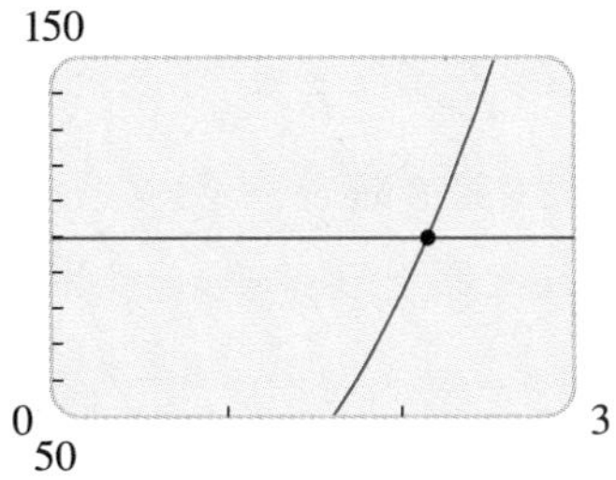

FIGURE 5
$y = \frac{4}{3}\pi x^3 + 4\pi x^2$ and $y = 100$

A negative solution for r would be meaningless in this physical situation, and by substitution we can verify that $r = 3$ leads to a tank that is over 226 ft^3 in volume, much larger than the required 100 ft^3. Thus, we know the correct radius lies somewhere between 0 and 3 ft, and so we use a viewing rectangle of $[0, 3]$ by $[50, 150]$ to graph the function $y = \frac{4}{3}\pi x^3 + 4\pi x^2$, as shown in Figure 5. Since we want the value of this function to be 100, we also graph the horizontal line $y = 100$ in the same viewing rectangle. The correct radius will be the x-coordinate of the point of intersection of the curve and the line. Using the cursor and zooming in, we see that at the point of intersection $x \approx 2.15$, correct to two decimal places. Thus, the tank has a radius of about 2.15 ft. ■

Note that we also could have solved the equation in Example 7 by first writing it as

$$\frac{4}{3}\pi r^3 + 4\pi r^2 - 100 = 0$$

and then finding the x-intercept of the function $y = \frac{4}{3}\pi x^3 + 4\pi x^2 - 100$.

5.3 EXERCISES

1–4 ■ List all possible rational zeros given by the Rational Zeros Theorem (but don't check to see which actually are zeros).

1. $P(x) = x^3 - 4x^2 + 3$

2. $Q(x) = x^4 - 3x^3 - 6x + 20$

3. $R(x) = 2x^5 + 3x^3 + 4x^2 - 8$

4. $S(x) = 6x^4 - x^2 + 2x + 12$

5–32 ■ Find all rational zeros of the polynomial.

5. $P(x) = x^3 + 3x^2 - 4$

6. $P(x) = x^3 - 7x^2 + 14x - 8$

7. $P(x) = x^3 - 3x - 2$

8. $P(x) = x^3 + 4x^2 - 3x - 18$

9. $P(x) = x^3 - 6x^2 + 12x - 8$

10. $P(x) = x^3 - x^2 - 8x + 12$

11. $P(x) = x^3 - 4x^2 + x + 6$

12. $P(x) = x^3 - 4x^2 - 7x + 10$

13. $P(x) = x^3 + 3x^2 + 6x + 4$

14. $P(x) = x^3 - 2x^2 - 2x - 3$

15. $P(x) = x^4 - 5x^2 + 4$

16. $P(x) = x^4 - 2x^3 - 3x^2 + 8x - 4$

17. $P(x) = x^4 + 6x^3 + 7x^2 - 6x - 8$

18. $P(x) = x^4 - x^3 - 23x^2 - 3x + 90$

19. $P(x) = 4x^4 - 25x^2 + 36$

20. $P(x) = x^4 - x^3 - 5x^2 + 3x + 6$

21. $P(x) = x^4 + 8x^3 + 24x^2 + 32x + 16$

22. $P(x) = 2x^3 + 7x^2 + 4x - 4$

23. $P(x) = 4x^3 + 4x^2 - x - 1$

24. $P(x) = 2x^3 - 3x^2 - 2x + 3$

25. $P(x) = 4x^3 - 7x + 3$

26. $P(x) = 8x^3 + 10x^2 - x - 3$

27. $P(x) = 2x^4 - 7x^3 + 3x^2 + 8x - 4$

28. $P(x) = 6x^4 - 7x^3 - 12x^2 + 3x + 2$

29. $P(x) = x^5 + 3x^4 - 9x^3 - 31x^2 + 36$

30. $P(x) = x^5 - 4x^4 - 3x^3 + 22x^2 - 4x - 24$

31. $P(x) = 3x^5 - 14x^4 - 14x^3 + 36x^2 + 43x + 10$

32. $P(x) = 2x^6 - 3x^5 - 13x^4 + 29x^3 - 27x^2 + 32x - 12$

33–42 ■ Find all the real zeros of the polynomial. Use the quadratic formula if necessary, as in Example 2.

33. $P(x) = x^3 + 4x^2 + 3x - 2$

34. $P(x) = x^3 - 5x^2 + 2x + 12$

35. $P(x) = x^4 - 6x^3 + 4x^2 + 15x + 4$

36. $P(x) = x^4 + 2x^3 - 2x^2 - 3x + 2$

37. $P(x) = x^4 - 7x^3 + 14x^2 - 3x - 9$

38. $P(x) = x^5 - 4x^4 - x^3 + 10x^2 + 2x - 4$

39. $P(x) = 4x^3 - 6x^2 + 1$

40. $P(x) = 3x^3 - 5x^2 - 8x - 2$

41. $P(x) = 2x^4 + 15x^3 + 17x^2 + 3x - 1$

42. $P(x) = 4x^5 - 18x^4 - 6x^3 + 91x^2 - 60x + 9$

43–50 ■ A polynomial P is given. **(a)** Find all the real zeros of P. **(b)** Sketch the graph of P.

43. $P(x) = x^3 - 3x^2 - 4x + 12$

44. $P(x) = -x^3 - 2x^2 + 5x + 6$

45. $P(x) = -x^3 + 2x^2 + 15x - 36$

46. $P(x) = 3x^3 + 17x^2 + 21x - 9$

47. $P(x) = x^4 - 5x^3 + 6x^2 + 4x - 8$

48. $P(x) = -x^4 + 10x^2 + 8x - 8$

49. $P(x) = x^5 - x^4 - 5x^3 + x^2 + 8x + 4$

50. $P(x) = x^5 - x^4 - 6x^3 + 14x^2 - 11x + 3$

51–56 ■ Use Descartes' Rule of Signs to determine how many positive and how many negative real zeros the polynomial can have. Then determine the possible total number of real zeros.

51. $P(x) = x^3 - x^2 - x - 3$

52. $P(x) = 2x^3 - x^2 + 4x - 7$

53. $P(x) = 2x^6 + 5x^4 - x^3 - 5x - 1$

54. $P(x) = x^4 + x^3 + x^2 + x + 12$

55. $P(x) = x^5 + 4x^3 - x^2 + 6x$

56. $P(x) = x^8 - x^5 + x^4 - x^3 + x^2 - x + 1$

57–60 ■ Show that the given values for a and b are lower and upper bounds for the real zeros of the polynomial.

57. $P(x) = 2x^3 + 5x^2 + x - 2; \quad a = -3, b = 1$

58. $P(x) = x^4 - 2x^3 - 9x^2 + 2x + 8; \quad a = -3, b = 5$

59. $P(x) = 8x^3 + 10x^2 - 39x + 9; \quad a = -3, b = 2$

60. $P(x) = 3x^4 - 17x^3 + 24x^2 - 9x + 1; \quad a = 0, b = 6$

61–64 ■ Find integers that are upper and lower bounds for the real zeros of the polynomial.

61. $P(x) = x^3 - 3x^2 + 4$

62. $P(x) = 2x^3 - 3x^2 - 8x + 12$

63. $P(x) = x^4 - 2x^3 + x^2 - 9x + 2$

64. $P(x) = x^5 - x^4 + 1$

65–70 ■ Find all rational zeros of the polynomial, and then find the irrational zeros, if any. Whenever appropriate, use the Rational Zeros Theorem, the Upper and Lower Bounds Theorem, Descartes' Rule of Signs, the quadratic formula, or other factoring techniques.

65. $P(x) = 2x^4 + 3x^3 - 4x^2 - 3x + 2$

66. $P(x) = 2x^4 + 15x^3 + 31x^2 + 20x + 4$

67. $P(x) = 4x^4 - 21x^2 + 5$

68. $P(x) = 6x^4 - 7x^3 - 8x^2 + 5x$

69. $P(x) = x^5 - 7x^4 + 9x^3 + 23x^2 - 50x + 24$

70. $P(x) = 8x^5 - 14x^4 - 22x^3 + 57x^2 - 35x + 6$

71–74 ■ Show that the polynomial does not have any rational zeros.

71. $P(x) = x^3 - x - 2$

72. $P(x) = 2x^4 - x^3 + x + 2$

73. $P(x) = 3x^3 - x^2 - 6x + 12$

74. $P(x) = x^{50} - 5x^{25} + x^2 - 1$

75–78 ■ The real solutions of the given equation are rational. List all possible rational roots using the Rational Zeros Theorem, and then graph the polynomial in the given viewing rectangle to determine which values are actually solutions. (All solutions can be seen in the given viewing rectangle.)

75. $x^3 - 3x^2 - 4x + 12 = 0; \quad [-4, 4]$ by $[-15, 15]$

76. $x^4 - 5x^2 + 4 = 0; \quad [-4, 4]$ by $[-30, 30]$

77. $2x^4 - 5x^3 - 14x^2 + 5x + 12 = 0; \quad [-2, 5]$ by $[-40, 40]$

78. $3x^3 + 8x^2 + 5x + 2 = 0; \quad [-3, 3]$ by $[-10, 10]$

79–82 ■ Use a graphing device to find all real solutions of the equation, correct to two decimal places.

79. $x^4 - x - 4 = 0$

80. $2x^3 - 8x^2 + 9x - 9 = 0$

81. $4.00x^4 + 4.00x^3 - 10.96x^2 - 5.88x + 9.09 = 0$

82. $x^5 + 2.00x^4 + 0.96x^3 + 5.00x^2 + 10.00x + 4.80 = 0$

83. A grain silo consists of a cylindrical main section and a hemispherical roof. If the total volume of the silo (including the part inside the roof section) is 15,000 ft^3 and the cylindrical part is 30 ft tall, what is the radius of the silo, correct to the nearest tenth of a foot?

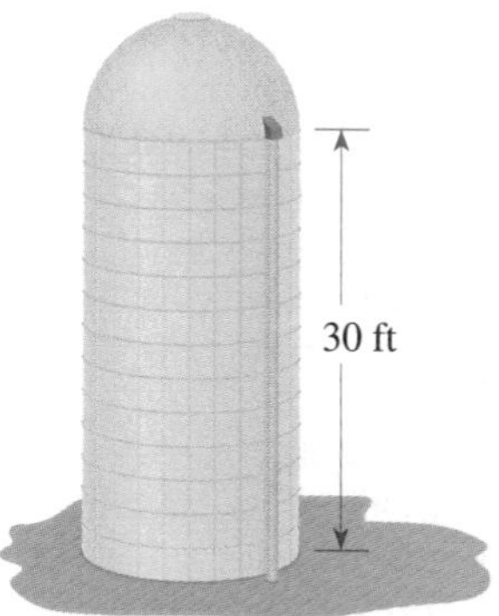

84. A rectangular parcel of land has an area of 5000 ft^2. A diagonal between opposite corners is measured to be 10 ft longer than one side of the parcel. What are the dimensions of the land, correct to the nearest foot?

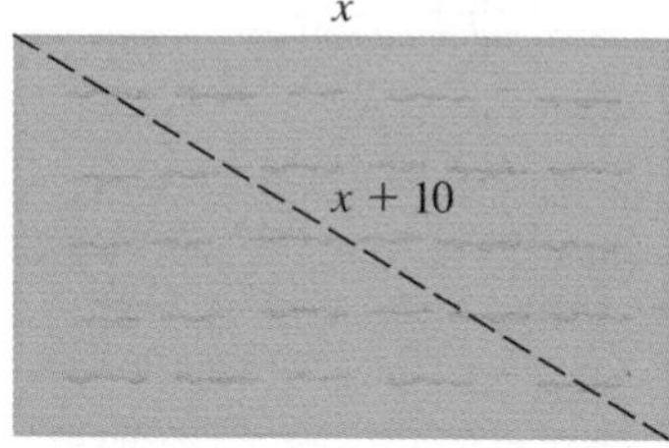

85. Snow began falling at noon on Sunday. The amount of snow on the ground at a certain location at time t was given by the function

$$h(t) = 11.60t - 12.41t^2 + 6.20t^3 - 1.58t^4 + 0.20t^5 - 0.01t^6$$

where t is measured in days from the start of the snowfall and $h(t)$ is the depth of snow in inches. Draw a graph of this function and use your graph to answer the following questions.

(a) What happened shortly after noon on Tuesday?

(b) Was there ever more than 5 in. of snow on the ground? If so, on what day(s)?

(c) On what day and at what time (to the nearest hour) did the snow disappear completely?

86. An open box with a volume of 1500 cm^3 is to be constructed by taking a piece of cardboard 20 cm by 40 cm, cutting squares of side length x cm from each corner, and folding up the sides. Show that this can be done in two different ways, and find the exact dimensions of the box in each case.

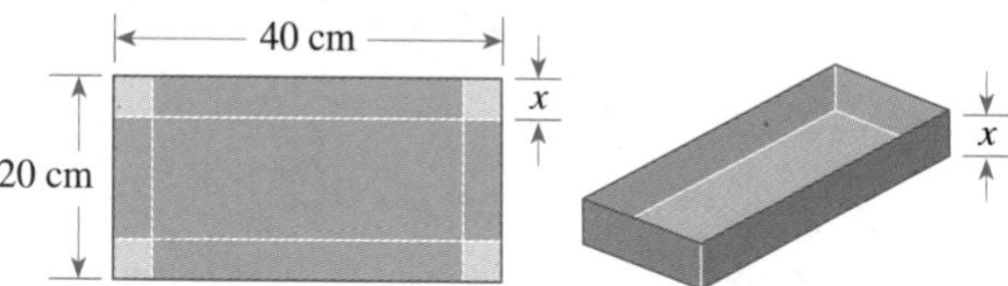

87. A rocket consists of a right circular cylinder of height 20 m surmounted by a cone whose height and diameter are equal and whose radius is the same as that of the cylindrical section. What should this radius be (correct to two decimal places) if the total volume is to be $500\pi/3$ m^3?

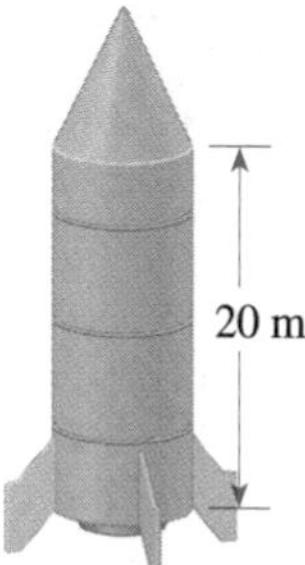

88. A rectangular box with a volume of $2\sqrt{2}$ ft^3 has a square base. The diagonal of the box (between a pair of opposite corners) is 1 ft longer than each side of the base.

(a) If the base has sides of length x feet, show that

$$x^6 - 2x^5 - x^4 + 8 = 0$$

(b) Show that two different boxes satisfy the given conditions. Find the dimensions in each case, correct to the nearest hundredth of a foot.

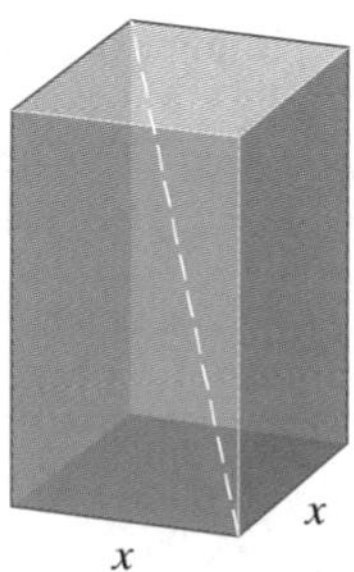

89. A box with a square base has length plus girth of 108 in. (Girth is the distance "around" the box.) What is the length of the box if its volume is 2200 in^3?

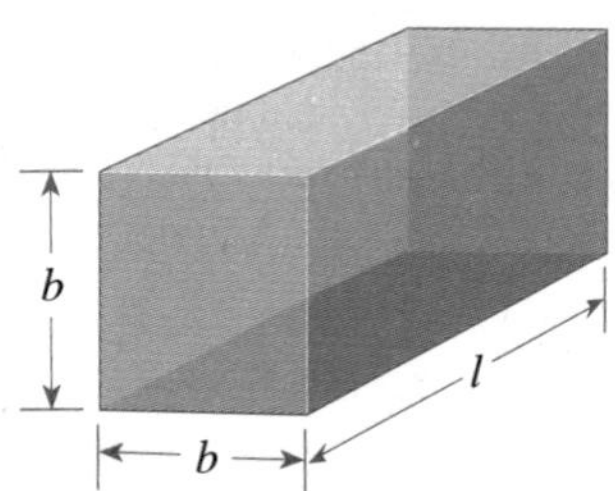

90. Let $P(x)$ be a polynomial with real coefficients and let $b > 0$. Use the Division Algorithm to write

$$P(x) = (x - b) \cdot Q(x) + r$$

Suppose that $r \geq 0$ and that all the coefficients in $Q(x)$ are nonnegative. Let $z > b$.

(a) Show that $P(z) > 0$.

(b) Prove the first part of the Upper and Lower Bounds Theorem.

91. Use the first part of the Upper and Lower Bounds Theorem to prove the second part. [*Hint:* Show that if $P(x)$ satisfies the second part of the theorem, then $P(-x)$ satisfies the first part.]

92. Show that the equation

$$x^5 - x^4 - x^3 - 5x^2 - 12x - 6 = 0$$

has exactly one rational root, and then prove that it must have either two or four irrational roots.

DISCOVERY • DISCUSSION

93. How Many Real Zeros Can a Polynomial Have? Give examples of polynomials that have the following properties, or explain why it is impossible to find such a polynomial.

(a) A polynomial of degree 3 that has no real zeros

(b) A polynomial of degree 4 that has no real zeros

(c) A polynomial of degree 3 that has three real zeros, only one of which is rational

(d) A polynomial of degree 4 that has four real zeros, none of which is rational

What must be true about the degree of a polynomial with integer coefficients if it has no real zeros?

94. The Depressed Cubic The most general cubic (third-degree) equation with rational coefficients can be written as $x^3 + ax^2 + bx + c = 0$.

(a) Show that if we replace x by $X - a/3$ and simplify, we end up with an equation that doesn't have an X^2 term, that is, an equation of the form $X^3 + pX + q = 0$. This is called a *depressed cubic*, because we have "depressed" the quadratic term.

(b) Use the procedure described in part (a) to depress the equation $x^3 + 6x^2 + 9x + 4 = 0$.

95. The Cubic Formula The quadratic formula can be used to solve any quadratic (or second-degree) equation. You may have wondered if similar formulas exist for cubic (third-degree), quartic (fourth-degree), and higher-degree equations. For the depressed cubic $x^3 + px + q = 0$, Cardano (page 351) found the following formula for one solution:

$$x = \sqrt[3]{\frac{-q}{2} + \sqrt{\frac{q^2}{4} + \frac{p^3}{27}}} + \sqrt[3]{\frac{-q}{2} - \sqrt{\frac{q^2}{4} + \frac{p^3}{27}}}$$

A formula for quartic equations was discovered by the Italian mathematician Ferrari in 1540. In 1824 the Norwegian mathematician Niels Henrik Abel proved that it is impossible to write a quintic formula, that is, a formula for fifth-degree equations. Finally, Galois (page 334) gave a criterion for determining which equations can be solved by a formula involving radicals.

Use the cubic formula to find a solution for the following equations. Then solve the equations using the methods you learned in this section. Which method is easier?

(a) $x^3 - 3x + 2 = 0$

(b) $x^3 - 27x - 54 = 0$

(c) $x^3 + 3x + 4 = 0$

ZEROING IN ON A ZERO

We have seen how to find the zeros of a polynomial algebraically and graphically. Let's work through a **numerical method** for finding the zeros. With this method we can find the value of any real zero to as many decimal places as we wish.

The Intermediate Value Theorem states: If P is a polynomial and if $P(a)$ and $P(b)$ are of opposite sign, then P has a zero between a and b. (See page 316.) The Intermediate Value Theorem is an example of an **existence theorem**—it tells us that a zero exists, but doesn't tell us exactly where it is. Nevertheless, we can use the theorem to zero in on the zero.

For example, consider the polynomial $P(x) = x^3 + 8x - 30$. Notice that $P(2) < 0$ and $P(3) > 0$. By the Intermediate Value Theorem P must have a zero between 2 and 3. To "trap" the zero in a smaller interval, we evaluate P at successive tenths between 2 and 3 until we find where P changes sign, as in Table 1. From the table we see that the zero we are looking for lies between 2.2 and 2.3, as shown in Figure 1.

x	$P(x)$	
2.1	−3.94	
2.2	−1.75	} change of sign
2.3	0.57	

TABLE 1

x	$P(x)$	
2.26	−0.38	
2.27	−0.14	} change of sign
2.28	0.09	

TABLE 2

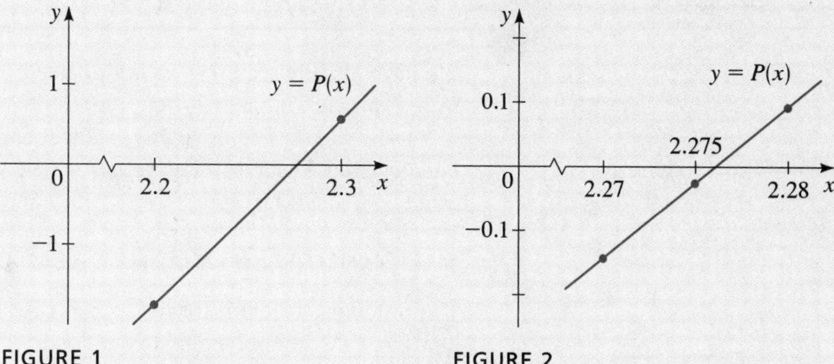

FIGURE 1

FIGURE 2

We can repeat this process by evaluating P at successive 100ths between 2.2 and 2.3, as in Table 2. By repeating this process over and over again, we can get a numerical value for the zero as accurately as we want. From Table 2 we see that the zero is between 2.27 and 2.28. To see whether it is closer to 2.27 or 2.28, we check the value of P halfway between these two numbers: $P(2.275) \approx -0.03$. Since this value is negative, the zero we are looking for lies between 2.275 and 2.28, as illustrated in Figure 2. Correct to the nearest 100th, the zero is 2.28.

1. (a) Show that $P(x) = x^2 - 2$ has a zero between 1 and 2.
   (b) Find the zero of P to the nearest tenth.
   (c) Find the zero of P to the nearest 100th.
   (d) Explain why the zero you found is an approximation to $\sqrt{2}$. Repeat the process several times to obtain $\sqrt{2}$ correct to three decimal places. Compare your results to $\sqrt{2}$obtained by a calculator.
2. Find a polynomial that has $\sqrt[3]{5}$ as a zero. Use the process described here to zero in on $\sqrt[3]{5}$ to four decimal places.
3. Show that the polynomial has a zero between the given integers, and then zero in on that zero, correct to two decimals.
   (a) $P(x) = x^3 + x - 7$; between 1 and 2
   (b) $P(x) = x^3 - x^2 - 5$; between 2 and 3
   (c) $P(x) = 2x^4 - 4x^2 + 1$; between 1 and 2
   (d) $P(x) = 2x^4 - 4x^2 + 1$; between -1 and 0
4. Find the indicated irrational zero, correct to two decimals.
   (a) The positive zero of $P(x) = x^4 + 2x^3 + x^2 - 1$
   (b) The negative zero of $P(x) = x^4 + 2x^3 + x^2 - 1$

5.4 THE FUNDAMENTAL THEOREM OF ALGEBRA

We have already seen that an nth-degree polynomial can have at most n real zeros. In the complex number system an nth-degree polynomial has *exactly* n zeros, and so can be factored into exactly n linear factors. This fact is a consequence of the Fundamental Theorem of Algebra, which was proved by the German mathematician C. F. Gauss in 1799.

THE FUNDAMENTAL THEOREM OF ALGEBRA

The following theorem forms the basis for much of our work in factoring polynomials and solving polynomial equations.

FUNDAMENTAL THEOREM OF ALGEBRA

Every polynomial

$$P(x) = a_n x^n + a_{n-1}x^{n-1} + \cdots + a_1 x + a_0 \qquad (n \geq 1,\ a_n \neq 0)$$

with complex coefficients has at least one complex zero.

Carl Friedrich Gauss (1777–1855) is considered the greatest mathematician of modern times. His contemporaries called him the "Prince of Mathematics." He was born into a poor family; his father made a living as a mason. As a very small child, Gauss found a calculation error in his father's accounts, the first of many incidents that gave evidence of his mathematical precocity. (See also page 593.) At 19 Gauss demonstrated that the regular 17-sided polygon can be constructed with straight-edge and compass alone. This was remarkable because, since the time of Euclid, it was thought that the only regular polygons constructible in this way were the triangle and pentagon. Because of this discovery Gauss decided to pursue a career in mathematics instead of languages, his other passion. In his doctoral dissertation, written at the age of 22, Gauss proved the Fundamental Theorem of Algebra: A polynomial of degree n with complex coefficients has n roots. His other accomplishments range over every branch of mathematics, as well as physics and astronomy.

Because any real number is also a complex number, the theorem applies to polynomials with real coefficients as well.

The Fundamental Theorem of Algebra and the Factor Theorem together show that a polynomial can be factored completely into linear factors, as we now prove.

COMPLETE FACTORIZATION THEOREM

If $P(x)$ is a polynomial of degree $n > 0$, then there exist complex numbers $a, c_1, c_2, \ldots, c_n$ (with $a \neq 0$) such that

$$P(x) = a(x - c_1)(x - c_2) \cdots (x - c_n)$$

■ **Proof** By the Fundamental Theorem of Algebra, P has at least one zero. Let's call it c_1. By the Factor Theorem, $P(x)$ can be factored as

$$P(x) = (x - c_1) \cdot Q_1(x)$$

where $Q_1(x)$ is of degree $n - 1$. Applying the Fundamental Theorem to the quotient $Q_1(x)$ gives us the factorization

$$P(x) = (x - c_1) \cdot (x - c_2) \cdot Q_2(x)$$

where $Q_2(x)$ is of degree $n - 2$ and c_2 is a zero of $Q_1(x)$. Continuing this process for n steps, we get a final quotient $Q_n(x)$ of degree 0, a nonzero constant that we will call a. This means that P has been factored as

$$P(x) = a(x - c_1)(x - c_2) \cdots (x - c_n)$$

□

In the Complete Factorization Theorem the numbers $c_1, c_2, \ldots, c_n$ are the zeros of P. These zeros need not all be different. If the factor $x - c$ appears k times in the complete factorization of $P(x)$, then we say that c is a zero of **multiplicity *k***. For example, the polynomial

$$P(x) = (x - 1)^3(x + 2)^2(x + 3)^5$$

has the following zeros:

$$1 \text{ (multiplicity 3)}, \qquad -2 \text{ (multiplicity 2)}, \qquad -3 \text{ (multiplicity 5)}$$

The polynomial P has the same number of zeros as its degree—it has degree 10 and has 10 zeros, provided we count multiplicities. This is true for all polynomials.

ZEROS THEOREM

Every polynomial of degree $n \geqslant 1$ has exactly n zeros, provided that a zero of multiplicity k is counted k times.

■ **Proof** Let P be a polynomial of degree n. By the Complete Factorization Theorem

$$P(x) = a(x - c_1)(x - c_2) \cdots (x - c_n)$$

Now suppose that c is a zero of P other than $c_1, c_2, \ldots, c_n$. Then

$$P(c) = a(c - c_1)(c - c_2) \cdots (c - c_n) = 0$$

Thus, by the Zero-Product Property one of the factors $c - c_i$ must be 0, so $c = c_i$ for some i. It follows that P has exactly the n zeros $c_1, c_2, \ldots, c_n$. □

EXAMPLE 1 ■ Factoring a Polynomial with Complex Zeros

Find the complete factorization and all five zeros of the polynomial

$$P(x) = 3x^5 + 24x^3 + 48x$$

SOLUTION

Since $3x$ is a common factor, we have

$$\begin{aligned} P(x) &= 3x(x^4 + 8x^2 + 16) \\ &= 3x(x^2 + 4)^2 \end{aligned}$$

To factor $x^2 + 4$, note that $2i$ and $-2i$ are zeros of this polynomial. Thus $x^2 + 4 = (x - 2i)(x + 2i)$, and so

$$\begin{aligned} P(x) &= 3x[(x - 2i)(x + 2i)]^2 \\ &= 3x(x - 2i)(x - 2i)(x + 2i)(x + 2i) \end{aligned}$$

The zeros of P are 0, $2i$, and $-2i$. Since the factors $x - 2i$ and $x + 2i$ each occur twice in the complete factorization of P, the zeros $2i$ and $-2i$ are of multiplicity 2 (or *double* zeros). So the total number of zeros is five. ■

The following table gives further examples of polynomials with their complete factorizations and zeros.

Degree	Polynomial	Zero(s)	Number of zeros
1	$P(x) = x - 4$	4	1
2	$P(x) = x^2 - 10x + 25$ $= (x - 5)(x - 5)$	5 (multiplicity 2)	2
3	$P(x) = x^3 + x$ $= x(x - i)(x + i)$	$0, i, -i$	3
4	$P(x) = x^4 + 8x^2 + 16$ $= (x - 2i)(x - 2i)(x + 2i)(x + 2i)$	$2i$ (multiplicity 2), $-2i$ (multiplicity 2)	4
5	$P(x) = x^5 - 2x^4 + x^3$ $= x^3(x - 1)^2$	0 (multiplicity 3), 1 (multiplicity 2)	5

EXAMPLE 2 ■ Finding Polynomials with Specified Zeros

Find a polynomial that satisfies the given description.

(a) A polynomial $P(x)$ of degree 4, with zeros i, $-i$, 2, and -2 and with $P(3) = 25$

(b) A polynomial $Q(x)$ of degree 4, with zeros -2 and 0, where -2 is a zero of multiplicity 3

SOLUTION

(a) The required polynomial has the form

$$\begin{aligned} P(x) &= a(x - i)(x - (-i))(x - 2)(x - (-2)) \\ &= a(x^2 + 1)(x^2 - 4) && \text{Difference of squares} \\ &= a(x^4 - 3x^2 - 4) && \text{Multiply} \end{aligned}$$

We know that $P(3) = a(3^4 - 3 \cdot 3^2 - 4) = 50a = 25$, so $a = \frac{1}{2}$. Thus

$$P(x) = \tfrac{1}{2}x^4 - \tfrac{3}{2}x^2 - 2$$

(b) We require

$$\begin{aligned} Q(x) &= a[x - (-2)]^3(x - 0) \\ &= a(x + 2)^3x \\ &= a(x^3 + 6x^2 + 12x + 8)x && \text{Product Formula 4 (Section 1.4)} \\ &= a(x^4 + 6x^3 + 12x^2 + 8x) \end{aligned}$$

Since we are given no information about Q other than its zeros and their multiplicity, we can choose any number for a. If we use $a = 1$, we get

$$Q(x) = x^4 + 6x^3 + 12x^2 + 8x$$

■

EXAMPLE 3 ■ Finding All the Zeros of a Polynomial

Find all four zeros of $P(x) = 3x^4 - 2x^3 - x^2 - 12x - 4$.

SOLUTION

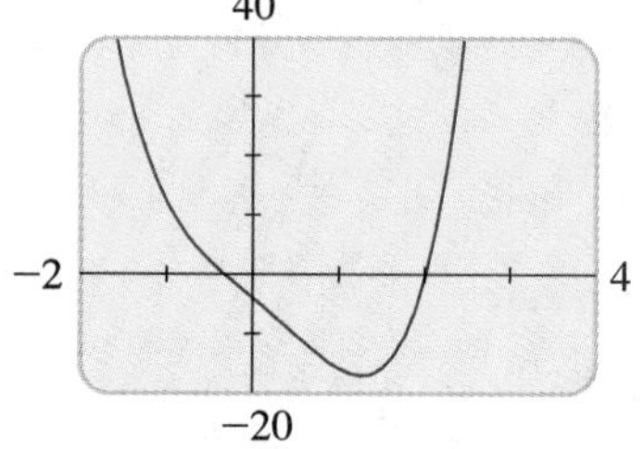

FIGURE 1

$P(x) = 3x^4 - 2x^3 - x^2 - 12x - 4$

Figure 1 shows the graph of the polynomial P in Example 3. The x-intercepts correspond to the real zeros of P. The imaginary zeros cannot be determined from the graph.

Using the Rational Zeros Theorem from Section 5.3, we obtain the following list of possible rational zeros: $\pm 1, \pm 2, \pm 4, \pm \frac{1}{3}, \pm \frac{2}{3}, \pm \frac{4}{3}$. Checking these using synthetic division, we find that 2 and $-\frac{1}{3}$ are zeros, and we get the following factorization.

$$\begin{aligned} P(x) &= 3x^4 - 2x^3 - x^2 - 12x - 4 \\ &= (x - 2)(3x^3 + 4x^2 + 7x + 2) && \text{Factor } x - 2 \\ &= (x - 2)\left(x + \tfrac{1}{3}\right)(3x^2 + 3x + 6) && \text{Factor } x + \tfrac{1}{3} \\ &= 3(x - 2)\left(x + \tfrac{1}{3}\right)(x^2 + x + 2) && \text{Factor 3} \end{aligned}$$

The zeros of the quadratic factor are

$$x = \frac{-1 \pm \sqrt{1 - 8}}{2} = -\frac{1}{2} \pm i\frac{\sqrt{7}}{2} \qquad \text{Quadratic formula}$$

so the zeros of $P(x)$ are

$$2, \quad -\frac{1}{3}, \quad -\frac{1}{2} + i\frac{\sqrt{7}}{2}, \quad \text{and} \quad -\frac{1}{2} - i\frac{\sqrt{7}}{2}$$

■

As you may have noticed from the examples so far, the complex zeros of polynomials with real coefficients come in pairs. Whenever $a + bi$ is a zero, its complex conjugate $a - bi$ is also a zero.

CONJUGATE ZEROS THEOREM

If the polynomial P has real coefficients, and if the complex number z is a zero of P, then its complex conjugate $\overline{z}$ is also a zero of P.

■ **Proof** Let

$$P(x) = a_n x^n + a_{n-1}x^{n-1} + \cdots + a_1 x + a_0$$

where each coefficient is real. Suppose that $P(z) = 0$. We must prove that $P(\overline{z}) = 0$.

We use the facts that the complex conjugate of a sum of two complex numbers is the sum of the conjugates and that the conjugate of a product is the product of the conjugates (see Exercises 61 and 62 in Section 3.4).

$$\begin{aligned} P(\overline{z}) &= a_n(\overline{z})^n + a_{n-1}(\overline{z})^{n-1} + \cdots + a_1\overline{z} + a_0 \\ &= \overline{a_n}\,\overline{z^n} + \overline{a_{n-1}}\,\overline{z^{n-1}} + \cdots + \overline{a_1}\,\overline{z} + \overline{a_0} && \text{Because the coefficients are real} \\ &= \overline{a_n z^n} + \overline{a_{n-1} z^{n-1}} + \cdots + \overline{a_1 z} + \overline{a_0} \\ &= \overline{a_n z^n + a_{n-1}z^{n-1} + \cdots + a_1 z + a_0} \\ &= \overline{P(z)} = \overline{0} = 0 \end{aligned}$$

This shows that $\overline{z}$ is also a zero of $P(x)$, which proves the theorem. □

EXAMPLE 4 ■ A Polynomial with a Specified Complex Zero

Find a polynomial $P(x)$ of degree 3 that has integer coefficients and zeros $\frac{1}{2}$ and $3 - i$.

SOLUTION

Since $3 - i$ is a zero, then so is $3 + i$ by the Conjugate Roots Theorem. This means that $P(x)$ has the form

$$\begin{aligned} P(x) &= a\left(x - \tfrac{1}{2}\right)[x - (3 - i)][x - (3 + i)] \\ &= a\left(x - \tfrac{1}{2}\right)[(x - 3) + i][(x - 3) - i] && \text{Regroup} \\ &= a\left(x - \tfrac{1}{2}\right)[(x - 3)^2 - i^2] && \text{Difference of squares formula} \\ &= a\left(x - \tfrac{1}{2}\right)(x^2 - 6x + 10) && \text{Expand} \\ &= a\left(x^3 - \tfrac{13}{2}x^2 + 13x - 5\right) && \text{Expand} \end{aligned}$$

To make all coefficients integers, we set $a = 2$ and get

$$P(x) = 2x^3 - 13x^2 + 26x - 10$$

Any other polynomial that satisfies the given requirements must be an integer multiple of this one. ■

We have seen that a polynomial factors completely into linear factors if we use complex numbers. If we do not use complex numbers, then a polynomial with real coefficients can always be factored into linear and quadratic factors. We will use this property in Section 7.8 when we study partial fractions. A quadratic polynomial with no real zeros is called **irreducible** over the real numbers. Such a polynomial cannot be factored without using complex numbers.

Gerolamo Cardano (1501–1576) is certainly one of the most colorful figures in the history of mathematics. He was the most renowned physician in Europe in his day, yet throughout his life he was plagued by numerous maladies, including ruptures, hemorrhoids, and an irrational fear of encountering rabid dogs. A doting father, his beloved sons broke his heart—his favorite was eventually beheaded for murdering his wife. Cardano was also a compulsive gambler; indeed, this vice may have driven him to write the *Book on Games of Chance*, the first study of probability from a mathematical point of view.

In Cardano's major mathematical work, the *Ars Magna*, he detailed the solution of the general third- and fourth-degree polynomial equations. At the time of its publication, mathematicians were uncomfortable even with negative numbers, but Cardano's formulas paved the way for the acceptance not just of negative numbers, but also of imaginary numbers, because they occurred naturally in solving polynomial equations. For example, one of his formulas gives the solution

$$x = \sqrt[3]{2 + \sqrt{-121}} + \sqrt[3]{2 - \sqrt{-121}}$$

for the cubic equation

$$x^3 - 15x - 4 = 0$$

This value for x actually turns out to be the *integer* 4, yet to find it Cardano had to use the imaginary number $\sqrt{-121} = 11i$.

LINEAR AND QUADRATIC FACTORS THEOREM

Every polynomial with real coefficients can be factored into a product of linear and irreducible quadratic factors with real coefficients.

■ **Proof** We first observe that if $c = a + bi$ is a complex number, then

$$\begin{aligned}(x - c)(x - \bar{c}) &= [x - (a + bi)][x - (a - bi)] \\ &= [(x - a) - bi][(x - a) + bi] \\ &= (x - a)^2 - (bi)^2 \\ &= x^2 - 2ax + (a^2 + b^2)\end{aligned}$$

The last expression is a quadratic with *real* coefficients.

Now, if P is a polynomial with real coefficients, then by the Complete Factorization Theorem

$$P(x) = a(x - c_1)(x - c_2) \cdots (x - c_n)$$

Since the complex roots occur in conjugate pairs, we can multiply the factors corresponding to each such pair to get a quadratic factor with real coefficients. This results in P being factored into linear and irreducible quadratic factors. □

EXAMPLE 5 ■ Factoring a Polynomial into Linear and Quadratic Factors

Let $P(x) = x^4 + 2x^2 - 8$.

(a) Factor P into linear and irreducible quadratic factors with real coefficients.
(b) Factor P completely into linear factors with complex coefficients.

SOLUTION

(a)
$$\begin{aligned}P(x) &= x^4 + 2x^2 - 8 \\ &= (x^2 - 2)(x^2 + 4) \\ &= (x - \sqrt{2})(x + \sqrt{2})(x^2 + 4)\end{aligned}$$

The factor $x^2 + 4$ is irreducible since it has only the imaginary zeros $\pm 2i$.

(b) To get the complete factorization, we factor the remaining quadratic factor.

$$\begin{aligned}P(x) &= (x - \sqrt{2})(x + \sqrt{2})(x^2 + 4) \\ &= (x - \sqrt{2})(x + \sqrt{2})(x - 2i)(x + 2i)\end{aligned}$$

■

5.4 EXERCISES

1–18 ■ Factor the polynomial completely and find all its zeros. State the multiplicity of each zero.

1. $P(x) = x^2 + 9$

2. $P(x) = 4x^2 + 25$

3. $Q(x) = x^2 + 2x + 2$

4. $Q(x) = x^2 - 8x + 17$

5. $P(x) = x^3 + 4x$

6. $P(x) = x^3 + x^2 + x$

7. $Q(x) = x^4 - 1$

8. $Q(x) = x^4 - 625$

9. $P(x) = 16x^4 - 81$

10. $P(x) = x^3 - 64$

11. $P(x) = x^3 + x^2 + 9x + 9$

12. $P(x) = x^6 - 729$

13. $Q(x) = x^4 + 2x^2 + 1$

14. $Q(x) = x^4 + 10x^2 + 25$

15. $P(x) = x^4 + 3x^2 - 4$

16. $P(x) = x^5 + 7x^3$

17. $P(x) = x^5 + 6x^3 + 9x$

18. $P(x) = x^6 + 16x^3 + 64$

19–28 ■ Find a polynomial with integer coefficients that satisfies the given conditions.

19. P has degree 2, and zeros $1 + i$ and $1 - i$.

20. P has degree 2, and zeros $1 + i\sqrt{2}$ and $1 - i\sqrt{2}$.

21. Q has degree 3, and zeros 3, $2i$, and $-2i$.

22. Q has degree 3, and zeros 0 and i.

23. P has degree 3, and zeros 2 and i.

24. Q has degree 3, and zeros -3 and $1 + i$.

25. R has degree 4, and zeros $1 - 2i$ and 1, with 1 a zero of multiplicity 2.

26. S has degree 4, and zeros $2i$ and $3i$.

27. T has degree 4, zeros i and $1 + i$, and constant coefficient 12.

28. U has degree 5, zeros $\frac{1}{2}$, -1, and $-i$, and leading coefficient 4; the zero -1 has multiplicity 2.

29–44 ■ Find all zeros of the polynomial.

29. $P(x) = x^3 + 2x^2 + 4x + 8$

30. $P(x) = x^3 - 7x^2 + 17x - 15$

31. $P(x) = x^3 - 2x^2 + 2x - 1$

32. $P(x) = x^3 + 7x^2 + 18x + 18$

33. $P(x) = x^3 - 3x^2 + 3x - 2$

34. $P(x) = x^3 - x - 6$

35. $P(x) = 2x^3 + 7x^2 + 12x + 9$

36. $P(x) = 2x^3 - 8x^2 + 9x - 9$

37. $P(x) = x^4 + x^3 + 7x^2 + 9x - 18$

38. $P(x) = x^4 - 2x^3 - 2x^2 - 2x - 3$

39. $P(x) = x^5 - x^4 + 7x^3 - 7x^2 + 12x - 12$

40. $P(x) = x^5 + x^3 + 8x^2 + 8$ [*Hint:* Factor by grouping.]

41. $P(x) = x^4 - 6x^3 + 13x^2 - 24x + 36$

42. $P(x) = x^4 - x^2 + 2x + 2$

43. $P(x) = x^5 - 3x^4 + 12x^3 - 28x^2 + 27x - 9$

44. $P(x) = x^5 - 2x^4 + 2x^3 - 4x^2 + x - 2$

45–50 ■ A polynomial P is given.

(a) Factor P into linear and irreducible quadratic factors with real coefficients.

(b) Factor P completely into linear factors with complex coefficients.

45. $P(x) = x^3 - 5x^2 + 4x - 20$

46. $P(x) = x^3 - 2x - 4$

47. $P(x) = x^4 + 8x^2 - 9$

48. $P(x) = x^4 + 8x^2 + 16$

49. $P(x) = x^6 - 64$

50. $P(x) = x^5 - 16x$

51. By the Zeros Theorem, every nth-degree polynomial equation has exactly n solutions (including possibly some that are repeated). Some of these may be real and some may be imaginary. Use a graphing device to determine how many real and imaginary solutions each equation has.

(a) $x^4 - 2x^3 - 11x^2 + 12x = 0$

(b) $x^4 - 2x^3 - 11x^2 + 12x - 5 = 0$

(c) $x^4 - 2x^3 - 11x^2 + 12x + 40 = 0$

52–54 ■ So far we have worked only with polynomials that have real coefficients. These exercises involve polynomials with real and imaginary coefficients.

52. Find all solutions of the equation.

(a) $2x + 4i = 1$

(b) $x^2 - ix = 0$

(c) $x^2 + 2ix - 1 = 0$

(d) $ix^2 - 2x + i = 0$

53. (a) Show that $2i$ and $1 - i$ are both solutions of the equation

$$x^2 - (1 + i)x + (2 + 2i) = 0$$

but that their complex conjugates $-2i$ and $1 + i$ are not.

(b) Explain why the result of part (a) does not violate the Conjugate Roots Theorem.

54. (a) Find the polynomial with *real* coefficients of the smallest possible degree for which i and $1 + i$ are zeros and in which the coefficient of the highest power is 1.

(b) Find the polynomial with *complex* coefficients of the smallest possible degree for which i and $1 + i$ are zeros and in which the coefficient of the highest power is 1.

DISCOVERY • DISCUSSION

55. Polynomials of Odd Degree The Conjugate Zeros Theorem says that the complex zeros of a polynomial with real coefficients occur in complex conjugate pairs. Explain how this fact proves that a polynomial with real coefficients and odd degree has at least one real zero.

56. Roots of Unity There are two square roots of 1, namely 1 and -1. These are the solutions of $x^2 = 1$. The fourth roots of 1 are the solutions of the equation $x^4 = 1$ or $x^4 - 1 = 0$. How many fourth roots of 1 are there? Find them. The cube roots of 1 are the solutions of the equation $x^3 = 1$ or $x^3 - 1 = 0$. How many cube roots of 1 are there? Find them. How would you find the sixth roots of 1? How many are there? Make a conjecture about the number of nth roots of 1.

5.5 RATIONAL FUNCTIONS

A rational function is a function of the form

$$r(x) = \frac{P(x)}{Q(x)}$$

where P and Q are polynomials. We assume that $P(x)$ and $Q(x)$ have no factor in common. Even though rational functions are constructed from polynomials, their graphs look quite different than the graphs of polynomial functions.

RATIONAL FUNCTIONS AND ASYMPTOTES

Rational functions are not defined for those values of x for which the denominator is zero. When graphing a rational function, we must pay special attention to the behavior of the graph near those x-values. We begin by graphing a very simple rational function.

EXAMPLE 1 ■ A Simple Rational Function

Sketch a graph of the rational function $r(x) = \dfrac{1}{x}$.

SOLUTION

The function r is not defined for $x = 0$. The following tables show that when x is close to zero, the value of $|r(x)|$ is large, and the closer x gets to zero, the larger $|r(x)|$ gets.

For positive real numbers,

$$\frac{1}{\text{BIG NUMBER}} = \text{small number}$$

$$\frac{1}{\text{small number}} = \text{BIG NUMBER}$$

x	$r(x)$
-0.1	-10
-0.01	-100
-0.00001	$-100{,}000$
↑ approaching 0^-	↑ approaching $-\infty$

x	$r(x)$
0.1	10
0.01	100
0.00001	100,000
↑ approaching 0^+	↑ approaching ∞

We describe this behavior in words and in symbols as follows. The first table shows that as x approaches 0 from the left, the values of $r(x)$ decrease without bound. In symbols,

$$r(x) \to -\infty \quad \text{as} \quad x \to 0^-$$

"y approaches negative infinity as x approaches 0 from the left"

The second table shows that as x approaches 0 from the right, the values of $r(x)$ increase without bound. In symbols,

$$r(x) \to \infty \quad \text{as} \quad x \to 0^+$$

"y approaches infinity as x approaches 0 from the right"

The next two tables show the behavior of the function r as $|x|$ becomes large.

x	$r(x)$
-10	-0.1
-100	-0.01
$-100{,}000$	-0.00001
↑ approaching $-\infty$	↑ approaching 0

x	$r(x)$
10	0.1
100	0.01
100,000	0.00001
↑ approaching ∞	↑ approaching 0

These tables show that as $|x|$ becomes large, the value of $r(x)$ gets closer and closer to zero. We describe this situation in symbols by writing

$$r(x) \to 0 \quad \text{as} \quad x \to -\infty \qquad \text{and} \qquad r(x) \to 0 \quad \text{as} \quad x \to \infty$$

Using the information in these tables and plotting a few additional points, we obtain the graph shown in Figure 1.

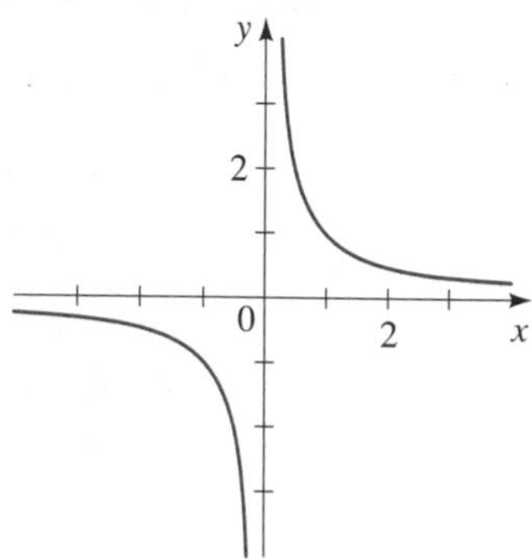

FIGURE 1

$r(x) = \dfrac{1}{x}$

■

In Example 1 we used the following arrow notation.

Symbol	Meaning
$x \to a^-$	x approaches a from the left
$x \to a^+$	x approaches a from the right
$x \to -\infty$	x goes to negative infinity; that is, x decreases without bound
$x \to \infty$	x goes to infinity; that is, x increases without bound

The line $x = 0$ is called a *vertical asymptote* of the graph in Figure 1, and the line $y = 0$ is a *horizontal asymptote*. Informally speaking, an asymptote of a function is a line that the graph of the function gets closer and closer to as one travels along that line in either direction.

DEFINITION OF ASYMPTOTES

1. The line $x = a$ is a **vertical asymptote** of the function $y = f(x)$ if

$$y \to \infty \quad \text{or} \quad y \to -\infty \qquad \text{as} \qquad x \to a^+ \quad \text{or} \quad x \to a^-$$

2. The line $y = b$ is a **horizontal asymptote** of the function $y = f(x)$ if

$$y \to b \qquad \text{as} \qquad x \to \infty \quad \text{or} \quad x \to -\infty$$

A rational function has vertical asymptotes where the function is undefined, that is, where the denominator is zero.

EXAMPLE 2 ■ Asymptotes of a Rational Function

Sketch a graph of the rational function $s(x) = \dfrac{x - 1}{x - 2}$.

SOLUTION 1

We find the asymptotes of s and use them to sketch the graph.

VERTICAL ASYMPTOTE The line $x = 2$ is a vertical asymptote because the denominator of s is 0 when $x = 2$.

BEHAVIOR NEAR VERTICAL ASYMPTOTE We make tables of values for x-values to the left and right of 2. From the tables shown at the top of the next page we see that

$$y \to -\infty \quad \text{as} \quad x \to 2^- \qquad \text{and} \qquad y \to \infty \quad \text{as} \quad x \to 2^+$$

$x \to 2^-$

x	y
1	0
1.5	−1
1.9	−9
1.95	−19
1.99	−99
1.999	−999
↑ approaching 2^-	↑ approaching $-\infty$

$x \to 2^+$

x	y
3	2
2.5	3
2.1	11
2.05	21
2.01	101
2.001	1001
↑ approaching 2^+	↑ approaching ∞

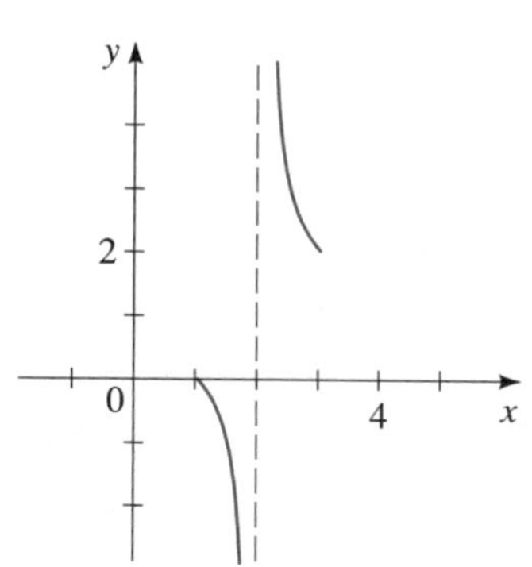

FIGURE 2

The graph of $y = r(x)$ therefore has the shape near $x = 2$ shown in Figure 2.

HORIZONTAL ASYMPTOTE The horizontal asymptote is the value that y approaches as $x \to \pm\infty$. To find this value divide both numerator and denominator by x:

$$y = s(x) = \frac{x-1}{x-2} \cdot \frac{\frac{1}{x}}{\frac{1}{x}} = \frac{1 - \frac{1}{x}}{1 - \frac{2}{x}}$$

The fractional expressions $\frac{1}{x}$ and $\frac{2}{x}$ both approach 0 as $x \to \pm\infty$ (see Exercise 60, Section 1.2). So as $x \to \pm\infty$, we have

$$y = \frac{1 - \frac{1}{x}}{1 - \frac{2}{x}} \quad \to \quad \frac{1-0}{1-0} = 1$$

Thus, the horizontal asymptote is the line $y = 1$.

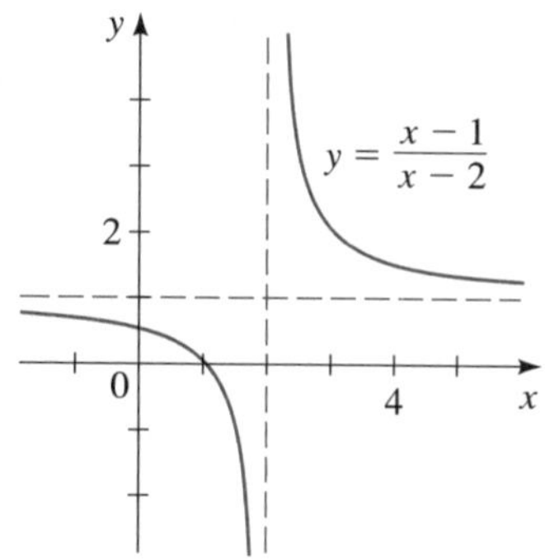

FIGURE 3

GRAPH Since the graph must approach the horizontal asymptote, we can complete it as shown in Figure 3.

SOLUTION 2

Using long division, we see that

$$s(x) = 1 + \frac{1}{x-2}$$

Shifting functions is discussed in Section 4.5.

This means that the graph of s is just the graph of the function r of Example 1 shifted upward 1 unit and to the right 2 units. Thus, we obtain the graph in Figure 3 from the graph in Figure 1 simply by shifting horizontally and then vertically. ■

GRAPHS OF RATIONAL FUNCTIONS

We have seen that asymptotes are important when graphing simple rational functions. In general, we use the following guidelines to graph rational functions.

SKETCHING GRAPHS OF RATIONAL FUNCTIONS

1. FACTOR. Factor the numerator and denominator.
2. INTERCEPTS. Find the x-intercepts by determining the zeros of the numerator, and the y-intercept from the value of the function at $x = 0$.
3. VERTICAL ASYMPTOTES. Find the vertical asymptotes by determining the zeros of the denominator, and then see if $y \to \infty$ or $y \to -\infty$ on each side of every vertical asymptote.
4. HORIZONTAL ASYMPTOTE. Find the horizontal asymptote (if any) by dividing both numerator and denominator by the highest power of x that appears in the denominator, and then letting $x \to \pm\infty$.
5. SKETCH THE GRAPH. Graph the information provided by the first four steps. Then plot as many additional points as needed to fill in the rest of the graph of the function.

EXAMPLE 3 ■ Graphing a Rational Function

Sketch a graph of the rational function $r(x) = \dfrac{2x^2 + 7x - 4}{x^2 + x - 2}$.

SOLUTION

We factor the numerator and denominator, find the intercepts and asymptotes, and sketch the graph.

FACTOR: $r(x) = \dfrac{(2x - 1)(x + 4)}{(x - 1)(x + 2)}$

A fraction is 0 if and only if its numerator is 0.

x-INTERCEPTS: The x-intercepts are the zeros of the numerator, $x = \frac{1}{2}$ and $x = -4$.

y-INTERCEPT: To find the y-intercept, we substitute $x = 0$ into the original form of the function:

$$r(0) = \frac{2(0)^2 + 7(0) - 4}{(0)^2 + (0) - 2} = \frac{-4}{-2} = 2$$

The y-intercept is 2.

HORIZONTAL ASYMPTOTE: The horizontal asymptote (if it exists) is the value that y approaches as $x \to \pm\infty$. To find this value, we divide both the numerator and the denominator by the highest power of x that appears in the denominator (in this case, x^2).

$$y = r(x) = \frac{2x^2 + 7x - 4}{x^2 + x - 2} \cdot \frac{\frac{1}{x^2}}{\frac{1}{x^2}} = \frac{2 + \frac{7}{x} - \frac{4}{x^2}}{1 + \frac{1}{x} - \frac{2}{x^2}}$$

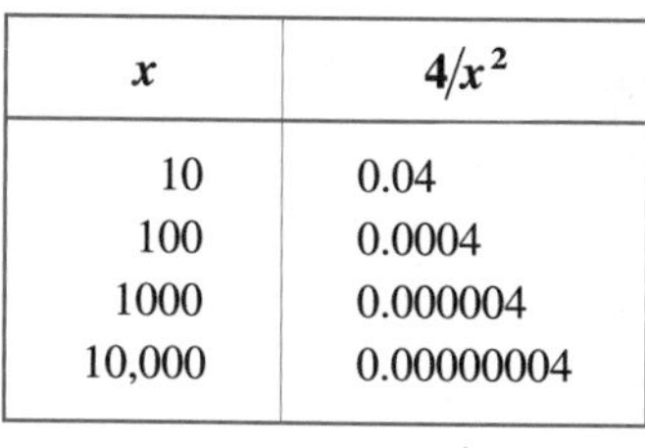

x	$4/x^2$
10	0.04
100	0.0004
1000	0.000004
10,000	0.00000004

↑ approaching ∞ ↑ approaching 0

Any expression of the form c/x^n approaches 0 as $x \to \pm\infty$ (if $n > 0$). The table in the margin illustrates this for the term $4/x^2$. So as $x \to \pm\infty$, we have

$$y = \frac{2 + \frac{7}{x} - \frac{4}{x^2}}{1 + \frac{1}{x} - \frac{2}{x^2}} \quad \to \quad \frac{2 + 0 - 0}{1 + 0 - 0} = 2$$

Thus, the horizontal asymptote is $y = 2$.

VERTICAL ASYMPTOTES: The vertical asymptotes occur where the denominator is 0, that is, where the function is undefined. From the factored form we see that the vertical asymptotes are $x = 1$ and $x = -2$.

BEHAVIOR NEAR VERTICAL ASYMPTOTES: We need to know whether $y \to \infty$ or $y \to -\infty$ on each side of each vertical asymptote. To determine the sign of y for x-values near the vertical asymptotes, we use test values. For instance, as $x \to 1^-$, we use a test value close to and to the left of 1 ($x = 0.9$, say) to check whether y is positive or negative to the left of $x = 1$:

$$y = \frac{(2(0.9) - 1)((0.9) + 4)}{((0.9) - 1)((0.9) + 2)} \qquad \text{whose sign is} \qquad \frac{(+)(+)}{(-)(+)} \quad \text{(negative)}$$

So $y \to -\infty$ as $x \to 1^-$. On the other hand, as $x \to 1^+$, we use a test value close to and to the right of 1 ($x = 1.1$, say), to get

$$y = \frac{(2(1.1) - 1)((1.1) + 4)}{((1.1) - 1)((1.1) + 2)} \qquad \text{whose sign is} \qquad \frac{(+)(+)}{(+)(+)} \quad \text{(positive)}$$

So $y \to \infty$ as $x \to 1^+$. The other entries in the following table are calculated similarly.

As $x \to$	-2^-	-2^+	1^-	1^+
the sign of $y = \frac{(2x-1)(x+4)}{(x-1)(x+2)}$ is	$\frac{(-)(+)}{(-)(-)}$	$\frac{(-)(+)}{(-)(+)}$	$\frac{(+)(+)}{(-)(+)}$	$\frac{(+)(+)}{(+)(+)}$
so $y \to$	$-\infty$	∞	$-\infty$	∞

ADDITIONAL VALUES:

x	y
−6	0.93
−3	−1.75
−1	4.50
1.5	6.29
2	4.50
3	3.50

GRAPH:

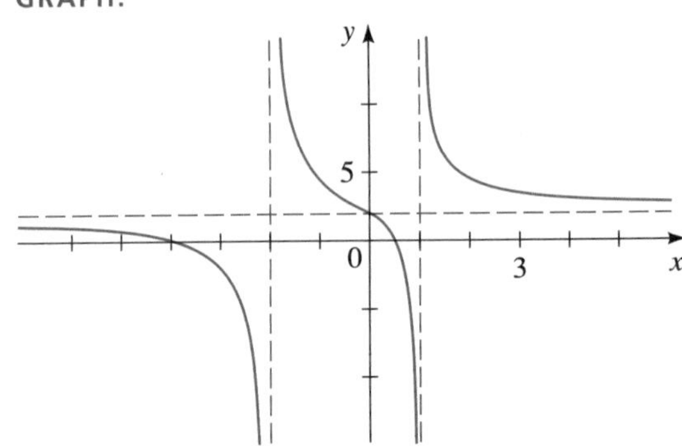

FIGURE 4

■

The Three Forms of A Rational Function

To find the **y-intercept** of a rational function, we use the *original form* of the function.

To find the **x-intercepts** and **vertical asymptotes**, we use the *factored form* of the function.

To find the **horizontal asymptote**, we use a *compound-fraction form* of the function.

Notice that we use three forms of the equation defining the rational function when we perform this analysis. In Example 3 the *original form*

$$r(x) = \frac{2x^2 + 7x - 4}{x^2 + x - 2}$$

easily gave us the y-intercept $r(0) = (-4)/(-2) = 2$. The *factored form*

$$r(x) = \frac{(2x - 1)(x + 4)}{(x - 1)(x + 2)}$$

gave us the x-intercepts $\frac{1}{2}$ and -4, and the vertical asymptotes $x = 1$ and $x = -2$. Finally, the *compound-fraction form*

$$r(x) = \frac{2 + \dfrac{7}{x} - \dfrac{4}{x^2}}{1 + \dfrac{1}{x} - \dfrac{2}{x^2}}$$

tells us that the horizontal asymptote is $y = \frac{2}{1} = 2$.

We can determine whether a rational function $r(x) = P(x)/Q(x)$ has a horizontal asymptote by considering the degrees of the numerator and denominator. If the degrees of P and Q are the same (both n, say), then dividing numerator and denominator by x^n shows that the horizontal asymptote is

$$y = \frac{\text{leading coefficient of } P}{\text{leading coefficient of } Q}$$

as we saw in Example 3. The following box summarizes the procedure for finding asymptotes.

ASYMPTOTES OF RATIONAL FUNCTIONS

Let r be the rational function

$$r(x) = \frac{a_n x^n + a_{n-1}x^{n-1} + \cdots + a_1 x + a_0}{b_m x^m + b_{m-1}x^{m-1} + \cdots + b_1 x + b_0}$$

1. The vertical asymptotes of r are the lines $x = a$, where a is a zero of the denominator.
2. (a) If $n < m$, then r has horizontal asymptote $y = 0$.

   (b) If $n = m$, then r has horizontal asymptote $y = \frac{a_n}{b_m}$.

   (c) If $n > m$, then r has no horizontal asymptote.

EXAMPLE 4 ■ Graphing a Rational Function

Sketch a graph of the rational function $r(x) = \dfrac{x-2}{x^2-1}$.

SOLUTION

FACTOR: $y = \dfrac{x-2}{(x-1)(x+1)}$

x-INTERCEPT: 2, from $x - 2 = 0$

y-INTERCEPT: 2, because $r(0) = \dfrac{0-2}{0^2-1} = 2$

HORIZONTAL ASYMPTOTE: $y = 0$, since degree of numerator < degree of denominator

VERTICAL ASYMPTOTES: $x = 1$ and $x = -1$, from the zeros of the denominator

BEHAVIOR NEAR VERTICAL ASYMPTOTES:

As $x \to$	1^+	1^-	-1^+	-1^-
the sign of $y = \dfrac{x-2}{(x-1)(x+1)}$ is	$\dfrac{(-)}{(+)(+)}$	$\dfrac{(-)}{(-)(+)}$	$\dfrac{(-)}{(-)(+)}$	$\dfrac{(-)}{(-)(-)}$
so $y \to$	$-\infty$	∞	∞	$-\infty$

ADDITIONAL VALUES:

x	y
-2	-1.33
-0.5	3.33
0.5	2
1.5	-0.4
3	0.125
4	0.133
5	0.125

GRAPH:

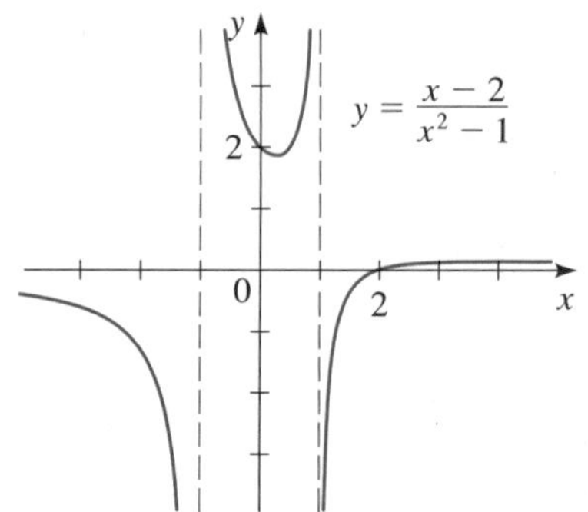

FIGURE 5 ■

From the graph in Figure 5, we see that, contrary to common misconception, a graph may cross a horizontal asymptote.

EXAMPLE 5 ■ Graphing a Rational Function

Sketch a graph of the rational function $r(x) = \dfrac{x^2 - 3x - 4}{2x^2 + 4x}$.

SOLUTION

FACTOR: $y = \dfrac{(x+1)(x-4)}{2x(x+2)}$

x-INTERCEPTS: -1 and 4, from $x + 1 = 0$ and $x - 4 = 0$

y-INTERCEPT: None, because $r(0)$ is undefined

HORIZONTAL ASYMPTOTE: $y = \frac{1}{2}$, because degree of numerator and denominator are the same and

$$\frac{\text{leading coefficient of numerator}}{\text{leading coefficient of denominator}} = \frac{1}{2}$$

VERTICAL ASYMPTOTES: $x = 0$ and $x = -2$, from the zeros of the denominator

BEHAVIOR NEAR VERTICAL ASYMPTOTES:

As $x \to$	-2^-	-2^+	0^-	0^+
the sign of $y = \dfrac{(x+1)(x-4)}{2x(x+2)}$ is	$\dfrac{(-)(-)}{(-)(-)}$	$\dfrac{(-)(-)}{(-)(+)}$	$\dfrac{(+)(-)}{(-)(+)}$	$\dfrac{(+)(-)}{(+)(+)}$
so $y \to$	∞	$-\infty$	∞	$-\infty$

ADDITIONAL VALUES:

x	y
-3	2.33
-2.5	3.90
-0.5	1.50
1	-1.00
3	-0.13
5	0.09

GRAPH:

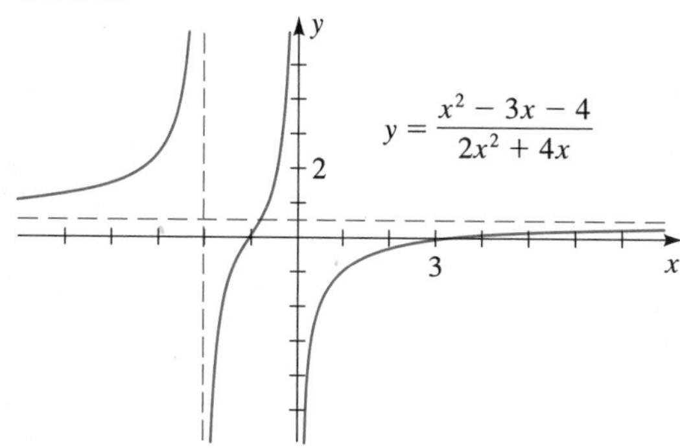

FIGURE 6

■

SLANT ASYMPTOTES AND END BEHAVIOR

If $r(x) = P(x)/Q(x)$ is a rational function in which the degree of the numerator is one more than the degree of the denominator, we can use the Division Algorithm to express the function in the form

$$r(x) = ax + b + \frac{R(x)}{Q(x)}$$

where the degree of R is less than the degree of Q and $a \neq 0$. This means that as $x \to \pm\infty$, $R(x)/Q(x) \to 0$, so for large values of $|x|$, the graph of $y = r(x)$ approaches the graph of the line $y = ax + b$. In this situation we say that $y = ax + b$ is a **slant asymptote**, or an **oblique asymptote**.

EXAMPLE 6 ■ A Rational Function with a Slant Asymptote

Sketch a graph of the rational function $r(x) = \dfrac{x^2 - 4x - 5}{x - 3}$.

SOLUTION

FACTOR: $y = \dfrac{(x + 1)(x - 5)}{x - 3}$

x-INTERCEPTS: -1 and 5, from $x + 1 = 0$ and $x - 5 = 0$

y-INTERCEPTS: $\frac{5}{3}$, because $r(0) = \dfrac{0^2 - 4 \cdot 0 - 5}{0 - 3} = \dfrac{5}{3}$

HORIZONTAL ASYMPTOTE: None, because degree of numerator is greater than degree of denominator

VERTICAL ASYMPTOTE: $x = 3$, from the zero of the denominator

BEHAVIOR NEAR VERTICAL ASYMPTOTE: $y \to \infty$ as $x \to 3^-$ and $y \to -\infty$ as $x \to 3^+$

$$\begin{array}{r} x - 1 \\ x-3\overline{)x^2 - 4x - 5} \\ \underline{x^2 - 3x} \\ -x - 5 \\ \underline{-x + 3} \\ -8 \end{array}$$

SLANT ASYMPTOTE: Since the degree of the numerator is one more than the degree of the denominator, the function has a slant asymptote. Dividing (see the margin), we obtain

$$r(x) = x - 1 + \frac{8}{x - 3}$$

Thus, $y = x - 1$ is the slant asymptote.

ADDITIONAL VALUES:

x	y
−2	−1.4
1	4
2	9
4	−5
6	2.33

GRAPH:

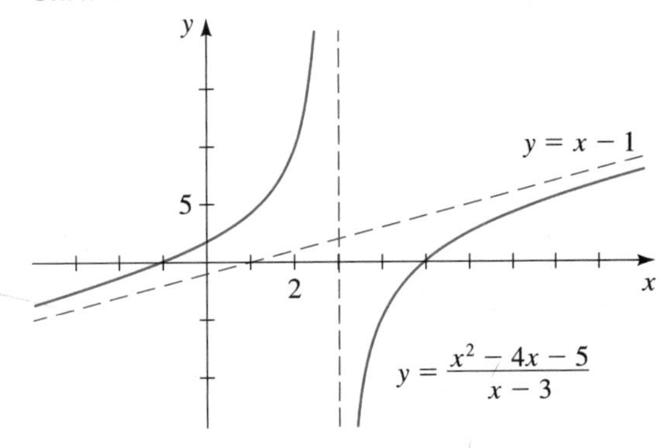

FIGURE 7

■

So far we have considered only horizontal and slant asymptotes as end behaviors for rational functions. In the next example we graph a function whose end behavior is like that of a parabola.

EXAMPLE 7 ■ End Behavior of a Rational Function

Graph the rational function

$$r(x) = \frac{x^3 - 2x^2 + 3}{x - 2}$$

and describe its end behavior.

SOLUTION

FACTOR: $y = \dfrac{(x + 1)(x^2 - 3x + 3)}{x - 2}$

x-INTERCEPTS: -1, from $x + 1 = 0$ (The other factor in the numerator has no real zeros.)

y-INTERCEPTS: $-\frac{3}{2}$, because $r(0) = \frac{0^3 - 2 \cdot 0^2 + 3}{0 - 2} = -\frac{3}{2}$

HORIZONTAL ASYMPTOTE: None, because degree of numerator is greater than degree of denominator

VERTICAL ASYMPTOTE: $x = 2$, from the zero of the denominator

BEHAVIOR NEAR VERTICAL ASYMPTOTE: $y \to -\infty$ as $x \to 2^-$ and $y \to \infty$ as $x \to 2^+$

$$\begin{array}{r} x^2 \\ x - 2 \overline{) x^3 - 2x^2 + 0x + 3} \\ \underline{x^3 - 2x^2 } \\ 3 \end{array}$$

END BEHAVIOR: Dividing (see the margin), we get

$$r(x) = x^2 + \frac{3}{x - 2}$$

This shows that the end behavior of r is like that of the parabola $y = x^2$ because $3/(x - 2)$ is small when $|x|$ is large. That is, $3/(x - 2) \to 0$ as $x \to \pm\infty$. This means that the graph of r will be close to the graph of $y = x^2$ for large $|x|$.

GRAPH: In Figure 8(a) we graph r in a small viewing rectangle; we can see the intercepts, the vertical asymptotes, and the local minimum. In Figure 8(b) we graph r in a larger viewing rectangle; here the graph looks almost like the graph of a parabola. In Figure 8(c) we graph both $y = r(x)$ and $y = x^2$; these graphs are very close to each other except near the vertical asymptote.

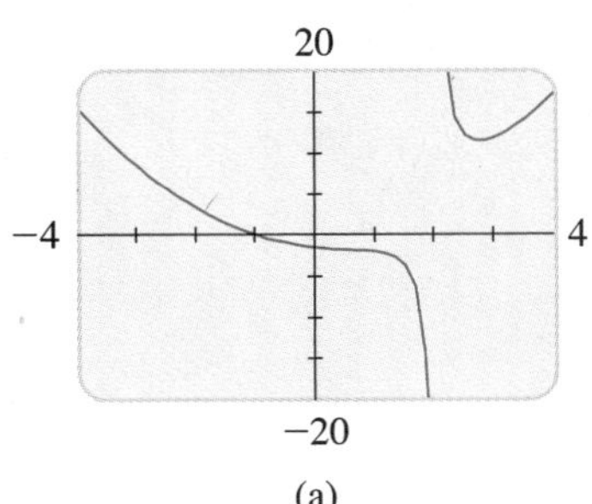

(a)

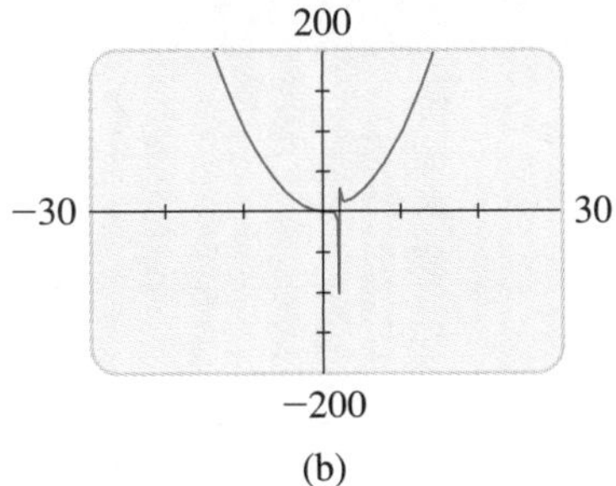

(b)

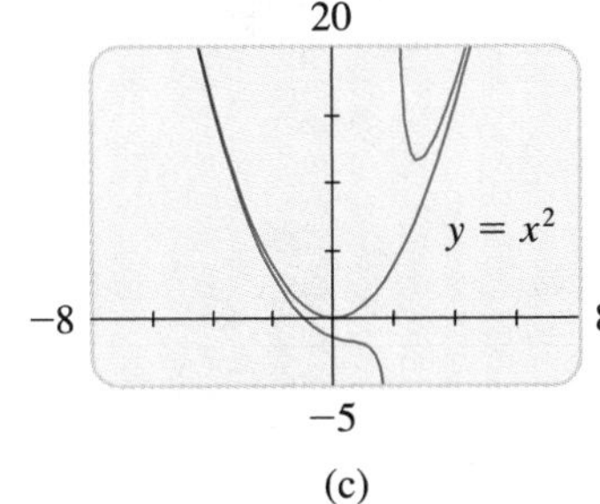

(c)

FIGURE 8
$r(x) = \frac{x^3 - 2x^2 + 3}{x - 2}$

■

Rational functions occur frequently in scientific applications of algebra. In the next example, we analyze the graph of a function from the theory of electricity.

EXAMPLE 8 ■ Electrical Resistance

When two resistors with resistances R_1 and R_2 are connected in parallel, their combined resistance R is given by the formula

$$R = \frac{R_1 R_2}{R_1 + R_2}$$

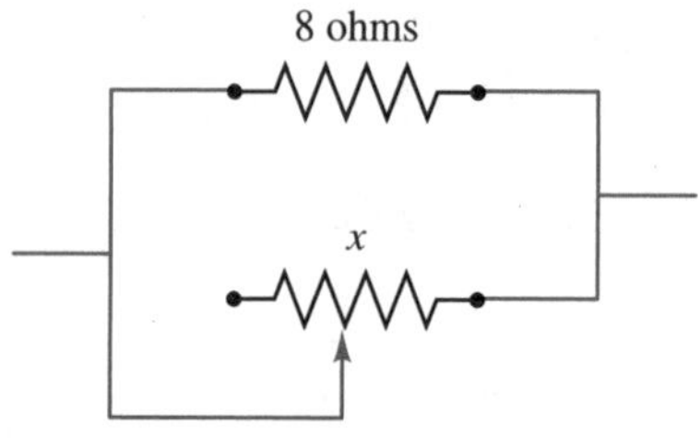

FIGURE 9

Suppose that a fixed 8-ohm resistor is connected in parallel with a variable resistor, as shown in Figure 9. If the resistance of the variable resistor is denoted by x, then the combined resistance R is a function of x. Graph R and give a physical interpretation of the graph.

SOLUTION

Substituting $R_1 = 8$ and $R_2 = x$ into the formula gives the function

$$R(x) = \frac{8x}{8 + x}$$

Since resistance cannot be negative, this function has physical meaning only when $x > 0$. The function is graphed in Figure 10(a) using the viewing rectangle $[0, 20]$ by $[0, 10]$. The function has no vertical asymptote when x is restricted to positive values. The combined resistance R increases as the variable resistance x increases. If we widen the viewing rectangle to $[0, 100]$ by $[0, 10]$, we obtain the graph in Figure 10(b). For large x, the combined resistance R levels off, getting closer and closer to the horizontal asymptote $R = 8$. No matter how large the variable resistance x, the combined resistance is never greater than 8 ohms.

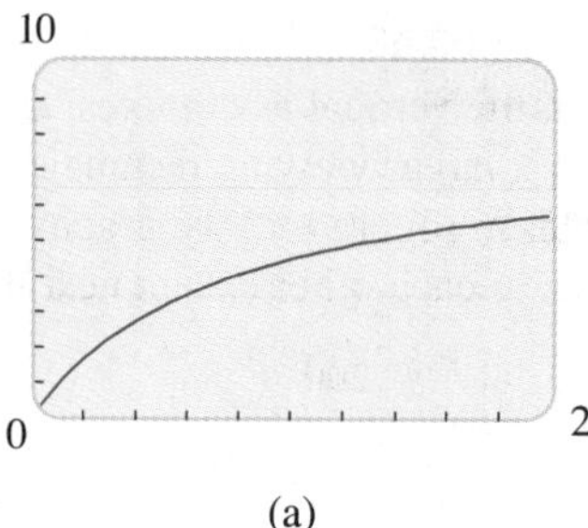

(a)

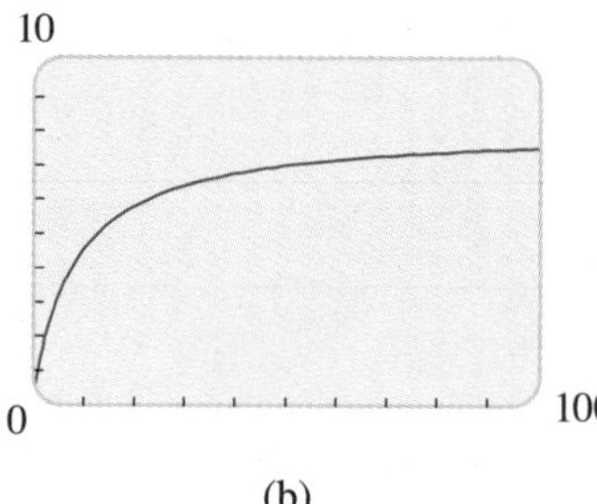

(b)

FIGURE 10 $R(x) = \frac{8x}{8 + x}$

■

5.5 EXERCISES

1–6 ■ Find the x- and y-intercepts of the rational function.

1. $r(x) = \frac{x - 2}{x + 3}$

2. $s(x) = \frac{2x}{3x + 5}$

3. $t(x) = \frac{x^2 - x - 2}{x - 6}$

4. $r(x) = \frac{2}{x^2 + 3x - 4}$

5. $r(x) = \frac{x^2 - 9}{x^2}$

6. $r(x) = \frac{x^3 + 8}{x^2 + 4}$

7–8 ■ From the graph, determine the x- and y-intercepts and the vertical and horizontal asymptotes.

7.

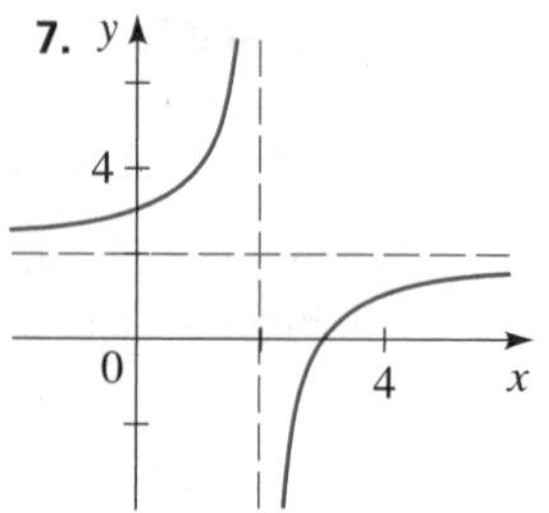

8.

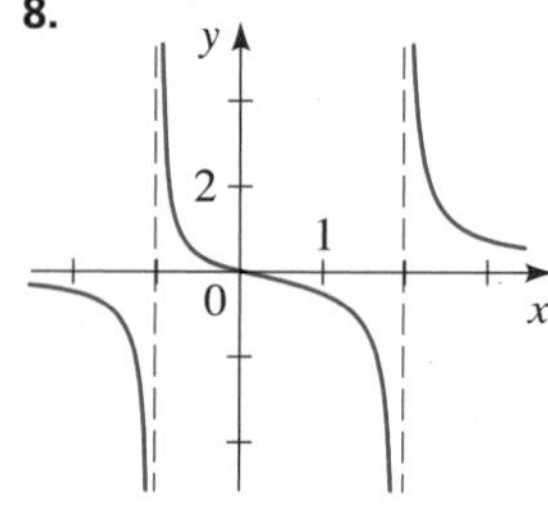

9–18 ■ Find all horizontal and vertical asymptotes (if any).

9. $r(x) = \dfrac{5}{x+3}$

10. $s(x) = \dfrac{3x+3}{x-3}$

11. $t(x) = \dfrac{x^2}{x^2 - x - 6}$

12. $r(x) = \dfrac{2x-4}{x^2+2x+1}$

13. $s(x) = \dfrac{6}{x^2+2}$

14. $t(x) = \dfrac{(x-1)(x-2)}{(x-3)(x-4)}$

15. $r(x) = \dfrac{6x-2}{x^2+5x-6}$

16. $s(x) = \dfrac{3x^2}{x^2+2x+5}$

17. $t(x) = \dfrac{x^2+2}{x-1}$

18. $r(x) = \dfrac{x^3+3x^2}{x^2-4}$

19–46 ■ Find the intercepts and asymptotes, and then sketch a graph of the rational function.

19. $r(x) = \dfrac{4}{x-2}$

20. $r(x) = \dfrac{9}{x+3}$

21. $r(x) = \dfrac{x-1}{x-2}$

22. $r(x) = \dfrac{x+9}{x-3}$

23. $r(x) = \dfrac{4x-4}{x+2}$

24. $r(x) = \dfrac{2x+6}{-6x+3}$

25. $s(x) = \dfrac{4-3x}{x+7}$

26. $s(x) = \dfrac{1-2x}{2x+3}$

27. $r(x) = \dfrac{18}{(x-3)^2}$

28. $r(x) = \dfrac{x-2}{(x+1)^2}$

29. $s(x) = \dfrac{4x-8}{(x-4)(x+1)}$

30. $s(x) = \dfrac{x+2}{(x+3)(x-1)}$

31. $s(x) = \dfrac{6}{x^2-5x-6}$

32. $s(x) = \dfrac{2x-4}{x^2+x-2}$

33. $t(x) = \dfrac{3x+6}{x^2+2x-8}$

34. $t(x) = \dfrac{x-2}{x^2-4x}$

35. $r(x) = \dfrac{(x-1)(x+2)}{(x+1)(x-3)}$

36. $r(x) = \dfrac{2x(x+4)}{(x-1)(x-2)}$

37. $r(x) = \dfrac{x^2-2x+1}{x^2+2x+1}$

38. $r(x) = \dfrac{4x^2}{x^2-2x-3}$

39. $r(x) = \dfrac{2x^2+10x-12}{x^2+x-6}$

40. $r(x) = \dfrac{2x^2+2x-4}{x^2+x}$

41. $r(x) = \dfrac{x^2-x-6}{x^2+3x}$

42. $r(x) = \dfrac{x^2+3x}{x^2-x-6}$

43. $r(x) = \dfrac{3x^2+6}{x^2-2x-3}$

44. $r(x) = \dfrac{5x^2+5}{x^2+4x+4}$

45. $s(x) = \dfrac{x^2-2x+1}{x^3-3x^2}$

46. $t(x) = \dfrac{x^3-x^2}{x^3-3x-2}$

47–54 ■ Find the slant asymptote, the vertical asymptotes, and sketch a graph of the function.

47. $r(x) = \dfrac{x^2}{x-2}$

48. $r(x) = \dfrac{x^2+2x}{x-1}$

49. $r(x) = \dfrac{x^2-2x-8}{x}$

50. $r(x) = \dfrac{3x-x^2}{2x-2}$

51. $r(x) = \dfrac{x^2+5x+4}{x-3}$

52. $r(x) = \dfrac{x^3+4}{2x^2+x-1}$

53. $r(x) = \dfrac{x^3+x^2}{x^2-4}$

54. $r(x) = \dfrac{2x^3+2x}{x^2-1}$

55–58 ■ Graph the rational function f and determine all vertical asymptotes from your graph. Then graph f and g in a sufficiently large viewing rectangle to show that they have the same end behavior.

55. $f(x) = \dfrac{2x^2+6x+6}{x+3}, \quad g(x) = 2x$

56. $f(x) = \dfrac{-x^3+6x^2-5}{x^2-2x}, \quad g(x) = -x+4$

57. $f(x) = \dfrac{x^3-2x^2+16}{x-2}, \quad g(x) = x^2$

58. $f(x) = \dfrac{-x^4+2x^3-2x}{(x-1)^2}, \quad g(x) = 1-x^2$

59–64 ■ Graph the rational function and find all vertical asymptotes, x- and y-intercepts, and local extrema, correct to the nearest decimal. Then use long division to find a polynomial that has the same end behavior as the rational function, and graph both functions in a sufficiently large viewing rectangle to verify that the end behaviors of the polynomial and the rational function are the same.

59. $y = \dfrac{2x^2-5x}{2x+3}$

60. $y = \dfrac{x^4-3x^3+x^2-3x+3}{x^2-3x}$

61. $y = \dfrac{x^5}{x^3-1}$

62. $y = \dfrac{x^4}{x^2-2}$

63. $r(x) = \dfrac{x^4 - 3x^3 + 6}{x - 3}$

64. $r(x) = \dfrac{4 + x^2 - x^4}{x^2 - 1}$

65. The rabbit population on Mr. Jenkins' farm follows the formula

$$p(t) = \frac{3000t}{t + 1}$$

where $t \geqslant 0$ is the time (in months) since the beginning of the year.

(a) Draw a graph of the rabbit population.

(b) What eventually happens to the rabbit population?

66. After a certain drug is injected into a patient, the concentration c of the drug in the bloodstream is monitored. At time $t \geqslant 0$ (in minutes since the injection), the concentration (in mg/L) is given by

$$c(t) = \frac{30t}{t^2 + 2}$$

(a) Draw a graph of the drug concentration.

(b) What eventually happens to the concentration of drug in the bloodstream?

67. A drug is administered to a patient and the concentration of the drug in the bloodstream is monitored. At time $t \geqslant 0$ (in hours since giving the drug), the concentration (in mg/L) is given by

$$c(t) = \frac{5t}{t^2 + 1}$$

Graph the function c with a graphing device.

(a) What is the highest concentration of drug that is reached in the patient's bloodstream?

(b) What happens to the drug concentration after a long period of time?

(c) How long does it take for the concentration to drop below 0.3 mg/L?

68. Suppose a rocket is fired upward from the surface of the earth with an initial velocity v (measured in m/s). Then the maximum height h (in meters) reached by the rocket is given by the function

$$h(v) = \frac{Rv^2}{2gR - v^2}$$

where $R = 6.4 \times 10^6$ m is the radius of the earth and $g = 9.8$ m/s^2 is the acceleration due to gravity. Use a graphing device to draw a graph of the function h. (Note that h and v must both be positive, so the viewing rectangle need not contain negative values.) What does the vertical asymptote represent physically?

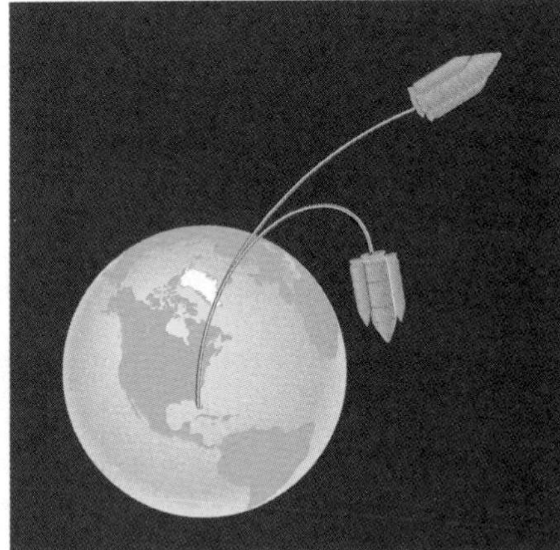

69. As a train moves toward an observer (see the figure), the pitch of its whistle sounds higher to the observer than it would if the train were at rest, because the crests of the sound waves are compressed closer together. This phenomenon is called the *Doppler effect*. The observed pitch P is a function of the speed v of the train and is given by

$$P(v) = P_0\left(\frac{s_0}{s_0 - v}\right)$$

where P_0 is the actual pitch of the whistle at the source and $s_0 = 332$ m/s is the speed of sound in air. Suppose that a train has a whistle pitched at $P_0 = 440$ Hz. Graph the function $y = P(v)$ using a graphing device. How can the vertical asymptote of this function be interpreted physically?

70. For a camera with a lens of fixed focal length F to focus on an object located a distance x from the lens, the film must be placed a distance y behind the lens, where F, x, and y are related by

$$\frac{1}{x} + \frac{1}{y} = \frac{1}{F}$$

(See the figure.) Suppose the camera has a 55-mm lens ($F = 55$).

(a) Express y as a function of x and graph the function.

(b) What happens to the focusing distance y as the object moves far away from the lens?

(c) What happens to the focusing distance y as the object moves close to the lens?

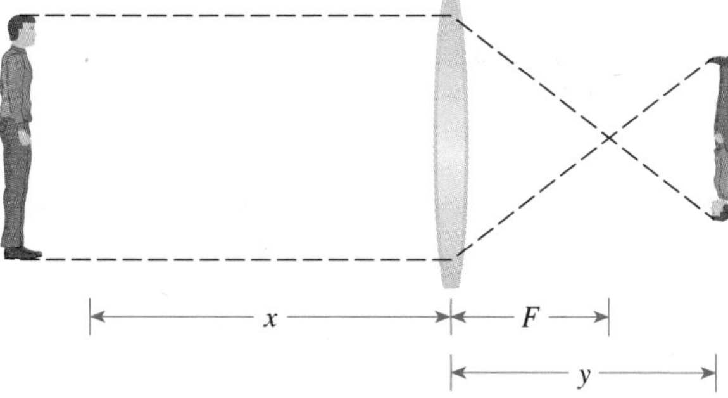

DISCOVERY • DISCUSSION

71. Constructing a Rational Function from Its Asymptotes Give an example of a rational function that has vertical asymptote $x = 3$. Now give an example of one that has vertical asymptote $x = 3$ *and* horizontal asymptote $y = 2$. Now give an example of a rational function with vertical asymptotes $x = 1$ and $x = -1$, horizontal asymptote $y = 0$, and x-intercept 4.

72. A Rational Function with No Asymptote Explain how you can tell (without graphing it) that the function

$$r(x) = \frac{x^6 + 10}{x^4 + 8x^2 + 15}$$

has no x-intercept and no horizontal, vertical, or slant asymptote. What is its end behavior?'

73. Graphs with Holes In this chapter we adopted the convention that in rational functions, the numerator and denominator don't share a common factor. In this exercise we consider the graph of a rational function that doesn't satisfy this rule.

(a) Show that the graph of

$$r(x) = \frac{3x^2 - 3x - 6}{x - 2}$$

is the line $y = 3x + 3$ with the point $(2, 9)$ removed. [*Hint:* Factor. What is the domain of r?]

(b) Graph the rational functions:

$$s(x) = \frac{x^2 + x - 20}{x + 5}$$

$$t(x) = \frac{2x^2 - x - 1}{x - 1}$$

$$u(x) = \frac{x - 2}{x^2 - 2x}$$

5 REVIEW

CONCEPT CHECK

1. (a) Write the defining equation for a polynomial P of degree n.
   (b) What does it mean to say that c is a zero of P?
2. Sketch graphs showing the possible end behaviors of polynomials of odd degree and of even degree.
3. What steps would you follow to graph a polynomial by hand?
4. (a) What is meant by a local maximum point or local minimum point of a polynomial?
   (b) How many local extrema can a polynomial of degree n have?
5. State the Division Algorithm and identify the dividend, divisor, quotient, and remainder.
6. How does synthetic division work?
7. (a) State the Remainder Theorem.
   (b) State the Factor Theorem.
8. (a) State the Rational Zeros Theorem.
   (b) What steps would you take to find the rational zeros of a polynomial?
9. State Descartes' Rule of Signs.
10. (a) What does it mean to say that a is a lower bound and b is an upper bound for the zeros of a polynomial?
    (b) State the Upper and Lower Bounds Theorem.
11. (a) State the Fundamental Theorem of Algebra.
    (b) State the Complete Factorization Theorem.
    (c) What does it mean to say that c is a zero of multiplicity k of a polynomial P?
    (d) State the Zeros Theorem.
    (e) State the Conjugate Roots Theorem.
12. (a) What is a rational function?
    (b) What does it mean to say that $x = a$ is a vertical asymptote of $y = f(x)$?
    (c) How do you locate a vertical asymptote?
    (d) What does it mean to say that $y = b$ is a horizontal asymptote of $y = f(x)$?
    (e) How do you locate a horizontal asymptote?
    (f) What steps do you follow to sketch the graph of a rational function by hand?
    (g) Under what circumstances does a rational function have a slant asymptote? If one exists, how do you find it?
    (h) How do you determine the end behavior of a rational function?

EXERCISES

1–6 ■ Graph the polynomial. Show clearly all x- and y-intercepts.

1. $P(x) = (x - 2)^3 + 8$
2. $P(x) = 32 - 2x^4$
3. $P(x) = x^3 - 9x$
4. $P(x) = x^3 - 5x^2 - 6x$
5. $P(x) = x^3 - 5x^2 - 4x + 20$
6. $P(x) = x^4 - 9x^2$

7–10 ■ Use a graphing device to graph the polynomial. Find the x- and y-intercepts and the coordinates of all local extrema, correct to the nearest decimal. Describe the end behavior of the function.

7. $P(x) = 2x^3 + x^2 - 18x - 9$
8. $P(x) = x^4 - 8x^2 + 16$
9. $P(x) = x^5 + x^2 - 5$
10. $P(x) = 3x^5 + x^4 - 4x$

11–18 ■ Find the quotient and remainder.

11. $\dfrac{x^3 - x^2 + x - 11}{x - 3}$
12. $\dfrac{x^4 + 30x + 12}{x + 3}$
13. $\dfrac{x^3 - x^2 - 11x + 6}{x^2 + 2x - 5}$
14. $\dfrac{x^5 - 3x^4 + 3x^3 + 20x - 6}{x^2 + 2x - 6}$
15. $\dfrac{x^4 - 25x^2 + 4x + 15}{x + 5}$
16. $\dfrac{2x^3 - x^2 - 5}{x - \frac{3}{2}}$
17. $\dfrac{x^4 + x^3 - 2x^2 - 3x - 1}{x - 2}$

18. $\dfrac{x^4 - 3x^2 + 1}{x + 1}$

19–20 ■ Find the indicated value of the polynomial using the Remainder Theorem.

19. $P(x) = 2x^3 - 9x^2 - 7x + 13$; find $P(5)$

20. $Q(x) = x^4 + 4x^3 + 7x^2 + 10x + 15$; find $Q(-3)$

21. Show that $\frac{1}{2}$ is a zero of the polynomial

$$P(x) = 2x^4 + x^3 - 5x^2 + 10x - 4$$

22. Use the Factor Theorem to show that $x + 4$ is a factor of the polynomial

$$P(x) = x^5 + 4x^4 - 7x^3 - 23x^2 + 23x + 12$$

23. What is the remainder when the polynomial

$$P(x) = x^{500} + 6x^{201} - x^2 - 2x + 4$$

is divided by $x - 1$?

24. What is the remainder when $x^{101} - x^4 + 2$ is divided by $x + 1$?

25–26 ■ **(a)** List all possible rational roots (without testing to see if they actually are roots). **(b)** Determine the possible number of positive and negative real roots using Descartes' Rule of Signs.

25. $x^5 - 6x^3 - x^2 + 2x + 18 = 0$

26. $6x^4 + 3x^3 + x^2 + 3x + 4 = 0$

27. Find a polynomial of degree 3 with constant coefficient 12 and zeros $-\frac{1}{2}$, 2, and 3.

28. Find a polynomial of degree 4 having integer coefficients and zeros $3i$ and 4, with 4 a double zero.

29. Does there exist a polynomial of degree 4 with integer coefficients that has zeros i, $2i$, $3i$, and $4i$? If so, find it. If not, explain why.

30. Prove that the equation $3x^4 + 5x^2 + 2 = 0$ has no real root.

31–40 ■ Find all rational, irrational, and complex zeros (and state their multiplicities). Use Descartes' Rule of Signs, the Upper and Lower Bounds Theorem, the quadratic formula, or other factoring techniques to help you whenever possible.

31. $P(x) = x^3 - 3x^2 - 13x + 15$

32. $P(x) = 2x^3 + 5x^2 - 6x - 9$

33. $P(x) = x^4 + 6x^3 + 17x^2 + 28x + 20$

34. $P(x) = x^4 + 7x^3 + 9x^2 - 17x - 20$

35. $P(x) = x^5 - 3x^4 - x^3 + 11x^2 - 12x + 4$

36. $P(x) = x^4 - 81$

37. $P(x) = x^6 - 64$

38. $P(x) = 18x^3 + 3x^2 - 4x - 1$

39. $P(x) = 6x^4 - 18x^3 + 6x^2 - 30x + 36$

40. $P(x) = x^4 + 15x^2 + 54$

41–44 ■ Use a graphing device to find all real solutions of the equation.

41. $2x^2 = 5x + 3$

42. $x^3 + x^2 - 14x - 24 = 0$

43. $x^4 - 3x^3 - 3x^2 - 9x - 2 = 0$

44. $x^5 = x + 3$

45–50 ■ Graph the rational function. Show clearly all x- and y-intercepts and asymptotes.

45. $r(x) = \dfrac{3x - 12}{x + 1}$

46. $r(x) = \dfrac{1}{(x + 2)^2}$

47. $r(x) = \dfrac{x - 2}{x^2 - 2x - 8}$

48. $r(x) = \dfrac{2x^2 - 6x - 7}{x - 4}$

49. $r(x) = \dfrac{x^2 - 9}{2x^2 + 1}$

50. $r(x) = \dfrac{x^3 + 27}{x + 4}$

51–54 ■ Use a graphing device to analyze the graph of the rational function. Find all x- and y-intercepts, and all vertical, horizontal, and slant asymptotes. If the function has no horizontal or slant asymptote, find a polynomial that has the same end behavior as the rational function.

51. $r(x) = \dfrac{x - 3}{2x + 6}$

52. $r(x) = \dfrac{2x - 7}{x^2 + 9}$

53. $r(x) = \dfrac{x^3 + 8}{x^2 - x - 2}$

54. $r(x) = \dfrac{2x^3 - x^2}{x + 1}$

55. Find the coordinates of all points of intersection of the graphs of

$$y = x^4 + x^2 + 24x \quad \text{and} \quad y = 6x^3 + 20$$

5 TEST

1. Graph the polynomial $P(x) = x^3 - x^2 - 9x + 9$, showing clearly all x- and y-intercepts.

2. (a) Use synthetic division to find the quotient and remainder when $x^4 - 4x^2 + 2x + 5$ is divided by $x - 2$.
   (b) Use long division to find the quotient and remainder when $2x^5 + 4x^4 - x^3 - x^2 + 7$ is divided by $2x^2 - 1$.

3. Let $P(x) = 2x^4 - 7x^3 + x^2 + 7x - 3$.
   (a) List all possible rational zeros of P.
   (b) Find the complete factorization of P.
   (c) Find the zeros of P.

4. Find all real and complex zeros of $P(x) = x^3 - x^2 - 4x - 6$.

5. Find the complete factorization of $P(x) = x^4 - 2x^3 + 5x^2 - 8x + 4$.

6. Find a fourth-degree polynomial with integer coefficients that has zeros $3i$ and -1, with -1 a zero of multiplicity 2.

7. Let $P(x) = 2x^4 - 7x^3 + x^2 - 18x + 3$.
   (a) Use Descartes' Rule of Signs to determine how many positive and how many negative real zeros P can have.
   (b) Show that 4 is an upper bound and -1 is a lower bound for the real zeros of P.

   (c) Draw a graph of P and use it to estimate the real zeros of P, correct to two decimal places.

8. Consider the following rational functions:

$$r(x) = \frac{2x - 1}{x^2 - x - 2} \qquad s(x) = \frac{x^3 + 27}{x^2 + 4} \qquad t(x) = \frac{x^3 - 9x}{x + 2} \qquad u(x) = \frac{x^2 + x - 6}{x^2 - 25}$$

   (a) Which of these rational functions has a horizontal asymptote?
   (b) Which of these functions has a slant asymptote?
   (c) Which of these functions has no vertical asymptote?
   (d) Graph $y = u(x)$, showing clearly any asymptotes and x- and y-intercepts the function may have.

   (e) Use long division to find a polynomial P that has the same end behavior as t. Graph both P and t on the same screen to verify that they have the same end behavior.

FOCUS ON PROBLEM SOLVING

It is often necessary to combine different problem-solving principles. The principles of **drawing a diagram** and **introducing something extra** are used in the two problems that follow. Euler's solution of the "Königsberg bridge problem" is so brilliant that it is the basis for a whole field of mathematics called *network theory*, which has practical applications to electric circuits and economics.

THE KÖNIGSBERG BRIDGE PROBLEM

The old city of Königsberg, now called Kaliningrad, is situated on both banks of the Pregel River and on two islands in the river. Seven bridges connect the parts of the city, as shown in Figure 1. After the last bridge was built, the Königsberg townspeople posed the problem of finding a route through the city that would cross all seven bridges in a continuous walk, without recrossing any bridge. No one was able to find such a route. When Euler heard of the problem, he realized that an interesting mathematical principle must be at work. To discover the principle, Euler first stripped away the nonessential parts of the problem by drawing a simpler diagram of the city in which he represented the land masses as points and the bridges as lines connecting the points (Figure 2).

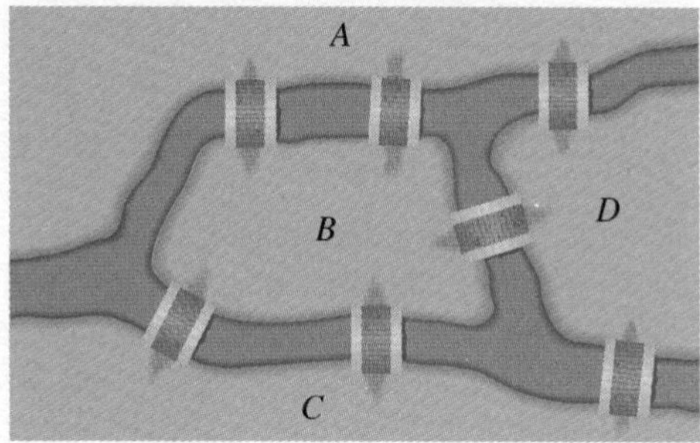

FIGURE 1

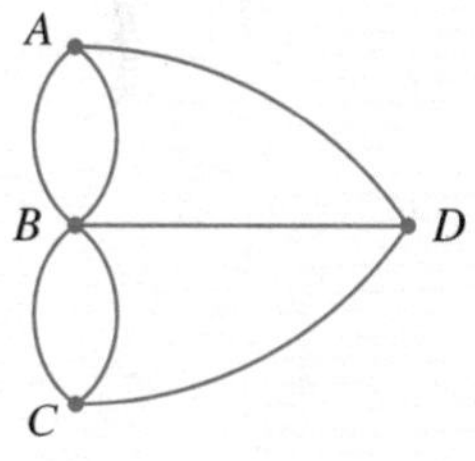

FIGURE 2

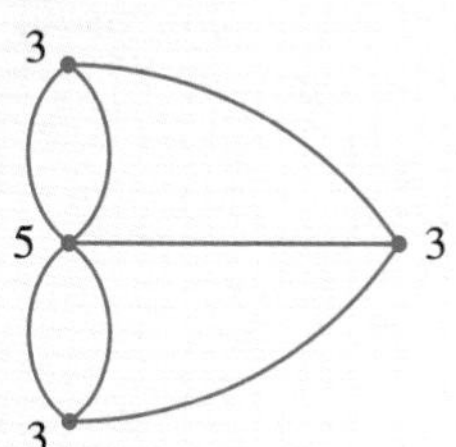

FIGURE 3

The problem now was to draw the diagram without lifting the pencil and without retracing any line. To show that this was impossible, Euler introduced something extra by labeling each point with the number of line segments that end at that point (Figure 3). He then argued that every pass through a point accounts for two line ends (one entering and one leaving). Thus for the walk to be possible, each point, except possibly the starting and ending points, must be labeled by an even number. Since more than two points are labeled by odd numbers, the walk is impossible!

The beauty of Euler's solution is that it is completely general. In modern language a diagram consisting of points and connecting lines is called a *network*, each point is a *vertex*, each line an *edge*, and the number of edges ending at a vertex is the *order* of the vertex. A network is *traversable* if it can be drawn in one stroke without retracing any edge. Euler's reasoning about the Königsberg bridge problem proved the following principle: *If a network has more than two vertices with odd order, then it is not traversable.*

Here is another clever use of the principle of **introducing something extra**.

THE IMPOSSIBLE MUSEUM TOUR

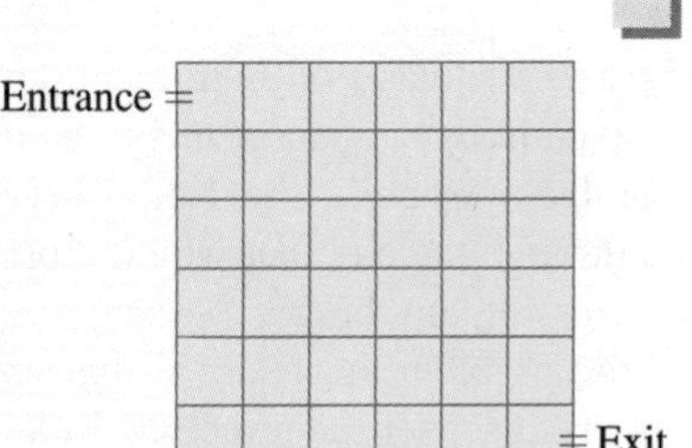

FIGURE 4

The floor plan of a museum is in the shape of a square with six square rooms to a side. Each room is connected to each adjacent room by a door. The museum entrance and exit are at diagonally opposite corners, as shown in Figure 4.

A visitor making a tour of this museum would like to visit each room exactly once and then exit. Can you find such a path? Here are examples of attempts that failed:

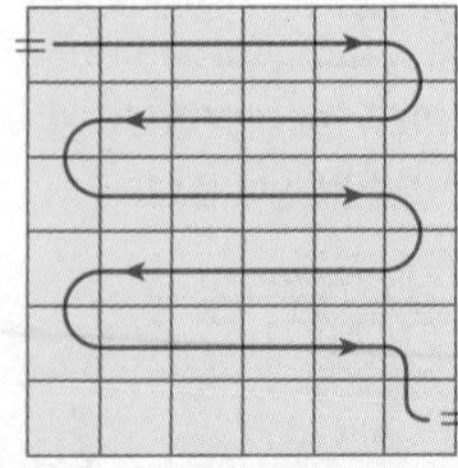

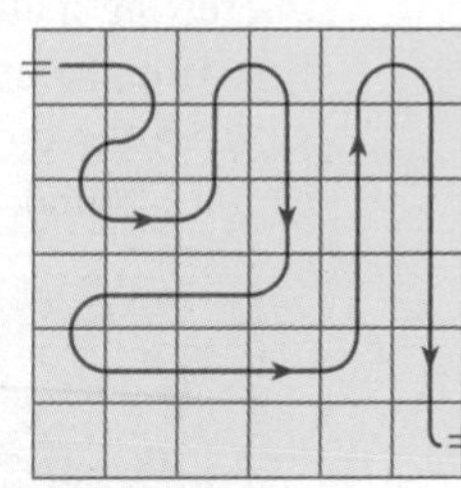

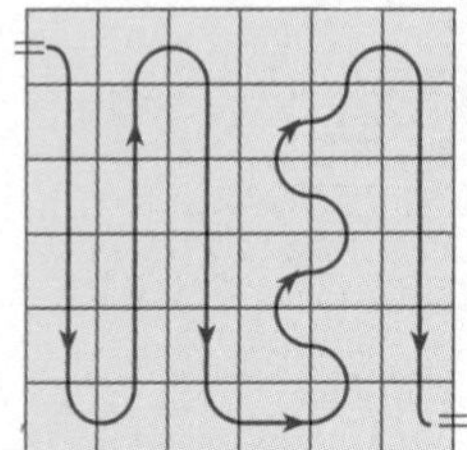

Trial and error will make it plausible to you that such a path is in fact not possible. But trial and error does not *prove* that our proposed tour of the museum is impossible. How do we give a convincing argument for this? Here is a clever idea: Imagine that the rooms are colored alternately black and white, like the checkerboard in Figure 5. So there are 18 black rooms and 18 white rooms. By coloring the rooms, we've introduced something extra that at first does not seem relevant. Indeed, the tour either can or cannot be done, regardless of the colors of the rooms. But notice how coloring the rooms in this fashion allows us to give a convincing argument that explains why this tour is impossible. We reason as follows.

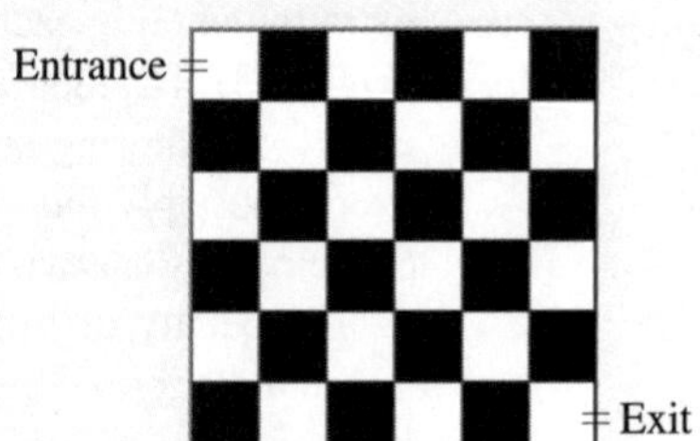

FIGURE 5

If a visitor is in a white room, he must next go to a black room, and if he is in a black room, he must next go to a white room. The entrance is in a white room, so the visitor's path, in terms of colors of rooms, is

W B W B W B ...

Since the museum contains an even number of rooms and the visitor starts in a white room, he must end his tour in a black room. But the exit is in a white room, so the proposed tour is impossible!

PROBLEMS

1–4 ■ Determine whether each network is traversable. If it is, find a path that traverses it.

1.

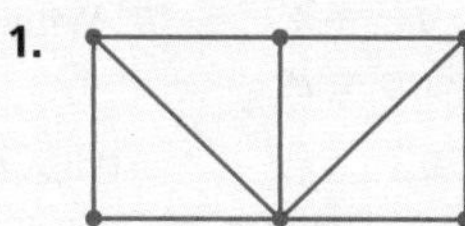

2.

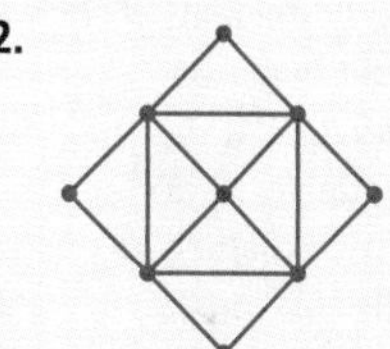

3.

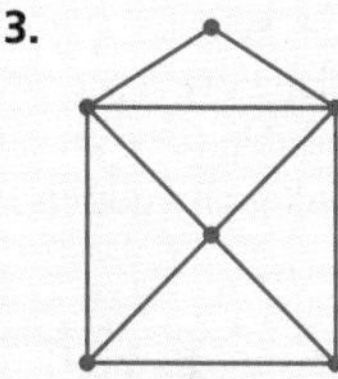

4.

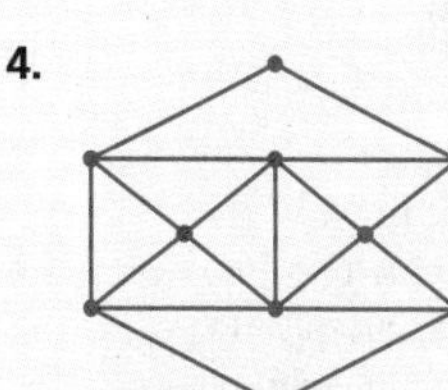

5. Is the museum tour possible if there are five rooms to a side instead of six? Find a general result about when the tour is possible for an $n \times n$ museum.

6. A Tibetan monk leaves the monastery at 7:00 A.M. and walks along his usual path to the top of the mountain, arriving at 7:00 P.M. The following morning, he starts at 7:00 A.M. at the top and takes the same path back, arriving at the monastery at 7:00 P.M. Show that there is exactly one point on the path that the monk will cross at exactly the same time of day on both trips.

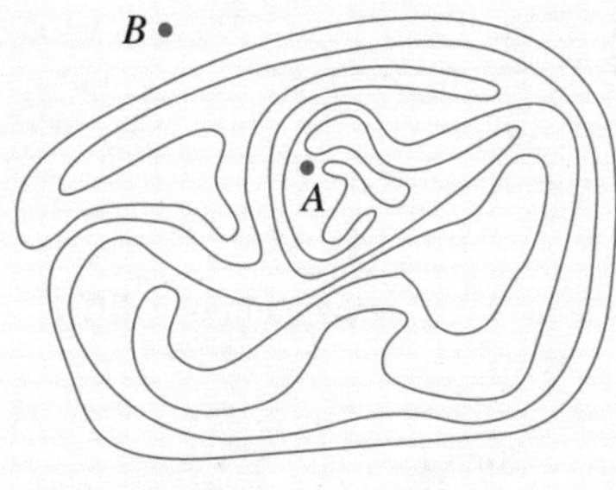

Jordan Curve

7. A curve that starts and ends at the same point and does not intersect itself is called a **Jordan curve**. Such a curve always has an "inside" and an "outside." Is the point A in the figure inside or outside this Jordan curve? Find an easy method of determining whether a point is inside or outside a Jordan cuve. [*Hint:* The point B is clearly outside the curve. Add a line segment connecting A and B.]

Srinivasa Ramanujan (1887–1920) was born into a poor family in the small town of Kumbakonam in India. Self-taught in mathematics, he worked in virtual isolation from other mathematicians. At the age of 25 he wrote a letter to G. H. Hardy, the leading British mathematician at the time, listing some of his discoveries. Hardy immediately recognized Ramanujan's genius and for the next six years the two worked together in London until Ramanujan fell ill and returned to his hometown in India, where he died a year later. Ramanujan was a genius with phenomenal ability to see hidden patterns in the properties of numbers. Most of his discoveries were written as complicated infinite series, the importance of which was not recognized until many years after his death. In the last year of his life he wrote 130 pages of mysterious formulas, many of which still defy proof. Hardy tells the story that when he visited Ramanujan in a hospital and arrived in a taxi, he remarked to Ramanujan that the cab's number,

(continued)

8. Consider the following sequence of polynomials:

$$P_1(x) = (x - a)$$
$$P_2(x) = (x - a)(x - b)$$
$$P_3(x) = (x - a)(x - b)(x - c)$$
$$\vdots$$

(a) Expand the polynomials P_2 and P_3.

(b) Using the pattern you observe from part (a), write the expansions of P_4 and P_5 without actually multiplying out these polynomials. The patterns you have found are called **Viète's relations** (see page 154).

(c) For a polynomial of degree n, express the coefficient of x^{n-1} and the constant coefficient in terms of the zeros of the polynomial.

(d) Let $P(x) = x^6 - 7x^5 + 2x^3 + 17$. Without finding the zeros, find the product and the sum of the zeros of P.

9. If the equation $x^4 + ax^2 + bx + c = 0$ has roots 1, 2, and 3, find c. [*Hint:* Use the Complete Factorization Theorem together with Problem 8(c).]

10. Find a polynomial of degree 3 with integer coefficients that has $2 - \sqrt[3]{2}$ as one of its zeros. How many other real zeros does the polynomial have?

11. You have an 8×8 grid with squares removed from two diagonally opposite corners, as shown in the figure. You also have a set of dominoes, each of which covers exactly two squares of the grid. Is it possible to cover all the remaining squares of the grid with dominoes, without overlapping any dominoes? [*Hint:* Color the grid as a checkerboard.]

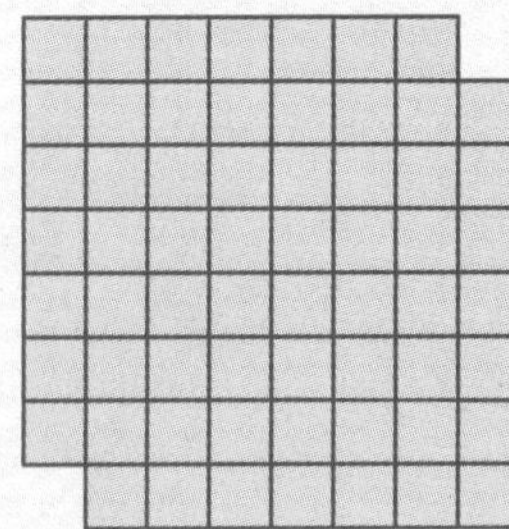

12. The number 1729 is the smallest positive integer that can be represented in two different ways as the sum of two cubes. What are the two ways?

13. If $f_0(x) = \dfrac{1}{1 - x}$ and $f_{n+1} = f_0 \circ f_n$ for $n = 0, 1, 2, \ldots$, find $f_{100}(3)$.

14. Sketch a graph of the function $g(x) = |x^2 - 1| - |x^2 - 4|$.

1729, was uninteresting. Ramanujan replied "No, it is a very interesting number. It is the smallest number expressible as the sum of two cubes in two different ways." (See Problem 12.)

15. Sketch a graph of the function $f(x) = x - [\![x]\!]$, where $[\![x]\!]$ is the greatest integer function defined on page 222.

16. Sketch the region in the plane defined by the equation

$$[\![x]\!]^2 + [\![y]\!]^2 = 1$$

17. (a) Show that if you multiply four consecutive integers and then add 1 to the result, you get a perfect square.

(b) Show that if you multiply three consecutive integers and then add the middle integer to the result, you get a perfect cube.

18. The equation $x^3 = 1$ has only one real solution, so it must have two imaginary solutions. Let w be one of the imaginary solutions. Show that the expression

$$(1 - w + w^2)(1 + w - w^2)$$

is a real number, and find its value.

19. A rectangle is divided into smaller rectangles in such a way that each of the smaller rectangles has at least one side of integer length. Show that one side of the original, large rectangle must have integer length. [*Hint:* Imagine the rectangle lying in the coordinate plane, which has been tiled into a checkerboard pattern of black and white squares, each $\frac{1}{2}$ unit by $\frac{1}{2}$ unit. Observe that a rectangle has at least one integer side if and only if its area is composed of equal amounts of black and white.]

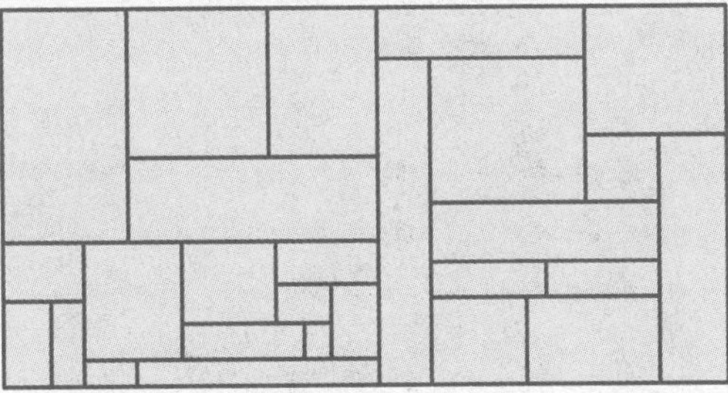

6 EXPONENTIAL AND LOGARITHMIC FUNCTIONS

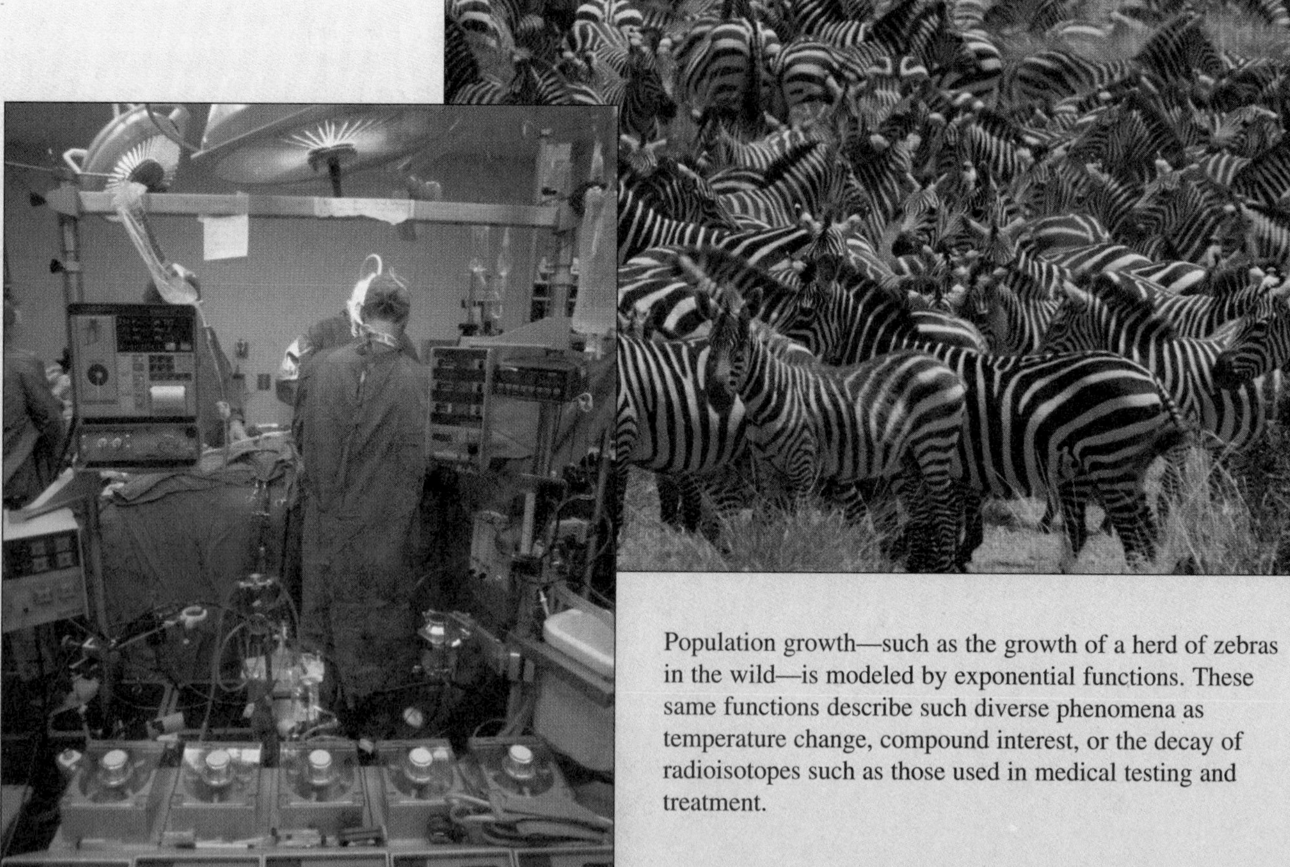

Population growth—such as the growth of a herd of zebras in the wild—is modeled by exponential functions. These same functions describe such diverse phenomena as temperature change, compound interest, or the decay of radioisotopes such as those used in medical testing and treatment.

Mathematics compares the most diverse phenomena, and discovers the secret analogies which unite them.

JOSEPH FOURIER

So far we have studied relatively simple functions such as polynomial and rational functions. We now turn our attention to two of the most important functions in mathematics, the *exponential function* and its inverse function, the *logarithmic function*. We use these functions to describe exponential growth in biology and economics and radioactive decay in physics and chemistry, and other phenomena.

6.1 EXPONENTIAL FUNCTIONS

In Section 1.3 we defined a^x if $a > 0$ and x is a rational number, but we have not yet defined irrational powers. For instance, what is meant by $2^{\sqrt{3}}$ or 5^{π}? To help us answer these questions, we first look at the graph of the function $f(x) = 2^x$, where x is rational. A very crude representation of this graph is shown in Figure 1. The graph has a hole wherever x is irrational, and we want to fill in these holes with a smooth curve. To do this, we must define irrational powers of 2 appropriately. For example, to define $2^{\sqrt{3}}$ we use rational approximations of $\sqrt{3}$. Since

$$\sqrt{3} \approx 1.73205 \ldots$$

we successively approximate $2^{\sqrt{3}}$ by the following rational powers:

$$2^{1.7}, \quad 2^{1.73}, \quad 2^{1.732}, \quad 2^{1.7320}, \quad 2^{1.73205}, \ldots$$

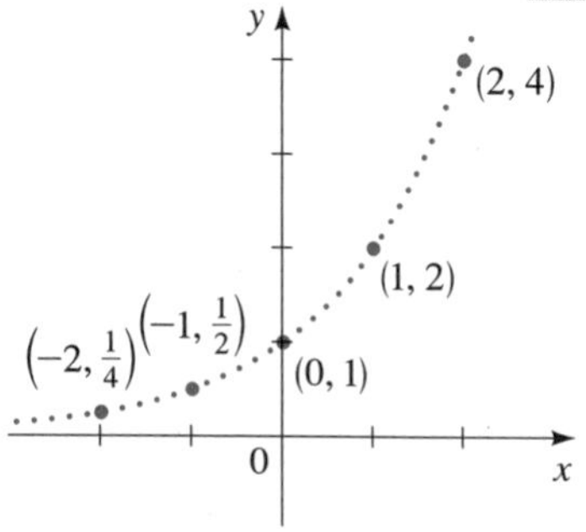

FIGURE 1

Representation of $f(x) = 2^x$ for x rational

Using advanced mathematics, it can be shown that there is exactly one number that these powers approach. We define $2^{\sqrt{3}}$ to be this number. Intuitively, these rational powers of 2 are getting closer and closer to $2^{\sqrt{3}}$. By this process we can approximate $2^{\sqrt{3}}$ to as many decimal places as we want:

$$2^{\sqrt{3}} \approx 3.321997 \ldots$$

Similarly, we can define 2^x (or a^x if $a > 0$) where x is any irrational number.

Using this definition of irrational powers, we can graph $f(x) = 2^x$ on all of $\mathbb{R}$, as shown in Figure 2. It is an increasing function and, in fact, it increases very rapidly when $x > 0$ (see the margin note).

To demonstrate just how quickly $f(x) = 2^x$ increases, let's perform the following thought experiment. Suppose we start with a piece of paper a thousandth of an inch thick, and we fold it in half 50 times. Each time we fold the paper, the thickness of the paper stack doubles, so the thickness of the resulting stack would be $2^{50}/1000$ inches. How thick do you think that is? It works out to be more than 17 million miles!

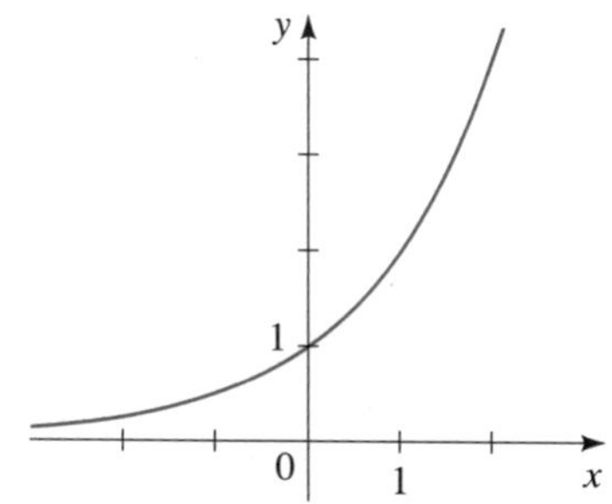

FIGURE 2

$f(x) = 2^x$ for x real

It can be proved that *the Laws of Exponents are still true when the exponents are real numbers.*

EXAMPLE 1 ■ Graphing Exponential Functions by Plotting Points

Draw the graph of each function.

(a) $f(x) = 3^x$ (b) $g(x) = \left(\dfrac{1}{3}\right)^x$

SOLUTION

We calculate values of $f(x)$ and $g(x)$ and plot points to sketch the graphs in Figure 3.

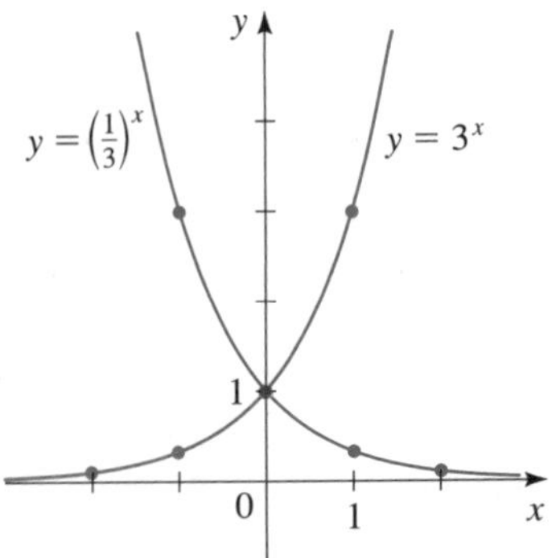

FIGURE 3

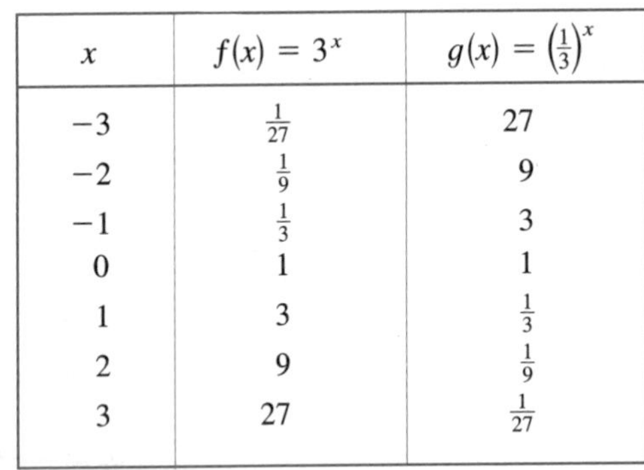

x	$f(x) = 3^x$	$g(x) = \left(\frac{1}{3}\right)^x$
-3	$\frac{1}{27}$	27
-2	$\frac{1}{9}$	9
-1	$\frac{1}{3}$	3
0	1	1
1	3	$\frac{1}{3}$
2	9	$\frac{1}{9}$
3	27	$\frac{1}{27}$

Notice that

$$g(x) = \left(\frac{1}{3}\right)^x = \frac{1}{3^x} = 3^{-x} = f(-x)$$

and so we could have obtained the graph of g from the graph of f by reflecting in the y-axis. ■

Reflecting graphs is explained in Section 4.5.

Figure 4 shows the graphs of the family of exponential functions $f(x) = a^x$ for various values of the base a. All of these graphs pass through the point $(0, 1)$

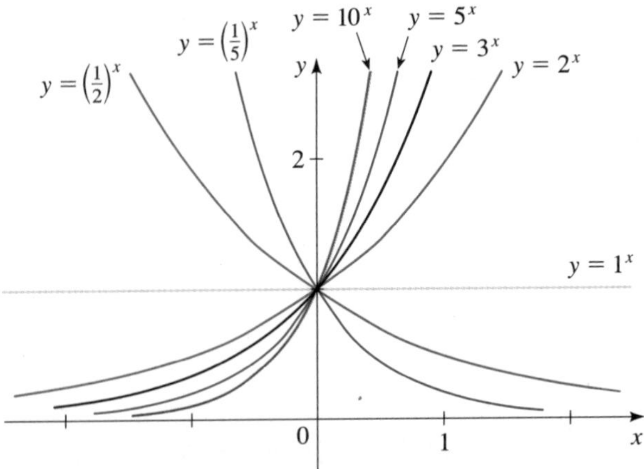

FIGURE 4
A family of exponential functions

because $a^0 = 1$ for $a \neq 0$. You can see from Figure 4 that there are three kinds of exponential functions $y = a^x$: If $0 < a < 1$, the exponential function decreases rapidly. If $a = 1$, it is constant. If $a > 1$, the function increases rapidly, and the larger the base, the more rapid the increase.

If $a > 1$, then the graph of $y = a^x$ approaches $y = 0$ as x decreases through negative values, and so the x-axis is a horizontal asymptote. If $0 < a < 1$, the graph approaches $y = 0$ as x increases indefinitely and, again, the x-axis is a horizontal asymptote. In either case the graph never touches the x-axis because $a^x > 0$ for all x. Thus, for $a \neq 1$, the exponential function $f(x) = a^x$ has domain $\mathbb{R}$ and range $(0, \infty)$. Let's summarize the preceding discussion.

EXPONENTIAL FUNCTIONS

For $a > 0$, the **exponential function with base a** is defined by

$$f(x) = a^x$$

For $a \neq 1$, the domain of f is $\mathbb{R}$, the range of f is $(0, \infty)$, and the graph of f has one of the following shapes:

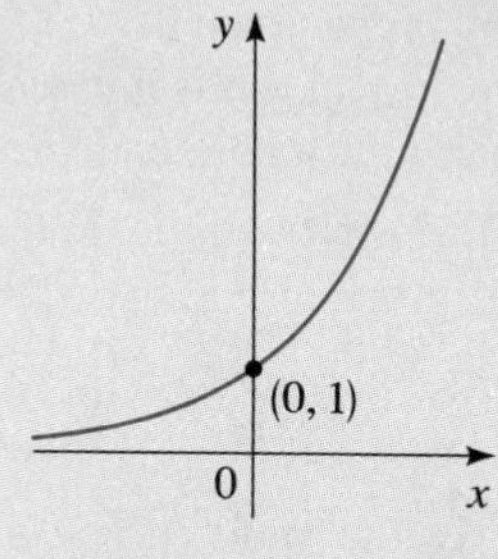

$f(x) = a^x$ for $a > 1$

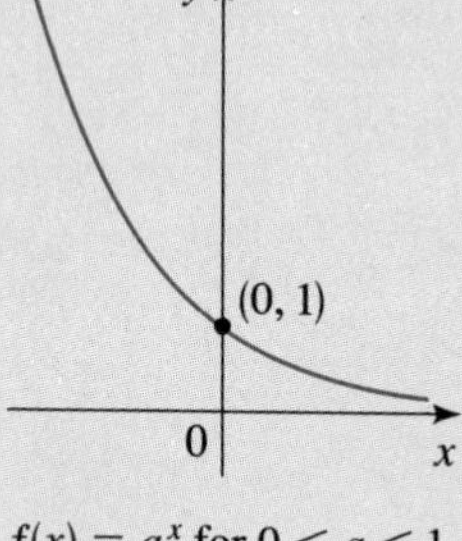

$f(x) = a^x$ for $0 < a < 1$

EXAMPLE 2 ■ Identifying Graphs of Exponential Functions

Find the exponential function $f(x) = a^x$ whose graph is given.

(a)

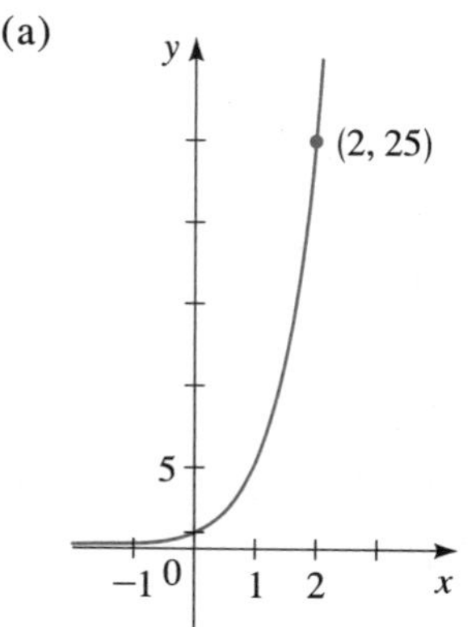

(b)

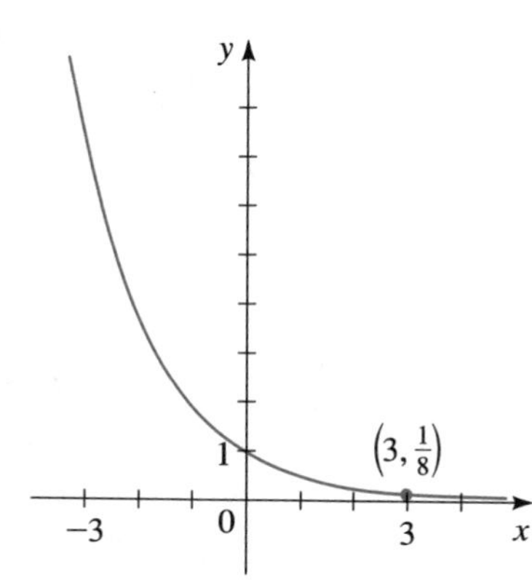

SOLUTION

(a) Since $f(2) = a^2 = 25$, we see by inspection that the base is $a = 5$. So $f(x) = 5^x$.

(b) Since $f(3) = a^3 = \frac{1}{8}$, we see by inspection that the base is $a = \frac{1}{2}$. So $f(x) = \left(\frac{1}{2}\right)^x$. ■

In the next two examples we see how to graph certain functions, not by plotting points, but by taking the basic graphs of the exponential functions in Figure 4 and applying the shifting and reflecting transformations of Section 4.5.

EXAMPLE 3 ■ Transformations of Exponential Functions

Use the graph $f(x) = 2^x$ to sketch the graph of each function.

(a) $g(x) = 1 + 2^x$ (b) $h(x) = -2^x$

SOLUTION

Vertical shifting and reflecting of graphs is explained in Section 4.5.

(a) To obtain the graph of $g(x) = 1 + 2^x$, we start with the graph of $f(x) = 2^x$ in Figure 5(a) and shift it upward 1 unit. Notice from Figure 5(b) that the line $y = 1$ is now a horizontal asymptote.

(b) Again we start with the graph of $f(x) = 2^x$, but here we reflect in the x-axis to get the graph of $h(x) = -2^x$ shown in Figure 5(c).

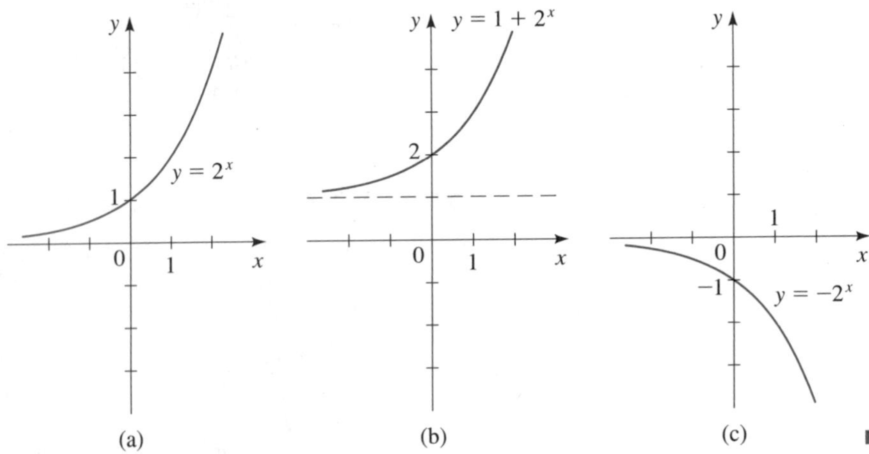

FIGURE 5 ■

EXAMPLE 4 ■ Transformations of Exponential Functions

(a) Use the graph of $f(x) = 10^x$ to sketch the graph of $g(x) = 10^{x-1} - 2$.

(b) State the asymptote, the domain, and the range of this function.

SOLUTION

Vertical and horizontal shifting of graphs is explained in Section 4.5.

(a) Recall from Section 4.5 that we get the graph of $y = f(x - 1)$ from the graph of $y = f(x)$ by shifting to the right 1 unit. Thus, to get the graph of $y = 10^{x-1} - 2$, we shift the graph of $y = 10^x$ to the right 1 unit and downward 2 units, as shown in Figure 6.

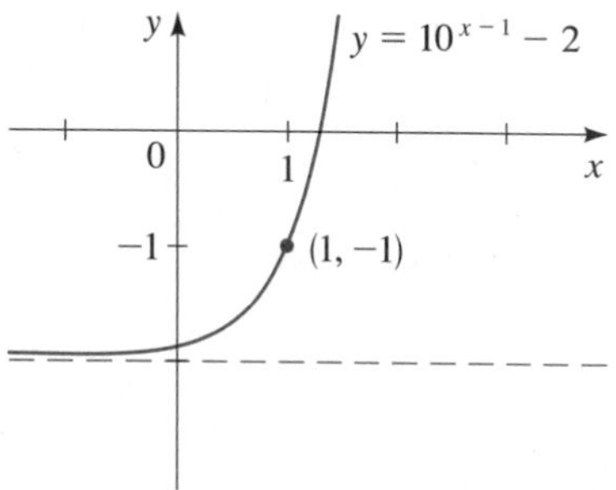

FIGURE 6

(b) The horizontal asymptote is $y = -2$, the domain is $\mathbb{R}$, and the range is

$$\{y \mid y > -2\} = (-2, \infty)$$

■

EXAMPLE 5 ■ Comparing Exponential and Power Functions

Compare the rates of growth of the exponential function $f(x) = 2^x$ and the power function $g(x) = x^2$ by drawing the graphs of both functions in the following viewing rectangles.

(a) $[0, 3]$ by $[0, 8]$ (b) $[0, 6]$ by $[0, 25]$ (c) $[0, 20]$ by $[0, 1000]$

SOLUTION

(a) Figure 7(a) shows that the graph of $g(x) = x^2$ catches up with, and becomes higher than, the graph of $f(x) = 2^x$ at $x = 2$.

(b) The larger viewing rectangle in Figure 7(b) shows that the graph of $f(x) = 2^x$ overtakes that of $g(x) = x^2$ when $x = 4$.

(c) Figure 7(c) gives a more global view and shows that, when x is large, $f(x) = 2^x$ is much larger than $g(x) = x^2$.

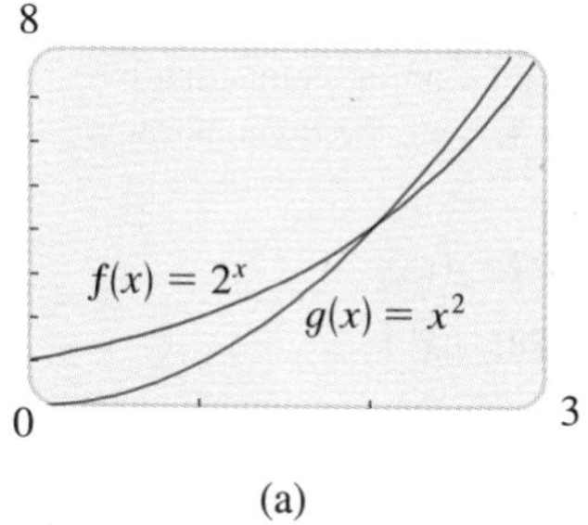

(a)

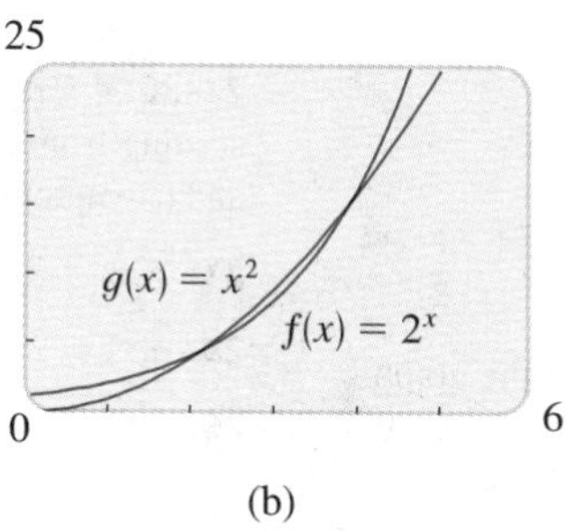

(b)

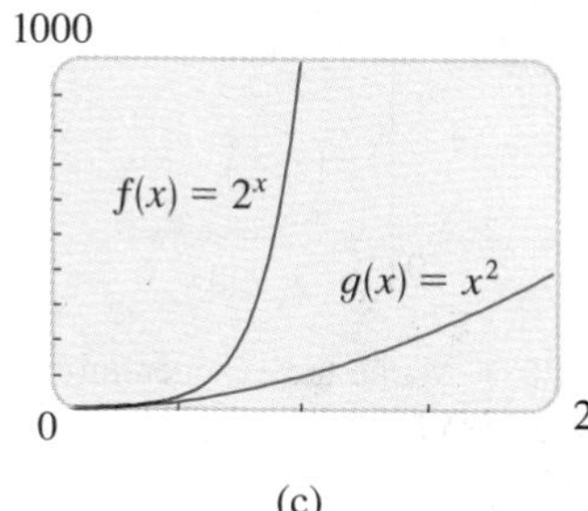

(c)

FIGURE 7

■

6.1 EXERCISES

1–8 ■ Sketch the graph of the function by making a table of values. Use a calculator if necessary.

1. $f(x) = 2^x$ **2.** $g(x) = 8^x$

3. $h(x) = 6^x$ **4.** $h(x) = (0.8)^x$

5. $f(x) = \left(\frac{1}{3}\right)^x$ **6.** $h(x) = (1.1)^x$

7. $g(x) = \left(\frac{1}{4}\right)^x$ **8.** $f(x) = \left(\frac{3}{2}\right)^x$

9–10 ■ Graph both functions on one set of axes.

9. $y = 4^x$ and $y = 7^x$

10. $y = \left(\frac{2}{3}\right)^x$ and $y = \left(\frac{4}{3}\right)^x$

11–14 ■ Find the exponential function $f(x) = a^x$ whose graph is given.

11.

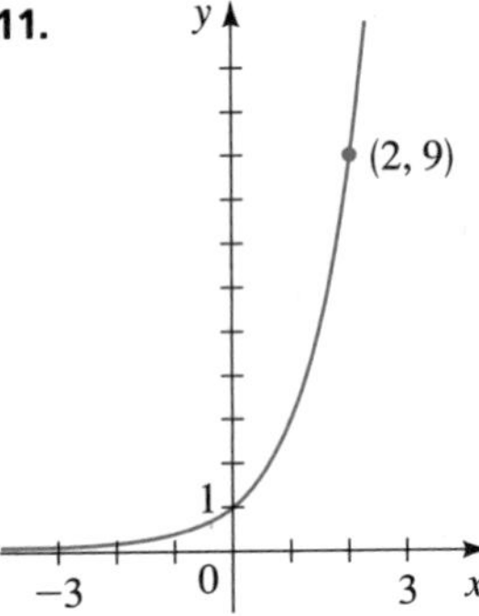

12.

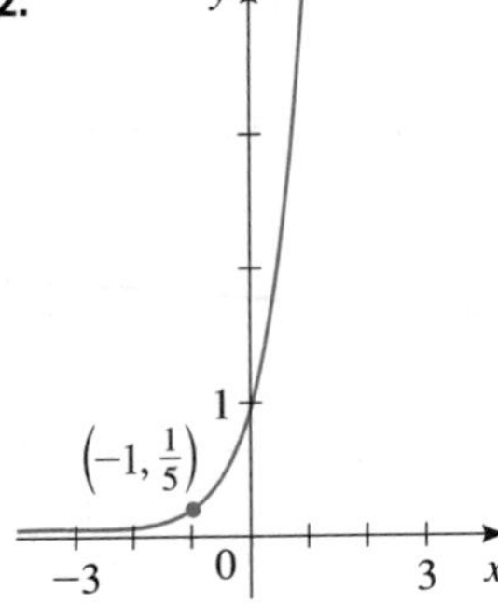

13.

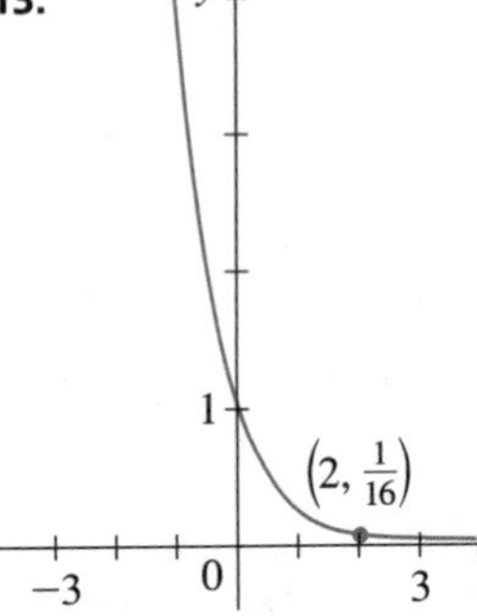

14.

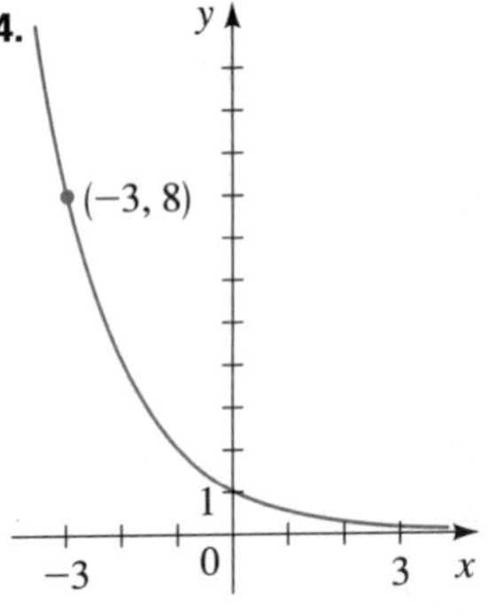

15–20 ■ Match the exponential function with one of the graphs labeled I–VI.

15. $f(x) = 5^x$ **16.** $f(x) = -5^x$

17. $f(x) = 5^{-x}$ **18.** $f(x) = 5^x + 3$

19. $f(x) = 5^{x-3}$

20. $f(x) = 5^{x+1} - 4$

I

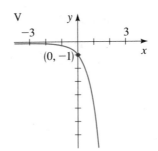

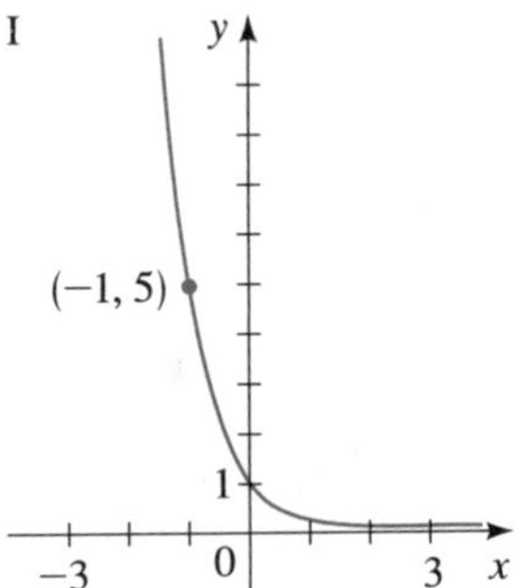

II

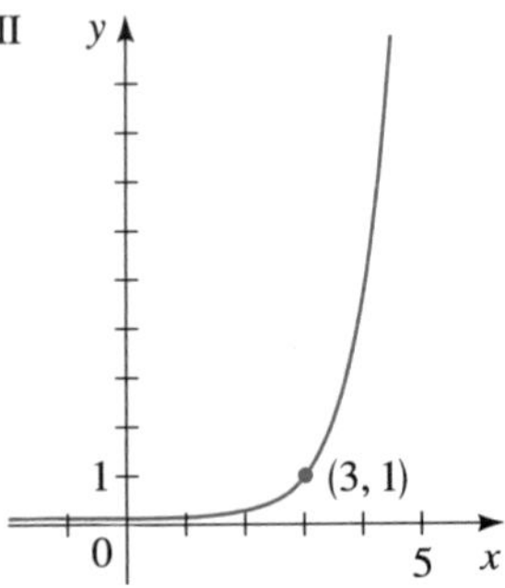

III

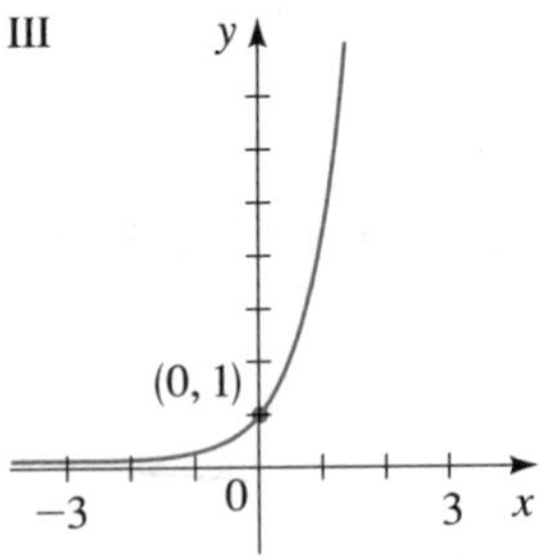

IV

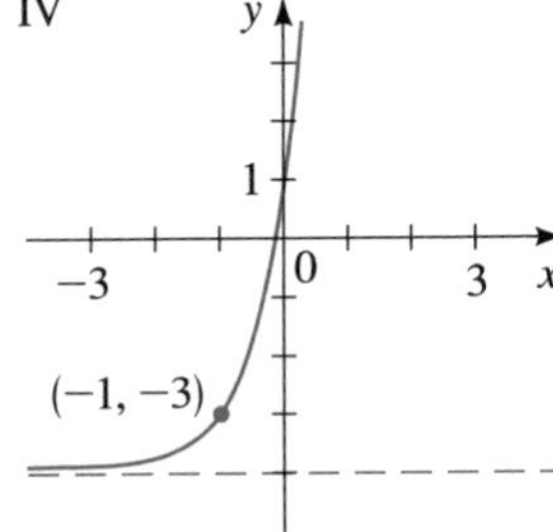

V

VI

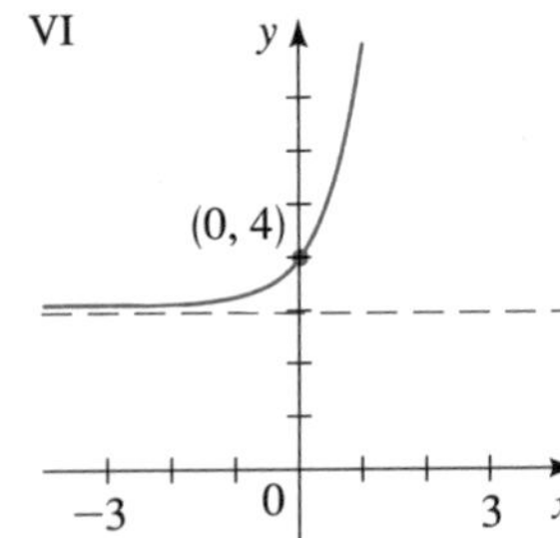

21–36 ■ Graph the function, not by plotting points, but by starting from the graphs in Figure 4. State the domain, range, and asymptote.

21. $f(x) = -3^x$ **22.** $f(x) = 10^{-x}$

23. $g(x) = 2^x - 3$ **24.** $g(x) = 2^{x-3}$

25. $h(x) = 4 + \left(\frac{1}{2}\right)^x$ **26.** $h(x) = 6 - 3^x$

27. $f(x) = 10^{x+3}$ **28.** $f(x) = -\left(\frac{1}{5}\right)^x$

29. $f(x) = -3^{-x}$

30. $f(x) = 10^{-x} - 4$

31. $y = 5^{-2x}$

32. $y = 1 + 2^{x+1}$

33. $f(x) = 5 - 2^{x-1}$

34. $f(x) = 1 - 2^{-x}$

35. $y = 2^{|x|}$

36. $y = 2^{-|x|}$

37–38 ■ Find the function of the form $f(x) = Ca^x$ whose graph is given.

37.

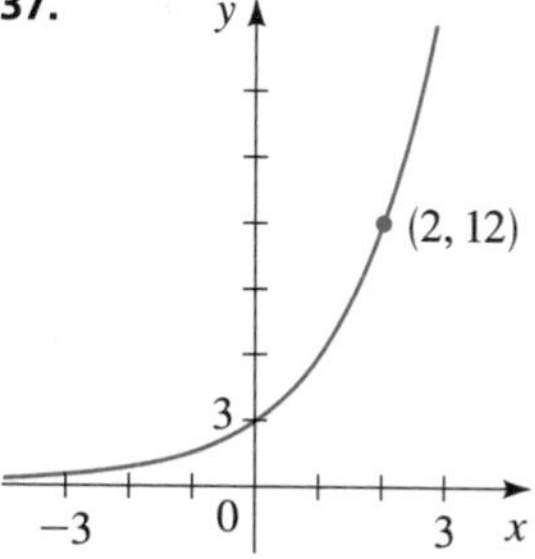

38.

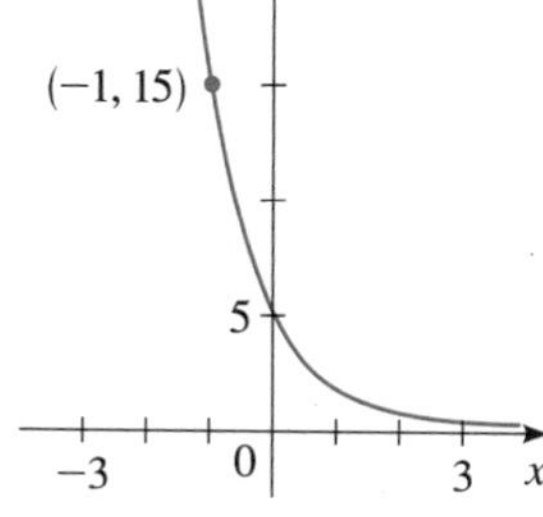

39. (a) Sketch the graphs of $f(x) = 2^x$ and $g(x) = 3(2^x)$.
(b) How are the graphs related?

40. (a) Sketch the graphs of $f(x) = 9^{x/2}$ and $g(x) = 3^x$.
(b) Use the Laws of Exponents to explain the relationship between these graphs.

41. If $f(x) = 10^x$, show that

$$\frac{f(x+h) - f(x)}{h} = 10^x\left(\frac{10^h - 1}{h}\right)$$

42. Compare the functions $f(x) = x^3$ and $g(x) = 3^x$ by evaluating both of them for $x = 0, 1, 2, 3, 4, 5, 6, 7, 8, 9, 10, 15,$ and 20. Then draw the graphs of f and g on the same set of axes.

43. (a) Compare the rates of growth of the functions $f(x) = 2^x$ and $g(x) = x^5$ by drawing the graphs of both functions in the following viewing rectangles.
(i) $[0, 5]$ by $[0, 20]$
(ii) $[0, 25]$ by $[0, 10^7]$
(iii) $[0, 50]$ by $[0, 10^8]$
(b) Find the solutions of the equation $2^x = x^5$, correct to one decimal place.

44. (a) Compare the rates of growth of the functions $f(x) = 3^x$ and $g(x) = x^4$ by drawing the graphs of both functions in the following viewing rectangles:
(i) $[-4, 4]$ by $[0, 20]$
(ii) $[0, 10]$ by $[0, 5000]$
(iii) $[0, 20]$ by $[0, 10^5]$
(b) Find the solutions of the equation $3^x = x^4$, correct to two decimal places.

45–46 ■ Draw graphs of the given family of functions for $c = 0.25, 0.5, 1, 2, 4$. How are the graphs related?

45. $f(x) = c2^x$ **46.** $f(x) = 2^{cx}$

47–48 ■ Graph the function and comment on vertical and horizontal asymptotes.

47. $y = 2^{1/x}$ **48.** $y = \frac{2^x}{x}$

49–50 ■ Find the local maximum and minimum values of the function and the value of x at which each occurs. State each answer correct to two decimal places.

49. $g(x) = x^x \quad (x > 0)$

50. $g(x) = \sqrt[x]{x} \quad (x > 0)$

51–52 ■ Find, correct to two decimal places, (a) the intervals on which the function is increasing or decreasing, and (b) the range of the function.

51. $y = 10^{x-x^2}$ **52.** $y = x2^x$

DISCOVERY • DISCUSSION

53. Growth of an Exponential Function Suppose you are offered a job that lasts one month, and you are to be very well paid. Which of the following methods of payment is more profitable for you?
(a) One million dollars at the end of the month
(b) Two cents on the first day of the month, 4 cents on the second day, 8 cents on the third day, and, in general, 2^n cents on the nth day

54. The Height of the Graph of an Exponential Function Your mathematics instructor asks you to sketch a graph of the exponential function

$$f(x) = 2^x$$

for x between 0 and 40, using a scale of 10 units to one inch. What are the dimensions of the sheet of paper you will need to sketch this graph?

EXPONENTIAL EXPLOSION

To see how difficult it is to comprehend exponential growth, let's try a thought experiment.

Suppose you put a penny in your piggy bank today, two pennies tomorrow, four pennies the next day, and so on, doubling the number of pennies you add to the bank each day (see the table). How many pennies will you put in your piggy bank on day 30? The answer is 2^{30} pennies. That's simple, but can you guess how many dollars that is? 2^{30} pennies is more than 10 million dollars!

Day	Pennies
0	1
1	2
2	4
3	8
4	16
⋮	⋮
n	2^n
⋮	⋮

As you can see, the exponential function $f(x) = 2^x$ grows extremely fast. This is the principle behind atomic explosions. An atom splits releasing two neutrons which cause two atoms to split, each releasing two neutrons, causing four atoms to split, and so on. At the nth stage 2^n atoms split—an exponential explosion!

As we know, populations grow exponentially. Let's see what this means for a type of bacteria that splits every minute. Suppose that at 12:00 noon a single bacterium colonizes a discarded food can. The bacterium and his descendants are all happy, but they fear the time when the can is completely full of bacteria—doomsday.

1. How many bacteria are in the can at 12:05? at 12:10?
2. The can is completely full of bacteria at 1:00 P.M. At what time was the can only half full of bacteria?
3. When the can is exactly half full, the president of the bacteria colony reassures his constituents that doomsday is far away—after all, there is as much room left in the can as has been used in the entire previous history of the colony. Is the president correct? How much time is left before doomsday?
4. When the can is one-quarter full, how much time remains till doomsday?
5. A wise bacterium decides to start a new colony in another can and slow down splitting time to 2 minutes. How much time does this new colony have?

6.2 THE NATURAL EXPONENTIAL FUNCTION

n	$\left(1+\dfrac{1}{n}\right)^n$
1	2.00000
5	2.48832
10	2.59374
100	2.70481
1000	2.71692
10,000	2.71815
100,000	2.71827
1,000,000	2.71828
10,000,000	2.71828

Any positive number can be used as the base for an exponential function, but some bases are used more frequently than others. We will see in the remaining sections of this chapter that the bases 2 and 10 are convenient for certain applications, but the most important base is the number denoted by the letter e.

The number e is defined as the value that $(1 + 1/n)^n$ approaches as n becomes large. (In calculus this idea is made more precise through the concept of a limit. See Exercise 41.) The table in the margin shows the values of the expression $(1 + 1/n)^n$ for increasingly large values of n. It appears that, correct to five decimal places, $e \approx 2.71828$; in fact, the approximate value to 20 decimal places is

$$e \approx 2.71828182845904523536$$

It can be shown that e is an irrational number, so we cannot write its exact value.

Why use such a strange base for an exponential function? It may seem at first that a base such as 10 is easier to work with. We will see, however, that in certain applications the number e is the best possible base. In this section we study how e occurs in the description of compound interest and population growth.

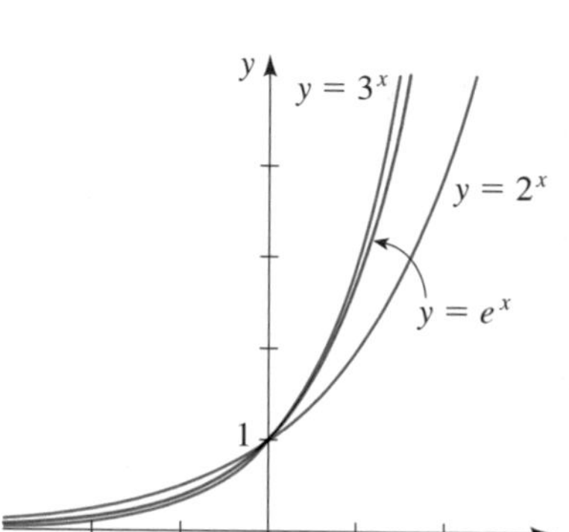

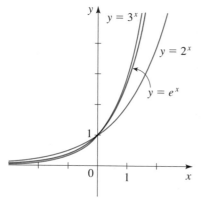

FIGURE 1
Graph of the natural exponential function

THE NATURAL EXPONENTIAL FUNCTION

The **natural exponential function** is the exponential function

$$f(x) = e^x$$

with base e. It is often referred to as *the* exponential function.

Since $2 < e < 3$, the graph of the natural exponential function lies between the graphs of $y = 2^x$ and $y = 3^x$, as shown in Figure 1.

Scientific calculators have a special key for the function $f(x) = e^x$. We use this key in the next example.

The notation e for the base of the natural exponential function was chosen by the Swiss mathematician Leonhard Euler (see page 165), probably because it is the first letter of the word *exponential*.

EXAMPLE 1 ■ Evaluating the Exponential Function

Evaluate each expression correct to five decimal places.

(a) e^3 (b) $2e^{-0.53}$ (c) $e^{4.8}$

SOLUTION

We use the $\boxed{e^x}$ key on a calculator to evaluate the exponential function.

(a) $e^3 \approx 20.08554$

(b) $2e^{-0.53} \approx 1.17721$

(c) $e^{4.8} \approx 121.51042$ ■

EXAMPLE 2 ■ Transformations of the Exponential Function

Sketch the graph of each function.

(a) $f(x) = e^{-x}$ (b) $g(x) = 3e^{0.5x}$

SOLUTION

Reflecting graphs is explained in Section 4.5.

(a) We start with the graph of $y = e^x$ and reflect in the y-axis to obtain the graph of $y = e^{-x}$ as in Figure 2.

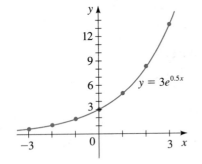

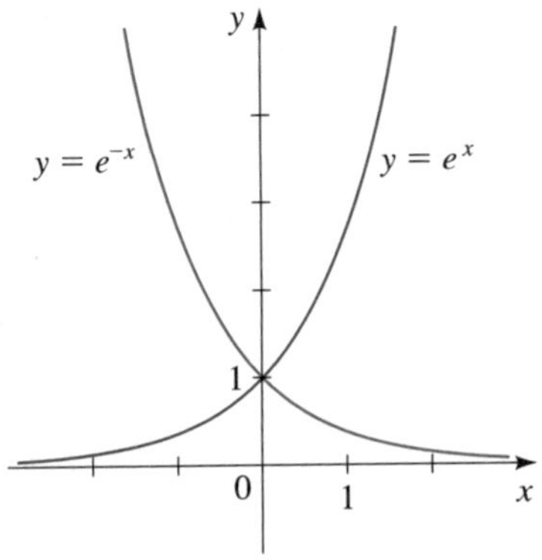

FIGURE 2

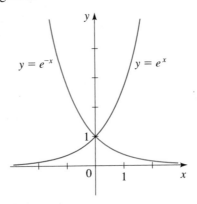

(b) We calculate several values, plot the resulting points, then connect the points with a smooth curve. The graph is shown in Figure 3.

x	$f(x) = 3e^{0.5x}$
−3	0.67
−2	1.10
−1	1.82
0	3.00
1	4.95
2	8.15
3	13.45

■

$y = 3e^{0.5x}$

FIGURE 3

COMPOUND INTEREST

The natural exponential function occurs in calculating *continuously* compounded interest. We begin by explaining compound interest.

If an amount of money P, called the **principal**, is invested at a simple interest rate i, then after one time period the interest is Pi, and the amount A of money is

$$A = P + Pi = P(1 + i)$$

If the interest is reinvested, then the new principal is $P(1 + i)$, and the amount after another time period is $A = P(1 + i)(1 + i) = P(1 + i)^2$. Similarly, after a third time period the amount is $A = P(1 + i)^3$. In general, after k periods the amount is

$$A = P(1 + i)^k$$

Notice that this is an exponential function with base $1 + i$.

If the annual interest rate is r and if interest is compounded n times per year, then in each time period the interest rate is $i = r/n$, and there are nt time periods in t years. This leads to the following formula for the amount after t years.

r is often referred to as the *nominal annual interest rate*.

COMPOUND INTEREST

Compound interest is calculated by the formula

$$A(t) = P\left(1 + \frac{r}{n}\right)^{nt}$$

where $A(t)$ = amount after t years

P = principal

r = interest rate

n = number of times interest is compounded per year

t = number of years

EXAMPLE 3 ■ Calculating Compound Interest

A sum of \$1000 is invested at an interest rate of 12% per year. Find the amounts in the account after 3 years if interest is compounded annually, semiannually, quarterly, monthly, and daily.

SOLUTION

We use the compound interest formula with $P = \$1000$, $r = 0.12$, and $t = 3$.

Compounding	n	Amount after 3 years
Annual	1	$1000\left(1 + \frac{0.12}{1}\right)^{1(3)} = \$1404.93$
Semiannual	2	$1000\left(1 + \frac{0.12}{2}\right)^{2(3)} = \$1418.52$
Quarterly	4	$1000\left(1 + \frac{0.12}{4}\right)^{4(3)} = \$1425.76$
Monthly	12	$1000\left(1 + \frac{0.12}{12}\right)^{12(3)} = \$1430.77$
Daily	365	$1000\left(1 + \frac{0.12}{365}\right)^{365(3)} = \$1433.24$

■

We see from Example 3 that the interest paid increases as the number of compounding periods n increases. Let's see what happens as n increases indefinitely. If we let $m = n/r$, then

$$A(t) = P\left(1 + \frac{r}{n}\right)^{nt} = P\left[\left(1 + \frac{r}{n}\right)^{n/r}\right]^{rt} = P\left[\left(1 + \frac{1}{m}\right)^{m}\right]^{rt}$$

Recall that as m becomes large, the quantity $(1 + 1/m)^m$ approaches the number e. Thus, the amount approaches $A = Pe^{rt}$. This expression gives the amount when the interest is compounded at "every instant."

CONTINUOUSLY COMPOUNDED INTEREST

Continuously compounded interest is calculated by the formula

$$A(t) = Pe^{rt}$$

where $A(t)$ = amount after t years

P = principal

r = interest rate

t = number of years

EXAMPLE 4 ■ Calculating Continuously Compounded Interest

Find the amount after 3 years if $1000 is invested at an interest rate of 12% per year, compounded continuously.

SOLUTION

We use the formula for continuously compounded interest with $P = \$1000$, $r = 0.12$, and $t = 3$ to get

$$A(3) = 1000e^{(0.12)3} = 1000e^{0.36} = \$1433.33$$

Compare this amount with the amounts in Example 3. ■

EXPONENTIAL MODELS OF POPULATION GROWTH

Biologists have observed that the population of a species doubles its size in a fixed period of time. For example, under ideal conditions a certain population of bacteria doubles in size every 3 hours. If the culture is started with 1000 bacteria, then after 3 hours there will be 2000 bacteria, after another 3 hours there will be 4000, and so on. If we let $n = n(t)$ be the number of bacteria after t hours, then

The **Gateway Arch** in St. Louis, Missouri, is shaped in the form of the graph of a combination of exponential functions (*not* a parabola, as it might first appear). Specifically, it is a **catenary**, which is the graph of an equation of the form

$$y = a(e^{bx} + e^{-bx})$$

(see Exercise 43). This shape was chosen because it is optimal for distributing the internal structural forces of the arch. Chains and cables suspended between two points (for example, the stretches of cable between pairs of telephone poles) hang in the shape of a catenary.

$$\begin{aligned} n(0) &= 1000 \\ n(3) &= 1000 \cdot 2 \\ n(6) &= (1000 \cdot 2) \cdot 2 = 1000 \cdot 2^2 \\ n(9) &= (1000 \cdot 2^2) \cdot 2 = 1000 \cdot 2^3 \\ n(12) &= (1000 \cdot 2^3) \cdot 2 = 1000 \cdot 2^4 \end{aligned}$$

From this pattern it appears that the number of bacteria after t hours is modeled by the function

$$n(t) = 1000 \cdot 2^{t/3}$$

In general, suppose that the initial size of a population is n_0 and the doubling period is a. Then the size of the population at time t is modeled by

$$n(t) = n_0 2^{ct}$$

where $c = 1/a$. If we knew the tripling time b, then the formula would be $n(t) = n_0 3^{ct}$ where $c = 1/b$. These formulas indicate that the growth of the bacteria is modeled by an exponential function. But what base should we use? The answer is e, because then it can be shown (using calculus) that the population is modeled by

$$n(t) = n_0 e^{rt}$$

where r is the *relative rate of growth of population, expressed as a proportion of the population at any time*. For instance, if $r = 0.02$, then at any time t the growth rate is 2% of the population at time t.

Notice that the formula for population growth is the same as that for continuously compounded interest. In fact, the same principle is at work in both cases: The growth of a population (or an investment) per time period is proportional to the size of the population (or the amount of the investment). A population of 1,000,000 will increase more in one year than a population of 1000; in exactly the same way, an investment of $1,000,000 will increase more in one year than an investment of $1000.

EXPONENTIAL GROWTH

A population that experiences **exponential growth** increases according to the model

$$n(t) = n_0 e^{rt}$$

where $n(t)$ = population at time t

n_0 = initial size of the population

r = relative rate of growth (expressed as a proportion of the population)

t = time

In the following examples we assume that the populations experience exponential growth.

EXAMPLE 5 ■ Predicting the Size of a Population

The initial bacterium count in a culture is 500. A biologist later makes a sample count of bacteria in the culture and finds that the relative rate of growth is 40% per hour.

(a) Find a function that models the number of bacteria after t hours.
(b) What is the estimated count after 10 hours?
(c) Sketch the graph of the function $n(t)$.

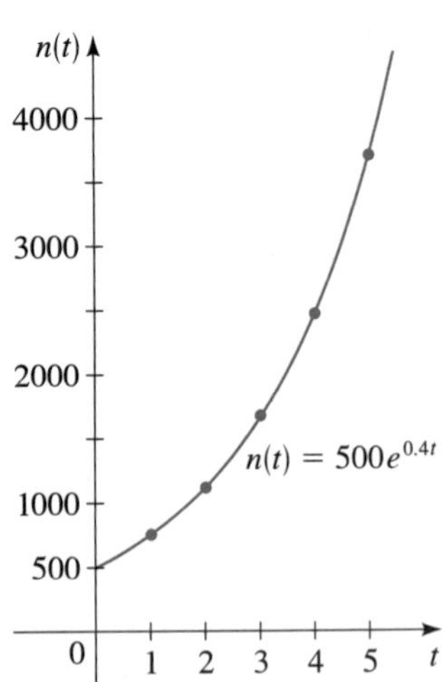

FIGURE 4

SOLUTION

(a) We use the exponential growth model with $n_0 = 500$ and $r = 0.4$ to get

$$n(t) = 500e^{0.4t}$$

where t is measured in hours.

(b) Using the function in part (a), we find that the bacterium count after 10 hours is

$$n(10) = 500e^{0.4(10)} = 500e^4 \approx 27{,}300$$

(c) We plot several points and sketch the graph in Figure 4. ■

EXAMPLE 6 ■ Comparing Effects of Different Rates of Population Growth

In 2000 the population of the world was 6.1 billion and the relative rate of growth was 1.4% per year. It is claimed that a rate of 1.0% per year would make a significant difference in the total population in just a few decades. Test this claim by estimating the population of the world in the year 2050 using a relative rate of growth of (a) 1.4% per year and (b) 1.0% per year.

Graph the population functions for the next 100 years for the two relative growth rates in the same viewing rectangle.

SOLUTION

(a) By the exponential growth model, we have

$$n(t) = 6.1e^{0.014t}$$

where $n(t)$ is measured in billions and t is measured in years since 2000. Because the year 2050 is 50 years after 2000, we find

$$n(50) = 6.1e^{0.014(50)} = 6.1e^{0.7} \approx 12.3$$

Thus, the estimated population in the year 2050 is about 12.3 billion.

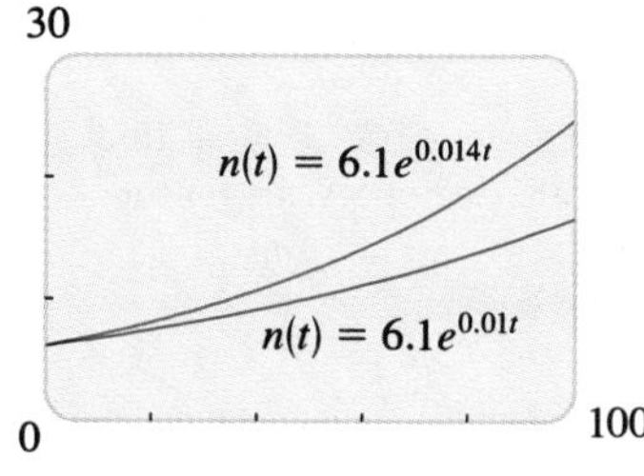

FIGURE 5

(b) We use the function

$$n(t) = 6.1e^{0.010t}$$

and find $\quad n(50) = 6.1e^{0.010(50)} = 6.1e^{0.50} \approx 10.1$

So, the estimated population in the year 2050 is about 10.1 billion.

The graphs in Figure 5 show that a small change in the relative rate of growth will, over time, make a large difference in population size. ■

EXAMPLE 7 ■ Finding the Initial Population

A certain breed of rabbit was introduced onto a small island about 8 years ago. The current rabbit population on the island is estimated to be 4100, with a relative growth rate of 55% per year.

(a) What was the initial size of the rabbit population?
(b) Estimate the population 12 years from now.

SOLUTION

(a) From the exponential growth model, we have

$$n(t) = n_0 e^{0.55t}$$

and we know that the population at time $t = 8$ is $n(8) = 4100$. We substitute what we know into the equation and solve for n_0:

$$4100 = n_0 e^{0.55(8)}$$

$$n_0 = \frac{4100}{e^{0.55(8)}} \approx \frac{4100}{81.45} \approx 50$$

Thus, we estimate that 50 rabbits were introduced onto the island.

(b) Now that we know n_0, we can write a formula for population growth:

$$n(t) = 50e^{0.55t}$$

Twelve years from now, $t = 20$ and

$$n(20) = 50e^{0.55(20)} \approx 2{,}993{,}707$$

So, we estimate that the rabbit population on the island 12 years from now will be about 3 million. ■

Another way to solve part (b) is to let t be the number of years from now. In this case, $n_0 = 4100$ (the current population). So, the population 12 years from now will be

$$n(12) = 4100e^{0.55(12)} \approx 3 \text{ million}$$

Can the rabbit population in Example 7(b) actually reach such a high number? In reality, as the island becomes overpopulated with rabbits, the rabbit population growth will be slowed due to food shortage and other factors. One model that takes into account such factors is the *logistic growth model* described in Exercises 31 and 32.

6.2 EXERCISES

1–2 ■ Complete the table and use it to graph the function.

1. $f(x) = 3e^x$

x	$f(x) = 3e^x$
-2	
-1.5	
-1	
-0.5	
0	
0.5	
1	
1.5	
2	

2. $g(x) = 2e^{-0.5x}$

x	$g(x) = 2e^{-0.5x}$
-4	
-3	
-2	
-1	
0	
1	
2	
3	
4	

3–8 ■ Graph the function, not by plotting points, but by starting from the graph of $y = e^x$ in Figure 1. State the domain, range, and asymptote.

3. $y = -e^x$

4. $y = 1 - e^x$

5. $y = e^{-x} - 1$

6. $y = -e^{-x}$

7. $y = e^{x-2}$

8. $y = e^{x-3} + 4$

9. If \$10,000 is invested at an interest rate of 10% per year, compounded semiannually, find the value of the investment after the given number of years.
(a) 5 years (b) 10 years (c) 15 years

10. If \$4000 is borrowed at a rate of 16% interest per year, compounded quarterly, find the amount due at the end of the given number of years.
(a) 4 years (b) 6 years (c) 8 years

11. If \$3000 is invested at an interest rate of 9% per year, find the amount of the investment at the end of 5 years for the following compounding methods.
(a) Annual (b) Semiannual
(c) Monthly (d) Weekly
(e) Daily (f) Hourly
(g) Continuously

12. If \$4000 is invested in an account for which interest is compounded quarterly, find the amount of the investment at the end of 5 years for the following interest rates.
(a) 6% (b) $6\frac{1}{2}$% (c) 7% (d) 8%

13. Which of the given interest rates and compounding periods would provide the best investment?
(i) $8\frac{1}{2}$% per year, compounded semiannually
(ii) $8\frac{1}{4}$% per year, compounded quarterly
(iii) 8% per year, compounded continuously

14. Which of the given interest rates and compounding periods would provide the better investment?
(i) $9\frac{1}{4}$% per year, compounded semiannually
(ii) 9% per year, compounded continuously

15–16 ■ The **present value** of a sum of money is the amount that must be invested now, at a given rate of interest, to produce the desired sum at a later date.

15. Find the present value of \$10,000 if interest is paid at a rate of 9% per year, compounded semiannually, for 3 years.

16. Find the present value of \$100,000 if interest is paid at a rate of 8% per year, compounded monthly, for 5 years.

17. The number of bacteria in a culture is given by the function

$$n(t) = 500e^{0.45t}$$

where t is measured in hours.
(a) What is the relative rate of growth of this bacterium population? Express your answer as a percentage.
(b) What is the initial population of the culture (at $t = 0$)?
(c) How many bacteria will the culture contain at time $t = 5$?

18. The rat population in New York City is given by the function

$$n(t) = 54e^{0.12t}$$

where t is measured in years since 1990 and $n(t)$ is measured in millions.
(a) What is the relative rate of growth of the rat population? Express your answer as a percentage.

(b) What was the rat population in 1990?
(c) What is the rat population in the year 2000?
(d) Sketch a graph of the population function.

19. The fox population in a certain region has a relative growth rate of 8% per year. It is estimated that the population in 2000 was 18,000.
(a) Find a function that models the population t years after 2000.
(b) Use the function from part (a) to estimate the fox population in the year 2008.
(c) Sketch a graph of the fox population function for the years 2000–2008.

20. The population of a certain city has a relative growth rate of 5% per year. The population in 1988 was 421,000. Find the projected population of the city for each year.
(a) 2000 (b) 2030

21. The population of a country has relative growth rate of 3% per year. The government is trying to reduce the growth rate to 2%. The population in 1995 was approximately 110 million. Find the projected population for the year 2020 for the following conditions.
(a) The relative growth rate remains at 3% per year.
(b) The relative growth rate is reduced to 2% per year.

22. The population of a certain city was 680,000 in the year 2000 and is growing at the relative growth rate of 12% per year.
(a) Find a function that models the population of this city t years after 2000.
(b) Estimate the population in 2008.

23. The population of the world in 1987 was 5 billion and the relative growth rate was estimated at 2% per year. Assuming that the world population follows an exponential growth model, find the projected world population in 1995. (Compare this with the actual world population of 5.7 billion in 1995.)

24. The relative growth rate for a certain bacteria population is 80% per hour. A small culture is formed, and 3 hours later a count shows approximately 21,500 bacteria in the culture.
(a) Find the initial number of bacteria in the culture.
(b) Estimate the number of bacteria 5 hours from the time the culture was started.

25. Under ideal conditions, a certain type of bacteria has a relative growth rate of 220% per hour. A number of these bacteria are introduced accidentally into a food product. Two hours after contamination, a bacterium count shows that there are about 40,000 bacteria in the food.
(a) Find the initial number of bacteria introduced into the food.
(b) Estimate the number of bacteria in the food 3 hours after contamination.

26. A radioactive substance decays in such a way that the amount of mass remaining after t days is given by the function

$$m(t) = 13e^{-0.015t}$$

where $m(t)$ is measured in kilograms.
(a) Find the mass at time $t = 0$.
(b) How much of the mass remains after 45 days?

27. Radioactive iodine is used by doctors as a tracer in diagnosing certain thyroid gland disorders. This type of iodine decays in such a way that the mass remaining after t days is given by the function

$$m(t) = 6e^{-0.087t}$$

where $m(t)$ is measured in grams.
(a) Find the mass at time $t = 0$.
(b) How much of the mass remains after 20 days?

28. A sky diver jumps from a reasonable height above the ground. The air resistance she experiences is proportional to her velocity, and the constant of proportionality is 0.2. It can be shown that the downward velocity of the sky diver at time t is given by

$$v(t) = 80(1 - e^{-0.2t})$$

where t is measured in seconds and $v(t)$ is measured in feet per second (ft/s).
(a) Find the initial velocity of the sky diver.
(b) Find the velocity after 5 s and after 10 s.
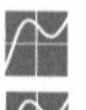
(c) Draw a graph of the velocity function $v(t)$.
(d) The maximum velocity of a falling object with wind resistance is called its *terminal velocity*. From the graph in part (c) find the terminal velocity of this sky diver.

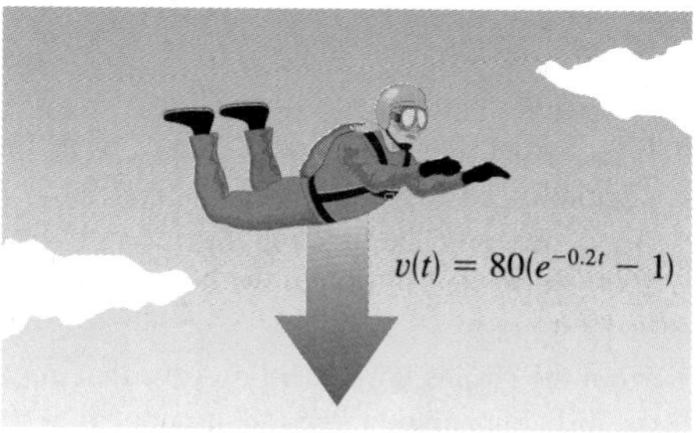

29. A 50-gallon barrel is filled completely with pure water. Salt water with a concentration of 0.3 lb/gal is then pumped into the barrel, and the resulting mixture overflows at the same rate. The amount of salt in the barrel at time t is given by

$$Q(t) = 15(1 - e^{-0.04t})$$

where t is measured in minutes and $Q(t)$ is measured in pounds.

(a) How much salt is in the barrel after 5 min?

(b) How much salt is in the barrel after 10 min?

(c) Draw a graph of the function $Q(t)$.

(d) Use the graph in part (c) to determine the value that the amount of salt in the barrel approaches as t becomes large. Is this what you would expect?

$Q(t) = 15(1 - e^{-0.04t})$

 **30.** A sum of $5000 is invested at an interest rate of 9% per year, compounded semiannually.

(a) Find the value $A(t)$ of the investment after t years.

(b) Draw a graph of $A(t)$.

(c) Use the graph of $A(t)$ to determine when this investment will amount to $25,000.

31. Assume that the rabbit population in Example 7 behaves according to the *logistic growth model*

$$n(t) = \frac{300}{0.05 + \left(\dfrac{300}{n_0} - 0.05\right)e^{-0.55t}}$$

where n_0 is the initial rabbit population.

(a) If the initial population is 50 rabbits, what will the population be after 12 years?

(b) Draw graphs of the function $n(t)$ for $n_0 = 50$, 500, 2000, 8000, and 12,000 in the viewing rectangle [0, 15] by [0, 12,000].

(c) From the graphs in part (b), observe that, regardless of the initial population, the rabbit population seems to approach a certain number as time goes on. What is that number? (This is the number of rabbits that the island can support.)

 **32.** The population of a certain species of bird is limited by the type of habitat required for nesting. The population behaves according to the *logistic growth model*

$$n(t) = \frac{5600}{0.5 + 27.5e^{-0.044t}}$$

where t is measured in years.

(a) Find the initial bird population.

(b) Draw a graph of the function $n(t)$.

(c) What size does the population approach as time goes on?

33. Use a calculator to help graph the function $f(x) = e^{-x^2}$ for $x \geqslant 0$. Then use the fact that f is an even function to draw the rest of the graph.

 **34.** (a) Compare the functions $f(x) = e^x$ and $g(x) = x^3$ by drawing their graphs in the following viewing rectangles:

(i) [0, 3] by [0, 15]

(ii) [0, 6] by [0, 120]

(iii) [0, 20] by [0, 10,000]

(b) Find the solutions of the equation $e^x = x^3$, correct to two decimal places.

35. The *hyperbolic cosine function* is defined by

$$\cosh(x) = \frac{e^x + e^{-x}}{2}$$

Sketch the graphs of the functions $y = \frac{1}{2}e^x$ and $y = \frac{1}{2}e^{-x}$ on the same axes and use graphical addition (see Section 4.7) to sketch the graph of $y = \cosh(x)$.

36. The *hyperbolic sine function* is defined by

$$\sinh(x) = \frac{e^x - e^{-x}}{2}$$

Sketch the graph of this function using graphical addition as in Exercise 35.

37–40 ■ Use the definitions in Exercises 35 and 36 to prove the identity.

37. $\cosh(-x) = \cosh(x)$

38. $\sinh(-x) = -\sinh(x)$

39. $[\cosh(x)]^2 - [\sinh(x)]^2 = 1$

40. $\sinh(x + y) = \sinh(x)\cosh(y) + \cosh(x)\sinh(y)$

 **41.** Illustrate the definition of the number e by graphing the curve $y = (1 + 1/x)^x$ and the line $y = e$ on the same screen using the viewing rectangle $[0, 40]$ by $[0, 4]$.

 **42.** Investigate the behavior of the function

$$f(x) = \left(1 - \frac{1}{x}\right)^x$$

as $x \to \infty$ by graphing f and the line $y = 1/e$ on the same screen using the viewing rectangle $[0, 20]$ by $[0, 1]$.

 **43.** (a) Draw the graphs of the family of functions

$$f(x) = \frac{a}{2}(e^{x/a} + e^{-x/a})$$

for $a = 0.5$, 1, 1.5, and 2.

(b) How does a larger value of a affect the graph?

 **44.** Graph the function $y = e^x/x$ and comment on vertical and horizontal asymptotes.

 **45–46** ■ Find the local maximum and minimum values of the function and the value of x at which each occurs. State each answer correct to two decimal places.

45. $f(x) = xe^{-x}$

46. $f(x) = e^x + e^{-3x}$

6.3 LOGARITHMIC FUNCTIONS

Every exponential function $f(x) = a^x$, with $a > 0$ and $a \neq 1$, is a one-to-one function by the Horizontal Line Test (see Figure 1 for the case $a > 1$) and therefore has an inverse function. The inverse function f^{-1} is called the *logarithmic function with base a* and is denoted by $\log_a$. Recall from Section 4.8 that f^{-1} is defined by

$$f^{-1}(x) = y \quad \Leftrightarrow \quad f(y) = x$$

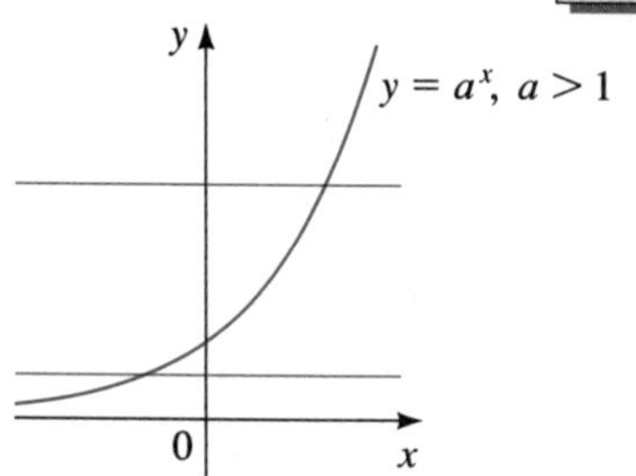

FIGURE 1
$f(x) = a^x$ is one-to-one

This leads to the following definition of the logarithmic function.

DEFINITION OF THE LOGARITHMIC FUNCTION

Let a be a positive number with $a \neq 1$. The **logarithmic function with base *a***, denoted by $\mathbf{log_a}$, is defined by

$$\log_a x = y \quad \Leftrightarrow \quad a^y = x$$

We read $\log_a x = y$ as "log base a of x is y."

In words, this says that

> $\log_a x$ is the exponent to which the base a must be raised to give x.

By tradition, the name of the logarithmic function is $\log_a$, not just a single letter. Also, we usually omit the parentheses in the function notation and write

$$\log_a(x) = \log_a x$$

When we use the definition of logarithms to switch back and forth between the **logarithmic form** $\log_a x = y$ and the **exponential form** $a^y = x$, it's helpful to notice that, in both forms, the base is the same:

The following examples illustrate how to change an equation from one of these forms to the other.

EXAMPLE 1 ■ Logarithmic and Exponential Forms

The logarithmic and exponential forms are equivalent equations—if one is true, then so is the other. So, we can switch from one form to the other as in the following illustrations.

Logarithmic form	Exponential form
$\log_{10} 100{,}000 = 5$	$10^5 = 100{,}000$
$\log_2 8 = 3$	$2^3 = 8$
$\log_2\left(\frac{1}{8}\right) = -3$	$2^{-3} = \frac{1}{8}$
$\log_5 s = r$	$5^r = s$

■

x	$\log_{10} x$
10	1
10^2	2
10^3	3
10^4	4
10^{-1}	-1
10^{-2}	-2
10^{-3}	-3
10^{-4}	-4

It's important to understand that $\log_a x$ is an *exponent*. For example, the numbers in the right column of the table in the margin are the logarithms (base 10) of the numbers in the left column. This is the case for all bases, as the following example illustrates.

EXAMPLE 2 ■ Evaluating Logarithms

(a) $\log_{10} 1000 = 3$ because $10^3 = 1000$
(b) $\log_2 32 = 5$ because $2^5 = 32$
(c) $\log_{10} 0.1 = -1$ because $10^{-1} = 0.1$
(d) $\log_{16} 4 = \frac{1}{2}$ because $16^{1/2} = 4$ ■

GRAPHS OF LOGARITHMIC FUNCTIONS

Recall that if a one-to-one function f has domain A and range B, then its inverse function f^{-1} has domain B and range A. Since the exponential function $f(x) = a^x$ with $a \neq 1$ has domain $\mathbb{R}$ and range $(0, \infty)$, we conclude that its inverse function, $f^{-1}(x) = \log_a x$, has domain $(0, \infty)$ and range $\mathbb{R}$.

The graph of $f^{-1}(x) = \log_a x$ is obtained by reflecting the graph of $f(x) = a^x$ in the line $y = x$. Figure 2 shows the case $a > 1$. The fact that $y = a^x$ (for $a > 1$) is a very rapidly increasing function for $x > 0$ implies that $y = \log_a x$ is a very slowly increasing function for $x > 1$ (see Exercise 74).

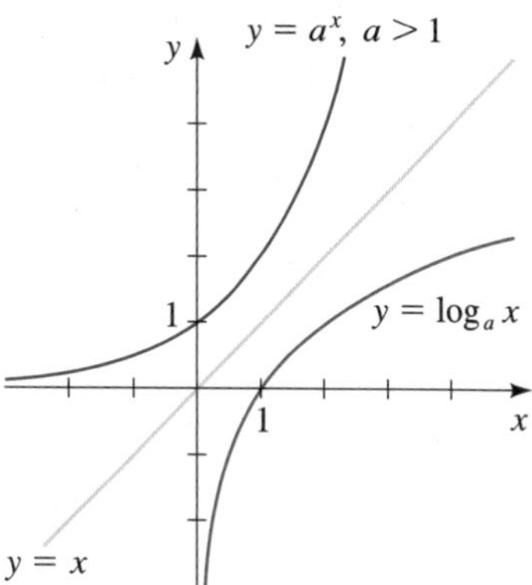

FIGURE 2
Graph of the logarithmic function $f(x) = \log_a x$

Notice that since $a^0 = 1$, we have

$$\log_a 1 = 0$$

and so the x-intercept of the function $y = \log_a x$ is 1. Notice also that because the x-axis is a horizontal asymptote of $y = a^x$, the y-axis is a vertical asymptote of $y = \log_a x$. In fact, using the notation of Section 5.5, we can write

$$\log_a x \to -\infty \quad \text{as} \quad x \to 0^+$$

for the case $a > 1$.

EXAMPLE 3 ■ Graphing a Logarithmic Function by Plotting Points

Sketch the graph of $f(x) = \log_2 x$.

SOLUTION

To make a table of values, we choose the x-values to be powers of 2 so that we can easily find their logarithms. We plot these points and connect them with a smooth curve as in Figure 3.

x	$\log_2 x$
1	0
2	1
2^2	2
2^3	3
2^4	4
2^{-1}	-1
2^{-2}	-2
2^{-3}	-3
2^{-4}	-4

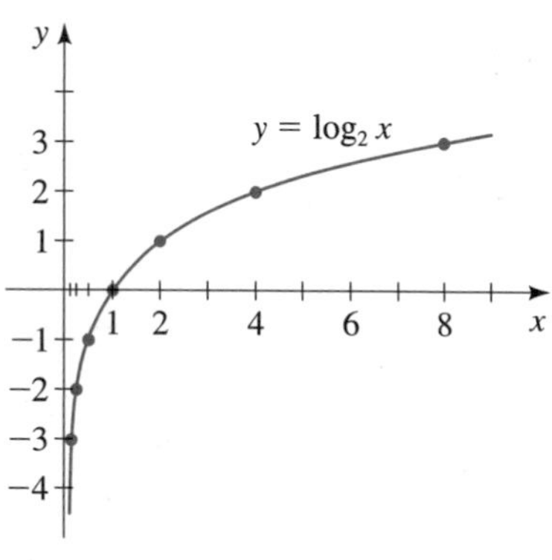

FIGURE 3

■

Figure 4 shows the graphs of the family of logarithmic functions with bases 2, 3, 5, and 10. These graphs are drawn by reflecting the graphs of $y = 2^x$, $y = 3^x$, $y = 5^x$, and $y = 10^x$ (see Figure 4 in Section 6.1) in the line $y = x$. We can also plot points as an aid to sketching these graphs, as illustrated in Example 3.

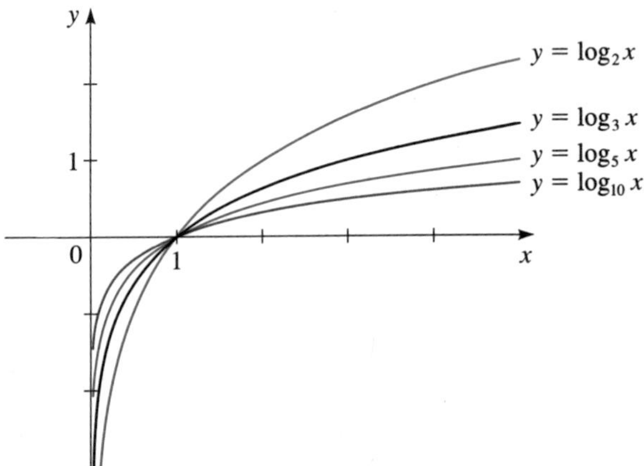

FIGURE 4
A family of logarithmic functions

In the next two examples we graph logarithmic functions by starting with the basic graphs in Figure 4 and using the transformations of Section 4.5.

EXAMPLE 4 ■ Reflecting Graphs of Logarithmic Functions

Sketch the graph of each function.

(a) $f(x) = -\log_2 x$ (b) $g(x) = \log_2(-x)$

SOLUTION

Reflecting graphs is explained in Section 4.5.

(a) We start with the graph of $y = \log_2 x$ in Figure 5(a) and reflect in the x-axis to get the graph of $f(x) = -\log_2 x$ in Figure 5(b).

(b) To obtain the graph of $g(x) = \log_2(-x)$, we reflect the graph of $y = \log_2 x$ in the y-axis. See Figure 5(c).

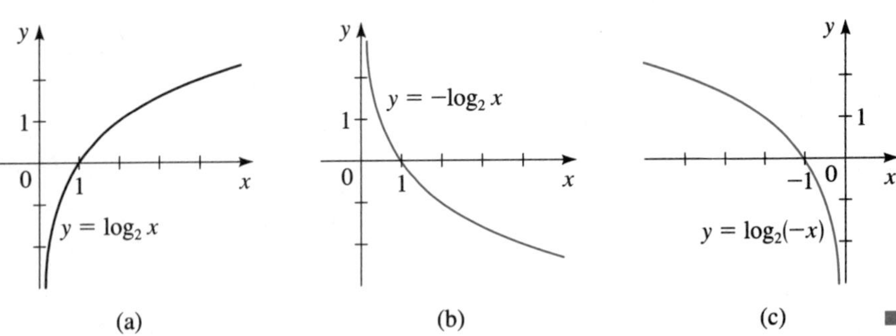

FIGURE 5 (a) (b) (c) ■

EXAMPLE 5 ■ Shifting Graphs of Logarithmic Functions

Find the domain of each function, and sketch the graph.

(a) $f(x) = 2 + \log_5 x$ (b) $g(x) = \log_{10}(x - 3)$

SOLUTION

(a) The graph of f is obtained from the graph of $y = \log_5 x$ (Figure 4) by shifting upward 2 units (see Figure 6). The domain of f is $(0, \infty)$.

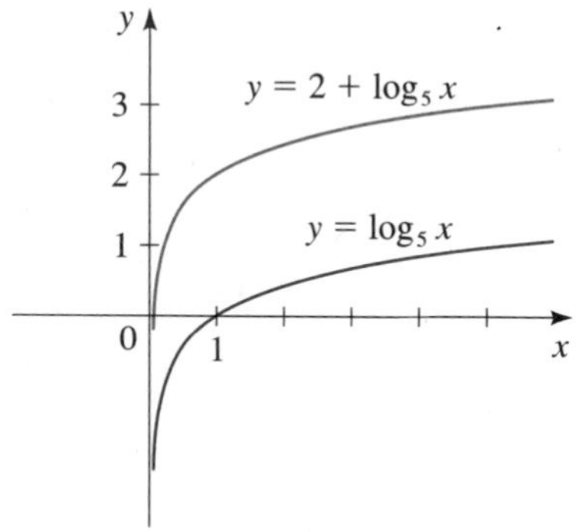

FIGURE 6

(b) The graph of g is obtained from the graph of $y = \log_{10} x$ (Figure 4) by shifting to the right 3 units (see Figure 7). The line $x = 3$ is a vertical asymptote. Since $\log_{10} x$ is defined only when $x > 0$, the domain of $g(x) = \log_{10}(x - 3)$ is

$$\{x \mid x - 3 > 0\} = \{x \mid x > 3\} = (3, \infty)$$

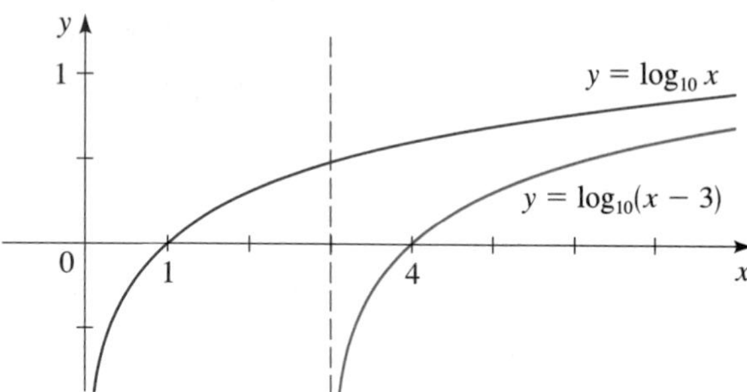

FIGURE 7

■

In Section 4.8 we saw that a function f and its inverse function f^{-1} satisfy the equations

$$f^{-1}(f(x)) = x \qquad \text{for } x \text{ in the domain of } f$$

$$f(f^{-1}(x)) = x \qquad \text{for } x \text{ in the domain of } f^{-1}$$

When applied to $f(x) = a^x$ and $f^{-1}(x) = \log_a x$, these equations become

$$\log_a(a^x) = x \qquad x \in \mathbb{R}$$

$$a^{\log_a x} = x \qquad x > 0$$

We list these and other properties of logarithms discussed in this section.

PROPERTIES OF LOGARITHMS

Property	Reason
1. $\log_a 1 = 0$	We must raise a to the power 0 to get 1.
2. $\log_a a = 1$	We must raise a to the power 1 to get a.
3. $\log_a a^x = x$	We must raise a to the power x to get a^x.
4. $a^{\log_a x} = x$	$\log_a x$ is the power to which a must be raised to get x.

EXAMPLE 6 ■ Applying Properties of Logarithms

We illustrate the properties of logarithms when the base is 5.

$\log_5 1 = 0$	Property 1	$\log_5 5 = 1$	Property 2
$\log_5 5^8 = 8$	Property 3	$5^{\log_5 12} = 12$	Property 4

■

COMMON LOGARITHMS

We now study logarithms with base 10.

COMMON LOGARITHM

The logarithm with base 10 is called the **common logarithm** and is denoted by omitting the base:

$$\log x = \log_{10} x$$

From the definition of logarithms we can easily find that

$$\log 10 = 1 \qquad \text{and} \qquad \log 100 = 2$$

But how do we find log 50? We need to find the exponent y such that $10^y = 50$. Clearly, 1 is too small and 2 is too large. So

$$1 < \log 50 < 2$$

To get a better approximation, we can experiment to find a power of 10 closer to 50. Fortunately, scientific calculators are equipped with a [log] key that directly gives values of common logarithms.

EXAMPLE 7 ■ Evaluating Common Logarithms

Use a calculator to find appropriate values of $f(x) = \log x$ and use the values to sketch the graph.

SOLUTION

We make a table of values, using a calculator to evaluate the function at those values of x that are not powers of 10. We plot those points and connect them by a smooth curve as in Figure 8.

x	$\log x$
0.01	−2
0.1	−1
0.5	−0.301
1	0
4	0.602
5	0.699
10	1
15	1.176

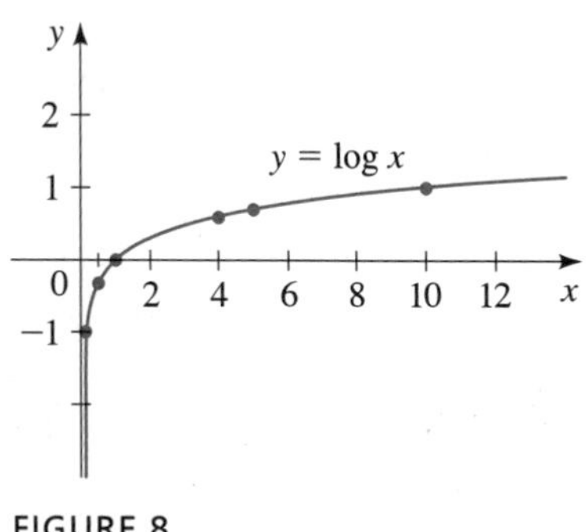

FIGURE 8 ■

NATURAL LOGARITHMS

Of all possible bases a for logarithms, it turns out that the most convenient choice for the purposes of calculus is the number e, which we defined in Section 6.2.

The notation ln is an abbreviation for the Latin name *logarithmus naturalis*.

NATURAL LOGARITHM

The logarithm with base e is called the **natural logarithm** and is denoted by **ln**:

$$\ln x = \log_e x$$

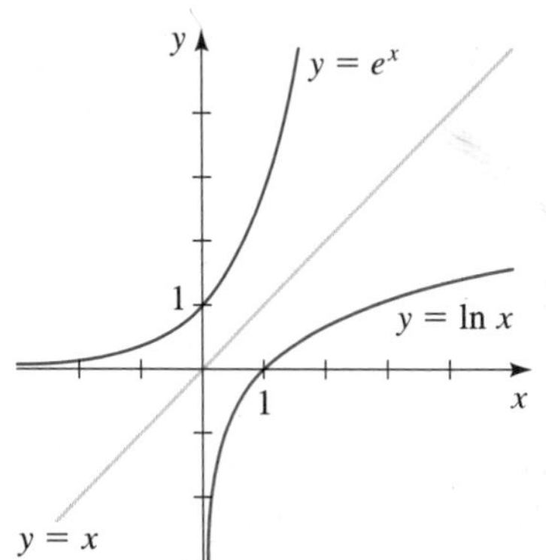

FIGURE 9
Graph of the natural logarithmic function

The natural logarithmic function $y = \ln x$ is the inverse function of the exponential function $y = e^x$. Both functions are graphed in Figure 9. By the definition of inverse functions we have

$$\ln x = y \quad \Leftrightarrow \quad e^y = x$$

If we substitute $a = e$ and write "ln" for "$\log_e$" in the properties of logarithms mentioned earlier, we obtain the following properties of natural logarithms.

PROPERTIES OF NATURAL LOGARITHMS

Property	Reason
1. $\ln 1 = 0$	We must raise e to the power 0 to get 1.
2. $\ln e = 1$	We must raise e to the power 1 to get e.
3. $\ln e^x = x$	We must raise e to the power x to get e^x.
4. $e^{\ln x} = x$	$\ln x$ is the power to which e must be raised to get x.

Calculators are equipped with an [ln] key that directly gives the values of natural logarithms.

EXAMPLE 8 ■ Evaluating the Natural Logarithm Function

(a) $\ln e^8 = 8$ — Definition of natural logarithm

(b) $\ln\left(\dfrac{1}{e^2}\right) = \ln e^{-2} = -2$ — Definition of natural logarithm

(c) $\ln 5 \approx 1.609$ — Use the [ln] key on a calculator ■

EXAMPLE 9 ■ Finding the Domain of a Logarithmic Function

Find the domain of the function $f(x) = \ln(4 - x^2)$.

SOLUTION

As with any logarithmic function, $\ln x$ is defined when $x > 0$. Thus, the domain of f is

$$\{x \mid 4 - x^2 > 0\} = \{x \mid x^2 < 4\} = \{x \mid |x| < 2\}$$
$$= \{x \mid -2 < x < 2\} = (-2, 2)$$ ■

EXAMPLE 10 ■ Drawing the Graph of a Logarithmic Function

Draw the graph of the function $y = x\ln(4 - x^2)$ and use it to find the asymptotes and local maximum and minimum values.

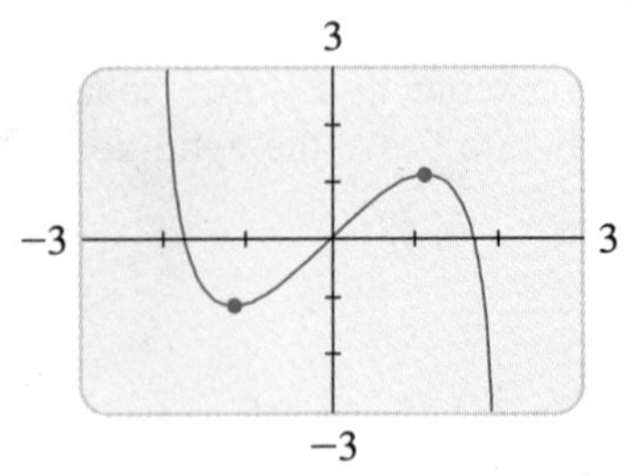

FIGURE 10
$y = x\ln(4 - x^2)$

SOLUTION

As in Example 9 the domain of this function is the interval $(-2, 2)$, so we choose the viewing rectangle $[-3, 3]$ by $[-3, 3]$. The graph is shown in Figure 10, and from it we see that the lines $x = -2$ and $x = 2$ are vertical asymptotes.

The function has a local maximum point to the right of $x = 1$ and a local minimum point to the left of $x = -1$. By zooming in and tracing along the graph with the cursor, we find that the local maximum value is approximately 1.13 and this occurs when $x \approx 1.15$. Similarly (or by noticing that the function is odd), we find that the local minimum value is about -1.13, and it occurs when $x \approx -1.15$. ■

6.3 EXERCISES

1–6 ■ Express the equation in exponential form.

1. (a) $\log_2 32 = 5$ (b) $\log_5 1 = 0$

2. (a) $\log_{10} 0.1 = -1$ (b) $\log_8 512 = 3$

3. (a) $\log_4 2 = \frac{1}{2}$ (b) $\log_2\left(\frac{1}{16}\right) = -4$

4. (a) $\log_3 81 = 4$ (b) $\log_8 4 = \frac{2}{3}$

5. (a) $\ln 5 = x$ (b) $\ln y = 5$

6. (a) $\ln(x + 1) = 2$ (b) $\ln(x - 1) = 4$

7–12 ■ Express the equation in logarithmic form.

7. (a) $2^3 = 8$ (b) $10^{-3} = 0.001$

8. (a) $10^4 = 10{,}000$ (b) $81^{1/2} = 9$

9. (a) $4^{-3/2} = 0.125$ (b) $7^3 = 343$

10. (a) $8^{-1} = \frac{1}{8}$ (b) $10^m = n$

11. (a) $e^x = 2$ (b) $e^3 = y$

12. (a) $e^{x+1} = 0.5$ (b) $e^{0.5x} = t$

13–22 ■ Evaluate the expression.

13. (a) $\log_5 5^4$ (b) $\log_4 64$ (c) $\log_9 9$

14. (a) $\log_3 3$ (b) $\log_3 1$ (c) $\log_3 3^2$

15. (a) $\log_8 64$ (b) $\log_7 49$ (c) $\log_7 7^{10}$

16. (a) $\log_2 32$ (b) $\log_8 8^{17}$ (c) $\log_6 1$

17. (a) $\log_3\left(\frac{1}{27}\right)$ (b) $\log_{10} \sqrt{10}$ (c) $\log_5 0.2$

18. (a) $\log_5 125$ (b) $\log_{49} 7$ (c) $\log_9 \sqrt{3}$

19. (a) $2^{\log_2 37}$ (b) $3^{\log_3 8}$ (c) $e^{\ln \sqrt{5}}$

20. (a) $e^{\ln \pi}$ (b) $10^{\log 5}$ (c) $10^{\log 87}$

21. (a) $\log_8 0.25$ (b) $\ln e^4$ (c) $\ln(1/e)$

22. (a) $\log_4 \sqrt{2}$ (b) $\log_4\left(\frac{1}{2}\right)$ (c) $\log_4 8$

23–28 ■ Use the definition of the logarithmic function to find x.

23. (a) $\log_2 x = 5$ (b) $\log_2 16 = x$

24. (a) $\log_5 x = 4$ (b) $\log_{10} 0.1 = x$

25. (a) $\log_{10} x = 2$ (b) $\log_5 x = 2$

26. (a) $\log_x 1000 = 3$ (b) $\log_x 25 = 2$

27. (a) $\log_x 16 = 4$ (b) $\log_x 8 = \frac{3}{2}$

28. (a) $\log_x 6 = \frac{1}{2}$ (b) $\log_x 3 = \frac{1}{3}$

29–32 ■ Use a calculator to evaluate the expression, correct to four decimal places.

29. (a) $\log 2$ (b) $\log 35.2$ (c) $\log\left(\frac{2}{3}\right)$

30. (a) $\log 50$ (b) $\log \sqrt{2}$ (c) $\log\left(3\sqrt{2}\right)$

31. (a) $\ln 5$ (b) $\ln 25.3$ (c) $\ln\left(1 + \sqrt{3}\right)$

32. (a) $\ln 27$ (b) $\ln 7.39$ (c) $\ln 54.6$

33–36 ■ Find the function of the form $y = \log_a x$ whose graph is given.

33.

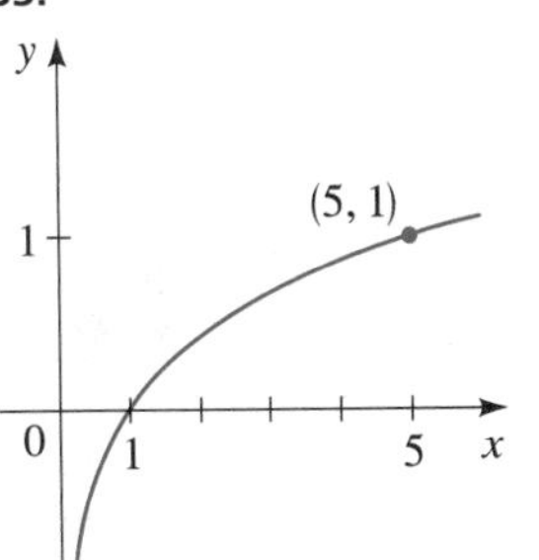

34.

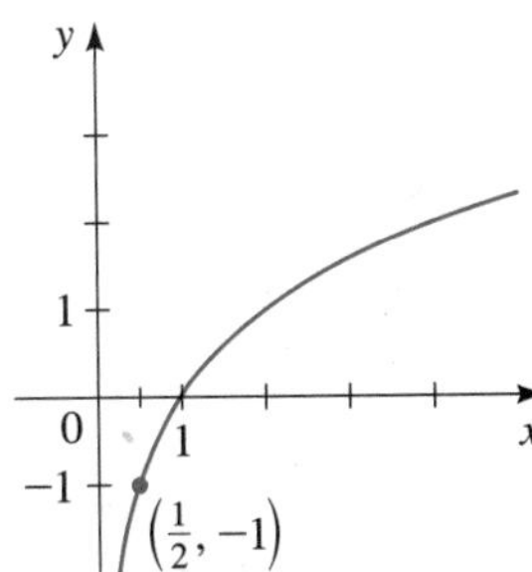

35.

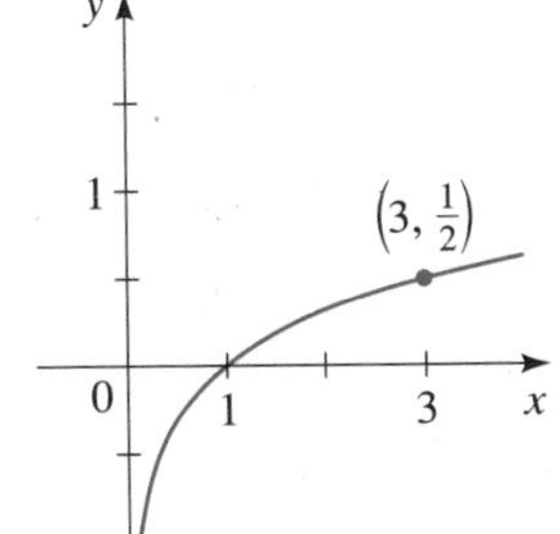

36.

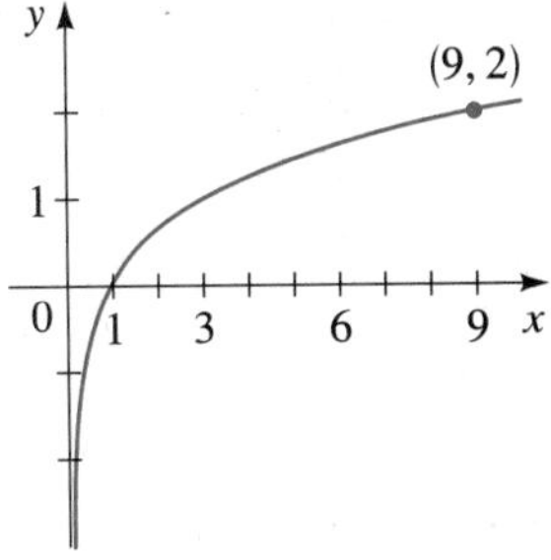

37–42 ■ Match the logarithmic function with one of the graphs labeled I–VI.

37. $f(x) = \ln x$ **38.** $f(x) = \ln(x - 2)$

39. $f(x) = 2 + \ln x$ **40.** $f(x) = \ln(-x)$

41. $f(x) = \ln(2 - x)$ **42.** $f(x) = -\ln(-x)$

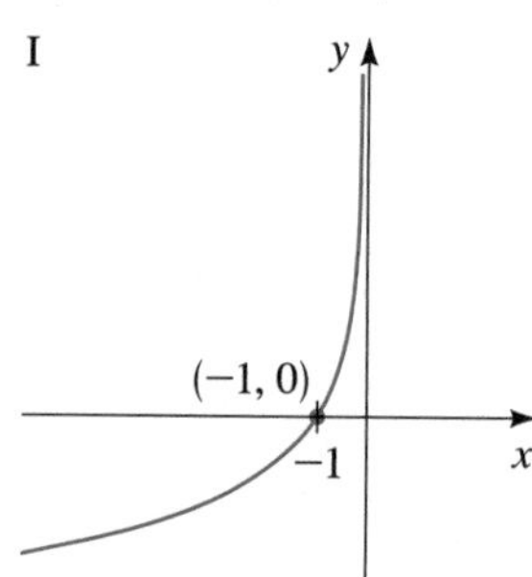

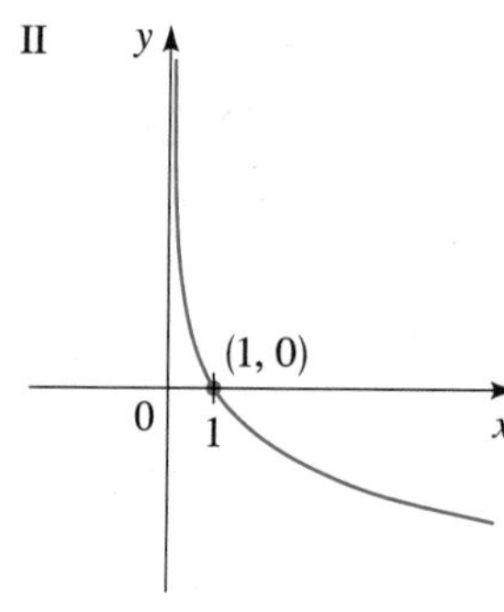

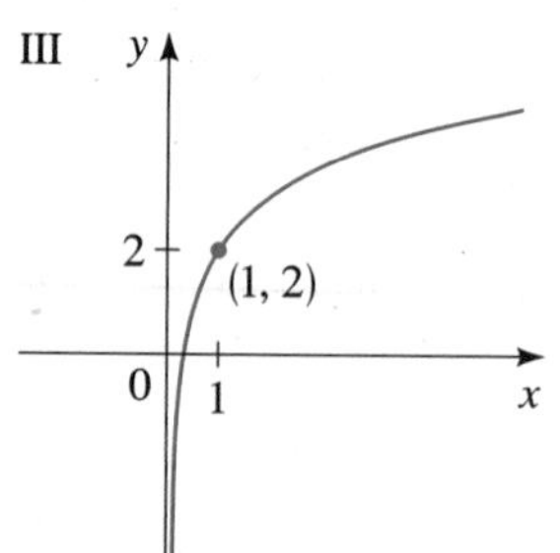

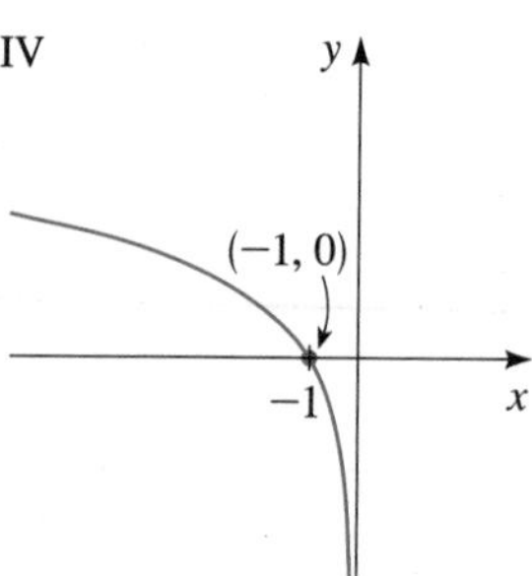

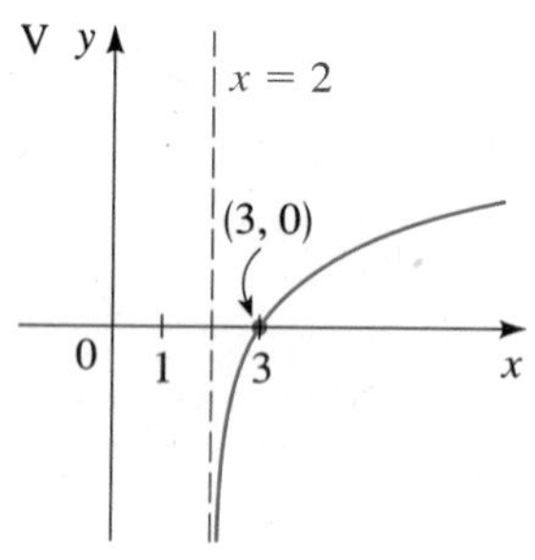

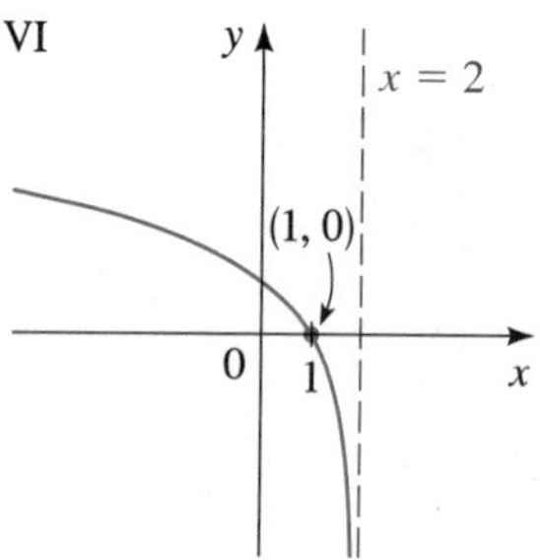

43. Draw the graph of $y = 4^x$, then use it to draw the graph of $y = \log_4 x$.

44. Draw the graph of $y = 3^x$, then use it to draw the graph of $y = \log_3 x$.

45–54 ■ Graph the function, not by plotting points, but by starting from the graphs in Figures 4 and 9. State the domain, range, and asymptote.

45. $f(x) = \log_2(x - 4)$ **46.** $f(x) = -\log_{10} x$

47. $g(x) = \log_5(-x)$ **48.** $g(x) = \ln(x + 2)$

49. $y = 2 + \log_3 x$ **50.** $y = \log_3(x - 1) - 2$

51. $y = 1 - \log_{10} x$ **52.** $y = 1 + \ln(-x)$

53. $y = |\ln x|$ **54.** $y = \ln |x|$

55–60 ■ Find the domain of the function.

55. $f(x) = \log_{10}(2 + 5x)$

56. $f(x) = \log_2(10 - 3x)$

57. $g(x) = \log_3(x^2 - 1)$

58. $g(x) = \ln(x - x^2)$

59. $h(x) = \ln x + \ln(2 - x)$

60. $h(x) = \sqrt{x - 2} - \log_5(10 - x)$

61–66 ■ Draw the graph of the function in a suitable viewing rectangle and use it to find the domain, the asymptotes, and the local maximum and minimum values.

61. $y = \log_{10}(1 - x^2)$ **62.** $y = \ln(x^2 - x)$

63. $y = x + \ln x$ **64.** $y = x(\ln x)^2$

65. $y = \dfrac{\ln x}{x}$ **66.** $y = x \log_{10}(x + 10)$

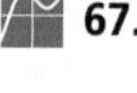

67. Compare the rates of growth of the functions $f(x) = \ln x$ and $g(x) = \sqrt{x}$ by drawing their graphs on a common screen using the viewing rectangle $[-1, 30]$ by $[-1, 6]$.

68. (a) By drawing the graphs of the functions

$$f(x) = 1 + \ln(1 + x) \quad \text{and} \quad g(x) = \sqrt{x}$$

in a suitable viewing rectangle, show that even when a logarithmic function starts out higher than a root function, it is ultimately overtaken by the root function.

(b) Find, correct to two decimal places, the solutions of the equation $\sqrt{x} = 1 + \ln(1 + x)$.

69–70 ■ A family of functions is given.

(a) Draw graphs of the family for $c = 1, 2, 3$, and 4.

(b) How are the graphs in part (a) related?

69. $f(x) = \log(cx)$

70. $f(x) = c \log x$

71–72 ■ A function $f(x)$ is given.

(a) Find the domain of the function f.

(b) Find the inverse function of f.

71. $f(x) = \log_2(\log_{10} x)$

72. $f(x) = \ln(\ln(\ln x))$

73. (a) Find the inverse of the function $f(x) = \dfrac{2^x}{1 + 2^x}$.

(b) What is the domain of the inverse function?

DISCOVERY • DISCUSSION

74. The Height of the Graph of a Logarithmic Function Suppose that the graph of $y = 2^x$ is drawn on a coordinate plane where the unit of measurement is an inch.

(a) Show that at a distance 2 ft to the right of the origin the height of the graph is about 265 mi.

(b) If the graph of $y = \log_2 x$ is drawn on the same set of axes, how far to the right of the origin do we have to go before the height of the curve reaches 2 ft?

75. The Googolplex A **googol** is 10^{100}, and a **googolplex** is 10^{googol}. Find

$$\log(\log(\text{googol})) \qquad \text{and} \qquad \log(\log(\log(\text{googolplex})))$$

76. Comparing Logarithms Which is larger, $\log_4 17$ or $\log_5 24$? Explain your reasoning.

77. The Number of Digits in an Integer Compare log 1000 to the number of digits in 1000. Do the same for 10,000. How many digits does any number between 1000 and 10,000 have? Between what two values must the common logarithm of such a number lie? Use your observations to explain why the number of digits in any positive integer x is $[\![\log x]\!] + 1$. (The symbol $[\![n]\!]$ is the greatest integer function defined in Section 4.2.)

6.4 LAWS OF LOGARITHMS

Since logarithms are exponents, the Laws of Exponents give rise to the Laws of Logarithms. These properties give logarithmic functions a wide range of applications, as we will see in Section 6.6.

LAWS OF LOGARITHMS

Let a be a positive number, with $a \neq 1$. Let $A > 0$, $B > 0$, and C be any real numbers.

Law	Description
1. $\log_a(AB) = \log_a A + \log_a B$	The logarithm of a product of numbers is the sum of the logarithms of the numbers.
2. $\log_a\left(\dfrac{A}{B}\right) = \log_a A - \log_a B$	The logarithm of a quotient of numbers is the difference of the logarithms of the numbers.
3. $\log_a(A^C) = C \log_a A$	The logarithm of a power of a number is the exponent times the logarithm of the number.

■ **Proof** We make use of the property $\log_a a^x = x$ from Section 6.3.

Law 1. Let $\qquad \log_a A = u \qquad \text{and} \qquad \log_a B = v$

When written in exponential form, these equations become

$$a^u = A \qquad \text{and} \qquad a^v = B$$

John Napier (1550–1617) was a Scottish landowner for whom mathematics was a hobby. We know him today because of his key invention—logarithms, which he published in 1614 under the title *A Description of the Marvelous Rule of Logarithms*. In Napier's time, logarithms were used exclusively for simplifying complicated calculations. For example, to multiply two large numbers we would write them as powers of 10. The exponents are simply the logarithms of the numbers. For instance,

$$\begin{aligned}4532 \times 57783 &\approx 10^{3.65629} \times 10^{4.76180}\\ &= 10^{8.41809}\\ &\approx 261{,}872{,}564\end{aligned}$$

The idea is that multiplying powers of 10 is easy (we simply add their exponents). Napier produced extensive tables giving the logarithms (or exponents) of numbers. Since the advent of calculators and computers, logarithms are no longer used for this purpose. The logarithmic functions, however, have found many applications, some of which are described in this chapter.

Napier wrote on many topics. One of his most colorful works is a book entitled *A Plaine Discovery of the Whole Revelation of Saint John*, in which he predicted that the world would end in the year 1700.

Thus
$$\begin{aligned}\log_a(AB) &= \log_a(a^u a^v) = \log_a(a^{u+v})\\ &= u + v = \log_a A + \log_a B\end{aligned}$$

Law 2. Using Law 1, we have

$$\log_a A = \log_a\left[\left(\frac{A}{B}\right)B\right] = \log_a\left(\frac{A}{B}\right) + \log_a B$$

so
$$\log_a\left(\frac{A}{B}\right) = \log_a A - \log_a B$$

Law 3. Let $\log_a A = u$. Then $a^u = A$, so

$$\log_a(A^C) = \log_a(a^u)^C = \log_a(a^{uC}) = uC = C\log_a A$$ □

As the following examples illustrate, these laws are used in both directions. Since the domain of any logarithmic function is the interval $(0, \infty)$, we assume that all quantities whose logarithms occur are positive.

EXAMPLE 1 ■ Using the Laws of Logarithms to Expand Expressions

Use the Laws of Logarithms to rewrite each expression.

(a) $\log_2(6x)$ (b) $\log\sqrt{5}$

(c) $\log_5(x^3y^6)$ (d) $\ln\left(\dfrac{ab}{\sqrt[3]{c}}\right)$

SOLUTION

(a) $\log_2(6x) = \log_2 6 + \log_2 x$ Law 1

(b) $\log\sqrt{5} = \log 5^{1/2} = \frac{1}{2}\log 5$ Law 3

(c) $\log_5(x^3y^6) = \log_5 x^3 + \log_5 y^6$ Law 1

$= 3\log_5 x + 6\log_5 y$ Law 3

(d) $\ln\left(\dfrac{ab}{\sqrt[3]{c}}\right) = \ln(ab) - \ln\sqrt[3]{c}$ Law 2

$= \ln a + \ln b - \ln c^{1/3}$ Law 1

$= \ln a + \ln b - \frac{1}{3}\ln c$ Law 3 ■

EXAMPLE 2 ■ Using the Laws of Logarithms to Evaluate Expressions

Evaluate each expression.

(a) $\log_4 2 + \log_4 32$ (b) $\log_2 80 - \log_2 5$ (c) $-\frac{1}{3}\log 8$

SOLUTION

(a) $\log_4 2 + \log_4 32 = \log_4(2 \cdot 32)$ Law 1

$= \log_4 64 = 3$ Because $4^3 = 64$

(b) $\log_2 80 - \log_2 5 = \log_2\left(\frac{80}{5}\right)$ Law 2

$= \log_2 16 = 4$ Because $2^4 = 16$

(c) $-\frac{1}{3}\log 8 = \log 8^{-1/3}$ Law 3

$= \log\left(\frac{1}{2}\right)$ Property of negative exponents

≈ -0.301 Use a calculator ■

EXAMPLE 3 ■ Writing an Expression as a Single Logarithm

Express $3 \log x + \frac{1}{2}\log(x + 1)$ as a single logarithm.

SOLUTION

$$3 \log x + \tfrac{1}{2}\log(x+1) = \log x^3 + \log(x+1)^{1/2} \quad \text{Law 3}$$

$$= \log(x^3(x+1)^{1/2}) \quad \text{Law 1}$$ ■

EXAMPLE 4 ■ Writing an Expression as a Single Logarithm

Express $3 \ln s + \frac{1}{2}\ln t - 4\ln(t^2 + 1)$ as a single logarithm.

SOLUTION

$$3 \ln s + \tfrac{1}{2}\ln t - 4\ln(t^2+1) = \ln s^3 + \ln t^{1/2} - \ln(t^2+1)^4 \quad \text{Law 3}$$

$$= \ln(s^3 t^{1/2}) - \ln(t^2+1)^4 \quad \text{Law 1}$$

$$= \ln\left(\frac{s^3\sqrt{t}}{(t^2+1)^4}\right) \quad \text{Law 2}$$ ■

WARNING Although the Laws of Logarithms tell us how to compute the logarithm of a product or a quotient, *there is no corresponding rule for the logarithm of a sum or a difference.* For instance,

$$\log_a(x + y) \neq \log_a x + \log_a y$$

In fact, we know that the right side is equal to $\log_a(xy)$.

Also, don't improperly simplify quotients or powers of logarithms. For instance,

$$\frac{\log 6}{\log 2} \neq \log\left(\frac{6}{2}\right)$$

$$(\log_2 x)^3 \neq 3\log_2 x$$

CHANGE OF BASE

For some purposes, we find it useful to change from logarithms in one base to logarithms in another base. Suppose we are given $\log_a x$ and want to find $\log_b x$. Let

$$y = \log_b x$$

We write this in exponential form and take the logarithm, with base a, of each side.

$$b^y = x \qquad \text{Exponential form}$$

$$\log_a(b^y) = \log_a x \qquad \text{Take } \log_a \text{ of each side}$$

$$y \log_a b = \log_a x \qquad \text{Law 3}$$

$$y = \frac{\log_a x}{\log_a b} \qquad \text{Divide by } \log_a b$$

This proves the following formula.

We may write the change of base formula as

$$\log_b x = \left(\frac{1}{\log_a b}\right)\log_a x$$

So, $\log_b x$ is just a constant multiple of $\log_a x$; the constant is $\dfrac{1}{\log_a b}$.

CHANGE OF BASE FORMULA

$$\log_b x = \frac{\log_a x}{\log_a b}$$

In particular, if we put $x = a$, then $\log_a a = 1$ and this formula becomes

$$\log_b a = \frac{1}{\log_a b}$$

We can now evaluate a logarithm to *any* base by using the Change of Base Formula to express the logarithm in terms of common logarithms or natural logarithms and then using a calculator.

EXAMPLE 5 ■ Using the Change of Base Formula to Evaluate Logarithms

Use the Change of Base Formula and common or natural logarithms to evaluate each logarithm, correct to five decimal places.

(a) $\log_8 5$ (b) $\log_9 20$

SOLUTION

(a) We use the Change of Base Formula with $b = 8$ and $a = 10$:

$$\log_8 5 = \frac{\log_{10} 5}{\log_{10} 8} \approx 0.77398$$

(b) We use the Change of Base Formula with $b = 9$ and $a = e$:

$$\log_9 20 = \frac{\ln 20}{\ln 9} \approx 1.36342$$

■

EXAMPLE 6 ■ Using the Change of Base Formula To Graph a Logarithmic Function

Use a graphing calculator to graph $f(x) = \log_6 x$.

SOLUTION

Calculators don't have a key for $\log_6$, so we use the Change of Base Formula to write

$$f(x) = \log_6 x = \frac{\ln x}{\ln 6}$$

Since calculators do have an ln key, we can enter this new form of the function and graph it. The graph is shown in Figure 1. ■

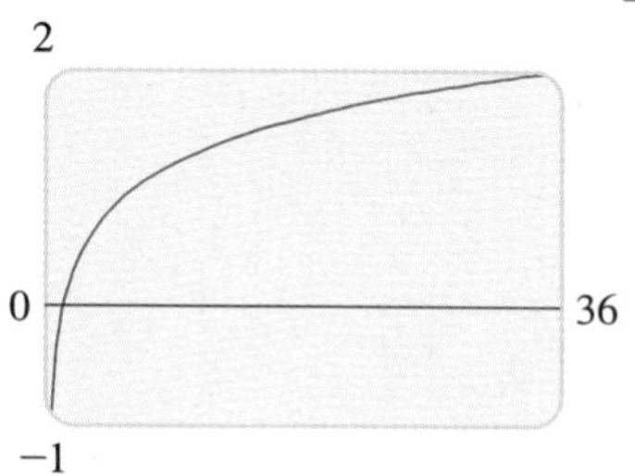

FIGURE 1

$f(x) = \log_6 x = \dfrac{\ln x}{\ln 6}$

6.4 EXERCISES

1–24 ■ Use the Laws of Logarithms to rewrite the expression in a form with no logarithm of a product, quotient, or power.

1. $\log_2(x(x-1))$

2. $\log_5\left(\frac{x}{2}\right)$

3. $\log 7^{23}$

4. $\ln(\pi x)$

5. $\log_2(AB^2)$

6. $\log_6 \sqrt[4]{17}$

7. $\log_3(x\sqrt{y})$

8. $\log_2(xy)^{10}$

9. $\log_5 \sqrt[3]{x^2+1}$

10. $\log_a\left(\frac{x^2}{yz^3}\right)$

11. $\ln \sqrt{ab}$

12. $\ln \sqrt[3]{3r^2s}$

13. $\log\left(\frac{x^3y^4}{z^6}\right)$

14. $\log\left(\frac{a^2}{b^4\sqrt{c}}\right)$

15. $\log_2\left(\frac{x(x^2+1)}{\sqrt{x^2-1}}\right)$

16. $\log_5\sqrt{\frac{x-1}{x+1}}$

17. $\ln\left(x\sqrt{\frac{y}{z}}\right)$

18. $\ln \frac{3x^2}{(x+1)^{10}}$

19. $\log \sqrt[4]{x^2+y^2}$

20. $\log\left(\frac{x}{\sqrt[3]{1-x}}\right)$

21. $\log\sqrt{\frac{x^2+4}{(x^2+1)(x^3-7)^2}}$

22. $\log\sqrt{x\sqrt{y\sqrt{z}}}$

23. $\ln\left(\frac{z^4\sqrt{x}}{\sqrt[3]{y^2+6y+17}}\right)$

24. $\log\left(\frac{10^x}{x(x^2+1)(x^4+2)}\right)$

25–34 ■ Evaluate the expression.

25. $\log_5\sqrt{125}$

26. $\log_2 112 - \log_2 7$

27. $\log 2 + \log 5$

28. $\log\sqrt{0.1}$

29. $\log_4 192 - \log_4 3$

30. $\log_{12} 9 + \log_{12} 16$

31. $\ln 6 - \ln 15 + \ln 20$

32. $e^{3\ln 5}$

33. $10^{2\log 4}$

34. $\log_2 8^{33}$

35–44 ■ Rewrite the expression as a single logarithm.

35. $\log_3 5 + 5\log_3 2$

36. $\log 12 + \frac{1}{2}\log 7 - \log 2$

37. $\log_2 A + \log_2 B - 2\log_2 C$

38. $\log_5(x^2-1) - \log_5(x-1)$

39. $4\log x - \frac{1}{3}\log(x^2+1) + 2\log(x-1)$

40. $\ln(a+b) + \ln(a-b) - 2\ln c$

41. $\ln 5 + 2\ln x + 3\ln(x^2+5)$

42. $2(\log_5 x + 2\log_5 y - 3\log_5 z)$

43. $\frac{1}{3}\log(2x+1) + \frac{1}{2}[\log(x-4) - \log(x^4-x^2-1)]$

44. $\log_a b + c \log_a d - r \log_a s$

45–52 ■ Use the Change of Base Formula and a calculator to evaluate the logarithm, correct to six decimal places. Use either natural or common logarithms.

45. $\log_2 7$ **46.** $\log_5 2$ **47.** $\log_3 11$

48. $\log_6 92$ **49.** $\log_7 3.58$ **50.** $\log_6 532$

51. $\log_4 322$ **52.** $\log_{12} 2.5$

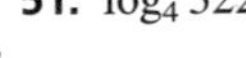

53. Use the Change of Base Formula to show that

$$\log_3 x = \frac{\ln x}{\ln 3}$$

Then use this fact to draw the graph of the function $f(x) = \log_3 x$.

54. Draw graphs of the family of functions $y = \log_a x$ for $a = 2, e, 5$, and 10 on the same screen, using the viewing rectangle $[0, 5]$ by $[-3, 3]$. How are these graphs related?

55. Use the Change of Base Formula to show that

$$\log e = \frac{1}{\ln 10}$$

56. Simplify: $(\log_2 5)(\log_5 7)$

57. Show that $-\ln\left(x - \sqrt{x^2 - 1}\right) = \ln\left(x + \sqrt{x^2 - 1}\right)$.

DISCOVERY • DISCUSSION

58. Is the Equation an Identity? Discuss each equation and determine whether it is true for all possible values of the variables. (Ignore values of the variables for which any term is undefined.)

(a) $\log\left(\frac{x}{y}\right) = \frac{\log x}{\log y}$

(b) $\log_2(x - y) = \log_2 x - \log_2 y$

(c) $\log_5\left(\frac{a}{b^2}\right) = \log_5 a - 2 \log_5 b$

(d) $\log 2^z = z \log 2$

(e) $(\log P)(\log Q) = \log P + \log Q$

(f) $\frac{\log a}{\log b} = \log a - \log b$

(g) $(\log_2 7)^x = x \log_2 7$

(h) $\log_a a^a = a$

(i) $\log(x - y) = \frac{\log x}{\log y}$

(j) $-\ln\left(\frac{1}{A}\right) = \ln A$

59. Find the Error What is wrong with the following argument?

$$\begin{aligned} \log 0.1 &< 2 \log 0.1 \\ &= \log(0.1)^2 \\ &= \log 0.01 \\ \log 0.1 &< \log 0.01 \\ 0.1 &< 0.01 \end{aligned}$$

60. Shifting, Shrinking, and Stretching Graphs of Functions Let $f(x) = x^2$. Show that $f(2x) = 4f(x)$, and explain how this shows that shrinking the graph of f horizontally has the same effect as stretching it vertically. Then use the identities $e^{2+x} = e^2 e^x$ and $\ln(2x) = \ln 2 + \ln x$ to show that for $g(x) = e^x$, a horizontal shift is the same as a vertical stretch and for $h(x) = \ln x$, a horizontal shrinking is the same as a vertical shift.

6.5 EXPONENTIAL AND LOGARITHMIC EQUATIONS

In this section we solve equations that involve exponential or logarithmic functions. The techniques we develop here will be used in the next section for solving applied problems.

EXPONENTIAL EQUATIONS

An exponential equation is one in which the variable occurs in the exponent. For example,

$$2^x = 7$$

The variable x presents a difficulty because it is in the exponent. To deal with this difficulty, we take the logarithm of each side and then use the Laws of Logarithms to "bring down x" from the exponent.

$$2^x = 7$$

$$\ln 2^x = \ln 7 \quad \text{Take ln of each side}$$

$$x \ln 2 = \ln 7 \quad \text{Law 3 (bring down the exponent)}$$

$$x = \frac{\ln 7}{\ln 2} \quad \text{Solve for } x$$

$$\approx 2.807 \quad \text{Use a calculator}$$

Recall that Law 3 of the Laws of Logarithms says that $\log_a A^C = C \log_a A$.

The method we used to solve $2^x = 7$ is typical of the methods we use to solve all exponential equations, and it can be summarized as follows.

GUIDELINES FOR SOLVING EXPONENTIAL EQUATIONS

1. Isolate the exponential expression on one side of the equation.
2. Take the logarithm of each side, then use the Laws of Logarithms to "bring down the exponent."
3. Solve for the variable.

EXAMPLE 1 ■ Solving an Exponential Equation

Find the solution of the equation $3^{x+2} = 7$, correct to six decimal places.

SOLUTION

We take the common logarithm of each side and use Law 3.

$$3^{x+2} = 7$$

$$\log(3^{x+2}) = \log 7 \quad \text{Take log of each side}$$

$$(x + 2) \log 3 = \log 7 \quad \text{Law 3 (bring down the exponent)}$$

$$x + 2 = \frac{\log 7}{\log 3} \quad \text{Divide by log 3}$$

$$x = \frac{\log 7}{\log 3} - 2 \quad \text{Subtract 2}$$

$$\approx -0.228756 \quad \text{Use a calculator}$$

We could have used natural logarithms instead of common logarithms. In fact, using the same steps, we get

$$x = \frac{\ln 7}{\ln 3} - 2 \approx -0.228756$$

CHECK YOUR ANSWER ■ Substituting $x = -0.228756$ into the original equation and using a calculator, we get

$$3^{(-0.228756)+2} \approx 7 \quad ✓$$

■

EXAMPLE 2 ■ Solving an Exponential Equation

Solve the equation $8e^{2x} = 20$.

SOLUTION

We first divide by 8 in order to isolate the exponential term on one side of the equation.

$$\begin{aligned} 8e^{2x} &= 20 \\ e^{2x} &= \tfrac{20}{8} && \text{Divide by 8} \\ \ln e^{2x} &= \ln 2.5 && \text{Take ln of each side} \\ 2x &= \ln 2.5 && \text{Property of ln} \\ x &= \frac{\ln 2.5}{2} && \text{Divide by 2} \\ &\approx 0.458 && \text{Use a calculator} \end{aligned}$$

CHECK YOUR ANSWER ■ Substituting $x = 0.458$ into the original equation and using a calculator, we get

$$8e^{2(0.458)} \approx 20 \quad ✓$$

■

An alternative method is to solve graphically (see Section 3.1). For the equation in Example 3 we draw the graphs of $y = e^{3-2x}$ and $y = 4$ in the same viewing rectangle. Zooming in on the point of intersection of the two graphs, we see that the solution is $x \approx 0.807$.

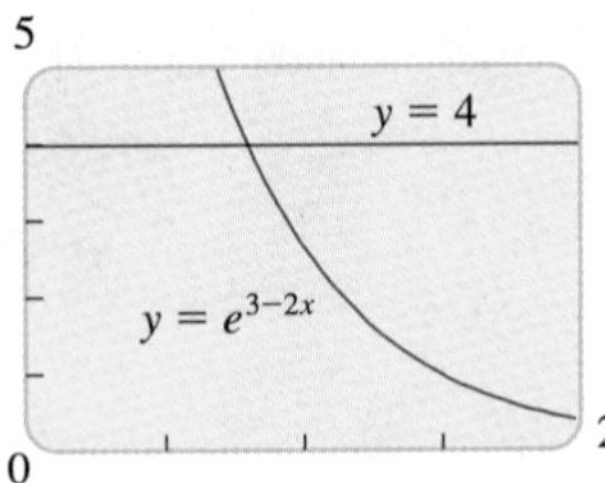

EXAMPLE 3 ■ Solving an Exponential Equation

Solve the equation $e^{3-2x} = 4$.

SOLUTION

Since the base of the exponential term is e, we use natural logarithms to solve this equation.

$$\begin{aligned} \ln(e^{3-2x}) &= \ln 4 && \text{Take ln of each side} \\ 3 - 2x &= \ln 4 && \text{Property of ln} \\ 2x &= 3 - \ln 4 \\ x &= \tfrac{1}{2}(3 - \ln 4) \approx 0.807 \end{aligned}$$

You should check that this answer satisfies the original equation. ■

EXAMPLE 4 ■ Solving an Exponential Equation

Solve the equation $e^{2x} - e^x - 6 = 0$.

SOLUTION

To isolate the exponential term, we factor.

$$e^{2x} - e^x - 6 = 0$$

$$(e^x)^2 - e^x - 6 = 0 \qquad \text{Law of Exponents}$$

$$(e^x - 3)(e^x + 2) = 0 \qquad \text{Factor (a quadratic in } e^x\text{)}$$

$$e^x - 3 = 0 \quad \text{or} \quad e^x + 2 = 0 \qquad \text{Zero-Product Property}$$

$$e^x = 3 \qquad\qquad e^x = -2$$

The equation $e^x = 3$ leads to $x = \ln 3$. But the equation $e^x = -2$ has no solution because $e^x > 0$ for all x. Thus, $x = \ln 3 \approx 1.0986$ is the only solution. You should check that this answer satisfies the original equation. ■

EXAMPLE 5 ■ Solving an Exponential Equation

Solve the equation $3x^2e^x + x^3e^x = 0$.

SOLUTION

First we factor the left side of the equation.

$$3x^2e^x + x^3e^x = 0$$

$$(3x^2 + x^3)e^x = 0 \qquad \text{Factor out } e^x$$

$$x^2(3 + x)e^x = 0 \qquad \text{Factor out } x^2$$

$$x^2 = 0 \quad \text{or} \quad 3 + x = 0 \quad \text{or} \quad e^x = 0 \qquad \text{Zero-Product Property}$$

$$x = 0 \qquad\qquad x = -3$$

CHECK YOUR ANSWERS

$x = 0$:

$3(0)^2e^0 + 0^3e^0 = 0$ ✓

$x = -3$:

$3(-3)^2e^{-3} + (-3)^3e^{-3}$

$= 27e^{-3} - 27e^{-3} = 0$ ✓

The equation $e^x = 0$ has no solution because $e^x > 0$ for all x. Thus, $x = 0$ and $x = -3$ are the only solutions. ■

LOGARITHMIC EQUATIONS

A *logarithmic equation* is one in which a logarithm of the variable occurs. For example,

$$\log_2(x + 2) = 5$$

Radiocarbon dating is a method archeologists use to determine the age of ancient objects. The carbon dioxide in the atmosphere always contains a fixed fraction of radioactive carbon, carbon-14 (^{14}C), with a half-life of about 5730 years. Plants absorb carbon dioxide from the atmosphere, which then makes its way to animals through the food chain. Thus, all living creatures contain the same fixed proportions of ^{14}C to nonradioactive ^{12}C as the atmosphere.

After an organism dies, it stops assimilating ^{14}C, and the amount of ^{14}C in it begins to decay exponentially. We can then determine the time elapsed since the death of the organism by measuring the amount of ^{14}C left in it.

For example, if a donkey bone contains 73% as much ^{14}C as a living donkey and it died t years ago, then by the formula for radioactive decay (Section 6.6),

$$0.73 = (1.00)e^{-(t \ln 2)/5730}$$

We solve this exponential equation to find $t \approx 2600$, so the bone is about 2600 years old.

To solve for x, we write the equation in exponential form.

$$x + 2 = 2^5 \quad \text{Exponential form}$$

$$x = 32 - 2 = 30 \quad \text{Solve for } x$$

Another way of looking at the first step is to raise the base, 2, to each side of the equation.

$$2^{\log_2(x+2)} = 2^5 \quad \text{Raise 2 to each side}$$

$$x + 2 = 2^5 \quad \text{Property of logarithms}$$

$$x = 32 - 2 = 30 \quad \text{Solve for } x$$

The methods used to solve this simple problem are typical. We summarize the steps as follows.

GUIDELINES FOR SOLVING LOGARITHMIC EQUATIONS

1. Isolate the logarithmic term on one side of the equation; you may need to first combine the logarithmic terms.
2. Write the equation in exponential form (or raise the base to each side of the equation).
3. Solve for the variable.

EXAMPLE 6 ■ Solving Logarithmic Equations

Solve each equation for x.

(a) $\ln x = 8$ (b) $\log_2(25 - x) = 3$

SOLUTION

(a)

$$\ln x = 8$$

$$x = e^8 \quad \text{Exponential form}$$

Therefore, $x = e^8 \approx 2981$.

We can also solve this problem another way:

$$\ln x = 8$$

$$e^{\ln x} = e^8 \quad \text{Raise } e \text{ to each side}$$

$$x = e^8 \quad \text{Property of ln}$$

(b) The first step is to rewrite the equation in exponential form.

$$\log_2(25 - x) = 3$$

$$25 - x = 2^3 \qquad \text{Exponential form (or raise 2 to each side)}$$

$$25 - x = 8$$

$$x = 25 - 8 = 17$$

■

CHECK YOUR ANSWER

If $x = 17$, we get

$\log_2(25 - 17) = \log_2 8 = 3$ ✓

EXAMPLE 7 ■ Solving Logarithmic Equations

Solve the equation $4 + 3\log(2x) = 16$.

SOLUTION

We first isolate the logarithmic term. This allows us to write the equation in exponential form.

$$4 + 3\log(2x) = 16$$

$$3\log(2x) = 12 \qquad \text{Subtract 4}$$

$$\log(2x) = 4 \qquad \text{Divide by 3}$$

$$2x = 10^4 \qquad \text{Exponential form (or raise 10 to each side)}$$

$$x = 5000 \qquad \text{Divide by 2}$$

■

CHECK YOUR ANSWER

If $x = 5000$, we get

$$\begin{aligned} 4 + 3\log 2(5000) &= 4 + 3\log 10{,}000 \\ &= 4 + 3(4) \\ &= 16 \quad ✓ \end{aligned}$$

EXAMPLE 8 ■ Solving Logarithmic Equations

Solve the equation $\log(x + 2) + \log(x - 1) = 1$.

SOLUTION

We first combine the logarithmic terms using the Laws of Logarithms.

$$\log[(x + 2)(x - 1)] = 1 \qquad \text{Law 1}$$

$$(x + 2)(x - 1) = 10 \qquad \text{Exponential form (or raise 10 to each side)}$$

$$x^2 + x - 2 = 10 \qquad \text{Expand left side}$$

$$x^2 + x - 12 = 0 \qquad \text{Subtract 10}$$

$$(x + 4)(x - 3) = 0 \qquad \text{Factor}$$

$$x = -4 \quad \text{or} \quad x = 3$$

We check these potential solutions in the original equation and find that $x = -4$ is not a solution (because logarithms of negative numbers are undefined), but $x = 3$ is a solution. (See *Check Your Answers*.) ■

CHECK YOUR ANSWERS

$x = -4$:

$$\begin{aligned} &\log(-4 + 2) + \log(-4 - 1) \\ &\quad = \log(-2) + \log(-5) \\ &\quad \text{undefined} \quad ✗ \end{aligned}$$

$x = 3$:

$$\begin{aligned} &\log(3 + 2) + \log(3 - 1) \\ &\quad = \log 5 + \log 2 = \log(5 \cdot 2) \\ &\quad = \log 10 = 1 \quad ✓ \end{aligned}$$

6.5 EXERCISES

1–24 ■ Find the solution of the exponential equation, correct to four decimal places.

1. $5^x = 16$

2. $10^{-x} = 2$

3. $2^{1-x} = 3$

4. $3^{2x-1} = 5$

5. $3e^x = 10$

6. $2e^{12x} = 17$

7. $e^{1-4x} = 2$

8. $4(1 + 10^{5x}) = 9$

9. $4 + 3^{5x} = 8$

10. $2^{3x} = 34$

11. $8^{0.4x} = 5$

12. $3^{x/14} = 0.1$

13. $5^{-x/100} = 2$

14. $e^{3-5x} = 16$

15. $e^{2x+1} = 200$

16. $\left(\frac{1}{4}\right)^x = 75$

17. $5^x = 4^{x+1}$

18. $10^{1-x} = 6^x$

19. $2^{3x+1} = 3^{x-2}$

20. $7^{x/2} = 5^{1-x}$

21. $\dfrac{50}{1 + e^{-x}} = 4$

22. $\dfrac{10}{1 + e^{-x}} = 2$

23. $100(1.04)^{2t} = 300$

24. $(1.00625)^{12t} = 2$

25–32 ■ Solve the equation.

25. $x^2 2^x - 2^x = 0$

26. $x^2 10^x - x10^x = 2(10^x)$

27. $4x^3e^{-3x} - 3x^4e^{-3x} = 0$

28. $x^2e^x + xe^x - e^x = 0$

29. $e^{2x} - 3e^x + 2 = 0$

30. $e^{2x} - e^x - 6 = 0$

31. $e^{4x} + 4e^{2x} - 21 = 0$

32. $e^x - 12e^{-x} - 1 = 0$

33–48 ■ Solve the logarithmic equation for x.

33. $\ln x = 10$

34. $\ln(2 + x) = 1$

35. $\log x = -2$

36. $\log(x - 4) = 3$

37. $\log(3x + 5) = 2$

38. $\log_3(2 - x) = 3$

39. $2 - \ln(3 - x) = 0$

40. $\log_2(x^2 - x - 2) = 2$

41. $\log_2 3 + \log_2 x = \log_2 5 + \log_2(x - 2)$

42. $2 \log x = \log 2 + \log(3x - 4)$

43. $\log x + \log(x - 1) = \log(4x)$

44. $\log_5 x + \log_5(x + 1) = \log_5 20$

45. $\log_5(x + 1) - \log_5(x - 1) = 2$

46. $\log x + \log(x - 3) = 1$

47. $\log_9(x - 5) + \log_9(x + 3) = 1$

48. $\ln(x - 1) + \ln(x + 2) = 1$

49. For what value of x is the following true?

$$\log(x + 3) = \log x + \log 3$$

50. For what value of x is it true that $(\log x)^3 = 3 \log x$?

51. Solve for x: $2^{2/\log_5 x} = \frac{1}{16}$

52. Solve for x: $\log_2(\log_3 x) = 4$

53. A 15-g sample of radioactive iodine decays in such a way that the mass remaining after t days is given by $m(t) = 15e^{-0.087t}$ where $m(t)$ is measured in grams. After how many days is there only 5 g remaining?

54. The velocity of a sky diver t seconds after jumping is given by $v(t) = 80(e^{-0.2t} - 1)$. After how many seconds is the velocity 70 ft/s?

55. The figure shows an electric circuit containing a battery producing a voltage of 60 volts (V), a resistor with a resistance of 13 ohms (Ω), and an inductor with an inductance of 5 henrys (H). Using calculus, it can be shown that the current $I = I(t)$ (in amperes, A) t seconds after the switch is closed is $I = \frac{60}{13}(1 - e^{-13t/5})$.

(a) Use this equation to express the time t as a function of the current I.

(b) After how many seconds is the current 2 A?

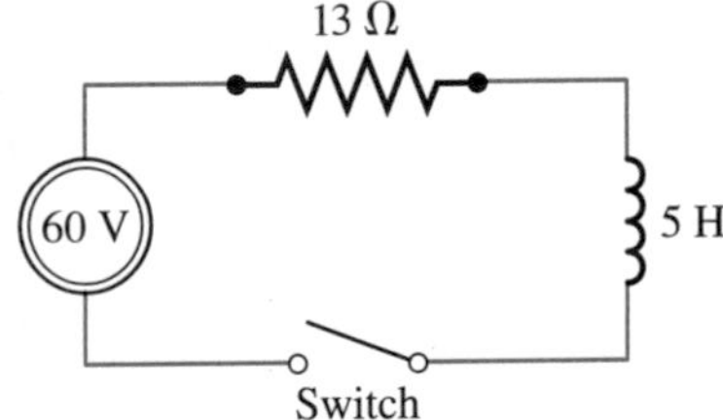

56. A *learning curve* is a graph of a function $P(t)$ that measures the performance of someone learning a skill as a function of the training time t. At first, the rate of learning is rapid. Then, as performance increases and approaches a maximal value M, the rate of learning decreases. It has been found that the function $P(t) = M - Ce^{-kt}$, where k and C are positive constants and $C < M$, is a reasonable model for learning.

(a) Express the learning time t as a function of the performance level P.

(b) For a pole-vaulter in training, the learning curve is given by $P(t) = 20 - 14e^{-0.024t}$, where $P(t)$ is the height he is able to pole-vault after t months. After how many months of training is he able to vault 12 ft?

(c) Draw a graph of the learning curve in part (b).

57–64 ■ Use a graphing device to find all solutions of the equation, correct to two decimal places.

57. $\ln x = 3 - x$

58. $\log x = x^2 - 2$

59. $x^3 - x = \log(x + 1)$

60. $x = \ln(4 - x^2)$

61. $e^x = -x$

62. $2^{-x} = x - 1$

63. $4^{-x} = \sqrt{x}$

64. $e^{x^2} - 2 = x^3 - x$

65–68 ■ Solve the inequality.

65. $\log(x - 2) + \log(9 - x) < 1$

66. $3 \leqslant \log_2 x \leqslant 4$

67. $2 < 10^x < 5$

68. $x^2e^x - 2e^x < 0$

69–71 ■ Solve the equation.

69. $(x - 1)^{\log(x-1)} = 100(x - 1)$

70. $\log_2 x + \log_4 x + \log_8 x = 11$

71. $4^x - 2^{x+1} = 3$
[*Hint:* First write the equation as a quadratic equation in 2^x.]

DISCOVERY • DISCUSSION

72. Estimating a Solution Without actually solving the equation, find two whole numbers between which the solution of $9^x = 20$ must lie. Do the same for $9^x = 100$. Explain how you reached your conclusions.

73. A Surprising Equation Take logarithms to show that the equation $x^{1/\log x} = 5$ has no solution. For what values of k does the equation $x^{1/\log x} = k$ have a solution? What does this tell us about the graph of the function $f(x) = x^{1/\log x}$? Confirm your answer using a graphing device.

6.6 APPLICATIONS OF EXPONENTIAL AND LOGARITHMIC FUNCTIONS

Logarithms were invented by John Napier (page 408) to eliminate the tedious calculations involved in multiplying, dividing, and taking powers and roots of the large numbers that occur in astronomy and other sciences. With the advent of computers and calculators, logarithms are no longer important for such calculations. However, logarithms arise in problems of exponential growth and decay because they are inverses of exponential functions. Because of the Laws of Logarithms, they also turn out to be useful in the measurement of the loudness of sounds, the intensity of earthquakes, and many other phenomena. In this section we study some of these applications.

COMPOUND INTEREST

Recall the formulas for interest that we found in Section 6.2.

> If a principal P is invested at an interest rate r for a period of t years, then the amount A of the investment is given by
>
> $A = P(1 + r)$ Simple interest (for one year)
>
> $A(t) = P\left(1 + \frac{r}{n}\right)^{nt}$ Interest compounded n times per year
>
> $A(t) = Pe^{rt}$ Interest compounded continuously

We can use logarithms to determine the time it takes for the principal to increase to a given amount.

EXAMPLE 1 ■ Finding the Term for an Investment to Double

A sum of \$5000 is invested at an interest rate of 9% per year. Find the time required for the money to double if the interest is compounded according to the following method.

(a) Semiannual (b) Continuous

SOLUTION

(a) We use the formula for compound interest with $P = \$5000$, $A(t) = \$10,000$, $r = 0.09$, $n = 2$, and solve the resulting exponential equation for t.

$$5000\left(1 + \frac{0.09}{2}\right)^{2t} = 10,000$$

$$(1.045)^{2t} = 2 \qquad \text{Divide by 5000}$$

$$\log 1.045^{2t} = \log 2 \qquad \text{Take log of each side}$$

$$2t \log 1.045 = \log 2 \qquad \text{Law 3 (bring down the exponent)}$$

$$t = \frac{\log 2}{2 \log 1.045} \qquad \text{Divide by 2 log 1.045}$$

$$t \approx 7.9 \qquad \text{Use a calculator}$$

The money will double in 7.9 years.

(b) We use the formula for continuously compounded interest with $P = \$5000$, $A(t) = \$10{,}000$, $r = 0.09$, and solve the resulting exponential equation for t.

$$5000e^{0.09t} = 10{,}000$$

$$e^{0.09t} = 2 \qquad \text{Divide by 5000}$$

$$\ln e^{0.09t} = \ln 2 \qquad \text{Take ln of each side}$$

$$0.09t = \ln 2 \qquad \text{Property of ln}$$

$$t = \frac{\ln 2}{0.09} \qquad \text{Divide by 0.09}$$

$$t \approx 7.702 \qquad \text{Use a calculator}$$

The money will double in 7.7 years. ■

If an investment earns compound interest, then the **annual percentage yield** (APY) is the *simple* interest rate that yields the same amount at the end of one year.

EXAMPLE 2 ■ Calculating the Annual Percentage Yield

Find the annual percentage yield for an investment that earns interest at a rate of 6% per year, compounded daily.

SOLUTION

After one year, a principal P will grow to the amount

$$A = P\left(1 + \frac{0.06}{365}\right)^{365} = P(1.06183)$$

The formula for simple interest is

$$A = P(1 + r)$$

Comparing, we see that $r = 0.06183$, so the annual percentage yield is 6.183%. ■

EXPONENTIAL GROWTH

In Section 6.2 we studied the formula for exponential growth, which models the growth of an animal or bacterium population.

If n_0 is the initial size of a population experiencing exponential growth, then the population $n(t)$ at time t is modeled by the function

$$n(t) = n_0 e^{rt}$$

where r is the relative rate of growth expressed as a fraction of the population.

Now that we are equipped with logarithms, we can answer questions concerning the time at which the population reaches a certain size.

EXAMPLE 3 ■ World Population Projections

The population of the world in 2000 was 6.1 billion, and the estimated relative growth rate was 1.4% per year. If the population continues to grow at this rate, when will it reach 122 billion?

SOLUTION

We use the population growth function with $n_0 = 6.1$ billion, $r = 0.014$, and $n(t) = 122$ billion. This leads to an exponential equation, which we solve for t.

$$\begin{aligned} 6.1e^{0.014t} &= 122 \\ e^{0.014t} &= 20 && \text{Divide by 6.1} \\ \ln e^{0.014t} &= \ln 20 && \text{Take ln of each side} \\ 0.014t &= \ln 20 && \text{Property of ln} \\ t &= \frac{\ln 20}{0.014} && \text{Divide by 0.014} \\ t &\approx 213.98 && \text{Use a calculator} \end{aligned}$$

Thus, the population will reach 122 billion in approximately 214 years, that is, in the year $2000 + 214 = 2214$. ■

STANDING ROOM ONLY

The population of the world was about 6.1 billion in 2000, and was increasing at 1.4% per year. Using the exponential model for population growth, and assuming that each person occupies an average of 4 ft^2 of the surface of the earth, we find that by the year 2801 there will be standing room only! (The total land surface area of the world is about 1.8×10^{15} ft^2.)

EXAMPLE 4 ■ The Number of Bacteria in a Culture

A culture starts with 10,000 bacteria, and the number doubles every 40 min.

(a) Find a function that models the number of bacteria at time t.
(b) Find the number of bacteria after one hour.
(c) After how many minutes will there be 50,000 bacteria?

(d) Sketch a graph of the number of bacteria at time t.

SOLUTION

(a) To find the function that models this population growth, we need to find the rate r. To do this, we use the formula for population growth with $n_0 = 10{,}000$, $t = 40$, and $n(t) = 20{,}000$, and then solve for r.

$$10{,}000 \cdot e^{r(40)} = 20{,}000$$

$$e^{40r} = 2 \qquad \text{Divide by 10,000}$$

$$\ln e^{40r} = \ln 2 \qquad \text{Take ln of each side}$$

$$40r = \ln 2 \qquad \text{Property of ln}$$

$$r = \frac{\ln 2}{40} \qquad \text{Divide by 40}$$

$$r \approx 0.01733 \qquad \text{Use a calculator}$$

Now that we know $r \approx 0.01733$, we can write the function for the population growth:

$$n(t) = 10{,}000 \cdot e^{0.01733t}$$

(b) Using the function we found in part (a) with $t = 60$ min (one hour), we get

$$n(60) = 10{,}000 \cdot e^{0.01733(60)} \approx 28{,}287$$

Thus, the number of bacteria after one hour is approximately 28,000.

(c) We use the function we found in part (a) with $n(t) = 50{,}000$ and solve the resulting exponential equation for t.

$$10{,}000 \cdot e^{0.01733t} = 50{,}000$$

$$e^{0.01733t} = 5 \qquad \text{Divide by 10,000}$$

$$\ln e^{0.01733t} = \ln 5 \qquad \text{Take ln of each side}$$

$$0.01733t = \ln 5 \qquad \text{Property of ln}$$

$$t = \frac{\ln 5}{0.01733} \qquad \text{Divide by 0.01733}$$

$$t \approx 92.9 \qquad \text{Use a calculator}$$

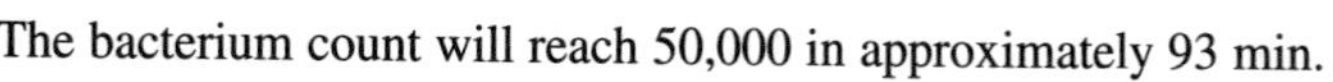

The bacterium count will reach 50,000 in approximately 93 min.

(d) We sketch a graph of the function $n(t) = 10{,}000 \cdot e^{0.01733t}$ in Figure 1. ■

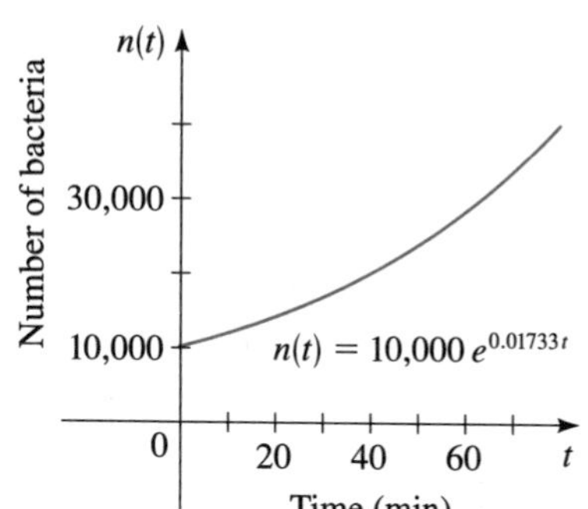

FIGURE 1

RADIOACTIVE DECAY

Radioactive substances decay by spontaneously emitting radiation. The rate of decay is directly proportional to the mass of the substance. This is analogous to population growth, except that the mass of radioactive material *decreases*. It can be shown that the mass $m(t)$ remaining at time t is modeled by the function

$$m(t) = m_0 e^{-rt}$$

where r is the rate of decay expressed as a proportion of the mass and m_0 is the initial mass. Physicists express the rate of decay in terms of **half-life**, the time required for half the mass to decay. We can obtain the rate r from this as follows. If h is the half-life, then a mass of 1 unit becomes $\frac{1}{2}$ unit when $t = h$. Substituting this into the model, we get

$$\tfrac{1}{2} = 1 \cdot e^{-rh}$$

$$\ln\left(\tfrac{1}{2}\right) = -rh \qquad \text{Take ln of each side}$$

$$r = -\frac{1}{h}\ln(2^{-1}) \qquad \text{Solve for } r$$

$$r = \frac{\ln 2}{h} \qquad \ln 2^{-1} = -\ln 2 \text{ by Law 3}$$

This last equation allows us to find the rate r from the half-life h.

> If m_0 is the initial mass of a radioactive substance with half-life h, then the mass remaining at time t is modeled by the function
>
> $$m(t) = m_0 e^{-rt}$$
>
> where $r = \dfrac{\ln 2}{h}$.

RADIOACTIVE WASTE

Harmful radioactive isotopes are produced whenever a nuclear reaction occurs, whether as the result of an atomic bomb test, a nuclear accident such as the one at Chernobyl in 1986, or the uneventful production of electricity at a nuclear power plant.

One radioactive material produced in atomic bombs is the isotope strontium-90 (^{90}Sr), with a half-life of 28 years. This is deposited like calcium in human bone tissue, where it can cause leukemia and other cancers. However, in the decades since atmospheric testing of nuclear weapons was halted, ^{90}Sr levels in the environment have fallen to a level that no longer poses a threat to health.

Nuclear power plants produce radioactive plutonium-239 (^{239}Pu), which has a half-life of 24,360 years. Because of its long half-life, ^{239}Pu could pose a threat to the environment for thousands of years. So, great care must be taken to dispose of it properly. The difficulty of ensuring the safety of the disposed radioactive waste is one reason that nuclear power plants remain controversial.

EXAMPLE 5 ■ Radioactive Decay

Polonium-210 (^{210}Po) has a half-life of 140 days. Suppose a sample of this substance has a mass of 300 mg.

(a) Find a function that models the amount of the sample remaining at time t.
(b) Find the mass remaining after one year.
(c) How long will it take for the sample to decay to a mass of 200 mg?
(d) Draw a graph of the sample mass as a function of time.

SOLUTION

(a) Using the model for radioactive decay with $m_0 = 300$ and $r = (\ln 2/140) \approx 0.00495$, we have

$$m(t) = 300e^{0.00495t}$$

(b) We use the function we found in part (a) with $t = 365$ (one year).

$$m(365) = 300e^{-0.00495(365)} \approx 49.256$$

Thus, approximately 49 mg of ^{210}Po remains after one year.

(c) We use the function we found in part (a) with $m(t) = 200$ and solve the resulting exponential equation for t.

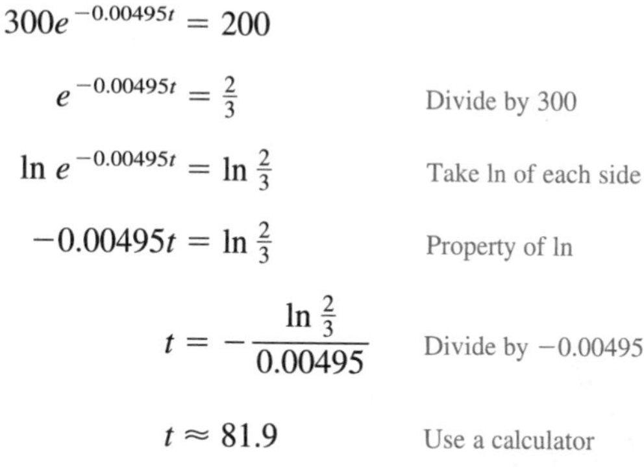

$$300e^{-0.00495t} = 200$$

$$e^{-0.00495t} = \frac{2}{3} \qquad \text{Divide by 300}$$

$$\ln e^{-0.00495t} = \ln \frac{2}{3} \qquad \text{Take ln of each side}$$

$$-0.00495t = \ln \frac{2}{3} \qquad \text{Property of ln}$$

$$t = -\frac{\ln \frac{2}{3}}{0.00495} \qquad \text{Divide by } -0.00495$$

$$t \approx 81.9 \qquad \text{Use a calculator}$$

The time required for the sample to decay to 200 mg is about 82 days.

(d) A graph of the function $m(t) = 300e^{-0.00495t}$ is shown in Figure 2. ■

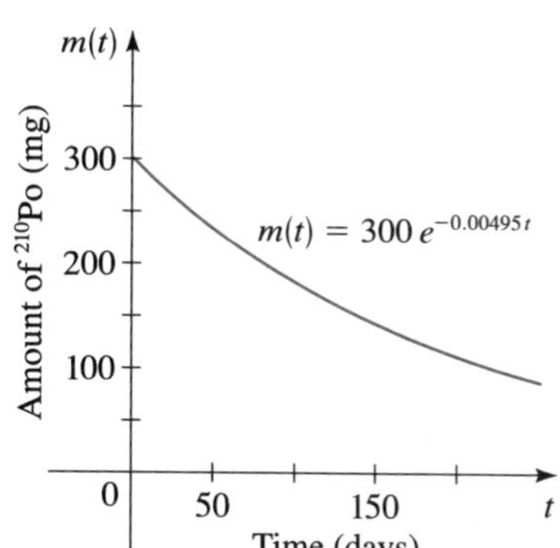

FIGURE 2

NEWTON'S LAW OF COOLING

Newton's Law of Cooling states that the rate of cooling of an object is proportional to the temperature difference between the object and its surroundings, provided that the temperature difference is not too large. Using calculus, the following model can be deduced from this law.

> If D_0 is the initial temperature difference between an object and its surroundings, and if its surroundings have temperature T_s, then the temperature of the object at time t is modeled by the function
>
> $$T(t) = T_s + D_0e^{-kt}$$
>
> where k is a positive constant that depends on the type of object.

EXAMPLE 6 ■ Newton's Law of Cooling

A cup of coffee has a temperature of 200°F and is placed in a room that has a temperature of 70°F. After 10 min the temperature of the coffee is 150°F.

(a) Find a function that models the temperature of the coffee at time t.

(b) Find the temperature of the coffee after 15 min.

(c) When will the coffee have cooled to 100°F?

(d) Illustrate by drawing a graph of the temperature function.

Half-lives of **radioactive elements** vary from very long to very short. Here are some examples.

Element	Half-life
Thorium-232	14.5 billion years
Uranium-235	4.5 billion years
Thorium-230	80,000 years
Plutonium-239	24,360 years
Carbon-14	5,730 years
Radium-226	1,600 years
Cesium-137	30 years
Strontium-90	28 years
Polonium-210	140 days
Thorium-234	25 days
Iodine-135	8 days
Radon-222	3.8 days
Lead-211	3.6 minutes
Krypton-91	10 seconds

SOLUTION

(a) The temperature of the room is $T_s = 70°\text{F}$, and the initial temperature difference is

$$D_0 = 200 - 70 = 130°\text{F}$$

So, by Newton's Law of Cooling, the temperature after t minutes is modeled by the function

$$T(t) = 70 + 130e^{-kt}$$

We need to find the constant k associated with this cup of coffee. To do this, we use the fact that when $t = 10$, the temperature is $T(10) = 150$. So we have

$$\begin{aligned} 70 + 130e^{-10k} &= 150 \\ 130e^{-10k} &= 80 && \text{Subtract 70} \\ e^{-10k} &= \tfrac{8}{13} && \text{Divide by 130} \\ -10k &= \ln \tfrac{8}{13} && \text{Take ln of each side} \\ k &= -\tfrac{1}{10} \ln \tfrac{8}{13} && \text{Divide by } -10 \\ k &\approx 0.04855 && \text{Use a calculator} \end{aligned}$$

Substituting this value of k into the expression for $T(t)$, we get

$$T(t) = 70 + 130e^{-0.04855t}$$

(b) We use the function we found in part (a) with $t = 15$.

$$T(15) = 70 + 130e^{-0.04855(15)} \approx 133°\text{F}$$

(c) We use the function we found in part (a) with $T(t) = 100$ and solve the resulting exponential equation for t.

$$\begin{aligned} 70 + 130e^{-0.04855t} &= 100 \\ 130e^{-0.04855t} &= 30 && \text{Subtract 70} \\ e^{-0.04855t} &= \tfrac{3}{13} && \text{Divide by 130} \\ -0.04855t &= \ln \tfrac{3}{13} && \text{Take ln of each side} \\ t &= \frac{\ln \tfrac{3}{13}}{0.04855} && \text{Divide by 0.04855} \\ t &\approx 30.2 && \text{Use a calculator} \end{aligned}$$

The coffee will have cooled to 100°F after about half an hour.

(d) The graph of the temperature function is sketched in Figure 3. Notice that the line $t = 70$ is a horizontal asymptote. (Why?)

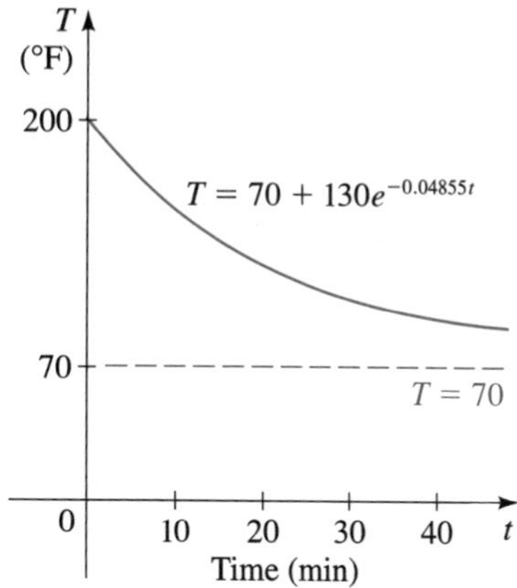

FIGURE 3
Temperature of coffee after t minutes

■

LOGARITHMIC SCALES

When a physical quantity varies over a very large range, it is often convenient to take its logarithm in order to have a more manageable set of numbers. We discuss three such situations: the pH scale, which measures acidity; the Richter scale, which measures the intensity of earthquakes; and the decibel scale, which measures the loudness of sounds. Other quantities that are measured on logarithmic scales are light intensity, information capacity, and radiation.

THE pH SCALE Chemists measured the acidity of a solution by giving its hydrogen ion concentration until Sorensen, in 1909, proposed a more convenient measure. He defined

$$\text{pH} = -\log[\text{H}^+]$$

where $[\text{H}^+]$ is the concentration of hydrogen ions measured in moles per liter (M). He did this to avoid very small numbers and negative exponents. For instance,

$$\text{if} \quad [\text{H}^+] = 10^{-4}\,\text{M}, \qquad \text{then} \qquad \text{pH} = -\log_{10}(10^{-4}) = -(-4) = 4$$

Solutions with a pH of 7 are defined as *neutral*, those with pH < 7 are *acidic*, and those with pH > 7 are *basic*. Notice that when the pH increases by one unit, $[\text{H}^+]$ decreases by a factor of 10.

EXAMPLE 7 ■ pH Scale and Hydrogen Ion Concentration

(a) The hydrogen ion concentration of a sample of human blood was measured to be $[H^+] = 3.16 \times 10^{-8}$ M. Find the pH and classify the blood as acidic or basic.

(b) The most acidic rainfall ever measured occurred in Scotland in 1974; its pH was 2.4. Find the hydrogen ion concentration.

SOLUTION

(a) A calculator gives

$$pH = -\log[H^+] = -\log(3.16 \times 10^{-8}) \approx 7.5$$

Since this is greater than 7, the blood is basic.

(b) To find the hydrogen ion concentration, we need to solve for $[H^+]$ in the logarithmic equation

$$\log[H^+] = -pH$$

So, we write it in exponential form.

$$[H^+] = 10^{-pH}$$

In this case, pH = 2.4, so

$$[H^+] = 10^{-2.4} \approx 4.0 \times 10^{-3} \text{ M}$$ ■

THE RICHTER SCALE In 1935 the American geologist Charles Richter (1900–1984) defined the magnitude M of an earthquake to be

$$M = \log \frac{I}{S}$$

where I is the intensity of the earthquake (measured by the amplitude of a seismograph reading taken 100 km from the epicenter of the earthquake) and S is the intensity of a "standard" earthquake (whose amplitude is 1 micron $= 10^{-4}$ cm). The magnitude of a standard earthquake is

$$M = \log \frac{S}{S} = \log 1 = 0$$

Richter studied many earthquakes that occurred between 1900 and 1950. The largest had magnitude 8.9 on the Richter scale, and the smallest had magnitude 0. This corresponds to a ratio of intensities of 800,000,000, so the Richter scale pro-

---

MATHEMATICS IN THE MODERN WORLD

LAW ENFORCEMENT

Mathematics aids law enforcement in numerous and surprising ways, from the reconstruction of bullet trajectories, to determining the time of death, to calculating the probability that a DNA sample is from a particular person. One interesting use is in the search for missing persons. If a person has been missing for several years that person may look quite different than their most recent available photograph. This is particularly true if the missing person is a child. Have you ever wondered what you will look like 5, 10, or 15 years from now?

Researchers have found that different parts of the body grow at different rates. For example, you no doubt noticed that a baby's head is much larger relative to its body than an adult's. As another example, the ratio of arm length to height is $\frac{1}{3}$ in a child but about $\frac{2}{5}$ in an adult. By collecting data and analyzing the graphs, researchers are able to determine the functions that model growth. As in all growth phenomena, exponential and logarithmic functions play a crucial role. For instance, the formula that relates arm length l to height h is $l = ae^{kh}$ where a and k are constants. By studying various

(continued)

physical characteristics of a person, mathematical biologists model each characteristic by a function that describes how it changes over time. Models of facial characteristics can be programmed into a computer to give a picture of how a person's appearance changes over time. These pictures aid law enforcement agencies in locating missing persons.

vides more manageable numbers to work with. For instance, an earthquake of magnitude 6 is ten times stronger than an earthquake of magnitude 5.

EXAMPLE 8 ■ Magnitude of Earthquakes

The 1906 earthquake in San Francisco had an estimated magnitude of 8.3 on the Richter scale. In the same year the strongest earthquake ever recorded occurred on the Colombia-Ecuador border and was four times as intense. What was the magnitude of the Colombia-Ecuador earthquake on the Richter scale?

SOLUTION

If I is the intensity of the San Francisco earthquake, then from the definition of magnitude we have

$$M = \log \frac{I}{S} = 8.3$$

The intensity of the Colombia-Ecuador earthquake was $4I$, so its magnitude was

$$M = \log \frac{4I}{S} = \log 4 + \log \frac{I}{S} = \log 4 + 8.3 \approx 8.9$$ ■

EXAMPLE 9 ■ Intensity of Earthquakes

The 1989 Loma Prieta earthquake that shook San Francisco had a magnitude of 7.1 on the Richter scale. How many times more intense was the 1906 earthquake (see Example 8) than the 1989 event?

SOLUTION

If I_1 and I_2 are the intensities of the 1906 and 1989 earthquakes, then we are required to find I_1/I_2. To relate this to the definition of magnitude, we divide numerator and denominator by S.

$$\log \frac{I_1}{I_2} = \log \frac{I_1/S}{I_2/S} \qquad \text{Divide numerator and denominator by } S$$

$$= \log \frac{I_1}{S} - \log \frac{I_2}{S} \qquad \text{Law 2 of Logarithms}$$

$$= 8.3 - 7.1 = 1.2 \qquad \text{Definition of earthquake magnitude}$$

Therefore

$$\frac{I_1}{I_2} = 10^{\log(I_1/I_2)} = 10^{1.2} \approx 16$$

The 1906 earthquake was about 16 times as intense as the 1989 earthquake. ■

THE DECIBEL SCALE The ear is sensitive to an extremely wide range of sound intensities. We take as a reference intensity $I_0 = 10^{-12}$ W/m^2 (watts per square meter) at a frequency of 1000 hertz, which measures a sound that is just barely audible (the threshold of hearing). The psychological sensation of loudness varies with the logarithm of the intensity (the Weber-Fechner Law) and so the **intensity level** β, measured in decibels (dB), is defined as

$$\beta = 10 \log \frac{I}{I_0}$$

The intensity level of the barely audible reference sound is

$$\beta = 10 \log \frac{I_0}{I_0} = 10 \log 1 = 0 \text{ dB}$$

EXAMPLE 10 ■ Sound Intensity of a Jet Takeoff

Find the decibel intensity level of a jet engine during takeoff if the intensity was measured at 100 W/m^2.

SOLUTION

From the definition of intensity level we see that

$$\beta = 10 \log \frac{I}{I_0} = 10 \log \frac{10^2}{10^{-12}} = 10 \log 10^{14} = 140 \text{ dB}$$

Thus, the intensity level is 140 dB. ■

Source of sound	β (dB)
Jet takeoff	140
Jackhammer	130
Rock concert	120
Subway	100
Heavy traffic	80
Ordinary traffic	70
Normal conversation	50
Whisper	30
Rustling leaves	10–20
Threshold of hearing	0

The table in the margin lists decibel intensity levels for some common sounds ranging from the threshold of hearing to the jet takeoff of Example 10. The threshold of pain is about 120 dB.

6.6 EXERCISES

1. A man invests \$10,000 in an account that pays 8.5% interest per year, compounded quarterly.
(a) Find the amount after 3 years.
(b) How long will it take for the investment to double?

2. A man invests \$6500 in an account that pays 6% interest per year, compounded continuously.
(a) What is the amount after 2 years?
(b) How long will it take for the amount to be \$8000?

3. Find the time required for an investment of \$5000 to grow to \$8000 at an interest rate of 9.5% per year, compounded quarterly.

4. Nancy wants to invest \$4000 in saving certificates that bear an interest rate of 9.75% per year, compounded semiannually. How long a time period should she choose in order to save an amount of \$5000?

5. How long will it take for an investment of \$1000 to double in value if the interest rate is 8.5% per year, compounded continuously?

6. A sum of \$1000 was invested for 4 years, and the interest was compounded semiannually. If this sum amounted to \$1435.77 in the given time, what was the interest rate?

7. Find the annual percentage yield for an investment that earns 8% per year, compounded monthly.

8. Find the annual percentage yield for an investment that earns $5\frac{1}{2}\%$ per year, compounded continuously.

9. The number of bacteria in a culture is modeled by the function

$$n(t) = 500e^{0.45t}$$

where t is measured in hours.

(a) What is the initial number of bacteria?
(b) What is the relative rate of growth of this bacterium population? Express your answer as a percentage.
(c) How many bacteria are in the culture after 3 hours?
(d) After how many hours will the number of bacteria reach 10,000?

10. The number of a certain species of fish is modeled by the function

$$n(t) = 12e^{0.012t}$$

where t is measured in years and $n(t)$ is measured in millions.

(a) What is the relative rate of growth of the fish population? Express your answer as a percentage.
(b) What will the fish population be after 5 years?
(c) After how many years will the number of fish reach 30 million?
(d) Sketch a graph of the fish population function $n(t)$.

11. The population of a certain city was 112,000 in 1994, and the observed relative growth rate is 4% per year.

(a) Find a function that models the population after t years.
(b) Find the projected population in the year 2000.
(c) In what year will the population reach 200,000?

12. The frog population in a small pond grows exponentially. The current population is 85 frogs, and the relative growth rate is 18% per year.

(a) Find a function that models the population after t years.
(b) Find the projected population after 3 years.
(c) Find the number of years required for the frog population to reach 600.

13. The graph shows the deer population in a Pennsylvania county between 1996 and 2000. Assume that the population grows exponentially.

(a) What was the deer population in 1996?
(b) Find a function that models the deer population t years after 1996.
(c) What is the projected deer population in 2004?
(d) In what year will the deer population reach 100,000?

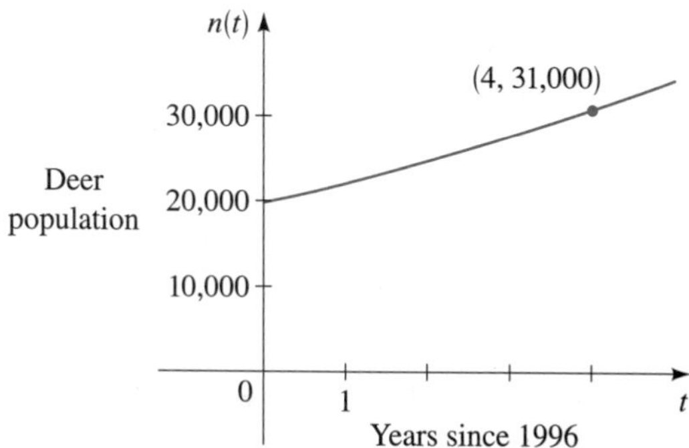

14. A culture contains 1500 bacteria initially and doubles every 30 min.

(a) Find a function that models the number of bacteria $n(t)$ after t minutes.
(b) Find the number of bacteria after 2 hours.
(c) After how many minutes will the culture contain 4000 bacteria?

15. A culture starts with 8600 bacteria. After one hour the count is 10,000.

(a) Find a function that models the number of bacteria $n(t)$ after t hours.
(b) Find the number of bacteria after 2 hours.
(c) After how many hours will the number of bacteria double?

16. The count in a culture of bacteria was 400 after 2 hours and 25,600 after 6 hours.

(a) What is the relative rate of growth of the bacterium population? Express your answer as a percentage.
(b) What was the initial size of the culture?
(c) Find a function that models the number of bacteria $n(t)$ after t hours.

(d) Find the number of bacteria after 4.5 hours.
(e) When will the number of bacteria be 50,000?

17. The population of the world was 5.7 billion in 1995 and the observed relative growth rate was 2% per year.
(a) By what year will the population have doubled?
(b) By what year will the population have tripled?

18. The population of California was 10,586,223 in 1950 and 23,668,562 in 1980. Assume the population grows exponentially.
(a) Find a function that models the population t years after 1950.
(b) Find the time required for the population to double.
(c) Use the function from part (a) to predict the population of California in the year 2000. Look up the actual California population in 2000, and compare.

19. An infectious strain of bacteria increases in number at a relative growth rate of 200% per hour. When a certain critical number of bacteria are present in the bloodstream, a person becomes ill. If a single bacterium infects a person, the critical level is reached in 24 hours. How long will it take for the critical level to be reached if the same person is infected with 10 bacteria?

20. The half-life of radium-226 is 1600 years. Suppose we have a 22-mg sample.
(a) Find a function that models the mass remaining after t years.
(b) How much of the sample will remain after 4000 years?
(c) After how long will only 18 mg of the sample remain?

21. The half-life of cesium-137 is 30 years. Suppose we have a 10-g sample.
(a) Find a function that models the mass remaining after t years.
(b) How much of the sample will remain after 80 years?
(c) After how long will only 2 g of the sample remain?

22. The mass $m(t)$ remaining after t days from a 40-g sample of thorium-234 is given by

$$m(t) = 40e^{-0.0277t}$$

(a) How much of the sample will remain after 60 days?
(b) After how long will only 10 g of the sample remain?
(c) Find the half-life of thorium-234.

23. The half-life of strontium-90 is 28 years. How long will it take a 50-mg sample to decay to a mass of 32 mg?

24. Radium-221 has a half-life of 30 s. How long will it take for 95% of a sample to decay?

25. If 250 mg of a radioactive element decays to 200 mg in 48 hours, find the half-life of the element.

26. After 3 days a sample of radon-222 has decayed to 58% of its original amount.
(a) What is the half-life of radon-222?
(b) How long will it take the sample to decay to 20% of its original amount?

27. A wooden artifact from an ancient tomb contains 65% of the carbon-14 that is present in living trees. How long ago was the artifact made? (The half-life of carbon-14 is 5730 years.)

28. The burial cloth of an Egyptian mummy is estimated to contain 59% of the carbon-14 it contained originally. How long ago was the mummy buried? (The half-life of carbon-14 is 5730 years.)

29. A hot bowl of soup is served at a dinner party. It starts to cool according to Newton's Law of Cooling so that its temperature at time t is given by

$$T(t) = 65 + 145e^{-0.05t}$$

where t is measured in minutes and T is measured in °F.
(a) What is the initial temperature of the soup?
(b) What is the temperature after 10 min?
(c) After how long will the temperature be 100°F?

30. Newton's Law of Cooling is used in homicide investigations to determine the time of death. The normal body temperature is 98.6°F. Immediately following death, the body begins to cool. It has been determined experimentally that the constant in Newton's Law of Cooling is approximately $k = 0.1947$, assuming time is measured in hours. Suppose that the temperature of the surroundings is 60°F.
(a) Find a function $T(t)$ that models the temperature t hours after death.
(b) If the temperature of the body is now 72°F, how long ago was the time of death?

31. A roasted turkey is taken from an oven when its temperature has reached 185°F and is placed on a table in a room where the temperature is 75°F.
(a) If the temperature of the turkey is 150°F after half an hour, what is its temperature after 45 min?
(b) When will the turkey cool to 100°F?

32. A kettle full of water is brought to a boil in a room with temperature 20°C. After 15 min the temperature of the water has decreased from 100°C to 75°C. Find the temperature after another 10 min. Illustrate by sketching a graph of the temperature function.

33. The hydrogen ion concentration of a sample of each substance is given. Calculate the pH of the substance.
(a) Lemon juice: $[H^+] = 5.0 \times 10^{-3}$ M
(b) Tomato juice: $[H^+] = 3.2 \times 10^{-4}$ M
(c) Seawater: $[H^+] = 5.0 \times 10^{-9}$ M

34. An unknown substance has a hydrogen ion concentration of $[H^+] = 3.1 \times 10^{-8}$ M. Find the pH and classify the substance as acidic or basic.

35. The pH reading of a sample of each substance is given. Calculate the hydrogen ion concentration of the substance.
(a) Vinegar: pH = 3.0
(b) Milk: pH = 6.5

36. The pH reading of a glass of liquid is given. Find the hydrogen ion concentration of the liquid.
(a) Beer: pH = 4.6
(b) Water: pH = 7.3

37. The hydrogen ion concentrations in cheeses range from 4.0×10^{-7} M to 1.6×10^{-5} M. Find the corresponding range of pH readings.

38. The pH readings for wines vary from 2.8 to 3.8. Find the corresponding range of hydrogen ion concentrations.

39. If one earthquake is 20 times as intense as another, how much larger is its magnitude on the Richter scale?

40. The 1906 earthquake in San Francisco had a magnitude of 8.3 on the Richter scale. At the same time in Japan an earthquake with magnitude 4.9 caused only minor damage. How many times more intense was the San Francisco earthquake than the Japanese earthquake?

41. The Alaska earthquake of 1964 had a magnitude of 8.6 on the Richter scale. How many times more intense was this than the 1906 San Francisco earthquake? (See Exercise 40.)

42. The Northridge, California, earthquake of 1994 had a magnitude of 6.8 on the Richter scale. A year later, a 7.2-magnitude earthquake struck Kobe, Japan. How many times more intense was the Kobe earthquake than the Northridge earthquake?

43. The 1985 Mexico City earthquake had a magnitude of 8.1 on the Richter scale. The 1976 earthquake in Tangshan, China, was 1.26 times as intense. What was the magnitude of the Tangshan earthquake?

44. The intensity of the sound of traffic at a busy intersection was measured at 2.0×10^{-5} W/m^2. Find the intensity level in decibels.

45. The intensity level of the sound of a subway train was measured at 98 dB. Find the intensity in W/m^2.

46. The noise from a power mower was measured at 106 dB. The noise level at a rock concert was measured at 120 dB. Find the ratio of the intensity of the rock music to that of the power mower.

47. A law of physics states that the intensity of sound is inversely proportional to the square of the distance d from the source:

$$I = \frac{k}{d^2}$$

(a) Use this and the equation

$$\beta = 10 \log \frac{I}{I_0}$$

(described in this section) to show that the decibel levels β_1 and β_2 at distances d_1 and d_2 from a sound source are related by the equation

$$\beta_2 = \beta_1 + 20 \log \frac{d_1}{d_2}$$

(b) The intensity level at a rock concert is 120 dB at a distance 2 m from the speakers. Find the intensity level at a distance of 10 m.

6 REVIEW

CONCEPT CHECK

1. (a) Write an equation that defines the exponential function with base a.
(b) What is the domain of this function?
(c) If $a \neq 1$, what is the range of this function?
(d) Sketch the general shape of the graph of the exponential function for each of the following cases.
(i) $a > 1$ (ii) $a = 1$
(iii) $a < 1$

2. If x is large, which function grows faster, $y = 2^x$ or $y = x^2$?

3. (a) How is the number e defined?
(b) What is the natural exponential function?

4. (a) How is the logarithmic function $y = \log_a x$ defined?
(b) What is the domain of this function?
(c) What is the range of this function?
(d) Sketch the general shape of the graph of the function $y = \log_a x$ if $a > 1$.
(e) What is the natural logarithm?
(f) What is the common logarithm?

5. State the three Laws of Logarithms.

6. State the Change of Base Formula.

7. (a) How do you solve an exponential equation?
(b) How do you solve a logarithmic equation?

8. Suppose an amount P is invested at an interest rate r and A is the amount after t years.
(a) Write an expression for A if the interest is compounded n times per year.
(b) Write an expression for A if the interest is compounded continuously?

9. If the initial size of a population is n_0 and the population grows exponentially with relative growth rate r, write an expression for the population $n(t)$ at time t.

10. (a) What is the half-life of a radioactive substance?
(b) If a radioactive substance has initial mass m_0 and half-life h, write an expression for the mass $m(t)$ remaining at time t.

11. What does Newton's Law of Cooling say?

12. What do the pH scale, the Richter scale, and the decibel scale have in common? What do they measure?

EXERCISES

1–12 ■ Sketch the graph of the function. State the domain, range, and asymptote.

1. $f(x) = \dfrac{1}{2^x}$
2. $g(x) = 3^{x-2}$
3. $y = 5 - 10^x$
4. $y = 1 + 5^{-x}$
5. $f(x) = \log_3(x - 1)$
6. $g(x) = \log(-x)$
7. $y = 2 - \log_2 x$
8. $y = 3 + \log_5(x + 4)$
9. $F(x) = e^x - 1$
10. $G(x) = \frac{1}{2} e^{x-1}$
11. $y = 2 \ln x$
12. $y = \ln(x^2)$

13–14 ■ Find the domain of the function.

13. $f(x) = 10^{x^2} + \log(1 - 2x)$

14. $g(x) = \ln(2 + x - x^2)$

15–18 ■ Write the equation in exponential form.

15. $\log_2 1024 = 10$
16. $\log_6 37 = x$
17. $\log x = y$
18. $\ln c = 17$

19–22 ■ Write the equation in logarithmic form.

19. $2^6 = 64$
20. $49^{-1/2} = \frac{1}{7}$
21. $10^x = 74$
22. $e^k = m$

23–38 ■ Evaluate the expression without using a calculator.

23. $\log_2 128$
24. $\log_8 1$
25. $10^{\log 45}$
26. $\log 0.000001$
27. $\ln(e^6)$
28. $\log_4 8$
29. $\log_3\left(\frac{1}{27}\right)$
30. $2^{\log_2 13}$

31. $\log_5 \sqrt{5}$

32. $e^{2\ln 7}$

33. $\log 25 + \log 4$

34. $\log_3 \sqrt{243}$

35. $\log_2 16^{23}$

36. $\log_5 250 - \log_5 2$

37. $\log_8 6 - \log_8 3 + \log_8 2$

38. $\log \log 10^{100}$

39–44 ■ Rewrite the expression in a form with no logarithms of products, quotients, or powers.

39. $\log(AB^2C^3)$

40. $\log_2\left(x\sqrt{x^2+1}\right)$

41. $\ln\sqrt{\dfrac{x^2-1}{x^2+1}}$

42. $\log\left(\dfrac{4x^3}{y^2(x-1)^5}\right)$

43. $\log_5\left(\dfrac{x^2(1-5x)^{3/2}}{\sqrt{x^3-x}}\right)$

44. $\ln\left(\dfrac{\sqrt[3]{x^4+12}}{(x+16)\sqrt{x-3}}\right)$

45–50 ■ Rewrite the expression as a single logarithm.

45. $\log 6 + 4\log 2$

46. $\log x + \log(x^2y) + 3\log y$

47. $\frac{3}{2}\log_2(x-y) - 2\log_2(x^2+y^2)$

48. $\log_5 2 + \log_5(x+1) - \frac{1}{3}\log_5(3x+7)$

49. $\log(x-2) + \log(x+2) - \frac{1}{2}\log(x^2+4)$

50. $\frac{1}{2}[\ln(x-4) + 5\ln(x^2+4x)]$

51–60 ■ Use a calculator to find the solution of the equation, correct to two decimal places.

51. $\log_2(1-x) = 4$

52. $2^{3x-5} = 7$

53. $5^{5-3x} = 26$

54. $\ln(2x-3) = 14$

55. $e^{3x/4} = 10$

56. $2^{1-x} = 3^{2x+5}$

57. $\log x + \log(x+1) = \log 12$

58. $\log_8(x+5) - \log_8(x-2) = 1$

59. $x^2e^{2x} + 2xe^{2x} = 8e^{2x}$

60. $2^{3^x} = 5$

61–64 ■ Use a calculator to find the solution of the equation, correct to six decimal places.

61. $5^{-2x/3} = 0.63$

62. $2^{3x-5} = 7$

63. $5^{2x+1} = 3^{4x-1}$

64. $e^{-15k} = 10{,}000$

65–68 ■ Draw a graph of the function and use it to determine the asymptotes and the local maximum and minimum values.

65. $y = e^{x/(x+2)}$

66. $y = 2x^2 - \ln x$

67. $y = \log(x^3 - x)$

68. $y = 10^x - 5^x$

69–70 ■ Find the solutions of the equation, correct to two decimal places.

69. $3\log x = 6 - 2x$

70. $4 - x^2 = e^{-2x}$

71–72 ■ Solve the inequality graphically.

71. $\ln x > x - 2$

72. $e^x < 4x^2$

73. Use a graph of $f(x) = e^x - 3e^{-x} - 4x$ to find, approximately, the intervals on which f is increasing and on which f is decreasing.

74. Find an equation of the line shown in the figure.

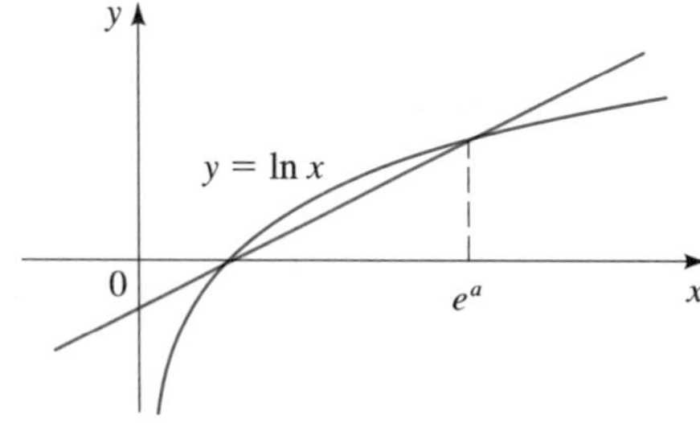

75. Evaluate $\log_4 15$, correct to six decimal places.

76. Solve the inequality: $0.2 \leq \log x < 2$

77. Which is larger, $\log_4 258$ or $\log_5 620$?

78. Find the inverse of the function $f(x) = 2^{3^x}$ and state its domain and range.

79. If \$12,000 is invested at an interest rate of 10% per year, find the amount of the investment at the end of 3 years for each compounding method.

(a) Semiannual
(b) Monthly
(c) Daily
(d) Continuous

80. A sum of \$5000 is invested at an interest rate of $8\frac{1}{2}\%$ per year, compounded semiannually.

(a) Find the amount of the investment after $1\frac{1}{2}$ years.
(b) After what period of time will the investment amount to \$7000?

81. The stray-cat population in a small town grows exponentially. In 1999, the town had 30 stray cats and the relative growth rate was 15% per year.

(a) Find a function that models the stray-cat population $n(t)$ after t years.
(b) Find the projected population after 4 years.

(c) Find the number of years required for the stray-cat population to reach 500.

82. A culture contains 10,000 bacteria initially. After an hour the bacteria count is 25,000.
(a) Find the doubling period.
(b) Find the number of bacteria after 3 hours.

83. Uranium-234 has a half-life of 2.7×10^5 years.
(a) Find the amount remaining from a 10-mg sample after a thousand years.
(b) How long will it take this sample to decompose until its mass is 7 mg?

84. A sample of bismuth-210 decayed to 33% of its original mass after 8 days.
(a) Find the half-life of this element.
(b) Find the mass remaining after 12 days.

85. The half-life of radium-226 is 1590 years.
(a) If a sample has a mass of 150 mg, find a function that models the mass that remains after t years.
(b) Find the mass that will remain after 1000 years.
(c) After how many years will only 50 mg remain?

86. The half-life of palladium-100 is 4 days. After 20 days a sample has been reduced to a mass of 0.375 g.
(a) What was the initial mass of the sample?
(b) Find a function that models the mass remaining after t days.
(c) What is the mass after 3 days?
(d) After how many days will only 0.15 g remain?

87. The graph shows the population of a rare species of bird, where t represents years since 1994 and $n(t)$ is measured in thousands.
(a) Find a function that models the bird population at time t in the form $n(t) = n_0e^{rt}$.
(b) What is the bird population expected to be in the year 2005?

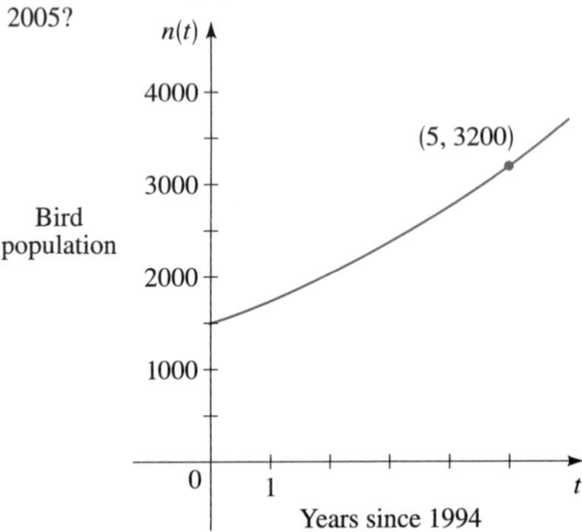

88. A car engine runs at a temperature of 190°F. When the engine is turned off, it cools according to Newton's Law of Cooling with constant $k = 0.0341$, where the time is measured in minutes. Find the time needed for the engine to cool to 90°F if the surrounding temperature is 60°F.

89. The hydrogen ion concentration of fresh egg whites was measured as

$$[H^+] = 1.3 \times 10^{-8}\ M$$

Find the pH, and classify the substance as acidic or basic.

90. The pH of lime juice is 1.9. Find the hydrogen ion concentration.

91. If one earthquake has magnitude 6.5 on the Richter scale, what is the magnitude of another quake that is 35 times as intense?

92. The drilling of a jackhammer was measured at 132 dB. The sound of whispering was measured at 28 dB. Find the ratio of the intensity of the drilling to that of the whispering.

6 TEST

1. Graph the functions $y = 4^x$ and $y = \log_4 x$ on the same axes.

2. Sketch the graph of the function $f(x) = \log(x + 2)$ and state the domain, range, and asymptote.

3. Evaluate each logarithmic expression.
   (a) $\log_3 \sqrt{27}$
   (b) $\log_2 56 - \log_2 7$
   (c) $\log_8 4$
   (d) $\log_6 4 + \log_6 9$

4. Use the Laws of Logarithms to rewrite the expression without logarithms of products, quotients, powers, or roots.

$$\log \sqrt{\frac{x^2 - 1}{x^3(y^2 + 1)^5}}$$

5. Write as a single logarithm: $\ln x - 2\ln(x^2 + 1) + \frac{1}{2}\ln(3 - x^4)$

6. Find the solution of the equation, correct to two decimal places.
   (a) $2^{x-1} = 10$
   (b) $5\ln(3 - x) = 4$
   (c) $10^{x+3} = 6^{2x}$
   (d) $\log_2(x + 2) + \log_2(x - 1) = 2$

7. The initial size of a culture of bacteria is 1000. After one hour the bacterium count is 8000.
   (a) Find a function that models the population after t hours.
   (b) Find the population after 1.5 hours.
   (c) When will the population reach 15,000?
   (d) Sketch the graph of the population function.

8. Suppose that \$12,000 is invested in a savings account paying 5.6% interest per year.
   (a) Write the formula for the amount in the account after t years if interest is compounded monthly.
   (b) Find the amount in the account after 3 years if interest is compounded daily.
   (c) How long will it take for the amount in the account to grow to \$20,000 if interest is compounded semiannually?

9. Let $f(x) = \dfrac{e^x}{x^3}$.
   (a) Graph f in an appropriate viewing rectangle.
   (b) State the asymptotes of f.
   (c) Find, correct to two decimal places, the local minimum value of f and the value of x at which it occurs.
   (d) Find the range of f.
   (e) Solve the equation $\dfrac{e^x}{x^3} = 2x + 1$. State each solution correct to two decimal places.

FOCUS ON MODELING

In *Principles of Modeling* (page 119), we learned how to construct linear models of data. Figure 1 shows some scatter plots of data; the first plot appears to be linear but the others are not. What do we do if the data we are studying are not linear? In this case, our model would be some other type of function that best fits the data. The type of function we choose is determined by the shape of the scatter plot or by some physical principle that underlies the data. In this section we learn how to construct exponential, power, and polynomial models.

FIGURE 1

EXPONENTIAL FUNCTIONS AS MODELS

Table 1 gives the population of the world in the 20th century. The scatter plot in Figure 2 shows that the population grows too quickly for a linear model to be appropriate. In fact, as we saw in Section 6.2, population tends to increase exponentially. So, we should seek an *exponential model* for population growth, that is, a function like

$$P = Ce^{kt}$$

TABLE 1 World population

Year (t)	World population (P, in millions)
1900	1650
1910	1750
1920	1860
1930	2070
1940	2300
1950	2520
1960	3020
1970	3700
1980	4450
1990	5300
2000	6060

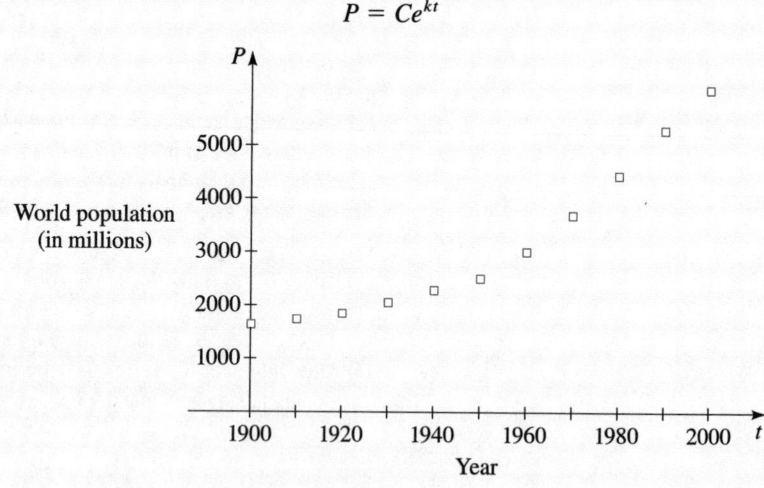

FIGURE 2 Scatter plot of world population

To find an appropriate exponential model of the data points

$$(t, P)$$

we first linearize or "straighten" the data by calculating $\ln P$ for each data point. Table 2 gives the linearized data

$$(t, \ln P)$$

[Because the population is given in millions, the first entry in Table 2 is $\ln(1{,}650{,}000{,}000) = 21.224$; subsequent entries are calculated in the same way.]

We expect the data $(t, \ln P)$ to be nearly linear, so we find a linear model. A calculator gives the regression line

$$\ln P = 0.013625t - 4.797$$

with correlation coefficient .98. Figure 3 shows the scatter plot and the regression line of the linearized data.

TABLE 2 World population data

t	Population P (in millions)	$\ln P$
1900	1650	21.224
1910	1750	21.283
1920	1860	21.344
1930	2070	21.451
1940	2300	21.556
1950	2520	21.648
1960	3020	21.829
1970	3700	22.032
1980	4450	22.216
1990	5300	22.391
2000	6060	22.525

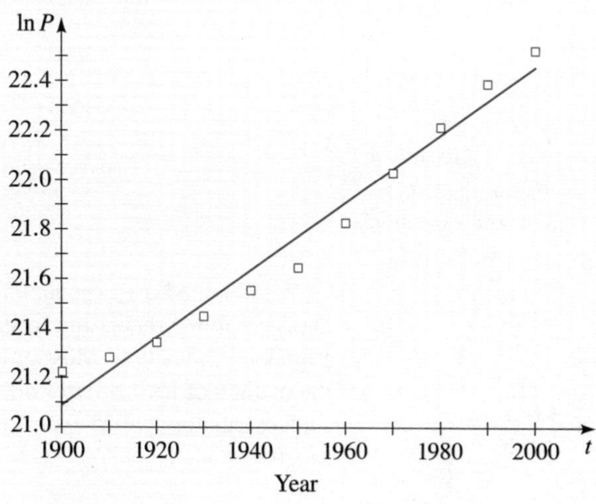

FIGURE 3

To find a model for the population P, we take the exponential of each side of the preceding equation:

$$e^{\ln P} = e^{0.013625t-4.797} \quad \text{Take exponential of each side}$$

$$P = e^{0.013625t-4.797} \quad \text{Property of logarithms}$$

$$P = e^{0.013625t}e^{-4.797} \quad \text{Law of exponentials}$$

$$P = 0.0082545e^{0.013625t} \quad \text{Use a calculator}$$

Thus, the exponential model is

$$\textbf{(1)} \qquad P(t) = 0.0082545e^{0.013625t}$$

Figure 4 at the top of page 440 shows the exponential model for population growth together with the original scatter plot. We see that the exponential curve fits the data reasonably well. The period of relatively slow population growth is explained by the depression of the 1930s and the two world wars.

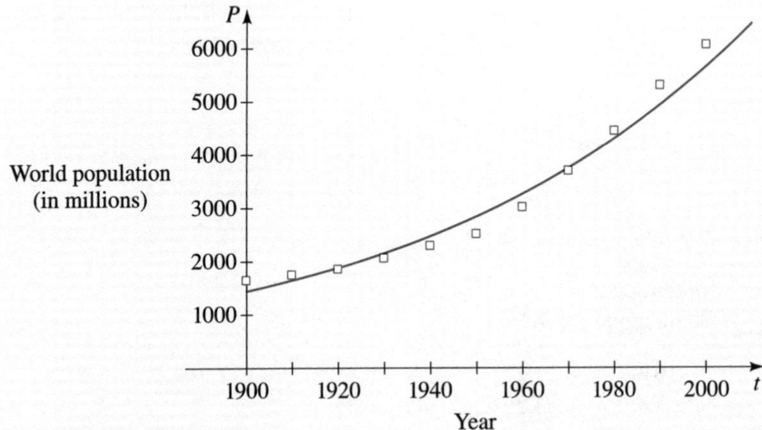

FIGURE 4
Exponential model
for world population

We obtained the exponential model for population growth by taking logarithms, applying linear regression to the resulting data, then solving for P by applying the exponential function. Most graphing calculators, however, are capable of applying the method of least squares *directly* to the data to obtain an exponential model. One such calculator applies exponential regression to the data in Table 1 and obtains the function

$$P(t) = (0.0082543) \cdot (1.0137186)^t \tag{2}$$

To compare the relative growth rates of different exponential functions, we need to use the same base. So, we convert to base e by writing

$$1.0137186 = e^k$$

$$k = \ln 1.0137186 \approx 0.013625 \qquad \text{Solve for } k$$

So $$P(t) = 0.0082543e^{0.013625t}$$

The slight discrepancy between Equations 1 and 2 is explained by the fact that applying least squares to the transformed data doesn't quite correspond to the least squares fit of the original data.

POWER FUNCTIONS AS MODELS

If the scatter plot of the data we are studying resembles the graph of $y = ax^2$, $y = ax^3$, or some other power function, then we seek a *power model*, that is, a function of the form

$$y = ax^n$$

where a is a positive constant and n is any real number. If we take the logarithm of each side of the preceding equation, we get

$$\begin{aligned} \ln y &= \ln ax^n && \text{Take ln of each side} \\ &= \ln a + \ln x^n && \text{Property of logarithms} \\ &= \ln a + n \ln x && \text{Property of logarithms} \end{aligned}$$

The last equation shows that $\ln y$ is a linear function of $\ln x$. To see this, write $X = \ln x$, $Y = \ln y$, and $A = \ln a$. The last equation becomes

$$Y = nX + A$$

and this is a linear equation. So, if we suspect that a power function is an appropriate model, we may linearize or "straighten" the data by graphing the points $(\ln x, \ln y)$. The graph of the points $(\ln x, \ln y)$ is called a **log-log plot**.

Now let's apply the preceding observations to a problem. Let's find the relationship between the distance of a planet from the sun and the period of the planet, that is, the time it takes the planet to make a complete revolution around the sun. Table 3 shows the mean distance d of each planet from the sun in astronomical units and its period T in years. An astronomical unit (AU) is the mean distance from the sun to the earth. We seek a power model of this data, so we calculate the points $(\ln d, \ln T)$ in Table 4. The resulting log-log plot in Figure 5 is nearly linear. Using a calculator, we find the regression line of the log-log data:

$$\ln T = 1.5 \ln d + 0.0004$$

This is the linear model for the data in Table 4.

TABLE 3 Distances and periods of the planets

Planet	d	T
Mercury	0.387	0.241
Venus	0.723	0.615
Earth	1.000	1.000
Mars	1.523	1.881
Jupiter	5.203	11.861
Saturn	9.541	29.457
Uranus	19.190	84.008
Neptune	30.086	164.784
Pluto	39.507	248.350

TABLE 4 Log-log table

$\ln d$	$\ln T$
−0.94933	−1.4230
−0.32435	−0.48613
0	0
0.42068	0.6318
1.6492	2.4733
2.2556	3.3829
2.9544	4.4309
3.4041	5.1046
3.6765	5.5148

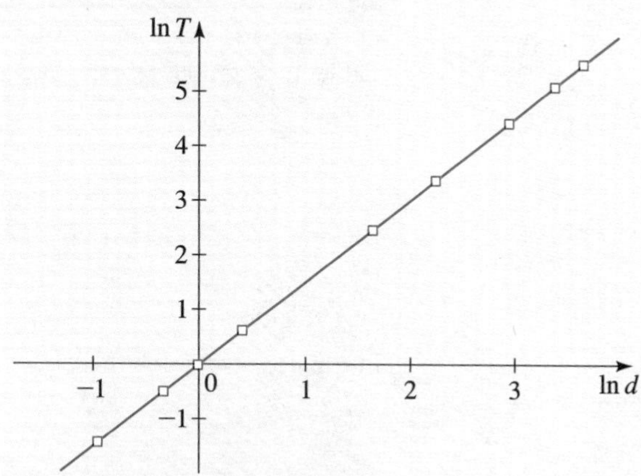

FIGURE 5 Log-log plot and regression line of data in Table 4

To find a model for the original data, we take exponentials of each side.

$$e^{\ln T} = e^{1.5 \ln d + 0.0004} \qquad \text{Take exponentials of each side}$$

$$T = e^{1.5 \ln d} e^{0.0004} \qquad \text{Law of Exponents}$$

$$= e^{\ln d^{1.5}} e^{0.0004} \qquad \text{Property of logarithms}$$

$$= 1.0004 d^{1.5} \qquad \text{Property of logarithms}$$

Since $1.0004 \approx 1$, the power model for this data is

$$T = d^{1.5}$$

This is the remarkable relationship discovered by Johannes Kepler (1571–1630). We can now use this model to calculate the period of an asteroid whose mean distance is five times as far from the sun as that of the earth. In this case, $d = 5$ AU, and so the period is $T = 5^{1.5} \approx 11.18$ years.

Graphing calculators are capable of applying the method of least squares *directly* to obtain a power model of data. One such calculator applies power regression to the data in Table 2 and gives the model

$$T = 1.0039 d^{1.49966}$$

Compare this with the model we obtained above.

POLYNOMIAL FUNCTIONS AS MODELS

Polynomial functions can also be used to model data. Most graphing calculators use the method of least squares to fit a polynomial of specified degree to the given data. Let's try, for instance, to fit a cubic polynomial to the world population data in Table 1. A graphing calculator using the method of least squares gives the cubic model

$$P = at^3 + bt^2 + ct + d$$

where $a = 1252.914$ $\qquad b = -6.817424 \times 10^6$

$c = 1.233675 \times 10^{10}$ $\qquad d = -7.420959 \times 10^{12}$

In Figure 6 we graph this cubic function and the data points of Table 1.

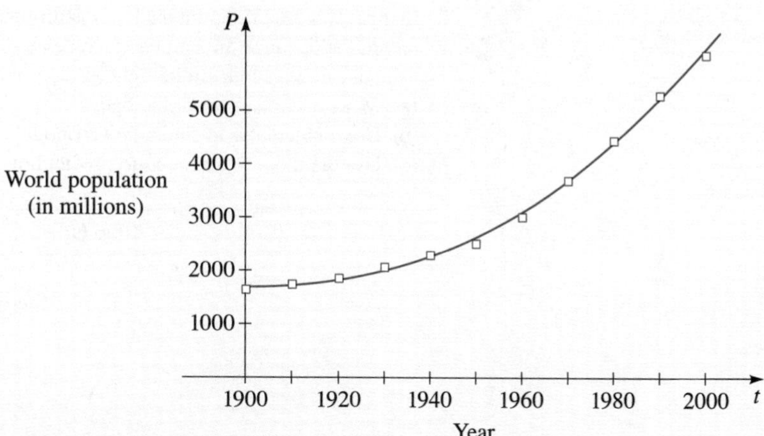

FIGURE 6
Cubic model for world population

We see from the graph that the cubic function models the world population of the 20th century very well. Perhaps surprisingly, it's quite a bit better than the exponential model and would be useful for estimating the world population in 1925 or 1985, for instance. For the purpose of predicing the population in 2050 or 2100, however, the cubic model would probably not be nearly as accurate, because we know that the physical mechanism of population growth is inherently exponential.

PROBLEMS

1. The U.S. Constitution requires a census every 10 years. The census data for 1790–1990 is given in the table.

(a) Make a scatter plot of the data.

(b) Use a calculator to find an exponential model for the data.

(c) Use your model to predict the population at the 2000 census.

(d) Use your model to estimate the population in 1965.

(e) Compare your answers from parts (c) and (d) to the values in the table. Do you think an exponential model is appropriate for these data?

Year	Population (in millions)	Year	Population (in millions)	Year	Population (in millions)
1790	3.9	1860	31.4	1930	123.2
1800	5.3	1870	38.6	1940	132.2
1810	7.2	1880	50.2	1950	151.3
1820	9.6	1890	63.0	1960	179.3
1830	12.9	1900	76.2	1970	203.3
1840	17.1	1910	92.2	1980	226.5
1850	23.2	1920	106.0	1990	248.7

2. In a physics experiment a lead ball is dropped from a height of 5 m. The students record the distance the ball has fallen every one-tenth of a second. (This can be done using a camera and a strobe light.)
(a) Make a scatter plot of the data.
(b) Use a calculator to find a power model.
(c) Use the model you found to predict how far a dropped ball would fall in 3 s.

Time (s)	Distance (m)
0.1	0.048
0.2	0.197
0.3	0.441
0.4	0.882
0.5	1.227
0.6	1.765
0.7	2.401
0.8	3.136
0.9	3.969
1.0	4.902

3. The U.S. health-care expenditures for 1960–1993 are given in the table, and a scatter plot of the data is shown in the figure.
(a) Does the scatter plot shown suggest an exponential model?
(b) Make a table of the values (t, ln E) and a scatter plot. Does the scatter plot appear to be linear?
(c) Find the regression line for the data in part (b).
(d) Use the results of part (b) to find an exponential model for the growth of health-care expenditures.
(e) Use the model you found in part (d) to predict the total health-care expenditures in 1996.

Year	Health expenditures (in billions of dollars)
1960	27.1
1970	74.3
1980	251.1
1985	434.5
1987	506.2
1989	623.9
1990	696.6
1991	755.8
1992	820.3
1993	884.2

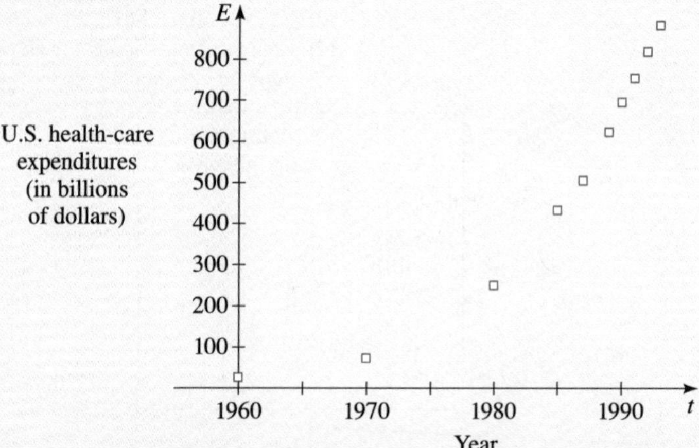

4. A student is trying to determine the half-life of radioactive iodine-131. He measures the amount of iodine-131 in a sample solution every 8 hours. His data are shown in the table.
   (a) Make a scatter plot of the data.
   (b) Use a calculator to find an exponential model.
   (c) Use your model to find the half-life of iodine-131.

Time (h)	Amount (g)
0	4.80
8	4.66
16	4.51
24	4.39
32	4.29
40	4.14
48	4.04

Year	Lead emissions
1970	199.1
1975	143.8
1980	68.0
1985	18.3
1988	5.9
1989	5.5
1990	5.1
1991	4.5
1992	4.7

5. The table gives U.S. lead emissions into the environment in millions of metric tons for 1970–1992.
   (a) Find an exponential model for these data.
   (b) Find a fourth-degree polynomial model for these data.
   (c) Which of these curves gives a better model for the data? Use graphs of the two models to decide.
   (d) Use each model to estimate the lead emissions in 1972 and 1982.

6. A study of the U.S. Office of Science and Technology in 1972 estimated the cost of reducing automobile emissions by certain percentages. Find a model that captures the "diminishing returns" trend of these data.

Reduction in emissions (%)	Cost per car ($)
50	45
55	55
60	62
65	70
70	80
75	90
80	100
85	200
90	375
95	600

7

SYSTEMS OF EQUATIONS AND INEQUALITIES

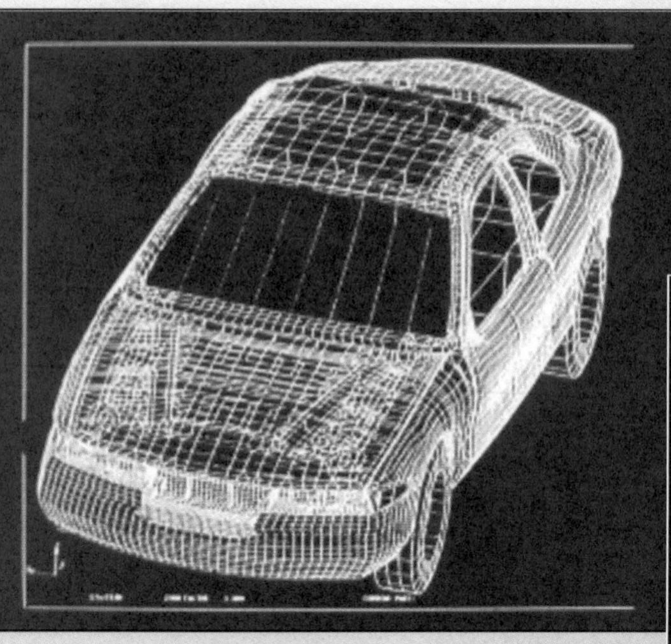

Systems of equations and inequalities are essential in modeling situations that involve many varying quantities. For instance, systems of equations are used in manipulating three-dimensional images on a computer monitor; systems of inequalities are used to find the most efficient routing for a telephone call.

Mathematics is the key and door to the sciences.

GALILEO GALILEI

Many of the problems to which we can apply the techniques of algebra give rise to sets of equations with several variables, rather than to just a single equation in a single variable. A set of equations with common variables is called a *system of equations*. In this chapter we develop techniques for finding simultaneous solutions of such systems. We first consider the simplest case: pairs of equations with two unknowns. To help us solve *linear* equations in an arbitrary number of variables, we study the algebra of matrices and determinants. We also study systems of inequalities, and in the *Focus on Modeling* section we examine linear programming, an optimization technique used widely in business and the social sciences.

7.1 SYSTEMS OF EQUATIONS

In this section we learn how to solve systems of two equations in two unknowns. Here is an example of how such systems arise.

A gas station in rural Wyoming sells two types of gas: regular at \$1.10 per gallon and premium at \$1.50 per gallon. At the end of a business day the cashier finds that receipts totaled \$340 and that 280 gallons of gasoline were sold. How many gallons of each type of gasoline were sold?

If we let x and y represent the number of gallons of regular and premium sold, respectively, we can express the problem as a system of equations:

$$\begin{cases} x + y = 280 & \text{Amount of gasoline sold is 280 gal} \\ 1.10x + 1.50y = 340 & \text{Total revenue is \$340} \end{cases}$$

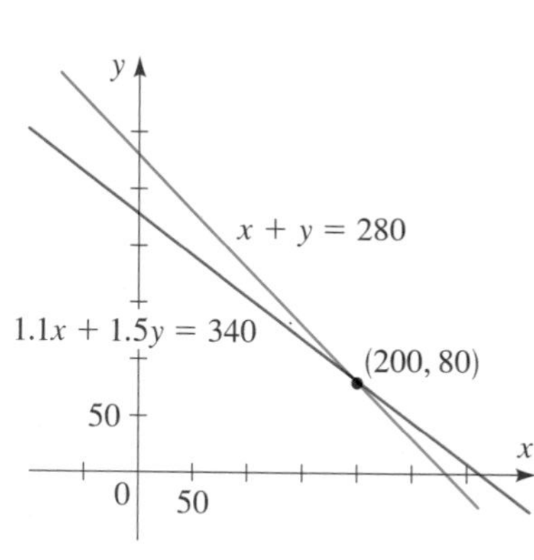

FIGURE 1

To solve this system, we must find values for x and y that satisfy both equations. A solution of this system is $x = 200$, $y = 80$ because

$$\begin{cases} (200) + (80) = 280 \\ 1.10(200) + 1.50(80) = 340 \end{cases}$$

We write the solution as the ordered pair (200, 80). This means that 200 gal of regular and 80 gal of premium were sold. Graphically, the solutions of such a system are the points where the graphs of the two equations intersect (see Figure 1).

We can solve a system of equations by substitution, by elimination, or graphically, as we show in this section.

SUBSTITUTION METHOD

In the **substitution method** we start with one equation in the system and solve for one variable in terms of the other variable. The following box describes the procedure.

> **SUBSTITUTION METHOD**
>
> 1. SOLVE FOR ONE VARIABLE. Choose one equation and solve for one variable in terms of the other variable.
>
> 2. SUBSTITUTE. Substitute the expression you found in Step 1 into the other equation to get an equation in one variable, then solve for that variable.
>
> 3. BACK-SUBSTITUTE. Substitute the value you found in Step 2 back into the expression found in Step 1 to solve for the remaining variable.

EXAMPLE 1 ■ Substitution Method

Find all solutions of the system.

$$\begin{cases} 2x + y = 1 \\ 3x + 4y = 14 \end{cases}$$

SOLUTION

We solve for y in the first equation.

$$y = 1 - 2x \qquad \text{Solve for } y \text{ in the first equation}$$

Now we substitute for y in the second equation and solve for x:

$$\begin{aligned} 3x + 4(1 - 2x) &= 14 && \text{Substitute } y = 1 - 2x \text{ into the second equation} \\ 3x + 4 - 8x &= 14 && \text{Expand} \\ -5x + 4 &= 14 && \text{Simplify} \\ -5x &= 10 && \text{Subtract 4} \\ x &= -2 && \text{Solve for } x \end{aligned}$$

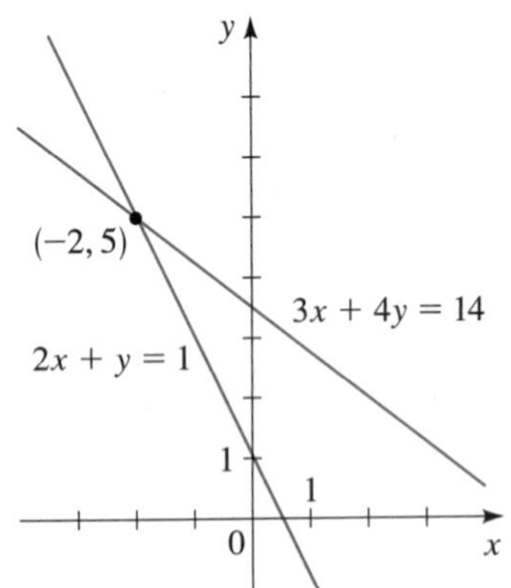

FIGURE 2

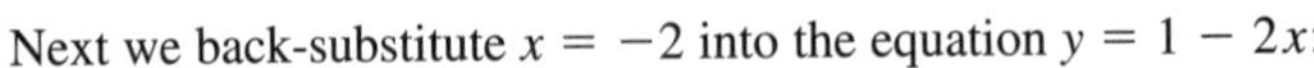

Next we back-substitute $x = -2$ into the equation $y = 1 - 2x$:

$$y = 1 - 2(-2) = 5 \qquad \text{Back-substitute}$$

Thus, $x = -2$ and $y = 5$, so the solution is the ordered pair $(-2, 5)$. Figure 2 shows that the graphs of the two equations intersect at the point $(-2, 5)$. ■

> CHECK YOUR ANSWER
>
> $x = -2, y = 5$:
>
> $$\begin{cases} 2(-2) + 5 = 1 \\ 3(-2) + 4(5) = 14 \end{cases} \quad ✓$$

EXAMPLE 2 ■ Substitution Method

Find all solutions of the system.

$$\begin{cases} x^2 + y^2 = 100 \\ 3x - y = 10 \end{cases}$$

SOLUTION

We start by solving for y in the second equation.

$$y = 3x - 10 \qquad \text{Solve for } y \text{ in the second equation}$$

Next we substitute for y in the first equation and solve for x:

$$x^2 + (3x - 10)^2 = 100 \qquad \text{Substitute } y = 3x - 10 \text{ into the first equation}$$

$$x^2 + (9x^2 - 60x + 100) = 100 \qquad \text{Expand}$$

$$10x^2 - 60x = 0 \qquad \text{Simplify}$$

$$10x(x - 6) = 0 \qquad \text{Factor}$$

$$x = 0 \quad \text{or} \quad x = 6 \qquad \text{Solve for } x$$

Now we back-substitute these values of x into the equation $y = 3x - 10$.

For $x = 0$: $y = 3(0) - 10 = -10$ Back-substitute

For $x = 6$: $y = 3(6) - 10 = 8$ Back-substitute

So we have two solutions: $(0, -10)$ and $(6, 8)$.

The graph of the first equation is a circle, and the graph of the second equation is a line; Figure 3 shows that the graphs intersect at the two points $(0, -10)$ and $(6, 8)$.

CHECK YOUR ANSWERS

$x = 0,\ y = -10$:

$$\begin{cases} (0)^2 + (-10)^2 = 100 \\ 3(0) - (-10) = 10 \end{cases} \checkmark$$

$x = 6,\ y = 8$:

$$\begin{cases} (6)^2 + (8)^2 = 36 + 64 = 100 \\ 3(6) - (8) = 18 - 8 = 10 \end{cases} \checkmark$$

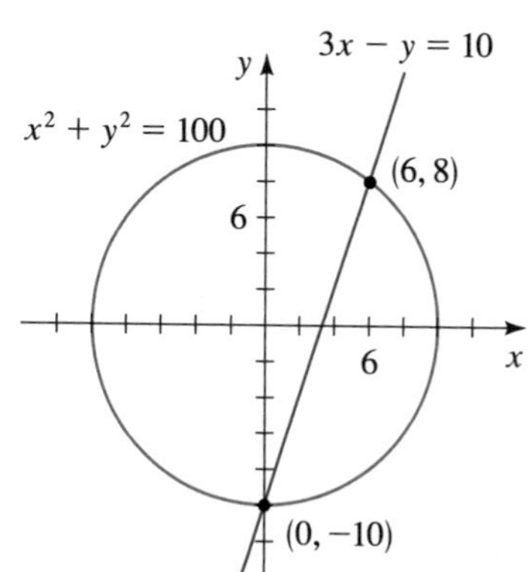

FIGURE 3

■

ELIMINATION METHOD

To solve a system using the **elimination method**, we try to combine the equations using sums or differences so as to eliminate one of the variables.

> **ELIMINATION METHOD**
>
> 1. ADJUST THE COEFFICIENTS. Multiply one or more of the equations by appropriate numbers so that the coefficient of one variable in the equation is the negative of its coefficient in the other equation.
>
> 2. ADD THE EQUATIONS. Add the two equations to eliminate one variable, then solve for the remaining variable.
>
> 3. BACK-SUBSTITUTE. Substitute the value you found in Step 2 back into one of the original equations, and solve for the remaining variable.

EXAMPLE 3 ■ Elimination Method

Find all solutions of the system.

$$\begin{cases} 3x + 2y = 14 \\ x - 2y = 2 \end{cases}$$

SOLUTION

Since the coefficients of the y-terms are negatives of each other, we can add the equations to eliminate y.

$$\begin{cases} 3x + 2y = 14 \\ x - 2y = 2 \end{cases} \qquad \text{System}$$

$$4x \qquad = 16 \qquad \text{Add}$$

$$x = 4 \qquad \text{Solve for } x$$

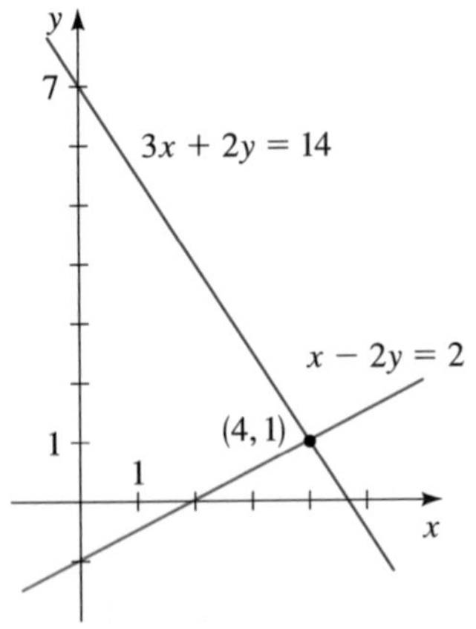

FIGURE 4

Now we back-substitute $x = 4$ into one of the original equations and solve for y. Let's choose the second equation because it looks simpler.

$$x - 2y = 2 \qquad \text{Second equation}$$

$$4 - 2y = 2 \qquad \text{Back-substitute } x = 4 \text{ into second equation}$$

$$-2y = -2 \qquad \text{Subtract 4}$$

$$y = 1 \qquad \text{Solve for } y$$

The solution is $(4, 1)$. Figure 4 shows that the graphs of the equations in the system intersect at the point $(4, 1)$. ■

EXAMPLE 4 ■ Elimination Method

Find all solutions of the system.

$$\begin{cases} 3x^2 + 2y = 26 \\ 5x^2 + 7y = 3 \end{cases}$$

SOLUTION

We choose to eliminate the x-term, so we multiply the first equation by 5 and the second equation by -3. Then we add the two equations and solve for y.

$$\begin{cases} 15x^2 + 10y = 130 \\ -15x^2 - 21y = -9 \end{cases} \qquad \begin{matrix}\text{First equation times 5} \\ \text{Second equation times } -3\end{matrix}$$

$$-11y = 121 \qquad \text{Add}$$

$$y = -11 \qquad \text{Solve for } y$$

Now we back-substitute $y = -11$ into one of the original equations, say $3x^2 + 2y = 26$, and solve for x:

$$3x^2 + 2(-11) = 26 \qquad \text{Back-substitute } y = -11 \text{ into the first equation}$$

$$3x^2 = 48 \qquad \text{Add 22}$$

$$x^2 = 16 \qquad \text{Divide by 3}$$

$$x = -4 \quad \text{or} \quad x = 4 \qquad \text{Solve for } x$$

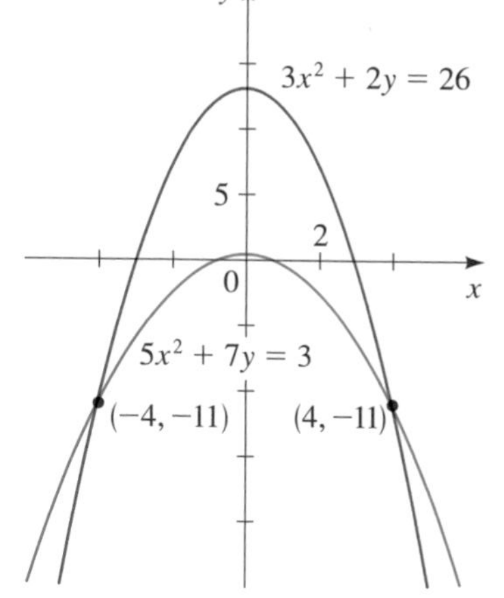

FIGURE 5

So we have two solutions: $(-4, -11)$ and $(4, -11)$.

The graphs of both equations are parabolas; Figure 5 shows that the graphs intersect at the two points $(-4, -11)$ and $(4, -11)$.

CHECK YOUR ANSWERS

$x = -4, y = -11$:

$$\begin{cases} 3(-4)^2 + 2(-11) = 26 \\ 5(-4)^2 + 7(-11) = 3 \end{cases} \checkmark$$

$x = 4, y = -11$:

$$\begin{cases} 3(4)^2 + 2(-11) = 26 \\ 5(4)^2 + 7(-11) = 3 \end{cases} \checkmark$$

■

GRAPHICAL METHOD

In the **graphical method** we use a graphing device to solve the system of equations. Note that with most graphing devices, any equation must first be expressed in terms of one or more functions of the form $y = f(x)$ before we can use the calculator to graph it. Not all equations can be readily expressed in this way, so not all systems can be solved by this method.

MATHEMATICS IN THE MODERN WORLD

GLOBAL POSITIONING SYSTEM (GPS)

On a cold foggy day in 1707, a British naval fleet was sailing at a fast clip directly home. Although the fleet was only a few yards from the rocky shores of England, the captain didn't know it. In the ensuing disaster the fleet was totally destroyed. This tragedy could have been avoided had the captain known his position accurately. In those days latitude was determined by the position of the north star (so this could only be done at night in good weather) and longitude by the position of the sun relative to where it would be in England *at that same time*. So navigation required an accurate method of telling time on ships. (Eventually, the solution was the invention of the spring loaded clock).

Since then, several different methods have been developed to determine position, and all rely heavily on mathematics (see LORAN, page 560). The latest method, called the Global Positioning System, uses triangulation. In

(continued)

GRAPHICAL METHOD

1. GRAPH EACH EQUATION. To express each equation in a form suitable for the graphing calculator, solve for y as a function of x. Graph the equations on the same screen.
2. FIND THE INTERSECTION POINTS. The solutions are the x- and y-coordinates of the points of intersection.

It may be more convenient to solve for x in terms of y in the equations. In that case, in Step 1 graph x as a function of y instead.

EXAMPLE 5 ■ Graphical Method

Find all solutions of the system.

$$\begin{cases} x^2 - y = 2 \\ 2x - y = -1 \end{cases}$$

SOLUTION

Solving for y in terms of x, we get the equivalent system

$$\begin{cases} y = x^2 - 2 \\ y = 2x + 1 \end{cases}$$

Figure 6 shows that the graphs of these equations intersect at two points. Zooming in, we see that the solutions are

$$(-1, -1) \qquad \text{and} \qquad (3, 7)$$

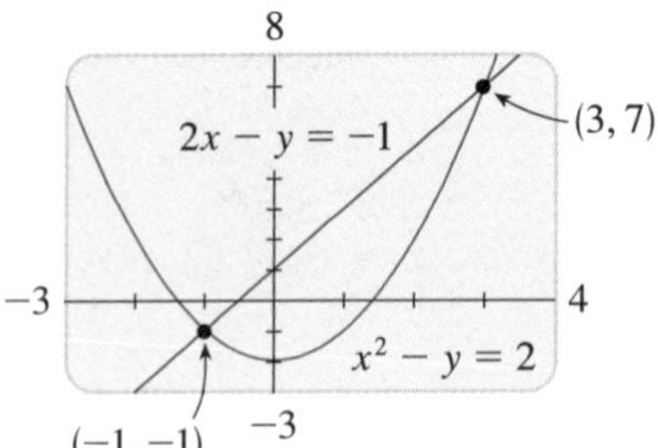

FIGURE 6

CHECK YOUR ANSWERS

$x = -1, y = -1$:

$$\begin{cases} (-1)^2 - (-1) = 2 \\ 2(-1) - (-1) = -1 \end{cases} \checkmark$$

$x = 3, y = 7$:

$$\begin{cases} 3^2 - 7 = 2 \\ 2(3) - 7 = -1 \end{cases} \checkmark$$

■

this system 24 satellites are strategically located above the surface of the earth. A hand-held GPS device measures distance from a satellite using the travel time of radio signals emitted from the satellite. Knowing the distance to three different satellites tells us that we are at the point of intersection of three different spheres. This uniquely determines our position (see Exercise 49, page 455).

EXAMPLE 6 ■ Solving a System of Equations with a Graphing Calculator

Find all solutions of the system, correct to one decimal place.

$$\begin{cases} \dfrac{x^2}{12} + \dfrac{y^2}{7} = 1 \\ y = 3x^2 - 6x + \frac{1}{2} \end{cases}$$

SOLUTION

Solving for y in terms of x, we get

$$\frac{y^2}{7} = 1 - \frac{x^2}{12} \qquad \text{Isolate } y\text{-term on LHS}$$

$$y^2 = 7\left(1 - \frac{x^2}{12}\right) \qquad \text{Solve for } y^2$$

$$y = \pm\sqrt{7\left(1 - \frac{x^2}{12}\right)} \qquad \text{Take square roots}$$

To graph the entire curve, we must graph both functions:

$$y = \sqrt{7[1 - (x^2/12)]} \qquad \text{and} \qquad y = -\sqrt{7[1 - (x^2/12)]}$$

In Figure 7 the first equation's graph is shown in red, and the second one is shown in blue. The graphs intersect in quadrants I and II. Zooming in, we see that their coordinates are $(-0.3, 2.6)$ and $(2.2, 2.0)$. There also appears to be an intersection point in quadrant IV. However, when we zoom in, we see that the curves come close to each other but don't intersect (see Figure 8). Thus, the system has only two solutions; correct to the nearest tenth, they are

$$(-0.3, 2.6) \qquad \text{and} \qquad (2.2, 2.0)$$

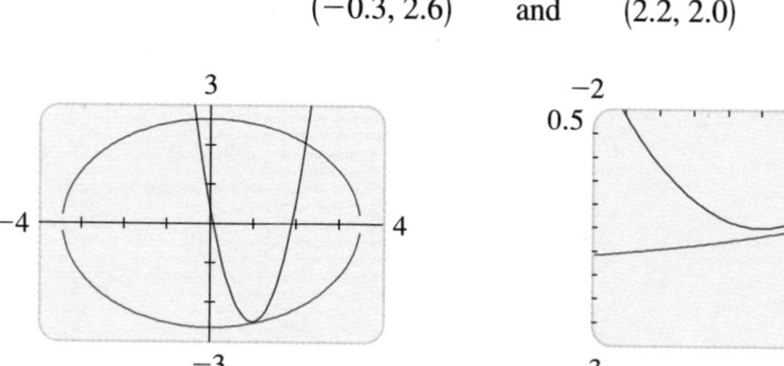

FIGURE 7

$\frac{x^2}{12} + \frac{y^2}{7} = 1, \ y = 3x^2 - 6x + \frac{1}{2}$

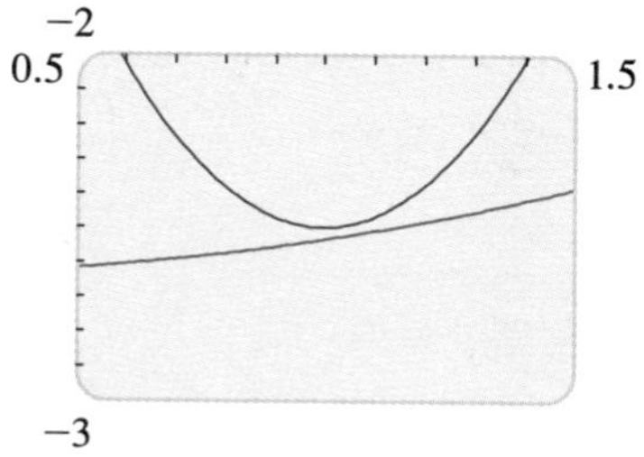

FIGURE 8

Zooming in

■

7.1 EXERCISES

1–6 ■ Use the substitution method to find all solutions of the system of equations.

1. $\begin{cases} x + y = 8 \\ x - 3y = 0 \end{cases}$

2. $\begin{cases} 2x + y = 7 \\ 3x - y = 13 \end{cases}$

3. $\begin{cases} y = x^2 \\ y = x + 6 \end{cases}$

4. $\begin{cases} x^2 + y^2 = 25 \\ y = \frac{3}{4}x \end{cases}$

5. $\begin{cases} x^2 + y^2 = 8 \\ x + y = 0 \end{cases}$

6. $\begin{cases} x^2 + y = 9 \\ x - y + 3 = 0 \end{cases}$

7–12 ■ Use the elimination method to find all solutions of the system of equations.

7. $\begin{cases} 5x + 2y = 2 \\ 7x + 3y = 6 \end{cases}$

8. $\begin{cases} 4x - 3y = 10 \\ 9x + 4y = 1 \end{cases}$

9. $\begin{cases} x^2 - 2y = 1 \\ x^2 + 5y = 29 \end{cases}$

10. $\begin{cases} 3x^2 + 4y = 17 \\ 2x^2 + 5y = 2 \end{cases}$

11. $\begin{cases} 3x^2 - y^2 = 11 \\ x^2 + 4y^2 = 8 \end{cases}$

12. $\begin{cases} 2x^2 + 4y = 13 \\ x^2 - y^2 = \frac{7}{2} \end{cases}$

13–26 ■ Find all solutions of the system of equations.

13. $\begin{cases} y + x^2 = 4x \\ y + 4x = 16 \end{cases}$

14. $\begin{cases} x - y^2 = 0 \\ y - x^2 = 0 \end{cases}$

15. $\begin{cases} x - 2y = 2 \\ y^2 - x^2 = 2x + 4 \end{cases}$

16. $\begin{cases} y = 4 - x^2 \\ y = x^2 - 4 \end{cases}$

17. $\begin{cases} x - y = 4 \\ xy = 12 \end{cases}$

18. $\begin{cases} xy = 24 \\ 2x^2 - y^2 + 4 = 0 \end{cases}$

19. $\begin{cases} x^2y = 16 \\ x^2 + 4y + 16 = 0 \end{cases}$

20. $\begin{cases} x + \sqrt{y} = 0 \\ y^2 - 4x^2 = 12 \end{cases}$

21. $\begin{cases} x^2 + y^2 = 9 \\ x^2 - y^2 = 1 \end{cases}$

22. $\begin{cases} x^2 + 2y^2 = 2 \\ 2x^2 - 3y = 15 \end{cases}$

23. $\begin{cases} 2x^2 - 8y^3 = 19 \\ 4x^2 + 16y^3 = 34 \end{cases}$

24. $\begin{cases} x^4 - y^3 = 17 \\ 3x^4 + 5y^3 = 53 \end{cases}$

25. $\begin{cases} \dfrac{2}{x} - \dfrac{3}{y} = 1 \\ -\dfrac{4}{x} + \dfrac{7}{y} = 1 \end{cases}$

26. $\begin{cases} \dfrac{4}{x^2} + \dfrac{6}{y^4} = \dfrac{7}{2} \\ \dfrac{1}{x^2} - \dfrac{2}{y^4} = 0 \end{cases}$

27–36 ■ Use the graphical method to find all solutions of the system of equations, correct to two decimal places.

27. $\begin{cases} y = 2x + 6 \\ y = -x + 5 \end{cases}$

28. $\begin{cases} y = x^2 \\ y = x + 3 \end{cases}$

29. $\begin{cases} y = x^3 \\ y = 1 - x^4 \end{cases}$

30. $\begin{cases} y = x^2 - 4x \\ 2x - y = 2 \end{cases}$

31. $\begin{cases} x^2 + y^2 = 25 \\ x + 3y = 2 \end{cases}$

32. $\begin{cases} x^2 + y^2 = 17 \\ x^2 - 2x + y^2 = 13 \end{cases}$

33. $\begin{cases} \dfrac{x^2}{9} + \dfrac{y^2}{18} = 1 \\ y = -x^2 + 6x - 2 \end{cases}$

34. $\begin{cases} x^2 - y^2 = 3 \\ y = x^2 - 2x - 8 \end{cases}$

35. $\begin{cases} x^4 + 16y^4 = 32 \\ x^2 + 2x + y = 0 \end{cases}$

36. $\begin{cases} y = e^x + e^{-x} \\ y = 5 - x^2 \end{cases}$

37. A rectangle has an area of 180 cm^2 and a perimeter of 54 cm. What are its dimensions?

38. A right triangle has an area of 84 ft^2 and a hypotenuse 25 ft long. What are the lengths of its other two sides?

39. The perimeter of a rectangle is 70 and its diagonal is 25. Find its length and width.

40. A circular piece of sheet metal has a diameter of 20 in. The edges are to be cut off to form a rectangle of area 160 in^2 (see the figure). What are the dimensions of the rectangle?

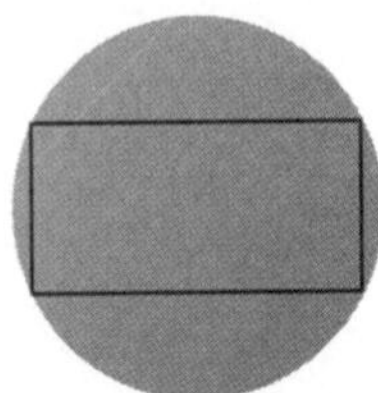

41. A hill is inclined so that its "slope" is $\frac{1}{2}$, as shown in the figure. We introduce a coordinate system with the origin at the base of the hill and with the scales on the axes measured in meters. A rocket is fired from the base of the hill in such a way that its trajectory is the parabola $y = -x^2 + 401x$. At what point does the rocket strike the hillside? How far is this point from the base of the hill (to the nearest cm)?

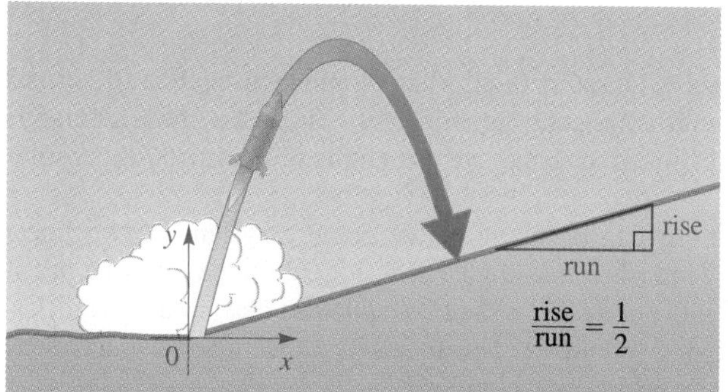

42. A rectangular piece of sheet metal with an area of 1200 in^2 is to be bent into a cylindrical length of stovepipe having a volume of 600 in^3. What are the dimensions of the sheet metal?

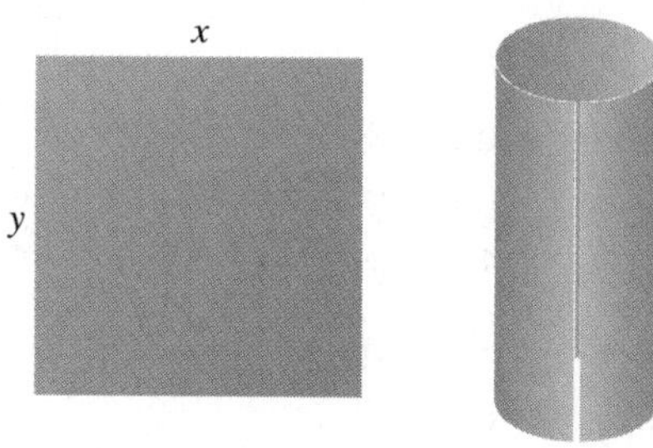

43. Find an equation for the line that passes through the points of intersection of the circles $x^2 + y^2 = 25$ and $x^2 - 3x + y^2 + y = 30$. [*Hint:* Try to eliminate the x^2- and y^2-terms.]

44–47 ■ Find all solutions of the system.

44. $\begin{cases} x - y = 3 \\ x^3 - y^3 = 387 \end{cases}$ [*Hint:* Factor the left side of the second equation.]

45. $\begin{cases} x^2 + xy = 1 \\ xy + y^2 = 3 \end{cases}$ [*Hint:* Add the equations and factor the result.]

46. $\begin{cases} 2^x + 2^y = 10 \\ 4^x + 4^y = 68 \end{cases}$ [*Hint:* Note that $4^x = 2^{2x} = (2^x)^2$.]

47. $\begin{cases} \log x + \log y = \frac{3}{2} \\ 2 \log x - \log y = 0 \end{cases}$

DISCOVERY • DISCUSSION

48. Intersection of a Parabola and a Line On a sheet of graph paper, or using a graphing calculator, draw the parabola $y = x^2$. Then draw the graphs of the linear equation $y = x + k$ on the same coordinate plane for various values of k. Try to choose values of k so that the line and the parabola intersect at two points for some of your k's, and not for others. For what value of k is there exactly one intersection point? Use the results of your experiment to make a conjecture about the values of k for which the following system has two solutions, one solution, and no solution. Prove your conjecture.

$$\begin{cases} y = x^2 \\ y = x + k \end{cases}$$

49. Global Positioning System (GPS) The Global Positioning System determines the location of an object from its distances to satellites in orbit around the earth. In the simplified, two-dimensional situation shown in the figure, determine the coordinates of P from the fact that P is 26 units from satellite A and 20 units from satellite B.

How many satellites do you think are needed to determine position in the real world (which is three-dimensional)?

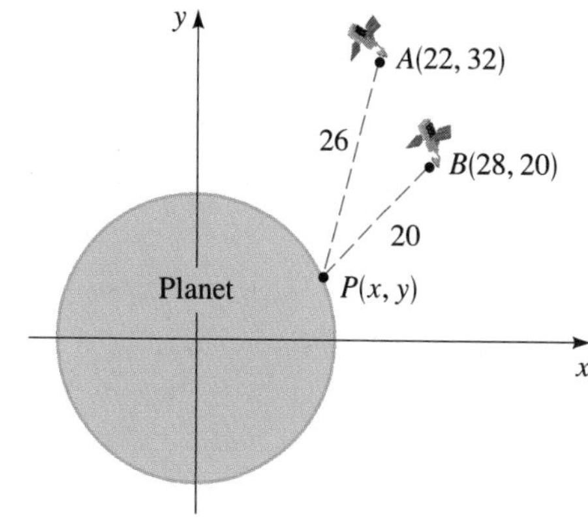

7.2 PAIRS OF LINES

In Section 2.4 we saw that the graph of any equation of the form

$$Ax + By = C$$

is a line. In this section we study systems of two such equations:

$$\begin{cases} ax + by = c \\ dx + ey = f \end{cases}$$

We can use either the substitution method or the elimination method (described in Section 7.1) to solve such systems algebraically. But since the elimination method is usually easier for linear systems, we use elimination instead of substitution in our examples.

In general, three situations can occur when we graph a system of two linear equations. The graphs may intersect at a single point (Figure 1), they may be parallel with no intersection point (Figure 2), or the two equations may simply be different equations for the same line (Figure 3). This means that the system can have one solution, no solution, or infinitely many solutions.

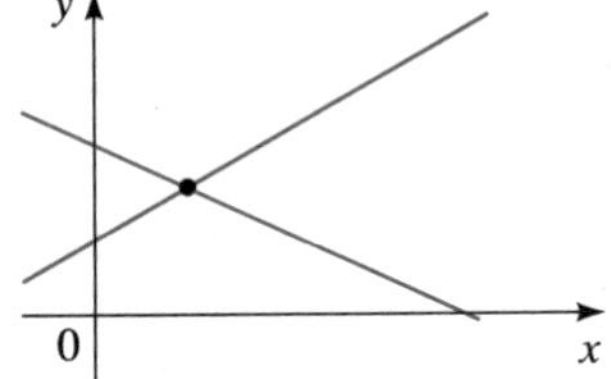

FIGURE 1
Linear system with one solution.
Lines intersect at a single point.

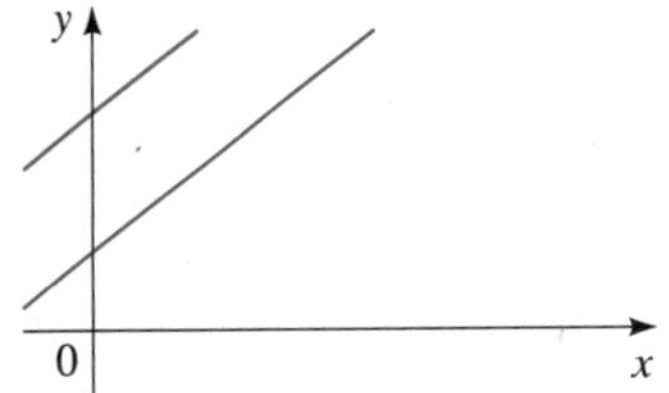

FIGURE 2
Linear system with no solution.
Lines are parallel—they do not intersect.

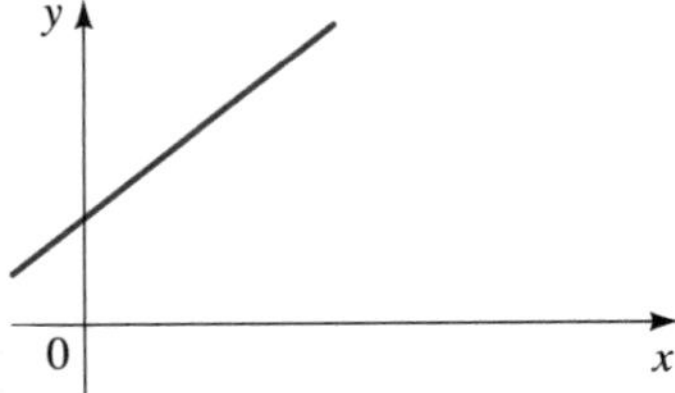

FIGURE 3
Linear system with infinitely many solutions.
Lines coincide—equations are for the same line.

EXAMPLE 1 ■ A Linear System with One Solution

Solve the system and graph the lines.

$$\begin{cases} 3x - y = 0 \\ 5x + 2y = 22 \end{cases}$$

SOLUTION

We eliminate y from the equations and solve for x.

$$\begin{cases} 6x - 2y = 0 & \text{First equation times 2} \\ 5x + 2y = 22 \end{cases}$$

$$11x \qquad = 22 \qquad \text{Add}$$

$$x = \tfrac{22}{11} = 2 \qquad \text{Solve for } x$$

Now we back-substitute into the first equation and solve for y:

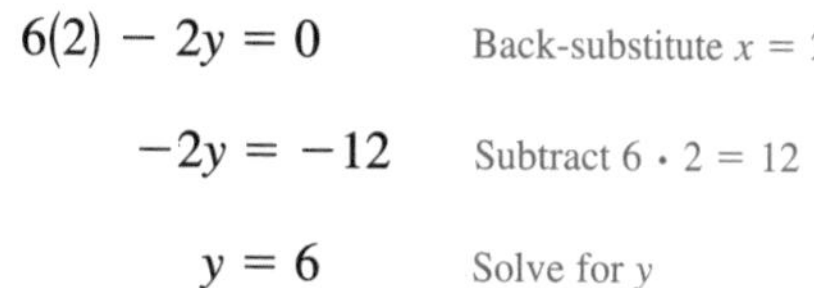

$$\begin{aligned} 6(2) - 2y &= 0 && \text{Back-substitute } x = 2 \\ -2y &= -12 && \text{Subtract } 6 \cdot 2 = 12 \\ y &= 6 && \text{Solve for } y \end{aligned}$$

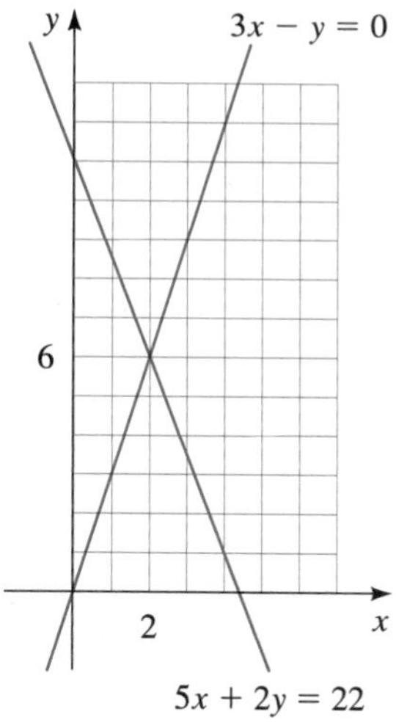

FIGURE 4

The solution of the system is the ordered pair $(2, 6)$, that is,

$$x = 2, \qquad y = 6$$

The graph in Figure 4 shows that the lines in the system intersect at the point $(2, 6)$.

CHECK YOUR ANSWER

$x = 2, y = 6$:

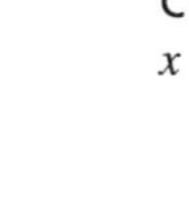

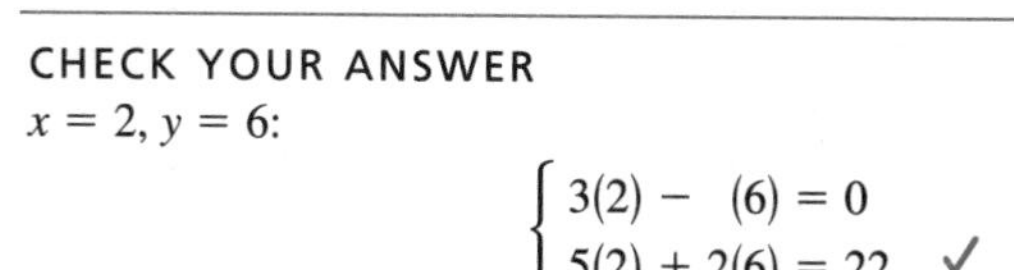

$$\begin{cases} 3(2) - (6) = 0 \\ 5(2) + 2(6) = 22 \end{cases} \quad ✓$$

■

EXAMPLE 2 ■ A Pair of Linear Equations with No Solution

Solve the system.

$$\begin{cases} 8x - 2y = 5 \\ -12x + 3y = 7 \end{cases}$$

SOLUTION

This time we try to find a suitable combination of the two equations to eliminate the variable y. Multiplying the first equation by 3 and the second by 2 gives

$$\begin{aligned} \begin{cases} 24x - 6y = 15 \\ -24x + 6y = 14 \end{cases} && \begin{matrix}\text{First equation times 3} \\ \text{Second equation times 2}\end{matrix} \\ 0 = 29 && \text{Add} \end{aligned}$$

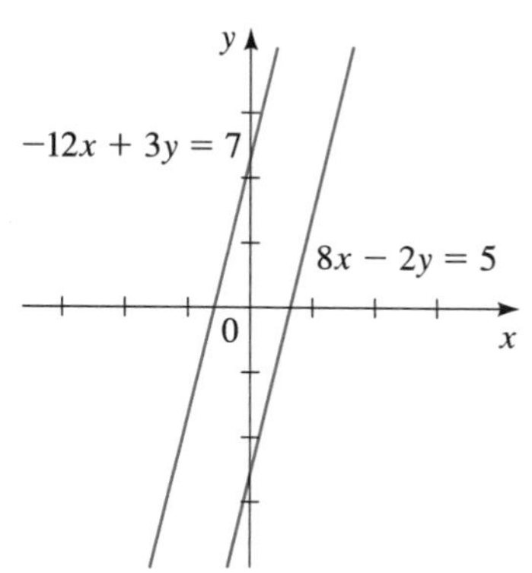

FIGURE 5

Adding the two equations eliminates *both x and y* in this case, and we end up with $0 = 29$, which is obviously false. No matter what values we assign to x and y, we cannot make this statement true, so the system has *no solution*. Figure 5 shows that the lines in the system are parallel and do not intersect. ■

A system that has no solution, like the one in Example 2, is said to be **inconsistent**.

EXAMPLE 3 ■ A Pair of Linear Equations with Infinitely Many Solutions

Solve the system.

$$\begin{cases} 3x - 6y = 12 \\ 4x - 8y = 16 \end{cases}$$

SOLUTION

We multiply the first equation by 4 and the second by 3 to prepare for subtracting the equations to eliminate x. The new equations are

$$\begin{cases} 12x - 24y = 48 & \text{First equation times 4} \\ 12x - 24y = 48 & \text{Second equation times 3} \end{cases}$$

We see that the two equations in the original system are simply different ways of expressing the equation of one single line. The coordinates of any point on this line give a solution of the system. Writing the equation in slope-intercept form, we have $y = \frac{1}{2}x - 2$. So, the solution of the system is

$$\left(x, \tfrac{1}{2}x - 2\right)$$

where x can be any real number. The system has infinitely many solutions. ■

MODELING WITH LINEAR SYSTEMS

Frequently, when we use equations to solve problems in the sciences or in other areas, we obtain systems like the ones we've been considering. When modeling with systems of equations, we use the following guidelines, similar to those in Section 3.2.

GUIDELINES FOR MODELING WITH SYSTEMS OF EQUATIONS

1. IDENTIFY THE VARIABLES. Identify the quantities the problem asks you to find. These are usually determined by a careful reading of the question posed at the end of the problem. Introduce notation for the variables (call them x and y or some other letters).

2. EXPRESS ALL UNKNOWN QUANTITIES IN TERMS OF THE VARIABLES. Read the problem again and express all the quantities mentioned in the problem in terms of the variables you defined in Step 1.

3. SET UP A SYSTEM OF EQUATIONS. Find the crucial facts in the problem that give the relationships between the expressions you found in Step 2. Set up a system of equations (or a model) that expresses these relationships.

4. SOLVE THE SYSTEM AND INTERPRET THE RESULTS. Solve the system you found in Step 3, check your solutions, and state your final answer as a sentence that answers the question posed in the problem.

The next two examples illustrate how to model with systems of equations.

EXAMPLE 4 ■ A Distance-Speed-Time Problem

A woman rows a boat upstream from one point on a river to another point 4 mi away in $1\frac{1}{2}$ hours. The return trip, traveling with the current, takes only 45 min. How fast does she row relative to the water, and at what speed is the current flowing?

SOLUTION

We are asked to find the rowing speed and the speed of the current, so we let

Identify the variables

$$x = \text{rowing speed (mi/h)}$$

$$y = \text{current speed (mi/h)}$$

The woman's speed when she rows upstream is her rowing speed minus the speed of the current; her speed downstream is her rowing speed plus the speed of the current. Now we translate this information into the language of algebra.

Express unknown quantities in terms of the variable

In Words	In Algebra
Rowing speed	x
Current speed	y
Speed upstream	$x - y$
Speed downstream	$x + y$

The distance upstream and downstream is 4 mi, so using the fact that speed × time = distance for both legs of the trip, we get the equations

Set up a system of equations

$$(x - y)\tfrac{3}{2} = 4$$

$$(x + y)\tfrac{3}{4} = 4$$

(The times have been converted to hours, since we are expressing the speeds in miles per *hour*.) If we multiply the equations by 2 and 4, respectively, to clear the denominators, we get the system

Solve the system

$$\begin{cases} 3x - 3y = 8 & \text{First equation times 2} \\ 3x + 3y = 16 & \text{Second equation times 4} \end{cases}$$

$$6x = 24 \quad \text{Add}$$

$$x = 4 \quad \text{Solve for } x$$

Back-substituting this value of x into the first equation (the second works just as well) and solving for y gives

$$3(4) - 3y = 8 \qquad \text{Back-substitute } x = 4$$

$$-3y = 8 - 12 \qquad \text{Subtract 12}$$

$$y = \tfrac{4}{3} \qquad \text{Solve for } y$$

The woman rows at 4 mi/h and the current flows at $1\frac{1}{3}$ mi/h.

CHECK YOUR ANSWER

Speed upstream is

$$\frac{\text{distance}}{\text{time}} = \frac{4 \text{ mi}}{1\frac{1}{2}\text{ h}} = 2\tfrac{2}{3} \text{ mi/h}$$

and this should equal

rowing speed − current flow

$$= 4 \text{ mi/h} - \tfrac{4}{3} \text{ mi/h} = 2\tfrac{2}{3} \text{ mi/h}$$

Speed downstream is

$$\frac{\text{distance}}{\text{time}} = \frac{4 \text{ mi}}{\frac{3}{4}\text{ h}} = 5\tfrac{1}{3} \text{ mi/h}$$

and this should equal

rowing speed + current flow

$$= 4 \text{ mi/h} + \tfrac{4}{3} \text{ mi/h} = 5\tfrac{1}{3} \text{ mi/h} \quad ✓$$

■

EXAMPLE 5 ■ A Mixture Problem

A vintner fortifies wine that contains 10% alcohol by adding 70% alcohol solution to it. The resulting mixture has an alcoholic strength of 16% and fills 1000 one-liter bottles. How many liters (L) of the wine and of the alcohol solution does he use?

SOLUTION

Since we are asked for the amounts of wine and alcohol, we let

Identify the variables

$$x = \text{amount of wine used (L)}$$

$$y = \text{amount of alcohol used (L)}$$

From the fact that the wine contains 10% alcohol and the solution 70% alcohol, we get the following.

Express unknown quantities in terms of the variable

In Words	In Algebra
Amount of wine used (L)	x
Amount of alcohol used (L)	y
Amount of alcohol in wine (L)	$0.10x$
Amount of alcohol in solution (L)	$0.70y$

The volume of the mixture must be the total of the two volumes the vintner is adding together, so

$$x + y = 1000$$

Also, the amount of alcohol in the mixture must be the total of the alcohol contributed by the wine and by the alcohol solution, that is

$$0.10x + 0.70y = (0.16)1000$$

$$0.10x + 0.70y = 160 \qquad \text{Simplify}$$

$$x + 7y = 1600 \qquad \text{Multiply by 10 to clear decimals}$$

Thus, we get the system

Set up a system of equations

$$\begin{cases} x + y = 1000 \\ x + 7y = 1600 \end{cases}$$

Subtracting the first equation from the second eliminates the variable x, and we get

Solve the system

$$6y = 600 \qquad \text{Subtract first equation from second}$$

$$y = 100 \qquad \text{Solve for } y$$

We now back-substitute $y = 100$ into the first equation and solve for x:

$$x + 100 = 1000 \qquad \text{Back-substitute } y = 100$$

$$x = 900 \qquad \text{Solve for } x$$

The vintner uses 900 L of wine and 100 L of the alcohol solution. ■

7.2 EXERCISES

1–6 ■ Graph each pair of lines on a single set of axes. Determine whether the lines are parallel, and if they are not parallel, estimate the coordinates of their point of intersection from your graph.

1. $\begin{cases} x + y = 3 \\ 2x - y = 0 \end{cases}$

2. $\begin{cases} 3x + 2y = 3 \\ -x + 5y = 16 \end{cases}$

3. $\begin{cases} 2x + 3y = 12 \\ x - y = 1 \end{cases}$

4. $\begin{cases} 3x + 5y = 15 \\ x + \frac{5}{3}y = 10 \end{cases}$

5. $\begin{cases} 2x + 5y = 15 \\ 4x + 10y = 20 \end{cases}$

6. $\begin{cases} -4x + 14y = 28 \\ 10x - 35y = 70 \end{cases}$

7–28 ■ Solve the system. If a system has infinitely many solutions, express them in the form given in Example 3.

7. $\begin{cases} -x + y = 2 \\ 4x - 3y = -3 \end{cases}$

8. $\begin{cases} 4x - 3y = 28 \\ 9x - y = -6 \end{cases}$

9. $\begin{cases} x + 2y = 7 \\ 5x - y = 2 \end{cases}$

10. $\begin{cases} -4x + 12y = 0 \\ 12x + 4y = 160 \end{cases}$

11. $\begin{cases} \frac{1}{2}x + \frac{1}{3}y = 2 \\ \frac{1}{5}x - \frac{2}{3}y = 8 \end{cases}$

12. $\begin{cases} 0.2x - 0.2y = -1.8 \\ -0.3x + 0.5y = 3.3 \end{cases}$

13. $\begin{cases} 3x + 2y = 8 \\ x - 2y = 0 \end{cases}$

14. $\begin{cases} 4x + 2y = 16 \\ x - 5y = 70 \end{cases}$

15. $\begin{cases} x + 4y = 8 \\ 3x + 12y = 2 \end{cases}$

16. $\begin{cases} -3x + 5y = 2 \\ 9x - 15y = 6 \end{cases}$

17. $\begin{cases} 2x - 6y = 10 \\ -3x + 9y = -15 \end{cases}$

18. $\begin{cases} 2x - 3y = -8 \\ 14x - 21y = 3 \end{cases}$

19. $\begin{cases} 6x + 4y = 12 \\ 9x + 6y = 18 \end{cases}$

20. $\begin{cases} 25x - 75y = 100 \\ -10x + 30y = -40 \end{cases}$

21. $\begin{cases} 8s - 3t = -3 \\ 5s - 2t = -1 \end{cases}$

22. $\begin{cases} u - 30v = -5 \\ -3u + 80v = 5 \end{cases}$

23. $\begin{cases} \frac{1}{2}x + \frac{3}{5}y = 3 \\ \frac{5}{3}x + 2y = 10 \end{cases}$

24. $\begin{cases} \frac{3}{2}x - \frac{1}{3}y = \frac{1}{2} \\ 2x - \frac{1}{2}y = -\frac{1}{2} \end{cases}$

25. $\begin{cases} \dfrac{2x-5}{3} + \dfrac{y-1}{6} = \dfrac{1}{2} \\ \dfrac{x}{5} + \dfrac{3y-6}{12} = 1 \end{cases}$

26. $\begin{cases} x - 3y = 4x - 6y - 10 \\ 2x = 12y + 10 \end{cases}$

27. $x - 2y = 2x + 2y = 1$

28. $x = 2x + y = 2y + 1$

29–32 ■ Find x and y in terms of a and b.

29. $\begin{cases} x + y = 0 \\ x + ay = 1 \end{cases}$ $(a \neq 1)$

30. $\begin{cases} ax + by = 0 \\ x + y = 1 \end{cases}$ $(a \neq b)$

31. $\begin{cases} ax + by = 1 \\ bx + ay = 1 \end{cases}$ $(a^2 - b^2 \neq 0)$

32. $\begin{cases} ax + by = 0 \\ a^2x + b^2y = 1 \end{cases}$ $(a \neq 0, b \neq 0, a \neq b)$

33. Find two numbers whose sum is 34 and whose difference is 10.

34. The sum of two numbers is twice their difference. The larger number is 6 more than twice the smaller. Find the numbers.

35. A man has 14 coins in his pocket, all of which are dimes and quarters. If the total value of his change is \$2.75, how many dimes and how many quarters does he have?

36. The admission fee at an amusement park is \$1.50 for children and \$4.00 for adults. On a certain day, 2200 people entered the park, and the admission fees collected totaled \$5050. How many children and how many adults were admitted?

37. A man flies a small airplane from Fargo to Bismarck, North Dakota—a distance of 180 mi. Because he is flying into a head wind, the trip takes him 2 hours. On the way back, the wind is still blowing at the same speed, so the return trip takes only 1 h 12 min. What is his speed in still air, and how fast is the wind blowing?

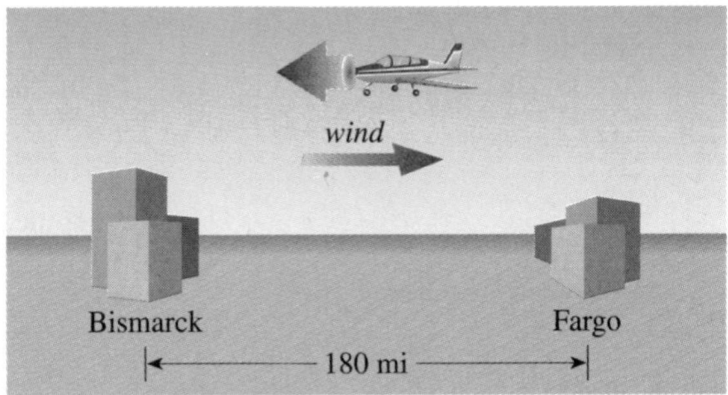

38. A boat on a river travels downstream between two points, 20 mi apart, in one hour. The return trip against the current takes $2\frac{1}{2}$ hours. What is the boat's speed, and how fast does the current in the river flow?

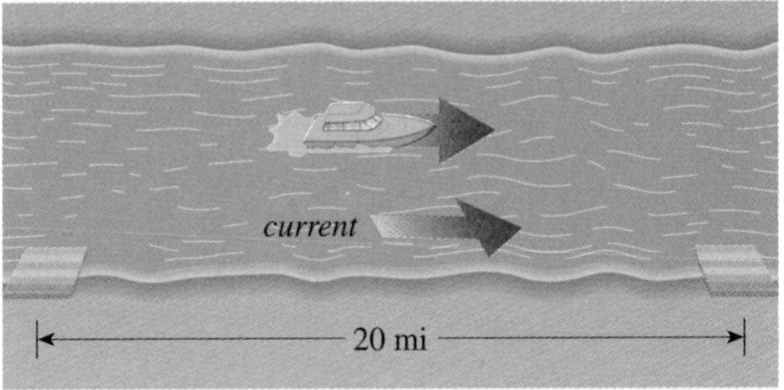

39. A woman keeps fit by bicycling and running every day. On Monday she spends $\frac{1}{2}$ hour at each activity, covering a total of $12\frac{1}{2}$ mi. On Tuesday, she runs for 12 min and cycles for 45 min, covering a total of 16 mi. Assuming her running and cycling speeds don't change from day to day, find these speeds.

40. A biologist has two brine solutions, one containing 5% salt and another containing 20% salt. How many milliliters of each solution should he mix to obtain 1 L of a solution that contains 14% salt?

41. A researcher performs an experiment to test a hypothesis that involves the nutrients niacin and retinol. She feeds one group of laboratory rats a daily diet of precisely 32 units of niacin and 22,000 units of retinol. She uses two types of commercial pellet foods. Food A contains 0.12 unit of niacin and 100 units of retinol per gram. Food B contains 0.20 unit of niacin and 50 units of retinol per gram. How many grams of each food does she feed this group of rats each day?

42. A customer in a coffee shop purchases a blend of two coffees: Kenyan, costing \$3.50 a pound, and Sri Lankan, costing \$5.60 a pound. He buys 3 lb of the blend, which costs him \$11.55. How many pounds of each kind went into the mixture?

43. A chemist has two large containers of sulfuric acid solution, with different concentrations of acid in each container. Blending 300 mL of the first solution and 600 mL of the second gives a mixture that is 15% acid, whereas 100 mL of the first mixed with 500 mL of the second gives a $12\frac{1}{2}\%$ acid mixture. What are the concentrations of sulfuric acid in the original containers?

44. A woman invests a total of \$20,000 in two accounts, one paying 5% and the other paying 8% simple interest per year. Her annual interest is \$1180. How much did she invest at each rate?

45. A man invests his savings in two accounts, one paying 6% and the other paying 10% simple interest per year. He puts twice as much in the lower-yielding account because it is less risky. His annual interest is \$3520. How much did he invest at each rate?

46. John and Mary leave their house at the same time and drive in opposite directions. John drives at 60 mi/h and travels 35 mi farther than Mary, who drives at 40 mi/h. Mary's trip takes 15 min longer than John's. For what length of time does each of them drive?

47. The sum of the digits of a two-digit number is 7. When the digits are reversed, the number is increased by 27. Find the number.

48. Find the area of the triangle that lies in the first quadrant (with its base on the x-axis) and that is bounded by the lines $y = 2x - 4$ and $y = -4x + 20$.

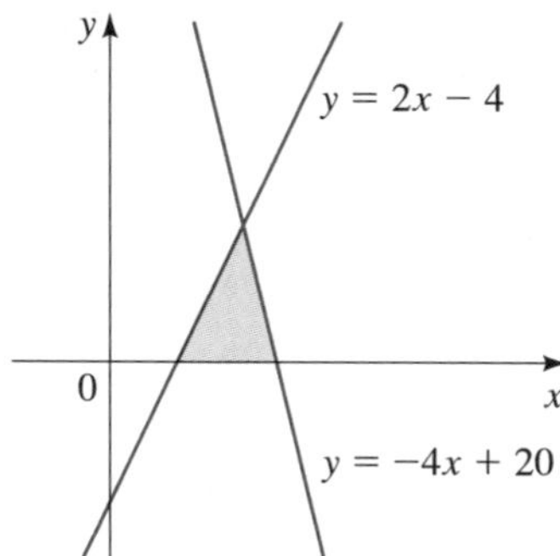

 **49–52** ■ Use a graphing device to graph both lines in the same viewing rectangle. (Note that you must solve for y in terms of x before graphing if you are using a graphing calculator.) Solve the system by zooming in to the point of intersection and then using the cursor to find its coordinates, correct to two decimal places.

49. $\begin{cases} 0.21x + 3.17y = 9.51 \\ 2.35x - 1.17y = 5.89 \end{cases}$

50. $\begin{cases} 18.72x - 14.91y = 12.33 \\ 6.21x - 12.92y = 17.82 \end{cases}$

51. $\begin{cases} 2371x - 6552y = 13{,}591 \\ 9815x + 992y = 618{,}555 \end{cases}$

52. $\begin{cases} -435x + 912y = 0 \\ 132x + 455y = 994 \end{cases}$

DISCOVERY • DISCUSSION

53. The Least Squares Line The *least squares* line or *regression* line is the line that best fits a set of points in the plane. We studied this line in *Principles of Modeling* (see page 121). Using calculus, it can be shown that the line that best fits the n data points $(x_1, y_1), (x_2, y_2), \ldots, (x_n, y_n)$ is the line $y = ax + b$, where the coefficients a and b satisfy the following pair of linear equations. [The notation $\Sigma_{k=1}^{n} x_k$ stands for the sum of all the x's. See Section 9.1 for a complete description of sigma (Σ) notation.]

$$\left(\sum_{k=1}^{n} x_k\right)a + nb = \sum_{k=1}^{n} y_k$$

$$\left(\sum_{k=1}^{n} x_k^2\right)a + \left(\sum_{k=1}^{n} x_k\right)b = \sum_{k=1}^{n} x_k y_k$$

Use these equations to find the least squares line for the following data points.

$$(1, 3), \quad (2, 5), \quad (3, 6), \quad (5, 6), \quad (7, 9)$$

Sketch the points and your line to confirm that the line fits these points well. If your calculator computes regression lines, see whether it gives you the same line as the formulas.

7.3 SYSTEMS OF LINEAR EQUATIONS

Linear equations

$6x_1 - 3x_2 + \sqrt{5}\,x_3 = 1000$

$x + y + z = 2w - \frac{1}{2}$

Nonlinear equations

$x^2 + 3y - \sqrt{z} = 5$

$x_1x_2 + 6x_3 = -6$

A **linear equation in n variables** is an equation that can be put in the form

$$a_1x_1 + a_2x_2 + \cdots + a_nx_n = c$$

where $a_1, a_2, \ldots, a_n$ and c are real numbers, and $x_1, x_2, \ldots, x_n$ are the variables. If we have no more than three or four variables, we generally use x, y, z, and w instead of x_1, x_2, x_3, and x_4. Such equations are called *linear* because if we have just two variables, the equation is

$$a_1x + a_2y = c$$

which is the equation of a line. Each term of a linear equation is either a constant or a constant multiple of one of the variables.

To solve a system of linear equations, we adapt the elimination method introduced in Section 7.1 to obtain an **equivalent system** (that is, a system with the same solution as the original system), which is easy to solve. For a system of three equations, the idea is to eliminate all but one variable from one equation, then eliminate all but two variables from another equation. This allows us to use back-substitution to find the solution.

EXAMPLE 1 ■ Solving a System Using the Elimination Method

Solve the system of linear equations.

$$\begin{cases} x - y + 3z = 4 \\ x + 2y - 2z = 10 \\ 3x - y + 5z = 14 \end{cases}$$

SOLUTION

We use elimination to obtain a system whose last equation involves only the variable z, and the next-to-last involves just y and z. Then we back-substitute to find first z, then y, and finally x.

We begin by eliminating the x-terms from the second and third equations. We multiply the first equation by -1 and add it to the second equation.

$$\begin{array}{rl} -x + y - 3z = -4 & \text{First equation times } -1 \\ x + 2y - 2z = 10 & \text{Second equation} \\ \hline 3y - 5z = 6 & \text{Add} \end{array}$$

The resulting equation does not contain the variable x. Similarly, we add -3 times the first equation to the third.

$$\begin{array}{rl} -3x + 3y - 9z = -12 & \text{First equation times } -3 \\ 3x - y + 5z = 14 & \text{Third equation} \\ \hline 2y - 4z = 2 & \text{Add} \end{array}$$

This equation also does not contain the variable x. Now we have a simpler system that is equivalent to the original system.

$$\begin{cases} x - y + 3z = 4 \\ 3y - 5z = 6 \\ 2y - 4z = 2 \end{cases} \quad \text{New system}$$

Since each term in the third equation has a common factor of 2, we multiply the equation by $\frac{1}{2}$.

$$\begin{cases} x - y + 3z = 4 \\ 3y - 5z = 6 \\ y - 2z = 1 \end{cases} \quad \text{Third equation times } \tfrac{1}{2}$$

We now eliminate the variable y from the second equation by adding -3 times the third equation to the second.

$$\begin{aligned} 3y - 5z &= 6 \\ -3y + 6z &= -3 \\ \hline z &= 3 \end{aligned}$$

$$\begin{cases} x - y + 3z = 4 \\ z = 3 \\ y - 2z = 1 \end{cases} \quad \text{Add } -3 \text{ times third equation to second}$$

To reach our goal, we interchange the second and third equations.

$$\begin{cases} x - y + 3z = 4 \\ y - 2z = 1 \\ z = 3 \end{cases} \quad \text{Interchange second and third equations}$$

Now we back-substitute $z = 3$ into the second equation and solve for y:

$$y - 2(3) = 1 \quad \text{Back-substitute } z = 3 \text{ into the second equation}$$

$$y = 7 \quad \text{Solve for } y$$

Finally, we back-substitute the values for y and z into the first equation and solve for x:

$$x - (7) + 3(3) = 4 \quad \text{Back-substitute } y = 7 \text{ and } z = 3 \text{ into the first equation}$$

$$x = 2 \quad \text{Solve for } x$$

Thus, the solution to the system is $x = 2$, $y = 7$, $z = 3$. The solution may be written as $(2, 7, 3)$. ■

CHECK YOUR ANSWER

$x = 2, y = 7, z = 3$:

$$\begin{cases} (2) - (7) + 3(3) = 4 \\ (2) + 2(7) - 2(3) = 10 \\ 3(2) - (7) + 5(3) = 14 \end{cases} \quad ✓$$

MATRIX NOTATION

If we examine the solution to Example 1, we see that the variables x, y, and z act simply as place-holders in our computations. It's only the coefficients of the variables and the constants that actually enter into the calculations. We use this fact to simplify our notation. Instead of writing out the equations of a system in full, we write only the coefficients and constants in a rectangular array, called the **matrix form** of the system. The matrix form of the system of Example 1 is as follows.

System of linear equations	Matrix form
$\begin{cases} x - y + 3z = 4 \\ x + 2y - 2z = 10 \\ 3x - y + 5z = 14 \end{cases}$	$\begin{bmatrix} 1 & -1 & 3 & 4 \\ 1 & 2 & -2 & 10 \\ 3 & -1 & 5 & 14 \end{bmatrix}$

Any rectangular array of numbers is called a **matrix**. The **rows** of a matrix are the horizontal lists of numbers in the array; the **columns** of a matrix are the vertical lists of numbers in the array. Each row of the matrix form of a system represents an equation. For example, the first row of the matrix in the preceding display is $[1 \quad -1 \quad 3 \quad 4]$; it represents the equation $x - y + 3z = 4$.

The operations we use on a system of equations (as in Example 1) correspond to operations on the matrix form of the system. These are called the **elementary row operations**.

ELEMENTARY ROW OPERATIONS

1. Add a multiple of one row to another.
2. Multiply a row by a nonzero constant.
3. Interchange two rows.

Note that performing any of these operations on the matrix form of a system does not change its solution. We use the following notation to describe the elementary row operations:

Symbol	Description
$R_i + kR_j \rightarrow R_i$	Change the ith row by adding k times row j to it, and then put the result back in row i.
kR_i	Multiply the ith row by k.
$R_i \leftrightarrow R_j$	Interchange the ith and jth rows.

In the next example we compare the two ways of writing systems of linear equations, using the system of Example 1.

EXAMPLE 2 ■ Using the Matrix Form to Solve a System

Solve the system using its matrix form.

$$\begin{cases} x - y + 3z = 4 \\ x + 2y - 2z = 10 \\ 3x - y + 5z = 14 \end{cases}$$

SOLUTION

This is the system in Example 1. As before, our goal is to eliminate the x-term from the second equation and the x- and y-terms from the third equation. For comparison, we write both the full form of the system and the shorthand matrix form.

System		Matrix form
$\begin{cases} x - y + 3z = 4 \\ x + 2y - 2z = 10 \\ 3x - y + 5z = 14 \end{cases}$		$\begin{bmatrix} 1 & -1 & 3 & 4 \\ 1 & 2 & -2 & 10 \\ 3 & -1 & 5 & 14 \end{bmatrix}$

Add -1 times the first equation to the second.

Add -3 times the first to the third.

$$\begin{cases} x - y + 3z = 4 \\ 3y - 5z = 6 \\ 2y - 4z = 2 \end{cases} \qquad \xrightarrow[R_3 - 3R_1 \to R_3]{R_2 - R_1 \to R_2} \qquad \begin{bmatrix} 1 & -1 & 3 & 4 \\ 0 & 3 & -5 & 6 \\ 0 & 2 & -4 & 2 \end{bmatrix}$$

Multiply the third equation by $\frac{1}{2}$.

$$\begin{cases} x - y + 3z = 4 \\ 3y - 5z = 6 \\ y - 2z = 1 \end{cases} \qquad \xrightarrow{\frac{1}{2}R_3} \qquad \begin{bmatrix} 1 & -1 & 3 & 4 \\ 0 & 3 & -5 & 6 \\ 0 & 1 & -2 & 1 \end{bmatrix}$$

Add -3 times the third equation to the second (to eliminate y from the second equation).

$$\begin{cases} x - y + 3z = 4 \\ z = 3 \\ y - 2z = 1 \end{cases} \qquad \xrightarrow{R_2 - 3R_3 \to R_2} \qquad \begin{bmatrix} 1 & -1 & 3 & 4 \\ 0 & 0 & 1 & 3 \\ 0 & 1 & -2 & 1 \end{bmatrix}$$

Interchange the second and third equations.

$$\begin{cases} x - y + 3z = 4 \\ y - 2z = 1 \\ z = 3 \end{cases} \qquad \xrightarrow{R_2 \leftrightarrow R_3} \qquad \begin{bmatrix} 1 & -1 & 3 & 4 \\ 0 & 1 & -2 & 1 \\ 0 & 0 & 1 & 3 \end{bmatrix}$$

At this point we arrive at the solution $(2, 7, 3)$ by back-substitution, as in Example 1. ■

In general, to solve a system of equations using a matrix, we use elementary row operations to arrive at a matrix in a certain form. This form is described in the following box.

ECHELON FORM AND REDUCED ECHELON FORM OF A MATRIX

A matrix is in **echelon form** if it satisfies the following conditions.

1. The first nonzero number in each row (reading from left to right) is 1. This is called the **leading entry**.
2. The leading entry in each row is to the right of the leading entry in the row immediately above it.
3. All rows consisting entirely of zeros are at the bottom of the matrix.

A matrix in echelon form is in **reduced echelon form** if it also satisfies the following condition.

4. Every number above and below each leading entry is a 0.

In the following matrices the first matrix is in reduced echelon form, but the second one is just in echelon form. The third matrix is not in echelon form.

Reduced echelon form

$$\begin{bmatrix} 1 & 3 & 0 & 0 & 0 \\ 0 & 0 & 1 & 0 & -3 \\ 0 & 0 & 0 & 1 & \frac{1}{2} \\ 0 & 0 & 0 & 0 & 0 \end{bmatrix}$$

Echelon form

$$\begin{bmatrix} 1 & 3 & -6 & 10 & 0 \\ 0 & 0 & 1 & 4 & -3 \\ 0 & 0 & 0 & 1 & \frac{1}{2} \\ 0 & 0 & 0 & 0 & 0 \end{bmatrix}$$

Not in echelon form

$$\begin{bmatrix} 0 & 1 & -\frac{1}{2} & 0 & 7 \\ 1 & 0 & 3 & 4 & -5 \\ 0 & 0 & 0 & 1 & 0.4 \\ 0 & 1 & 1 & 0 & 0 \end{bmatrix}$$

To solve a system in matrix form, we use the elementary row operations to arrive at a matrix in echelon form or reduced echelon form. The technique of using elementary row operations to get an echelon form matrix is called **Gaussian elimination**, in honor of the German mathematician C. F. Gauss (see page 347). The process of arriving at a matrix in reduced echelon form is called **Gauss-Jordan elimination**.

EXAMPLE 3 ■ Solving a System Using Echelon Form

Solve the system of linear equations using Gaussian elimination.

$$\begin{cases} 4x + 8y - 4z = 4 \\ 3x + 6y + 5z = -13 \\ -2x + y + 12z = -17 \end{cases}$$

SOLUTION

We first write the system in matrix form, and then use elementary row operations to put it in echelon form.

$$\begin{bmatrix} 4 & 8 & -4 & 4 \\ 3 & 6 & 5 & -13 \\ -2 & 1 & 12 & -17 \end{bmatrix} \xrightarrow{\frac{1}{4}R_1} \begin{bmatrix} 1 & 2 & -1 & 1 \\ 3 & 6 & 5 & -13 \\ -2 & 1 & 12 & -17 \end{bmatrix} \xrightarrow[R_3 + 2R_1 \to R_3]{R_2 - 3R_1 \to R_2} \begin{bmatrix} 1 & 2 & -1 & 1 \\ 0 & 0 & 8 & -16 \\ 0 & 5 & 10 & -15 \end{bmatrix}$$

$$\xrightarrow{R_2 \leftrightarrow R_3} \begin{bmatrix} 1 & 2 & -1 & 1 \\ 0 & 5 & 10 & -15 \\ 0 & 0 & 8 & -16 \end{bmatrix} \xrightarrow[\frac{1}{8}R_3]{\frac{1}{5}R_2} \begin{bmatrix} 1 & 2 & -1 & 1 \\ 0 & 1 & 2 & -3 \\ 0 & 0 & 1 & -2 \end{bmatrix}$$

We now have an equivalent matrix in echelon form, and the corresponding system of equations is

$$\begin{cases} x + 2y - z = 1 \\ y + 2z = -3 \\ z = -2 \end{cases}$$

We use back-substitution to solve the system.

$$y + 2(-2) = -3 \qquad \text{Back-substitute } z = -2 \text{ into the second equation}$$

$$y = 1 \qquad \text{Solve for } y$$

$$x + 2(1) - (-2) = 1 \qquad \text{Back-substitute } y = 1 \text{ and } z = -2 \text{ into the first equation}$$

$$x = -3 \qquad \text{Solve for } x$$

So the solution of the system is $(-3, 1, -2)$. ■

The advantage of using reduced echelon form is that we don't need to use back-substitution when solving a system in this form, as we see in the next example.

EXAMPLE 4 ■ Solving a System Using Reduced Echelon Form

Solve the system of linear equations, using Gauss-Jordan elimination.

$$\begin{cases} 4x + 8y - 4z = 4 \\ 3x + 6y + 5z = -13 \\ -2x + y + 12z = -17 \end{cases}$$

SOLUTION

In Example 3 we used Gaussian elimination on the matrix form of this system of equations to arrive at an equivalent matrix in echelon form. We continue using elementary row operations on the last matrix in Example 3 to arrive at an equivalent matrix in reduced echelon form.

$$\begin{bmatrix} 1 & 2 & -1 & 1 \\ 0 & 1 & 2 & -3 \\ 0 & 0 & 1 & -2 \end{bmatrix} \xrightarrow[\text{R}_1 + \text{R}_3 \to \text{R}_1]{\text{R}_2 - 2\text{R}_3 \to \text{R}_2} \begin{bmatrix} 1 & 2 & 0 & -1 \\ 0 & 1 & 0 & 1 \\ 0 & 0 & 1 & -2 \end{bmatrix} \xrightarrow{\text{R}_1 - 2\text{R}_2 \to \text{R}_1} \begin{bmatrix} 1 & 0 & 0 & -3 \\ 0 & 1 & 0 & 1 \\ 0 & 0 & 1 & -2 \end{bmatrix}$$

We now have an equivalent matrix in reduced echelon form, and the corresponding system of equations is

Since the system is in reduced echelon form, back-substitution is not required to get the solution.

$$\begin{cases} x = -3 \\ y = 1 \\ z = -2 \end{cases}$$

Hence we immediately arrive at the solution $(-3, 1, -2)$. ■

INCONSISTENT AND DEPENDENT SYSTEMS

When you study calculus or linear algebra, you will learn that the graph of a linear equation in three variables is a *plane* in a three-dimensional coordinate system. For a system of three equations in three variables, the following situations arise:

1. The three planes intersect in a single point.
   The system has a unique solution.

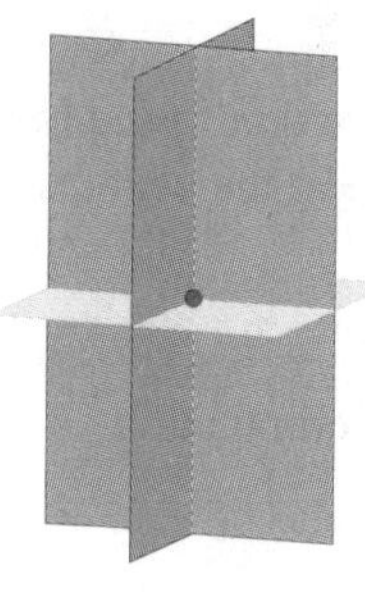

2. The three planes intersect in more than one point.
   The system has infinitely many solutions.

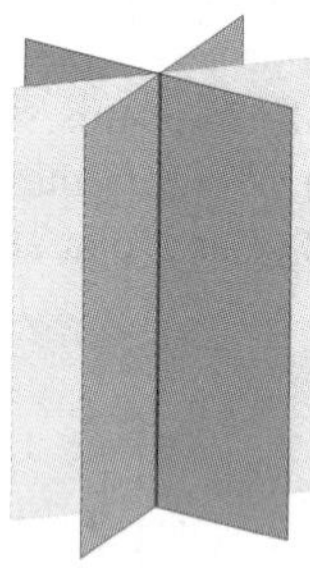

3. The three planes have no point in common.
   The system has no solution.

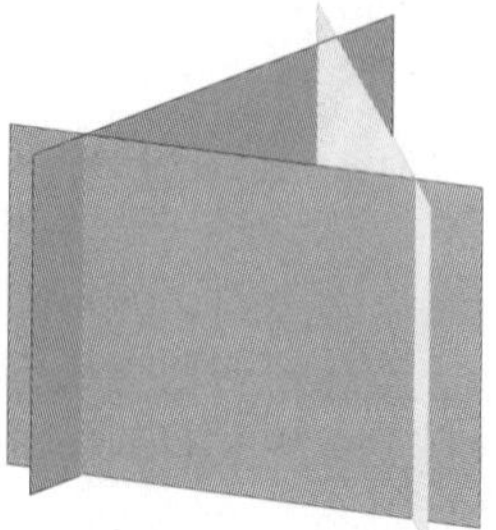

The systems of linear equations that we considered in Examples 1–4 had one solution for each unknown. But as we saw in Section 7.2, a system of two linear equations in two variables can have one solution, no solution, or infinitely many solutions. The same cases arise when we study linear systems with more equations and more variables.

A system that has no solution is said to be **inconsistent**. If we use Gaussian elimination to change a system to echelon form, and if one of the equations we arrive at is false, then the system is inconsistent. (The false equation will always have the form $0 = c$, where c is not zero.)

A system that has infinitely many solutions is called **dependent**. If we use Gaussian elimination to change a system to echelon form and there are fewer nonzero rows than variables, then the system is dependent.

In the following box we give a procedure for solving a system in matrix form. A **leading variable** corresponds to a leading entry in the echelon form matrix of a system.

SOLVING A SYSTEM IN MATRIX FORM

Suppose the matrix form of a system of linear equations has been transformed by Gaussian elimination into echelon form.

1. NO SOLUTION. If the echelon form contains a row that represents the equation $0 = c$, where c is not zero, then the system has *no solution*.

2. ONE SOLUTION. If each variable in the echelon form is a leading variable, then the system has *exactly one solution*, which we find using back-substitution or Gauss-Jordan elimination.

3. INFINITELY MANY SOLUTIONS. If the variables in the echelon form are not all leading variables, then the system has *infinitely many solutions*, which we find by using Gauss-Jordan elimination to convert to reduced echelon form and then expressing the leading variables in terms of the nonleading variables. The nonleading variables take on any real numbers as their values.

EXAMPLE 5 ■ A System with No Solution

Solve the system.

$$\begin{cases} x - 3y + 2z = 12 \\ 2x - 5y + 5z = 14 \\ x - 2y + 3z = 20 \end{cases}$$

SOLUTION

We transform the system into echelon form.

$$\begin{bmatrix} 1 & -3 & 2 & 12 \\ 2 & -5 & 5 & 14 \\ 1 & -2 & 3 & 20 \end{bmatrix} \xrightarrow[\text{R}_3 - \text{R}_1 \to \text{R}_3]{\text{R}_2 - 2\text{R}_1 \to \text{R}_2} \begin{bmatrix} 1 & -3 & 2 & 12 \\ 0 & 1 & 1 & -10 \\ 0 & 1 & 1 & 8 \end{bmatrix}$$

$$\xrightarrow{\text{R}_3 - \text{R}_2 \to \text{R}_3} \begin{bmatrix} 1 & -3 & 2 & 12 \\ 0 & 1 & 1 & -10 \\ 0 & 0 & 0 & 18 \end{bmatrix} \xrightarrow{\frac{1}{18}\text{R}_3} \begin{bmatrix} 1 & -3 & 2 & 12 \\ 0 & 1 & 1 & -10 \\ 0 & 0 & 0 & 1 \end{bmatrix}$$

This last matrix is in echelon form, so we may stop the Gaussian elimination process. Now if we translate the last row back into equation form, we get $0x + 0y + 0z = 1$, or $0 = 1$, which is false. No matter what values we pick for x, y, and z, the last equation will never be a true statement. This means the system *has no solution.* ■

EXAMPLE 6 ■ A System with Infinitely Many Solutions

Find the complete solution of the system.

$$\begin{cases} -3x - 5y + 36z = 10 \\ -x \qquad\quad + 7z = 5 \\ x + y - 10z = -4 \end{cases}$$

SOLUTION

We transform the system into echelon form.

$$\begin{bmatrix} -3 & -5 & 36 & 10 \\ -1 & 0 & 7 & 5 \\ 1 & 1 & -10 & -4 \end{bmatrix} \xrightarrow{\text{R}_1 \leftrightarrow \text{R}_3} \begin{bmatrix} 1 & 1 & -10 & -4 \\ -1 & 0 & 7 & 5 \\ -3 & -5 & 36 & 10 \end{bmatrix}$$

$$\xrightarrow[\text{R}_3 + 3\text{R}_1 \to \text{R}_3]{\text{R}_2 + \text{R}_1 \to \text{R}_2} \begin{bmatrix} 1 & 1 & -10 & -4 \\ 0 & 1 & -3 & 1 \\ 0 & -2 & 6 & -2 \end{bmatrix} \xrightarrow{\text{R}_3 + 2\text{R}_2 \to \text{R}_3} \begin{bmatrix} 1 & 1 & -10 & -4 \\ 0 & 1 & -3 & 1 \\ 0 & 0 & 0 & 0 \end{bmatrix}$$

$$\xrightarrow{\text{R}_1 - \text{R}_2 \to \text{R}_1} \begin{bmatrix} 1 & 0 & -7 & -5 \\ 0 & 1 & -3 & 1 \\ 0 & 0 & 0 & 0 \end{bmatrix}$$

The third row corresponds to the equation $0 = 0$. This equation is always true, no matter what values are used for x, y, and z. Since the equation adds no new

MATHEMATICS IN THE MODERN WORLD

MATHEMATICAL ECONOMICS

The health of the global economy is determined by such interrelated factors as supply, demand, production, consumption, pricing, distribution, and thousands of other factors. These factors are in turn determined by economic decisions (for example, whether or not you buy a certain brand of toothpaste) made by billions of different individuals each day. How will today's creation and distribution of goods affect tomorrow's economy? Such questions are tackled by mathematicians who work on mathematical models of the economy. One pioneer in this area is the mathematician Wassily Leontief who in the 1940s created a model consisting of thousands of equations that describe how different sectors of the economy such as the oil industry, transportation, communication, and so on, interact with each other. A different approach to economic models, one dealing with individuals in the economy as opposed to large sectors, was pioneered by John Nash in the 1950s. In his model, which uses *Game Theory*, the economy is a game where individual players make decisions that often lead to mutual gain. Leontief and Nash were awarded the Nobel Prize in Economics in 1973 and 1994, respectively. Economic theory continues to be a major area of mathematical research.

information about the variables, we can drop it from the system. So the last matrix corresponds to the system

Leading variables

$$\begin{cases} x \qquad -7z = -5 \\ \qquad y - 3z = \;\;\, 1 \end{cases}$$

We solve for the leading variables x and y in terms of the nonleading variable z:

$$x = 7z - 5 \qquad \text{Solve for } x \text{ in the first equation}$$

$$y = 3z + 1 \qquad \text{Solve for } y \text{ in the second equation}$$

To obtain the complete solution, we let z be any real number:

$$x = 7z - 5$$

$$y = 3z + 1$$

$$z = \text{any real number}$$

■

In Example 6, to get specific solutions we give a specific value to z. For example, if $z = 1$, then

$$x = 7(1) - 5 = 2$$

$$y = 3(1) + 1 = 4$$

Thus, $(2, 4, 1)$ is a solution to the system. We would get a different solution if we let $z = 2$, because we would then have

$$x = 7(2) - 5 = 9$$

$$y = 3(2) + 1 = 7$$

So $(9, 7, 2)$ is also a solution. There are infinitely many choices for z, so there are infinitely many solutions to the system.

EXAMPLE 7 ■ A System with Infinitely Many Solutions

Find the complete solution of the system.

$$\begin{cases} x + 2y - 3z - 4w = 10 \\ x + 3y - 3z - 4w = 15 \\ 2x + 2y - 6z - 8w = 10 \end{cases}$$

SOLUTION

We transform the system into reduced echelon form.

$$\begin{bmatrix} 1 & 2 & -3 & -4 & 10 \\ 1 & 3 & -3 & -4 & 15 \\ 2 & 2 & -6 & -8 & 10 \end{bmatrix} \xrightarrow[\mathrm{R}_3 - 2\mathrm{R}_1 \to \mathrm{R}_3]{\mathrm{R}_2 - \mathrm{R}_1 \to \mathrm{R}_2} \begin{bmatrix} 1 & 2 & -3 & -4 & 10 \\ 0 & 1 & 0 & 0 & 5 \\ 0 & -2 & 0 & 0 & -10 \end{bmatrix}$$

$$\xrightarrow{\mathrm{R}_3 + 2\mathrm{R}_2 \to \mathrm{R}_3} \begin{bmatrix} 1 & 2 & -3 & -4 & 10 \\ 0 & 1 & 0 & 0 & 5 \\ 0 & 0 & 0 & 0 & 0 \end{bmatrix} \xrightarrow{\mathrm{R}_1 - 2\mathrm{R}_2 \to \mathrm{R}_1} \begin{bmatrix} 1 & 0 & -3 & -4 & 0 \\ 0 & 1 & 0 & 0 & 5 \\ 0 & 0 & 0 & 0 & 0 \end{bmatrix}$$

This is in reduced echelon form. Since the last row represents the equation $0 = 0$, we may discard it. So the last matrix corresponds to the system

Leading variables

$$\begin{cases} x \qquad -3z - 4w = 0 \\ \quad y \qquad\qquad\quad = 5 \end{cases}$$

To obtain the complete solution, we solve for the leading variables x and y in terms of the nonleading variables z and w, and we let z and w be any real numbers. Thus, the complete solution is

$$x = 3z + 4w$$

$$y = 5$$

$$z = \text{any real number}$$

$$w = \text{any real number}$$ ■

Note that z and w do *not* necessarily have to be the *same* real number in the solution for Example 7. We can choose arbitrary values for each if we wish to construct a specific solution to the system. For example, if we let $z = 1$ and $w = 2$, then we get the solution $(11, 5, 1, 2)$. You should check that this does indeed satisfy all three of the original equations in Example 7.

Examples 6 and 7 illustrate this general fact: If a system in echelon form has n equations in m variables $(m > n)$, then the complete solution will have $m - n$ nonleading variables. For instance, in Example 6 we arrived at two equations in the three variables x, y, and z with $3 - 2 = 1$ nonleading variable.

APPLICATIONS OF LINEAR SYSTEMS

Linear equations, often containing hundreds or even thousands of variables, occur frequently in the applications of algebra to the sciences and to other fields. For now, let's consider an example that involves only three variables.

EXAMPLE 8 ■ Nutritional Analysis Using a System of Linear Equations

A nutritionist is performing an experiment on student volunteers. He wishes to feed one of his subjects a daily diet that consists of a combination of three commercial diet foods: MiniCal, SloStarve, and SlimQuick. For the experiment it's important that the subject consume exactly 500 mg of potassium, 75 g of protein, and 1150 units of vitamin D every day. The amounts of these nutrients in one ounce of each food are given in the table. How many ounces of each food should the subject eat every day to satisfy the nutrient requirements exactly?

	MiniCal	SloStarve	SlimQuick
Potassium (mg)	50	75	10
Protein (g)	5	10	3
Vitamin D (units)	90	100	50

SOLUTION

Let x, y, and z represent the number of ounces of MiniCal, SloStarve, and SlimQuick, respectively, that the subject should eat every day. This means that he will get $50x$ mg of potassium from MiniCal, $75y$ mg from SloStarve, and $10z$ mg from SlimQuick, for a total of $50x + 75y + 10z$ mg potassium in all. Since the potassium requirement is 500 mg, we get the first equation below. Similar reasoning for the protein and vitamin D requirements leads to the system

$$\begin{cases} 50x + 75y + 10z = 500 & \text{Potassium} \\ 5x + 10y + 3z = 75 & \text{Protein} \\ 90x + 100y + 50z = 1150 & \text{Vitamin D} \end{cases}$$

Dividing the first equation by 5 and the third one by 10 gives the system

$$\begin{cases} 10x + 15y + 2z = 100 \\ 5x + 10y + 3z = 75 \\ 9x + 10y + 5z = 115 \end{cases}$$

We solve this system using Gaussian elimination.

$$\begin{bmatrix} 10 & 15 & 2 & 100 \\ 5 & 10 & 3 & 75 \\ 9 & 10 & 5 & 115 \end{bmatrix} \xrightarrow{R_1 - R_3 \to R_1} \begin{bmatrix} 1 & 5 & -3 & -15 \\ 5 & 10 & 3 & 75 \\ 9 & 10 & 5 & 115 \end{bmatrix}$$

$$\xrightarrow[R_3 - 9R_1 \to R_3]{R_2 - 5R_1 \to R_2} \begin{bmatrix} 1 & 5 & -3 & -15 \\ 0 & -15 & 18 & 150 \\ 0 & -35 & 32 & 250 \end{bmatrix} \xrightarrow{-\frac{1}{3}R_2} \begin{bmatrix} 1 & 5 & -3 & -15 \\ 0 & 5 & -6 & -50 \\ 0 & -35 & 32 & 250 \end{bmatrix}$$

$$\xrightarrow{R_3 + 7R_2 \to R_3} \begin{bmatrix} 1 & 5 & -3 & -15 \\ 0 & 5 & -6 & -50 \\ 0 & 0 & -10 & -100 \end{bmatrix} \xrightarrow{-\frac{1}{10}R_3} \begin{bmatrix} 1 & 5 & -3 & -15 \\ 0 & 5 & -6 & -50 \\ 0 & 0 & 1 & 10 \end{bmatrix}$$

CHECK YOUR ANSWER

$x = 5, y = 2, z = 10$:

$$\begin{cases} 10(5) + 15(2) + 2(10) = 100 \\ 5(5) + 10(2) + 3(10) = 75 \\ 9(5) + 10(2) + 5(10) = 115 \end{cases} \checkmark$$

Now we back-substitute to get $z = 10$, $y = 2$, and $x = 5$. The subject should be fed 5 oz of MiniCal, 2 oz of SloStarve, and 10 oz of SlimQuick every day. ■

A more practical application might involve dozens of foods and nutrients, rather than just three. As you can imagine, such a problem is almost impossible to solve without a computer. Many graphing calculators, including the TI-82, TI-83, TI-85, and TI-86 are capable of performing elementary row operations on matrices, so you can use them to solve more complicated systems.

7.3 EXERCISES

1–4 ■ State whether the equation or system of equations is linear.

1. $6x - 3y + 1000z - w = \sqrt{13}$

2. $x_1^2 + x_2^2 + x_3^2 = 36$

3. $e^2 x_1 + \pi x_2 - \sqrt{5} = x_3 - \frac{1}{2}x_4$

4. $\begin{cases} x - 3xy + 5y = 0 \\ 12x + 321y = 123 \end{cases}$

5–10 ■ **(a)** Determine whether the matrix is in echelon form. **(b)** Determine whether the matrix is in reduced echelon form. **(c)** Write the system of equations that corresponds to the matrix.

5. $\begin{bmatrix} 1 & 0 & -3 \\ 0 & 1 & 5 \end{bmatrix}$

6. $\begin{bmatrix} 1 & 3 & -3 \\ 0 & 1 & 5 \end{bmatrix}$

7. $\begin{bmatrix} 1 & 2 & 8 & 0 \\ 0 & 1 & 3 & 2 \\ 0 & 0 & 0 & 0 \end{bmatrix}$

8. $\begin{bmatrix} 1 & 0 & -7 & 0 \\ 0 & 1 & 3 & 0 \\ 0 & 0 & 0 & 1 \end{bmatrix}$

9. $\begin{bmatrix} 1 & 0 & 0 & 0 \\ 0 & 0 & 0 & 0 \\ 0 & 1 & 5 & 1 \end{bmatrix}$

10. $\begin{bmatrix} 1 & 0 & 0 & 1 \\ 0 & 1 & 0 & 2 \\ 0 & 0 & 1 & 3 \end{bmatrix}$

11–20 ■ The system of linear equations has a unique solution. Find the solution using Gaussian elimination or Gauss-Jordan elimination.

11. $\begin{cases} x - 2y + z = 1 \\ y + 2z = 5 \\ x + y + 3z = 8 \end{cases}$

12. $\begin{cases} x + y + 6z = 3 \\ x + y + 3z = 3 \\ x + 2y + 4z = 7 \end{cases}$

13. $\begin{cases} x + y + z = 2 \\ 2x - 3y + 2z = 4 \\ 4x + y - 3z = 1 \end{cases}$

14. $\begin{cases} x + y + z = 4 \\ -x + 2y + 3z = 17 \\ 2x - y = -7 \end{cases}$

15. $\begin{cases} x + 2y - z = -2 \\ x + z = 0 \\ 2x - y - z = -3 \end{cases}$

16. $\begin{cases} 2y + z = 4 \\ x + y = 4 \\ 3x + 3y - z = 10 \end{cases}$

17. $\begin{cases} x_1 + 2x_2 - x_3 = 9 \\ 2x_1 - x_3 = -2 \\ 3x_1 + 5x_2 + 2x_3 = 22 \end{cases}$

18. $\begin{cases} 2x_1 + x_2 = 7 \\ 2x_1 - x_2 + x_3 = 6 \\ 3x_1 - 2x_2 + 4x_3 = 11 \end{cases}$

19. $\begin{cases} 2x - 3y - z = 13 \\ -x + 2y - 5z = 6 \\ 5x - y - z = 49 \end{cases}$

20. $\begin{cases} 10x + 10y - 20z = 60 \\ 15x + 20y + 30z = -25 \\ -5x + 30y - 10z = 45 \end{cases}$

21–30 ■ Determine whether the system of linear equations is inconsistent or dependent. If it is dependent, find the complete solution.

21. $\begin{cases} x + y + z = 2 \\ y - 3z = 1 \\ 2x + y + 5z = 0 \end{cases}$

22. $\begin{cases} x + 3z = 3 \\ 2x + y - 2z = 5 \\ -y + 8z = 8 \end{cases}$

23. $\begin{cases} 2x - 3y - 9z = -5 \\ x + 3z = 2 \\ -3x + y - 4z = -3 \end{cases}$

24. $\begin{cases} x - 2y + 5z = 3 \\ -2x + 6y - 11z = 1 \\ 3x - 16y + 20z = -26 \end{cases}$

25. $\begin{cases} x - y + 3z = 3 \\ 4x - 8y + 32z = 24 \\ 2x - 3y + 11z = 4 \end{cases}$

26. $\begin{cases} -2x + 6y - 2z = -12 \\ x - 3y + 2z = 10 \\ -x + 3y + 2z = 6 \end{cases}$

27. $\begin{cases} x + 4y - 2z = -3 \\ 2x - y + 5z = 12 \\ 8x + 5y + 11z = 30 \end{cases}$

28. $\begin{cases} 3r + 2s - 3t = 10 \\ r - s - t = -5 \\ r + 4s - t = 20 \end{cases}$

29. $\begin{cases} 2x + y - 2z = 12 \\ -x - \frac{1}{2}y + z = -6 \\ 3x + \frac{3}{2}y - 3z = 18 \end{cases}$

30. $\begin{cases} y - 5z = 7 \\ 3x + 2y = 12 \\ 3x + 10z = 80 \end{cases}$

31–42 ■ Solve the system of linear equations.

31. $\begin{cases} 4x - 3y + z = -8 \\ -2x + y - 3z = -4 \\ x - y + 2z = 3 \end{cases}$

32. $\begin{cases} 2x - 3y + 5z = 14 \\ 4x - y - 2z = -17 \\ -x - y + z = 3 \end{cases}$

33. $\begin{cases} x + 2y - 3z = -5 \\ -2x - 4y - 6z = 10 \\ 3x + 7y - 2z = -13 \end{cases}$

34. $\begin{cases} 3x - y + 2z = -1 \\ 4x - 2y + z = -7 \\ -x + 3y - 2z = -1 \end{cases}$

35. $\begin{cases} -x + 2y + z - 3w = 3 \\ 3x - 4y + z + w = 9 \\ -x - y + z + w = 0 \\ 2x + y + 4z - 2w = 3 \end{cases}$

36. $\begin{cases} x + y - z - w = 6 \\ 2x + z - 3w = 8 \\ x - y + 4w = -10 \\ 3x + 5y - z - w = 20 \end{cases}$

37. $\begin{cases} x + y + 2z - w = -2 \\ 3y + z + 2w = 2 \\ x + y + 3w = 2 \\ -3x + z + 2w = 5 \end{cases}$

38. $\begin{cases} x - 3y + 2z + w = -2 \\ x - 2y - 2w = -10 \\ z + 5w = 15 \\ 3x + 2z + w = -3 \end{cases}$

39. $\begin{cases} x + z + w = 4 \\ y - z = -4 \\ x - 2y + 3z + w = 12 \\ 2x - 2z + 5w = -1 \end{cases}$

40. $\begin{cases} y - z + 2w = 0 \\ 3x + 2y + w = 0 \\ 2x + 4w = 12 \\ -2x - 2z + 5w = 6 \end{cases}$

41. $\begin{cases} x - y + w = 0 \\ 3x - z + 2w = 0 \\ x - 4y + z + 2w = 0 \end{cases}$

42. $\begin{cases} 2x - y + 2z + w = 5 \\ -x + y + 4z - w = 3 \\ 3x - 2y - z = 0 \end{cases}$

43. A doctor recommends that a patient take 50 mg each of niacin, riboflavin, and thiamin daily to alleviate a vitamin deficiency. In his medicine chest at home, the patient finds three brands of vitamin pills. The amounts of the relevant vitamins per pill are given in the table. How many pills of each type should he take every day to get 50 mg of each vitamin?

	VitaMax	Vitron	VitaPlus
Niacin (mg)	5	10	15
Riboflavin (mg)	15	20	0
Thiamin (mg)	10	10	10

44. A chemist has three acid solutions at various concentrations. The first is 10% acid, the second is 20%, and the third is 40%. How many milliliters of each should he use to make 100 mL of 18% solution, if he has to use four times as much of the 10% solution as the 40% solution?

45. Amanda, Bryce, and Corey enter a race in which they have to run, swim, and cycle over a marked course. Their average speeds are given in the table. Corey finishes first with a total time of 1 h 45 min. Amanda comes in second with a time of 2 h 30 min. Bryce finishes last with a time of 3 h. Find the distance (in miles) for each part of the race.

	Average Speed (mi/h)		
	Running	**Swimming**	**Cycling**
Amanda	10	4	20
Bryce	$7\frac{1}{2}$	6	15
Corey	15	3	40

46. A small school has 100 students who occupy three classrooms: A, B, and C. After the first period of the school day, half the students in room A move to room B, one-fifth of the students in room B move to room C, and one-third of the students in room C move to room A. Nevertheless, the total number of students in each room is the same for both periods. How many students occupy each room?

47. A furniture factory makes wooden tables, chairs, and armoires. Each piece of furniture requires three operations: cutting the wood, assembling, and finishing. Each operation requires the number of hours (h) given in the table. The workers in the factory can provide 300 hours of cutting, 400 hours of assembling, and 590 hours of finishing each work week. How many tables, chairs, and armoires should be produced so that all available labor-hours are used? Or is this impossible?

	Table	**Chair**	**Armoire**
Cutting (h)	$\frac{1}{2}$	1	1
Assembling (h)	$\frac{1}{2}$	$1\frac{1}{2}$	1
Finishing (h)	1	$1\frac{1}{2}$	2

48. A diagram of a section of a city's street network is shown in the figure. The arrows indicate one-way streets, and the numbers show how many cars enter or leave this section of the city via the indicated street in a certain one-hour period. The variables x, y, z, and w represent the number of cars that travel along the portions of First, Second, Avocado, and Birch Streets during this period. Find x, y, z, and w, assuming that none of the cars stop or park on any of the streets shown in the figure.

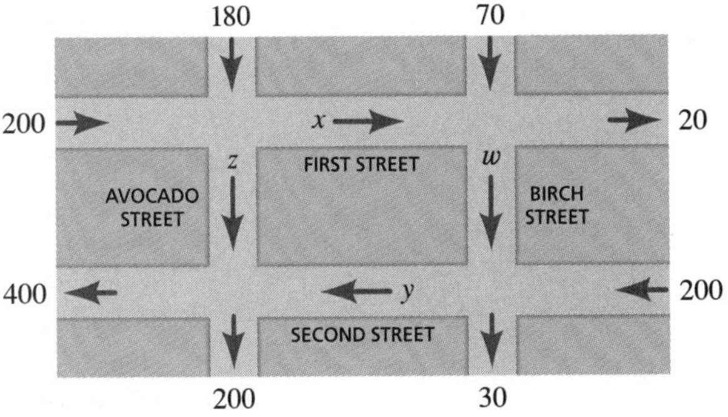

49. (a) Suppose that (x_0, y_0, z_0) and (x_1, y_1, z_1) are solutions of the system

$$\begin{cases} a_1x + b_1y + c_1z = d_1 \\ a_2x + b_2y + c_2z = d_2 \\ a_3x + b_3y + c_3z = d_3 \end{cases}$$

Show that $\left(\frac{x_0 + x_1}{2}, \frac{y_0 + y_1}{2}, \frac{z_0 + z_1}{2}\right)$ is also a solution.

(b) Use the result of part (a) to prove that if the system has two different solutions, then it has infinitely many solutions.

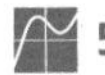

DISCOVERY • DISCUSSION

50. Polynomials Determined by a Set of Points We all know that two points uniquely determine a line $y = ax + b$ in the coordinate plane. Similarly, three points uniquely determine a quadratic (second-degree) polynomial $y = ax^2 + bx + c$, four points uniquely determine a cubic (third-degree) polynomial $y = ax^3 + bx^2 + cx + d$, and so on. (Some exceptions to this rule are if the three points actually lie on a line, or the four points lie on a quadratic or line, and so on.) For the following set of four points, find the line that contains the first two points, the quadratic that contains the first three points, and the cubic that contains all four points.

$$(0, 0), \quad (1, 12), \quad (3, 6), \quad (-1, -14)$$

Graph the points and functions in the same viewing rectangle using a graphing device.

7.4 THE ALGEBRA OF MATRICES

Thus far we've used matrices simply for notational convenience. Matrices have many other uses in mathematics and the sciences, and for most of these applications a knowledge of matrix algebra is essential. Like numbers, matrices can be added, subtracted, multiplied, and divided. In this section we learn how to perform these algebraic operations on matrices.

Recall that a matrix is simply a rectangular array of numbers enclosed between brackets. For example, let A be the matrix

$$A = \begin{bmatrix} -1 & 4 & 7 & 0 \\ 0 & 2 & 13 & 14 \\ \frac{1}{2} & 22 & 8 & -2 \end{bmatrix}$$

The **dimension** of a matrix is a pair of numbers that indicates how many **rows** and **columns** a matrix has. The matrix A is a 3×4 matrix because it has 3 horizontal rows and 4 vertical columns. The individual numbers that make up a matrix are called its **entries**, and they are specified by their row and column position. In the matrix A, the number 13 is the $(2, 3)$ entry, since it is in the second row and the third column. If the name of a matrix is A, we often use the symbol a_{ij} to denote the (i, j) entry of the matrix. Thus, for the preceding matrix, we have $a_{24} = 14$ and $a_{32} = 22$.

Equal matrices

$$\begin{bmatrix} \sqrt{4} & 2^2 & e^0 \\ 0.5 & 1 & 1-1 \end{bmatrix} = \begin{bmatrix} 2 & 4 & 1 \\ \frac{1}{2} & \frac{2}{2} & 0 \end{bmatrix}$$

Unequal matrices

$$\begin{bmatrix} 1 & 2 \\ 3 & 4 \\ 5 & 6 \end{bmatrix} \neq \begin{bmatrix} 1 & 3 & 5 \\ 2 & 4 & 6 \end{bmatrix}$$

Two matrices are **equal** if they have the same dimension and their corresponding entries are equal. So

> $A = B$ if and only if both A and B have the same dimension $m \times n$, and $a_{ij} = b_{ij}$ for $i = 1, 2, \ldots, m$ and $j = 1, 2, \ldots, n$.

ALGEBRAIC OPERATIONS ON MATRICES

Two matrices can be added or subtracted if they have the same dimension. (Otherwise, their sum or difference is undefined.) We add or subtract the matrices by adding or subtracting corresponding entries. To multiply a matrix by a number, we multiply every element of the matrix by that number. This is called the *scalar product*.

SUM, DIFFERENCE, AND SCALAR PRODUCT OF MATRICES

If A and B are matrices of the same dimension and if k is any real number, then

1. The **sum** $A + B$ is a matrix of the same dimension as A and B whose (i, j) entry is $a_{ij} + b_{ij}$.
2. The **difference** $A - B$ is a matrix of the same dimension as A and B whose (i, j) entry is $a_{ij} - b_{ij}$.
3. The **scalar product** kA is a matrix of the same dimension as A whose (i, j) entry is ka_{ij}.

Julia Robinson (1919–1985) was born in St. Louis, Missouri, and grew up at Point Loma, California. Due to an illness, Robinson missed two years of school but later, with the aid of a tutor, she completed fifth, sixth, seventh, and eighth grades, all in one year. At San Diego State University, she found mathematics especially interesting after reading biographies of mathematicians in the book *Men of Mathematics* by E. T. Bell. She said, "I cannot overemphasize the importance of such books . . . in the intellectual life of a student." Robinson is famous for her work on Hilbert's tenth problem (page 498), which asks for a general procedure for determining whether an equation has integer solutions. Her ideas led to a complete answer to the problem. Interestingly, the answer involved certain properties of the Fibonacci numbers (page 584) discovered by the then 22-year-old Russian mathematician Yuri Matijasevič. As a result of her brilliant work on Hilbert's tenth problem, Robinson was offered a professorship at the University of California, Berkeley, and became the first woman mathematician elected to the National Academy of Sciences. She also served as president of the American Mathematical Society.

EXAMPLE 1 ■ Performing Algebraic Operations on Matrices

Let

$$A = \begin{bmatrix} 2 & -3 \\ 0 & 5 \\ 7 & -\frac{1}{2} \end{bmatrix} \qquad B = \begin{bmatrix} 1 & 0 \\ -3 & 1 \\ 2 & 2 \end{bmatrix}$$

$$C = \begin{bmatrix} 7 & -3 & 0 \\ 0 & 1 & 5 \end{bmatrix} \qquad D = \begin{bmatrix} 6 & 0 & -6 \\ 8 & 1 & 9 \end{bmatrix}$$

Carry out each indicated operation, or explain why it cannot be performed.

(a) $A + B$ (b) $C - D$ (c) $C + A$ (d) $5A$

SOLUTION

(a) $$A + B = \begin{bmatrix} 2 & -3 \\ 0 & 5 \\ 7 & -\frac{1}{2} \end{bmatrix} + \begin{bmatrix} 1 & 0 \\ -3 & 0 \\ 2 & 2 \end{bmatrix} = \begin{bmatrix} 3 & -3 \\ -3 & 0 \\ 9 & \frac{3}{2} \end{bmatrix}$$

(b) $$C - D = \begin{bmatrix} 7 & -3 & 0 \\ 0 & 1 & 5 \end{bmatrix} - \begin{bmatrix} 6 & 0 & -6 \\ 8 & 1 & 9 \end{bmatrix} = \begin{bmatrix} 1 & -3 & 6 \\ -8 & 0 & -4 \end{bmatrix}$$

(c) $C + A$ is undefined because we can't add matrices of different dimensions.

(d) $$5A = 5\begin{bmatrix} 2 & -3 \\ 0 & 5 \\ 7 & -\frac{1}{2} \end{bmatrix} = \begin{bmatrix} 10 & -15 \\ 0 & 25 \\ 35 & -\frac{5}{2} \end{bmatrix}$$

■

MATRIX MULTIPLICATION

Multiplication of two matrices is more difficult to describe than other matrix operations. In later examples we will see why taking the matrix product involves a rather complex procedure, which we now describe.

First, the product AB (or $A \cdot B$) of two matrices A and B is defined only when the number of columns in A is equal to the number of rows in B. This means that if we write their dimensions side by side, the two inner numbers must match:

Matrices	A	B
Dimensions	$m \times n$	$n \times k$
	↑ columns in A	↑ rows in B

If the dimensions of A and B match in this fashion, then the product AB will have dimension $m \times k$. Before describing the procedure for obtaining the elements of AB, we define the *inner product* of a row of A and a column of B.

If $[a_1 \quad a_2 \quad \cdots \quad a_n]$ is a row of A, and if $\begin{bmatrix} b_1 \\ b_2 \\ \vdots \\ b_n \end{bmatrix}$ is a column of B, then

their **inner product** is the number $a_1b_1 + a_2b_2 + \cdots + a_nb_n$.

For example,

$$[2 \quad -1 \quad 0 \quad 4] \cdot \begin{bmatrix} 5 \\ 4 \\ -3 \\ \frac{1}{2} \end{bmatrix} = 2 \cdot 5 + (-1) \cdot 4 + 0 \cdot (-3) + 4 \cdot \tfrac{1}{2} = 8$$

We now define the **product** AB of two matrices.

THE PRODUCT OF TWO MATRICES

Suppose that A is an $m \times n$ matrix and B an $n \times k$ matrix. Then $C = AB$ is an $m \times k$ matrix, where c_{ij} is the inner product of the ith row of A and the jth column of B.

EXAMPLE 2 ■ Multiplying Matrices

Let

$$A = \begin{bmatrix} 1 & 3 \\ -1 & 0 \end{bmatrix} \quad \text{and} \quad B = \begin{bmatrix} -1 & 5 & 2 \\ 0 & 4 & 7 \end{bmatrix}$$

Calculate, if possible, the products of AB and BA.

SOLUTION

$A \cdot B$

$2 \times 2 \quad 2 \times 3$

Inner numbers match, so product is defined.
Outer numbers give dimension of product: 2×3

Since A has dimension 2×2 and B has dimension 2×3, the product AB is defined and has dimension 2×3. We can thus write

$$AB = \begin{bmatrix} 1 & 3 \\ -1 & 0 \end{bmatrix} \begin{bmatrix} -1 & 5 & 2 \\ 0 & 4 & 7 \end{bmatrix} = \begin{bmatrix} ? & ? & ? \\ ? & ? & ? \end{bmatrix}$$

where the question marks must be filled in using the rule defining the entries of a matrix product. The (1, 1) entry will be the inner product of the first row of A and the first column of B:

$$\begin{bmatrix} 1 & 3 \\ -1 & 0 \end{bmatrix} \begin{bmatrix} -1 & 5 & 2 \\ 0 & 4 & 7 \end{bmatrix} \qquad 1 \cdot (-1) + 3 \cdot 0 = -1$$

Similarly, we calculate the remaining entries as follows.

Entry	Inner product of:	Value	Product matrix
(1, 2)	$\begin{bmatrix} 1 & 3 \\ -1 & 0 \end{bmatrix}\begin{bmatrix} -1 & 5 & 2 \\ 0 & 4 & 7 \end{bmatrix}$	$1 \cdot 5 + 3 \cdot 4 = 17$	$\begin{bmatrix} -1 & 17 & \\ & & \end{bmatrix}$
(1, 3)	$\begin{bmatrix} 1 & 3 \\ -1 & 0 \end{bmatrix}\begin{bmatrix} -1 & 5 & 2 \\ 0 & 4 & 7 \end{bmatrix}$	$1 \cdot 2 + 3 \cdot 7 = 23$	$\begin{bmatrix} -1 & 17 & 23 \\ & & \end{bmatrix}$
(2, 1)	$\begin{bmatrix} 1 & 3 \\ -1 & 0 \end{bmatrix}\begin{bmatrix} -1 & 5 & 2 \\ 0 & 4 & 7 \end{bmatrix}$	$(-1) \cdot (-1) + 0 \cdot 0 = 1$	$\begin{bmatrix} -1 & 17 & 23 \\ 1 & & \end{bmatrix}$
(2, 2)	$\begin{bmatrix} 1 & 3 \\ -1 & 0 \end{bmatrix}\begin{bmatrix} -1 & 5 & 2 \\ 0 & 4 & 7 \end{bmatrix}$	$(-1) \cdot 5 + 0 \cdot 4 = -5$	$\begin{bmatrix} -1 & 17 & 23 \\ 1 & -5 & \end{bmatrix}$
(2, 3)	$\begin{bmatrix} 1 & 3 \\ -1 & 0 \end{bmatrix}\begin{bmatrix} -1 & 5 & 2 \\ 0 & 4 & 7 \end{bmatrix}$	$(-1) \cdot 2 + 0 \cdot 7 = -2$	$\begin{bmatrix} -1 & 17 & 23 \\ 1 & -5 & -2 \end{bmatrix}$

Thus, we have

$$AB = \begin{bmatrix} -1 & 17 & 23 \\ 1 & -5 & -2 \end{bmatrix}$$

The product BA is not defined, however, because the dimensions are

$$2 \times 3 \qquad \text{and} \qquad 2 \times 2$$

The inner two numbers are not the same, so the rows and columns won't match up when we try to calculate the product. ■

The next example shows that even when both AB and BA are defined, they aren't necessarily equal. This result proves that matrix multiplication is *not* commutative.

EXAMPLE 3 ■ Matrix Multiplication Is Not Commutative

Let

$$A = \begin{bmatrix} 5 & 7 \\ -3 & 0 \end{bmatrix} \qquad \text{and} \qquad B = \begin{bmatrix} 1 & 2 \\ 9 & -1 \end{bmatrix}$$

Calculate the products AB and BA.

SOLUTION

Since both matrices A and B have dimension 2×2, both products AB and BA are defined, and each product is also a 2×2 matrix.

$$AB = \begin{bmatrix} 5 & 7 \\ -3 & 0 \end{bmatrix}\begin{bmatrix} 1 & 2 \\ 9 & -1 \end{bmatrix} = \begin{bmatrix} 5 \cdot 1 + 7 \cdot 9 & 5 \cdot 2 + 7 \cdot (-1) \\ (-3) \cdot 1 + 0 \cdot 9 & (-3) \cdot 2 + 0 \cdot (-1) \end{bmatrix} = \begin{bmatrix} 68 & 3 \\ -3 & -6 \end{bmatrix}$$

$$BA = \begin{bmatrix} 1 & 2 \\ 9 & -1 \end{bmatrix}\begin{bmatrix} 5 & 7 \\ -3 & 0 \end{bmatrix} = \begin{bmatrix} 1 \cdot 5 + 2 \cdot (-3) & 1 \cdot 7 + 2 \cdot 0 \\ 9 \cdot 5 + (-1) \cdot (-3) & 9 \cdot 7 + (-1) \cdot 0 \end{bmatrix} = \begin{bmatrix} -1 & 7 \\ 48 & 63 \end{bmatrix}$$

This shows that, in general, $AB \neq BA$. In fact, in this example AB and BA don't even have an entry in common. ■

Although matrix multiplication is not commutative, it does obey the Associative and Distributive Properties.

PROPERTIES OF MATRIX MULTIPLICATION

Let A, B, and C be matrices for which the following products are defined. Then

$A(BC) = (AB)C$	Associative Property
$A(B + C) = AB + AC$ $(B + C)A = BA + CA$	Distributive Property

The next two examples give some indication of why mathematicians chose to define the matrix product in such an apparently bizarre fashion.

EXAMPLE 4 ■ Writing a System of Equations as a Matrix Equation

Matrix equations are described in more detail on page 492.

Show that the following matrix equation is equivalent to the system of equations in Example 1 of Section 7.3.

$$\begin{bmatrix} 1 & -1 & 3 \\ 1 & 2 & -2 \\ 3 & -1 & 5 \end{bmatrix}\begin{bmatrix} x \\ y \\ z \end{bmatrix} = \begin{bmatrix} 4 \\ 10 \\ 14 \end{bmatrix}$$

SOLUTION

If we perform the matrix multiplication on the left side of the given equation, we get

$$\begin{bmatrix} x - y + 3z \\ x + 2y - 2z \\ 3x - y + 5z \end{bmatrix} = \begin{bmatrix} 4 \\ 10 \\ 14 \end{bmatrix}$$

Because two matrices are equal only if their corresponding entries are equal, this matrix equation means that

$$\begin{cases} x - y + 3z = 4 \\ x + 2y - 2z = 10 \\ 3x - y + 5z = 14 \end{cases}$$

This is exactly the system of equations in Example 1 of Section 7.3. ■

Olga Taussky-Todd (1906–1995) was instrumental in developing applications of Matrix Theory. Described as "in love with anything matrices can do," she successfully applied matrices to aerodynamics, a field used in the design of airplanes and rockets. Taussky-Todd was also famous for her work in Number Theory, which deals with prime numbers and divisibility. Although Number Theory has often been called the least applicable branch of mathematics, it is now used in significant ways throughout the computer industry.

Taussky-Todd studied mathematics at a time when young women rarely aspired to be mathematicians. She said, "When I entered university I had no idea what it meant to study mathematics." One of the most respected mathematicians of her day, she was for many years a professor of mathematics at Caltech in Pasadena.

The preceding example shows that our definition of matrix product allows us to express a system of linear equations as a single matrix equation.

EXAMPLE 5 ■ Representing Demographic Data in Terms of Matrices

In a certain city the proportion of voters in each age group who are registered as Democrats, Republicans, or Independents is given by the following matrix.

	Age		
	18–30	31–50	Over 50
Democrat	0.30	0.60	0.50
Republican	0.50	0.35	0.25
Independent	0.20	0.05	0.25

$$\begin{bmatrix} 0.30 & 0.60 & 0.50 \\ 0.50 & 0.35 & 0.25 \\ 0.20 & 0.05 & 0.25 \end{bmatrix} = A$$

The next matrix gives the distribution, by age and sex, of the voting population of this city.

Age	Male	Female
18–30	5,000	6,000
31–50	10,000	12,000
Over 50	12,000	15,000

$$\begin{bmatrix} 5{,}000 & 6{,}000 \\ 10{,}000 & 12{,}000 \\ 12{,}000 & 15{,}000 \end{bmatrix} = B$$

For this problem, let's make the (highly unrealistic) assumption that within each age group, political preference is not related to gender. That is, the percentage of Democrat males in the 18–30 group, for example, is the same as the percentage of Democrat females in this group.

(a) Calculate the product AB.
(b) How many males are registered as Democrats in this city?
(c) How many females are registered as Republicans?

SOLUTION

$$\text{(a) } AB = \begin{bmatrix} 0.30 & 0.60 & 0.50 \\ 0.50 & 0.35 & 0.25 \\ 0.20 & 0.05 & 0.25 \end{bmatrix} \begin{bmatrix} 5{,}000 & 6{,}000 \\ 10{,}000 & 12{,}000 \\ 12{,}000 & 15{,}000 \end{bmatrix} = \begin{bmatrix} 13{,}500 & 16{,}500 \\ 9{,}000 & 10{,}950 \\ 4{,}500 & 5{,}550 \end{bmatrix}$$

(b) When we take the inner product of a row in A with a column in B, we are adding the number of people in each age group who belong to the category in question. For example, the (2, 1) entry of AB (the 9,000) is obtained by taking the inner product of the Republican row in A with the Male column in B. This number is therefore the total number of male Republicans in this city. We can label the rows and columns of AB as follows.

	Male	Female
Democrat	13,500	16,500
Republican	9,000	10,950
Independent	4,500	5,550

$= AB$

Thus, 13,500 males are registered as Democrats in this city.

(c) There are 10,950 females registered as Republicans. ■

If we add the entries in the columns of matrix A in Example 5, we see that in each case the sum is 1. (Can you see why this has to be true, given what the matrix describes?) A matrix with this property is called **stochastic**. Stochastic matrices are used extensively in statistics, where they arise frequently in situations like the one described here.

7.4 EXERCISES

1–21 ■ The matrices A, B, C, D, E, F, and G are defined as follows.

$$A = \begin{bmatrix} 2 & -5 \\ 0 & 7 \end{bmatrix} \quad B = \begin{bmatrix} 3 & \frac{1}{2} & 5 \\ 1 & -1 & 3 \end{bmatrix} \quad C = \begin{bmatrix} 2 & -\frac{5}{2} & 0 \\ 0 & 2 & -3 \end{bmatrix}$$

$$D = \begin{bmatrix} 7 & 3 \end{bmatrix} \quad E = \begin{bmatrix} 1 \\ 2 \\ 0 \end{bmatrix} \quad F = \begin{bmatrix} 1 & 0 & 0 \\ 0 & 1 & 0 \\ 0 & 0 & 1 \end{bmatrix}$$

$$G = \begin{bmatrix} 5 & -3 & 10 \\ 6 & 1 & 0 \\ -5 & 2 & 2 \end{bmatrix}$$

Carry out the indicated algebraic operation, or explain why it cannot be performed.

1. $B + C$ **2.** $B + F$ **3.** $C - B$

4. $5A$ **5.** $3B + 2C$ **6.** $C - 5A$

7. $2C - 6B$ **8.** DA **9.** AD

10. BC **11.** BF **12.** GF

13. $(DA)B$ **14.** $D(AB)$ **15.** GE

16. A^2 **17.** A^3 **18.** $DB + DC$

19. B^2 **20.** F^2 **21.** $BF + FE$

22. What must be true about the dimensions of the matrices A and B if both products AB and BA are defined?

23–26 ■ Write the system of equations as a matrix equation (see Example 4).

23. $\begin{cases} 2x - 5y = 7 \\ 3x + 2y = 4 \end{cases}$

24. $\begin{cases} 6x - y + z = 12 \\ 2x + z = 7 \\ y - 2z = 4 \end{cases}$

25. $\begin{cases} 3x_1 + 2x_2 - x_3 + x_4 = 0 \\ x_1 - x_3 = 5 \\ 3x_2 + x_3 - x_4 = 4 \end{cases}$

26. $\begin{cases} x - y + z = 2 \\ 4x - 2y - z = 2 \\ x + y + 5z = 2 \\ -x - y - z = 2 \end{cases}$

27–28 ■ Solve for x and y.

27. $\begin{bmatrix} 6 & x \\ 1 & 0 \end{bmatrix}\begin{bmatrix} y & 2 \\ -1 & 2 \end{bmatrix} = \begin{bmatrix} 4 & 16 \\ 1 & 2 \end{bmatrix}$

28. $\begin{bmatrix} 2 & 5 \\ 1 & 3 \end{bmatrix}\begin{bmatrix} x \\ y \end{bmatrix} = \begin{bmatrix} -2 \\ -3 \end{bmatrix}$

29–32 ■ Solve the matrix equation for the unknown matrix X, or explain why no solution exists.

$$A = \begin{bmatrix} 4 & 6 \\ 1 & 3 \end{bmatrix} \qquad B = \begin{bmatrix} 2 & 5 \\ 3 & 7 \end{bmatrix}$$

$$C = \begin{bmatrix} 2 & 3 \\ 1 & 0 \\ 0 & 2 \end{bmatrix} \qquad D = \begin{bmatrix} 10 & 20 \\ 30 & 20 \\ 10 & 0 \end{bmatrix}$$

29. $2X - A = B$

30. $5(X - C) = D$

31. $3X + B = C$

32. $A + D = 3X$

33. A small fast-food chain with restaurants in Santa Monica, Long Beach, and Anaheim sells only hamburgers, hot dogs, and milk shakes. On a certain day, sales were distributed according to the following matrix.

	Number of items sold		
	Santa Monica	Long Beach	Anaheim
Hamburgers	4000	1000	3500
Hot dogs	400	300	200
Milk shakes	700	500	9000

$= A$

The price of each item is given by the following matrix.

Hamburger	Hot dog	Milk shake
[$0.90	$0.80	$1.10]

$= B$

(a) Calculate the product BA.

(b) Interpret the entries in the product matrix BA.

34. A specialty-car manufacturer has plants in Auburn, Biloxi, and Chattanooga. Three models are produced, with daily production given in the following matrix.

	Cars produced each day		
	Model K	Model R	Model W
Auburn	12	10	0
Biloxi	4	4	20
Chattanooga	8	9	12

$= A$

Because of a wage increase, February profits are less than January profits. The profit per car is tabulated by model in the following matrix.

	January	February
Model K	$1000	$500
Model R	$2000	$1200
Model W	$1500	$1000

$= B$

(a) Calculate AB.

(b) Assuming all cars produced were sold, what was the daily profit in January from the Biloxi plant?

(c) What was the total daily profit (from all three plants) in February?

35. Let

$$A = \begin{bmatrix} 1 & 0 & 6 & -1 \\ 2 & \frac{1}{2} & 4 & 0 \end{bmatrix} \qquad C = \begin{bmatrix} 1 \\ 0 \\ -1 \\ -2 \end{bmatrix}$$

$$B = \begin{bmatrix} 1 & 7 & -9 & 2 \end{bmatrix}$$

Determine which of the following products are defined, and calculate the ones that are:

$$ABC \qquad ACB \qquad BAC$$

$$BCA \qquad CAB \qquad CBA$$

36. Prove that if A and B are 2×2 matrices, then

$$(A + B)^2 = A^2 + AB + BA + B^2$$

37. If A and B are 2×2 matrices, is it necessarily true that

$$(A + B)^2 \stackrel{?}{=} A^2 + 2AB + B^2$$

DISCOVERY • DISCUSSION

38. Powers of a Matrix Let

$$A = \begin{bmatrix} 1 & 1 \\ 0 & 1 \end{bmatrix}$$

Calculate $A^2, A^3, A^4, \ldots$ until you detect a pattern. Write a general formula for A^n.

39. Powers of a Matrix Let $A = \begin{bmatrix} 1 & 1 \\ 1 & 1 \end{bmatrix}$. Calculate A^2, $A^3, A^4, \ldots$ until you detect a pattern. Write a general formula for A^n.

40. Square Roots of Matrices A **square root** of a matrix B is a matrix A with the property that $A^2 = B$. (This is the same definition as for the square root of a number.) Find as many square roots as you can of each matrix:

$$\begin{bmatrix} 4 & 0 \\ 0 & 9 \end{bmatrix} \qquad \begin{bmatrix} 1 & 5 \\ 0 & 9 \end{bmatrix}$$

[*Hint:* If $A = \begin{bmatrix} a & b \\ c & d \end{bmatrix}$, write the equations that a, b, c, and d would have to satisfy if A is the square root of the given matrix.]

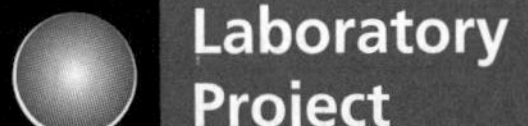

Laboratory Project

WILL THE SPECIES SURVIVE?

To study the survival of animal species, mathematicians model their populations by observing the different stages in their life. They consider, for example, the stage at which the animal is fertile, the proportion of the population that reproduces, and the proportion of the young that survive each year. For a certain species, there are three stages: immature, juvenile, and adult. An animal is considered immature for the first year of its life, juvenile for the second year, and an adult from then on. Conservation biologists have collected the following field data for this species:

$$A = \begin{bmatrix} 0 & 0 & .4 \\ .1 & 0 & 0 \\ 0 & .3 & .8 \end{bmatrix} \begin{matrix} \text{immature} \\ \text{juvenile} \\ \text{adult} \end{matrix} \qquad X_0 = \begin{bmatrix} 600 \\ 400 \\ 3500 \end{bmatrix} \begin{matrix} \text{immature} \\ \text{juvenile} \\ \text{adult} \end{matrix}$$

(Columns of A: immature, juvenile, adult)

The entries in the matrix A indicate the proportion of the population that survives *to the next year*. For example, the first column describes what happens to the immature population: None remain immature, 10% survive to become juveniles, and of course none become adults. The second column describes what happens to the juvenile population: None become immature or remain juvenile, and 30% survive to adulthood. The third column describes the adult population: The number of their new offspring is 40% of the adult population, no adults become juveniles, and 80% survive to live another year. The entries in the population matrix X_0 indicate the current population (year 0) of immature, juvenile, and adult animals.

Let $X_1 = AX_0$, $X_2 = AX_1$, $X_3 = AX_2$, and so on.

1. Explain why X_1 gives the population in year 1, X_2 the population in year 2, and so on.
2. Find the population matrix for years 1, 2, 3, and 4. (Round fractional entries to the nearest whole number.) Do you see any trend?
3. Show that $X_2 = A^2X_0$, $X_3 = A^3X_0$, and so on.
4. Find the population after 50 years—that is, find X_{50}. (Use part 3, and a graphing calculator.) Does it appear that the species will survive?

5. Suppose the environment is improved so that the proportion of immatures that become junveniles each year is increased to .3 from .1, the proportion of juveniles that become adults is increased to .7 from .3, and the proportion of adults that survives to the next year is increased to .9. Find the population after 50 years with the new matrix A. Does it appear that the species will survive under these new conditions?

6. The matrix A in the example is called a **transition matrix**. Such matrices occur in many applications of matrix algebra. The following transition matrix T predicts the calculus grades of a class of students who must take a four-semester sequence of calculus courses. The first column of the matrix, for instance, indicates that of those students who get an A in one course, 70% will get an A in the following course, 15% will get a B, and 10% will get a C. (Students who receive D or F are not permitted to go on to the next course, so are not included in the matrix.) The entries in the matrix Y_0 give the number of incoming students who got A, B, and C, respectively, in their final high school mathematics course.

Let $Y_1 = TY_0$, $Y_2 = TY_1$, $Y_3 = TY_2$, and $Y_4 = TY_3$. Calculate and interpret the entries of Y_1, Y_2, Y_3, and Y_4.

$$T = \begin{array}{c} \begin{matrix} \text{A} & \text{B} & \text{C} \end{matrix} \\ \begin{bmatrix} .70 & .25 & .05 \\ .15 & .50 & .25 \\ .10 & .15 & .45 \end{bmatrix} \end{array} \begin{matrix} \text{A} \\ \text{B} \\ \text{C} \end{matrix} \qquad Y_0 = \begin{bmatrix} 140 \\ 320 \\ 400 \end{bmatrix} \begin{matrix} \text{A} \\ \text{B} \\ \text{C} \end{matrix}$$

7.5 INVERSES OF MATRICES AND MATRIX EQUATIONS

In the preceding section we saw that, when the dimensions are appropriate, matrices can be added, subtracted, and multiplied. In this section we investigate division of matrices. With this operation we can solve equations that involve matrices.

First, we define *identity matrices*, which play the same role for matrix multiplication as the number 1 does for ordinary multiplication of numbers; that is, $1 \cdot a = a \cdot 1 = a$ for all numbers a. In the following definition the term **main diagonal** refers to the entries of a square matrix whose row and column numbers are the same. These entries stretch diagonally down the matrix, from top left to bottom right.

The **identity matrix** I_n is the $n \times n$ matrix for which each main diagonal entry is a 1 and for which all other entries are 0.

Thus, the 2×2, 3×3, and 4×4 identity matrices are, respectively,

$$I_2 = \begin{bmatrix} 1 & 0 \\ 0 & 1 \end{bmatrix} \qquad I_3 = \begin{bmatrix} 1 & 0 & 0 \\ 0 & 1 & 0 \\ 0 & 0 & 1 \end{bmatrix} \qquad I_4 = \begin{bmatrix} 1 & 0 & 0 & 0 \\ 0 & 1 & 0 & 0 \\ 0 & 0 & 1 & 0 \\ 0 & 0 & 0 & 1 \end{bmatrix}$$

Identity matrices behave like the number 1 in the sense that

$$A \cdot I_n = A \qquad \text{and} \qquad I_n \cdot B = B$$

whenever these products are defined. Thus, multiplication by an identity of the appropriate size leaves a matrix unchanged. For example, we can verify by direct calculation that

$$\begin{bmatrix} 1 & 0 \\ 0 & 1 \end{bmatrix} \begin{bmatrix} 3 & 5 & 6 \\ -1 & 2 & 7 \end{bmatrix} = \begin{bmatrix} 3 & 5 & 6 \\ -1 & 2 & 7 \end{bmatrix}$$

or that

$$\begin{bmatrix} -1 & 7 & \frac{1}{2} \\ 12 & 1 & 3 \\ -2 & 0 & 7 \end{bmatrix} \begin{bmatrix} 1 & 0 & 0 \\ 0 & 1 & 0 \\ 0 & 0 & 1 \end{bmatrix} = \begin{bmatrix} -1 & 7 & \frac{1}{2} \\ 12 & 1 & 3 \\ -2 & 0 & 7 \end{bmatrix}$$

If A and B are $n \times n$ matrices, and if $AB = BA = I_n$, then we say that B is the *inverse* of A, and we write $B = A^{-1}$. The concept of the inverse of a matrix is analogous to that of the reciprocal of a real number.

INVERSE OF A MATRIX

Let A be a square $n \times n$ matrix. If there exists an $n \times n$ matrix A^{-1} with the property that

$$AA^{-1} = A^{-1}A = I_n$$

then we say that A^{-1} is the **inverse** of A.

Not every square matrix has an inverse. The following rule provides a simple way for calculating the inverse of a 2×2 matrix, when it exists. For larger matrices, there's a more general procedure for finding inverses, which we consider later in this section.

INVERSE OF A 2 × 2 MATRIX

$$\text{If } A = \begin{bmatrix} a & b \\ c & d \end{bmatrix} \qquad \text{then} \qquad A^{-1} = \frac{1}{ad - bc} \begin{bmatrix} d & -b \\ -c & a \end{bmatrix}.$$

Arthur Cayley (1821–1895) was an English mathematician who invented matrices and developed Matrix Theory. He practiced law until the age of 42, but his primary interest from adolescence was mathematics, and he published almost 200 articles on the subject in his spare time. In 1863 he accepted the offer of a professorship in mathematics at Cambridge, where he taught until his death. Cayley's work on matrices was of purely theoretical interest in his day, but in the 20th century many of his results have found applications in physics, the social sciences, business, and other fields.

EXAMPLE 1 ■ Finding the Inverse of a 2×2 Matrix

Let

$$A = \begin{bmatrix} 4 & 5 \\ 2 & 3 \end{bmatrix}$$

Find A^{-1} and verify that $AA^{-1} = A^{-1}A = I_2$.

SOLUTION

Using the rule, we get

$$A^{-1} = \frac{1}{4 \cdot 3 - 5 \cdot 2}\begin{bmatrix} 3 & -5 \\ -2 & 4 \end{bmatrix} = \frac{1}{2}\begin{bmatrix} 3 & -5 \\ -2 & 4 \end{bmatrix} = \begin{bmatrix} \frac{3}{2} & -\frac{5}{2} \\ -1 & 2 \end{bmatrix}$$

To verify that this is indeed the inverse of A, we calculate AA^{-1} and $A^{-1}A$:

$$AA^{-1} = \begin{bmatrix} 4 & 5 \\ 2 & 3 \end{bmatrix}\begin{bmatrix} \frac{3}{2} & -\frac{5}{2} \\ -1 & 2 \end{bmatrix} = \begin{bmatrix} 4 \cdot \frac{3}{2} + 5(-1) & 4\left(-\frac{5}{2}\right) + 5 \cdot 2 \\ 2 \cdot \frac{3}{2} + 3(-1) & 2\left(-\frac{5}{2}\right) + 3 \cdot 2 \end{bmatrix} = \begin{bmatrix} 1 & 0 \\ 0 & 1 \end{bmatrix}$$

$$A^{-1}A = \begin{bmatrix} \frac{3}{2} & -\frac{5}{2} \\ -1 & 2 \end{bmatrix}\begin{bmatrix} 4 & 5 \\ 2 & 3 \end{bmatrix} = \begin{bmatrix} \frac{3}{2} \cdot 4 + \left(-\frac{5}{2}\right)2 & \frac{3}{2} \cdot 5 + \left(-\frac{5}{2}\right)3 \\ (-1)4 + 2 \cdot 2 & (-1)5 + 2 \cdot 3 \end{bmatrix} = \begin{bmatrix} 1 & 0 \\ 0 & 1 \end{bmatrix}$$

■

The quantity $ad - bc$ that appears in the rule for calculating the inverse is called the **determinant** of the matrix. If the determinant is 0, then the matrix does not have an inverse (since we cannot divide by 0). In the next section we learn how to calculate the determinant of a square matrix of any size, and how to use determinants to solve systems of equations.

INVERSES OF $n \times n$ MATRICES

For 3×3 and larger square matrices, the following technique provides the most efficient way to calculate their inverses. If A is an $n \times n$ matrix, we first construct the $n \times 2n$ matrix that has the entries of A on the left and of the identity matrix I_n on the right:

$$\left[\begin{array}{cccc|cccc} a_{11} & a_{12} & \cdots & a_{1n} & 1 & 0 & \cdots & 0 \\ a_{21} & a_{22} & \cdots & a_{2n} & 0 & 1 & \cdots & 0 \\ \vdots & \vdots & \ddots & \vdots & \vdots & \vdots & \ddots & \vdots \\ a_{n1} & a_{n2} & \cdots & a_{nn} & 0 & 0 & \cdots & 1 \end{array}\right]$$

We then use the elementary row operations on this new large matrix to change the left side into the identity matrix. The right side is transformed automatically into A^{-1}. (We omit the proof of this fact.)

MATHEMATICS IN THE MODERN WORLD

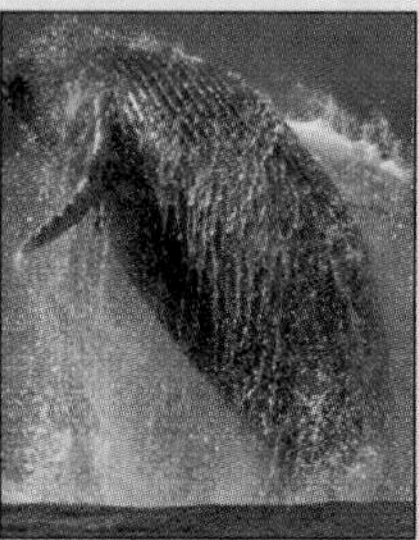

MATHEMATICAL ECOLOGY

In the 1970s humpback whales became a center of controversy. Environmentalists believed that whaling threatened the whales with imminent extinction; whalers saw their livelihood threatened by any attempt to stop whaling. Are whales really threatened to extinction by whaling? What level of whaling is safe to guarantee survival of the whales? These questions motivated mathematicians to study population patterns of whales and other species more closely.

As early as the 1920s Lotka and Volterra had founded the field of mathematical biology by creating predator-prey models. Their models, which draw on a branch of mathematics called differential equations, take into account the rates at which predator eats prey and the rates of growth of each population. Notice that as predator eats prey, the prey population decreases; this means less food supply for the predators, so their population begins to decrease; with fewer predators the prey population begins to increase, and so on. Normally, a state of equilibrium develops, and the two populations alternate between a minimum and a maximum. Notice that if the

(continued)

EXAMPLE 2 ■ Finding the Inverse of a 3 × 3 Matrix

Let A be the matrix

$$A = \begin{bmatrix} 1 & -2 & -4 \\ 2 & -3 & -6 \\ -3 & 6 & 15 \end{bmatrix}$$

(a) Find A^{-1}.
(b) Verify that $AA^{-1} = A^{-1}A = I_3$.

SOLUTION

(a) We begin with the 3×6 matrix whose left half is A and whose right half is the identity matrix.

$$\left[\begin{array}{rrr|rrr} 1 & -2 & -4 & 1 & 0 & 0 \\ 2 & -3 & -6 & 0 & 1 & 0 \\ -3 & 6 & 15 & 0 & 0 & 1 \end{array}\right]$$

We then transform the left half of this new matrix into the identity matrix by performing the following sequence of elementary row operations on the *entire* new matrix:

$$\xrightarrow[\text{R}_3 + 3\text{R}_1 \to \text{R}_3]{\text{R}_2 - 2\text{R}_1 \to \text{R}_2} \left[\begin{array}{rrr|rrr} 1 & -2 & -4 & 1 & 0 & 0 \\ 0 & 1 & 2 & -2 & 1 & 0 \\ 0 & 0 & 3 & 3 & 0 & 1 \end{array}\right]$$

$$\xrightarrow{\frac{1}{3}\text{R}_3} \left[\begin{array}{rrr|rrr} 1 & -2 & -4 & 1 & 0 & 0 \\ 0 & 1 & 2 & -2 & 1 & 0 \\ 0 & 0 & 1 & 1 & 0 & \frac{1}{3} \end{array}\right]$$

$$\xrightarrow{\text{R}_1 + 2\text{R}_2 \to \text{R}_1} \left[\begin{array}{rrr|rrr} 1 & 0 & 0 & -3 & 2 & 0 \\ 0 & 1 & 2 & -2 & 1 & 0 \\ 0 & 0 & 1 & 1 & 0 & \frac{1}{3} \end{array}\right]$$

$$\xrightarrow{\text{R}_2 - 2\text{R}_3 \to \text{R}_2} \left[\begin{array}{rrr|rrr} 1 & 0 & 0 & -3 & 2 & 0 \\ 0 & 1 & 0 & -4 & 1 & -\frac{2}{3} \\ 0 & 0 & 1 & 1 & 0 & \frac{1}{3} \end{array}\right]$$

We have now transformed the left half of this matrix into an identity matrix. (This means we've put the entire matrix in reduced echelon form.) Note that to do this in as systematic a fashion as possible, we first changed the elements below the main diagonal to zeros, just as we would if we were using Gaussian elimination. We then changed each main diagonal element to a 1 by multiplying by the appropriate constant(s). Finally, we completed the process predators eat the prey too fast they will be left without food and ensure their own extinction.

Since Lotka and Volterra's time, more detailed mathematical models of animal populations have been developed. For many species the population is divided into several stages—immature, juvenile, adult, and so on. The proportion of each stage that survives or reproduces in a given time period is entered into a matrix (called a transition matrix); matrix multiplication is then used to predict the population in succeeding time periods. (See Laboratory Project, page 486.) As you can see, the power of mathematics to model and predict is an invaluable tool in the ongoing debate over the environment.

by changing the remaining entries on the left side to zeros. The right half is now A^{-1}.

$$A^{-1} = \begin{bmatrix} -3 & 2 & 0 \\ -4 & 1 & -\frac{2}{3} \\ 1 & 0 & \frac{1}{3} \end{bmatrix}$$

(b) We calculate AA^{-1} and $A^{-1}A$, and verify that both products give the identity matrix I_3.

$$AA^{-1} = \begin{bmatrix} 1 & -2 & -4 \\ 2 & -3 & -6 \\ -3 & 6 & 15 \end{bmatrix} \begin{bmatrix} -3 & 2 & 0 \\ -4 & 1 & -\frac{2}{3} \\ 1 & 0 & \frac{1}{3} \end{bmatrix} = \begin{bmatrix} 1 & 0 & 0 \\ 0 & 1 & 0 \\ 0 & 0 & 1 \end{bmatrix}$$

$$A^{-1}A = \begin{bmatrix} -3 & 2 & 0 \\ -4 & 1 & -\frac{2}{3} \\ 1 & 0 & \frac{1}{3} \end{bmatrix} \begin{bmatrix} 1 & -2 & -4 \\ 2 & -3 & -6 \\ -3 & 6 & 15 \end{bmatrix} = \begin{bmatrix} 1 & 0 & 0 \\ 0 & 1 & 0 \\ 0 & 0 & 1 \end{bmatrix}$$

■

The next example shows that not every square matrix has an inverse.

EXAMPLE 3 ■ A Matrix That Does Not Have an Inverse

Find the inverse of the matrix.

$$\begin{bmatrix} 2 & -3 & -7 \\ 1 & 2 & 7 \\ 1 & 1 & 4 \end{bmatrix}$$

SOLUTION

We proceed as follows.

$$\left[\begin{array}{ccc|ccc} 2 & -3 & -7 & 1 & 0 & 0 \\ 1 & 2 & 7 & 0 & 1 & 0 \\ 1 & 2 & 4 & 0 & 0 & 1 \end{array}\right] \xrightarrow{R_1 \leftrightarrow R_2} \left[\begin{array}{ccc|ccc} 1 & 2 & 7 & 0 & 1 & 0 \\ 2 & -3 & -7 & 1 & 0 & 0 \\ 1 & 1 & 4 & 0 & 0 & 1 \end{array}\right]$$

$$\xrightarrow[R_3 - R_1 \rightarrow R_3]{R_2 - 2R_1 \rightarrow R_2} \left[\begin{array}{ccc|ccc} 1 & 2 & 7 & 0 & 1 & 0 \\ 0 & -7 & -21 & 1 & -2 & 0 \\ 0 & -1 & -3 & 0 & -1 & 1 \end{array}\right]$$

$$\xrightarrow{-\frac{1}{7}R_2} \left[\begin{array}{ccc|ccc} 1 & 2 & 7 & 0 & 1 & 0 \\ 0 & 1 & 3 & -\frac{1}{7} & \frac{2}{7} & 0 \\ 0 & -1 & -3 & 0 & -1 & 1 \end{array}\right]$$

$$\xrightarrow[R_1 - 2R_2 \rightarrow R_1]{R_3 + R_2 \rightarrow R_3} \left[\begin{array}{ccc|ccc} 1 & 0 & 1 & \frac{2}{7} & \frac{3}{7} & 0 \\ 0 & 1 & 3 & -\frac{1}{7} & \frac{2}{7} & 0 \\ 0 & 0 & 0 & -\frac{1}{7} & -\frac{5}{7} & 1 \end{array}\right]$$

At this point, we would like to change the 0 in the (3, 3) position of this matrix to a 1, without changing the zeros in the (3, 1) and (3, 2) positions. But there is no way to accomplish this, because no matter what multiple of rows 1 and/or 2 we add to row 3, we can't change the third zero in row 3 without changing the first or second as well. Thus, we cannot change the left half to the identity matrix. The original matrix doesn't have an inverse. ■

If we encounter a row of zeros on the left when trying to find an inverse, as in Example 3, then the original matrix does not have an inverse.

MATRIX EQUATIONS

We saw in Section 7.4 that a system of linear equations can be written as a single matrix equation. For example, the system

$$\begin{cases} x - 2y - 4z = 7 \\ 2x - 3y - 6z = 5 \\ -3x + 6y + 15z = 0 \end{cases}$$

is equivalent to the matrix equation

$$\begin{bmatrix} 1 & -2 & -4 \\ 2 & -3 & -6 \\ -3 & 6 & 15 \end{bmatrix} \begin{bmatrix} x \\ y \\ z \end{bmatrix} = \begin{bmatrix} 7 \\ 5 \\ 0 \end{bmatrix}$$

If we let

The matrix A is called the *coefficient matrix.*

$$A = \begin{bmatrix} 1 & -2 & -4 \\ 2 & -3 & -6 \\ -3 & 6 & 15 \end{bmatrix} \qquad X = \begin{bmatrix} x \\ y \\ z \end{bmatrix} \qquad B = \begin{bmatrix} 7 \\ 5 \\ 0 \end{bmatrix}$$

then this matrix equation can be written as

$$AX = B$$

We solve this matrix equation by multiplying each side by the inverse of A (provided this inverse exists):

Solving the matrix equation $AX = B$ is very similar to solving the simple real-number equation

$$3x = 12$$

which we solve by multiplying each side by the reciprocal (or inverse) of 3:

$$\tfrac{1}{3}(3x) = \tfrac{1}{3}(12)$$

$$x = 4$$

$$\begin{aligned} AX &= B \\ A^{-1}(AX) &= A^{-1}B && \text{Multiply on left by } A^{-1} \\ (A^{-1}A)X &= A^{-1}B && \text{Associative property} \\ I_3X &= A^{-1}B && \text{Property of inverses} \\ X &= A^{-1}B && \text{Property of identity matrix} \end{aligned}$$

In Example 2, we showed that

$$A^{-1} = \begin{bmatrix} -3 & 2 & 0 \\ -4 & 1 & -\frac{2}{3} \\ 1 & 0 & \frac{1}{3} \end{bmatrix}$$

So, from $X = A^{-1}B$ we have

$$\begin{bmatrix} x \\ y \\ z \end{bmatrix} = \begin{bmatrix} -3 & 2 & 0 \\ -4 & 1 & -\frac{2}{3} \\ 1 & 0 & \frac{1}{3} \end{bmatrix} \begin{bmatrix} 7 \\ 5 \\ 0 \end{bmatrix} = \begin{bmatrix} -11 \\ -23 \\ 7 \end{bmatrix}$$

Thus, $x = -11$, $y = -23$, $z = 7$ is the solution of the original system.

SOLVING A MATRIX EQUATION

If A is a square $n \times n$ matrix that has an inverse A^{-1}, and if X is a variable matrix and B a known matrix, both with n rows, then the solution of the matrix equation

$$AX = B$$

is given by

$$X = A^{-1}B$$

EXAMPLE 4 ■ Solving a System Using the Matrix Inverse

Solve the system of equations.

$$\begin{cases} 2x - 5y = 15 \\ 3x - 6y = 36 \end{cases}$$

SOLUTION

We first convert this to a matrix equation of the form $AX = B$:

$AX = B$

$$\begin{bmatrix} 2 & -5 \\ 3 & -6 \end{bmatrix} \begin{bmatrix} x \\ y \end{bmatrix} = \begin{bmatrix} 15 \\ 36 \end{bmatrix}$$

Using the rule for calculating the inverse of a 2×2 matrix, we get

Find A^{-1}.

$$\begin{bmatrix} 2 & -5 \\ 3 & -6 \end{bmatrix}^{-1} = \frac{1}{2(-6) - (-5)3} \begin{bmatrix} -6 & -(-5) \\ -3 & 2 \end{bmatrix} = \frac{1}{3} \begin{bmatrix} -6 & 5 \\ -3 & 2 \end{bmatrix}$$

Multiplying each side of the matrix equation by this inverse matrix, we get

$X = A^{-1}B$

$$\begin{bmatrix} x \\ y \end{bmatrix} = \frac{1}{3} \begin{bmatrix} -6 & 5 \\ -3 & 2 \end{bmatrix} \begin{bmatrix} 15 \\ 36 \end{bmatrix} = \begin{bmatrix} 30 \\ 9 \end{bmatrix}$$

So $x = 30$ and $y = 9$. ■

EXAMPLE 5 ■ An Application of Matrix Equations

A pet-store owner feeds his hamsters and gerbils different mixtures of three types of rodent food pellets, which we will call brands A, B, and C. He wishes to feed his animals the correct amount of each brand to satisfy their daily requirements for protein, fat, and carbohydrates exactly. Suppose that hamsters require 340 mg of protein, 280 mg of fat, and 440 mg of carbohydrates, and gerbils need 480 mg of protein, 360 mg of fat, and 680 mg of carbohydrates each day. The amount of each nutrient in one gram of each brand is given in the following table. How many grams of each food should the storekeeper feed his hamsters and gerbils daily to satisfy their nutrient requirements?

	Brand A	Brand B	Brand C
Protein (mg)	10	0	20
Fat (mg)	10	20	10
Carbohydrates (mg)	5	10	30

SOLUTION

On the TI-82 and TI-83 calculators, matrices are stored in memory using names such as [A], [B], [C], ... To find the inverse of [A], we key in

[A] x^{-1} ENTER

If we let x_1, x_2, and x_3 be the grams of brands A, B, and C, respectively, that the hamsters should eat, and if we let y_1, y_2, and y_3 be the corresponding amounts for the gerbils, then we want to solve the matrix equations

$$\begin{bmatrix} 10 & 0 & 20 \\ 10 & 20 & 10 \\ 5 & 10 & 30 \end{bmatrix}\begin{bmatrix} x_1 \\ x_2 \\ x_3 \end{bmatrix} = \begin{bmatrix} 340 \\ 280 \\ 440 \end{bmatrix} \qquad \text{Hamster equation}$$

$$\begin{bmatrix} 10 & 0 & 20 \\ 10 & 20 & 10 \\ 5 & 10 & 30 \end{bmatrix}\begin{bmatrix} y_1 \\ y_2 \\ y_3 \end{bmatrix} = \begin{bmatrix} 480 \\ 360 \\ 680 \end{bmatrix} \qquad \text{Gerbil equation}$$

The coefficient matrix is the same in both equations, so we only need to find its inverse once. Using a calculator with matrix functions, we get

$$\begin{bmatrix} 10 & 0 & 20 \\ 10 & 20 & 10 \\ 5 & 10 & 30 \end{bmatrix}^{-1} = \begin{bmatrix} 0.10 & 0.04 & -0.08 \\ -0.05 & 0.04 & 0.02 \\ 0 & -0.02 & 0.04 \end{bmatrix} = \frac{1}{100}\begin{bmatrix} 10 & 4 & -8 \\ -5 & 4 & 2 \\ 0 & -2 & 4 \end{bmatrix}$$

We now multiply each side of our matrix equations by this inverse matrix, to get

$$\begin{bmatrix} x_1 \\ x_2 \\ x_3 \end{bmatrix} = \frac{1}{100}\begin{bmatrix} 10 & 4 & -8 \\ -5 & 4 & 2 \\ 0 & -2 & 4 \end{bmatrix}\begin{bmatrix} 340 \\ 280 \\ 440 \end{bmatrix} = \begin{bmatrix} 10 \\ 3 \\ 12 \end{bmatrix} \qquad \text{Hamster solution matrix}$$

$$\begin{bmatrix} y_1 \\ y_2 \\ y_3 \end{bmatrix} = \frac{1}{100}\begin{bmatrix} 10 & 4 & -8 \\ -5 & 4 & 2 \\ 0 & -2 & 4 \end{bmatrix}\begin{bmatrix} 480 \\ 360 \\ 680 \end{bmatrix} = \begin{bmatrix} 8 \\ 4 \\ 20 \end{bmatrix} \qquad \text{Gerbil solution matrix}$$

Thus, each hamster should be fed 10 g of brand A, 3 g of brand B, and 12 g of brand C, and each gerbil should be fed 8 g of brand A, 4 g of brand B, and 20 g of brand C daily. ■

Since finding the inverse of a 3×3 or larger matrix usually involves a lot of work, the method used in Example 5 is really useful only when we are solving several systems of equations with the same coefficient matrix. However, if we have access to a calculator or computer program that calculates matrix inverses, then this becomes the preferred method in all circumstances. Most graphing calculators can compute matrix inverses.

7.5 EXERCISES

1–2 ■ Find the inverse of the matrix and verify that $A^{-1}A = AA^{-1} = I_2$ and $B^{-1}B = BB^{-1} = I_3$.

1. $A = \begin{bmatrix} 7 & 4 \\ 3 & 2 \end{bmatrix}$

2. $B = \begin{bmatrix} 1 & 3 & 2 \\ 0 & 2 & 2 \\ -2 & -1 & 0 \end{bmatrix}$

3–18 ■ Find the inverse of the matrix if it exists.

3. $\begin{bmatrix} 5 & 3 \\ 3 & 2 \end{bmatrix}$

4. $\begin{bmatrix} 3 & 4 \\ 7 & 9 \end{bmatrix}$

5. $\begin{bmatrix} 2 & 5 \\ -5 & -13 \end{bmatrix}$

6. $\begin{bmatrix} -7 & 4 \\ 8 & -5 \end{bmatrix}$

7. $\begin{bmatrix} 6 & -3 \\ -8 & 4 \end{bmatrix}$

8. $\begin{bmatrix} \frac{1}{2} & \frac{1}{3} \\ 5 & 4 \end{bmatrix}$

9. $\begin{bmatrix} 0.4 & -1.2 \\ 0.3 & 0.6 \end{bmatrix}$

10. $\begin{bmatrix} 4 & 2 & 3 \\ 3 & 3 & 2 \\ 1 & 0 & 1 \end{bmatrix}$

11. $\begin{bmatrix} 2 & 4 & 1 \\ -1 & 1 & -1 \\ 1 & 4 & 0 \end{bmatrix}$

12. $\begin{bmatrix} 5 & 7 & 4 \\ 3 & -1 & 3 \\ 6 & 7 & 5 \end{bmatrix}$

13. $\begin{bmatrix} 1 & 2 & 3 \\ 4 & 5 & -1 \\ 1 & -1 & -10 \end{bmatrix}$

14. $\begin{bmatrix} 2 & 1 & 0 \\ 1 & 1 & 4 \\ 2 & 1 & 2 \end{bmatrix}$

15. $\begin{bmatrix} 0 & -2 & 2 \\ 3 & 1 & 3 \\ 1 & -2 & 3 \end{bmatrix}$

16. $\begin{bmatrix} 3 & -2 & 0 \\ 5 & 1 & 1 \\ 2 & -2 & 0 \end{bmatrix}$

17. $\begin{bmatrix} 1 & 2 & 0 & 3 \\ 0 & 1 & 1 & 1 \\ 0 & 1 & 0 & 1 \\ 1 & 2 & 0 & 2 \end{bmatrix}$

18. $\begin{bmatrix} 1 & 0 & 1 & 0 \\ 0 & 1 & 0 & 1 \\ 1 & 1 & 1 & 0 \\ 1 & 1 & 1 & 1 \end{bmatrix}$

19–26 ■ Solve the system of equations by converting to a matrix equation and using the inverse of the coefficient matrix, as in Example 4. Use the inverses from Exercises 3–6, 11, 12, 15, and 17.

19. $\begin{cases} 5x + 3y = 4 \\ 3x + 2y = 0 \end{cases}$

20. $\begin{cases} 3x + 4y = 10 \\ 7x + 9y = 20 \end{cases}$

21. $\begin{cases} 2x + 5y = 2 \\ -5x - 13y = 20 \end{cases}$

22. $\begin{cases} -7x + 4y = 0 \\ 8x - 5y = 100 \end{cases}$

23. $\begin{cases} 2x + 4y + z = 7 \\ -x + y - z = 0 \\ x + 4y = -2 \end{cases}$

24. $\begin{cases} 5x + 7y + 4z = 1 \\ 3x - y + 3z = 1 \\ 6x + 7y + 5z = 1 \end{cases}$

25. $\begin{cases} -2y + 2z = 12 \\ 3x + y + 3z = -2 \\ x - 2y + 3z = 8 \end{cases}$

26. $\begin{cases} x + 2y + 3w = 0 \\ y + z + w = 1 \\ y + w = 2 \\ x + 2y + 2w = 3 \end{cases}$

27–28 ■ Solve the matrix equation by multiplying each side by the appropriate inverse matrix.

27. $\begin{bmatrix} 3 & -2 \\ -4 & 3 \end{bmatrix}\begin{bmatrix} x & y & z \\ u & v & w \end{bmatrix} = \begin{bmatrix} 1 & 0 & -1 \\ 2 & 1 & 3 \end{bmatrix}$

28. $\begin{bmatrix} 0 & -2 & 2 \\ 3 & 1 & 3 \\ 1 & -2 & 3 \end{bmatrix}\begin{bmatrix} x & u \\ y & v \\ z & w \end{bmatrix} \begin{bmatrix} 3 & 6 \\ 6 & 12 \\ 0 & 0 \end{bmatrix}$

29. A nutritionist is studying the effects of the nutrients folic acid, choline, and inositol. He has three types of food available, and each type contains the following amounts of these nutrients per ounce:

	Type A	Type B	Type C
Folic acid (mg)	3	1	3
Choline (mg)	4	2	4
Inositol (mg)	3	2	4

(a) Find the inverse of the matrix

$$\begin{bmatrix} 3 & 1 & 3 \\ 4 & 2 & 4 \\ 3 & 2 & 4 \end{bmatrix}$$

and use it to solve the remaining parts of this problem.

(b) How many ounces of each food should the nutritionist feed his laboratory rats if he wants their daily diet to contain 10 mg of folic acid, 14 mg of choline, and 13 mg of inositol?

(c) How much of each food is needed to supply 9 mg of folic acid, 12 mg of choline, and 10 mg of inositol?

(d) Will any combination of these foods supply 2 mg of folic acid, 4 mg of choline, and 11 mg of inositol?

30. Refer to Exercise 29. Suppose food type C has been improperly labeled, and it actually contains 4 mg of folic acid, 6 mg of choline, and 5 mg of inositol per ounce. Would it still be possible to use matrix inversion to solve parts (b), (c), and (d) of Exercise 29? Why or why not?

31. An encyclopedia saleswoman works for a company that offers three different grades of bindings for its encyclopedias: standard, deluxe, and leather. For each set she sells, she earns a commission that is based on the set's binding grade. One week she sells one standard, one deluxe, and two leather sets and makes \$675 in commission. The next week she sells two standard, one deluxe, and one leather set for a \$600 commission. The third week she sells one standard, two deluxe, and one leather set, earning \$625 in commission.

(a) Let x, y, and z represent the commission she earns on standard, deluxe, and leather sets, respectively. Translate the given information into a system of equations in x, y, and z.

(b) Express the system of equations you found in part (a) as a matrix equation of the form $AX = B$.

(c) Find the inverse of the coefficient matrix A and use it to solve the matrix equation in part (b). How much commission does the saleswoman earn on a set of encyclopedias in each grade of binding?

32–35 ■ Find the inverse of the matrix. For what value(s) of x, if any, does the matrix have no inverse?

32. $\begin{bmatrix} x & 1 \\ -1 & 1/x \end{bmatrix}$

33. $\begin{bmatrix} e^x & -e^{2x} \\ e^{2x} & e^{3x} \end{bmatrix}$

34. $\begin{bmatrix} 1 & e^x & 0 \\ e^x & -e^{2x} & 0 \\ 0 & 0 & 2 \end{bmatrix}$

35. $\begin{bmatrix} 1 & x & 1 \\ 1 & 0 & 1 \\ 1 & 2x & 2 \end{bmatrix}$

36. Find the inverse of the following matrix, where $abcd \neq 0$:

$$\begin{bmatrix} a & 0 & 0 & 0 \\ 0 & b & 0 & 0 \\ 0 & 0 & c & 0 \\ 0 & 0 & 0 & d \end{bmatrix}$$

DISCOVERY • DISCUSSION

37. No Zero-Product Property for Matrices We have used the Zero-Product Property to solve algebraic equations. Matrices do *not* have this property. Let O represent the **2×2 zero matrix**:

$$O = \begin{bmatrix} 0 & 0 \\ 0 & 0 \end{bmatrix}$$

Find 2×2 matrices $A \neq O$ and $B \neq O$ such that $AB = O$. Can you find a matrix $A \neq O$ such that $A^2 = O$?

7.6 DETERMINANTS AND CRAMER'S RULE

If a matrix is **square** (that is, if it has the same number of rows as columns), then we can assign to it a number called its **determinant**. Determinants can be used to solve matrix equations, as we will see later in this section. They are also useful in determining whether a matrix has an inverse.

We denote the determinant of a square matrix A by the symbol $|A|$, and we begin by defining $|A|$ for the simplest case. If A is a 1×1 matrix, then it has only one entry, and we define its determinant to be the value of that entry; that is, if $A = [a]$, then $|A| = a$. If A is a 2×2 matrix,

$$A = \begin{bmatrix} a & b \\ c & d \end{bmatrix}$$

then the determinant of A is defined by

$$|A| = \begin{vmatrix} a & b \\ c & d \end{vmatrix} = ad - bc$$

EXAMPLE 1 ■ Determinant of a 2 × 2 Matrix

Evaluate $|A|$ for $A = \begin{bmatrix} 6 & -3 \\ 2 & 3 \end{bmatrix}$.

SOLUTION

$$\begin{vmatrix} 6 & -3 \\ 2 & 3 \end{vmatrix} = 6 \cdot 3 - (-3)2 = 18 - (-6) = 24$$

■

We can think of the evaluation of a 2×2 determinant as a "cross-product" operation. We take the product of the diagonal from top left to bottom right, and subtract the product from top right to bottom left.

To define the concept of determinant for an arbitrary $n \times n$ matrix, we need the following terminology.

Let A be an $n \times n$ matrix.

1. The **minor** M_{ij} of the element a_{ij} is the determinant of the matrix obtained by deleting the ith row and jth column of A.

2. The **cofactor** A_{ij} of the element a_{ij} is

$$A_{ij} = (-1)^{i+j} M_{ij}$$

David Hilbert (1862–1943) was born in Königsberg, Germany, and became a professor at Göttingen University. He is considered by many to be the greatest mathematician of the 20th century. At the International Congress of Mathematicians held in Paris in 1900, Hilbert set the direction of mathematics for the about-to-dawn 20th century by posing 23 problems he believed to be of crucial importance. He said that "these are problems whose solutions we expect from the future." Most of Hilbert's problems have now been solved (see Julia Robinson, page 479 and Alan Turing, page 95), and their solutions have led to important new areas of mathematical research. Yet as we enter the new millennium, some of his problems remain unsolved. In his work, Hilbert emphasized structure, logic, and the foundations of mathematics. Part of his genius lay in his ability to see the most general possible statement of a problem. For instance, Euler proved that every whole number is the sum of four squares; Hilbert proved a similar statement for all powers of positive integers.

For example, if A is the matrix

$$\begin{bmatrix} 2 & 3 & -1 \\ 0 & 2 & 4 \\ -2 & 5 & 6 \end{bmatrix}$$

then the minor M_{12} is the determinant of the matrix obtained by deleting the first row and second column from A. Thus

$$M_{12} = \begin{vmatrix} 2 & 3 & -1 \\ 0 & 2 & 4 \\ -2 & 5 & 6 \end{vmatrix} = \begin{vmatrix} 0 & 4 \\ -2 & 6 \end{vmatrix} = 0(6) - 4(-2) = 8$$

So, the cofactor $A_{12} = (-1)^{1+2}M_{12} = -8$. Similarly

$$M_{33} = \begin{vmatrix} 2 & 3 & -1 \\ 0 & 2 & 4 \\ -2 & 5 & 6 \end{vmatrix} = \begin{vmatrix} 2 & 3 \\ 0 & 2 \end{vmatrix} = 2 \cdot 2 - 3 \cdot 0 = 4$$

So, $A_{33} = (-1)^{3+3}M_{33} = 4$.

Note that the cofactor of a_{ij} is simply the minor of a_{ij} multiplied by either 1 or -1, depending on whether $i + j$ is even or odd. Thus, in a 3×3 matrix we obtain the cofactor of any element by prefixing its minor with the sign obtained from the following checkerboard pattern:

$$\begin{bmatrix} + & - & + \\ - & + & - \\ + & - & + \end{bmatrix}$$

We are now ready to define the determinant of any square matrix.

THE DETERMINANT OF A SQUARE MATRIX

If A is an $n \times n$ matrix, then the **determinant** of A is obtained by multiplying each element of the first row by its cofactor, and then adding the results. In symbols,

$$|A| = \begin{vmatrix} a_{11} & a_{12} & \cdots & a_{1n} \\ a_{21} & a_{22} & \cdots & a_{2n} \\ \vdots & \vdots & \ddots & \vdots \\ a_{n1} & a_{n2} & \cdots & a_{nn} \end{vmatrix} = a_{11}A_{11} + a_{12}A_{12} + \cdots + a_{1n}A_{1n}$$

EXAMPLE 2 ■ Determinant of a 3×3 Matrix

Evaluate the determinant of the matrix.

$$A = \begin{bmatrix} 2 & 3 & -1 \\ 0 & 2 & 4 \\ -2 & 5 & 6 \end{bmatrix}$$

SOLUTION

$$|A| = \begin{vmatrix} 2 & 3 & -1 \\ 0 & 2 & 4 \\ -2 & 5 & 6 \end{vmatrix} = 2\begin{vmatrix} 2 & 4 \\ 5 & 6 \end{vmatrix} - 3\begin{vmatrix} 0 & 4 \\ -2 & 6 \end{vmatrix} + (-1)\begin{vmatrix} 0 & 2 \\ -2 & 5 \end{vmatrix}$$

$$= 2(2 \cdot 6 - 4 \cdot 5) - 3[0 \cdot 6 - 4(-2)] - [0 \cdot 5 - 2(-2)]$$

$$= -16 - 24 - 4$$

$$= -44$$

■

In our definition of the determinant we used the cofactors of elements in the first row only. This is called **expanding the determinant by the first row**. In fact, *we can expand the determinant by any row or column in the same way, and obtain the same result in each case* (although we won't prove this). The next example illustrates this principle.

EXAMPLE 3 ■ Expanding a Determinant about a Row and a Column

Expand the determinant of the matrix A in Example 2 by the second row and by the third column, and show that the value obtained is the same in each case.

SOLUTION

Expanding by the second row, we get

$$|A| = \begin{vmatrix} 2 & 3 & -1 \\ 0 & 2 & 4 \\ -2 & 5 & 6 \end{vmatrix} = -0\begin{vmatrix} 3 & -1 \\ 5 & 6 \end{vmatrix} + 2\begin{vmatrix} 2 & -1 \\ -2 & 6 \end{vmatrix} - 4\begin{vmatrix} 2 & 3 \\ -2 & 5 \end{vmatrix}$$

$$= 0 + 2[2 \cdot 6 - (-1)(-2)] - 4[2 \cdot 5 - 3(-2)]$$

$$= 0 + 20 - 64$$

$$= -44$$

Emmy Noether (1882–1935) was one of the foremost mathematicians of the early 20th century. Her groundbreaking work in abstract algebra provided much of the foundation for this field, and her work in Invariant Theory was essential in the development of Einstein's theory of general relativity. Although women weren't allowed to study at German universities at that time, she audited courses unofficially and went on to receive a doctorate at Erlangen *summa cum laude*, despite the opposition of the academic senate, which declared that women students would "overthrow all academic order." She subsequently taught mathematics at Göttingen, Moscow, and Frankfurt. In 1933 she left Germany to escape Nazi persecution, accepting a position at Bryn Mawr College in suburban Philadelphia. She lectured there and at the Institute for Advanced Study in Princeton, New Jersey, until her untimely death in 1935.

Expanding by the third column gives

$$\begin{aligned}|A| &= \begin{vmatrix} 2 & 3 & -1 \\ 0 & 2 & 4 \\ -2 & 5 & 6 \end{vmatrix} \\ &= -1\begin{vmatrix} 0 & 2 \\ -2 & 5 \end{vmatrix} - 4\begin{vmatrix} 2 & 3 \\ -2 & 5 \end{vmatrix} + 6\begin{vmatrix} 2 & 3 \\ 0 & 2 \end{vmatrix} \\ &= -[0 \cdot 5 - 2(-2)] - 4[2 \cdot 5 - 3(-2)] + 6(2 \cdot 2 - 3 \cdot 0) \\ &= -4 - 64 + 24 \\ &= -44\end{aligned}$$

In both cases, we obtain the same value for the determinant as when we expanded by the first row in Example 2. ■

The following principle allows us to determine whether a square matrix has an inverse without actually calculating the inverse. This is one of the most important uses of the determinant in matrix algebra, and it is the reason for the name *determinant*.

INVERTIBILITY CRITERION

If A is a square matrix, then A has an inverse if and only if $|A| \neq 0$.

Although we won't prove this fact, we have already seen (in the preceding section) why it is true in the case of 2×2 matrices.

EXAMPLE 4 ■ Using the Determinant to Show That a Matrix Is Not Invertible

Show that the matrix A has no inverse.

$$A = \begin{bmatrix} 1 & 2 & 0 & 4 \\ 0 & 0 & 0 & 3 \\ 5 & 6 & 2 & 6 \\ 2 & 4 & 0 & 9 \end{bmatrix}$$

SOLUTION

We begin by calculating the determinant of A. Since all but one of the elements of the second row is zero, we expand the determinant by the second row. If we do this, we see from the following equation that only the cofactor A_{24} will have to be calculated.

$$
\begin{aligned}
|A| &= -0 \cdot A_{21} + 0 \cdot A_{22} - 0 \cdot A_{23} + 3 \cdot A_{24} = 3A_{24} \\
&= 3\begin{vmatrix} 1 & 2 & 0 \\ 5 & 6 & 2 \\ 2 & 4 & 0 \end{vmatrix} \qquad \text{Expand this by the third column} \\
&= 3(-2)\begin{vmatrix} 1 & 2 \\ 2 & 4 \end{vmatrix} \\
&= 3(-2)(1 \cdot 4 - 2 \cdot 2) = 0
\end{aligned}
$$

Since the determinant of A is zero, A cannot have an inverse, by the Invertibility Criterion. ■

The preceding example shows that if we expand a determinant about a row or column that contains many zeros, our work is reduced considerably because we don't have to evaluate the cofactors of the elements that are zero. The following principle often enables us to simplify the process of finding a determinant by introducing zeros into it without changing its value.

ROW AND COLUMN TRANSFORMATIONS OF A DETERMINANT

If A is a square matrix, and if the matrix B is obtained from A by adding a multiple of one row to another, or a multiple of one column to another, then $|A| = |B|$.

EXAMPLE 5 ■ Using Row and Column Transformations to Calculate a Determinant

Find the determinant of the matrix A. Does it have an inverse?

$$
A = \begin{bmatrix} 8 & 2 & -1 & -4 \\ 3 & 5 & -3 & 11 \\ 24 & 6 & 1 & -12 \\ 2 & 2 & 7 & -1 \end{bmatrix}
$$

SOLUTION

If we add -3 times row 1 to row 3, we change all but one element of row 3 to zeros:

$$
\begin{bmatrix} 8 & 2 & -1 & -4 \\ 3 & 5 & -3 & 11 \\ 0 & 0 & 4 & 0 \\ 2 & 2 & 7 & -1 \end{bmatrix}
$$

This new matrix has the same determinant as A, and if we expand its determinant by the third row, we get

$$|A| = 4\begin{vmatrix} 8 & 2 & -4 \\ 3 & 5 & 11 \\ 2 & 2 & -1 \end{vmatrix}$$

Now, adding 2 times column 3 to column 1 in this determinant gives us

$$|A| = 4\begin{vmatrix} 0 & 2 & -4 \\ 25 & 5 & 11 \\ 0 & 2 & -1 \end{vmatrix} \qquad \text{Expand this by the first column}$$

$$= 4(-25)\begin{vmatrix} 2 & -4 \\ 2 & -1 \end{vmatrix}$$

$$= 4(-25)[2(-1) - (-4)2] = -600$$

Since the determinant of A is not zero, A does have an inverse. ■

CRAMER'S RULE

The solutions of linear equations can sometimes be expressed using determinants. To illustrate, let's solve the following pair of linear equations for the variable x.

$$\begin{cases} ax + by = r \\ cx + dy = s \end{cases}$$

To eliminate the variable y, we multiply the first equation by d and the second by b, and subtract.

$$\begin{array}{l} adx + bdy = rd \\ \underline{bcx + bdy = bs} \\ adx - bcx = rd - bs \end{array}$$

Factoring the left-hand side, we get $(ad - bc)x = rd - bs$. Assuming that $ad - bc \neq 0$, we can now solve this equation for x, obtaining

$$x = \frac{rd - bs}{ad - bc}$$

The numerator and denominator of this fraction look like the determinants of 2×2 matrices. In fact, we can write the solution of the system as

$$x = \frac{\begin{vmatrix} r & b \\ s & d \end{vmatrix}}{\begin{vmatrix} a & b \\ c & d \end{vmatrix}} \qquad y = \frac{\begin{vmatrix} a & r \\ c & s \end{vmatrix}}{\begin{vmatrix} a & b \\ c & d \end{vmatrix}}$$

where the solution for y was obtained using the same sort of technique as the solution for x. Notice that the denominator in each case is the determinant of the coefficient matrix, which we will call D. The numerator in the solution for x is the determinant of the matrix obtained from D by replacing the first column, the coefficients of x, by r and s, respectively. Similarly, in the solution for y the numerator is the determinant of the matrix obtained from D by replacing the second column, the coefficients of y, by r and s. Thus, if we define

$$D = \begin{bmatrix} a & b \\ c & d \end{bmatrix} \qquad D_x = \begin{bmatrix} r & b \\ s & d \end{bmatrix} \qquad D_y = \begin{bmatrix} a & r \\ c & s \end{bmatrix}$$

then we can write the solution of the system as

$$x = \frac{|D_x|}{|D|} \quad \text{and} \quad y = \frac{|D_y|}{|D|}$$

This pair of formulas is known as **Cramer's Rule**, and it can be used to solve any pair of linear equations in two unknowns in which the determinant of the coefficient matrix is not zero.

EXAMPLE 6 ■ Using Cramer's Rule to Solve a System with Two Variables

Use Cramer's Rule to solve the system.

$$\begin{cases} 2x + 6y = -1 \\ x + 8y = 2 \end{cases}$$

SOLUTION

For this system we have

$$|D| = \begin{vmatrix} 2 & 6 \\ 1 & 8 \end{vmatrix} = 2 \cdot 8 - 6 \cdot 1 = 10$$

$$|D_x| = \begin{vmatrix} -1 & 6 \\ 2 & 8 \end{vmatrix} = (-1)8 - 6 \cdot 2 = -20$$

$$|D_y| = \begin{vmatrix} 2 & -1 \\ 1 & 2 \end{vmatrix} = 2 \cdot 2 - (-1)1 = 5$$

The solution is

$$x = \frac{|D_x|}{|D|} = \frac{-20}{10} = -2$$

$$y = \frac{|D_y|}{|D|} = \frac{5}{10} = \frac{1}{2}$$

■

Cramer's Rule can be extended to apply to any system of n linear equations in n variables in which the determinant of the coefficient matrix is not zero. As we saw in the preceding section, any such system can be written in matrix form as

$$\begin{bmatrix} a_{11} & a_{12} & \cdots & a_{1n} \\ a_{21} & a_{22} & \cdots & a_{2n} \\ \vdots & \vdots & \ddots & \vdots \\ a_{n1} & a_{n2} & \cdots & a_{nn} \end{bmatrix} \begin{bmatrix} x_1 \\ x_2 \\ \vdots \\ x_n \end{bmatrix} = \begin{bmatrix} b_1 \\ b_2 \\ \vdots \\ b_n \end{bmatrix}$$

By analogy with our derivation of Cramer's Rule in the case of two equations in two unknowns, we let D be the coefficient matrix in this system, and D_{x_i} be the matrix obtained by replacing the ith column of D by the numbers $b_1, b_2, \ldots, b_n$ that appear to the right of the equal sign. The solution of the system is then given by the following rule.

CRAMER'S RULE

If a system of n linear equations in the n variables $x_1, x_2, \ldots, x_n$ is equivalent to the matrix equation $DX = B$, and if $|D| \neq 0$, then its solutions are

$$x_1 = \frac{|D_{x_1}|}{|D|}, \quad x_2 = \frac{|D_{x_2}|}{|D|}, \quad \ldots, \quad x_n = \frac{|D_{x_n}|}{|D|}$$

where D_{x_i} is the matrix obtained by replacing the ith column of D by the $n \times 1$ matrix B.

EXAMPLE 7 ■ Using Cramer's Rule to Solve a System with Three Variables

Use Cramer's Rule to solve the system.

$$\begin{cases} 2x - 3y + 4z = 1 \\ x + 6z = 0 \\ 3x - 2y = 5 \end{cases}$$

SOLUTION

First, we evaluate the determinants that appear in Cramer's Rule.

$$|D| = \begin{vmatrix} 2 & -3 & 4 \\ 1 & 0 & 6 \\ 3 & -2 & 0 \end{vmatrix} = -38 \qquad |D_x| = \begin{vmatrix} 1 & -3 & 4 \\ 0 & 0 & 6 \\ 5 & -2 & 0 \end{vmatrix} = -78$$

$$|D_y| = \begin{vmatrix} 2 & 1 & 4 \\ 1 & 0 & 6 \\ 3 & 5 & 0 \end{vmatrix} = -22 \qquad |D_z| = \begin{vmatrix} 2 & -3 & 1 \\ 1 & 0 & 0 \\ 3 & -2 & 5 \end{vmatrix} = 13$$

Now we use Cramer's Rule to get the solution:

$$x = \frac{|D_x|}{|D|} = \frac{-78}{-38} = \frac{39}{19}$$

$$y = \frac{|D_y|}{|D|} = \frac{-22}{-38} = \frac{11}{19}$$

$$z = \frac{|D_z|}{|D|} = \frac{13}{-38} = -\frac{13}{38}$$

■

Solving the system in Example 7 using Gaussian elimination would involve matrices whose elements are fractions with fairly large denominators. Thus, in cases like Examples 6 and 7, Cramer's Rule provides an efficient method for solving systems of linear equations. But in systems with more than three equations, evaluating the various determinants involved is usually a long and tedious task. Moreover, the rule doesn't apply if $|D| = 0$ or if D is not a square matrix. So, Cramer's Rule is a useful alternative to Gaussian elimination, but only in some situations.

Most graphing calculators can evaluate determinants.

7.6 EXERCISES

1–8 ■ Find the determinant of the matrix, if it exists.

1. $[3]$

2. $[0]$

3. $\begin{bmatrix} 4 & 5 \\ 0 & -1 \end{bmatrix}$

4. $\begin{bmatrix} -2 & 1 \\ 3 & -2 \end{bmatrix}$

5. $[2 \quad 5]$

6. $\begin{bmatrix} 3 \\ 0 \end{bmatrix}$

7. $\begin{bmatrix} \frac{1}{2} & \frac{1}{8} \\ 1 & \frac{1}{2} \end{bmatrix}$

8. $\begin{bmatrix} 2.2 & -1.4 \\ 0.5 & 1.0 \end{bmatrix}$

9–14 ■ Evaluate the minor and cofactor using the matrix A.

$$A = \begin{bmatrix} 1 & 0 & \frac{1}{2} \\ -3 & 5 & 2 \\ 0 & 0 & 4 \end{bmatrix}$$

9. M_{11}, A_{11}

10. M_{33}, A_{33}

11. M_{12}, A_{12}

12. M_{13}, A_{13}

13. M_{23}, A_{23}

14. M_{32}, A_{32}

15–20 ■ Find the determinant of the matrix. Determine whether the matrix has an inverse, but don't calculate the inverse.

15. $\begin{bmatrix} 1 & 3 & 7 \\ 2 & 0 & -1 \\ 0 & 2 & 6 \end{bmatrix}$

16. $\begin{bmatrix} -2 & -\frac{3}{2} & \frac{1}{2} \\ 2 & 4 & 0 \\ \frac{1}{2} & 2 & 1 \end{bmatrix}$

17. $\begin{bmatrix} 30 & 0 & 20 \\ 0 & -10 & -20 \\ 40 & 0 & 10 \end{bmatrix}$

18. $\begin{bmatrix} 1 & 2 & 5 \\ -2 & -3 & 2 \\ 3 & 5 & 3 \end{bmatrix}$

19. $\begin{bmatrix} 1 & 3 & 3 & 0 \\ 0 & 2 & 0 & 1 \\ -1 & 0 & 0 & 2 \\ 1 & 6 & 4 & 1 \end{bmatrix}$

20. $\begin{bmatrix} 1 & 2 & 0 & 2 \\ 3 & -4 & 0 & 4 \\ 0 & 1 & 6 & 0 \\ 1 & 0 & 2 & 0 \end{bmatrix}$

21–24 ■ Evaluate the determinant, using row or column operations whenever possible to simplify your work.

21. $\begin{vmatrix} 0 & 0 & 4 & 6 \\ 2 & 1 & 1 & 3 \\ 2 & 1 & 2 & 3 \\ 3 & 0 & 1 & 7 \end{vmatrix}$

22. $\begin{vmatrix} -2 & 3 & -1 & 7 \\ 4 & 6 & -2 & 3 \\ 7 & 7 & 0 & 5 \\ 3 & -12 & 4 & 0 \end{vmatrix}$

23. $\begin{vmatrix} 1 & 2 & 3 & 4 & 5 \\ 0 & 2 & 4 & 6 & 8 \\ 0 & 0 & 3 & 6 & 9 \\ 0 & 0 & 0 & 4 & 8 \\ 0 & 0 & 0 & 0 & 5 \end{vmatrix}$

24. $\begin{vmatrix} 2 & -1 & 6 & 4 \\ 7 & 2 & -2 & 5 \\ 4 & -2 & 10 & 8 \\ 6 & 1 & 1 & 4 \end{vmatrix}$

25. Let

$$B = \begin{bmatrix} 4 & 1 & 0 \\ -2 & -1 & 1 \\ 4 & 0 & 3 \end{bmatrix}$$

(a) Evaluate $|B|$ by expanding by the second row.
(b) Evaluate $|B|$ by expanding by the third column.
(c) Do your results in parts (a) and (b) agree?

26. Consider the system

$$\begin{cases} x + 2y + 6z = 5 \\ -3x - 6y + 5z = 8 \\ 2x + 6y + 9z = 7 \end{cases}$$

(a) Verify that $x = -1$, $y = 0$, $z = 1$ is a solution of the system.
(b) Find the determinant of the coefficient matrix.
(c) Without solving the system, determine whether there are any other solutions.
(d) Can Cramer's Rule be used to solve this system? Why or why not?

27–42 ■ Use Cramer's Rule to solve the system.

27. $\begin{cases} 2x - y = -9 \\ x + 2y = 8 \end{cases}$

28. $\begin{cases} 6x + 12y = 33 \\ 4x + 7y = 20 \end{cases}$

29. $\begin{cases} x - 6y = 3 \\ 3x + 2y = 1 \end{cases}$

30. $\begin{cases} \frac{1}{2}x + \frac{1}{3}y = 1 \\ \frac{1}{4}x - \frac{1}{6}y = -\frac{3}{2} \end{cases}$

31. $\begin{cases} 0.4x + 1.2y = 0.4 \\ 1.2x + 1.6y = 3.2 \end{cases}$

32. $\begin{cases} 10x - 17y = 21 \\ 20x - 31y = 39 \end{cases}$

33. $\begin{cases} x - y + 2z = 0 \\ 3x + z = 11 \\ -x + 2y = 0 \end{cases}$

34. $\begin{cases} 5x - 3y + z = 6 \\ 4y - 6z = 22 \\ 7x + 10y = -13 \end{cases}$

35. $\begin{cases} 2x_1 + 3x_2 - 5x_3 = 1 \\ x_1 + x_2 - x_3 = 2 \\ 2x_2 + x_3 = 8 \end{cases}$

36. $\begin{cases} -2a + c = 2 \\ a + 2b - c = 9 \\ 3a + 5b + 2c = 22 \end{cases}$

37. $\begin{cases} \frac{1}{3}x - \frac{1}{5}y + \frac{1}{2}z = \frac{7}{10} \\ -\frac{2}{3}x + \frac{2}{5}y + \frac{3}{2}z = \frac{11}{10} \\ x - \frac{4}{5}y + z = \frac{9}{5} \end{cases}$

38. $\begin{cases} 2x - y = 5 \\ 5x + 3z = 19 \\ 4y + 7z = 17 \end{cases}$

39. $\begin{cases} 3y + 5z = 4 \\ 2x - z = 10 \\ 4x + 7y = 0 \end{cases}$

40. $\begin{cases} 2x - 5y = 4 \\ x + y - z = 8 \\ 3x + 5z = 0 \end{cases}$

41. $\begin{cases} x + y + z + w = 0 \\ 2x + w = 0 \\ y - z = 0 \\ x + 2z = 1 \end{cases}$

42. $\begin{cases} x + y = 1 \\ y + z = 2 \\ z + w = 3 \\ w - x = 4 \end{cases}$

43. (a) Show that the equation

$$\begin{vmatrix} x_1 & y_1 & 1 \\ x_2 & y_2 & 1 \\ x & y & 1 \end{vmatrix} = 0$$

is an equation for the line that passes through the points (x_1, y_1) and (x_2, y_2).

(b) Use the result of part (a) to find an equation for the line that passes through the points $(20, 50)$ and $(-10, 25)$.

44. Evaluate the following determinant.

$$\begin{vmatrix} a & a & a & a & a \\ 0 & a & a & a & a \\ 0 & 0 & a & a & a \\ 0 & 0 & 0 & a & a \\ 0 & 0 & 0 & 0 & a \end{vmatrix}$$

45–48 ■ Solve for x.

45. $\begin{vmatrix} x & 12 & 13 \\ 0 & x - 1 & 23 \\ 0 & 0 & x - 2 \end{vmatrix} = 0$

46. $\begin{vmatrix} x & 1 & 1 \\ 1 & 1 & x \\ x & 1 & x \end{vmatrix} = 0$

47. $\begin{vmatrix} 1 & 0 & x \\ x^2 & 1 & 0 \\ x & 0 & 1 \end{vmatrix} = 0$

48. $\begin{vmatrix} a & b & x - a \\ x & x + b & x \\ 0 & 1 & 1 \end{vmatrix} = 0$

DISCOVERY • DISCUSSION

49. Matrices with Determinant Zero Use the definition of the determinant and the elementary row and column operations to explain why matrices of the following types have determinant 0.

(a) A matrix with a row or column consisting entirely of zeros

(b) A matrix with two rows the same or two columns the same

(c) A matrix in which one row is a multiple of another row, or one column is a multiple of another column

50. Solving Linear Systems Suppose you had to solve a linear system with five equations and five variables without the assistance of a calculator or computer. Which method would you prefer: Cramer's Rule or Gaussian elimination? Write a short paragraph explaining the reasons for your answer.

7.7 SYSTEMS OF INEQUALITIES

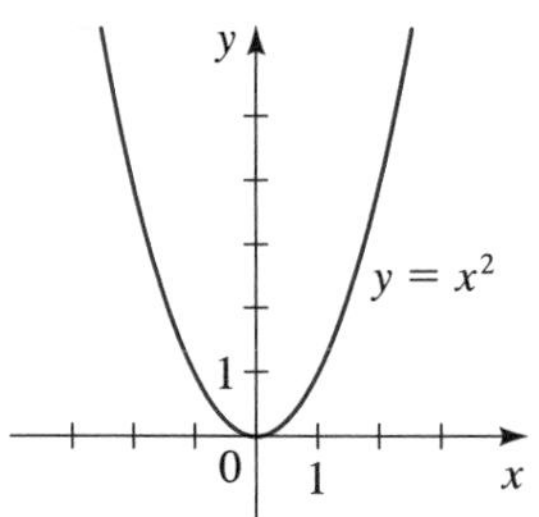

FIGURE 1

In this section we study systems of inequalities in two variables from a graphical point of view. First we consider the graph of a single inequality. We already know that the graph of $y = x^2$, for example, is the *parabola* in Figure 1. If we replace the equal sign by the symbol $\geqslant$, we obtain the *inequality*

$$y \geqslant x^2$$

Its graph consists of not just the parabola in Figure 1, but also every point whose y-coordinate is *larger* than x^2. We indicate the solution in Figure 2 by shading the points *above* the parabola.

Similarly, the graph of $y \leqslant x^2$ in Figure 3 consists of all points on and *below* the parabola, whereas the graphs of $y > x^2$ and $y < x^2$ don't include the points on the parabola itself, as indicated by the dashed curves in Figures 4 and 5.

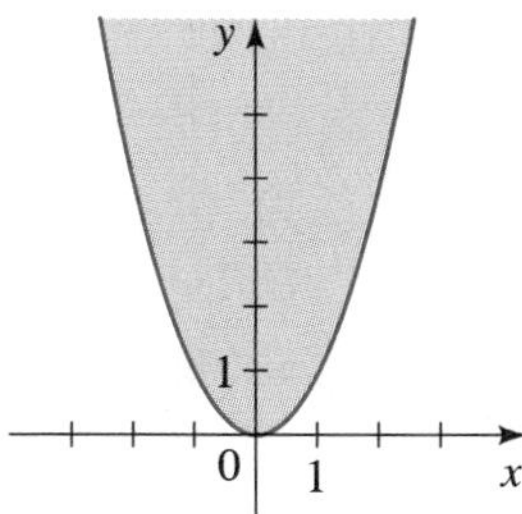

FIGURE 2
$y \geqslant x^2$

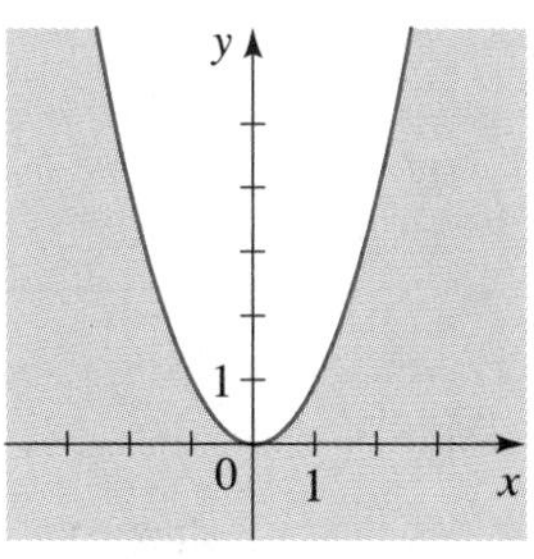

FIGURE 3
$y \leqslant x^2$

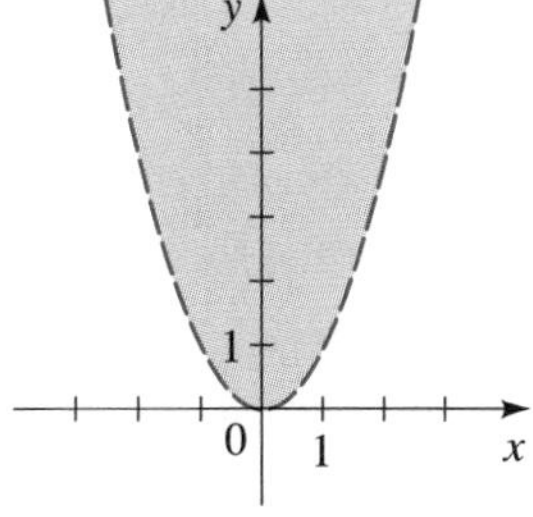

FIGURE 4
$y > x^2$

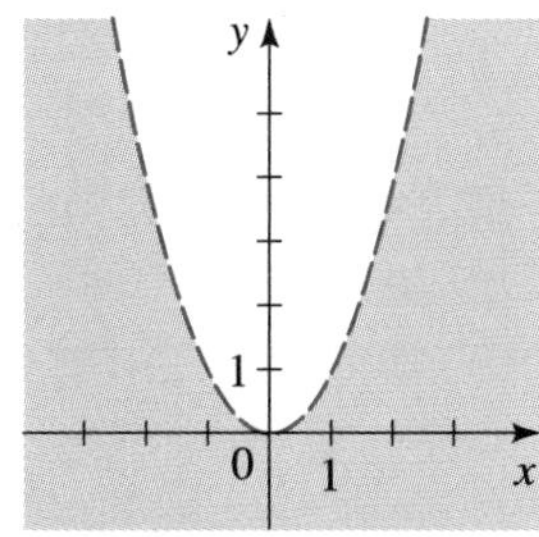

FIGURE 5
$y < x^2$

The graph of an inequality, in general, consists of a region in the plane whose boundary is the graph of the equation obtained by replacing the inequality sign ($\geqslant$, $\leqslant$, $>$, or $<$) with an equal sign. To determine which side of the graph gives the solution set of the inequality, we need only check **test points**.

> **GRAPHING INEQUALITIES**
>
> To graph an inequality, we carry out the following steps.
>
> 1. GRAPH EQUATION. Graph the equation corresponding to the inequality. Use a dashed curve for $>$ or $<$, or a solid curve for $\leq$ or $\geq$.
>
> 2. TEST POINTS. Test one point in each of the regions formed by the graph in Step 1. If the point satisfies the inequality, then all the points in that region satisfy the inequality. (In that case, shade the region to indicate it is part of the graph.) If the test point does not satisfy the inequality, then the region isn't part of the graph.

EXAMPLE 1 ■ Graphs of Nonlinear Inequalities

Graph each inequality.

(a) $x^2 + y^2 < 25$ (b) $x + 2y \geq 5$

SOLUTION

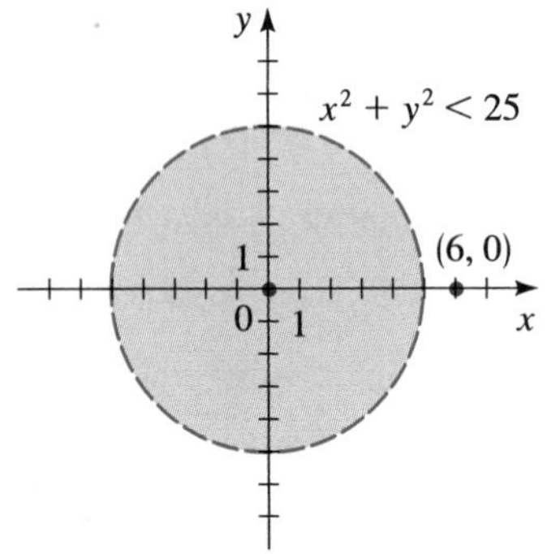

FIGURE 6

(a) The graph of $x^2 + y^2 = 25$ is a circle of radius 5 centered at the origin. The points on the circle itself do not satisfy the inequality because it is of the form $<$, so we graph the circle with a dashed curve, as shown in Figure 6.

To determine whether the inside or the outside of the circle satisfies the inequality, we use the test points $(0, 0)$ on the inside and $(6, 0)$ on the outside. To do this, we substitute the coordinates of each point into the inequality and determine whether the result satisfies the inequality. (Note that *any* point inside or outside the circle can serve as a test point. We have chosen these points for simplicity.)

Test point	$x^2 + y^2 < 25$	Conclusion
(0, 0)	$0^2 + 0^2 = 0 < 25$	Part of graph
(6, 0)	$6^2 + 0^2 = 36 \not< 25$	Not part of graph

Thus the graph of $x^2 + y^2 < 25$ is the set of all points *inside* the circle (see Figure 6).

(b) The graph of $x + 2y = 5$ is the line shown in Figure 7. We use the test points $(0, 0)$ and $(5, 5)$ on opposite sides of the line.

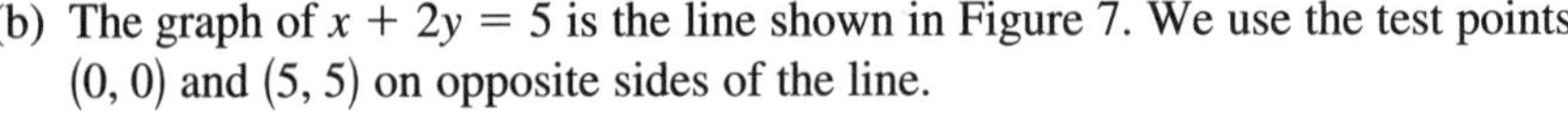

Test point	$x + 2y \geq 5$	Conclusion
(0, 0)	$0 + 2(0) = 0 \not\geq 5$	Not part of graph
(5, 5)	$5 + 2(5) = 15 \geq 5$	Part of graph

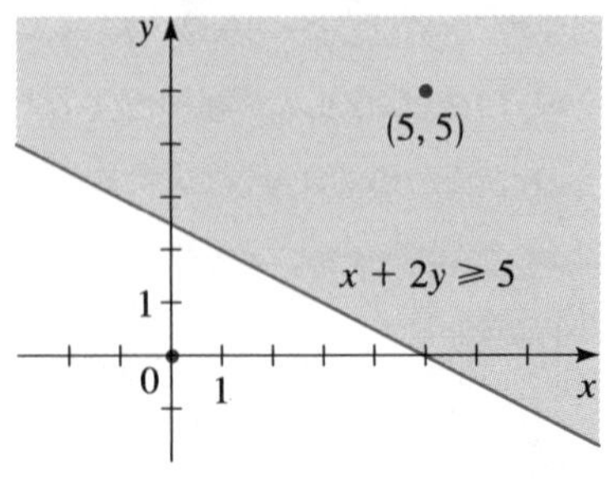

FIGURE 7

Our check shows that the points *above* the line satisfy the inequality.

Alternatively, we could put the inequality into slope-intercept form and graph it directly:

$$x + 2y \geqslant 5$$
$$2y \geqslant -x + 5$$
$$y \geqslant -\tfrac{1}{2}x + \tfrac{5}{2}$$

From this form we see that the graph includes all points whose y-coordinates are *greater* than those on the line $y = -\frac{1}{2}x + \frac{5}{2}$; that is, the graph consists of the points *on or above* this line, as shown in Figure 7. ■

EXAMPLE 2 ■ A System of Two Inequalities

Graph the solution set of the system of inequalities.

$$\begin{cases} x^2 + y^2 < 25 \\ x + 2y \geqslant 5 \end{cases}$$

SOLUTION

These are the two inequalities of Example 1. In this example we wish to graph only those points that simultaneously satisfy both inequalities. The solution consists of the intersection of the graphs in Example 1. In Figure 8(a) we show the two regions on the same coordinate plane (in different colors), and in Figure 8(b) we show their intersection.

Vertices: The points $(-3, 4)$ and $(5, 0)$ in Figure 8(b) are the **vertices** of the solution set. They are obtained by solving the system of *equations*

$$\begin{cases} x^2 + y^2 = 25 \\ x + 2y = 5 \end{cases}$$

We solve this system of equations by substitution. Solving for x in the second equation gives $x = 5 - 2y$, and substituting this into the first equation gives

$$(5 - 2y)^2 + y^2 = 25 \qquad \text{Substitute } x = 5 - 2y$$
$$(25 - 20y + 4y^2) + y^2 = 25 \qquad \text{Expand}$$
$$-20y + 5y^2 = 0 \qquad \text{Simplify}$$
$$-5y(4 - y) = 0 \qquad \text{Factor}$$

Thus, $y = 0$ or $y = 4$. When $y = 0$, we have $x = 5 - 2(0) = 5$, and when $y = 4$, we have $x = 5 - 2(4) = -3$. So the points of intersection of these curves are $(5, 0)$ and $(-3, 4)$.

Note that in this case the vertices are not part of the solution set, since they don't satisfy the inequality $x^2 + y^2 < 25$ (and so they are graphed as open circles in the figure). They simply show where the "corners" of the solution set lie. ■

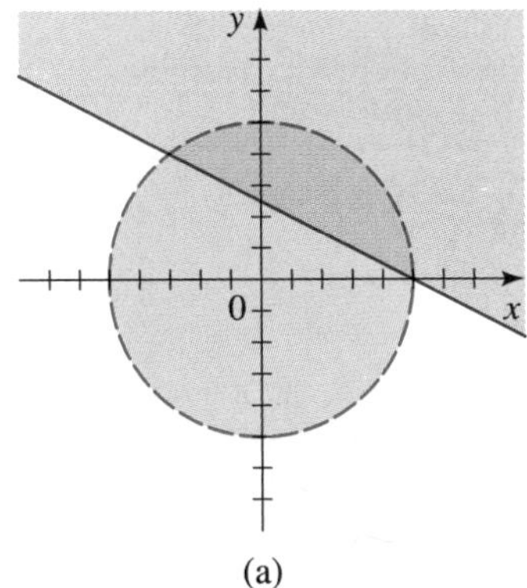

(a)

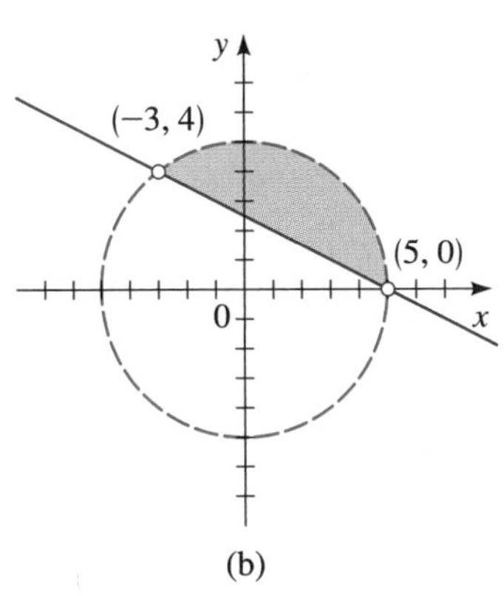

(b)

FIGURE 8
$\begin{cases} x^2 + y^2 < 25 \\ x + 2y \geqslant 5 \end{cases}$

An inequality is **linear** if it can be put into one of the following forms:

$$ax + by \geq c \qquad ax + by \leq c \qquad ax + by > c \qquad ax + by < c$$

In the next example we graph the solution set of a system of linear inequalities.

EXAMPLE 3 ■ A System of Four Linear Inequalities

Graph the solution set of the system, and label its vertices.

$$\begin{cases} x + 3y \leq 12 \\ x + y \leq 8 \\ x \geq 0 \\ y \geq 0 \end{cases}$$

SOLUTION

In Figure 9 we first graph the lines given by the equations that correspond to each inequality. To determine the graphs of the linear inequalities, we only need to check one test point. For simplicity let's use the point (0, 0).

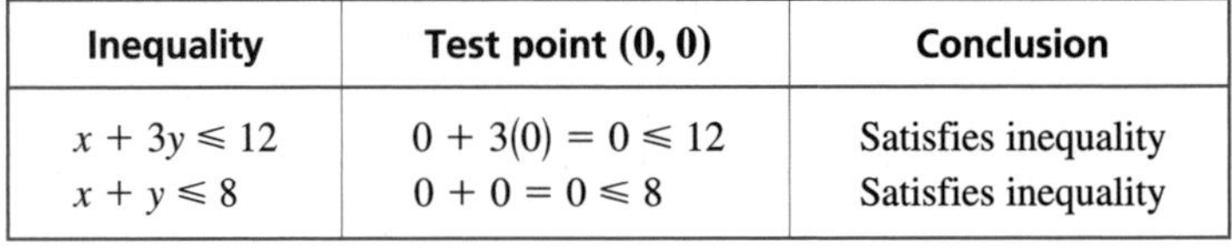

Inequality	Test point (0, 0)	Conclusion
$x + 3y \leq 12$	$0 + 3(0) = 0 \leq 12$	Satisfies inequality
$x + y \leq 8$	$0 + 0 = 0 \leq 8$	Satisfies inequality

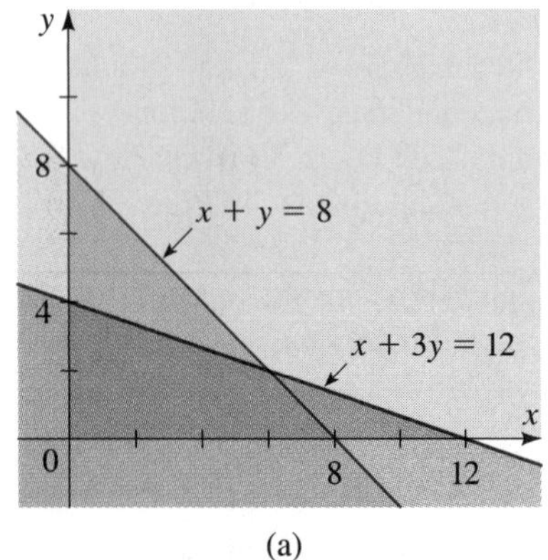

(a)

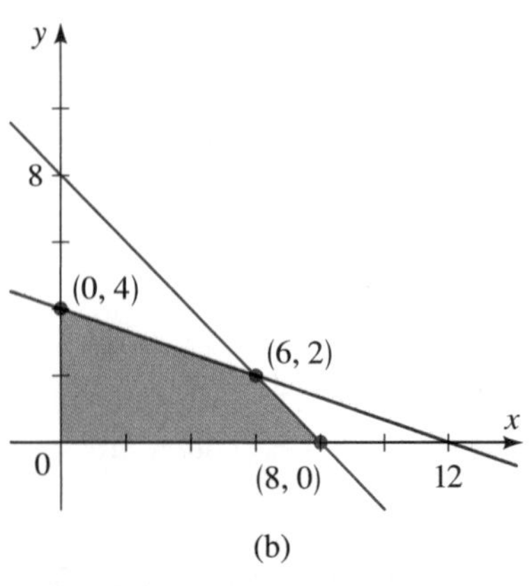

(b)

FIGURE 9

Since (0, 0) is below the line $x + 3y = 12$, our check shows that the region on or below the line must satisfy the inequality. Likewise, since (0, 0) is below the line $x + y = 8$, our check shows that the region on or below this line must satisfy the inequality. The inequalities $x \geq 0$ and $y \geq 0$ say that x and y are nonnegative. These regions are sketched in Figure 9(a), and the intersection—the solution set—is sketched in Figure 9(b).

Vertices: The coordinates of each vertex are obtained by simultaneously solving the equations of the lines that intersect at that vertex. From the system

$$\begin{cases} x + 3y = 12 \\ x + y = 8 \end{cases}$$

we get the vertex (6, 2). The other two vertices are at the x- and y-intercepts of the corresponding lines: (8, 0) and (0, 4). In this case, all the vertices *are* part of the solution set. ■

When a region in the plane can be covered by a (sufficiently large) circle, it's said to be **bounded**. A region that is not bounded is called **unbounded**. For example, the regions graphed in Figure 6, 8(b), and 9(b) are bounded, whereas those in Figures 2–5 and 7 are unbounded. An unbounded region cannot be "fenced in"—it extends infinitely far in at least one direction.

7.7 EXERCISES

1–12 ■ Graph the inequality.

1. $x \leqslant 2$

2. $y > -3$

3. $y > x$

4. $y < x + 2$

5. $y \leqslant 2x + 2$

6. $y < -x + 5$

7. $2x - y \leqslant 8$

8. $3x + 4y + 12 > 0$

9. $4x + 5y < 25$

10. $-x^2 + y \geqslant 10$

11. $y > x^2 + 1$

12. $x^2 + y^2 \geqslant 5$

13–34 ■ Graph the solution of the system of inequalities. In each case, find the coordinates of all vertices, and determine whether the solution set is bounded.

13. $\begin{cases} x + y \leqslant 4 \\ y \geqslant x \end{cases}$

14. $\begin{cases} 2x + 3y > 12 \\ 3x - y < 21 \end{cases}$

15. $\begin{cases} y < \frac{1}{4}x + 2 \\ y \geqslant 2x - 5 \end{cases}$

16. $\begin{cases} x - y > 0 \\ 4 + y \leqslant 2x \end{cases}$

17. $\begin{cases} x \geqslant 0 \\ y \geqslant 0 \\ 3x + 5y \leqslant 15 \\ 3x + 2y \leqslant 9 \end{cases}$

18. $\begin{cases} x > 2 \\ y < 12 \\ 2x - 4y > 8 \end{cases}$

19. $\begin{cases} y < 9 - x^2 \\ y \geqslant x + 3 \end{cases}$

20. $\begin{cases} y \geqslant x^2 \\ x + y \geqslant 6 \end{cases}$

21. $\begin{cases} x^2 + y^2 \leqslant 4 \\ x - y > 0 \end{cases}$

22. $\begin{cases} x > 0 \\ y > 0 \\ x + y < 10 \\ x^2 + y^2 > 9 \end{cases}$

23. $\begin{cases} x^2 - y \leqslant 0 \\ 2x^2 + y \leqslant 12 \end{cases}$

24. $\begin{cases} x^2 + y^2 < 9 \\ 2x + y^2 \geqslant 1 \end{cases}$

25. $\begin{cases} x + 2y \leqslant 14 \\ 3x - y \geqslant 0 \\ x - y \geqslant 2 \end{cases}$

26. $\begin{cases} y < x + 6 \\ 3x + 2y \geqslant 12 \\ x - 2y \leqslant 2 \end{cases}$

27. $\begin{cases} x \geqslant 0 \\ y \geqslant 0 \\ x \leqslant 5 \\ x + y \leqslant 7 \end{cases}$

28. $\begin{cases} x \geqslant 0 \\ y \geqslant 0 \\ y \leqslant 4 \\ 2x + y \leqslant 8 \end{cases}$

29. $\begin{cases} y > x + 1 \\ x + 2y \leqslant 12 \\ x + 1 > 0 \end{cases}$

30. $\begin{cases} x + y > 12 \\ y < \frac{1}{2}x - 6 \\ 3x + y < 6 \end{cases}$

31. $\begin{cases} x^2 + y^2 \leqslant 8 \\ x \geqslant 2 \\ y \geqslant 0 \end{cases}$

32. $\begin{cases} x^2 - y \geqslant 0 \\ x + y < 6 \\ x - y < 6 \end{cases}$

33. $\begin{cases} x^2 + y^2 < 9 \\ x + y > 0 \\ x \leqslant 0 \end{cases}$

34. $\begin{cases} y \geqslant x^3 \\ y \leqslant 2x + 4 \\ x + y \geqslant 0 \end{cases}$

35. A publishing company publishes a total of no more than 100 books every year. At least 20 of these are nonfiction, but the company always publishes at least as much fiction as nonfiction. Find a system of inequalities that describes the possible numbers of fiction and nonfiction books that the company can produce each year consistent with these policies. Graph the solution set.

36. A man and his daughter manufacture unfinished tables and chairs. Each table requires 3 hours of sawing and 1 hour of assembly. Each chair requires 2 hours of sawing and 2 hours of assembly. The two of them can put in up to 12 hours of sawing and 8 hours of assembly work each day. Find a system of inequalities that describes all possible combinations of tables and chairs that they can make daily. Graph the solution set.

DISCOVERY • DISCUSSION

37. Shading Unwanted Regions To graph the solution of a system of inequalities, we have shaded the solution of each inequality in a different color; the solution of the system is the region where all the shaded parts overlap. Here is a different method: for each inequality, shade the region that does *not* satisfy the inequality. Explain why the part of the plane that is left unshaded is the solution of the system. Solve the following system by both methods. Which do you prefer?

$$\begin{cases} x + 2y > 4 \\ -x + y < 1 \\ x + 3y < 9 \\ x < 3 \end{cases}$$

7.8 PARTIAL FRACTIONS

To write a sum or difference of fractional expressions as a single fraction, we bring them to a common denominator. For example,

$$\frac{1}{x-1} + \frac{1}{2x+1} = \frac{(2x+1)+(x-1)}{(x-1)(2x+1)} = \frac{3x}{2x^2-x-1}$$

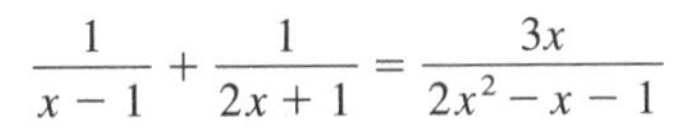

Common denominator
$\frac{1}{x-1} + \frac{1}{2x+1} = \frac{3x}{2x^2-x-1}$
Partial fractions

But for some applications of algebra to calculus, we must reverse this process—that is, we must express a fraction such as $3x/(2x^2 - x - 1)$ as the sum of the simpler fractions $1/(x-1)$ and $1/(2x+1)$. These simpler fractions are called *partial fractions*; we learn how to find them in this section.

Let r be the rational function

$$r(x) = \frac{P(x)}{Q(x)}$$

where the degree of P is less than the degree of Q. By the Linear and Quadratic Factors Theorem in Section 5.4, every polynomial with real coefficients can be factored completely into linear and irreducible quadratic factors, that is, factors of the form $ax + b$ and $ax^2 + bx + c$, where a, b, and c are real numbers. For instance,

$$x^4 - 1 = (x^2-1)(x^2+1) = (x-1)(x+1)(x^2+1)$$

After we have completely factored the denominator Q of r, we can express $r(x)$ as a sum of **partial fractions** of the form

$$\frac{A}{(ax+b)^i} \quad \text{and} \quad \frac{Ax+B}{(ax^2+bx+c)^j}$$

This sum is called the **partial fraction decomposition** of r. Let's examine the details of the four possible cases.

CASE 1: The Denominator Is a Product of Distinct Linear Factors

This means that we can write

$$Q(x) = (a_1x+b_1)(a_2x+b_2)\cdots(a_nx+b_n)$$

with no factor repeated. In this case the partial fraction decomposition of r takes the form

$$r(x) = \frac{P(x)}{Q(x)} = \frac{A_1}{a_1x+b_1} + \frac{A_2}{a_2x+b_2} + \cdots + \frac{A_n}{a_nx+b_n}$$

where the constants, $A_1, A_2, \ldots, A_n$ are determined as in the following example.

The Rhind papyrus is the oldest known mathematical document. It is an Egyptian scroll written in 1650 B.C. by the scribe Ahmes, who explains that it is an exact copy of a scroll written 200 years earlier. Ahmes claims that his papyrus contains "a thorough study of all things, insight into all that exists, knowledge of all obscure secrets." Actually, the document contains rules for doing arithmetic, including multiplication and division of fractions and several exercises with solutions. The exercise shown below reads: A heap and its seventh make nineteen; how large is the heap? In solving problems of this sort, the Egyptians used partial fractions because their number system required all fractions to be written as sums of reciprocals of whole numbers. For example, $\frac{7}{12}$ would be written as $\frac{1}{3} + \frac{1}{4}$.

The papyrus gives a correct formula for the volume of a truncated pyramid (page 72). It also gives the formula $A = \left(\frac{8}{9}d\right)^2$ for the area of a circle with diameter d. How close is this to the actual area?

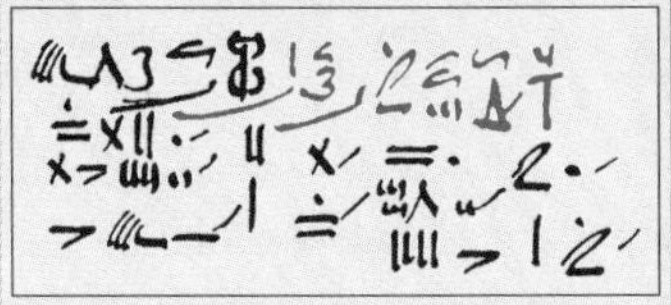

EXAMPLE 1 ■ Distinct Linear Factors

Find the partial fraction decomposition of $\dfrac{5x+7}{x^3+2x^2-x-2}$.

SOLUTION

The denominator factors as follows:

$$\begin{aligned} x^3+2x^2-x-2 &= x^2(x+2)-(x+2) = (x^2-1)(x+2) \\ &= (x-1)(x+1)(x+2) \end{aligned}$$

This gives us the partial fraction decomposition

$$\frac{5x+7}{x^3+2x^2-x-2} = \frac{A}{x-1} + \frac{B}{x+1} + \frac{C}{x+2}$$

Multiplying each side by the common denominator, $(x-1)(x+1)(x+2)$, we get

$$\begin{aligned} 5x+7 &= A(x+1)(x+2) + B(x-1)(x+2) + C(x-1)(x+1) \\ &= A(x^2+3x+2) + B(x^2+x-2) + C(x^2-1) && \text{Expand} \\ &= (A+B+C)x^2 + (3A+B)x + (2A-2B-C) && \text{Combine like terms} \end{aligned}$$

If two polynomials are equal, then their coefficients are equal. Thus, since $5x+7$ has no x^2-term, we have $A+B+C=0$. Similarly, by comparing the coefficients of x, we see that $3A+B=5$, and by comparing constant terms, we get $2A-2B-C=7$. This leads to the following system of linear equations for A, B, and C.

$$\begin{cases} A + B + C = 0 \\ 3A + B = 5 \\ 2A - 2B - C = 7 \end{cases}$$

We use the methods developed in Section 7.3 to solve this system.

$$\begin{bmatrix} 1 & 1 & 1 & 0 \\ 3 & 1 & 0 & 5 \\ 2 & -2 & -1 & 7 \end{bmatrix} \xrightarrow[\mathrm{R}_3 - 2\mathrm{R}_1 \to \mathrm{R}_3]{\mathrm{R}_2 - 3\mathrm{R}_1 \to \mathrm{R}_2} \begin{bmatrix} 1 & 1 & 1 & 0 \\ 0 & -2 & -3 & 5 \\ 0 & -4 & -3 & 7 \end{bmatrix} \xrightarrow[-\mathrm{R}_2]{\mathrm{R}_3 - 2\mathrm{R}_2 \to \mathrm{R}_3}$$

$$\begin{bmatrix} 1 & 1 & 1 & 0 \\ 0 & 2 & 3 & -5 \\ 0 & 0 & 3 & -3 \end{bmatrix} \xrightarrow[\frac{1}{3}\mathrm{R}_3]{\mathrm{R}_2 - \mathrm{R}_3 \to \mathrm{R}_2} \begin{bmatrix} 1 & 1 & 1 & 0 \\ 0 & 2 & 0 & -2 \\ 0 & 0 & 1 & -1 \end{bmatrix}$$

Thus, $C = -1$, $B = -1$, and $A = 2$, so the required partial fraction decomposition is

$$\frac{5x + 7}{x^3 + 2x^2 - x - 2} = \frac{2}{x - 1} + \frac{-1}{x + 1} + \frac{-1}{x + 2}$$

■

The same method of attack works in the remaining cases. We set up the partial fraction decomposition with the unknown constants, $A, B, C, \ldots$. Then we multiply each side of the resulting equation by the common denominator, simplify the right-hand side of the equation, and equate coefficients. This gives a set of linear equations that will always have a unique solution (provided that the partial fraction decomposition has been set up correctly).

CASE 2: The Denominator Is a Product of Linear Factors, Some of Which Are Repeated

Suppose the complete factorization of $Q(x)$ contains the linear factor $ax + b$ repeated k times; that is, $(ax + b)^k$ is a factor of $Q(x)$. Then, corresponding to each such factor, the partial fraction decomposition for $P(x)/Q(x)$ contains

$$\frac{A_1}{ax + b} + \frac{A_2}{(ax + b)^2} + \cdots + \frac{A_k}{(ax + b)^k}$$

EXAMPLE 2 ■ Repeated Linear Factors

Find the partial fraction decomposition of $\dfrac{x^2 + 1}{x(x - 1)^3}$.

SOLUTION

Because the factor $x - 1$ is repeated three times in the denominator, the partial fraction decomposition has the form

$$\frac{x^2 + 1}{x(x - 1)^3} = \frac{A}{x} + \frac{B}{x - 1} + \frac{C}{(x - 1)^2} + \frac{D}{(x - 1)^3}$$

Multiplying each side by the common denominator, $x(x - 1)^3$, gives

$$\begin{aligned} x^2 + 1 &= A(x - 1)^3 + Bx(x - 1)^2 + Cx(x - 1) + Dx \\ &= A(x^3 - 3x^2 + 3x - 1) + B(x^3 - 2x^2 + x) + C(x^2 - x) + Dx && \text{Expand} \\ &= (A + B)x^3 + (-3A - 2B + C)x^2 + (3A + B - C + D)x - A && \text{Combine like terms} \end{aligned}$$

Equating coefficients, we get the equations

$$\begin{cases} A + B = 0 \\ -3A - 2B + C = 1 \\ 3A + B - C + D = 0 \\ -A = 1 \end{cases}$$

If we rearrange these equations by putting the last one in the first position, we can easily see (using substitution) that the solution to the system is $A = -1$, $B = 1$, $C = 0$, $D = 2$, and so the partial fraction decomposition is

$$\frac{x^2+1}{x(x-1)^3} = \frac{-1}{x} + \frac{1}{x-1} + \frac{2}{(x-1)^3}$$

■

CASE 3: The Denominator Has Irreducible Quadratic Factors, None of Which Is Repeated

If the complete factorization of the denominator $Q(x)$ contains the quadratic factor $ax^2 + bx + c$ (which can't be factored further), then corresponding to this the partial fraction decomposition of $P(x)/Q(x)$ will have a term of the form

$$\frac{Ax+B}{ax^2+bx+c}$$

EXAMPLE 3 ■ Distinct Quadratic Factors

Find the partial fraction decomposition of $\dfrac{2x^2-x+4}{x^3+4x}$.

SOLUTION

Since $x^3 + 4x = x(x^2 + 4)$, which can't be factored further, we write

$$\frac{2x^2-x+4}{x^3+4x} = \frac{A}{x} + \frac{Bx+C}{x^2+4}$$

Multiplying by $x(x^2 + 4)$, we get

$$\begin{aligned} 2x^2 - x + 4 &= A(x^2+4) + (Bx+C)x \\ &= (A+B)x^2 + Cx + 4A \end{aligned}$$

Equating coefficients gives us the equations

$$\begin{cases} A + B = 2 \\ \quad\;\; C = -1 \\ \quad 4A = 4 \end{cases}$$

and so $A = 1$, $B = 1$, and $C = -1$. The required partial fraction decomposition is

$$\frac{2x^2-x+4}{x^3+4x} = \frac{1}{x} + \frac{x-1}{x^2+4}$$

■

CASE 4: The Denominator Has a Repeated Irreducible Quadratic Factor

If the complete factorization of $Q(x)$ contains the factor $(ax^2 + bx + c)^k$, where $ax^2 + bx + c$ can't be factored further, then the partial fraction decomposition of $P(x)/Q(x)$ will have the terms

$$\frac{A_1x + B_1}{ax^2 + bx + c} + \frac{A_2x + B_2}{(ax^2 + bx + c)^2} + \cdots + \frac{A_kx + B_k}{(ax^2 + bx + c)^k}$$

EXAMPLE 4 ■ Repeated Quadratic Factors

Write the form of the partial fraction decomposition of

$$\frac{x^5 - 3x^2 + 12x - 1}{x^3(x^2 + x + 1)(x^2 + 2)^3}$$

SOLUTION

$$\frac{x^5 - 3x^2 + 12x - 1}{x^3(x^2 + x + 1)(x^2 + 2)^3}$$

$$= \frac{A}{x} + \frac{B}{x^2} + \frac{C}{x^3} + \frac{Dx + E}{x^2 + x + 1} + \frac{Fx + G}{x^2 + 2} + \frac{Hx + I}{(x^2 + 2)^2} + \frac{Jx + K}{(x^2 + 2)^3}$$ ■

To find the values of A, B, C, D, E, F, G, H, I, J, and K in Example 4, we would have to solve a system of 11 linear equations. Although possible, this would certainly involve a great deal of work!

The techniques we have described in this section apply only to rational functions $P(x)/Q(x)$ in which the degree of P is less than the degree of Q. If this isn't the case, we must first use long division to divide Q into P.

EXAMPLE 5 ■ Using Long Division to Prepare for Partial Fractions

Find the partial fraction decomposition of

$$\frac{2x^4 + 4x^3 - 2x^2 + x + 7}{x^3 + 2x^2 - x - 2}$$

SOLUTION

Since the degree of the numerator is larger than the degree of the denominator, we use long division to obtain

$$\begin{array}{r} 2x \\ x^3 + 2x^2 - x - 2 \overline{\smash{)}\, 2x^4 + 4x^3 - 2x^2 + x + 7} \\ \underline{2x^4 + 4x^3 - 2x^2 - 4x } \\ 5x + 7 \end{array}$$

$$\frac{2x^4 + 4x^3 - 2x^2 + x + 7}{x^3 + 2x^2 - x - 2} = 2x + \frac{5x + 7}{x^3 + 2x^2 - x - 2}$$

The remainder term now satisfies the requirement that the degree of the numerator is less than the degree of the denominator. At this point we proceed as in Example 1 to obtain the decomposition

$$\frac{2x^4 + 4x^3 - 2x^2 + x + 7}{x^3 + 2x^2 - x - 2} = 2x + \frac{2}{x - 1} + \frac{-1}{x + 1} + \frac{-1}{x + 2}$$ ■

7.8 EXERCISES

1–10 ■ Write the form of the partial fraction decomposition of the function (as in Example 4). Do not determine the numerical values of the coefficients.

1. $\dfrac{1}{(x - 1)(x + 2)}$

2. $\dfrac{x}{x^2 + 3x - 4}$

3. $\dfrac{x^2 - 3x + 5}{(x - 2)^2(x + 4)}$

4. $\dfrac{1}{x^4 - x^3}$

5. $\dfrac{x^2}{(x - 3)(x^2 + 4)}$

6. $\dfrac{1}{x^4 - 1}$

7. $\dfrac{x^3 - 4x^2 + 2}{(x^2 + 1)(x^2 + 2)}$

8. $\dfrac{x^4 + x^2 + 1}{x^2(x^2 + 4)^2}$

9. $\dfrac{x^3 + x + 1}{x(2x - 5)^3(x^2 + 2x + 5)^2}$

10. $\dfrac{1}{(x^6 - 1)(x^4 - 1)}$

11–40 ■ Find the partial fraction decomposition of the rational function.

11. $\dfrac{5}{(x - 1)(x + 4)}$

12. $\dfrac{x + 6}{x(x + 3)}$

13. $\dfrac{12}{x^2 - 9}$

14. $\dfrac{x - 12}{x^2 - 4x}$

15. $\dfrac{4}{x^2 - 4}$

16. $\dfrac{2x + 1}{x^2 + x - 2}$

17. $\dfrac{x + 14}{x^2 - 2x - 8}$

18. $\dfrac{8x - 3}{2x^2 - x}$

19. $\dfrac{x}{8x^2 - 10x + 3}$

20. $\dfrac{7x - 3}{x^3 + 2x^2 - 3x}$

21. $\dfrac{9x^2 - 9x + 6}{2x^3 - x^2 - 8x + 4}$

22. $\dfrac{-3x^2 - 3x + 27}{(x + 2)(2x^2 + 3x - 9)}$

23. $\dfrac{x^2 + 1}{x^3 + x^2}$

24. $\dfrac{3x^2 + 5x - 13}{(3x + 2)(x^2 - 4x + 4)}$

25. $\dfrac{2x}{4x^2 + 12x + 9}$

26. $\dfrac{x - 4}{(2x - 5)^2}$

27. $\dfrac{4x^2 - x - 2}{x^4 + 2x^3}$

28. $\dfrac{x^3 - 2x^2 - 4x + 3}{x^4}$

29. $\dfrac{-10x^2 + 27x - 14}{(x - 1)^3(x + 2)}$

30. $\dfrac{-2x^2 + 5x - 1}{x^4 - 2x^3 + 2x - 1}$

31. $\dfrac{3x^3 + 22x^2 + 53x + 41}{(x + 2)^2(x + 3)^2}$

32. $\dfrac{3x^2 + 12x - 20}{x^4 - 8x^2 + 16}$

33. $\dfrac{x - 3}{x^3 + 3x}$

34. $\dfrac{3x^2 - 2x + 8}{x^3 - x^2 + 2x - 2}$

35. $\dfrac{2x^3 + 7x + 5}{(x^2 + x + 2)(x^2 + 1)}$

36. $\dfrac{x^2 + x + 1}{2x^4 + 3x^2 + 1}$

37. $\dfrac{x^4 + x^3 + x^2 - x + 1}{x(x^2 + 1)^2}$

38. $\dfrac{2x^2 - x + 8}{(x^2 + 4)^2}$

39. $\dfrac{x^5 - 2x^4 + x^3 + x + 5}{x^3 - 2x^2 + x - 2}$

40. $\dfrac{x^5 - 3x^4 + 3x^3 - 4x^2 + 4x + 12}{(x - 2)^2(x^2 + 2)}$

41. Determine A and B in terms of a and b:

$$\frac{ax + b}{x^2 - 1} = \frac{A}{x - 1} + \frac{B}{x + 1}$$

42. Determine A, B, C, and D in terms of a and b:

$$\frac{ax^3 + bx^2}{(x^2 + 1)^2} = \frac{Ax + B}{x^2 + 1} + \frac{Cx + D}{(x^2 + 1)^2}$$

DISCOVERY • DISCUSSION

43. Recognizing Partial Fraction Decompositions For each expression, determine whether it is already a partial fraction decomposition, or whether it can be decomposed further.

(a) $\dfrac{x}{x^2+1}+\dfrac{1}{x+1}$ (b) $\dfrac{x}{(x+1)^2}$

(c) $\dfrac{1}{x+1}+\dfrac{2}{(x+1)^2}$ (d) $\dfrac{x+2}{(x^2+1)^2}$

44. Assembling and Disassembling Partial Fractions The following expression is a partial fraction decomposition:

$$\frac{2}{x-1}+\frac{1}{(x-1)^2}+\frac{1}{x+1}$$

Use a common denominator to combine the terms into one fraction. Then use the techniques of this section to find its partial fraction decomposition. Did you get back the original expression?

7 REVIEW

CONCEPT CHECK

1. Suppose you are asked to solve a system of two equations (not necessarily linear) in two variables. Explain how you would solve them
   (a) by the substitution method
   (b) by the elimination method
   (c) graphically

2. Suppose you are asked to solve a system of two *linear* equations in two variables.
   (a) Would you prefer to use substitution or elimination?
   (b) How many solutions are possible? Draw diagrams to illustrate the possibilities.

3. Explain how Gaussian elimination works. Your explanation should include a discussion of matrices, elementary row operations, echelon form, back-substitution, and leading variables.

4. (a) What is meant by an inconsistent system?
   (b) What is meant by a dependent system? What is a leading variable?

5. Suppose you have used Gaussian elimination to transform a matrix that represents a system of linear equations into echelon form. How can you tell if the system has
   (a) no solution?
   (b) exactly one solution?
   (c) infinitely many solutions?

6. How can you tell if a matrix is in reduced echelon form? What is the advantage of using a matrix in reduced echelon form?

7. (a) What does it mean to say that A is a matrix with dimension $m \times n$?
   (b) If A and B have the same dimension and k is a real number, how do you find $A + B$, $A - B$, and kA?

8. (a) Under what conditions is it possible to find the product AB of two matrices?
   (b) If those conditions are satisfied, how do you calculate the product AB?

9. (a) What is the identity matrix I_n?
   (b) If A is a square $n \times n$ matrix, what is its inverse matrix?
   (c) Write a formula for the inverse of a 2×2 matrix.
   (d) Explain how you would find the inverse of a 3×3 matrix.

10. Suppose A is an $n \times n$ matrix.
    (a) What is the minor M_{ij} of the element a_{ij}?
    (b) What is the cofactor A_{ij}?
    (c) How do you find the determinant of A?
    (d) How can you tell if A has an inverse?

11. State Cramer's Rule for solving a system of linear equations in terms of determinants. Do you prefer to use Cramer's Rule or Gaussian elimination? Explain.

12. Explain the procedure for solving a system of inequalities.

13. Explain how to find the partial fraction decomposition of a fractional expression. Your explanation should discuss each of the four cases that arise.

EXERCISES

1–6 ■ Solve the system of equations and graph the lines.

1. $\begin{cases} 3x - y = 5 \\ 2x + y = 5 \end{cases}$

2. $\begin{cases} y = 2x + 6 \\ y = -x + 3 \end{cases}$

3. $\begin{cases} 2x - 7y = 28 \\ y = \frac{2}{7}x - 4 \end{cases}$

4. $\begin{cases} 6x - 8y = 15 \\ -\frac{3}{2}x + 2y = -4 \end{cases}$

5. $\begin{cases} 2x - y = 1 \\ x + 3y = 10 \\ 3x + 4y = 15 \end{cases}$

6. $\begin{cases} 2x + 5y = 9 \\ -x + 3y = 1 \\ 7x - 2y = 14 \end{cases}$

7–10 ■ Solve the system of equations.

7. $\begin{cases} y = x^2 + 2x \\ y = 6 + x \end{cases}$

8. $\begin{cases} x^2 + y^2 = 8 \\ y = x + 2 \end{cases}$

9. $\begin{cases} 3x + \dfrac{4}{y} = 6 \\ x - \dfrac{8}{y} = 4 \end{cases}$

10. $\begin{cases} x^2 + y^2 = 10 \\ x^2 + 2y^2 - 7y = 0 \end{cases}$

11–18 ■ Find the complete solution of the system using Gaussian elimination, or show that the system has no solution.

11. $\begin{cases} x + y + 2z = 6 \\ 2x + 5z = 12 \\ x + 2y + 3z = 9 \end{cases}$

12. $\begin{cases} x - 2y + 3z = 1 \\ x - 3y - z = 0 \\ 2x - 6z = 6 \end{cases}$

13. $\begin{cases} x - 2y + 3z = 1 \\ 2x - y + z = 3 \\ 2x - 7y + 11z = 2 \end{cases}$

14. $\begin{cases} x + y + z + w = 2 \\ 2x - 3z = 5 \\ x - 2y + 4w = 9 \\ x + y + 2z + 3w = 5 \end{cases}$

15. $\begin{cases} x - 3y + z = 4 \\ 4x - y + 15z = 5 \end{cases}$

16. $\begin{cases} 2x - 3y + 4z = 3 \\ 4x - 5y + 9z = 13 \\ 2x + 7z = 0 \end{cases}$

17. $\begin{cases} -x + 4y + z = 8 \\ 2x - 6y + z = -9 \\ x - 6y - 4z = -15 \end{cases}$

18. $\begin{cases} x - z + w = 2 \\ 2x + y - 2w = 12 \\ 3y + z + w = 4 \\ x + y - z = 10 \end{cases}$

19. A man invests his savings in two accounts, one paying 6% interest per year and the other paying 7%. He has twice as much invested in the 7% account as in the 6% account, and his annual interest income is \$600. How much is invested in each account?

20. Find the values of a, b, and c if the parabola

$$y = ax^2 + bx + c$$

passes through the points $(1, 0)$, $(-1, -4)$, and $(2, 11)$.

21–32 ■ Let

$$A = \begin{bmatrix} 2 & 0 & -1 \end{bmatrix} \qquad B = \begin{bmatrix} 1 & 2 & 4 \\ -2 & 1 & 0 \end{bmatrix}$$

$$C = \begin{bmatrix} \frac{1}{2} & 3 \\ 2 & \frac{3}{2} \\ -2 & 1 \end{bmatrix} \qquad D = \begin{bmatrix} 1 & 4 \\ 0 & -1 \\ 2 & 0 \end{bmatrix}$$

$$E = \begin{bmatrix} 2 & -1 \\ -\frac{1}{2} & 1 \end{bmatrix} \qquad F = \begin{bmatrix} 4 & 0 & 2 \\ -1 & 1 & 0 \\ 7 & 5 & 0 \end{bmatrix}$$

$$G = \begin{bmatrix} 5 \end{bmatrix}$$

Carry out the indicated operation, or explain why it cannot be performed.

21. $A + B$ **22.** $C - D$ **23.** $2C + 3D$

24. $5B - 2C$ **25.** GA **26.** AG

27. BC **28.** CB **29.** BF

30. FC **31.** $(C + D)E$ **32.** $F(2C - D)$

33–38 ■ Find the determinant and, if possible, the inverse of the matrix.

33. $\begin{bmatrix} 1 & 4 \\ 2 & 9 \end{bmatrix}$

34. $\begin{bmatrix} 2 & 2 \\ 1 & -3 \end{bmatrix}$

35. $\begin{bmatrix} 4 & -12 \\ -2 & 6 \end{bmatrix}$

36. $\begin{bmatrix} 2 & 4 & 0 \\ -1 & 1 & 2 \\ 0 & 3 & 2 \end{bmatrix}$

37. $\begin{bmatrix} 3 & 0 & 1 \\ 2 & -3 & 0 \\ 4 & -2 & 1 \end{bmatrix}$

38. $\begin{bmatrix} 1 & 0 & 0 & 1 \\ 0 & 2 & 0 & 2 \\ 0 & 0 & 3 & 3 \\ 0 & 0 & 0 & 4 \end{bmatrix}$

39–40 ■ Express the system of linear equations as a matrix equation. Then solve the matrix equation by multiplying each side by the inverse of the coefficient matrix.

39. $\begin{cases} 12x - 5y = 10 \\ 5x - 2y = 17 \end{cases}$

40. $\begin{cases} 2x + y + 5z = \frac{1}{3} \\ x + 2y + 2z = \frac{1}{4} \\ x + 3z = \frac{1}{6} \end{cases}$

41–44 ■ Solve the system using Cramer's Rule.

41. $\begin{cases} 2x + 7y = 13 \\ 6x + 16y = 30 \end{cases}$

42. $\begin{cases} 12x - 11y = 140 \\ 7x + 9y = 20 \end{cases}$

43. $\begin{cases} 2x - y + 5z = 0 \\ -x + 7y = 9 \\ 5x + 4y + 3z = -9 \end{cases}$

44. $\begin{cases} 3x + 4y - z = 10 \\ x - 4z = 20 \\ 2x + y + 5z = 30 \end{cases}$

45–48 ■ Graph the solution set of the system of inequalities. Find the coordinates of all vertices, and determine whether the solution set is bounded or unbounded.

45. $\begin{cases} x^2 + y^2 < 9 \\ x + y < 0 \end{cases}$

46. $\begin{cases} y - x^2 \geqslant 4 \\ y < 20 \end{cases}$

47. $\begin{cases} x \geqslant 0, \quad y \geqslant 0 \\ x + 2y \leqslant 12 \\ y \leqslant x + 4 \end{cases}$

48. $\begin{cases} x \geqslant 4 \\ x + y \geqslant 24 \\ x \leqslant 2y + 12 \end{cases}$

49–50 ■ Solve for x, y, and z in terms of a, b, and c.

49. $\begin{cases} -x + y + z = a \\ x - y + z = b \\ x + y - z = c \end{cases}$

50. $\begin{cases} ax + by + cz = a - b + c \\ bx + by + cz = c \\ cx + cy + cz = c \end{cases}$ $\quad (a \neq b, b \neq c, c \neq 0)$

51. For what values of k do the following three lines have a common point of intersection?

$$x + y = 12$$
$$kx - y = 0$$
$$y - x = 2k$$

52. For what value of k does the following system have infinitely many solutions?

$$\begin{cases} kx + y + z = 0 \\ x + 2y + kz = 0 \\ -x + 2y + 3z = 0 \end{cases}$$

53–56 ■ Use a graphing device to solve the system, correct to the nearest hundredth.

53. $\begin{cases} 0.32x + 0.43y = 0 \\ 7x - 12y = 341 \end{cases}$

54. $\begin{cases} \sqrt{12}\,x - 3\sqrt{2}\,y = 660 \\ 7137x + 3931y = 20{,}000 \end{cases}$

55. $\begin{cases} x - y^2 = 10 \\ x = \frac{1}{22}y + 12 \end{cases}$

56. $\begin{cases} y = 5^x + x \\ y = x^5 + 5 \end{cases}$

57–60 ■ Find the partial fraction decomposition of the rational function.

57. $\dfrac{3x + 1}{x^2 - 2x - 15}$

58. $\dfrac{8}{x^3 - 4x}$

59. $\dfrac{2x - 4}{x(x - 1)^2}$

60. $\dfrac{x + 6}{x^3 - 2x^2 + 4x - 8}$

7 TEST

1. In $2\frac{1}{2}$ hours an airplane travels 600 km against the wind. It takes 50 min to travel 300 km with the wind. Find the speed of the wind and the speed of the airplane in still air.

2–5 ■ Find all solutions of the system. Determine whether the system is linear or nonlinear. If it is linear, state whether it is inconsistent, dependent, or neither.

2. $\begin{cases} 3x - y = 10 \\ 2x + 5y = 1 \end{cases}$

3. $\begin{cases} x - y + 9z = -8 \\ x \qquad - 4z = 7 \\ 3x - y + z = 5 \end{cases}$

4. $\begin{cases} 2x - y + z = 0 \\ 3x + 2y - 3z = 1 \\ x - 4y + 5z = -1 \end{cases}$

5. $\begin{cases} 2x^2 + y^2 = 6 \\ 3x^2 - 4y = 11 \end{cases}$

6–13 ■ Let

$$A = \begin{bmatrix} 2 & 3 \\ 2 & 4 \end{bmatrix} \qquad B = \begin{bmatrix} 2 & 4 \\ -1 & 1 \\ 3 & 0 \end{bmatrix} \qquad C = \begin{bmatrix} 1 & 0 & 4 \\ -1 & 1 & 2 \\ 0 & 1 & 3 \end{bmatrix}$$

Carry out the indicated operation, or explain why it cannot be performed.

6. $A + B$
7. AB
8. $BA - 3B$
9. CBA
10. A^{-1}
11. B^{-1}
12. $|B|$
13. $|C|$

14. Write a matrix equation equivalent to the following system.

$$\begin{cases} 4x - 3y = 10 \\ 3x - 2y = 30 \end{cases}$$

Find the inverse of the coefficient matrix, and use it to solve the system.

15. Solve using Cramer's Rule:

$$\begin{cases} 2x \qquad - z = 14 \\ 3x - y + 5z = 0 \\ 4x + 2y + 3z = -2 \end{cases}$$

16. Only one of the following matrices has an inverse. Find the determinant of each matrix, and use the determinants to identify the one that has an inverse. Then find the inverse.

$$A = \begin{bmatrix} 1 & 4 & 1 \\ 0 & 2 & 0 \\ 1 & 0 & 1 \end{bmatrix} \qquad B = \begin{bmatrix} 1 & 4 & 0 \\ 0 & 2 & 0 \\ -3 & 0 & 1 \end{bmatrix}$$

17. Graph the following system of inequalities, and state the coordinates of the vertices.

$$\begin{cases} x^2 - y + 5 \leqslant 0 \\ y \leqslant 5 + 2x \end{cases}$$

18. Find the partial fraction decomposition of the following rational function.

$$\frac{4x - 1}{(x - 1)^2(x + 2)}$$

19. Using a graphing device to find all solutions of the following system, correct to two decimal places.

$$\begin{cases} 2x^2 + y^2 = 16 \\ y = x^4 - 4x^3 + 6x^2 - 4x \end{cases}$$

FOCUS ON MODELING

Linear programming is a modeling technique used to determine the optimal allocation of resources in business, the military, and other areas of human endeavor. For example, a manufacturer who makes several different products from the same raw materials can use linear programming to determine how much of each product should be produced to maximize the profit. This modeling technique is probably the most important practical application of systems of linear inequalities. In 1975 Leonid Kantorovich and T. C. Koopmans won the Nobel Prize in economics for their work in the development of this technique.

Although linear programming can be applied to very complex problems with hundreds or even thousands of variables, we consider only a few simple examples to which the graphical methods of Section 7.7 can be applied. (For large numbers of variables, a linear programming method based on matrices is used.) Let's examine a typical problem.

EXAMPLE 1 ■ Manufacturing for Maximum Profit

A small shoe manufacturer makes two styles of shoes: oxfords and loafers. Two machines are used in the process: a cutting machine and a sewing machine. Each type of shoe requires 15 min per pair on the cutting machine. Oxfords require 10 min of sewing per pair, and loafers require 20 min of sewing per pair. Because the manufacturer can hire only one operator for each machine, each process is available for just 8 hours per day. If the profit is \$15 on each pair of oxfords and \$20 on each pair of loafers, how many pairs of each type should be produced per day for maximum profit?

Because loafers produce more profit, it would seem best to manufacture only loafers. Surprisingly, this does not turn out to be the most profitable solution.

SOLUTION

First we organize the given information into a table. To be consistent, let's convert all times to hours.

	Oxfords	**Loafers**	**Time available**
Time on cutting machine (h)	$\frac{1}{4}$	$\frac{1}{4}$	8
Time on sewing machine (h)	$\frac{1}{6}$	$\frac{1}{3}$	8
Profit	\$15	\$20	

We describe the model and solve the problem in four steps.

Linear programming helps the telephone industry determine the most efficient way to route telephone calls. The computerized routing decisions must be made very rapidly so callers are not kept waiting for connections. Since the database of customers and routes is huge, an extremely fast method for solving linear programming problems is essential. In 1984 the 28-year-old mathematician **Narendra Karmarkar**, working at Bell Labs in Murray Hill, New Jersey, discovered just such a method. His idea is so ingenious and his method so fast that the discovery caused a sensation in the mathematical world. Although mathematical discoveries rarely make the news, this one was reported in *Time*, on December 3, 1984. Today airlines routinely use Karmarkar's technique to minimize costs in scheduling passengers, flight personnel, fuel, baggage, and maintenance workers.

CHOOSING THE VARIABLES To make a mathematical model, we first give names to the variable quantities. For this problem we let

$$x = \text{number of pairs of oxfords made daily.}$$

$$y = \text{number of pairs of loafers made daily}$$

FINDING THE OBJECTIVE FUNCTION The objective of this problem is to determine which values for x and y give maximum profit. Since each pair of oxfords provides \$15 profit and each pair of loafers \$20, the total profit is given by

$$P = 15x + 20y$$

This function is called the *objective function.*

GRAPHING THE FEASIBLE REGION The larger x and y are, the greater the profit. But we cannot choose arbitrarily large values for these variables, because of the restrictions, or *constraints*, in the problem. Each restriction is an inequality in the variables.

In this problem the total number of cutting hours needed is $\frac{1}{4}x + \frac{1}{4}y$. Since only 8 hours are available on the cutting machine, we have

$$\tfrac{1}{4}x + \tfrac{1}{4}y \leqslant 8$$

Similarly, by considering the amount of time needed and available on the sewing machine, we get

$$\tfrac{1}{6}x + \tfrac{1}{3}y \leqslant 8$$

We cannot produce a negative number of shoes, so we also have

$$x \geqslant 0 \qquad \text{and} \qquad y \geqslant 0$$

Thus, x and y must satisfy the constraints

$$\begin{cases} \tfrac{1}{4}x + \tfrac{1}{4}y \leqslant 8 \\ \tfrac{1}{6}x + \tfrac{1}{3}y \leqslant 8 \\ x \geqslant 0 \\ y \geqslant 0 \end{cases}$$

If we multiply the first inequality by 4 and the second by 6, we obtain the simplified system

$$\begin{cases} x + y \leqslant 32 \\ x + 2y \leqslant 48 \\ x \geqslant 0 \\ y \geqslant 0 \end{cases}$$

The solution of this sytem (with vertices labeled) is sketched in Figure 1. The only values that satisfy the restrictions of the problem are the ones that correspond to points of the shaded region in Figure 1. This is called the *feasible region* for the problem.

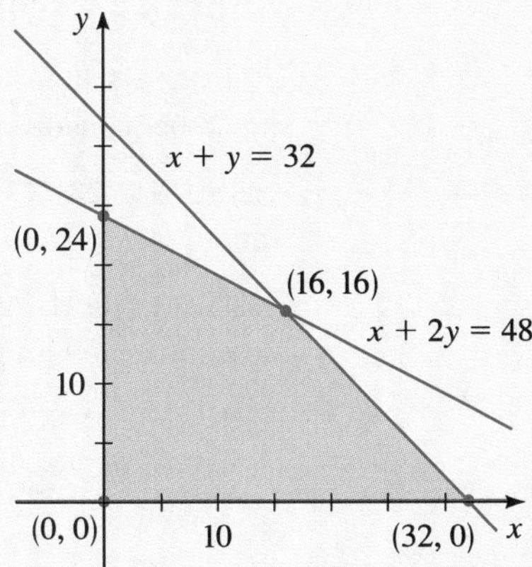

FIGURE 1

FINDING MAXIMUM PROFIT As x or y increases, profit will increase as well. Thus, it seems reasonable that the maximum profit will occur at a point on one of the outside edges of the feasible region, where it's impossible to increase x or y without going outside the region. In fact, it can be shown that the maximum value occurs at a vertex. This means that we need to check the profit only at the vertices.

Vertex	$P = 15x + 20y$	
(0, 0)	0	
(0, 24)	15(0) + 20(24) = \$480	
(16, 16)	15(16) + 20(16) = \$560	← maximum profit
(32, 0)	15(32) + 20(0) = \$480	

The largest value of P occurs at the point $(16, 16)$, where $P = \$560$. Thus, the manufacturer should make 16 pairs of oxfords and 16 pairs of loafers, for a maximum daily profit of \$560. ■

The linear programming problems that we consider all follow the pattern of Example 1. Each problem involves two variables. The problem describes restrictions, called **constraints**, that lead to a system of linear inequalities whose solution is called the **feasible region**. The function we wish to maximize or minimize is called the **objective function**. This function always attains its largest and smallest values at the **vertices** of the feasible region. This modeling technique involves four steps, summarized in the following box.

GUIDELINES FOR LINEAR PROGRAMMING

1. CHOOSE THE VARIABLES. Decide what variable quantities in the problem should be named x and y.

2. FIND THE OBJECTIVE FUNCTION. Write an expression for the function we want to maximize or minimize.

3. GRAPH THE FEASIBLE REGION. Express the constraints as a system of inequalities and graph the solution of this system (the feasible region).

4. FIND THE MAXIMUM OR MINIMUM. Evaluate the objective function at the vertices of the feasible region to determine its maximum or minimum value.

EXAMPLE 2 ■ A Shipping Problem

A car dealer has warehouses in Millville and Trenton and dealerships in Camden and Atlantic City. Every car sold at the dealerships must be delivered from one of the warehouses. On a certain day the Camden dealers sell 10 cars, and the Atlantic City dealers sell 12. The Millville warehouse has 15 cars available, and the Trenton warehouse has 10. The cost of shipping one car is \$50 from Millville to Camden, \$40 from Millville to Atlantic City, \$60 from Trenton to Camden, and \$55 from Trenton to Atlantic City. How many cars should be moved from each warehouse to each dealership to fill the orders at minimum cost?

SOLUTION

Our first step is to organize the given information. Rather than construct a table, we draw a diagram to show the flow of cars from the warehouses to the dealerships (see Figure 2). The diagram shows the number of cars available at each warehouse or required at each dealership and the cost of shipping between these locations.

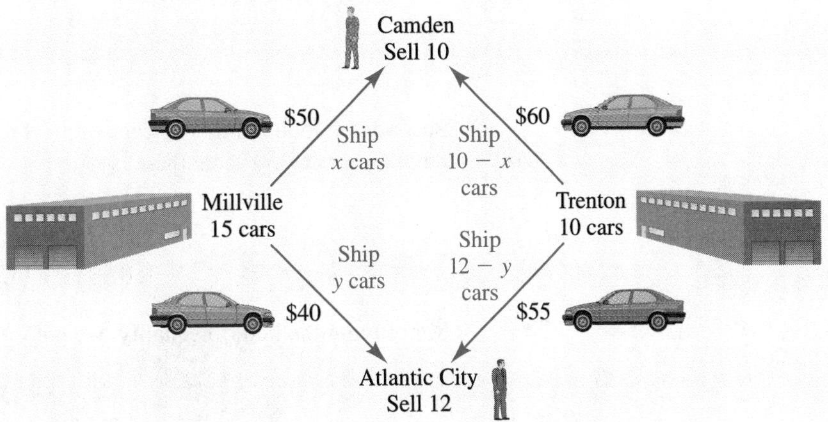

FIGURE 2

CHOOSING THE VARIABLES The arrows in Figure 2 indicate four possible routes, so the problem seems to involve four variables. But we let

x = number of cars to be shipped from Millville to Camden

y = number of cars to be shipped from Millville to Atlantic City

To fill the orders, we must have

$10 - x$ = number of cars shipped from Trenton to Camden

$12 - y$ = number of cars shipped from Trenton to Atlantic City

So the only variables in the problem are x and y.

FINDING THE OBJECTIVE FUNCTION The objective of this problem is to minimize cost. From Figure 2 we see that the total cost of shipping the cars is

$$\begin{aligned} C &= 50x + 40y + 60(10 - x) + 55(12 - y) \\ &= 50x + 40y + 600 - 60x + 660 - 55y \\ &= 1260 - 10x - 15y \end{aligned}$$

This is the objective function.

GRAPHING THE FEASIBLE REGION Now we derive the constraint inequalities that define the feasible region. First, the number of cars shipped on each route can't be negative, so we have

$$x \geqslant 0 \qquad\qquad y \geqslant 0$$

$$10 - x \geqslant 0 \qquad\qquad 12 - y \geqslant 0$$

Second, the total number of cars shipped from each warehouse can't exceed the number of cars available there, so

$$x + y \leqslant 15$$

$$(10 - x) + (12 - y) \leqslant 10$$

Simplifying the latter inequality, we get

$$22 - x - y \leqslant 10$$

$$-x - y \leqslant -12$$

$$x + y \geqslant 12$$

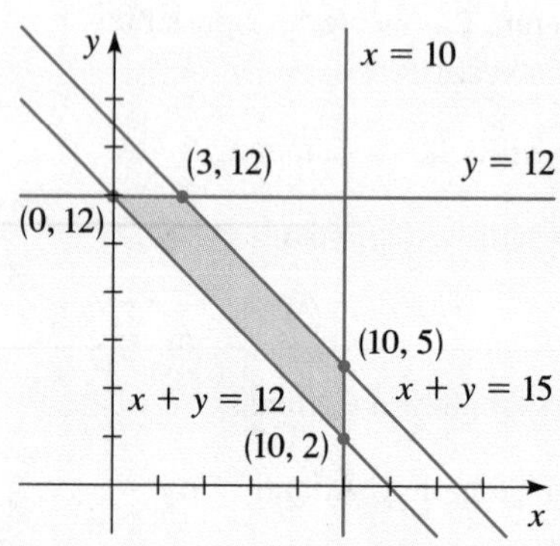

FIGURE 3

The inequalities $10 - x \geqslant 0$ and $12 - y \geqslant 0$ can be rewritten as $x \leqslant 10$ and $y \leqslant 12$, respectively. Thus, the feasible region is described by the constraints

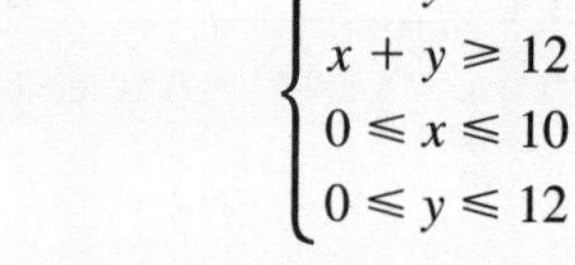

$$\begin{cases} x + y \leqslant 15 \\ x + y \geqslant 12 \\ 0 \leqslant x \leqslant 10 \\ 0 \leqslant y \leqslant 12 \end{cases}$$

The feasible region is graphed in Figure 3.

FINDING MINIMUM COST We check the value of the objective function at each vertex of the feasible region.

Vertex	$C = 1260 - 10x - 15y$	
(0, 12)	$1260 - 10(0) - 15(12) = \$1080$	
(3, 12)	$1260 - 10(3) - 15(12) = \$1050$	← minimum cost
(10, 5)	$1260 - 10(10) - 15(5) = \$1085$	
(10, 2)	$1260 - 10(10) - 15(2) = \$1130$	

The lowest cost is incurred at the point (3, 12). Thus, the dealer should ship

3 cars from Millville to Camden
12 cars from Millville to Atlantic City
7 cars from Trenton to Camden
0 cars from Trenton to Atlantic City ■

In the 1940s mathematicians developed matrix methods for solving linear programming problems that involve more than two variables. These methods were first used by the Allies in World War II to solve supply problems similar to (but, of course, much more complicated than) Example 2. Improving these matrix methods is an active and exciting area of current mathematical research.

PROBLEMS

1–4 ■ Find the maximum and minimum values of the given objective function on the indicated feasible region.

1. $M = 200 - x - y$

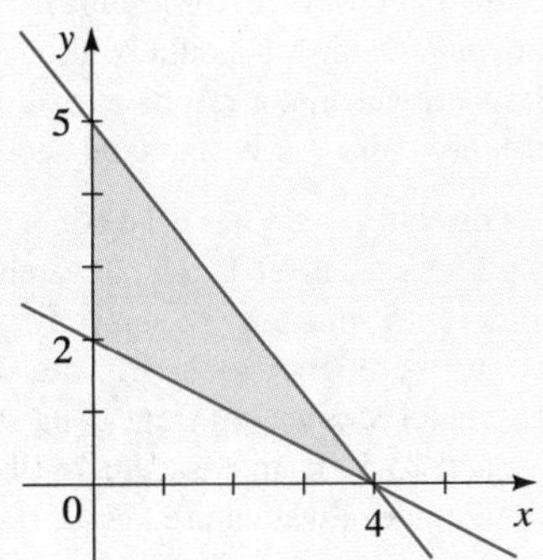

2. $N = \frac{1}{2}x + \frac{1}{4}y + 40$

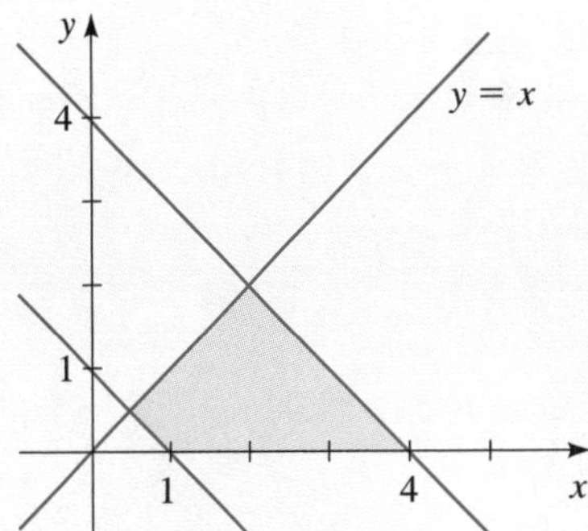

3. $P = 140 - x + 3y$

$$\begin{cases} x \geqslant 0, \quad y \geqslant 0 \\ 2x + y \leqslant 10 \\ 2x + 4y \leqslant 28 \end{cases}$$

4. $Q = 70x + 82y$

$$\begin{cases} x \geqslant 0, \quad y \geqslant 0 \\ x \leqslant 10, \quad y \leqslant 20 \\ x + y \geqslant 5 \\ x + 2y \leqslant 18 \end{cases}$$

5. A furniture manufacturer makes wooden tables and chairs. The production process involves two basic types of labor: carpentry and finishing. A table requires 2 hours of carpentry and 1 hour of finishing, whereas a chair requires 3 hours of carpentry and $\frac{1}{2}$ hour of finishing. The profit is \$35 per table and \$20 per chair. The manufacturer's employees can supply a maximum of 108 hours of carpentry work and 20 hours of finishing work per day. How many tables and chairs should be made each day to maximize profit?

6. A housing contractor has subdivided a farm into 100 building lots. He has designed two types of homes for these lots: colonial and ranch style. A colonial requires $30,000 of capital and produces a profit of $4000 when sold. A ranch-style house requires $40,000 of capital and provides an $8000 profit. If he has $3.6 million of capital on hand, how many houses of each type should he build for maximum profit? Will any of the lots be left vacant?

7. A trucker hauls citrus fruit from Florida to Montreal. Each crate of oranges is 4 ft^3 in volume and weighs 80 lb. Each crate of grapefruit has a volume of 6 ft^3 and weighs 100 lb. Her truck has a maximum capacity of 300 ft^3 and can carry no more than 5600 lb. Moreover, she is not permitted to carry more crates of grapefruit than crates of oranges. If she makes a profit of $2.50 on each crate of oranges and $4 on each crate of grapefruit, how many crates of each fruit should she carry for maximum profit?

8. A manufacturer of calculators produces two models: standard and scientific. Long-term demand for the two models mandates that the company manufacture at least 100 standard and 80 scientific calculators each day. However, because of limitations on production capacity, no more than 200 standard and 170 scientific calculators can be made daily. To satisfy a shipping contract, a total of at least 200 calculators must be shipped every day.
   (a) If the production cost is $5 for a standard calculator and $7 for a scientific one, how many of each model should be produced daily to minimize this cost?
   (b) If each standard calculator results in a $2 loss but each scientific one produces a $5 profit, how many of each model should be made daily to maximize profit?

9. An electronics discount chain has a sale on a certain brand of stereo. The chain has stores in Santa Monica and El Toro and warehouses in Long Beach and Pasadena. To satisfy rush orders, 15 sets must be shipped from the warehouses to the Santa Monica store, and 19 must be shipped to the El Toro store. The cost of shipping a set is $5 from Long Beach to Santa Monica, $6 from Long Beach to El Toro, $4 from Pasadena to Santa Monica, and $5.50 from Pasadena to El Toro. If the Long Beach warehouse has 24 sets and the Pasadena warehouse has 18 sets in stock, how many sets should be shipped from each warehouse to each store to fill the orders at a minimum shipping cost?

10. A man owns two building supply stores, one on the east side and one on the west side of a city. Two customers order some $\frac{1}{2}$-inch plywood. Customer A needs 50 sheets and customer B needs 70 sheets. The east-side store has 80 sheets and the west-side store has 45 sheets of this plywood in stock. Delivery cost per sheet is $0.50 from the east-side store to customer A, $0.60 from the east-side store to customer B, $0.40 from the west-side store to A, and $0.55 from the west-side store to B. How many sheets should be shipped from each store to each customer to minimize delivery costs?

11. A confectioner sells two types of nut mixtures. The standard-mixture package contains 100 g of cashews and 200 g of peanuts and sells for $1.95. The deluxe-mixture package contains 150 g of cashews and 50 g of peanuts and sells for $2.25. The confectioner has 15 kg of cashews and 20 kg of peanuts available. Based on past sales, he needs to have

at least as many standard as deluxe packages available. How many bags of each mixture should he package to maximize his revenue?

12. A biologist wishes to feed laboratory rabbits a mixture of two types of foods. Type I contains 8 g of fat, 12 g of carbohydrate, and 2 g of protein per ounce, whereas type II contains 12 g of fat, 12 g of carbohydrate, and 1 g of protein per ounce. Type I costs \$0.20 per ounce and type II cost \$0.30 per ounce. The rabbits each receive a daily minimum of 24 g of fat, 36 g of carbohydrate, and 4 g of protein, but get no more than 5 oz of food per day. How many ounces of each food type should be fed to each rabbit daily to satisfy the dietary requirements at minimum cost?

13. A woman wishes to invest \$12,000 in three types of bonds: municipal bonds paying 7% interest per year, bank investment certificates paying 8%, and high-risk bonds paying 12%. For tax reasons, she wants the amount invested in municipal bonds to be at least three times the amount invested in bank certificates. To keep her level of risk manageable, she will invest no more than \$2000 in high-risk bonds. How much should she invest in each type of bond to maximize her annual interest yield? [*Hint:* Let x = amount in municipal bonds and y = amount in bank certificates. Then the amount in high-risk bonds will be $12{,}000 - x - y$.]

14. Refer to Problem 13. Suppose the investor decides to increase to \$3000 the maximum amount she will allow to be invested in high-risk bonds but leaves the other conditions unchanged. By how much will her maximum possible interest yield increase?

15. A small software company publishes computer games and educational and utility software. Their policy is to market a total of 36 new programs each year, with at least four of these being games. The number of utility programs published is never more than twice the number of educational programs. On average, the company makes an annual profit of \$5000 on each computer game, \$8000 on each educational program, and \$6000 on each utility program. How many of each type of software should they publish annually for maximum profit?

16. All parts of this problem refer to the following feasible region and objective function.

$$\begin{cases} x \geqslant 0 \\ x \geqslant y \\ x + 2y \leqslant 12 \\ x + y \leqslant 10 \end{cases}$$

$$P = x + 4y$$

(a) Graph the feasible region.

(b) On your graph from part (a), sketch the graphs of the linear equations obtained by setting P equal to 40, 36, 32, and 28.

(c) If we continue to decrease the value of P, at which vertex of the feasible region will these lines first touch the feasible region?

(d) Verify that the maximum value of P on the feasible region occurs at the vertex you chose in part (c).

8 CONIC SECTIONS

The path of an object moving under the gravitational pull of another is a conic section. For instance, the trajectory of a basketball is a parabola; the orbit of a planet around the sun is an ellipse.

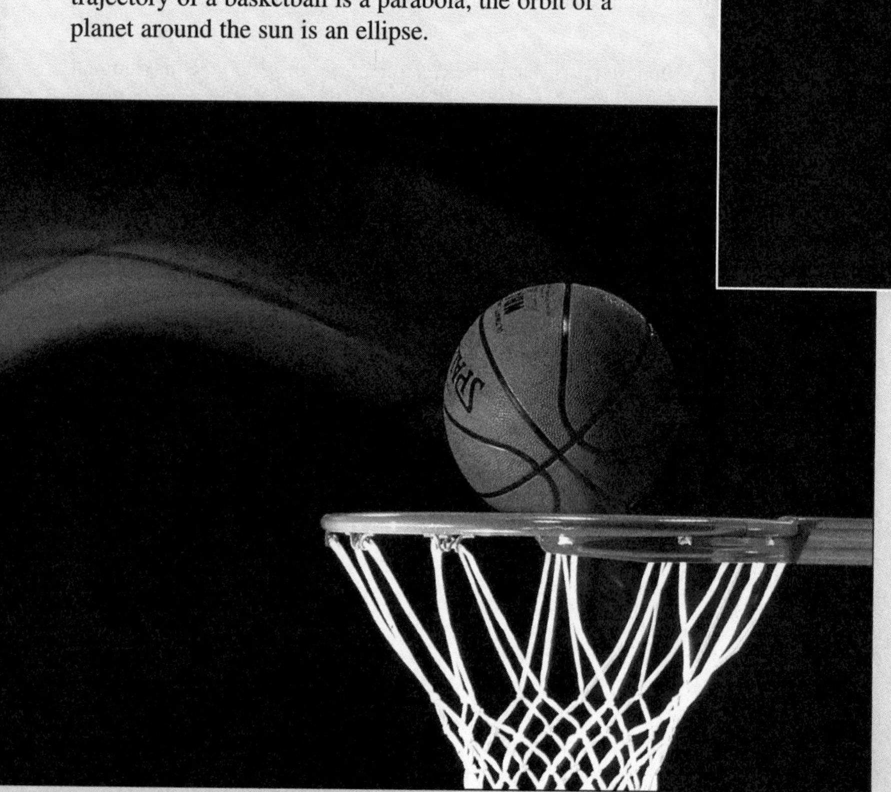

In studying the procedures of geometric thought we may hope to reach what is most essential in the human mind.

HENRI POINCARÉ

In this chapter we study the geometry of **conic sections** (or simply **conics**). Conic sections are the curves formed by the intersection of a plane with a pair of circular cones. These curves have four basic shapes, called **circles**, **ellipses**, **parabolas**, and **hyperbolas**, as illustrated in the figure.

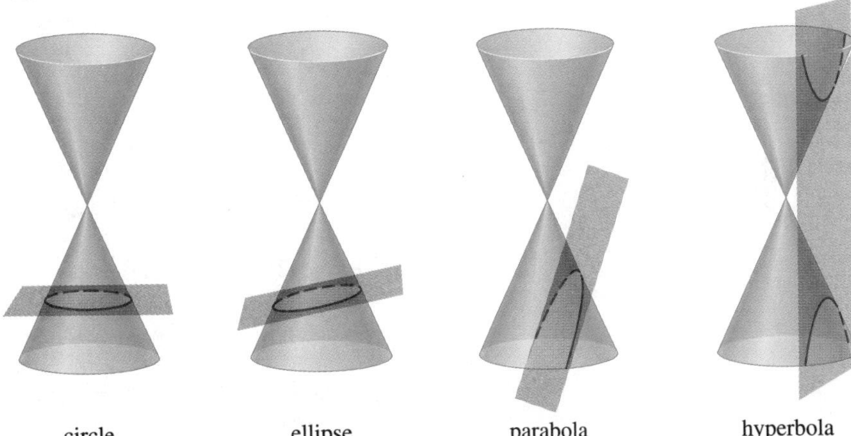

The ancient Greeks studied these curves because they considered the geometry of conic sections to be very beautiful. The mathematician Apollonius (262–190 B.C.) wrote a definitive eight-volume work on the subject. In more modern times, conics have been found to be useful as well as beautiful. Galileo discovered in 1590 that the path of a missile shot upward at an angle is a parabola. In 1609 Kepler found that the planets move in elliptical orbits around the sun. In 1668 Newton built the first reflecting telescope, whose principle is based on the properties of parabolas and hyperbolas. In this century, many further applications of conic sections have been developed. One important application is the LORAN radio navigation system, which uses the intersection points of hyperbolas to pinpoint the location of ships and aircraft. Another application is the medical procedure of lithotripsy, a method of removing kidney stones without surgery that uses a property of ellipses. These and other applications of conic sections will be considered in this chapter.

8.1 PARABOLAS

We saw in Section 4.6 that the graph of the equation $y = ax^2 + bx + c$ is a U-shaped curve called a *parabola* that opens either upward or downward, depending on whether the sign of a is positive or negative (see Figure 1). The lowest or highest point of the parabola is called the *vertex*, and the parabola is symmetric about its *axis*.

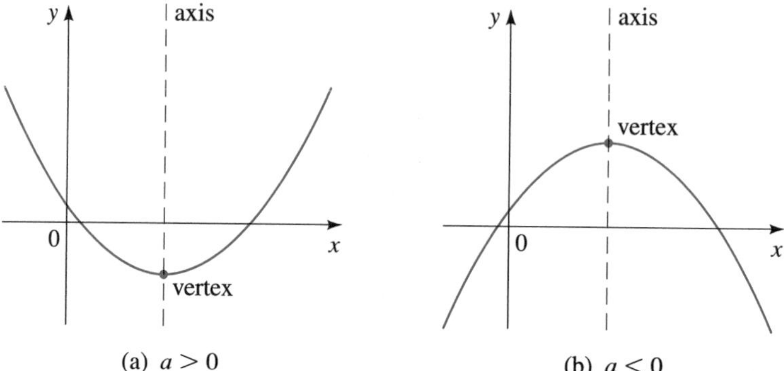

FIGURE 1
$y = ax^2 + bx + c$

In this section we study parabolas from a geometric rather than an algebraic point of view. We begin with the geometric definition of a parabola and show how this leads to the algebraic formula that we are already familiar with.

GEOMETRIC DEFINITION OF A PARABOLA

A **parabola** is the set of points in the plane equidistant from a fixed point F (called the **focus**) and a fixed line l (called the **directrix**).

This definition is illustrated in Figure 2. Note that the vertex V of the parabola lies halfway between the focus and the directrix and that the axis of symmetry is the line that runs through the focus perpendicular to the directrix.

In this section we restrict our attention to parabolas that are situated with the vertex at the origin and that have a vertical or horizontal axis of symmetry. (Parabolas in more general positions will be considered in Section 8.4.) If the focus of such a parabola is the point $F(0, p)$, then the axis of symmetry must be vertical and the directrix has the equation $y = -p$. Figure 3 illustrates the case $p > 0$.

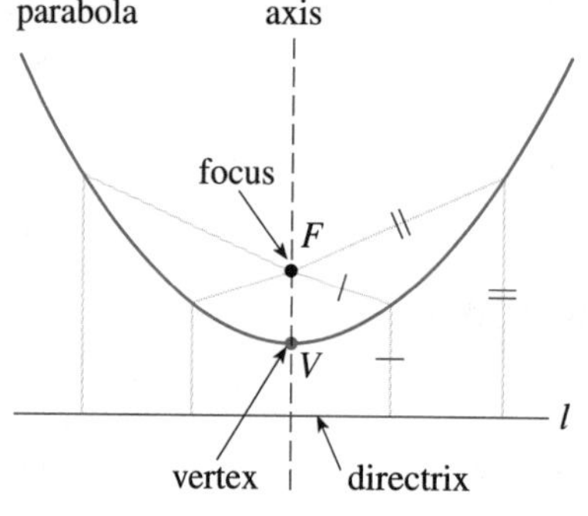

FIGURE 2

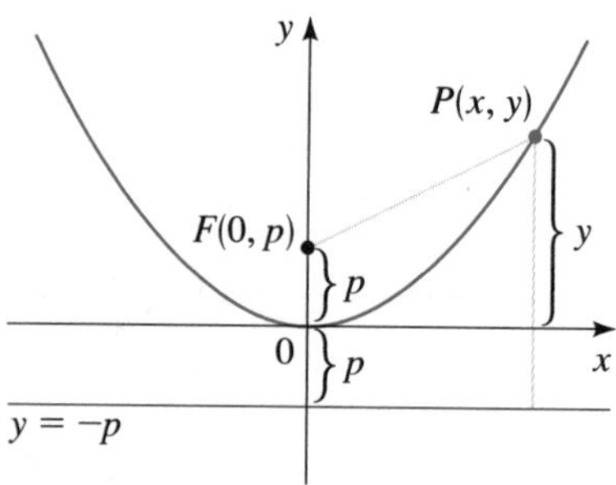

FIGURE 3

If $P(x, y)$ is any point on the parabola, then the distance from P to the focus F (using the Distance Formula) is

$$\sqrt{x^2 + (y - p)^2}$$

The distance from P to the directrix is

$$|y - (-p)| = |y + p|$$

By the definition of a parabola, these two distances must be equal:

$$\sqrt{x^2 + (y - p)^2} = |y + p|$$

$$x^2 + (y - p)^2 = |y + p|^2 = (y + p)^2 \qquad \text{Square both sides}$$

$$x^2 + y^2 - 2py + p^2 = y^2 + 2py + p^2 \qquad \text{Expand}$$

$$x^2 - 2py = 2py \qquad \text{Simplify}$$

$$x^2 = 4py$$

If $p > 0$, then the parabola opens upward, but if $p < 0$, it opens downward. When x is replaced by $-x$, the equation remains unchanged, so the graph is symmetric about the y-axis. We summarize what we have proved in the following box.

PARABOLA WITH VERTICAL AXIS

The graph of the equation

$$x^2 = 4py$$

is a parabola with the following properties.

VERTEX $V(0, 0)$

FOCUS $F(0, p)$

DIRECTRIX $y = -p$

The parabola opens upward if $p > 0$ or downward if $p < 0$.

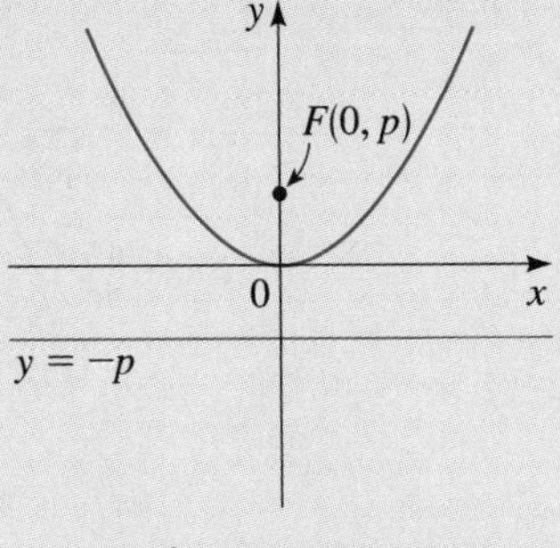

(a) $x^2 = 4py$ with $p > 0$

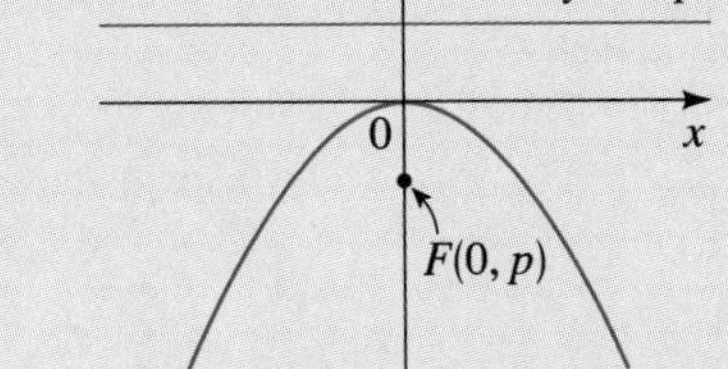

(b) $x^2 = 4py$ with $p < 0$

EXAMPLE 1 ■ Finding the Equation of a Parabola

Find the equation of the parabola with vertex $V(0, 0)$ and focus $F(0, 2)$, and sketch its graph.

SOLUTION

Since the focus is $F(0, 2)$, we conclude that $p = 2$ (and so the directrix is $y = -2$). Thus, the equation of the parabola is

$$x^2 = 4(2)y \qquad x^2 = 4py \text{ with } p = 2$$

$$x^2 = 8y$$

Since $p = 2 > 0$, the parabola opens upward. See Figure 4.

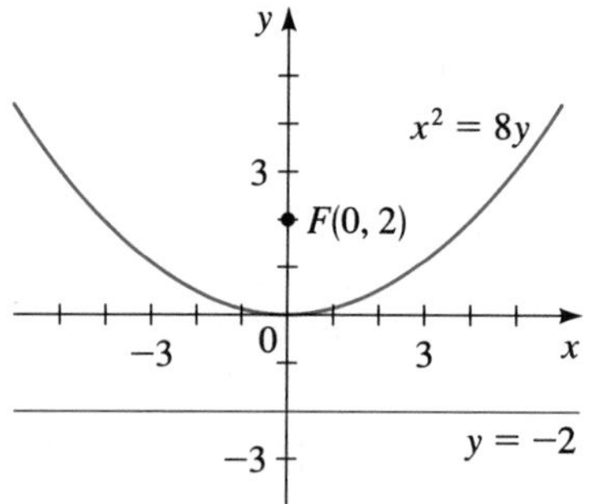

FIGURE 4

■

EXAMPLE 2 ■ Finding the Focus and Directrix of a Parabola from Its Equation

Find the focus and directrix of the parabola $y = -x^2$, and sketch its graph.

SOLUTION

Comparing the equation $y = -x^2$ with the general equation $x^2 = 4py$, we see that $4p = -1$, so $p = -\frac{1}{4}$. Thus, the focus is $F\left(0, -\frac{1}{4}\right)$ and the directrix is $y = \frac{1}{4}$. The graph is shown in Figure 5.

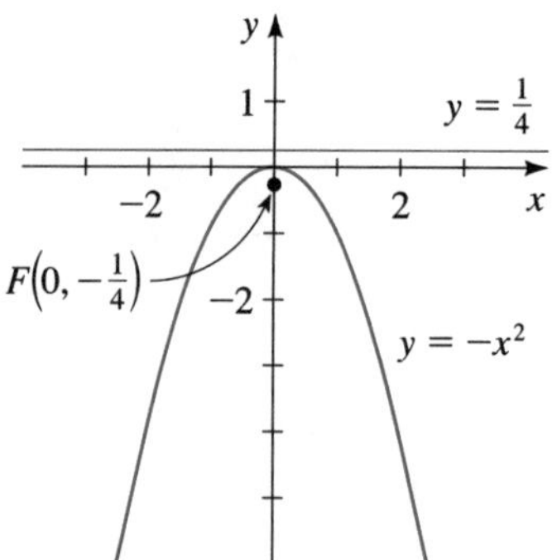

FIGURE 5

■

Reflecting the graph in Figure 3 about the diagonal line $y = x$ has the effect of interchanging the roles of x and y. This results in a parabola with horizontal axis. By the same method as before, we can prove the following properties.

PARABOLA WITH HORIZONTAL AXIS

The graph of the equation

$$y^2 = 4px$$

is a parabola with the following properties.

VERTEX	$V(0, 0)$
FOCUS	$F(p, 0)$
DIRECTRIX	$x = -p$

The parabola opens to the right if $p > 0$ or to the left if $p < 0$.

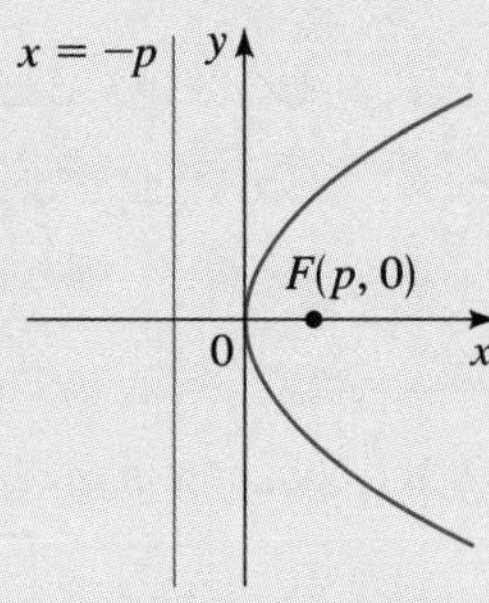

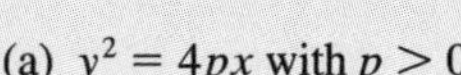
(a) $y^2 = 4px$ with $p > 0$

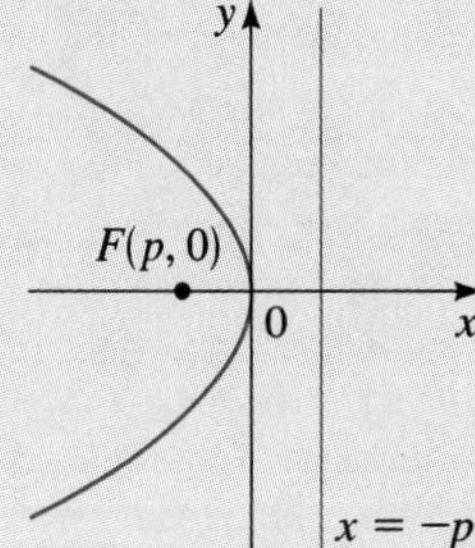

(b) $y^2 = 4px$ with $p < 0$

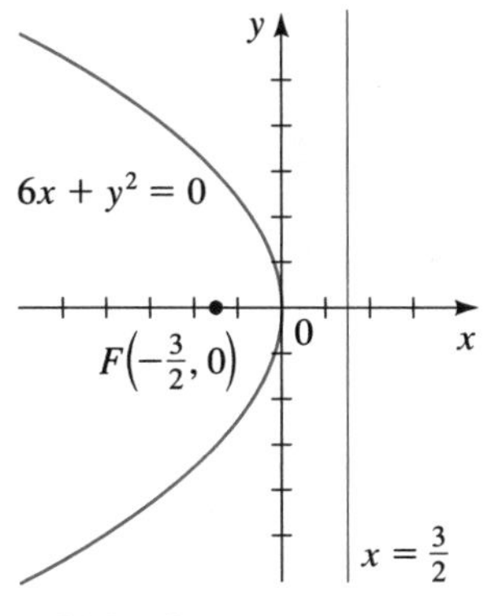

FIGURE 6

EXAMPLE 3 ■ A Parabola with Horizontal Axis

Find the focus and directrix of the parabola $6x + y^2 = 0$, and sketch its graph.

SOLUTION

We first write the equation as $y^2 = -6x$. Comparing this with the general equation $y^2 = 4px$, we see that $-6 = 4p$, so $p = -\frac{3}{2}$. Thus, the focus is $\left(-\frac{3}{2}, 0\right)$ and the directrix is $x = \frac{3}{2}$.

Since $p = -\frac{3}{2} < 0$, the parabola opens to the left. See Figure 6. ■

We can use the coordinates of the focus to estimate the "width" of a parabola when sketching its graph. The line segment that runs through the focus perpendicular to the axis, with endpoints on the parabola, is called the **latus rectum**, and its

Archimedes (287–212 B.C.) was the greatest mathematician of the ancient world. He was born in Syracuse, a Greek colony on Sicily, a generation after Euclid (see page 55). Renowned as a mechanical genius for his many engineering inventions, he designed pulleys for lifting heavy ships and the spiral screw for transporting water to higher levels. He is said to have used parabolic mirrors to concentrate the rays of the sun to set fire to Roman ships attacking Syracuse. King Hieron II of Syracuse once suspected a goldsmith of keeping part of the gold intended for the king's crown and replacing it with an equal amount of silver. The king asked Archimedes for advice. While in deep thought at a public bath, Archimedes discovered the solution to the king's problem when he noticed that his body's volume was the same as the volume of water it displaced from the tub. As the story is told, he ran home naked, shouting "Eureka, eureka!" ("I have found it, I have found it!") This incident attests to his enormous powers of concentration. In spite of his engineering prowess, Archimedes was most proud of his mathematical discoveries. These included the formulas for the volume of a sphere, $V = \frac{4}{3}\pi r^3$; the surface area of a sphere, $S = 4\pi r^2$; and a careful analysis of the properties of parabolas and other conics.

length is the **focal diameter** of the parabola. From Figure 7 we can see that the distance from an endpoint Q of the latus rectum to the directrix is $|2p|$. Thus, the distance from Q to the focus must be $|2p|$ as well (by the definition of a parabola), and so the focal diameter is $|4p|$. In the next example, we use the focal diameter to determine the "width" of a parabola when graphing it.

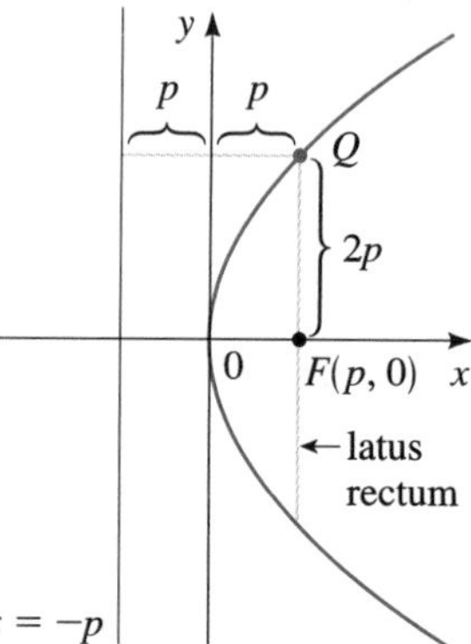

FIGURE 7

EXAMPLE 4 ■ Using the Focal Diameter to Sketch a Parabola

Find the focus, directrix, and focal diameter of the parabola $y = \frac{1}{2}x^2$, and sketch its graph.

SOLUTION

We first put the equation in the form $x^2 = 4py$.

$$y = \tfrac{1}{2}x^2$$

$$x^2 = 2y \qquad \text{Multiply each side by 2}$$

From this equation we see that $4p = 2$, so the focal diameter is 2. Solving for p gives $p = \frac{1}{2}$, so the focus is $\left(0, \frac{1}{2}\right)$ and the directrix is $y = -\frac{1}{2}$. Since the focal diameter is 2, the latus rectum extends 1 unit to the left and 1 unit to the right of the focus. This enables us to sketch the graph in Figure 8.

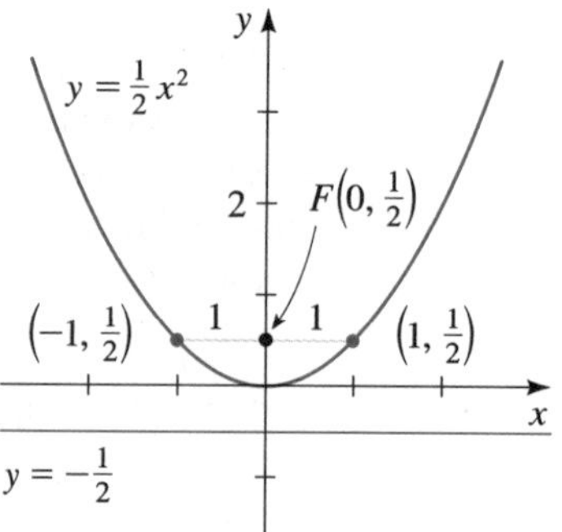

FIGURE 8

■

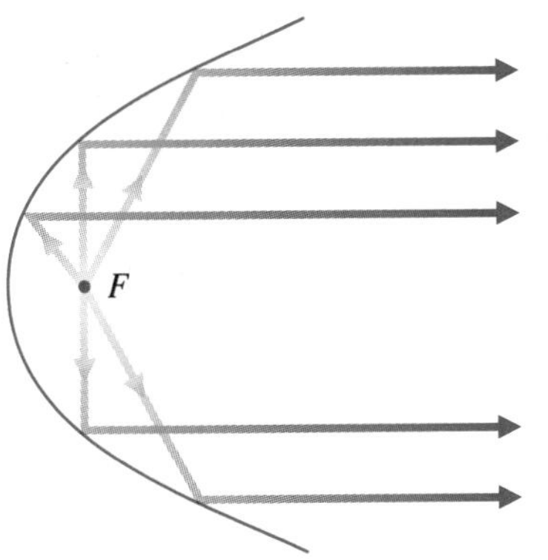

FIGURE 9
Parabolic reflector

Parabolas have an important property that makes them useful as reflectors for lamps and telescopes. Light from a source placed at the focus of a surface with parabolic cross section will be reflected in such a way that it travels parallel to the axis of the parabola (see Figure 9). Thus, a parabolic mirror reflects the light into a beam of parallel rays. Conversely, light approaching the reflector in rays parallel to its axis of symmetry is concentrated to the focus. This *reflection property*, which can be proved using calculus, is used in the construction of reflecting telescopes.

EXAMPLE 5 ■ Finding the Focal Point of a Searchlight Reflector

A searchlight has a parabolic reflector that forms a "bowl," which is 12 in. wide from rim to rim and 8 in. deep, as shown in Figure 10. If the filament of the light bulb is located at the focus, how far from the vertex of the reflector is it?

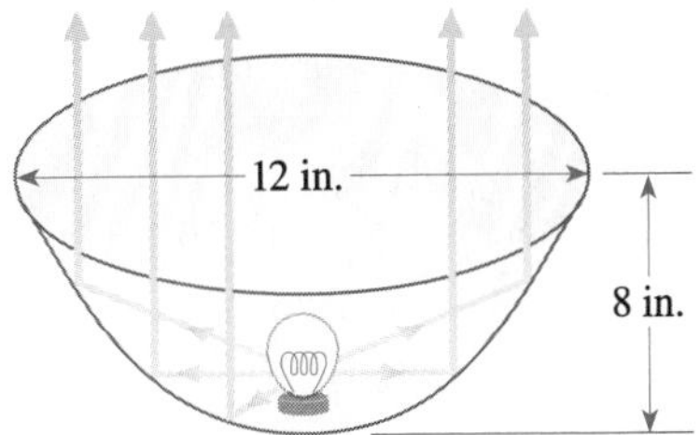

FIGURE 10
A parabolic reflector

SOLUTION

We introduce a coordinate system and place a parabolic cross section of the reflector so that its vertex is at the origin and its axis is vertical (see Figure 11). Then the equation of this parabola has the form $x^2 = 4py$. From Figure 11 we see that the point (6, 8) lies on the parabola. We use this to find p.

$$6^2 = 4p(8) \qquad \text{The point } (6, 8) \text{ satisfies the equation } x^2 = 4py$$

$$36 = 32p$$

$$p = \tfrac{9}{8}$$

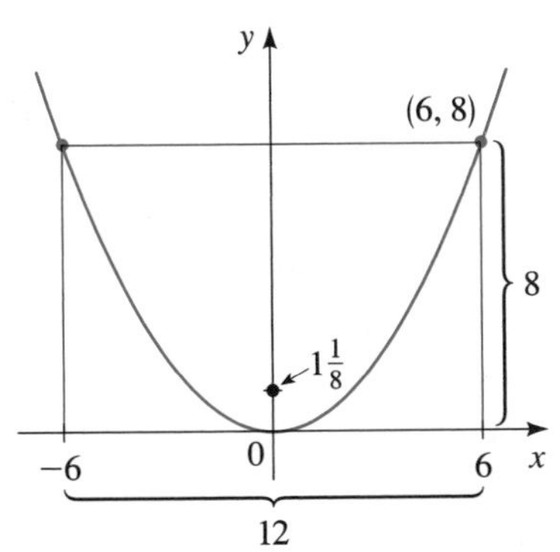

FIGURE 11

The focus is $F\left(0, \frac{9}{8}\right)$, so the distance between the vertex and the focus is $\frac{9}{8} = 1\frac{1}{8}$ in. Because the filament is positioned at the focus, it is located $1\frac{1}{8}$ in. from the vertex of the reflector. ■

In the next example we graph a family of parabolas, to show how changing the distance between the focus and the vertex affects the "width" of a parabola.

EXAMPLE 6 ■ A Family of Parabolas

(a) Find equations for the parabolas with vertex at the origin and foci $F_1\left(0, \frac{1}{8}\right)$, $F_2\left(0, \frac{1}{2}\right)$, $F_3(0, 1)$, and $F_4(0, 4)$.

(b) Draw the graphs of the parabolas in part (a). What do you conclude?

SOLUTION

(a) Since the foci are on the positive y-axis, the parabolas open upward and have equations of the form $x^2 = 4py$. This leads to the following equations.

Focus	p	Equation $x^2 = 4py$	Form of the equation for graphing calculator
$F_1\left(0, \frac{1}{8}\right)$	$p = \frac{1}{8}$	$x^2 = \frac{1}{2}y$	$y = 2x^2$
$F_2\left(0, \frac{1}{2}\right)$	$p = \frac{1}{2}$	$x^2 = 2y$	$y = 0.5x^2$
$F_3(0, 1)$	$p = 1$	$x^2 = 4y$	$y = 0.25x^2$
$F_4(0, 4)$	$p = 4$	$x^2 = 16y$	$y = 0.0625x^2$

(b) The graphs are drawn in Figure 12. We see that the closer the focus to the vertex, the narrower the parabola.

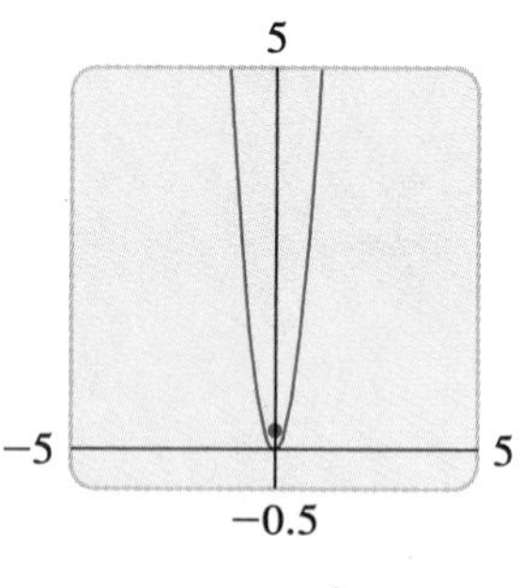

$y = 2x^2$

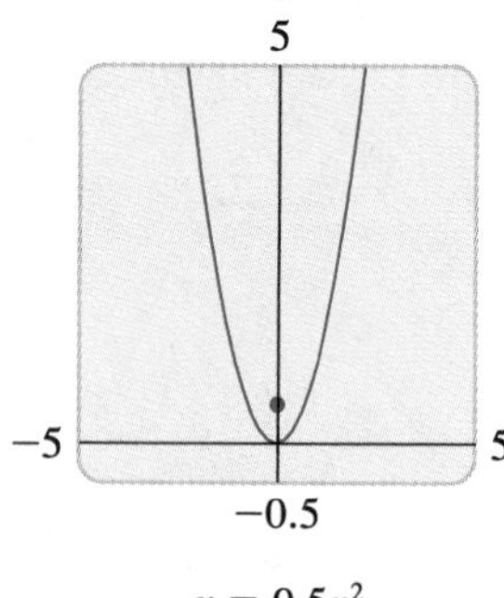

$y = 0.5x^2$

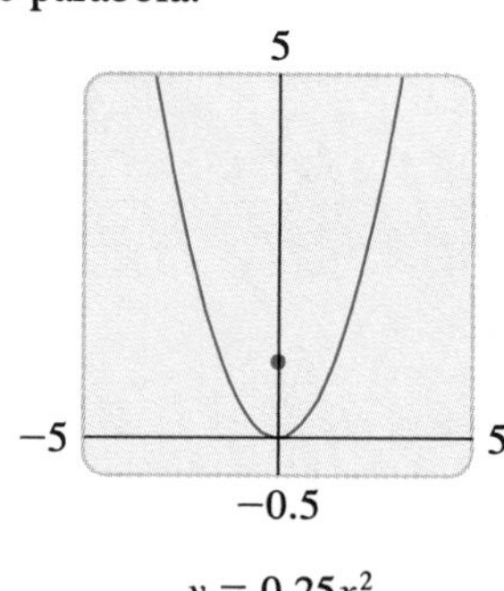

$y = 0.25x^2$

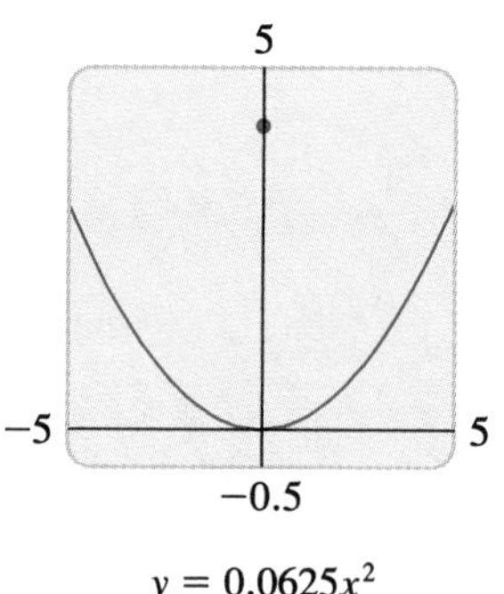

$y = 0.0625x^2$

FIGURE 12 A family of parabolas

■

8.1 EXERCISES

1–12 ■ Find the focus, directrix, and focal diameter of the parabola, and sketch its graph.

1. $y^2 = 4x$ **2.** $x^2 = y$ **3.** $x^2 = 9y$

4. $y^2 = 3x$ **5.** $y = 5x^2$ **6.** $y = -2x^2$

7. $x = -8y^2$ **8.** $x = \frac{1}{2}y^2$ **9.** $x^2 + 6y = 0$

10. $x - 7y^2 = 0$ **11.** $5x + 3y^2 = 0$

12. $8x^2 + 12y = 0$

13–24 ■ Find an equation for the parabola that has its vertex at the origin and satisfies the given condition(s).

13. Focus $F(0, 2)$ **14.** Focus $F\left(0, -\frac{1}{2}\right)$

15. Focus $F(-8, 0)$ **16.** Focus $F(5, 0)$

17. Directrix $x = 2$ **18.** Directrix $y = 6$

19. Directrix $y = -10$ **20.** Directrix $x = -\frac{1}{8}$

21. Focus on the positive x-axis, 2 units away from the directrix

22. Directrix has y-intercept 6

23. Opens upward with focus 5 units from the vertex

24. Focal diameter 8 and focus on the negative y-axis

25–32 ■ Find an equation of the parabola whose graph is shown.

25.

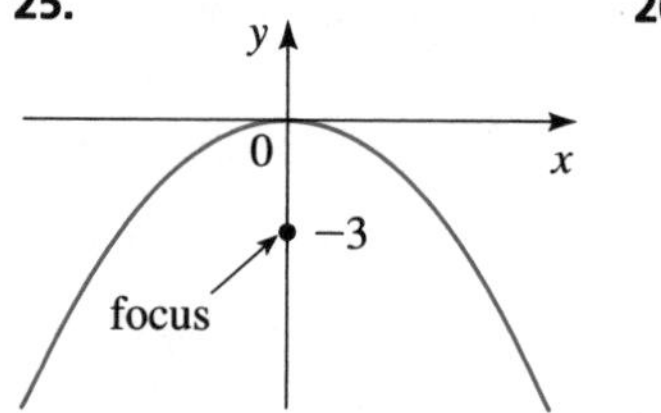

26.

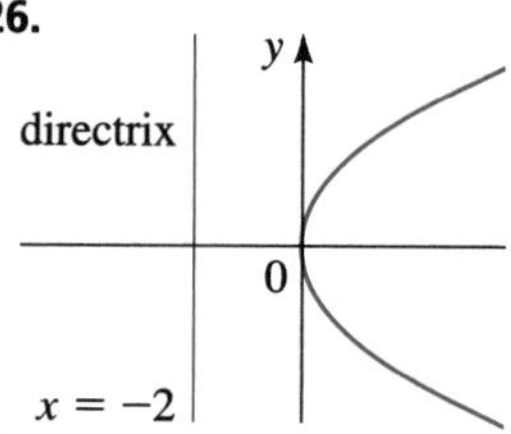

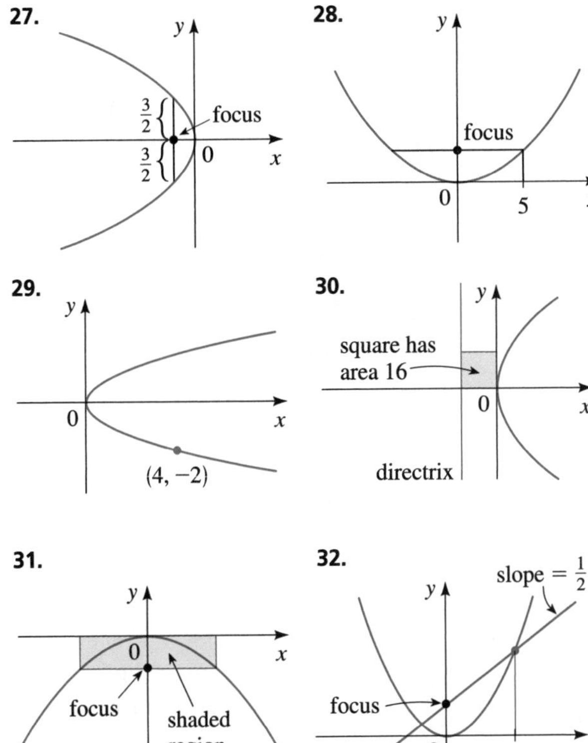

33. A lamp with a parabolic reflector is shown in the figure. The bulb is placed at the focus and the focal diameter is 12 cm.

(a) Find an equation of the parabola.

(b) Find the diameter $d(C, D)$ of the opening, 20 cm from the vertex.

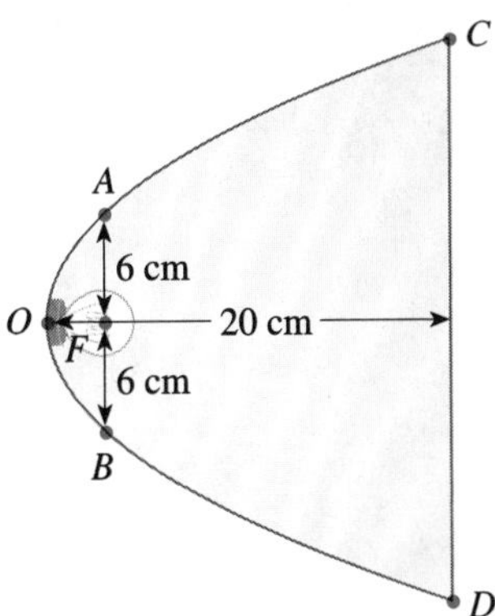

34. A reflector for a satellite dish is parabolic in cross section, with the receiver at the focus F. The reflector is 1 ft deep and 20 ft wide from rim to rim (see the figure). How far is the receiver from the vertex of the parabolic reflector?

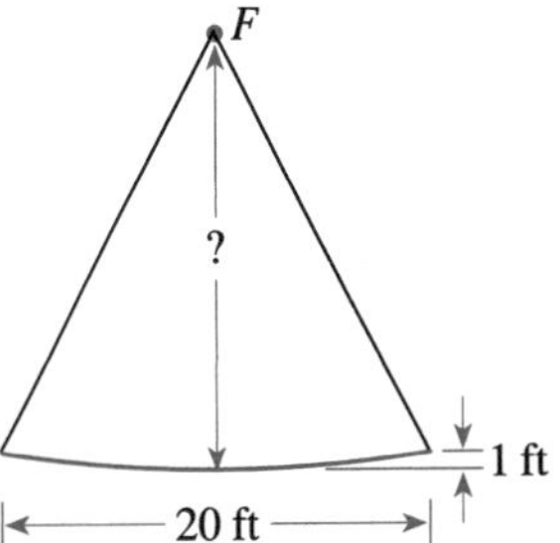

35. In a suspension bridge the shape of the suspension cables is parabolic. The bridge shown in the figure has towers that are 600 m apart, and the lowest point of the suspension cables is 150 m below the top of the towers. Find the equation of the parabolic part of the cables, placing the origin of the coordinate system at the vertex.

NOTE: This equation is used to find the length of cable needed in the construction of the bridge.

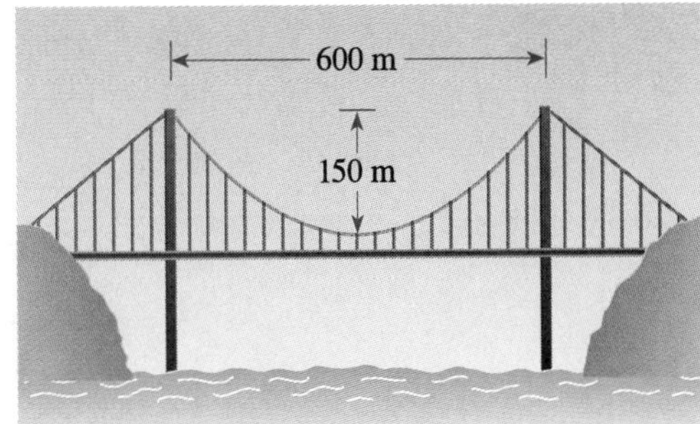

36. The Hale telescope at the Mount Palomar Observatory has a 200-in. mirror, as shown. The mirror is constructed in a parabolic shape that collects light from the stars and focuses it at the **prime focus**, that is, the focus of the parabola. The mirror is 3.79 in. deep at its center. Find the **focal length** of this parabolic mirror, that is, the distance from the vertex to the focus.

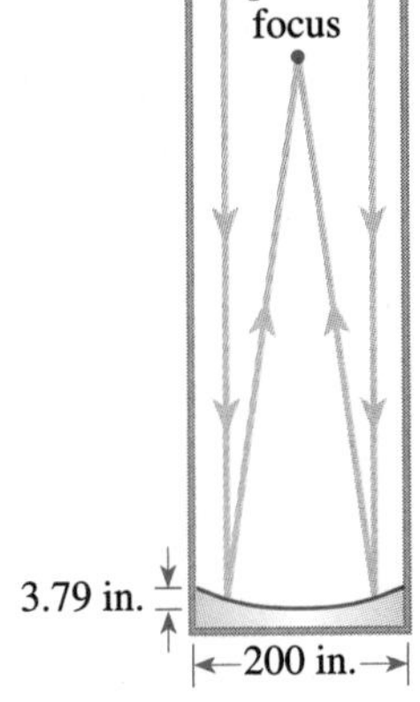

37. (a) Find equations for the family of parabolas with vertex at the origin and with directrixes $y = \frac{1}{2}$, $y = 1$, $y = 4$, and $y = 8$.

(b) Draw the graphs. What do you conclude?

DISCOVERY • DISCUSSION

38. Parabolas in the Real World Several examples of the uses of parabolas are given in the text. Find other situations in real life where parabolas occur. Consult a scientific encyclopedia in the reference section of your library, or search the Internet.

39. Light Cone from a Flashlight When a flashlight is held horizontally, a lighted area forms on the ground, as shown in the figure. Is it possible to angle the flashlight in such a way that the boundary of the lighted area is a parabola? Explain your answer.

Laboratory Project

ROLLING DOWN A RAMP

Galileo was the first to show that the motion of a falling ball can be modeled by a parabola (see page 566). To accomplish this, he rolled balls down a ramp and used his acute sense of timing to study the motion of the ball. In this project you will perform a similar experiment, using modern equipment. You will need the following:

- A wide board, about 6–8 ft long, to be used as a ramp
- A large ball, the size of a soccer or volleyball
- A calculator-based motion detector, such as the Texas Instruments CBR system (Calculator Based Ranger)

Use books to prop up the ramp at an angle of about 15°, clamp the motion detector to the top of the ramp, and connect it to the calculator, as shown in the figure. (Read the manual carefully to make sure you have set up the calculator and motion detector correctly.)

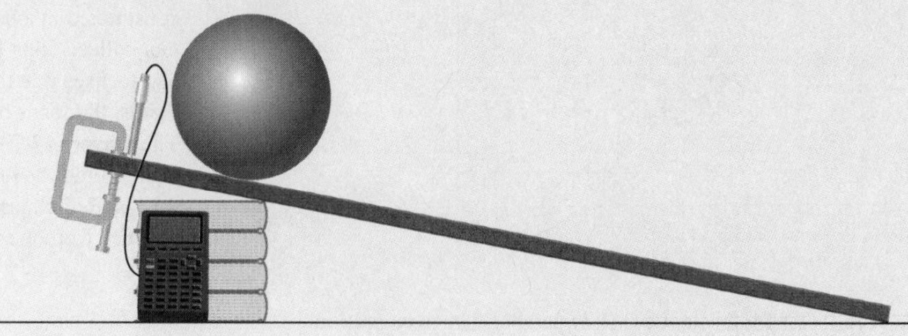

1. Mark a spot on the ramp about 2 ft from the top. Place the ball on the mark and release it to see how many seconds it takes to roll to the bottom. Adjust the motion detector to record the ball's position every 0.05 s over this length of time. Now let the ball roll down the ramp again, this time with the motion detector running. The calculator should record at least 50 data points that indicate the distance between the ball and the motion detector every 0.05 s.
2. Make a scatter plot of your data, plotting time on the x-axis and distance on the y-axis. Does the plot look like half of a parabola?
3. Use the quadratic regression command on your calculator (called QuadReg on the TI-83) to find the parabola equation $y = ax^2 + bx + c$ that best fits the data. Graph this equation on your scatter plot to see how well it fits. Do the data points really form part of a parabola?
4. Repeat the experiment with the ramp inclined at a shallower angle. How does reducing the angle of the ramp affect the shape of the parabola?
5. Try rolling the ball up the ramp from the bottom to the spot you marked in Step 1, so that it rolls up and then back down again. If you perform the experiment this way instead of just letting the ball roll down, how does your graph in Step 3 change?

8.2 ELLIPSES

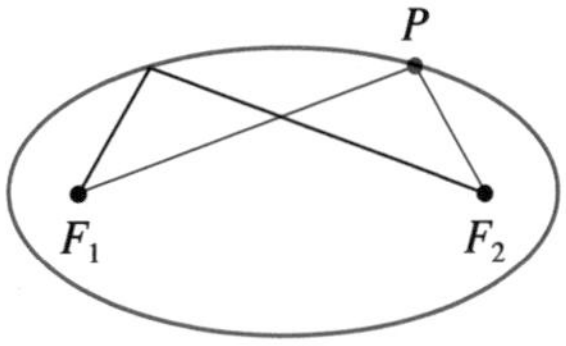

An ellipse is an oval curve that looks like an elongated circle. More precisely, we have the following definition.

FIGURE 1

GEOMETRIC DEFINITION OF AN ELLIPSE

An **ellipse** is the set of all points in the plane the sum of whose distances from two fixed points F_1 and F_2 is a constant. (See Figure 1.) These two fixed points are the **foci** (plural of **focus**) of the ellipse.

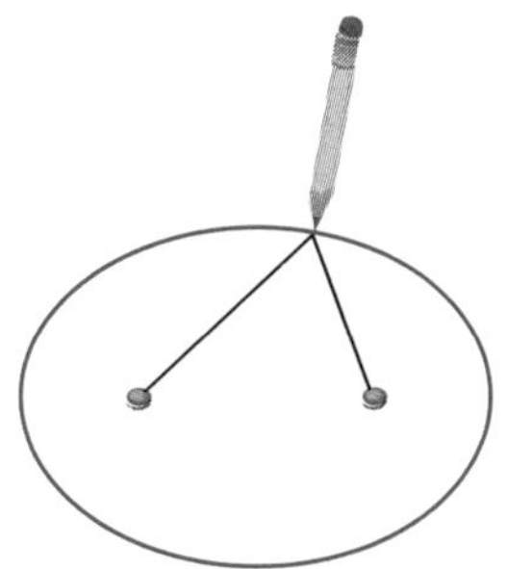

FIGURE 2

The geometric definition suggests a simple method for drawing an ellipse. Place a sheet of paper on a drawing board and insert thumbtacks at the two points that are to be the foci of the ellipse. Attach the ends of a string to the tacks, as shown in Figure 2. With the point of a pencil, hold the string taut. Then carefully move the pencil around the foci, keeping the string taut at all times. The pencil will trace out an ellipse, because the sum of the distances from the point of the pencil to the foci will always equal the length of the string, which is constant.

If the string is only slightly longer than the distance between the foci, then the ellipse traced out will be elongated in shape as in Figure 3(a), but if the foci are close together relative to the length of the string, the ellipse will be almost circular, as shown in Figure 3(b).

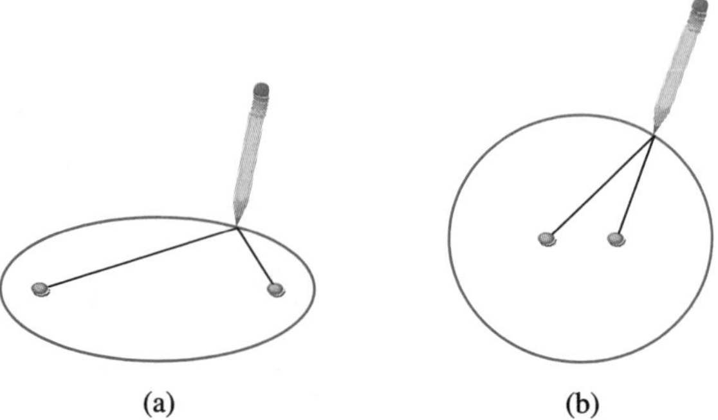

FIGURE 3 (a) (b)

To obtain the simplest equation for an ellipse, we place the foci on the x-axis at $F_1(-c, 0)$ and $F_2(c, 0)$, so that the origin is halfway between them (see Figure 4).

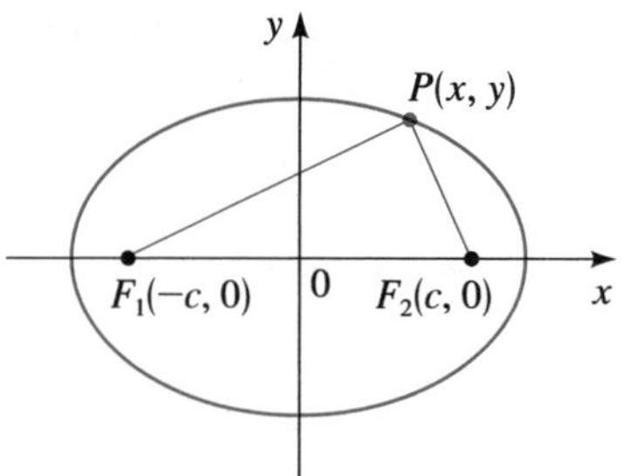

FIGURE 4

For later convenience we let the sum of the distances from a point on the ellipse to the foci be $2a$. Then if $P(x, y)$ is any point on the ellipse, we have

$$d(P, F_1) + d(P, F_2) = 2a$$

So, from the Distance Formula

$$\sqrt{(x+c)^2 + y^2} + \sqrt{(x-c)^2 + y^2} = 2a$$

or

$$\sqrt{(x-c)^2 + y^2} = 2a - \sqrt{(x+c)^2 + y^2}$$

Squaring each side and expanding, we get

$$x^2 - 2cx + c^2 + y^2 = 4a^2 - 4a\sqrt{(x+c)^2 + y^2} + (x^2 + 2cx + c^2 + y^2)$$

which simplifies to

$$4a\sqrt{(x+c)^2 + y^2} = 4a^2 + 4cx$$

Dividing each side by 4 and squaring again, we get

$$a^2[(x + c)^2 + y^2] = (a^2 + cx)^2$$

$$a^2x^2 + 2a^2cx + a^2c^2 + a^2y^2 = a^4 + 2a^2cx + c^2x^2$$

$$(a^2 - c^2)x^2 + a^2y^2 = a^2(a^2 - c^2)$$

Since the sum of the distances from P to the foci must be larger than the distance between the foci, we have that $2a > 2c$, or $a > c$. Thus, $a^2 - c^2 > 0$, and we can divide each side of the preceding equation by $a^2(a^2 - c^2)$ to get

$$\frac{x^2}{a^2} + \frac{y^2}{a^2 - c^2} = 1$$

For convenience let $b^2 = a^2 - c^2$ (with $b > 0$). Since $b^2 < a^2$, it follows that $b < a$. The preceding equation then becomes

$$\frac{x^2}{a^2} + \frac{y^2}{b^2} = 1 \qquad \text{with } a > b$$

This is the equation of the ellipse. To graph it, we need to know the x- and y-intercepts. Setting $y = 0$, we get

$$\frac{x^2}{a^2} = 1$$

so $x^2 = a^2$, or $x = \pm a$. Thus, the ellipse crosses the x-axis at $(a, 0)$ and $(-a, 0)$. These points are called the **vertices** of the ellipse, and the segment that joins them is called the **major axis**. Its length is $2a$.

Similarly, if we set $x = 0$, we get $y = \pm b$, so the ellipse crosses the y-axis at $(0, b)$ and $(0, -b)$. The segment that joins these points is called the **minor axis**, and it has length $2b$. Note that $2a > 2b$, so the major axis is longer than the minor axis.

In Section 2.2 we studied several tests that detect symmetry in a graph. If we replace x by $-x$ or y by $-y$ in the ellipse equation, it remains unchanged. Thus, the ellipse is symmetric about both the x- and y-axes, and hence about the origin as well. For this reason the origin is called the **center** of the ellipse. The complete graph is shown in Figure 5.

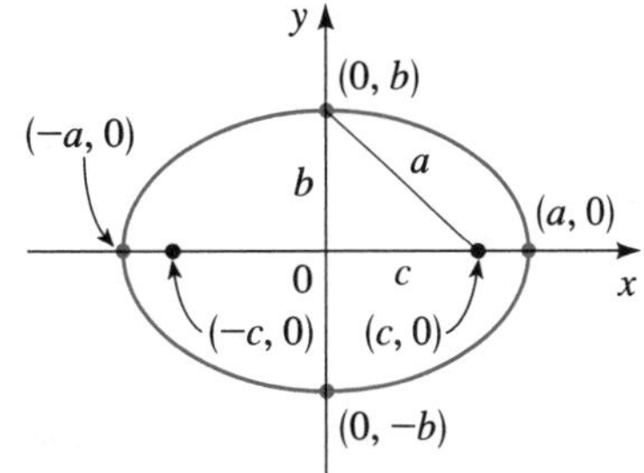

FIGURE 5

$\frac{x^2}{a^2} + \frac{y^2}{b^2} = 1$ with $a > b$

If the foci of the ellipse are placed on the y-axis at $(0, \pm c)$ rather than on the x-axis, then the roles of x and y are reversed in the preceding discussion, and we get a vertical ellipse. Thus, we have the following description of ellipses.

ELLIPSE WITH CENTER AT THE ORIGIN

The graph of each of the following equations is an ellipse with center at the origin and having the given properties.

EQUATION	$\frac{x^2}{a^2} + \frac{y^2}{b^2} = 1$	$\frac{x^2}{b^2} + \frac{y^2}{a^2} = 1$
	$a > b > 0$	$a > b > 0$
VERTICES	$(\pm a, 0)$	$(0, \pm a)$
MAJOR AXIS	Horizontal, length $2a$	Vertical, length $2a$
MINOR AXIS	Vertical, length $2b$	Horizontal, length $2b$
FOCI	$(\pm c, 0)$, $c^2 = a^2 - b^2$	$(0, \pm c)$, $c^2 = a^2 - b^2$
GRAPH		

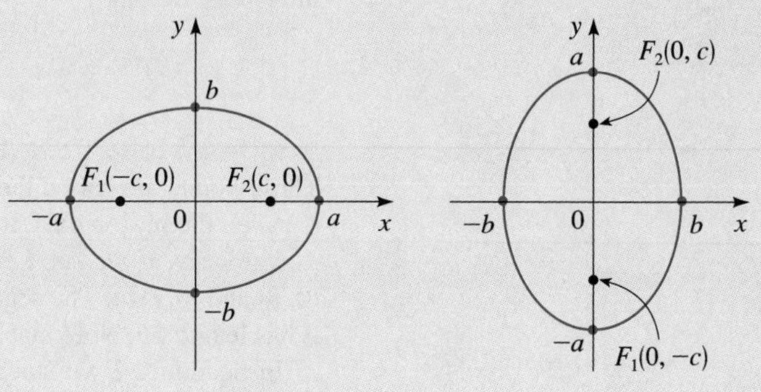

In the standard equation for an ellipse, a^2 is the *larger* denominator and b^2 is the *smaller*. To find c^2, we subtract: larger denominator minus smaller denominator.

EXAMPLE 1 ■ Sketching an Ellipse

Find the foci, vertices, and the lengths of the major and minor axes for the following ellipse, and sketch its graph.

$$\frac{x^2}{9} + \frac{y^2}{4} = 1$$

SOLUTION

Since the denominator of x^2 is larger, the ellipse has horizontal major axis. This gives $a^2 = 9$ and $b^2 = 4$, so $c^2 = a^2 - b^2 = 9 - 4 = 5$. Thus, $a = 3$, $b = 2$,

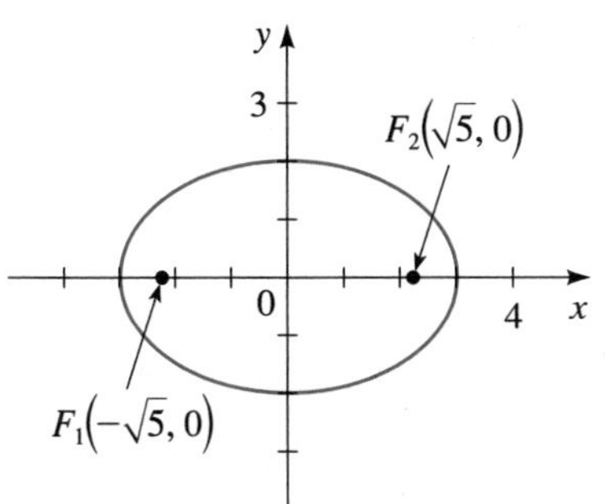

FIGURE 6
$\frac{x^2}{9} + \frac{y^2}{4} = 1$

and $c = \sqrt{5}$.

FOCI $(\pm\sqrt{5}, 0)$

VERTICES $(\pm 3, 0)$

LENGTH OF MAJOR AXIS 6

LENGTH OF MINOR AXIS 4

The graph is shown in Figure 6. ■

EXAMPLE 2 ■ Finding the Equation of an Ellipse

The vertices of an ellipse are $(\pm 4, 0)$ and the foci are $(\pm 2, 0)$. Find its equation and sketch the graph.

SOLUTION

Since the vertices are $(\pm 4, 0)$, we have $a = 4$. The foci are $(\pm 2, 0)$, so $c = 2$. To write the equation, we need to find b. Since $c^2 = a^2 - b^2$, we have

$$2^2 = 4^2 - b^2$$

$$b^2 = 16 - 4 = 12$$

Thus, the equation of the ellipse is

$$\frac{x^2}{16} + \frac{y^2}{12} = 1$$

The graph is shown in Figure 7.

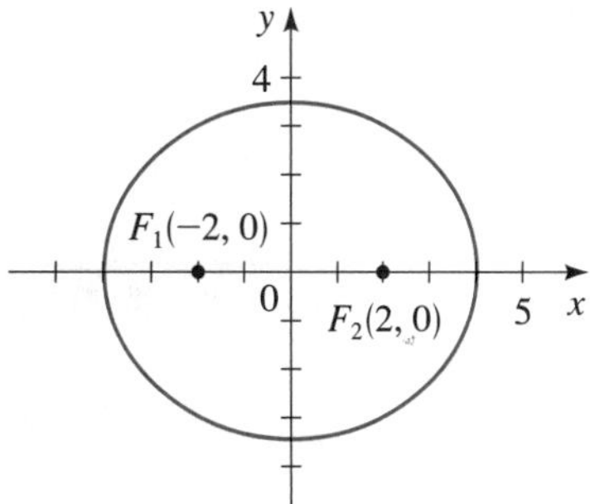

FIGURE 7
$\frac{x^2}{16} + \frac{y^2}{12} = 1$

■

EXAMPLE 3 ■ Finding the Foci of an Ellipse

Find the foci of the ellipse $16x^2 + 9y^2 = 144$, and sketch its graph.

SOLUTION

First we put the equation in standard form. Dividing by 144, we get

$$\frac{x^2}{9} + \frac{y^2}{16} = 1$$

Since $16 > 9$, this is an ellipse with its foci on the y-axis, and with $a = 4$ and $b = 3$. We have

$$c^2 = a^2 - b^2 = 16 - 9 = 7$$

$$c = \sqrt{7}$$

Thus, the foci are $(0, \pm\sqrt{7})$. The graph is shown in Figure 8.

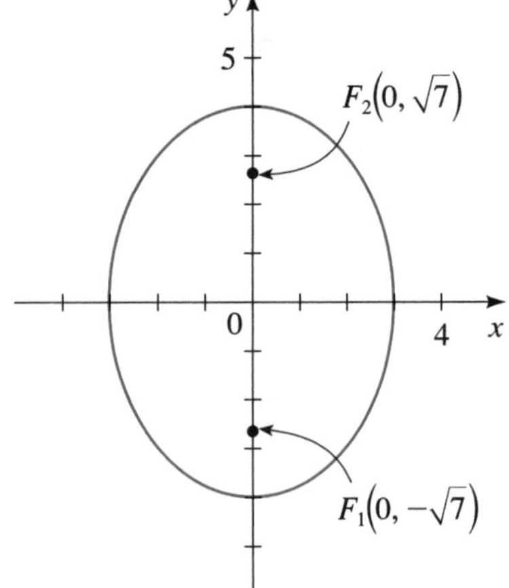

FIGURE 8
$16x^2 + 9y^2 = 144$

■

ECCENTRICITIES OF THE ORBITS OF THE PLANETS

The orbits of the planets are ellipses with the sun at one focus. For most planets these ellipses have very small eccentricity, so they are nearly circular. However, Mercury and Pluto, the innermost and outermost known planets, have visibly elliptical orbits.

Planet	Eccentricity
Mercury	0.206
Venus	0.007
Earth	0.017
Mars	0.093
Jupiter	0.048
Saturn	0.056
Uranus	0.046
Neptune	0.010
Pluto	0.248

We saw earlier in this section (Figure 3) that if $2a$ is only slightly greater than $2c$, the ellipse is long and thin, whereas if $2a$ is much greater than $2c$, the ellipse is almost circular. We measure the deviation of an ellipse from being circular by the ratio of a and c.

DEFINITION OF ECCENTRICITY

For the ellipse $\dfrac{x^2}{a^2} + \dfrac{y^2}{b^2} = 1$ or $\dfrac{x^2}{b^2} + \dfrac{y^2}{a^2} = 1$ (with $a > b > 0$), the **eccentricity** e is the number

$$e = \frac{c}{a}$$

where $c = \sqrt{a^2 - b^2}$. The eccentricity of every ellipse satisfies $0 < e < 1$.

Johannes Kepler (1571–1630) was the first to give a correct description of the motion of the planets. The cosmology of his time postulated complicated systems of circles moving on circles to describe these motions. Kepler sought a simpler and more harmonious description. As official astronomer at the imperial court in Prague, he studied the astronomical observations of the Danish astronomer Tycho Brahe, whose data were the most accurate available at the time. After numerous attempts to find a theory, Kepler made the momentous discovery that the orbits of the planets are elliptical. His three great laws of planetary motion are

1. The orbit of each planet is an ellipse with the sun at one focus.
2. The line segment that joins the sun to a planet sweeps out equal areas in equal time (see the figure).
3. The square of the period of revolution of a planet is proportional to the cube of the length of the major axis of its orbit.

His formulation of these laws is perhaps the most impressive deduction from empirical data in the history of science.

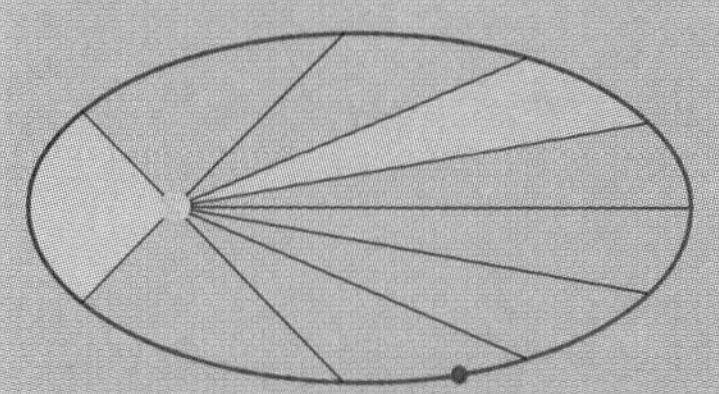

Thus, if e is close to 1, then c is almost equal to a, and the ellipse is elongated in shape, but if e is close to 0, then the ellipse is close to a circle in shape. The eccentricity is a measure of how "stretched" the ellipse is.

In Figure 9 we show a number of ellipses to demonstrate the effect of varying the eccentricity e.

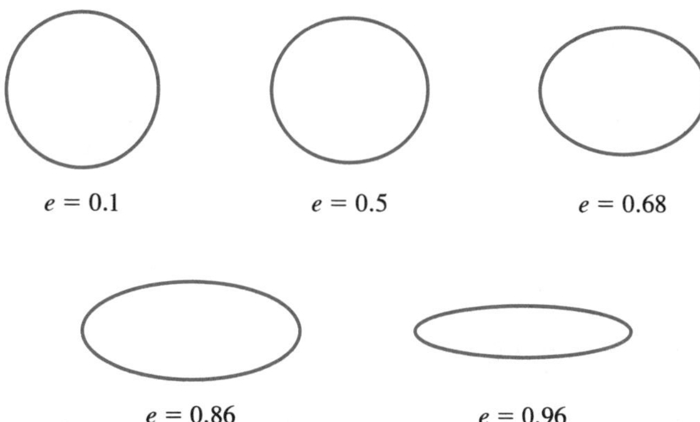

FIGURE 9 Ellipses with various eccentricities

EXAMPLE 4 ■ Finding the Equation of an Ellipse from Its Eccentricity and Foci

Find the equation of the ellipse with foci $(0, \pm 8)$ and eccentricity $e = \frac{4}{5}$, and sketch its graph.

SOLUTION

We are given $e = \frac{4}{5}$ and $c = 8$. Thus

$$\frac{4}{5} = \frac{8}{a} \qquad \text{Eccentricity } e = \frac{c}{a}$$

$$4a = 40 \qquad \text{Cross multiply}$$

$$a = 10$$

To find b, we use the fact that $c^2 = a^2 - b^2$.

$$8^2 = 10^2 - b^2$$

$$b^2 = 10^2 - 8^2 = 36$$

$$b = 6$$

Thus, the equation of the ellipse is

$$\frac{x^2}{36} + \frac{y^2}{100} = 1$$

Because the foci are on the y-axis, the ellipse is oriented vertically. To sketch the ellipse, we find the intercepts: The x-intercepts are ± 6 and the y-intercepts are ± 10. The graph is sketched in Figure 10.

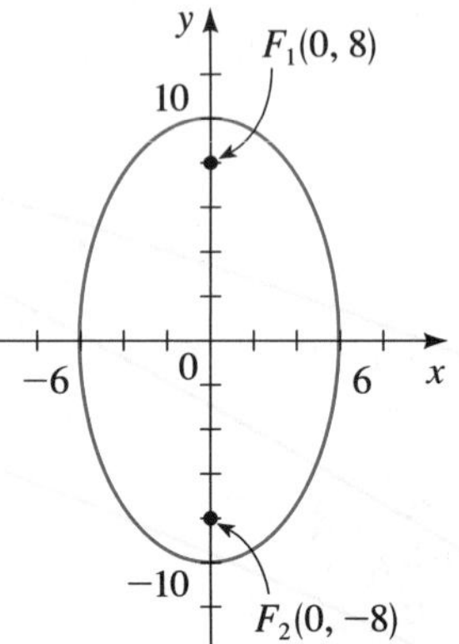

FIGURE 10
$\dfrac{x^2}{36} + \dfrac{y^2}{100} = 1$

■

Gravitational attraction causes the planets to move in elliptical orbits around the sun with the sun at one focus. This remarkable property was first observed by Johannes Kepler and was later deduced by Isaac Newton from his inverse square law of gravity, using calculus. The orbits of the planets have different eccentricities, but most are nearly circular (see the margin note on page 548).

Ellipses, like parabolas, have an interesting *reflection property* that leads to a number of practical applications. If a light source is placed at one focus of a reflecting surface with elliptical cross sections, then all the light will be reflected off the surface to the other focus, as shown in Figure 11. This principle, which works for sound waves as well as for light, is used in *lithotripsy*, a treatment for kidney stones. The patient is placed in a tub of water with elliptical cross sections in such a way that the kidney stone is accurately located at one focus. High-intensity sound waves generated at the other focus are reflected to the stone and destroy it with minimal damage to surrounding tissue. The patient is spared the trauma of surgery and recovers within days instead of weeks.

The reflection property of ellipses is also used in the construction of *whispering galleries*. Sound coming from one focus bounces off the walls and ceiling of an elliptical room and passes through the other focus. In these rooms even quiet whispers spoken at one focus can be heard clearly at the other. Famous whispering galleries include the National Statuary Gallery of the U.S. Capitol in Washington, D.C., and the Mormon Tabernacle in Salt Lake City, Utah.

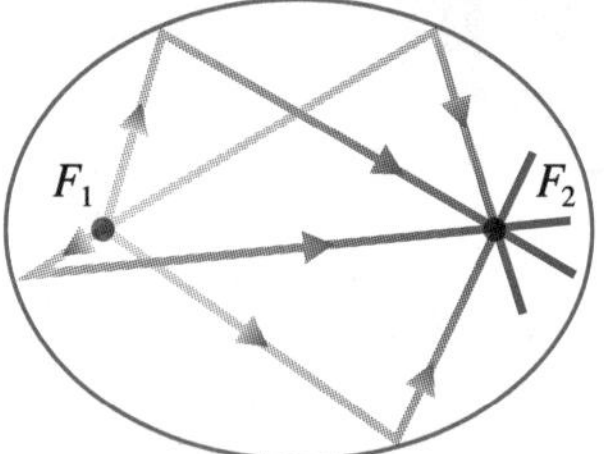

FIGURE 11

8.2 EXERCISES

1–14 ■ Find the vertices, foci, and eccentricity of the ellipse. Determine the lengths of the major and minor axes, and sketch the graph.

1. $\frac{x^2}{25} + \frac{y^2}{9} = 1$

2. $\frac{x^2}{16} + \frac{y^2}{25} = 1$

3. $9x^2 + 4y^2 = 36$

4. $4x^2 + 25y^2 = 100$

5. $x^2 + 4y^2 = 16$

6. $4x^2 + y^2 = 16$

7. $2x^2 + y^2 = 3$

8. $5x^2 + 6y^2 = 30$

9. $x^2 + 4y^2 = 1$

10. $9x^2 + 4y^2 = 1$

11. $\frac{1}{2}x^2 + \frac{1}{8}y^2 = \frac{1}{4}$

12. $x^2 = 4 - 2y^2$

13. $y^2 = 1 - 2x^2$

14. $20x^2 + 4y^2 = 5$

15–18 ■ Find an equation for the ellipse whose graph is shown.

15.

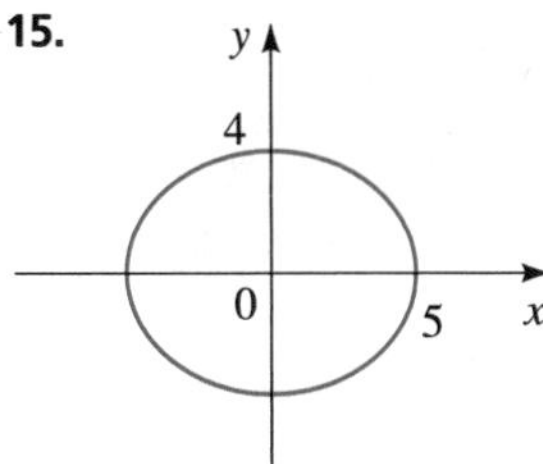

16.

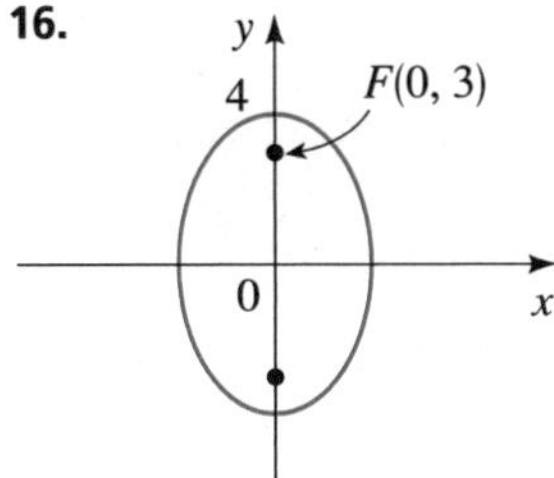

17.

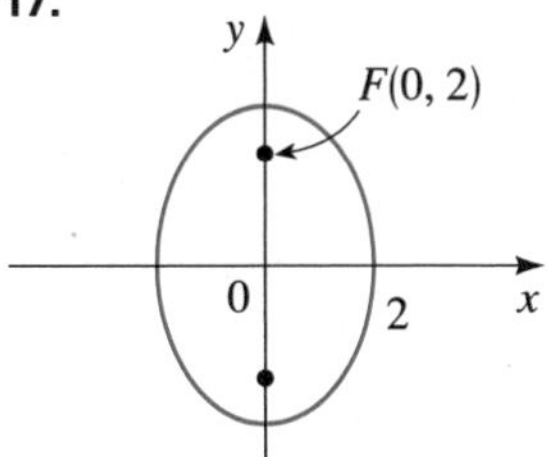

18.

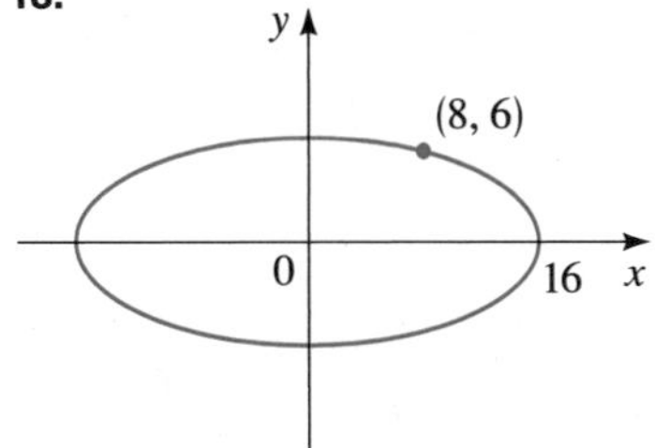

19–30 ■ Find an equation for the ellipse that satisfies the given conditions.

19. Foci $(\pm 4, 0)$, vertices $(\pm 5, 0)$

20. Foci $(0, \pm 3)$, vertices $(0, \pm 5)$

21. Length of major axis 4, length of minor axis 2, foci on y-axis

22. Length of major axis 6, length of minor axis 4, foci on x-axis

23. Foci $(0, \pm 2)$, length of minor axis 6

24. Foci $(\pm 5, 0)$, length of major axis 12

25. Endpoints of major axis $(\pm 10, 0)$, distance between foci 6

26. Endpoints of minor axis $(0, \pm 3)$, distance between foci 8

27. Length of major axis 10, foci on x-axis, ellipse passes through the point $(\sqrt{5}, 2)$

28. Eccentricity $\frac{1}{9}$, foci $(0, \pm 2)$

29. Eccentricity 0.8, foci $(\pm 1.5, 0)$

30. Eccentricity $\sqrt{3}/2$, foci on y-axis, length of major axis 4

31–32 ■ Find the intersection points of the pair of ellipses. Sketch the graphs of each pair of equations on the same coordinate axes and label the points of intersection.

31. $\begin{cases} 4x^2 + y^2 = 4 \\ 4x^2 + 9y^2 = 36 \end{cases}$

32. $\begin{cases} \dfrac{x^2}{16} + \dfrac{y^2}{9} = 1 \\ \dfrac{x^2}{9} + \dfrac{y^2}{16} = 1 \end{cases}$

33. The planets move around the sun in elliptical orbits with the sun at one focus. The point in the orbit at which the planet is closest to the sun is called **perihelion**, and the point at which it is farthest is called **aphelion**. These points are the vertices of the orbit. The earth's distance from the sun is 147,000,000 km at perihelion and 153,000,000 km at aphelion. Find an equation for the earth's orbit. (Place the origin at the center of the orbit with the sun on the x-axis.)

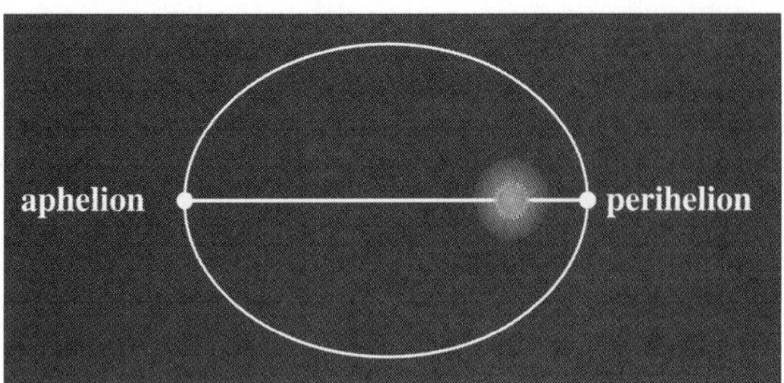

34. With an eccentricity of 0.25, Pluto's orbit is the most eccentric in the solar system. The length of the minor axis of its orbit is approximately 10,000,000,000 km. Find the distance between Pluto and the sun at perihelion and at aphelion. (See Exercise 33.)

35. For an object in an elliptical orbit around the moon, the points in the orbit that are closest to and farthest from the center of the moon are called **perilune** and **apolune**, respectively. These are the vertices of the orbit. The center of the moon is at one focus of the orbit. The *Apollo 11* spacecraft was placed in a lunar orbit with perilune at 68 mi and apolune at 195 mi above the surface of the moon. Assuming the moon is a sphere of radius 1075 mi, find an equation for the orbit of *Apollo 11*. (Place the coordinate axes so that the origin is at the center of the orbit and the foci are located on the x-axis.)

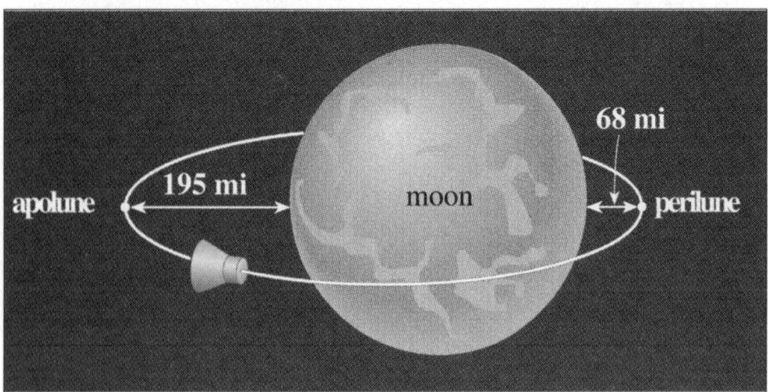

36. A carpenter wishes to construct an elliptical table top from a sheet of plywood, 4 ft by 8 ft. He will trace out the ellipse using the "thumbtack and string" method illustrated in Figures 2 and 3. What length of string should he use, and how far apart should the tacks be located, if the ellipse is to be the largest possible that can be cut out of the plywood sheet?

37. A "sunburst" window above a doorway is constructed in the shape of the top half of an ellipse, as shown in the figure. The window is 20 in. tall at its highest point and 80 in. wide at the bottom. Find the height of the window 25 in. from the center of the base.

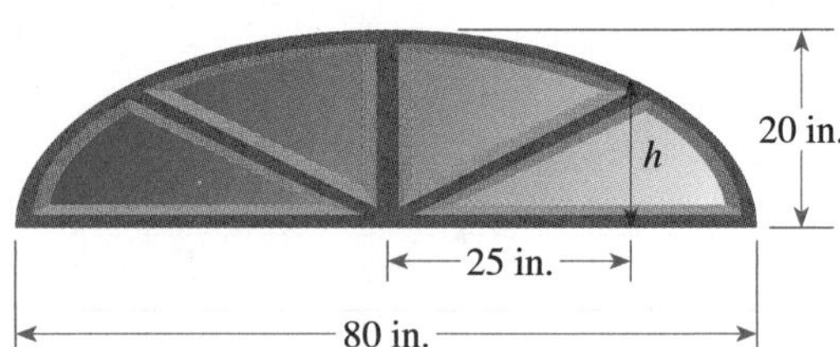

38. The **ancillary circle** of an ellipse is the circle with radius equal to half the length of the minor axis and center the same as the ellipse (see the figure). The ancillary circle is thus the largest circle that can fit within an ellipse.

(a) Find an equation for the ancillary circle of the ellipse $x^2 + 4y^2 = 16$.

(b) For the ellipse and ancillary circle of part (a), show that if (s, t) is a point on the ancillary circle, then $(2s, t)$ is a point on the ellipse.

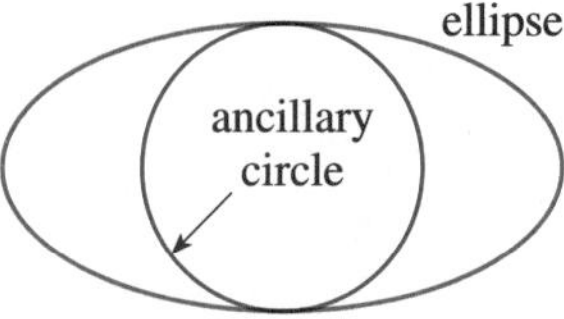

39. A **latus rectum** for an ellipse is a line segment perpendicular to the major axis at a focus, with endpoints on the ellipse, as shown. Show that the length of a latus rectum is $2b^2/a$ for the ellipse

$$\frac{x^2}{a^2} + \frac{y^2}{b^2} = 1 \qquad \text{with } a > b$$

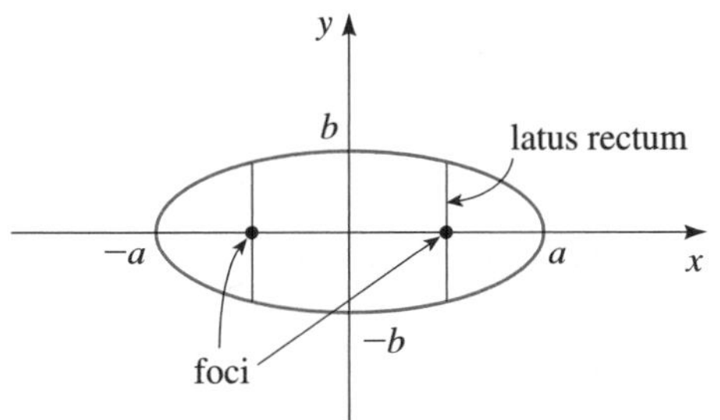

40. If $k > 0$, the following equation represents an ellipse:

$$\frac{x^2}{k} + \frac{y^2}{4 + k} = 1$$

Show that all the ellipses represented by this equation have the same foci, no matter what the value of k.

41. Use a graphing device to draw the following ellipses by solving for y and graphing both solutions.

(a) $\dfrac{x^2}{25} + \dfrac{y^2}{20} = 1$ (b) $6x^2 + y^2 = 36$

42. (a) Use a graphing device to sketch the top half (the portion in the first and second quadrants) of the family of ellipses $x^2 + ky^2 = 100$ for $k = 4, 10, 25$, and 50.

(b) What do the members of this family of ellipses have in common? How do they differ?

DISCOVERY • DISCUSSION

43. Drawing an Ellipse on a Blackboard Try drawing an ellipse as accurately as possible on a blackboard. How would a piece of string and two friends help this process?

44. Light Cone from a Flashlight A flashlight shines on a wall, as shown in the figure. What is the shape of the boundary of the lighted area? Explain your answer.

45. Is It an Ellipse? A piece of paper is wrapped around a cylindrical bottle, and then a compass is used to draw a circle on the paper, as shown in the figure. When the paper is laid flat, is the shape drawn on the paper an ellipse? (You don't need to prove your answer, but you may want to do the experiment and see what you get.)

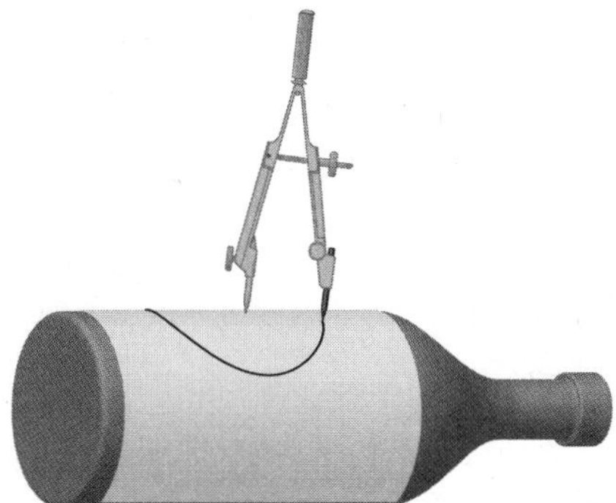

8.3 HYPERBOLAS

Although ellipses and hyperbolas have completely different shapes, their definitions and equations are similar. Instead of using the *sum* of distances from two fixed foci, as in the case of an ellipse, we use the *difference* to define a hyperbola.

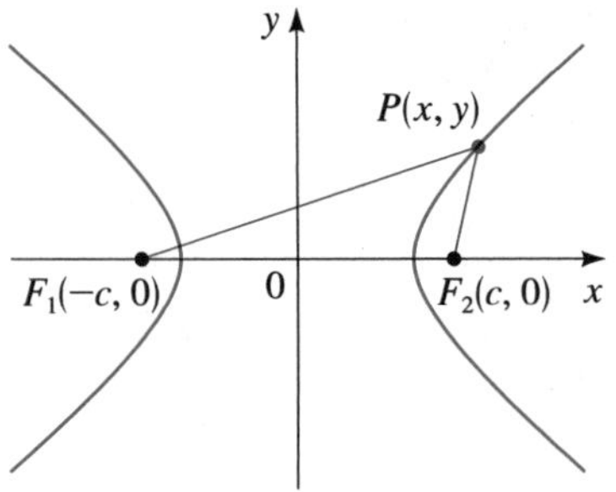

FIGURE 1

P is on the hyperbola if $|d(P, F_1) - d(P, F_2)| = 2a$

GEOMETRIC DEFINITION OF A HYPERBOLA

A **hyperbola** is the set of all points in the plane, the difference of whose distances from two fixed points F_1 and F_2 is a constant. (See Figure 1.) These two fixed points are the **foci** of the hyperbola.

As in the case of the ellipse, we get the simplest equation for the hyperbola by placing the foci on the x-axis at $(\pm c, 0)$, as shown in Figure 1. By definition, if $P(x, y)$ lies on the hyperbola, then either $d(P, F_1) - d(P, F_2)$ or $d(P, F_2) - d(P, F_1)$ must equal some positive constant, which we call $2a$. Thus, we have

$$d(P, F_1) - d(P, F_2) = \pm 2a$$

or

$$\sqrt{(x + c)^2 + y^2} - \sqrt{(x - c)^2 + y^2} = \pm 2a$$

Proceeding as we did in the case of the ellipse (Section 8.2), we simplify this to

$$(c^2 - a^2)x^2 - a^2y^2 = a^2(c^2 - a^2)$$

From triangle PF_1F_2 in Figure 1 we see that $|d(P, F_1) - d(P, F_2)| < 2c$. It follows that $2a < 2c$, or $a < c$. Thus, $c^2 - a^2 > 0$, so we can set $b^2 = c^2 - a^2$. We then simplify the last displayed equation to get

$$\frac{x^2}{a^2} - \frac{y^2}{b^2} = 1$$

This is the *equation of the hyperbola*. If we replace x by $-x$ or y by $-y$ in this equation, it remains unchanged, so the hyperbola is symmetric about both the x- and y-axes and about the origin. The x-intercepts are $\pm a$, and the points $(a, 0)$ and $(-a, 0)$ are the **vertices** of the hyperbola. There is no y-intercept, because setting $x = 0$ in the equation of the hyperbola leads to $-y^2 = b^2$, which has no real solution. Furthermore, the equation of the hyperbola implies that

$$\frac{x^2}{a^2} = \frac{y^2}{b^2} + 1 \geqslant 1$$

so $x^2/a^2 \geqslant 1$; thus, $x^2 \geqslant a^2$, and hence $x \geqslant a$ or $x \leqslant -a$. This means that the hyperbola consists of two parts, called its **branches**. The segment joining the two vertices on the separate branches is the **transverse axis** of the hyperbola, and the origin is called its **center**.

If we place the foci of the hyperbola on the y-axis rather than on the x-axis, then this has the effect of reversing the roles of x and y in the derivation of the equation of the hyperbola. This leads to a hyperbola with a vertical transverse axis.

The main properties of hyperbolas are listed in the following box.

HYPERBOLA WITH CENTER AT THE ORIGIN

The graph of each of the following equations is a hyperbola with center at the origin and having the given properties.

EQUATION	$\dfrac{x^2}{a^2} - \dfrac{y^2}{b^2} = 1 \quad (a > 0, b > 0)$	$\dfrac{y^2}{a^2} - \dfrac{x^2}{b^2} = 1 \quad (a > 0, b > 0)$
VERTICES	$(\pm a, 0)$	$(0, \pm a)$
TRANSVERSE AXIS	Horizontal, length $2a$	Vertical, length $2a$
ASYMPTOTES	$y = \pm \dfrac{b}{a} x$	$y = \pm \dfrac{a}{b} x$
FOCI	$(\pm c, 0), \quad c^2 = a^2 + b^2$	$(0, \pm c), \quad c^2 = a^2 + b^2$
GRAPH		

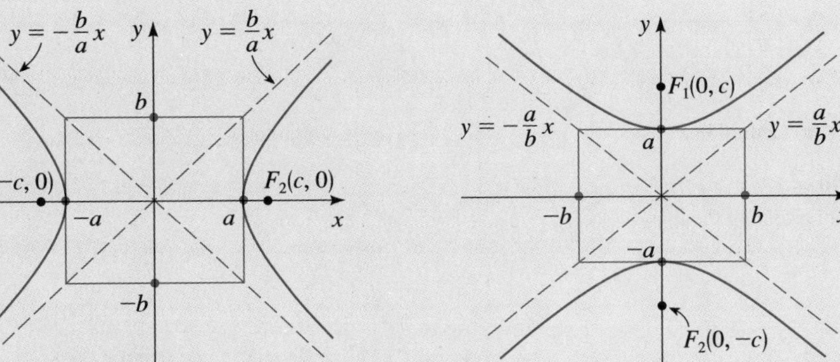

Asymptotes of rational functions are discussed in Section 5.5.

The *asymptotes* mentioned in this box are lines that the hyperbola approaches for large values of x and y. To find the asymptotes in the first case in the box, we solve the equation for y to get

$$y = \pm \frac{b}{a} \sqrt{x^2 - a^2}$$

$$= \pm \frac{b}{a} x \sqrt{1 - \frac{a^2}{x^2}}$$

As x gets large, a^2/x^2 gets closer to zero. In other words, as $x \to \infty$ we have $a^2/x^2 \to 0$. So, for large x the value of y can be approximated as $y = \pm(b/a)x$. This shows that these lines are asymptotes of the hyperbola.

Asymptotes are an essential aid for graphing a hyperbola; they help us determine its shape. A convenient way to find the asymptotes, for a parabola with horizontal transverse axis, is to first plot the points $(a, 0)$, $(-a, 0)$, $(0, b)$, and $(0, -b)$. Then sketch horizontal and vertical segments through these points to construct a rectangle, as shown in Figure 2(a). We call this rectangle the **central box** of the hyperbola. The slopes of the diagonals of the central box are $\pm b/a$, so by extending them we obtain the asymptotes $y = \pm(b/a)x$, as sketched in part (b) of the figure. Finally, we plot the vertices and use the asymptotes as a guide in sketching the hyperbola shown in part (c). (A similar procedure applies to graphing a hyperbola that has a vertical transverse axis.)

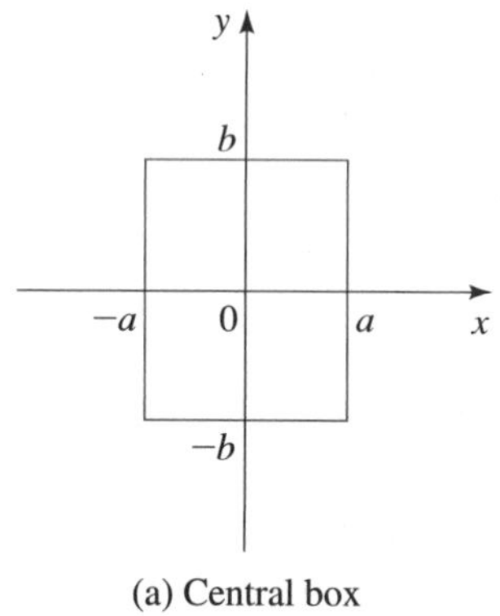

(a) Central box

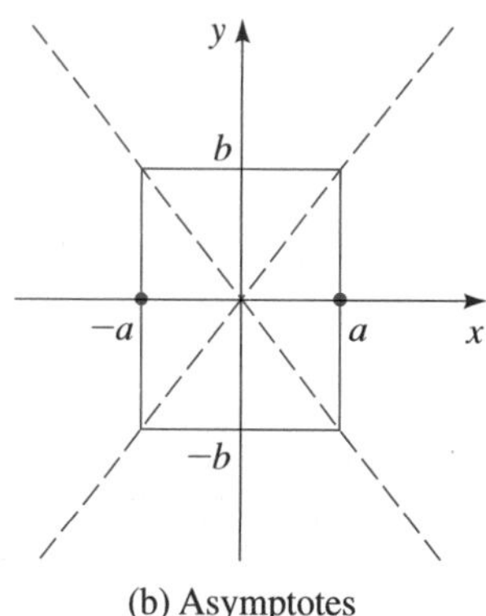

(b) Asymptotes

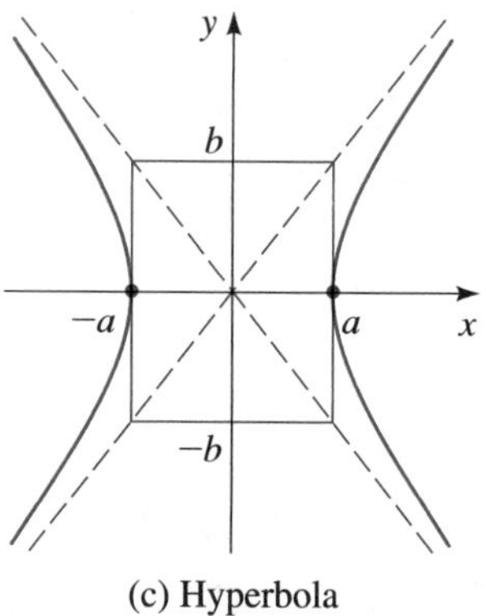

(c) Hyperbola

FIGURE 2
Steps in graphing the hyperbola $\dfrac{x^2}{a^2} - \dfrac{y^2}{b^2} = 1$

HOW TO SKETCH A HYPERBOLA

1. SKETCH THE CENTRAL BOX. This is the rectangle centered at the origin, with sides parallel to the axes, that crosses one axis at $\pm a$, the other at $\pm b$.
2. SKETCH THE ASYMPTOTES. These are the lines obtained by extending the diagonals of the central box.
3. PLOT THE VERTICES. These are the two x-intercepts or the two y-intercepts.
4. SKETCH THE HYPERBOLA. Start at a vertex and sketch a branch of the hyperbola, approaching the asymptotes. Sketch the other branch in the same way.

EXAMPLE 1 ■ A Hyperbola with Horizontal Transverse Axis

Find the vertices, foci, and asymptotes of the hyperbola, and sketch its graph.

$$9x^2 - 16y^2 = 144$$

PATHS OF COMETS

The path of a comet is an ellipse, a parabola, or a hyperbola with the sun at a focus. This fact can be proved using calculus and Newton's laws of motion.* If the path is a parabola or a hyperbola, the comet will never return. If the path is an ellipse, it can be determined precisely when and where the comet can be seen again. Halley's comet has an elliptical path and returns every 75 years; it was last seen in 1987. The brightest comet of the 20th century was comet Hale-Bopp, seen in 1997. Its orbit is also elliptical.

*James Stewart, *Calculus,* 4th ed. (Pacific Grove, CA: Brooks/Cole, 1999), pp. 897–899.

SOLUTION

First we divide both sides of the equation by 144 to put it into standard form:

$$\frac{x^2}{16} - \frac{y^2}{9} = 1$$

Because the x^2-term is positive, the hyperbola has a horizontal transverse axis; its vertices and foci are on the x-axis. Since $a^2 = 16$ and $b^2 = 9$, we get $a = 4$, $b = 3$, and $c = \sqrt{16 + 9} = 5$. Thus, we have

VERTICES	$(\pm 4, 0)$
FOCI	$(\pm 5, 0)$
ASYMPTOTES	$y = \pm \frac{3}{4}x$

After sketching the central box and asymptotes, we complete the sketch of the hyperbola as in Figure 3.

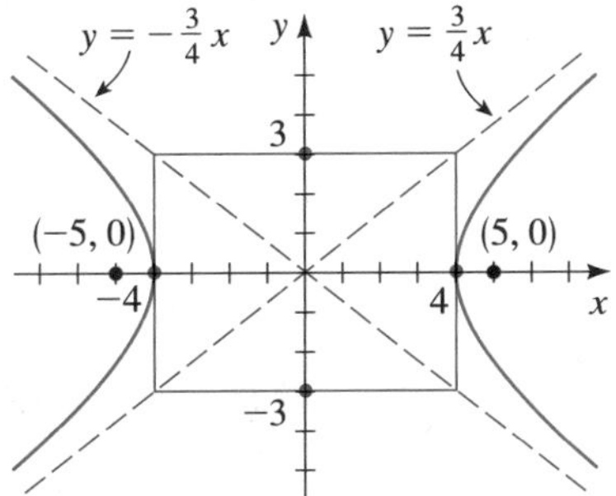

FIGURE 3
$9x^2 - 16y^2 = 144$

■

EXAMPLE 2 ■ A Hyperbola with Vertical Transverse Axis

Find the vertices, foci, and asymptotes of the hyperbola, and sketch its graph.

$$x^2 - 9y^2 + 9 = 0$$

SOLUTION

We begin by writing the equation in the standard form for a hyperbola.

$$x^2 - 9y^2 = -9$$

$$y^2 - \frac{x^2}{9} = 1 \qquad \text{Divide by } -9$$

Because the y^2-term is positive, the hyperbola has a vertical transverse axis; its foci and vertices are on the y-axis. Since $a^2 = 1$ and $b^2 = 9$, we get $a = 1$, $b = 3$, and $c = \sqrt{1 + 9} = \sqrt{10}$. Thus, we have

VERTICES	$(0, \pm 1)$
FOCI	$(0, \pm\sqrt{10})$
ASYMPTOTES	$y = \pm\frac{1}{3}x$

We sketch the central box and asymptotes, then complete the graph, as shown in Figure 4.

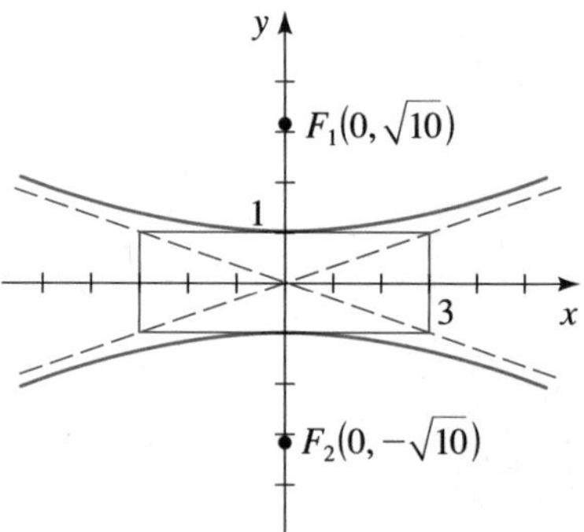

FIGURE 4
$x^2 - 9y^2 + 9 = 0$

■

EXAMPLE 3 ■ **Finding the Equation of a Hyperbola from Its Vertices and Foci**

Find the equation of the hyperbola with vertices $(\pm 3, 0)$ and foci $(\pm 4, 0)$. Sketch the graph.

SOLUTION

Since the vertices are on the x-axis, the hyperbola has a horizontal transverse axis. Its equation is of the form

$$\frac{x^2}{3^2} - \frac{y^2}{b^2} = 1$$

We have $a = 3$ and $c = 4$. To find b, we use the relation $a^2 + b^2 = c^2$:

$$3^2 + b^2 = 4^2$$
$$b^2 = 4^2 - 3^2 = 7$$
$$b = \sqrt{7}$$

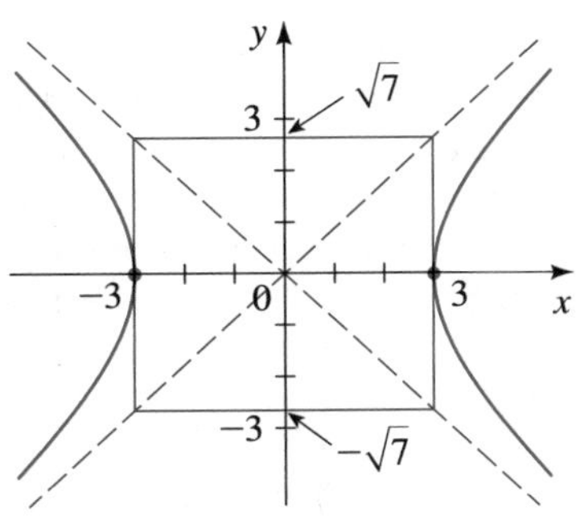

FIGURE 5
$\dfrac{x^2}{9} - \dfrac{y^2}{7} = 1$

Thus, the equation of the hyperbola is

$$\frac{x^2}{9} - \frac{y^2}{7} = 1$$

The graph is shown in Figure 5. ■

EXAMPLE 4 ■ Finding the Equation of a Hyperbola from Its Vertices and Asymptotes

Find the equation and the foci of the hyperbola with vertices $(0, \pm 2)$ and asymptotes $y = \pm 2x$. Sketch the graph.

SOLUTION

Since the vertices are on the y-axis, the hyperbola has a vertical transverse axis with $a = 2$. From the asymptote equation we see that

$$\frac{a}{b} = 2$$

$$\frac{2}{b} = 2 \qquad \text{Since } a = 2$$

$$b = \frac{2}{2} = 1$$

Thus, the equation of the hyperbola is

$$\frac{y^2}{4} - x^2 = 1$$

To find the foci, we calculate $c^2 = a^2 + b^2 = 2^2 + 1^2 = 5$, so $c = \sqrt{5}$. Thus, the foci are $(0, \pm\sqrt{5})$. The graph is shown in Figure 6.

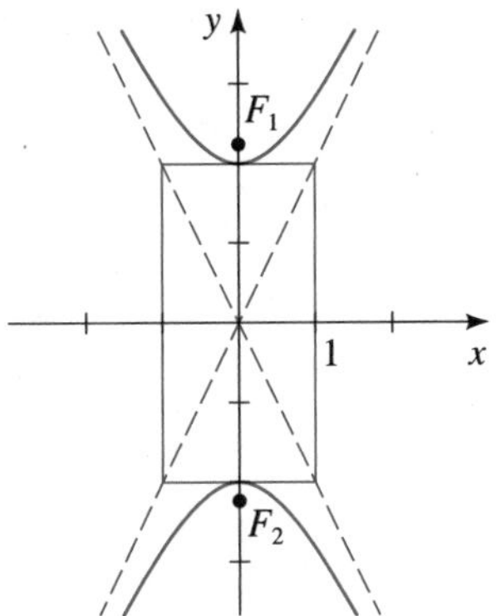

FIGURE 6
$\dfrac{y^2}{4} - x^2 = 1$

■

Like parabolas and ellipses, hyperbolas have an interesting *reflection property*. Light aimed at one focus of a hyperbolic mirror is reflected toward the other focus, as shown in Figure 7. This property is used in the construction of Cassegrain-type telescopes. A hyperbolic mirror is placed in the telescope tube so that light reflected from the primary parabolic reflector is aimed at one focus of the hyperbolic mirror. The light is then refocused at a more accessible point below the primary reflector (Figure 8).

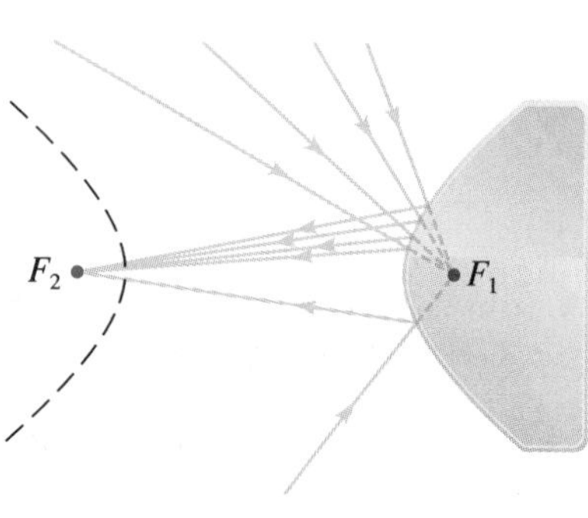

FIGURE 7
Reflection property of hyperbolas

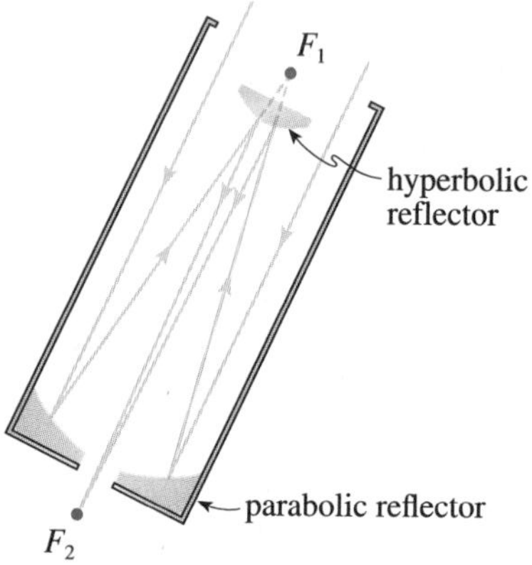

FIGURE 8
Cassegrain-type telescope

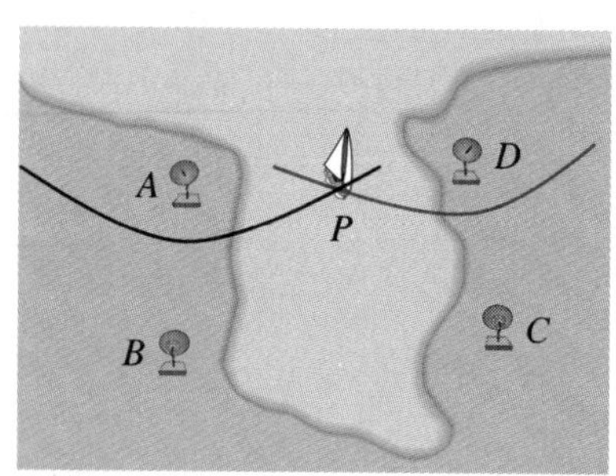

FIGURE 9
LORAN system for finding the location of a ship

The LORAN (LOng RAnge Navigation) system was used until the early 1990s; it has now been superseded by the GPS system (see page 452). In the LORAN system, hyperbolas are used onboard a ship to determine its location. In Figure 9 radio stations at A and B transmit signals simultaneously for reception by the ship at P. The onboard computer converts the time difference in reception of these signals into a distance difference $d(P, A) - d(P, B)$. From the definition of a hyperbola this locates the ship on one branch of a hyperbola with foci at A and B (sketched in black in the figure). The same procedure is carried out with two other radio stations at C and D, and this locates the ship on a second hyperbola (shown in red in the figure). (In practice only three stations are needed because one station can be used as a focus for both hyperbolas.) The coordinates of the intersection point of these two hyperbolas, which can be calculated precisely by the computer, give the location of P.

8.3 EXERCISES

1–12 ■ Find the vertices, foci, and asymptotes of the hyperbola, and sketch its graph.

1. $\dfrac{x^2}{4} - \dfrac{y^2}{16} = 1$

2. $\dfrac{y^2}{9} - \dfrac{x^2}{16} = 1$

3. $y^2 - \dfrac{x^2}{25} = 1$

4. $\dfrac{x^2}{2} - y^2 = 1$

5. $x^2 - y^2 = 1$

6. $9x^2 - 4y^2 = 36$

7. $25y^2 - 9x^2 = 225$

8. $x^2 - y^2 + 4 = 0$

9. $x^2 - 4y^2 - 8 = 0$

10. $x^2 - 2y^2 = 3$

11. $4y^2 - x^2 = 1$

12. $9x^2 - 16y^2 = 1$

13–16 ■ Find the equation for the hyperbola whose graph is shown.

13.

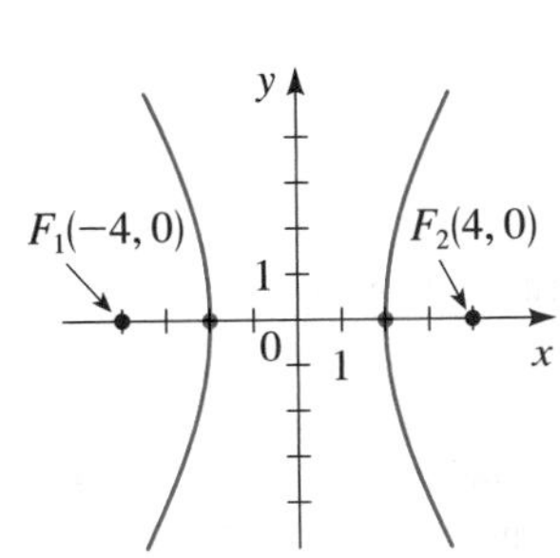

14.

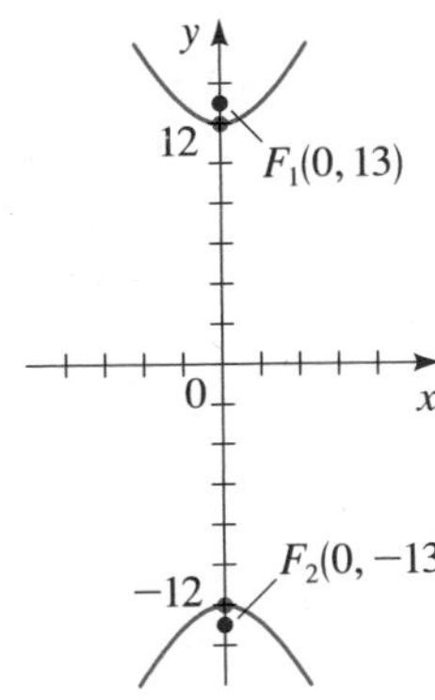

15.

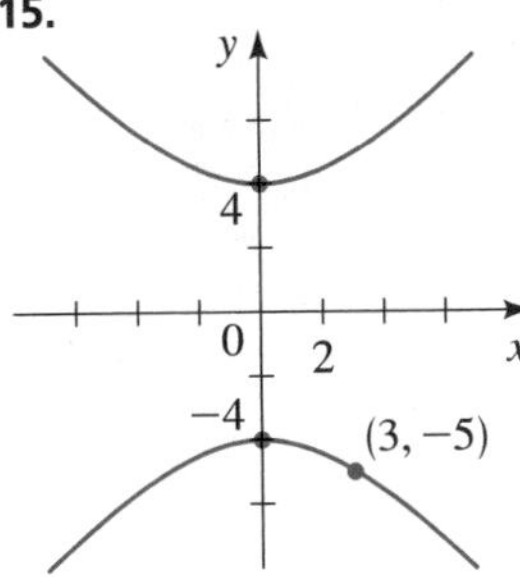

16.

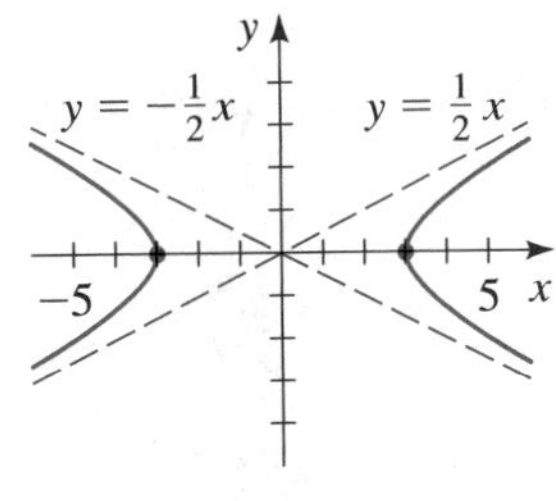

17–28 ■ Find an equation for the hyperbola that satisfies the given conditions.

17. Foci $(\pm 5, 0)$, vertices $(\pm 3, 0)$

18. Foci $(0, \pm 10)$, vertices $(0, \pm 8)$

19. Foci $(0, \pm 2)$, vertices $(0, \pm 1)$

20. Foci $(\pm 6, 0)$, vertices $(\pm 2, 0)$

21. Vertices $(\pm 1, 0)$, asymptotes $y = \pm 5x$

22. Vertices $(0, \pm 6)$, asymptotes $y = \pm \frac{1}{3}x$

23. Foci $(0, \pm 8)$, asymptotes $y = \pm \frac{1}{2}x$

24. Vertices $(0, \pm 6)$, hyperbola passes through $(-5, 9)$

25. Asymptotes $y = \pm x$, hyperbola passes through $(5, 3)$

26. Foci $(\pm 3, 0)$, hyperbola passes through $(4, 1)$

27. Foci $(\pm 5, 0)$, length of transverse axis 6

28. Foci $(0, \pm 1)$, length of transverse axis 1

29. (a) Show that the asymptotes of the hyperbola $x^2 - y^2 = 5$ are perpendicular to each other.

(b) Find an equation for the hyperbola with foci $(\pm c, 0)$ and with asymptotes perpendicular to each other.

30. The hyperbolas

$$\frac{x^2}{a^2} - \frac{y^2}{b^2} = 1 \qquad \text{and} \qquad \frac{x^2}{a^2} - \frac{y^2}{b^2} = -1$$

are said to be **conjugate** to each other.

(a) Show that the hyperbolas

$$x^2 - 4y^2 + 16 = 0 \qquad \text{and} \qquad 4y^2 - x^2 + 16 = 0$$

are conjugate to each other, and sketch their graphs on the same coordinate axes.

(b) What do the hyperbolas of part (a) have in common?

(c) Show that any pair of conjugate hyperbolas have the relationship you discovered in part (b).

31. In the derivation of the equation of the hyperbola at the beginning of this section, we said that the equation

$$\sqrt{(x + c)^2 + y^2} - \sqrt{(x - c)^2 + y^2} = \pm 2a$$

simplifies to

$$(c^2 - a^2)x^2 - a^2y^2 = a^2(c^2 - a^2)$$

Supply the steps needed to show this.

32. (a) For the hyperbola

$$\frac{x^2}{9} - \frac{y^2}{16} = 1$$

determine the values of a, b, and c, and find the coordinates of the foci F_1 and F_2.

(b) Show that the point $P(5, \frac{16}{3})$ lies on this hyperbola.

(c) Find $d(P, F_1)$ and $d(P, F_2)$.

(d) Verify that the difference between $d(P, F_1)$ and $d(P, F_2)$ is $2a$.

33. Refer to Figure 9 in the text. Suppose that the radio stations at A and B are 500 mi apart, and that the ship at P receives Station A's signal 2640 microseconds (μs) before it receives the signal from B.

(a) Assuming that radio signals travel at 980 ft/μs, find $d(P, A) - d(P, B)$.

(b) Find an equation for the branch of the hyperbola indicated in black in the figure. (Place A and B on the y-axis with the origin halfway between them. Use miles as the unit of distance.)

(c) If A is due north of B, and if P is due east of A, how far is P from A?

34. Some comets, such as Halley's comet, are a permanent part of the solar system, traveling in elliptical orbits around the sun. Others pass through the solar system only once, following a hyperbolic path with the sun at a focus. The figure shows the path of such a comet. Find an equation for the path, assuming that the closest the comet comes to the sun is 2×10^9 mi and that the path the comet was taking before it neared the solar system is at a right angle to the path it continues on after leaving the solar system.

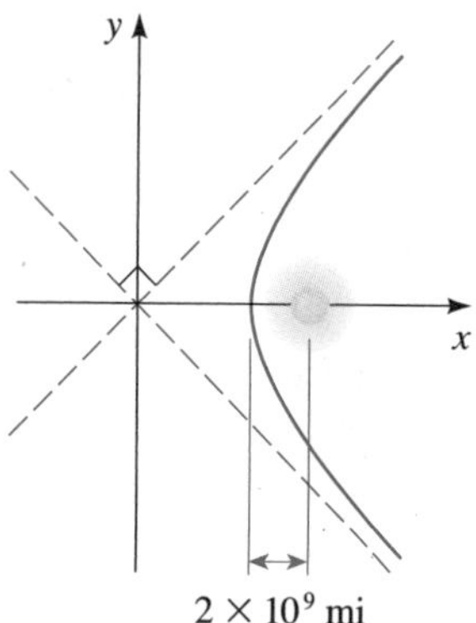

35. Hyperbolas are called **confocal** if they have the same foci.

(a) Show that the hyperbolas

$$\frac{y^2}{k} - \frac{x^2}{16-k} = 1 \qquad \text{with } 0 < k < 16$$

are confocal.

(b) Use a graphing device to draw the top branches of the family of hyperbolas in part (a) for $k = 1, 4, 8$, and 12. How does the shape of the graph change as k increases?

DISCOVERY • DISCUSSION

36. Hyperbolas in the Real World Several examples of the uses of hyperbolas are given in the text. Find other situations in real life where hyperbolas occur. Consult a scientific encyclopedia in the reference section of your library, or search the Internet.

37. Light from a Lamp The light from a lamp forms a lighted area on a wall, as shown in the figure. Why is the boundary of this lighted area a hyperbola? How can one hold a flashlight so that its beam forms a hyperbola on the ground?

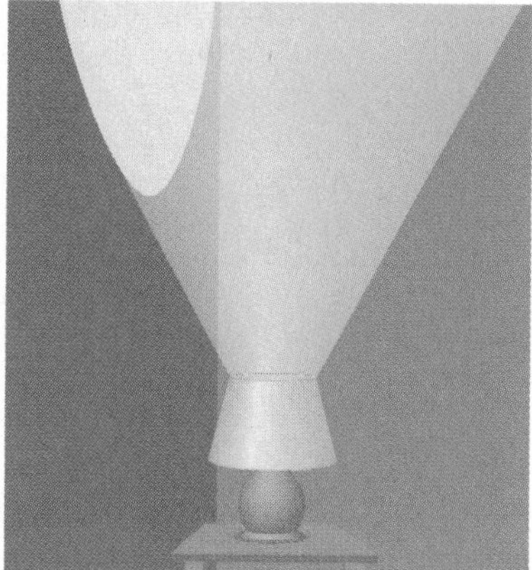

8.4 SHIFTED CONICS

In the preceding sections we studied parabolas with vertices at the origin and ellipses and hyperbolas with centers at the origin. We restricted ourselves to these cases because these equations have the simplest form. In this section we consider conics whose vertices and centers are not necessarily at the origin, and we determine how this affects their equations.

In Section 4.5 we studied transformations of functions that have the effect of shifting their graphs. In general, for any equation in x and y, if we replace x by $x - h$ or by $x + h$, the graph of the new equation is simply the old graph shifted horizontally; if y is replaced by $y - k$ or by $y + k$, the graph is shifted vertically. The following box gives the details.

SHIFTING GRAPHS OF EQUATIONS

If h and k are positive real numbers, then replacing x by $x - h$ or by $x + h$ and replacing y by $y - k$ or by $y + k$ has the following effect(s) on the graph of any equation in x and y.

Replacement	How the graph is shifted
1. x replaced by $x - h$	Right h units
2. x replaced by $x + h$	Left h units
3. y replaced by $y - k$	Upward k units
4. y replaced by $y + k$	Downward k units

For example, consider the ellipse with equation

$$\frac{x^2}{a^2} + \frac{y^2}{b^2} = 1$$

which is shown in Figure 1. If we shift it so that its center is at the point (h, k) instead of at the origin, then its equation becomes

$$\frac{(x - h)^2}{a^2} + \frac{(y - k)^2}{b^2} = 1$$

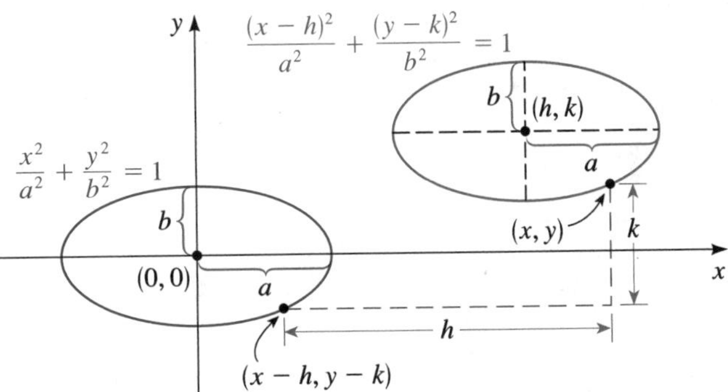

FIGURE 1
Shifted ellipse

EXAMPLE 1 ■ Sketching the Graph of a Shifted Ellipse

Sketch the graph of the ellipse

$$\frac{(x + 1)^2}{4} + \frac{(y - 2)^2}{9} = 1$$

and determine the coordinates of the foci.

SOLUTION

The ellipse

$$\frac{(x + 1)^2}{4} + \frac{(y - 2)^2}{9} = 1 \quad \text{Shifted ellipse}$$

is shifted so that its center is at $(-1, 2)$. It is obtained from the ellipse

$$\frac{x^2}{4} + \frac{y^2}{9} = 1 \qquad \text{Ellipse with center at origin}$$

by shifting it left 1 unit and upward 2 units. The endpoints of the minor and major axes of the unshifted ellipse are $(2, 0)$, $(-2, 0)$, $(0, 3)$, $(0, -3)$. We apply the required shifts to these points to obtain the corresponding points on the shifted ellipse:

$$(2, 0) \rightarrow (2 - 1, 0 + 2) = (1, 2)$$
$$(-2, 0) \rightarrow (-2 - 1, 0 + 2) = (-3, 2)$$
$$(0, 3) \rightarrow (0 - 1, 3 + 2) = (-1, 5)$$
$$(0, -3) \rightarrow (0 - 1, -3 + 2) = (-1, -1)$$

This helps us sketch the graph in Figure 2.

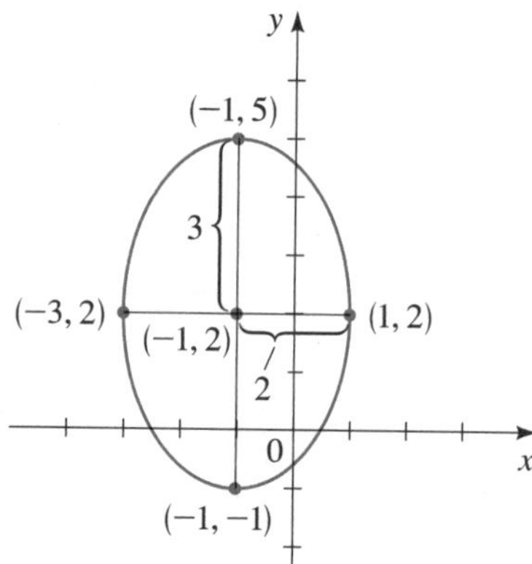

FIGURE 2
$$\frac{(x+1)^2}{4} + \frac{(y-2)^2}{9} = 1$$

To find the foci of the shifted ellipse, we first find the foci of the ellipse with center at the origin. Since $a^2 = 9$ and $b^2 = 4$, we have $c^2 = 9 - 4 = 5$, so $c = \sqrt{5}$. So the foci are $\left(0, \pm\sqrt{5}\right)$. Shifting left 1 unit and upward 2 units, we get

$$(0, \sqrt{5}) \rightarrow (0 - 1, \sqrt{5} + 2) = \left(-1, 2 + \sqrt{5}\right)$$
$$(0, -\sqrt{5}) \rightarrow (0 - 1, -\sqrt{5} + 2) = \left(-1, 2 - \sqrt{5}\right)$$

Thus, the foci of the shifted ellipse are

$$\left(-1, 2 + \sqrt{5}\right) \quad \text{and} \quad \left(-1, 2 - \sqrt{5}\right)$$

■

Applying shifts to parabolas and hyperbolas leads to the equations and graphs shown in Figures 3 and 4.

MATHEMATICS IN THE MODERN WORLD

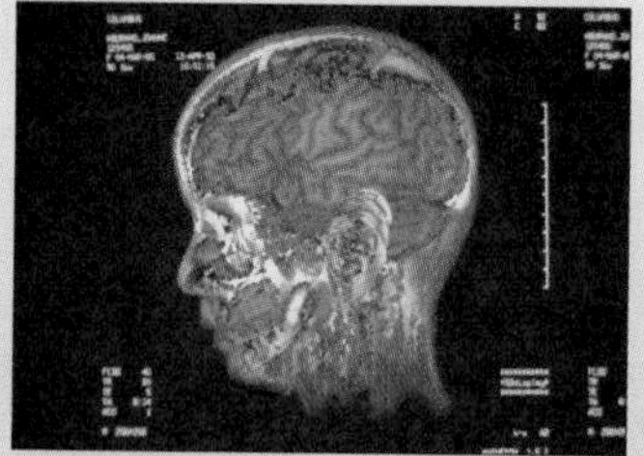

LOOKING INSIDE YOUR HEAD

How would you like to take a look inside your head? The idea may not be especially appealing to you, but doctors often need to do just that. If they can look without invasive surgery, all the better. An X-ray doesn't really give a look inside, it simply gives a "graph" of the density of tissue the X-rays must pass through. So an X-ray is a "flattened" view in one direction. Suppose you get an X-ray view from many different directions—can these "graphs" be used to reconstruct the three-dimensional inside view? This is a purely mathematical problem and was solved by mathematicians a long time ago. However, reconstructing the inside view requires thousands of tedious computations. Today, mathematics and high-speed computers make it possible to "look inside" by a process called Computer Aided Tomography (or CAT scan). Mathematicians continue to search for better ways of using mathematics to reconstruct images. One of the latest techniques, called magnetic resonance imaging (MRI), combines molecular biology and mathematics for a clear "look inside."

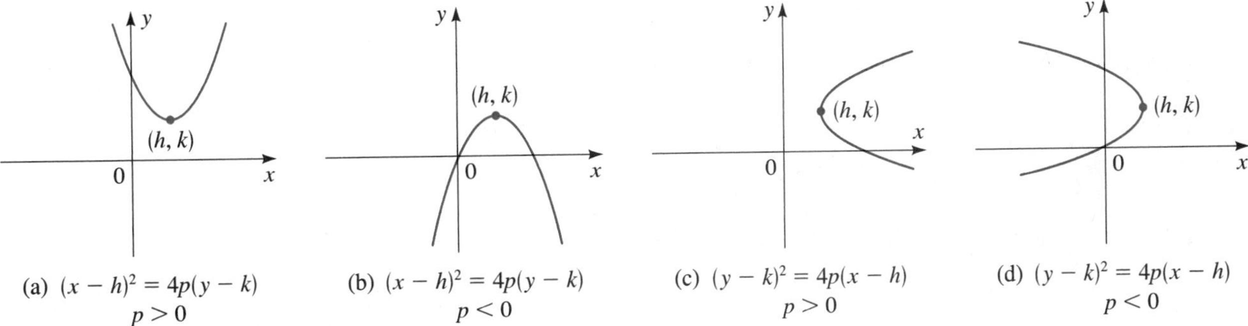

FIGURE 3 Shifted parabolas

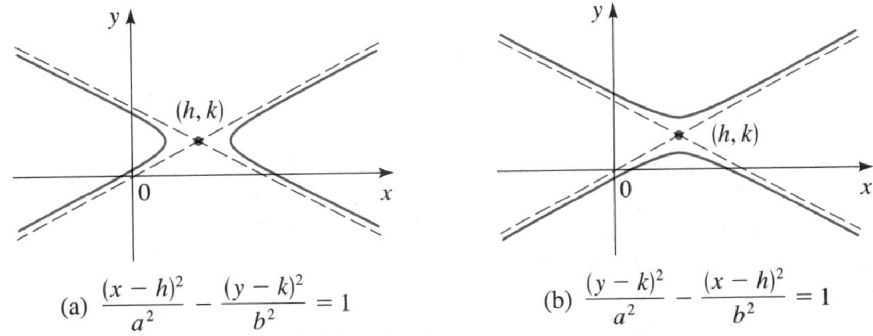

FIGURE 4
Shifted hyperbolas

EXAMPLE 2 ■ Graphing a Shifted Parabola

Determine the vertex, focus, and directrix and sketch the graph of the parabola.

$$x^2 - 4x = 8y - 28$$

SOLUTION

We complete the square in x to put this equation into one of the forms in Figure 3.

$$x^2 - 4x + 4 = 8y - 28 + 4 \quad \text{Add 4 to complete the square}$$

$$(x - 2)^2 = 8y - 24$$

$$(x - 2)^2 = 8(y - 3) \quad \text{Shifted parabola}$$

This parabola opens upward with vertex at $(2, 3)$. It is obtained from the parabola

$$x^2 = 8y \quad \text{Parabola with vertex at origin}$$

by shifting right 2 units and upward 3 units. Since $4p = 8$, we have $p = 2$, so the focus is 2 units above the vertex and the directrix is 2 units below the vertex. Thus, the focus is $(2, 5)$ and the directrix is $y = 1$. The graph is shown in Figure 5. ■

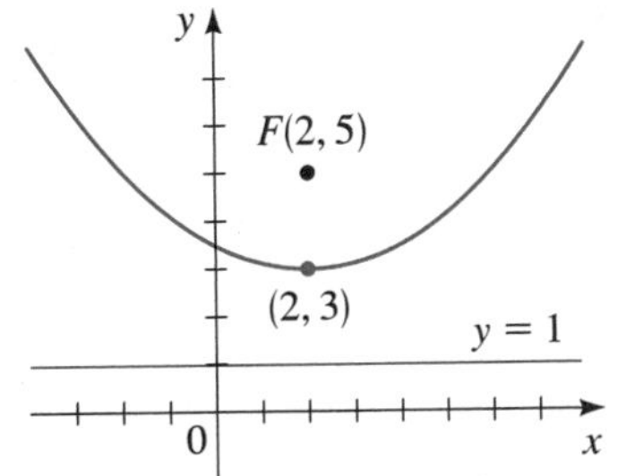

FIGURE 5
$x^2 - 4x = 8y - 28$

Galileo Galilei (1564–1642) was born in Pisa, Italy. He studied medicine, but later abandoned this in favor of science and mathematics. At the age of 25 he demonstrated that light objects fall at the same rate as heavier ones, by dropping cannonballs of various sizes from the Leaning Tower of Pisa. This contradicted the then-accepted view of Aristotle that heavier objects fall more quickly. He also showed that the distance an object falls is proportional to the square of the time it has been falling, and from this was able to prove that the path of a projectile is a parabola.

Galileo constructed the first telescope, and using it, discovered the moons of Jupiter. His advocacy of the Copernican view that the earth revolves around the sun (rather than being stationary) led to his being called before the Inquisition. By then an old man, he was forced to recant his views, but he is said to have muttered under his breath "the earth nevertheless does move." Galileo revolutionized science by expressing scientific principles in the language of mathematics. He said, "The great book of nature is written in mathematical symbols."

EXAMPLE 3 ■ Graphing a Shifted Hyperbola

Show that the equation represents a hyperbola.

$$9x^2 - 72x - 16y^2 - 32y = 16$$

Find its center, vertices, foci, and asymptotes, and sketch its graph.

SOLUTION

We first complete the squares in both x and y:

$$9(x^2 - 8x \qquad) - 16(y^2 + 2y \qquad) = 16$$

$$9(x^2 - 8x + 16) - 16(y^2 + 2y + 1) = 16 + 9 \cdot 16 - 16 \cdot 1 \qquad \text{Complete the squares}$$

$$9(x - 4)^2 - 16(y + 1)^2 = 144 \qquad \text{Divide this by 144}$$

$$\frac{(x - 4)^2}{16} - \frac{(y - 1)^2}{9} = 1 \qquad \text{Shifted hyperbola}$$

This hyperbola has center $(4, -1)$ and a horizontal transverse axis.

CENTER $(4, -1)$

Its graph will have the same shape as the unshifted hyperbola

$$\frac{x^2}{16} - \frac{y^2}{9} = 1 \qquad \text{Hyperbola with center at origin}$$

Since $a^2 = 16$ and $b^2 = 9$, we have $a = 4$, $b = 3$, and $c = \sqrt{a^2 + b^2} = \sqrt{16 + 9} = 5$. Thus, the foci lie 5 units to the left and to the right of the center, and the vertices lie 4 units to either side of the center.

FOCI $(-1, -1)$ and $(9, -1)$

VERTICES $(0, -1)$ and $(8, -1)$

The asymptotes of the unshifted hyperbola are $y = \pm\frac{3}{4}x$, so the asymptotes of the shifted parabola are found as follows.

ASYMPTOTES $(y + 1) = \pm\frac{3}{4}(x - 4)$

$$y + 1 = \pm\tfrac{3}{4}x \mp 3$$

$$y = \tfrac{3}{4}x - 4 \qquad \text{and} \qquad y = -\tfrac{3}{4}x + 2$$

To help us sketch the hyperbola, we draw the central box; it extends 4 units left and right from the center and 3 units upward and downward from the center. We then draw the asymptotes and complete the graph of the shifted hyperbola as shown in Figure 6.

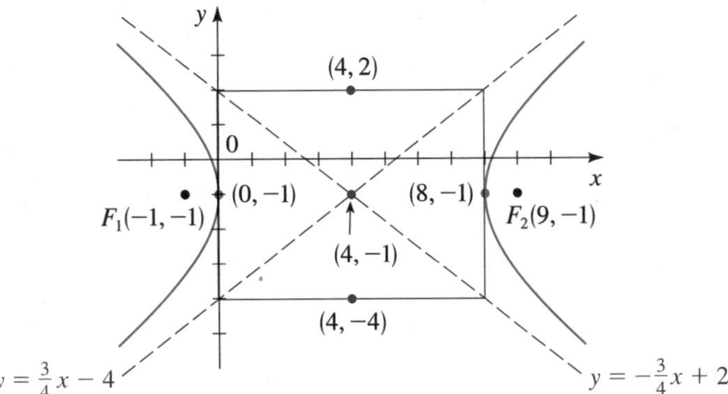

FIGURE 6
$9x^2 - 72x - 16y^2 - 32y = 16$

■

If we expand and simplify the equations of any of the shifted conics illustrated in Figures 1, 3, and 4, then we will always obtain an equation of the form

$$Ax^2 + Cy^2 + Dx + Ey + F = 0$$

where A and C are not both 0. Conversely, if we begin with an equation of this form, then we can complete the square in x and y to see which type of conic section the equation represents. In some cases, the graph of the equation turns out to be just a pair of lines, a single point, or there may be no graph at all. These cases are called **degenerate conics**. The next example illustrates such a case.

EXAMPLE 4 ■ An Equation That Leads to a Degenerate Conic

Sketch the graph of the equation

$$9x^2 - y^2 + 18x + 6y = 0$$

SOLUTION

Because the coefficients of x^2 and y^2 are of opposite sign, this equation looks as if it should represent a hyperbola (like the equation of Example 3). To see whether this is in fact the case, we complete the squares:

$$9(x^2 + 2x \qquad) - (y^2 - 6y \qquad) = 0$$

$$9(x^2 + 2x + 1) - (y^2 - 6y + 9) = 0 + 9 - 9$$

$$9(x + 1)^2 - (y - 3)^2 = 0$$

$$(x + 1)^2 - \frac{(y - 3)^2}{9} = 0 \qquad \text{Divide by 9}$$

For this to fit the form of the equation of a hyperbola, we would need a nonzero

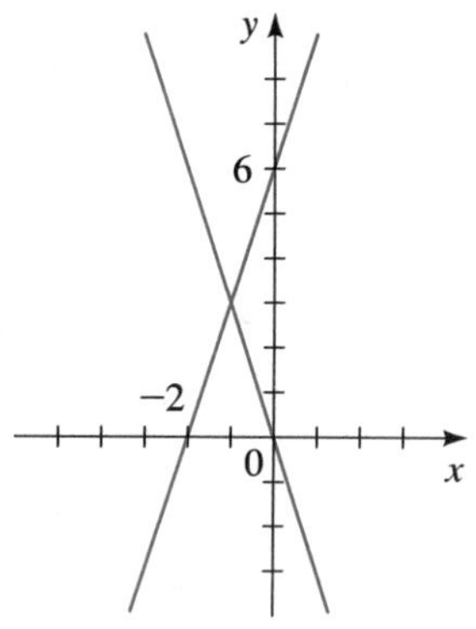

FIGURE 7
$9x^2 - y^2 + 18x + 6y = 0$

constant to the right of the equal sign. In fact, further analysis shows that this is the equation of a pair of intersecting lines:

$$(y-3)^2 = 9(x+1)^2$$

$$y - 3 = \pm 3(x+1) \qquad \text{Take square roots}$$

$$y = 3(x+1) + 3 \qquad \text{or} \qquad y = -3(x+1) + 3$$

$$y = 3x + 6 \qquad\qquad y = -3x$$

These lines are graphed in Figure 7. ■

Because the equation in Example 4 looked at first glance like the equation of a hyperbola but, in fact, turned out to represent simply a pair of lines,we refer to its graph as a **degenerate hyperbola**. Degenerate ellipses and parabolas can also arise when we complete the square(s) in an equation that seems to represent a conic. For example, the equation

$$4x^2 + y^2 - 8x + 2y + 6 = 0$$

looks as if it should represent an ellipse, because the coefficients of x^2 and y^2 have the same sign. But completing the squares leads to

$$(x-1)^2 + \frac{(y+1)^2}{4} = -\frac{1}{4}$$

which has no solution at all (since the sum of two squares cannot be negative). This equation is therefore degenerate.

To summarize, we have the following theorem.

GENERAL EQUATION OF A SHIFTED CONIC

The graph of the equation

$$Ax^2 + Cy^2 + Dx + Ey + F = 0$$

where A and C are not both 0, is a conic or a degenerate conic. In the nondegenerate cases, the graph is

1. a parabola if A or C is 0.
2. an ellipse if A and C have the same sign (or a circle if $A = C$).
3. a hyperbola if A and C have opposite signs.

8.4 EXERCISES

1–4 ■ Find the center, foci, and vertices of the ellipse, and determine the lengths of the major and minor axes. Then sketch the graph.

1. $\dfrac{(x-2)^2}{9} + \dfrac{(y-1)^2}{4} = 1$

2. $\dfrac{(x-3)^2}{16} + (y+3)^2 = 1$

3. $\dfrac{x^2}{9} + \dfrac{(y+5)^2}{25} = 1$

4. $\dfrac{(x+2)^2}{4} + y^2 = 1$

5–8 ■ Find the vertex, focus, and directrix of the parabola, and sketch the graph.

5. $(x-3)^2 = 8(y+1)$

6. $(y+5)^2 = -6x + 12$

7. $-4\left(x + \frac{1}{2}\right)^2 = y$

8. $y^2 = 16x - 8$

9–12 ■ Find the center, foci, vertices, and asymptotes of the hyperbola. Then sketch the graph.

9. $\dfrac{(x+1)^2}{9} - \dfrac{(y-3)^2}{16} = 1$

10. $(x-8)^2 - (y+6)^2 = 1$

11. $y^2 - \dfrac{(x+1)^2}{4} = 1$

12. $\dfrac{(y-1)^2}{25} - (x+3)^2 = 1$

13–18 ■ Find an equation for the conic whose graph is shown.

13.

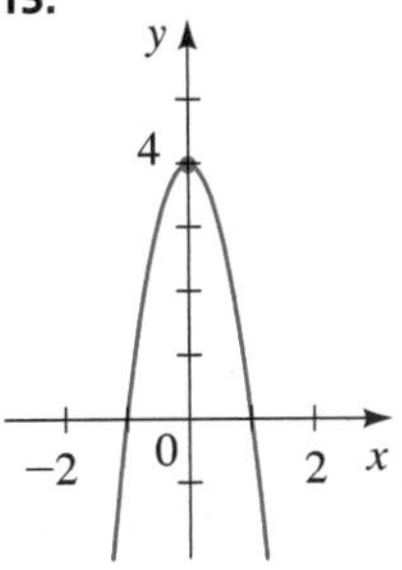

14.

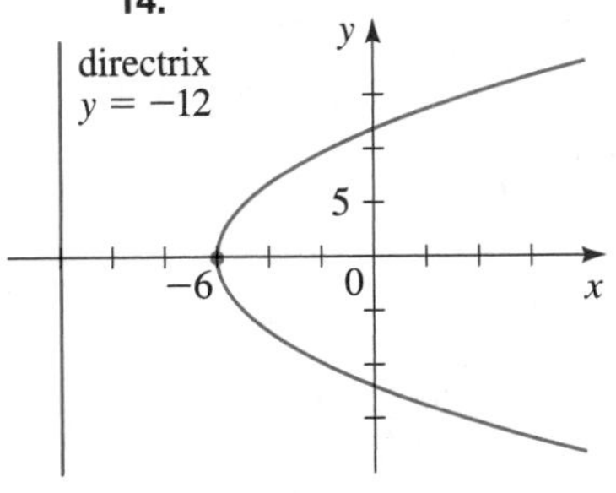

15.

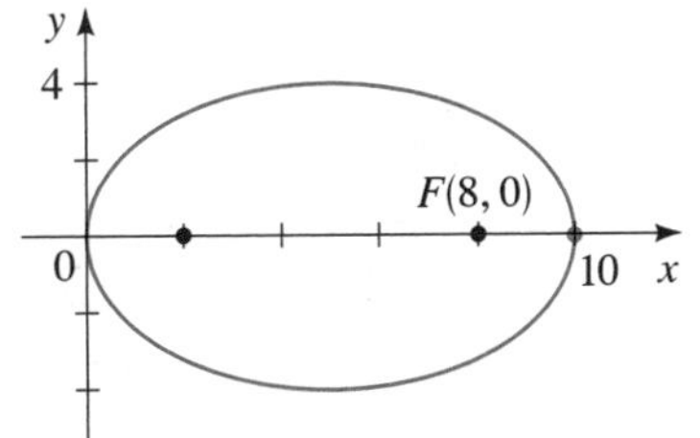

16.

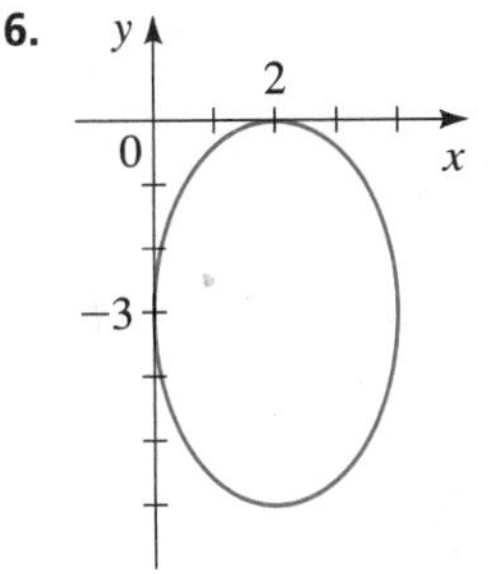

17.

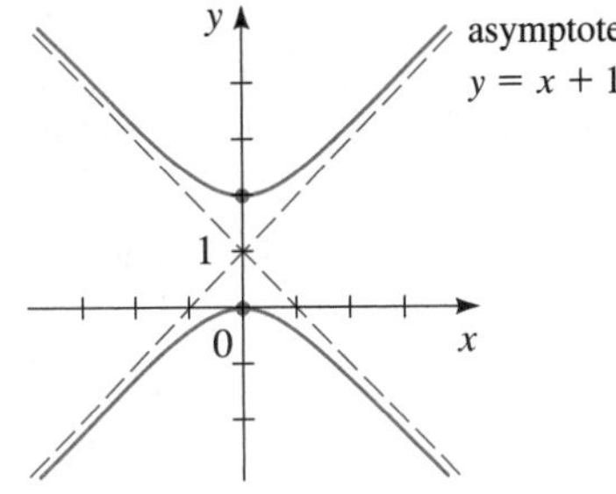

18.

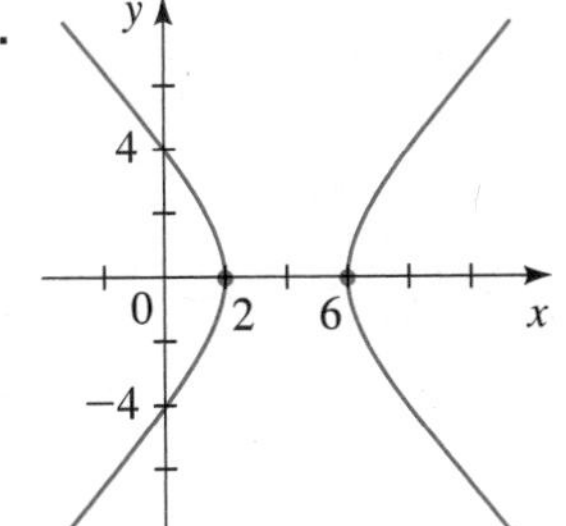

19–30 ■ Complete the square to determine whether the equation represents an ellipse, a parabola, a hyperbola, or a degenerate conic. If the graph is an ellipse, find the center, foci, vertices, and lengths of the major and minor axes. If it is a parabola, find the vertex, focus, and directrix. If it is a hyperbola, find the center, foci, vertices, and asymptotes. Then sketch the graph of the equation. If the equation has no graph, explain why.

19. $9x^2 - 36x + 4y^2 = 0$

20. $y^2 = 4(x + 2y)$

21. $x^2 - 4y^2 - 2x + 16y = 20$

22. $x^2 + 6x + 12y + 9 = 0$

23. $4x^2 + 25y^2 - 24x + 250y + 561 = 0$

24. $2x^2 + y^2 = 2y + 1$

25. $16x^2 - 9y^2 - 96x + 288 = 0$

26. $4x^2 - 4x - 8y + 9 = 0$

27. $x^2 + 16 = 4(y^2 + 2x)$

28. $x^2 - y^2 = 10(x - y) + 1$

29. $3x^2 + 4y^2 - 6x - 24y + 39 = 0$

30. $x^2 + 4y^2 + 20x - 40y + 300 = 0$

31. Determine what the value of F must be if the graph of the equation

$$4x^2 + y^2 + 4(x - 2y) + F = 0$$

is (a) an ellipse, (b) a single point, or (c) the empty set.

32. Find an equation for the ellipse that shares a vertex and a focus with the parabola $x^2 + y = 100$ and has its other focus at the origin.

33. This exercise deals with **confocal parabolas**, that is, families of parabolas that have the same focus.
   (a) Draw graphs of the family of parabolas

$$x^2 = 4p(y + p)$$

for $p = -2, -\frac{3}{2}, -1, -\frac{1}{2}, \frac{1}{2}, 1, \frac{3}{2}, 2$.

   (b) Show that each parabola in this family has its focus at the origin.
   (c) Describe the effect on the graph of moving the vertex closer to the origin.

DISCOVERY • DISCUSSION

34. **A Family of Confocal Conics** Conics that share a focus are called **confocal**. Consider the family of conics that have a focus at $(0, 1)$ and a vertex at the origin (see the figure).
   (a) Find equations of two different ellipses that have these properties.
   (b) Find equations of two different hyperbolas that have these properties.
   (c) Explain why only one parabola satisfies these properties. Find its equation.
   (d) Sketch the conics you found in parts (a), (b), and (c) on the same coordinate axes (for the hyperbolas, sketch the top branches only).
   (e) How are the ellipses and hyperbolas related to the parabola?

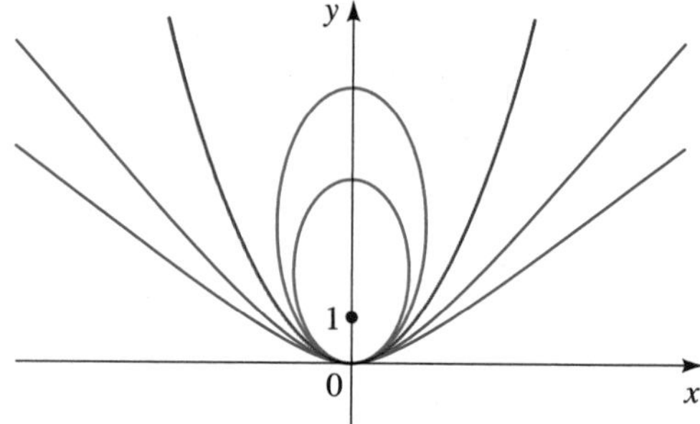

8 REVIEW

CONCEPT CHECK

1. (a) Give the geometric definition of a parabola. What are the focus and directrix of the parabola?
   (b) Sketch the parabola $x^2 = 4py$ for the case $p > 0$. Identify on your diagram the vertex, focus, and directrix. What happens if $p < 0$?
   (c) Sketch the parabola $y^2 = 4px$, together with its vertex, focus, and directrix, for the case $p > 0$. What happens if $p < 0$?

2. (a) Give the geometric definition of an ellipse. What are the foci of the ellipse?
   (b) For the ellipse with equation

$$\frac{x^2}{a^2} + \frac{y^2}{b^2} = 1$$

where $a > b > 0$, what are the coordinates of the vertices and the foci? What are the major and minor axes? Illustrate with a graph.

(c) Give an expression for the eccentricity of the ellipse in part (b).

(d) State the equation of an ellipse with foci on the y-axis.

3. (a) Give the geometric definition of a hyperbola. What are the foci of the hyperbola?

(b) For the hyperbola with equation

$$\frac{x^2}{a^2} - \frac{y^2}{b^2} = 1$$

what are the coordinates of the vertices and foci? What are the equations of the asymptotes? What is the transverse axis? Illustrate with a graph.

(c) State the equation of a hyperbola with foci on the y-axis.

(d) What steps would you take to sketch the hyperbola in part (b)?

4. Suppose h and k are positive numbers. What is the effect on the graph of an equation in x and y if

(a) x is replaced by $x - h$? By $x + h$?

(b) y is replaced by $y - k$? By $y + k$?

5. How can you tell whether the following nondegenerate conic is a parabola, an ellipse, or a hyperbola?

$$Ax^2 + Cy^2 + Dx + Ey + F = 0$$

EXERCISES

1–4 ■ Find the vertex, focus, and directrix of the parabola, and sketch the graph.

1. $x^2 + 8y = 0$

2. $2x - y^2 = 0$

3. $x - y^2 + 4y - 2 = 0$

4. $2x^2 + 6x + 5y + 10 = 0$

5–8 ■ Find the center, vertices, foci, and the lengths of the major and minor axes of the ellipse, and sketch the graph.

5. $x^2 + 4y^2 = 16$

6. $9x^2 + 4y^2 = 1$

7. $4x^2 + 9y^2 = 36y$

8. $2x^2 + y^2 = 2 + 4(x - y)$

9–12 ■ Find the center, vertices, foci, and asymptotes of the hyperbola, and sketch the graph.

9. $x^2 - 2y^2 = 16$

10. $x^2 - 4y^2 + 16 = 0$

11. $9y^2 + 18y = x^2 + 6x + 18$

12. $y^2 = x^2 + 6y$

13–18 ■ Find an equation for the conic whose graph is shown.

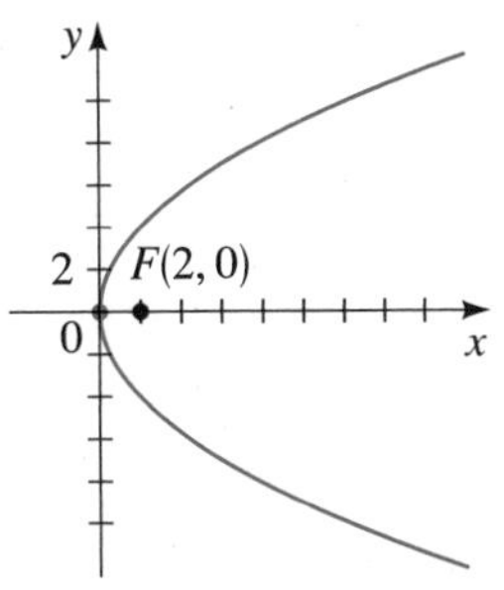

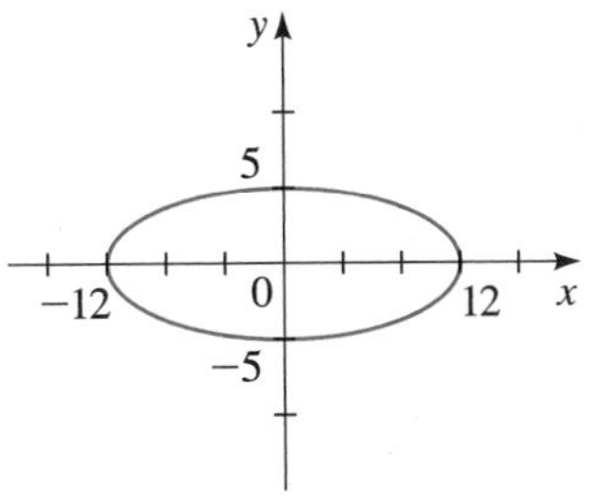

15.

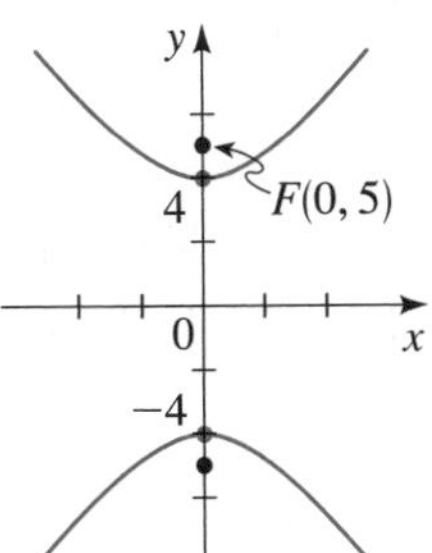

16.

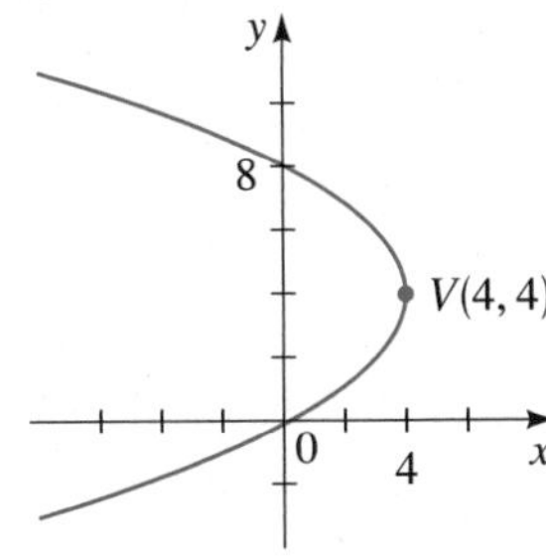

17.

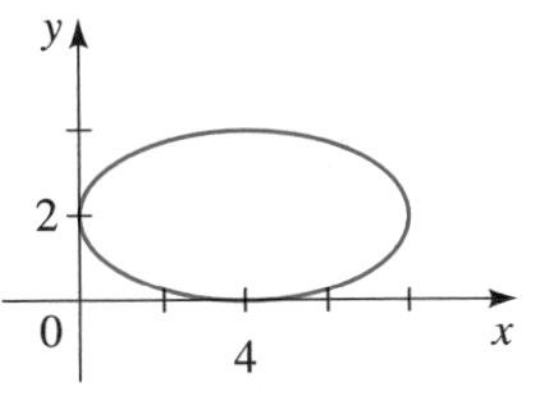

18.

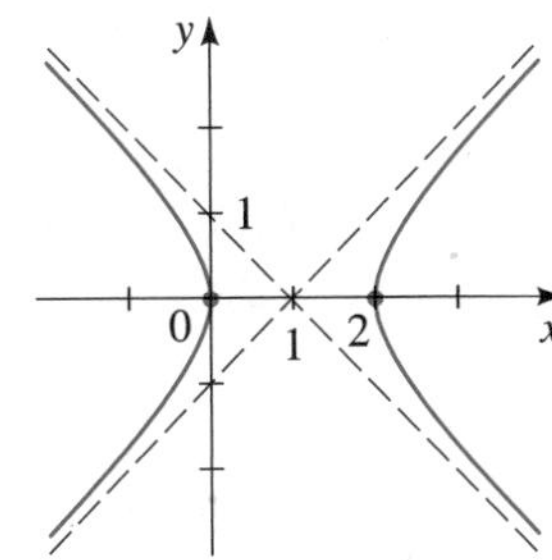

19–30 ■ Determine the type of curve represented by the equation. Find the foci and vertices (if any), and sketch the graph.

19. $\dfrac{x^2}{12} + y = 1$

20. $\dfrac{x^2}{12} + \dfrac{y^2}{144} = \dfrac{y}{12}$

21. $x^2 - y^2 + 144 = 0$

22. $x^2 + 6x = 9y^2$

23. $4x^2 + y^2 = 8(x + y)$

24. $3x^2 - 6(x + y) = 10$

25. $x = y^2 - 16y$

26. $2x^2 + 4 = 4x + y^2$

27. $2x^2 - 12x + y^2 + 6y + 26 = 0$

28. $36x^2 - 4y^2 - 36x - 8y = 31$

29. $9x^2 + 8y^2 - 15x + 8y + 27 = 0$

30. $x^2 + 4y^2 = 4x + 8$

31–38 ■ Find an equation for the conic section with the given properties.

31. The parabola with focus $F(0, 1)$ and directrix $y = -1$

32. The ellipse with center $C(0, 4)$, foci $F_1(0, 0)$ and $F_2(0, 8)$, and major axis of length 10

33. The hyperbola with vertices $V(0, \pm 2)$ and asymptotes $y = \pm\frac{1}{2}x$

34. The hyperbola with center $C(2, 4)$, foci $F_1(2, 1)$ and $F_2(2, 7)$, and vertices $V_1(2, 6)$ and $V_2(2, 2)$

35. The ellipse with foci $F_1(1, 1)$ and $F_2(1, 3)$, and with one vertex on the x-axis

36. The parabola with vertex $V(5, 5)$ and directrix the y-axis

37. The ellipse with vertices $V_1(7, 12)$ and $V_2(7, -8)$, and passing through the point $P(1, 8)$

38. The parabola with vertex $V(-1, 0)$ and horizontal axis of symmetry, and crossing the y-axis at $y = 2$

39. A cannon fires a cannonball as shown in the figure. The path of the cannonball is a parabola with vertex at the highest point of the path. If the cannonball lands 1600 ft from the cannon and the highest point it reaches is 3200 ft above the ground, find an equation for the path of the cannonball. Place the origin at the location of the cannon.

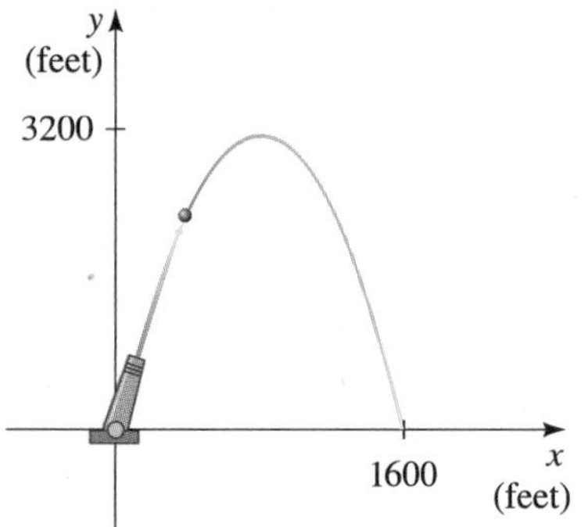

40. A satellite is in an elliptical orbit around the earth with the center of the earth at one focus. The height of the satellite above the earth varies between 140 mi and 440 mi. Assume the earth is a sphere with radius 3960 mi. Find an equation for the path of the satellite with the origin at the center of the earth.

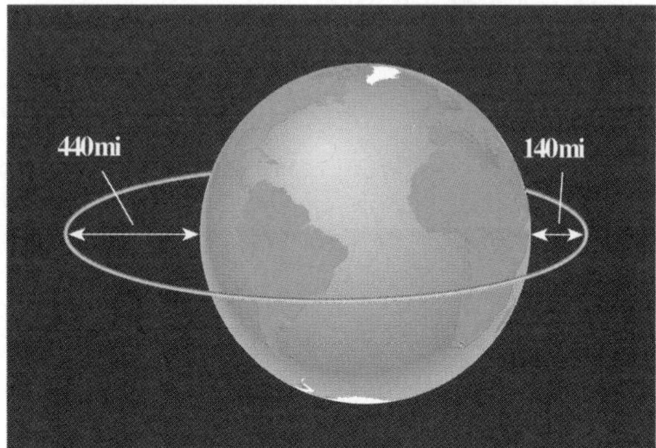

41. The path of the earth around the sun is an ellipse with the sun at one focus. The ellipse has major axis 186,000,000 mi and eccentricity 0.017. Find the distance between the earth and the sun when the earth is **(a)** closest to the sun and **(b)** farthest from the sun.

42. A ship is located 40 mi from a straight shoreline. LORAN stations A and B are located on the shoreline, 300 mi apart. From the LORAN signals, the captain determines that his ship is 80 mi closer to A than to B. Find the location of the ship. (Place A and B on the y-axis with the x-axis halfway between them. Find the x- and y-coordinates of the ship.)

43. **(a)** Draw graphs of the following family of ellipses for $k = 1, 2, 4$, and 8.

$$\frac{x^2}{16 + k^2} + \frac{y^2}{k^2} = 1$$

(b) Prove that all the ellipses in part (a) have the same foci.

44. **(a)** Draw graphs of the following family of parabolas for $k = \frac{1}{2}, 1, 2$, and 4.

$$y = kx^2$$

(b) Find the foci of the parabolas in part (a).

(c) How does the location of the focus change as k increases?

8 TEST

1. Find the focus and directrix of the parabola $x^2 = -12y$, and sketch its graph.

2. Find the vertices, foci, and the lengths of the major and minor axes for the following ellipse. Then sketch its graph.

$$\frac{x^2}{16} + \frac{y^2}{4} = 1$$

3. Find the vertices, foci, and asymptotes of the following hyperbola. Then sketch its graph

$$\frac{y^2}{9} - \frac{x^2}{16} = 1$$

4–6 ■ Find an equation for the conic whose graph is shown.

4.

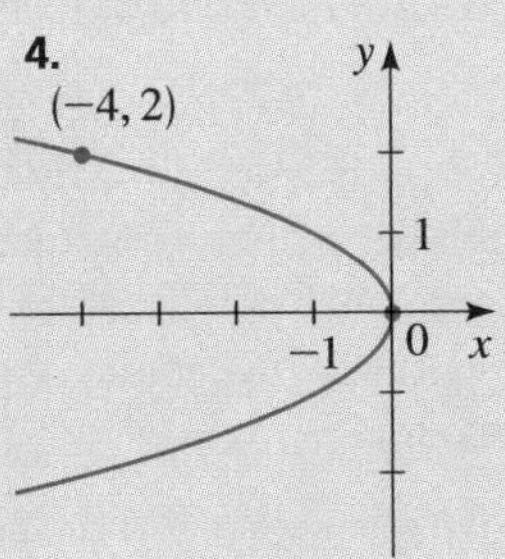

5.

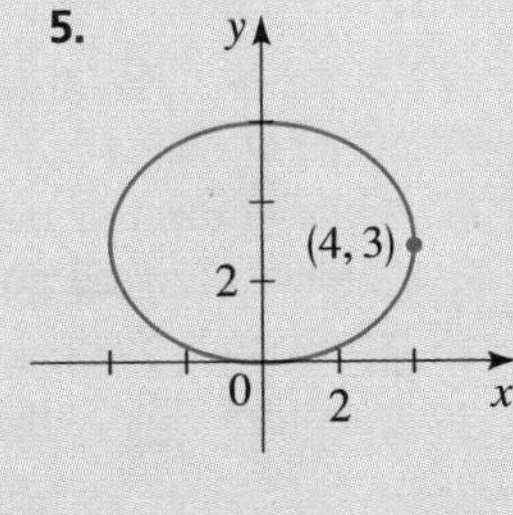

6.

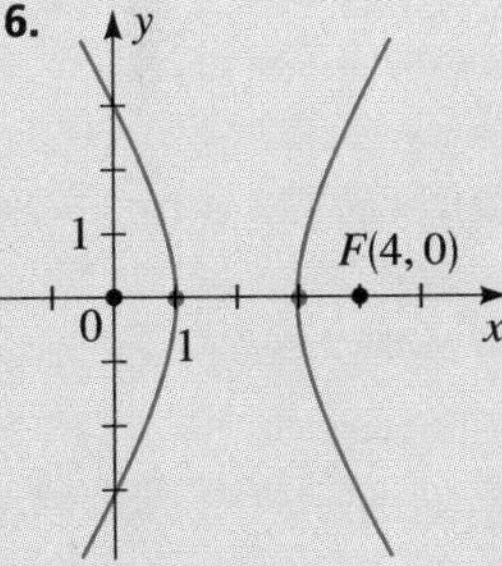

7–9 ■ Sketch the graph of the equation. Identify the vertices, foci, and asymptotes, if any.

7. $16x^2 + 36y^2 - 96x + 36y + 9 = 0$

8. $9x^2 - 8y^2 + 36x + 64y = 92$

9. $2x + y^2 + 8y + 8 = 0$

10. Find an equation for the hyperbola with foci $(0, \pm 5)$ and with asymptotes $y = \pm\frac{3}{4}x$.

11. Find an equation for the parabola with focus $(2, 4)$ and directrix the x-axis.

12. A parabolic reflector for a car headlight forms a bowl shape that is 6 in. wide at its opening and 3 in. deep, as shown in the figure. How far from the vertex should the bulb be placed if it is to be located at the focus?

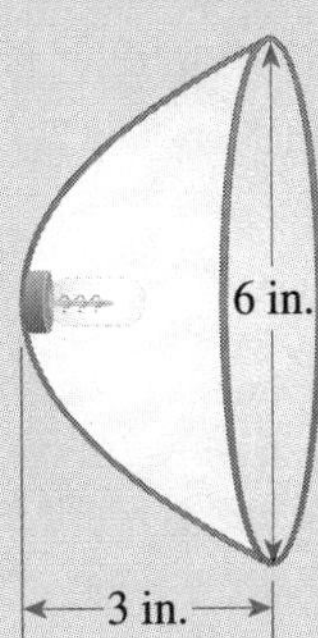

FOCUS ON PROBLEM SOLVING

One important problem-solving technique is the strategy of **taking cases**. A classic use of this strategy is in the classification of the regular polyhedra. These are called *Platonic solids* because they were mentioned in the writings of Plato.

A *regular polygon* is one in which all sides and all angles are equal. There are infinitely many regular polygons, as indicated in Figure 1.

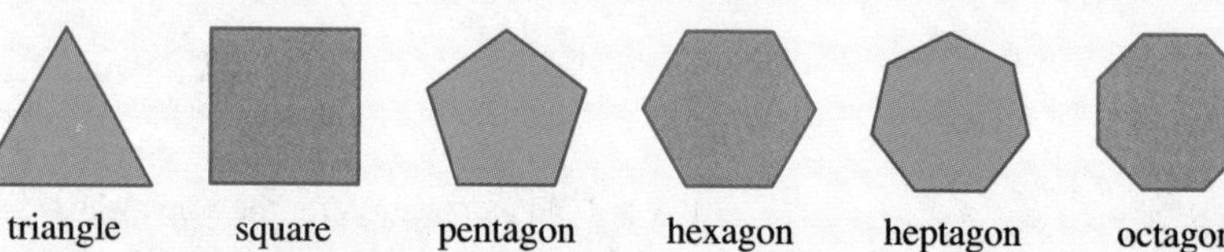

FIGURE 1
The regular polygons

For three-dimensional shapes, the analogous concept is that of regular polyhedra. A *regular polyhedron* is a solid in which all faces are congruent regular polygons, and the same number of polygons meet at each corner, or *vertex*. We would like to find all possible regular polyhedra. It might seem at first that there should be infinitely many, just as for regular polygons. But we will see that there are just a finite number of polyhedra.

CLASSIFYING THE REGULAR POLYHEDRA

We now prove that there are exactly five regular polyhedra (see Figure 2). To show this, let's consider all possible cases.

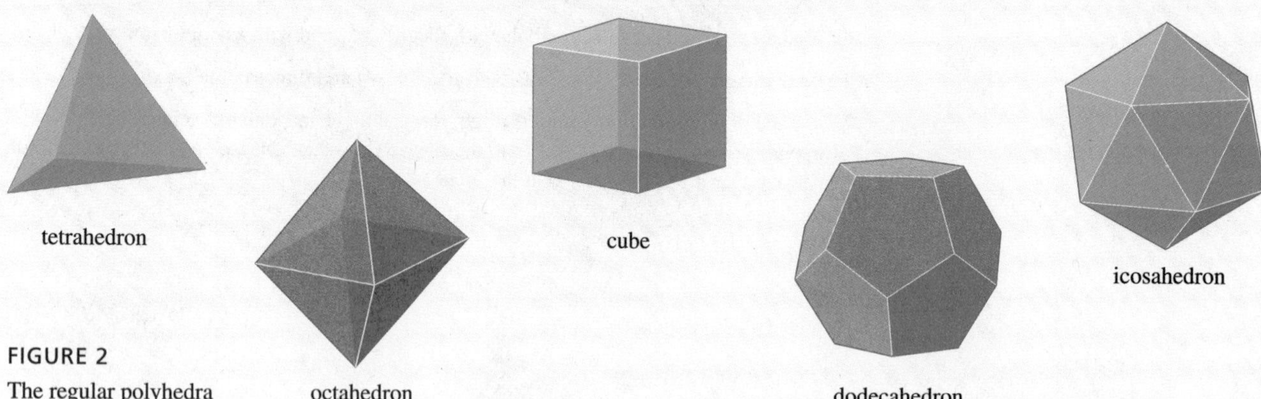

FIGURE 2
The regular polyhedra

Three, four, and five equilateral triangles can be folded up to make a corner, but six such triangles lie flat.

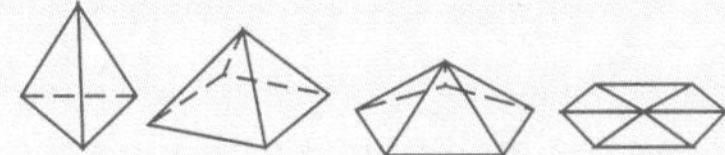

■ **CASE 1** Suppose the faces of a regular polyhedron are equilateral triangles. How many such triangles can meet at a corner? To make a corner there must be at least three triangles. We can also have four or five. But six equilateral triangles cannot meet at a point to make a corner. (Why?) If three triangles meet at each vertex, we can complete the polyhedron by adding one more triangle to make a *tetrahedron*. If four triangles meet at each vertex, we have an *octahedron*. If five triangles meet at each vertex, the resulting regular polyhedron is an *icosahedron*. Thus, we have found all the regular polyhedra with triangular faces.

■ **CASE 2** Suppose the faces of a regular polygon are squares. If three squares meet at each point, then the polyhedron is a *cube*. It's impossible for four or more squares to meet at a point to make a corner. (Why?) Thus, the only regular polyhedron with square faces is the cube.

■ **CASE 3** Suppose the faces of a regular polygon are pentagons. If three pentagons meet at each vertex, the resulting polyhedron is a *dodecahedron*. Since the angles of a regular pentagon are 108°, it's impossible for more than three regular pentagons to meet at a vertex. Thus, the only regular polyhedron with pentagonal faces is the dodecahedron.

■ **CASE 4** Is it possible for the faces of a regular polygon to be regular hexagons? Since the angle of a regular hexagon is 120°, when three such hexagons meet at a point they do not form a corner, so it's impossible for a regular polyhedron to have hexagonal faces. The same reasoning shows that no other regular polygon can be the face of a regular polyhedron.

Since these four cases account for all the possibilities, we have shown that there are exactly five regular polyhedra.

EULER'S FORMULA

How many faces, edges, and vertices does a regular polyhedron have? In the 18th century, Euler observed that

$$F - E + V = 2 \qquad \text{Euler's Formula}$$

where F is the number of faces, E is the number of edges, and V is the number of vertices. We can use Euler's Formula to answer the question. For example, the icosahedron is assembled from F equilateral triangles:

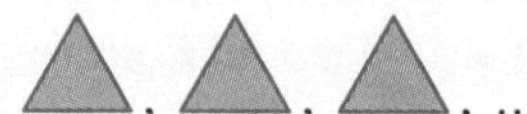

In these triangles the total number of sides is $3F$, and the total number of angles is also $3F$. In an icosahedron, five angles of these triangles meet to form a vertex, so the total number of vertices must be

$$V = \frac{3F}{5}$$

Since two sides of adjacent triangles meet to form one edge of the polyhedron, the number of edges must be

$$E = \frac{3F}{2}$$

It's interesting to note that there are infinitely many 2-dimensional regular polygons, but only five 3-dimensional regular polyhedra. Although it's impossible to draw the 4-dimensional regular polyhedra, mathematicians have shown that there are six of them. Surprisingly, in all higher dimensions there are exactly three regular polyhedra—the n-dimensional cube, tetrahedron, and octahedron.

Substituting into Euler's Formula gives

$$F - \frac{3F}{2} + \frac{3F}{5} = 2$$

Solving gives $F = 20$. Substituting this value of F into the formulas for edges and faces gives $E = 30$ and $V = 12$. Thus, for the icosahedron, $F = 20$, $E = 30$, and $V = 12$. Using similar reasoning we can find the number of faces, edges, and vertices for the other regular polyhedra.

PROBLEMS

1. Find the number of faces, edges, and vertices for each of the regular polyhedra, not by counting them, but by using Euler's Formula.
2. Describe the polyhedron whose edges are the line segments joining the centers of the faces of an octahedron, as shown in the figure below. Do the same for the other Platonic solids.

THE FOUR-COLOR PROBLEM

A famous example of a problem that was solved by taking cases is the *four-color problem*: Prove that every map can be colored using no more than four colors. The requirement is that any two regions that share a common border should have different colors. (If two regions meet only at a single point, as do Utah and New Mexico, for example, then they may have the same color.) The four-color problem was first stated as a conjecture in 1852, but it wasn't until 1976 that Kenneth Appel and Wolfgang Haken were able to prove it. In their proof they considered thousands of cases corresponding to different types of maps, and most of those cases were so complicated that they had to use a high-speed computer for the verification.

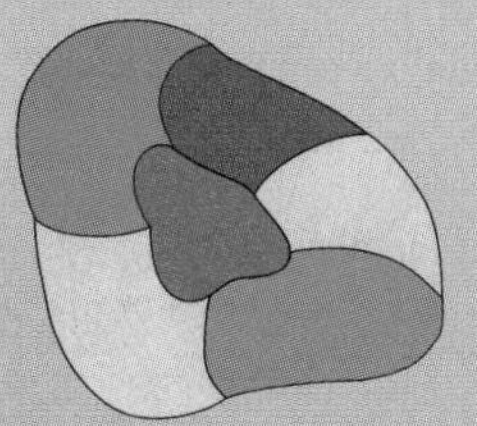

3. As the following figures indicate, it's possible to *tile* the plane (that is, completely cover it) with equilateral triangles and with squares. Find all other regular polygons that tile the plane. Prove your answer.

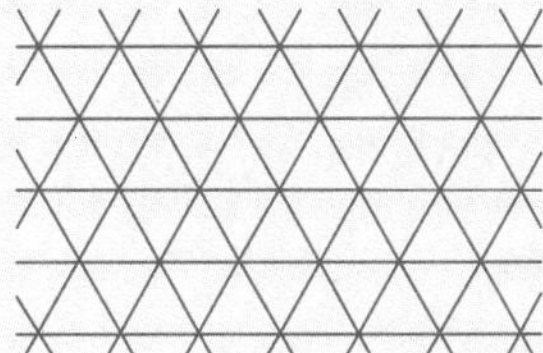

4. A group of n pulleys, all of radius 1, is fixed so that their centers form a convex n-gon of perimeter P. (The figure shows the case $n = 4$.) Find the length of the belt that fits around the pulleys. [*Hint:* Try the simplest case $n = 2$ first. Fit together the sectors of the pulleys that touch the belt.]

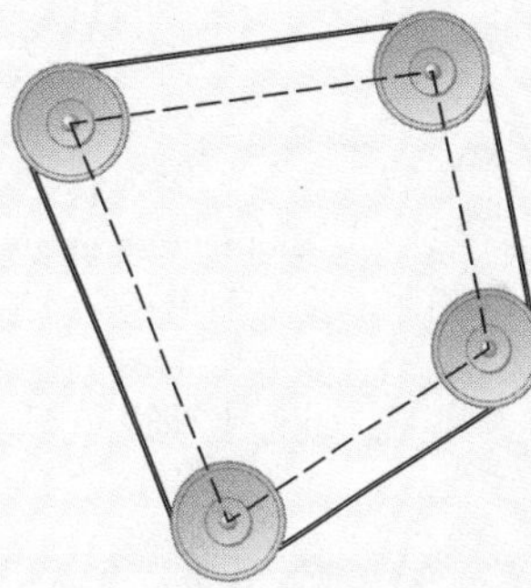

5. A bug is sitting at point A in one corner of a room and wants to crawl to point B in the corner diagonally opposite (as shown in the figure). Find the shortest path for the bug.

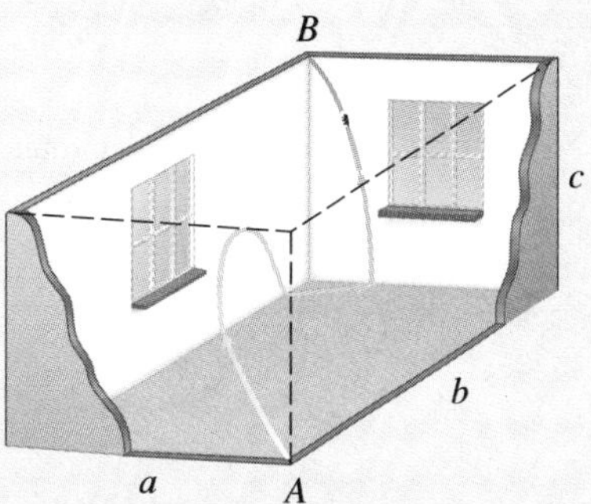

Pierre de Fermat (1601–1665) was a French lawyer who became interested in mathematics at the age of 30. Fermat's job as a magistrate left him little time to write complete proofs of his discoveries, so he often wrote them in the margin of whatever book he was reading at the time. After his death, his copy of Diophantus' *Arithmetica* (see page 45) was found to contain a particularly tantalizing comment. Where Diophantus discusses the solutions of $x^2 + y^2 = z^2$ (for example, $x = 3$, $y = 4$, $z = 5$), Fermat states in the margin that for $n \geqslant 3$ there are no natural number solutions to the equation $x^n + y^n = z^n$. In other words, it's impossible for a cube to equal the sum of two cubes, a fourth power to equal the sum of two fourth powers, and so on. Fermat writes "I have discovered a truly wonderful proof for this but the margin is too small to contain it." All the other margin comments in Fermat's copy of *Arithmetica* have been proved. This one, however, long remained

(continued)

6. For what values of the real number k does the circle $(x - k)^2 + y^2 = 4$ intersect the following ellipse in exactly 0, 1, 2, 3, 4, or 5 points?

$$x^2 + \frac{y^2}{9} = 1$$

7. On the parabola $x^2 = 4py$, let P and Q be two points with the property that $\angle POQ$ is a right angle. Show that the y-intercept of the segment PQ is the same for any such pair of points P and Q.

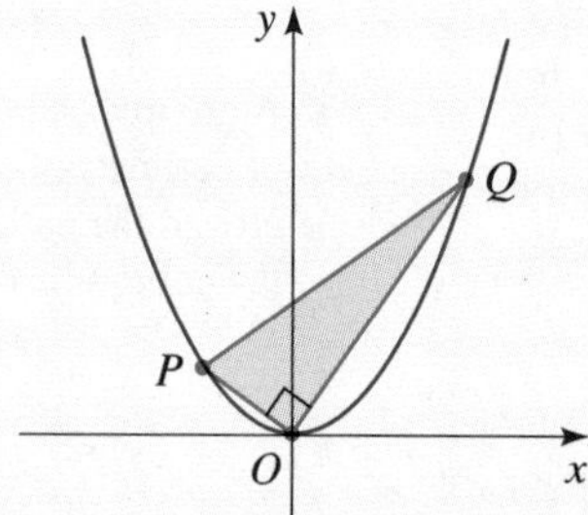

8. In a right triangle, the hypotenuse has length 5 cm and another side has length 3 cm. What is the length of the altitude that is perpendicular to the hypotenuse?

9. The perimeter of a right triangle is 60 cm, and the altitude perpendicular to the hypotenuse is 12 cm. Find the lengths of the three sides.

10. A point P is located in the interior of a rectangle so that the distance from P to one corner is 5 cm, from P to the opposite corner is 14 cm, and from P to a third corner is 10 cm. What is the distance from P to the fourth corner?

11. Find every positive integer that gives a perfect square if 132 is added to it and another perfect square if 200 is added to it.

12. Prove that every prime number is the leg of exactly one right triangle with integer sides. (This problem was first stated by Fermat.)

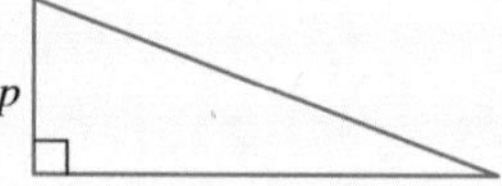

unproved, and it came to be known as "Fermat's Last Theorem."

In 1994 Andrew Wiles of Princeton University announced a proof of Fermat's Last Theorem, an astounding 350 years after it was conjectured. His proof is one of the most widely reported mathematical results in the popular press.

13. Show that the equation $x^2 + y^2 = 4z + 3$ has no solution in integers. [*Hint:* Recall that an even number is of the form $2n$ and an odd number is of the form $2n + 1$. Consider all possible cases for x and y even or odd.]

14. (a) Find all prime numbers p such that $2p + 1$ is a perfect square. [*Hint:* Write the equation $2p + 1 = n^2$ as $2p = n^2 - 1$ and factor. Then consider cases.]
(b) Find all prime numbers p such that $2p + 1$ is a perfect cube.

15. (a) Write 13 as the sum of two squares. Then do the same for 41.
(b) Verify that $(a^2 + b^2)(c^2 + d^2) = (ac + bd)^2 + (ad - bc)^2$.
(c) Express 533 as the sum of two squares in two different ways. [*Hint:* Factor 533 and use parts (a) and (b).]

9

SEQUENCES AND SERIES

Sequences and series abound in nature; the Fibonacci sequence, for example, is hidden in the intricate structure of the nautilus. Sequences are also used in everyday tasks, such as calculating the monthly mortgage payment on your dream home.

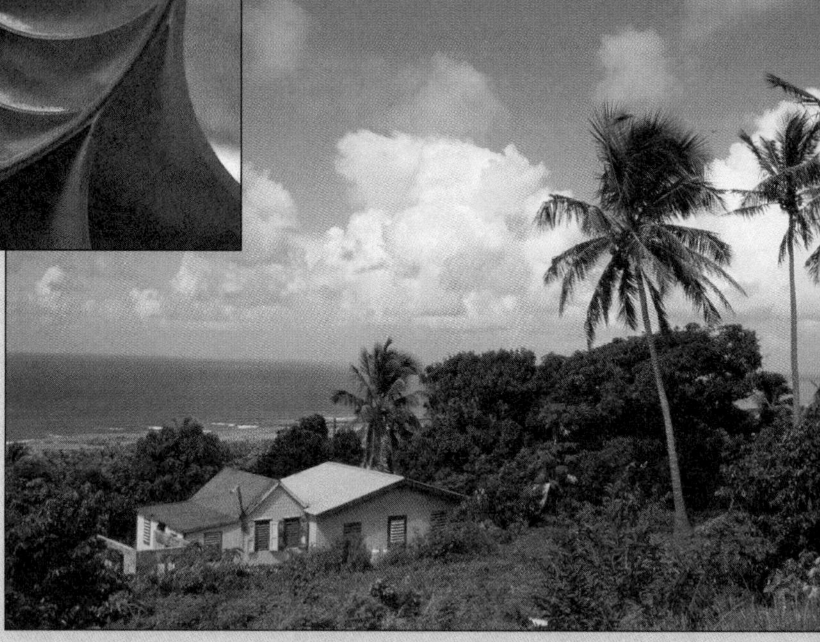

A mathematician, like a painter or poet, is a maker of patterns.

G. H. Hardy

In this chapter we study sequences and series of numbers. Roughly speaking, a *sequence* is a list of numbers written in a specific order, and a *series* is what one gets by adding the numbers in a sequence. Sequences and series have many theoretical and practical uses. Among other applications, we consider how series are used to calculate the value of an annuity.

Section 9.5 introduces a special kind of proof called *mathematical induction*. In Section 9.6 we use mathematical induction to prove a formula for expanding $(a + b)^n$ for any natural number n.

9.1 SEQUENCES AND SUMMATION NOTATION

A *sequence* is a set of numbers written in a specific order:

$$a_1, a_2, a_3, a_4, \ldots, a_n, \ldots$$

The number a_1 is called the *first term*, a_2 is the *second term*, and in general a_n is the *nth term*. Since for every natural number n there is a corresponding number a_n, we can define a sequence as a function.

DEFINITION OF A SEQUENCE

A **sequence** is a function f whose domain is the set of natural numbers. The values $f(1), f(2), f(3), \ldots$ are called the **terms** of the sequence.

We usually write a_n instead of the function notation $f(n)$ for the value of the function at the number n.

Here is a simple example of a sequence:

$$2, 4, 6, 8, 10, \ldots$$

The dots indicate that the sequence continues indefinitely. We can write a sequence in this way when it's clear what the subsequent terms of the sequence are. This sequence consists of even numbers. To be more accurate, however, we need to specify a procedure for finding *all* the terms of the sequence. This can be done by giving a formula for the nth term a_n of the sequence. In this case,

$$a_n = 2n$$

Another way to write this sequence is to use function notation:

$$a(n) = 2n$$

so $a(1) = 2$, $a(2) = 4$, $a(3) = 6, \ldots$

and the sequence can be written as

$$\underset{\substack{\uparrow \\ \text{1st} \\ \text{term}}}{2,} \quad \underset{\substack{\uparrow \\ \text{2nd} \\ \text{term}}}{4,} \quad \underset{\substack{\uparrow \\ \text{3rd} \\ \text{term}}}{6,} \quad \underset{\substack{\uparrow \\ \text{4th} \\ \text{term}}}{8,} \quad \ldots, \quad \underset{\substack{\uparrow \\ n\text{th} \\ \text{term}}}{2n,} \quad \ldots$$

Notice how the formula $a_n = 2n$ gives all the terms of the sequence. For instance, substituting 1, 2, 3, and 4 for n gives the first four terms:

$$a_1 = 2 \cdot 1 = 2 \qquad a_2 = 2 \cdot 2 = 4$$

$$a_3 = 2 \cdot 3 = 6 \qquad a_4 = 2 \cdot 4 = 8$$

To find the 103rd term of this sequence, we use $n = 103$ to get

$$a_{103} = 2 \cdot 103 = 206$$

EXAMPLE 1 ■ Finding the Terms of a Sequence

Find the first five terms and the 100th term of the sequence defined by each formula.

(a) $a_n = 2n - 1$ (b) $c_n = n^2 - 1$

(c) $t_n = \dfrac{n}{n+1}$ (d) $r_n = \dfrac{(-1)^n}{2^n}$

SOLUTION

To find the first five terms, we substitute $n = 1, 2, 3, 4$, and 5 in the formula for the nth term. To find the 100th term, we substitute $n = 100$. This gives the following.

nth term	First five terms	100th term
(a) $2n - 1$	1, 3, 5, 7, 9	199
(b) $n^2 - 1$	0, 3, 8, 15, 24	9999
(c) $\dfrac{n}{n+1}$	$\dfrac{1}{2}, \dfrac{2}{3}, \dfrac{3}{4}, \dfrac{4}{5}, \dfrac{5}{6}$	$\dfrac{100}{101}$
(d) $\dfrac{(-1)^n}{2^n}$	$-\dfrac{1}{2}, \dfrac{1}{4}, -\dfrac{1}{8}, \dfrac{1}{16}, -\dfrac{1}{32}$	$\dfrac{1}{2^{100}}$

■

In Example 1(d) the presence of $(-1)^n$ in the sequence has the effect of making successive terms alternately negative and positive.

It is often useful to picture a sequence by sketching its graph. Since a sequence is a function whose domain is the natural numbers, we can draw its graph in the Cartesian plane. For instance, the graph of the sequence

$$1, \frac{1}{2}, \frac{1}{3}, \frac{1}{4}, \frac{1}{5}, \frac{1}{6}, \ldots, \frac{1}{n}, \ldots$$

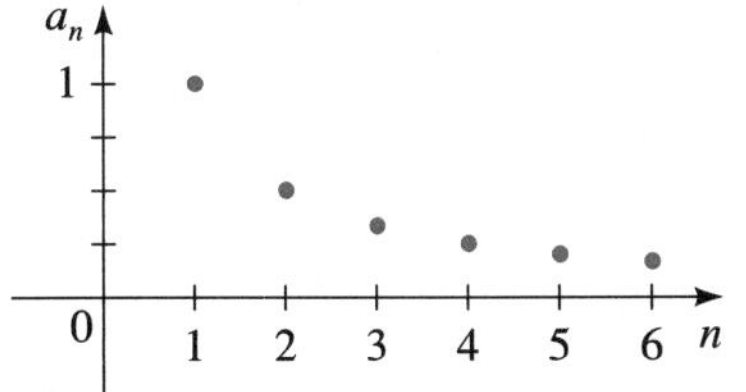

FIGURE 1

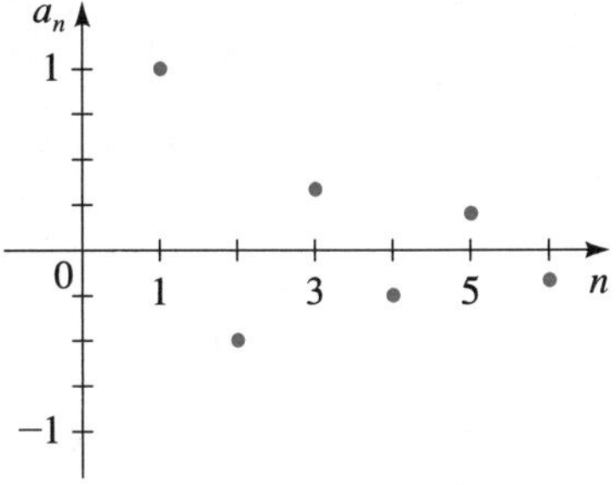

FIGURE 2

is shown in Figure 1. Compare this to the graph of

$$1, -\frac{1}{2}, \frac{1}{3}, -\frac{1}{4}, \frac{1}{5}, -\frac{1}{6}, \ldots, \frac{(-1)^{n+1}}{n}, \ldots$$

shown in Figure 2. The graph of every sequence consists of isolated points that are *not* connected.

Some sequences do not have simple defining formulas like those of the preceding example. The nth term of a sequence may depend on some or all of the terms preceding it. A sequence defined in this way is called **recursive**. Here are two examples.

EXAMPLE 2 ■ Finding the Terms of a Recursive Sequence

Find the first five terms of the sequence defined recursively by $a_1 = 1$ and

$$a_n = 3(a_{n-1} + 2)$$

SOLUTION

The defining formula for this sequence is recursive. It allows us to find the nth term a_n if we know the preceding term a_{n-1}. Thus, we can find the second term from the first term, the third term from the second term, the fourth term from the third term, and so on. Since we are given the first term $a_1 = 1$, we can proceed as follows.

$$a_2 = 3(a_1 + 2) = 3(1 + 2) = 9$$

$$a_3 = 3(a_2 + 2) = 3(9 + 2) = 33$$

$$a_4 = 3(a_3 + 2) = 3(33 + 2) = 105$$

$$a_5 = 3(a_4 + 2) = 3(105 + 2) = 321$$

Thus, the first five terms of this sequence are

$$1, 9, 33, 105, 321, \ldots$$

■

Note that in order to find the 100th term of the sequence in Example 2 we must first find all 99 preceding terms.

EXAMPLE 3 ■ The Fibonacci Sequence

Find the first 11 terms of the sequence defined recursively by $F_1 = 1$, $F_2 = 1$ and

$$F_n = F_{n-1} + F_{n-2}$$

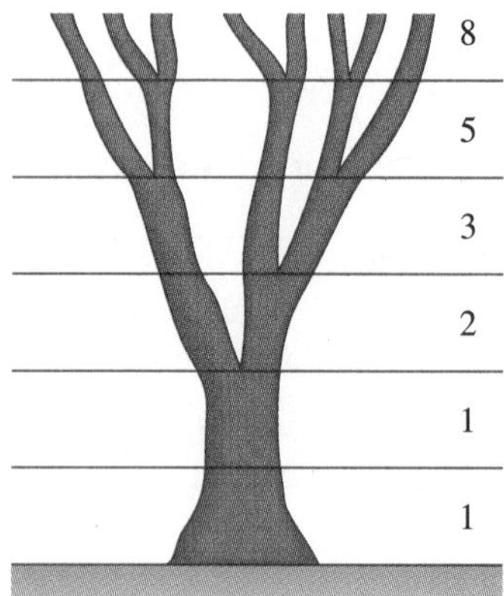

FIGURE 3
The Fibonacci sequence in the branching of a tree

SOLUTION

To find F_n we need to find the two preceding terms F_{n-1} and F_{n-2}. Since we are given F_1 and F_2, we proceed as follows.

$$F_3 = F_2 + F_1 = 1 + 1 = 2$$

$$F_4 = F_3 + F_2 = 2 + 1 = 3$$

$$F_5 = F_4 + F_3 = 3 + 2 = 5$$

It's clear what is happening here. Each term is simply the sum of the two terms that precede it, so we can easily write down as many terms as we please. Here are the first 11 terms:

$$1, 1, 2, 3, 5, 8, 13, 21, 34, 55, 89, \ldots$$ ■

The sequence in Example 3 is called the **Fibonacci sequence**, named after the 13th-century Italian mathematician who used it to solve a problem about the breeding of rabbits (see Exercise 56). The sequence also occurs in numerous other applications in nature. (See Figures 3 and 4.) In fact, so many phenomena behave like the Fibonacci sequence that one mathematical journal, the *Fibonacci Quarterly*, is devoted entirely to its properties.

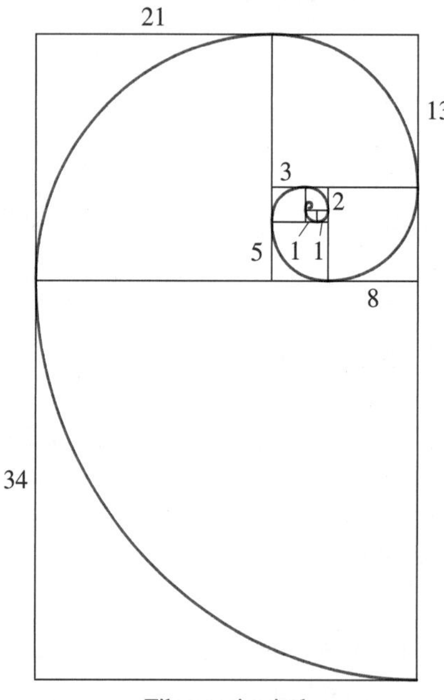

Fibonacci spiral

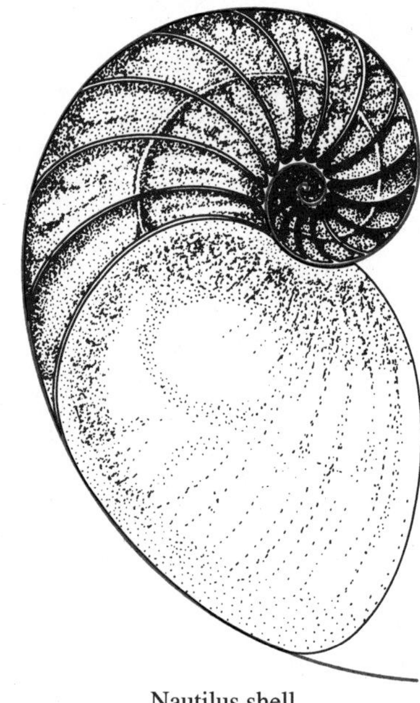
Nautilus shell

FIGURE 4

The sequences we have considered so far are defined by either a formula or a recursive procedure. But not all sequences can be defined in this way. For example,

A *prime number* is a natural number with no factor other than 1 and itself. By convention, 1 is not considered prime.

there is no known formula that produces the sequence of prime numbers:

$$2, 3, 5, 7, 11, 13, 17, 19, 23, \ldots$$

If we let a_n be the nth digit in the decimal expansion of the number π, we get the sequence

$$3, 1, 4, 1, 5, 9, 2, 6, 5, 4, \ldots$$

Again, no simple formula can be given for finding the terms of this sequence.

Finding patterns is an important part of mathematics. Consider a sequence that begins

$$1, 4, 9, 16, \ldots$$

Can you detect a pattern in these numbers? In other words, can you define a sequence whose first four terms are these numbers? The answer to this question seems easy; these numbers are the squares of the numbers 1, 2, 3, 4. Thus, the sequence we are looking for is defined by $a_n = n^2$. However, this is not the *only* sequence whose first four terms are 1, 4, 9, 16. In other words, the answer to our problem is not unique (see Exercise 57). In the next example we are interested in finding an *obvious* sequence whose first few terms agree with the given ones.

Fibonacci (1175–1250) was born in Pisa, Italy, and educated in North Africa. He traveled widely in the Mediterranean area and learned the various methods then in use for writing numbers. On returning to Pisa in 1202, Fibonacci advocated the use of the Hindu-Arabic decimal system, the one we use today, over the Roman numeral system used in Europe in his time. His most famous book *Liber Abaci* expounds on the advantages of the Hindu-Arabic numerals. In fact, multiplication and division were so complicated using Roman numerals that a college degree was necessary to master these skills. Interestingly, in 1299 the city of Florence outlawed the use of the decimal system for merchants and businesses, requiring numbers to be written in Roman numerals or words. One can only speculate about the reasons for this law.

EXAMPLE 4 ■ Finding the nth Term of a Sequence

Find the nth term of a sequence whose first several terms are given.

(a) $\frac{1}{2}, \frac{3}{4}, \frac{5}{6}, \frac{7}{8}, \ldots$ (b) $-2, 4, -8, 16, -32, \ldots$

SOLUTION

(a) We notice that the numerators of these fractions are the odd numbers and the denominators are the even numbers. Even numbers are of the form $2n$, and odd numbers are of the form $2n - 1$ (an odd number differs from an even number by 1). So, a sequence that has these numbers for its first four terms is given by

$$a_n = \frac{2n - 1}{2n}$$

(b) These numbers are powers of 2 and they alternate in sign, so a sequence that agrees with these terms is given by

$$a_n = (-1)^n 2^n$$

You should check that these formulas do indeed generate the given terms. ■

THE PARTIAL SUMS OF A SEQUENCE

In calculus we are often interested in adding the terms of a sequence. This leads to the following definition.

THE PARTIAL SUMS OF A SEQUENCE

For the sequence

$$a_1, \quad a_2, \quad a_3, \quad a_4, \quad \ldots, \quad a_n, \quad \ldots$$

the **partial sums** are

$$\begin{aligned} S_1 &= a_1 \\ S_2 &= a_1 + a_2 \\ S_3 &= a_1 + a_2 + a_3 \\ S_4 &= a_1 + a_2 + a_3 + a_4 \\ &\vdots \\ S_n &= a_1 + a_2 + a_3 + \cdots + a_n \\ &\vdots \end{aligned}$$

S_1 is called the **first partial sum**, S_2 is the **second partial sum**, and so on. S_n is called the ***n*th partial sum**. The sequence $S_1, S_2, S_3, \ldots, S_n, \ldots$ is called the **sequence of partial sums**.

EXAMPLE 5 ■ Finding the Partial Sums of a Sequence

Find the first four partial sums and the nth partial sum of the sequence given by $a_n = 1/2^n$.

SOLUTION

The terms of the sequence are

$$\frac{1}{2}, \frac{1}{4}, \frac{1}{8}, \ldots$$

The first four partial sums are

$$\begin{aligned} S_1 &= \frac{1}{2} && = \frac{1}{2} \\ S_2 &= \frac{1}{2} + \frac{1}{4} && = \frac{3}{4} \\ S_3 &= \frac{1}{2} + \frac{1}{4} + \frac{1}{8} && = \frac{7}{8} \\ S_4 &= \frac{1}{2} + \frac{1}{4} + \frac{1}{8} + \frac{1}{16} && = \frac{15}{16} \end{aligned}$$

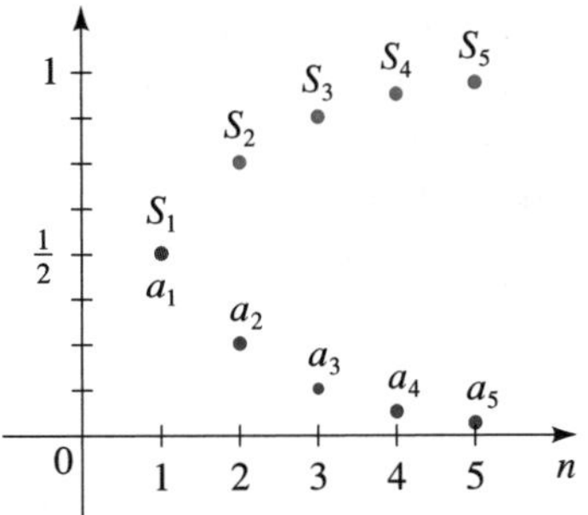

FIGURE 5
Graph of the sequence a_n and the sequence of partial sums S_n

Notice that in the value of each partial sum the denominator is a power of 2 and the numerator is one less than the denominator. In general, the nth partial sum is

$$S_n = \frac{2^n - 1}{2^n} = 1 - \frac{1}{2^n}$$

The first five terms of a_n and S_n are graphed in Figure 5. ■

EXAMPLE 6 ■ Finding the Partial Sums of a Sequence

Find the first four partial sums and the nth partial sum of the sequence given by

$$a_n = \frac{1}{n} - \frac{1}{n+1}$$

SOLUTION

The first four partial sums are

$$S_1 = \left(1 - \frac{1}{2}\right) = 1 - \frac{1}{2}$$

$$S_2 = \left(1 - \frac{1}{2}\right) + \left(\frac{1}{2} - \frac{1}{3}\right) = 1 - \frac{1}{3}$$

$$S_3 = \left(1 - \frac{1}{2}\right) + \left(\frac{1}{2} - \frac{1}{3}\right) + \left(\frac{1}{3} - \frac{1}{4}\right) = 1 - \frac{1}{4}$$

$$S_4 = \left(1 - \frac{1}{2}\right) + \left(\frac{1}{2} - \frac{1}{3}\right) + \left(\frac{1}{3} - \frac{1}{4}\right) + \left(\frac{1}{4} - \frac{1}{5}\right) = 1 - \frac{1}{5}$$

Do you detect a pattern here? Of course. The nth partial sum is

$$S_n = 1 - \frac{1}{n+1}$$ ■

SIGMA NOTATION

Given a sequence

$$a_1, a_2, a_3, a_4, \ldots$$

we can write the sum of the first n terms using **summation notation**, or **sigma notation**. This notation derives its name from the Greek letter Σ (capital sigma, corresponding to our S for "sum"). Sigma notation is used as follows:

$$\sum_{k=1}^{n} a_k = a_1 + a_2 + a_3 + a_4 + \cdots + a_n$$

The ancient Greeks considered a line segment to be divided into the **golden ratio** if the ratio of the shorter part to the longer part is the same as the ratio of the longer part to the whole segment.

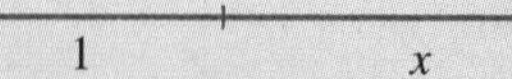

Thus, the segment shown is divided into the golden ratio if

$$\frac{1}{x} = \frac{x}{1 + x}$$

This leads to a quadratic equation whose positive solution is

$$x = \frac{1 + \sqrt{5}}{2} \approx 1.618$$

This ratio occurs naturally in many places. For instance, psychological experiments show that the most pleasing shape of rectangle is one whose sides are in golden ratio. The ancient Greeks agreed with this and built their temples in this ratio.

The golden ratio is related to the Fibonacci numbers (see page 585). In fact, it can be shown using calculus* that the ratio of two successive Fibonacci numbers

$$\frac{F_{n+1}}{F_n}$$

gets closer to the golden ratio the larger the value of n. Try finding this ratio for $n = 10$.

*James Stewart, *Calculus*, 4th ed. (Pacific Grove, CA: Brooks/Cole, 1999) p. 737.

The left side of this expression is read "The sum of a_k from $k = 1$ to $k = n$." The letter k is called the **index of summation**, or the **summation variable**, and the idea is to replace k in the expression after the sigma by the integers $1, 2, 3, \ldots, n$, and add the resulting expressions, arriving at the right side of the equation.

EXAMPLE 7 ■ Sigma Notation

Find each sum.

(a) $\sum_{k=1}^{5} k^2$ (b) $\sum_{j=3}^{5} \frac{1}{j}$ (c) $\sum_{i=5}^{10} i$ (d) $\sum_{i=1}^{6} 2$

SOLUTION

(a) $\sum_{k=1}^{5} k^2 = 1^2 + 2^2 + 3^2 + 4^2 + 5^2 = 55$

(b) $\sum_{j=3}^{5} \frac{1}{j} = \frac{1}{3} + \frac{1}{4} + \frac{1}{5} = \frac{47}{60}$

(c) $\sum_{i=5}^{10} i = 5 + 6 + 7 + 8 + 9 + 10 = 45$

(d) $\sum_{i=1}^{6} 2 = 2 + 2 + 2 + 2 + 2 + 2 = 12$ ■

EXAMPLE 8 ■ Writing Sums in Sigma Notation

Write each sum using sigma notation.

(a) $1^3 + 2^3 + 3^3 + 4^3 + 5^3 + 6^3 + 7^3$

(b) $\sqrt{3} + \sqrt{4} + \sqrt{5} + \cdots + \sqrt{77}$

SOLUTION

(a) We can write

$$1^3 + 2^3 + 3^3 + 4^3 + 5^3 + 6^3 + 7^3 = \sum_{k=1}^{7} k^3$$

(b) A natural way to write this sum is

$$\sqrt{3} + \sqrt{4} + \sqrt{5} + \cdots + \sqrt{77} = \sum_{k=3}^{77} \sqrt{k}$$

However, there is no unique way of writing a sum in sigma notation. We also could write this sum as

$$\sqrt{3} + \sqrt{4} + \sqrt{5} + \cdots + \sqrt{77} = \sum_{k=0}^{74} \sqrt{k + 3}$$

or

$$\sqrt{3} + \sqrt{4} + \sqrt{5} + \cdots + \sqrt{77} = \sum_{k=1}^{75} \sqrt{k + 2}$$ ■

The following properties of sums are natural consequences of properties of the real numbers.

PROPERTIES OF SUMS

Let $a_1, a_2, a_3, a_4, \ldots$ and $b_1, b_2, b_3, b_4, \ldots$ be sequences. Then for every positive integer n and any real number c, the following properties hold.

1. $\displaystyle\sum_{k=1}^{n} (a_k + b_k) = \sum_{k=1}^{n} a_k + \sum_{k=1}^{n} b_k$

2. $\displaystyle\sum_{k=1}^{n} (a_k - b_k) = \sum_{k=1}^{n} a_k - \sum_{k=1}^{n} b_k$

3. $\displaystyle\sum_{k=1}^{n} ca_k = c\left(\sum_{k=1}^{n} a_k\right)$

■ **Proof** To prove Property 1, we write out the left side of the equation to get

$$\sum_{k=1}^{n} (a_k + b_k) = (a_1 + b_1) + (a_2 + b_2) + (a_3 + b_3) + \cdots + (a_n + b_n)$$

Because addition is commutative and associative, we can rearrange the terms on the right side to read

$$\sum_{k=1}^{n} (a_k + b_k) = (a_1 + a_2 + a_3 + \cdots + a_n) + (b_1 + b_2 + b_3 + \cdots + b_n)$$

Rewriting the right side using sigma notation gives Property 1. Property 2 is proved in a similar manner. To prove Property 3, we use the Distributive Property:

$$\begin{aligned}\sum_{k=1}^{n} ca_k &= ca_1 + ca_2 + ca_3 + \cdots + ca_n \\ &= c(a_1 + a_2 + a_3 + \cdots + a_n) = c\left(\sum_{k=1}^{n} a_k\right)\end{aligned}$$

□

9.1 EXERCISES

1–10 ■ Find the first four terms and the 1000th term of the sequence.

1. $a_n = n + 1$

2. $a_n = 2n + 3$

3. $a_n = \dfrac{1}{n + 1}$

4. $a_n = n^2 + 1$

5. $a_n = \dfrac{(-1)^n}{n^2}$

6. $a_n = \dfrac{1}{n^2}$

7. $a_n = 1 + (-1)^n$

8. $a_n = (-1)^{n+1} \dfrac{n}{n + 1}$

9. $a_n = n^n$

10. $a_n = 3$

11–16 ■ Find the first five terms of the given recursively defined sequence.

11. $a_n = 2(a_{n-1} - 2)$ and $a_1 = 3$

12. $a_n = \dfrac{a_{n-1}}{2}$ and $a_1 = -8$

13. $a_n = 2a_{n-1} + 1$ and $a_1 = 1$

14. $a_n = \dfrac{1}{1 + a_{n-1}}$ and $a_1 = 1$

15. $a_n = a_{n-1} + a_{n-2}$ and $a_1 = 1, a_2 = 2$

16. $a_n = a_{n-1} + a_{n-2} + a_{n-3}$ and $a_1 = a_2 = a_3 = 1$

17–24 ■ Find the nth term of a sequence whose first several terms are given.

17. $2, 4, 8, 16, \ldots$

18. $-\frac{1}{3}, \frac{1}{9}, -\frac{1}{27}, \frac{1}{81}, \ldots$

19. $1, 4, 7, 10, \ldots$

20. $5, -25, 125, -625, \ldots$

21. $1, \frac{3}{4}, \frac{5}{9}, \frac{7}{16}, \frac{9}{25}, \ldots$

22. $\frac{3}{4}, \frac{4}{5}, \frac{5}{6}, \frac{6}{7}, \ldots$

23. $0, 2, 0, 2, 0, 2, \ldots$

24. $1, \frac{1}{2}, 3, \frac{1}{4}, 5, \frac{1}{6}, \ldots$

25–28 ■ Find the first six partial sums $S_1, S_2, S_3, S_4, S_5, S_6$ of the sequence.

25. $1, 3, 5, 7, \ldots$

26. $1^2, 2^2, 3^2, 4^2, \ldots$

27. $\dfrac{1}{3}, \dfrac{1}{3^2}, \dfrac{1}{3^3}, \dfrac{1}{3^4}, \ldots$

28. $-1, 1, -1, 1, \ldots$

29–32 ■ Find the first four partial sums and the nth partial sum of the sequence a_n.

29. $a_n = \dfrac{2}{3^n}$

30. $a_n = \dfrac{1}{n+1} - \dfrac{1}{n+2}$

31. $a_n = \sqrt{n} - \sqrt{n+1}$

32. $a_n = \log\left(\dfrac{n}{n+1}\right)$ [*Hint:* Use a property of logarithms to write the nth term as a difference.]

33–40 ■ Find the sum.

33. $\sum_{k=1}^{4} k$

34. $\sum_{k=1}^{4} k^2$

35. $\sum_{k=1}^{3} \dfrac{1}{k}$

36. $\sum_{j=1}^{100} (-1)^j$

37. $\sum_{i=1}^{8} [1 + (-1)^i]$

38. $\sum_{i=4}^{12} 10$

39. $\sum_{k=1}^{5} 2^{k-1}$

40. $\sum_{i=1}^{3} i2^i$

41–46 ■ Write the sum without using sigma notation.

41. $\sum_{k=1}^{5} \sqrt{k}$

42. $\sum_{i=0}^{4} \dfrac{2i-1}{2i+1}$

43. $\sum_{k=0}^{6} \sqrt{k+4}$

44. $\sum_{k=6}^{9} k(k+3)$

45. $\sum_{k=3}^{100} x^k$

46. $\sum_{j=1}^{n} (-1)^{j+1} x^j$

47–54 ■ Write the sum using sigma notation.

47. $1 + 2 + 3 + 4 + \cdots + 100$

48. $2 + 4 + 6 + \cdots + 20$

49. $1^2 + 2^2 + 3^2 + \cdots + 10^2$

50. $\dfrac{1}{2 \ln 2} - \dfrac{1}{3 \ln 3} + \dfrac{1}{4 \ln 4} - \dfrac{1}{5 \ln 5} + \cdots + \dfrac{1}{100 \ln 100}$

51. $\dfrac{1}{1 \cdot 2} + \dfrac{1}{2 \cdot 3} + \dfrac{1}{3 \cdot 4} + \cdots + \dfrac{1}{999 \cdot 1000}$

52. $\dfrac{\sqrt{1}}{1^2} + \dfrac{\sqrt{2}}{2^2} + \dfrac{\sqrt{3}}{3^2} + \cdots + \dfrac{\sqrt{n}}{n^2}$

53. $1 + x + x^2 + x^3 + \cdots + x^{100}$

54. $1 - 2x + 3x^2 - 4x^3 + 5x^4 + \cdots - 100x^{99}$

55. Find a formula for the nth term of the sequence

$$\sqrt{2},\quad \sqrt{2\sqrt{2}},\quad \sqrt{2\sqrt{2\sqrt{2}}},\quad \sqrt{2\sqrt{2\sqrt{2\sqrt{2}}}},\quad \ldots$$

[*Hint:* Write each term as a power of 2.]

56. Fibonacci posed the following problem: Suppose that rabbits live forever and that every month each pair produces a new pair that becomes productive at age 2 months. If we start with one newborn pair, how many pairs of rabbits will we have in the nth month? Show that the answer is F_n, where F_n is the nth term of the Fibonacci sequence.

DISCOVERY • DISCUSSION

57. Different Sequences That Start the Same

(a) Show that the first four terms of the sequence $a_n = n^2$ are

$$1, 4, 9, 16, \ldots$$

(b) Show that the first four terms of the sequence $a_n = n^2 + (n-1)(n-2)(n-3)(n-4)$ are also

$$1, 4, 9, 16, \ldots$$

(c) Find a sequence whose first six terms are the same as those of $a_n = n^2$ but whose succeeding terms differ from this sequence.

(d) Find two different sequences that begin

$$2, 4, 8, 16, \ldots$$

58. A Recursive Sequence Find the first 40 terms of the sequence defined by

$$a_{n+1} = \begin{cases} \dfrac{a_n}{2} & \text{if } a_n \text{ is an even number} \\ 3a_n + 1 & \text{if } a_n \text{ is an odd number} \end{cases}$$

and $a_1 = 11$. Do the same if $a_1 = 25$. Make a conjecture about this type of sequence. Try several other values for a_1, to test your conjecture.

59. A Different Type of Recursion Find the first 10 terms of the sequence defined by

$$a_n = a_{n-a_{n-1}} + a_{n-a_{n-2}}$$

with

$$a_1 = 1 \text{ and } a_2 = 1$$

How is this recursive sequence different from the others in this section?

9.2 ARITHMETIC SEQUENCES

Perhaps the simplest way to generate a sequence is to start with a number a and add to it a fixed constant d, over and over again.

DEFINITION OF AN ARITHMETIC SEQUENCE

An **arithmetic sequence** is a sequence of the form

$$a, a + d, a + 2d, a + 3d, a + 4d, \ldots$$

The number a is the **first term**, and d is the **common difference** of the sequence. The ***n*th term** of an arithmetic sequence is given by

$$a_n = a + (n - 1)d$$

The number d is called the common difference because any two consecutive terms of an arithmetic sequence differ by d.

EXAMPLE 1 ■ Arithmetic Sequences

(a) If $a = 2$ and $d = 3$, then we have the arithmetic sequence

$$2, 2 + 3, 2 + 6, 2 + 9, \ldots$$

or

$$2, 5, 8, 11, \ldots$$

Any two consecutive terms of this sequence differ by $d = 3$. The nth term is $a_n = 2 + 3(n - 1)$.

(b) Consider the arithmetic sequence

$$9, 4, -1, -6, -11, \ldots$$

Here the common difference is $d = -5$. The terms of an arithmetic sequence decrease if the common difference is negative. The nth term is $a_n = 9 - 5(n - 1)$. ■

An arithmetic sequence is determined completely by the first term a and the common difference d. Thus, if we know the first two terms of an arithmetic sequence, then we can find a formula for the nth term, as the next example shows.

EXAMPLE 2 ■ Finding Terms of an Arithmetic Sequence

Find the first six terms and the 300th term of the arithmetic sequence

$$13, 7, \ldots$$

SOLUTION

Since the first term is 13, we have $a = 13$. The common difference is $d = 7 - 13 = -6$. Thus, the nth term of this sequence is

$$a_n = 13 - 6(n - 1)$$

From this we find the first six terms:

$$13, 7, 1, -5, -11, -17, \ldots$$

The 300th term is $a_{300} = 13 - 6(299) = -1781$. ■

The next example shows that an arithmetic sequence is determined completely by *any* two of its terms.

EXAMPLE 3 ■ Finding Terms of an Arithmetic Sequence

The 11th term of an arithmetic sequence is 52, and the 19th term is 92. Find the 1000th term.

SOLUTION

To find the nth term of this sequence, we need to find a and d in the formula

$$a_n = a + (n - 1)d$$

From this formula we get

$$a_{11} = a + (11 - 1)d = a + 10d$$

$$a_{19} = a + (19 - 1)d = a + 18d$$

Since $a_{11} = 52$ and $a_{19} = 92$, we get the two equations:

$$\begin{cases} 52 = a + 10d \\ 92 = a + 18d \end{cases}$$

MATHEMATICS IN THE MODERN WORLD

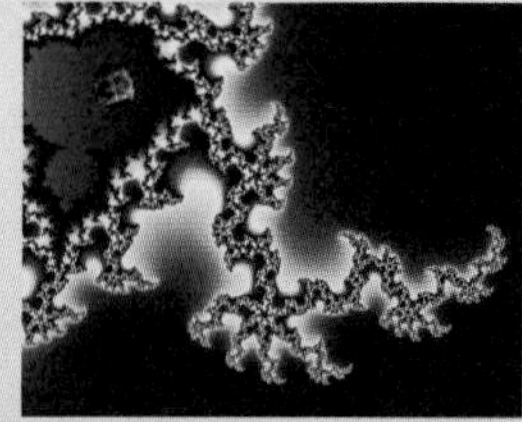

FRACTALS

Many of the things we model in this book have regular predictable shapes. But recent advances in mathematics have made it possible to mathematically model such seemingly random or even chaotic shapes as those of a cloud, a flickering flame, a mountain, or a jagged coastline. The basic tools in this type of modeling are the fractals invented by the mathematician Benoit Mandelbrot. A *fractal* is a geometric shape built up from a simple basic shape by scaling and repeating the shape indefinitely according to a given rule (an example is the snowflake curve on page 638). Fractals have infinite detail; this means the closer you look, the more you see. They are also *self-similar*; that is, zooming in on a portion of the fractal yields the same detail as the original shape. Because of their beautiful shapes, fractals are used by movie makers to create fictional landscapes or exotic backgrounds.

Although a fractal is a complex shape, it is produced according to very simple rules. This property of fractals is exploited in a process of storing pictures on a computer called *fractal image compression.* In this process a picture is stored as a simple basic shape and a rule;

(continued)

repeating the shape according to the rule produces the original picture. This is an extremely efficient method of storage; that's how 5000 color pictures can be put on a single compact disc.

Solving this system for a and d, we get $a = 2$ and $d = 5$. (Verify this.) Thus, the nth term of this sequence is

$$a_n = 2 + 5(n - 1)$$

The 1000th term is $a_{1000} = 2 + 5(999) = 4997$. ■

PARTIAL SUMS OF ARITHMETIC SEQUENCES

Suppose we want to find the sum of the numbers 1, 2, 3, 4, . . . , 100, that is,

$$\sum_{k=1}^{100} k$$

When the famous mathematician C. F. Gauss was a schoolboy, his teacher posed this problem to the class and expected that it would keep the students busy for a long time. But Gauss answered the question almost immediately. His idea was this: Since we are adding numbers produced according to a fixed pattern, there must also be a pattern (or formula) for finding the sum. He started by writing the numbers from 1 to 100 and below them the same numbers in reverse order. Writing S for the sum and adding corresponding terms gives

$$\begin{aligned} S &= 1 + 2 + 3 + \cdots + 98 + 99 + 100 \\ S &= 100 + 99 + 98 + \cdots + 3 + 2 + 1 \\ \hline 2S &= 101 + 101 + 101 + \cdots + 101 + 101 + 101 \end{aligned}$$

It follows that $2S = 100(101) = 10{,}100$ and so $S = 5050$.

Of course, the sequence of natural numbers 1, 2, 3, . . . is an arithmetic sequence (with $a = 1$ and $d = 1$), and the method for summing the first 100 terms of this sequence can be used to find a formula for the nth partial sum of any arithmetic sequence. We want to find the sum of the first n terms of the arithmetic sequence whose terms are $a_k = a + (k - 1)d$; that is, we want to find

$$\begin{aligned} S_n &= \sum_{k=1}^{n} [a + (k - 1)d] \\ &= a + (a + d) + (a + 2d) + (a + 3d) + \cdots + [a + (n - 1)d] \end{aligned}$$

Using Gauss's method, we write

$$\begin{aligned} S_n &= a + (a + d) + \cdots + [a + (n - 2)d] + [a + (n - 1)d] \\ S_n &= [a + (n - 1)d] + [a + (n - 2)d] + \cdots + (a + d) + a \\ \hline 2S_n &= [2a + (n - 1)d] + [2a + (n - 1)d] + \cdots + [2a + (n - 1)d] + [2a + (n - 1)d] \end{aligned}$$

There are n identical terms on the right side of this equation, so

$$2S_n = n[2a + (n - 1)d]$$

$$S_n = \frac{n}{2}[2a + (n - 1)d]$$

Notice that $a_n = a + (n - 1)d$ is the nth term of this sequence. So, we can write

$$S_n = \frac{n}{2}[a + a + (n - 1)d] = n\left(\frac{a + a_n}{2}\right)$$

This last formula says that the sum of the first n terms of an arithmetic sequence is the average of the first and nth terms multiplied by n, the number of terms in the sum. We now summarize this result.

PARTIAL SUMS OF AN ARITHMETIC SEQUENCE

For the arithmetic sequence $a_n = a + (n - 1)d$, the **nth partial sum**

$$S_n = a + (a + d) + (a + 2d) + (a + 3d) + \cdots + [a + (n - 1)d]$$

is given by either of the following formulas.

1. $S_n = \frac{n}{2}[2a + (n - 1)d]$ **2.** $S_n = n\left(\frac{a + a_n}{2}\right)$

EXAMPLE 4 ■ Finding a Partial Sum of an Arithmetic Sequence

Find the sum of the first 40 terms of the arithmetic sequence

$$3, 7, 11, 15, \ldots$$

SOLUTION

For this arithmetic sequence, $a = 3$ and $d = 4$. Using Formula 1 for the partial sum of an arithmetic sequence, we get

$$S_{40} = \tfrac{40}{2}[2(3) + (40 - 1)4] = 20(6 + 156) = 3240$$ ■

EXAMPLE 5 ■ Finding a Partial Sum of an Arithmetic Sequence

Find the sum of the first 50 odd numbers.

SOLUTION

The odd numbers form an arithmetic sequence with $a = 1$ and $d = 2$. The nth term is $a_n = 1 + 2(n - 1) = 2n - 1$, so the 50th odd number is $a_{50} = 2(50) - 1 = 99$.

Substituting in Formula 2 for the partial sum of an arithmetic sequence, we get

$$S_{50} = 50\left(\frac{a + a_{50}}{2}\right) = 50\left(\frac{1 + 99}{2}\right) = 50 \cdot 50 = 2500$$ ■

EXAMPLE 6 ■ Finding the Seating Capacity of an Amphitheater

An amphitheater has 50 rows of seats with 30 seats in the first row, 32 in the second, 34 in the third, and so on. Find the total number of seats.

SOLUTION

The numbers of seats in the rows form an arithmetic sequence with $a = 30$ and $d = 2$. Since there are 50 rows, the total number of seats is the sum

$$\begin{aligned} S_{50} &= \tfrac{50}{2}[2(30) + 49(2)] \qquad \text{Formula 1} \\ &= 3950 \end{aligned}$$

Thus, the amphitheater has 3950 seats. ■

EXAMPLE 7 ■ Finding the Number of Terms in a Partial Sum

How many terms of the arithmetic sequences 5, 7, 9, . . . must be added to get 572?

SOLUTION

We are asked to find n when $S_n = 572$. Substituting $a = 5$, $d = 2$, and $S_n = 572$ in Formula 1 for the partial sum of an arithmetic sequence, we get

$$\begin{aligned} 572 &= \frac{n}{2}[2 \cdot 5 + (n - 1)2] \qquad \text{Formula 1} \\ 572 &= 5n + n(n - 1) \\ 0 &= n^2 + 4n - 572 \\ 0 &= (n - 22)(n + 26) \end{aligned}$$

This gives $n = 22$ or $n = -26$. But since n is the *number* of terms in this partial sum, we must have $n = 22$. ■

9.2 EXERCISES

1–6 ■ Determine whether the sequence is arithmetic. If it is arithmetic, find the common difference.

1. 5, 8, 11, 14, . . .

2. 3, 6, 9, 13, . . .

3. 2, 4, 8, 16, . . .

4. 2, 4, 6, 8, . . .

5. $3, \frac{3}{2}, 0, -\frac{3}{2}, \ldots$

6. ln 2, ln 4, ln 8, ln 16, . . .

7–16 ■ Determine the common difference, the fifth term, the nth term, and the 100th term of the arithmetic sequence.

7. $2, 5, 8, 11, \ldots$

8. $1, 5, 9, 13, \ldots$

9. $4, 9, 14, 19, \ldots$

10. $11, 8, 5, 2, \ldots$

11. $-12, -8, -4, 0, \ldots$

12. $\frac{7}{6}, \frac{5}{3}, \frac{13}{6}, \frac{8}{3}, \ldots$

13. $25, 26.5, 28, 29.5, \ldots$

14. $15, 12.3, 9.6, 6.9, \ldots$

15. $2, 2 + s, 2 + 2s, 2 + 3s, \ldots$

16. $-t, -t + 3, -t + 6, -t + 9, \ldots$

17. The tenth term of an arithmetic sequence is $\frac{55}{2}$, and the second term is $\frac{7}{2}$. Find the first term.

18. The 12th term of an arithmetic sequence is 32, and the fifth term is 18. Find the 20th term.

19. The 100th term of an arithmetic sequence is 98, and the common difference is 2. Find the first three terms.

20. The 20th term of an arithmetic sequence is 101, and the common difference is 3. Find a formula for the nth term.

21. Which term of the arithmetic sequence $1, 4, 7, \ldots$ is 88?

22. The first term of an arithmetic sequence is 1, and the common difference is 4. Is 11,937 a term of this sequence? If so, which term is it?

23–28 ■ Find the partial sum S_n of the arithmetic sequence that satisfies the given conditions.

23. $a = 1, d = 2, n = 10$

24. $a = 3, d = 2, n = 12$

25. $a = 4, d = 2, n = 20$

26. $a = 100, d = -5, n = 8$

27. $a_1 = 55, d = 12, n = 10$

28. $a_2 = 8, a_5 = 9.5, n = 15$

29–34 ■ A partial sum of an arithmetic sequence is given. Find the sum.

29. $1 + 5 + 9 + \cdots + 401$

30. $-3 + \left(-\frac{3}{2}\right) + 0 + \frac{3}{2} + 3 + \cdots + 30$

31. $0.7 + 2.7 + 4.7 + \cdots + 56.7$

32. $-10 - 9.9 - 9.8 - \cdots - 0.1$

33. $\sum_{k=0}^{10} (3 + 0.25k)$

34. $\sum_{n=0}^{20} (1 - 2n)$

35. The purchase value of an office computer is \$12,500. Its annual depreciation is \$1875. Find the value of the computer after 6 years.

36. Telephone poles are stored in a pile with 25 poles in the first layer, 24 in the second, and so on. If there are 12 layers, how many telephone poles does the pile contain?

37. A man gets a job with a salary of \$30,000 a year. He is promised a \$2300 raise each subsequent year. Find his total earnings for a 10-year period.

38. A drive-in theater has spaces for 20 cars in the first parking row, 22 in the second, 24 in the third, and so on. If there are 21 rows in the theater, find the number of cars that can be parked.

39. An architect designs a theater with 15 seats in the first row, 18 in the second, 21 in the third, and so on. If the theater is to have a seating capacity of 870, how many rows must the architect use in his design?

40. An arithmetic sequence has first term $a = 5$ and common difference $d = 2$. How many terms of this sequence must be added to get 2700?

41. When an object is allowed to fall freely near the surface of the earth, the gravitational pull is such that the object falls 16 ft in the first second, 48 ft in the next second, 80 ft in the next second, and so on.

(a) Find the total distance a ball falls in 6 s.

(b) Find a formula for the total distance a ball falls in n seconds.

42. In the well-known song "The Twelve Days of Christmas," a person gives his sweetheart k gifts on the kth day for each of the 12 days of Christmas. The person also repeats each gift identically on each subsequent day. Thus, on the 12th day the sweetheart receives a gift for the first day, 2 gifts for the second, 3 gifts for the third, and so on. Show that the number of gifts received on the 12th day is a partial sum of an arithmetic sequence. Find this sum.

43. Show that a right triangle whose sides are in arithmetic progression is similar to a 3–4–5 triangle.

44. Find the product of the numbers

$$10^{1/10}, 10^{2/10}, 10^{3/10}, 10^{4/10}, \ldots, 10^{19/10}$$

45. A sequence is **harmonic** if the reciprocals of the terms of the sequence form an arithmetic sequence. Determine whether the following sequence is harmonic:

$$1, \tfrac{3}{5}, \tfrac{3}{7}, \tfrac{1}{3}, \ldots$$

46. The **harmonic mean** of two numbers is the reciprocal of the average of the reciprocals of the two numbers. Find the harmonic mean of 3 and 5.

DISCOVERY • DISCUSSION

47. Arithmetic Means The **arithmetic mean** (or average) of two numbers a and b is

$$m = \frac{a+b}{2}$$

Note that m is the same distance from a as from b, so a, m, b is an arithmetic sequence. In general, if $m_1, m_2, \ldots, m_k$ are equally spaced between a and b so that

$$a, m_1, m_2, \ldots, m_k, b$$

is an arithmetic sequence, then $m_1, m_2, \ldots, m_k$ are called k arithmetic means between a and b.

(a) Insert two arithmetic means between 10 and 18.

(b) Insert three arithmetic means between 10 and 18.

(c) Suppose a doctor needs to increase a patient's dosage of a certain medicine from 100 mg to 300 mg per day in five equal steps. How many arithmetic means must be inserted between 100 and 300 to give the progression of daily doses, and what are these means?

9.3 GEOMETRIC SEQUENCES

Another simple way of generating a sequence is to start with a number a and repeatedly multiply by a fixed nonzero constant r.

DEFINITION OF A GEOMETRIC SEQUENCE

A **geometric sequence** is a sequence of the form

$$a, ar, ar^2, ar^3, ar^4, \ldots$$

The number a is the **first term**, and r is the **common ratio** of the sequence. The ***n*th term** of a geometric sequence is given by

$$a_n = ar^{n-1}$$

The number r is called the common ratio because the ratio of any two consecutive terms of the sequence is r.

EXAMPLE 1 ■ Geometric Sequences

(a) If $a = 3$ and $r = 2$, then we have the geometric sequence

$$3, \quad 3 \cdot 2, \quad 3 \cdot 2^2, \quad 3 \cdot 2^3, \quad 3 \cdot 2^4, \quad \ldots$$

or

$$3, 6, 12, 24, 48, \ldots$$

Notice that the ratio of any two consecutive terms is $r = 2$. The nth term is $a_n = 3(2)^{n-1}$.

(b) The sequence

$$2, -10, 50, -250, 1250, \ldots$$

is a geometric sequence with $a = 2$ and $r = -5$. When r is negative, the terms of the sequence alternate in sign. The nth term is $a_n = 2(-5)^{n-1}$.

(c) The sequence

$$1, \frac{1}{3}, \frac{1}{9}, \frac{1}{27}, \frac{1}{81}, \ldots$$

is a geometric sequence with $a = 1$ and $r = \frac{1}{3}$. The nth term is $a_n = 1\left(\frac{1}{3}\right)^{n-1}$. If $0 < r < 1$, then the terms of the sequence decrease, but if $r > 1$, then the terms increase. (What happens if $r = 1$?) ■

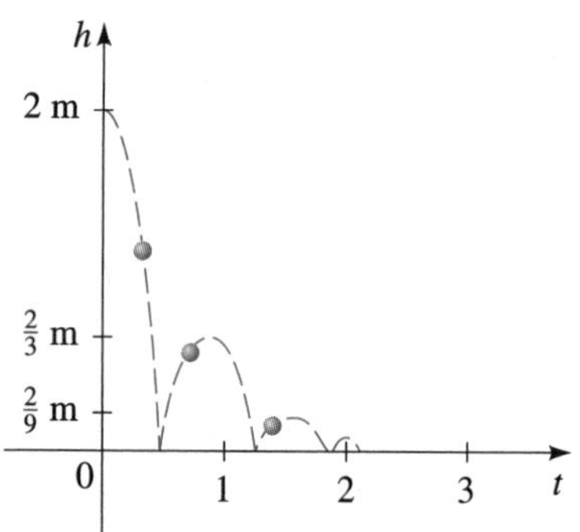

FIGURE 1

Geometric sequences occur naturally. Here is a simple example. Suppose a ball has elasticity such that when it is dropped it bounces up one-third of the distance it has fallen. If this ball is dropped from a height of 2 m, then it bounces up to a height of $2\left(\frac{1}{3}\right) = \frac{2}{3}$ m. On its second bounce, it returns to a height of $\left(\frac{2}{3}\right)\left(\frac{1}{3}\right) = \frac{2}{9}$ m, and so on (see Figure 1). Thus, the height h_n that the ball reaches on its nth bounce is given by the geometric sequence

$$h_n = \frac{2}{3}\left(\frac{1}{3}\right)^{n-1} = 2\left(\frac{1}{3}\right)^n$$

We can find the nth term of a geometric sequence if we know any two terms, as the following examples show.

EXAMPLE 2 ■ Finding Terms of a Geometric Sequence

Find the eighth term of the geometric sequence 5, 15, 45, ...

SOLUTION

To find a formula for the nth term of this sequence, we need to find a and r. Clearly, $a = 5$. To find r, we find the ratio of any two consecutive terms. For instance, $r = \frac{45}{15} = 3$. Thus

$$a_n = 5(3)^{n-1}$$

The eighth term is $a_8 = 5(3)^{8-1} = 5(3)^7 = 10{,}935$. ■

EXAMPLE 3 ■ Finding Terms of a Geometric Sequence

The third term of a geometric series is $\frac{63}{4}$, and the sixth term is $\frac{1701}{32}$. Find the fifth term.

SOLUTION

Since this series is geometric, its nth term is given by the formula $a_n = ar^{n-1}$. Thus

$$a_3 = ar^{3-1} = ar^2$$

$$a_6 = ar^{6-1} = ar^5$$

From the values we are given for these two terms, we get the following system of equations:

$$\begin{cases} \frac{63}{4} = ar^2 \\ \frac{1701}{32} = ar^5 \end{cases}$$

We can solve this system by dividing.

$$\frac{ar^5}{ar^2} = \frac{\frac{1701}{32}}{\frac{63}{4}}$$

$$r^3 = \frac{27}{8} \qquad \text{Simplify}$$

$$r = \frac{3}{2} \qquad \text{Take the cube root of each side}$$

Substituting for r in the first equation, $\frac{63}{4} = ar^2$, gives

$$\frac{63}{4} = a\left(\frac{3}{2}\right)^2$$

$$a = 7 \qquad \text{Solve for } a$$

It follows that the nth term of this sequence is

$$a_n = 7\left(\frac{3}{2}\right)^{n-1}$$

Thus, the fifth term is

$$a_5 = 7\left(\frac{3}{2}\right)^{5-1} = 7\left(\frac{3}{2}\right)^4 = \frac{567}{16}$$

■

PARTIAL SUMS OF GEOMETRIC SEQUENCES

For the geometric sequence $a, ar, ar^2, ar^3, ar^4, \ldots, ar^{n-1}, \ldots$, the nth partial sum is

$$S_n = \sum_{k=1}^{n} ar^{k-1} = a + ar + ar^2 + ar^3 + ar^4 + \cdots + ar^{n-1}$$

To find a formula for S_n, we multiply S_n by r and subtract from S_n to get

$$S_n = a + ar + ar^2 + ar^3 + ar^4 + \cdots + ar^{n-1}$$

$$rS_n = \quad ar + ar^2 + ar^3 + ar^4 + \cdots + ar^{n-1} + ar^n$$

$$S_n - rS_n = a - ar^n$$

So, $$S_n(1 - r) = a(1 - r^n)$$

$$S_n = \frac{a(1 - r^n)}{1 - r} \qquad (r \neq 1)$$

We summarize this result.

PARTIAL SUMS OF A GEOMETRIC SEQUENCE

For the geometric sequence $a_n = ar^{n-1}$, the **nth partial sum**

$$S_n = a + ar + ar^2 + ar^3 + ar^4 + \cdots + ar^{n-1} \qquad (r \neq 1)$$

is given by

$$S_n = a\frac{1 - r^n}{1 - r}$$

EXAMPLE 4 ■ Finding a Partial Sum of a Geometric Sequence

Find the sum of the first five terms of the geometric sequence

$$1, 0.7, 0.49, 0.343, \ldots$$

SOLUTION

The required sum is the sum of the first five terms of a geometric sequence with $a = 1$ and $r = 0.7$. Using the formula for S_n with $n = 5$, we get

$$S_5 = 1 \cdot \frac{1 - (0.7)^5}{1 - 0.7} = 2.7731$$

Thus, the sum of the first five terms of this sequence is 2.7731. ■

EXAMPLE 5 ■ Finding a Partial Sum of a Geometric Sequence

Find the sum $\sum_{k=1}^{5} 7\left(-\frac{2}{3}\right)^k$.

SOLUTION

The given sum is the fifth partial sum of a geometric sequence with first term $a = 7\left(-\frac{2}{3}\right) = -\frac{14}{3}$ and common ratio $r = -\frac{2}{3}$. Thus, by the formula for S_n, we have

$$S_5 = -\frac{14}{3} \cdot \frac{1 - \left(-\frac{2}{3}\right)^5}{1 - \left(-\frac{2}{3}\right)} = -\frac{14}{3} \cdot \frac{1 + \frac{32}{243}}{\frac{5}{3}} = -\frac{770}{243}$$

■

WHAT IS AN INFINITE SERIES?

An expression of the form

$$a_1 + a_2 + a_3 + a_4 + \cdots$$

is called an **infinite series**. The dots mean that we are to continue the addition indefinitely. What meaning can we attach to the sum of infinitely many numbers? It seems at first that it is not possible to add infinitely many numbers and arrive at a finite number. But consider the following problem. You have a cake and you want to eat it by first eating half the cake, then eating half of what remains, then again eating half of what remains. This process can continue indefinitely because at each stage some of the cake remains. (See Figure 2.)

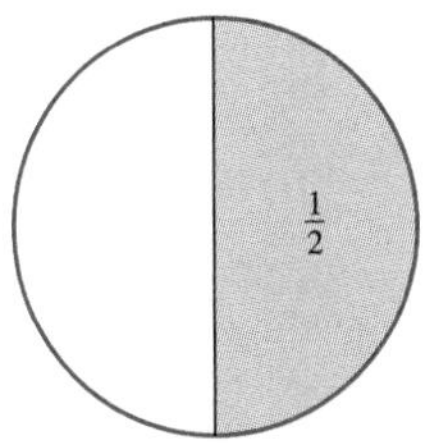

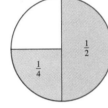

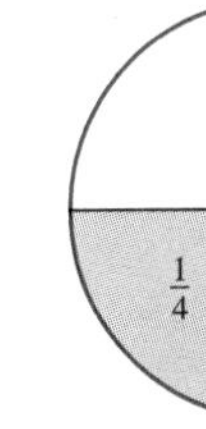

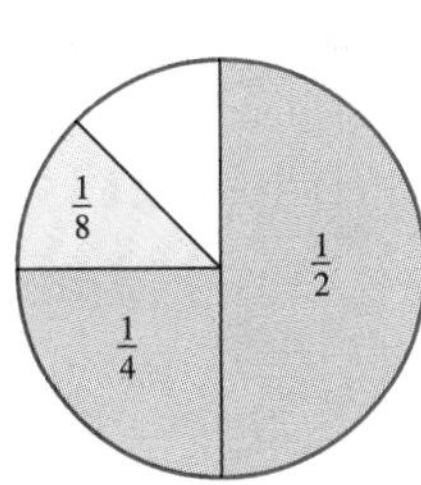

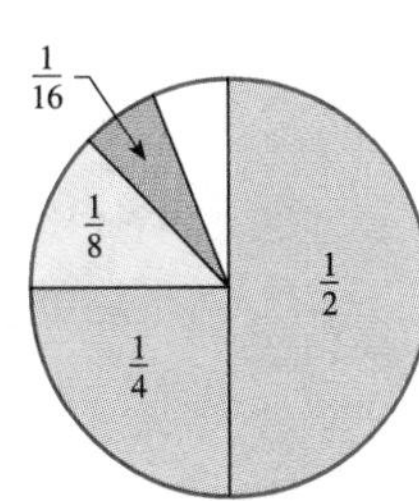

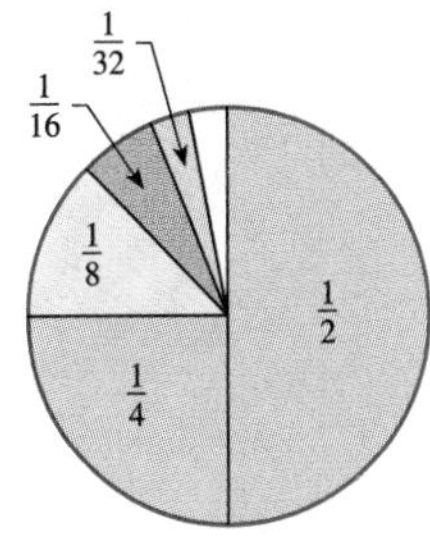

FIGURE 2

Does this mean that it's impossible to eat all of the cake? Of course not. Let's write down what you have eaten from this cake:

$$\frac{1}{2} + \frac{1}{4} + \frac{1}{8} + \frac{1}{16} + \cdots + \frac{1}{2^n} + \cdots$$

This is an infinite series, and we note two things about it: First, from Figure 2 it's clear that no matter how many terms of this series we add, the total will never exceed 1. Second, the more terms of this series we add, the closer the sum is to 1 (see Figure 2). This suggests that the number 1 can be written as the sum of infinitely many smaller numbers:

$$1 = \frac{1}{2} + \frac{1}{4} + \frac{1}{8} + \frac{1}{16} + \cdots + \frac{1}{2^n} + \cdots$$

To make this more precise, let's look at the partial sums of this series:

$$S_1 = \frac{1}{2} = \frac{1}{2}$$

$$S_2 = \frac{1}{2} + \frac{1}{4} = \frac{3}{4}$$

$$S_3 = \frac{1}{2} + \frac{1}{4} + \frac{1}{8} = \frac{7}{8}$$

$$S_4 = \frac{1}{2} + \frac{1}{4} + \frac{1}{8} + \frac{1}{16} = \frac{15}{16}$$

and, in general (see Example 5 of Section 9.1),

$$S_n = 1 - \frac{1}{2^n}$$

As n gets larger and larger, we are adding more and more of the terms of this series. Intuitively, as n gets larger, S_n gets closer to the sum of the series. Now notice that as n gets large, $1/2^n$ gets closer and closer to 0. Thus, S_n gets close to $1 - 0 = 1$. Using the notation of Section 5.5, we can write

$$S_n \to 1 \qquad \text{as} \qquad n \to \infty$$

In general, if S_n gets close to a finite number S as n gets large, we say that S is the **sum of the infinite series.**

INFINITE GEOMETRIC SERIES

An **infinite geometric series** is a series of the form

$$a + ar + ar^2 + ar^3 + ar^4 + \cdots + ar^{n-1} + \cdots$$

We can apply the reasoning used earlier to find the sum of an infinite geometric series. The nth partial sum of such a series is given by the formula

$$S_n = a\frac{1 - r^n}{1 - r} \qquad (r \neq 1)$$

It can be shown that if $|r| < 1$, then r^n gets close to 0 as n gets large (you can easily convince yourself of this using a calculator). It follows that S_n gets close to $a/(1 - r)$ as n gets large, or

$$S_n \to \frac{a}{1 - r} \qquad \text{as} \qquad n \to \infty$$

Thus, the sum of this infinite geometric series is $a/(1 - r)$.

Here is another way to arrive at the formula for the sum of an infinite geometric series:

$$\begin{aligned} S &= a + ar + ar^2 + ar^3 + \cdots \\ &= a + r(a + ar + ar^2 + \cdots) \\ &= a + rS \end{aligned}$$

Solve the equation $S = a + rS$ for S to get

$$\begin{aligned} S - rS &= a \\ (1 - r)S &= a \\ S &= \frac{a}{1 - r} \end{aligned}$$

SUM OF AN INFINITE GEOMETRIC SERIES

If $|r| < 1$, then the infinite geometric series

$$a + ar + ar^2 + ar^3 + ar^4 + \cdots + ar^{n-1} + \cdots$$

has the sum

$$S = \frac{a}{1 - r}$$

EXAMPLE 6 ■ Finding the Sum of an Infinite Geometric Series

Find the sum of the infinite geometric series

$$2 + \frac{2}{5} + \frac{2}{25} + \frac{2}{125} + \cdots + \frac{2}{5^n} + \cdots$$

SOLUTION

We use the formula for the sum of an infinite geometric series. In this case, $a = 2$ and $r = \frac{1}{5}$. Thus, the sum of this infinite series is

$$S = \frac{2}{1 - \frac{1}{5}} = \frac{5}{2}$$

■

EXAMPLE 7 ■ Writing a Repeated Decimal as a Fraction

Find the fraction that represents the rational number $2.3\overline{51}$.

SOLUTION

This repeating decimal can be written as a series:

$$\frac{23}{10} + \frac{51}{1000} + \frac{51}{100{,}000} + \frac{51}{10{,}000{,}000} + \frac{51}{1{,}000{,}000{,}000} + \cdots$$

After the first term, the terms of this series form an infinite geometric series with

$$a = \frac{51}{1000} \qquad \text{and} \qquad r = \frac{1}{100}$$

Thus, the sum of this part of the series is

$$S = \frac{\frac{51}{1000}}{1 - \frac{1}{100}} = \frac{\frac{51}{1000}}{\frac{99}{100}} = \frac{51}{1000} \cdot \frac{100}{99} = \frac{51}{990}$$

So,

$$2.3\overline{51} = \frac{23}{10} + \frac{51}{990} = \frac{2328}{990} = \frac{388}{165}$$

■

9.3 EXERCISES

1–6 ■ Determine whether the sequence is geometric. If it is geometric, find the common ratio.

1. $2, 4, 8, 16, \ldots$

2. $2, 6, 18, 36, \ldots$

3. $3, \frac{3}{2}, \frac{3}{4}, \frac{3}{8}, \ldots$

4. $27, -9, 3, -1, \ldots$

5. $\frac{1}{2}, \frac{1}{3}, \frac{1}{4}, \frac{1}{5}, \ldots$

6. $e^2, e^4, e^6, e^8, \ldots$

7–16 ■ Determine the common ratio, the fifth term, and the nth term of the geometric sequence.

7. $2, 6, 18, 54, \ldots$

8. $7, \frac{14}{3}, \frac{28}{9}, \frac{56}{27}, \ldots$

9. $0.3, -0.09, 0.027, -0.0081, \ldots$

10. $1, \sqrt{2}, 2, 2\sqrt{2}, \ldots$

11. $144, -12, 1, -\frac{1}{12}, \ldots$

12. $-8, -2, -\frac{1}{2}, -\frac{1}{8}, \ldots$

13. $3, 3^{5/3}, 3^{7/3}, 27, \ldots$

14. $t, \dfrac{t^2}{2}, \dfrac{t^3}{4}, \dfrac{t^4}{8}, \ldots$

15. $1, s^{2/7}, s^{4/7}, s^{6/7}, \ldots$

16. $5, 5^{c+1}, 5^{2c+1}, 5^{3c+1}, \ldots$

17. The first term of a geometric sequence is 8, and the second term is 4. Find the fifth term.

18. The first term of a geometric sequence is 3, and the third term is $\frac{4}{3}$. Find the fifth term.

19. The common ratio in a geometric sequence is $\frac{2}{5}$, and the fourth term is $\frac{5}{2}$. Find the third term.

20. The common ratio in a geometric sequence is $\frac{3}{2}$, and the fifth term is 1. Find the first three terms.

21. Which term of the geometric sequence 2, 6, 18, . . . is 118,098?

22. The second and the fifth terms of a geometric sequence are 10 and 1250, respectively. Is 31,250 a term of this sequence? If so, which term is it?

23–26 ■ Find the partial sum S_n of the geometric sequence that satisifies the given conditions.

23. $a = 5, \quad r = 2, \quad n = 6$

24. $a = \frac{2}{3}, \quad r = \frac{1}{3}, \quad n = 4$

25. $a_3 = 28, \quad a_6 = 224, \quad n = 6$

26. $a_2 = 0.12, \quad a_5 = 0.00096, \quad n = 4$

27–30 ■ Find the sum.

27. $1 + 3 + 9 + \cdots + 2187$

28. $1 - \frac{1}{2} + \frac{1}{4} - \frac{1}{8} + \cdots - \frac{1}{512}$

29. $\displaystyle\sum_{k=0}^{10} 3\left(\tfrac{1}{2}\right)^k$

30. $\displaystyle\sum_{j=0}^{5} 7\left(\tfrac{3}{2}\right)^j$

31. A ball is dropped from a height of 80 ft. The elasticity of this ball is such that it rebounds three-fourths of the distance it has fallen. How high does the ball rebound on the fifth bounce? Find a formula for how high the ball rebounds on the nth bounce.

32. A culture initially has 5000 bacteria, and its size increases by 8% every hour. How many bacteria are present at the end of 5 hours? Find a formula for the number of bacteria present after n hours.

33. A truck radiator holds 5 gal and is filled with water. A gallon of water is removed from the radiator and replaced with a gallon of antifreeze; then, a gallon of the mixture is removed from the radiator and again replaced by a gallon of antifreeze. This process is repeated indefinitely. How much water remains in the tank after this process is repeated 3 times? 5 times? n times?

34. A very patient woman wishes to become a billionaire. She decides to follow a simple scheme: She puts aside 1 cent the first day, 2 cents the second day, 4 cents the third day, and so on, doubling the number of cents each day. How much money will she have at the end of 30 days? How many days will it take this woman to realize her wish?

35. A ball is dropped from a height of 9 ft. The elasticity of the ball is such that it always bounces up one-third the distance it has fallen.
(a) Find the total distance the ball has traveled at the instant it hits the ground the fifth time.
(b) Find a formula for the total distance the ball has traveled at the instant it hits the ground the nth time.

36. The following is a well-known children's rhyme:

As I was going to St. Ives
I met a man with seven wives;
Every wife had seven sacks;
Every sack had seven cats;
Every cat had seven kits;
Kits, cats, sacks, and wives,
How many were going to St. Ives?

Assuming that the entire group is actually going to St. Ives, show that the answer to the question in the rhyme is a partial sum of a geometric sequence, and find the sum.

37–44 ■ Find the sum of the infinite geometric series.

37. $1 + \dfrac{1}{3} + \dfrac{1}{9} + \dfrac{1}{27} + \cdots$

38. $1 - \dfrac{1}{2} + \dfrac{1}{4} - \dfrac{1}{8} + \cdots$

39. $1 - \dfrac{1}{3} + \dfrac{1}{9} - \dfrac{1}{27} + \cdots$

40. $\dfrac{2}{5} + \dfrac{4}{25} + \dfrac{8}{125} + \cdots$

41. $\dfrac{1}{3^6} + \dfrac{1}{3^8} + \dfrac{1}{3^{10}} + \dfrac{1}{3^{12}} + \cdots$

42. $3 - \dfrac{3}{2} + \dfrac{3}{4} - \dfrac{3}{8} + \cdots$

43. $-\frac{100}{9} + \frac{10}{3} - 1 + \frac{3}{10} - \cdots$

44. $\frac{1}{\sqrt{2}} + \frac{1}{2} + \frac{1}{2\sqrt{2}} + \frac{1}{4} + \cdots$

45–50 ■ Express the repeating decimal as a fraction.

45. $0.777\ldots$

46. $0.2\overline{53}$

47. $0.030303\ldots$

48. $2.11\overline{25}$

49. $0.\overline{112}$

50. $0.123123123\ldots$

51. A certain ball rebounds to half the height from which it is dropped. Use an infinite geometric series to approximate the total distance the ball travels, after being dropped from 1 m above the ground, until it comes to rest.

52. If the ball in Exercise 51 is dropped from a height of 8 ft, then 1 s is required for its first complete bounce—from the instant it first touches the ground until it next touches the ground. Each subsequent complete bounce requires $1/\sqrt{2}$ as long as the preceding complete bounce. Use an infinite geometric series to estimate the time period from the instant the ball first touches the ground until it stops bouncing.

53. The midpoints of the sides of a square of side 1 are joined to form a new square. This procedure is repeated for each new square. (See the figure.)

(a) Find the sum of the areas of all the squares.

(b) Find the sum of the perimeters of all the squares.

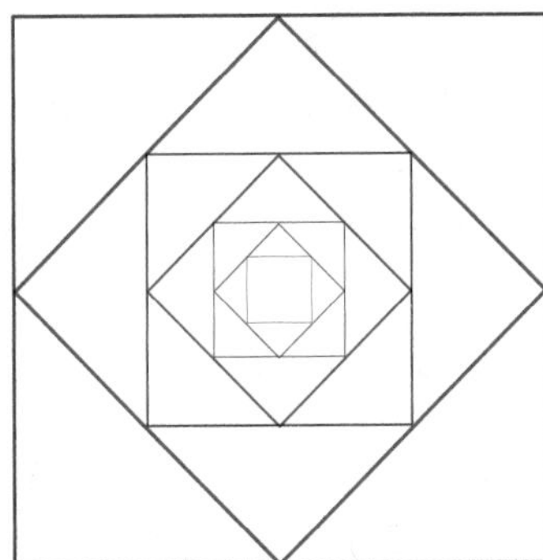

54. A circular disk of radius R is cut out of paper, as shown in figure (a). Two disks of radius $\frac{1}{2}R$ are cut out of paper and placed on top of the first disk, as in figure (b), and then four disks of radius $\frac{1}{4}R$ are placed on these two disks [figure (c)]. Assuming that this process can be repeated indefinitely, find the total area of all the disks.

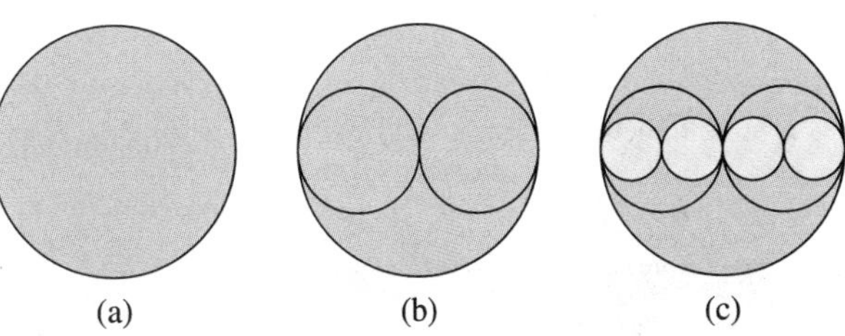

55. A yellow square of side 1 is divided into nine smaller squares, and the middle square is colored blue as shown in the figure. Each of the smaller yellow squares is in turn divided into nine squares, and each middle square is colored blue. If this process is continued indefinitely, what is the total area colored blue?

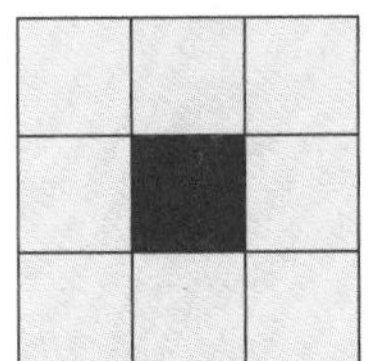
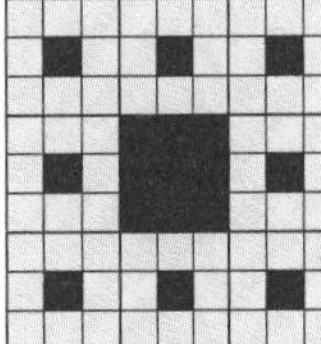

56. If the numbers $a_1, a_2, \ldots, a_n$ form a geometric sequence, then $a_2, a_3, \ldots, a_{n-1}$ are **geometric means** between a_1 and a_n. Insert three geometric means between 5 and 80.

57. Find the sum of the first ten terms of the sequence

$$a + b, a^2 + 2b, a^3 + 3b, a^4 + 4b, \ldots$$

 DISCOVERY • DISCUSSION

58. Arithmetic or Geometric? The first four terms of a sequence are given. Determine whether these terms can be the terms of an arithmetic sequence, a geometric sequence, or neither. Find the next term if the sequence is arithmetic or geometric.

(a) $5, -3, 5, -3, \ldots$

(b) $\frac{1}{3}, 1, \frac{5}{3}, \frac{7}{3}, \ldots$

(c) $\sqrt{3}, 3, 3\sqrt{3}, 9, \ldots$

(d) $1, -1, 1, -1, \ldots$

(e) $2, -1, \frac{1}{2}, 2, \ldots$

(f) $x - 1, x, x + 1, x + 2, \ldots$

(g) $-3, -\frac{3}{2}, 0, \frac{3}{2}, \ldots$

(h) $\sqrt{5}, \sqrt[3]{5}, \sqrt[6]{5}, 1, \ldots$

59. Reciprocals of a Geometric Sequence If $a_1, a_2, a_3, \ldots$ is a geometric sequence with common ratio r, show that the sequence

$$\frac{1}{a_1}, \frac{1}{a_2}, \frac{1}{a_3}, \ldots$$

is also a geometric sequence, and find the common ratio.

60. Logarithms of a Geometric Sequence If $a_1, a_2, a_3, \ldots$ is a geometric sequence with a common ratio $r > 0$ and $a_1 > 0$, show that the sequence

$$\log a_1, \log a_2, \log a_3, \ldots$$

is an arithmetic sequence, and find the common difference.

61. Exponentials of an Arithmetic Sequence If $a_1, a_2, a_3, \ldots$ is an arithmetic sequence with common difference d, show that the sequence

$$10^{a_1}, 10^{a_2}, 10^{a_3}, \ldots$$

is a geometric sequence, and find the common ratio.

FINDING PATTERNS

The ancient Greeks studied triangular numbers, square numbers, pentagonal numbers, and other **polygonal numbers**, like those shown in the figure.

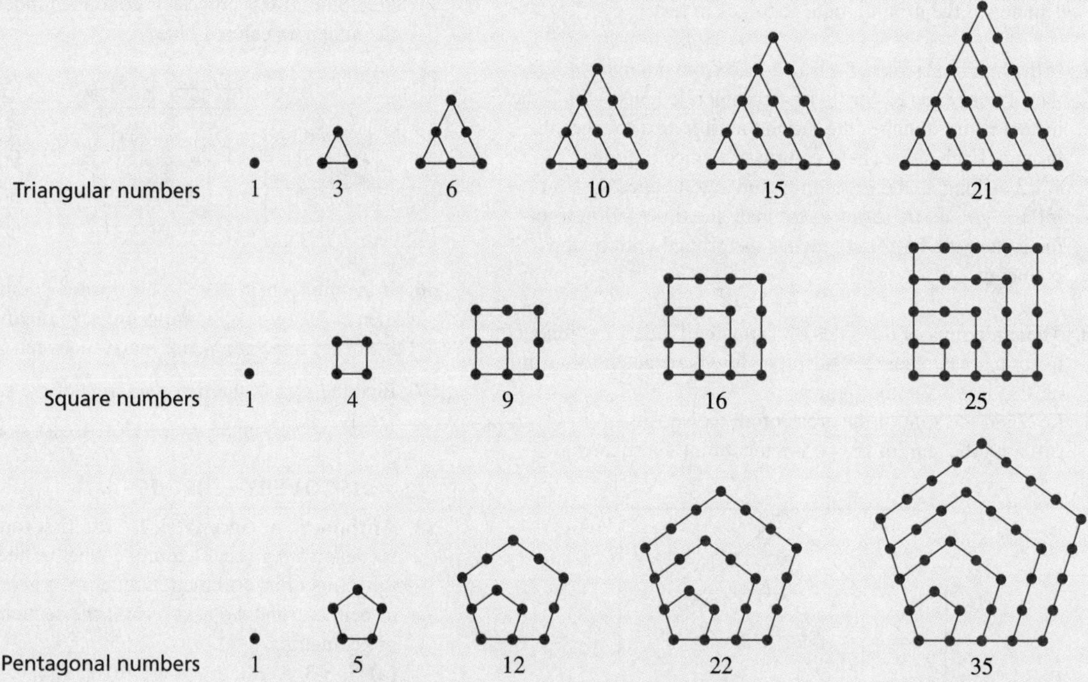

To find a pattern for such numbers, we construct a **first difference sequence** by taking differences of successive terms; we repeat the process to get a **second difference sequence**, **third difference sequence**, and so on. For the sequence of triangular numbers T_n we get the following **difference table**:

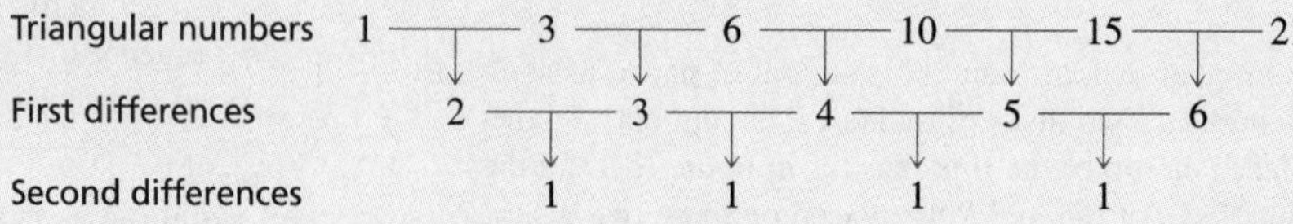

The formula for the nth triangular number can be found using the formula for the sum of the first n whole numbers (Example 2, Section 9.5). From the definition of T_n we have

$$T_n = 1 + 2 + \cdots + n = \frac{n(n+1)}{2} = \frac{1}{2}n^2 + \frac{1}{2}n$$

We stop at the second difference sequence because it's a constant sequence. Assuming that this sequence will continue to have constant value 1, we can work backward from the bottom row to find more terms of the first difference sequence, and from these, more triangular numbers.

If a sequence is given by a polynomial function and if we calculate the first differences, the second differences, the third differences, and so on, then eventually we get a constant sequence. For example, the triangular numbers are given by the polynomial $T_n = \frac{1}{2}n^2 + \frac{1}{2}n$ (see the margin note); the second difference sequence is the constant sequence 1, 1, 1, ...

1. Construct a difference table for the square numbers and the pentagonal numbers. Use the table to find the tenth pentagonal number.
2. From the patterns you've observed so far, what do you think the second difference would be for the *hexagonal numbers*? Use this, together with the fact that the first two hexagonal numbers are 1 and 6, to find the first eight hexagonal numbers.
3. Construct difference tables for $C_n = n^3$. Which difference sequence is constant? Do the same for $F_n = n^4$.
4. Make up a polynomial of degree 5 and construct a difference table. Which difference sequence is constant?
5. The first few terms of a polynomial sequence are 1, 2, 4, 8, 16, 31, 57, ... Construct a difference table and use it to find four more terms of this sequence.

9.4 ANNUITIES AND INSTALLMENT BUYING

Many financial transactions involve payments that are made at regular intervals. For example, if you deposit \$100 each month in an interest-bearing account, what will the value of the account be at the end of 5 years? If you borrow \$100,000 to buy a house, how much must the monthly payments be in order to pay off the loan in 30 years? Each of these questions involves the sum of a sequence of numbers; we use the results of the preceding section to answer them here.

THE AMOUNT OF AN ANNUITY

An **annuity** is a sum of money that is paid in regular equal payments. Although the word *annuity* suggests annual (or yearly) payments, they can be made semi-annually, quarterly, monthly, or at some other regular interval. Payments are usually made at the end of the payment interval. The **amount of an annuity** is the sum of all the individual payments from the time of the first payment until the last payment is made, together with all the interest. We denote this sum by A_f (the subscript f here is used to denote *final* amount).

EXAMPLE 1 ■ Calculating the Amount of an Annuity

An investor deposits \$400 every December 15 and June 15 for 10 years in an account that earns interest at the rate of 8% per year, compounded semiannually. How much will be in the account immediately after the last payment?

SOLUTION

When using interest rates in calculators, remember to convert percentages to decimals. For example, 8% is 0.08.

We need to find the amount of an annuity consisting of 20 semiannual payments of \$400 each. Since the interest rate is 8% per year, compounded semiannually, the interest rate per time period is $i = 0.08/2 = 0.04$. The first payment is in the account for 19 time periods, the second for 18 time periods, and so on. The last payment receives no interest. The situation can be illustrated by the time line in Figure 1.

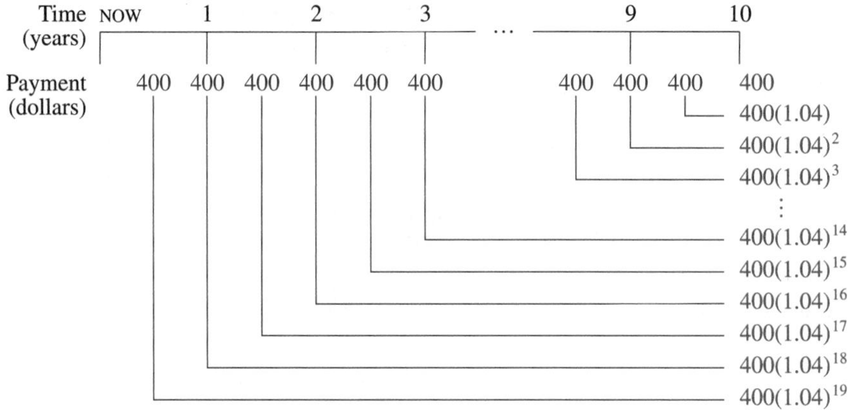

FIGURE 1

The amount A_f of the annuity is the sum of these 20 amounts. Thus

$$A_f = 400 + 400(1.04) + 400(1.04)^2 + \cdots + 400(1.04)^{19}$$

But this is a geometric series with $a = 400$, $r = 1.04$, and $n = 20$, so

$$A_f = 400\,\frac{1-(1.04)^{20}}{1-1.04} \approx 11{,}911.23$$

Thus, the amount in the account after the last payment is \$11,911.23. ■

In general, the regular annuity payment is called the **periodic rent** and is denoted by R. We also let i denote the interest rate per time period and n the number of payments. *We always assume that the time period in which interest is compounded is equal to the time between payments.* By the same reasoning as in

Example 1, we see that the amount A_f of an annuity is

$$A_f = R + R(1 + i) + R(1 + i)^2 + \cdots + R(1 + i)^{n-1}$$

Since this is the nth partial sum of a geometric sequence with $a = R$ and $r = 1 + i$, the formula for the partial sum gives

$$A_f = R\,\frac{1 - (1 + i)^n}{1 - (1 + i)} = R\,\frac{1 - (1 + i)^n}{-i} = R\,\frac{(1 + i)^n - 1}{i}$$

AMOUNT OF AN ANNUITY

The amount A_f of an annuity consisting of n regular equal payments of size R with interest rate i per time period is given by

$$A_f = R\,\frac{(1 + i)^n - 1}{i}$$

EXAMPLE 2 ■ Calculating the Amount of an Annuity

How much money should be invested every month at 12% per year, compounded monthly, in order to have \$4000 in 18 months?

SOLUTION

In this problem $i = 0.12/12 = 0.01$, $A_f = 4000$, and $n = 18$. We need to find the amount R of each payment. By the formula for the amount of an annuity.

$$4000 = R\,\frac{(1 + 0.01)^{18} - 1}{0.01}$$

Solving for R, we get

$$R = \frac{4000(0.01)}{(1 + 0.01)^{18} - 1} \approx 203.928$$

Thus, the monthly investment should be \$203.93. ■

THE PRESENT VALUE OF AN ANNUITY

If you were to receive \$10,000 five years from now, it would be worth much less than getting \$10,000 right now. This is because of the interest you could accumulate during the next five years if you invested the money now. What smaller amount would you be willing to accept *now* instead of receiving \$10,000 in five years? This is the amount of money that, together with interest, would be worth

\$10,000 in five years. The amount we are looking for here is called the *discounted value* or *present value*. If the interest rate is 8% per year, compounded quarterly, then the interest per time period is $i = 0.08/4 = 0.02$, and there are $4 \times 5 = 20$ time periods. If we let PV denote the present value, then by the formula for compound interest (Section 6.2) we have

$$10{,}000 = PV(1 + i)^n = PV(1 + 0.02)^{20}$$

so

$$PV = 10{,}000(1 + 0.02)^{-20} \approx 6729.713$$

Thus, in this situation, the present value of \$10,000 is \$6729.71. This reasoning leads to a general formula for present value:

$$PV = A(1 + i)^{-n}$$

Similarly, the **present value of an annuity** is the amount A_p that must be invested now at the interest rate i per time period in order to provide n payments, each of amount R. Clearly, A_p is the sum of the present values of each individual payment (see Exercise 22). Another way of finding A_p is to note that A_p is the present value of A_f:

$$A_p = A_f(1 + i)^{-n} = R\,\frac{(1 + i)^n - 1}{i}\,(1 + i)^{-n} = R\,\frac{1 - (1 + i)^{-n}}{i}$$

THE PRESENT VALUE OF AN ANNUITY

The **present value** A_p of an annuity consisting of n regular equal payments of size R and interest rate i per time period is given by

$$A_p = R\,\frac{1 - (1 + i)^{-n}}{i}$$

EXAMPLE 3 ■ Calculating the Present Value of an Annuity

A person wins \$10,000,000 in the California lottery, and the amount is paid in yearly installments of half a million dollars each for 20 years. What is the present value of his winnings? Assume that he can earn 10% interest, compounded annually.

SOLUTION

Since the amount won is paid as an annuity, we need to find its present value. Here $i = 0.1$, $R = \$500{,}000$, and $n = 20$. Thus

$$A_p = 500{,}000\,\frac{1 - (1 + 0.1)^{-20}}{0.1} \approx 4{,}256{,}781.859$$

This means that the winner really won only \$4,256,781.86 if it were paid immediately. ■

INSTALLMENT BUYING

When you buy a house or a car by installment, the payments you make are an annuity whose present value is the amount of the loan.

EXAMPLE 4 ■ The Amount of a Loan

A student wishes to buy a car. He can afford to pay \$200 per month but has no money for a down payment. If he can make these payments for four years and the interest rate is 12%, what purchase price can he afford?

SOLUTION

The payments the student makes constitute an annuity whose present value is the price of the car (which is also the amount of the loan, in this case). Here we have $i = 0.12/12 = 0.01$, $R = 200$, $n = 12 \times 4 = 48$, so

$$A_p = R\,\frac{1 - (1+i)^{-n}}{i} = 200\,\frac{1 - (1+0.01)^{-48}}{0.01} \approx 7594.792$$

Thus, the student can buy a car priced at \$7594.79. ■

When a bank makes a loan that is to be repaid with regular equal payments R, then the payments form an annuity whose present value A_p is the amount of the loan. So, to find the size of the payments, we solve for R in the formula for the amount of an annuity. This gives the following formula for R.

INSTALLMENT BUYING

If a loan A_p is to be repaid in n regular equal payments with interest rate i per time period, then the size R of each payment is given by

$$R = \frac{iA_p}{1 - (1+i)^{-n}}$$

EXAMPLE 5 ■ Calculating Monthly Mortgage Payments

A couple borrows \$100,000 at 9% interest as a mortage loan on a house. They expect to make monthly payments for 30 years to repay the loan. What is the size of each payment?

SOLUTION

The mortgage payments form an annuity whose present value is $A_p = \$100{,}000$. Also, $i = 0.09/12 = 0.0075$, and $n = 12 \times 30 = 360$. We are looking for the amount R of each payment. From the formula for installment buying, we get

$$\begin{aligned} R &= \frac{iA_p}{1-(1+i)^{-n}} \\ &= \frac{(0.0075)(100{,}000)}{1-(1+0.0075)^{-360}} \approx 804.623 \end{aligned}$$

Thus, the monthly payments are \$804.62. ■

We now illustrate the use of graphing devices in solving problems related to installment buying.

EXAMPLE 6 ■ Calculating the Interest Rate from the Size of Monthly Payments

A car dealer sells a new car for \$18,000. He offers the buyer payments of \$405 per month for 5 years. What interest rate is this car dealer charging?

SOLUTION

The payments form an annuity with present value $A_p = \$18{,}000$, $R = 405$, and $n = 12 \times 5 = 60$. To find the interest rate, we must solve for i in the equation

$$R = \frac{iA_p}{1-(1+i)^{-n}}$$

A little experimentation will convince you that it's not possible to algebraically solve this equation for i. So, to find i we use a graphing device to graph R as a function of the interest rate x, and we then use the graph to find the interest rate corresponding to the value of R we want (\$405 in this case). Since $i = x/12$, we graph the function

$$R(x) = \frac{\dfrac{x}{12}(18{,}000)}{1-\left(1+\dfrac{x}{12}\right)^{-60}}$$

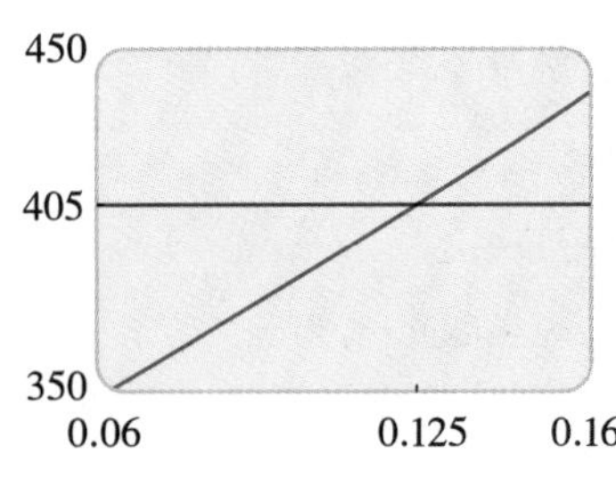

FIGURE 2

in the viewing rectangle $[0.06, 0.16] \times [350, 450]$, as shown in Figure 2. We also graph the horizontal line $R(x) = 405$ in the same viewing rectangle. Then, by moving the cursor to the point of intersection of the two graphs, we find that the corresponding x-value is approximately 0.125. Thus, the interest rate is about $12\frac{1}{2}\%$. ■

9.4 EXERCISES

1. Find the amount of an annuity that consists of 10 annual payments of \$1000 each into an account that pays 6% interest per year.

2. Find the amount of an annuity that consists of 24 monthly payments of \$500 each into an account that pays 8% interest per year, compounded monthly.

3. Find the amount of an annuity that consists of 20 annual payments of \$5000 each into an account that pays interest of 12% per year.

4. Find the amount of an annuity that consists of 20 semiannual payments of \$500 each into an account that pays 6% interest per year, compounded semiannually.

5. Find the amount of an annuity that consists of 16 quarterly payments of \$300 each into an account that pays 8% interest per year, compounded quarterly.

6. How much money should be invested every quarter at 10% per year, compounded quarterly, in order to have \$5000 in 2 years?

7. How much money should be invested monthly at 6% per year, compounded monthly, in order to have \$2000 in 8 months?

8. What is the present value of an annuity that consists of 20 semiannual payments of \$1000 at the interest rate of 9% per year, compounded semiannually?

9. How much money must be invested now at 9% per year, compounded semiannually, to fund an annuity of 20 payments of \$200 each, paid every 6 months, the first payment being 6 months from now?

10. A 55-year-old man deposits \$50,000 to fund an annuity with an insurance company. The money will be invested at 8% per year, compounded semiannually. He is to draw semiannual payments until he reaches age 65. What is the amount of each payment?

11. A woman wants to borrow \$12,000 in order to buy a car. She wants to repay the loan by monthly installments for 4 years. If the interest rate on this loan is $10\frac{1}{2}\%$ per year, compounded monthly, what is the amount of each payment?

12. What is the monthly payment on a 30-year mortgage of \$80,000 at 9% interest? What is the monthly payment on this same mortgage if it is to be repaid over a 15-year period?

13. What is the monthly payment on a 30-year mortgage of \$100,000 at 8% interest per year, compounded monthly? What is the total amount paid on this loan over the 30-year period?

14. A couple can afford to make a monthly mortgage payment of \$650. If the mortgage rate is 9% and the couple intends to secure a 30-year mortgage, how much can they borrow?

15. A couple secures a 30-year loan of \$100,000 at $9\frac{3}{4}\%$ per year, compounded monthly, to buy a house.
    (a) What is the amount of their monthly payment?
    (b) What total amount will they pay over the 30-year period?
    (c) If, instead of taking the loan, the couple deposits the monthly payments in an account that pays $9\frac{3}{4}\%$ interest per year, compounded monthly, how much will be in the account at the end of the 30-year period?

16. Jane agrees to buy a car for a down payment of \$2000 and payments of \$220 per month for 3 years. If the interest rate is 8% per year, compounded monthly, what is the actual purchase price of her car?

17. Mike buys a ring for his fiancee by paying \$30 per month for one year. If the interest rate is 10% per year, compounded monthly, what is the price of the ring?

18. Janet's payments on her \$12,500 car are \$420 a month for 3 years. Assuming that interest is compounded monthly, what interest rate is she paying on the car loan?

19. John buys a stereo system for \$640. He agrees to pay \$32 per month for 2 years. Assuming that interest is compounded monthly, what interest rate is he paying?

20. A man purchases a \$2000 diamond ring for a down payment of \$200 and monthly installments of \$88 for 2 years. Assuming that interest is compounded monthly, what interest rate is he paying?

21. An item at a department store is priced at \$189.99 and can be bought by making 20 payments of \$10.50. Find the interest rate, assuming that interest is compounded monthly.

22. (a) Draw a time line as in Example 1 to show that the present value of an annuity is the sum of the present values of each payment, that is,

$$A_p = \frac{R}{1+i} + \frac{R}{(1+i)^2} + \frac{R}{(1+i)^3} + \cdots + \frac{R}{(1+i)^n}$$

(b) Use part (a) to derive the formula for A_p given in the text.

DISCOVERY • DISCUSSION

23. **An Annuity That Lasts Forever** An **annuity in perpetuity** is one that continues forever. Such annuities are useful in setting up scholarship funds to ensure that the award continues.

(a) Draw a time line (as in Example 1) to show that to set up an annuity in perpetuity of amount R per time period, the amount that must be invested now is

$$A_p = \frac{R}{1+i} + \frac{R}{(1+i)^2} + \frac{R}{(1+i)^3} + \cdots + \frac{R}{(1+i)^n} + \cdots$$

where i is the interest rate per time period.

(b) Find the sum of the infinite series in part (a) to show that

$$A_p = \frac{R}{i}$$

(c) How much money must be invested now at 10% per year, compounded annually, to provide an annuity in perpetuity of \$5000 per year? The first payment is due in one year.

(d) How much money must be invested now at 8% per year, compounded quarterly, to provide an annuity in perpetuity of \$3000 per year? The first payment is due in one year.

9.5 MATHEMATICAL INDUCTION

There are two aspects to mathematics—discovery and proof—and both are of equal importance. We must discover something before we can attempt to prove it, and we can only be certain of its truth once it has been proved. In this section we examine the relationship between these two key components of mathematics more closely.

CONJECTURE AND PROOF

Let's try a simple experiment. We add more and more of the odd numbers as follows:

$$1 = 1$$

$$1 + 3 = 4$$

$$1 + 3 + 5 = 9$$

$$1 + 3 + 5 + 7 = 16$$

$$1 + 3 + 5 + 7 + 9 = 25$$

What do you notice about the numbers on the right side of these equations? They are in fact all perfect squares. These equations say the following:

The sum of the first 1 odd number is 1^2.

The sum of the first 2 odd numbers is 2^2.

The sum of the first 3 odd numbers is 3^2.

The sum of the first 4 odd numbers is 4^2.

The sum of the first 5 odd numbers is 5^2.

Consider the polynomial

$$p(n) = n^2 - n + 41$$

Here are some values of $p(n)$:

$$p(1) = 41 \qquad p(2) = 43$$
$$p(3) = 47 \qquad p(4) = 53$$
$$p(5) = 61 \qquad p(6) = 71$$
$$p(7) = 83 \qquad p(8) = 97$$

All the values so far are prime numbers. In fact, if you keep going, you will find that $p(n)$ is prime for all natural numbers up to $n = 40$. It may seem reasonable at this point to conjecture that $p(n)$ is prime for *every* natural number n. But our conjecture would be too hasty, because it is easily seen that $p(41)$ is *not* prime. This illustrates that we cannot be certain of the truth of a statement no matter how many special cases we check. We need a convincing argument—a *proof*—to determine the truth of a statement.

This leads naturally to the following question: Is it true that for every natural number n, the sum of the first n odd numbers is n^2? Could this remarkable property be true? We could try a few more numbers and find that the pattern persists for the first 6, 7, 8, 9, and 10 odd numbers. At this point, we feel quite sure that this is always true, so we make a *conjecture*:

The sum of the first n odd numbers is n^2.

Since we know that the nth odd number is $2n - 1$, we can write this statement more precisely as

$$1 + 3 + 5 + \cdots + (2n - 1) = n^2$$

It's important to realize that this is still a conjecture. We cannot conclude by checking a finite number of cases that a property is true for all numbers (there are infinitely many). To see this more clearly, suppose someone tells us he has added up the first trillion odd numbers and found that they do *not* add up to 1 trillion squared. What would you tell this person? It would be silly to say that you're sure it's true because you've already checked the first five cases. You could, however, take out paper and pencil and start checking it yourself, but this task would probably take the rest of your life. The tragedy would be that after completing this task you would still not be sure of the truth of the conjecture. Do you see why?

Herein lies the power of mathematical proof. A **proof** is a clear argument that demonstrates the truth of a statement beyond doubt.

MATHEMATICAL INDUCTION

Let's consider a special kind of proof called **mathematical induction**. Here is how it works: Suppose we have a statement that says something about all natural numbers n. Let's call this statement P. For example, we could consider the statement

P: For every natural number n, the sum of the first n odd numbers is n^2.

Since this statement is about *all* natural numbers, it contains infinitely many statements; we will call them $P(1), P(2), \ldots$.

$P(1)$: The sum of the first 1 odd number is 1^2.

$P(2)$: The sum of the first 2 odd numbers is 2^2.

$P(3)$: The sum of the first 3 odd numbers is 3^2.

$\vdots \qquad \vdots$

How can we prove all of these statements at once? Mathematical induction is a clever way of doing just that.

The crux of the idea is this: Suppose we can prove that whenever one of these statements is true, then the one following it in the list is also true. In other words,

For every k, if $P(k)$ is true, then $P(k + 1)$ is true.

This is called the **induction step** because it leads us from the truth of one statement to the next. Now, suppose that we can also prove that

$$P(1) \text{ is true.}$$

The induction step now leads us through the following chain of statements:

$$P(1) \text{ is true, so } P(2) \text{ is true.}$$
$$P(2) \text{ is true, so } P(3) \text{ is true.}$$
$$P(3) \text{ is true, so } P(4) \text{ is true.}$$
$$\vdots \qquad\qquad \vdots$$

So we see that if both the induction step and $P(1)$ are proved, then statement P is proved for all n. Here is a summary of this important method of proof.

PRINCIPLE OF MATHEMATICAL INDUCTION

For each natural number n, let $P(n)$ be a statement depending on n. Suppose that the following two conditions are satisfied.

1. $P(1)$ is true.
2. For every natural number k, if $P(k)$ is true, then $P(k+1)$ is true.

Then $P(n)$ is true for all natural numbers n.

To apply this principle, there are two steps:

Step 1 Prove that $P(1)$ is true.

Step 2 Assume that $P(k)$ is true and use this assumption to prove that $P(k+1)$ is true.

Notice that in Step 2 we do not prove that $P(k)$ is true. We only show that *if* $P(k)$ is true, *then* $P(k+1)$ is also true. The assumption that $P(k)$ is true is called the **induction hypothesis**.

We now use mathematical induction to prove that the conjecture we made at the beginning of this section is true.

EXAMPLE 1 ■ A Proof by Mathematical Induction

Prove that for all natural numbers n,

$$1 + 3 + 5 + \cdots + (2n - 1) = n^2$$

SOLUTION

Let $P(n)$ denote the statement $1 + 3 + 5 + \cdots + (2n - 1) = n^2$.

Step 1 We need to show that $P(1)$ is true. But $P(1)$ is simply the statement that $1 = 1^2$, which is of course true.

Step 2 We assume that $P(k)$ is true. Thus, our induction hypothesis is

$$1 + 3 + 5 + \cdots + (2k - 1) = k^2$$

We want to use this to show that $P(k + 1)$ is true, that is,

$$1 + 3 + 5 + \cdots + (2k - 1) + [2(k + 1) - 1] = (k + 1)^2$$

[Note that we get $P(k + 1)$ by substituting $k + 1$ for each n in the statement $P(n)$.] We start with the left side and use the induction hypothesis to obtain the right side of the equation:

$$\begin{aligned}
&1 + 3 + 5 + \cdots + (2k - 1) + [2(k + 1) - 1] \\
&\quad = [1 + 3 + 5 + \cdots + (2k - 1)] + [2(k + 1) - 1] && \text{Group the first } k \text{ terms} \\
&\quad = k^2 + [2(k + 1) - 1] && \text{Induction hypothesis} \\
&\quad = k^2 + [2k + 2 - 1] && \text{Distributive Property} \\
&\quad = k^2 + 2k + 1 && \text{Simplify} \\
&\quad = (k + 1)^2 && \text{Factor}
\end{aligned}$$

Thus, $P(k + 1)$ follows from $P(k)$ and this completes the induction step.

Having proved Steps 1 and 2, we conclude by the Principle of Mathematical Induction that $P(n)$ is true for all natural numbers n. ■

EXAMPLE 2 ■ A Proof by Mathematical Induction

Prove that for every natural number n,

$$1 + 2 + 3 + \cdots + n = \frac{n(n + 1)}{2}$$

Blaise Pascal (1623–1662) is considered one of the most versatile minds in modern history. He was a writer and philosopher as well as a gifted mathematician and physicist. Among his contributions that appear in this book are the theory of probability, Pascal's triangle, and the Principle of Mathematical Induction.

Pascal's father, himself a mathematician, believed that his son should not study mathematics until he was 15 or 16. But at age 12, Blaise insisted on learning geometry, and proved most of its elementary theorems himself. At 19, he invented the first mechanical adding machine. In 1647, after writing a major treatise on the conic sections, he abruptly abandoned mathematics because he felt his intense studies were contributing to his ill health. He devoted himself instead to frivolous recreations such as gambling, but this only served to pique his interest in probability. In 1654 he miraculously survived a carriage accident in which his horses ran off a bridge. Taking this to be a sign from God, he entered a monastery, where he pursued theology and philosophy, writing his famous *Pensées*. He also continued his

(continued)

SOLUTION

Let $P(n)$ be the statement $1 + 2 + 3 + \cdots + n = n(n + 1)/2$. We want to show that $P(n)$ is true for all natural numbers n.

Step 1 We need to show that $P(1)$ is true. But $P(1)$ says that

$$1 = \frac{1(1 + 1)}{2}$$

and this statement is clearly true.

Step 2 Assume that $P(k)$ is true. Thus, our induction hypothesis is

$$1 + 2 + 3 + \cdots + k = \frac{k(k + 1)}{2}$$

We want to use this to show that $P(k + 1)$ is true, that is,

$$1 + 2 + 3 + \cdots + k + (k + 1) = \frac{(k + 1)[(k + 1) + 1]}{2}$$

So, we start with the left side and use the induction hypothesis to obtain the right side:

$$
\begin{aligned}
&1 + 2 + 3 + \cdots + k + (k + 1) \\
&\quad = [1 + 2 + 3 + \cdots + k] + (k + 1) && \text{Group the first } k \text{ terms} \\
&\quad = \frac{k(k + 1)}{2} + (k + 1) && \text{Induction hypothesis} \\
&\quad = (k + 1)\left(\frac{k}{2} + 1\right) && \text{Factor } k + 1 \\
&\quad = (k + 1)\left(\frac{k + 2}{2}\right) && \text{Common denominator} \\
&\quad = \frac{(k + 1)[(k + 1) + 1]}{2} && \text{Write } k + 2 \text{ as } k + 1 + 1
\end{aligned}
$$

Thus, $P(k + 1)$ follows from $P(k)$ and this completes the induction step.

Having proved Steps 1 and 2, we conclude by the Principle of Mathematical Induction that $P(n)$ is true for all natural numbers n. ■

mathematical research. He valued faith and intuition more than reason as the source of truth, declaring that "the heart has its own reasons, which reason cannot know."

It might happen that a statement $P(n)$ is false for the first few natural numbers, but true from some number on. For example, we may want to prove that $P(n)$ is true for $n \geq 5$. Notice that if we prove the $P(5)$ is true, then this fact, together with the induction step, would imply the truth of $P(5)$, $P(6)$, $P(7)$, ... The next example illustrates this point.

EXAMPLE 3 ■ Proving an Inequality by Mathematical Induction

Prove that $4n < 2^n$ for all $n \geq 5$.

SOLUTION

Let $P(n)$ denote the statement $4n < 2^n$.

Step 1 $P(5)$ is the statement that $4 \cdot 5 < 2^5$, or $20 < 32$, which is true.

Step 2 Assume that $P(k)$ is true. Thus, our induction hypothesis is

$$4k < 2^k$$

We want to use this to show that $P(k+1)$ is true, that is,

$$4(k+1) < 2^{k+1}$$

We get $P(k+1)$ by replacing k by $k+1$ in the statement $P(k)$.

So, we start with the left side of the inequality and use the induction hypothesis to show that it is less than the right side. For $k \geq 5$, we have

$$\begin{aligned} 4(k+1) &= 4k + 4 \\ &< 2^k + 4 && \text{Induction hypothesis} \\ &< 2^k + 4k && \text{Because } 4 < 4k \\ &< 2^k + 2^k && \text{Induction hypothesis} \\ &= 2 \cdot 2^k \\ &= 2^{k+1} && \text{Property of exponents} \end{aligned}$$

Thus, $P(k+1)$ follows from $P(k)$ and this completes the induction step.

Having proved Steps 1 and 2, we conclude by the Principle of Mathematical Induction that $P(n)$ is true for all natural numbers $n \geq 5$. ■

9.5 EXERCISES

1–12 ■ Use mathematical induction to prove that the formula is true for all natural numbers n.

1. $2 + 4 + 6 + \cdots + 2n = n(n + 1)$

2. $1 + 4 + 7 + \cdots + 3(n - 2) = \dfrac{n(3n - 1)}{2}$

3. $5 + 8 + 11 + \cdots + (3n + 2) = \dfrac{n(3n + 7)}{2}$

4. $1^2 + 2^2 + 3^2 + \cdots + n^2 = \dfrac{n(n + 1)(2n + 1)}{6}$

5. $1 \cdot 2 + 2 \cdot 3 + 3 \cdot 4 + \cdots + n(n + 1) = \dfrac{n(n + 1)(n + 2)}{3}$

6. $1 \cdot 3 + 2 \cdot 4 + 3 \cdot 5 + \cdots + n(n + 2) = \dfrac{n(n + 1)(2n + 7)}{6}$

7. $1^3 + 2^3 + 3^3 + \cdots + n^3 = \dfrac{n^2(n + 1)^2}{4}$

8. $1^3 + 3^3 + 5^3 + \cdots + (2n - 1)^3 = n^2(2n^2 - 1)$

9. $2^3 + 4^3 + 6^3 + \cdots + (2n)^3 = 2n^2(n + 1)^2$

10. $\dfrac{1}{1 \cdot 2} + \dfrac{1}{2 \cdot 3} + \dfrac{1}{3 \cdot 4} + \cdots + \dfrac{1}{n(n + 1)} = \dfrac{n}{(n + 1)}$

11. $1 \cdot 2 + 2 \cdot 2^2 + 3 \cdot 2^3 + 4 \cdot 2^4 + \cdots + n \cdot 2^n = 2[1 + (n - 1)2^n]$

12. $1 + 2 + 2^2 + \cdots + 2^{n-1} = 2^n - 1$

13. Show that $n^2 + n$ is divisible by 2 for all natural numbers n.

14. Show that $5^n - 1$ is divisible by 4 for all natural numbers n.

15. Show that $n^2 - n + 41$ is odd for all natural numbers n.

16. Show that $n^3 - n + 3$ is divisible by 3 for all natural numbers n.

17. Show that $8^n - 3^n$ is divisible by 5 for all natural numbers n.

18. Show that $3^{2n} - 1$ is divisible by 8 for all natural numbers n.

19. Prove that $n < 2^n$ for all natural numbers n.

20. Prove that $(n + 1)^2 < 2n^2$ for all natural numbers $n \geq 3$.

21. Prove that if $x > -1$, then $(1 + x)^n \geq 1 + nx$ for all natural numbers n.

22. Show that $100n \leq n^2$ for all $n \geq 100$.

23. Let $a_{n+1} = 3a_n$ and $a_1 = 5$. Show that $a_n = 5 \cdot 3^{n-1}$ for all natural numbers n.

24. A sequence is defined recursively by $a_{n+1} = 3a_n - 8$ and $a_1 = 4$. Find an explicit formula for a_n and then use mathematical induction to prove that the formula you found is true.

25. Show that $x - y$ is a factor of $x^n - y^n$ for all natural numbers n.
[*Hint:* $x^{k+1} - y^{k+1} = x^k(x - y) + (x^k - y^k)y$]

26. Show that $x + y$ is a factor of $x^{2n-1} + y^{2n-1}$ for all natural numbers n.

27–31 ■ F_n denotes the nth term of the Fibonacci sequence discussed in Section 9.1. Use mathematical induction to prove the statement.

27. F_{3n} is even for all natural numbers n.

28. $F_1 + F_2 + F_3 + \cdots + F_n = F_{n+2} - 1$

29. $F_1^2 + F_2^2 + F_3^2 + \cdots + F_n^2 = F_n F_{n+1}$

30. If $a_{n+2} = a_{n+1} \cdot a_n$ and $a_1 = a_2 = 2$, then $a_n = 2^{F_n}$ for all natural numbers n.

31. For all $n \geq 2$,

$$\begin{bmatrix} 1 & 1 \\ 1 & 0 \end{bmatrix}^n = \begin{bmatrix} F_{n+1} & F_n \\ F_n & F_{n-1} \end{bmatrix}$$

32. Let a_n be the nth term of the sequence defined recursively by

$$a_{n+1} = \frac{1}{1 + a_n}$$

and $a_1 = 1$. Find a formula for a_n in terms of the Fibonacci numbers F_n. Prove that the formula you found is valid for all natural numbers n.

33. Let F_n be the nth term of the Fibonacci sequence. Find and prove an inequality relating n and F_n for natural numbers n.

34. Find and prove an inequality relating $100n$ and n^3.

DISCOVERY • DISCUSSION

35. **True or False?** Determine whether each statement is true or false. If you think the statement is true, prove it. If you think it is false, give an example where it fails.
    (a) $p(n) = n^2 - n + 11$ is prime for all n.
    (b) $n^2 > n$ for all $n \geqslant 2$.
    (c) $2^{2n+1} + 1$ is divisible by 3 for all $n \geqslant 1$.
    (d) $n^3 \geqslant (n + 1)^2$ for all $n \geqslant 2$.
    (e) $n^3 - n$ is divisible by 3 for all $n \geqslant 2$.
    (f) $n^3 - 6n^2 + 11n$ is divisible by 6 for all $n \geqslant 1$.

36. **All Cats Are Black?** What is wrong with the following "proof" by mathematical induction that all cats are black? Let $P(n)$ denote the statement: In any group of n cats, if one is black, then they are all black.

    Step 1 The statement is clearly true for $n = 1$.

    Step 2 Suppose that $P(k)$ is true. We show that $P(k + 1)$ is true. Suppose we have a group of $k + 1$ cats, one of whom is black; call this cat "Midnight." Remove some other cat (call it "Sparky") from the group. We are left with k cats, one of whom (Midnight) is black, so by the induction hypothesis, all k of these are black. Now put Sparky back in the group and take out Midnight. We again have a group of k cats, all of whom—except possibly Sparky—are black. Then by the induction hyothesis, Sparky must be black, too. So all $k + 1$ cats in the original group are black.

    Thus, by induction $P(n)$ is true for all n. Since everyone has seen at least one black cat, it follows that all cats are black.

9.6 THE BINOMIAL THEOREM

An expression of the form $a + b$ is called a **binomial**. Although in principle it's easy to raise $a + b$ to any power, raising it to a very high power would be a tedious task. In this section we find a formula that gives the expansion of $(a + b)^n$ for any natural number n and then prove it using mathematical induction. We discover this formula by finding a pattern for the successive powers of $a + b$. Let's first look at some special cases:

$$\begin{aligned}
(a + b)^1 &= a + b \\
(a + b)^2 &= a^2 + 2ab + b^2 \\
(a + b)^3 &= a^3 + 3a^2b + 3ab^2 + b^3 \\
(a + b)^4 &= a^4 + 4a^3b + 6a^2b^2 + 4ab^3 + b^4 \\
(a + b)^5 &= a^5 + 5a^4b + 10a^3b^2 + 10a^2b^3 + 5ab^4 + b^5 \\
&\vdots
\end{aligned}$$

The following simple patterns emerge for the expansion of $(a + b)^n$:

1. There are $n + 1$ terms, the first being a^n and the last b^n.
2. The exponents of a decrease by 1 from term to term while the exponents of b increase by 1.
3. The sum of the exponents of a and b in each term is n.

Pascal's triangle appears in this Chinese document by Chu Shi-kie, dated 1303. The title reads "The Old Method Chart of the Seven Multiplying Squares." The triangle was rediscovered by Pascal (see page 618).

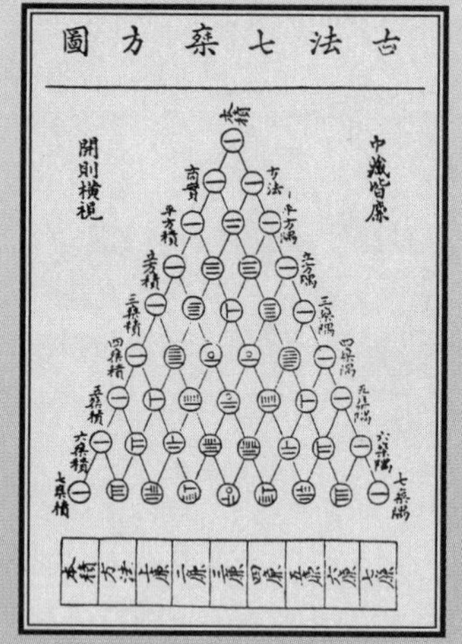

For instance, notice how the exponents of a and b behave in the expansion of $(a + b)^5$.

The exponents of *a* decrease:

$$(a + b)^5 = a^{\textcircled{5}} + 5a^{\textcircled{4}}b^1 + 10a^{\textcircled{3}}b^2 + 10a^{\textcircled{2}}b^3 + 5a^{\textcircled{1}}b^4 + b^5$$

The exponents of *b* increase:

$$(a + b)^5 = a^5 + 5a^4b^{\textcircled{1}} + 10a^3b^{\textcircled{2}} + 10a^2b^{\textcircled{3}} + 5a^1b^{\textcircled{4}} + b^{\textcircled{5}}$$

With these observations we can write the form of the expansion of $(a + b)^n$ for any natural number n. For example, writing a question mark for the missing coefficients, we have

$$(a + b)^8 = a^8 + ?a^7b + ?a^6b^2 + ?a^5b^3 + ?a^4b^4 + ?a^3b^5 + ?a^2b^6 + ?ab^7 + b^8$$

To complete the expansion, we need to determine these coefficients. To find a pattern, let's write the coefficients in the expansion of $(a + b)^n$ for the first few values of n in a triangular array as shown in the following array, which is called **Pascal's triangle**.

$$\begin{array}{lccccccccccc}
(a+b)^0 & & & & & & 1 & & & & & \\
(a+b)^1 & & & & & 1 & & 1 & & & & \\
(a+b)^2 & & & & 1 & & 2 & & 1 & & & \\
(a+b)^3 & & & \textcircled{1} & & \textcircled{3} & & 3 & & 1 & & \\
(a+b)^4 & & 1 & & \textcircled{4} & & \textcircled{6} & & \textcircled{4} & & 1 & \\
(a+b)^5 & 1 & & 5 & & 10 & & \textcircled{10} & & 5 & & 1
\end{array}$$

The row corresponding to $(a + b)^0$ is called the zeroth row and is included to show the symmetry of the array. The key observation about Pascal's triangle is the following property.

KEY PROPERTY OF PASCAL'S TRIANGLE

Every entry (other than a 1) is the sum of the two entries diagonally above it.

From this property it's easy to find any row of Pascal's triangle from the row above it. For instance, we find the sixth and seventh rows, starting with the fifth row:

$$\begin{array}{lccccccccccccccc}
(a+b)^5 & & & 1 & & 5 & & 10 & & 10 & & 5 & & 1 & & \\
(a+b)^6 & & 1 & & 6 & & 15 & & 20 & & 15 & & 6 & & 1 & \\
(a+b)^7 & 1 & & 7 & & 21 & & 35 & & 35 & & 21 & & 7 & & 1
\end{array}$$

To see why this property holds, let's consider the following expansions:

$$(a + b)^5 = a^5 + 5a^4b + 10a^3b^2 + \boxed{10a^2b^3} + \boxed{5ab^4} + b^5$$

$$(a + b)^6 = a^6 + 6a^5b + 15a^4b^2 + 20a^3b^3 + \boxed{15a^2b^4} + 6ab^5 + b^6$$

We arrive at the expansion of $(a + b)^6$ by multiplying $(a + b)^5$ by $(a + b)$. Notice, for instance, that the circled term in the expansion of $(a + b)^6$ is obtained via this multiplication from the two circled terms above it. We get this term when the two terms above it are multiplied by b and a, respectively. Thus, its coefficient is the sum of the coefficients of these two terms. We will use this observation at the end of this section when we prove the Binomial Theorem.

Having found these patterns, we can now easily obtain the expansion of any binomial, at least to relatively small powers.

EXAMPLE 1 ■ Expanding a Binomial Using Pascal's Triangle

Find the expansion of $(a + b)^7$ using Pascal's triangle.

SOLUTION

The first term in the expansion is a^7, and the last term is b^7. Using the fact that the exponent of a decreases by 1 from term to term and that of b increases by 1 from term to term, we have

$$(a + b)^7 = a^7 + ?a^6b + ?a^5b^2 + ?a^4b^3 + ?a^3b^4 + ?a^2b^5 + ?ab^6 + b^7$$

The appropriate coefficients appear in the seventh row of Pascal's triangle. Thus

$$(a + b)^7 = a^7 + 7a^6b + 21a^5b^2 + 35a^4b^3 + 35a^3b^4 + 21a^2b^5 + 7ab^6 + b^7 \quad ■$$

EXAMPLE 2 ■ Expanding a Binomial Using Pascal's Triangle

Use Pascal's triangle to expand $(2 - 3x)^5$.

SOLUTION

We find the expansion of $(a + b)^5$ and then substitute 2 for a and $-3x$ for b. Using Pascal's triangle for the coefficients, we get

$$(a + b)^5 = a^5 + 5a^4b + 10a^3b^2 + 10a^2b^3 + 5ab^4 + b^5$$

Substituting $a = 2$ and $b = -3x$ gives

$$\begin{aligned}(2 - 3x)^5 &= (2)^5 + 5(2)^4(-3x) + 10(2)^3(-3x)^2 + 10(2)^2(-3x)^3 + 5(2)(-3x)^4 + (-3x)^5 \\ &= 32 - 240x + 720x^2 - 1080x^3 + 810x^4 - 243x^5\end{aligned}$$

■

THE BINOMIAL COEFFICIENTS AND PASCAL'S TRIANGLE

Although Pascal's triangle is useful in finding the binomial expansion for reasonably small values of n, it isn't practical for finding $(a + b)^n$ for large values of n. The reason is that the method we use for finding the successive rows of Pascal's triangle is recursive. Thus, to find the 100th row of this triangle, we must first find the preceding 99 rows.

We need to examine the pattern in the coefficients more carefully to develop a formula that allows us to calculate directly any coefficient in the binomial expansion. Such a formula exists, and the rest of this section is devoted to finding and proving it. However, to state this formula we need some notation.

The product of the first n natural numbers is denoted by $\boldsymbol{n!}$ and is called ***n*** **factorial**:

$4! = 1 \cdot 2 \cdot 3 \cdot 4 = 24$

$7! = 1 \cdot 2 \cdot 3 \cdot 4 \cdot 5 \cdot 6 \cdot 7 = 5040$

$10! = 1 \cdot 2 \cdot 3 \cdot 4 \cdot 5 \cdot 6 \cdot 7 \cdot 8 \cdot 9 \cdot 10 = 3{,}628{,}800$

$$n! = 1 \cdot 2 \cdot 3 \cdot \cdots \cdot (n-1) \cdot n$$

We also define 0! as follows:

$$0! = 1$$

This definition of 0! makes many formulas involving factorials shorter and easier to write.

THE BINOMIAL COEFFICIENT

Let n and r be nonnegative integers with $r \leq n$. The **binomial coefficient** is denoted by $\binom{n}{r}$ and is defined by

$$\binom{n}{r} = \frac{n!}{r!(n-r)!}$$

EXAMPLE 3 ■ Calculating Binomial Coefficients

(a) $$\binom{9}{4} = \frac{9!}{4!(9-4)!} = \frac{9!}{4!5!} = \frac{\cancel{1 \cdot 2 \cdot 3 \cdot 4 \cdot 5} \cdot 6 \cdot 7 \cdot 8 \cdot 9}{(1 \cdot 2 \cdot 3 \cdot 4)\cancel{(1 \cdot 2 \cdot 3 \cdot 4 \cdot 5)}}$$

$$= \frac{6 \cdot 7 \cdot 8 \cdot 9}{1 \cdot 2 \cdot 3 \cdot 4} = 126$$

(b) $$\binom{100}{3} = \frac{100!}{3!(100-3)!} = \frac{\cancel{1 \cdot 2 \cdot 3 \cdot \cdots \cdot 97} \cdot 98 \cdot 99 \cdot 100}{(1 \cdot 2 \cdot 3)\cancel{(1 \cdot 2 \cdot 3 \cdot \cdots \cdot 97)}}$$

$$= \frac{98 \cdot 99 \cdot 100}{1 \cdot 2 \cdot 3} = 161{,}700$$

(c) $\binom{100}{97} = \dfrac{100!}{97!(100-97)!} = \dfrac{\cancel{1 \cdot 2 \cdot 3 \cdot \cdots \cdot 97} \cdot 98 \cdot 99 \cdot 100}{(\cancel{1 \cdot 2 \cdot 3 \cdot \cdots \cdot 97})(1 \cdot 2 \cdot 3)}$

$$= \frac{98 \cdot 99 \cdot 100}{1 \cdot 2 \cdot 3} = 161{,}700$$

■

Although the binomial coefficient $\binom{n}{r}$ is defined in terms of a fraction, all the results of Example 3 are natural numbers. In fact, $\binom{n}{r}$ is always a natural number (see Exercise 50). Notice that the binomial coefficients in parts (b) and (c) of Example 3 are equal. This is a special case of the following relation, which you are asked to prove in Exercise 48.

$$\binom{n}{r} = \binom{n}{n-r}$$

To see the connection between the binomial coefficients and the binomial expansion of $(a + b)^n$, let's calculate the following binomial coefficients:

$$\binom{5}{2} = \frac{5!}{2!(5-2)!} = 10$$

$$\binom{5}{0} = 1 \quad \binom{5}{1} = 5 \quad \binom{5}{2} = 10 \quad \binom{5}{3} = 10 \quad \binom{5}{4} = 5 \quad \binom{5}{5} = 1$$

These are precisely the entries in the fifth row of Pascal's triangle. In fact, we can write Pascal's triangle as follows.

$$\binom{0}{0}$$

$$\binom{1}{0} \quad \binom{1}{1}$$

$$\binom{2}{0} \quad \binom{2}{1} \quad \binom{2}{2}$$

$$\binom{3}{0} \quad \binom{3}{1} \quad \binom{3}{2} \quad \binom{3}{3}$$

$$\binom{4}{0} \quad \binom{4}{1} \quad \binom{4}{2} \quad \binom{4}{3} \quad \binom{4}{4}$$

$$\binom{5}{0} \quad \binom{5}{1} \quad \binom{5}{2} \quad \binom{5}{3} \quad \binom{5}{4} \quad \binom{5}{5}$$

$$\cdot \quad \cdot \quad \cdot \quad \cdot \quad \cdot \quad \cdot \quad \cdot$$

$$\binom{n}{0} \quad \binom{n}{1} \quad \binom{n}{2} \quad \cdot \quad \cdot \quad \cdot \quad \binom{n}{n-1} \quad \binom{n}{n}$$

To demonstrate that this pattern holds, we need to show that any entry in this version of Pascal's triangle is the sum of the two entries diagonally above it. In other words, we must show that each entry satisfies the key property of Pascal's triangle. We now state this property in terms of the binomial coefficients.

KEY PROPERTY OF THE BINOMIAL COEFFICIENTS

For any nonnegative integers r and k with $r \leq k$,

$$\binom{k}{r-1} + \binom{k}{r} = \binom{k+1}{r}$$

Notice that the two terms on the left side of this equation are adjacent entries in the kth row of Pascal's triangle and the term on the right side is the entry diagonally below them, in the $(k + 1)$st row. Thus, this equation is a restatement of the key property of Pascal's triangle in terms of the binomial coefficients. A proof of this formula is outlined in Exercise 49.

THE BINOMIAL THEOREM

We are now ready to state the Binomial Theorem.

THE BINOMIAL THEOREM

$$(a+b)^n = \binom{n}{0}a^n + \binom{n}{1}a^{n-1}b + \binom{n}{2}a^{n-2}b^2 + \cdots + \binom{n}{n-1}ab^{n-1} + \binom{n}{n}b^n$$

We prove this theorem at the end of this section. First, let's look at some of its applications.

EXAMPLE 4 ■ Expanding a Binomial Using the Binomial Theorem

Use the Binomial Theorem to expand $(x + y)^4$.

SOLUTION

By the Binomial Theorem,

$$(x+y)^4 = \binom{4}{0}x^4 + \binom{4}{1}x^3y + \binom{4}{2}x^2y^2 + \binom{4}{3}xy^3 + \binom{4}{4}y^4$$

Verify that

$$\binom{4}{0} = 1 \qquad \binom{4}{1} = 4 \qquad \binom{4}{2} = 6 \qquad \binom{4}{3} = 4 \qquad \binom{4}{4} = 1$$

It follows that

$$(x + y)^4 = x^4 + 4x^3y + 6x^2y^2 + 4xy^3 + y^4$$

■

EXAMPLE 5 ■ Expanding a Binomial Using the Binomial Theorem

Use the Binomial Theorem to expand $(\sqrt{x} - 1)^8$.

SOLUTION

We first find the expansion of $(a + b)^8$ and then substitute $\sqrt{x}$ for a and -1 for b. Using the Binomial Theorem, we have

$$\begin{aligned}(a + b)^8 = \binom{8}{0}a^8 + \binom{8}{1}a^7b + \binom{8}{2}a^6b^2 + \binom{8}{3}a^5b^3 + \binom{8}{4}a^4b^4 \\ + \binom{8}{5}a^3b^5 + \binom{8}{6}a^2b^6 + \binom{8}{7}ab^7 + \binom{8}{8}b^8\end{aligned}$$

Verify that

$$\binom{8}{0} = 1 \qquad \binom{8}{1} = 8 \qquad \binom{8}{2} = 28 \qquad \binom{8}{3} = 56 \qquad \binom{8}{4} = 70$$

$$\binom{8}{5} = 56 \qquad \binom{8}{6} = 28 \qquad \binom{8}{7} = 8 \qquad \binom{8}{8} = 1$$

So

$$\begin{aligned}(a + b)^8 = a^8 + 8a^7b + 28a^6b^2 + 56a^5b^3 + 70a^4b^4 + 56a^3b^5 \\ + 28a^2b^6 + 8ab^7 + b^8\end{aligned}$$

Performing the substitutions $a = x^{1/2}$ and $b = -1$ gives

$$\begin{aligned}(\sqrt{x} - 1)^8 = (x^{1/2})^8 + 8(x^{1/2})^7(-1) + 28(x^{1/2})^6(-1)^2 + 56(x^{1/2})^5(-1)^3 \\ + 70(x^{1/2})^4(-1)^4 + 56(x^{1/2})^3(-1)^5 + 28(x^{1/2})^2(-1)^6 \\ + 8(x^{1/2})(-1)^7 + (-1)^8\end{aligned}$$

This simplifies to

$$(\sqrt{x} - 1)^8 = x^4 - 8x^{7/2} + 28x^3 - 56x^{5/2} + 70x^2 - 56x^{3/2} + 28x - 8x^{1/2} + 1$$

■

The Binomial Theorem can be used to find a particular term of a binomial expansion without having to find the entire expansion.

Sir Isaac Newton (1642–1727) is universally regarded as one of the giants of physics and mathematics. He is well known for discovering the laws of motion and gravity and for inventing the calculus, but he also proved the Binomial Theorem and the laws of optics, and developed methods for solving polynomial equations to any desired accuracy. He was born on Christmas Day, a few months after the death of his father. After an unhappy childhood, he entered Cambridge University, where he learned mathematics by studying the writings of Euclid and Descartes.

During the plague years of 1665 and 1666, when the university was closed, Newton thought and wrote about ideas that, once published, instantly revolutionized the sciences. Imbued with a pathological fear of criticism, he published these writings only after many years of encouragement from Edmund Halley (who discovered the now-famous comet) and other colleagues. Newton's works brought him enormous fame and prestige. Even poets were moved to praise;

(*continued*)

GENERAL TERM OF THE BINOMIAL EXPANSION

The term that contains a^r in the expansion of $(a + b)^n$ is

$$\binom{n}{n-r}a^r b^{n-r}$$

EXAMPLE 6 ■ Finding a Particular Term in a Binomial Expansion

Find the term that contains x^5 in the expansion of $(2x + y)^{20}$.

SOLUTION

The term that contains x^5 is given by the formula for the general term with $a = 2x$, $b = y$, $n = 20$, and $r = 5$. So, this term is

$$\binom{20}{15}a^5b^{15} = \frac{20!}{15!(20-15)!}(2x)^5y^{15} = \frac{20!}{15!5!}32x^5y^{15} = 496{,}128x^5y^{15} \quad ■$$

EXAMPLE 7 ■ Finding a Particular Term in a Binomial Expansion

Find the coefficient of x^8 in the expansion of $\left(x^2 + \dfrac{1}{x}\right)^{10}$.

SOLUTION

Both x^2 and $1/x$ are powers of x, so the power of x in each term of the expansion is determined by both terms of the binomial. To find the required coefficient, we first find the general term in the expansion. By the formula we have $a = x^2$, $b = 1/x$, and $n = 10$, so the general term is

$$\binom{10}{10-r}(x^2)^r\left(\frac{1}{x}\right)^{10-r} = \binom{10}{10-r}x^{2r}(x^{-1})^{10-r} = \binom{10}{10-r}x^{3r-10}$$

Thus, the term that contains x^8 is the term in which

$$3r - 10 = 8$$

$$r = 6$$

So the required coefficient is

$$\binom{10}{10-6} = \binom{10}{4} = 210 \quad ■$$

PROOF OF THE BINOMIAL THEOREM

We now give a proof of the Binomial Theorem using mathematical induction.

Alexander Pope wrote:

Nature and Nature's Laws
lay hid in Night.
God said, "Let Newton be"
and all was Light.

Newton was far more modest about his accomplishments. He said, "I seem to have been only like a boy playing on the seashore . . . while the great ocean of truth lay all undiscovered before me." Newton was knighted by Queen Anne in 1705 and was buried with great honor in Westminster Abbey.

■ **Proof** Let $P(n)$ denote the statement

$$(a+b)^n = \binom{n}{0}a^n + \binom{n}{1}a^{n-1}b + \binom{n}{2}a^{n-2}b^2 + \cdots + \binom{n}{n-1}ab^{n-1} + \binom{n}{n}b^n$$

Step 1 We show that $P(1)$ is true. But $P(1)$ is just the statement

$$(a+b)^1 = \binom{1}{0}a^1 + \binom{1}{1}b^1 = 1a + 1b = a + b$$

which is certainly true.

Step 2 We assume that $P(k)$ is true. Thus, our induction hypothesis is

$$(a+b)^k = \binom{k}{0}a^k + \binom{k}{1}a^{k-1}b + \binom{k}{2}a^{k-2}b^2 + \cdots + \binom{k}{k-1}ab^{k-1} + \binom{k}{k}b^k$$

We use this to show that $P(k+1)$ is true.

$$\begin{aligned}
(a+b)^{k+1} &= (a+b)[(a+b)^k] \\
&= (a+b)\left[\binom{k}{0}a^k + \binom{k}{1}a^{k-1}b + \binom{k}{2}a^{k-2}b^2 + \cdots + \binom{k}{k-1}ab^{k-1} + \binom{k}{k}b^k\right] \quad \text{Induction hypothesis} \\
&= a\left[\binom{k}{0}a^k + \binom{k}{1}a^{k-1}b + \binom{k}{2}a^{k-2}b^2 + \cdots + \binom{k}{k-1}ab^{k-1} + \binom{k}{k}b^k\right] \\
&\quad + b\left[\binom{k}{0}a^k + \binom{k}{1}a^{k-1}b + \binom{k}{2}a^{k-2}b^2 + \cdots + \binom{k}{k-1}ab^{k-1} + \binom{k}{k}b^k\right] \\
&= \binom{k}{0}a^{k+1} + \binom{k}{1}a^kb + \binom{k}{2}a^{k-1}b^2 + \cdots + \binom{k}{k-1}a^2b^{k-1} + \binom{k}{k}ab^k \\
&\quad + \binom{k}{0}a^kb + \binom{k}{1}a^{k-1}b^2 + \binom{k}{2}a^{k-2}b^3 + \cdots + \binom{k}{k-1}ab^k + \binom{k}{k}b^{k+1} \\
&= \binom{k}{0}a^{k+1} + \left[\binom{k}{0} + \binom{k}{1}\right]a^kb + \left[\binom{k}{1} + \binom{k}{2}\right]a^{k-1}b^2 + \cdots + \left[\binom{k}{k-1} + \binom{k}{k}\right]ab^k + \binom{k}{k}b^{k+1}
\end{aligned}$$

Using the key property of the binomial coefficients, we can write each of the expressions in square brackets as a single binomial coefficient. Also, writing the first and last coefficients as $\binom{k+1}{0}$ and $\binom{k+1}{k+1}$ (these are equal to 1 by Exercise 46) gives

$$(a+b)^{k+1} = \binom{k+1}{0}a^{k+1} + \binom{k+1}{1}a^kb + \binom{k+1}{2}a^{k-1}b^2 + \cdots + \binom{k+1}{k}ab^k + \binom{k+1}{k+1}b^{k+1}$$

But this last equation is precisely $P(k+1)$, and this completes the induction step.

Having proved Steps 1 and 2, we conclude by the Principle of Mathematical Induction that the theorem is true for all natural numbers n. □

9.6 EXERCISES

1–12 ■ Use Pascal's triangle to expand the expression.

1. $(x+y)^6$ **2.** $(2x+1)^4$ **3.** $\left(x+\frac{1}{x}\right)^4$

4. $(x-y)^5$ **5.** $(x-1)^5$ **6.** $\left(\sqrt{a}+\sqrt{b}\right)^6$

7. $(x^2y-1)^5$ **8.** $\left(1+\sqrt{2}\right)^6$ **9.** $(2x-3y)^3$

10. $(1+x^3)^3$ **11.** $\left(\frac{1}{x}-\sqrt{x}\right)^5$ **12.** $\left(2+\frac{x}{2}\right)^5$

13–20 ■ Evaluate the expression.

13. $\binom{6}{4}$ **14.** $\binom{8}{3}$ **15.** $\binom{100}{98}$

16. $\binom{10}{5}$ **17.** $\binom{3}{1}\binom{4}{2}$ **18.** $\binom{5}{2}\binom{5}{3}$

19. $\binom{5}{0}+\binom{5}{1}+\binom{5}{2}+\binom{5}{3}+\binom{5}{4}+\binom{5}{5}$

20. $\binom{5}{0}-\binom{5}{1}+\binom{5}{2}-\binom{5}{3}+\binom{5}{4}-\binom{5}{5}$

21–24 ■ Use the Binomial Theorem to expand the expression.

21. $(x+2y)^4$ **22.** $(1-x)^5$

23. $\left(1+\frac{1}{x}\right)^6$ **24.** $(2A+B^2)^4$

25. Find the first three terms in the expansion of $(x+2y)^{20}$.

26. Find the first four terms in the expansion of $(x^{1/2}+1)^{30}$.

27. Find the last two terms in the expansion of $(a^{2/3}+a^{1/3})^{25}$.

28. Find the first three terms in the expansion of

$$\left(x+\frac{1}{x}\right)^{40}$$

29. Find the middle term in the expansion of $(x^2+1)^{18}$.

30. Find the fifth term in the expansion of $(ab-1)^{20}$.

31. Find the 24th term in the expansion of $(a+b)^{25}$.

32. Find the 28th term in the expansion of $(A-B)^{30}$.

33. Find the 100th term in the expansion of $(1+y)^{100}$.

34. Find the second term in the expansion of

$$\left(x^2-\frac{1}{x}\right)^{25}$$

35. Find the term containing x^4 in the expansion of $(x+2y)^{10}$.

36. Find the term containing y^3 in the expansion of $\left(\sqrt{2}+y\right)^{12}$.

37. Find the term containing b^8 in the expansion of $(a+b^2)^{12}$.

38. Find the term that does not contain x in the expansion of

$$\left(8x+\frac{1}{2x}\right)^8$$

39–42 ■ Factor using the Binomial Theorem.

39. $x^4+4x^3y+6x^2y^2+4xy^3+y^4$

40. $(x-1)^5+5(x-1)^4+10(x-1)^3+10(x-1)^2 + 5(x-1)+1$

41. $8a^3+12a^2b+6ab^2+b^3$

42. $x^8+4x^6y+6x^4y^2+4x^2y^3+y^4$

43–44 ■ Simplify using the Binomial Theorem.

43. $\dfrac{(x+h)^3-x^3}{h}$

44. $\dfrac{(x+h)^4-x^4}{h}$

45. Show that $(1.01)^{100}>2$.
[*Hint:* Note that $(1.01)^{100}=(1+0.01)^{100}$ and use the Binomial Theorem to show that the sum of the first two terms of the expansion is greater than 2.]

46. Show that $\binom{n}{0}=1$ and $\binom{n}{n}=1$.

47. Show that $\binom{n}{1}=\binom{n}{n-1}=n$.

48. Show that $\binom{n}{r} = \binom{n}{n-r}$ for $0 \leq r \leq n$.

49. In this exercise we prove the identity

$$\binom{n}{r-1} + \binom{n}{r} = \binom{n+1}{r}$$

(a) Write the left side of this equation as the sum of two fractions.

(b) Show that a common denominator of the expression you found in part (a) is $r!(n-r+1)!$.

(c) Add the two fractions using the common denominator in part (b), simplify the numerator, and note that the resulting expression is equal to the right side of the equation.

50. Prove that $\binom{n}{r}$ is an integer for all n and for $0 \leq r \leq n$. [*Suggestion:* Use induction to show that the statement is true for all n, and use Exercise 49 for the induction step.]

DISCOVERY • DISCUSSION

51. Powers of Factorials Which is larger, $(100!)^{101}$ or $(101!)^{100}$? [*Hint:* Try factoring the expressions. Do they have any common factors?]

52. Sums of Binomial Coefficients Add each of the first five rows of Pascal's triangle, as indicated. Do you see a pattern?

$$1 + 1 = ?$$
$$1 + 2 + 1 = ?$$
$$1 + 3 + 3 + 1 = ?$$
$$1 + 4 + 6 + 4 + 1 = ?$$
$$1 + 5 + 10 + 10 + 5 + 1 = ?$$

Based on the pattern you have found, find the sum of the nth row:

$$\binom{n}{0} + \binom{n}{1} + \binom{n}{2} + \cdots + \binom{n}{n}$$

Prove your result by expanding $(1 + 1)^n$ using the Binomial Theorem.

53. Alternating Sums of Binomial Coefficients Find the sum

$$\binom{n}{0} - \binom{n}{1} + \binom{n}{2} - \cdots + (-1)^n\binom{n}{n}$$

by finding a pattern as in Exercise 52. Prove your result by expanding $(1 - 1)^n$ using the Binomial Theorem.

9 REVIEW

CONCEPT CHECK

1. (a) What is a sequence?

(b) What is an arithmetic sequence? Write an expression for the nth term of an arithmetic sequence.

(c) What is a geometric sequence? Write an expression for the nth term of a geometric sequence.

2. (a) What is a recursive sequence?

(b) What is the Fibonacci sequence?

3. (a) What is meant by the partial sums of a sequence?

(b) If an arithmetic sequence has first term a and common difference d, write an expression for the sum of its first n terms.

(c) If a geometric sequence has first term a and common ratio r, write an expression for the sum of its first n terms.

(d) Write an expression for the sum of an infinite geometric series with first term a and common ratio r. For what values of r is your formula valid?

4. (a) Write the sum $\sum_{k=1}^{n} a_k$ without using Σ-notation.

(b) Write $b_1 + b_2 + b_3 + \cdots + b_n$ using Σ-notation.

5. Write an expression for the amount A_f of an annuity consisting of n regular equal payments of size R with interest rate i per time period.

6. State the Principle of Mathematical Induction.

7. Write the first five rows of Pascal's Triangle. How are the entries related to each other?

8. (a) What does the symbol $n!$ mean?

(b) Write an expression for the binomial coefficient $\binom{n}{r}$.

(c) State the Binomial Theorem.

(d) Write the term that contains a^r in the expansion of $(a + b)^n$.

EXERCISES

1–6 ■ Find the first four terms as well as the tenth term of the sequence with the given nth term.

1. $a_n = \dfrac{n^2}{n+1}$

2. $a_n = (-1)^n \dfrac{2^n}{n}$

3. $a_n = \dfrac{(-1)^n + 1}{n^3}$

4. $a_n = \dfrac{n(n+1)}{2}$

5. $a_n = \dfrac{(2n)!}{2^n n!}$

6. $a_n = \dbinom{n+1}{2}$

7–10 ■ A sequence is defined recursively. Find the first seven terms of the sequence.

7. $a_n = a_{n-1} + 2n - 1, \quad a_1 = 1$

8. $a_n = \dfrac{a_{n-1}}{n}, \quad a_1 = 1$

9. $a_n = a_{n-1} + 2a_{n-2}, \quad a_1 = 1, a_2 = 3$

10. $a_n = \sqrt{3a_{n-1}}, \quad a_1 = \sqrt{3}$

11–18 ■ The first four terms of a sequence are given. Determine whether they can be the terms of an arithmetic sequence, a geometric sequence, or neither. If the sequence is arithmetic or geometric, find the fifth term.

11. $5, 5.5, 6, 6.5, \ldots$

12. $1, -\frac{3}{2}, 2, -\frac{5}{2}, \ldots$

13. $\sqrt{2}, 2\sqrt{2}, 3\sqrt{2}, 4\sqrt{2}, \ldots$

14. $\sqrt{2}, 2, 2\sqrt{2}, 4, \ldots$

15. $t - 3, t - 2, t - 1, t, \ldots$

16. $t^3, t^2, t, 1, \ldots$

17. $\dfrac{3}{4}, \dfrac{1}{2}, \dfrac{1}{3}, \dfrac{2}{9}, \ldots$

18. $a, 1, \dfrac{1}{a}, \dfrac{1}{a^2}, \ldots$

19. Show that $3, 6i, -12, -24i, \ldots$ is a geometric sequence, and find the common ratio. (Here $i = \sqrt{-1}$.)

20. Find the nth term of the geometric sequence $2, 2 + 2i, 4i, -4 + 4i, -8, \ldots$ (Here $i = \sqrt{-1}$.)

21. The sixth term of an arithmetic sequence is 17, and the fourth term is 11. Find the second term.

22. The 20th term of an arithmetic sequence is 96, and the common difference is 5. Find the nth term.

23. The third term of a geometric sequence is 9, and the common ratio is $\frac{3}{2}$. Find the fifth term.

24. The second term of a geometric sequence is 10, and the fifth term is $\frac{1250}{27}$. Find the nth term.

25. The frequencies of musical notes (measured in cycles per second) form a geometric sequence. Middle C has a frequency of 256, and C, an octave higher, has a frequency of 512. Find the frequency of C two octaves below middle C.

26. A person has two parents, four grandparents, eight great-grandparents, and so on. How many ancestors does a person have 15 generations back?

27. A certain type of bacteria divides every 5 s. If three of these bacteria are put into a petri dish, how many bacteria are in the dish at the end of 1 min?

28. If $a_1, a_2, a_3, \ldots$ and $b_1, b_2, b_3, \ldots$ are arithmetic sequences, show that $a_1 + b_1, a_2 + b_2, a_3 + b_3, \ldots$ is also an arithmetic sequence.

29. If $a_1, a_2, a_3, \ldots$ and $b_1, b_2, b_3, \ldots$ are geometric sequences, show that $a_1b_1, a_2b_2, a_3b_3, \ldots$ is also a geometric sequence.

30. (a) If $a_1, a_2, a_3, \ldots$ is an arithmetic sequence, is the sequence $a_1 + 2, a_2 + 2, a_3 + 2, \ldots$ arithmetic?
(b) If $a_1, a_2, a_3, \ldots$ is a geometric sequence, is the sequence $5a_1, 5a_2, 5a_3, \ldots$ geometric?

31. Find the values of x for which the sequence $6, x, 12, \ldots$ is
(a) arithmetic (b) geometric

32. Find the values of x and y for which the sequence $2, x, y, 17, \ldots$ is
(a) arithmetic (b) geometric

33–36 ■ Find the sum.

33. $\displaystyle\sum_{k=3}^{6} (k+1)^2$

34. $\displaystyle\sum_{i=1}^{4} \frac{2i}{2i-1}$

35. $\displaystyle\sum_{k=1}^{6} (k+1)2^{k-1}$

36. $\displaystyle\sum_{m=1}^{5} 3^{m-2}$

37–40 ■ Write the sum without using sigma notation. Do not evaluate.

37. $\displaystyle\sum_{k=1}^{10} (k-1)^2$

38. $\displaystyle\sum_{j=2}^{100} \frac{1}{j-1}$

39. $\displaystyle\sum_{k=1}^{50} \frac{3^k}{2^{k+1}}$

40. $\displaystyle\sum_{n=1}^{10} n^2 2^n$

41–44 ■ Write the sum using sigma notation. Do not evaluate.

41. $3 + 6 + 9 + 12 + \cdots + 99$

42. $1^2 + 2^2 + 3^2 + \cdots + 100^2$

43. $1 \cdot 2^3 + 2 \cdot 2^4 + 3 \cdot 2^5 + 4 \cdot 2^6 + \cdots + 100 \cdot 2^{102}$

44. $\frac{1}{1 \cdot 2} + \frac{1}{2 \cdot 3} + \frac{1}{3 \cdot 4} + \cdots + \frac{1}{999 \cdot 1000}$

45–50 ■ Determine whether the expression is a partial sum of an arithmetic or geometric sequence. Then find the sum.

45. $1 + 0.9 + (0.9)^2 + \cdots + (0.9)^5$

46. $3 + 3.7 + 4.4 + \cdots + 10$

47. $\sqrt{5} + 2\sqrt{5} + 3\sqrt{5} + \cdots + 100\sqrt{5}$

48. $\frac{1}{3} + \frac{2}{3} + 1 + \frac{4}{3} + \cdots + 33$

49. $\sum_{n=0}^{6} 3(-4)^n$

50. $\sum_{k=0}^{8} 7(5)^{k/2}$

51. The first term of an arithmetic sequence is $a = 7$, and the common difference is $d = 3$. How many terms of this sequence must be added to obtain 325?

52. The sum of the first three terms of a geometric series is 52, and the common ratio is $r = 3$. Find the first term.

53. Refer to Exercise 26. What is the total number of a person's ancestors in 15 generations?

54. Find the amount of an annuity consisting of 16 annual payments of \$1000 each into an account that pays 8% interest per year, compounded annually.

55. How much money should be invested every quarter at 12% per year, compounded quarterly, in order to have \$10,000 in one year?

56. What are the monthly payments on a mortgage of \$60,000 at 9% interest if the loan is to be repaid in
(a) 30 years? (b) 15 years?

57–60 ■ Find the sum of the infinite geometric series.

57. $1 - \frac{2}{5} + \frac{4}{25} - \frac{8}{125} + \cdots$

58. $0.1 + 0.01 + 0.001 + 0.0001 + \cdots$

59. $1 + \frac{1}{3^{1/2}} + \frac{1}{3} + \frac{1}{3^{3/2}} + \cdots$

60. $a + ab^2 + ab^4 + ab^6 + \cdots$

61–63 ■ Use mathematical induction to prove that the formula is true for all natural numbers n.

61. $1 + 4 + 7 + \cdots + (3n - 2) = \frac{n(3n - 1)}{2}$

62. $\frac{1}{1 \cdot 3} + \frac{1}{3 \cdot 5} + \frac{1}{5 \cdot 7} + \cdots + \frac{1}{(2n - 1)(2n + 1)} = \frac{n}{2n + 1}$

63. $\left(1 + \frac{1}{1}\right)\left(1 + \frac{1}{2}\right)\left(1 + \frac{1}{3}\right) \cdots \left(1 + \frac{1}{n}\right) = n + 1$

64. Show that $7^n - 1$ is divisible by 6 for all natural numbers n.

65. Let $a_{n+1} = 3a_n + 4$ and $a_1 = 4$. Show that $a_n = 2 \cdot 3^n - 2$ for all natural numbers n.

66. Prove that the Fibonacci number F_{4n} is divisible by 3 for all natural numbers n.

67. Find and prove an inequality that relates 2^n and $n!$.

68–71 ■ Evaluate the expression.

68. $\binom{5}{2}\binom{5}{3}$

69. $\binom{10}{2} + \binom{10}{6}$

70. $\sum_{k=0}^{5} \binom{5}{k}$

71. $\sum_{k=0}^{8} \binom{8}{k}\binom{8}{8-k}$

72–73 ■ Expand the expression.

72. $(1 - x^2)^6$

73. $(2x + y)^4$

74. Find the 20th term in the expansion of $(a + b)^{22}$.

75. Find the first three terms in the expansion of $(b^{-2/3} + b^{1/3})^{20}$.

76. Find the term containing A^6 in the expansion of $(A + 3B)^{10}$.

9 TEST

1. Find the first four terms and the tenth term of the sequence whose nth term is

$$a_n = n^2 - 1$$

2. Find the common difference, the fourth term, and the nth term in the arithmetic sequence 80, 76, 72, ...

3. The first term of a geometric sequence is 25, and the fourth term is $\frac{1}{5}$. Find the common ratio and the fifth term.

4. Determine whether each statement is true or false. If it is true, prove it. If it is false, give an example where it fails.
   (a) If $a_1, a_2, a_3, \ldots$ is an arithmetic sequence, then the sequence $a_1^2, a_2^2, a_3^2, \ldots$ is also arithmetic.
   (b) If $a_1, a_2, a_3, \ldots$ is a geometric sequence, then the sequence $a_1^2, a_2^2, a_3^2, \ldots$ is also geometric.

5. (a) Write the formula for the nth partial sum of an arithmetic sequence.
   (b) The first term of an arithmetic sequence is 10, and the tenth term is 2. Find the partial sum of the first ten terms.
   (c) Find the common difference and the 100th term of the sequence in part (b).

6. (a) Write the formula for the nth partial sum of a geometric sequence.
   (b) Find the sum

$$\frac{1}{3} + \frac{2}{3^2} + \frac{2^2}{3^3} + \frac{2^3}{3^4} + \cdots + \frac{2^9}{3^{10}}$$

7. Find the sum of the infinite geometric series $1 + \dfrac{1}{2^{1/2}} + \dfrac{1}{2} + \dfrac{1}{2^{3/2}} + \cdots$.

8. Use mathematical induction to prove that, for all natural numbers n,

$$1^2 + 2^2 + 3^2 + \cdots + n^2 = \frac{n(n+1)(2n+1)}{6}$$

9. Write the expression without using sigma notation, and then find the sum.
   (a) $\displaystyle\sum_{n=1}^{5} (1 - n^2)$ (b) $\displaystyle\sum_{n=3}^{6} (-1)^n 2^{n-2}$

10. Expand $(2x + y^2)^5$.

11. A sequence is defined recursively by

$$a_{n+2} = (a_n)^2 - a_{n+1}$$

If $a_1 = 1$ and $a_2 = 1$, find a_5.

FOCUS ON PROBLEM SOLVING

The solutions to many problems in mathematics involve **finding patterns**. The algebraic formulas we've found in this book are compact ways of describing patterns. For example, the familiar equation $(a + b)^2 = a^2 + 2ab + b^2$ gives the pattern for squaring the sum of two numbers. Another pattern we have encountered is the pattern for the sum of the first n natural numbers:

$$1 + 2 + 3 + \cdots + n = \frac{n(n + 1)}{2}$$

How do we discover patterns? In many cases, a good way to start is to experiment with the problem. To prove that a pattern always holds, we can often use mathematical induction, but other proofs are possible. A geometrical method is used in the problem we give here. It is attributed to the 11th-century mathematician Abu Bekr Mohammed ibn Al Husain Al Karchi.

THE GNOMONS OF AL KARCHI

We prove the beautiful formula

$$1^3 + 2^3 + 3^3 + \cdots + n^3 = (1 + 2 + 3 + \cdots + n)^2$$

But first, let's see how this formula was discovered. The sum of the first n natural numbers is called a *triangular number*. The name "triangular" comes from the following figures:

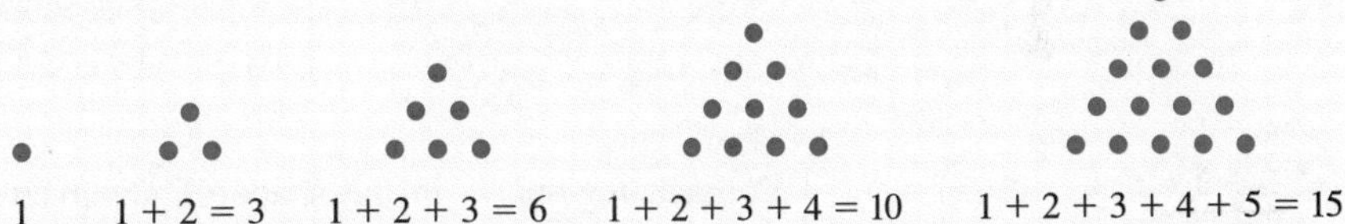

$1 \qquad 1 + 2 = 3 \qquad 1 + 2 + 3 = 6 \qquad 1 + 2 + 3 + 4 = 10 \qquad 1 + 2 + 3 + 4 + 5 = 15$

The first few triangular numbers are

$$1, 3, 6, 10, 15, \ldots$$

Now, let's look at the sums of the cubes:

$$1^3 = 1$$

$$1^3 + 2^3 = 9$$

$$1^3 + 2^3 + 3^3 = 36$$

$$1^3 + 2^3 + 3^3 + 4^3 = 100$$

$$1^3 + 2^3 + 3^3 + 4^3 + 5^3 = 225$$

We get the sequence

$$1, 9, 36, 100, 225, \ldots$$

It doesn't take long to notice that these are the squares of the triangular numbers. It appears that the sum of the first n cubes equals the square of the sum of the first n numbers.

To show that this pattern always holds, Al Karchi sketches the following diagram:

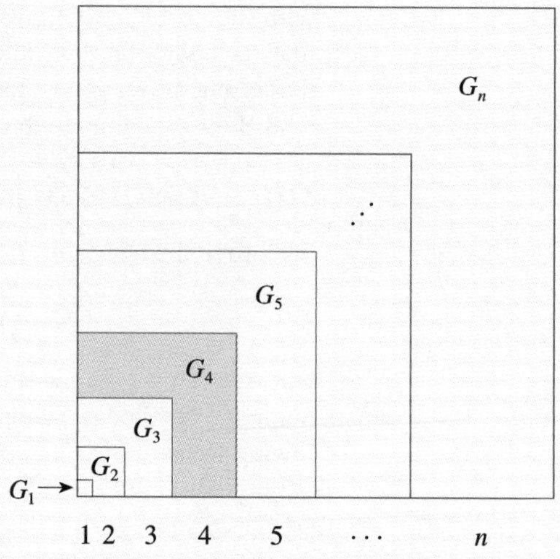

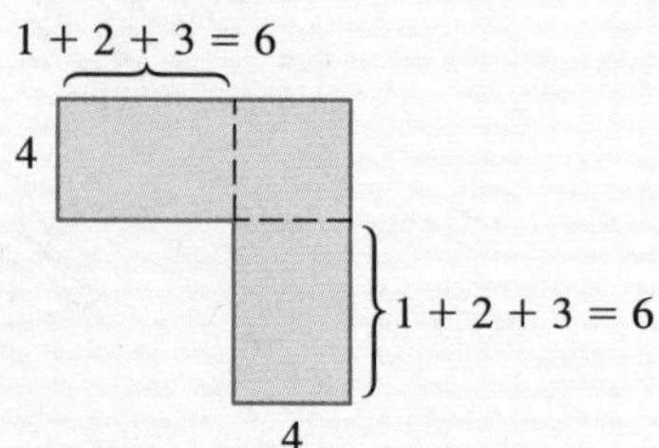

Each region G_n in the shape of an inverted L is called a *gnomon*. The area of gnomon G_4 is

$$\text{area} = 4^2 + 2[4 \times (1 + 2 + 3)] = 64$$

Similarly, the areas of gnomons G_1, G_2, G_3, G_4, $G_5, \ldots$ are 1, 8, 27, 64, 125, ... These are the cubes of the natural numbers. The pattern persists here also, because the area of the nth gnomon G_n is n^3, as the following calculation shows:

$$n^2 + 2(n \times [1 + 2 + \cdots + (n - 1)]) = n^2 + 2n\frac{(n-1)n}{2}$$

$$= n^2 + n^3 - n^2 = n^3$$

Now comes Al Karchi's punchline: The first n gnomons form a square of side $1 + 2 + 3 + \cdots + n$ and so the sum of the areas of these gnomons equals the area of the square, that is, $1^3 + 2^3 + \cdots + n^3 = (1 + 2 + \cdots + n)^2$.

PROBLEMS

1. Use the diagram below to find and prove a formula for the sum of the first n odd numbers.

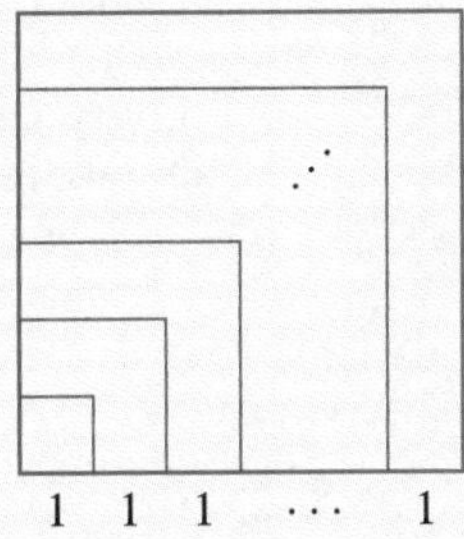

2. Use the figure to find a simple formula for $F_1^2 + F_2^2 + \cdots + F_n^2$, where F_k is the kth Fibonacci number.

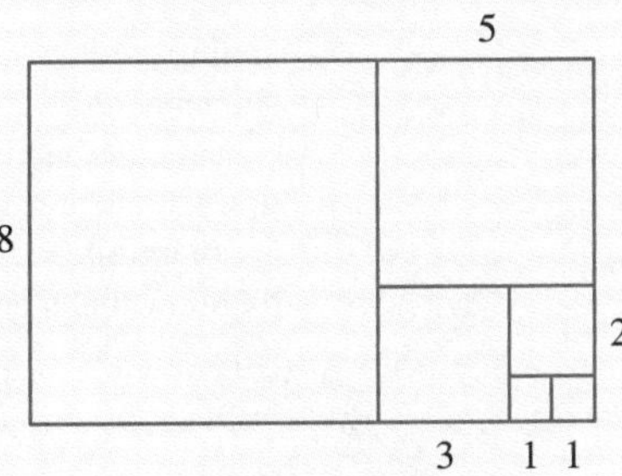

3. Show that every perfect cube is a difference of two squares. [*Hint:* Use the formula for the sum of cubes that we have just proved.]

4. **(a)** Find the product $\left(1 - \frac{1}{2}\right)\left(1 - \frac{1}{3}\right)\left(1 - \frac{1}{4}\right)\cdots\left(1 - \frac{1}{200}\right)$.

(b) Find the product $\left(1 - \frac{1}{4}\right)\left(1 - \frac{1}{9}\right)\left(1 - \frac{1}{16}\right)\cdots\left(1 - \frac{1}{n^2}\right)$.

The earliest known **magic square** is a 3 × 3 magic square from ancient China. According to Chinese legend, it appeared on the back of a turtle emerging from the river Lo.

4	9	2
3	5	7
8	1	6

Here is an example of a 4 × 4 magic square:

16	3	2	13
5	10	11	8
9	6	7	12
4	15	14	1

5. An $n \times n$ *magic square* consists of the numbers from 1 to n^2 arranged in a square in such a way that the sum of the numbers in any row, column, or diagonal is the same. Let's call this constant sum S. Find a formula for S for an $n \times n$ magic square.

6. Choose n different points on the circumference of a circle, and then connect them with line segments. We are interested in the number of regions into which these segments divide the circle. Let's see what happens for $n = 1, 2, 3$, and 4.

(a) How many regions result from five points on the circle?

(b) Based on the pattern you have observed, how many regions do you think would arise from n different points on the circle?

(c) Draw a diagram of the regions that result from six points on the circle. (Do this very carefully.) Does this result fit with the pattern you conjectured in part (b)?

(d) Take the sequence of numbers of regions for $n = 1, 2, 3, 4, 5, 6$, and construct the difference table (see page 606). Use the difference table to determine the number of regions for $n = 7, 8, 9$, and 10.

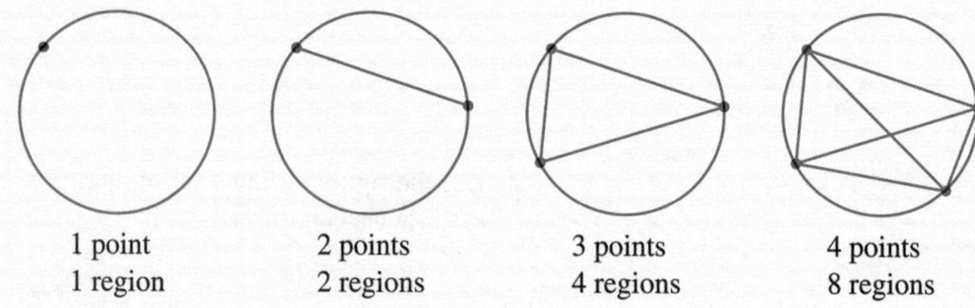

7. Starting with an equilateral triangle of side 1, we successively construct new figures with more and more sides as shown. The *snowflake curve* is the result of repeating this process indefinitely.

(a) Find the area enclosed by the snowflake curve. [*Hint:* This area is the sum of an infinite series.]

(b) Show that the length of the snowflake curve is infinite.

Parts (a) and (b) show that if your lawn has the shape of a snowflake curve, you could easily mow it, because it has finite area, but you could never put a fence along its boundary.

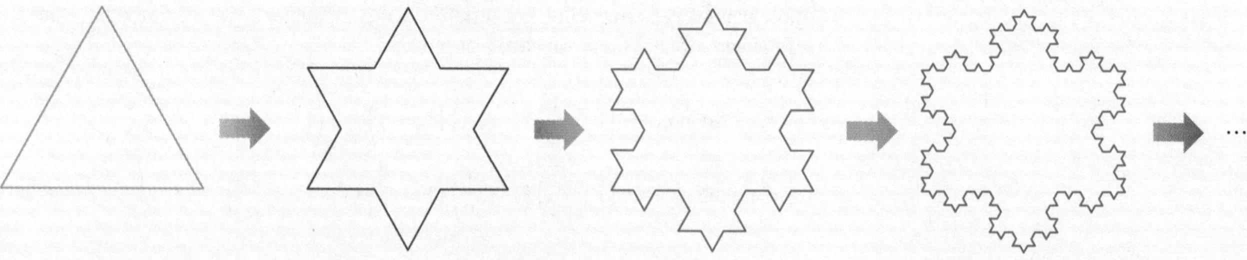

8. Prove that the number of people who have shaken hands an odd number of times is an even number.

9. Consider a 6×6 grid as shown in the figure.

(a) Find the number of squares of all sizes in this grid. Generalize your result to an $n \times n$ grid.

(b) Find the number of rectangles of all sizes in this grid. Generalize your result to an $n \times n$ grid.

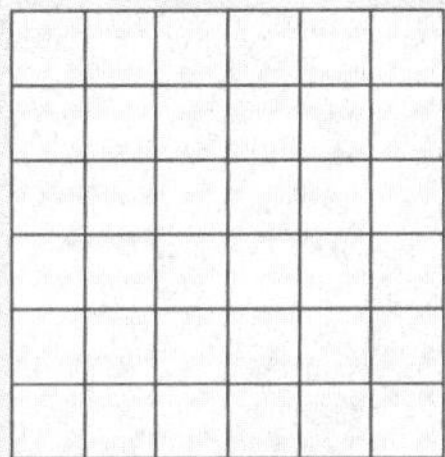

10. Let p be a prime number.

(a) Show that $\binom{p}{k}$ is divisible by p for $1 \leqslant k \leqslant p$.

(b) Use mathematical induction to prove that $n^p - n$ is divisible by p for all n. [*Hint:* For the induction step, use the Binomial Theorem to expand $(n + 1)^p$, and then use part (a).] This fact is known as Fermat's Little Theorem. It is used in constructing public-key codes (see page 38).

10

COUNTING AND PROBABILITY

Probability is the mathematical study of chance and random processes. The laws of probability are essential for understanding genetics, opinion polls, setting odds in horseracing and games of chance, and many other fields.

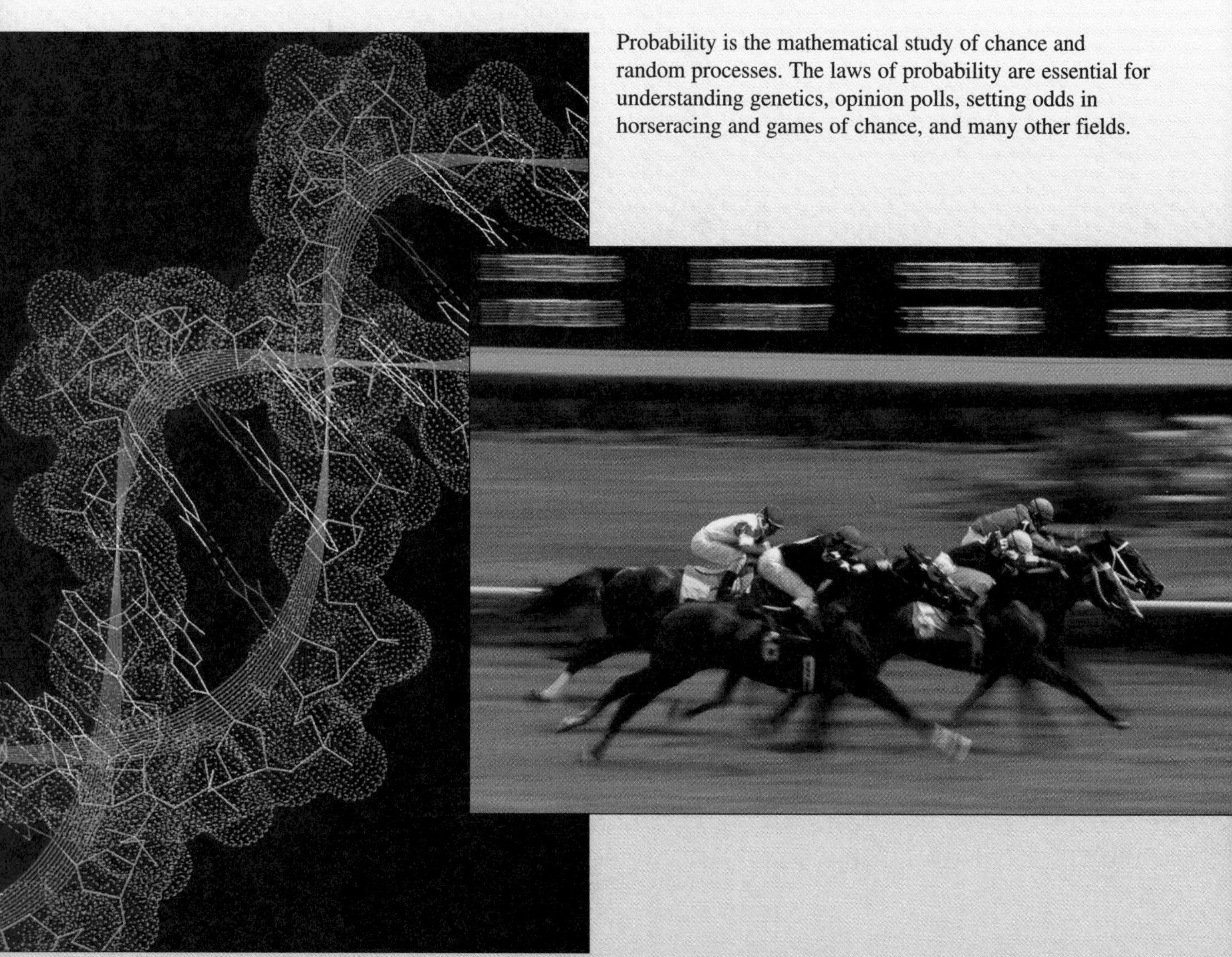

When it is not in our power to determine what is true, we ought to follow what is most probable.

RENÉ DESCARTES

Many questions in mathematics involve *counting*. For example, in how many ways can a committee of two men and three women be chosen from a group of 35 men and 40 women? How many different license plates can be made using three letters followed by three digits? How many different poker hands are possible?

Closely related to the problem of counting is that of *probability*. We consider questions such as these: If a committee of five people is chosen randomly from a group of 35 men and 40 women, what are the chances that no women will be chosen for the committee? What is the likelihood of getting a straight flush in a poker game? In studying probability we give precise mathematical meaning to phrases such as "what are the chances . . . ?" and "what is the likelihood . . . ?"

10.1 COUNTING PRINCIPLES

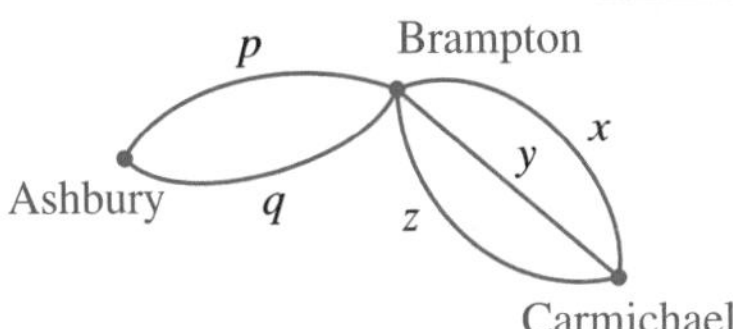

Suppose that three towns, Ashbury, Brampton, and Carmichael, are located in such a way that two roads connect Ashbury to Brampton and three roads connect Brampton to Carmichael. How many different routes can one take to travel from Ashbury to Carmichael via Brampton? The key idea in answering this question is to consider the problem in stages. At the first stage—from Ashbury to Brampton—there are two choices. For each of these choices, there are three choices to make at the second stage—from Brampton to Carmichael. Thus, the number of different routes is $2 \times 3 = 6$. These routes are conveniently enumerated by a *tree diagram* as in Figure 1.

The method used to solve this problem leads to the following principle.

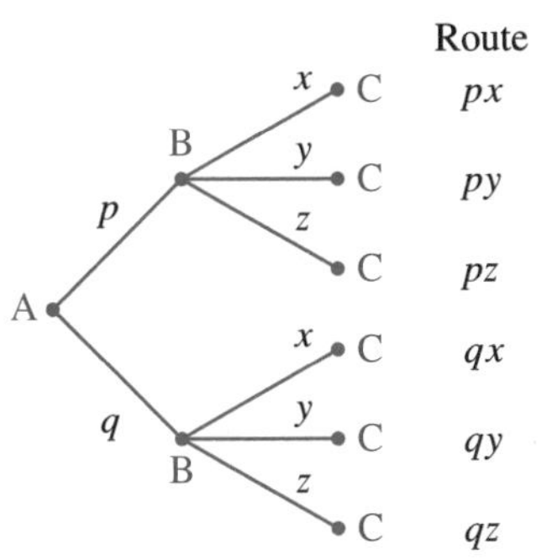

FIGURE 1
Tree diagram

FUNDAMENTAL COUNTING PRINCIPLE

Suppose that two events occur in order. If the first can occur in m ways and the second in n ways (after the first has occurred), then the two events can occur in order in $m \times n$ ways.

There is an immediate consequence of this principle for any number of events: If $E_1, E_2, \ldots, E_k$ are events that occur in order and if E_1 can occur in n_1 ways, E_2 in n_2 ways, and so on, then the events can occur in order in $n_1 \times n_2 \times \cdots \times n_k$ ways.

Persi Diaconis (b. 1945) is currently professor of statistics at Stanford University in California. He was born in New York City into a musical family and studied violin until the age of 14. At that time he left home to become a magician. He was a magician (apprentice and master) for ten years. Magic is still his major passion, and if there were a professorship for magic, he would certainly qualify for such a post! His interest in card tricks led him to a study of probability and statistics. He is now one of the leading statisticians in the world. With his background he approaches mathematics with an undeniable flair. He says "Statistics is the physics of numbers. Numbers seem to arise in the world in an orderly fashion. When we examine the world, the same regularities seem to appear again and again." Among his many original contributions to mathematics is a probabilistic study of the perfect card shuffle.

EXAMPLE 1 ■ Using the Fundamental Counting Principle

An ice-cream store offers three types of cones and 31 flavors. How many different single-scoop ice-cream cones is it possible to buy at this store?

SOLUTION

There are two choices: type of cone and flavor of ice cream. At the first stage we choose a type of cone, and at the second stage we choose a flavor. We can think of the different stages as boxes:

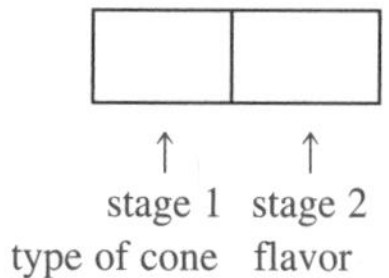

The first box can be filled in three ways and the second in 31 ways:

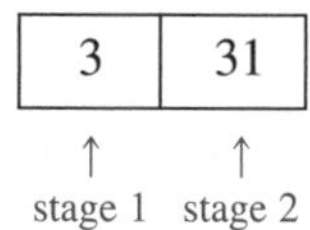

Thus, by the Fundamental Counting Principle, there are $3 \times 31 = 93$ ways of choosing a single-scoop ice-cream cone at this store. ■

EXAMPLE 2 ■ Using the Fundamental Counting Principle

In a certain state, automobile license plates display three letters followed by three digits. How many such plates are possible if repetition of the letters

(a) is allowed? (b) is not allowed?

SOLUTION

(a) There are six choices, one for each letter or digit on the license plate. As in the preceding example, we sketch a box for each stage:

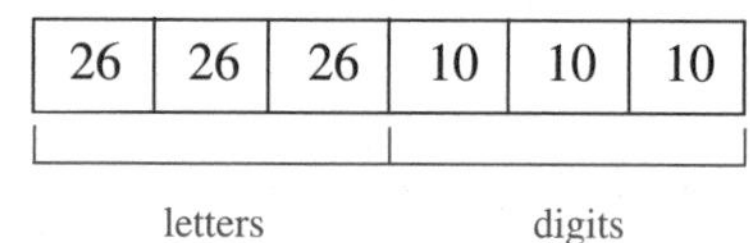

At the first stage, we choose a letter (from 26 possible choices); at the second stage, another letter (again from 26 choices); at the third stage, another letter (26 choices); at the fourth stage, a digit (from 10 possible choices); at the fifth stage, a digit (again from 10 choices); and at the sixth stage, another digit (10 choices). By the Fundamental Counting Principle, the number of possible license plates is

$$26 \times 26 \times 26 \times 10 \times 10 \times 10 = 17{,}576{,}000$$

(b) If repetition of letters is not allowed, then we can arrange the choices as follows:

26	25	24	10	10	10
letters			digits		

At the first stage, we have 26 letters to choose from, but once the first letter is chosen, there are only 25 letters to choose from at the second stage. Once the first two letters are chosen, 24 letters are left to choose from for the third stage. The digits are chosen as before. Thus, the number of possible license plates in this case is

$$26 \times 25 \times 24 \times 10 \times 10 \times 10 = 15{,}600{,}000$$ ■

EXAMPLE 3 ■ Using Factorial Notation

In how many different ways can a race with six runners be completed? Assume there is no tie.

SOLUTION

There are six possible choices for first place, five choices for second place (since only five runners are left after first place has been decided), four choices for third place, and so on. So, by the Fundamental Counting Principle, the number of different ways this race can be completed is

Factorial notation is explained on page 624.

$$6 \times 5 \times 4 \times 3 \times 2 \times 1 = 6! = 720$$ ■

10.1 EXERCISES

1. A vendor sells ice cream from a cart on the boardwalk. He offers vanilla, chocolate, strawberry, and pistachio ice cream, served on either a waffle, sugar, or plain cone. How many different single-scoop ice-cream cones can you buy from this vendor?

2. How many three-letter "words" (strings of letters) can be formed using the 26 letters of the alphabet if repetition of letters
(a) is allowed? (b) is not allowed?

3. How many three-letter "words" (strings of letters) can be formed using the letters *WXYZ* if repetition of letters
(a) is allowed? (b) is not allowed?

4. Eight horses are entered in a race.
(a) How many different orders are possible for completing the race?
(b) In how many different ways can first, second, and third places be decided? (Assume there is no tie.)

5. A multiple-choice test has five questions with four choices for each question. In how many different ways can the test be completed?

6. Telephone numbers consist of seven digits; the first digit cannot be 0 or 1. How many telephone numbers are possible?

7. In how many different ways can a race with five runners be completed? (Assume there is no tie.)

8. In how many ways can five people be seated in a row of five seats?

9. A restaurant offers six different main courses, eight types of drinks, and three kinds of desserts. How many different meals consisting of a main course, a drink, and a dessert does the restaurant offer?

10. In how many ways can five different mathematics books be placed next to each other on a shelf?

11. Towns A, B, C, and D are located in such a way that there are four roads from A to B, five roads from B to C, and six roads from C to D. How many routes are there from town A to town D via towns B and C?

12. In a family of four children, how many different boy-girl birth-order combinations are possible? (The birth orders *BBBG* and *BBGB* are different.)

13. A coin is flipped five times, and the resulting sequence of heads and tails is recorded. How many such sequences are possible?

14. A red die and a white die are rolled, and the numbers showing are recorded. How many different outcomes are possible? (The singular form of the word *dice* is *die*.)

15. A red die, a blue die, and a white die are rolled, and the numbers showing are recorded. How many different outcomes are possible?

16. Two cards are chosen in order from a deck. In how many ways can this be done if

(a) the first card must be a spade and the second must be a heart?

(b) both cards must be spades?

17. A girl has 5 skirts, 8 blouses, and 12 pairs of shoes. How many different skirt-blouse-shoe outfits can she wear? (Assume that each item matches all the others, so she is willing to wear any combination.)

18. A company's employee ID number system consists of one letter followed by three digits. How many different ID numbers are possible with this system?

19. A company has 2844 employees. Each employee is to be given an ID number that consists of one letter followed by two digits. Is it possible to give each employee a different ID number using this scheme? Explain.

20. An all-star baseball team has a roster of seven pitchers and three catchers. How many pitcher-catcher pairs can the manager select from this roster?

21. Standard automobile license plates in California display a nonzero digit, followed by three letters, followed by three digits. How many different standard plates are possible in this system?

22. A combination lock has 60 different positions. To open the lock, the dial is turned to a certain number in the clockwise direction, then to a number in the counterclockwise direction, and finally to a third number in the clockwise direction. If successive numbers in the combination cannot be the same, how many different combinations are possible?

23. A true-false test contains ten questions. In how many different ways can this test be completed?

24. An automobile dealer offers five models. Each model comes in a choice of four colors, three types of stereo equipment, with or without air conditioning, and with or without a sunroof. In how many different ways can a customer order an auto from this dealer?

25. The registrar at a certain university classifies students according to a major, minor, year (1, 2, 3, 4), and sex (M, F). Each student must choose one major and either one or no minor from the 32 fields taught at this university. How many different student classifications are possible?

26. How many monograms consisting of three initials are possible?

27. A state has registered 8 million automobiles. To simplify the license plate system, a state employee suggests that each plate display only two letters followed by three digits. Will this system create enough different license plates for all the vehicles registered?

28. A state license plate design has six places. Each plate begins with a fixed number of letters, and the remaining places are filled with digits. (For example, one letter followed by five digits, two letters followed by four digits, and so on.) The state has 17 million registered vehicles.
(a) The state decides to change to a system consisting of one letter followed by five digits. Will this design allow for enough different plates to accommodate all the vehicles registered?
(b) Find a system that will be sufficient if the smallest possible number of letters is to be used.

29. In how many ways can a president, vice president, and secretary be chosen from a class of 30 students?

30. In how many ways can a president, vice president, and secretary be chosen from a class of 20 females and 30 males if the president must be a female and the vice president a male?

31. A senate subcommittee consists of ten Democrats and seven Republicans. In how many ways can a chairman, vice chairman, and secretary be chosen if the chairman must be a Democrat and the vice chairman must be a Republican?

32. Social Security numbers consist of nine digits, with the first digit between 0 and 6, inclusive. How many Social Security numbers are possible?

33. Five-letter "words" are formed using the letters A, B, C, D, E, F, G. How many such words are possible for each of the following conditions?
(a) No condition is imposed.
(b) No letter can be repeated in a word.
(c) Each word must begin with the letter A.
(d) The letter C must be in the middle.
(e) The middle letter must be a vowel.

34. How many five-letter palindromes are possible? (A *palindrome* is a string of letters that reads the same backward and forward, such as the string *XCZCX*.)

35. A certain computer programming language allows names of variables to consist of two characters, the first being any letter and the second any letter or digit. How many names of variables are possible?

36. How many different three-character code words consisting of letters or digits are possible for the following code designs?
(a) The first entry must be a letter.
(b) The first entry cannot be zero.

37. In how many ways can four men and four women be seated in a row of eight seats for the following situations?
(a) The women are to be seated together, and the men are to be seated together.
(b) They are to be seated alternately by gender.

38. In how many ways can five different mathematics books be placed on a shelf if the two algebra books are to be placed next to each other?

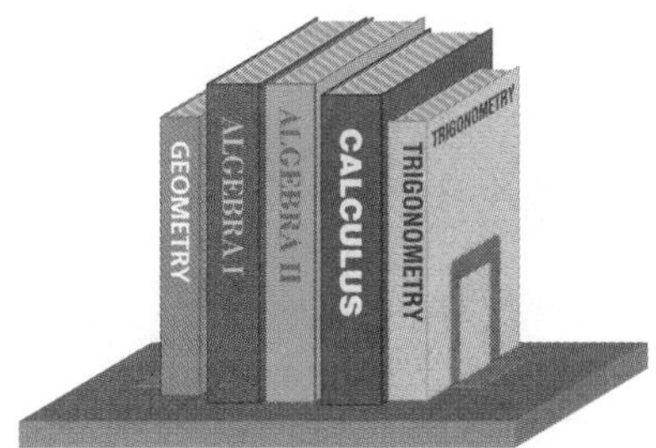

39. Eight mathematics books and three chemistry books are to be placed on a shelf. In how many ways can this be done if the mathematics books are next to each other and the chemistry books are next to each other?

40. Three-digit numbers are formed using the digits 2, 4, 5, and 7, with repetition of digits allowed. How many such numbers can be formed if
(a) the numbers are less than 700?
(b) the numbers are even?
(c) the numbers are divisible by 5?

41. How many three-digit odd numbers can be formed using the digits 1, 2, 4, and 6 if repetition of digits is not allowed?

DISCOVERY • DISCUSSION

42. Pairs of Initials Explain why in any group of 677 people, at least two people must have the same pair of initials.

43. Area Codes Until recently, telephone area codes in the United States, Canada, and the Caribbean islands were chosen according to the following rules: (i) The first digit *cannot* be 0 or a 1, and (ii) the second digit *must* be a 0 or a 1. But in 1995, the second rule was abandoned when the area code 360 was introduced in parts of western Washington State. Since then, many other new area codes that violate Rule (ii) have come into use, although Rule (i) still remains in effect.

(a) How many area code + telephone number combinations were possible under the old rules? (See Exercise 6 for a description of local telephone numbers.)

(b) How many area code + telephone number combinations are now possible under the new rules?

(c) Why do you think it was necessary to make this change?

(d) Look at the area code listings in a recent telephone directory and find five or more area codes that violate Rule (ii).

10.2 PERMUTATIONS AND COMBINATIONS

In this section we single out two important special cases of the Fundamental Counting Principle—permutations and combinations.

PERMUTATIONS

A **permutation** of a set of distinct objects is an ordering of these objects. For example, some permutations of the letters *ABCDWXYZ* are

XAYBZWCD *ZAYBCDWX* *DBWAZXYC* *YDXAWCZB*

Permutations of three colored squares

How many such permutations are possible? Since there are eight choices for the first position, seven for the second (after the first has been chosen), six for the third (after the first two have been chosen), and so on, the Fundamental Counting Principle tells us that the number of possible permutations is

$$8 \times 7 \times 6 \times 5 \times 4 \times 3 \times 2 \times 1 = 40{,}320$$

This same reasoning with 8 replaced by n leads to the following observation.

> The number of permutations of n objects is $n!$.

How many permutations consisting of five letters can be made from these same eight letters? Some of these permutations are

XYZWC *AZDWX* *AZXYB* *WDXZB*

Again, there are eight choices for the first position, seven for the second, six for the third, five for the fourth, and four for the fifth. By the Fundamental Counting Principle, the number of such permutations is

$$8 \times 7 \times 6 \times 5 \times 4 = 6720$$

In general, if a set has n elements, then the number of ways of ordering r elements from the set is denoted by $P(n, r)$ and is called **the number of permutations of n objects taken r at a time**.

We have just shown that $P(8, 5) = 6720$. The same reasoning used to find $P(8, 5)$ will help us find a general formula for $P(n, r)$. Indeed, there are n objects and r positions to place them in. Thus, there are n choices for the first position, $n - 1$ choices for the second, $n - 2$ choices for the third, and so on. The last position can be filled in $n - r + 1$ ways. By the Fundamental Counting Principle,

$$P(n, r) = n(n-1)(n-2)\cdots(n-r+1)$$

This formula can be written more compactly using factorial notation:

$$\begin{aligned}P(n, r) &= n(n-1)(n-2)\cdots(n-r+1)\\ &= \frac{n(n-1)(n-2)\cdots(n-r+1)(n-r)\cdots 3\cdot 2\cdot 1}{(n-r)\cdots 3\cdot 2\cdot 1} = \frac{n!}{(n-r)!}\end{aligned}$$

PERMUTATIONS OF n OBJECTS TAKEN r AT A TIME

The number of permutations of n objects taken r at a time is

$$P(n, r) = \frac{n!}{(n-r)!}$$

EXAMPLE 1 ■ Finding the Number of Permutations

A club has nine members. In how many ways can a president, vice president, and secretary be chosen from the members of this club?

SOLUTION

We need the number of ways of selecting three members, *in order*, for the positions of president, vice president, and secretary from the nine club members. This number is

$$P(9, 3) = \frac{9!}{(9-3)!} = \frac{9!}{6!} = 9 \times 8 \times 7 = 504$$

■

EXAMPLE 2 ■ Finding the Number of Permutations

From 20 raffle tickets in a hat, four tickets are to be selected in order. The holder of the first ticket wins a car, the second a motorcycle, the third a bicycle, and the fourth a skateboard. In how many different ways can these prizes be awarded?

SOLUTION

The order in which the tickets are chosen determines who wins each prize. So, we need to find the number of ways of selecting four objects, *in order*, from 20 objects (the tickets). This number is

$$P(20, 4) = \frac{20!}{(20-4)!} = \frac{20!}{16!} = 20 \times 19 \times 18 \times 17 = 116{,}280$$

■

DISTINGUISHABLE PERMUTATIONS

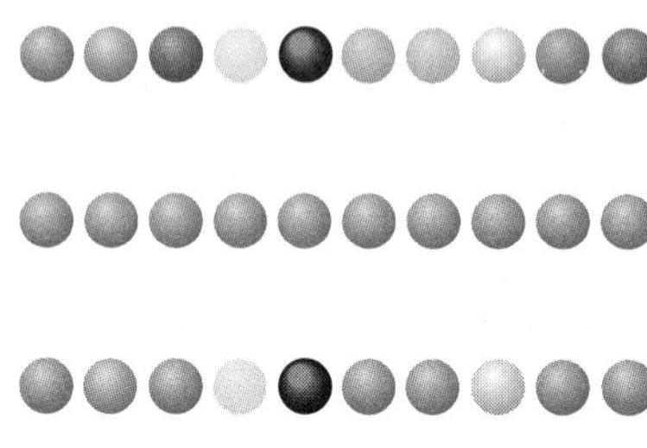

If we have a collection of ten balls, each a different color, then the number of permutations of these balls is $P(10, 10) = 10!$. If all ten balls are red, then we have just one distinguishable permutation because all the ways of ordering these balls look exactly the same. In general, when considering a set of objects, some of which are of the same kind, then two permutations are **distinguishable** if one cannot be obtained from the other by interchanging the positions of elements of the same kind. For example, if we have ten balls, of which six are red and the other four are each a different color, then how many distinguishable permutations are possible? The key point here is that balls of the same color are not distinguishable. So each rearrangement of the red balls, keeping all the other balls fixed, gives essentially the same permutation. Since there are 6! rearrangements of the red balls for each fixed position of the other balls, the total number of distinguishable permutations is $10!/6!$. The same type of reasoning gives the following general rule.

DISTINGUISHABLE PERMUTATIONS

If a set of n objects consists of k different kinds of objects with n_1 objects of the first kind, n_2 objects of the second kind, n_3 objects of the third kind, and so on, where $n_1 + n_2 + \cdots + n_k = n$, then the number of distinguishable permutations of these objects is

$$\frac{n!}{n_1!n_2!n_3!\cdots n_k!}$$

EXAMPLE 3 ■ Finding the Number of Distinguishable Permutations

Find the number of different ways of placing 15 balls in a row given that 4 are red, 3 are yellow, 6 are black, and 2 are blue.

SOLUTION

We want to find the number of distinguishable permutations of these balls. By the formula, this number is

$$\frac{15!}{4!\,3!\,6!\,2!} = 6{,}306{,}300$$

■

Suppose we have 15 wooden balls in a row and four colors of paint: red, yellow, black, and blue. In how many different ways can the 15 balls be painted in such a way that we have 4 red, 3 yellow, 6 black, and 2 blue balls? A little thought will show that this number is exactly the same as that calculated in Example 3. This way of looking at the problem is somewhat different, however. Here we think of the number of ways to **partition** the balls into four groups, each containing 4, 3, 6, and 2 balls to be painted red, yellow, black, and blue, respectively. The next example shows how this reasoning is used.

EXAMPLE 4 ■ Finding the Number of Partitions

Fourteen construction workers are to be assigned to three different tasks. Seven workers are needed for mixing cement, five for laying bricks, and two for carrying the bricks to the brick layers. In how many different ways can the workers be assigned to these tasks?

SOLUTION

We need to partition the workers into three groups containing 7, 5, and 2 workers, respectively. This number is

$$\frac{14!}{7!\,5!\,2!} = 72{,}072$$

■

COMBINATIONS

When finding permutations, we are interested in the number of ways of ordering elements of a set. In many counting problems, however, order is *not* important. For example, a poker hand is the same hand, regardless of how it is ordered. A poker player interested in the number of possible hands wants to know the number of ways of drawing five cards from 52 cards, without regard to the order in which the cards of a given hand are dealt. In this section we develop a formula for counting in situations such as this, where order doesn't matter.

A **combination** of r elements of a set is any subset of r elements from the set (without regard to order). If the set has n elements, then the number of combinations of r elements is denoted by $\boldsymbol{C(n, r)}$ and is called the **number of combinations of $\boldsymbol{n}$ elements taken $\boldsymbol{r}$ at a time**.

For example, consider a set with the four elements, A, B, C, and D. The combinations of these four elements taken three at a time are

$$ABC \qquad ABD \qquad ACD \qquad BCD$$

The permutations of these elements taken three at a time are

ABC	ABD	ACD	BCD
ACB	ADB	ADC	BDC
BAC	BAD	CAD	CBD
BCA	BDA	CDA	CDB
CAB	DAB	DAC	DBC
CBA	DBA	DCA	DCB

We notice that the number of combinations is a lot fewer than the number of permutations. In fact, each combination of three elements generates 3! permutations. So

$$C(4, 3) = \frac{P(4, 3)}{3!} = \frac{4!}{3!\,(4-3)!} = 4$$

In general, each combination of r objects gives rise to $r!$ permutations of these objects. Thus

$$C(n, r) = \frac{P(n, r)}{r!} = \frac{n!}{r!\,(n-r)!}$$

In Section 9.6 we denoted $C(n, r)$ by $\binom{n}{r}$, but it is customary to use the notation $C(n, r)$ in the context of counting. For an explanation of why these are the same, see Exercise 66.

COMBINATIONS OF n OBJECTS TAKEN r AT A TIME

The number of combinations of n objects taken r at a time is

$$C(n, r) = \frac{n!}{r!\,(n-r)!}$$

The key difference between permutations and combinations is *order*. If we are interested in ordered arrangements, then we are counting permutations; but if we are concerned with subsets without regard to order, then we are counting combinations. Compare Examples 5 and 6 below (where order doesn't matter) with Examples 1 and 2 (where order does matter).

EXAMPLE 5 ■ Finding the Number of Combinations

A club has nine members. In how many ways can a committee of three be chosen from the members of this club?

SOLUTION

We need the number of ways of choosing three of the nine members. Order is not important here, because the committee is the same no matter how its members are ordered. So, we want the number of combinations of nine objects (the club members) taken three at a time. This number is

$$C(9, 3) = \frac{9!}{3!(9-3)!} = \frac{9!}{3!\,6!} = \frac{9 \times 8 \times 7}{3 \times 2 \times 1} = 84$$

■

EXAMPLE 6 ■ Finding the Number of Combinations

From 20 raffle tickets in a hat, four tickets are to be chosen at random. The holders of the winning tickets are to be awarded free trips to the Bahamas. In how many ways can the four winners be chosen?

Ronald Graham was born in Taft, California, in 1935. He has been described as the world's leading mathematician in the field of combinatorics, the branch of mathematics that deals with counting. For many years Graham was head of the Mathematical Studies Center at Bell Laboratories in Murray Hill, New Jersey. There he solved numerous problems that arose in the telephone industry. During the *Apollo* program, NASA needed to evaluate mission schedules so that the three astronauts aboard the spacecraft could find the time to perform all the necessary tasks. The number of ways to allot these tasks were astronomical—too vast for even a computer to sort out. Graham, using his knowledge of combinatorics, was able to reassure NASA that there were easy ways of solving their problem that were not too far from the theoretically best possible solution. Besides being a prolific mathematician, Graham is an accomplished juggler and has been president of the International Jugglers Association. He has written several research papers that study the mathematical aspects of juggling.

SOLUTION

We need to find the number of ways of choosing four winners from 20 entries. The order in which the tickets are chosen doesn't matter, because the same prize is awarded to each of the four winners. So, we want the number of combinations of 20 objects (the tickets) taken four at a time. This number is

$$C(20, 4) = \frac{20!}{4!\,(20-4)!} = \frac{20!}{4!\,16!} = \frac{20 \times 19 \times 18 \times 17}{4 \times 3 \times 2 \times 1} = 4845$$ ■

If a set S has n elements, then $C(n, k)$ is the number of ways of choosing k elements from S, that is, the number of k-element subsets of S. Thus, the number of subsets of S of all possible sizes is given by the sum

$$C(n, 0) + C(n, 1) + C(n, 2) + \cdots + C(n, n) = 2^n$$

(See Section 9.6, Exercise 52, where this sum is discussed.)

> A set with n elements has 2^n subsets.

EXAMPLE 7 ■ Finding the Number of Subsets of a Set

A pizza parlor offers the basic cheese pizza and a choice of 16 toppings. How many different kinds of pizza can be ordered at this pizza parlor?

SOLUTION

We need the number of possible subsets of the 16 toppings (including the empty set, which corresponds to a plain cheese pizza). Thus, $2^{16} = 65{,}536$ different pizzas can be ordered. ■

The crucial step in solving counting problems is deciding whether to use permutations, combinations, or the Fundamental Counting Principle.

GUIDELINES FOR USING PERMUTATIONS AND COMBINATIONS

When we want to find the number of ways of picking r objects from n objects, we need to ask ourselves: Does the order of the objects matter?

If the order matters, we use permutations.

If the order doesn't matter, we use combinations.

EXAMPLE 8 ■ A Problem Involving Permutations and Combinations

A committee of seven—consisting of a chairman, a vice chairman, a secretary, and four other members—is to be chosen from a class of 20 students. In how many ways can this committee be chosen?

SOLUTION

In choosing the three officers, order is important. So, the number of ways of choosing them is

$$P(20, 3) = 6840$$

Next, we need to choose four other students from the 17 remaining. Since order doesn't matter in this case, the number of ways of doing this is

$$C(17, 4) = 2380$$

Thus, by the Fundamental Counting Principle, the number of ways of choosing this committee is

$$P(20, 3) \times C(17, 4) = 6840 \times 2380 = 16{,}279{,}200$$

■

We could have first chosen the four unordered members of the committee—in $C(20, 4)$ ways—and then the three officers from the remaining 16 members, in $P(16, 3)$ ways. Check that this gives the same answer.

10.2 EXERCISES

1–6 ■ Evaluate the expression.

1. $P(8, 3)$ **2.** $P(9, 2)$ **3.** $P(11, 4)$

4. $P(10, 5)$ **5.** $P(100, 1)$ **6.** $P(99, 3)$

7. In how many different ways can a president, vice president, and secretary be chosen from a class of 15 students?

8. In how many different ways can first, second, and third prizes be awarded in a game with eight contestants?

9. In how many different ways can six of ten people be seated in a row of six chairs?

10. In how many different ways can six people be seated in a row of six chairs?

11. How many three-letter "words" can be made from the letters *FGHIJK*? (Letters may not be repeated.)

12. How many permutations are possible from the letters of the word *LOVE*?

13. How many different three-digit whole numbers can be formed using the digits 1, 3, 5, and 7 if no repetition of digits is allowed?

14. A pianist plans to play eight pieces at a recital. In how many ways can she arrange these pieces in the program?

15. In how many different ways can a race with nine runners be completed, assuming there is no tie?

16. A ship carries five signal flags of different colors. How many different signals can be sent by hoisting exactly three of the five flags on the ship's flagpole in different orders?

17. In how many ways can first, second, and third prizes be awarded in a contest with 1000 contestants?

18. In how many ways can a president, vice president, secretary, and treasurer be chosen from a class of 30 students?

19. In how many ways can five students be seated in a row of five chairs if Jack insists on sitting in the first chair?

20. In how many ways can the students in Exercise 19 be seated if Jack insists on sitting in the middle chair?

21–24 ■ Find the number of distinguishable permutations of the given letters.

21. *AAABBC* **22.** *AAABBBCCC*

23. *AABCD* **24.** *ABCDDDEE*

25. In how many ways can two blue marbles and four red marbles be arranged in a row?

26. In how many different ways can five red balls, two white balls, and seven blue balls be arranged in a row?

27. In how many different ways can four pennies, three nickels, two dimes, and three quarters be arranged in a row?

28. In how many different ways can the letters of the word *ELEEMOSYNARY* be arranged?

29. A man bought three vanilla ice-cream cones, two chocolate cones, four strawberry cones, and five butterscotch cones for his 14 chidren. In how many ways can he distribute the cones among his children?

30. When seven students take a trip, they find a hotel with three rooms available—a room for one person, a room for two people, and a room for three people. In how many different ways can the students be assigned to these rooms? (One student has to sleep in the car.)

31. Eight workers are cleaning a large house. Five are needed to clean windows, two to clean the carpets, and one to clean the rest of the house. In how many different ways can these tasks be assigned to the eight workers?

32. A jogger jogs every morning to his health club, which is eight blocks east and five blocks north of his home. He always takes a route that is as short as possible, but he likes to vary it (see the figure). How many different routes can he take? [*Hint:* The route shown can be thought of as *ENNEEENENEENE*, where *E* is East and *N* is North.]

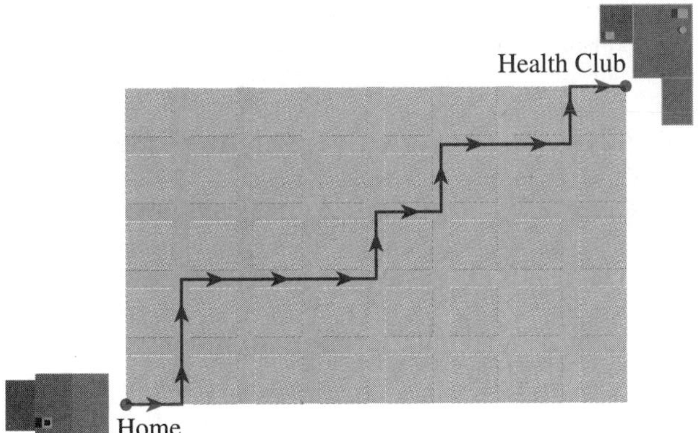

33–38 ■ Evaluate the expression.

33. $C(8, 3)$ **34.** $C(9, 2)$ **35.** $C(11, 4)$

36. $C(10, 5)$ **37.** $C(100, 1)$ **38.** $C(99, 3)$

39. In how many ways can three books be chosen from a group of six?

40. In how many ways can three pizza toppings be chosen from 12 available toppings?

41. In how many ways can six people be chosen from a group of ten?

42. In how many ways can a committee of three members be chosen from a club of 25 members?

43. How many different five-card hands can be dealt from a deck of 52 cards?

44. How many different seven-card hands can be picked from a deck of 52 cards?

45. A student must answer seven of the ten questions on an exam. In how many ways can she choose the seven questions?

46. A pizza parlor offers a choice of 16 different toppings. How many three-topping pizzas are possible?

47. A violinist has practiced 12 pieces. In how many ways can he choose eight of these pieces for a recital?

48. If a woman has eight skirts, in how many ways can she choose five of these to take on a weekend trip?

49. In how many ways can seven students from a class of 30 be chosen for a field trip?

50. In how many ways can the seven students in Exercise 49 be chosen if Jack must go on the field trip?

51. In how many ways can the seven students in Exercise 49 be chosen if Jack is not allowed to go on the field trip?

52. In the 6/49 lottery game, a player picks six numbers from 1 to 49. How many different choices does the player have?

53. In the California Lotto game, a player chooses six numbers from 1 to 53. It costs $1 to play this game. How much would it cost to buy every possible combination of six numbers to ensure picking the winning six numbers?

YOU CAN WIN $1,000,000

•••LOTTO•••

1	2	3	4	5	6	7	8	9	10
11	12	13	14	15	16	17	18	19	20
21	22	23	24	25	26	27	28	29	30
31	32	33	34	35	36	37	38	39	40
41	42	43	44	45	46	47	48	49	50
51	52	53							

54. A class has 20 students, of which 12 are females and 8 are males. In how many ways can a committee of five students be picked from this class under each condition?

(a) No restriction is placed on the number of males or females on the committee.

(b) No males are to be included on the committee.

(c) The committee must have three females and two males.

55. A set has eight elements.

(a) How many subsets containing five elements does this set have?

(b) How many subsets does this set have?

56. A travel agency has limited numbers of eight different free brochures about Australia. The agent tells you to take any that you like, but no more than one of any kind. How many different ways can you choose brochures (including not choosing any)?

57. A hamburger chain gives their customers a choice of ten different hamburger toppings. In how many different ways can a customer order a hamburger?

58. Each of 20 shoppers in a shopping mall chooses to enter or not to enter the Dressfastic clothing store. How many different outcomes of their decisions are possible?

59. From a group of ten male and ten female tennis players, two men and two women are to face each other in a men-versus-women doubles match. In how many different ways can this match be arranged?

60. A school dance committee is to consist of two freshmen, three sophomores, four juniors, and five seniors. If six freshmen, eight sophomores, twelve juniors, and ten seniors are eligible to be on the committee, in how many ways can the committee be chosen?

61. A group of 22 aspiring thespians contains ten men and twelve women. For the next play the director wants to choose a leading man, a leading lady, a supporting male role, a supporting female role, and eight extras—three women and five men. In how many ways can the cast be chosen?

62. A hockey team has 20 players of which twelve play forward, six play defense, and two are goalies. In how many ways can the coach pick a starting lineup consisting of three forwards, two defense players, and one goalie?

63. A pizza parlor offers four sizes of pizza (small, medium, large, and colossus), two types of crust (thick and thin), and 14 different toppings. How many different pizzas can be made with these choices?

DISCOVERY • DISCUSSION

64. Complementary Combinations Without performing any calculations, explain in words why the number of ways of choosing two objects from ten objects is the same as the number of ways of choosing eight objects from ten objects. In general, explain why $C(n, r) = C(n, n - r)$.

65. An Identity Involving Combinations Kevin has ten different marbles, and he wants to give three of them to Luke and two to Mark. How many ways can he choose to do this? There are two ways of analyzing this problem: He could first pick three for Luke and then two for Mark, or he could first pick two for Mark and then three for Luke. Explain how these two viewpoints show that $C(10, 3) \cdot C(7, 2) = C(10, 2) \cdot C(8, 3)$. In general, explain why

$$C(n, r) \cdot C(n - r, k) = C(n, k) \cdot C(n - k, r)$$

66. Why is $\binom{n}{r}$ the Same as $C(n, r)$? This exercise explains why the binomial coefficients $\binom{n}{r}$ that appear in the expansion of $(x + y)^n$ are the same as $C(n, r)$, the number of ways of choosing r objects from n objects. First, note that expanding a binomial using only the Distributive Property gives

$$\begin{aligned}(x + y)^2 &= (x + y)(x + y)\\ &= (x + y)x + (x + y)y\\ &= xx + xy + yx + yy\\ (x + y)^3 &= (x + y)(xx + xy + yx + yy)\\ &= xxx + xxy + xyx + xyy + yxx\\ &\quad + yxy + yyx + yyy\end{aligned}$$

(a) Expand $(x + y)^5$ using only the Distributive Property.
(b) Write all the terms that represent x^2y^3 together. These are all the terms that contain two x's and three y's.
(c) Note that the two x's appear in all possible positions. Conclude that the number of terms that represent x^2y^3 is $C(5, 2)$.
(d) In general, explain why $\binom{n}{r}$ in the Binomial Theorem is the same as $C(n, r)$.

10.3 PROBABILITY

If you roll a pair of dice, what are the chances of rolling a double six? What is the likelihood of winning a state lottery? The subject of probability was invented to give precise answers to questions like these. It is now an indispensable tool for making decisions in such diverse areas as business, manufacturing, psychology, genetics, and in many of the sciences. Probability is used to determine the effectiveness of new medicines, assess a fair price for an insurance policy, decide on the likelihood of a candidate winning an election, determine the opinion of many people on a certain topic (without interviewing everyone), and answer many other questions that involve a measure of uncertainty.

The mathematical theory of probability began in 1654 in a series of letters between Pascal (see page 618) and Fermat (see page 578). Their correspondence was prompted by a question raised by the experienced gambler the Chevalier de Méré. The Chevalier was interested in the equitable distribution of the stakes of an interrupted gambling game (see Problem 3, page 677).

To discuss probability, let's begin by defining some terms. An **experiment** is a process, such as tossing a coin or rolling a die, that gives definite results, called the **outcomes** of the experiment. For tossing a coin, the possible outcomes are "heads" and "tails"; for rolling a die, the outcomes are 1, 2, 3, 4, 5, and 6. The **sample space** of an experiment is the set of all possible outcomes. If we let H stand for heads and T for tails, then the sample space of the coin-tossing experiment is

$$S = \{H, T\}$$

The sample space for rolling a die is

$$S = \{1, 2, 3, 4, 5, 6\}$$

We will be concerned only with experiments for which all the outcomes are "equally likely." We already have an intuitive feeling for what this means. When tossing a perfectly balanced coin, heads and tails are equally likely outcomes in the sense that if this experiment is repeated many times, we expect that about half the results will be heads and half will be tails.

In any given experiment we are often concerned with a particular set of outcomes. We might be interested in a die showing an even number or in picking an ace from a deck of cards. Any particular set of outcomes is a subset of the sample space. This leads to the following definition.

DEFINITION OF AN EVENT

If S is the sample space of an experiment, then an **event** is any subset of the sample space.

MATHEMATICS IN THE MODERN WORLD

FAIR VOTING METHODS

The methods of mathematics have recently been applied to problems in the social sciences. For example, how do we find fair voting methods? You may ask, what is the problem with how we vote in elections? Well, suppose candidates A, B, and C are running for president. The final vote tally is as follows: A gets 40%, B gets 39%, and C gets 21%. So candidate A wins. But 60% of the voters *didn't* want A. Moreover, you voted for C, but you dislike A so much that you would have been willing to change your vote to B to avoid having A win. Most of the voters who voted for C feel the same way you do, so we have a situation where most of the voters prefer B over A, but A wins. So is that fair?

In the 1950s Kenneth Arrow showed mathematically that no democratic method of voting can be completely fair, and later won a Nobel Prize for his work. Mathematicians continue to work on finding fairer voting systems. The system most often used in federal, state, and local elections is called *plurality voting* (the candidate with the most votes wins). Other systems include *majority voting* (if no candidate gets a majority, a runoff is held between the top two vote-getters), *approval voting* (each voter can vote for as many candidates as he or she approves of), *preference voting* (each voter orders the candidates according to his or her preference), and *cumulative voting* (each voter gets as

(continued)

EXAMPLE 1 ■ Events in a Sample Space

If an experiment consists of tossing a coin three times and recording the results in order, the sample space is

$$S = \{HHH, HHT, HTH, THH, TTH, THT, HTT, TTT\}$$

The event E of showing "exactly two heads" is the subset of S that consists of all outcomes with two heads. Thus

$$E = \{HHT, HTH, THH\}$$

The event F of showing "at least two heads" is

$$F = \{HHH, HHT, HTH, THH\}$$

and the event of showing "no heads" is $G = \{TTT\}$. ■

We are now ready to define the notion of probability. Intuitively, we know that rolling a die may result in any of six equally likely outcomes, so the chance of any particular outcome occuring is $\frac{1}{6}$. What is the chance of showing an even number? Of the six equally likely outcomes possible, three are even numbers. So, it's reasonable to say that the chance of showing an even number is $\frac{3}{6} = \frac{1}{2}$. This reasoning is the intuitive basis for the following definition of probability.

DEFINITION OF PROBABILITY

Let S be the sample space of an experiment and E an event. The probability of E, written $P(E)$, is

$$P(E) = \frac{n(E)}{n(S)} = \frac{\text{number of elements in } E}{\text{number of elements in } S}$$

Notice that $0 \leqslant n(E) \leqslant n(S)$, so the probability $P(E)$ of an event is a number between 0 and 1, that is,

$$0 \leqslant P(E) \leqslant 1$$

The closer the probability of an event is to 1, the more likely the event is to happen; the closer to 0, the less likely. If $P(E) = 1$, then E is called the **certain event** and if $P(E) = 0$, then E is called the **impossible event**.

EXAMPLE 2 ■ Finding the Probability of an Event

A coin is tossed three times and the results are recorded. What is the probability of getting exactly two heads? At least two heads? No heads?

many votes as there are candidates and can give all his or her votes to one candidate or distribute them among the candidates as he or she sees fit). This last system is often used to select corporate boards of directors. Each system of voting has its advantages as well as disadvantages.

SOLUTION

By the results of Example 1, the sample space S of this experiment contains eight outcomes and the event E of getting "exactly two heads" contains three outcomes, $\{HHT\ HTH,\ THH\}$, so by the definition of probability,

$$P(E) = \frac{n(E)}{n(S)} = \frac{3}{8}$$

Similarly, the event F of getting "at least two heads" has four outcomes, $\{HHH,\ HHT,\ HTH,\ THH\}$, and so

$$P(F) = \frac{n(F)}{n(S)} = \frac{4}{8} = \frac{1}{2}$$

The event G of getting "no heads" has one element, so

$$P(G) = \frac{n(G)}{n(S)} = \frac{1}{8}$$

■

To find the probability of an event, we do not need to list all the elements in the sample space and the event. All we do need is the *number* of elements in these sets. The counting techniques we've learned in the preceding sections will be very useful here.

EXAMPLE 3 ■ Finding the Probability of an Event

A five-card poker hand is drawn from a standard deck of 52 cards. What is the probability that all five cards are spades?

SOLUTION

The experiment here consists of choosing five cards from the deck, and the sample space S consists of all possible five-card hands. Thus, the number of elements in the sample space is

$$n(S) = C(52, 5) = \frac{52!}{5!\,(52-5)!} = 2{,}598{,}960$$

The event E we are interested in consists of choosing five spades. Since the deck contains only 13 spades, the number of ways of choosing five spades is

$$n(E) = C(13, 5) = \frac{13!}{5!\,(13-5)!} = 1287$$

Thus, the probability of drawing five spades is

It's traditional to write probabilities without a leading zero before the decimal point.

$$P(E) = \frac{n(E)}{n(S)} = \frac{1287}{2{,}598{,}960} \approx .0005$$

■

What does the answer to Example 3 tell us? Since $.0005 = \frac{1}{2000}$, this means that if you play poker many, many times, on average you will be dealt a hand consisting of only spades about once in every 2000 hands.

EXAMPLE 4 ■ Finding the Probability of an Event

A bag contains 20 tennis balls, of which four are defective. If two balls are selected at random from the bag, what is the probability that both are defective?

SOLUTION

The experiment consists of choosing two balls from 20, so the number of elements in the sample space S is $C(20, 2)$. Since there are four defective balls, the number of ways of picking two defective balls is $C(4, 2)$. Thus, the probability of the event E of picking two defective balls is

$$P(E) = \frac{n(E)}{n(S)} = \frac{C(4, 2)}{C(20, 2)} = \frac{6}{190} \approx .032$$

■

The **complement** of an event E is the set of outcomes in the sample space that is not in E. We denote the complement of an event E by E'. We can calculate the probability of E' using the definition and the fact that $n(E') = n(S) - n(E)$:

$$P(E') = \frac{n(E')}{n(S)} = \frac{n(S) - n(E)}{n(S)} = \frac{n(S)}{n(S)} - \frac{n(E)}{n(S)} = 1 - P(E)$$

PROBABILITY OF THE COMPLEMENT OF AN EVENT

Let S be the sample space of an experiment and E an event. Then

$$P(E') = 1 - P(E)$$

This is an extremely useful result, since it is often difficult to calculate the probability of an event E but easy to find the probability of E', from which $P(E)$ can be calculated immediately using this formula.

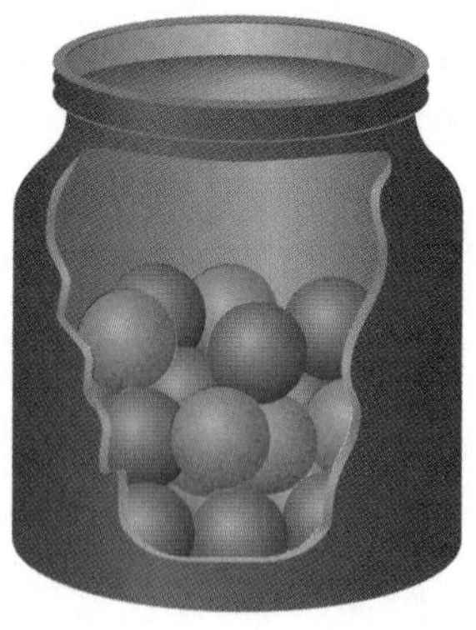

EXAMPLE 5 ■ Finding the Probability of the Complement of an Event

An urn contains 10 red balls and 15 blue balls. Six balls are drawn at random from the urn. What is the probability that at least one ball is red?

SOLUTION

Let E be the event that at least one red ball is drawn. It is tedious to count all the possible ways in which one or more of the balls drawn are red. So, let's consider E', the complement of this event—namely, that none of the balls chosen is red. The number of ways of choosing 6 blue balls from the 15 blue balls is $C(15, 6)$;

the number of ways of choosing 6 balls from the 25 balls is $C(25, 6)$. Thus

$$P(E') = \frac{n(E')}{n(S)} = \frac{C(15, 6)}{C(25, 6)} = \frac{5005}{177{,}100} = \frac{13}{460}$$

Since

$$P(E') = 1 - P(E)$$

we have

$$P(E) = 1 - P(E')$$

By the formula for the complement of an event, we have

$$P(E) = 1 - P(E') = 1 - \frac{13}{460} = \frac{447}{460} \approx .97$$

■

MUTUALLY EXCLUSIVE EVENTS

Two events that have no outcome in common are said to be **mutually exclusive** (see Figure 1). For example, in drawing a card from a deck, the events

E: The card is an ace

F: The card is a queen

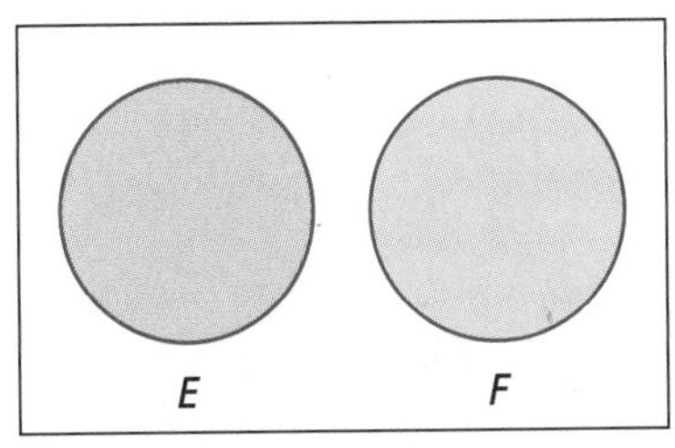

FIGURE 1

are mutually exclusive, because a card cannot be both an ace and a queen.

If E and F are mutually exclusive events, what is the probability that E or F occurs? The word *or* indicates that we want the probability of the *union* of these events, that is, $E \cup F$. Since E and F have no element in common,

$$n(E \cup F) = n(E) + n(F)$$

Thus

$$P(E \cup F) = \frac{n(E \cup F)}{n(S)} = \frac{n(E) + n(F)}{n(S)} = \frac{n(E)}{n(S)} + \frac{n(F)}{n(S)} = P(E) + P(F)$$

We have proved the following formula.

PROBABILITY OF THE UNION OF MUTUALLY EXCLUSIVE EVENTS

If E and F are mutually exclusive events in a sample space S, then the probability of E *or* F is

$$P(E \cup F) = P(E) + P(F)$$

There is a natural extension of this formula for any number of mutually exclusive events: If $E_1, E_2, \ldots, E_n$ are pairwise mutually exclusive, then

$$P(E_1 \cup E_2 \cup \cdots \cup E_n) = P(E_1) + P(E_2) + \cdots + P(E_n)$$

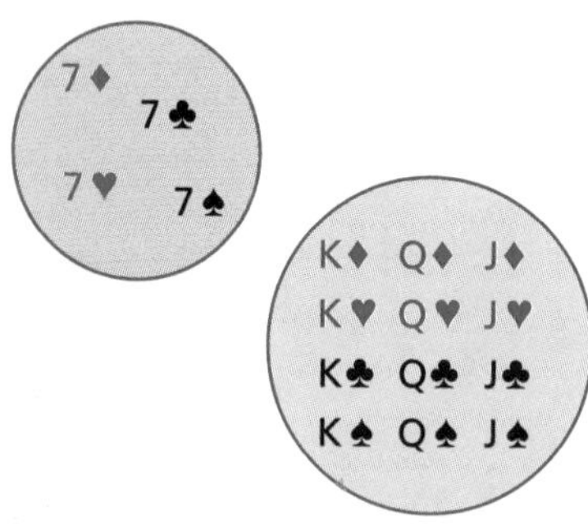

EXAMPLE 6 ■ The Probability of Mutually Exclusive Events

A card is drawn at random from a standard deck of 52 cards. What is the probability that the card is either a seven or a face card?

SOLUTION

Let E and F denote the following events.

E: The card is a seven

F: The card is a face card

Since a card cannot be both a seven and a face card, the events are mutually exclusive. We want the probability of E *or* F; in other words, the probability of $E \cup F$. By the formula,

$$P(E \cup F) = P(E) + P(F) = \frac{4}{52} + \frac{12}{52} = \frac{4}{13}$$

■

THE PROBABILITY OF THE UNION OF TWO EVENTS

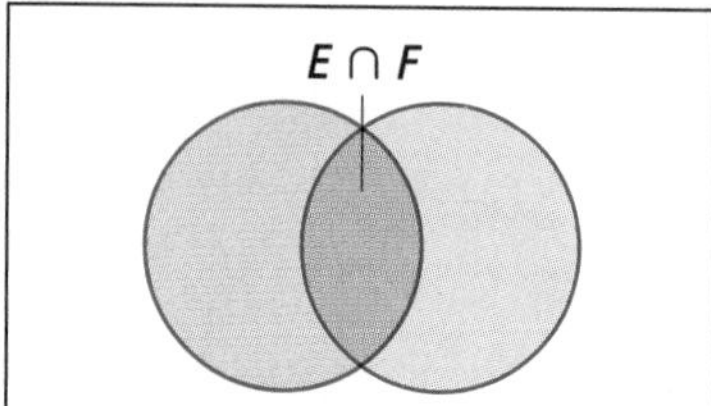

FIGURE 2

If two events E and F are not mutually exclusive, then they share outcomes in common. The situation is described graphically in Figure 2. The overlap of the two sets is their intersection, that is, $E \cap F$. Again, we are interested in the event E *or* F, so we must count the elements in $E \cup F$. If we simply added the number of elements in E to the number of elements in F, then we would be counting the elements in the overlap twice—once in E and once in F. So, to get the correct total, we must subtract the number of elements in $E \cap F$. Thus

$$n(E \cup F) = n(E) + n(F) - n(E \cap F)$$

Using the formula for probability, we get

$$\begin{aligned} P(E \cup F) &= \frac{n(E \cup F)}{n(S)} = \frac{n(E) + n(F) - n(E \cap F)}{n(S)} \\ &= \frac{n(E)}{n(S)} + \frac{n(F)}{n(S)} - \frac{n(E \cap F)}{n(S)} \\ &= P(E) + P(F) - P(E \cap F) \end{aligned}$$

We have proved the following.

PROBABILITY OF THE UNION OF TWO EVENTS

If E and F are events in a sample space S, then the probability of E *or* F is

$$P(E \cup F) = P(E) + P(F) - P(E \cap F)$$

EXAMPLE 7 ■ The Probability of the Union of Events

What is the probability that a card drawn at random from a standard 52-card deck is either a face card or a spade?

SOLUTION

We let E and F denote the following events:

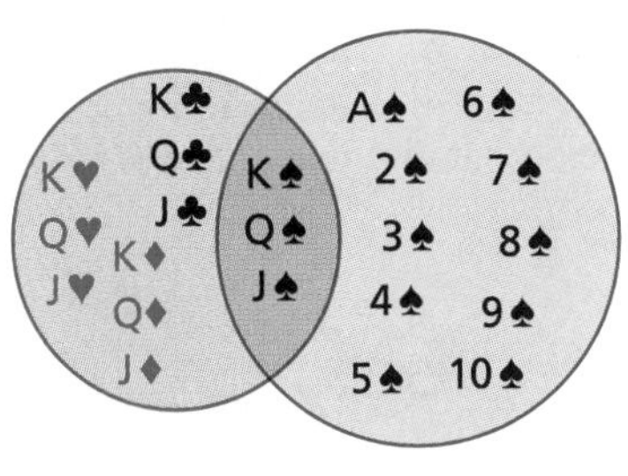

E: The card is a face card

F: The card is a spade

There are 12 face cards and 13 spades in a 52-card deck, so

$$P(E) = \frac{12}{52} \qquad \text{and} \qquad P(F) = \frac{13}{52}$$

Since 3 cards are simultaneously face cards and spades, we have

$$P(E \cap F) = \frac{3}{52}$$

Thus, by the formula for the probability of the union of two events, we have

$$\begin{aligned} P(E \cup F) &= P(E) + P(F) - P(E \cap F) \\ &= \frac{12}{52} + \frac{13}{52} - \frac{3}{52} = \frac{11}{26} \end{aligned}$$

■

THE INTERSECTION OF INDEPENDENT EVENTS

We have considered the probability of events joined by the word *or*, that is, the union of events. Now we study the probability of events joined by the word *and*—in other words, the intersection of events.

When the occurrence of one event does not affect the probability of another event, we say that the events are **independent**. For instance, if a balanced coin is tossed, the probability of showing heads on the second toss is $\frac{1}{2}$, regardless of the outcome of the first toss. So, any two tosses of a coin are independent.

PROBABILITY OF THE INTERSECTION OF INDEPENDENT EVENTS

If E and F are independent events in a sample space S, then

$$P(E \cap F) = P(E)P(F)$$

EXAMPLE 8 ■ The Probability of Independent Events

A jar contains five red balls and four black balls. A ball is drawn at random from the jar and then replaced; then another ball is picked. What is the probability that both balls are red?

SOLUTION

The events are independent. The probability that the first ball is red is $\frac{5}{9}$. The probability that the second is red is also $\frac{5}{9}$. Thus, the probability that both balls are red is

$$\frac{5}{9} \times \frac{5}{9} = \frac{25}{81} \approx .31$$

■

EXAMPLE 9 ■ The Birthday Problem

What is the probability that in a class of 35 students, at least two have the same birthday?

SOLUTION

It's reasonable to assume that the 35 birthdays are independent and that each day of the 365 days in a year is equally likely as a date of birth. (We ignore February 29.)

Let E be the event that two of the students have the same birthday. It is tedious to list all the possible ways in which at least two of the students have matching birthdays. So, we consider the complementary event E', that is, that *no* two students have the same birthday. To find this probability, we consider the students one at a time. The probability that the first student has a birthday is 1, the probability that the second has a birthday different from the first is $\frac{364}{365}$, the probability that the third has a birthday different from the first two is $\frac{363}{365}$, and so on. Thus

$$P(E') = 1 \cdot \tfrac{364}{365} \cdot \tfrac{363}{365} \cdot \tfrac{362}{365} \cdot \cdots \cdot \tfrac{331}{365} \approx .186$$

So $$P(E) = 1 - P(E') \approx 1 - .186 = .814$$

■

Number of people in a group	Probability that at least two have the same birthday
5	.02714
10	.11695
15	.25290
20	.41144
22	.47569
23	.50730
24	.53834
25	.56870
30	.70631
35	.81438
40	.89123
50	.97037

Most people are surprised that the probability in Example 9 is so high. For this reason, this problem is sometimes called the "birthday paradox." The table in the margin gives the probability that two people in a group will share the same birthday for groups of various sizes.

10.3 EXERCISES

1. An experiment consists of tossing a coin twice.
   (a) Find the sample space.
   (b) Find the probability of getting heads exactly two times.
   (c) Find the probability of getting heads at least one time.
   (d) Find the probability of getting heads exactly one time.

2. An experiment consists of tossing a coin and rolling a die.
   (a) Find the sample space.
   (b) Find the probability of getting heads and an even number.
   (c) Find the probability of getting heads and a number greater than 4.
   (d) Find the probability of getting tails and an odd number.

3–4 ■ A die is rolled. Find the probability of the given event.

3. (a) The number showing is a six.
   (b) The number showing is an even number.
   (c) The number showing is greater than 5.

4. (a) The number showing is a two or a three.
   (b) The number showing is an odd number.
   (c) The number showing is a number divisible by 3.

5–6 ■ A card is drawn randomly from a standard 52-card deck. Find the probability of the given event.

5. (a) The card drawn is a king.
   (b) The card drawn is a face card.
   (c) The card drawn is not a face card.

6. (a) The card drawn is a heart.
   (b) The card drawn is either a heart or a spade.
   (c) The card drawn is a heart, a diamond, or a spade.

7–8 ■ A ball is drawn randomly from a jar that contains five red balls, two white balls, and one yellow ball. Find the probability of the given event.

7. (a) A red ball is drawn.
   (b) The ball drawn is not yellow.
   (c) A black ball is drawn.

8. (a) Neither a white nor yellow ball is drawn.
   (b) A red, white, or yellow ball is drawn.
   (c) The ball drawn is not white.

9. A drawer contains an unorganized collection of 18 socks—three pairs are red, two pairs are white, and four pairs are black.
   (a) If one sock is drawn at random from the drawer, what is the probability that it is red?
   (b) Once a sock is drawn and discovered to be red, what is the probability of drawing another red sock to make a matching pair?

10. A child's game has a spinner as shown in the figure. Find the probability of the given event.
    (a) The spinner stops on an even number.
    (b) The spinner stops on an odd number or a number greater than 3.

11. A letter is chosen at random from the word *EXTRATERRESTRIAL*. Find the probability of the given event.
    (a) The letter T is chosen.
    (b) The letter chosen is a vowel.
    (c) The letter chosen is a consonant.

12–15 ■ A poker hand, consisting of five cards, is dealt from a standard deck of 52 cards. Find the probability that the hand contains the cards described.

12. Five hearts

13. Five cards of the same suit

14. Five face cards

15. An ace, king, queen, jack, and 10 of the same suit (royal flush)

16. A pair of dice is rolled, and the numbers showing are observed.
    (a) List the sample space of this experiment.
    (b) Find the probability of getting a sum of 7.
    (c) Find the probability of getting a sum of 9.
    (d) Find the probability that the two dice show doubles (the same number).

(e) Find the probability that the two dice show different numbers.
(f) Find the probability of getting a sum of 9 or higher.

17. A couple intends to have four children. Assume that having a boy or a girl is an equally likely event.
(a) List the sample space of this experiment.
(b) Find the probability that the couple has only boys.
(c) Find the probability that the couple has two boys and two girls.
(d) Find the probability that the couple has four children of the same sex.
(e) Find the probability that the couple has at least two girls.

18. What is the probability that a 13-card bridge hand consists of all cards from the same suit?

19. An American roulette wheel has 38 slots; two slots are numbered 0 and 00, and the remaining slots are numbered from 1 to 36. Find the probability that the ball lands in an odd-numbered slot.

20. A toddler has wooden blocks showing the letters *C*, *E*, *F*, *H*, *N*, and *R*. Find the probability that the child arranges the letters in the indicated order.
(a) In the order *FRENCH*
(b) In alphabetical order

21. In the 6/49 lottery game, a player selects six numbers from 1 to 49. What is the probability of picking the six winning numbers?

22. The president of a large company selects six employees to receive a special bonus. He claims that the six employees are chosen randomly from among the 30 employees, of which 19 are women and 11 are men. What is the probability that no woman is chosen?

23. An exam has ten true-false questions. A student who has not studied answers all ten questions by just guessing. Find the probability that the student correctly answers the given number of questions.
(a) All ten questions
(b) Exactly seven questions

24. To control the quality of their product, the Bright-Light Company inspects three light bulbs out of each batch of ten bulbs manufactured. If a defective bulb is found, the batch is discarded. Suppose a batch contains two defective bulbs. What is the probability that the batch will be discarded?

25. An often-quoted example of an event of extremely low probability is that a monkey types Shakespeare's entire play *Hamlet* by randomly striking keys on a typewriter. Assume that the typewriter has 48 keys (including the space bar) and that the monkey is equally likely to hit any key.
(a) Find the probability that such a monkey will actually correctly type just the title of the play as his first word.
(b) What is the probability that the monkey will type the phrase "To be or not to be" as his first words?

26. A monkey is trained to arrange wooden blocks in a straight line. He is then given six blocks showing the letters *A*, *E*, *H*, *L*, *M*, *T*. What is the probability that he will arrange them to spell the word *HAMLET*?

27. A monkey is trained to arrange wooden blocks in a straight line. He is then given 11 blocks showing the letters *A*, *B*, *B*, *I*, *I*, *L*, *O*, *P*, *R*, *T*, *Y*. What is the probability that the monkey will arrange the blocks to spell the word *PROBABILITY*?

28. Eight horses are entered in a race. You randomly predict a particular order for the horses to complete the race. What is the probability that your prediction is correct?

29. Many genetic traits are controlled by two genes, one dominant and one recessive. In Gregor Mendel's original experiments with peas, the genes controlling the height of the plant are denoted by T (tall) and t (short). The gene T is dominant, so a plant with the genotype (genetic makeup) TT or Tt is tall, whereas one with genotype tt is short. By a statistical analysis of the offspring in his experiments, Mendel concluded that offspring inherit one gene from each parent, and each possible combination of the two genes is equally likely. If each parent has the genotype Tt, then the following chart gives the possible genotypes of the offspring:

		Parent 2 T	Parent 2 t
Parent 1	T	TT	Tt
	t	Tt	tt

Find the probability that a given offspring of these parents will be (a) tall or (b) short.

30. Refer to Exercise 29. Make a chart of the possible genotypes of the offspring if one parent has genotype Tt and the other tt. Find the probability that a given offspring will be (a) tall or (b) short.

31–32 ■ Determine whether the events E and F in the given experiment are mutually exclusive.

31. The experiment consists of selecting a person at random.
(a) E: The person is male
F: The person is female
(b) E: The person is tall
F: The person is blond

32. The experiment consists of choosing at random a student from your class.
(a) E: The student is female
F: The student wears glasses
(b) E: The student has long hair
F: The student is male

33–34 ■ A die is rolled and the number showing is observed. Determine whether the events E and F are mutually exclusive. Then find the probability of the event $E \cup F$.

33. (a) E: The number is even
F: The number is odd
(b) E: The number is even
F: The number is greater than 4

34. (a) E: The number is greater than 3
F: The number is less than 5
(b) E: The number is divisible by 3
F: The number is less than 3

35–36 ■ A card is drawn at random from a standard 52-card deck. Determine whether the events E and F are mutually exclusive. Then find the probability of the event $E \cup F$.

35. (a) E: The card is a face card
F: The card is a spade
(b) E: The card is a heart
F: The card is a spade

36. (a) E: The card is a club
F: The card is a king
(b) E: The card is an ace
F: The card is a spade

37–38 ■ Refer to the spinner shown in the figure. Find the probability of the given event.

37. (a) The spinner stops on red.
(b) The spinner stops on an even number.
(c) The spinner stops on red or an even number.

38. (a) The spinner stops on blue.
(b) The spinner stops on an odd number.
(c) The spinner stops on blue or an odd number.

39. An American roulette wheel has 38 slots: two of the slots are numbered 0 and 00, and the rest are numbered from 1 to 36. Find the probability that the ball lands in an odd-numbered slot or in a slot with a number higher than 31.

40. A monkey is trained to arrange wooden blocks in a straight line. He is then given blocks with the letters *A, E, H, L, M, T*. What is the probability that he will arrange them to spell one of the words *HAMLET* or *THELMA*?

41. A committee of five is chosen randomly from a group of six males and eight females. What is the probability that the committee includes either all males or all females?

42. In the 6/49 lottery game a player selects six numbers from 1 to 49. What is the probability of selecting at least five of the six winning numbers?

43. A jar contains six red marbles numbered 1 to 6 and ten blue marbles numbered 1 to 10. A marble is drawn at random from the jar. Find the probability that the given event occurs.
(a) The marble is red.
(b) The marble is odd-numbered.
(c) The marble is red or odd-numbered.
(d) The marble is blue or even-numbered.

44. A coin is tossed twice. Let E be the event "the first toss shows heads" and F the event "the second toss shows heads."
(a) Are the events E and F independent?
(b) Find the probability of showing heads on both tosses.

45. A die is rolled twice. Let E be the event "the first roll shows a six" and F the event "the second roll shows a six."
(a) Are the events E and F independent?
(b) Find the probability of showing a six on both rolls.

46–47 ■ Spinners A and B shown in the figure are spun at the same time.

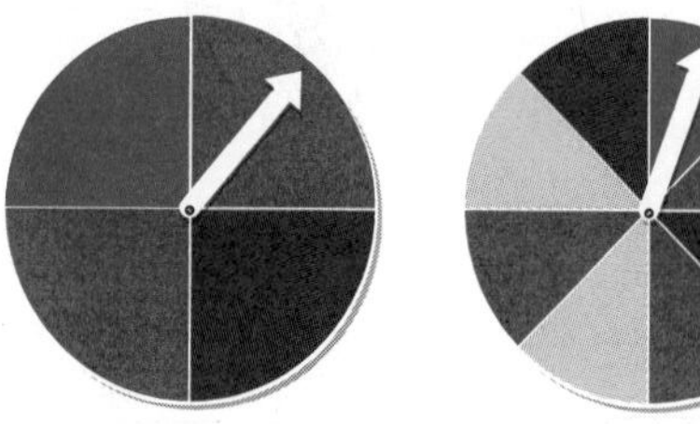

Spinner A Spinner B

46. (a) Are the events "spinner A stops on red" and "spinner B stops on yellow" independent?
(b) Find the probability that spinner A stops on red and spinner B stops on yellow.

47. (a) Find the probability that both spinners stop on purple.
(b) Find the probability that both spinners stop on blue.

48. A die is rolled twice. What is the probability of showing a one on both rolls?

49. A die is rolled twice. What is the probability of showing a one on the first roll and an even number on the second roll?

50. A card is drawn from a deck and replaced, and then a second card is drawn.
(a) What is the probability that both cards are aces?
(b) What is the probability that the first is an ace and the second a spade?

51. A roulette wheel has 38 slots: Two slots are numbered 0 and 00, and the rest are numbered 1 to 36. A player places a bet on a number between 1 and 36 and wins if a ball thrown into the spinning roulette wheel lands in the slot with the same number. Find the probability of winning on two consecutive spins of the roulette wheel.

52. A researcher claims that she has taught a monkey to spell the word *MONKEY* using the five wooden letters *E, O, K, M, N, Y*. If the monkey has not actually learned anything and is merely arranging the blocks randomly, what is the probability that he will spell the word correctly three consecutive times?

53. What is the probability of rolling "snake eyes" (double ones) three times in a row with a pair of dice?

54. In the 6/49 lottery game, a player selects six numbers from 1 to 49 and wins if he selects the winning six numbers. What is the probability of winning the lottery two times in a row?

55. Jar A contains three red balls and four white balls. Jar B contains five red balls and two white balls. Which one of the following ways of randomly selecting balls gives the greatest probability of drawing two red balls?
(i) Draw two balls from jar B.
(ii) Draw one ball from each jar.
(iii) Put all the balls in one jar, and then draw two balls.

56. A slot machine has three wheels: Each wheel has 11 positions—a bar and the digits 0, 1, 2, . . . , 9. When the handle is pulled, the three wheels spin independently before coming to rest. Find the probability that the wheels stop on the following positions.
(a) Three bars
(b) The same number on each wheel
(c) At least one bar

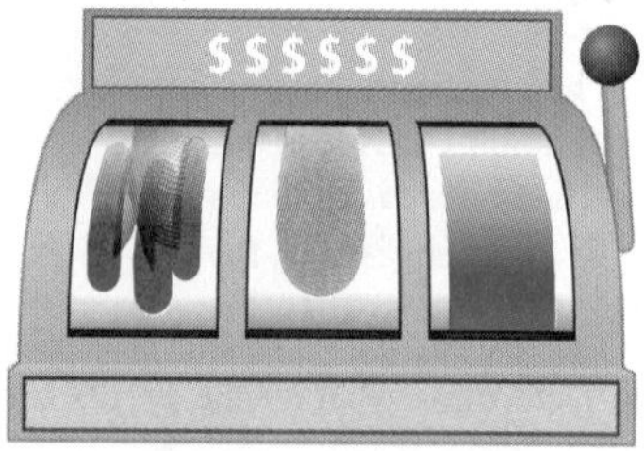

57. Find the probability that in a group of eight students at least two people have the same birthday.

58. What is the probability that in a group of six students at least two have birthdays in the same month?

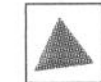

DISCOVERY • DISCUSSION

59. The "Second Son" Paradox Mrs. Smith says, "I have two children—the older one is named William." Mrs. Jones replies, "One of my two children is also named William." For each woman, list the sample space for the genders of her children, and calculate the probability that her other child is also a son. Explain why these two probabilities are different.

60. The "Oldest Son or Daughter" Phenomenon Poll your class to determine how many of your male classmates are the oldest sons in their families, and how many of your female classmates are the oldest daughters in their families. You will most likely find that they form a majority of the class. Explain why a randomly selected individual has a high probability of being the oldest son or daughter in his or her family.

Laboratory Project

SMALL SAMPLES, BIG RESULTS

A national poll finds that voter preference for presidential candidates is as follows:

Candidate A	57%
Candidate B	43%

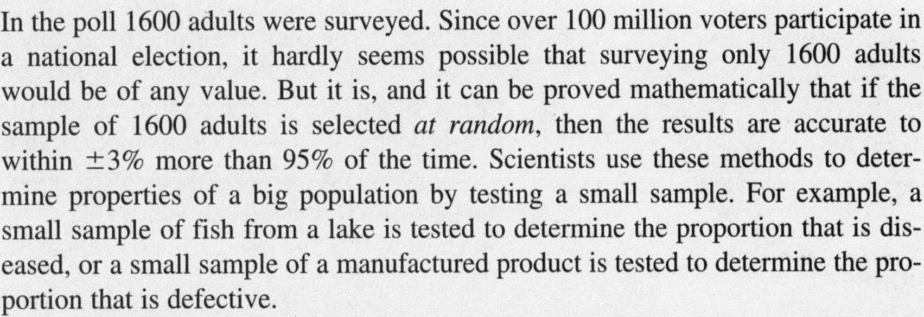

In the poll 1600 adults were surveyed. Since over 100 million voters participate in a national election, it hardly seems possible that surveying only 1600 adults would be of any value. But it is, and it can be proved mathematically that if the sample of 1600 adults is selected *at random*, then the results are accurate to within $\pm 3\%$ more than 95% of the time. Scientists use these methods to determine properties of a big population by testing a small sample. For example, a small sample of fish from a lake is tested to determine the proportion that is diseased, or a small sample of a manufactured product is tested to determine the proportion that is defective.

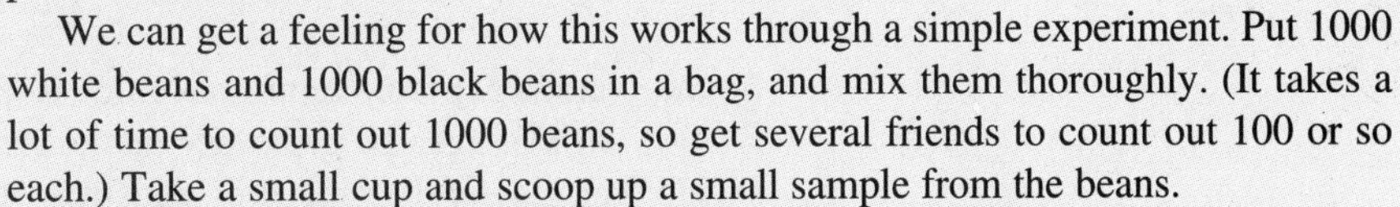

We can get a feeling for how this works through a simple experiment. Put 1000 white beans and 1000 black beans in a bag, and mix them thoroughly. (It takes a lot of time to count out 1000 beans, so get several friends to count out 100 or so each.) Take a small cup and scoop up a small sample from the beans.

1. Record the proportion of black (or white) beans in the sample. How closely does the proportion in the sample compare with the actual proportion in the bag?

2. Take several samples and record the proportion of black (or white) beans in each sample.
(a) Graph your results.
(b) Average your results. How close is your average to .5?

3. Try the experiment again but with 500 black beans and 1500 white beans. What proportion of black (or white) beans would you expect in this sample?

4. Have some friends mix black and white beans without telling you the number of each. Estimate the proportion of black (or white) beans by taking a few small samples.

5. This bean experiment can be simulated on your graphing calculator. Let's designate the numbers in the interval $[0, .25]$ as "black beans" and those in $(.25, 1]$ as "white beans." Now use the random number generator in your calculator to randomly pick a sample of numbers between 0 and 1. Try samples of size 100, 400, 1600, and larger. Determine the proportion of these that are "black beans." What would you expect this proportion to be? Do your results improve when you use a larger sample? (The TI-85 program in the margin takes a sample of 100 random numbers.)

```
PROGRAM: SAMPLE
:100 → N
:0 → C
:For (J, 1, N)
:rand → B
:(B ≤ 0.25) + C → C
:End
:Disp "PROPORTION="
:Disp C/N
```

10.4 EXPECTED VALUE

In the game shown in Figure 1, you pay \$1 to spin the arrow. If the arrow stops in a red region, you get \$3 (the dollar you paid plus \$2); otherwise, you lose the dollar you paid. If you play this game many times, how much would you expect to win? Or lose? To answer these questions, let's consider the probabilities of winning and losing. Since three of the regions are red, the probability of winning is $\frac{3}{10} = .3$ and that of losing is $\frac{7}{10} = .7$. Remember, this means that if you play this game many times, you expect to win "on average" three out of ten times. So, suppose you play the game 1000 times. Then you would expect to win 300 times and lose 700 times. Since we win \$2 or lose \$1 in each game, our expected payoff in 1000 games is

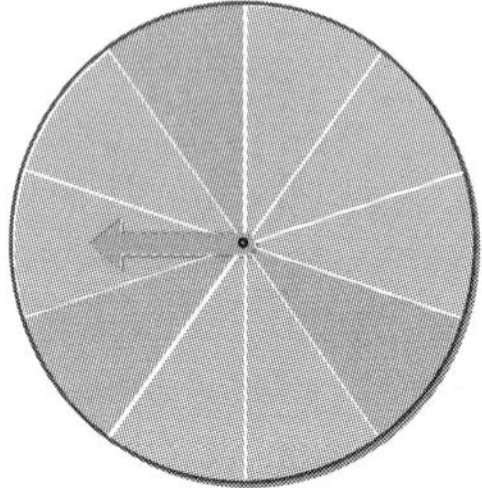

FIGURE 1

$$2(300) + (-1)(700) = -100$$

So the average expected return per game is $\frac{-100}{1000} = -0.1$. In other words, we expect to lose, on average, 10 cents per game. Another way to view this average is to divide each side of the preceding equation by 1000. Writing E for the result, we get

$$\begin{aligned} E &= \frac{2(300) + (-1)(700)}{1000} \\ &= 2\left(\frac{300}{1000}\right) + (-1)\frac{700}{1000} \\ &= 2(.3) + (-1)(.7) \end{aligned}$$

Thus, the expected return, or *expected value*, per game is

$$E = a_1 p_1 + a_2 p_2$$

where a_1 is the payoff that occurs with probability p_1 and a_2 is the payoff that occurs with probability p_2. This example leads us to the following definition of expected value.

DEFINITION OF EXPECTED VALUE

A game gives payoffs $a_1, a_2, \ldots, a_n$ with probabilities $p_1, p_2, \ldots, p_n$. The **expected value** (or **expectation**) E of this game is

$$E = a_1p_1 + a_2p_2 + \cdots + a_np_n$$

The expected value is an average expectation per game if the game is played many times. In general, E need not be one of the possible payoffs. In the preceding example the expected value is -10 cents, but it's impossible to lose exactly 10 cents in any given trial of the game.

EXAMPLE 1 ■ Finding an Expected Value

A die is rolled, and you receive \$1 for each point that shows. What is your expectation?

SOLUTION

Each face of the die has probability $\frac{1}{6}$ of showing. So you get \$1 with probability $\frac{1}{6}$, \$2 with probability $\frac{1}{6}$, \$3 with probability $\frac{1}{6}$, and so on. Thus, the expected value is

$$E = 1\left(\frac{1}{6}\right) + 2\left(\frac{1}{6}\right) + 3\left(\frac{1}{6}\right) + 4\left(\frac{1}{6}\right) + 5\left(\frac{1}{6}\right) + 6\left(\frac{1}{6}\right) = \frac{21}{6} = 3.5$$

This means that if you play this game many times, you will make, on average, \$3.50 per game. ■

EXAMPLE 2 ■ Finding an Expected Value

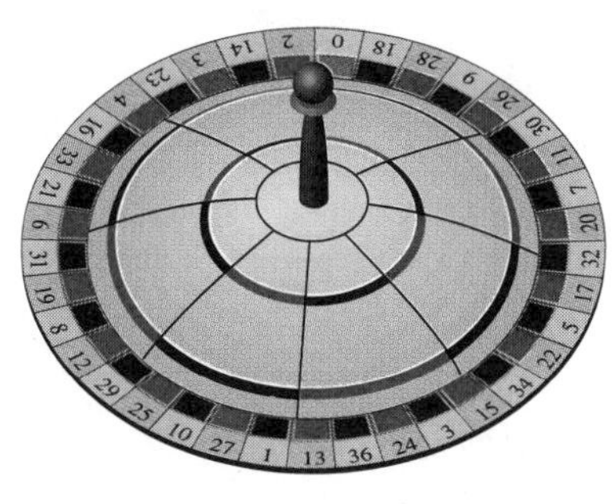

In Monte Carlo, the game of roulette is played on a wheel with slots numbered 0, 1, 2, . . . , 36. The wheel is spun, and a ball dropped in the wheel is equally likely to end up in any one of the slots. To play the game, you bet \$1 on any number other than zero. (For example, you may bet \$1 on number 23.) If the ball stops in your slot, you get \$36 (the \$1 you bet plus \$35). Find the expected value of this game.

SOLUTION

You gain \$35 with probability $\frac{1}{37}$, and you lose \$1 with probability $\frac{36}{37}$. Thus

$$E = (35)\,\frac{1}{37} + (-1)\,\frac{36}{37} \approx -0.027$$

In other words, if you play this game many times, you would expect to lose 2.7 cents on every dollar you bet (on average). Consequently, the house expects to gain 2.7 cents on every dollar that is bet. This expected value is what makes gambling very profitable for the gaming house and very unprofitable for the gambler. ■

10.4 EXERCISES

1–10 ■ Find the expected value (or expectation) of the games described.

1. Mike wins \$2 if a coin toss shows heads and \$1 if it shows tails.

2. Jane wins \$10 if a die roll shows a six, and she loses \$1 otherwise.

3. The game consists of drawing a card from a deck. You win \$100 if you draw the ace of spades or lose \$1 if you draw any other card.

4. Tim wins \$3 if a coin toss shows heads or \$2 if it shows tails.

5. Carol wins \$3 if a die roll shows a six, and she wins \$0.50 otherwise.

6. A coin is tossed twice. Albert wins \$2 for each heads and must pay \$1 for each tails.

7. A die is rolled. Tom wins \$2 if the die shows an even number and he pays \$2 otherwise.

8. A card is drawn from a deck. You win \$104 if the card is an ace, \$26 if it is a face card, and \$13 if it is the 8 of clubs.

9. A bag contains two silver dollars and eight slugs. You pay 50 cents to reach into the bag and take a coin, which you get to keep.

10. A bag contains eight white balls and two black balls. John picks two balls at random from the bag, and he wins \$5 if he does not pick a black ball.

11. In the game of roulette as played in Las Vegas, the wheel has 38 slots: Two slots are numbered 0 and 00, and the rest are numbered 1 to 36. A \$1 bet on any number other than 0 or 00 wins \$36 (\$35 plus the \$1 bet). Find the expected value of this game.

12. A sweepstakes offers a first prize of \$1,000,000, second prize of \$100,000, and third prize of \$10,000. Suppose that two million people enter the contest and three names are drawn randomly for the three prizes.
(a) Find the expected winnings for a person participating in this contest.
(b) Is it worth paying a dollar to enter this sweepstakes?

13. A box contains 100 envelopes. Ten envelopes contain \$10 each, ten contain \$5 each, two are "unlucky," and the rest are empty. A player draws an envelope from the box and keeps whatever is in it. If a person draws an unlucky envelope, however, he must pay \$100. What is the expectation of a person playing this game?

14. A safe containing \$1,000,000 is locked with a combination lock. You pay \$1 for one guess at the six-digit combination. If you open the lock, you get to keep the million dollars. What is your expectation?

15. An investor buys 1000 shares of a risky stock for \$5 a share. She estimates that the probability the stock will rise in value to \$20 a share is .1 and the probability that it will fall to \$1 a share is .9. If the only criterion for her decision to buy this stock was the expected value of her profit, did she make a wise investment?

16. A slot machine has three wheels, and each wheel has 11 positions—the digits 0, 1, 2, . . . , 9 and the picture of a watermelon. When a quarter is placed in the machine and the handle is pulled, the three wheels spin independently and come to rest. When three watermelons show, the payout is \$5; otherwise, nothing is paid. What is the expected value of this game?

17. In a 6/49 lottery game, a player pays \$1 and selects six numbers from 1 to 49. Any player who has chosen the six winning numbers wins \$1,000,000. Assuming this is the only way to win, what is the expected value of this game?

18. A bag contains two silver dollars and six slugs. A game consists of reaching into the bag and drawing a coin, which you get to keep. Determine the "fair price" of playing this game, that is, the price at which the player can be expected to break even if he plays the game many times (in other words, the price at which his expectation is zero).

19. A game consists of drawing a card from a deck. You win \$13 if you draw an ace. What is a "fair price" to pay to play this game? (See Exercise 18.)

DISCOVERY • DISCUSSION

20. **The Expected Value of a Sweepstakes Contest** A magazine clearinghouse holds a sweepstakes contest to sell subscriptions. If you return the winning number, you win \$1,000,000. You have a 1-in-20-million chance of winning, but your only cost to enter the contest is a first-class stamp to mail the entry. Use the current price of a first-class stamp to calculate your expected net winnings if you enter this contest. Is it worth entering the sweepstakes?

10 REVIEW

CONCEPT CHECK

1. What does the Fundamental Counting Principle say?

2. (a) What is a permutation of a set of distinct objects?
   (b) How many permutations are there of n objects?
   (c) How many permutations are there of n objects taken r at a time?
   (d) What is the number of distinguishable permutations of n objects if there are k different kinds of objects with n_1 objects of the first kind, n_2 objects of the second kind, and so on?

3. (a) What is a combination of r elements of a set?
   (b) How many combinations are there of n elements taken r at a time?
   (c) How many subsets does a set with n elements have?

4. In solving a problem involving picking r objects from n objects, how do you know whether to use permutations or combinations?

5. (a) What is meant by the sample space of an experiment?
   (b) What is an event?
   (c) Define the probability of an event E in a sample space S.
   (d) What is the probability of the complement of E?

6. (a) What are mutually exclusive events?
   (b) If E and F are mutually exclusive events, what is the probability of the union of E and F? What if E and F are not mutually exclusive?

7. (a) What are independent events?
   (b) If E and F are independent events, what is the probability of the intersection of E and F?

8. Suppose that a game gives payoffs $a_1, a_2, \ldots, a_n$ with probabilities $p_1, p_2, \ldots, p_n$. What is the expected value of the game? What is the significance of the expected value?

EXERCISES

1. A coin is tossed, a die is rolled, and a card is drawn from a deck. How many possible outcomes does this experiment have?

2. How many three-digit numbers can be formed using the digits 1, 2, 3, 4, 5, 6 if repetition of digits
   (a) is allowed?
   (b) is not allowed?

3. (a) How many different two-element subsets does the set $\{A, E, I, O, U\}$ have?
   (b) How many different two-letter "words" can be made using the letters from the set in part (a)?

4. An airline company overbooks a particular flight and seven passengers are "bumped" from the flight. If 120 passengers are booked on this flight, in how many ways can the airline choose the seven passengers to be bumped?

5. A quiz has ten true-false questions. How many different ways is it possible to earn a score of exactly 70% on this quiz?

6. A test has ten true-false questions and five multiple-choice questions with four choices for each. In how many ways can this test be completed?

7. If you must answer only eight of ten questions on a test, how many ways do you have of choosing the questions you will omit?

8. An ice-cream store offers 15 flavors of ice cream. The specialty is a banana split with four scoops of ice cream. If each scoop must be a different flavor, how many different banana splits may be ordered?

9. A company uses a different three-letter security code for each of its employees. What is the maximum number of codes this security system can generate?

10. A group of students determines that they can stand in a row for their class picture in 120 different ways. How many students are in this class?

11. A coin is tossed ten times. In how many different ways can the result be three heads and seven tails?

12. The Yukon Territory in Canada uses a license-plate system for automobiles that consists of two letters followed by three numbers. Explain how we can know that fewer than 700,000 autos are licensed in the Yukon.

13. A group of friends have reserved a tennis court. They find that there are ten different ways in which two of them can play a singles game on this court. How many friends are in this group?

14. A pizza parlor advertises that they prepare 2048 different types of pizza. How many toppings does this parlor offer?

15. In Morse code, each letter is represented by a sequence of dots and dashes, with repetition allowed. How many letters can be represented using Morse code if three or fewer symbols are used?

16. The genetic code is based on the four nucleotides adenine (A), cytosine (C), guanine (G), and thymine (T). These are connected in long strings to form DNA molecules. For example, a sequence in the DNA may look like CAGTGGTACC . . . . The code uses "words," all the same length, that are composed of the nucleotides A, C, G, and T. It is known that at least 20 different words exist. What is the minimum word length necessary to generate 20 words?

17. Given 16 subjects from which to choose, in how many ways can a student select fields of study as follows?
(a) A major and a minor
(b) A major, a first minor, and a second minor
(c) A major and two minors

18. (a) How many three-digit numbers can be formed using the digits 0, 1, . . . , 9? (Remember, a three-digit number cannot have 0 as the leftmost digit.)
(b) If a number is chosen randomly from the set $\{0, 1, 2, \ldots, 1000\}$, what is the probability that the number chosen is a three-digit number?

19–20 ■ An **anagram** of a word is a permutation of the letters of that word. For example, anagrams of the word *triangle* include *griantle*, *integral*, and *tenalgir*.

19. How many anagrams of the word *TRIANGLE* are possible?

20. How many anagrams are possible from the word *MISSISSIPPI*?

21. A shelf has ten books: two mysteries, four romance novels, and four mathematics textbooks. If you select a book at random to take to the beach, what is the probability that it turns out to be a mathematics text?

22. A jar contains ten red balls labeled 0, 1, 2, . . . , 9 and five white balls labeled 0, 1, 2, 3, 4. If a ball is drawn from the jar, find the probability of the given event.
(a) The ball is red.
(b) The ball is even-numbered.
(c) The ball is white and odd-numbered.
(d) The ball is red or odd-numbered.

23. A coin is tossed three times in a row, and the outcomes of each toss are observed.
(a) Find the sample space for this experiment.
(b) Find the probability of getting three heads.
(c) Find the probability of getting two or more heads.
(d) Find the probability of getting tails on the first toss.

24. A die is rolled and a card is selected from a standard 52-card deck. What is the probability that both the die and the card show a six?

25. Find the probability that the indicated card is drawn at random from a 52-card deck.
(a) An ace
(b) An ace or a jack
(c) An ace or a spade
(d) A red ace

26. A card is drawn from a 52-card deck, a die is rolled, and a coin is tossed. Find the probability of each outcome.
(a) The ace of spades, a six, and heads
(b) A spade, a six, and heads
(c) A face card, a number greater than 3, and heads

27. Two dice are rolled. Find the probability of each outcome.
(a) The dice show the same number.
(b) The dice show different numbers.

28. Four cards are dealt from a standard 52-card deck. Find the probability that the cards are
(a) all kings (b) all spades (c) all the same color

29. In the "numbers game" lottery, a player picks a three-digit number (from 000 to 999), and if the number is selected in the drawing, the player wins \$500. If another number with the same digits (in any order) is drawn, the player wins \$50. John plays the number 159.
(a) What is the probability that he will win \$500?
(b) What is the probability that he will win \$50?

30. In a television game show, a contestant is given five cards with a different digit on each and is asked to arrange them to match the price of a brand-new car. If she gets the price right, she wins the car. What is the probability that she wins, assuming that she knows the first digit but must guess the remaining four?

31. Two dice are rolled. John gets \$5 if they show the same number, or he pays \$1 if they show different numbers. What is the expected value of this game?

32. Three dice are rolled. John gets \$5 if they all show the same number; he pays \$1 otherwise. What is the expected value of this game?

33. Mary will win \$1,000,000 if she can name the 13 original states in the order in which they ratified the U.S. Constitution. Mary has no knowledge of this order, so she makes a guess. What is her expectation?

34. A pizza parlor offers 12 different toppings, one of which is anchovies. If a pizza is ordered at random, what is the probability that anchovies is one of the toppings?

35. A drawer contains an unorganized collection of 50 socks—20 are red and 30 are blue. Suppose the lights go out so Kathy can't distinguish the color of the socks.
(a) What is the minimum number of socks Kathy must take out of the drawer to be sure of getting a matching pair?
(b) If two socks are taken at random from the drawer, what is the probability that they make a matching pair?

36. A volleyball team has nine players. In how many ways can a starting lineup be chosen if it consists of two forward players and three defense players?

37. Zip codes consist of five digits.
(a) How many different zip codes are possible?
(b) How many different zip codes can be read when the envelope is turned upside down? (An upside down 9 is a 6; and 0, 1, and 8 are the same when read upside down.)
(c) What is the probability that a randomly chosen zip code can be read upside down?
(d) How many zip codes read the same upside down as right side up?

38. In the Zip+4 postal code system, zip codes consist of nine digits.
(a) How many different Zip+4 codes are possible?
(b) How many different Zip+4 codes are palindromes? (A *palindrome* is a number that reads the same from left to right as right to left.)
(c) What is the probability that a randomly chosen Zip+4 code is a palindrome?

39. Let $N = 3{,}600{,}000$. (Note that $N = 2^7 3^2 5^5$.)
(a) How many divisors does N have?
(b) How many even divisors does N have?
(c) How many divisors of N are multiples of 6?
(d) What is the probability that a randomly chosen divisor of N is even?

40. The U.S. Senate has two senators from each of the 50 states. In how many ways can a committee of five senators be chosen if no state is to have two members on the committee?

10 TEST

1. How many five-letter "words" can be made from the letters $A, B, C, D, E, F, G, H, I, J$ if repetition (a) is allowed or (b) is not allowed?

2. A restaurant offers five main courses, three types of desserts, and four kinds of drinks. In how many ways can a customer order a meal consisting of one choice from each category?

3. A board of directors consisting of eight members is to be chosen from a pool of 30 candidates. The board is to have a chairman, a treasurer, a secretary, and five other members. In how many ways can the board of directors be chosen?

4. A commuter must travel from Ajax to Barrie and back every day. Four roads join the two cities. The commuter likes to vary the trip as much as possible, so he always leaves and returns by different roads. In how many different ways can he make the round-trip?

5. A pizza parlor offers four sizes of pizza and 14 different toppings. A customer may choose any number of toppings (or no topping at all). How many different pizzas does this parlor offer?

6. An *anagram* of a word is a rearrangement of the letters of the word.

(a) How many anagrams of the word *LOVE* are possible?

(b) How many different anagrams of the word *KISSES* are possible?

7. Three people are chosen at random from a group of five men and ten women. What is the probability that all three are men?

8. Two dice are rolled. What is the probability of getting doubles?

9. One card is drawn from a deck. Find the probability of the given event.

(a) The card is red.

(b) The card is a king.

(c) The card is a red king.

10. A jar contains five red balls, numbered 1 to 5, and eight white balls, numbered 1 to 8. A ball is chosen at random from the jar. Find the probability of the given event.

(a) The ball is red.

(b) The ball is even-numbered.

(c) The ball is red or even-numbered.

11. You are to draw one card from a deck. If it is an ace, you win \$10; if it is a face card, you win \$1; otherwise, you lose \$0.50. What is the expected value of this game?

12. In a group of four students, what is the probability that at least two have the same astrological sign?

FOCUS ON MODELING

A good way to familiarize ourselves with a fact is to experiment with it. For instance, to convince ourselves that the earth is a sphere (which was considered a major paradox at one time), we could go up in a space shuttle to see that it is so; to see whether a given equation is an identity, we might try some special cases to make sure there are no obvious counterexamples. In problems involving probability, we can perform an experiment many times and use the results to estimate the probability in question. In fact, we often model the experiment on a computer, thereby making it feasible to perform the experiment a large number of times. This technique is called the **Monte Carlo method**, named after the famous gambling casino in Monaco.

EXAMPLE 1 ■ The Contestant's Dilemma

In a TV game show, a contestant chooses one of three doors. Behind one of them is a valuable prize—the other two doors have nothing behind them. After the contestant has made her choice, the host opens one of the other two doors, one that he knows does not conceal a prize, and then gives her the opportunity to change her choice.

Should the contestant switch, stay, or does it matter? In other words, by switching doors, does she increase, decrease, or leave unchanged her probability of winning? At first, it may seem that switching doors doesn't make any difference. After all, two doors are left—one with the prize and one without—so it seems reasonable that the contestant has an equal chance of winning or losing. But if you play this game many times, you will find that by switching doors you actually win about $\frac{2}{3}$ of the time.

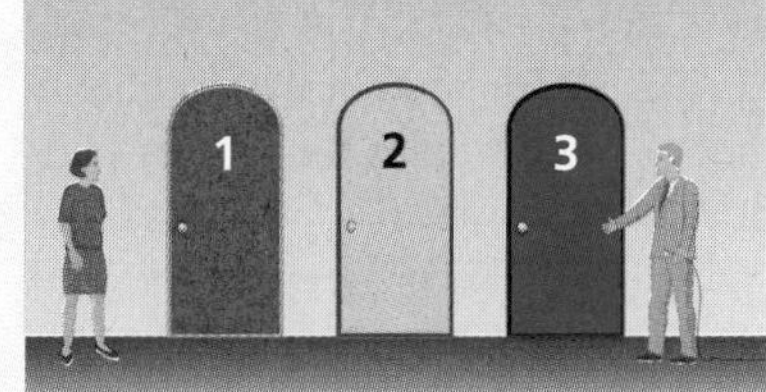

Contestant: "I choose door number 2."

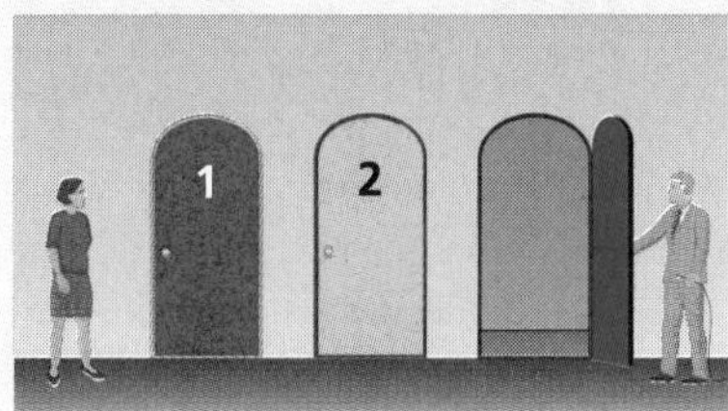

Contestant: "Oh no, what should I do?"

The authors modeled this game on a computer and found that in one million games the simulated contestant (who always switches) won 667,049 times—very close to $\frac{2}{3}$ of the time. Thus, it seems that switching doors does make a difference: Switching increases the contestant's chances of winning. This experiment forces us to reexamine our reasoning. Here is why switching doors is the correct strategy:

1. When the contestant first made her choice, she had a $\frac{1}{3}$ chance of winning. If she doesn't switch, no matter what the host does, her probability of winning remains $\frac{1}{3}$.

2. If the contestant decides to switch, she will switch to the winning door if she had initially chosen a losing one, or to a losing door if she had initially chosen the winning one. Since the probability of having initially selected a losing door is $\frac{2}{3}$, by switching the probability of winning then becomes $\frac{2}{3}$.

We conclude that the contestant should switch, because her probability of winning is $\frac{2}{3}$ if she switches and $\frac{1}{3}$ if she doesn't. Put simply, there is a much

The "contestant's dilemma" problem discussed here is an example of how subtle probability can be. This problem was posed in a nationally syndicated column in *Parade* magazine in 1990. The correct solution was presented in the column, but it generated considerable controversy, with thousands of letters arguing that the solution was wrong. This shows how problems in probability can be quite tricky. Without a lot of experience in probabilistic thinking, it's easy to make a mistake. Even great mathematicians like D'Alembert and Roberval (see Problems 2 and 3) made mistakes in probability. Professor David Burton writes in his book *The History of Mathematics*, "Probability theory abounds in paradoxes that wrench the common sense and trip the unwary."

greater chance that she initially chose a losing door (since there are more of these), so she should switch. ■

An experiment can be modeled using any computer language or programmable calculator that has random-number generator. This is a command or function (usually called RND or RAND) that returns a randomly chosen number x with $0 \leq x < 1$. In the next example we see how to use this to model a simple experiment.

EXAMPLE 2 ■ Monte Carlo Model of a Coin Toss

When a balanced coin is tossed, each outcome—"heads" or "tails"—has probability $\frac{1}{2}$. This doesn't mean that if we toss a coin several times, we will necessarily get exactly half heads and half tails. We would expect, however, the proportion of heads and of tails to get closer and closer to $\frac{1}{2}$ as the number of tosses increases. To test this hypothesis, we could toss a coin a very large number of times and keep track of the results. But this is a very tedious process, so we will use the Monte Carlo method to model this process.

To model a coin toss with a calculator or computer, we use the random-number generator to get a random number x such that $0 \leq x < 1$. Because the number is chosen randomly, the probability that it lies in the first half of this interval $(0 \leq x < \frac{1}{2})$ is the same as the probability that it lies in the second half $(\frac{1}{2} \leq x < 1)$. Thus, we could model the outcome "heads" by the event that $0 \leq x < \frac{1}{2}$ and the outcome "tails" by the event that $\frac{1}{2} \leq x < 1$.

An easier way to keep track of heads and tails is to note that if $0 \leq x < 1$, then $0 \leq 2x < 2$, and so $[\![2x]\!]$, the integer part of $2x$, is either 0 or 1, each with probability $\frac{1}{2}$. (On most programmable calculators, the function INT gives the integer part of a number.) Thus, we could model "heads" with the outcome "0" and "tails" with the outcome "1" when we take the integer part of $2x$. The program in the margin models 100 tosses of a coin on the TI-82 calculator. The graph in Figure 1 shows what proportion p of the tosses have come up "heads" after n tosses. As you can see, this proportion settles down near .5 as the number n of tosses increases—just as we hypothesized. ■

```
PROGRAM: HEADTAIL
:0→J:0→K
:For(N, 1, 100)
:rand→X
:int(2X)→Y
:J+(1−Y)→J
:K+Y→K
:END
:Disp "HEADS=",J
:Disp "TAILS=",K
```

In general, if a process has n equally likely outcomes, then we can model the process using a random-number generator as follows: If our program or calculator produces the random number x, with $0 \leq x < 1$, then the integer part of nx will be a random choice from the n integers $0, 1, 2, \ldots, n - 1$. Thus, we can use the outcomes $0, 1, 2, \ldots, n - 1$ as models for the outcomes of the actual experiment.

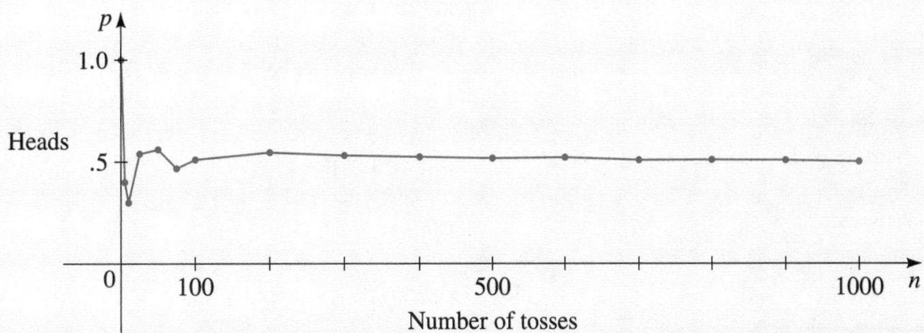

FIGURE 1
Relative frequency of "heads"

PROBLEMS

1. In a game show like the one described in Example 1, a prize is concealed behind one of ten doors. After the contestant chooses a door, the host opens eight losing doors and then gives the contestant the opportunity to switch to the other unopened door.
   (a) Play this tame with a friend 30 or more times, using the strategy of switching doors each time. Count the number of times you win, and estimate the probability of winning with this strategy.
   (b) Calculate the probability of winning with the "switching" strategy using reasoning similar to that in Example 1. Compare with your result from part (a).

2. A couple intend to have two children. What is the probability that they will have one child of each sex? The French mathematician D'Alembert analyzed this problem (incorrectly) by reasoning that three outcomes are possible: two boys, or two girls, or one child of each sex. He concluded that the probability of having one of each sex is $\frac{1}{3}$, mistakenly assuming that the three outcomes are "equally likely."
   (a) Model this problem with a pair of coins (using "heads" for boys and "tails" for girls), or write a program to model the problem. Perform the experiment 40 or more times, counting the number of boy-girl combinations. Estimate the probability of having one child of each sex.
   (b) Calculate the correct probability of having one child of each sex, and compare this with your result from part (a).

3. A game between two players consists of tossing a coin. Player A gets a point if the coin shows heads, and player B gets a point if it shows tails. The first player to get six points wins an \$8000 jackpot. As it happens, the police raid the place when player A has five points and B has three points. After everyone has calmed down, how should the jackpot be divided between the two players? In other words, what is the probability of A winning (and that of B winning) if the game were to continue?

   The French mathematicians Pascal and Fermat corresponded about this problem, and both came to the same correct conclusion (though by very different reasonings). Their friend Roberval disagreed with both of them. He argued that player A has probability $\frac{3}{4}$ of winning, because the game can end in the four ways H, TH, TTH, TTT, and in three of these, A wins. Roberval's reasoning was wrong.

(a) Continue the game from the point at which it was interrupted, using either a coin or a modeling program. Perform this experiment 80 or more times, and estimate the probability that player A wins.

(b) Calculate the probability that player A wins. Compare with your estimate from part (a).

4. In the World Series, the top teams in the National League and the American League play a best-of-seven series; that is, they play until one team has won four games. (No tie is allowed, so this results in a maximum of seven games.) Suppose the teams are evenly matched, so that the probability that either team wins a given game is $\frac{1}{2}$.

(a) Use a coin or a modeling program to model a World Series, where "heads" represents a win by Team A and "tails" a win by Team B. Perform this experiment at least 80 times, keeping track of how many games are needed to decide each series. Estimate the probability that an evenly matched series will end in four games. Do the same for five, six, and seven games.

(b) What is the probability that the series will end in four games? Five games? Six games? Seven games? Compare with your estimates from part (a).

(c) Find the expected value for the number of games until the series ends. [*Hint:* This will be $P(\text{four games}) \times 4 + P(\text{five}) \times 5 + P(\text{six}) \times 6 + P(\text{seven}) \times 7$.]

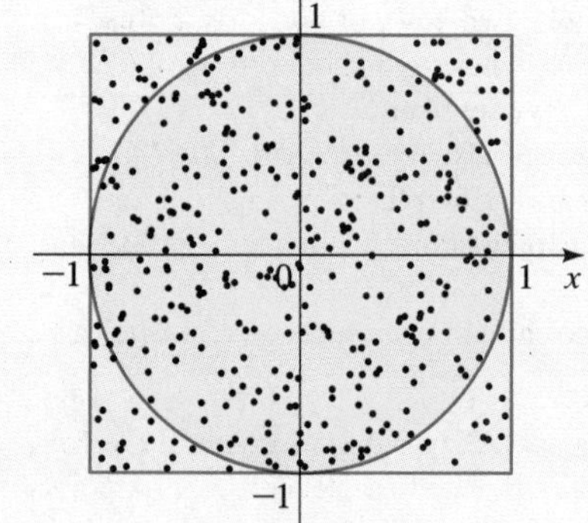

5. In this problem we use the Monte Carlo method to estimate the value of π. The circle in the figure has radius 1, so its area is π and the square has area 4. If we choose a point at random from the square, the probability that it lies inside the circle will be

$$\frac{\text{area of circle}}{\text{area of square}} = \frac{\pi}{4}$$

The Monte Carlo method involves choosing many points inside the square. Then we have

$$\frac{\text{number of hits inside circle}}{\text{number of hits inside square}} \approx \frac{\pi}{4}$$

Thus, 4 times this ratio will give us an approximation for π.

```
PROGRAM: PI
:0→P
:For(N, 1, 1000)
:rand→X:rand→Y
:P+((X²+Y²)<1)→P
:End
:Disp "PI IS APPROX",4*P/N
```

To implement this method, we use a random-number generator to obtain the coordinates (x, y) of a random point in the square, and then check to see if it lies inside the circle (that is, we check if $x^2 + y^2 < 1$). Note that we only need to use points in the first quadrant, since the ratio of areas is the same in each quadrant. The program in the margin shows a way of doing this on the TI-82 calculator for 1000 randomly selected points.

Carry out this Monte Carlo simulation for as many points as you can. How do your results compare with the actual value of π? Do you think this is a reasonable way to get a good approximation for π?

6. Choose two numbers at random from the interval $[0, 1)$. What is the probability that the sum of the two numbers is less than 1?

(a) Use a Monte Carlo model to estimate the probability.

(b) Calculate the exact value of the probability. [*Hint:* Call the numbers x and y. Choosing these numbers is the same as choosing an ordered pair (x, y) in the unit square $\{(x, y) \mid 0 \leq x < 1, 0 \leq y < 1\}$. What proportion of the points in this square corresponds to $x + y$ being less than 1?]

ANSWERS TO ODD-NUMBERED EXERCISES AND CHAPTER TESTS

CHAPTER 1

Section 1.1 ■ page 7

1. $ab = ba$ **3.** $(ab)c = a(bc)$ **5.** $(a - b)(a + b) = a^2 - b^2$
7. $(ab)^2 = a^2b^2$ **9.** $A = (a + b)/2$ **11.** $S = x + 2x^2$
13. $d = 7w$ **15.** $P = n(n + 1)$ **17.** $S = x^2 + y^2$
19. $t = d/r$ **21.** $A = x^2$ **23.** $V = 2x^3$
25. $S = 2lw + 2lh + 2wh$ **27.** $L = 2x + 2\pi r$ **29.** $A = h^2$
31. (a) 84.5 (b) $A = (a + b + 2f)/4$ (c) 89
33. (a) 4500 ft^2 (b) $A = 150T\ \text{ft}^2$ (c) 64 min
(d) 160 ft^2/min
35. (a) $\text{GPA} = (4a + 3b + 2c + d)/(a + b + c + d + f)$
(b) 2.84

Section 1.2 ■ page 18

1. Commutative Property for addition
3. Associative Property for addition **5.** Distributive Property
7. $3x + 3y$ **9.** $8m$ **11.** $-5x + 10y$
13. (a) $\frac{7}{13}$ (b) $\frac{23}{30}$ **15.** (a) $\frac{4}{9}$ (b) $\frac{15}{2}$
17. (a) False (b) True **19.** (a) False (b) True
21. (a) $x > 0$ (b) $t < 4$ (c) $a \geqslant \pi$ (d) $-5 < x < \frac{1}{3}$
(e) $|p - 3| \leqslant 5$
23. (a) $\{1, 2, 3, 4, 5, 6, 7, 8\}$ (b) $\{2, 4, 6\}$
25. (a) $\{1, 2, 3, 4, 5, 6, 7, 8, 9, 10\}$ (b) $\varnothing$
27. (a) $\{x \mid x \leqslant 5\}$ (b) $\{x \mid -1 < x < 4\}$
29. $-3 < x < 0$
31. $2 \leqslant x < 8$
33. $x \geqslant 2$
35. $(-\infty, 1]$
37. $(-2, 1]$ 
39. $(-1, \infty)$
41. 
43.
45. 
47. (a) 100 (b) 73 **49.** (a) 2 (b) -1
51. (a) 12 (b) 5 **53.** (a) 15 (b) 24 (c) $\frac{67}{40}$
55. (a) $\frac{7}{9}$ (b) $\frac{13}{45}$ (c) $\frac{19}{33}$ **57.** (a) Yes (b) 170

Section 1.3 ■ page 30

1. (a) 16 (b) -16 (c) 1
3. (a) $\frac{16}{25}$ (b) 1000 (c) 1024
5. (a) $\frac{2}{3}$ (b) 4 (c) $\frac{1}{2}$ **7.** (a) 6 (b) 4 (c) $\frac{3}{5}$
9. (a) $\frac{3}{2}$ (b) $\frac{9}{4}$ (c) $\frac{125}{512}$ **11.** $\sqrt[3]{4}$ **13.** $2\sqrt{5}$
15. a^4 **17.** $6x^7y^5$ **19.** $16x^{10}$ **21.** $4/b^2$
23. $64r^7s$ **25.** $648y^7$ **27.** $\dfrac{x^3}{y}$ **29.** $\dfrac{y^2z^9}{x^5}$
31. $\dfrac{s^3}{q^7r^6}$ **33.** $x^{13/15}$ **35.** $16b^{9/10}$ **37.** $\dfrac{1}{c^{2/3}d}$
39. $y^{1/2}$ **41.** $\dfrac{32x^{12}}{y^{16/15}}$ **43.** $\dfrac{x^{15}}{y^{15/2}}$ **45.** $\dfrac{4a^2}{3b^{1/3}}$
47. $\dfrac{3t^{25/6}}{s^{1/2}}$ **49.** $|x|$ **51.** $x\sqrt[3]{y}$ **53.** $ab\sqrt[5]{ab^2}$
55. $2|x|$ **57.** (a) $\dfrac{\sqrt{6}}{6}$ (b) $\dfrac{\sqrt{3xy}}{3y}$ (c) $\dfrac{\sqrt{15}}{10}$

59. (a) $\frac{\sqrt[3]{x^2}}{x}$ (b) $\frac{\sqrt[5]{x^3}}{x}$ (c) $\frac{\sqrt[7]{x^4}}{x}$

61. (a) 6.93×10^7 (b) 2.8536×10^{-5} (c) 1.2954×10^8

63. (a) 319,000 (b) 0.00000002670 (c) 710,000,000,000,000

65. (a) 5.9×10^{12} mi (b) 4×10^{-13} cm (c) 3.3×10^{19} molecules

67. 1.3×10^{-20} **69.** 1.429×10^{19} **71.** 7.4×10^{-14}

73. $8\frac{1}{3}$ min **75.** 41.3 mi

Section 1.4 ■ page 41

1. $5x^2 - 2x - 4$ **3.** $x^3 + 3x^2 - 6x + 11$
5. $9x + 103$ **7.** $-t^4 + t^3 - t^2 - 10t + 5$
9. $x^{3/2} - x$ **11.** $y^{7/3} - y^{1/3}$
13. $21t^2 - 29t + 10$ **15.** $3x^2 + 5xy - 2y^2$
17. $1 - 4y + 4y^2$ **19.** $2x^3 - 7x^2 + 7x - 5$
21. $x^3 + x^2 - 2x$ **23.** $4x^4 + 12x^2y^2 + 9y^4$
25. $x^4 - a^4$ **27.** $1 + 3a^3 + 3a^6 + a^9$ **29.** $a - 1/b^2$
31. $x^5 + x^4 - 3x^3 + 3x - 2$ **33.** $1 - x^{2/3} + x^{4/3} - x^2$
35. $1 - 2b^2 + b^4$ **37.** $3x^4y^4 + 7x^3y^5 - 6x^2y^3 - 14xy^4$
39. $2x(1 + 6x^2)$ **41.** $3y^3(2y - 5)$ **43.** $(x + 6)(x + 1)$
45. $(x - 4)(x + 2)$ **47.** $(y - 3)(y - 5)$
49. $(2x + 3)(x + 1)$ **51.** $9(x - 2)(x + 2)$
53. $(3x + 2)(2x - 3)$ **55.** $3(x - 1)(x + 2)$
57. $y^4(y + 2)^3(y + 1)^2$ **59.** $(a + 1)(a - 1)(b + 2)(b - 2)$
61. $(t + 1)(t^2 - t + 1)$ **63.** $(2t - 3)^2$ **65.** $x(x + 1)^2$
67. $(2x + y)^2$ **69.** $x^2(x + 3)(x - 1)$
71. $(2x - 5)(4x^2 + 10x + 25)$ **73.** $(x^2 + 2)(x + 1)(x - 1)$
75. $(y + 2)(y - 2)(y - 3)$ **77.** $(2x^2 + 1)(x + 2)$
79. $4(2 + x^2)$ **81.** $x^{1/2}(x + 1)(x - 1)$ **83.** $x^{-3/2}(x + 1)^2$
85. $(x^2 + 3)(x^2 + 1)^{-1/2}$ **87.** $(a + 2)(a - 2)(a + 1)(a - 1)$
89. $16x^2(x - 3)(5x - 9)$ **91.** $(2x - 1)^2(x + 3)^{-1/2}\left(7x + \frac{35}{2}\right)$

Section 1.5 ■ page 50

1. $\frac{1}{x + 2}$ **3.** $\frac{x + 2}{x + 1}$ **5.** $\frac{y}{y - 1}$ **7.** $\frac{x(2x + 3)}{2x - 3}$

9. $\frac{1}{t^2 + 9}$ **11.** $\frac{x + 4}{x + 1}$ **13.** $\frac{(2x + 1)(2x - 1)}{(x + 5)^2}$

15. $x^2(x + 1)$ **17.** $\frac{x}{yz}$ **19.** $\frac{3x + 7}{(x - 3)(x + 5)}$

21. $\frac{1}{(x + 1)(x + 2)}$ **23.** $\frac{3x + 2}{(x + 1)^2}$ **25.** $\frac{u^2 + 3u + 1}{u + 1}$

27. $\frac{2x + 1}{x^2(x + 1)}$ **29.** $\frac{2x + 7}{(x + 3)(x + 4)}$ **31.** $\frac{x - 2}{(x + 3)(x - 3)}$

33. $\frac{5x - 6}{x(x - 1)}$ **35.** $\frac{-5}{(x + 1)(x + 2)(x - 3)}$ **37.** $-xy$

39. $\frac{c}{c - 2}$ **41.** $\frac{3x + 7}{x^2 + 2x - 1}$ **43.** $\frac{y - x}{xy}$ **45.** 1

47. $\frac{-1}{a(a + h)}$ **49.** $\frac{-3}{(2 + x)(2 + x + h)}$ **51.** $\frac{1}{\sqrt{1 - x^2}}$

53. $\frac{(x + 2)^2(x - 13)}{(x - 3)^3}$ **55.** $\frac{x + 2}{(x + 1)^{3/2}}$

57. $\frac{2x + 3}{(x + 1)^{4/3}}$ **59.** $\frac{3 - \sqrt{5}}{2}$ **61.** $\frac{2(\sqrt{7} - \sqrt{2})}{5}$

63. $\frac{-4}{3(1 + \sqrt{5})}$ **65.** $\frac{r - 2}{5(\sqrt{r} - \sqrt{2})}$ **67.** $\frac{1}{\sqrt{x^2 + 1} + x}$

69. True **71.** False **73.** False **75.** True **77.** False

79. (a) $\frac{R_1R_2}{R_1 + R_2}$ (b) $\frac{20}{3} \approx 6.7$ ohms

Section 1.6 ■ page 60

1. (a) Yes (b) No **3.** (a) No (b) Yes
5. (a) No (b) Yes **7.** 4 **9.** -9 **11.** -3 **13.** 12
15. $-\frac{3}{4}$ **17.** $\frac{32}{9}$ **19.** $-\frac{1}{3}$ **21.** -20 **23.** 3 **25.** $-\frac{1}{2}$
27. $-\frac{4}{9}$ **29.** $\frac{13}{3}$ **31.** $\frac{29}{2}$ **33.** 30 **35.** $\frac{3}{97}$ **37.** 2
39. No solution **41.** No solution **43.** ± 7 **45.** $\pm 2\sqrt{6}$
47. $\pm 2\sqrt{2}$ **49.** No solution **51.** $-4, 0$ **53.** 3
55. ± 2 **57.** No solution **59.** $-5, 1$ **61.** 8 **63.** 125
65. -8 **67.** $x \approx 5.06$ **69.** $x \approx 43.66$ **71.** $R = \frac{PV}{nT}$

73. $R_1 = \frac{RR_2}{R_2 - R}$ **75.** $x = \frac{2d - b}{a - 2c}$ **77.** $x = \frac{1 - a}{a^2 - a - 1}$

79. $r = \pm\sqrt{\frac{3V}{\pi h}}$ **81.** $b = \pm\sqrt{c^2 - a^2}$ **83.** $r = \sqrt[3]{\frac{3V}{4\pi}}$

85. (a) Yes (b) 3292 ft^2

Chapter 1 Review ■ page 63

1. Commutative Property for addition
3. Distributive Property
5. $-1 < x \leq 3$ [number line: open circle at -1, closed dot at 3]
7. $(2, \infty)$ [number line: open circle at 2, arrow to the right]

9. 6 **11.** $\frac{1}{72}$ **13.** $\frac{1}{6}$ **15.** 11 **17.** 4 **19.** x^{-2}
21. x^{4m+2} **23.** x^{a+b+c} **25.** x^{5c-1} **27.** $12x^5y^4$
29. $9x^3$ **31.** x^2y^2 **33.** $\frac{x(2 - \sqrt{x})}{4 - x}$ **35.** $\frac{4r^{5/2}}{s^7}$
37. 7.825×10^{10} **39.** 1.65×10^{-32}
41. $3xy^2(4xy^2 - y^3 + 3x^2)$ **43.** $(x - 2)(x + 5)$
45. $(4t + 3)(t - 4)$ **47.** $(5 - 4t)(5 + 4t)$
49. $(x - 1)(x^2 + x + 1)(x + 1)(x^2 - x + 1)$

51. $x^{-1/2}(x-1)^2$ **53.** $(x-2)(4x^2+3)$
55. $\sqrt{x^2+2}\,(x^2+x+2)^2$ **57.** $y(a+b)(a-b)$ **59.** x^2
61. $6x^2-21x+3$ **63.** $4a^4-4a^2b+b^2$
65. $x^3-6x^2+11x-6$ **67.** $2x^{3/2}+x-x^{1/2}$
69. $2x^3-6x^2+4x$ **71.** $\dfrac{x-3}{2x+3}$ **73.** $\dfrac{3(x+3)}{x+4}$
75. $\dfrac{x+1}{x-4}$ **77.** $\dfrac{x+1}{(x-1)(x^2+1)}$ **79.** $\dfrac{1}{x+1}$ **81.** $-\dfrac{1}{2x}$
83. $6x+3h-5$ **85.** 4 **87.** 5 **89.** $\frac{15}{2}$ **91.** -6
93. 0 **95.** ± 12 **97.** $\sqrt[3]{3}$ **99.** -5 **101.** -27 **103.** 625
105. No **107.** Yes **109.** No **111.** No

Chapter 1 Test ■ page 66

1. (a) $H=24n$ (b) $P=n^3-n$
2. (a)

(b) $(-\infty,5)$; $[-2,1]$ (c) 53
3. (a) 81 (b) $\frac{1}{16}$ (c) $5^6=15{,}625$
4. (a) $\frac{3}{2}$ (b) 2 (c) $\frac{1}{8}$ **5.** $x^{b/2}$
6. (a) $8\sqrt{2}$ (b) $54a^6b^{14}$ (c) $\dfrac{y^{32}}{x^8}$ (d) $\dfrac{8x^{1/4}}{y}$
7. (a) $\dfrac{x+2}{x-2}$ (b) $\dfrac{1}{x-2}$ (c) $-x-y$
8. (a) 3.25×10^{11} (b) 8.931×10^{-6}
9. (a) $-3-7x$ (b) $2x^2-7x-15$ (c) $x-y$
(d) $9t^2+24t+16$ (e) $8-12x^2+6x^4-x^6$
10. (a) $(3x-5)(3x+5)$ (b) $(3x+5)(2x-1)$
(c) $(x^2-3)(x-4)$ (d) $x(x+3)(x^2-3x+9)$
(e) $3x^{-1/2}(x-1)(x-2)$ (f) $xy(x+2)(x-2)$
11. $5\sqrt{2}+2\sqrt{10}$
12. (a) -10 (b) 1 (c) $2\pm\dfrac{4\sqrt{3}}{3}$ (d) -3 (e) $8,-8$
13. $w=\dfrac{V}{2h(2h+1)}$

PRINCIPLES OF PROBLEM SOLVING ■ page 71

1. 37.5 mi/h **3.** 150 mi **5.** 2 **7.** 57 min
9. No, not necessarily **11.** The same amount
13. (a) Uncle (b) Father (d) No **15.** 8.49 **17.** 427
19. 40 **21.** $\frac{11}{2}$

CHAPTER 2

Section 2.1 ■ page 81

1.

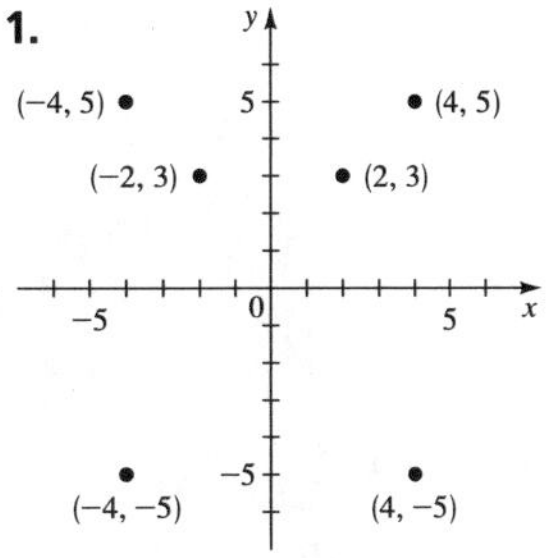

3. (a)

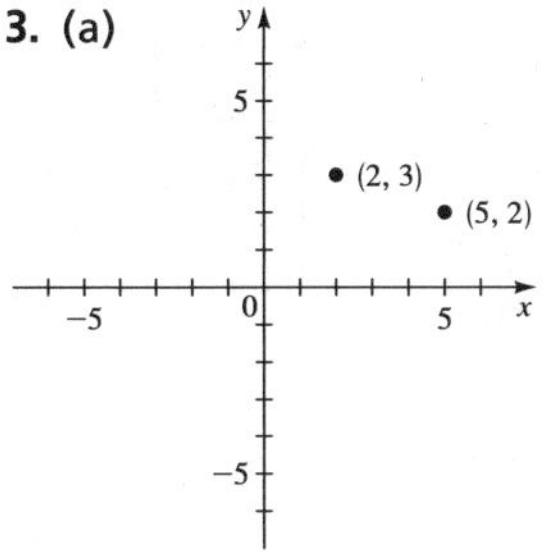

(b) $\sqrt{10}$
(c) $\left(\frac{7}{2},\frac{5}{2}\right)$

5. (a)

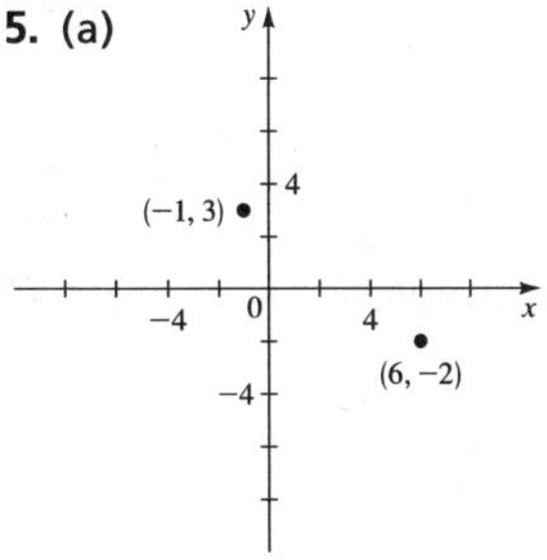

(b) $\sqrt{74}$
(c) $\left(\frac{5}{2},\frac{1}{2}\right)$

7. (a)

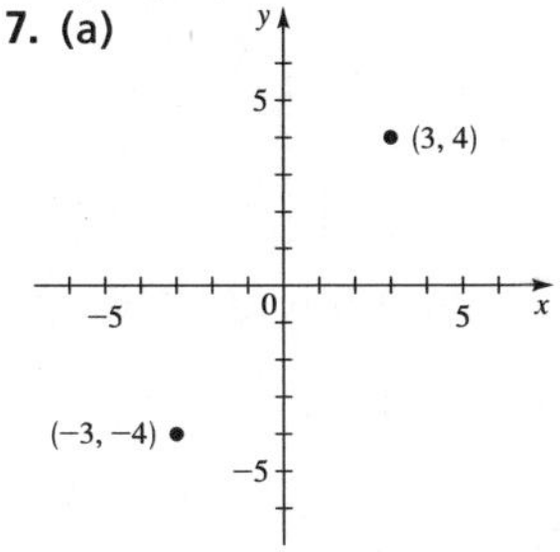

(b) 10
(c) $(0,0)$

9. 24

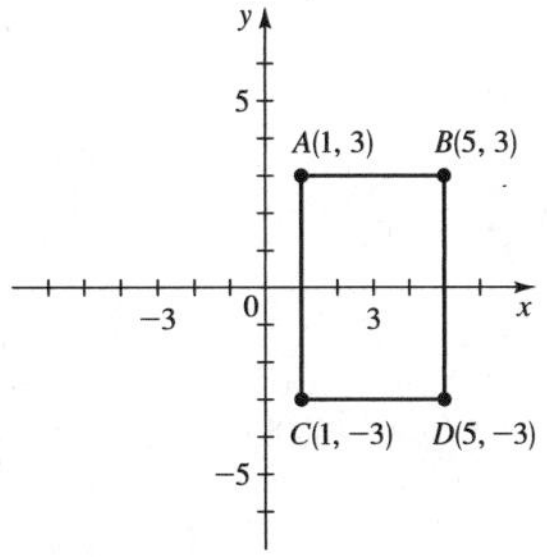

11. Trapezoid, area = 9

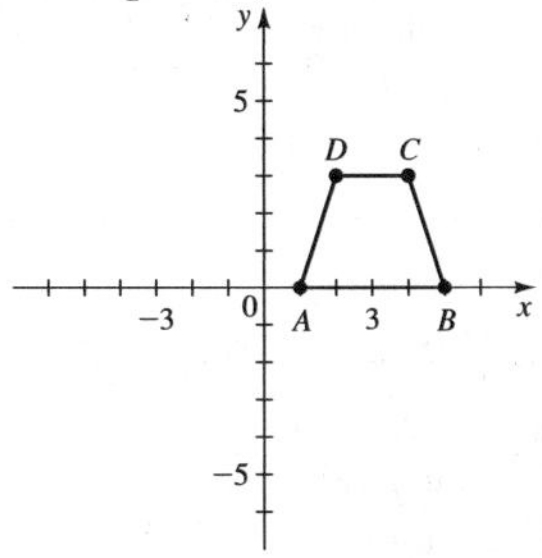

13.

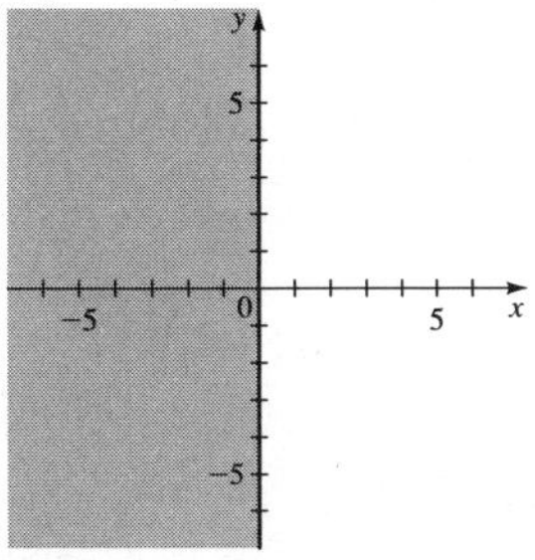

15.

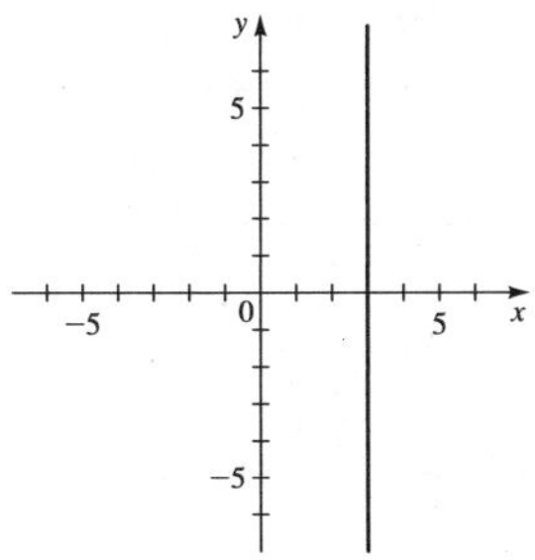

17.

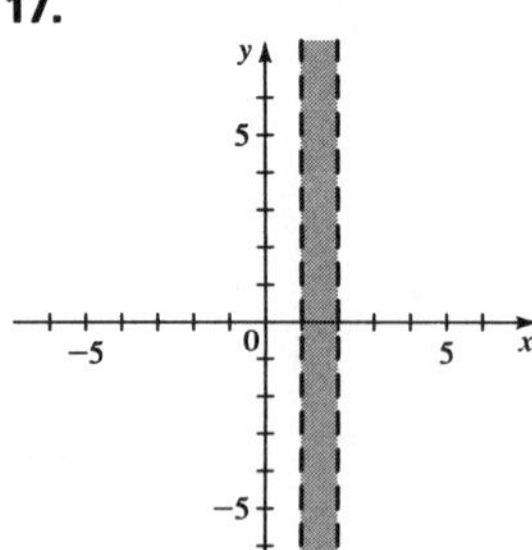

19.

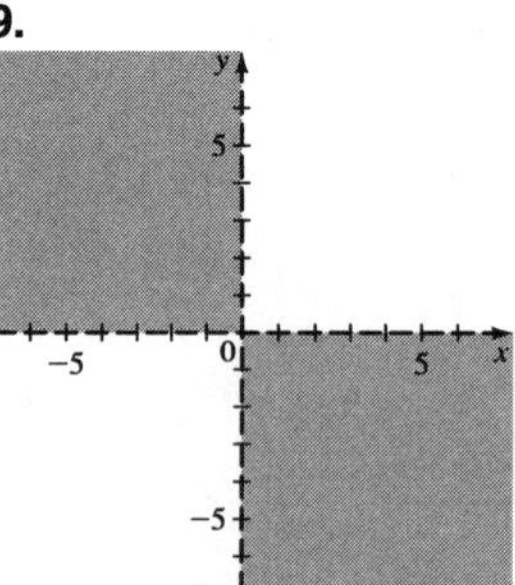

21.

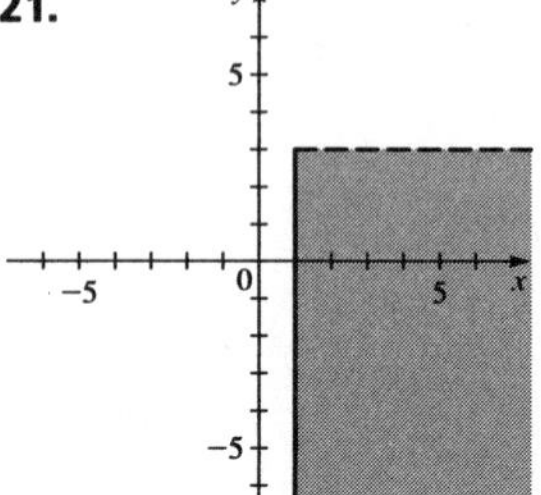

23.

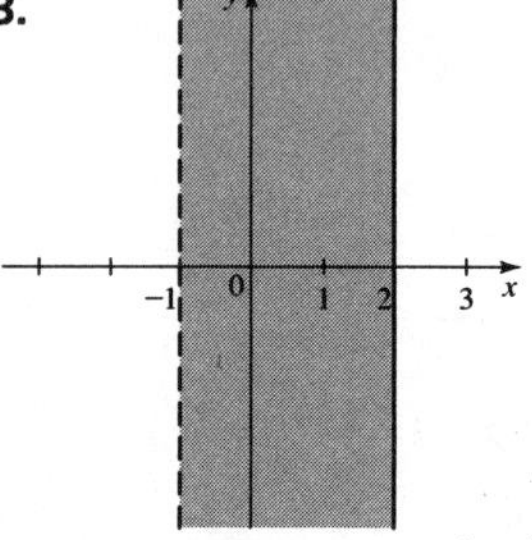

25. $A(6, 7)$ **27.** $Q(-1, 3)$ **31.** (b) 10 **35.** $(0, -4)$ **37.** $\left(1, \frac{7}{2}\right)$

39. $(2, -3)$

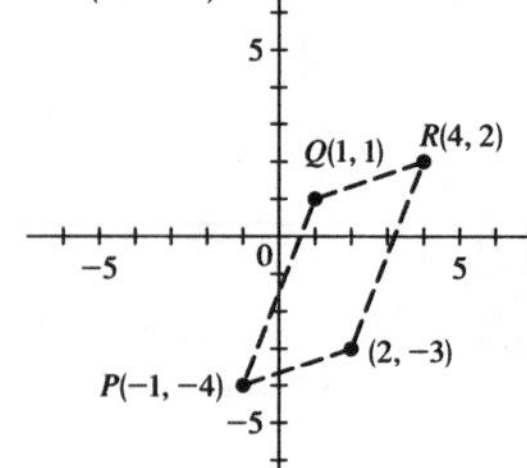

41. (a)

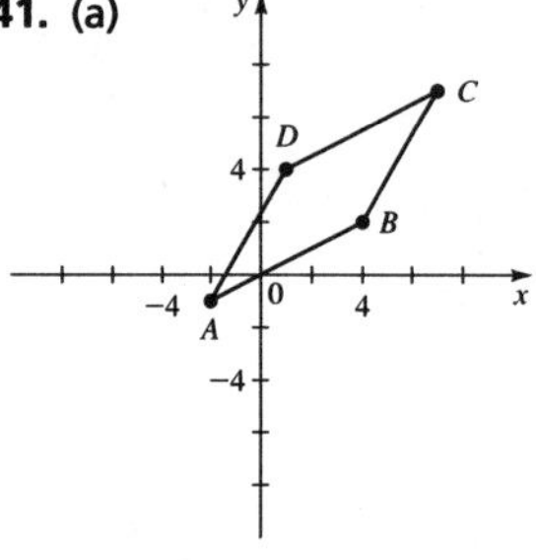

(b) $\left(\frac{5}{2}, 3\right), \left(\frac{5}{2}, 3\right)$

Section 2.2 ■ page 92

1. No, yes, no **3.** No, yes, yes **5.** Yes, no, yes

7. x-intercept 3, y-intercept -3

9. x-intercepts ± 3, y-intercept -9

11. x-intercepts ± 2, y-intercepts ± 2 **13.** None

15. x-intercept 0, y-intercept 0, symmetry about origin

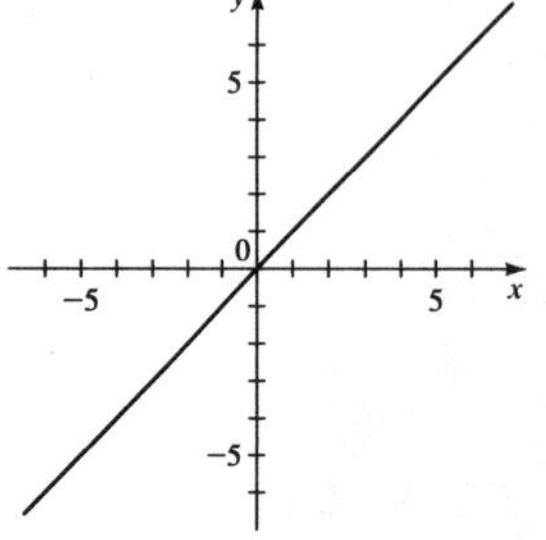

17. x-intercept 1, y-intercept -1, no symmetry

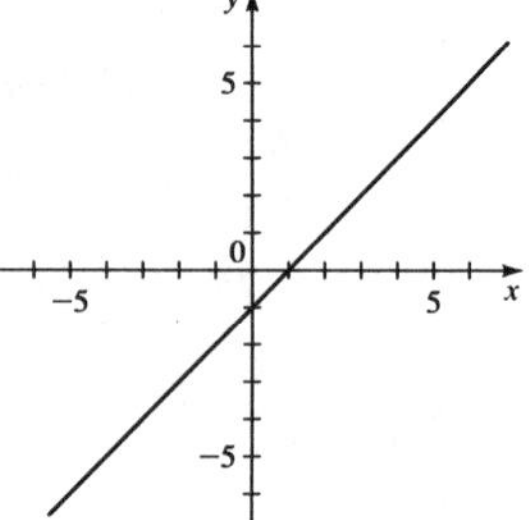

19. x-intercept $\frac{5}{3}$, y-intercept -5, no symmetry

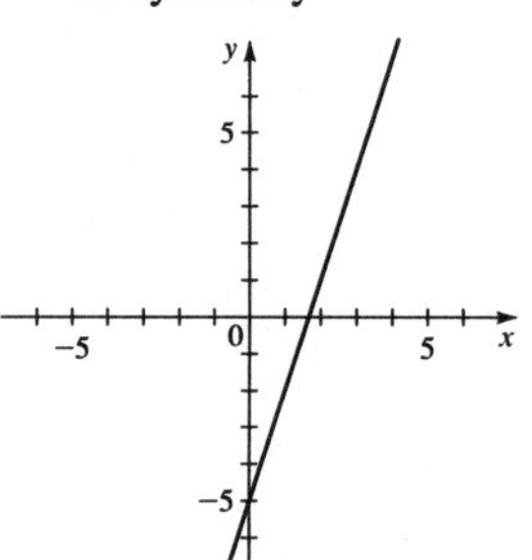

21. x-intercepts ± 1, y-intercept 1, symmetry about y-axis

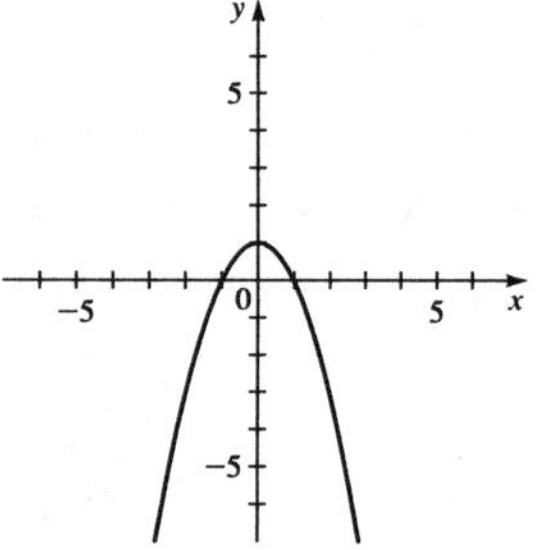

23. x-intercept 0, y-intercept 0, symmetry about y-axis

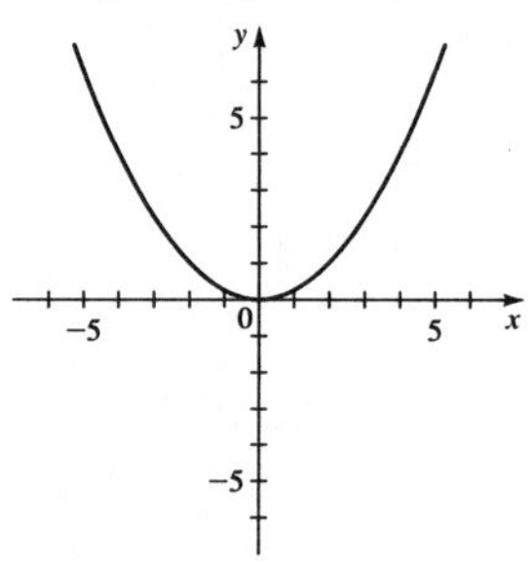

25. x-intercepts ± 3, y-intercept -9, symmetry about y-axis

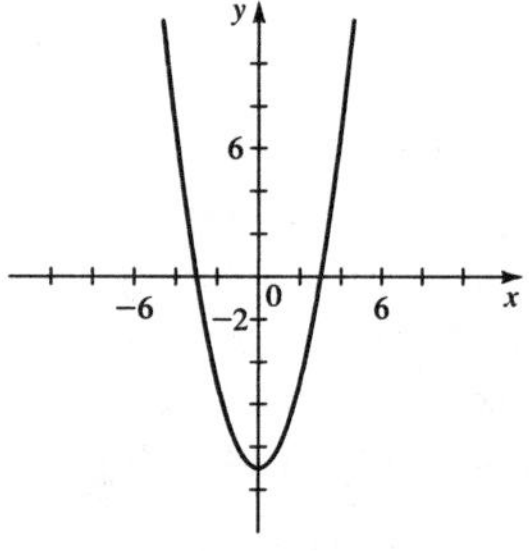

27. No intercepts, symmetry about origin

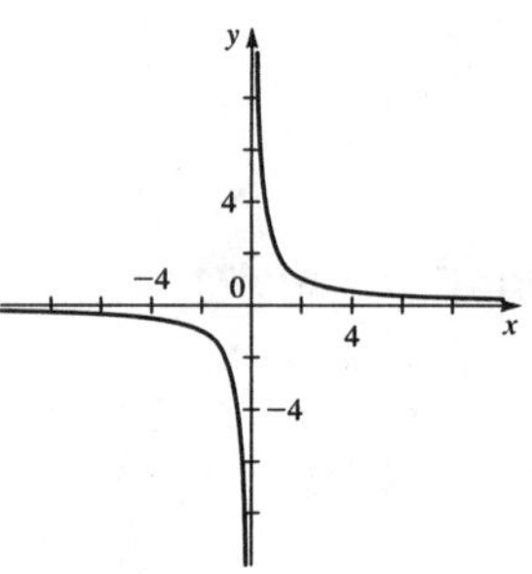

29. x-intercept 0, y-intercept 0, no symmetry

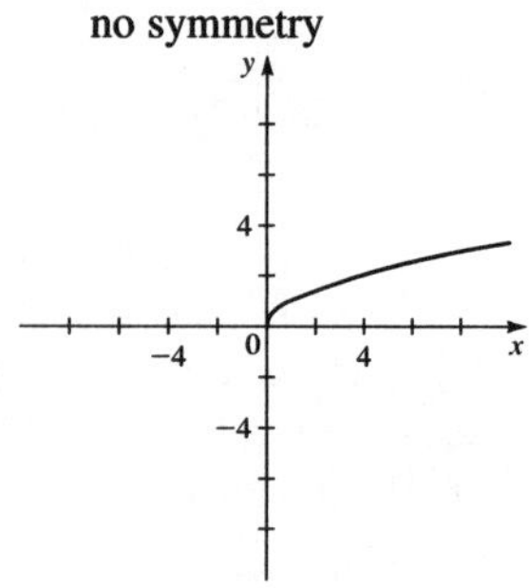

31. x-intercepts ± 2, y-intercept 2, symmetry about y-axis

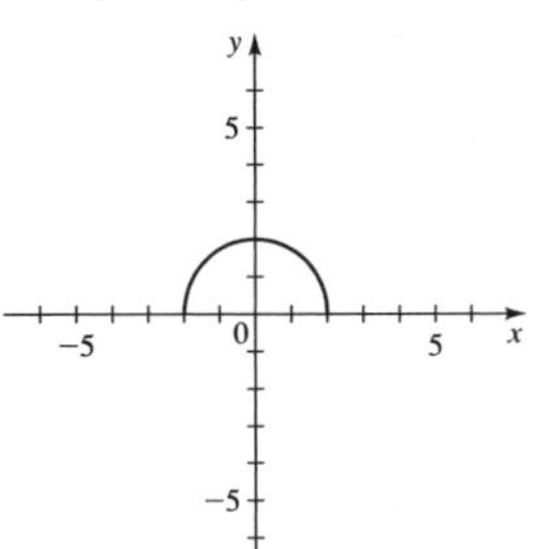

33. x-intercept 0, y-intercept 0, symmetry about y-axis

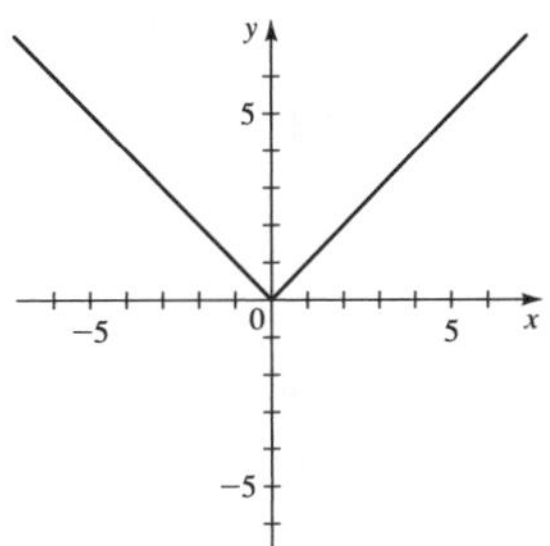

35. x-intercepts ± 4, y-intercept 4, symmetry about y-axis

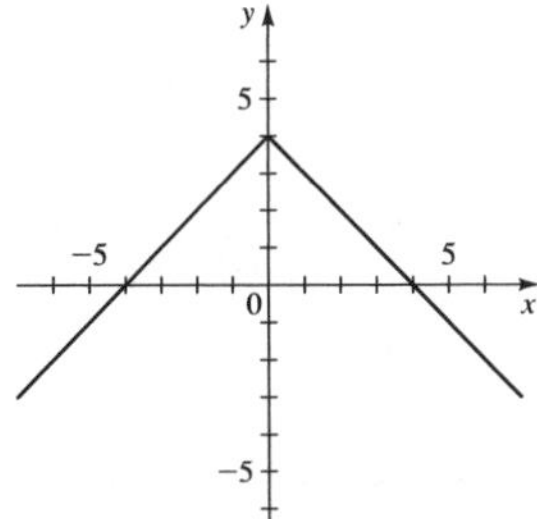

37. x-intercept 0, y-intercept 0, symmetry about origin

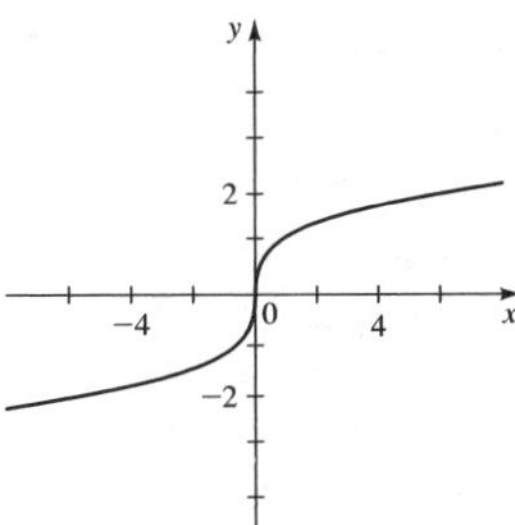

39. x-intercept 0, y-intercept 0, symmetry about y-axis

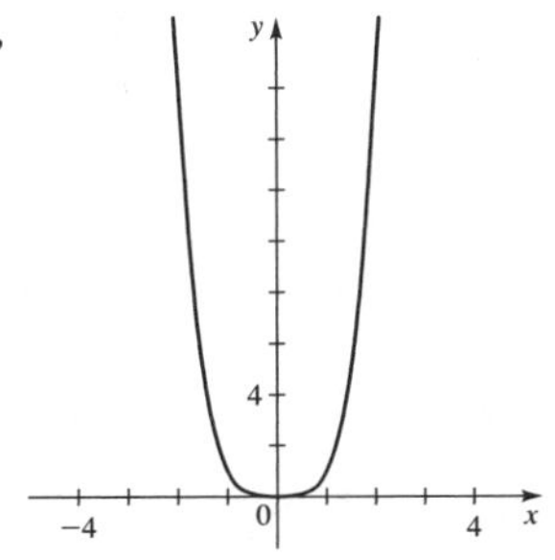
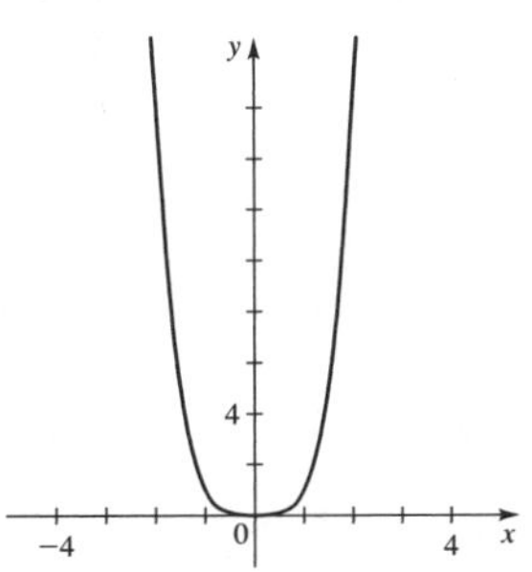

41. Symmetry about y-axis

43. Symmetry about origin

45. Symmetry about origin

47.

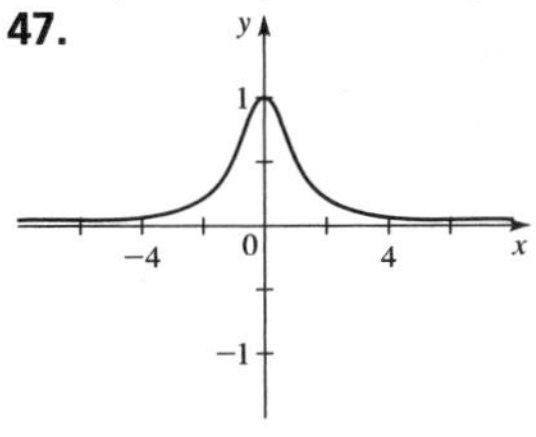

49.

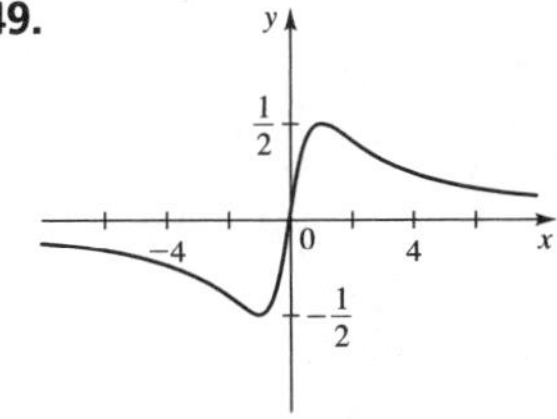

51. $(x-2)^2 + (y+1)^2 = 9$

53. $x^2 + y^2 = 65$

55. $(x-2)^2 + (y-3)^2 = 13$

57. $(x-7)^2 + (y+3)^2 = 9$

59. $(x+2)^2 + (y-2)^2 = 4$

61. $(1, -2)$, 2

63. $(2, -5)$, 4 **65.** $\left(-\frac{1}{2}, 0\right)$, $\frac{1}{2}$ **67.** $\left(\frac{1}{4}, -\frac{1}{4}\right)$, $\frac{1}{2}$

69.

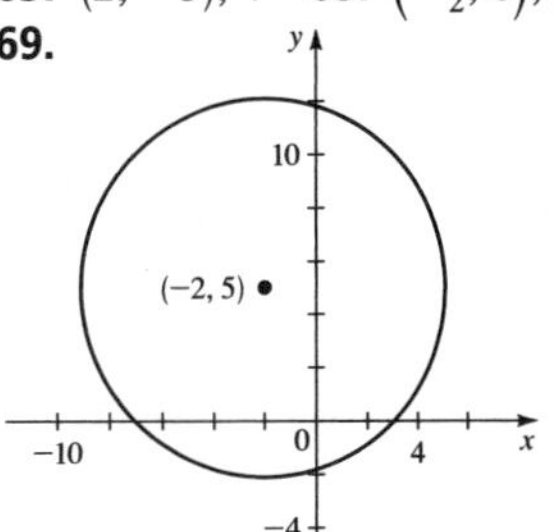

71.

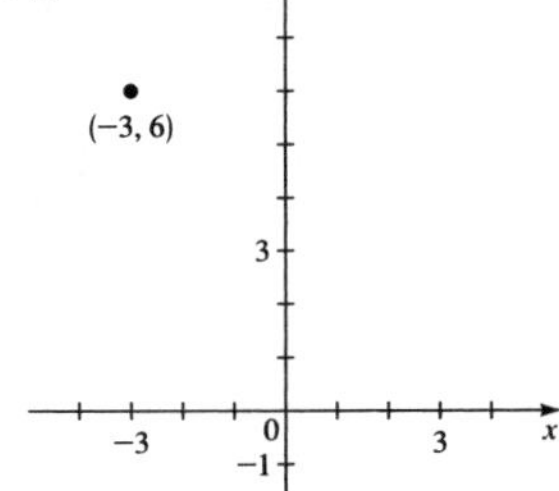

73.

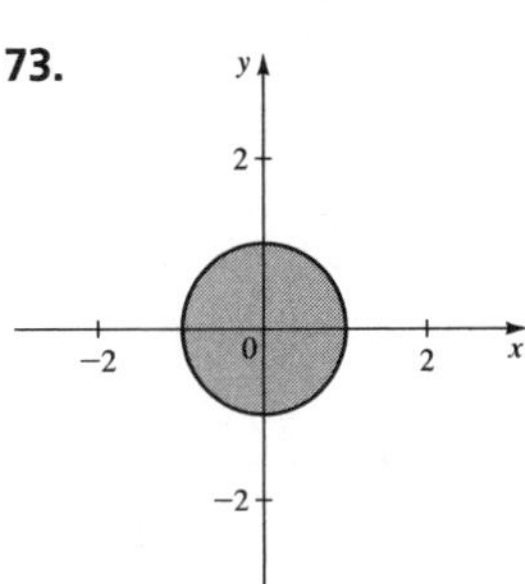

75.

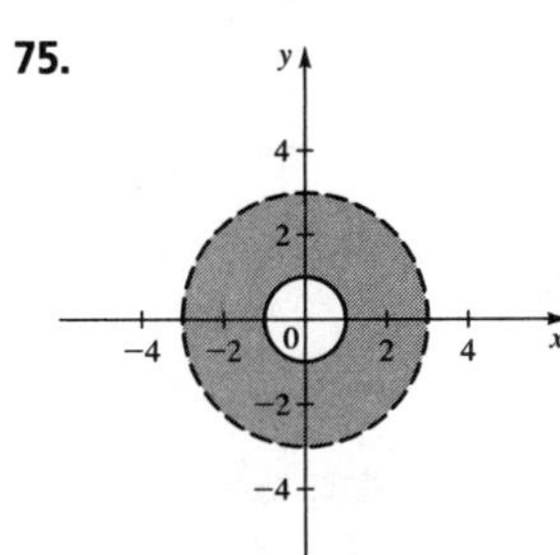

77. 12π

Section 2.3 ■ page 99

1. (b) **3.** (c) **5.** (c) **7.** (c)

9.

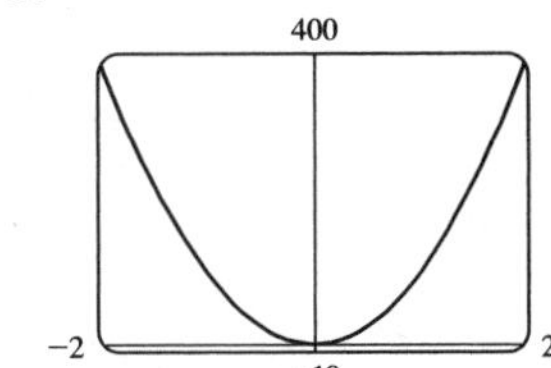

11.

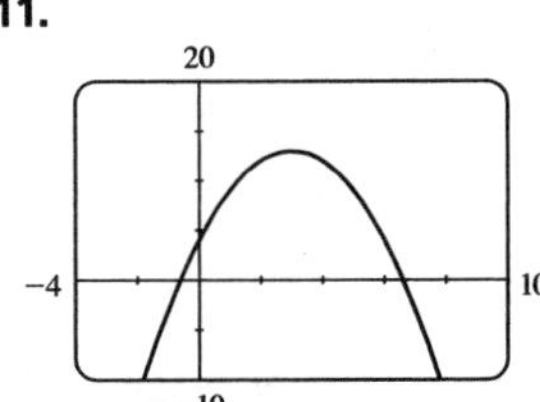

13.

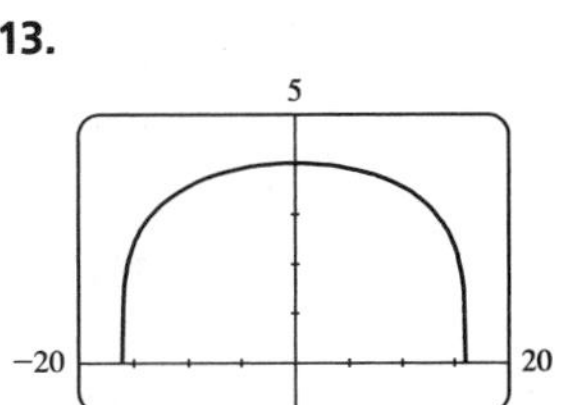

15.

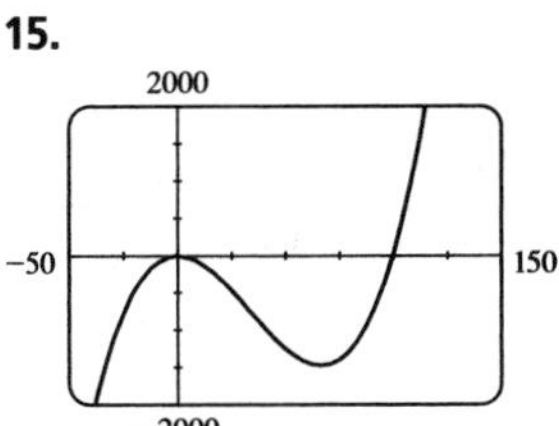

17.

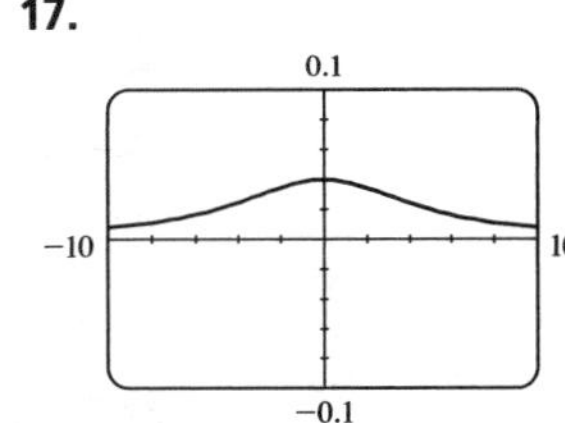

19.

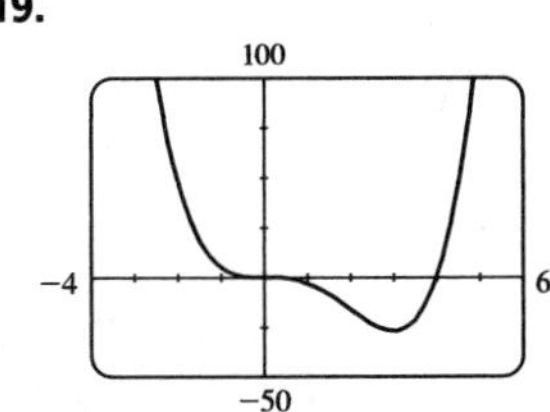

21.

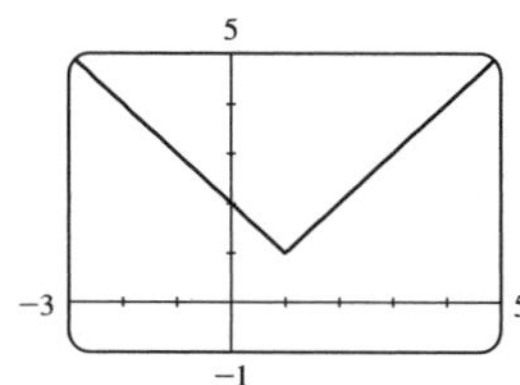

23.

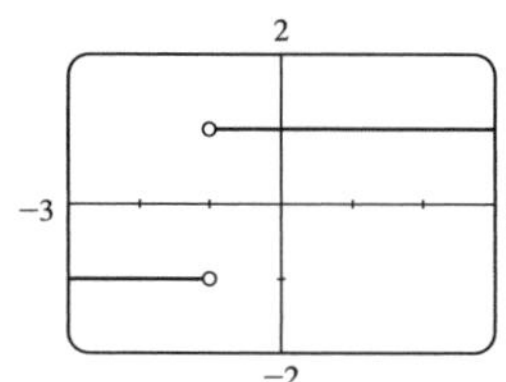

25.

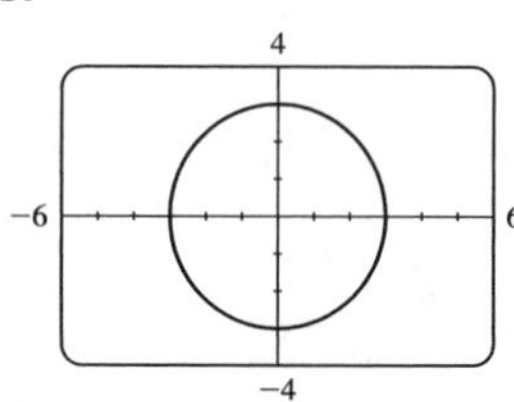

27.

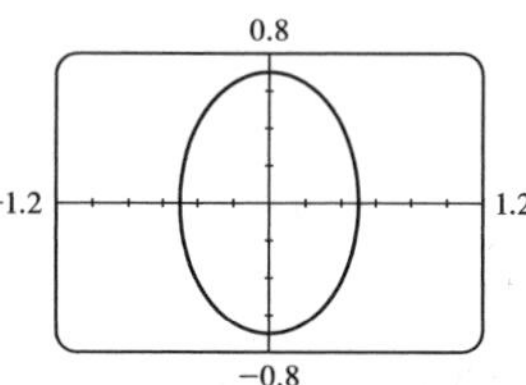

29. No

31. Yes, 2

Section 2.4 ■ page 111

1. 2 **3.** $-\frac{1}{4}$ **5.** 4 **7.** $-\frac{9}{2}$

9. $-2, \frac{1}{2}, 3, -\frac{1}{4}$ **11.** $x + y - 4 = 0$ **13.** $3x - 2y - 6 = 0$

15. $x - y + 1 = 0$ **17.** $2x - 3y + 19 = 0$

19. $5x + y - 11 = 0$ **21.** $3x - y - 2 = 0$

23. $3x - y - 3 = 0$ **25.** $y = 5$ **27.** $x + 2y + 11 = 0$

29. $x = -1$ **31.** $5x - 2y + 1 = 0$ **33.** $x - y + 6 = 0$

35. (a)

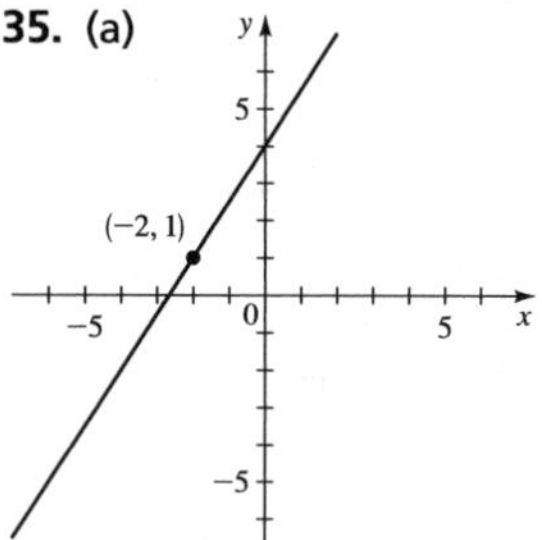

(b) $3x - 2y + 8 = 0$

37. All lines pass through (3, 2)

39. $-1, 3$

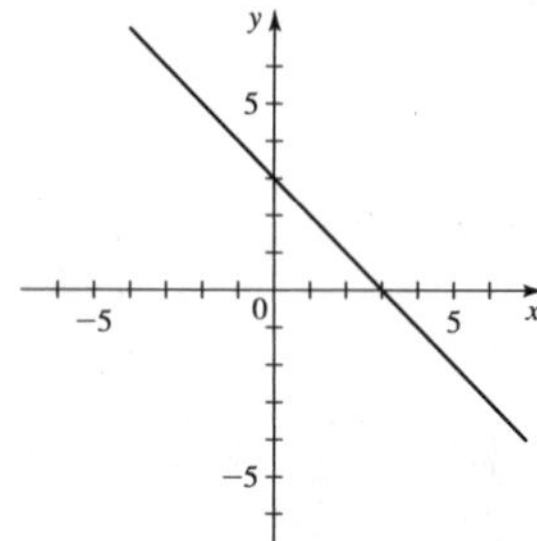

41. $-\frac{1}{3}, 0$

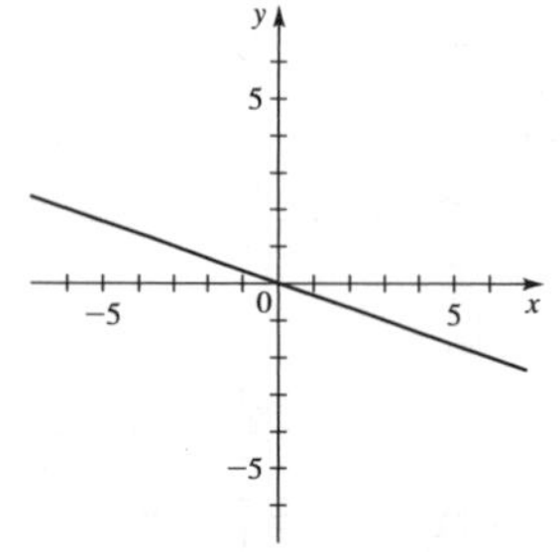

43. $\frac{3}{2}, 3$

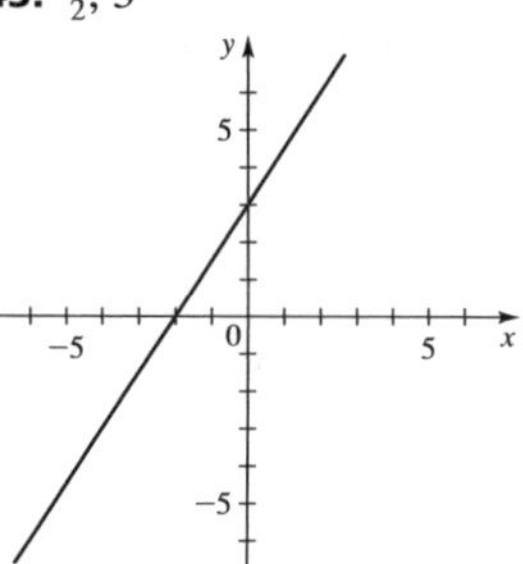

45. 0, 4

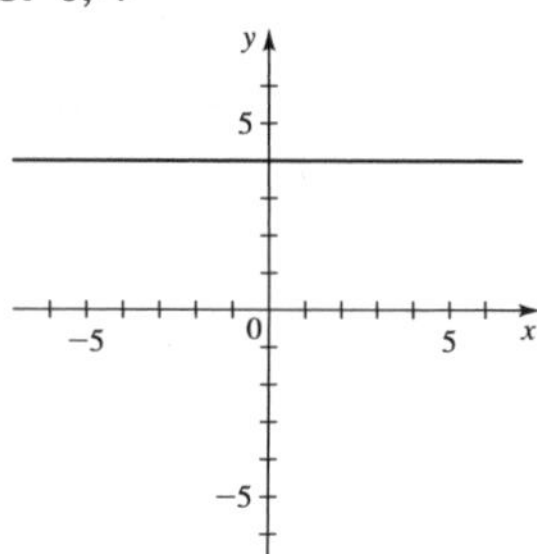

47. $\frac{3}{4}, -3$

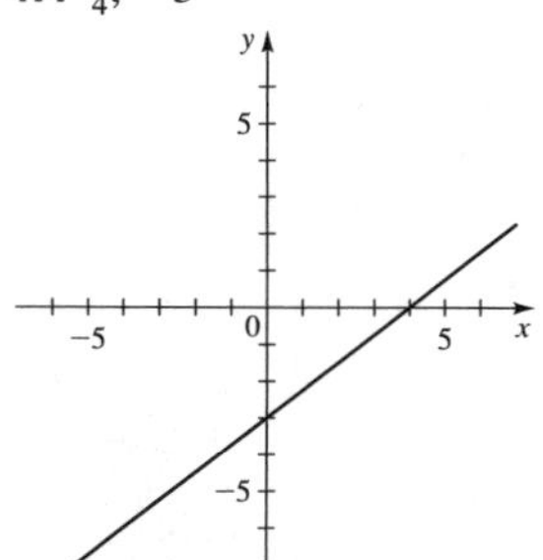

49. $-\frac{3}{4}, \frac{1}{4}$

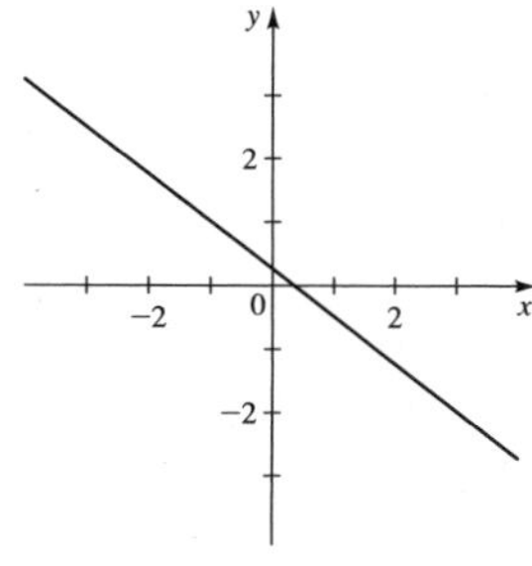

55. $x - y - 3 = 0$ **57.** (b) $4x - 3y - 24 = 0$

59. 16,667 ft

61. (a) 8.34; the slope represents the increase in dosage for a one-year increase in age.
(b) 8.34 mg

63. (a)

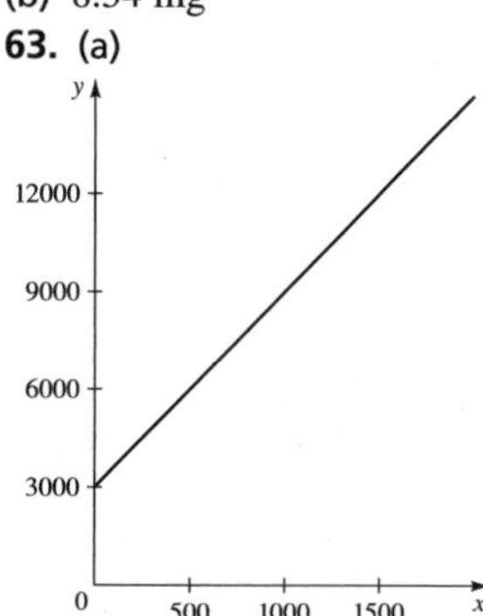

(b) The slope represents production cost per toaster; the y-intercept represents monthly fixed cost.

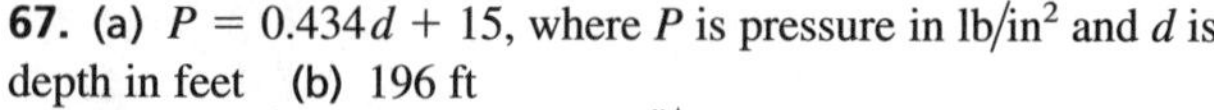

65. (a) $t = \frac{5}{24}n + 45$ (b) 76°F

67. (a) $P = 0.434d + 15$, where P is pressure in lb/in^2 and d is depth in feet (b) 196 ft

69. (a) $C = \frac{1}{4}d + 260$
(b) \$635
(c) The slope represents cost per mile.
(d) The y-intercept represents annual fixed cost.

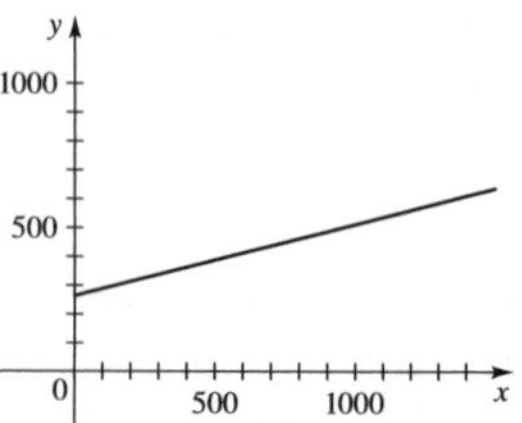

71. (a)

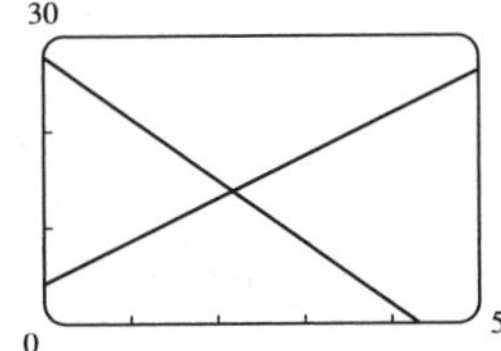

(b) (21.82, 13.82)
(c) Price \$21.82, amount 13.82

Chapter 2 Review ■ page 115

1. (a)

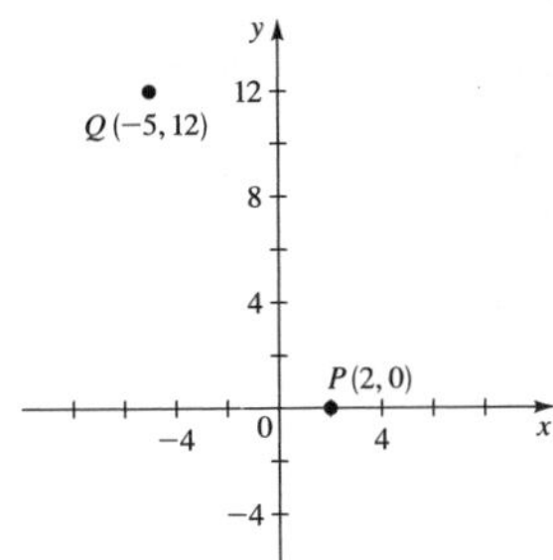

(b) $\sqrt{193}$
(c) $\left(-\frac{3}{2}, 6\right)$
(d) $y = -\frac{12}{7}x + \frac{24}{7}$
(e) $(x - 2)^2 + y^2 = 193$

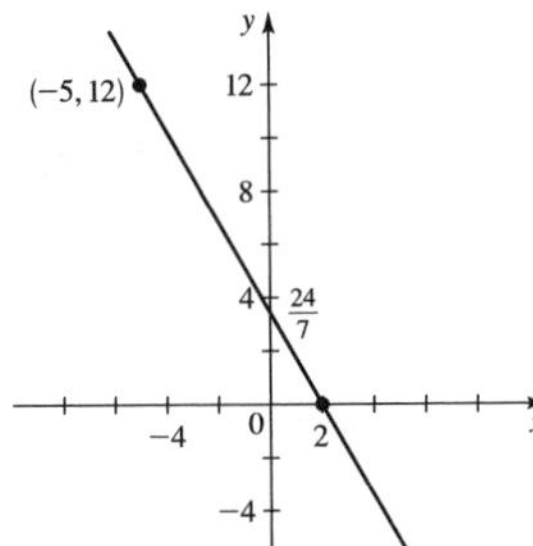

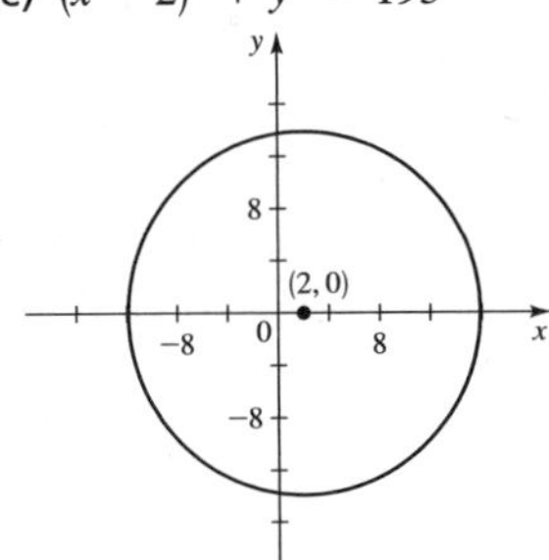

3.

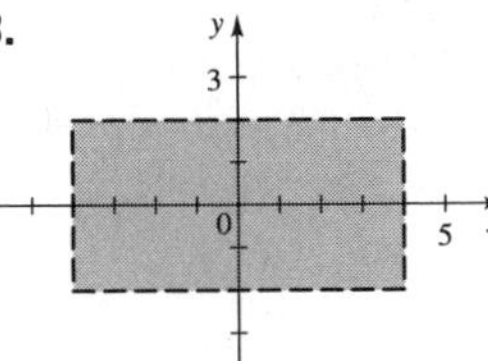

5. B

7. $(x + 5)^2 + (y + 1)^2 = 26$

9. Circle, center $(-1, 3)$, radius 1

11. No graph

13. No symmetry

15. No symmetry

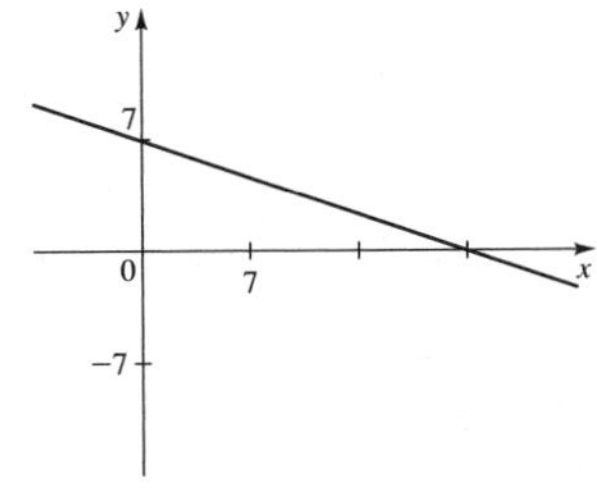

17. No symmetry

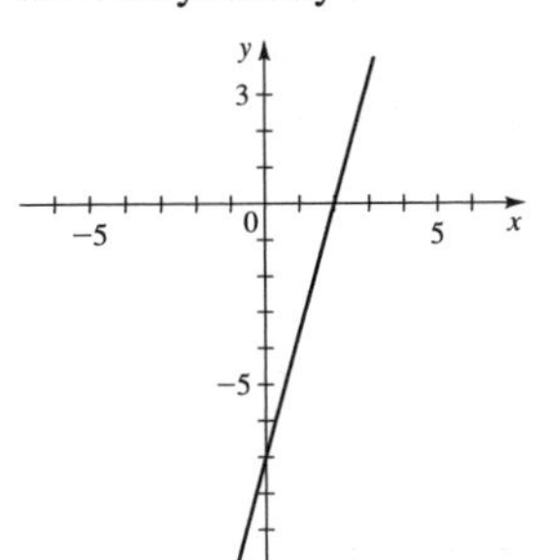

19. Symmetry about y-axis

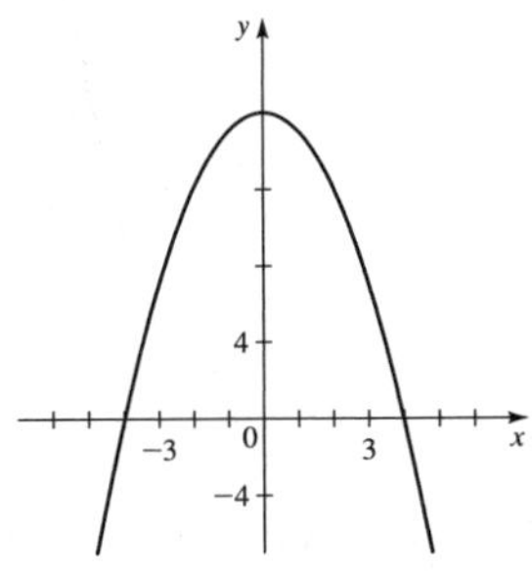

21. No symmetry

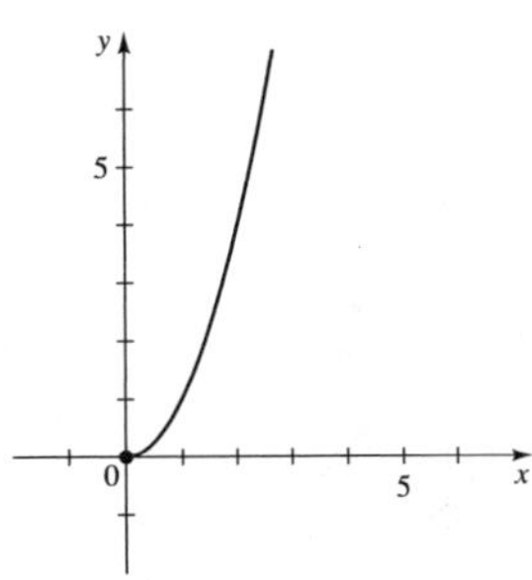

23.

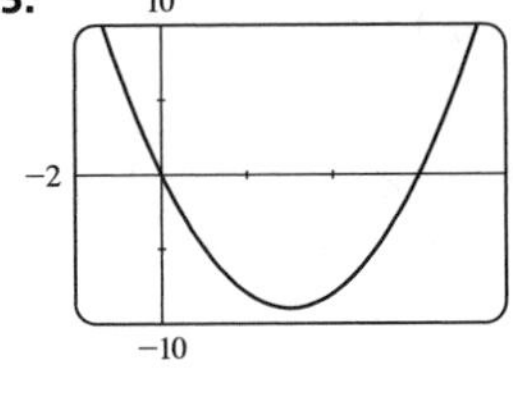

25.

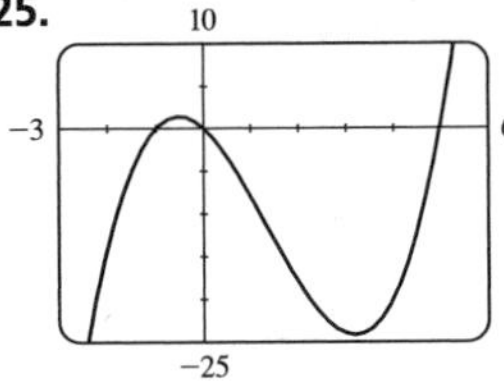

27. $2x - 3y - 16 = 0$

29. $3x + y - 12 = 0$ **31.** $x + 5y = 0$

33. (a) The slope represents the amount the spring lengthens for a one-pound increase in weight. The S-intercept represents the unstretched length of the spring.
(b) 4 in.

35. $x^2 + y^2 = 169$, $5x - 12y + 169 = 0$

Chapter 2 Test ■ page 118

1. (a) $S(3, 6)$

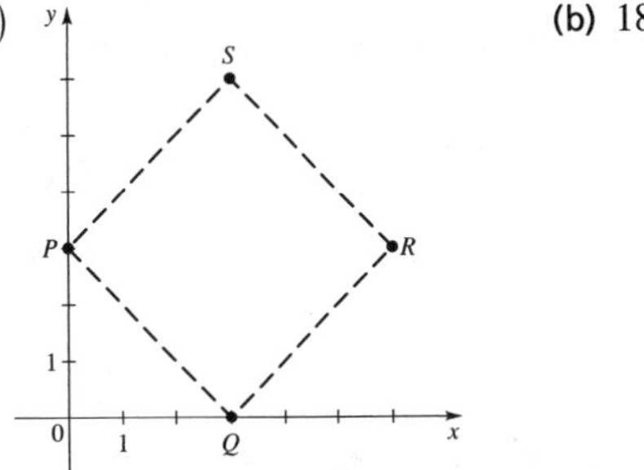

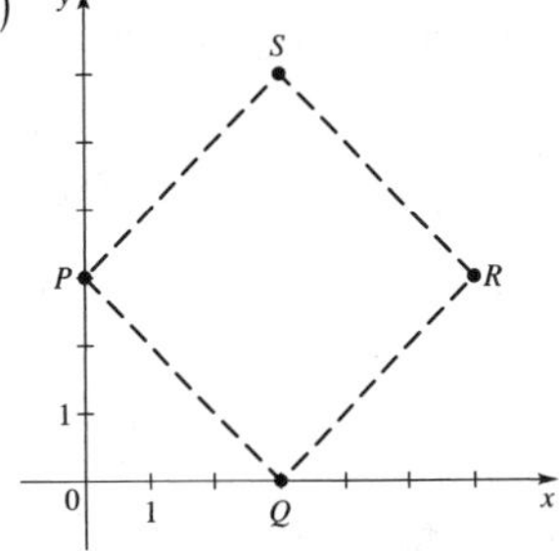

(b) 18

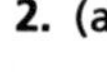

2. (a)

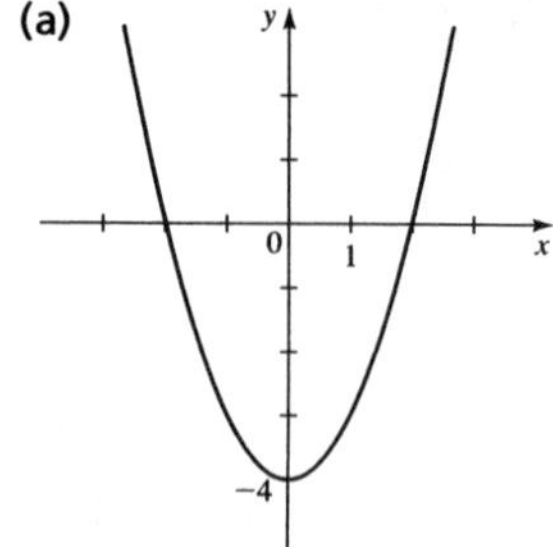

(b) x-intercepts $-2, 2$
y-intercept -4
(c) Symmetric about y-axis

3. (a) 20 **(b)** $(-1, -4)$ **(c)** $4x + 3y + 16 = 0$
(d) $3x - 4y - 13 = 0$ **(e)** $(x + 1)^2 + (y + 4)^2 = 100$
4. (a) Center $(0, 4)$; radius 3 **(b)** Center $(3, -5)$; radius 5

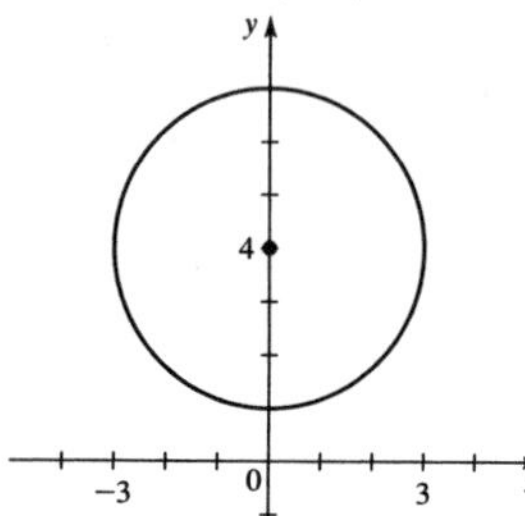

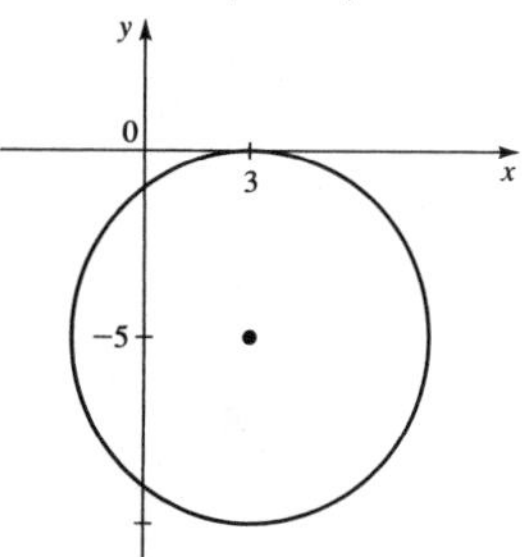

5. $y = \frac{3}{4}x + 6$, slope $\frac{3}{4}$, y-intercept 6

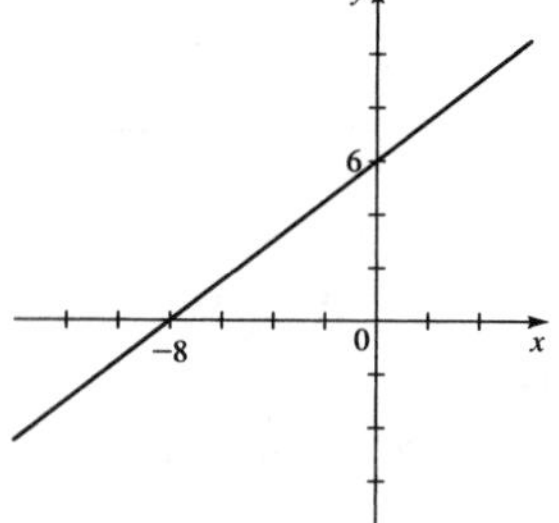

6. (a) $x + 3y - 7 = 0$ **(b)** $4x - y + 12 = 0$
7. (a) 35,000
(b) The slope represents the cost of producing one more blender. The C-intercept represents the fixed costs.

PRINCIPLES OF MODELING ■ page 123

1. (a)

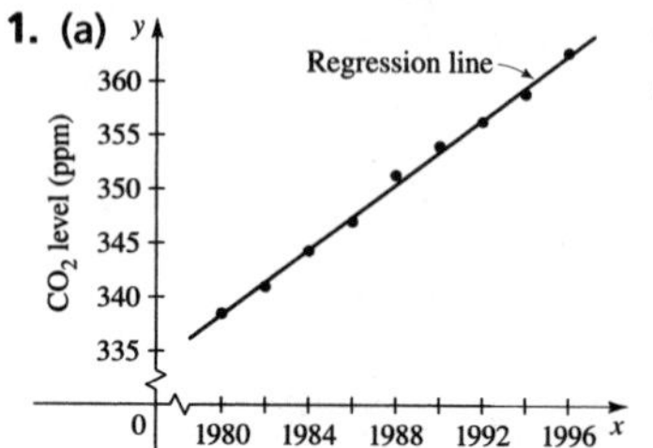

(b) $y = 1.5125x - 2656.4$
(c) 365.6 ppm

3. (a)

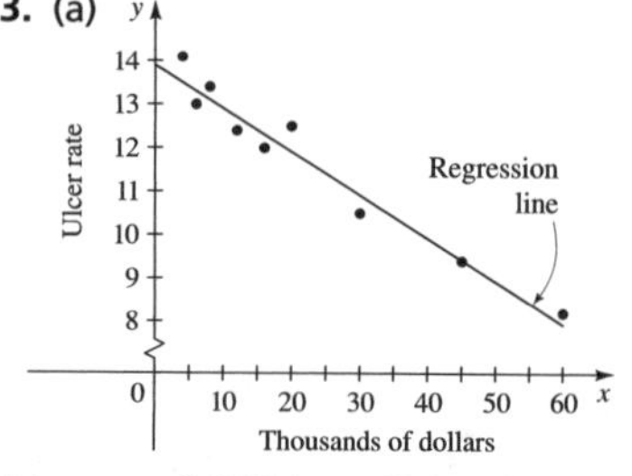

(b) $y = -0.0995x + 13.9$, x in thousands of dollars
(c) 11.4 per 100 population
(d) 5.8 per 100 population

5. (a)

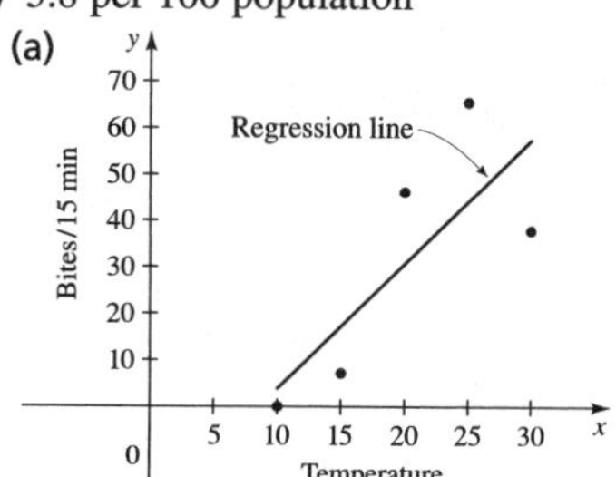

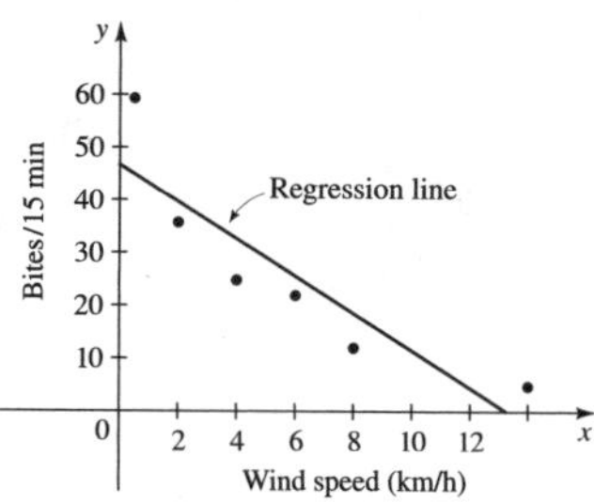

(b) $y = 2.6768x - 22.19$, $y = -3.5336x + 46.73$
(c) 0.77 is the correlation coefficient for the temperature-bites data, so a linear model is not appropriate; -0.89 is the correlation coefficient for the wind/bites data, so a linear model is more suitable in this case.

7. (a)

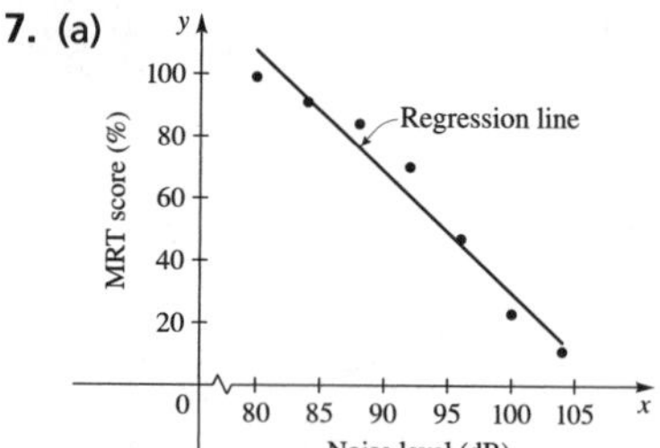

(b) $y = -3.9018x + 419.7$
(c) The correlation coefficient is -0.98, so a linear model is appropriate.
(d) 53%

9. (a)

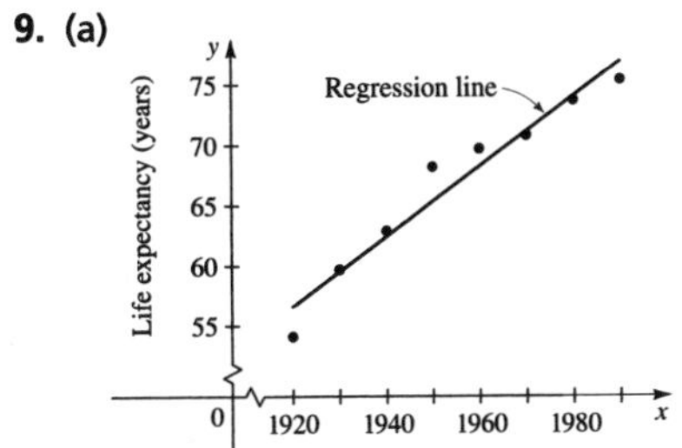

(b) $y = 0.29083x - 501.8$
(c) 79.9 years

11. (a)

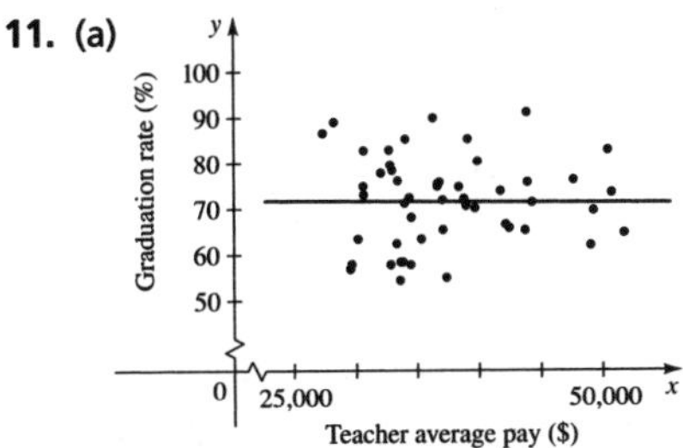

(b) $y = -0.000012x + 72.0$
(c) $r = -.00812$; very little correlation; it appears that teachers' salaries do not affect graduation rates.

CHAPTER 3

Section 3.1 ■ page 135

1. 7 **3.** −4 **5.** −6 **7.** $\frac{5}{14}$ **9.** $4\sqrt{2} \approx 5.7, -4\sqrt{2} \approx -5.7$ **11.** 2.5, −2.5 **13.** $5 + 2\sqrt[4]{5} \approx 7.99, 5 - 2\sqrt[4]{5} \approx 2.01$ **15.** 3.00, 4.00 **17.** 1.00, 2.00, 3.00 **19.** 1.62 **21.** −1.00, 0.00, 1.00 **23.** 2.55 **25.** −2.05, 0, 1.05 **27.** 0, 0.01

Section 3.2 ■ page 146

1. $3n + 3$ **3.** $n^2 + (n+1)^2$ **5.** $\dfrac{78 + 82 + s}{3}$ **7.** $50w$ **9.** $3s + 15$ **11.** $\dfrac{3a - 8}{3}$ **13.** \$7,400 **15.** \$45,000 **17.** 7 years **19.** 111, 112, 113 **21.** 16 ft **23.** 45 ft **25.** \$9000 at $4\frac{1}{2}$% and \$3000 at 4% **27.** Plumber, 70 h; assistant, 35 h **29.** 5 quarters, 10 dimes, 15 nickels **31.** 4 in. **33.** \$3000 **35.** 8 min 20 s **37.** 6.4 ft from the fulcrum **39.** 240 ft **41.** 9 cm **43.** 10 m **45.** 200 mL **47.** 18 g **49.** 0.6 L **51.** 37 min 20 s **53.** 3 h **55.** $1\frac{1}{2}$ h **57.** 450 mi **59.** 4 h **61.** 500 mi/h **63.** 120 ft **65.** 35 yards **67.** 4.55 ft

Section 3.3 ■ page 161

1. −2, 3 **3.** 2 **5.** $-3, -\frac{1}{2}$ **7.** $-\frac{4}{3}, \frac{1}{2}$ **9.** −20, 25 **11.** $-1 \pm \sqrt{3}$ **13.** $3 \pm 3\sqrt{2}$ **15.** −21, −1 **17.** $-2 \pm \dfrac{\sqrt{14}}{2}$ **19.** $0, \frac{1}{4}$ **21.** −2, 4 **23.** $-6 \pm 3\sqrt{7}$ **25.** $\dfrac{-3 \pm 2\sqrt{6}}{3}$ **27.** $\dfrac{1 \pm \sqrt{5}}{4}$ **29.** $-\frac{9}{2}, \frac{1}{2}$ **31.** $\dfrac{-5 \pm \sqrt{13}}{2}$ **33.** $\dfrac{\sqrt{5} \pm 1}{2}$ **35.** $\dfrac{8 \pm \sqrt{14}}{10}$ **37.** No real solution **39.** $1, 1 + \dfrac{1}{a}$ **41.** −0.248, 0.259 **43.** 1.200, 1.250 **45.** $t = \dfrac{v_0 \pm \sqrt{v_0^2 + 2gh}}{g}$ **47.** $r = \dfrac{-\pi h \pm \sqrt{\pi^2 h^2 + 2\pi A}}{2\pi}$ **49.** −10.05, 0.55 **51.** −0.80, 0.61 **53.** 2 **55.** 1 **57.** 2 **59.** $k = \pm 20$ **61.** 19 and 36 **63.** 25 ft by 35 ft **65.** 60 ft by 40 ft **67.** 48 **69.** 13 in. by 13 in. **71.** 120 ft by 126 ft **73.** 4.24 s **75. (a)** After 1 s and $1\frac{1}{2}$ s **(b)** Never **(c)** 25 ft **(d)** After $1\frac{1}{4}$ s **(e)** After $2\frac{1}{2}$ s **77. (a)** After 17 yr, on Jan. 1, 2009 **(b)** After 18.612 yr, on Aug. 12, 2010 **79.** 50 mi/h (or 240 mi/h) **81.** 6 km/h **83.** Irene 3 h, Henry $4\frac{1}{2}$ h **85.** 215,000 mi

Section 3.4 ■ page 170

1. Real part 3, imaginary part −5
3. Real part 0, imaginary part 6
5. Real part $\sqrt{2}$, imaginary part $\sqrt{3}$
7. $9 + i$ **9.** $12 + i$ **11.** $-19 + 4i$ **13.** $-4 + 8i$ **15.** $30 + 10i$ **17.** $-33 - 56i$ **19.** $27 - 8i$ **21.** $-i$ **23.** $\frac{8}{5} + \frac{1}{5}i$ **25.** $-5 + 12i$ **27.** $-4 + 2i$ **29.** $2 - \frac{4}{3}i$ **31.** $-i$ **33.** $-i$ **35.** 1 **37.** $5i$ **39.** −6 **41.** $(3 + \sqrt{5}) + (3 - \sqrt{5})i$ **43.** 2 **45.** $-i\sqrt{2}$ **47.** $\pm 3i$ **49.** $2 \pm i$ **51.** $-\dfrac{1}{2} \pm \dfrac{\sqrt{3}}{2}i$ **53.** $\frac{1}{2} \pm \frac{1}{2}i$ **55.** $-\dfrac{3}{2} \pm \dfrac{\sqrt{3}}{2}i$ **57.** $\dfrac{-6 \pm \sqrt{6}\,i}{6}$ **59.** $1 \pm 3i$

Section 3.5 ■ page 178

1. $0, \pm 8$ **3.** $0, \pm\sqrt{6}$ **5.** 0, 1, 2 **7.** $0, -2 \pm \sqrt{2}$ **9.** $\pm\sqrt{2}, 5$ **11.** 2 **13.** −1.4, 2 **15.** −50, 100 **17.** −4 **19.** No real solution **21.** −7, 0 **23.** $-\frac{3}{2}, -\frac{3}{4}$ **25.** $\pm 2\sqrt{2}, \pm\sqrt{5}$ **27.** No real solution **29.** $\pm 3\sqrt{3}, \pm 2\sqrt{2}$ **31.** −1, 0, 3 **33.** 27, 729 **35.** $-\frac{1}{2}$ **37.** 4 **39.** 4

41. 21 **43.** $-3, \frac{1 \pm \sqrt{13}}{2}$ **45.** 2 **47.** $1, \frac{-1 \pm i\sqrt{3}}{2}$

49. $0, \frac{-1 \pm i\sqrt{3}}{2}$ **51.** $\pm\sqrt{2}, \pm 2$

53. $1, 2, \frac{-1 \pm i\sqrt{3}}{2}, -1 \pm i\sqrt{3}$ **55.** $\pm 3i$

57. 50 **59.** 89 days **61.** 2 ft by 6 ft by 15 ft
63. 16 mi; no **65.** 7.52 ft **67.** 49 ft, 168 ft, and 175 ft
69. 132.6 ft **71.** $-\frac{b}{a}, \frac{b \pm \sqrt{3}\,bi}{2a}$ **73.** No solution

Section 3.6 ■ page 185

1. $\{-1, 0, \frac{1}{2}, \sqrt{2}, 2\}$ **3.** $\{0, \frac{1}{2}, \sqrt{2}, 2\}$ **5.** $\{-1, 2\}$
7. $(-\infty, 4]$ **9.** $(-\infty, -5)$ 
11. $(4, \infty)$ **13.** $(-\infty, 2]$ 
15. $(-\infty, -\frac{1}{2})$  **17.** $[1, \infty)$
19. $[-1, \infty)$ **21.** $(\frac{16}{3}, \infty)$
23. $[\frac{2}{3}, \infty)$ **25.** $(-\infty, -1]$ 
27. $[-3, -1)$ **29.** $(2, 6)$
31. $(0, 1]$  **33.** $[\frac{9}{2}, 5)$
35. $(\frac{5}{2}, \frac{11}{2}]$

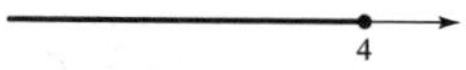

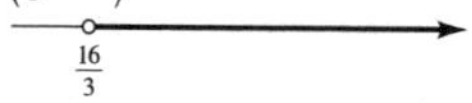

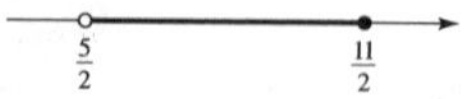

37. More than 200 mi **39.** $68 \leqslant F \leqslant 86$
41. (a) $-\frac{1}{3}P + \frac{560}{3}$ (b) From \$215 to \$290
43. Between 12,000 mi and 14,000 mi
45. $x \geqslant \frac{c(a + b)}{ab}$ **47.** $x > \frac{c - b}{a}$

Section 3.7 ■ page 194

1. $(-\infty, 2) \cup (5, \infty)$ **3.** $[-3, 6]$ 
5. $(-\infty, -1] \cup [\frac{1}{2}, \infty)$ **7.** $(-1, 4)$
9. $(-\infty, -3) \cup (6, \infty)$  **11.** $(-2, 2)$ 
13. 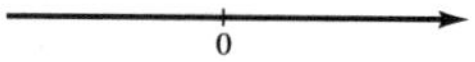 $(-\infty, \infty)$ **15.** $[-2, 0] \cup [2, \infty)$
17. $(-\infty, -1) \cup [3, \infty)$ **19.** $(-\infty, -\frac{3}{2})$
21. $(-\infty, 5) \cup [16, \infty)$ **23.** $(-2, 0) \cup (2, \infty)$
25. $(-\infty, -2] \cup [2, \infty)$ **27.** $[-2, -1) \cup (0, 1]$ 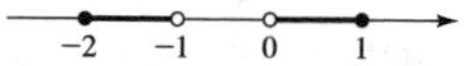
29. $(0, \frac{3}{4}] \cup (1, \infty)$ 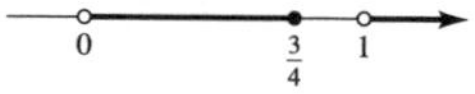 **31.** $(-\infty, -3) \cup (6, \infty)$ 
33. $[-8, -\frac{5}{2})$ **35.** $[-2, 0) \cup (1, 3]$ 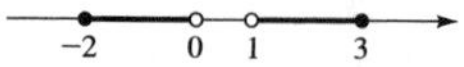
37. $(-3, -\frac{1}{2}) \cup (2, \infty)$ 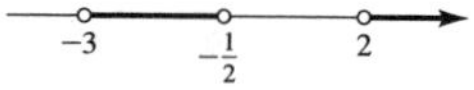 **39.** $(-\infty, -1) \cup (1, \infty)$

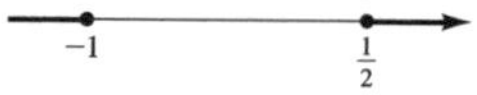

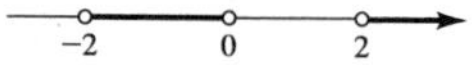

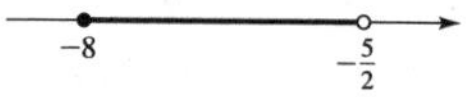

41. $[-2.0, 5.0]$ **43.** $(-\infty, 1.0] \cup [2.0, 3.0]$
45. $(-1.0, 0) \cup (1.0, \infty)$ **47.** $(-\infty, 0)$
49. From 0 s to 3 s **51.** Distances greater than 30 m
53. Between 0 and 60 mi/h **55.** Between 20 ft and 40 ft
57. $-\frac{4}{3} \leqslant x \leqslant \frac{4}{3}$ **59.** $(-\infty, -2) \cup (7, \infty)$
61. $(-\infty, -c) \cup [-a, b]$

Section 3.8 ■ page 199

1. $-5, 5$ **3.** $1, 5$ **5.** $-4.5, -3.5$ **7.** $-4, \frac{1}{2}$
9. $-3, -1$ **11.** $-8, -2$ **13.** $-\frac{25}{2}, \frac{35}{2}$ **15.** $-\frac{3}{2}, -\frac{1}{4}$
17. $(-2, 2)$ **19.** $[2, 8]$ **21.** $(-\infty, -2] \cup [0, \infty)$

23. $(-\infty, -7] \cup [-3, \infty)$ **25.** $[1.3, 1.7]$ **27.** $(-4, 8)$
29. $(-6.001, -5.999)$ **31.** $(-6, 2)$ **33.** $\left[-\frac{1}{2}, \frac{3}{2}\right]$
35. $\left(-\infty, -\frac{1}{2}\right) \cup \left(\frac{1}{3}, \infty\right)$ **37.** $[-4, -1] \cup [1, 4]$
39. $\left(-\frac{15}{2}, -7\right) \cup \left(-7, -\frac{13}{2}\right)$ **41.** $-8, 4$
43. $-4.45, 0.45, 0.59, 3.41$ **45.** $[1, 2.3]$
47. $\left(-\infty, \frac{1}{2}\right) \cup \left(\frac{1}{2}, \frac{3}{2}\right) \cup \left(\frac{3}{2}, \infty\right)$

Chapter 3 Review ■ page 200

1. 5 **3.** ± 3.5 **5.** No solution **7.** 2, 7 **9.** $-1, \frac{1}{2}$
11. $0, \pm\frac{5}{2}$ **13.** $\dfrac{-2 \pm \sqrt{7}}{3}$ **15.** $\dfrac{3 \pm \sqrt{6}}{3}$ **17.** ± 3
19. 1 **21.** 3, 11 **23.** $-2, 7$ **25.** $-1, 7$
27. $-2.72, -1.15, 1.00, 2.87$ **29.** 20 lb raisins, 30 lb nuts
31. $\frac{1}{4}(\sqrt{329} - 3) \approx 3.78$ mi/h **33.** 12 cm, 16 cm
35. 23 ft by 46 ft by 8 ft **37.** $-3 - 9i$ **39.** $19 + 40i$
41. $\dfrac{-5 - 12i}{13}$ **43.** i **45.** $\left(2 + 2\sqrt{3}\right) + \left(2 - 2\sqrt{3}\right)i$
47. $\pm 4i$ **49.** $-3 \pm i$ **51.** $\pm 4, \pm 4i$ **53.** $\pm\dfrac{\sqrt{3}}{3}i$

55. $(-3, \infty)$
−3

57. $(-3, -1]$
−3 −1

59. $(-\infty, -6) \cup (2, \infty)$
−6 2

61. $[-4, -1)$
−4 −1

63. $(-\infty, -2) \cup (2, 4]$
−2 2 4

65. $[2, 8]$
2 8

67. $(-\infty, -1] \cup [0, \infty)$
−1 0

69. $[1, 3]$ **71.** $(-1.85, -0.60) \cup (0.45, 2.00)$
73. (a) $\left[-3, \frac{8}{3}\right]$ (b) $(0, 1)$

Chapter 3 Test ■ page 202

1. (a) $\frac{14}{5}$ (b) 2.80 **2.** 120 mi
3. (a) $-1 - \frac{3}{2}i$ (b) $5 + i$ (c) $-1 + 2i$ (d) -1
(e) $6\sqrt{2}$
4. (a) $-3, 4$ (b) $\dfrac{-2 \pm i\sqrt{2}}{2}$ (c) No solution
(d) 1, 16 (e) $0, \pm 4$ (f) $\frac{2}{3}, \frac{22}{3}$
5. $-2.94, -0.11, 3.05$ **6.** 50 ft by 120 ft

7. (a) $\left(-\frac{5}{2}, 3\right]$ (b) $(0, 1) \cup (2, \infty)$ (c) $(1, 5)$ (d) $[-4, -1)$
8. $[-1.62, 2.30]$ **9.** 41°F to 50°F **10.** $0 \leq x \leq 4$

FOCUS ON PROBLEM SOLVING ■ page 204

3. 3 ft by 6 ft, 4 ft by 4 ft
5. *Hint:* Try proof by contradiction.
7. *Hint:* First divide the coins into two groups of four.
13.

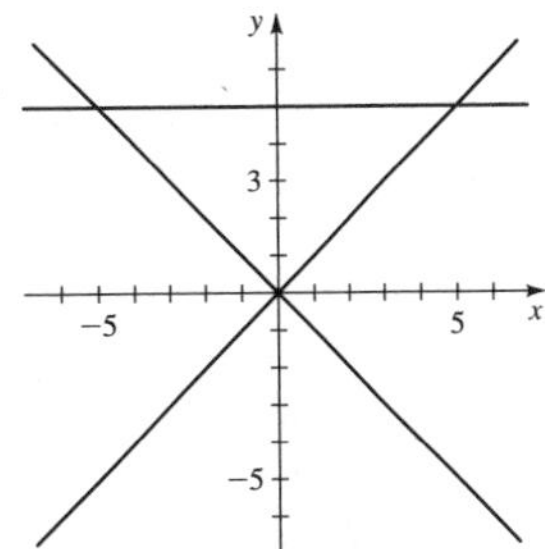

15. (a) A power of 2

CHAPTER 4

Section 4.1 ■ page 217

1. $f(x) = 3x + 1$ **3.** $f(x) = (x + 2)^2$
5. Divide by 3, then subtract 5
7. Square, multiply by 2, then subtract 3
9.

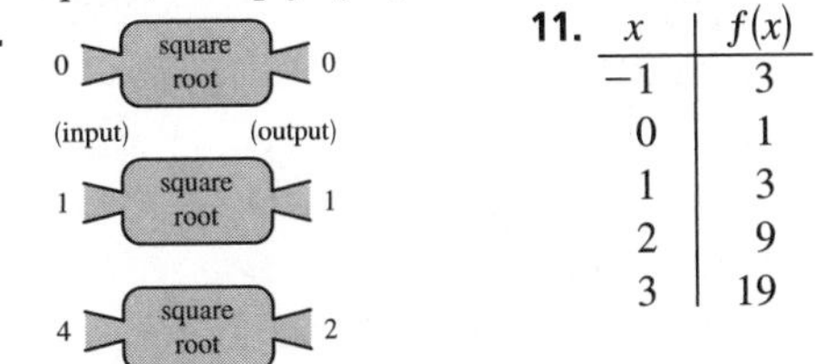

11.

x	$f(x)$
−1	3
0	1
1	3
2	9
3	19

13. $3, -3, 2, 2a + 1, -2a + 1, 2a + 2b + 1$
15. $-\dfrac{1}{3}, -3, \dfrac{1}{3}, \dfrac{1 - a}{1 + a}, \dfrac{2 - a}{a}$, undefined
17. $-4, 10, -2, 3\sqrt{2}, 2x^2 + 7x + 1, 2x^2 - 3x - 4$
19. $6, 2, 1, 2, 2|x|, 2(x^2 + 1)$ **21.** 4, 1, 1, 2, 3
23. $8, -\frac{3}{4}, -1, 0, 1$ **25.** $x^2 + 4x + 5, x^2 + 6$
27. $x^2 + 4, x^2 + 8x + 16$ **29.** $3a + 2, 3(a + h) + 2, 3$
31. 5, 5, 0
33. $3 - 5a + 4a^2$,
$3 - 5a - 5h + 4a^2 + 8ah + 4h^2, -5 + 8a + 4h$
35. $(-\infty, \infty)$ **37.** $[-1, 5]$ **39.** $\{x \mid x \neq 3\}$
41. $\{x \mid x \neq \pm 1\}$ **43.** $[5, \infty)$ **45.** $(-\infty, \infty)$ **47.** $\left[\frac{5}{2}, \infty\right)$
49. $[-2, 3) \cup (3, \infty)$ **51.** $(-\infty, 0] \cup [6, \infty)$

53. This person's weight increases as he grows, then continues to increase; the person then goes on a crash diet (possibly) at age 30, then gains weight again, the weight gain eventually leveling off.

55.

57.

59.

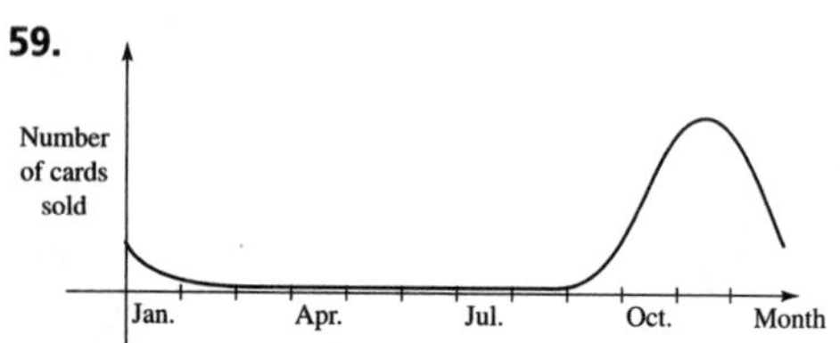

61. **(a)** 500 MW, 725 MW **(b)** Between 3:00 A.M. and 4:00 A.M. **(c)** Just before noon

63.

Section 4.2 ■ page 228

1. **(a)** $1, -1, 3, 4$ **(b)** Domain $[-3, 4]$, range $[-1, 4]$
3. **(a)** $f(0)$ **(b)** $g(-3)$ **(c)** $-2, 2$
5. **(a)** Yes **(b)** No **(c)** Yes **(d)** No
7. Function, domain $[-3, 2]$, range $[-2, 2]$
9. Not a function
11. **(a)**

(b) $(-\infty, \infty)$

13. **(a)**

(b) $(-\infty, \infty)$

15. **(a)**

(b) $(-\infty, 9]$

17. **(a)**

(b) $[-4, 4]$

19.

21.

23.

25.

27.

29.

31.

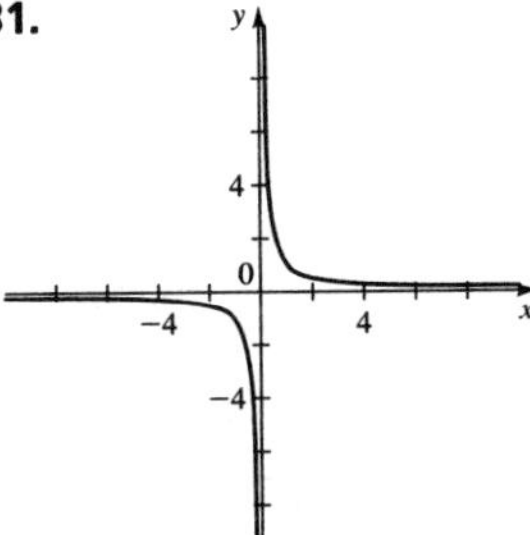

33.

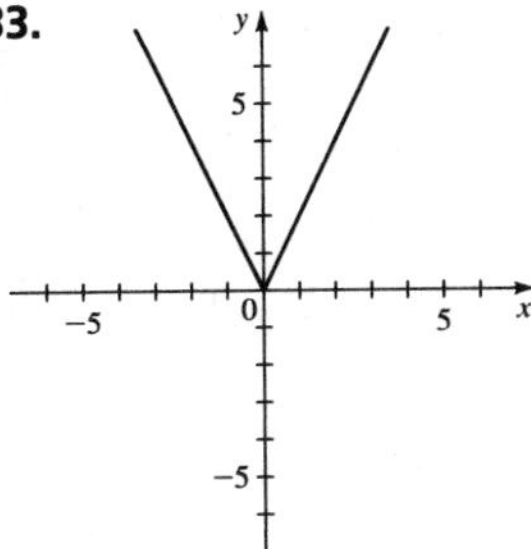

35.

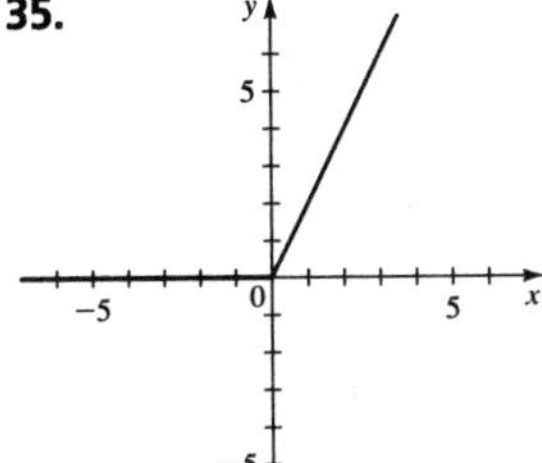

37.

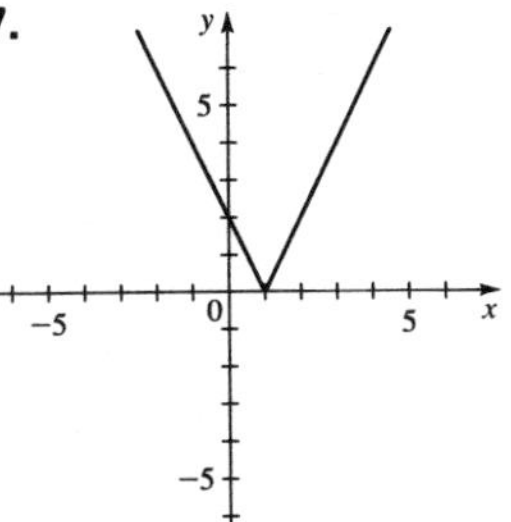

39.

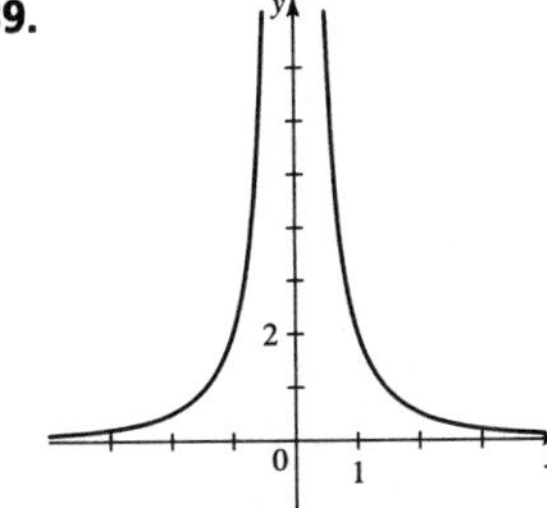

41. Yes
43. No
45. No
47. Yes
49. Yes
51. Yes

53. (a)

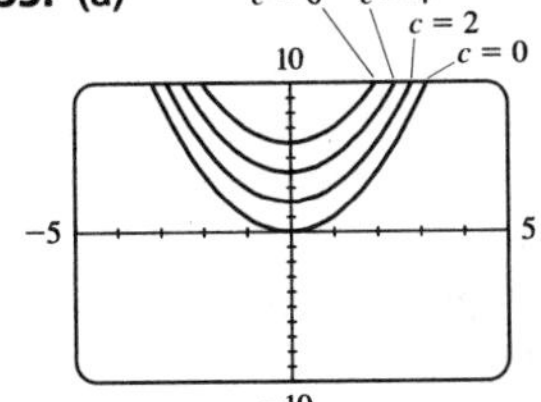

(b)

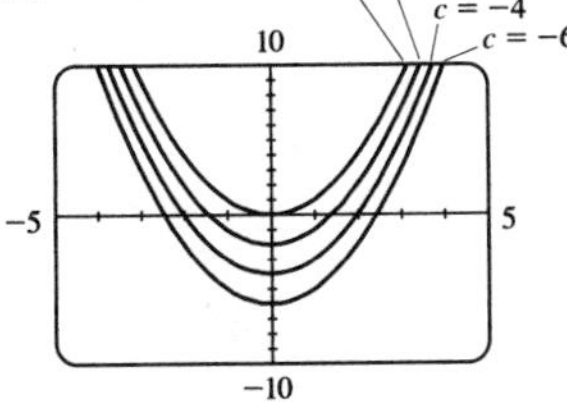

(c) If $c > 0$, then the graph of $f(x) = x^2 + c$ is the same as the graph of $y = x^2$ shifted upward c units. If $c < 0$, then the graph of $f(x) = x^2 + c$ is the same as the graph of $y = x^2$ shifted downward c units.

55. (a)

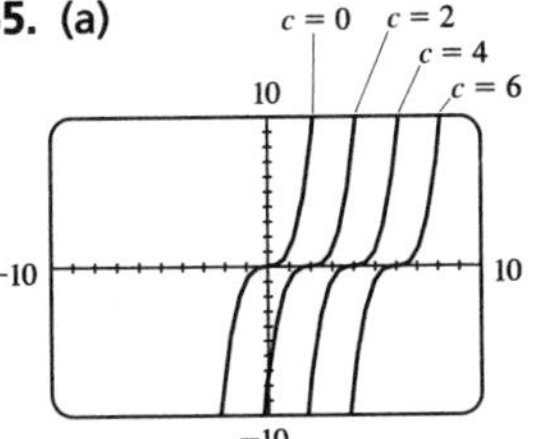

(b)

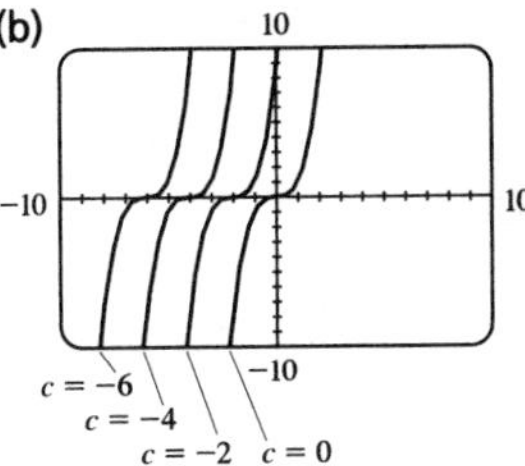

(c) If $c > 0$, then the graph of $f(x) = (x - c)^3$ is the same as the graph of $y = x^3$ shifted right c units. If $c < 0$, then the graph of $f(x) = (x - c)^3$ is the same as the graph of $y = x^3$ shifted left c units.

57. (a) (b)

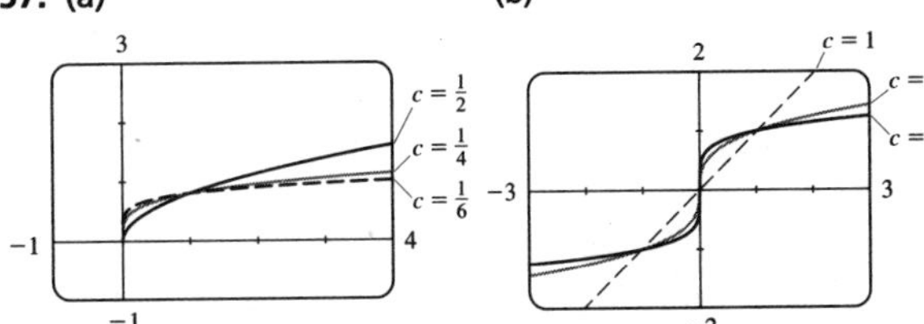

(c) Graphs of even roots are similar to $\sqrt{x}$; graphs of odd roots are similar to $\sqrt[3]{x}$. As c increases, the graph of $y = \sqrt[c]{x}$ becomes steeper near 0 and flatter when $x > 1$.

59.

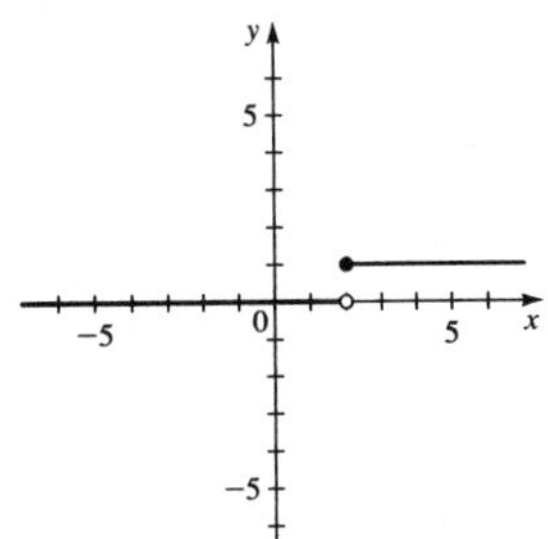

61.

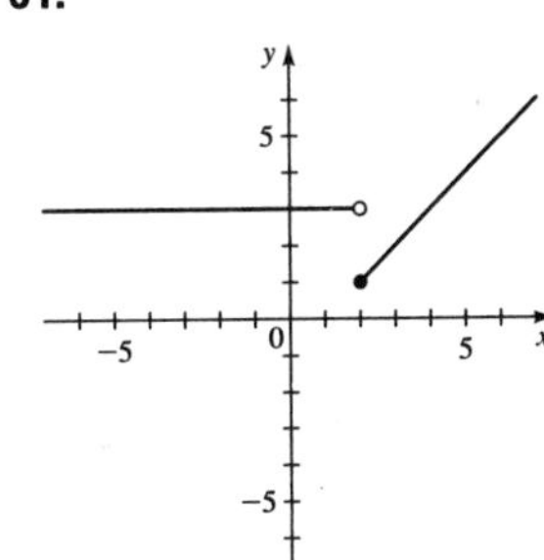

63.

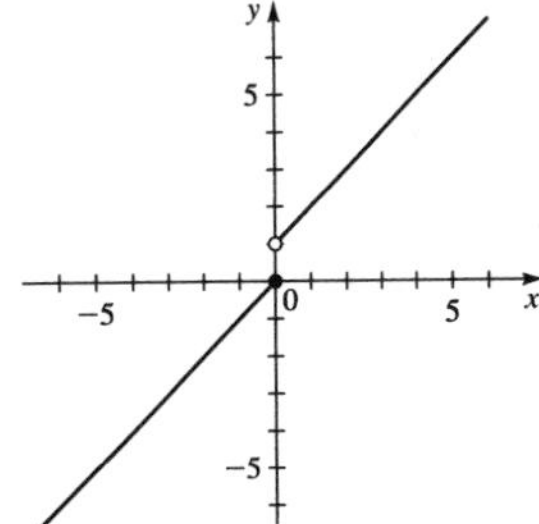

65.

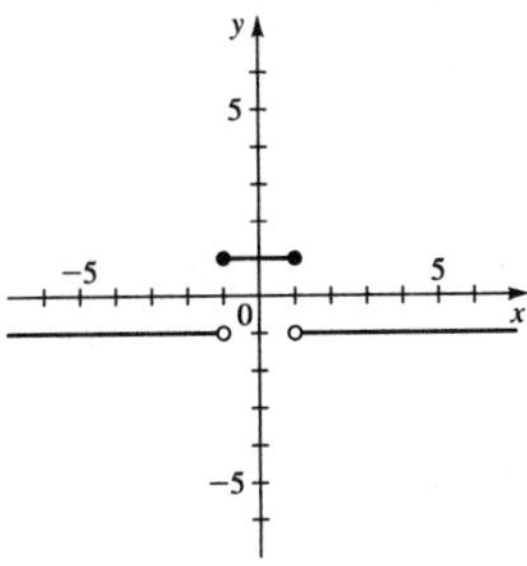

67.

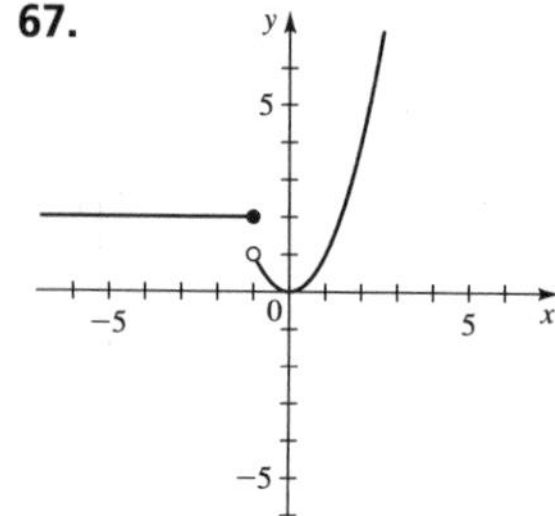

69.

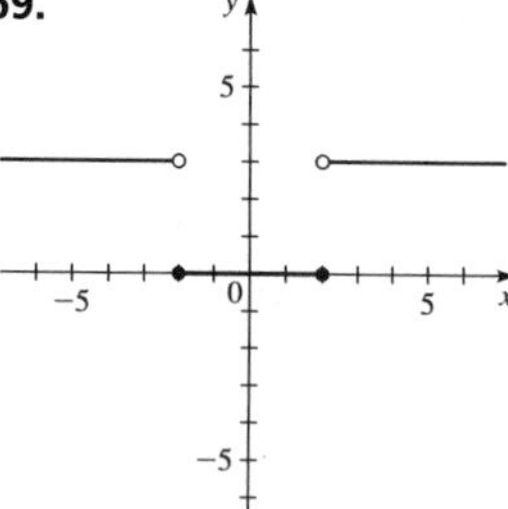

71.

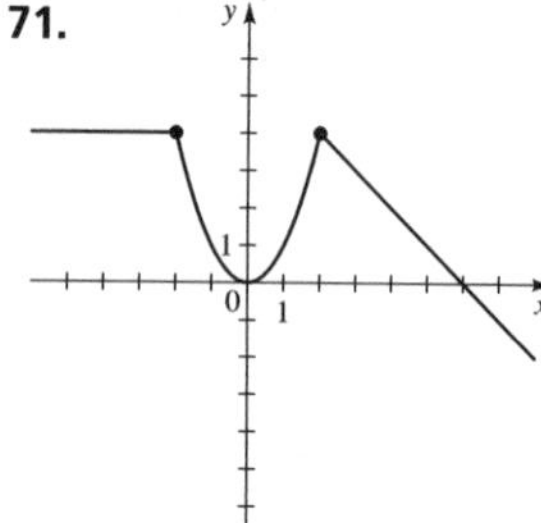

73.

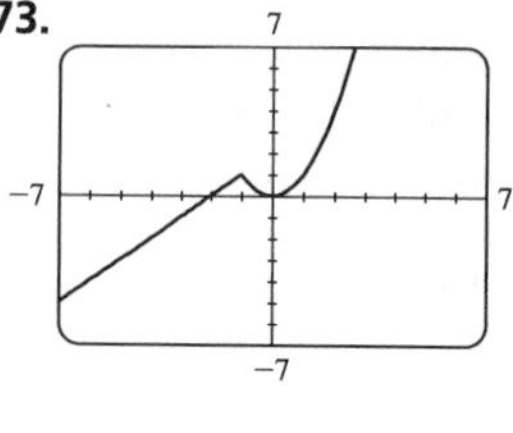

75.

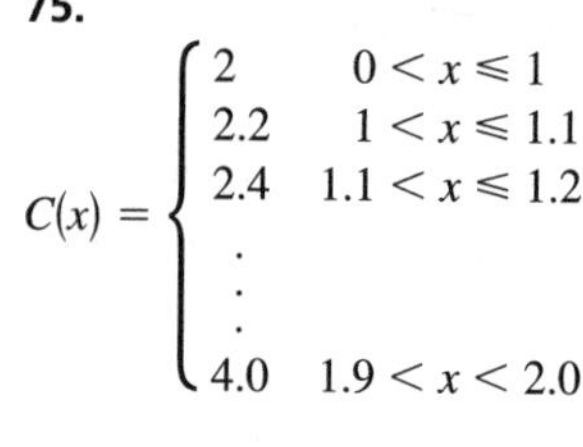

$$C(x) = \begin{cases} 2 & 0 < x \leqslant 1 \\ 2.2 & 1 < x \leqslant 1.1 \\ 2.4 & 1.1 < x \leqslant 1.2 \\ \vdots & \\ 4.0 & 1.9 < x < 2.0 \end{cases}$$

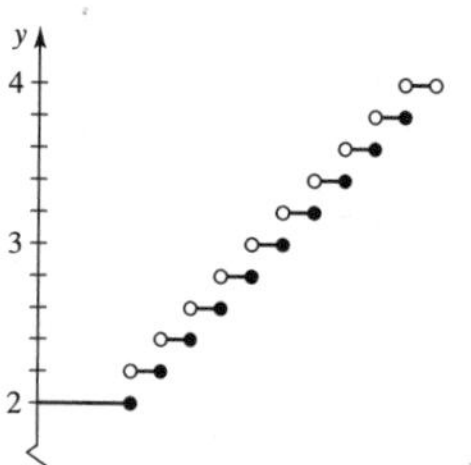

77. $f(x) = -\frac{7}{6}x - \frac{4}{3}, -2 \leqslant x \leqslant 4$ **79.** $f(x) = 1 - \sqrt{-x}$

Section 4.3 ■ page 235

1. $R = kt$ **3.** $v = k/z$ **5.** $y = ks/t$ **7.** $z = k\sqrt{y}$

9. $V = klwh$ **11.** $R = k\dfrac{i}{Pt}$ **13.** $y = 18x$

15. $M = 15x/y$ **17.** $W = 360/r^2$ **19.** $C = 16lwh$

21. $s = 500/\sqrt{t}$ **23.** (a) $F = kx$ (b) 8 (c) 32 N

25. (a) $C = kpm$ (b) 0.125 (c) \$57,500

27. (a) $R = kL/d^2$ (b) $0.00291\overline{6}$ (c) $R \approx 137\ \Omega$

29. \$3.75 **31.** \$51,200

Section 4.4 ■ page 245

1. 3 **3.** 5 **5.** 60 **7.** $12 + 3h$ **9.** $-1/a$ **11.** $\dfrac{-2}{a(a+h)}$

13. (a) $\frac{1}{2}$ **15.** $\frac{2}{3}$ **17.** $-\frac{4}{5}$

19. (a) 245 persons/yr (b) −328.5 persons/yr
(c) 1990–1994 (d) 1994–1999

21. (a) 7.2 units/yr (b) 8 units/yr (c) −55 units/yr
(d) 1996–1997, 1997–1998

23. Increasing on $[-1, 1]$, $[2, 4]$; decreasing on $[1, 2]$

25. Increasing on $[-2, -1]$, $[1, 2]$; decreasing on $[-3, -2]$, $[-1, 1]$, $[2, 3]$

27. (a)

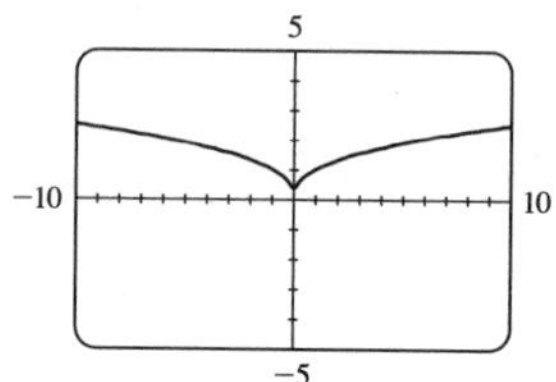

(b) Increasing on $[0, \infty)$; decreasing on $(-\infty, 0]$

29. (a)

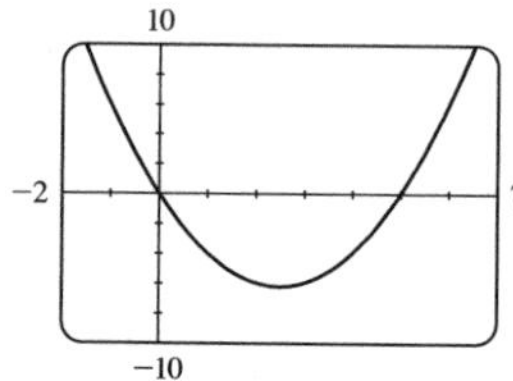

(b) Increasing on $[2.5, \infty)$; decreasing on $(-\infty, 2.5]$

31. (a)

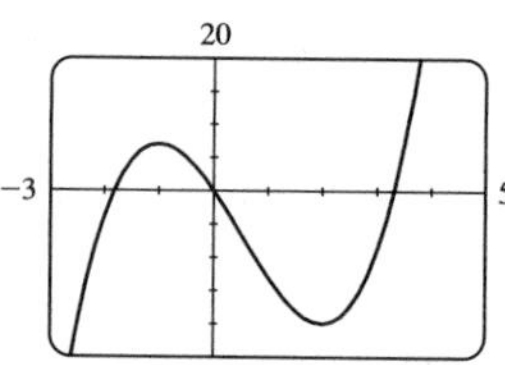

(b) Increasing on $(-\infty, -1]$, $[2, \infty)$; decreasing on $[-1, 2]$

33. (a)

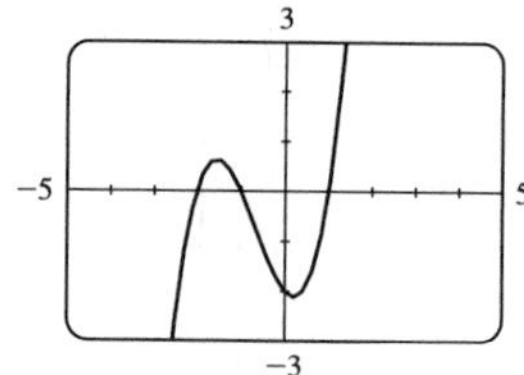

(b) Increasing on $(-\infty, -1.55]$, $[0.22, \infty)$; decreasing on $[-1.55, 0.22]$

Section 4.5 ■ page 256

1. (a) Shift downward 4 units (b) Shift right 4 units

3. (a) Stretch vertically by a factor of 3
(b) Shrink vertically by a factor of $\frac{1}{3}$

5. (a) Reflect in the x-axis and shift upward 5 units
(b) Reflect in the y-axis and shift upward 5 units

7. (a) Shift right 2 units and downward 3 units
(b) Shift right 3 units and stretch vertically by a factor of 2

9. (a)

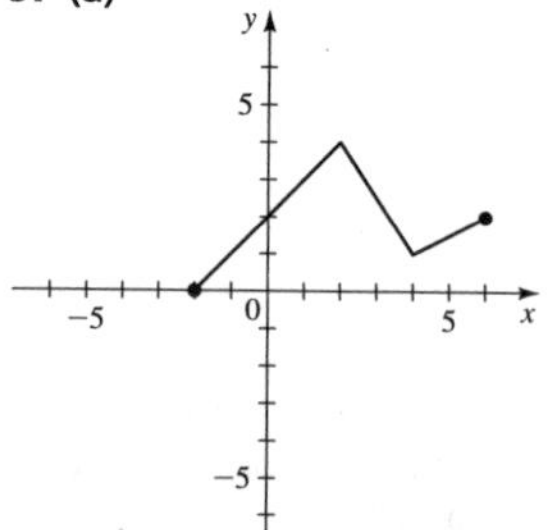

(b)

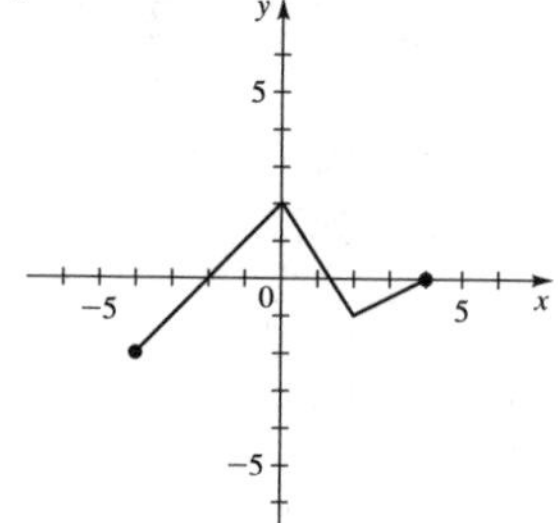

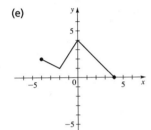

(c)

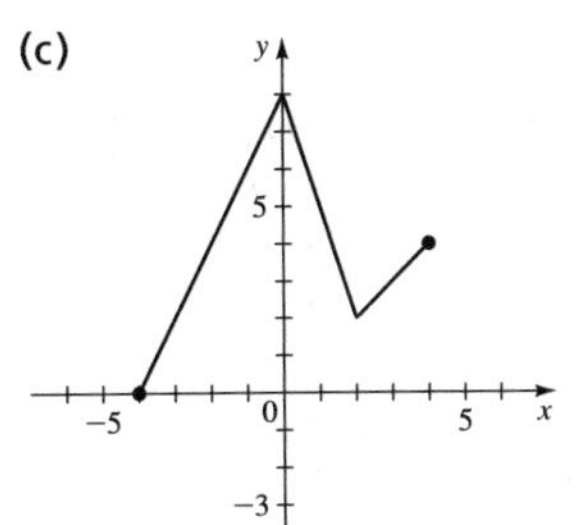

(d)

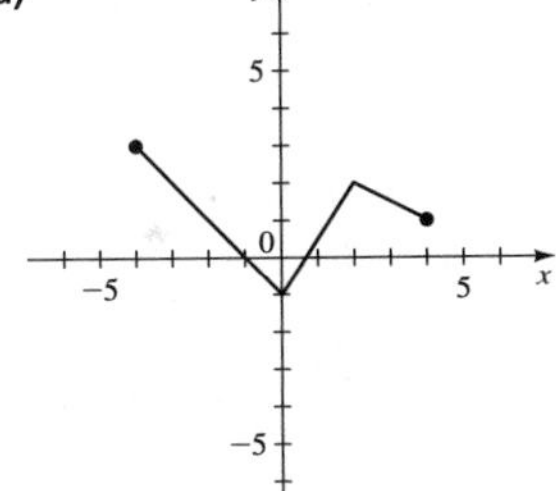

(e)

(f)

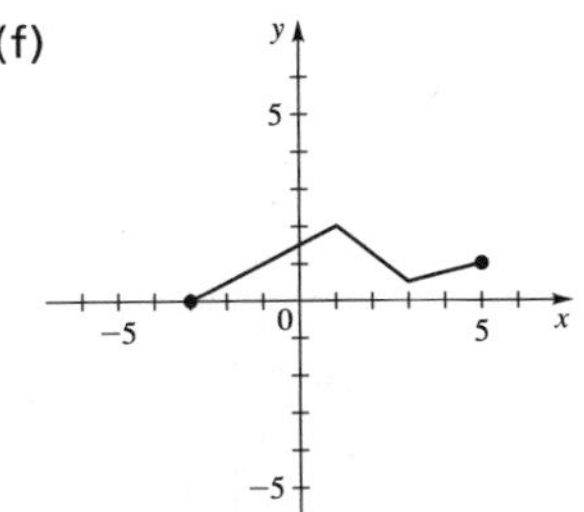

11. (a)

(b) (i)

(ii)

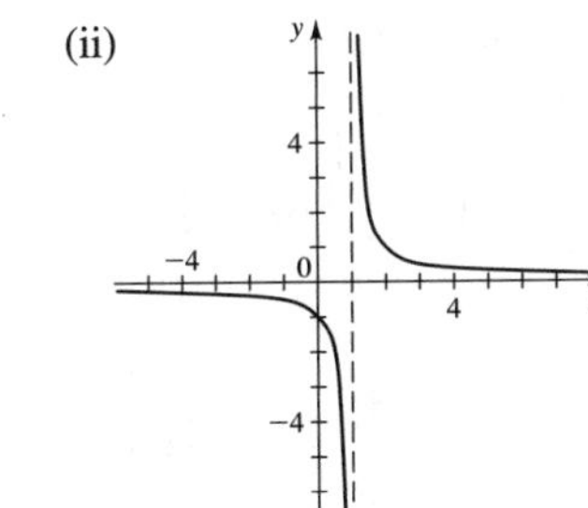

(iii)

(iv)

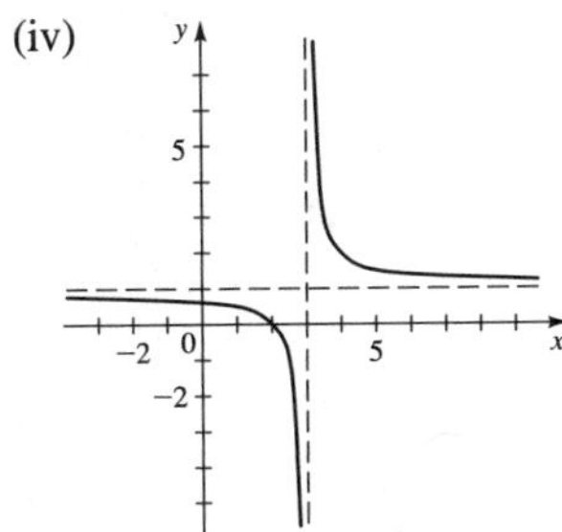

13. (a) Shift left 2 units
(b) Shift up 2 units

15. (a) Stretch vertically by a factor of 2
(b) Shift right 2 units, then shrink vertically by a factor of $\frac{1}{2}$

17.

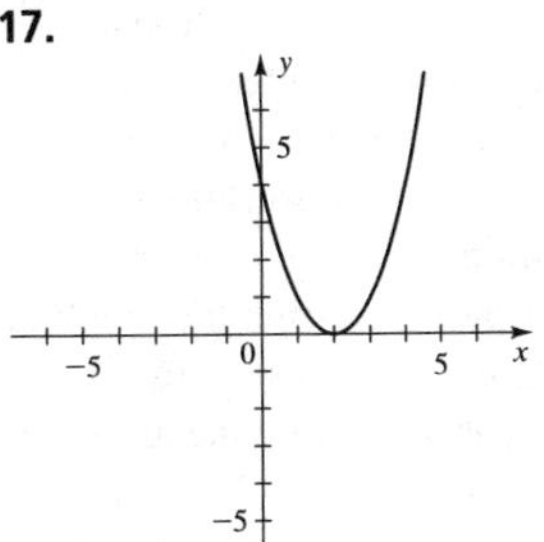

19.

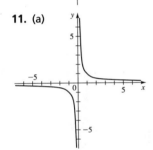

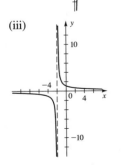

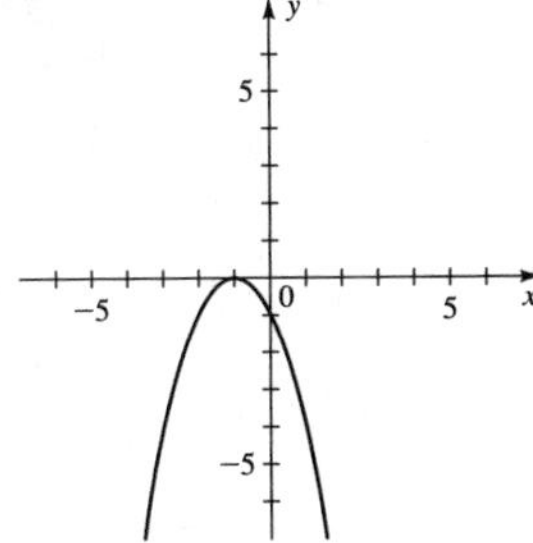

21.

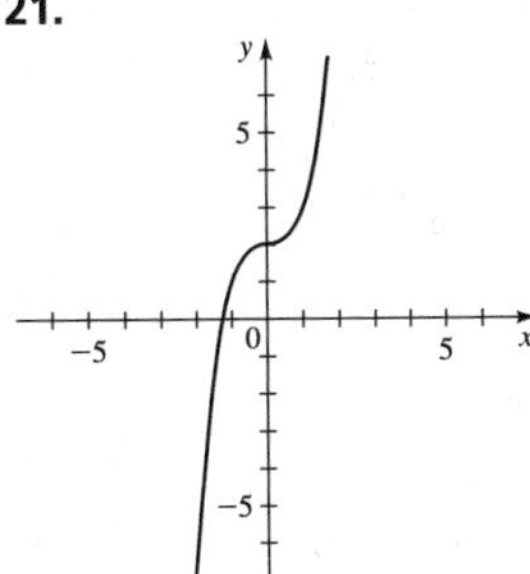

23.

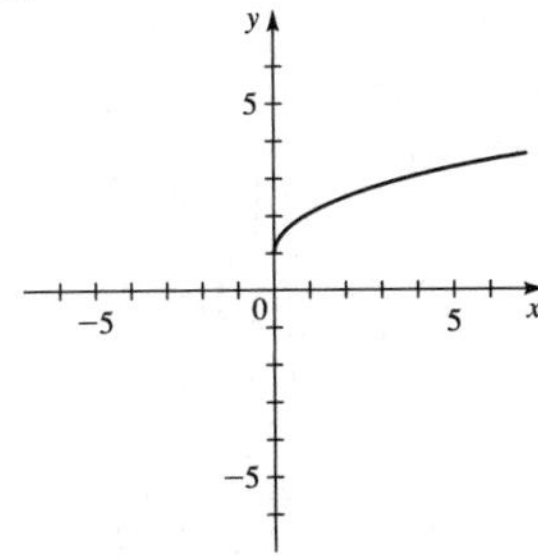

25.

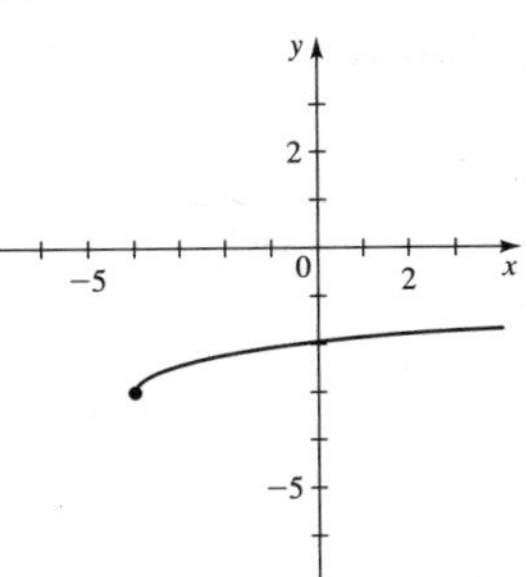

27.

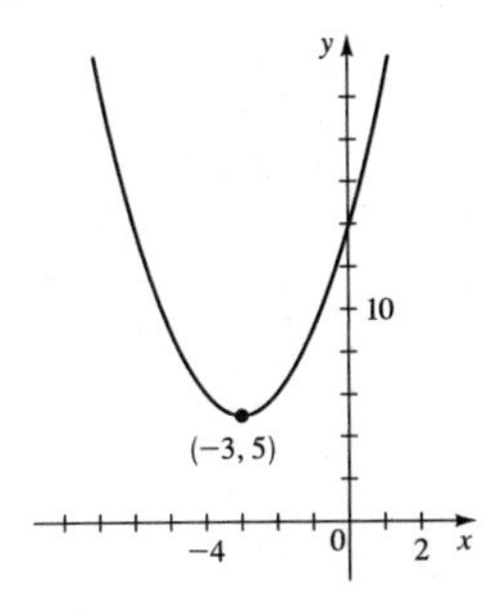

29.

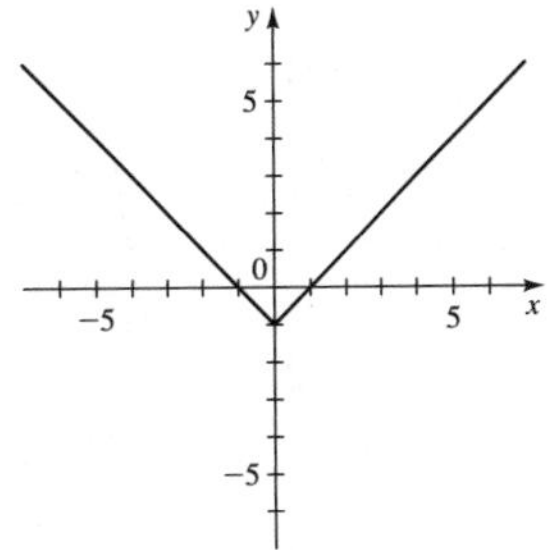

31.

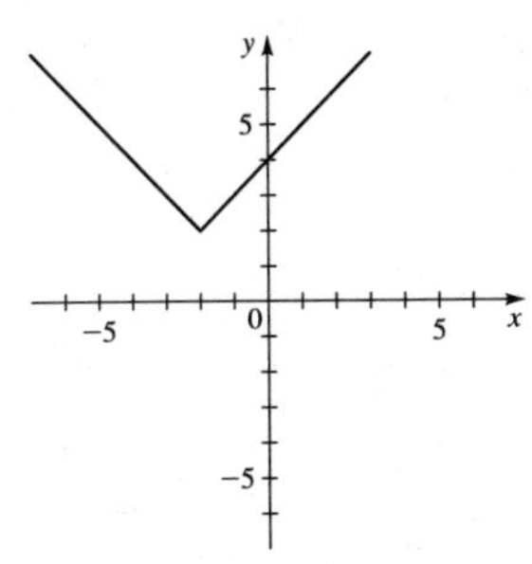

33. 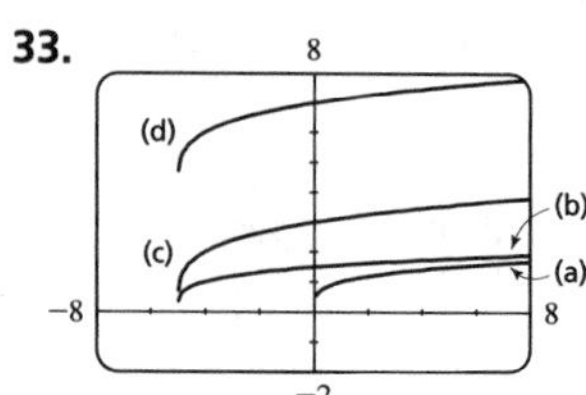

For part (b) shift the graph in (a) left 5 units; for part (c) shift the graph in (a) left 5 units and stretch vertically by a factor of 2; for part (d) shift the graph in (a) left 5 units, stretch vertically by a factor of 2, and then shift upward 4 units.

35. 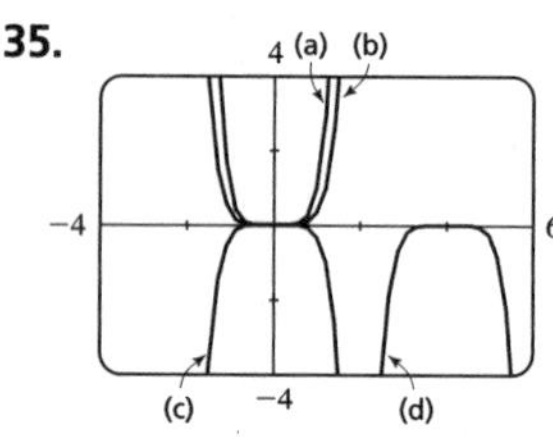

For part (b) shrink the graph in (a) vertically by a factor of $\frac{1}{3}$; for part (c) shrink the graph in (a) vertically by a factor of $\frac{1}{3}$ and reflect in the x-axis; for part (d) shift the graph in (a) right 4 units, shrink vertically by a factor of $\frac{1}{3}$, and then reflect in the x-axis.

37. Even

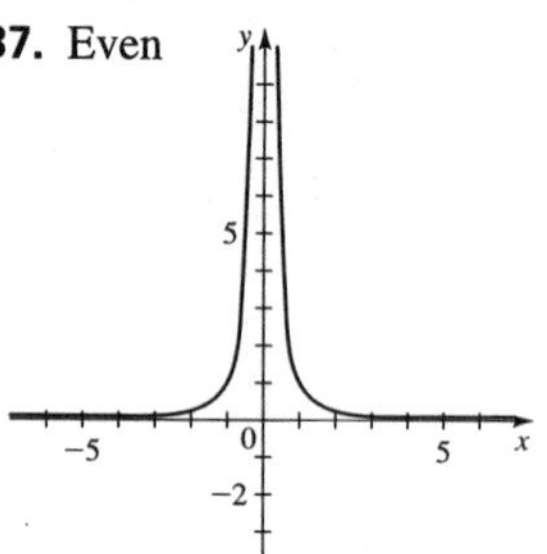

39. Neither

41. Odd

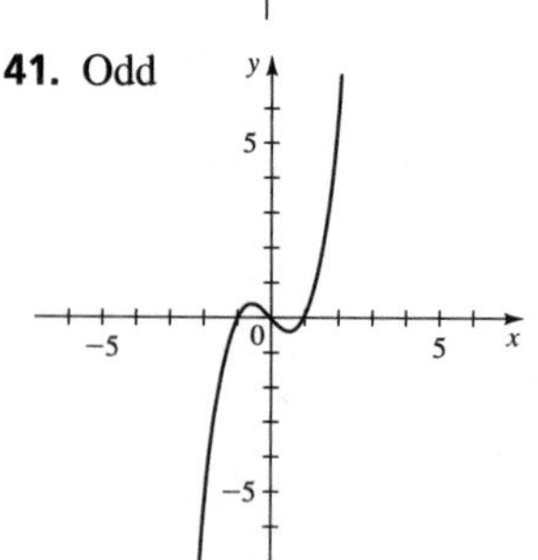

43. Neither

45. To obtain the graph of g, reflect in the x-axis the part of the graph of f that is below the x-axis.

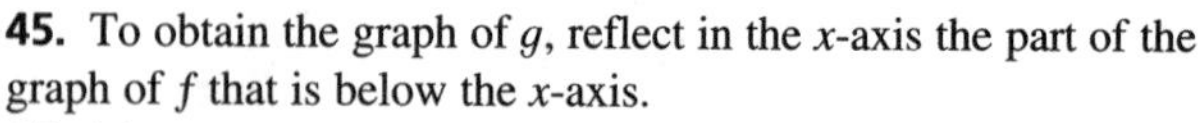

47. (a)

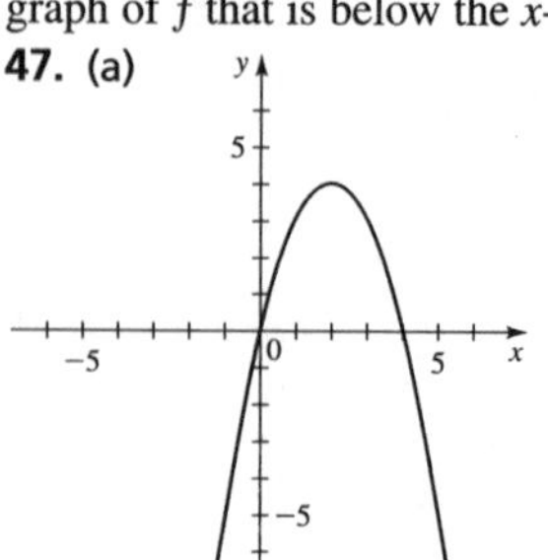

(b)

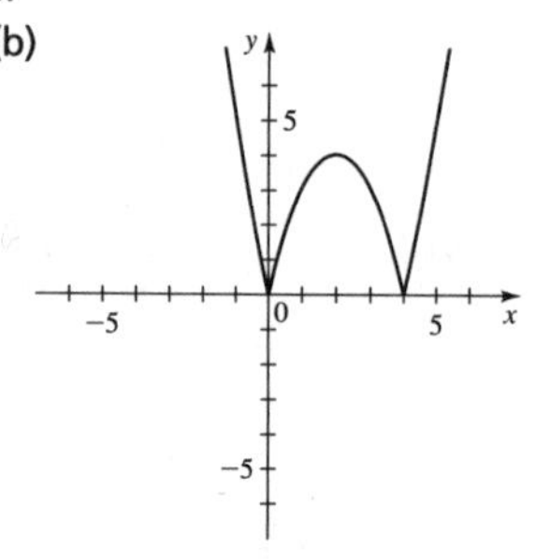

Section 4.6 ■ page 266

1. Vertex $(4, -16)$
x-intercepts 0, 8
y-intercept 0

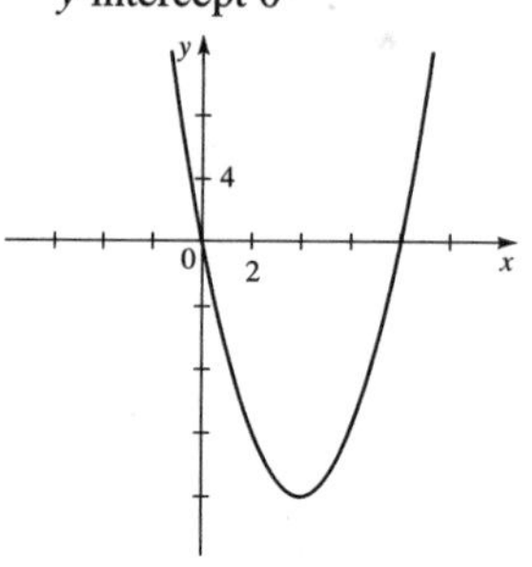

3. Vertex $\left(\frac{3}{2}, -\frac{9}{2}\right)$
x-intercepts 0, 3
y-intercept 0

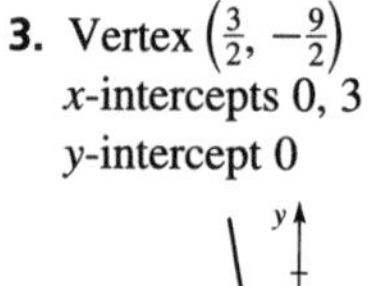

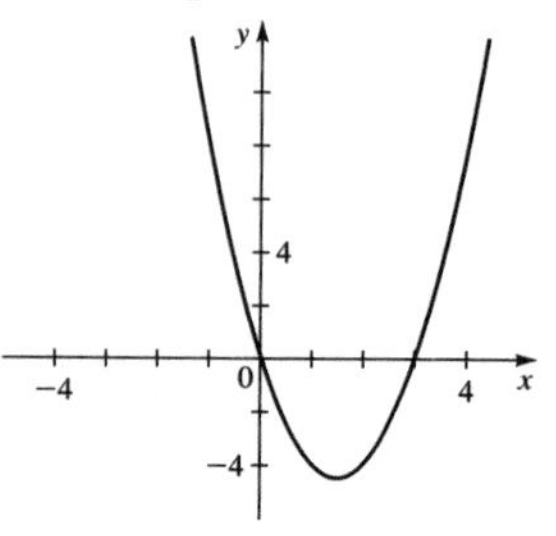

5. Vertex $(-2, -3)$
x-intercepts $-2 \pm \sqrt{3}$
y-intercept 1

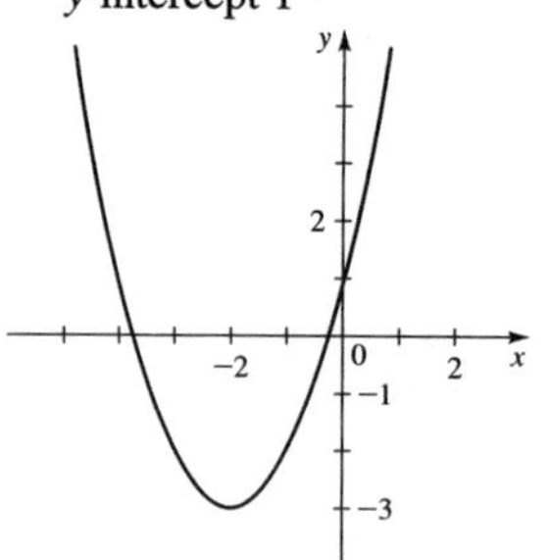

7. Vertex $(-3, -1)$
x-intercepts $-4, -2$
y-intercept 8

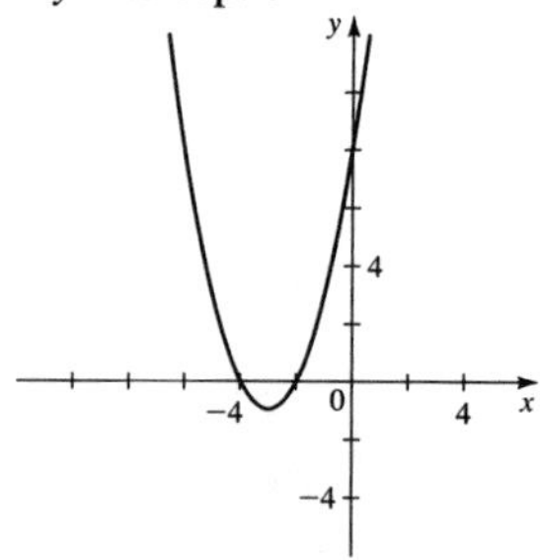

9. Vertex $(-1, 1)$
no x-intercept
y-intercept 3

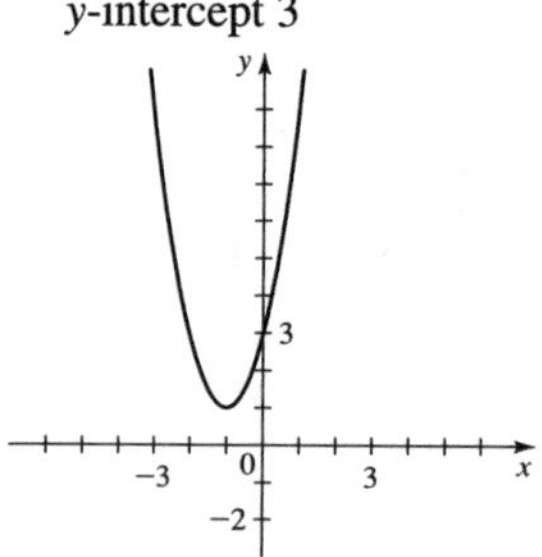

11. Vertex $(5, 7)$
no x-intercept
y-intercept 57

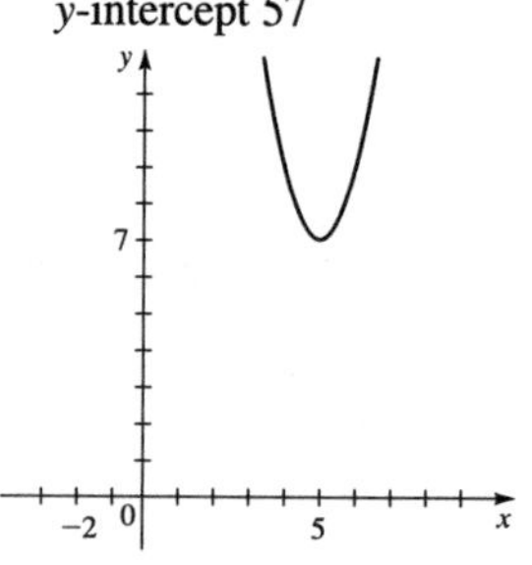

13. Vertex $(-2, 19)$
x-intercepts $-2 \pm \frac{1}{2}\sqrt{19}$
y-intercept 3

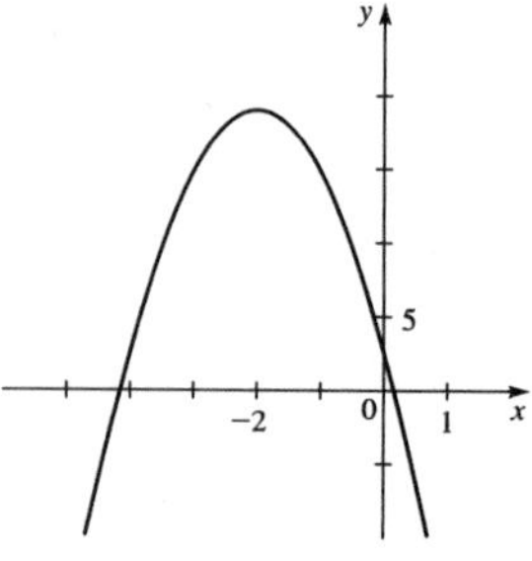

15. **(a)** $f(x) = -(x - 1)^2 + 1$
(b)

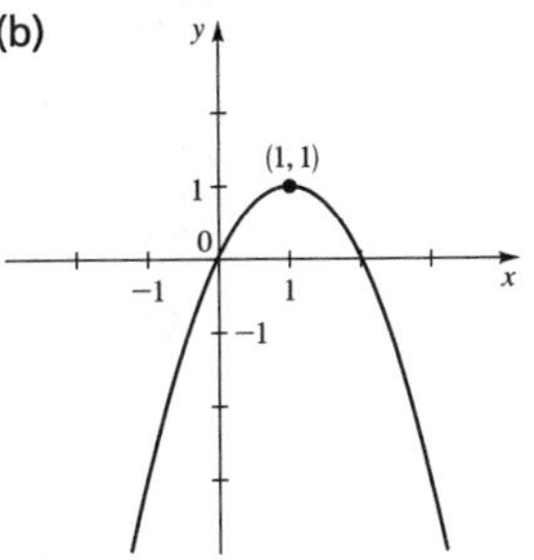

(c) Maximum $f(1) = 1$

17. **(a)** $(x + 1)^2 - 2$
(b)

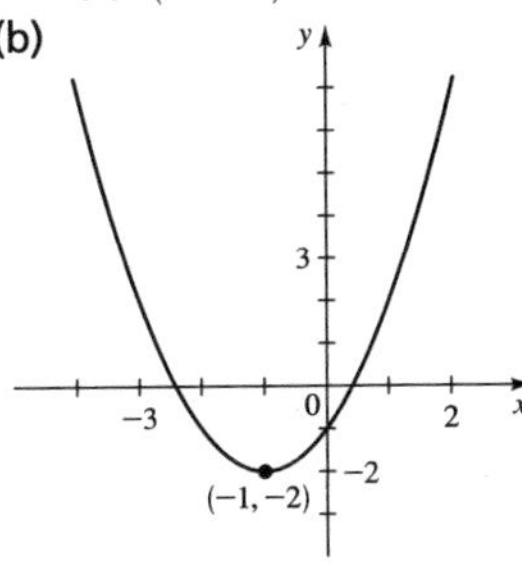

(c) Minimum $f(-1) = -2$

19. **(a)** $f(x) = -\left(x + \frac{3}{2}\right)^2 + \frac{21}{4}$
(b)

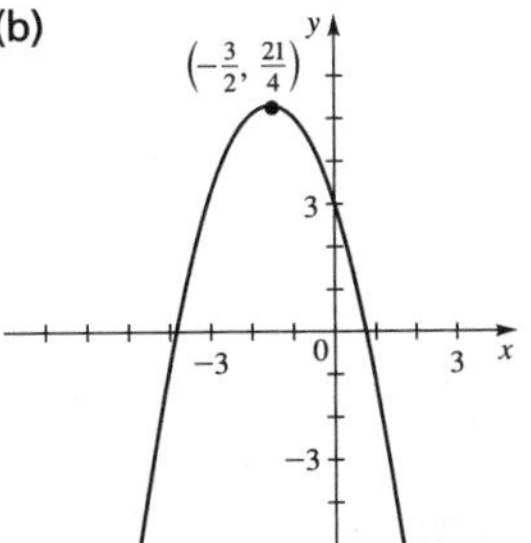

(c) Maximum $f\left(-\frac{3}{2}\right) = \frac{21}{4}$

21. **(a)** $g(x) = 3(x - 2)^2 + 1$
(b)

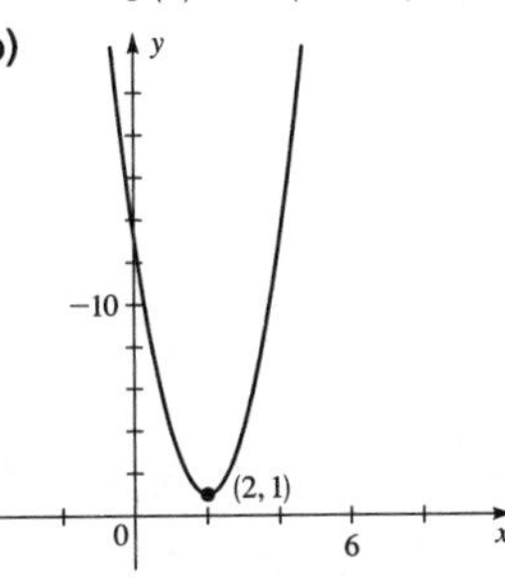

(c) Minimum $g(2) = 1$

23. **(a)** $h(x) = -\left(x + \frac{1}{2}\right)^2 + \frac{5}{4}$
(b)

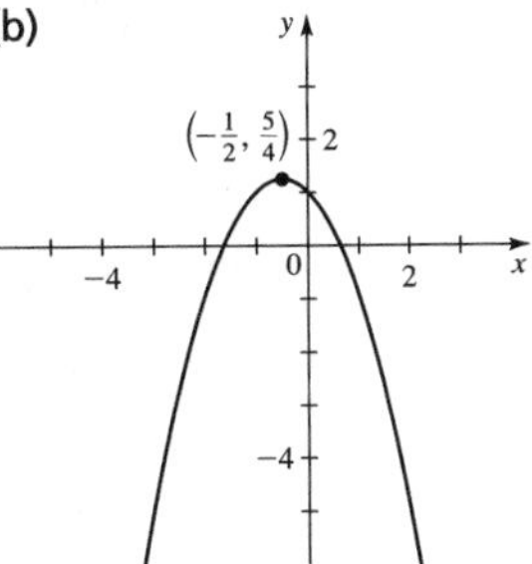

(c) Maximum $h\left(-\frac{1}{2}\right) = \frac{5}{4}$

25. Minimum $f\left(-\frac{1}{2}\right) = \frac{3}{4}$
27. Maximum $f(-3.5) = 185.75$
29. Minimum $f(0.6) = 15.64$
31. Minimum $h(-2) = -8$
33. Maximum $f(-1) = \frac{7}{2}$
35. $f(x) = 2x^2 - 4x$ **37.** $(-\infty, \infty), (-\infty, 1]$ **39.** 25 ft
41. \$4000, 100 units **43.** 30 times
45. **(a)** -4.01 **(b)** -4.011025
47. Local maximum ≈ 0.38 when $x \approx -0.58$; local minimum ≈ -0.38 when $x \approx 0.58$
49. Local maximum 0 when $x = 0$; local minimum ≈ -13.61 when $x \approx -1.71$; local minimum ≈ -73.32 when $x \approx 3.21$
51. Local maximum ≈ 5.66 when $x \approx 4.00$
53. Local maximum ≈ 0.38 when $x \approx -1.73$; local minimum ≈ -0.38 when $x \approx 1.73$
55. 7.5 mi/h **57.** 3.96°C **59.** 7

Section 4.7 ■ page 275

1. $(f + g)(x) = x^2 + 5, (-\infty, \infty)$;
$(f - g)(x) = x^2 - 2x - 5, (-\infty, \infty)$;
$(fg)(x) = x^3 + 4x^2 - 5x, (-\infty, \infty)$;
$(f/g)(x) = (x^2 - x)/(x + 5), (-\infty, -5) \cup (-5, \infty)$
3. $(f + g)(x) = \sqrt{1 + x} + \sqrt{1 - x}, [-1, 1]$;
$(f - g)(x) = \sqrt{1 + x} - \sqrt{1 - x}, [-1, 1]$;
$(fg)(x) = \sqrt{1 - x^2}, [-1, 1]$;
$(f/g)(x) = \sqrt{(1 + x)/(1 - x)}, [-1, 1)$
5. $(f + g)(x) = 8/(x(x + 4)), x \neq -4, x \neq 0$;
$(f - g)(x) = 4(x + 2)/(x(x + 4)), x \neq -4, x \neq 0$;
$(fg)(x) = -4/(x(x + 4)), x \neq -4, x \neq 0$;
$(f/g)(x) = -(x + 4)/x, x \neq -4, x \neq 0$
7. $[0, 1]$ **9.** $[4, \infty)$
11.

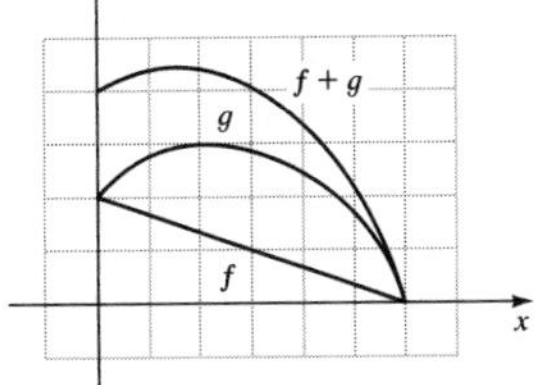

13.

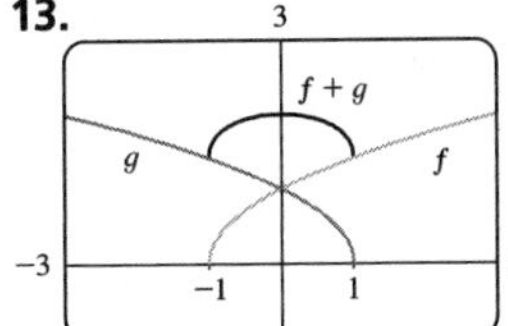

15.

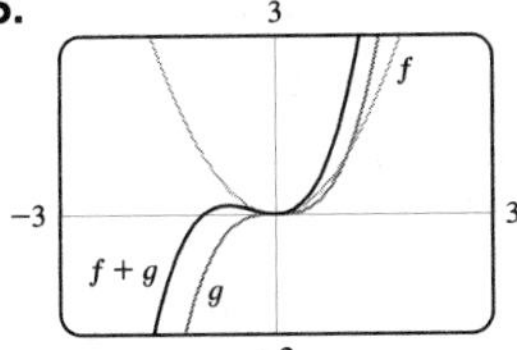

17. **(a)** 1 **(b)** -23 **19.** **(a)** -11 **(b)** -119
21. **(a)** $-3x^2 + 1$ **(b)** $-9x^2 + 30x - 23$
23. 4 **25.** 5 **27.** 4
29. $(f \circ g)(x) = 8x + 1, (-\infty, \infty)$;
$(g \circ f)(x) = 8x + 11, (-\infty, \infty)$;
$(f \circ f)(x) = 4x + 9, (-\infty, \infty)$;
$(g \circ g)(x) = 16x - 5, (-\infty, \infty)$
31. $(f \circ g)(x) = (x + 1)^2, (-\infty, \infty)$
$(g \circ f)(x) = x^2 + 1, (-\infty, \infty)$
$(f \circ f)(x) = x^4, (-\infty, \infty)$
$(g \circ g)(x) = x + 2, (-\infty, \infty)$

33. $(f \circ g)(x) = \dfrac{1}{2x + 4}, x \neq -2$
$(g \circ f)(x) = \dfrac{2}{x} + 4, x \neq 0$
$(f \circ f)(x) = x, x \neq 0$
$(g \circ g)(x) = 4x + 12, (-\infty, \infty)$
35. $(f \circ g)(x) = |2x + 3|, (-\infty, \infty)$
$(g \circ f)(x) = 2|x| + 3, (-\infty, \infty)$
$(f \circ f)(x) = |x|, (-\infty, \infty)$
$(g \circ g)(x) = 4x + 9, (-\infty, \infty)$
37. $(f \circ g)(x) = \dfrac{2x - 1}{2x}, x \neq 0$
$(g \circ f)(x) = \dfrac{2x}{x + 1} - 1, x \neq -1$
$(f \circ f)(x) = \dfrac{x}{2x + 1}, x \neq -1, x \neq -\frac{1}{2}$
$(g \circ g)(x) = 4x - 3, (-\infty, \infty)$
39. $(f \circ g)(x) = \sqrt[12]{x}, [0, \infty)$
$(g \circ f)(x) = \sqrt[12]{x}, [0, \infty)$
$(f \circ f)(x) = \sqrt[9]{x}, (-\infty, \infty)$
$(g \circ g)(x) = \sqrt[16]{x}, [0, \infty)$
41. $(f \circ g \circ h)(x) = \sqrt{x - 1} - 1$
43. $(f \circ g \circ h)(x) = (\sqrt{x} - 5)^4 + 1$
45. $g(x) = x - 9, f(x) = x^5$
47. $g(x) = x^2, f(x) = x/(x + 4)$
49. $g(x) = 1 - x^3, f(x) = |x|$
51. $h(x) = x^2, g(x) = x + 1, f(x) = 1/x$
53. $h(x) = \sqrt[3]{x}, g(x) = 4 + x, f(x) = x^9$
55. $A(t) = 3600\pi t^2$
57. (a) $s = f(d) = \sqrt{1 + d^2}$
(b) $d = g(t) = 350t$
(c) $s = f(g(t)) = \sqrt{1 + 122{,}500t^2}$

Section 4.8 ■ page 287

1. No **3.** Yes **5.** No **7.** Yes **9.** Yes **11.** Yes **13.** No
15. No **17.** (a) 2 (b) 3 **19.** 1
31. $f^{-1}(x) = \frac{1}{2}(x - 1)$
33. $f^{-1}(x) = \frac{1}{4}(x - 7)$
35. $f^{-1}(x) = 2x$
37. $f^{-1}(x) = (1/x) - 2$
39. $f^{-1}(x) = (5x - 1)/(2x + 3)$
41. $f^{-1}(x) = \frac{1}{5}(x^2 - 2), x \geq 0$
43. $f^{-1}(x) = \sqrt{4 - x}$
45. $f^{-1}(x) = (x - 4)^3$
47. $f^{-1}(x) = x^2 - 2x, x \geq 1$
49. $f^{-1}(x) = \sqrt[4]{x}$

51. (a)

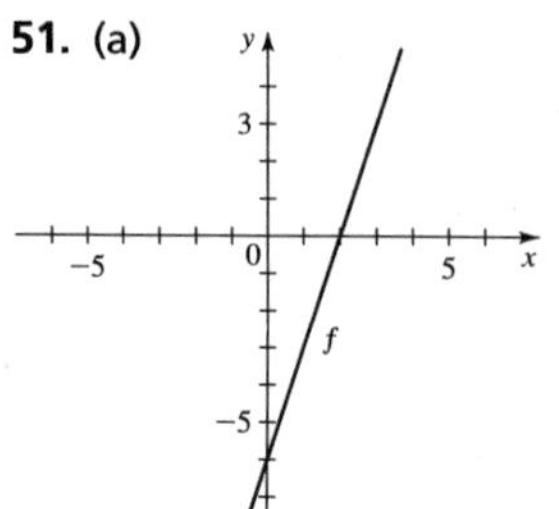

(b)

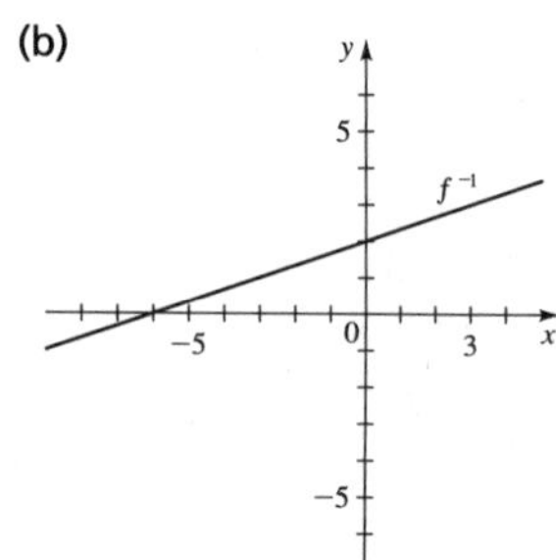

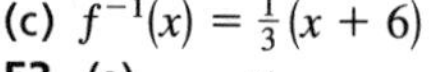

(c) $f^{-1}(x) = \frac{1}{3}(x + 6)$

53. (a)

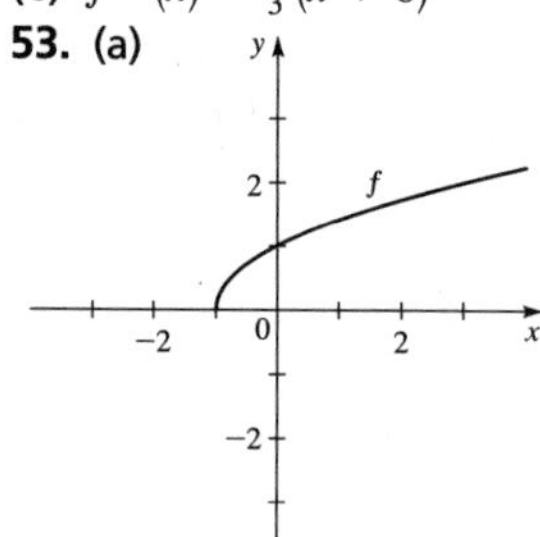

(b)

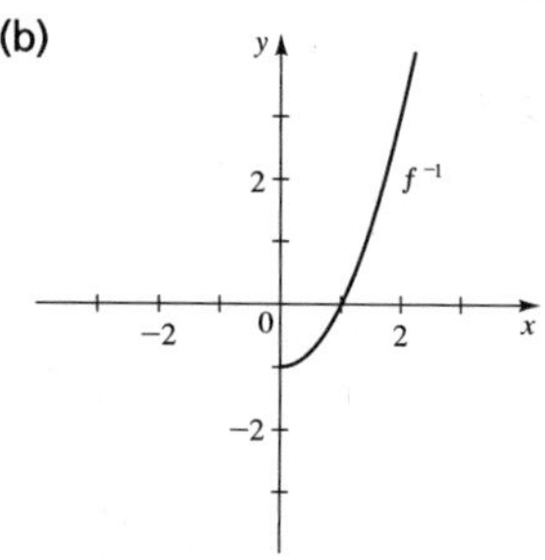

(c) $f^{-1}(x) = x^2 - 1, x \geq 0$

55. Not one-to-one

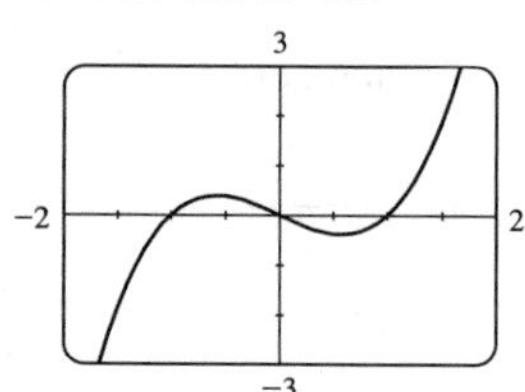

57. One-to-one

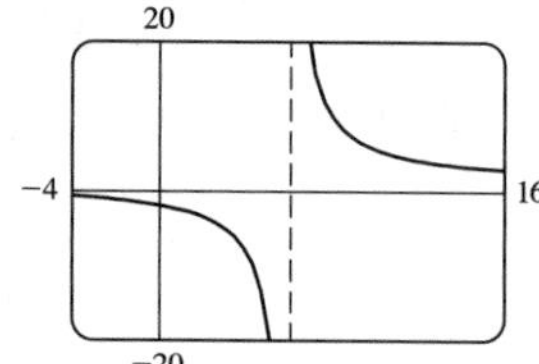

59. Not one-to-one

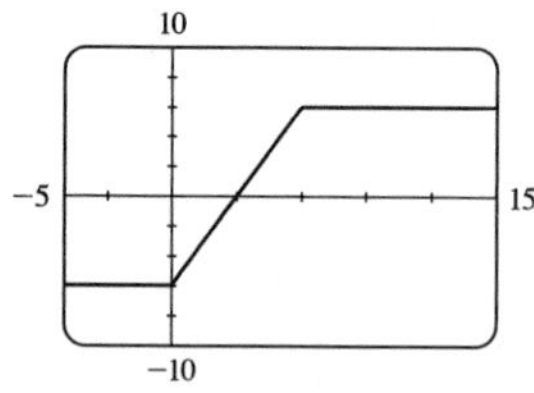

61. $x \geq 0, f^{-1}(x) = \sqrt{4 - x}$
63. $x \geq -2, h^{-1}(x) = \sqrt{x} - 2$
65.

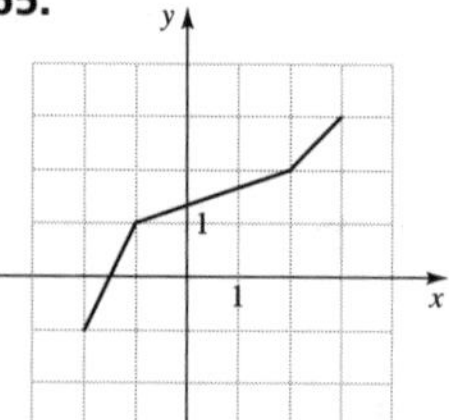

Chapter 4 Review ■ page 289

1. $1, 3, 7, a^2 - a + 1, a^2 + a + 1, x^2 + x + 1, 4x^2 - 2x + 1, 2x^2 - 2x$

3. (a) $-1, 2$ (b) $[-4, 5]$ (c) $[-4, 4]$
(d) Increasing on $[-4, -2]$ and $[-1, 4]$; decreasing on $[-2, -1]$ and $[4, 5]$ (e) No

5. Domain $[-3, \infty)$, range $[0, \infty)$ **7.** $(-\infty, \infty)$

9. $[-4, \infty)$ **11.** $\{x \mid x \neq -2, -1, 0\}$

13. $(-\infty, -1] \cup [1, 4]$

15.

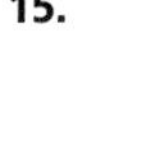

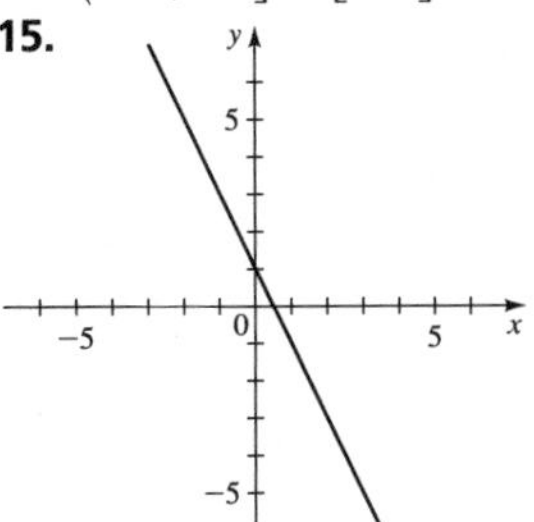

17.

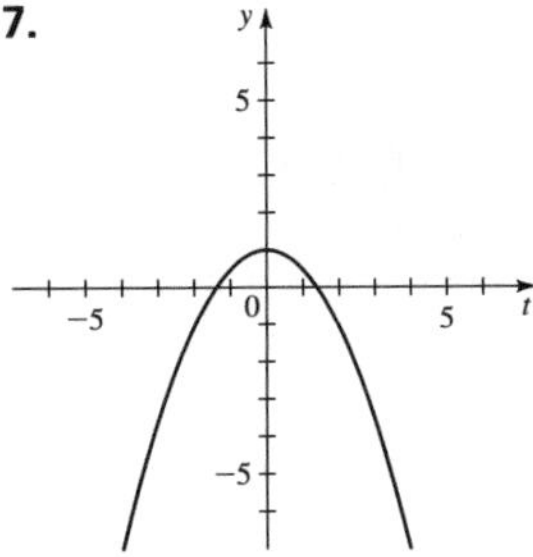

19.

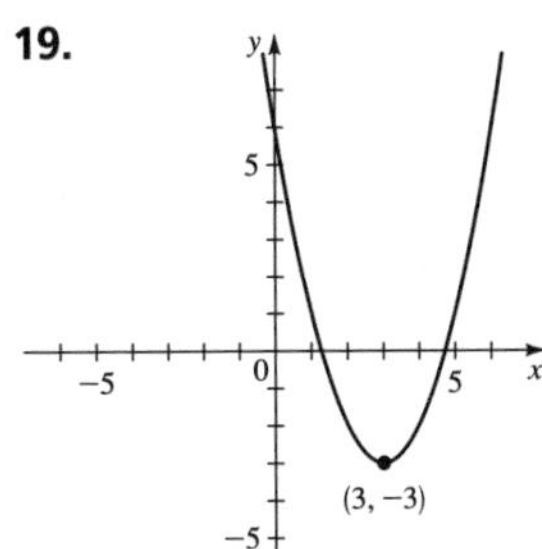

21.

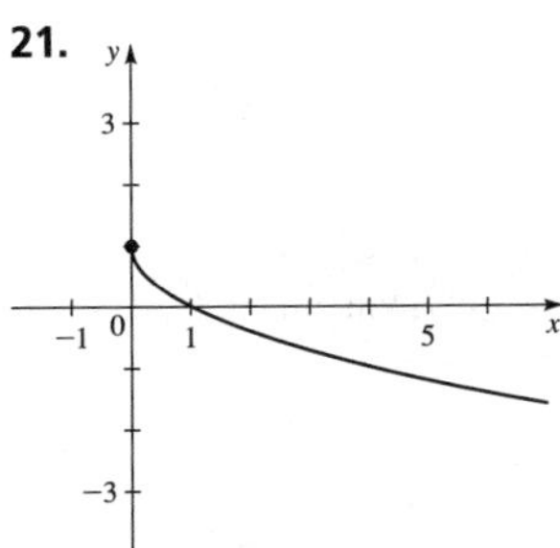

23.

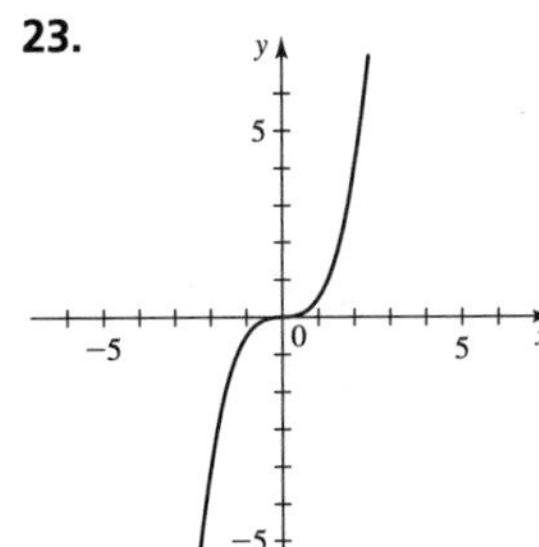

25.

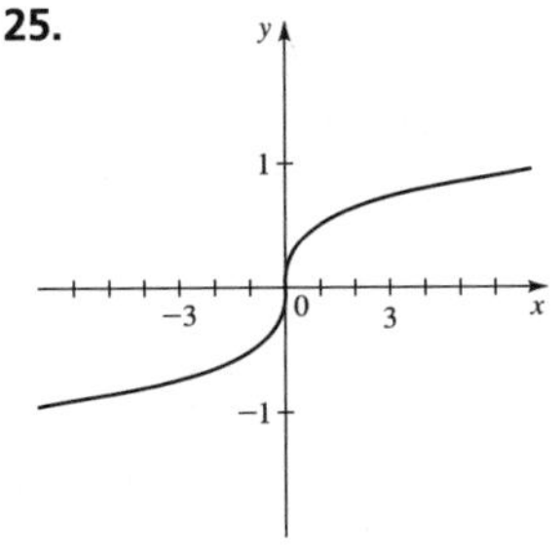

27.

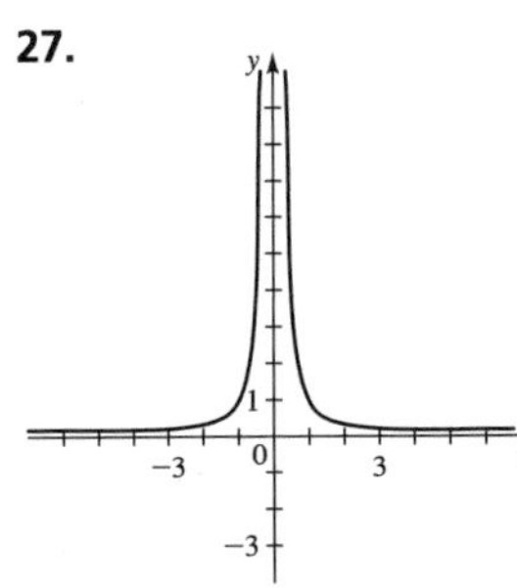

29.

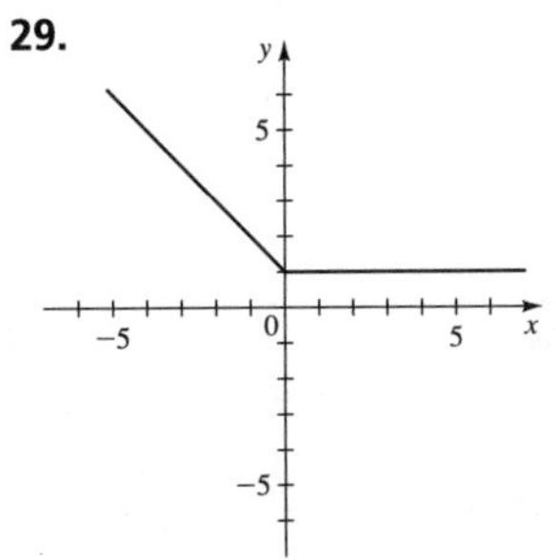

31.

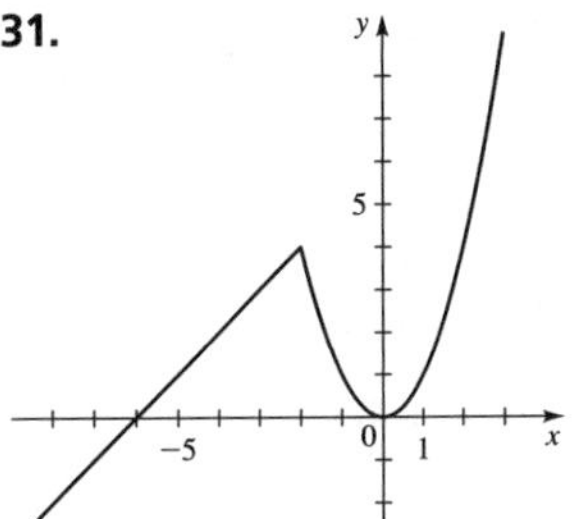

33. (iii)

35. **37.**

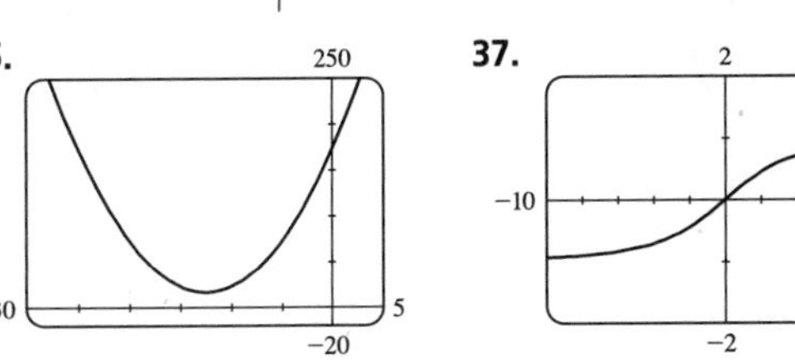

39. $[-2.1, 0.2] \cup [1.9, \infty)$

41. $M = 8z$ **43.** (a) $I = k/d^2$ (b) 64,000 (c) 160 candles

45. 11.0 mi/h **47.** 5 **49.** $\dfrac{-1}{3(3 + h)}$

51.

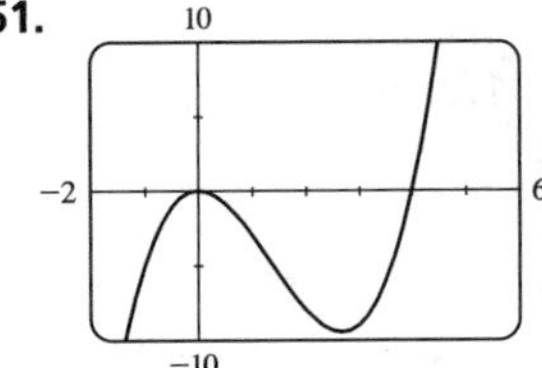

Increasing on $(-\infty, 0]$, $[2.67, \infty)$; decreasing on $[0, 2.67]$

53. (a) Shift upward 8 units (b) Shift left 8 units
(c) Stretch vertically by a factor of 2, then shift upward 1 unit
(d) Shift right 2 units and downward 2 units
(e) Reflect in y-axis
(f) Reflect in y-axis, then in x-axis
(g) Reflect in x-axis (h) Reflect in line $y = x$

55. (a) Neither (b) Odd (c) Even (d) Neither

57. $f(x) = (x + 2)^2 - 3$ **59.** $g(-1) = -7$ **61.** 68 ft

63. Local maximum ≈ 3.79 when $x \approx 0.46$; local minimum ≈ 2.81 when $x \approx -0.46$

65. (a) $(f + g)(x) = x^2 - 6x + 6$
(b) $(f - g)(x) = x^2 - 2$
(c) $(fg)(x) = -3x^3 + 13x^2 - 18x + 8$
(d) $(f/g)(x) = (x^2 - 3x + 2)/(4 - 3x)$
(e) $(f \circ g)(x) = 9x^2 - 15x + 6$
(f) $(g \circ f)(x) = -3x^2 + 9x - 2$

67. $(f \circ g)(x) = -3x^2 + 6x - 1, (-\infty, \infty)$;
$(g \circ f)(x) = -9x^2 + 12x - 3, (-\infty, \infty)$;
$(f \circ f)(x) = 9x - 4, (-\infty, \infty)$;
$(g \circ g)(x) = -x^4 + 4x^3 - 6x^2 + 4x, (-\infty, \infty)$

69. $(f \circ g \circ h)(x) = 1 + \sqrt{x}$ **71.** Yes **73.** No **75.** No

77. $f^{-1}(x) = \dfrac{x + 2}{3}$ **79.** $f^{-1}(x) = \sqrt[3]{x} - 1$

81. (a), (b)

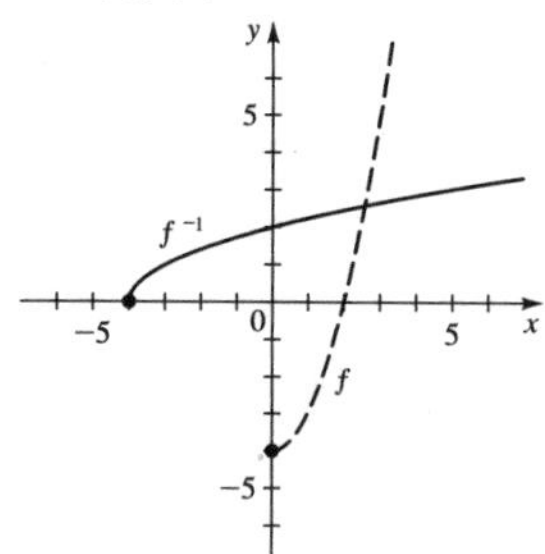

(c) $f^{-1}(x) = \sqrt{x + 4}$

Chapter 4 Test ■ page 294

1. (a) and (b) are graphs of functions, (a) is one-to-one

2. (a) $f(4) = \frac{2}{3}$, $f(6) = \frac{\sqrt{6}}{5}$, $f(a + 1) = \frac{\sqrt{a + 1}}{a}$

(b) $[0, 1) \cup (1, \infty)$

3. (a) $M = kwh^2/L$ (b) 400 (c) 12,000 lb **4.** $h + 7$

5. (a) (b)

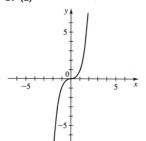

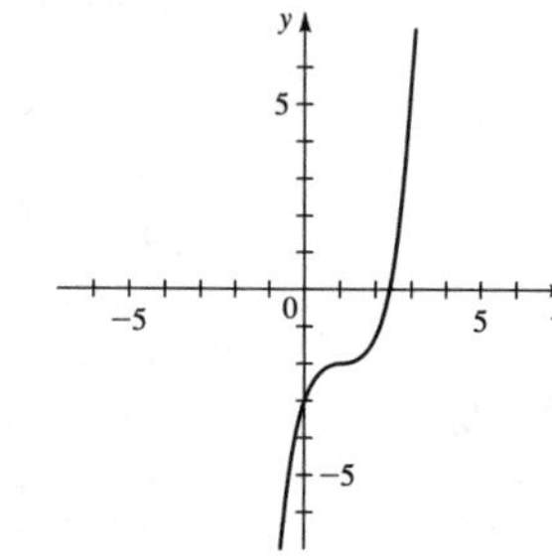

6. (a) Shift right 3 units, then shift upward 2 units

(b) Reflect in y-axis

7. (a) $f(x) = 2(x - 2)^2 + 5$

(b) $f(2) = 5$

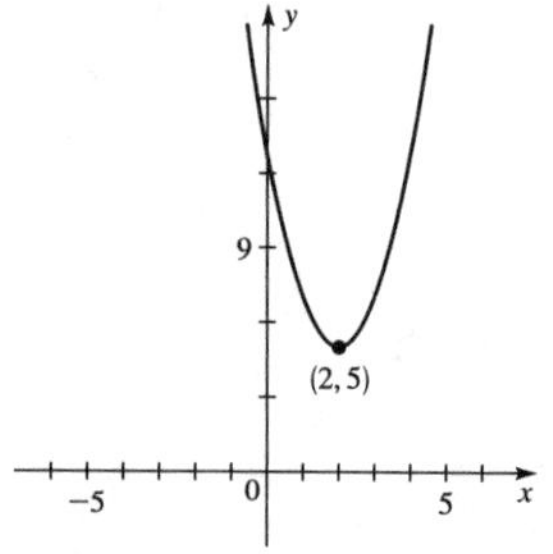

8. (a) $-3, 3$ (b)

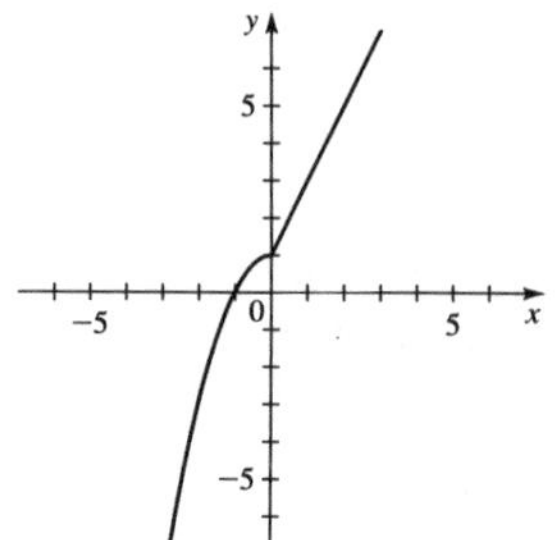

9. (a) $(f \circ g)(x) = 4x^2 - 8x + 2$

(b) $(g \circ f)(x) = 2x^2 + 4x - 5$ (c) 2 (d) 11

(e) $(g \circ g \circ g)(x) = 8x - 21$

10. (a) $f^{-1}(x) = 3 - x^2, x \geqslant 0$ (b)

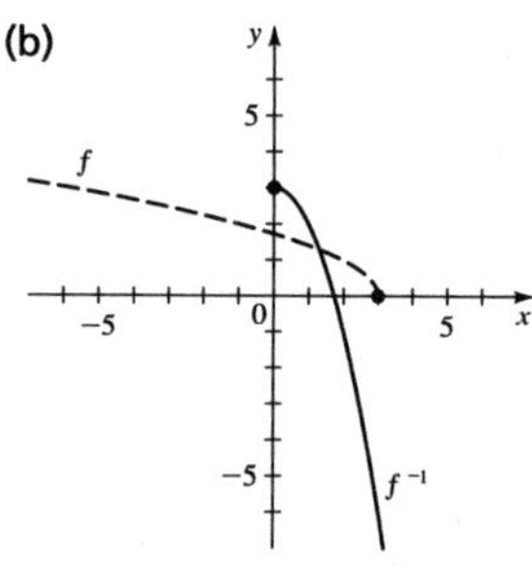

11. (a) Domain $[0, 6]$, range $[1, 7]$

(b)

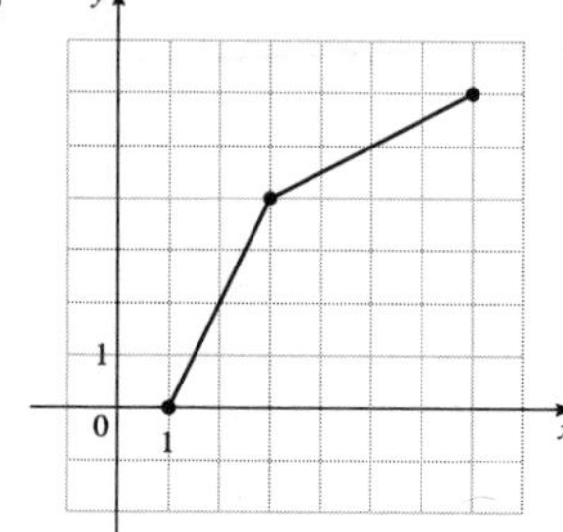

(c) $\frac{5}{4}$

12. (a)

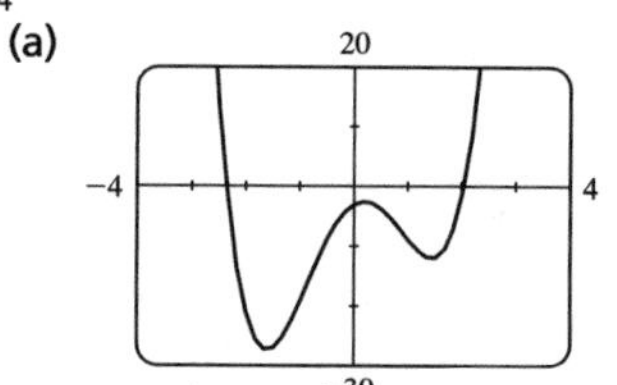

(b) No

(c) Local minimum ≈ -27.18 when $x \approx -1.61$; local maximum ≈ -2.55 when $x \approx 0.18$; local minimum ≈ -11.93 when $x \approx 1.43$

(d) $[-27.18, \infty)$

(e) Increasing on $[-1.61, 0.18] \cup [1.43, \infty)$; decreasing on $(-\infty, -1.61] \cup [0.18, 1.43]$

FOCUS ON MODELING ■ page 304

1. $A(w) = 3w^2, w > 0$ **3.** $V(w) = \frac{1}{2}w^3, w > 0$
7. $A(x) = 10x - x^2, 0 < x < 10$ **7.** $A(x) = (\sqrt{3}/4)x^2, x > 0$
9. $r(A) = \sqrt{A/\pi}, A > 0$ **11.** $A(x) = 2x^2 + 240/x, x > 0$
13. $D(t) = 25t, t \geqslant 0$ **15.** $A(x) = x\sqrt{4 - x}, 0 < x < 4$
17. $A(h) = 2h\sqrt{100 - h^2}, 0 < h < 10$
19. (b) $p(x) = x(19 - x)$ (c) 9.5, 9.5
21. $-12, -12$
23. (b) $A(x) = x(2400 - 2x)$ (c) 600 ft by 1200 ft
25. Width along road $15\sqrt{2} \approx 21.2$ ft, length $20\sqrt{2} \approx 28.3$ ft
27. \$9.50
29. Width ≈ 8.40 ft, height of rectangular part ≈ 4.20 ft
31. Height ≈ 1.44 ft, width of base ≈ 2.88 ft
33. 10 m by 10 m
35. To point C, 5.1 mi from point B

CHAPTER 5

Section 5.1 ■ page 323

1.

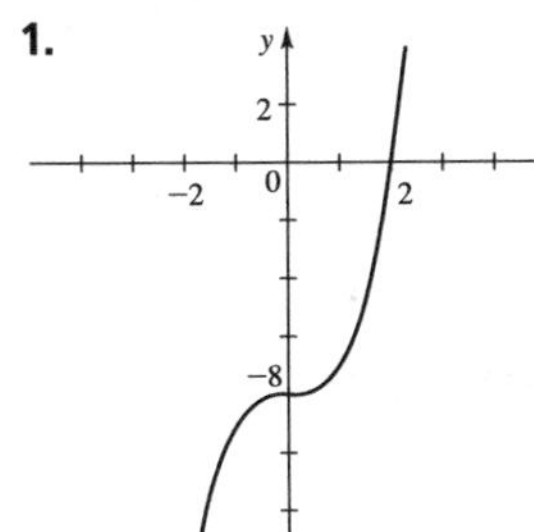

3.

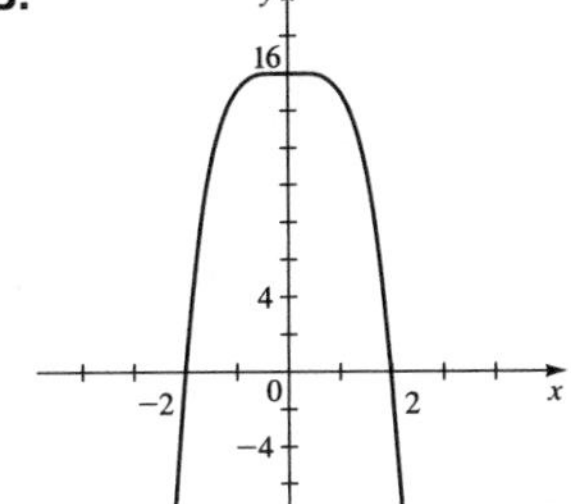

5.

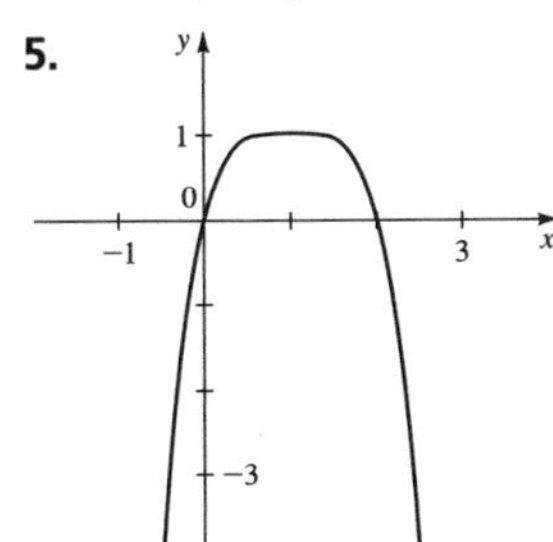

7.

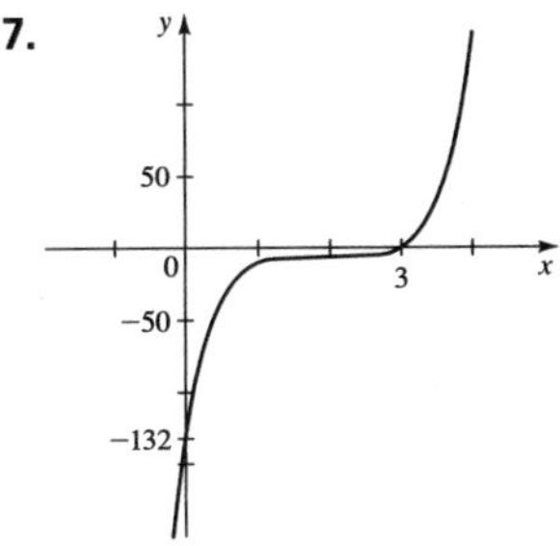

9.

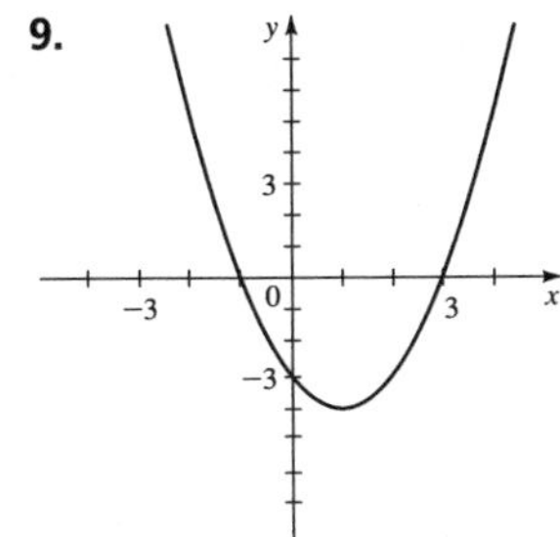

11.

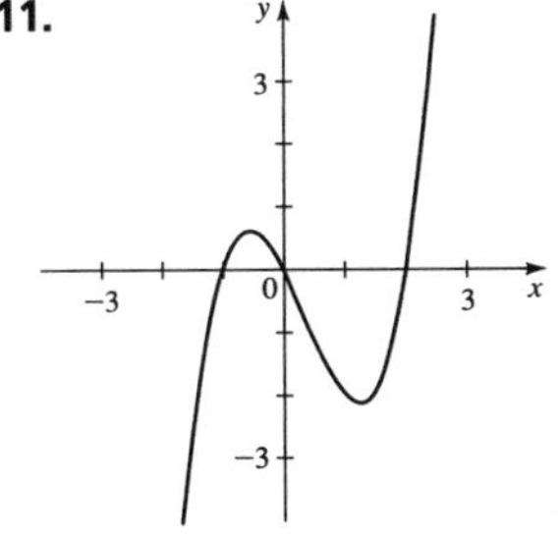

13.

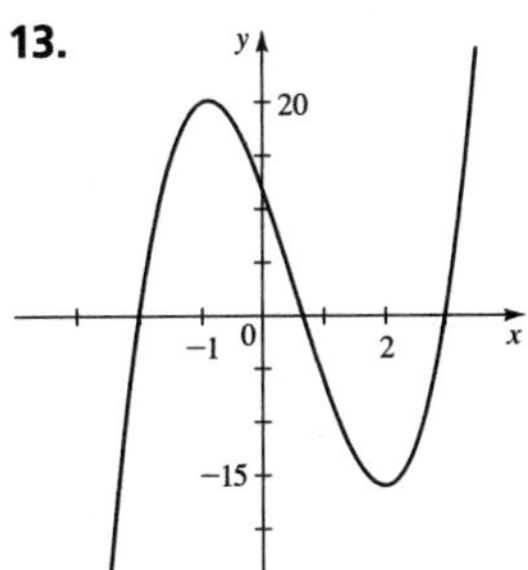

15.

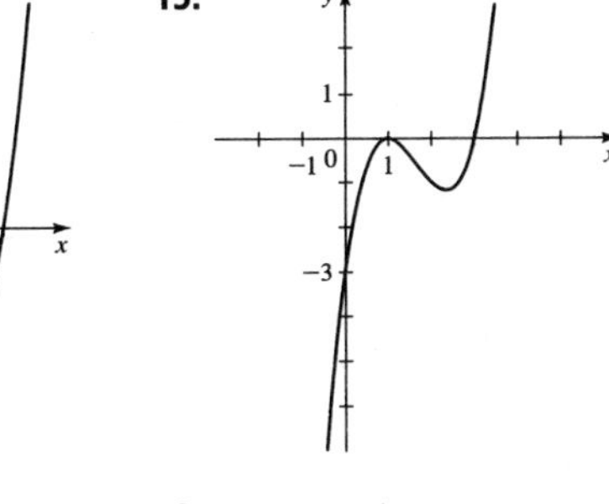

17.

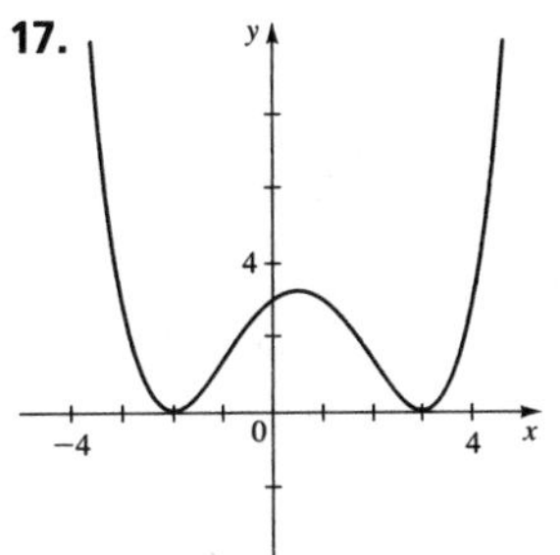

19.

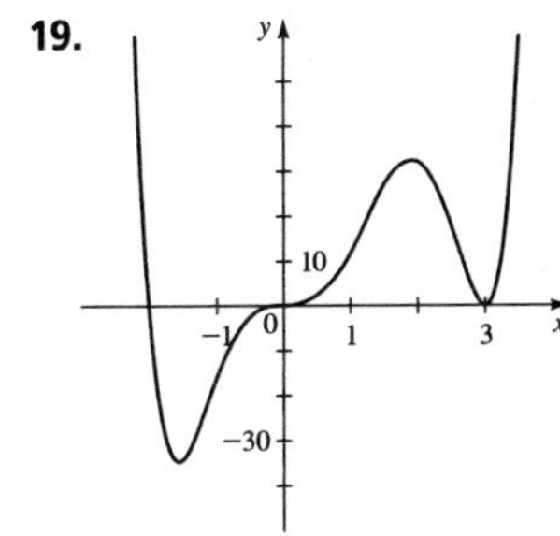

21.

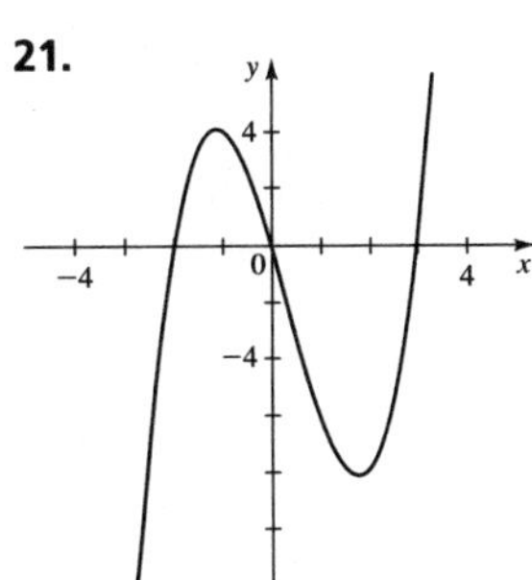

23.

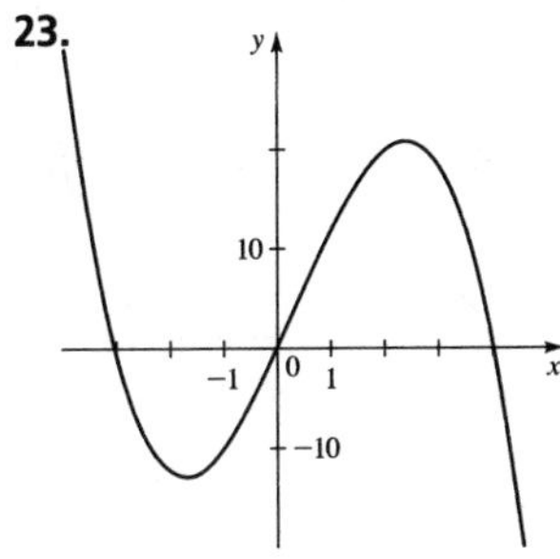

25.

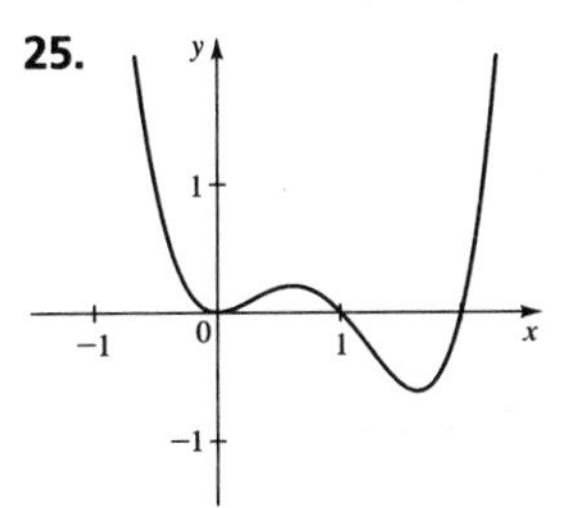

27.

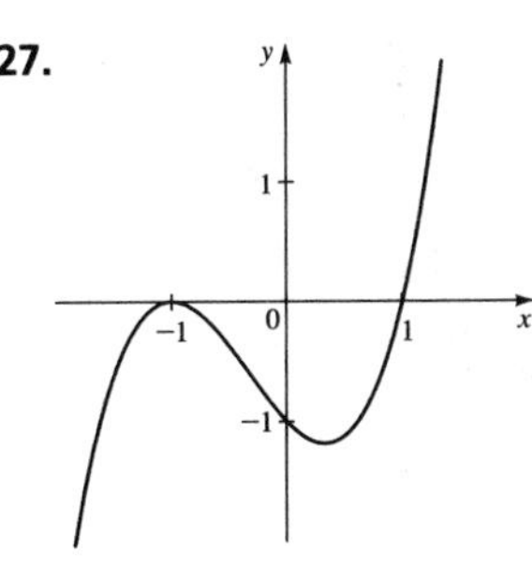

29.

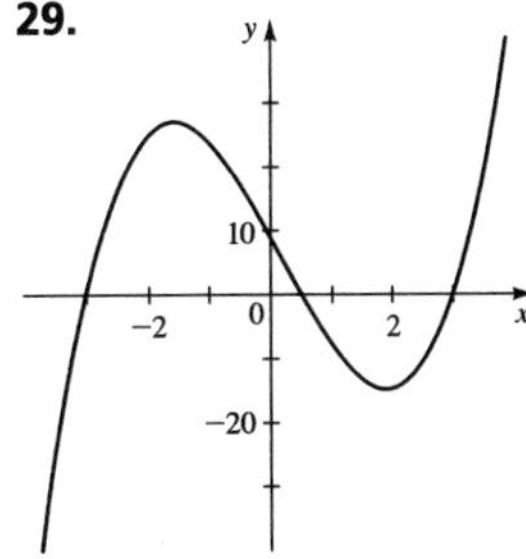

31. 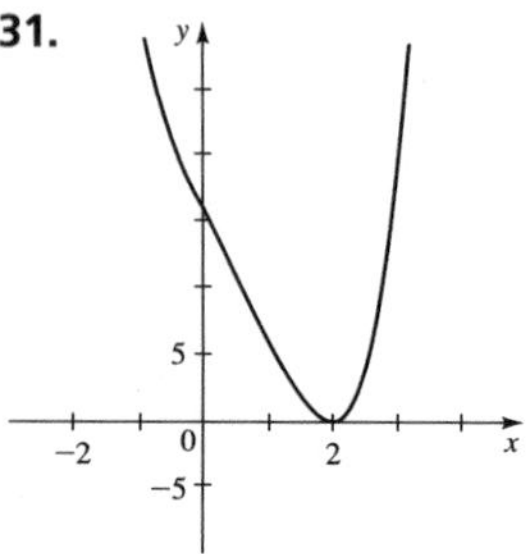

33. 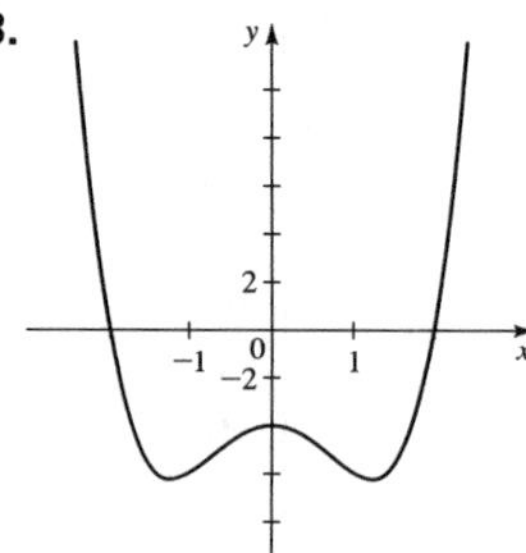

35. III

37. V

39. VI

41. $y \to \infty$ as $x \to \infty$, $y \to -\infty$ as $x \to -\infty$

43. $y \to \infty$ as $x \to \pm\infty$

45. $y \to \infty$ as $x \to \infty$, $y \to -\infty$ as $x \to -\infty$

47. 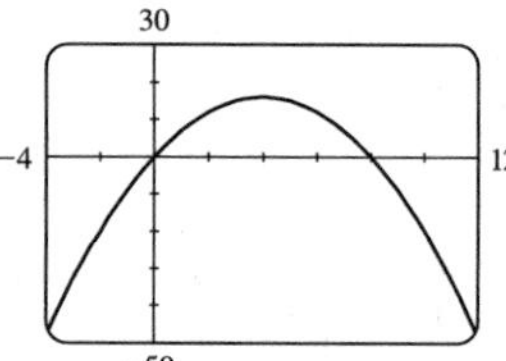

local maximum $(4, 16)$

49. 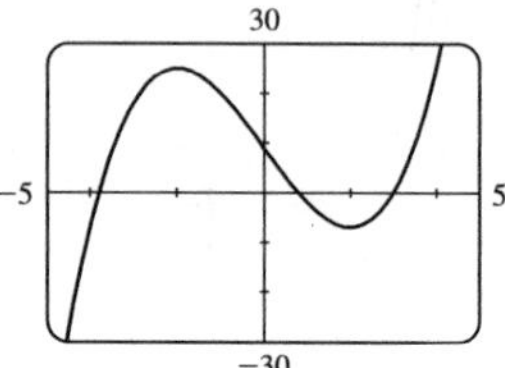

local maximum $(-2, 25)$
local minimum $(2, -7)$

51. 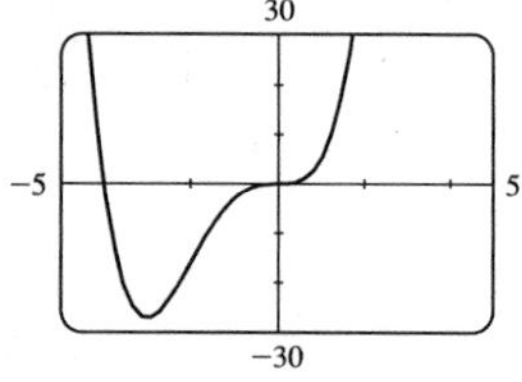

local minimum $(-3, -27)$

53. 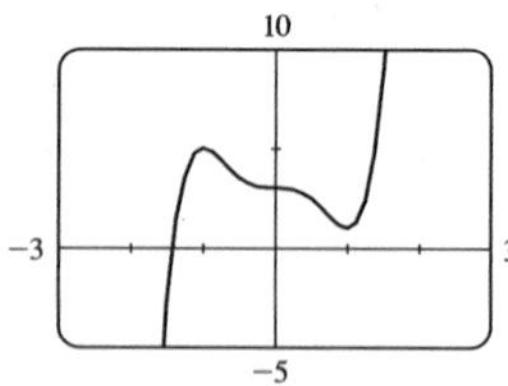

local maximum $(-1, 5)$
local minimum $(1, 1)$

55. One local maximum, no local minimum

57. One local maximum, one local minimum

59. One local maximum, two local minima

61. No local extrema

63. One local maximum, two local minima

65. 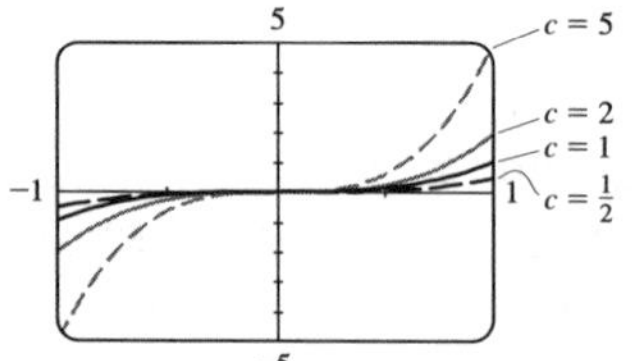

Increasing the value of c stretches the graph vertically.

67. 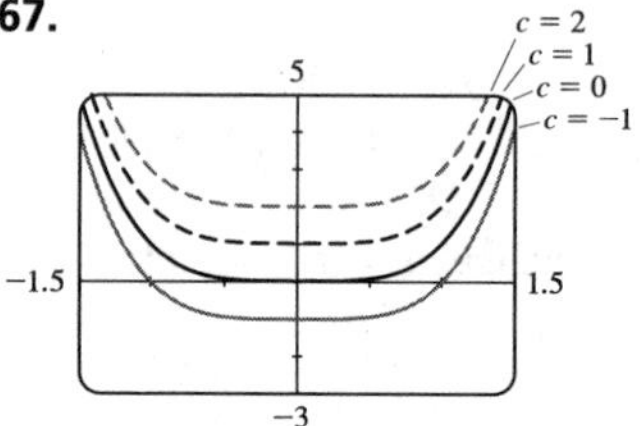

Increasing the value of c moves the graph up.

69. 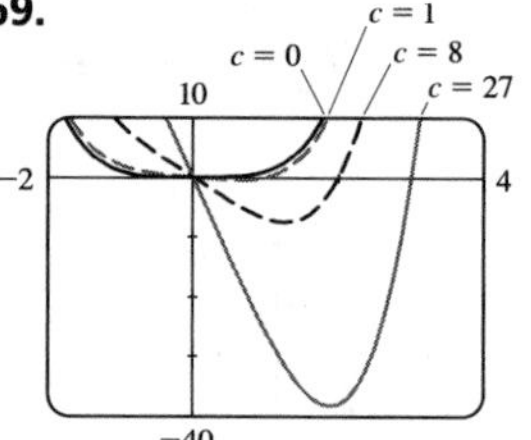

Increasing the value of c causes a deeper dip in the graph in the fourth quadrant and moves the positive x-intercept to the right.

71. **(a)** 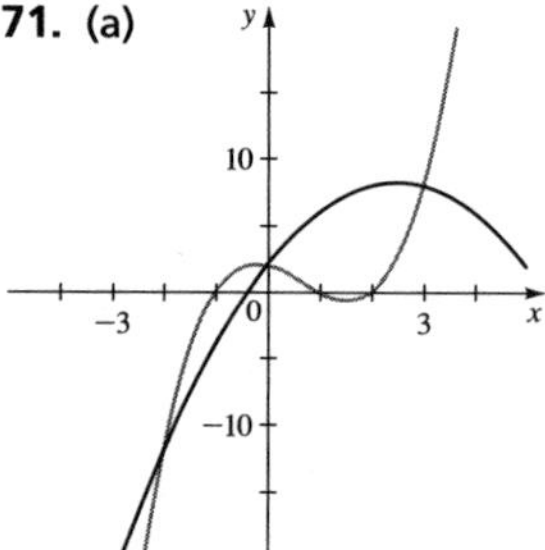

(b) Three

(c) $(0, 2)$, $(3, 8)$, $(-2, -12)$

73. **(g)** $P(x) = P_O(x) + P_E(x)$, where $P_O(x) = x^5 + 6x^3 - 2x$ and $P_E(x) = -x^2 + 5$

75. **(b)** Domain $\{x \mid 0 < x < 18\}$

(c) 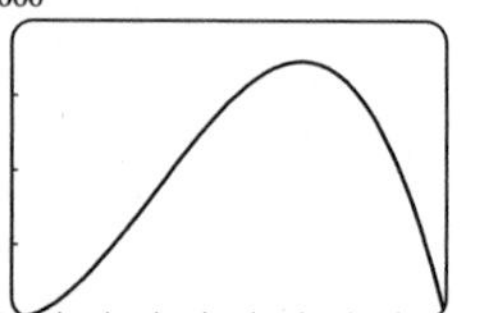

1728 in^3

77. (a) 1380 rabbits, after 4.2 months
(b) After 8.4 months
79. (a) Three x-intercepts, two local extrema
(b) One x-intercept, no local extrema

Section 5.2 ■ page 332

In answers 1–23, the first polynomial given is the quotient and the second is the remainder.

1. $x + 1, \quad -11$ **3.** $x^2 + 4x + 22, \quad 93$
5. $x + 2, \quad 8x - 1$ **7.** $3x + 1, \quad 7x - 5$ **9.** $x^4 + 1, \quad 0$
11. $x - 2, \quad -2$ **13.** $3x + 23, \quad 138$ **15.** $x^2 + 2, \quad -3$
17. $x^2 - 3x + 1, \quad -1$ **19.** $x^4 + x^3 + 4x^2 + 4x + 4, \quad -2$
21. $2x^2 + 4x, \quad 1$ **23.** $x^2 + 3x + 9, \quad 0$
25. -3 **27.** 12 **29.** -7 **31.** -483 **33.** 2159 **35.** $\frac{7}{3}$
37. -8.279 **43.** $-1 \pm \sqrt{6}$ **45.** $x^3 - 3x^2 - x + 3$
47. $x^4 - 8x^3 + 14x^2 + 8x - 15$
49. $-\frac{3}{2}x^3 + 3x^2 + \frac{15}{2}x - 9$

Section 5.3 ■ page 341

1. $\pm 1, \pm 3$ **3.** $\pm 1, \pm 2, \pm 4, \pm 8, \pm\frac{1}{2}$ **5.** $-2, 1$ **7.** $-1, 2$
9. 2 **11.** $-1, 2, 3$ **13.** -1 **15.** $\pm 1, \pm 2$
17. $1, -1, -2, -4$ **19.** $\pm 2, \pm\frac{3}{2}$ **21.** -2
23. $-1, -\frac{1}{2}, \frac{1}{2}$ **25.** $-\frac{3}{2}, \frac{1}{2}, 1$ **27.** $-1, \frac{1}{2}, 2$
29. $-3, -2, 1, 3$ **31.** $-1, -\frac{1}{3}, 2, 5$ **33.** $-2, -1 \pm \sqrt{2}$
35. $-1, 4, \dfrac{3 \pm \sqrt{13}}{2}$ **37.** $3, \dfrac{1 \pm \sqrt{5}}{2}$ **39.** $\frac{1}{2}, \dfrac{1 \pm \sqrt{3}}{2}$
41. $-1, -\frac{1}{2}, -3 \pm \sqrt{10}$
43. (a) $-2, 2, 3$ (b)

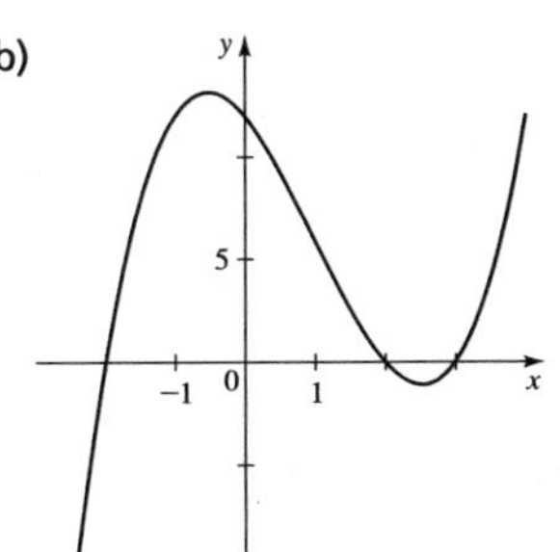

45. (a) $-4, 3$ (b)

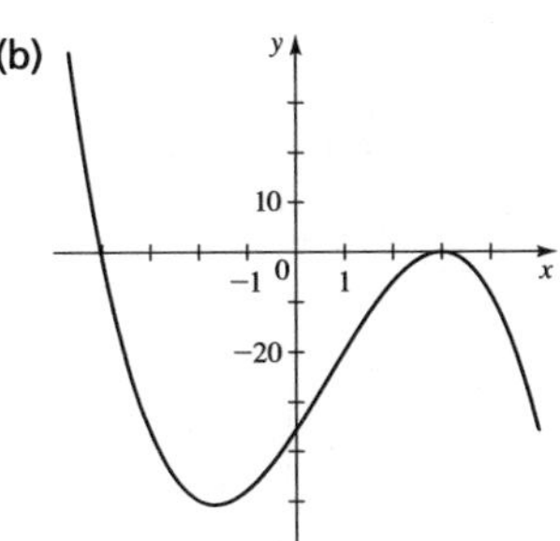

47. (a) $-1, 2$ (b)

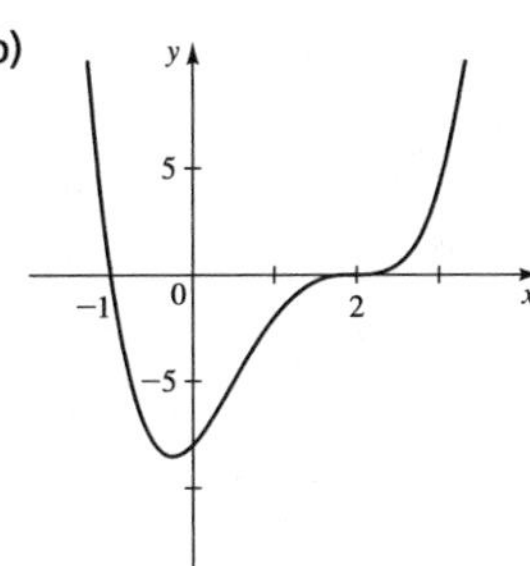

49. (a) $-1, 2$ (b)

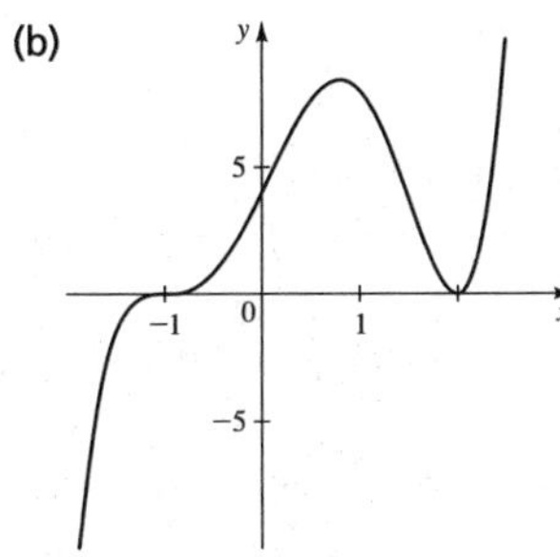

51. 1 positive, 2 or 0 negative; 3 or 1 real
53. 1 positive, 1 negative; 2 real
55. 2 or 0 positive, 0 negative; 3 or 1 real (since 0 is a zero but is neither positive nor negative)
61. $3, -2$ **63.** $3, -1$ **65.** $-2, \frac{1}{2}, \pm 1$ **67.** $\pm\frac{1}{2}, \pm\sqrt{5}$
69. $-2, 1, 3, 4$ **75.** $-2, 2, 3$ **77.** $-\frac{3}{2}, -1, 1, 4$
79. $-1.28, 1.53$ **81.** -1.50 **83.** 11.3 ft
85. (a) It began to snow again. (b) No
(c) Just before midnight on Saturday night
87. 2.76 m **89.** 88 in. (or 3.21 in.)

Section 5.4 ■ page 353

In answers 1–17, the factored form is given first, then the zeros are listed with the multiplicity of each in parentheses.

1. $(x - 3i)(x + 3i)$; $3i$ (1), $-3i$ (1)
3. $[x - (-1 + i)][x - (-1 - i)]$; $-1 + i$ (1), $-1 - i$ (1)
5. $x(x - 2i)(x + 2i)$; 0 (1), $2i$ (1), $-2i$ (1)
7. $(x - 1)(x + 1)(x - i)(x + i)$; 1 (1), -1 (1), i (1), $-i$ (1)
9. $16\left(x - \frac{3}{2}\right)\left(x + \frac{3}{2}\right)\left(x - \frac{3}{2}i\right)\left(x + \frac{3}{2}i\right)$; $\frac{3}{2}$ (1), $-\frac{3}{2}$ (1), $\frac{3}{2}i$ (1), $-\frac{3}{2}i$ (1)
11. $(x + 1)(x - 3i)(x + 3i)$; -1 (1), $3i$ (1), $-3i$ (1)
13. $(x - i)^2(x + i)^2$; i (2), $-i$ (2)
15. $(x - 1)(x + 1)(x - 2i)(x + 2i)$; 1 (1), -1 (1), $2i$ (1), $-2i$ (1)
17. $x(x - i\sqrt{3})^2(x + i\sqrt{3})^2$; 0 (1), $i\sqrt{3}$ (2), $-i\sqrt{3}$ (2)
19. $P(x) = x^2 - 2x + 2$

21. $Q(x) = x^3 - 3x^2 + 4x - 12$
23. $P(x) = x^3 - 2x^2 + x - 2$
25. $R(x) = x^4 - 4x^3 + 10x^2 - 12x + 5$
27. $T(x) = 6x^4 - 12x^3 + 18x^2 - 12x + 12$
29. $-2, \pm 2i$ **31.** $1, \dfrac{1 \pm i\sqrt{3}}{2}$ **33.** $2, \dfrac{1 + i\sqrt{3}}{2}$

35. $-\frac{3}{2}, -1 \pm i\sqrt{2}$ **37.** $-2, 1, \pm 3i$ **39.** $1, \pm 2i, \pm i\sqrt{3}$

41. 3 (multiplicity 2), $\pm 2i$ **43.** 1 (multiplicity 3), $\pm 3i$
45. **(a)** $(x - 5)(x^2 + 4)$ **(b)** $(x - 5)(x - 2i)(x + 2i)$
47. **(a)** $(x - 1)(x + 1)(x^2 + 9)$
(b) $(x - 1)(x + 1)(x - 3i)(x + 3i)$
49. **(a)** $(x - 2)(x + 2)(x^2 - 2x + 4)(x^2 + 2x + 4)$
(b) $(x - 2)(x + 2)[x - (1 + i\sqrt{3})][x - (1 - i\sqrt{3})]$
$[x + (1 + i\sqrt{3})][x + (1 - i\sqrt{3})]$
51. **(a)** 4 real **(b)** 2 real, 2 imaginary **(c)** 4 imaginary

Section 5.5 ■ page 366

1. x-intercept 2; y-intercept $-\frac{2}{3}$
3. x-intercepts $-1, 2$; y-intercept $\frac{1}{3}$
5. x-intercepts $-3, 3$; no y-intercept
7. x-intercept 3, y-intercept 3, vertical $x = 2$; horizontal $y = 2$
9. Vertical $x = -3$; horizontal $y = 0$
11. Vertical $x = 3, x = -2$; horizontal $y = 1$
13. Horizontal $y = 0$
15. Vertical $x = -6, x = 1$; horizontal $y = 0$
17. Vertical $x = 1$

19.

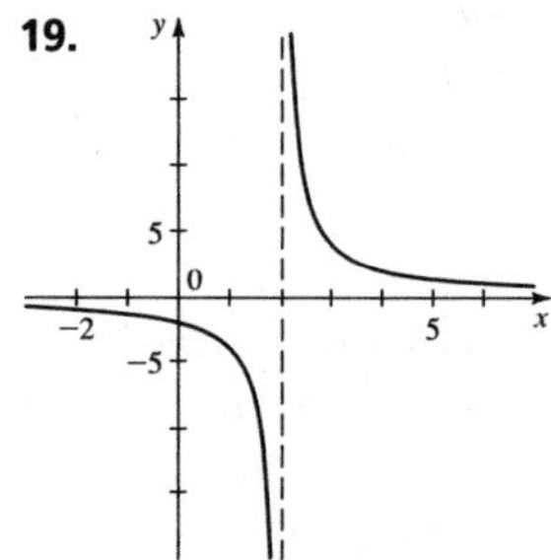

21.

23.

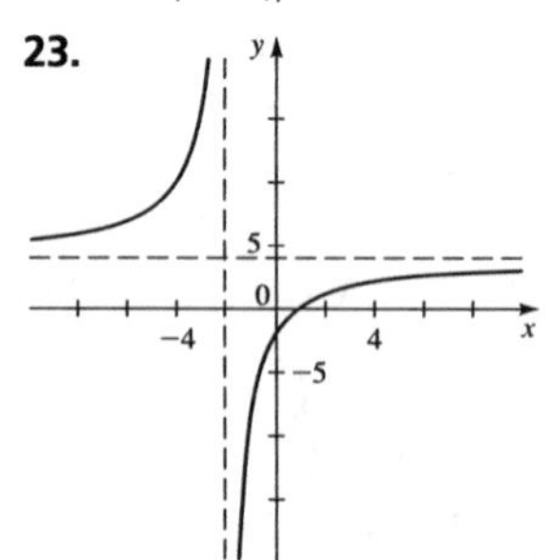

25.

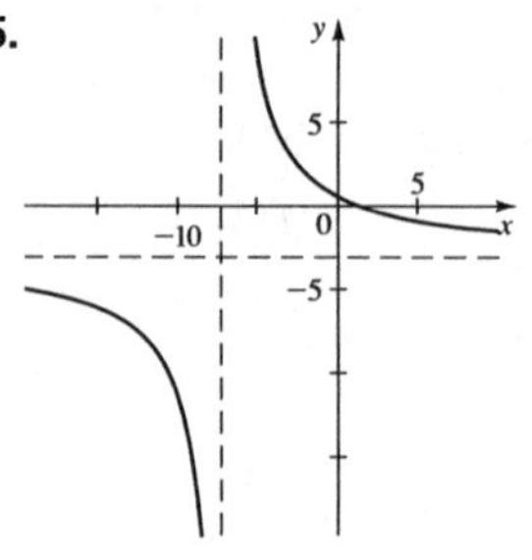

27.

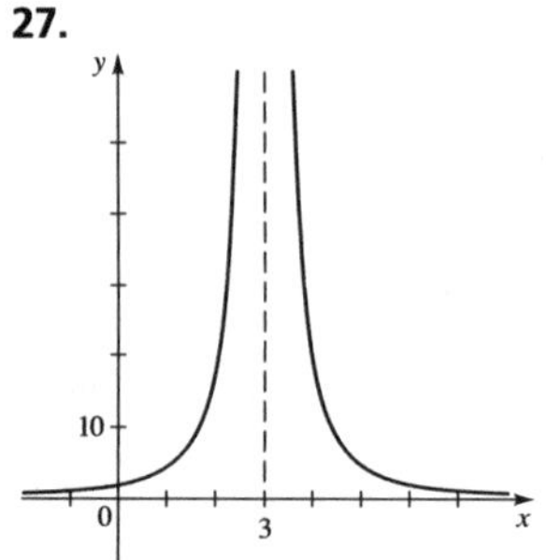

29.

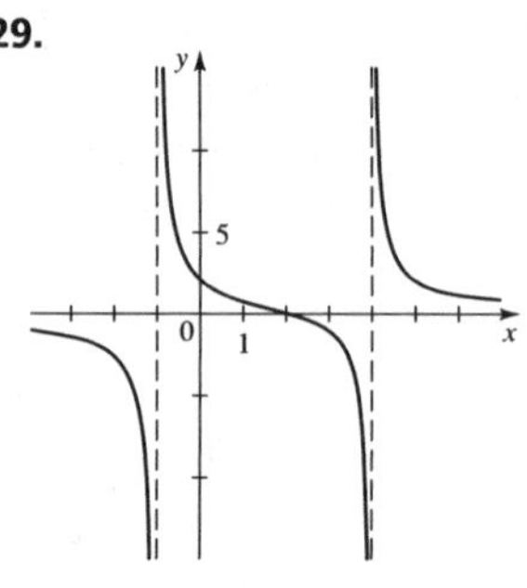

31.

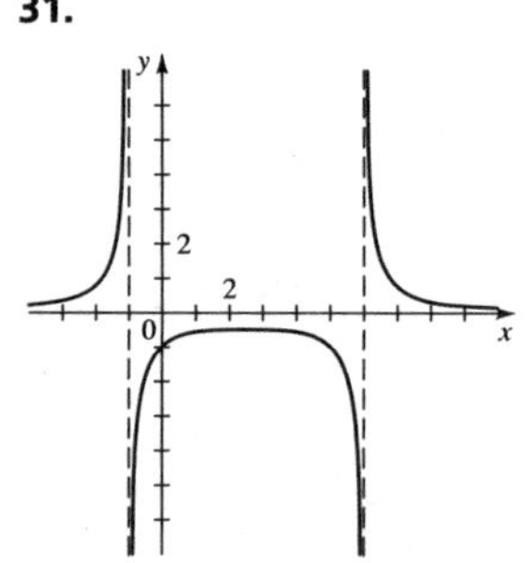

33.

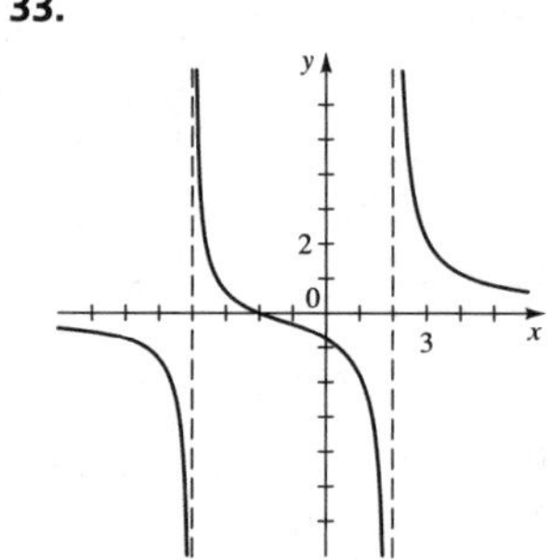

35.

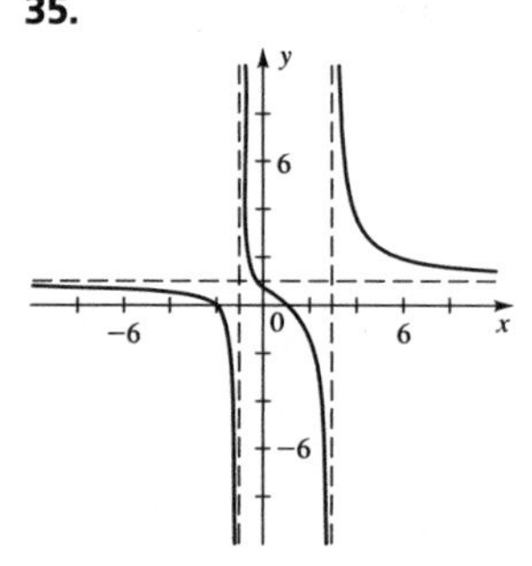

37.

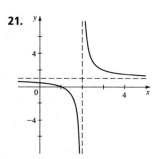

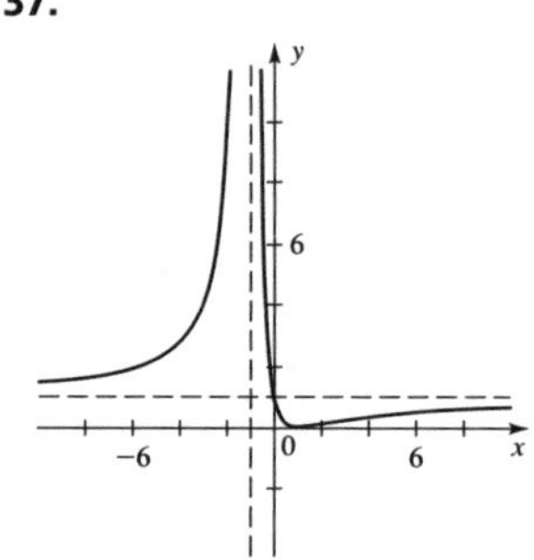

39.

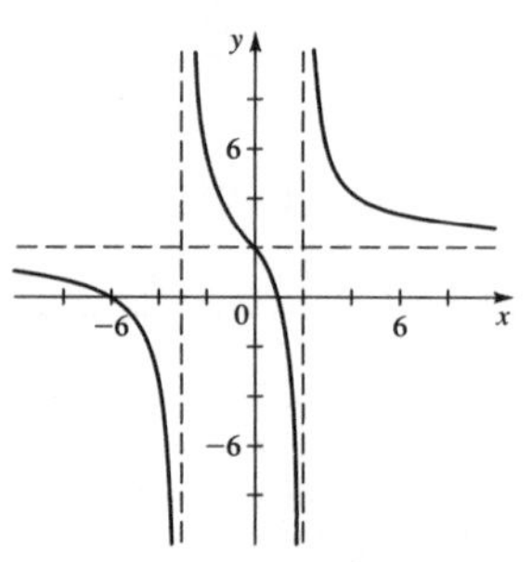

41.

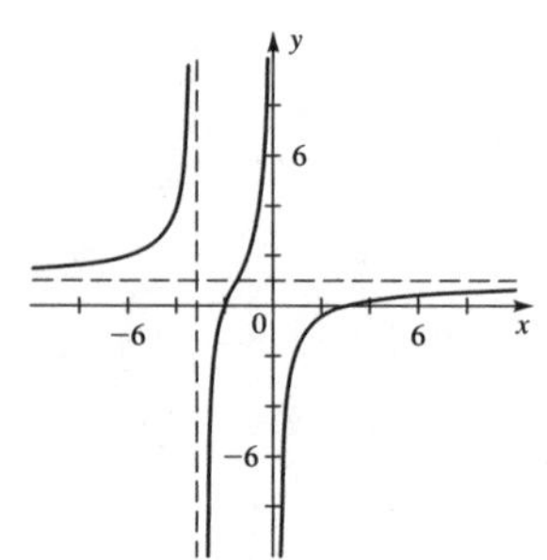

43.

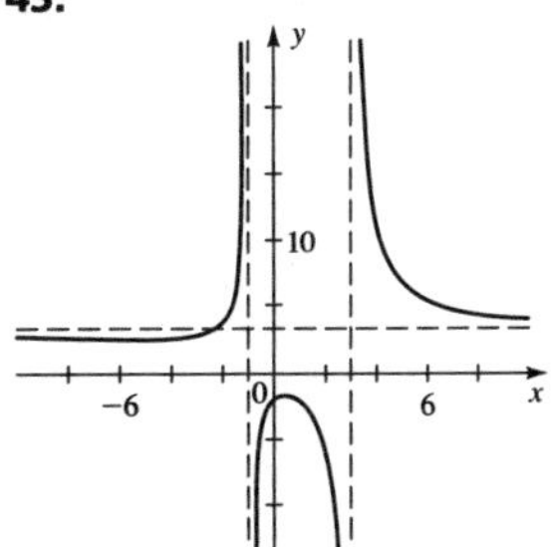

45.

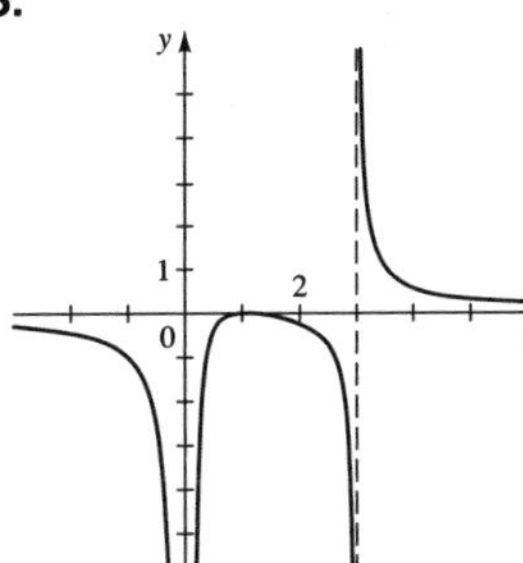

47.

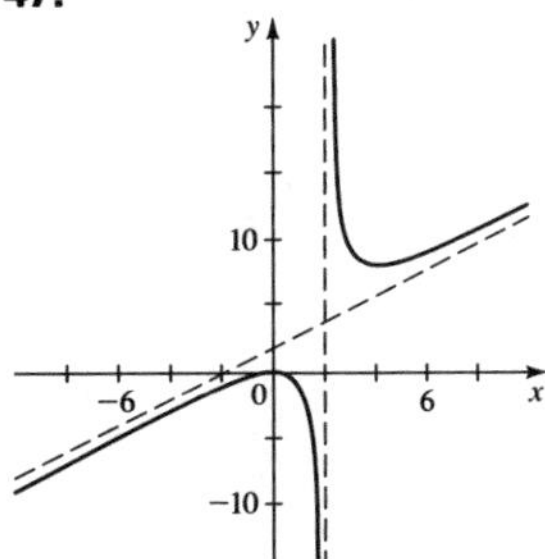

49.

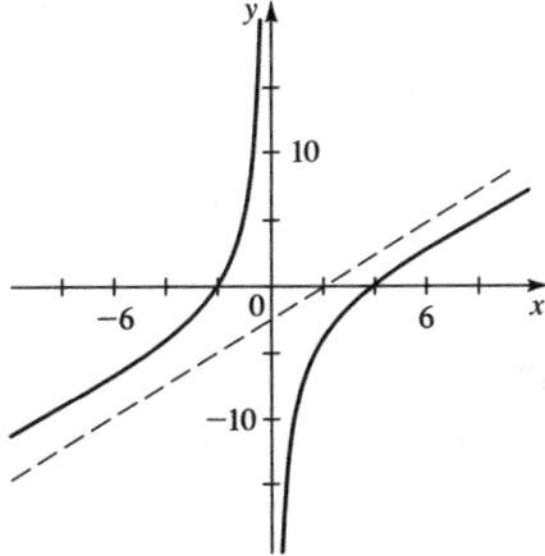

51.

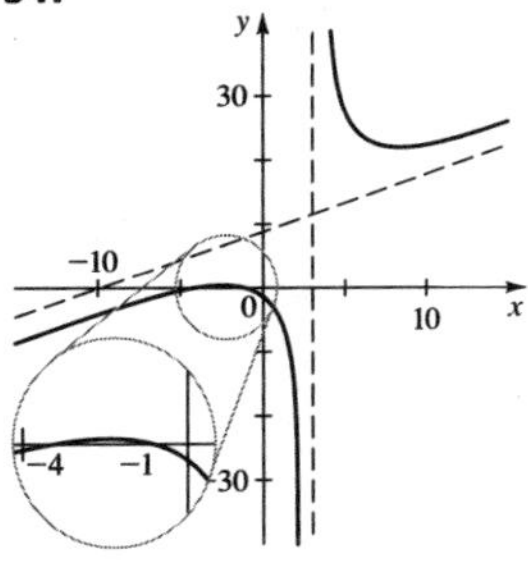

53.

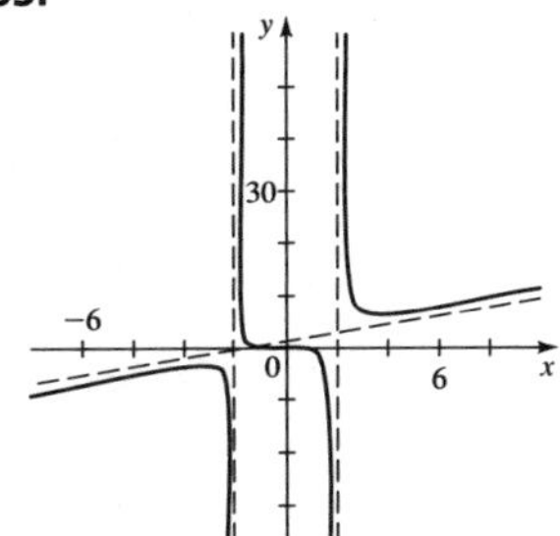

55.

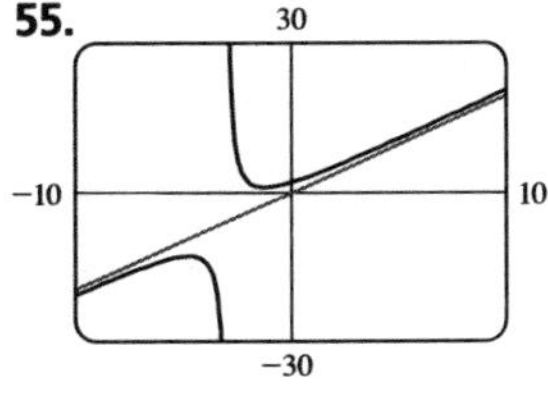

vertical $x = -3$

57.

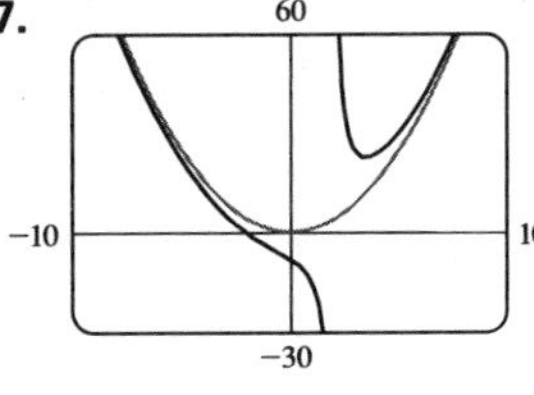

vertical $x = 2$

59.

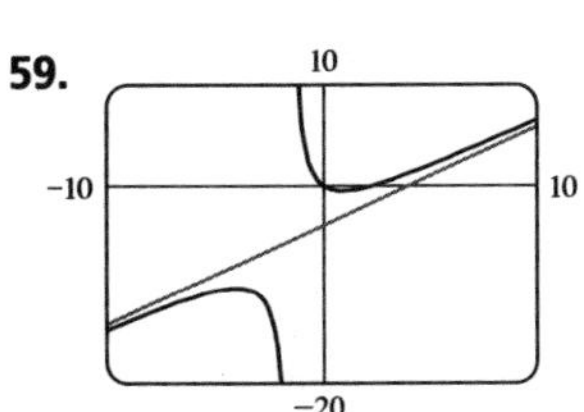

vertical $x = -1.5$
x-intercepts 0, 2.5
y-intercept 0
local maximum $(-3.9, -10.4)$
local minimum $(0.9, -0.6)$
end behavior: $y = x - 4$

61.

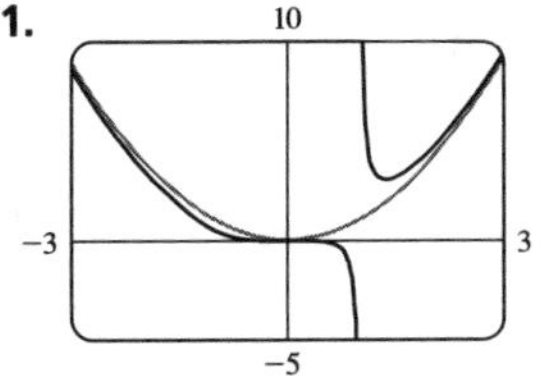

vertical $x = 1$
x-intercept 0
y-intercept 0
local minimum $(1.4, 3.1)$
end behavior: $y = x^2$

63.

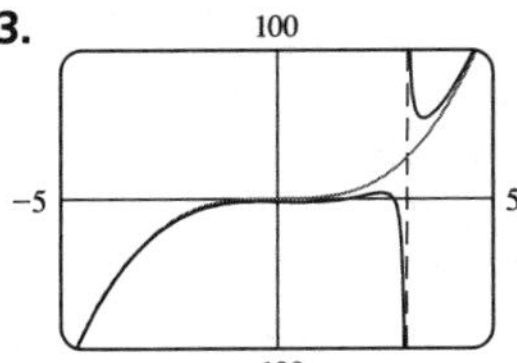

vertical $x = 3$
x-intercepts 1.6, 2.7
y-intercept -2
local maxima $(-0.4, -1.8)$, $(2.4, 3.8)$
local minima $(0.6, -2.3)$ $(3.4, 54.3)$
end behavior $y = x^3$

65. (a)

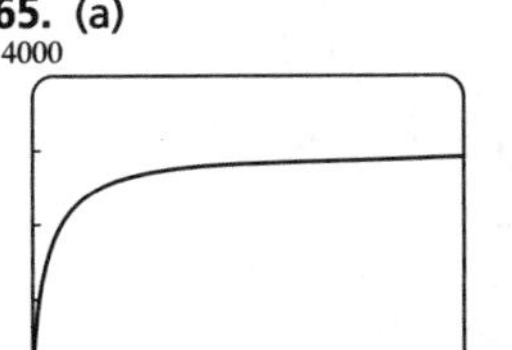

(b) It levels off at 3000.

67. (a) 2.50 mg/L (b) It decreases to 0. (c) 16.61 h

69.

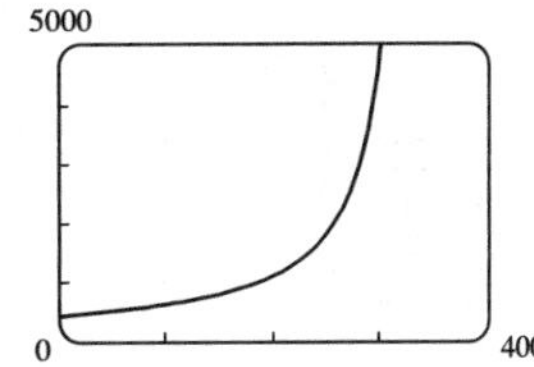

If the speed of the train approaches the speed of sound, then the pitch increases indefinitely (a sonic boom).

Chapter 5 Review ■ page 370

1.

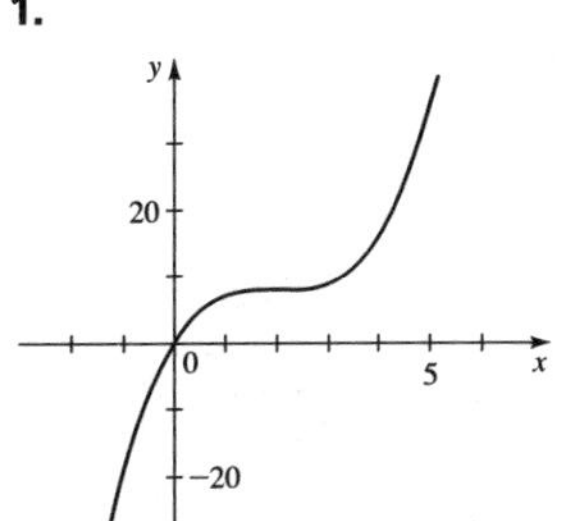

3.

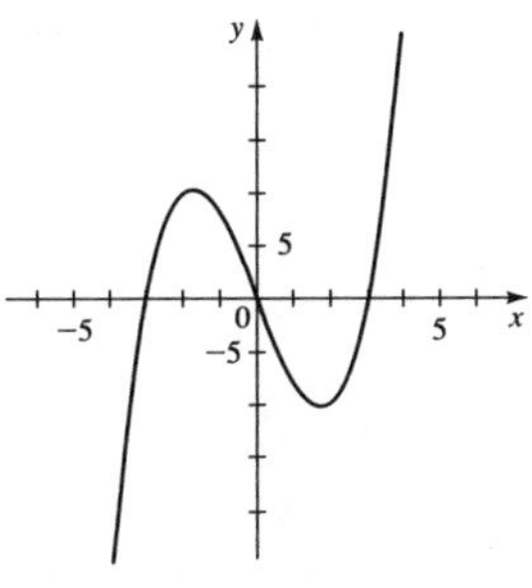

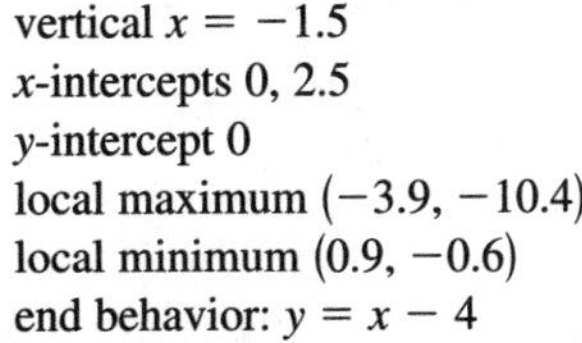

5.

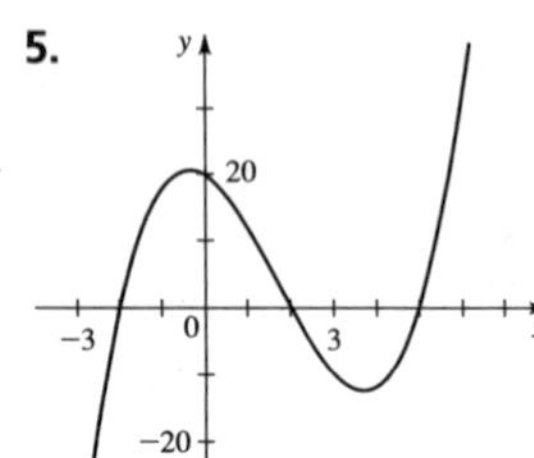

7.

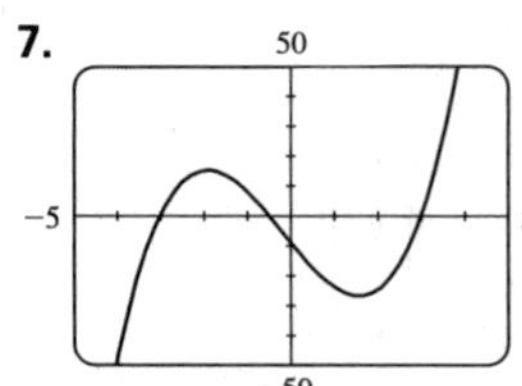

x-intercepts $-3, -0.5, 3$
y-intercept -9
local maximum: $(-1.9, 15.1)$
local minimum: $(1.6, -27.1)$
$y \to \infty$ as $x \to \infty$
$y \to -\infty$ as $x \to -\infty$

9.

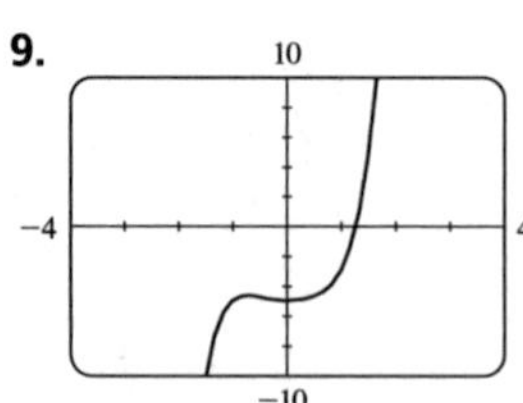

x-intercept 1.3
y-intercept -5
local maximum $(-0.7, -4.7)$
local minimum $(0, -5)$
$y \to \infty$ as $x \to \infty$
$y \to -\infty$ as $x \to -\infty$

In answers 11–17, the first polynomial given is the quotient and the second is the remainder.

11. $x^2 + 2x + 7, 10$ **13.** $x - 3, -9$ **15.** $x^3 - 5x^2 + 4, -5$
17. $x^3 + 3x^2 + 4x + 5, 9$ **19.** 3 **23.** 8
25. (a) $\pm 1, \pm 2, \pm 3, \pm 6, \pm 9, \pm 18$
(b) 2 or 0 positive, 3 or 1 negative
27. $4x^3 - 18x^2 + 14x + 12$
29. No; since the complex conjugates of imaginary zeros will also be zeros, the polynomial would have 8 zeros, contradicting the requirement that it have degree 4.
31. $-3, 1, 5$ **33.** $-1 \pm 2i, -2$ (multiplicity 2)
35. $\pm 2, 1$ (multiplicity 3) **37.** $\pm 2, \pm 1 \pm i\sqrt{3}$
39. $1, 3, \dfrac{-1 \pm i\sqrt{7}}{2}$
41. $x = -0.5, 3$ **43.** $x \approx -0.24, 4.24$

45.

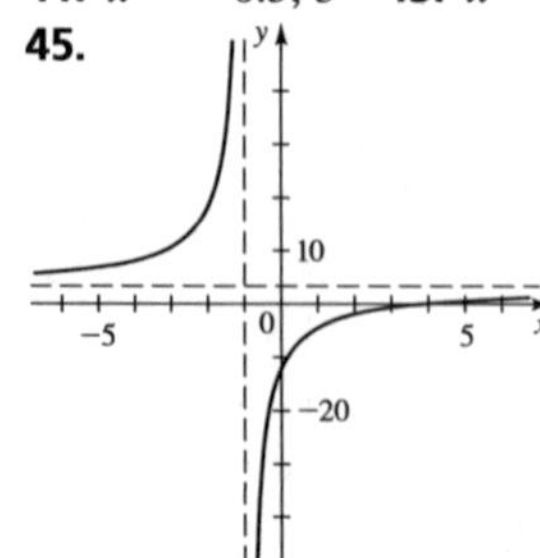

47.

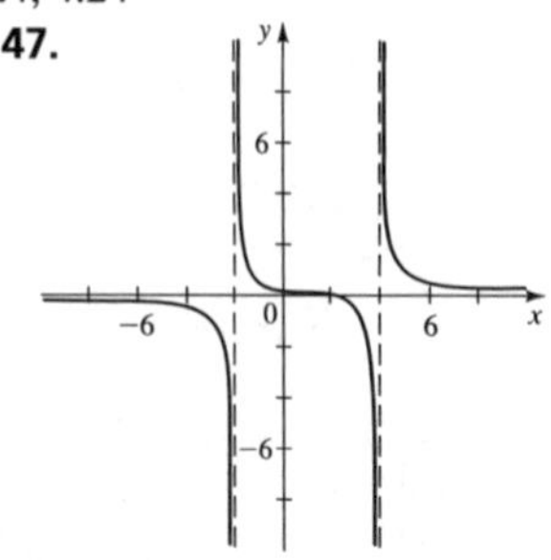

49.

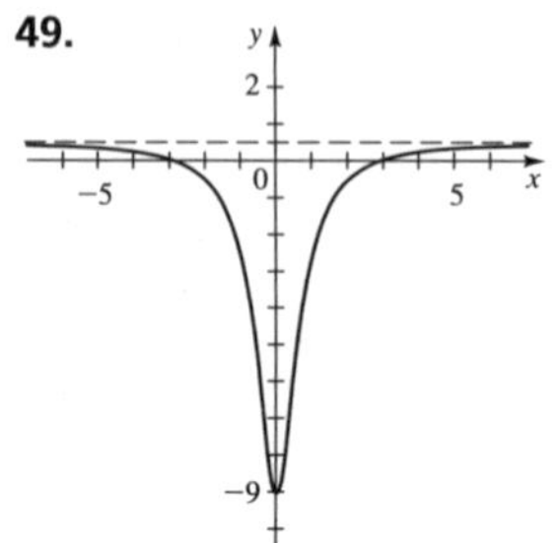

51.

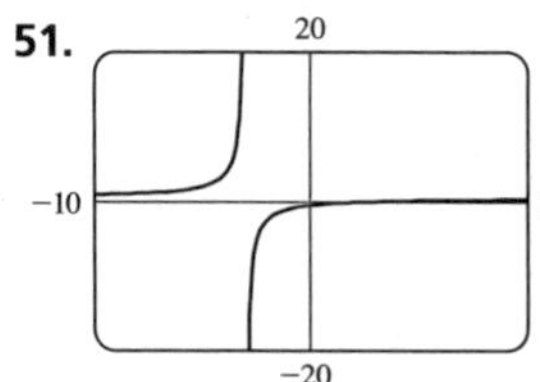

x-intercept 3
y-intercept -0.5
vertical $x = -3$
horizontal $y = 0.5$
no local extrema

53.

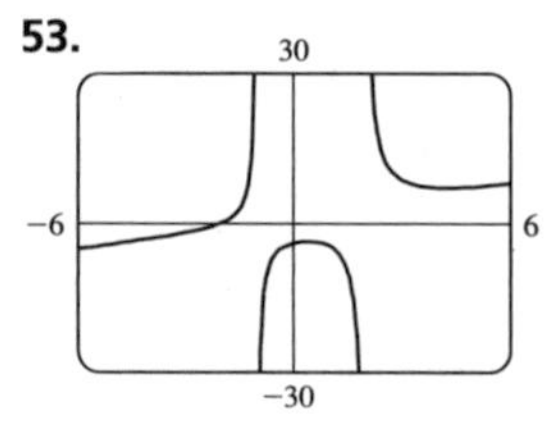

x-intercept -2
y-intercept -4
vertical $x = -1, x = 2$
slant $y = x + 1$
local maximum $(0.425, -3.599)$
local minimum $(4.216, 7.175)$

55. $(-2, -28), (1, 26), (2, 68), (5, 770)$

Chapter 5 Test ■ page 372

1.

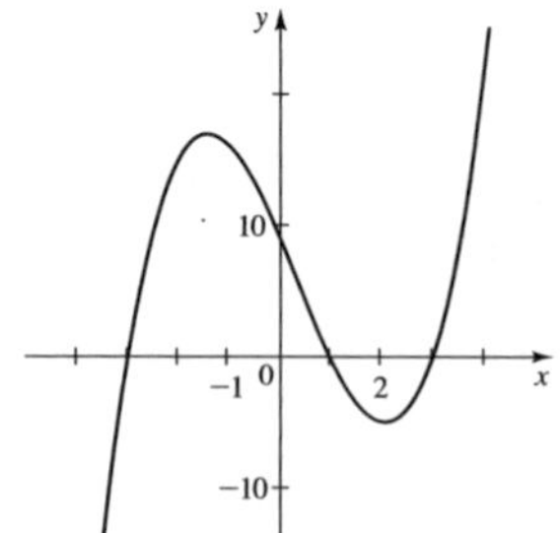

2. (a) $x^3 + 2x^2 + 2, 9$ (b) $x^3 + 2x^2 + \frac{1}{2}, \frac{15}{2}$
3. (a) $\pm 1, \pm 3, \pm\frac{1}{2}, \pm\frac{3}{2}$
(b) $2(x + 1)\left(x - \frac{1}{2}\right)(x - 1)(x - 3)$
(c) $-1, \frac{1}{2}, 1, 3$
4. $3, -1 \pm i$ **5.** $(x - 1)^2(x - 2i)(x + 2i)$
6. $x^4 + 2x^3 + 10x^2 + 18x + 9$
7. (a) 4, 2, or 0 positive; 0 negative (c) 0.17, 3.93

8. (a) r, u (b) s (c) s

(d)

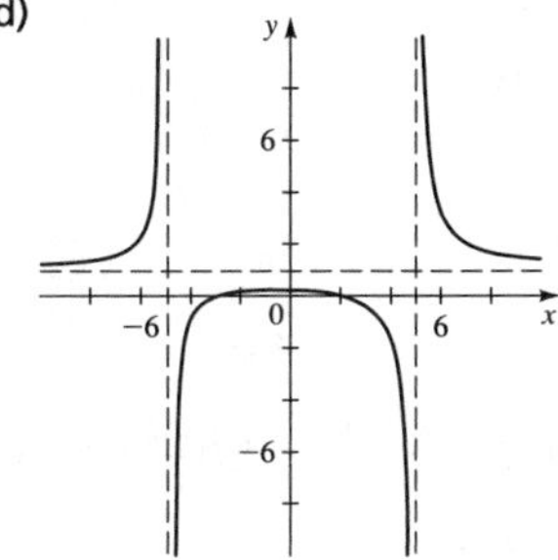

(e) $x^2 - 2x - 5$

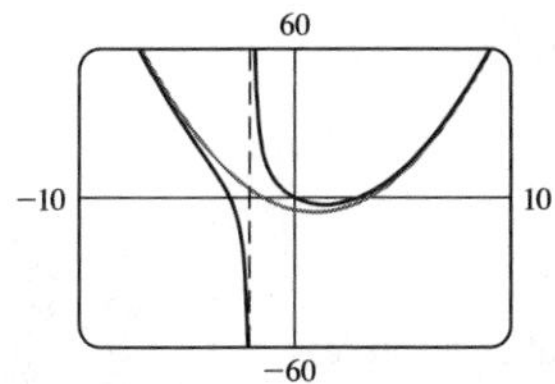

FOCUS ON PROBLEM SOLVING ■ page 375

1. Not traversable **3.** Traversable

5. Yes; tour is possible for odd n. **7.** Inside

9. -36 **11.** No **13.** $\frac{2}{3}$

15.

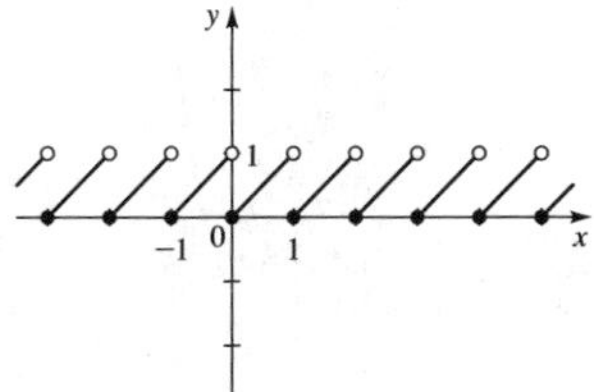

CHAPTER 6

Section 6.1 ■ page 384

1.

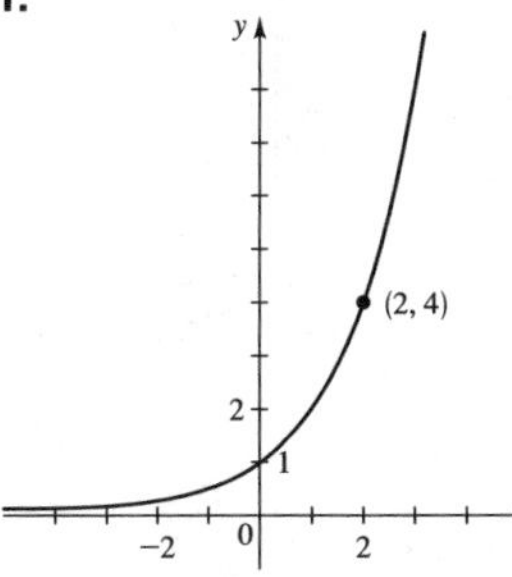

3.

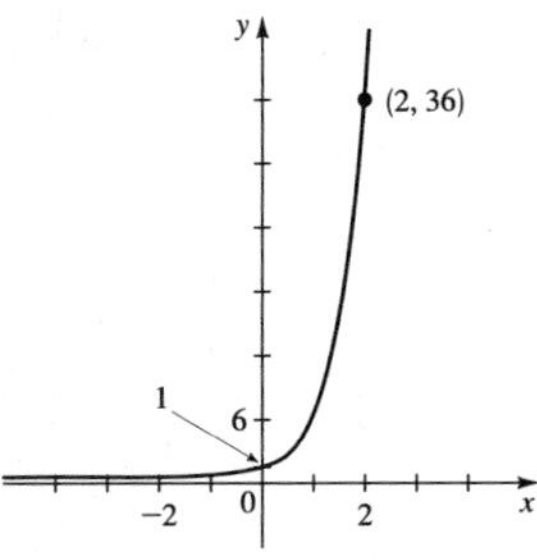

5.

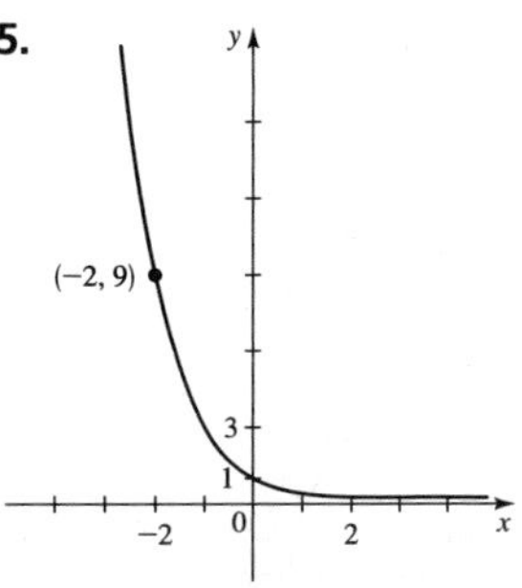

7.

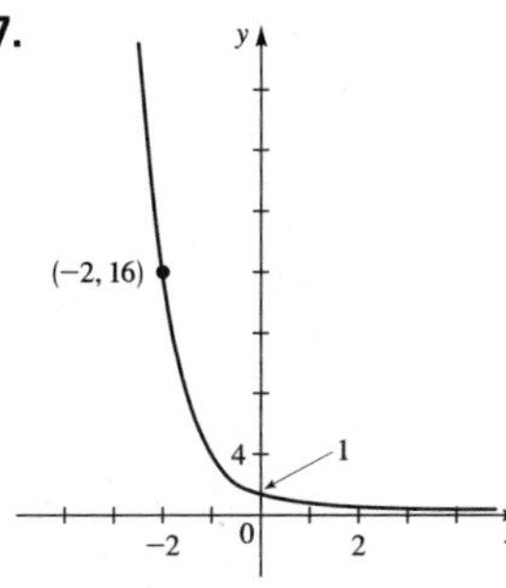

9.

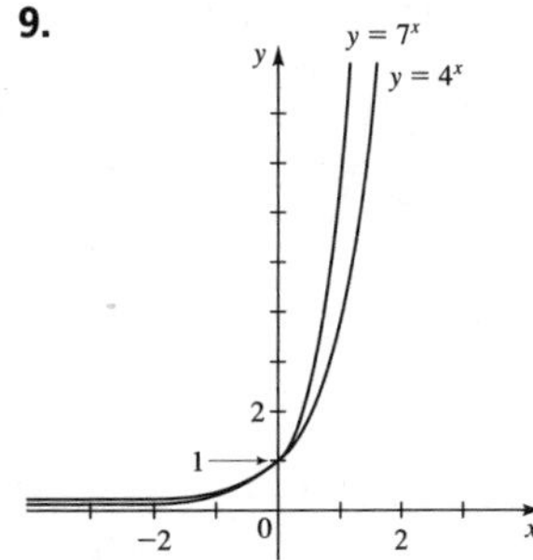

11. $f(x) = 3^x$ **13.** $f(x) = \left(\frac{1}{4}\right)^x$

15. III **17.** I **19.** II

21. $\mathbb{R}, (-\infty, 0), y = 0$

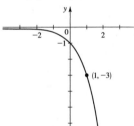

23. $\mathbb{R}, (-3, \infty), y = -3$

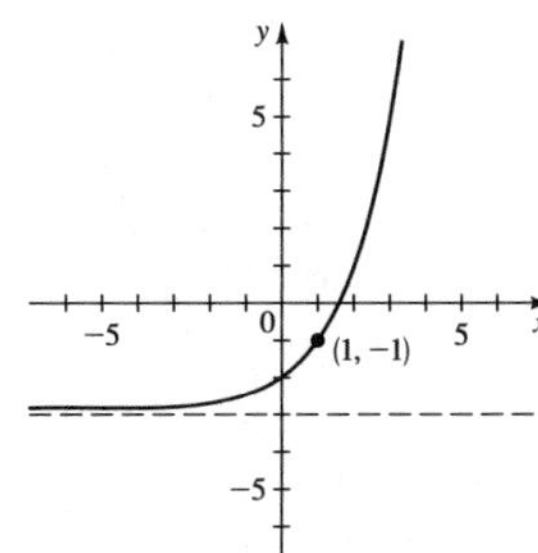

25. $\mathbb{R}, (4, \infty), y = 4$

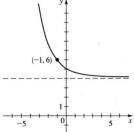

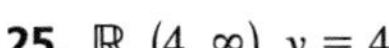

27. $\mathbb{R}, (0, \infty), y = 0$

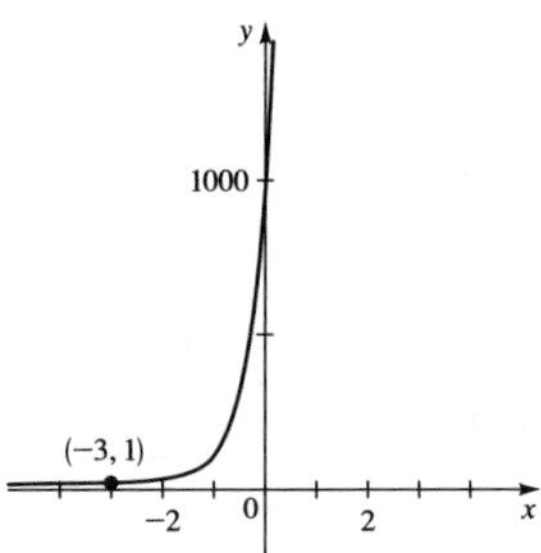

29. $\mathbb{R}$, $(-\infty, 0)$, $y = 0$

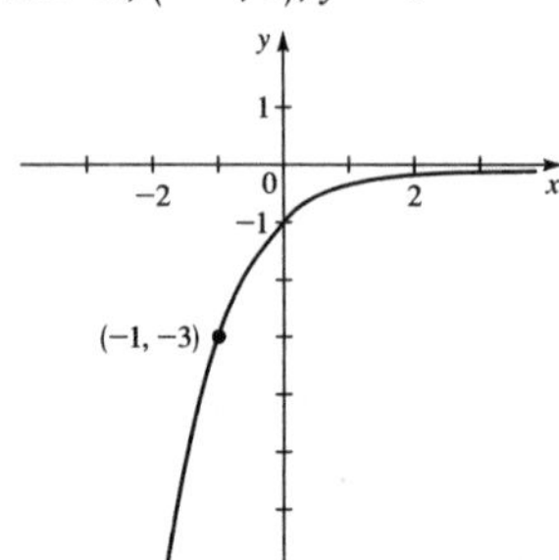

31. $\mathbb{R}$, $(0, \infty)$, $y = 0$

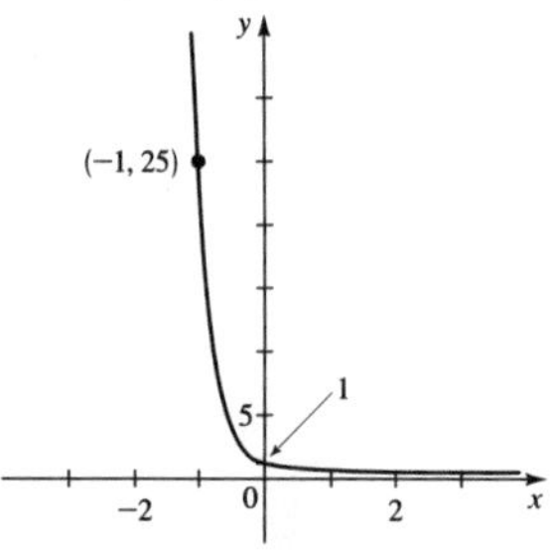

33. $\mathbb{R}$, $(-\infty, 5)$, $y = 5$

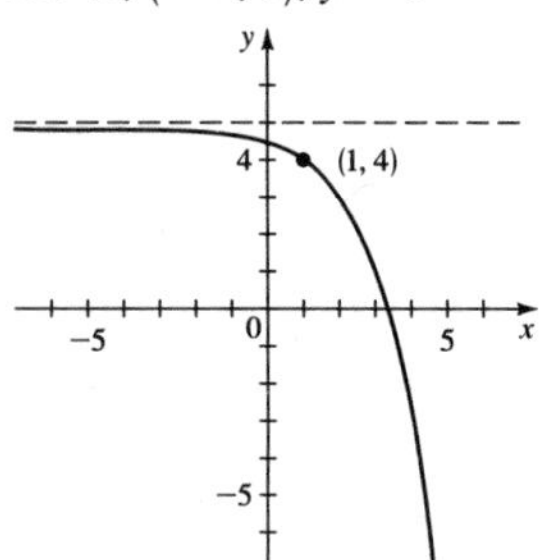

35. $\mathbb{R}$, $[1, \infty)$, no asymptote

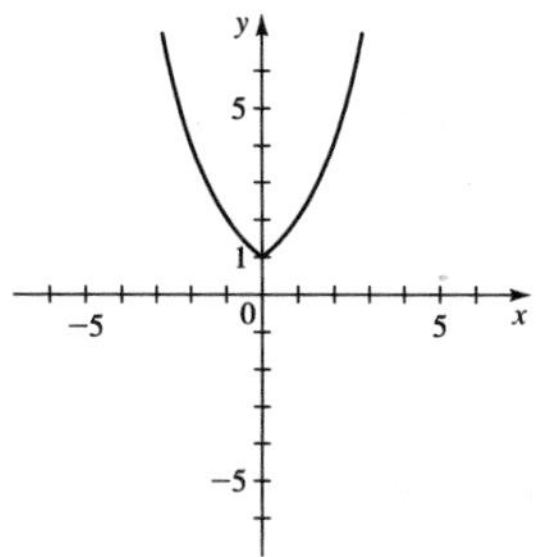

37. $y = 3(2^x)$

39. **(a)**

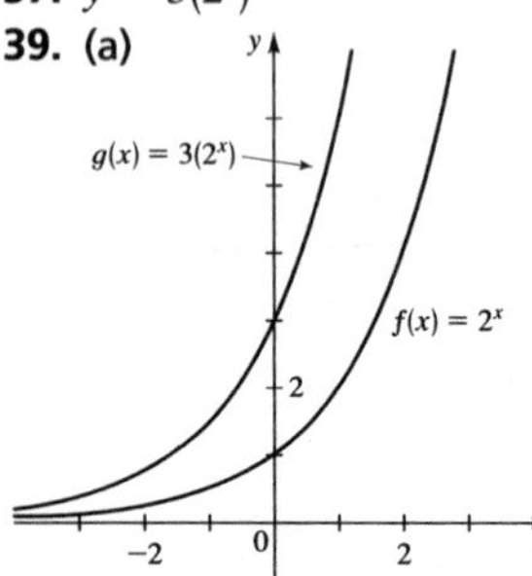

(b) The graph of g is steeper than that of f.

43. **(a)**

(i)

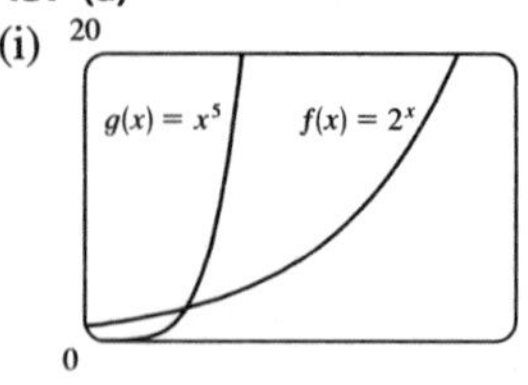

(ii)

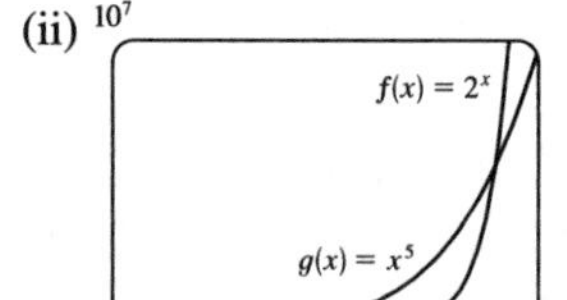
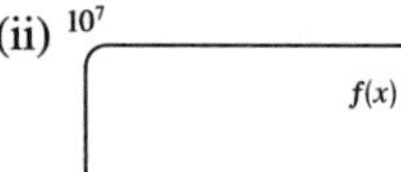
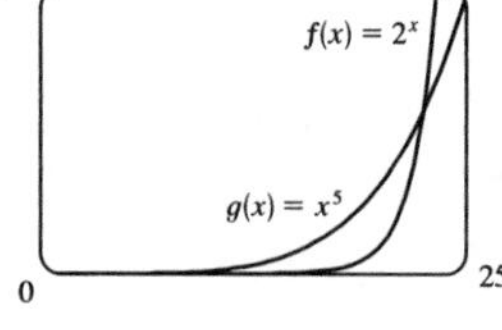

(iii)

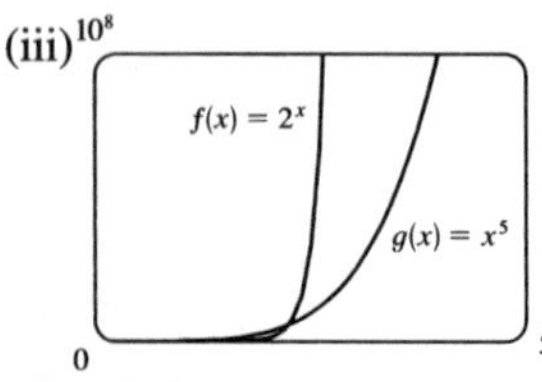

The graph of f ultimately increases much more quickly than g.

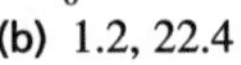

(b) 1.2, 22.4

45.

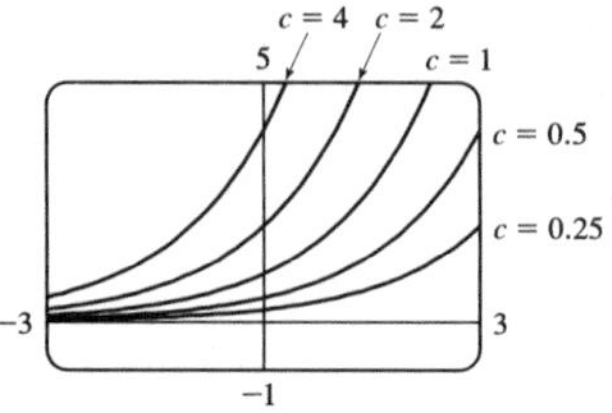

The larger the value of c, the more rapidly the graph increases.

47.

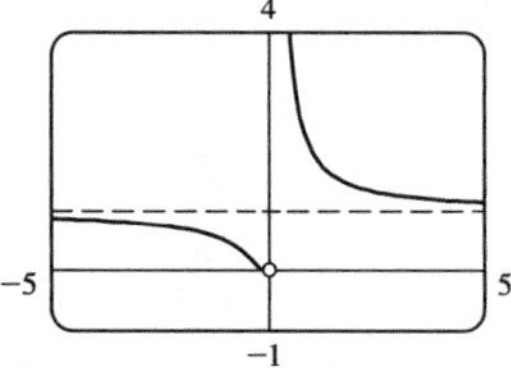

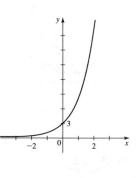

Vertical asymptote $x = 0$, horizontal asymptote $y = 1$

49. Local minimum $\approx (0.37, 0.69)$

51. **(a)** Increasing on $(-\infty, 0.50]$, decreasing on $[0.50, \infty)$
(b) $(0, 1.78]$

Section 6.2 ■ page 394

1.

x	$f(x) = 3e^x$
−2	0.4
−1.5	0.7
−1	1.1
−0.5	1.8
0	3
0.5	4.9
1	8.2
1.5	13.4
2	22.2

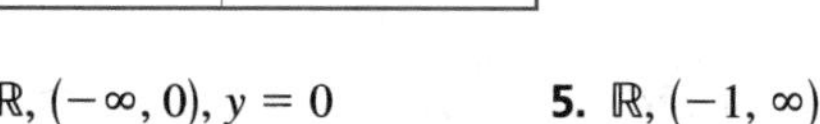

3. $\mathbb{R}$, $(-\infty, 0)$, $y = 0$

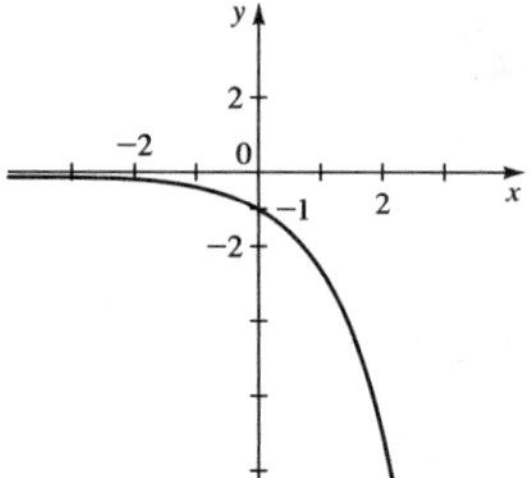

5. $\mathbb{R}$, $(-1, \infty)$, $y = -1$

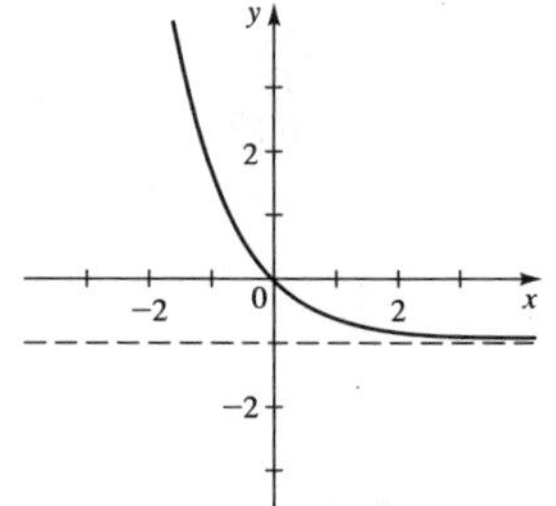

7. $\mathbb{R}$, $(0, \infty)$, $y = 0$

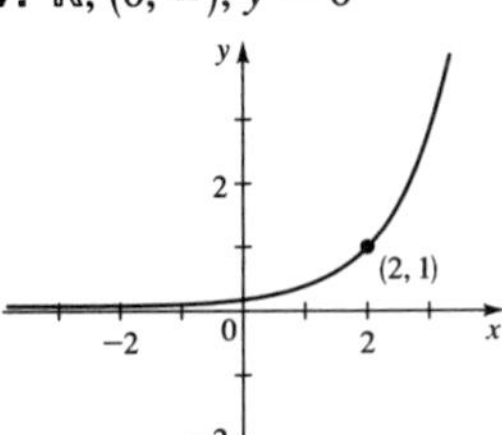

9. (a) \$16,288.95 **(b)** \$26,532.98 **(c)** \$43,219.42

11. (a) \$4,615.87 **(b)** \$4,658.91 **(c)** \$4,697.04 **(d)** \$4,703.11 **(e)** \$4,704.68 **(f)** \$4,704.93 **(g)** \$4,704.94

13. (i) **15.** \$7,678.96 **17. (a)** 45% **(b)** 500 **(c)** 4744

19. (a) $n(t) = 18{,}000e^{0.08t}$ **(b)** 34,137

(c)

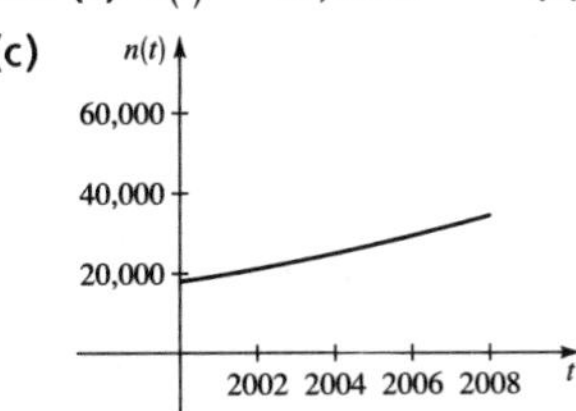

21. (a) 233 million **(b)** 181 million

23. 5.87 billion **25. (a)** About 500 **(b)** 361,000

27. (a) 6 g **(b)** 1 g

29. (a) 2.7 lb **(b)** 4.9 lb

(c)

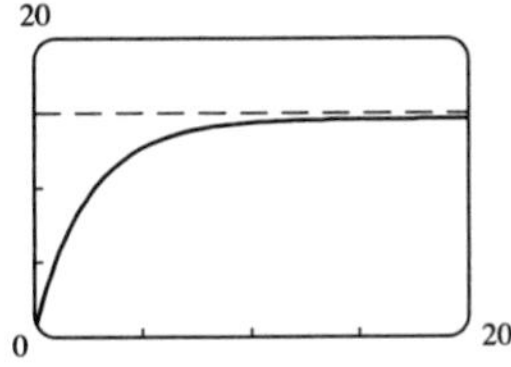

(d) 15 lb

31. (a) 5164 **(b)**

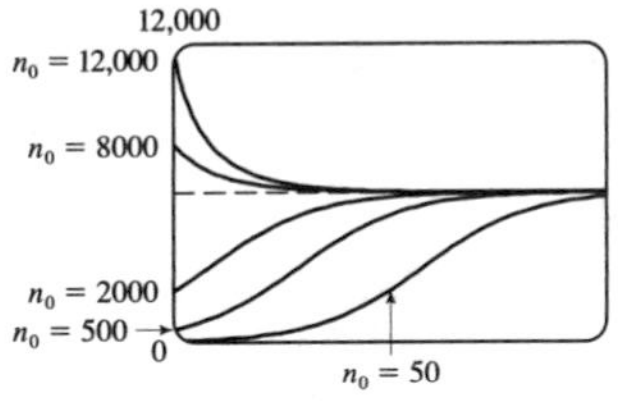

(c) 6000

33.

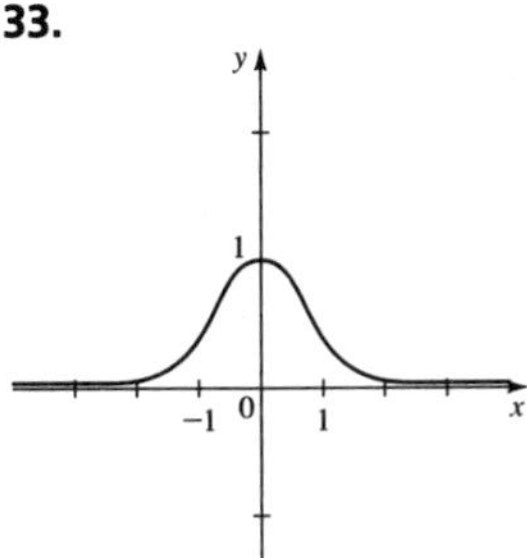

35.

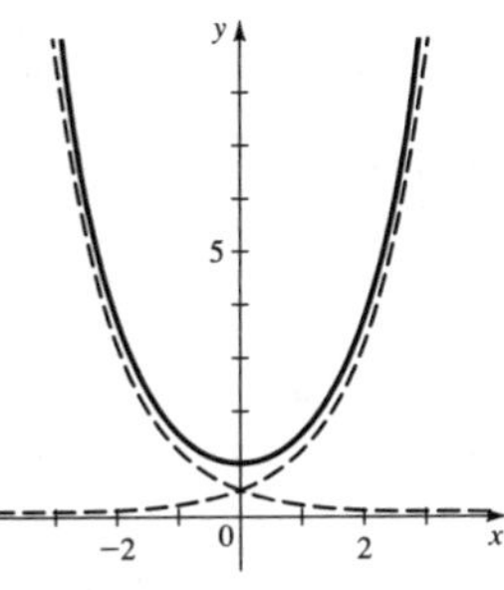

41.

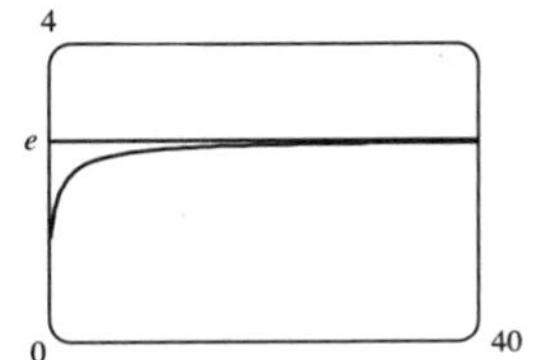

43. (a)

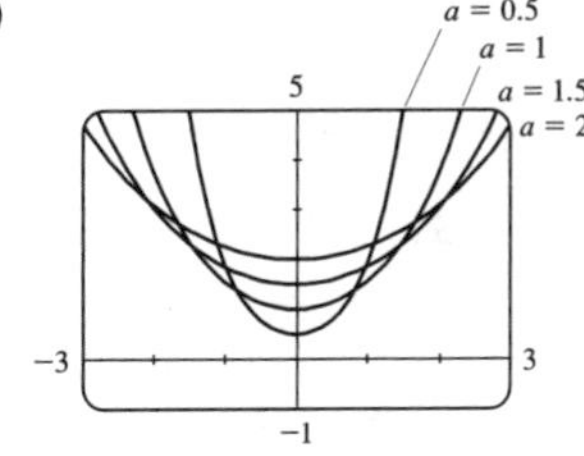

(b) The larger the value of a, the less steep the graph and the higher its minimum value.

45. Local maximum $\approx (1.00, 0.37)$

Section 6.3 ■ page 405

1. (a) $2^5 = 32$ **(b)** $5^0 = 1$

3. (a) $4^{1/2} = 2$ **(b)** $2^{-4} = \frac{1}{16}$

5. (a) $e^x = 5$ **(b)** $e^5 = y$

7. (a) $\log_2 8 = 3$ **(b)** $\log_{10} 0.001 = -3$

9. (a) $\log_4 0.125 = -\frac{3}{2}$ **(b)** $\log_7 343 = 3$

11. (a) $\ln 2 = x$ **(b)** $\ln y = 3$

13. (a) 4 **(b)** 3 **(c)** 1 **15. (a)** 2 **(b)** 2 **(c)** 10

17. (a) -3 **(b)** $\frac{1}{2}$ **(c)** -1

19. (a) 37 **(b)** 8 **(c)** $\sqrt{5}$

21. (a) $-\frac{2}{3}$ **(b)** 4 **(c)** -1 **23. (a)** 32 **(b)** 4

25. (a) 100 **(b)** 25 **27. (a)** 2 **(b)** 4

29. (a) 0.3010 **(b)** 1.5465 **(c)** -0.1761

31. (a) 1.6094 **(b)** 3.2308 **(c)** 1.0051

33. $y = \log_5 x$ **35.** $y = \log_9 x$

37. II **39.** III **41.** VI

43.

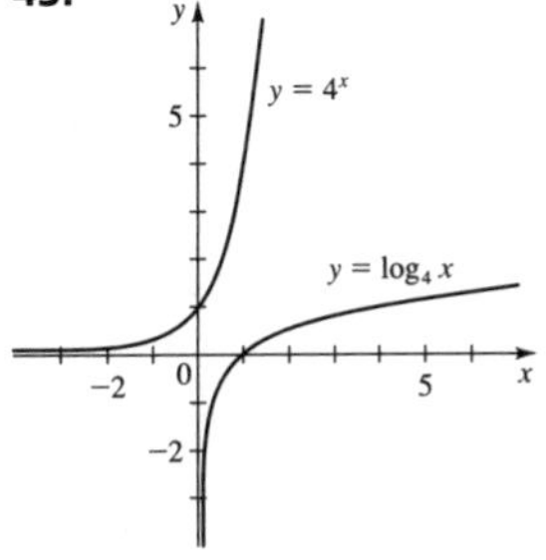

45. $(4, \infty), \mathbb{R}, x = 4$

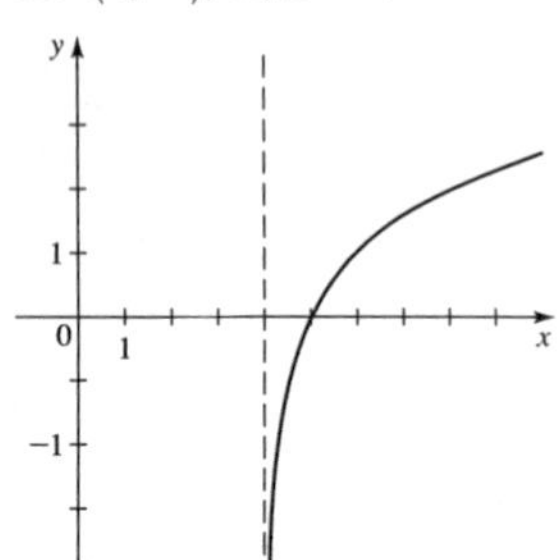

47. $(-\infty, 0), \mathbb{R}, x = 0$

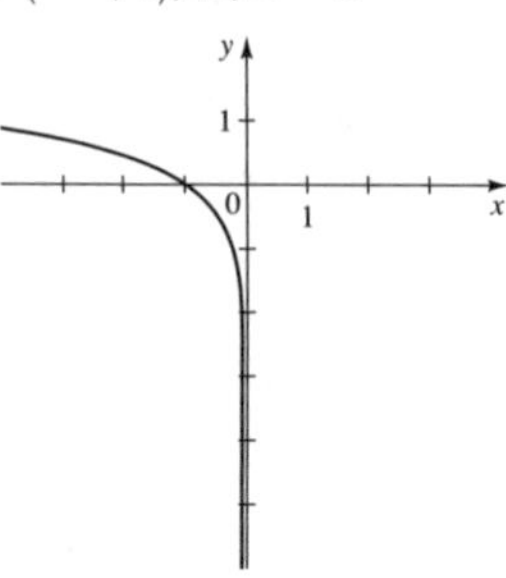

49. $(0, \infty), \mathbb{R}, x = 0$

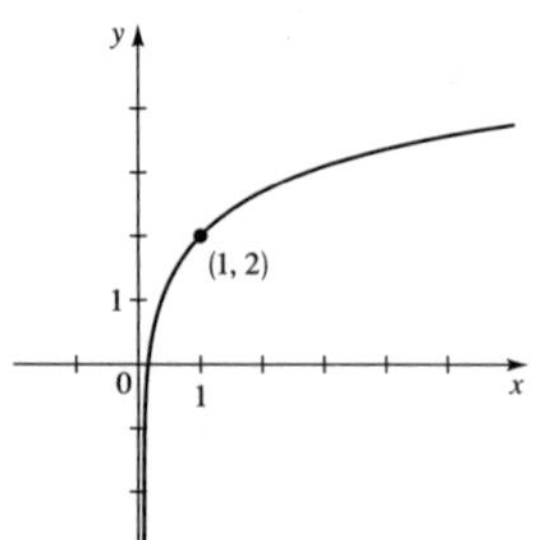

51. $(0, \infty), \mathbb{R}, x = 0$

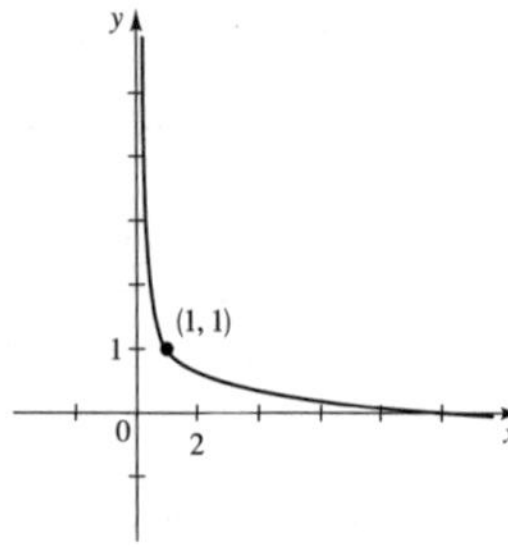

53. $(0, \infty), [0, \infty), x = 0$

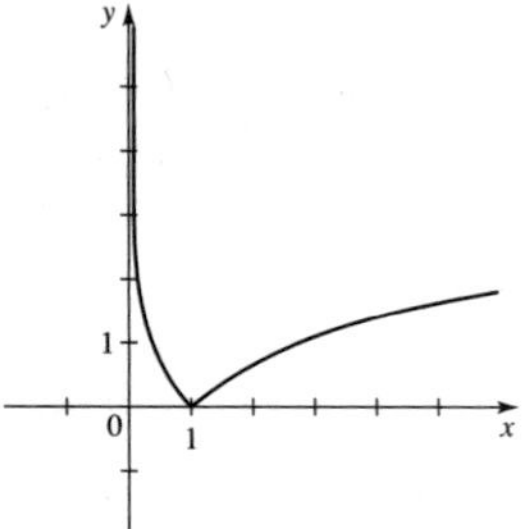

55. $\left(-\frac{2}{5}, \infty\right)$ **57.** $(-\infty, -1) \cup (1, \infty)$ **59.** $(0, 2)$

61.

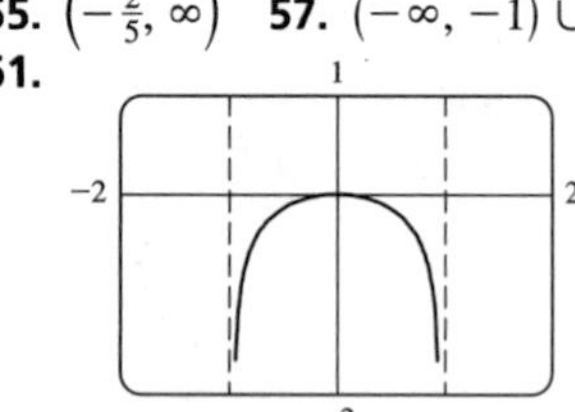

domain $= (-1, 1)$
vertical asymptotes
$x = 1, x = -1$
local maximum $(0, 0)$

63.

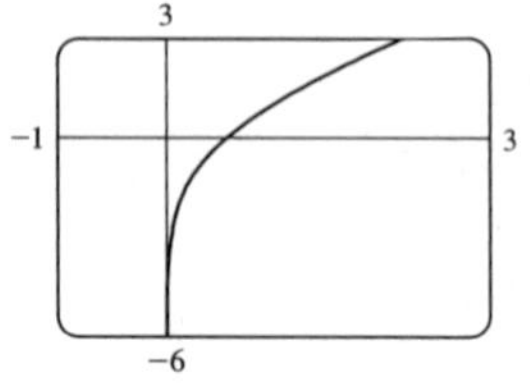

domain $= (0, \infty)$
vertical asymptote $x = 0$
no maximum or minimum

65.

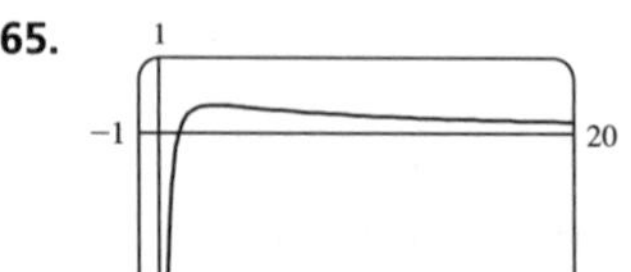

domain $= (0, \infty)$
vertical asymptote $x = 0$
horizontal asymptote $y = 0$
local maximum
$\approx (2.72, 0.37)$

67. The graph of f grows more slowly than g.

69. (a)

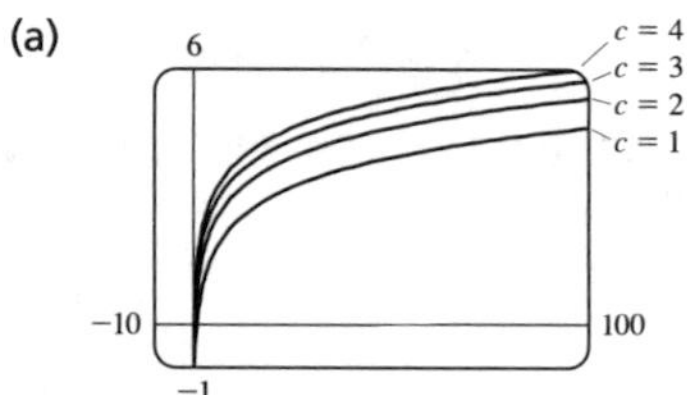

(b) The graph of $f(x) = \log(cx)$ is the graph of $f(x) = \log x$ shifted upward $\log c$ units.

71. (a) $(1, \infty)$ (b) $f^{-1}(x) = 10^{2^x}$

73. (a) $f^{-1}(x) = \log_2\left(\dfrac{x}{1-x}\right)$ (b) $(0, 1)$

Section 6.4 ■ page 411

1. $\log_2 x + \log_2(x - 1)$ **3.** $23 \log 7$
5. $\log_2 A + 2 \log_2 B$ **7.** $\log_3 x + \frac{1}{2} \log_3 y$
9. $\frac{1}{3} \log_5(x^2 + 1)$ **11.** $\frac{1}{2}(\ln a + \ln b)$
13. $3 \log x + 4 \log y - 6 \log z$
15. $\log_2 x + \log_2(x^2 + 1) - \frac{1}{2} \log_2(x^2 - 1)$
17. $\ln x + \frac{1}{2}(\ln y - \ln z)$ **19.** $\frac{1}{4} \log(x^2 + y^2)$
21. $\frac{1}{2}[\log(x^2 + 4) - \log(x^2 + 1) - 2 \log(x^3 - 7)]$
23. $\frac{1}{2} \ln x + 4 \ln z - \frac{1}{3} \ln(y^2 + 6y + 17)$
25. $\frac{3}{2}$ **27.** 1 **29.** 3 **31.** $\ln 8$ **33.** 16
35. $\log_3 160$ **37.** $\log_2(AB/C^2)$ **39.** $\log\left[\dfrac{x^4(x-1)^2}{\sqrt[3]{x^2+1}}\right]$
41. $\ln[5x^2(x^2 + 5)^3]$
43. $\log\left[\sqrt[3]{2x+1}\sqrt{(x-4)/(x^4 - x^2 - 1)}\right]$
45. 2.807355 **47.** 2.182658 **49.** 0.655407
51. 4.165458
53.

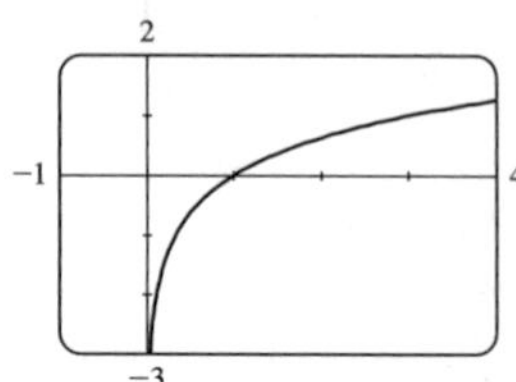

Section 6.5 ■ page 418

1. 1.7227 **3.** -0.5850 **5.** 1.2040 **7.** 0.0767 **9.** 0.2524 **11.** 1.9349 **13.** -43.0677 **15.** 2.1492 **17.** 6.2126 **19.** -2.9469 **21.** -2.4423 **23.** 14.0055 **25.** ± 1 **27.** $0, \frac{4}{3}$ **29.** $\ln 2 \approx 0.6931, 0$ **31.** $\frac{1}{2}\ln 3 \approx 0.5493$ **33.** $e^{10} \approx 22026$ **35.** 0.01 **37.** $\frac{95}{3}$ **39.** $3 - e^2 \approx -4.3891$ **41.** 5 **43.** 5 **45.** $\frac{13}{12}$ **47.** 6 **49.** $\frac{3}{2}$ **51.** $1/\sqrt{5} \approx 0.4472$ **53.** 13 days **55.** (a) $t = -\frac{5}{13}\ln\left(1 - \frac{13}{60}I\right)$ (b) 0.218 s **57.** 2.21 **59.** 0.00, 1.14 **61.** -0.57 **63.** 0.36 **65.** $2 < x < 4$ or $7 < x < 9$ **67.** $\log 2 < x < \log 5$ **69.** 101, 1.1 **71.** $\log_2 3 \approx 1.58$

Section 6.6 ■ page 430

1. (a) \$12,870.19 (b) 8.24 yr **3.** 5 yr **5.** 8.15 yr **7.** 8.30% **9.** (a) 500 (b) 45% (c) 1929 (d) 6.66 h **11.** (a) $n(t) = 112{,}000e^{0.04t}$ (b) About 142,000 (c) 2008 **13.** (a) 20,000 (b) $n(t) = 20{,}000e^{0.1096t}$ (c) About 48,000 (d) 2010 **15.** (a) $n(t) = 8600e^{0.1508t}$ (b) About 11,600 (c) 4.6 h **17.** (a) 2029 (b) 2049 **19.** 22.85 h **21.** (a) $n(t) = 10e^{-0.0231t}$ (b) 1.6 g (c) 70 yr **23.** 18 yr **25.** 149 h **27.** 3560 yr **29.** (a) 210°F (b) 153°F (c) 28 min **31.** (a) 137°F (b) 116 min **33.** (a) 2.3 (b) 3.5 (c) 8.3 **35.** (a) 10^{-3} M (b) 3.2×10^{-7} M **37.** $4.8 \leq \text{pH} \leq 6.4$ **39.** $\log 20 \approx 1.3$ **41.** Twice as intense **43.** 8.2 **45.** 6.3×10^{-3} W/m² **47.** (b) 106 dB

Chapter 6 Review ■ page 434

1. $\mathbb{R}, (0, \infty), y = 0$

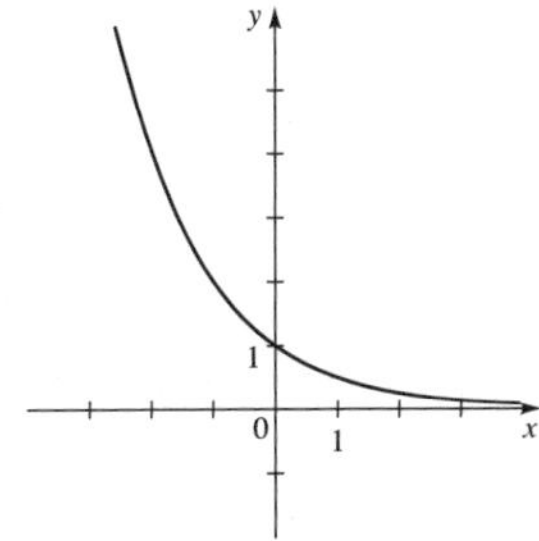

3. $\mathbb{R}, (-\infty, 5), y = 5$

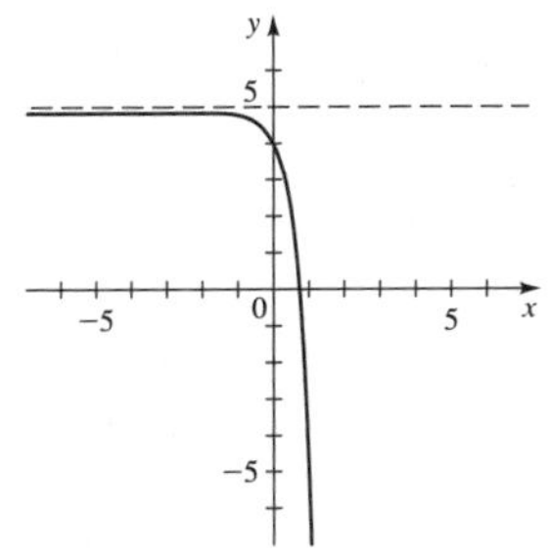

5. $(1, \infty), \mathbb{R}, x = 1$

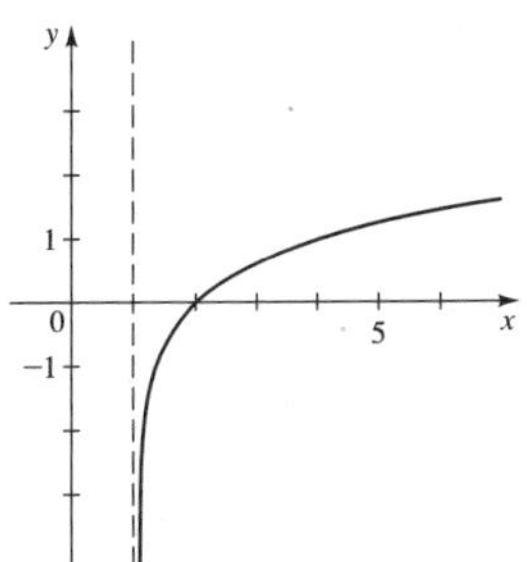

7. $(0, \infty), \mathbb{R}, x = 0$

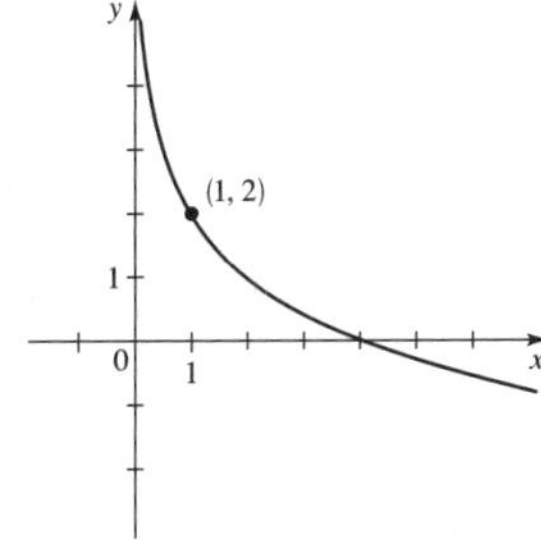

9. $\mathbb{R}, (-1, \infty), y = -1$

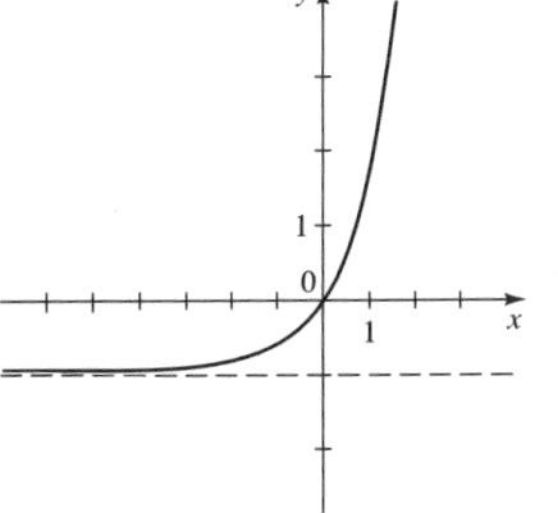

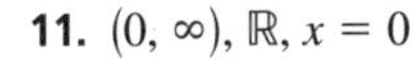

11. $(0, \infty), \mathbb{R}, x = 0$

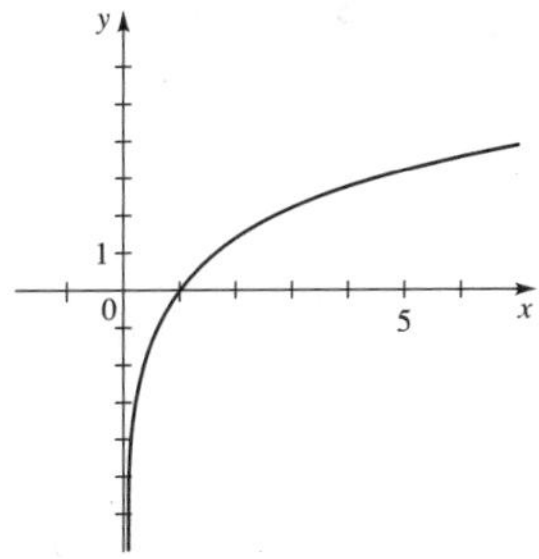

13. $\left(-\infty, \frac{1}{2}\right)$ **15.** $2^{10} = 1024$ **17.** $10^y = x$ **19.** $\log_2 64 = 6$ **21.** $\log 74 = x$ **23.** 7 **25.** 45 **27.** 6 **29.** -3 **31.** $\frac{1}{2}$ **33.** 2 **35.** 92 **37.** $\frac{2}{3}$ **39.** $\log A + 2\log B + 3\log C$ **41.** $\frac{1}{2}[\ln(x^2 - 1) - \ln(x^2 + 1)]$ **43.** $2\log_5 x + \frac{3}{2}\log_5(1 - 5x) - \frac{1}{2}\log_5(x^3 - x)$ **45.** $\log 96$ **47.** $\log_2\left[\frac{(x - y)^{3/2}}{(x^2 + y^2)^2}\right]$ **49.** $\log\left(\frac{x^2 - 4}{\sqrt{x^2 + 4}}\right)$ **51.** -15 **53.** $\frac{1}{3}(5 - \log_5 26) \approx 0.99$ **55.** $\frac{4}{3}\ln 10 \approx 3.07$ **57.** 3 **59.** $-4, 2$ **61.** 0.430618 **63.** 2.303600

65.

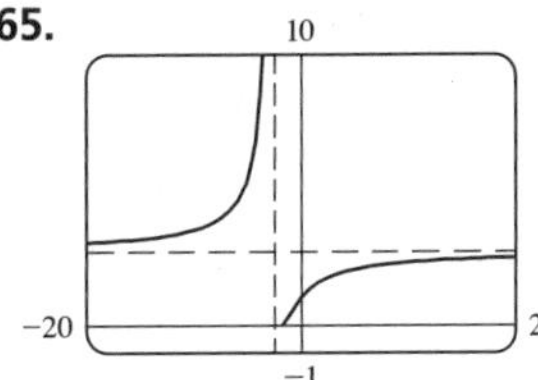

vertical asymptote $x = -2$
horizontal asymptote $y = 2.72$
no maximum or minimum

67.

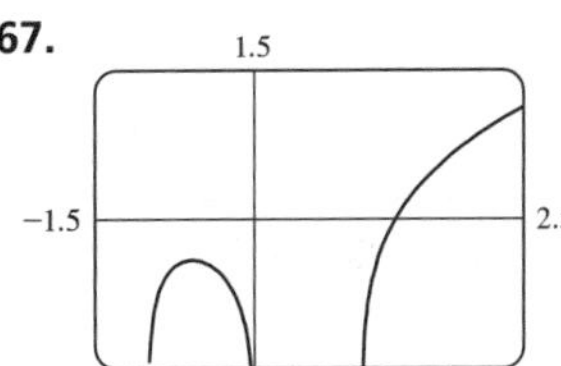

vertical asymptotes $x = -1, x = 0, x = 1$
local maximum $\approx (-0.58, -0.41)$

69. 2.42 **71.** $0.16 < x < 3.15$
73. Increasing on $(-\infty, 0]$ and $[1.10, \infty)$, decreasing on $[0, 1.10]$
75. 1.953445 **77.** $\log_4 258$
79. (a) \$16,081.15 (b) \$16,178.18 (c) \$16,197.64 (d) \$16,198.31
81. (a) $n(t) = 30e^{0.15t}$ (b) 55 (c) 19 yr
83. (a) 9.97 mg (b) 1.39×10^5 yr
85. (a) $n(t) = 150e^{-0.0004359t}$ (b) 97.0 mg (c) 2520 yr
87. (a) $n(t) = 1500e^{0.1515t}$ (b) 7940 **89.** 7.9, basic
91. 8.0

Chapter 6 Test ■ page 437

1.

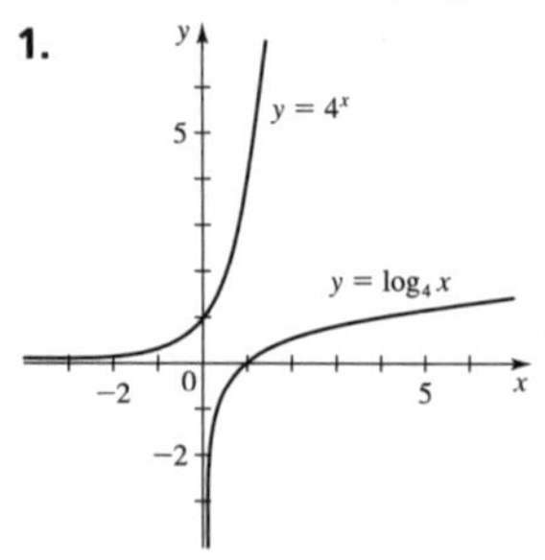

2.

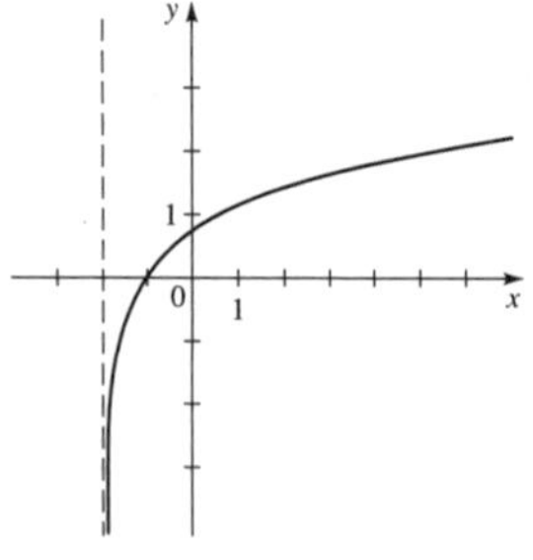

$(-2, \infty)$, $\mathbb{R}$, $x = -2$

3. (a) $\frac{3}{2}$ (b) 3 (c) $\frac{2}{3}$ (d) 2
4. $\frac{1}{2}[\log(x^2 - 1) - 3\log x - 5\log(y^2 + 1)]$
5. $\ln\left[\frac{x\sqrt{3 - x^4}}{(x^2 + 1)^2}\right]$ **6.** (a) 4.32 (b) 0.77 (c) 5.39 (d) 2
7. (a) $n(t) = 1000e^{2.07944t}$ (b) 22,627 (c) 1.3 h
(d)

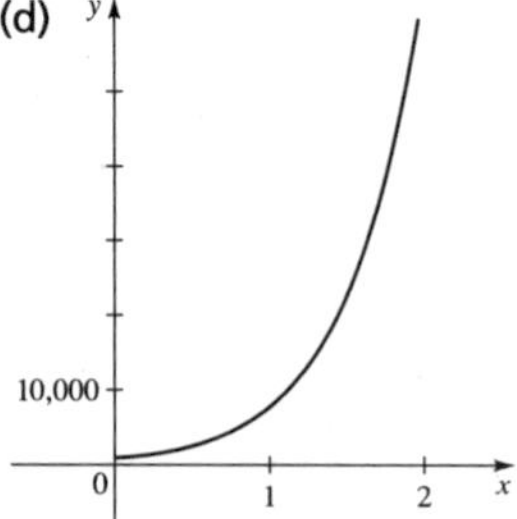

8. (a) $A(t) = 12{,}000\left(1 + \frac{0.056}{12}\right)^{12t}$
(b) \$14,195.06 (c) 9.249 yr
9. (a)

5
-5 10
-5

(b) $x = 0, y = 0$
(c) Local minimum $\approx (3.00, 0.74)$
(d) $(-\infty, 0) \cup [0.74, \infty)$
(e) $-0.85, 0.96, 9.92$

FOCUS ON MODELING ■ page 443

1. (a)

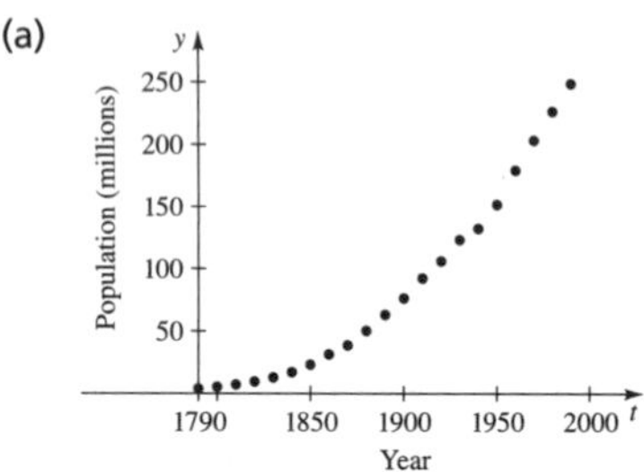

(b) $y = ab^t$, where $a = 4.041807 \times 10^{-16}$ and $b = 1.021003194$, and y is the population in millions in the year t.
(c) 457.9 million (d) 221.2 million (e) No
3. (a) Yes
(b) Yes, the scatter plot appears linear.

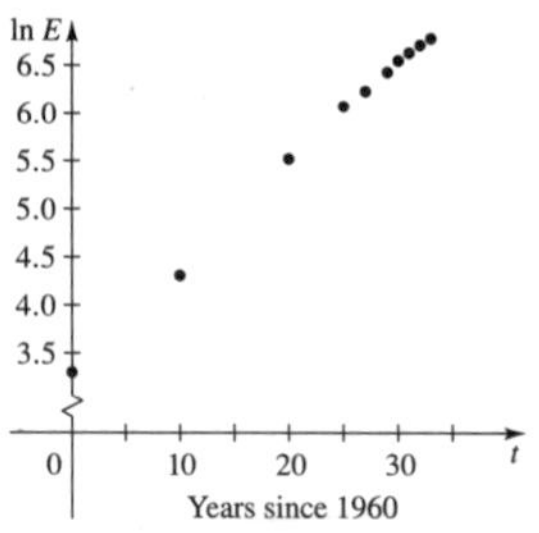

(c) $\ln E = 3.30161 \times 0.10769t$, where t is years since 1960 and E is expenditure in billions of dollars.
(d) $E = 27.15633e^{0.10769t}$
(e) 1310.9 billion dollars
5. (a) $y = ab^t$, where $a = 301.813054$, $b = 0.819745$, and t is the number of years since 1970
(b) $y = at^4 + bt^3 + ct^2 + dt + e$, where $a = -0.002430$, $b = 0.135159$, $c = -2.014322$, $d = -4.055294$, $e = 199.092227$, and t is the number of years since 1970
(c) From the graphs we see that the fourth-degree polynomial is a better model.

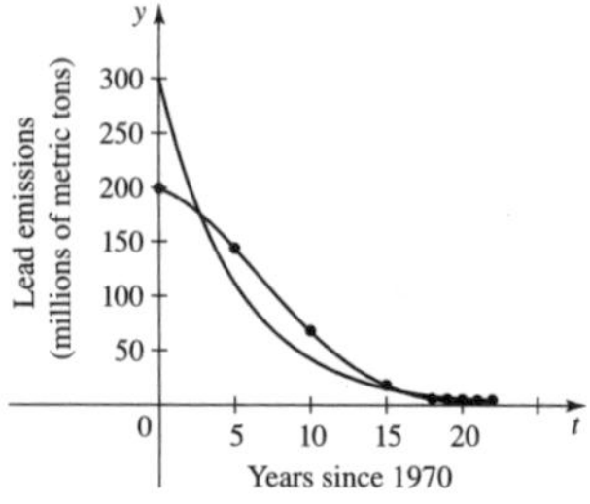

(d) 202.8, 27.8; 184.0, 43.5

CHAPTER 7

Section 7.1 ■ page 447

1. $(6, 2)$ **3.** $(-2, 4), (3, 9)$ **5.** $(2, -2), (-2, 2)$
7. $(-6, 16)$ **9.** $(-3, 4), (3, 4)$
11. $(-2, -1), (-2, 1), (2, -1), (2, 1)$ **13.** $(4, 0)$
15. $(-2, -2)$ **17.** $(6, 2), (-2, -6)$
19. No solution
21. $(\sqrt{5}, 2), (\sqrt{5}, -2), (-\sqrt{5}, 2), (-\sqrt{5}, -2)$
23. $\left(3, -\frac{1}{2}\right), \left(-3, -\frac{1}{2}\right)$ **25.** $\left(\frac{1}{5}, \frac{1}{3}\right)$ **27.** $(-0.33, 5.33)$
29. $(-1.38, -2.63), (0.82, 0.55)$
31. $(-4.51, 2.17), (4.91, -0.97)$
33. $(1.23, 3.87), (-0{,}35, -4.21)$
35. $(-2.30, -0.70), (0.48, -1.19)$
37. 12 cm by 15 cm **39.** 15, 20
41. $(400.50, 200.25)$, 447.77 m **43.** $y = 3x + 5$
45. $\left(-\frac{1}{2}, -\frac{3}{2}\right), \left(\frac{1}{2}, \frac{3}{2}\right)$ **47.** $(\sqrt{10}, 10)$

Section 7.2 ■ page 456

1. $(1, 2)$

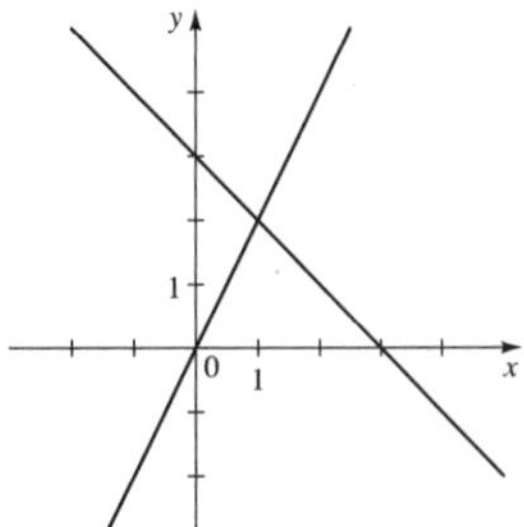

3. $(3, 2)$

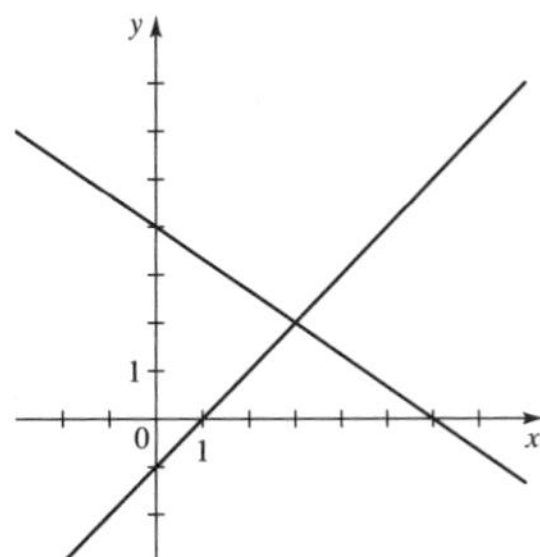

5. Parallel

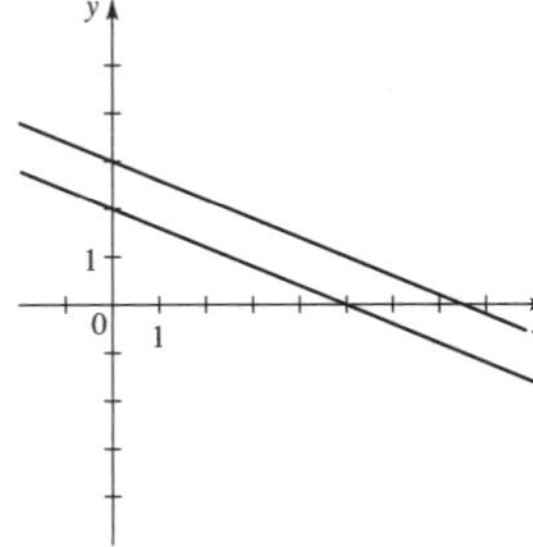

7. $(3, 5)$ **9.** $(1, 3)$ **11.** $(10, -9)$ **13.** $(2, 1)$
15. No solution **17.** $\left(x, \frac{1}{3}x - \frac{5}{3}\right)$ **19.** $\left(x, 3 - \frac{3}{2}x\right)$
21. $(-3, -7)$ **23.** $\left(x, 5 - \frac{5}{6}x\right)$ **25.** $\left(\frac{5}{2}, 4\right)$
27. $\left(\frac{2}{3}, -\frac{1}{6}\right)$ **29.** $\left(-\dfrac{1}{a-1}, \dfrac{1}{a-1}\right)$
31. $\left(\dfrac{1}{a+b}, \dfrac{1}{a+b}\right)$ **33.** 22, 12 **35.** 5 dimes, 9 quarters
37. Plane's speed 120 mi/h, wind speed 30 mi/h
39. Run 5 mi/h, cycle 20 mi/h
41. 200 g of A, 40 g of B **43.** 25%, 10%
45. \$16,000 at 10%, \$32,000 at 6% **47.** 25
49. $(3.87, 2.74)$ **51.** $(61.00, 20.00)$

Section 7.3 ■ page 464

1. Linear **3.** Linear
5. (a) Yes (b) Yes (c) $\begin{cases} x = -3 \\ y = 5 \end{cases}$
7. (a) Yes (b) No (c) $\begin{cases} x + 2y + 8z = 0 \\ \quad\; y + 3z = 0 \end{cases}$
9. (a) No (b) No (c) $\begin{cases} x \qquad\; = 0 \\ \qquad 0 = 0 \\ y + 5z = 1 \end{cases}$
11. $(1, 1, 2)$ **13.** $(1, 0, 1)$ **15.** $(-1, 0, 1)$
17. $(-1, 5, 0)$ **19.** $(10, 3, -2)$ **21.** No solution
23. $x = 2 - 3z$, $y = 3 - 5z$, $z =$ any number
25. No solution
27. $x = -2z + 5$, $y = z - 2$, $z =$ any number
29. $x = -\frac{1}{2}y + z + 6$, $y =$ any number, $z =$ any number
31. $(-2, 1, 3)$ **33.** $(-9, 2, 0)$
35. $(0, -3, 0, -3)$ **37.** $(-1, 0, 0, 1)$
39. $x = \frac{7}{4} - \frac{7}{4}w$, $y = -\frac{7}{4} + \frac{3}{4}w$, $z = \frac{9}{4} + \frac{3}{4}w$, $w =$ any number
41. $x = \frac{1}{3}z - \frac{2}{3}w$, $y = \frac{1}{3}z + \frac{1}{3}w$, $z =$ any number, $w =$ any number
43. 2 VitaMax, 1 Vitron, 2 VitaPlus
45. 5 mile run, 2 mile swim, 30 mile cycle
47. Impossible

Section 7.4 ■ page 478

1. $\begin{bmatrix} 5 & -2 & 5 \\ 1 & 1 & 0 \end{bmatrix}$ **3.** $\begin{bmatrix} -1 & -3 & -5 \\ -1 & 3 & -6 \end{bmatrix}$ **5.** $\begin{bmatrix} 13 & -\frac{7}{2} & 15 \\ 3 & 1 & 3 \end{bmatrix}$
7. $\begin{bmatrix} -14 & -8 & -30 \\ -6 & 10 & -24 \end{bmatrix}$ **9.** Impossible **11.** $\begin{bmatrix} 3 & \frac{1}{2} & 5 \\ 1 & -1 & 3 \end{bmatrix}$
13. $\begin{bmatrix} 28 & 21 & 28 \end{bmatrix}$ **15.** $\begin{bmatrix} -1 \\ 8 \\ -1 \end{bmatrix}$ **17.** $\begin{bmatrix} 8 & -335 \\ 0 & 343 \end{bmatrix}$
19. Impossible **21.** Impossible
23. $\begin{bmatrix} 2 & -5 \\ 3 & 2 \end{bmatrix} \begin{bmatrix} x \\ y \end{bmatrix} = \begin{bmatrix} 7 \\ 4 \end{bmatrix}$

25. $\begin{bmatrix} 3 & 2 & -1 & 1 \\ 1 & 0 & -1 & 0 \\ 0 & 3 & 1 & -1 \end{bmatrix} \begin{bmatrix} x_1 \\ x_2 \\ x_3 \\ x_4 \end{bmatrix} = \begin{bmatrix} 0 \\ 5 \\ 4 \end{bmatrix}$ **27.** $x = 2, y = 1$

29. $\begin{bmatrix} 3 & \frac{11}{2} \\ 2 & 5 \end{bmatrix}$ **31.** Impossible

33. (a) $[4{,}690 \quad 1{,}690 \quad 13{,}210]$
(b) Total revenue in Santa Monica, Long Beach, and Anaheim, respectively

35. Only ACB is defined. $\begin{bmatrix} -3 & -21 & 27 & -6 \\ -2 & -14 & 18 & -4 \end{bmatrix}$ **37.** No

Section 7.5 ■ page 487

1. $\begin{bmatrix} 1 & -2 \\ -\frac{3}{2} & \frac{7}{2} \end{bmatrix}$ **3.** $\begin{bmatrix} 2 & -3 \\ -3 & 5 \end{bmatrix}$ **5.** $\begin{bmatrix} 13 & 5 \\ -5 & -2 \end{bmatrix}$

7. No inverse **9.** $\begin{bmatrix} 1 & 2 \\ -\frac{1}{2} & \frac{2}{3} \end{bmatrix}$ **11.** $\begin{bmatrix} -4 & -4 & 5 \\ 1 & 1 & -1 \\ 5 & 4 & -6 \end{bmatrix}$

13. No inverse **15.** $\begin{bmatrix} -\frac{9}{2} & -1 & 4 \\ 3 & 1 & -3 \\ \frac{7}{2} & 1 & -3 \end{bmatrix}$

17. $\begin{bmatrix} 0 & 0 & -2 & 1 \\ -1 & 0 & 1 & 1 \\ 0 & 1 & -1 & 0 \\ 1 & 0 & 0 & -1 \end{bmatrix}$ **19.** $x = 8, y = -12$

21. $x = 126, y = -50$ **23.** $x = -38, y = 9, z = 47$

25. $x = -20, y = 10, z = 16$ **27.** $\begin{bmatrix} 7 & 2 & 3 \\ 10 & 3 & 5 \end{bmatrix}$

29. (a) $\begin{bmatrix} 0 & 1 & -1 \\ -2 & \frac{3}{2} & 0 \\ 1 & -\frac{3}{2} & 1 \end{bmatrix}$ (b) 1 oz A, 1 oz B, 2 oz C
(c) 2 oz A, 0 oz B, 1 oz C (d) No

31. (a) $\begin{cases} x + y + 2z = 675 \\ 2x + y + z = 600 \\ x + 2y + z = 625 \end{cases}$

(b) $\begin{bmatrix} 1 & 1 & 2 \\ 2 & 1 & 1 \\ 1 & 2 & 1 \end{bmatrix} \begin{bmatrix} x \\ y \\ z \end{bmatrix} = \begin{bmatrix} 675 \\ 600 \\ 625 \end{bmatrix}$

(c) $A^{-1} = \begin{bmatrix} -\frac{1}{4} & \frac{3}{4} & -\frac{1}{4} \\ -\frac{1}{4} & -\frac{1}{4} & \frac{3}{4} \\ \frac{3}{4} & -\frac{1}{4} & -\frac{1}{4} \end{bmatrix}$

He earns \$125 on a standard set, \$150 on a deluxe set, and \$200 on a leather-bound set.

33. $\frac{1}{2}\begin{bmatrix} e^{-x} & e^{-2x} \\ -e^{-2x} & e^{-3x} \end{bmatrix}$

35. $\begin{bmatrix} 2 & 0 & -1 \\ 1/x & -1/x & 0 \\ -2 & 1 & 1 \end{bmatrix}$

Section 7.6 ■ page 497

1. 3 **3.** −4 **5.** Does not exist **7.** $\frac{1}{8}$ **9.** 20, 20
11. −12, 12 **13.** 0, 0 **15.** −6, has an inverse
17. 5000, has an inverse **19.** −4, has an inverse
21. −18 **23.** 120
25. (a) −2 (b) −2 (c) Yes
27. $(-2, 5)$ **29.** $(0.6, -0.4)$ **31.** $(4, -1)$
33. $(4, 2, -1)$ **35.** $(1, 3, 2)$ **37.** $(0, -1, 1)$
39. $\left(\frac{189}{29}, -\frac{108}{29}, \frac{88}{29}\right)$ **41.** $\left(\frac{1}{2}, \frac{1}{4}, \frac{1}{4}, -1\right)$
43. (b) $5x - 6y = -200$ **45.** 0, 1, 2 **47.** 1, −1

Section 7.7 ■ page 507

1.

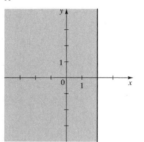

3.

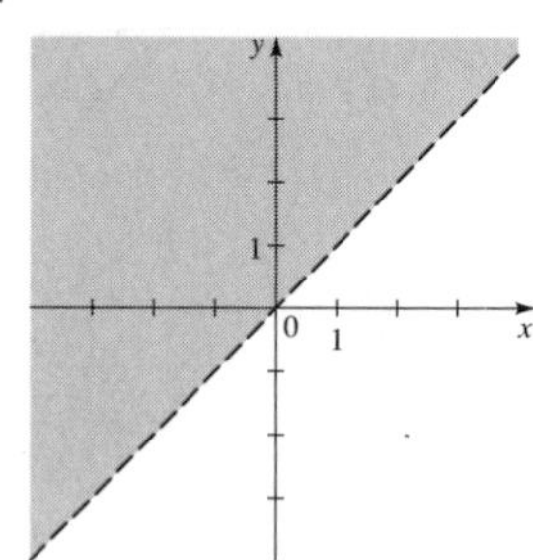

5.

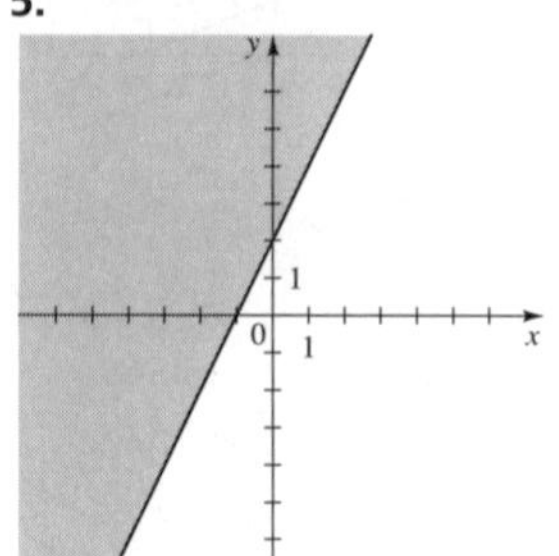

7.

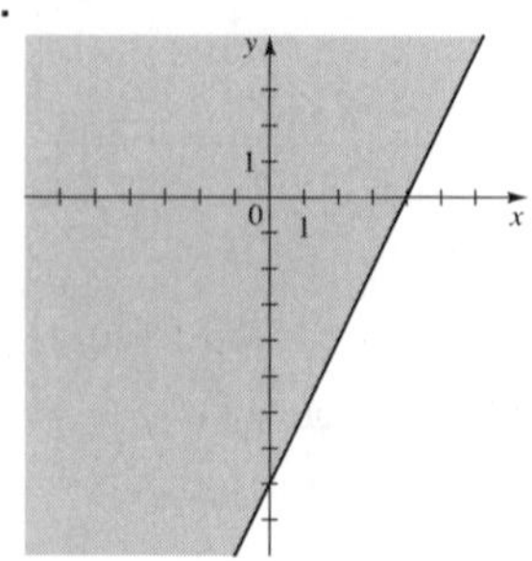

9.

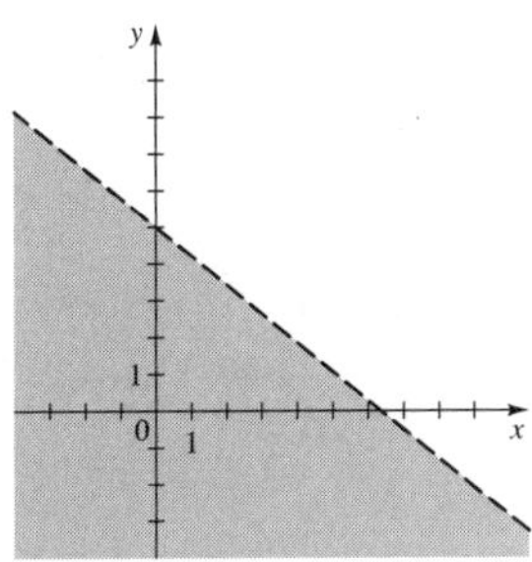

11.

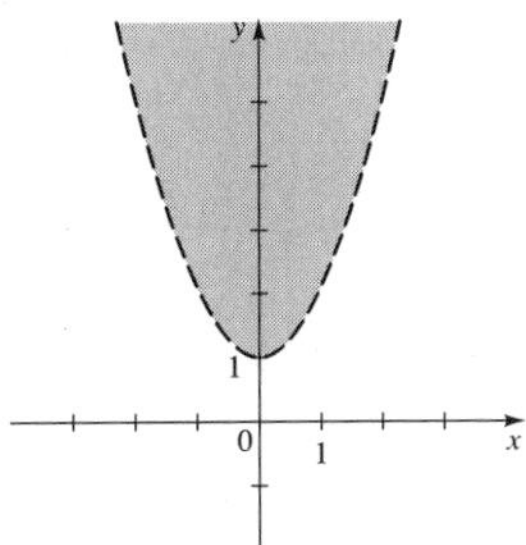

25.

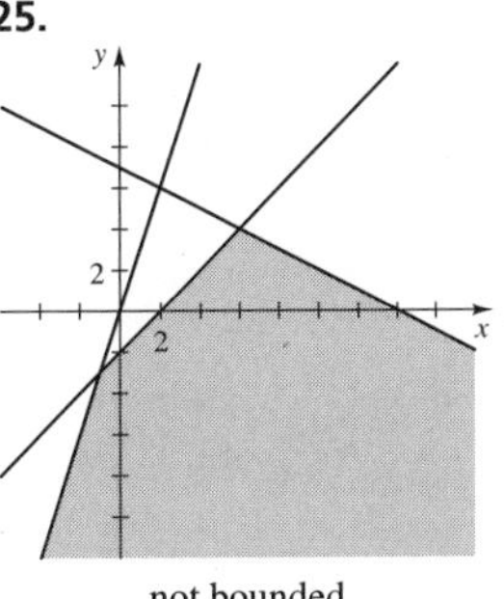

not bounded

27.

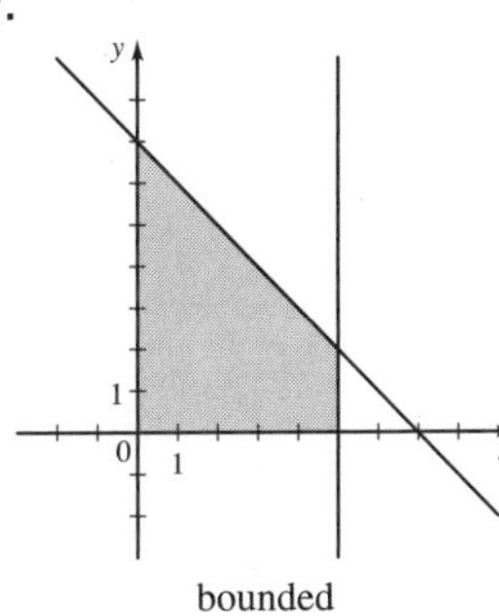

bounded

13.

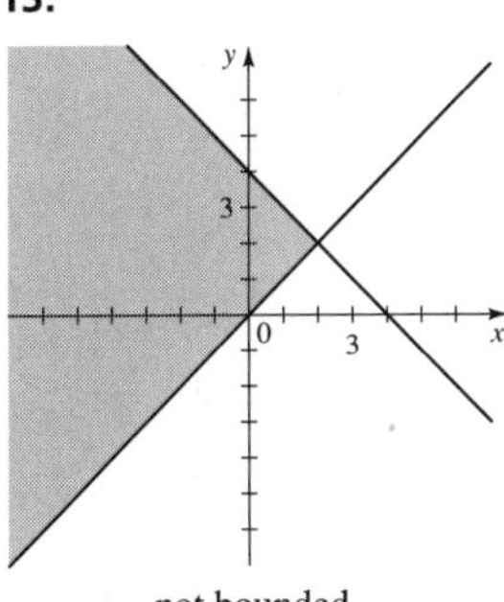

not bounded

15.

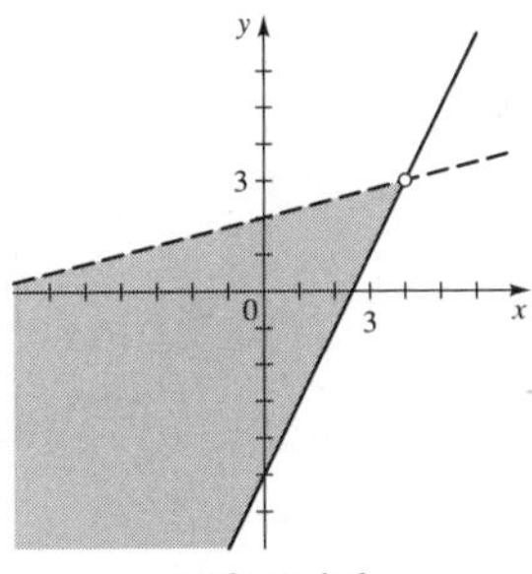

not bounded

29.

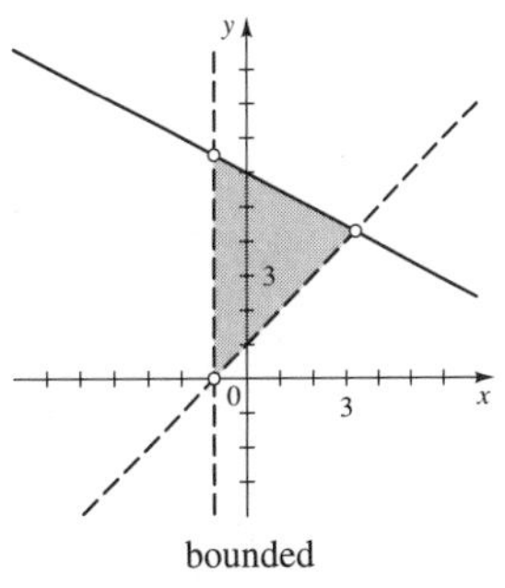

bounded

31.

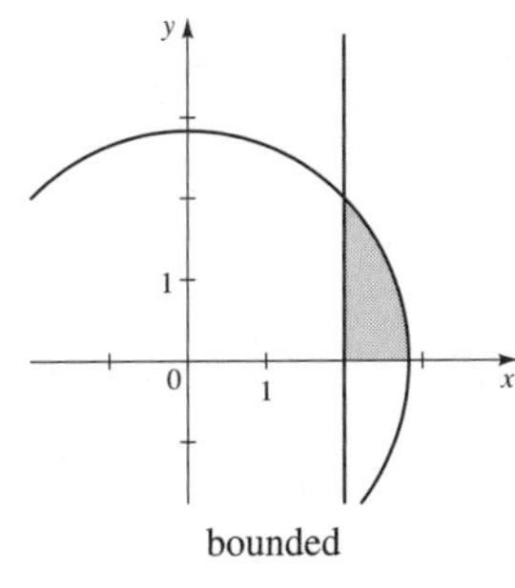

bounded

17.

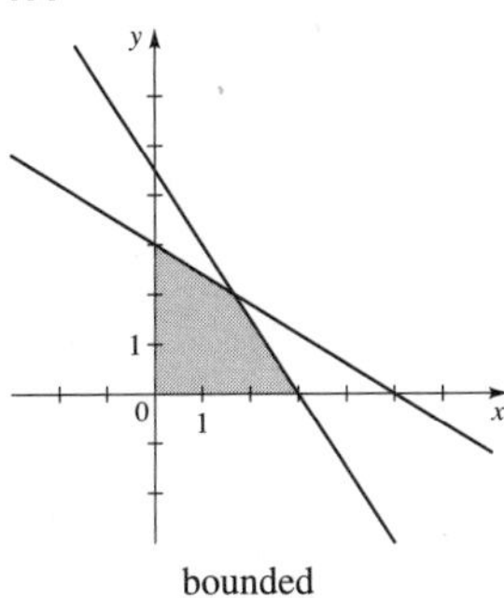

bounded

19.

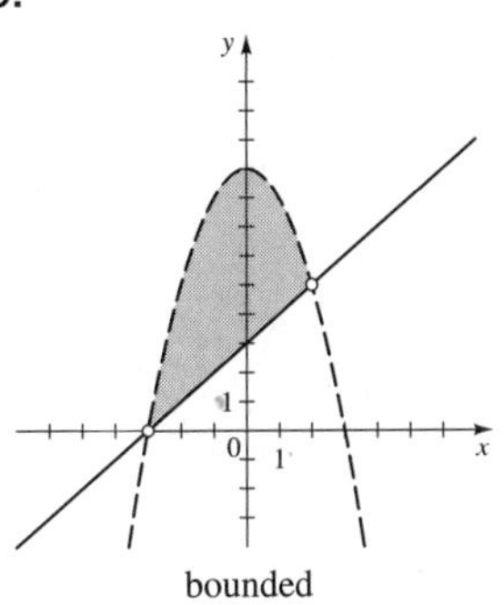

bounded

33.

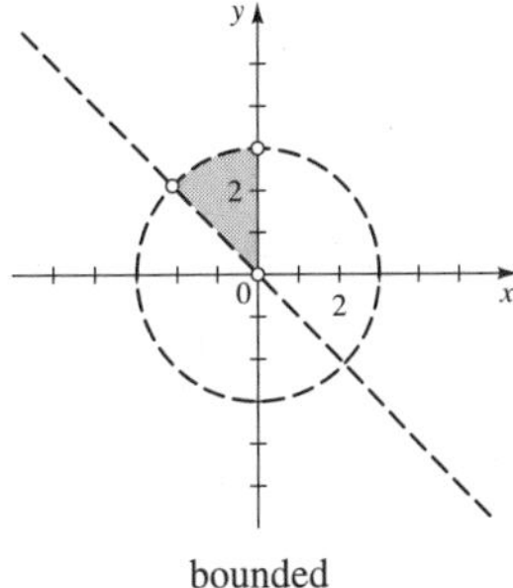

bounded

21.

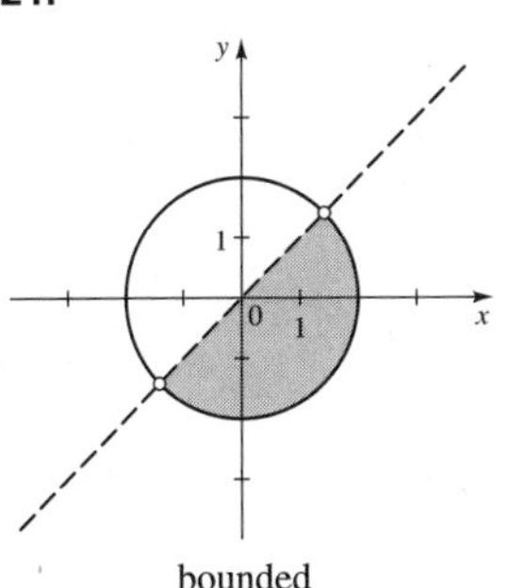

bounded

23.

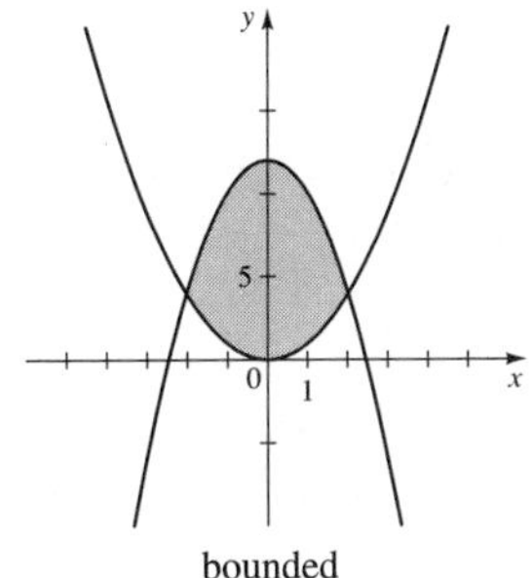

bounded

35. x = number of fiction books
y = number of nonfiction books

$$\begin{cases} x + y \leqslant 100 \\ 20 \leqslant y,\ x \geqslant y \\ x \geqslant 0,\ y \geqslant 0 \end{cases}$$

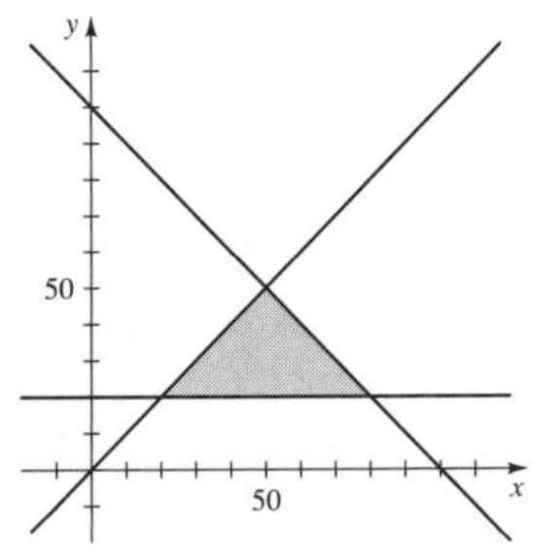

Section 7.8 ■ page 512

1. $\dfrac{A}{x-1}+\dfrac{B}{x+2}$ **3.** $\dfrac{A}{x-2}+\dfrac{B}{(x-2)^2}+\dfrac{C}{x+4}$

5. $\dfrac{A}{x-3}+\dfrac{Bx+C}{x^2+4}$ **7.** $\dfrac{Ax+B}{x^2+1}+\dfrac{Cx+D}{x^2+2}$

9. $\dfrac{A}{x}+\dfrac{B}{2x-5}+\dfrac{C}{(2x-5)^2}+\dfrac{D}{(2x-5)^3}+\dfrac{Ex+F}{x^2+2x+5}+\dfrac{Gx+H}{(x^2+2x+5)^2}$

11. $\dfrac{1}{x-1}-\dfrac{1}{x+4}$ **13.** $\dfrac{2}{x-3}-\dfrac{2}{x+3}$

15. $\dfrac{1}{x-2}-\dfrac{1}{x+2}$ **17.** $\dfrac{3}{x-4}-\dfrac{2}{x+2}$

19. $\dfrac{-\frac{1}{2}}{2x-1}+\dfrac{\frac{3}{2}}{4x-3}$ **21.** $\dfrac{2}{x-2}+\dfrac{3}{x+2}-\dfrac{1}{2x-1}$

23. $\dfrac{2}{x+1}-\dfrac{1}{x}+\dfrac{1}{x^2}$ **25.** $\dfrac{1}{2x+3}-\dfrac{3}{(2x+3)^2}$

27. $\dfrac{2}{x}-\dfrac{1}{x^3}-\dfrac{2}{x+2}$

29. $\dfrac{4}{x+2}-\dfrac{4}{x-1}+\dfrac{2}{(x-1)^2}+\dfrac{1}{(x-1)^3}$

31. $\dfrac{3}{x+2}-\dfrac{1}{(x+2)^2}-\dfrac{1}{(x+3)^2}$

33. $\dfrac{x+1}{x^2+3}-\dfrac{1}{x}$ **35.** $\dfrac{2x-5}{x^2+x+2}+\dfrac{5}{x^2+1}$

37. $\dfrac{1}{x^2+1}-\dfrac{x+2}{(x^2+1)^2}+\dfrac{1}{x}$

39. $x^2+\dfrac{3}{x-2}-\dfrac{x+1}{x^2+1}$ **41.** $A=\dfrac{a+b}{2}, B=\dfrac{a-b}{2}$

Chapter 7 Review ■ page 518

1. (2, 1)

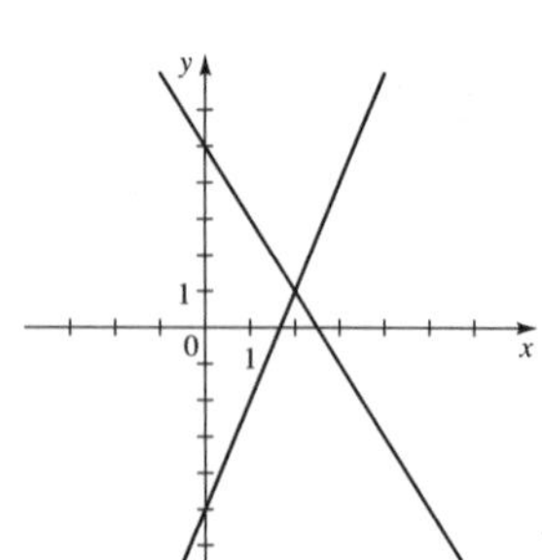

3. x = any number
$y=\frac{2}{7}x-4$

5. No solution

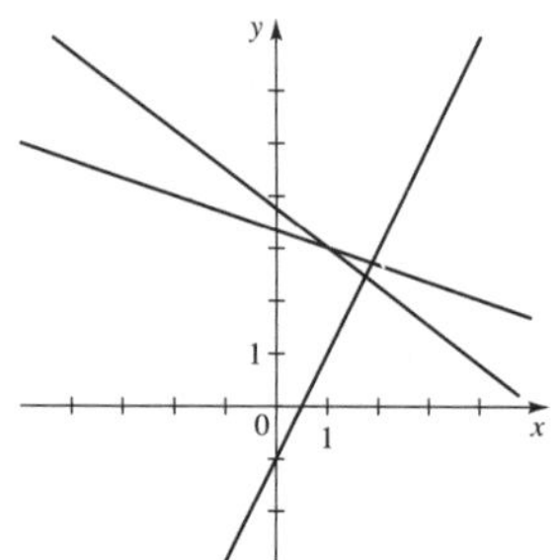

7. $(-3, 3), (2, 8)$ **9.** $\left(\frac{16}{7}, -\frac{14}{3}\right)$ **11.** (1, 1, 2) **13.** No solution **15.** $x=-4z+1$, $y=-z-1$, z = any number **17.** $x=6-5z$, $y=\frac{1}{2}(7-3z)$, z = any number

19. \$3000 at 6%, \$6000 at 7% **21.** Impossible

23. $\begin{bmatrix}4 & 18\\4 & 0\\2 & 2\end{bmatrix}$ **25.** $[10 \quad 0 \quad -5]$ **27.** $\begin{bmatrix}-\frac{7}{2} & 10\\1 & -\frac{9}{2}\end{bmatrix}$

29. $\begin{bmatrix}30 & 22 & 2\\-9 & 1 & -4\end{bmatrix}$ **31.** $\begin{bmatrix}-\frac{1}{2} & \frac{11}{2}\\\frac{15}{4} & -\frac{3}{2}\\-\frac{1}{2} & 1\end{bmatrix}$ **33.** $1, \begin{bmatrix}9 & -4\\-2 & 1\end{bmatrix}$

35. 0, no inverse **37.** $-1, \begin{bmatrix}3 & 2 & -3\\2 & 1 & -2\\-8 & -6 & 9\end{bmatrix}$ **39.** (65, 154)

41. $\left(\frac{1}{5}, \frac{9}{5}\right)$ **43.** $\left(-\frac{87}{26}, \frac{21}{26}, \frac{3}{2}\right)$

45.

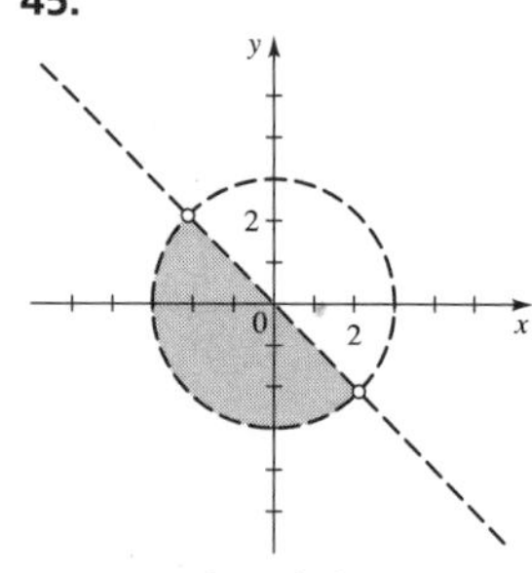

bounded

47.

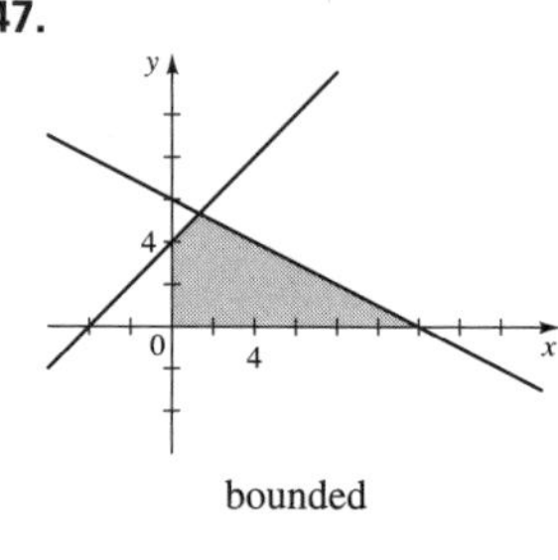

bounded

49. $x=\dfrac{b+c}{2}, y=\dfrac{a+c}{2}, z=\dfrac{a+b}{2}$ **51.** 2, 3

53. $(21.41, -15.93)$ **55.** $(11.94, -1.39), (12.07, 1.44)$

57. $\dfrac{2}{x-5}+\dfrac{1}{x+3}$ **59.** $\dfrac{-4}{x}+\dfrac{4}{x-1}+\dfrac{-2}{(x-1)^2}$

Chapter 7 Test ■ page 521

1. Wind speed 60 km/h, airplane 300 km/h
2. $(3, -1)$; linear, neither inconsistent nor dependent
3. No solution; linear, inconsistent

4. $x = \frac{1}{7}(z + 1)$, $y = \frac{1}{7}(9z + 2)$, $z =$ any number; linear, dependent
5. $(\pm 1, -2)$, $\left(\pm\frac{5}{3}, -\frac{2}{3}\right)$; nonlinear
6. Incompatible dimensions **7.** Incompatible dimensions

8. $\begin{bmatrix} 6 & 10 \\ 3 & -2 \\ -3 & 9 \end{bmatrix}$ **9.** $\begin{bmatrix} 36 & 58 \\ 0 & -3 \\ 18 & 28 \end{bmatrix}$ **10.** $\begin{bmatrix} 2 & -\frac{3}{2} \\ -1 & 1 \end{bmatrix}$

11. B is not square **12.** B is not square **13.** -3
14. $(70, 90)$ **15.** $(5, -5, -4)$

16. $|A| = 0$, $|B| = 2$, $B^{-1} = \begin{bmatrix} 1 & -2 & 0 \\ 0 & \frac{1}{2} & 0 \\ 3 & -6 & 1 \end{bmatrix}$

17.

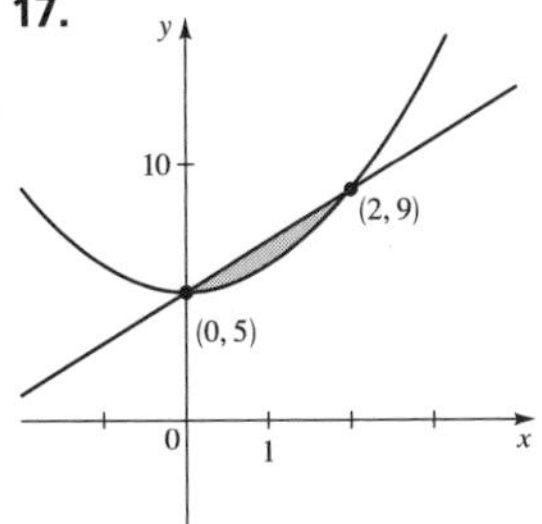

18. $\dfrac{1}{x-1} + \dfrac{1}{(x-1)^2} - \dfrac{1}{x+2}$

19. $(-0.49, 3.93)$, $(2.34, 2.24)$

FOCUS ON MODELING ■ page 523

1. 198, 195
3.

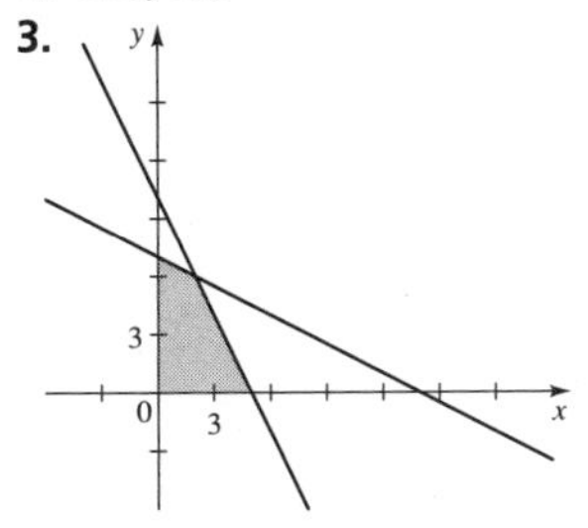

maximum 161
minimum 135

5. 3 tables, 34 chairs **7.** 30 grapefruit crates, 30 orange crates
9. 15 Pasadena to Santa Monica, 3 Pasadena to El Toro, 0 Long Beach to Santa Monica, 16 Long Beach to El Toro
11. 90 standard, 40 deluxe
13. $7500 in municipal bonds, $2500 in bank certificates, $2000 in high-risk bonds
15. 4 games, 32 educational, 0 utility

CHAPTER 8

Section 8.1 ■ page 533

Order of answers: focus; directrix; focal diameter

1. $F(1, 0)$; $x = -1$; 4 **3.** $F\left(0, \frac{9}{4}\right)$; $y = -\frac{9}{4}$; 9

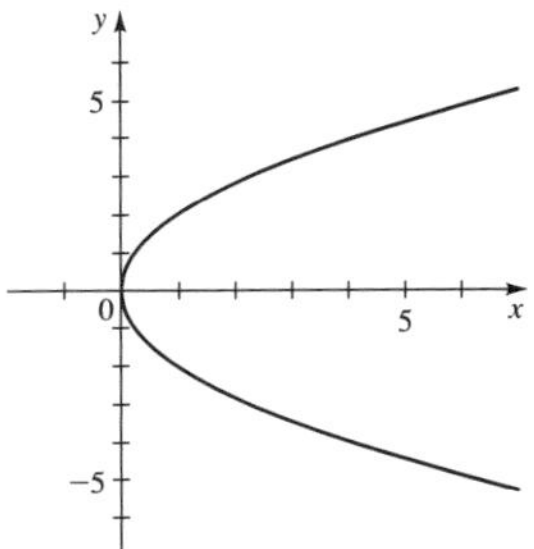

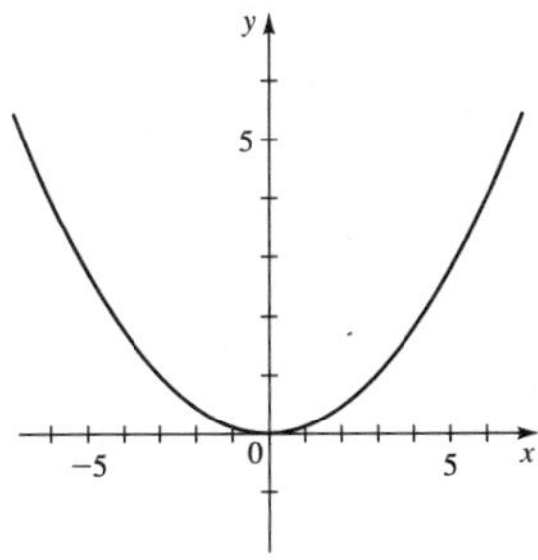

5. $F\left(0, \frac{1}{20}\right)$; $y = -\frac{1}{20}$; $\frac{1}{5}$

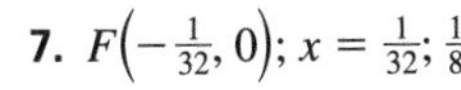
7. $F\left(-\frac{1}{32}, 0\right)$; $x = \frac{1}{32}$; $\frac{1}{8}$

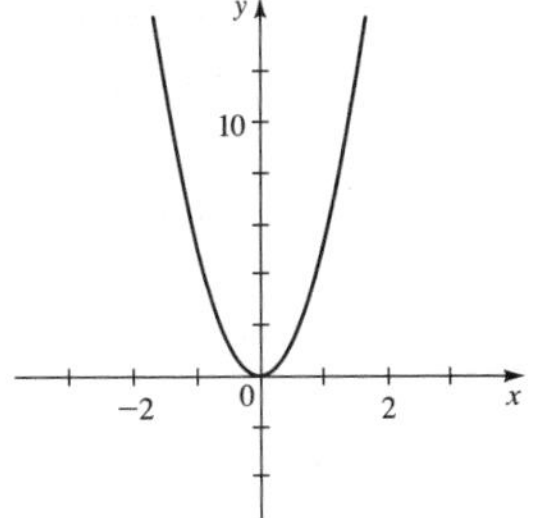

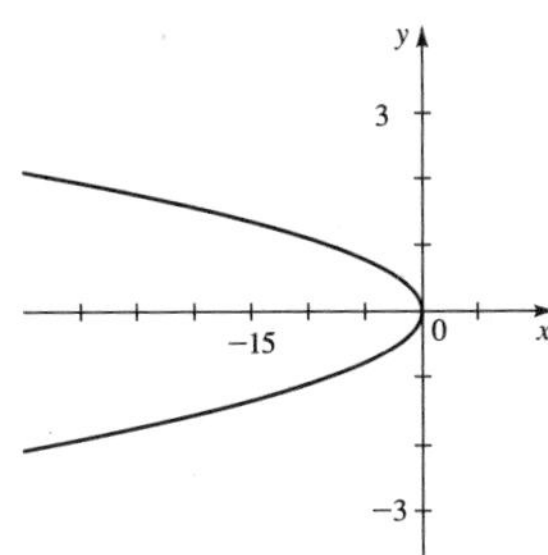

9. $F\left(0, -\frac{3}{2}\right)$; $y = \frac{3}{2}$; 6

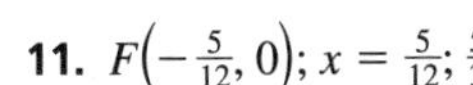
11. $F\left(-\frac{5}{12}, 0\right)$; $x = \frac{5}{12}$; $\frac{5}{3}$

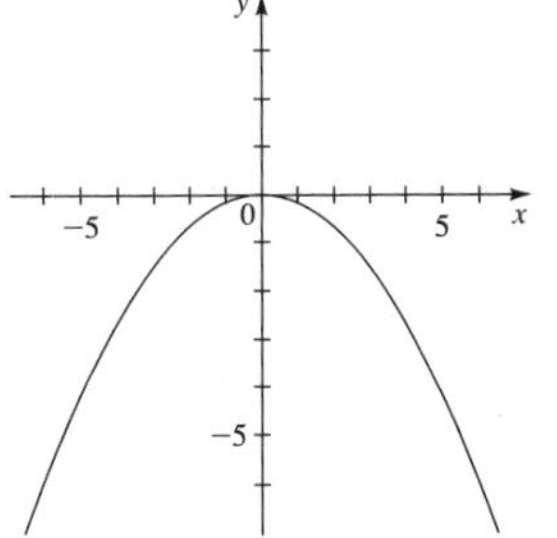

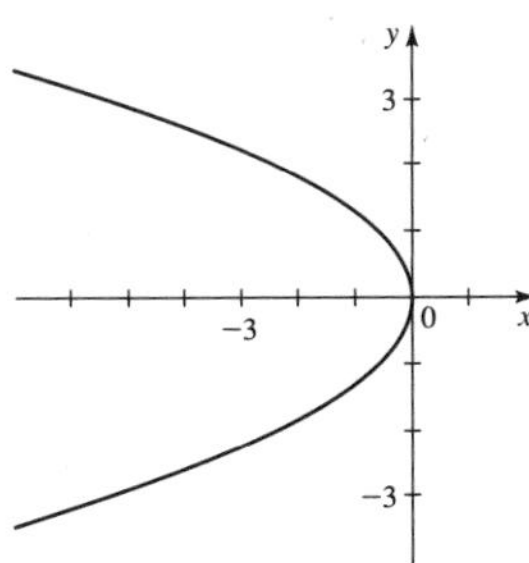

13. $x^2 = 8y$ **15.** $y^2 = -32x$
17. $y^2 = -8x$ **19.** $x^2 = 40y$
21. $y^2 = 4x$ **23.** $x^2 = 20y$
25. $x^2 = -12y$ **27.** $y^2 = -3x$
29. $x = y^2$ **31.** $x^2 = -4\sqrt{2}\,y$
33. (a) $y^2 = 12x$ (b) $8\sqrt{15} \approx 31$ cm
35. $x^2 = 600y$

37. (a) $x^2 = -4py$, $p = \frac{1}{2}$, 1, 4, and 8
(b) The closer the directrix to the vertex, the steeper the parabola.

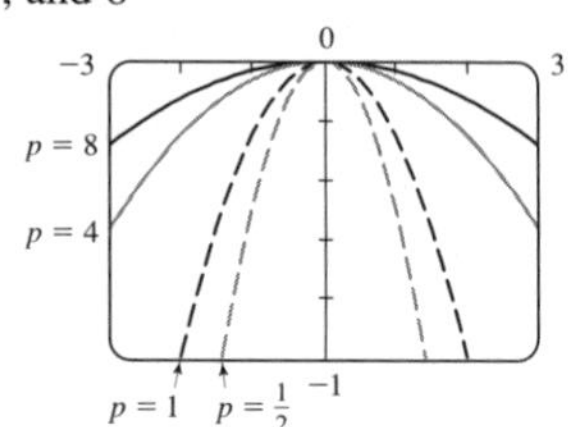

Section 8.2 ■ page 543

Order of answers: vertices; foci; eccentricity; major axis and minor axis

1. $V(\pm 5, 0)$; $F(\pm 4, 0)$; $\frac{4}{5}$; 10, 6

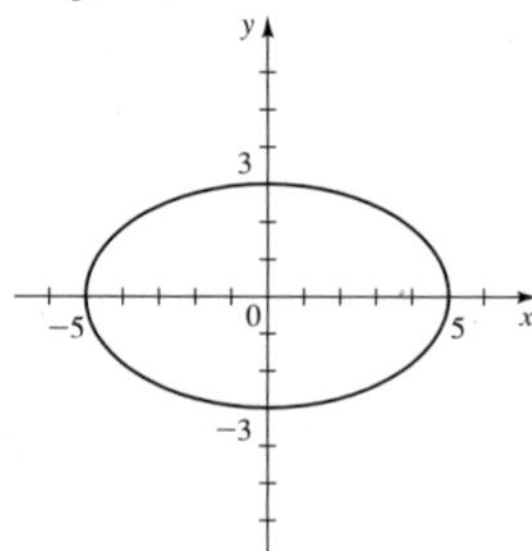

3. $V(0, \pm 3)$; $F(0, \pm\sqrt{5})$; $\sqrt{5}/3$; 6, 4

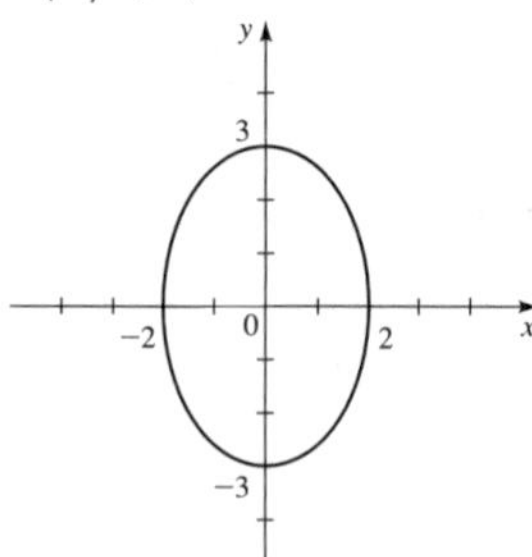

5. $V(\pm 4, 0)$; $F(\pm 2\sqrt{3}, 0)$; $\sqrt{3}/2$; 8, 4

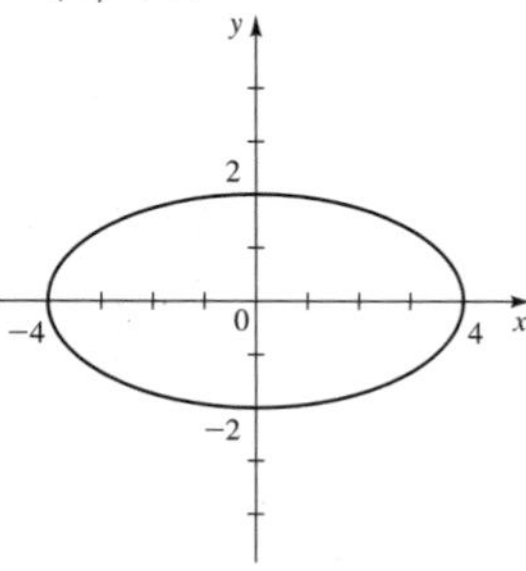

7. $V(0, \pm\sqrt{3})$; $F(0, \pm\sqrt{3/2})$; $1/\sqrt{2}$; $2\sqrt{3}$, $\sqrt{6}$

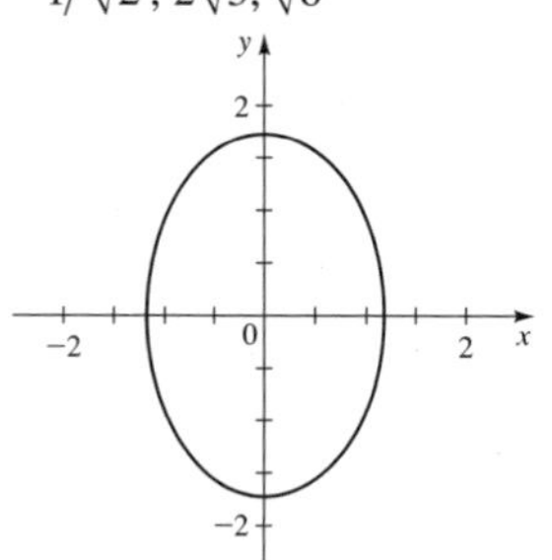

9. $V(\pm 1, 0)$; $F(\pm\sqrt{3}/2, 0)$; $\sqrt{3}/2$; 2, 1

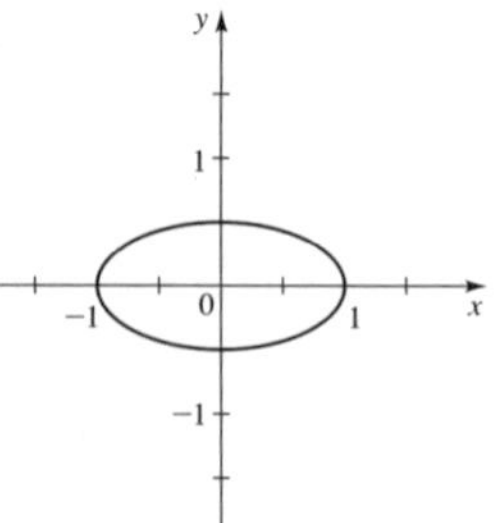

11. $V(0, \pm\sqrt{2})$; $F(0, \pm\sqrt{3/2})$; $\sqrt{3}/2$; $2\sqrt{2}$, $\sqrt{2}$

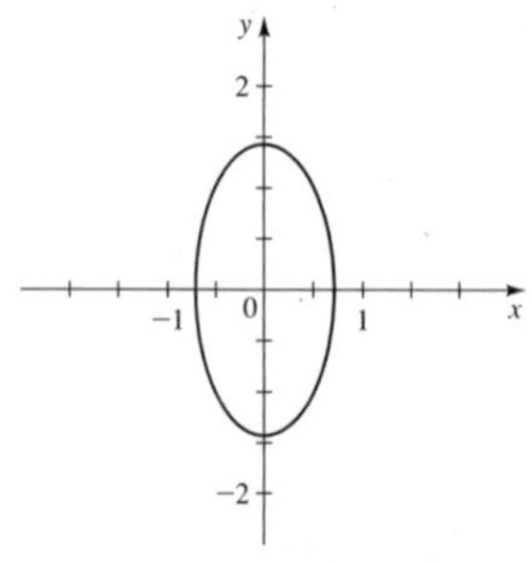

13. $V(0, \pm 1)$; $F(0, \pm 1/\sqrt{2})$; $1/\sqrt{2}$; 2, $\sqrt{2}$

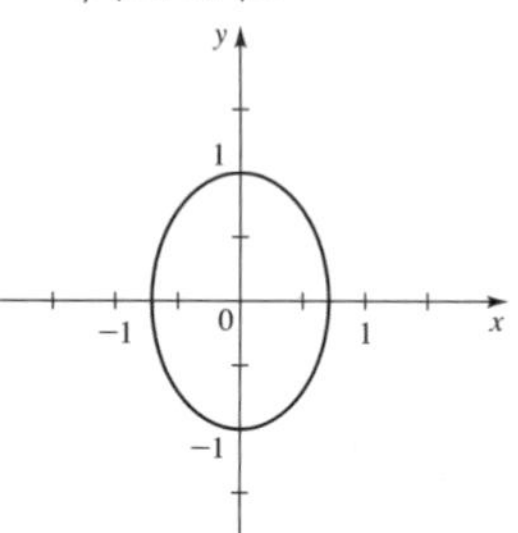

15. $\dfrac{x^2}{25} + \dfrac{y^2}{16} = 1$ **17.** $\dfrac{x^2}{4} + \dfrac{y^2}{8} = 1$

19. $\dfrac{x^2}{25} + \dfrac{y^2}{9} = 1$ **21.** $x^2 + \dfrac{y^2}{4} = 1$

23. $\dfrac{x^2}{9} + \dfrac{y^2}{13} = 1$ **25.** $\dfrac{x^2}{100} + \dfrac{y^2}{91} = 1$

27. $\dfrac{x^2}{25} + \dfrac{y^2}{5} = 1$ **29.** $\dfrac{64x^2}{225} + \dfrac{64y^2}{81} = 1$

31. $(0, \pm 2)$

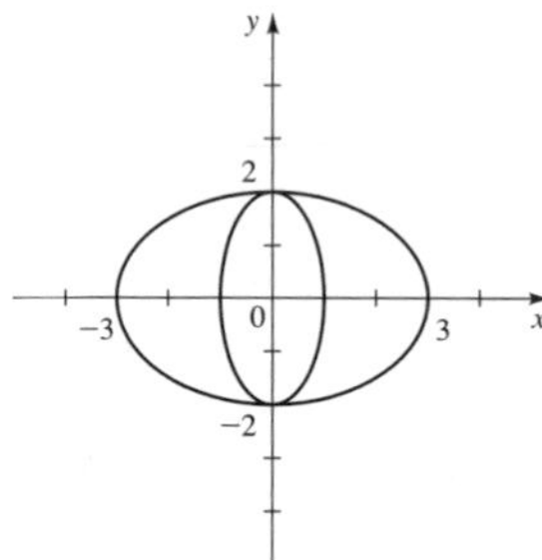

33. $\dfrac{x^2}{2.2500 \times 10^{16}} + \dfrac{y^2}{2.2491 \times 10^{16}} = 1$

35. $\dfrac{x^2}{1,455,642} + \dfrac{y^2}{1,451,610} = 1$

37. $5\sqrt{39}/2 \approx 15.6$ in.

41. (a)

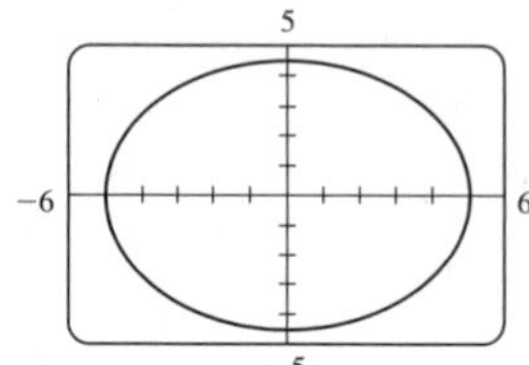

(b)

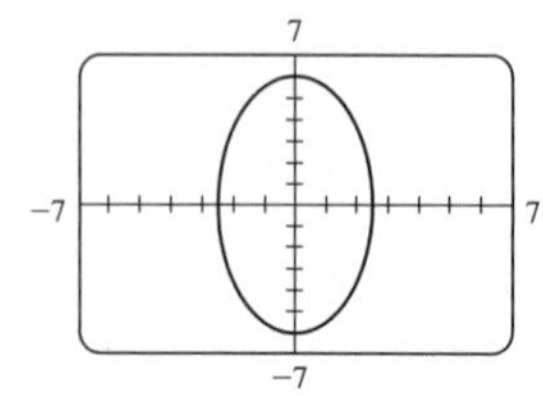

Section 8.3 ■ page 553

Order of answers: vertices; foci; asymptotes

1. $V(\pm 2, 0)$; $F(\pm 2\sqrt{5}, 0)$; $y = \pm 2x$

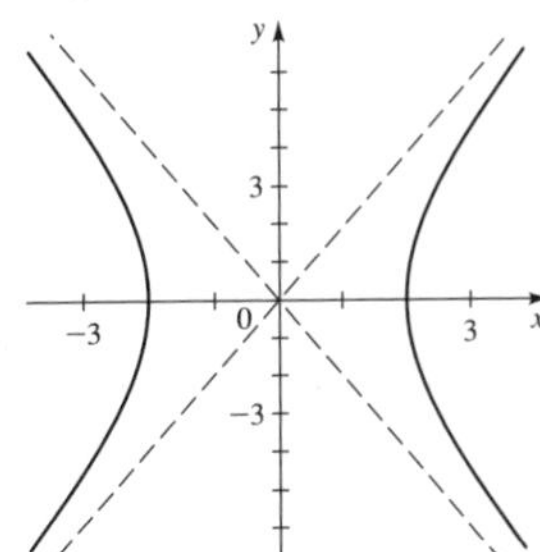

3. $V(0, \pm 1)$; $F(0, \pm\sqrt{26})$; $y = \pm\frac{1}{5}x$

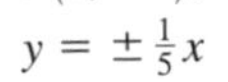

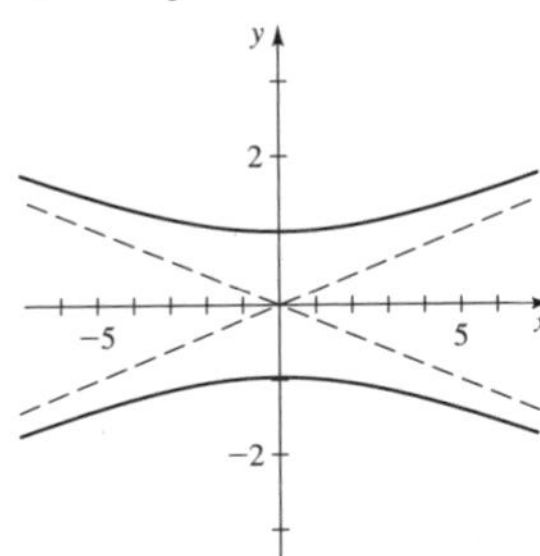

5. $V(\pm 1, 0)$; $F(\pm\sqrt{2}, 0)$; $y = \pm x$

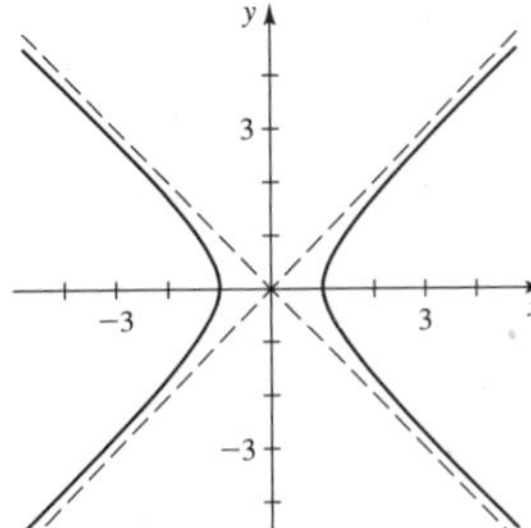

7. $V(0, \pm 3)$; $F(0, \pm\sqrt{34})$; $y = \pm\frac{3}{5}x$

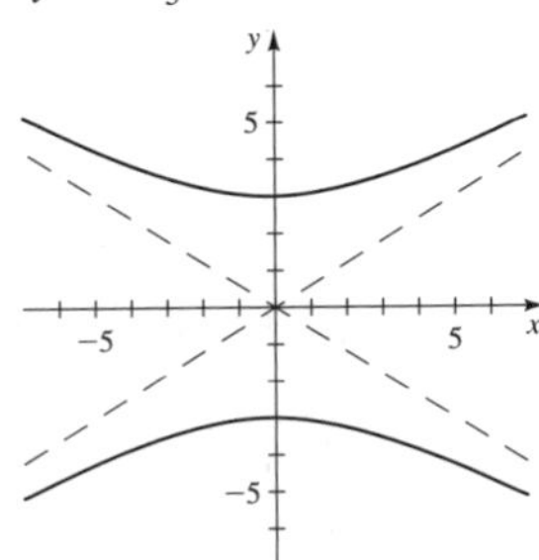

9. $V(\pm 2\sqrt{2}, 0)$; $F(\pm\sqrt{10}, 0)$; $y = \pm\frac{1}{2}x$

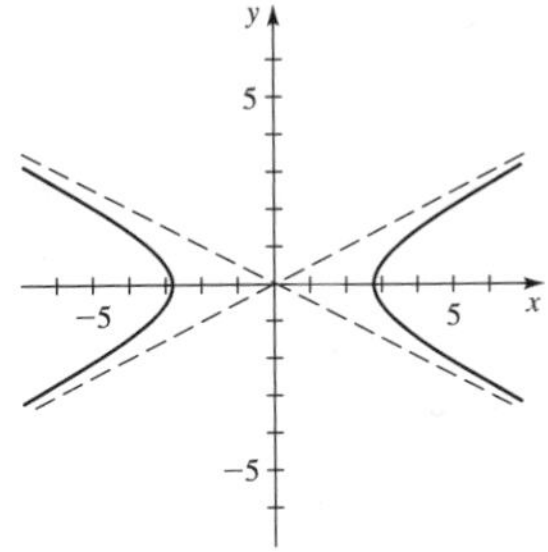

11. $V\left(0, \pm\frac{1}{2}\right)$; $F(0, \pm\sqrt{5}/2)$; $y = \pm\frac{1}{2}x$

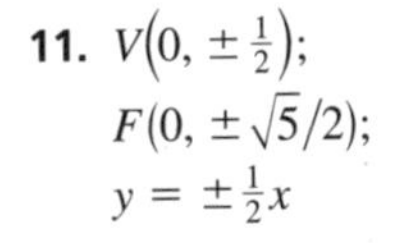

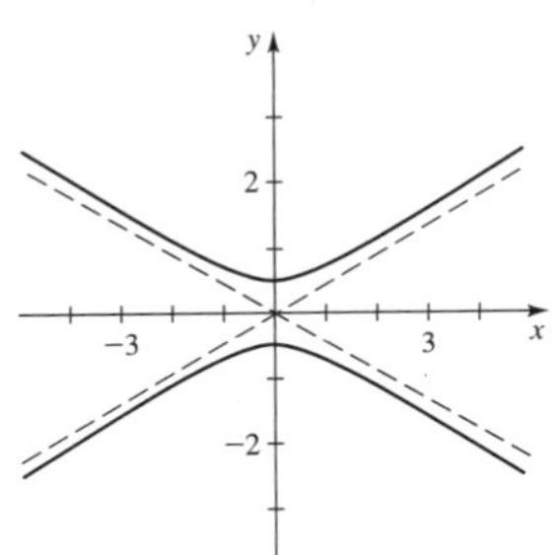

13. $\dfrac{x^2}{4} - \dfrac{y^2}{12} = 1$ **15.** $\dfrac{y^2}{16} - \dfrac{x^2}{16} = 1$

17. $\dfrac{x^2}{9} - \dfrac{y^2}{16} = 1$ **19.** $y^2 - \dfrac{x^2}{3} = 1$

21. $x^2 - \dfrac{y^2}{25} = 1$ **23.** $\dfrac{5y^2}{64} - \dfrac{5x^2}{256} = 1$

25. $\dfrac{x^2}{16} - \dfrac{y^2}{16} = 1$ **27.** $\dfrac{x^2}{9} - \dfrac{y^2}{16} = 1$

29. (a) $x^2 - y^2 = c^2/2$

33. (a) 490 mi (b) $\dfrac{y^2}{60{,}025} - \dfrac{x^2}{2475} = 1$ (c) 10.1 mi

35. (b)

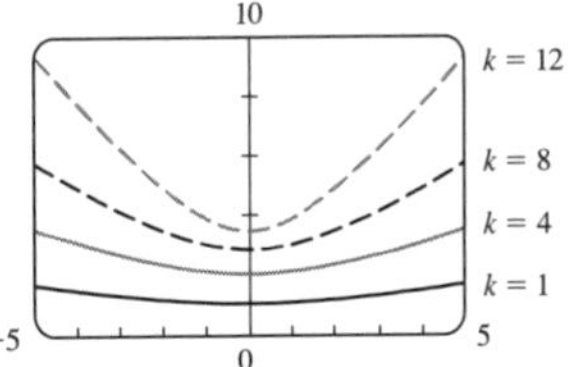

Section 8.4 ■ page 562

1. Center $C(2, 1)$; foci $F(2 \pm \sqrt{5}, 1)$; vertices $V_1(-1, 1)$, $V_2(5, 1)$; major axis 6, minor axis 4

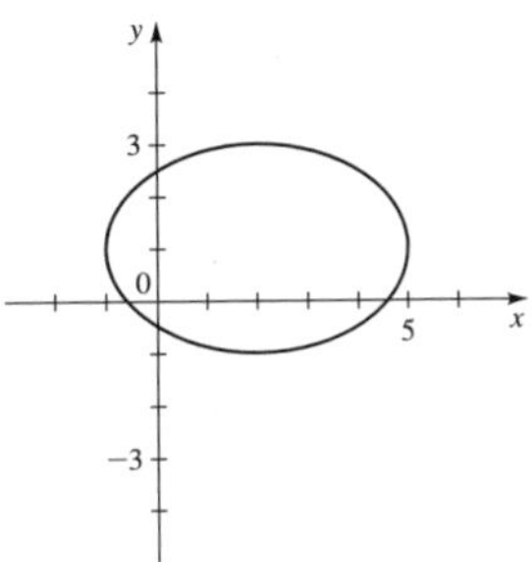

3. Center $C(0, -5)$; foci $F_1(0, -1)$, $F_2(0, -9)$; vertices $V_1(0, 0)$, $V_2(0, -10)$; major axis 10, minor axis 6

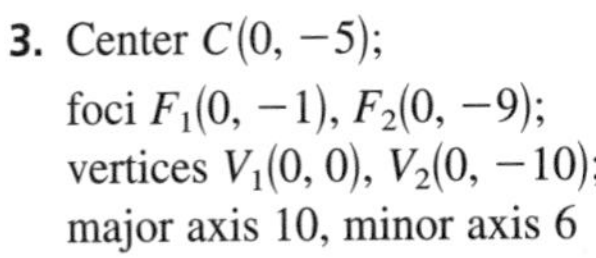

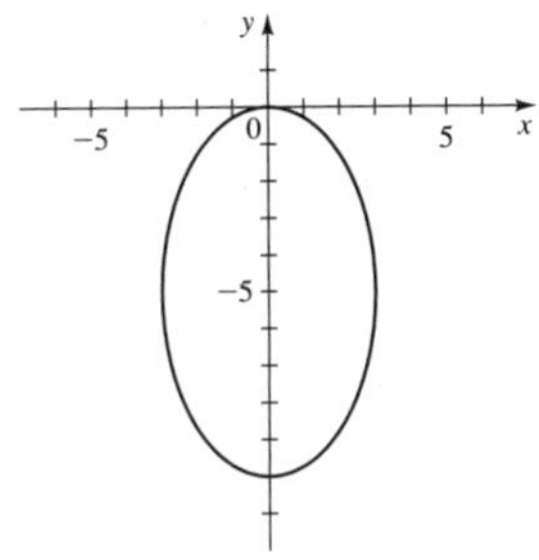

5. Vertex $V(3, -1)$; focus $F(3, 1)$; directrix $y = -3$

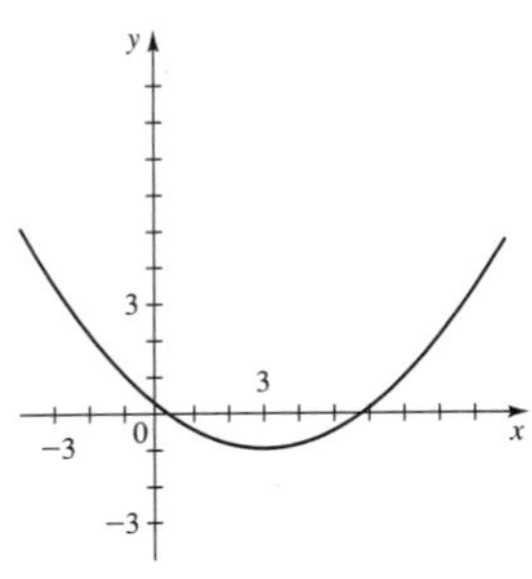

7. Vertex $V\left(-\frac{1}{2}, 0\right)$; focus $F\left(-\frac{1}{2}, -\frac{1}{16}\right)$; directrix $y = \frac{1}{16}$

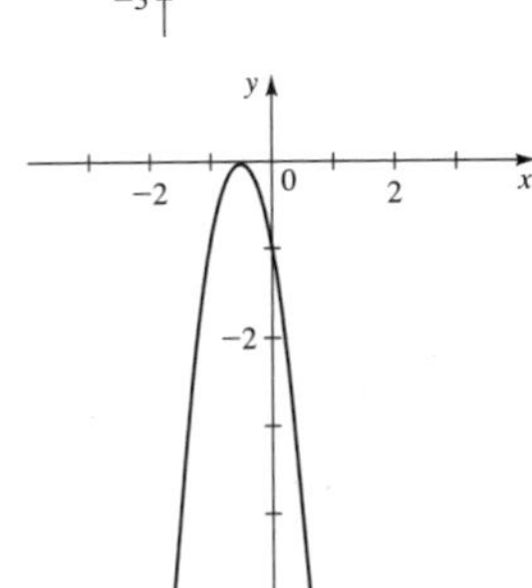

9. Center $C(-1, 3)$; foci $F_1(-6, 3)$, $F_2(4, 3)$; vertices $V_1(-4, 3)$, $V_2(2, 3)$; asymptotes

$$y = \pm\tfrac{4}{3}(x + 1) + 3$$

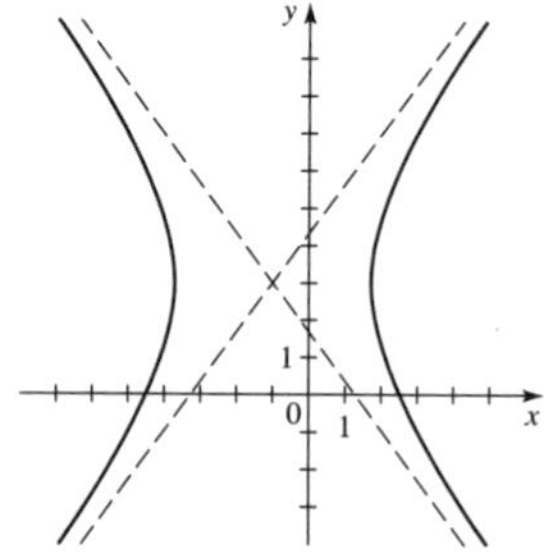

11. Center $C(-1, 0)$; foci $F(-1, \pm\sqrt{5})$; vertices $V(-1, \pm 1)$; asymptotes

$$y = \pm\tfrac{1}{2}(x + 1)$$

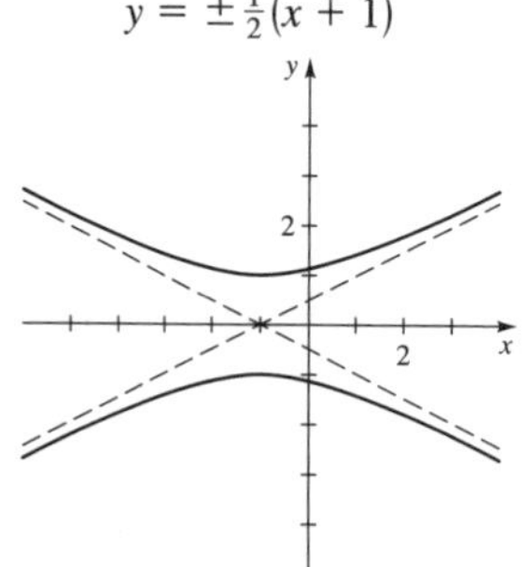

13. $x^2 = -\frac{1}{4}(y - 4)$

15. $\dfrac{(x - 5)^2}{25} + \dfrac{y^2}{16} = 1$

17. $(y - 1)^2 - x^2 = 1$

19. Ellipse; $C(2, 0)$; $F(2, \pm\sqrt{5})$; $V(2, \pm 3)$; major axis 6, minor axis 4

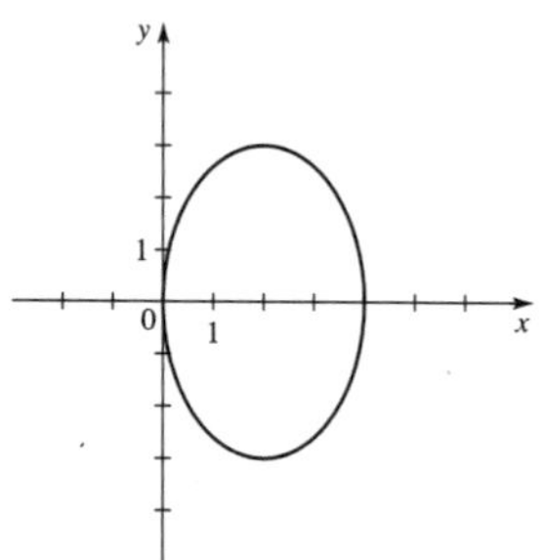

21. Hyperbola; $C(1, 2)$; $F_1\left(-\frac{3}{2}, 2\right)$, $F_2\left(\frac{7}{2}, 2\right)$; $V(1 \pm \sqrt{5}, 2)$; asymptotes

$$y = \pm\tfrac{1}{2}(x - 1) + 2$$

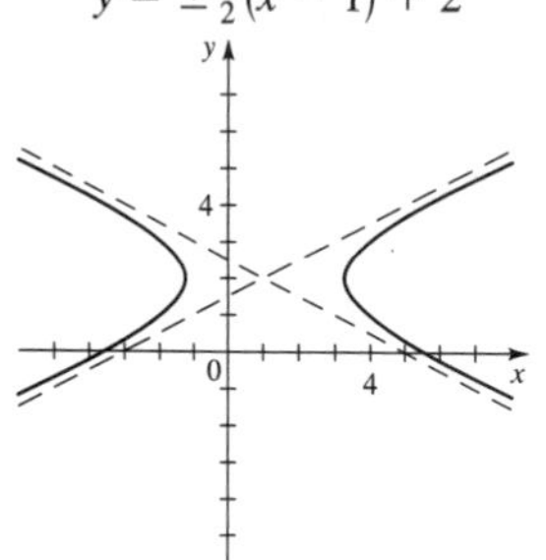

23. Ellipse; $C(3, -5)$; $F(3 \pm \sqrt{21}, -5)$; $V_1(-2, -5)$, $V_1(8, -5)$; major axis 10, minor axis 4

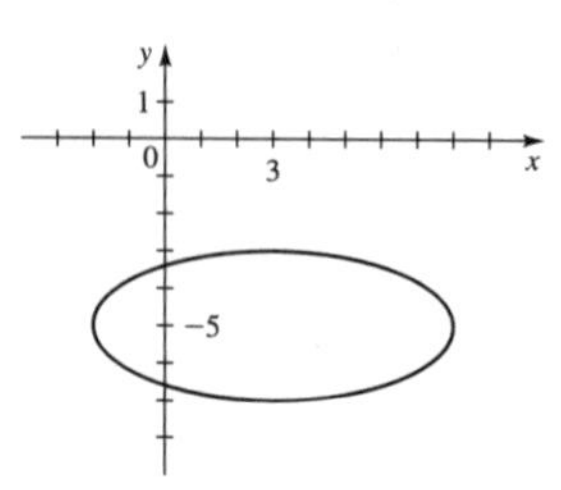

25. Hyperbola; $C(3, 0)$; $F(3, \pm 5)$; $V(3, \pm 4)$; asymptotes

$$y = \pm\tfrac{4}{3}(x - 3)$$

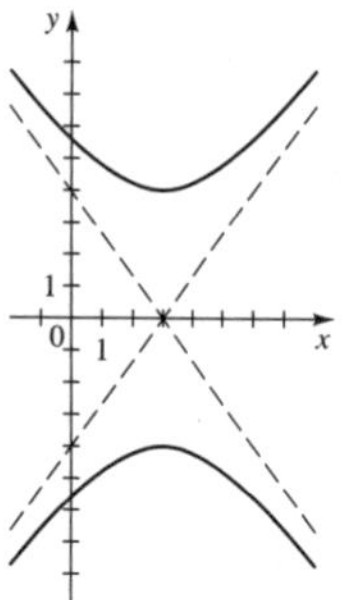

27. Degenerate conic (pairs of lines), $y = \pm\frac{1}{2}(x - 4)$

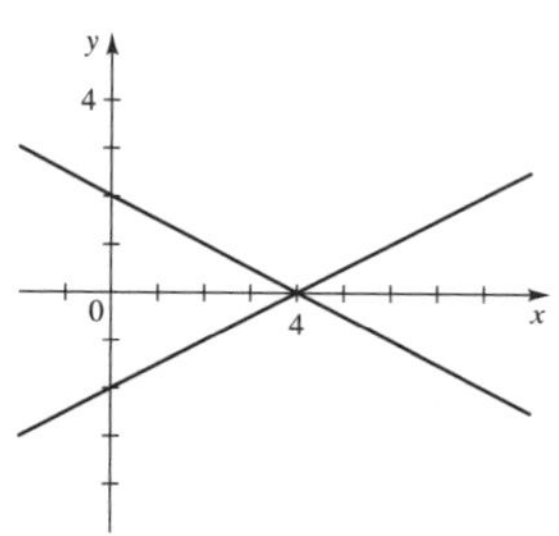

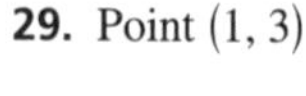

29. Point $(1, 3)$

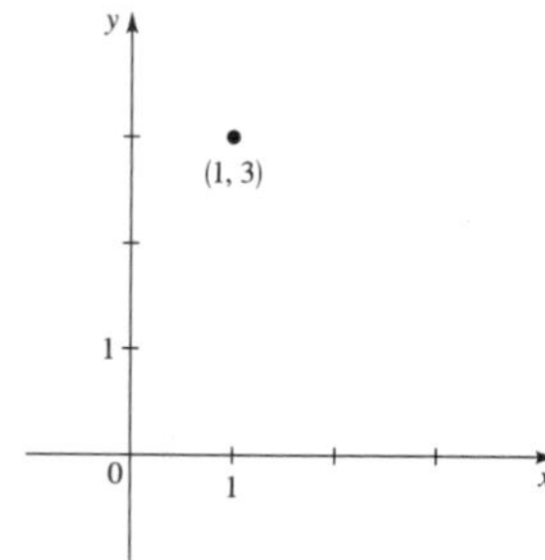

31. (a) $F < 17$ (b) $F = 17$ (c) $F > 17$

33. (a)

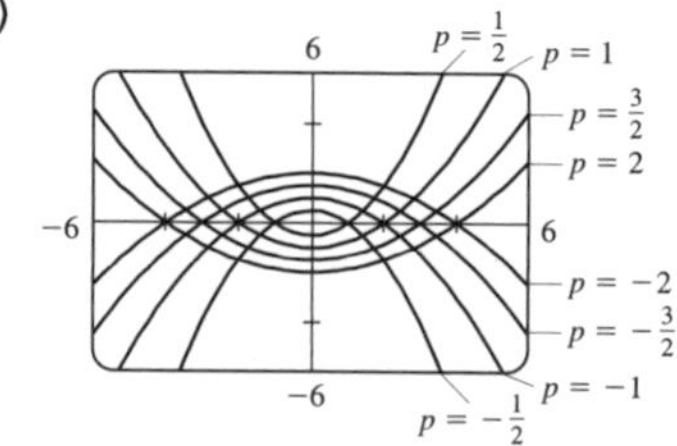

(c) The parabolas become narrower.

Chapter 8 Review ■ page 570

1. $V(0, 0)$; $F(0, -2)$; $y = 2$

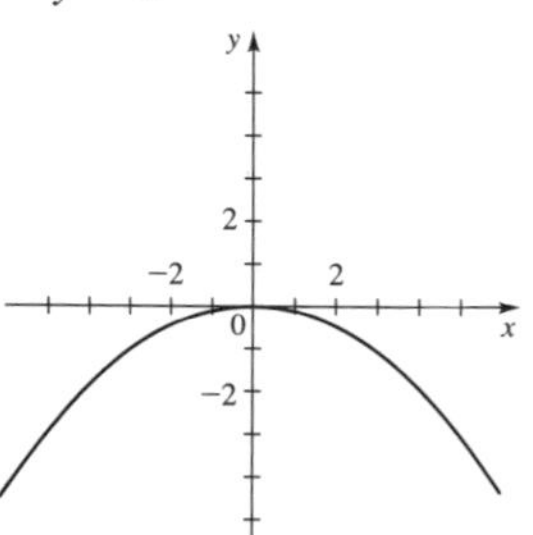

3. $V(-2, 2)$; $F\left(-\frac{7}{4}, 2\right)$; $x = -\frac{9}{4}$

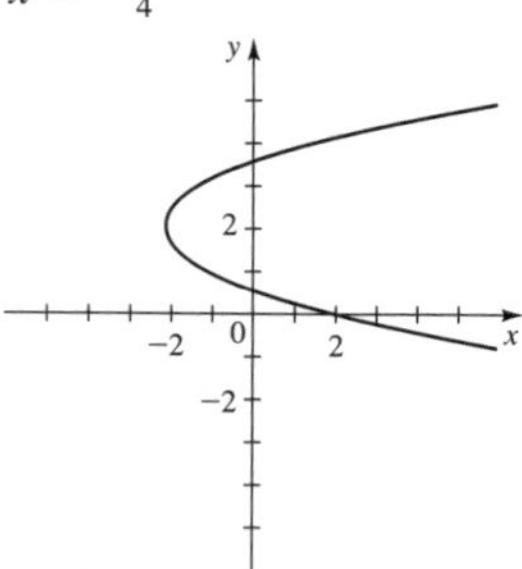

5. $C(0, 0)$; $V(\pm 4, 0)$; $F(\pm 2\sqrt{3}, 0)$; axes 8, 4

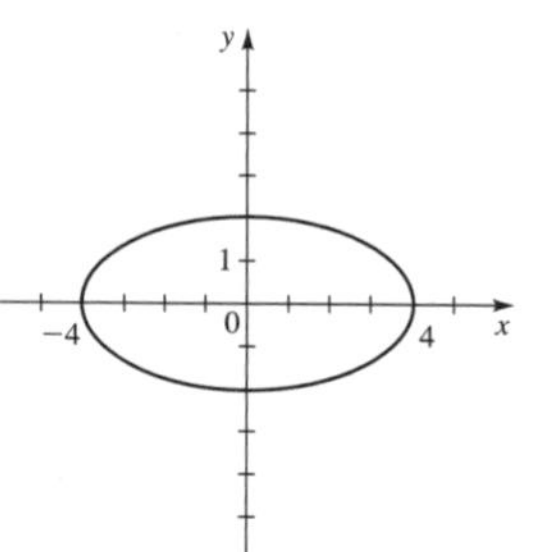

7. $C(0, 2)$; $V(\pm 3, 2)$; $F(\pm\sqrt{5}, 2)$; axes 6, 4

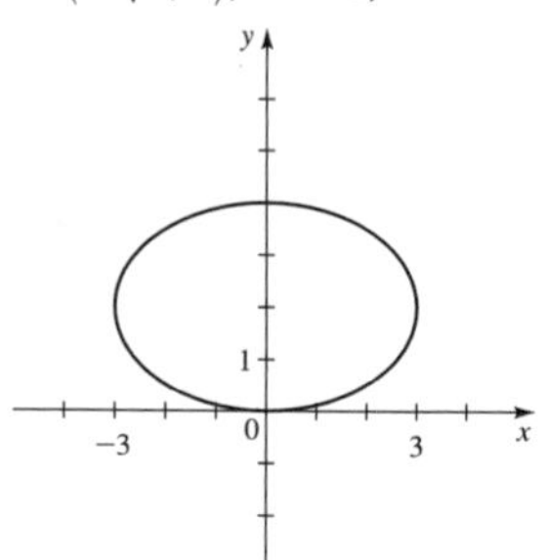

9. $C(0, 0)$; $V(\pm 4, 0)$; $F(\pm 2\sqrt{6}, 0)$; asymptotes $y = \pm \dfrac{1}{\sqrt{2}}x$

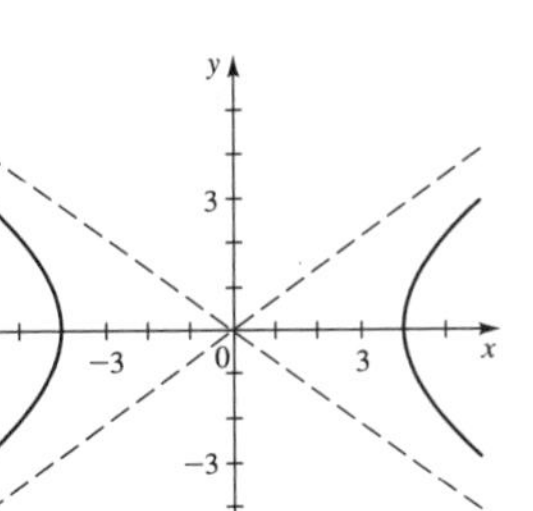

11. $C(-3, -1)$; $V(-3, -1 \pm \sqrt{2})$; $F(-3, -1 \pm 2\sqrt{5})$; asymptotes $y = \frac{1}{3}x$, $y = -\frac{1}{3}x - 2$

13. $y^2 = 8x$

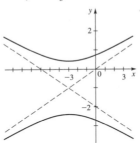

15. $\dfrac{y^2}{16} - \dfrac{x^2}{9} = 1$

17. $\dfrac{(x-4)^2}{16} + \dfrac{(y-2)^2}{4} = 1$

19. Parabola; $F(0, -2)$; $V(0, 1)$

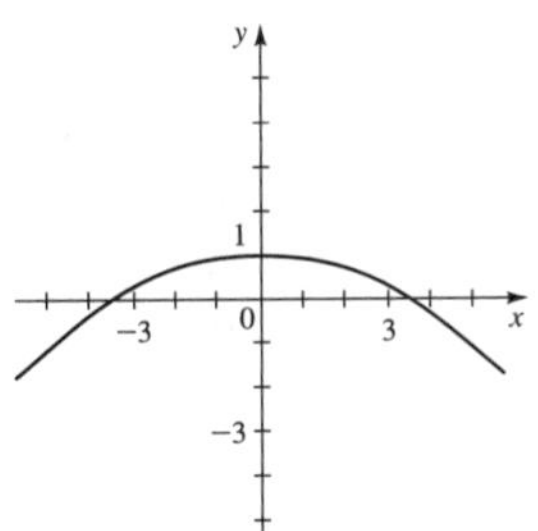

21. Hyperbola; $F(0, \pm 12\sqrt{2})$; $V(0, \pm 12)$

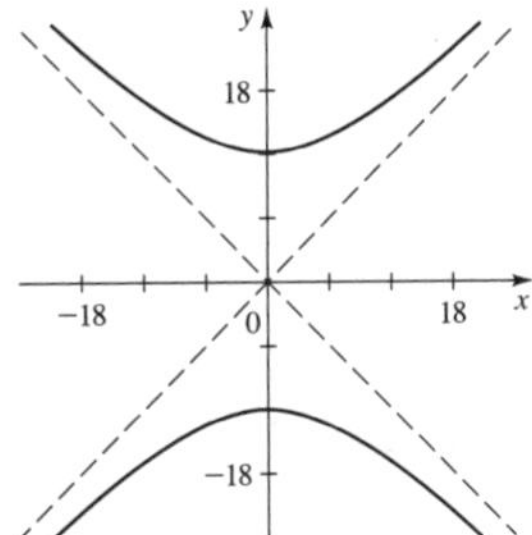

23. Ellipse; $F(1, 4 \pm \sqrt{15})$; $V(1, 4 \pm 2\sqrt{5})$

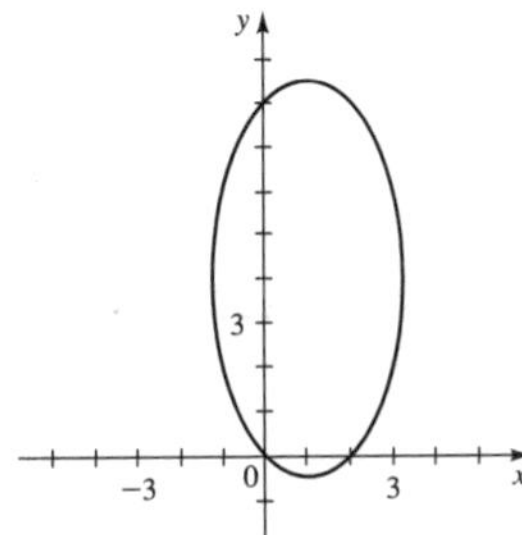

25. Parabola; $F\left(-\frac{255}{4}, 8\right)$; $V(-64, 8)$

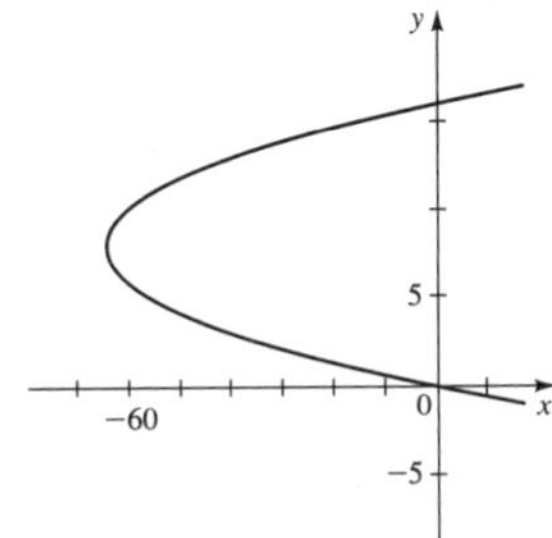

27. Ellipse; $F(3, -3 \pm 1/\sqrt{2})$; $V_1(3, -4)$, $V_2(3, -2)$

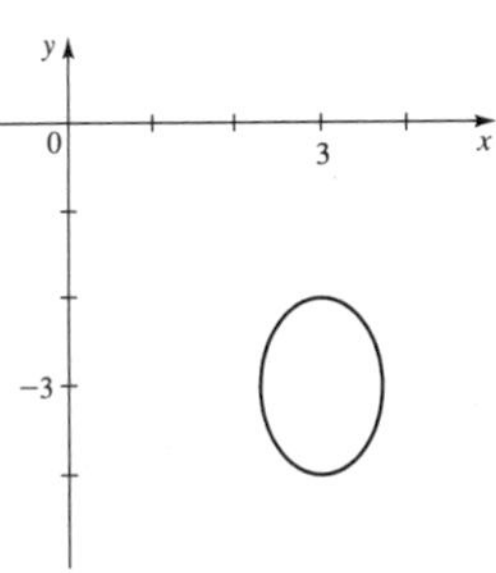

29. Has no graph **31.** $x^2 = 4y$ **33.** $\dfrac{y^2}{4} - \dfrac{x^2}{16} = 1$

35. $\dfrac{(x-1)^2}{3} + \dfrac{(y-2)^2}{4} = 1$

37. $\dfrac{4(x-7)^2}{225} + \dfrac{(y-2)^2}{100} = 1$

39. $(x - 800)^2 = -200(y - 3200)$

41. (a) 91,419,000 mi (b) 94,581,000 mi

43. (a)

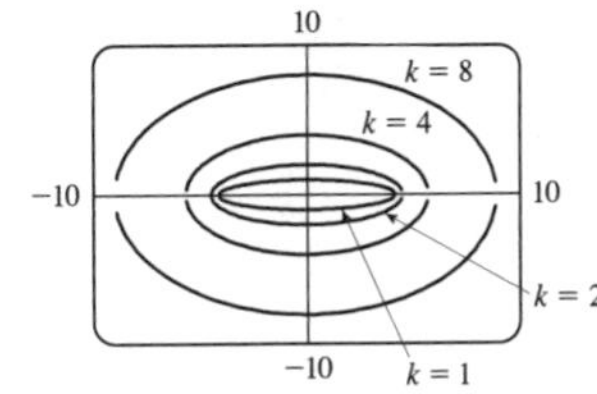

Chapter 8 Test ■ page 573

1. $F(0, -3)$, $y = 3$

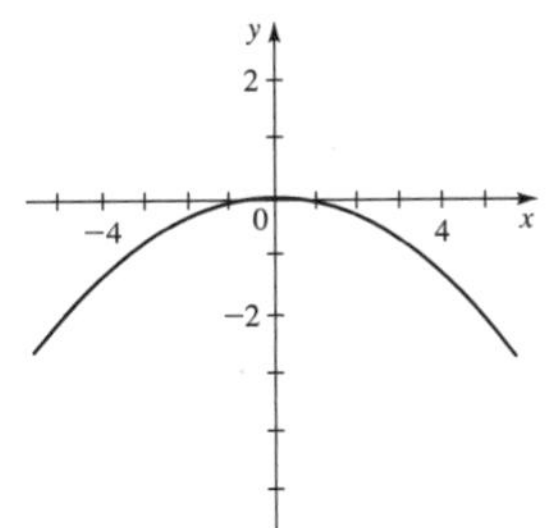

2. $V(\pm 4, 0)$; $F(\pm 2\sqrt{3}, 0)$; 8, 4

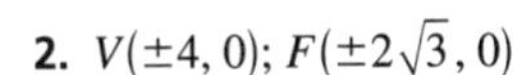

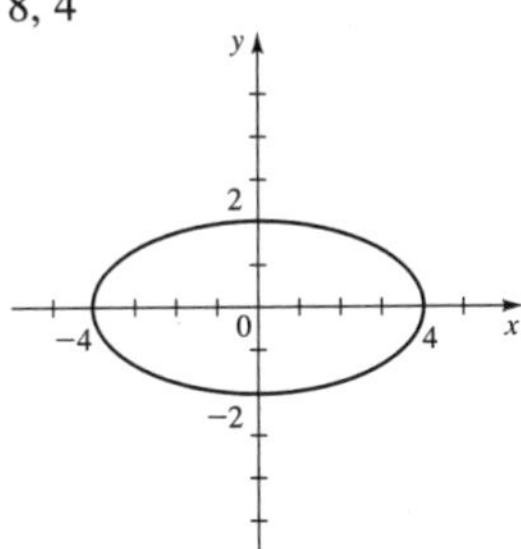

3. $V(0, \pm 3)$; $F(0, \pm 5)$; $y = \pm \frac{3}{4}x$

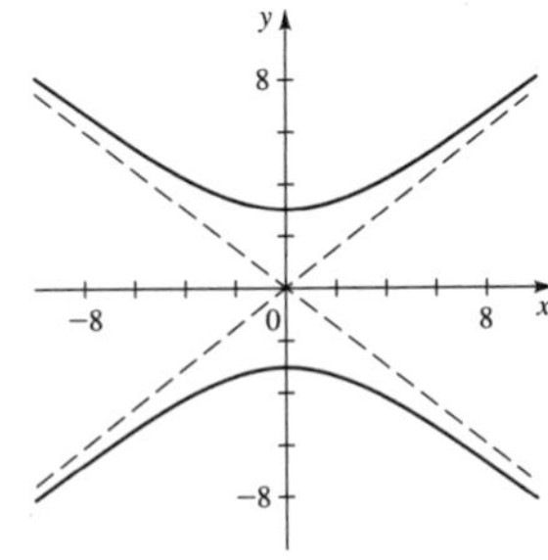

4. $y^2 = -x$ **5.** $\dfrac{x^2}{16} + \dfrac{(y-3)^2}{9} = 1$

6. $(x-2)^2 - \dfrac{y^2}{3} = 1$

7. $\dfrac{(x-3)^2}{9} + \dfrac{\left(y+\frac{1}{2}\right)^2}{4} = 1$

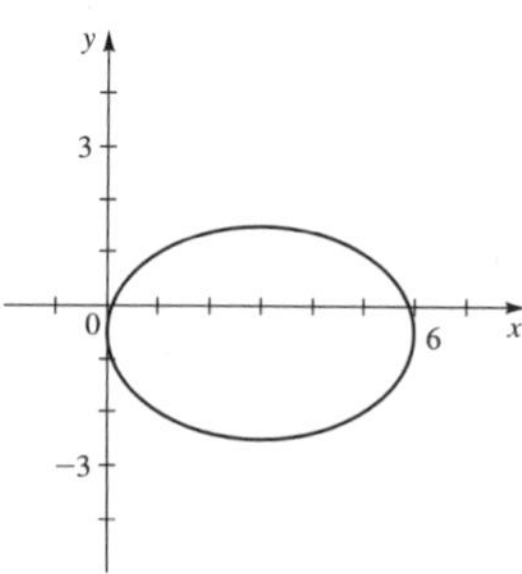

Vertices $\left(0, -\frac{1}{2}\right), \left(6, -\frac{1}{2}\right)$; foci $\left(3+\sqrt{5}, -\frac{1}{2}\right), \left(3-\sqrt{5}, -\frac{1}{2}\right)$

8. $9(x+2)^2 - 8(y-4)^2 = 0$

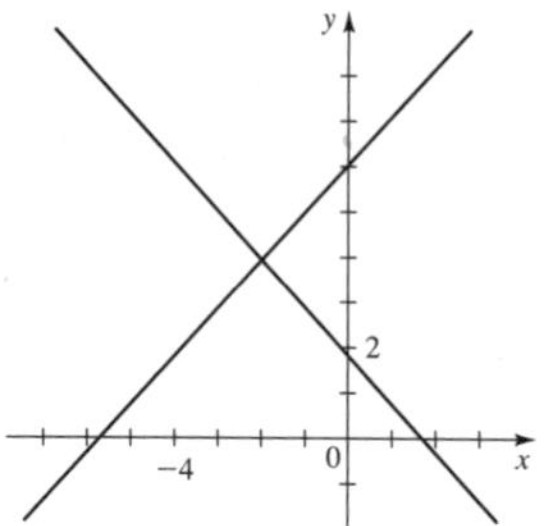

No vertices, foci, or asymptotes. Degenerate conic.

9. $(y+4)^2 = -2(x-4)]$

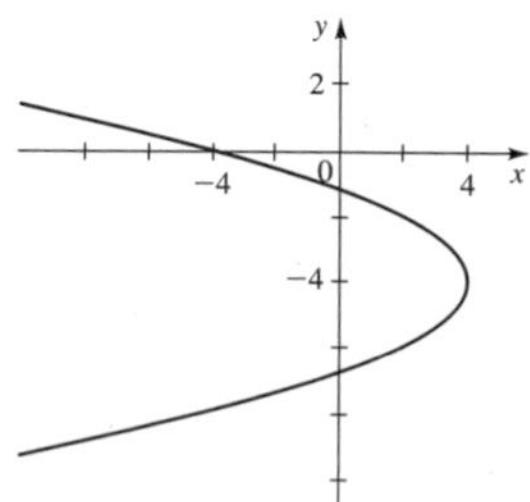

Vertex $(4, -4)$; focus $\left(\frac{7}{2}, -4\right)$

10. $\dfrac{y^2}{9} - \dfrac{x^2}{16} = 1$ **11.** $x^2 - 4x - 8y + 20 = 0$

12. $\frac{3}{4}$ in.

FOCUS ON PROBLEM SOLVING ■ page 574

1. Tetrahedron 4, 6, 4; octahedron 8, 12, 6; cube 6, 12, 8; dodecahedron 12, 30, 20; icosahedron 20, 30, 12
3. Hexagons **5.** $\sqrt{a^2 + (b+c)^2}$
9. 15, 20, 25 **11.** 124
15. (a) $13 = 2^2 + 3^2, 41 = 4^2 + 5^2$
(c) $533 = 2^2 + 23^2 = 7^2 + 22^2$

CHAPTER 9

Section 9.1 ■ page 581

1. 2, 3, 4, 5; 1001 **3.** $\frac{1}{2}, \frac{1}{3}, \frac{1}{4}, \frac{1}{5}; \frac{1}{1001}$
5. $-1, \frac{1}{4}, -\frac{1}{9}, \frac{1}{16}; \frac{1}{1{,}000{,}000}$ **7.** 0, 2, 0, 2; 2
9. 1, 4, 27, 256; 1000^{1000} **11.** 3, 2, 0, −4, −12
13. 1, 3, 7, 15, 31 **15.** 1, 2, 3, 5, 8 **17.** 2^n
19. $3n - 2$ **21.** $(2n-1)/n^2$ **23.** $1 + (-1)^n$
25. 1, 4, 9, 16, 25, 36 **27.** $\frac{1}{3}, \frac{4}{9}, \frac{13}{27}, \frac{40}{81}, \frac{121}{243}, \frac{364}{729}$
29. $\frac{2}{3}, \frac{8}{9}, \frac{26}{27}, \frac{80}{81}; S_n = 1 - \dfrac{1}{3^n}$
31. $1-\sqrt{2}, 1-\sqrt{3}, -1, 1-\sqrt{5}; S_n = 1 - \sqrt{n+1}$
33. 10 **35.** $\frac{11}{6}$ **37.** 8 **39.** 31
41. $\sqrt{1} + \sqrt{2} + \sqrt{3} + \sqrt{4} + \sqrt{5}$
43. $\sqrt{4} + \sqrt{5} + \sqrt{6} + \sqrt{7} + \sqrt{8} + \sqrt{9} + \sqrt{10}$
45. $x^3 + x^4 + \cdots + x^{100}$ **47.** $\sum_{k=1}^{100} k$ **49.** $\sum_{k=1}^{10} k^2$
51. $\sum_{k=1}^{999} \dfrac{1}{k(k+1)}$ **53.** $\sum_{k=0}^{100} x^k$ **55.** $2^{(2^n-1)/2^n}$

Section 9.2 ■ page 591

1. Arithmetic, 3 **3.** Not arithmetic **5.** Arithmetic, $-\frac{3}{2}$
7. 3, 14, $2 + 3(n-1)$, 299 **9.** 5, 24, $4 + 5(n-1)$, 499
11. 4, 4, $-12 + 4(n-1)$, 384
13. 1.5, 31, $25 + 1.5(n-1)$, 173.5
15. s, $2 + 4s$, $2 + (n-1)s$, $2 + 99s$ **17.** $\frac{1}{2}$
19. −100, −98, −96 **21.** 30th **23.** 100 **25.** 460
27. 1090 **29.** 20,301 **31.** 832.3 **33.** 46.75
35. \$1250 **37.** \$403,500 **39.** 20
41. (a) 576 ft (b) $16n^2$ ft **45.** Yes

Section 9.3 ■ page 597

1. Geometric, 2 **3.** Geometric, $\frac{1}{2}$ **5.** Not geometric
7. 3, 162, $2 \cdot 3^{n-1}$ **9.** −0.3, 0.00243, $(0.3)(-0.3)^{n-1}$
11. $-\frac{1}{12}, \frac{1}{144}, 144\left(-\frac{1}{12}\right)^{n-1}$ **13.** $3^{2/3}, 3^{11/3}, 3^{(2n+1)/3}$

15. $s^{2/7}, s^{8/7}, s^{2(n-1)/7}$ **17.** $\frac{1}{2}$ **19.** $\frac{25}{4}$ **21.** 11th
23. 315 **25.** 441 **27.** 3280 **29.** $\frac{6141}{1024}$
31. 19 ft, $80\left(\frac{3}{4}\right)^n$ **33.** $\frac{64}{25}, \frac{1024}{625}, 5\left(\frac{4}{5}\right)^n$
35. (a) $17\frac{8}{9}$ ft (b) $18 - \left(\frac{1}{3}\right)^{n-3}$ **37.** $\frac{3}{2}$ **39.** $\frac{3}{4}$ **41.** $\frac{1}{648}$
43. $-\frac{1000}{117}$ **45.** $\frac{7}{9}$ **47.** $\frac{1}{33}$ **49.** $\frac{112}{999}$ **51.** 3 m
53. (a) 2 (b) $8 + 4\sqrt{2}$ **55.** 1
57. $a \cdot \dfrac{1 - a^9}{1 - a} + 55b$

Section 9.4 ■ page 607

1. \$13,180.79 **3.** \$360,262.21 **5.** \$5,591.79
7. \$245.66 **9.** \$2,601.59 **11.** \$307.24
13. \$733.76, \$264,153.60
15. (a) \$859.15 (b) \$309,294.00 (c) \$1,841,519.29
17. \$341.24 **19.** 18.16% **21.** 11.68%

Section 9.6 ■ page 630

1. $x^6 + 6x^5y + 15x^4y^2 + 20x^3y^3 + 15x^2y^4 + 6xy^5 + y^6$
3. $x^4 + 4x^2 + 6 + \dfrac{4}{x^2} + \dfrac{1}{x^4}$
5. $x^5 - 5x^4 + 10x^3 - 10x^2 + 5x - 1$
7. $x^{10}y^5 - 5x^8y^4 + 10x^6y^3 - 10x^4y^2 + 5x^2y - 1$
9. $8x^3 - 36x^2y + 54xy^2 - 27y^3$
11. $\dfrac{1}{x^5} - \dfrac{5}{x^{7/2}} + \dfrac{10}{x^2} - \dfrac{10}{x^{1/2}} + 5x - x^{5/2}$
13. 15 **15.** 4950 **17.** 18 **19.** 32
21. $x^4 + 8x^3y + 24x^2y^2 + 32xy^3 + 16y^4$
23. $1 + \dfrac{6}{x} + \dfrac{15}{x^2} + \dfrac{20}{x^3} + \dfrac{15}{x^4} + \dfrac{6}{x^5} + \dfrac{1}{x^6}$
25. $x^{20} + 40x^{19}y + 760x^{18}y^2$ **27.** $25a^{26/3} + a^{25/3}$
29. $48{,}620x^{18}$ **31.** $300a^2b^{23}$ **33.** $100y^{99}$
35. $13{,}440x^4y^6$ **37.** $495a^8b^8$ **39.** $(x + y)^4$
41. $(2a + b)^3$ **43.** $3x^2 + 3xh + h^2$

Chapter 9 Review ■ page 631

1. $\frac{1}{2}, \frac{4}{3}, \frac{9}{4}, \frac{16}{5}; \frac{100}{11}$ **3.** $0, \frac{1}{4}, 0, \frac{1}{32}; \frac{1}{500}$
5. 1, 3, 15, 105; 654, 729, 075 **7.** 1, 4, 9, 16, 25, 36, 49
9. 1, 3, 5, 11, 21, 43, 85 **11.** Arithmetic, 7
13. Arithmetic, $5\sqrt{2}$ **15.** Arithmetic, $t + 1$
17. Geometric, $\frac{4}{27}$ **19.** $2i$ **21.** 5 **23.** $\frac{81}{4}$
25. 64 **27.** 12,288 **31.** (a) 9 (b) $\pm 6\sqrt{2}$
33. 126 **35.** 384 **37.** $0^2 + 1^2 + 2^2 + \cdots + 9^2$
39. $\dfrac{3}{2^2} + \dfrac{3^2}{2^3} + \dfrac{3^3}{2^4} + \cdots + \dfrac{3^{50}}{2^{51}}$ **41.** $\sum_{k=1}^{33} 3k$
43. $\sum_{k=1}^{100} k\, 2^{k+2}$ **45.** Geometric; 4.68559
47. Arithmetic, $5050\sqrt{5}$ **49.** Geometric, 9831
51. 13 **53.** 65,534 **55.** \$2390.27 **57.** $\frac{5}{7}$
59. $\frac{1}{2}(3 + \sqrt{3})$ **67.** $2^n < n!$ for $n \geq 4$ **69.** 255
71. 12,870 **73.** $16x^4 + 32x^3y + 24x^2y^2 + 8xy^3 + y^4$
75. $b^{-40/3} + 20b^{-37/3} + 190b^{-34/3}$

Chapter 9 Test ■ page 634

1. 0, 3, 8, 15; 99 **2.** -4; 68, $a_n = 80 - 4(n - 1)$
3. $\frac{1}{5}, \frac{1}{25}$ **4.** (a) False (b) True
5. (a) $S_n = \dfrac{n}{2}[2a + (n - 1)d]$ or $S_n = n\left(\dfrac{a + a_n}{2}\right)$
(b) 60 (c) $-\frac{8}{9}, -78$
6. (a) $S_n = \dfrac{a(1 - r^n)}{1 - r}$ (b) 58,025/59,049
7. $2 + \sqrt{2}$
9. (a) $(1 - 1^2) + (1 - 2^2) + (1 - 3^2) + (1 - 4^2) + (1 - 5^2) = -50$
(b) 10
10. $32x^5 + 80x^4y^2 + 80x^3y^4 + 40x^2y^6 + 10xy^8 + y^{10}$
11. -1

FOCUS ON PROBLEM SOLVING ■ page 635

1. $1 + 3 + 5 + \cdots + (2n - 1) = n^2$ **5.** $\dfrac{n(n^2 + 1)}{2}$
7. (a) $\dfrac{2\sqrt{3}}{5}$
9. (a) 91; $\dfrac{n(n + 1)(2n + 1)}{6}$ (b) 441; $\left[\dfrac{n(n + 1)}{2}\right]^2$

CHAPTER 10

Section 10.1 ■ page 641

1. 12 **3.** (a) 64 (b) 24 **5.** 1024 **7.** 120
9. 144 **11.** 120 **13.** 32 **15.** 216 **17.** 480
19. No **21.** 158,184,000 **23.** 1024 **25.** 8192
27. No **29.** 24,360 **31.** 1050
33. (a) 16,807 (b) 2520 (c) 2401 (d) 2401 (e) 4802
35. 936 **37.** (a) 1152 (b) 1152 **39.** 483,840 **41.** 6

Section 10.2 ■ page 646

1. 336 **3.** 7920 **5.** 100 **7.** 2730 **9.** 151,200
11. 120 **13.** 24 **15.** 362,880 **17.** 997,002,000

19. 24 **21.** 60 **23.** 60 **25.** 15 **27.** 277,200
29. 2,522,520 **31.** 168 **33.** 56 **35.** 330
37. 100 **39.** 20 **41.** 210 **43.** 2,598,960
45. 120 **47.** 495 **49.** 2,035,800 **51.** 1,560,780
53. \$22,957,480 **55.** (a) 56 (b) 256 **57.** 1024
59. 2025 **61.** 79,833,600 **63.** 131,072

Section 10.3 ■ page 655

1. (a) $S = \{HH, HT, TH, TT\}$ (b) $\frac{1}{4}$ (c) $\frac{3}{4}$ (d) $\frac{1}{2}$
3. (a) $\frac{1}{6}$ (b) $\frac{1}{2}$ (c) $\frac{1}{6}$ **5.** (a) $\frac{1}{13}$ (b) $\frac{3}{13}$ (c) $\frac{10}{13}$
7. (a) $\frac{5}{8}$ (b) $\frac{7}{8}$ (c) 0 **9.** (a) $\frac{1}{3}$ (b) $\frac{5}{17}$
11. (a) $\frac{3}{16}$ (b) $\frac{3}{8}$ (c) $\frac{5}{8}$
13. $4C(13, 5)/C(52, 5) \approx .00198$
15. $4/C(52, 5) \approx 1.53908 \times 10^{-6}$
17. (b) $\frac{1}{16}$ (c) $\frac{3}{8}$ (d) $\frac{1}{8}$ (e) $\frac{11}{16}$ **19.** $\frac{9}{19}$
21. $1/C(49, 6) \approx 7.15 \times 10^{-8}$ **23.** (a) $\frac{1}{1024}$ (b) $\frac{15}{128}$
25. (a) $1/48^6 \approx 8.18 \times 10^{-11}$ (b) $1/48^{18} \approx 5.47 \times 10^{-31}$
27. $4/11! \approx 1.00 \times 10^{-7}$ **29.** (a) $\frac{3}{4}$ (b) $\frac{1}{4}$
31. (a) Yes (b) No
33. (a) Mutually exclusive; 1
(b) Not mutually exclusive; $\frac{2}{3}$
35. (a) Not mutually exclusive; $\frac{11}{26}$
(b) Mutually exclusive; $\frac{1}{2}$
37. (a) $\frac{3}{4}$ (b) $\frac{1}{2}$ (c) 1 **39.** $\frac{21}{38}$ **41.** $\frac{31}{1001}$
43. (a) $\frac{3}{8}$ (b) $\frac{1}{2}$ (c) $\frac{11}{16}$ (d) $\frac{13}{16}$
45. (a) Yes (b) $\frac{1}{36}$ **47.** (a) $\frac{1}{16}$ (b) $\frac{1}{32}$
49. $\frac{1}{12}$ **51.** $\frac{1}{1444}$ **53.** $\frac{1}{36^3} \approx 2.14 \times 10^{-5}$
55. (i) **57.** $1 - \frac{P(365, 8)}{365^8} \approx 0.07434$

Section 10.4 ■ page 668

1. \$1.50 **3.** \$0.94 **5.** \$0.92 **7.** 0 **9.** −\$0.30
11. −\$0.0526 **13.** −\$0.50
15. No, she should expect to lose \$2.10 per stock.
17. −\$0.93 **19.** \$1

Chapter 10 Review ■ page 671

1. 624 **3.** (a) 10 (b) 20 **5.** 120 **7.** 45
9. 17,576 **11.** 120 **13.** 5 **15.** 14
17. (a) 240 (b) 3360 (c) 1680
19. 40,320 **21.** $\frac{2}{5}$
23. (a) $S = \{HHH, HHT, HTH, HTT, THH, THT, TTH, TTT\}$
(b) $\frac{1}{8}$ (c) $\frac{1}{2}$ (d) $\frac{1}{2}$
25. (a) $\frac{1}{13}$ (b) $\frac{2}{13}$ (c) $\frac{4}{13}$ (d) $\frac{1}{26}$
27. (a) $\frac{1}{6}$ (b) $\frac{5}{6}$ **29.** (a) $\frac{1}{1000}$ (b) $\frac{1}{200}$ **31.** 0
33. \$0.00016 **35.** (a) 3 (b) .51
37. (a) 10^5 (b) 5^5 (c) $\frac{1}{32}$ (d) 75
39. (a) 144 (b) 126 (c) 84 (d) $\frac{7}{8}$

Chapter 10 Test ■ page 674

1. (a) 10^5 (b) 30,240 **2.** 60
3. $30 \cdot 29 \cdot 28 \cdot C(27, 5) = 1{,}966{,}582{,}800$
4. 12 **5.** $4 \cdot 2^{14} = 65{,}536$
6. (a) $4! = 24$ (b) $6!/3! = 120$
7. $C(5, 3)/C(15, 3) \approx .022$ **8.** $\frac{1}{6}$
9. (a) $\frac{1}{2}$ (b) $\frac{1}{13}$ (c) $\frac{1}{26}$
10. (a) $\frac{5}{13}$ (b) $\frac{6}{13}$ (c) $\frac{9}{13}$
11. \$0.65 **12.** $1 - 1 \cdot \frac{11}{12} \cdot \frac{10}{12} \cdot \frac{9}{12} \approx .427$

FOCUS ON MODELING ■ page 675

1. (b) $\frac{9}{10}$ **3.** (b) $\frac{7}{8}$

INDEX

PHOTO CREDITS

CHAPTER 1

2 Jim Pickerell/Stock Boston; Corbis
68 Stanford University News Service
71 Corbis

CHAPTER 2

74 J.L. Amos, Superstock; Courtesy, NASA
95 Bassano Studios
120 R. Morris/Corbis
121 Eric & David Hosking/Corbis
126 Phil Schemeister/Corbis

CHAPTER 3

130 Lori Adamski Peck/Tony Stone Images; Gar Ladd/Photo Researchers
174 Robert Frerck/Woodfin Camp/PNI

CHAPTER 4

208 Sarah Putnam/The Picture Cube; Staffan Widstrand/Corbis, Kevin R. Morris/Corbis
225 Stanford University News Service
255 The Granger Collection

CHAPTER 5

310 Brian Yarvin/Photo Researchers; Dilip Mehta/Contact Press Images/PNI
318 Edward Moylan/Ford Motor Company
347 Smithsonian Institute
376 Michael Ng/NYT Pictures

CHAPTER 6

378 Art Wolfe/AllStock/PNI; Owen Franken/Corbis
391 Garry McMichael/Photo Researchers
428 Herman J. Kokojan/Black Star/PNI; Archive Photos/PNI

CHAPTER 7

446 Tamara Munzer/Stanford University, Eric Hoffman/Ipsilon Networks, K. Claffy/NLANR, and Bill Fenner/XeroxPARC; Edward Moylan/Ford Motor Company
452 Corbis
479 The Archives of the National Academy of Sciences
483 Courtesy of CalTech
486 Ron Sanford/AllStock/PNI
488 The Granger Collection
490 Ron Sanford/AllStock/PNI
498 Corbis
500 The Granger Collection

CHAPTER 8

532 TSM/Lew Long, 1996; Corbis
566 The Granger Collection
578 No Credit Listed

CHAPTER 9

580 Susan Van Etten/Photo Edit; Corbis
585 The Granger Collection
589 Corbis
618 Smithsonian Institute
628 Bill Sanderson/Science Photo Library/Photo Researchers

CHAPTER 10

640 Will and Denni McIntyre/Photo Researchers; Kevin R. Morris/Corbis
642 Stanford University News Service
651 Ken Regan/Camera 5